Mathematics for Physicists

This textbook is a comprehensive introduction to the key disciplines of mathematics – linear algebra, calculus and geometry – needed in the undergraduate physics curriculum. Its leitmotiv is that success in learning these subjects depends on a good balance between theory and practice. Reflecting this belief, mathematical foundations are explained in pedagogical depth, and computational methods are introduced from a physicist's perspective and in a timely manner. This original approach presents concepts and methods as inseparable entities, facilitating in-depth understanding and making even advanced mathematics tangible.

The book guides the reader from high-school level to advanced subjects such as tensor algebra, complex functions and differential geometry. It contains numerous worked examples, info sections providing context, biographical boxes, several detailed case studies, over 300 problems and fully worked solutions for all odd-numbered problems. An online solutions manual for all even-numbered problems will be made available to instructors.

Alexander Altland is Professor of Theoretical Physics at the University of Cologne. His areas of specialization include quantum field theory and the physics of disordered and chaotic systems. He is co-author of the hugely successful textbook *Condensed Matter Field Theory* (2nd edition published by Cambridge University Press, 2010). He received the Albertus Magnus Teaching Award of the faculty of mathematics and natural sciences of Cologne University.

Jan von Delft is Professor of Theoretical Physics at the Arnold Sommerfeld Center for Theoretical Physics at the Ludwig–Maximilians–Universität in Munich. His research is focused on mesoscopic physics and strongly interacting electron systems. For his engagement in teaching, utilizing electronic chalk and "example+practice" problem sheets including example problems with detailed solutions, he received a Golden Sommerfeld teaching award.

Mathematics for Physicists

Introductory Concepts and Methods

ALEXANDER ALTLAND

Universität zu Köln

JAN VON DELFT

Ludwig–Maximilians–Universität München

CAMBRIDGE
UNIVERSITY PRESS

CAMBRIDGE
UNIVERSITY PRESS

University Printing House, Cambridge CB2 8BS, United Kingdom

One Liberty Plaza, 20th Floor, New York, NY 10006, USA

477 Williamstown Road, Port Melbourne, VIC 3207, Australia

314–321, 3rd Floor, Plot 3, Splendor Forum, Jasola District Centre, New Delhi – 110025, India

79 Anson Road, #06–04/06, Singapore 079906

Cambridge University Press is part of the University of Cambridge.

It furthers the University's mission by disseminating knowledge in the pursuit of education, learning, and research at the highest international levels of excellence.

www.cambridge.org
Information on this title: www.cambridge.org/9781108471220
DOI: 10.1017/9781108557917

First published 2019

Printed in the United Kingdom by TJ International Ltd. Padstow Cornwall

A catalogue record for this publication is available from the British Library.

Library of Congress Cataloging-in-Publication Data
Names: Altland, Alexander, 1965– author. | Delft, Jan von, 1967– author.
Title: Mathematics for physicists introductory concepts and methods /
Alexander Altland (Universität zu Köln), Jan von Delft
(Ludwig-Maximilians-Universität München).
Description: Cambridge ; New York, NY : Cambridge University Press, 2019.
Identifiers: LCCN 2018043275 | ISBN 9781108471220
Subjects: LCSH: Mathematical physics. | Physics.
Classification: LCC QC20 .A4345 2019 | DDC 530.15–dc23
LC record available at https://lccn.loc.gov/2018043275

ISBN 978-1-108-47122-0 Hardback

Additional resources for this publication at www.cambridge.org/altland-vondelft.

Brief Contents

Contents

Preface

> The miracle of the appropriateness of the language of mathematics for the formulation of the laws of physics is a wonderful gift which we neither understand nor deserve.
>
> Eugene Paul Wigner

This text is an introduction to mathematics for beginner physics students. It contains all the material required in the undergraduate curriculum. The main feature distinguishing it from the large number of available books on the subject is that mathematical *concepts* and *methods* are presented in unison and on an equal footing. Let us explain what is meant by this statement.

Physicists teaching mathematics often focus on the *training of methods*. They provide recipes for the algebraic manipulation of vectors and matrices, the differentiation of functions, the computation of integrals, etc. Such pragmatic approaches are often justified by time pressure: physics courses require advanced mathematical methodology and students have to learn it as quickly as possible.

However, knowledge of computational methods alone will not carry a student through the physics curriculum. Equally important, she needs to understand the mathematical principles and *concepts* behind the machinery. For example, the methodological knowledge that the derivative of x^2 equals $2x$ remains hollow, unless the conceptual meaning of that $2x$ as a local linear approximation of a parabola is fully appreciated. Similar things can be said about any of the advanced mathematical methods required in academic physics teaching.

Recognizing this point, physics curricula often include lecture courses in pure mathematics – who would be better authorized to teach mathematical concepts than mathematicians themselves? However, there is a catch: mathematicians approach the conceptual framework of their science from a perspective different from that of physicists. Rigorous proofs and existence theorems stand in the foreground and are more important than the communication of concepts relevant to the understanding of structures in physics. This tendency is pervasive, including when mathematicians teach "mathematics for physicists".

For these reasons, the traditional division – physics courses focusing on methods, mathematics courses on proofs – is not ideal.

Pedagogical strategy – unified presentation of concepts and methods

This book aims to bridge the divide. It contains a *unified presentation* of concepts and methods, written from the perspective of theoretical physicists. Mathematical structures are motivated and introduced as an orienting framework for the understanding of methods. Although less emphasis is put on formal proofs, the text maintains a fair level of mathematical hygiene and generally does present material in a formally consistent manner. Importantly, it does not operate on a higher level of technicality or abstraction than is standard in physics.

As an example, consider *vectors*. First-time introductions often focus on three-dimensional vectors, visualized as arrows and described by three components. To many students this picture is familiar from high school, and it suffices to follow introductory mechanics courses in university. However, only one year later quantum mechanics is on the agenda. The mathematics of quantum mechanics is all about vectors, however these now live in a more abstract (Hilbert) space which is hard to visualize. This can be frustratingly difficult for students conditioned to a narrow understanding of vectors. In this text, we take the different approach of introducing vector spaces in full generality at a very early stage. Experience shows that beginner students have no difficulty in absorbing the concept. Once familiar with it, the categorization even of very abstract objects as vectors feels natural and does not present any difficulty. In this way, the later mathematics of quantum mechanics becomes much easier to comprehend.

Does the enhanced emphasis on concepts come at the expense of methodological training? The answer is an emphatic "no!" – a solid conceptual understanding of mathematics leads to greatly improved practical and methodological skills. These statements are backed by experience. The book is based on a course taught more than ten times to first-year students at the University of Cologne and Ludwig–Maximilians–Universität (LMU) Munich. Building on this text, these courses introduce mathematical methods at a pace compatible with standard physics curricula and at load levels manageable for average students. The introduction of this new teaching concept has significantly enhanced the students' performance and confidence. Its emphasis on the motivation of mathematical concepts also provides welcome tail wind in the understanding of concurrent courses in pure mathematics.

Organization and scope

The book is organized into *three parts*:

▷ Linear Algebra (L),

▷ Calculus (C),

▷ Vector Calculus (V).

Starting at high-school level, each part covers the material required in a standard Bachelor curriculum and reaches out somewhat beyond that. In fact, the whole text has been written with an eye on modern developments in physics research. This becomes apparent in the final chapters which include introductions to multilinear algebra, complex calculus, and differential forms, formulated in the language used in contemporary research. However, the early chapters are already formulated in ways which anticipate these developments and occasionally employ language and notation slightly different from (but never incompatible with) that of traditional teaching. Generally, the writing style of each part gradually changes from moderately paced and colloquial at the beginning to somewhat more concise and "scientific" towards the final chapters. Due to its modular structure, the text can also serve as a reference covering all elements of linear algebra, calculus and vector calculus encountered in a Bachelor physics curriculum.

The reading order of parts L, C, V is not fixed and can be varied according to individual taste and/or time constraints. A good way to start is to first read a few chapters of each of parts L and C and then move into V. Where later chapters draw connections between fields, initial *remarks* state the required background so that there is no risk of accidentally missing out on something essential. For concreteness, Table 1 details the organization of a one-semester course at LMU München. Table 2 is the outline of a more in-depth two-semester course at Cologne University where the first and second semesters focus on calculus and linear algebra, respectively.

Pedagogical features

Many beginning physics students struggle with mathematics. When confronted with abstract material they ask the "what for?" question or even perceive mathematics as a hostile subject. By contrast, the authors of the present text love mathematics and know that the symbiotic relationship between the disciplines is a gift. They have tried to convey as much as possible of this positive attitude in the text.

Examples, info sections, case studies, biographical boxes

The text includes numerous *examples* showing the application of general concepts or methods in physically motivated contexts. It also contains more than a hundred *info sections* addressing the background or relevance of mathematical material in physics. For example, the info sections on pp. 52 and 113 put general material of linear algebra into the context of Einstein's theory of relativity. A few *case studies*, more expansive in scope than the info sections, illustrate how mathematical concepts find applications in physics. For example, quantum mechanics is mentioned repeatedly throughout Part L. All these references are put into context in a case study (Section L8.4) discussing how the principles of quantum mechanics are naturally articulated in the language of linear algebra.

Almost all info sections and case studies can be read without further background in physics. However, it should be emphasized that this text is not an introduction to physics and that the added material only serves illustrative purposes. It puts mathematical material into a physics context but remains optional and can be skipped if time pressure is high and priorities have to be set.

Finally, abstract mathematical material often feels less alien if the actual person responsible for its creation is visible. Therefore numerous *biographical boxes* portray some of the great minds behind mathematical or physical invention.

Problems

Solving *problems* is an essential part of learning mathematics. About one third of the book is devoted to problems, more than 300 in number, all tried and tested in Munich and Cologne. In this text, there is an important distinction between odd- and even-numbered problems: the odd-numbered *example problems* include detailed solutions serving as efficient and streamlined templates for the handling of a technical task. They can be used for self-study or for discussion in tutorials. These exemplar problems prepare the reader for the subsequent even-numbered *practice problems*, which are of similar structure but should be solved independently.

To avoid disruption of the text flow, all problems are assembled in three separate chapters, one at the end of each part. Individual problems are referenced from the text location where they first become relevant. For example, $\rightarrow$ L5.5.1-2, referenced in Section L5.5 on *general linear maps and matrices*, points to an example problem on two-dimensional rotation matrices, followed by a practice problem on three-dimensional ones. Three chapters at the very end of the book contain the solutions to the odd-numbered problems. A password-protected manual containing solutions of all even-numbered problems will be made available to instructors.

Index, margin, hyperlinks and English language

Keywords appearing in the *index* are highlighted in **bold** throughout the text. Likewise in bold, we have added a large number of *margin keywords*. Margin keywords often duplicate index keywords for extra visibility. More generally they represent topical name tags providing an at-a-glance overview of what is going on on a page. *Slanted font* in the text is used for emphasis or to indicate topical structure.

The electronic version of this book is extensively hyperlinked. Clicking on a page number cited in the text causes a jump to that page, and similarly for citations of equations, chapters, sections, problems and index keywords. Likewise, clicking on the title of an odd-numbered problem jumps to its solution, and vice versa.

Finally, a word of encouragement for readers whose mother tongue isn't English: Learning to communicate in English – the lingua franca of science – at the earliest possible stage is more important then ever. This is why we have written this text in English and not in our own native language. Beginners will find that technical texts like this one are much easier to read than prose and that learning scientific English is easier than expected.

Some remarks for lecturers

We mentioned above that the present text deviates in some points from standard teaching in physics. None of these changes are drastic, and most amount to a slightly different accentuation of material. We already mentioned that we put emphasis on the general understanding of vectors. Students conditioned to "seeing vectors everywhere" have no difficulties in understanding the concept of spherical harmonics as a complete set of functions on the sphere, interpreting the Fourier transform as a basis change in function space, or thinking of a Green function as the inverse of a linear operator. We know from experience that once this way of thinking has become second nature the mathematics of quantum mechanics and of other advanced disciplines becomes much easier to comprehend.

On a related note, the physics community has the habit of regarding every object comprising components as either a vector or a matrix. However, only a fraction of the index-carrying objects encountered in physics are genuine vectors or matrices.[1] Equally important are dual vectors, bilinear forms, alternating forms, or tensors. Depending on the field one is working in, the "everything-is-a-vector" attitude can be tolerable or a notorious

---

[1] For example, a magnetic field "vector" does not change sign under a reflection of space. It therefore cannot be a true vector, which always causes confusion in teaching.

source of confusion. The latter is the case in fields such as particle physics and relativity, and in emerging areas such as quantum information or topological condensed matter physics. Linear algebra as introduced in this text naturally accommodates non-vectorial and non-matrix objects, first examples including the cross product of vectors and the metric of vector spaces. In the later parts of the text we introduce tensors and differential forms, and illustrate the potency of these concepts in a case study on electromagnetism (Chapter V7).

One of the less conventional aspects of this text is the use of *covariant notation* (indices of vector components upstairs, those of vectors downstairs). Covariant notation has numerous pedagogical advantages, both pragmatic and conceptual. For instance, it is very efficient as an error tracking device. Consistent summations extend over pairs of contravariant superscript and covariant subscript indices, and violations of this rule either indicate an error (a useful consistency check) or the hidden presence of "non-vectorial structures" (the latter occurring in connection with, e.g., the cross product). In all such cases, we explain what is going on either right away, or somewhat later in the text. In our teaching experience, the covariant approach is generally well received by students. As added value, it naturally prepares them for fields such as relativity or particle physics, where it is mainstream. (Readers consulting this text as a secondary reference and for which covariance does not feel natural are free to ignore it – just read all indices in a traditional way as subscripts.)

Acknowledgments

This project would not have been possible without the continued support of many people. Specifically, we thank Thomas Franosch for kindly making his lecture notes available to us when we first started teaching this material. We thank Stefan Kehrein, Volker Meden, Frank Wilhelm and Martin Zirnbauer for helpful feedback, Florian Bauer, Benedikt Bruognolo, Vitaly Golovach, Bernhard Emmer, Alessandro Fasse, Olga Goulko, Fanny Groll, Sebastian Huber, Michael Kogan, Björn Kuballa, Fabian Kugler, Jan Manousakis, Pouria Mazloumi, Dmitri Pimenov, Dennis Schimmel, Frauke Schwarz, Enrique Solano, Katharina Stadler, Valentin Thoss, Hong-Hao Tu, Elias Walter and Lukas Weidinger for valuable help in the formulation and debugging of exercise problems, and Martin Dupont and Abhiram Kidambi for help in translating them from German to English. We thank Nicholas Gibbons, Martina Markus and Ilaria Tassistro for help and advice in matters of design and layout, Frances Nex for careful copy-editing, Richard Hutchinson and Mairi Sutherland for meticulous proofreading and Rosie Crawley for managing production. Finally, we are deeply grateful to our families for accompanying the seemingly infinite process of writing this book with truly infinite patience and support.

Alexander Altland and Jan von Delft

Table 1   Outline of a moderately paced (top) or fast-paced (bottom) one-semester course based on this text. Each row refers to a 90-minute lecture.

	L	C	V	Topic
1	1.1-2			Basic concepts I: sets, maps and groups
2	1.3			Basic concepts II: fields and complex numbers
3		1		Differentiation of one-dimensional functions
4		2		Integration of one-dimensional functions
5	2.1-3			Vector spaces: standard vector space, general definition
6	2.4-5			Vector spaces: basis and dimension
7	3.1-2			Euclidean spaces I: scalar product, norm, orthogonality
8	3.3-4			Euclidean spaces II: metric, complex inner product
9	4			Vector product: Levi–Civita symbol, various identities
10			1	Curves, line integrals
11		3		Partial differentiation
12		4.1		Multidimensional integration I: Cartesian
13			2.1-3	Curvilinear coordinates: polar, cylindrical, spherical
14		4.2-3		Multidimensional integration II: curvilinear coordinates
15			3.1-2	Scalar fields and gradient
16			3.4	Vector fields: gradient fields, nabla operator
17	5.1-3			Linear maps I: matrices, matrix multiplication
18	5.4-6			Linear maps II: inverse, basis transformations
19	6		2.5	Determinants: definition, properties
20	7			Diagonalization: eigenvalues, eigenvectors
21		5.1-2		Taylor series: definition, complex Taylor series
22		7.1-3		Differential equations I: separable DEQs, linear first-order DEQs
23		7.4-5		Differential equations II: systems of linear DEQs
24		6.1-2		Fourier calculus I: Dirac delta function, Fourier series
25		6.3		Fourier calculus II: Fourier transforms
26		4.4-5	3.5	Integration in arbitrary dimensions; flux integrals
27			3.5	Sources of vector fields, Gauss's theorem
28			3.6-7	Circulation of vector fields, Stokes's theorem
1	1			Basic concepts I: sets, maps, groups, fields and complex numbers
2		1,2		Differentiation and integration of one-dimensional functions
3	2			Vector spaces: definition, examples, basis and dimension
4	3			Euclidean spaces: scalar product, norm, orthogonality, metric
5	4			Vector product: Levi–Civita symbol, various identities
6			1	Curves, line integrals
7		3, 4.1		Partial differentiation, multidimensional integration: Cartesian
8			2.1-3	Curvilinear coordinates: polar, cylindrical, spherical
9		4.2-4		Multidimensionl integration, curvilinear coordinates
10			3.1-2	Scalar fields and gradient
11			3.3-4	Extrema of functions with constraints, gradient fields
12	5.1-3			Linear maps I: matrices, matrix multiplication
13	5.4-6			Linear maps II: inverse, basis transformations
14	6	4.5	2.5	Determinants: definition, properties, applications
15	7			Diagonalization: eigenvalues, eigenvectors
16	8			Orthogonal, unitary, symmetric and Hermitian matrices
17		5.1-2		Taylor series: definition, complex Taylor series
18		7.1-3		Differential equations I: separable DEQs, linear first-order DEQs
19		7.4-5		Differential equations II: systems of linear DEQs
20		5.3-5		Perturbation expansions; higher-dimensional Taylor series
21		6.1-2		Fourier calculus I: Dirac delta function, Fourier series
22		6.3		Fourier calculus II: Fourier transforms
23		6.3,7.5		Fourier series for periodic functions; Green functions
24		7.6-7		Differential equations III: general nth-order DEQs, linearization
25			3.5	Sources of vector fields, Gauss's theorem
26			3.6-7	Circulation of vector fields, Stokes's theorem
27		9.1-2		Holomorphic functions, complex integration, Cauchy's theorem
28		9.3-5		Singularities, residue theorem, essential singularities

Table 2 Outline of a more in-depth two-semester course based on this text.

	L	C	V	Topic
1	1.1-2			Basic concepts I: sets, maps and groups
2	1.3-4			Basic concepts II: fields and complex numbers
3	2.1-3			Vector spaces I: standard vector space and general definition
4	2.4-5			Vector spaces II: basis and dimension
5	3			Euclidean geometry: scalar product, norm, orthogonality
6	4			Vector product
7		1		Differentiation of one-dimensional functions
8		2		Integration of one-dimensional functions
9		3		Partial differentiation
10		4.1		Multidimensional integration in Cartesian coordinates
11			1.1-2	Curves
12			1.3-4	Curve length and line integrals
13			2.1-2	Curvilinear coordinates I: polar coordinates, general concept
14			2.3-4	Curvilinear coordinates II: cylindrical, spherical
15			3.1-2	Scalar fields and gradient
16			3.4	Gradient fields
17		4.2-3		Curvilinear integration in two and three dimensions
18		4.4		Curvilinear surface integrals
19		5.1-2		Taylor series: definition, complex Taylor series
20		6.1		Fourier calculus I: Dirac delta function
21		6.2		Fourier calculus II: Fourier series
22		6.3		Fourier calculus III: Fourier transform
23		6.3-4		Fourier transform applications
24			3.5	Flux integrals of vector fields
25			3.5	Sources of vector fields, Gauss's theorem
26			3.6-7	Circulation of vector fields, Stokes's theorem
1	5.1-2			Linear maps and matrices
2	5.3-4			Matrix multiplication, and inverse
3	5.4			Dimension formula, linear systems of equations
4	5.5-6			Basis transformations I
5	5.6			Basis transformations II
6	6.1			Determinants I
7	6.1			Determinants II
8		4.2-5	2.5	Integration in arbitrary dimensions, revisited
9	7.1-3			Eigenvalues, eigenvectors, characteristic polynomial
10	7.4-5			Diagonalization of matrices
11	7.5	7.1		Differential equations (DEQ): motivation
12		7.2-3		DEQs: typology and linear first-order equations
13		7.3-4		Systems of first-order DEQs
14		7.4		Green functions
15		7.5		General first-order DEQs
16		7.6		nth-order differential equations
17		7.7-9		Linearization, fixed points, partial differential equations
18	8.1-2			Linear maps: unitary and orthogonality
19	8.3			Linear maps: Hermiticity and symmetry
20	8.4			Case study: linear algebra in quantum mechanics
21		9.1		Holomorphic functions
22		9.2		Complex integration, Cauchy's theorem
23		9.3-4		Singularities, residue theorem
24	9.1-2			Linear algebra in function spaces I
25	9.3-4			Linear algebra in function spaces II

L

LINEAR ALGEBRA

The first part of this book is an introduction to linear algebra, the mathematical discipline of structures that are, in a sense to be discussed, "straight". No previous knowledge of the subject is assumed. We start with an introduction to various basic structures in mathematics: sets, groups, fields, different types of "numbers", and finally vectors. This is followed by a discussion of elementary geometric operations involving vectors, the computation of lengths, angles, areas, volumes, etc. We then explain how to describe relations between vectorial objects via so-called linear maps, how to represent linear maps in terms of matrices, and how to work with these operations in practice. Part L concludes with two chapters on advanced material. The first introduces the interpretation of functions as vectors (a view of essential importance to quantum mechanics). In the second, we discuss linear algebra in vector spaces containing a high level of intrinsic structure, so-called tensor spaces, which appear in disciplines such as relativity theory, fluid mechanics and quantum information theory.

Mathematics before numbers

Many people believe that numbers are the most basic elements of mathematics. This, however, is an outside view which does not reflect the way mathematics itself treats numbers. Numbers can be added, subtracted, multiplied and divided by, which means that they possess a considerable degree of complexity.[1] Metaphorically speaking, they are high up in the evolutionary tree of mathematics, and beneath them there exist numerous structures of lesser complexity. Much as a basic understanding of evolutionary heritage is important in understanding life – reptiles, vs. birds, vs. mammals, etc. – the evolutionary ancestry of numbers is a key element in the understanding of mathematics, and physics. We take this as motivation to start with a synopsis of various pre-numerical structures which we will later see play a fundamental role throughout the text.

symmetry operations

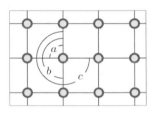

EXAMPLE Consider a two-dimensional square lattice that is invariant under rotations by 90 degrees (i.e. if you rotate the lattice by 90 deg[2] it looks the same as before, see figure). Then rotations by 0, 90, 180 or 270 deg are "*symmetry operations*" that map the lattice onto itself. Let us denote these operations by e, a, b and c, respectively. Two successive rotations by 90 deg are equivalent to one by 180 deg, a fact we may express as $a \cdot a = b$. Similarly, $b \cdot b = e$ (viewing a 360 deg rotation as equivalent to one by 0 deg). These operations are examples of mathematical objects which can be "combined" with each other, but not "divided" by one another. Together, they form a pre-number structure, soon to be identified as a "group". Generically groups have less structure than numbers and yet are very important in physics.

L1.1 Sets and maps

When we work with a complex systems of objects of *any* kind we need ways to categorize and store them. At the very least, we require containers capable of storing objects (think of the situation in a repair shop). On top of that one may want to establish connections between the objects of different containers (such as a tabular list indicating which screw

---

[1] At the end of the nineteenth century mathematicians became increasingly aware of gaps in the logical foundations of their science. It became understood that the self-consistent definition even of natural numbers $(1, 2, 3, \dots)$ was more complex than was previously thought. For an excellent account of the ensuing crisis of mathematics, including its social dimensions, we refer to the graphic novel *Logicomix*, A. Doxiadis, Bloomsbury Publishing, 2009.

[2] In this text we use the standard abbreviation "deg" for degrees.

in the screw-box matches which screwdriver in the screwdriver rack). In the terminology of mathematics, containers are called "sets", and the connections between them are established by "maps". In this section we define these two fundamental structures and introduce various concepts pertaining to them.

Sets

set   Perhaps the most basic mathematical structure is that of a **set**. (The question whether there are categories even more fundamental than sets is in fact a subject of current research.) As indicated above, one may think of a set as a container holding objects. In mathematical terminology, the objects contained in a set are called its **elements**. Unlike the containers in a repair shop, mathematical sets are not "physical" but simply serve to group objects according to certain categories (which implies that one object may be an element of different sets). For example, consider the set of all your relatives. Your mother is an element of that set, and at the same time one of the much larger set of all females on the planet, etc. More formally, the notation $a \in A$ indicates that a is an element of the set A, and $A = \{a, b, c, \dots\}$ denotes the full set.

notation   INFO   Be careful to exercise *precision in matters of notation.* For example, denoting a set by $(a, b, c, \dots)$ would be incompatible with the standard curly bracket format $\{a, b, c, \dots\}$ and an abuse of notation. Insistence on clean notation has nothing to do with pedantry and serves multiple important purposes. For example, the notation $B = \{1, 2, 3\}$ is understood by every mathematically educated person on the planet, meaning that standardized mathematical notation makes for the most international idiom there is. At the same time, uncertainties in matters of notation often indicate a lack of understanding of a concept. For example, $a \in \{a\}$ is correct notation indicating that a is an element of the set $\{a\}$ containing just this one element. However, it would be incorrect to write $a = \{a\}$. The element a and the one-element set $\{a\}$ are different objects. Uncertainty in matters of notation is a sure and general indicator of a problem in one's understanding and should always be considered a warning sign – stop and rethink.

The definition of sets and elements motivates a number of generally useful secondary definitions:

empty set   ▷   An **empty set** is a set containing no elements at all and denoted by $A = \{\}$, or $A = \emptyset$.

     ▷   A **subset** of A, denoted by $B \subset A$, contains some of the elements of A, for example, $\{a, b\} \subset \{a, b, c, d\}$. The notation $B \subseteq A$ indicates that the subset B may actually be equal to A. On the other hand, $B \subsetneq A$ means that this is certainly not the case.

     ▷   The **union** of two sets is denoted by $\cup$, for example, $\{a, b, c\} \cup \{c, d\} = \{a, b, c, d\}$. The **intersection** is denoted by $\cap$, for example, $\{a, b, c\} \cap \{c, d\} = \{c\}$.

     ▷   The removal of a subset $B \subset A$ from a set A results in the **difference**, denoted by $A \backslash B$. For example, $\{a, b, c, d\} \backslash \{c\} = \{a, b, d\}$.

     ▷   We will often define sets by *conditional rules.* The standard notation for this is set $= \{\text{elements} \,|\, \text{rule}\}$. For example, with $A = \{1, 2, 3, 4, 5, 6, 7, 8, 9, 10\}$ the set of all even integers up to 10 could be defined as $B = \{a \in A \,|\, a/2 \in A\} = \{2, 4, 6, 8, 10\}$.

Cartesian product

▷ Given two sets A and B, the **Cartesian product**,[3]

$$A \times B \equiv \{(a, b) \,|\, a \in A, b \in B\}, \tag{L1}$$

is a set containing all pairs (a, b) formed by elements of A and B.

The number of elements of a set is called its **cardinality**. The cardinality can be finite (the set of all your relatives) or infinite (the set of all natural numbers). Among the infinite sets one distinguishes between "countable" and "uncountable" sets. A set is **countable** if one can come up with a way to number its elements. For example, the set of even integers $A = \{0, 2, 4, \ldots\}$ is countable. The real numbers (see Section L1.3) form an uncountable set.

equivalence classes

It is often useful to organize sets in **equivalence classes** expressing the equality, $a \sim b$, of two elements relative to a certain criterion, R. For example, let A be the set of relatives and let the distinguishing criterion, R, be their sex. The notation Victoria $\sim$ Erna then indicates that the two relatives are equivalent in the sense that they are female. An equivalence relation has the following defining properties:

▷ reflexivity: $a \sim a$, every element is equivalent to itself,

▷ symmetry: $a \sim b$ implies $b \sim a$ and vice versa,

▷ transitivity: $a \sim b$ and $b \sim c$ implies $a \sim c$.

The subset of all elements equivalent to a given reference element a is called an *equivalence class* and denoted $[a] \subset A$. In the example of relatives and their sex, there are two such subsets, for example $A = [\text{Herbert}] \cup [\text{Erna}]$. The label used for an equivalence class is not unique; for example, one might relabel $[\text{Erna}] = [\text{Victoria}]$. The set of all equivalence classes relative to a criterion R is called its **quotient** and is denoted by A/R. In the example of relatives (A) and their sex (R), the quotient set $A/R = \{[\text{Herbert}], [\text{Victoria}]\}$ would have two elements, the class of males and that of females.

EXAMPLE   Consider the set of integers, and pick some integer q. Now view any two integers as equivalent if they have the same *remainder under division by q*. For example, $q = 4$ defines $0 \sim 4 \sim 8$, $1 \sim 5 \sim 9$. In this case there are four equivalence classes, denotable by $[0]$, $[1]$, $[2]$ and $[3]$. In general, the remainder of p divided by q is denoted by $p \bmod q$ (spoken "p-modulo-q", or just "p-mod-q"), e.g., $8 \bmod 4 = 0$, $6 \bmod 4 = 2$, or $-5 \bmod 4 = 3$ (by definition, remainders are taken to be positive). The equivalence class of all integers with the same remainder r under division by q is the set $[r] = \{p \in \mathbb{Z} \,|\, p \bmod q = r\}$. There are q such equivalence classes, and the set of these classes is denoted by $\mathbb{Z}_q \equiv \mathbb{Z}/q\mathbb{Z} = \{[0], [1], \ldots, [q-1]\}$.

Maps

Consider two sets, A and B, plus a rule, F, assigning to each element a of A an element b

map

of B. Such a rule, written as $F(a) \equiv b \in B$, is called a **map**. In mathematics and physics, maps are specified by the following standard notation:

---

[3]   We follow a widespread convention whereby $\square \equiv \triangle$ means "$\square$ is defined by $\triangle$". In the German literature, the alternative notation $\square := \triangle$ is frequently used.

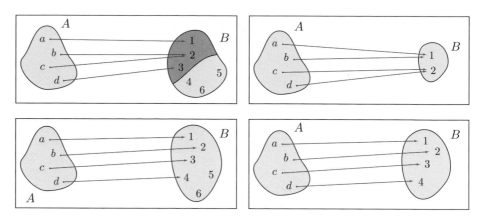

Fig. L1   Different types of maps. Top left: a generic map, top right: surjective map, bottom left: injective map, bottom right: bijective map.

$$F : A \to B, \qquad a \mapsto F(a). \tag{L2}$$

domain   The set A is called the **domain** of the map and B is its **codomain**.[4] An element $a \in A$ fed into the map is called an **argument** and $F(a)$ is its **value** or **image**. Note that different types of arrows are used for "domain $\to$ codomain" and "argument $\mapsto$ image".

image   The **image** of A under F, denoted by $F(A)$, is the set containing all image elements of F: $F(A) = \{F(a) | a \in A\} \subseteq B$ (see dark shaded area in the top left panel of Fig. L1). A map is called **surjective** (top right panel) if its image covers all of B, $F(A) = B$, i.e. if any element of the codomain is the image of at least one element of the domain. It is called **injective** (bottom left) if every element of the codomain is the image of at most one element of the domain. The map is **bijective** if it is both surjective and injective (bottom right panel), i.e. if every element $b \in B$ of the codomain is the image of precisely one element $a \in A$ of the domain. Bijective maps establish an unambiguous relation between the elements of the sets A and B. The one-to-one nature of this assignment means that it can be inverted: there exists an **inverse map**, $F^{-1} : B \to A$, such that $F^{-1}(F(a)) = a$ for every $a \in A$.

inverse map

composition of maps   Given two maps, $F : A \to B$ and $G : B \to C$, their **composition** is defined by substituting the image element of the first as an argument into the second:

$$G \circ F : A \to C, \qquad a \mapsto G(F(a)). \tag{L3}$$

For example, the above statement about bijective maps means that the composition of a bijective map F with its inverse, F^{-1}, yields the identity map: $F^{-1} \circ F : A \to A$, with $a \mapsto F^{-1}(F(a)) = a$.

Finally, a map F defined on a Cartesian product set, $A \times B$, is denoted as

$$F : A \times B \to C, \qquad (a, b) \mapsto c = F(a, b).$$

This map assigns to every pair (a, b) an element of C. For example, the shape of a sand dune can be described by a map, $h : \mathbb{R} \times \mathbb{R} \to \mathbb{R}$, $(x, y) \mapsto h(x, y)$, where for each

---

[4] The designation "codomain" is standard in mathematics, but not in physics. Oddly, physics does not seem to have an established designation for the "target set" of a map.

point (x, y) in the plane, the function $h(x, y)$ gives the height of the dune above that point. ($\rightarrow$ L1.1.1-2)

L1.2 Groups

Sets as such are just passive containers storing elements. Often, however, the elements of a set are introduced with the purpose of doing something with them. As an example, consider the set of 90 deg rotations, $R \equiv \{e, a, b, c\}$, introduced on p. 3. A two-fold rotation by 180 deg is equivalent to a non-rotation and this fact may be described as $b \cdot b = e$. Or we may say that $a \cdot b = c$, meaning that a 90 degree rotation following one by 180 degrees equals one by 270 degrees, etc. In this section, we define groups as the simplest category of sets endowed with an "active" operation on their elements.

Definition of groups

group The minimal structure[5] which brings a set to life in terms of operations between its elements is called a **group**. Let A be a set and consider an operation, "$\cdot$", equivalently called a **group law** or **composition rule**, assigning to every pair of elements a and b in A another element, $a \cdot b$:

$$\cdot : A \times A \to A, \qquad (a, b) \mapsto a \cdot b. \tag{L4}$$

group axioms This map defines a group operation provided that the following four **group axioms** are satisfied:[6]

(i) **Closure**: for all a and b in A the result of the operation $a \cdot b$ is again in A. (Although this condition is already implied by the definition (L4), it is generally counted as one of the group axioms.)

(ii) **Associativity**: for all a, b and c in A we have $(a \cdot b) \cdot c = a \cdot (b \cdot c)$.

(iii) **Neutral element**: there exists an element e in A such that for every a in A, the equation $e \cdot a = a \cdot e = a$ holds.

Depending on context, the neutral element is also called the **identity element** or **null element**.

(iv) **Inverse element**: For each a in A there exists an element b in A such that $a \cdot b = b \cdot a = e$.

---

[5] This statement is not fully accurate. There is a structure even more basic than a group, the **semigroup**. A semigroup need not have a neutral element, nor inverse elements to each element. In physics, semigroups play a less prominent role than groups, hence we will not discuss them further.

[6] Mathematicians often formulate statements of this type in a more compact notation. Frequently used symbols include $\forall$, abbreviating **for all**, and $\exists$, for **there exists**. Expressed in terms of these, the group axioms read: (i) $\forall a, b \in A, a \cdot b \in A$. (ii) $\forall a, b, c \in A, a \cdot (b \cdot c) = (a \cdot b) \cdot c$. (iii) $\exists e \in A$ such that $\forall a \in A, a \cdot e = e \cdot a = a$. (iv) $\forall a \in A, \exists b \in A$ such that $a \cdot b = b \cdot a = e$. Although this notation is less frequently used in physics texts, it is very convenient and we will use it at times.

Under these conditions, A and "·" define a group as $G \equiv (A, \cdot)$. A group should always be considered a "double", comprising a set and an operation. It is important to treat the operation as an integral part of the group definition: there are numerous examples of sets, A, which admit two different group operations, "·"

Nils Henrik Abel 1802–1829
Norwegian mathematician who made breakthrough contributions to several fields of mathematics before dying at a young age. Abel is considered the inventor (together with but independently from Galois) of group theory. He also worked on various types of special functions, and on the solution theory of algebraic equations.

and "$*$". The doubles $G = (A, \cdot)$ and $G' = (A, *)$ are then different groups. We finally note that in some cases it can be more natural to denote the group operation by a different symbol, "+", or "$*$", or "∘", . . .

EXAMPLE Here are a few first examples of groups.

▷ The simplest group of all, $G = (\{e\}, \cdot)$, contains just one element, its neutral element. Nothing much to discuss.

▷ The introductory example of 90 deg rotations, $R \equiv \{e, a, b, c\}$, defines a group of cardinality four. Its neutral element is e and for each element we have an inverse, for example $a \cdot c = e$. (Set up a "multiplication table" specifying the group operation for all elements of $R \times R$.) The same group, i.e. a set of four elements with the same group law, can be realized in different contexts. For example, for the quotient set $\mathbb{Z}_4 = \{[0], [1], [2], [3]\}$ defined on p. 5, a group operation may be defined as "addition modulo 4". This means that the addition of a number with remainder 1 mod 4 to one with remainder 3 mod 4 yields a number with remainder 0 mod 4, for example $[1] + [3] = [0]$. Set up the full group operation table for this group and show that it is identical to that of the group of 90 deg rotations discussed previously. This implies that $(\mathbb{Z}_4, +)$ and $(R, \cdot)$ define the same group. Explain in intuitive terms why this is so. The concept of different realizations of the same group is very important in both physics and mathematics. We will see many more examples of such correspondences throughout the text. ($\rightarrow$ L1.2.2)

group $\mathbb{Z}_2$ ▷ The simplest nontrivial group, which nevertheless has many important applications, contains just two elements, $\mathbb{Z}_2 = \{e, a\}$, with $a \cdot a = e$. The **group** $\mathbb{Z}_2$ can be realized by rotations by 180 deg, as the group of integers with addition mod 2 ($\rightarrow$ L1.2.1), and in many different ways. It plays a very important role in modern physics. For example, in information science, $\mathbb{Z}_2$ is the mathematical structure used to describe "**bits**", objects that can assume only one of two values, "on" and "off", or "0" and "1".

▷ The integers, $\mathbb{Z} \equiv \{\ldots, -2, -1, 0, 1, 2, \ldots\}$, with group operation "+" = "addition" (e.g. $2 + 4 = 6$) are an example of a group of infinite cardinality. $(\mathbb{Z}, +)$ has neutral element 0 and the inverse of a is $-a$, i.e. $a + (-a) = 0$. Why are the integers $(\mathbb{Z}, \cdot)$ with multiplicative composition ($2 \cdot 3 = 6$) *not* a group?

▷ Another important example of a discrete group is the translation group on a lattice. ($\rightarrow$ L1.2.3-4)

abelian group If the group operation is *commutative* in the sense that it satisfies $a \cdot b = b \cdot a$ for all elements the group is called an **abelian group**. All examples mentioned so far have this property. **Non-abelian groups** possess at least some elements for which $a \cdot b \neq b \cdot a$. An important example is the group formed by all *rotations of three-dimensional space*. This group can

be given a concrete realization by fixing three perpendicular coordinate axes in space. Any rotation can then be represented (see figure) as a succession of rotations around the

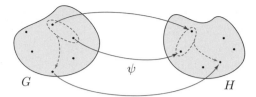

Fig. L2   The concept of a group homomorphism: a map between two groups that is compatible with the group operations (dashed) in that the image of the composition of two elements in the domain group (left) equals the composition of the image elements in the target group (right).

coordinate axes and the set of all these rotations forms a group where the group operation is the successive application of rotations. For example, $R_2 \cdot R_1$ is the rotation obtained by performing first R_1, then R_2. This concatenation is not commutative. For example, a rotation first around the x-axis and then around the z-axis is different from the operation in reverse order.

role in physics   **INFO** Groups play an important *role in physics*. This is because many classes of physical operations effectively carry a group structure. Simple examples include *rotations* or *translations* in space or time. These operations define groups because they can be applied in succession ("composed"), are associative, possess a neutral element (nothing is done), and can be inverted (undone). The translation and rotation groups play crucial roles in the description of momentum and angular momentum, respectively, both in classical and quantum mechanics. While continuous translations and rotations define groups of infinite cardinality, the physics of crystalline structures is frequently described in terms of finite restrictions. We mentioned the group $\mathbb{Z}_4$ of rotations by 90 deg around one axis as an example. In the late 1960s, group theory became important as a cornerstone of the *standard model* describing the fundamental structure of matter in terms of quarks and other elementary particles.

Despite the deceptive simplicity of the group axioms, the theory of groups is of great depth and beauty, and it remains a field of active research in modern mathematics.

Group homomorphism

Above, we have seen that the same group structure can be "realized" in different ways. For example, the group $\mathbb{Z}_2$ can be realized as the group of rotations by 180 deg, or as addition in $\mathbb{Z}$ mod 2. Identifications of this type frequently appear in physics and mathematics, and it is worthwhile to formulate them in a precise language. To this end, consider two groups, $(G, \cdot)$ and $(H, \cdot)$ with *a priori* independent group operations. Let $\psi : G \to H$ be a map from G to H. If this map is such that for all $a, b \in G$ the equality $\psi(a \cdot b) = \psi(a) \cdot \psi(b)$ **group homo-** holds, then ψ is called a **group homomorphism** (see Fig. L2). The defining feature of **morphism** a group homomorphism is its compatibility with the group law. As an example consider $G = H = (\mathbb{Z}, +)$, the addition of integers. Now assign to each integer its doubled value, $n \mapsto \psi(n) = 2n$. This map is a group homomorphism because $\psi(n + m) = 2(n + m) = 2n + 2m = \psi(n) + \psi(m)$. However, the map ϕ assigning to each integer its square, $n \mapsto \phi(n) = n^2$, is not a group homomorphism, because $\phi(1) + \phi(2) = 1^2 + 2^2 \neq \phi(1 + 2) = \phi(3) = 3^2$. As another example, consider the map $\psi : \mathbb{Z} \to \mathbb{Z}_2, n \mapsto \psi(n) = n \bmod 2$,

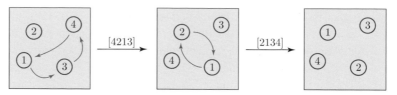

Fig. L3   Two permutations of four objects performed in succession.

assigning the number 0 or 1 to the integers, depending on whether n is even or odd. This is a homomorphism between the infinite group $(\mathbb{Z}, +)$ and the two-element group $\mathbb{Z}_2$.

    A perfect identification between two groups G and H is obtained if there exists a *bijective* **group** homomorphism between the two, a so-called **group isomorphism**. In this case, we write **isomorphism** $G \cong H$. Mathematicians tend to not even distinguish between isomorphic groups, a view that can be confusing to physicists. The identification $\mathbb{Z}_2 \cong (\mathbb{Z} \bmod 2) \cong$ (rotations group by 180 deg) discussed above is a group isomorphism.

> **EXERCISE**   Consider $\mathbb{Z}_n \equiv (\mathbb{Z} \bmod n, +)$, $n \in \mathbb{Z}$. Show that it defines a group of cardinality n. Show that $\mathbb{Z}_n$ is isomorphic to the group of rotations by $360/n$ deg around a fixed axis. ($\rightarrow$ L1.2.2)

Permutation group

The permutations of n objects define one of the most important finite groups, the permutation group, S_n. Consider n arbitrary but distinguishable objects. For definiteness it may be **permutation** useful to think of a set of n numbered billiard balls (see Fig. L3 for $n = 4$). A **permutation** is a rearrangement of these objects into a different order. For example, the reordering of four objects indicated in the left panel of the figure leads to the new arrangement shown in **factorial** the middle. There are **n-factorial**, $n! \equiv n(n-1)(n-2)\ldots 1$, different arrangements or permutations,[7] and we consider the set, S_n, of cardinality $n!$ containing all of them.

    Rearrangements can be composed, i.e. performed in succession. For example, the exchange in the middle panel of the figure leads to the final arrangement shown in the right panel. The group operation in S_n is this composition of permutations. Evidently, there is a trivial permutation (the one that leaves sequences unaltered), the composition of permutations is associative, and each permutation can be undone, such that there exists an inverse. This shows that S_n forms a group, the **permutation** **permutation group** or **symmetric group** of n objects. It is easy to verify that **group** the permutation group is non-abelian. (Invent examples of permutations proving the point.)

    Although the permutation group is easily defined, its mathematical structure is rather rich. (For example, the solution of a Rubik's cube amounts to a permutation of the 54 differently colored squares covering the six faces of the cube, and the solution algorithms

---

[7]   One way to understand this number is to notice that the first of n objects can be put in any of n places. This leaves $n-1$ options for the second object (one position is already occupied by the first object), $n-2$ for the third, etc. The total number of rearrangements is obtained as the product of the number of options for objects $1, 2, \ldots, n$, i.e. $n(n-1)(n-2)\ldots 1 = n!$.

reflect the mathematics of the permutation group S_{54}.) Below, we will frequently work with permutations and it will be useful to have a good notation for them. One popular labeling system denotes the permutation shown in the left part of the figure by [4213]. This notation logs the final configuration of the objects, i.e. the image of the map $(1, 2, 3, 4) \mapsto (4, 2, 1, 3)$, as a list in square brackets, where the map describes the permutation which replaces 1 by 4, 2 by 2, 3 by 1 and 4 by 3. The second permutation is thus denoted as [2134], and the composition of the two becomes [2134] ∘ [4213] = [4123].

EXERCISE Check that the permutation group of three objects can be represented as (→ L1.2.5)

$$S_3 = \{[123], [213], [321], [231], [312], [132]\}.$$

Alternatively, a permutation may be identified with a map $P : \mathbb{N}_n \to \mathbb{N}_n, j \mapsto P(j)$, where $\mathbb{N}_n = \{1, 2, \ldots, n\}$ is the set of n integers, and $P(j) \in \mathbb{N}_n$ the number by which j is replaced. Sometimes, the shorthand notation $Pj \equiv P(j)$ is used instead. In this language, [4213] is represented as $P1 = 4, P2 = 2, P3 = 1, P4 = 3$.

Note that each permutation can be reduced to a product of **pair permutations**, i.e. permutations which exchange just two objects at a time. This statement is easy to understand: any reordering of n objects can be achieved manually (with one's own two hands) by sequentially swapping pairs of objects. For example, the permutation [4213] of the figure can be effected by first exchanging $1 \leftrightarrow 3$, and then $3 \leftrightarrow 4$. For any permutation $P \in S_n$ we then have two options: the number of pair permutations needed to arrive at P may be even or odd (determine the even/odd attribute for the six permutations of S_3).[8] In the former/latter case, we call P an **even/odd permutation** and define

even/odd
permutation

$$\mathrm{sgn}(P) = \begin{cases} +1, & P \text{ even}, \\ -1, & P \text{ odd}, \end{cases} \tag{L5}$$

signum as the **signum** of a permutation. Note that $\mathrm{sgn}(P) = \mathrm{sgn}(P^{-1})$ (why?). (→ L1.2.6)

EXERCISE Define a map $S_n \to \mathbb{Z}_2, P \mapsto \mathrm{sgn}(P)$, where the two signs, $\mathrm{sgn}(P) = \pm 1$, are identified with the two elements of $\mathbb{Z}_2 = \{+1, -1\}$. Show that this is a group homomorphism between S_n and $\mathbb{Z}_2$, i.e. that $\mathrm{sgn}(P \cdot Q) = \mathrm{sgn}(P) \cdot \mathrm{sgn}(Q)$, where the multiplication on the right is that of the numbers ± 1. Understand this as the educated formulation of the statement that the product of two odd permutations is even, that of an even and an odd is odd, etc.

In this text, the signum of permutations will appear frequently and once more a good notation is called for. A convenient way to track this quantity is provided by the **Levi–Civita symbol**,

Levi–Civita
symbol

$$\epsilon_{i_1, i_2, \ldots, i_n}^{j_1, j_2, \ldots, j_n} \equiv \mathrm{sgn}\left(P_{i_1, i_2, \ldots, i_n}^{j_1, j_2, \ldots, j_n}\right), \tag{L6}$$

where $P_{i_1, i_2, \ldots, i_n}^{j_1, j_2, \ldots, j_n}$ denotes the permutation $P(j_l) = i_l$ which replaces the sequence $(j_1, j_2, \ldots, j_n)$ by $(i_1, i_2, \ldots, i_n)$. For example, $\epsilon_{321}^{231} = -1$, because a single pair

---

[8] Notice that the even/odd attribute is not entirely innocent: there are different ways of realizing a given P by a sequence of pair permutations. However, the "parity", i.e. the even- or oddness of the number of pair permutations, is an invariant. This makes the function sgn well defined.

permutation transmutes $(2, 3, 1)$ into $(3, 2, 1)$. For the same reason, the Levi–Civita symbol is **antisymmetric** under the exchange of any two indices, e.g. $\epsilon^{4321}_{2341} = -\epsilon^{4321}_{3241}$.

In applications, situations often arise where one of the involved permutations is the ordered one, $(j_1, j_2, \ldots, j_n) = (1, 2, \ldots, n)$. In such cases, it is customary to suppress the ordered sequence in the notation and just write $\epsilon_{i_1, i_2, \ldots, i_n} \equiv \epsilon^{1,2,\ldots,n}_{i_1, i_2, \ldots, i_n}$, and similarly $\epsilon^{j_1, j_2, \ldots, j_n} \equiv \epsilon^{j_1, j_2, \ldots, j_n}_{1,2,\ldots,n}$. For example, $\epsilon_{321} = 1$ because reordering $(1, 2, 3)$ to $(3, 2, 1)$ requires two pair permutations.

The Levi–Civita symbol is frequently used in contexts where two or more of its indices can be equal. In that case, its value is defined to be zero, e.g. $\epsilon_{112} = 0$. This is the only value consistent with the condition that the symbol change sign under the inconsequential exchange of two identical indices. To summarize,

$$\epsilon_{i_1, i_2, \ldots, i_n} \equiv \epsilon^{i_1, i_2, \ldots, i_n} \equiv \begin{cases} \text{sgn}\left(P^{1,2,\ldots n}_{i_1, i_2, \ldots, i_n}\right), & \text{if } (i_1, \ldots, i_n) \text{ is a permutation of } (1, \ldots, n), \\ 0, & \text{otherwise.} \end{cases} \tag{L7}$$

L1.3 Fields

Numbers are mathematical objects that can be added, subtracted, multiplied and divided. Seen as composition rules, **addition**, $a + b$, and **multiplication**, $a \cdot b$, have several features in common (associativity, commutativity, neutral element exists, inverse elements exist). A set for which both addition *and* multiplication are defined as separate operations **field** is called a (number) **field**. Referring to lecture courses in mathematics for a more rigorous approach, we here introduce the concept of fields in a quick and informal manner: a field is a *"triple"*, $F \equiv (A, +, \cdot)$, comprising a set A and *two* composition rules, addition and multiplication. Addition and multiplication each define their own *abelian* group structure, the neutral elements being denoted by 0 and 1, respectively, i.e. $a + 0 = a$ and $a \cdot 1 = a$.

The inverse element of a under addition is denoted by $-a$, i.e. $a + (-a) = 0$, and the **subtraction** addition of the inverse is called **subtraction**. Likewise, the inverse element of a under **division** multiplication is called a^{-1}, and multiplication by the inverse is called **division** (alternatively denoted by $b \cdot a^{-1} \equiv b/a \equiv \frac{b}{a}$). The group structures defined by addition and multiplication are independent, except for two points: (i) the neutral element of addition, 0, does not have a multiplicative inverse. In other words, 0^{-1} does not exist and "division" by zero is not allowed. (ii) Multiplication is distributive over addition in the sense that $a \cdot (b + c) = a \cdot b + a \cdot c$.

It is possible to construct fields with a finite number of elements, the **Galois fields** ($\rightarrow$ L1.3.7). However, most fields of relevance to physics are infinite, and the most important ones – the rational, the real and the complex numbers – are introduced below.

Rational and real numbers

rational
numbers

real numbers

irrational
numbers

The integers, $\mathbb{Z}$, do *not* form a field because the operation of multiplication does not have an inverse in $\mathbb{Z}$. For example, the multiplicative inverse of $3 \in \mathbb{Z}$ does not exist in $\mathbb{Z}$. There is no integer number that can be multiplied with three to obtain unity.[9]

The most elementary example of an infinite field is the **rational numbers**, $\mathbb{Q} \equiv \{\frac{q}{p} | q, p \in \mathbb{Z}, p \neq 0\}$, i.e. the set of all ratios of integers. The rational numbers $\mathbb{Q} \subset \mathbb{R}$ are contained in a larger number field, the **real numbers**. Heuristically, the set of real numbers may be imagined as a continuous line extending from $-\infty$ to $+\infty$. Each rational number can be positioned on a continuous line of numbers, which explains the "embedding" of the rationals in the reals. However, the field of real numbers also contains **irrational numbers**, $r \notin \mathbb{Q}$. Irrational numbers may be approximated to arbitrary precision by rational numbers but are not rational themselves. For example, $\sqrt{2}$ has rational approximations as $\sqrt{2} \simeq 1.4142 = \frac{14142}{10000}$, etc., but $\sqrt{2}$ itself can not be written as a ratio of two fixed integers and therefore is not rational. In mathematics courses one learns how the reals can be defined as the union of the rational numbers with the set of all limits of rational numbers (for example, $\sqrt{2}$ can be viewed as the limit of an infinitely refined approximation in terms of rational fractions). However, we do not discuss this formalization here.

The understanding of the reals as a continuous "line" implies that there are more real numbers than rational numbers. The precise understanding of what is meant by "more" is far from trivial. Pioneering work by Georg Cantor on the comparison of infinite sets of different size (1874) showed where the conceptual problems lie and caused consternation among his contemporaries.

**Georg Ferdinand Ludwig Philipp Cantor
1845–1918**
German mathematician who did pioneering work in set and number theory. Cantor proved that there are "more" real numbers than integer numbers. His work raised questions of philosophical significance and eventually triggered a crisis in the understanding of the logical foundations of mathematics.

The intuitive picture is that the rationals form a subset of infinitely many points embedded into the continuous line of the reals. The "discreteness" of these points means one can come up with a systematic numbering scheme that lists them all; much like the integer numbers, the rationals form a *countable set* as defined on p. 5. Between any two rational numbers there are gaps corresponding to irrational numbers (see figure, where each dot represents a rational number). Although the set of rationals is "dense" in the reals, in the sense that any real number can be rationally approximated to any desired accuracy, there is no way to count the real numbers

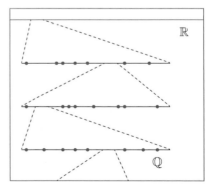

---

[9] A structure allowing addition, multiplication, subtraction but not division is called a **ring**.

lying between them; the reals are uncountable in the sense of the definition of p. 5. For a more substantial discussion of these aspects we refer to lecture courses in mathematics.

The complex numbers

The set of real numbers is large enough to accommodate operations which cannot be performed in the rationals, such as taking the square root of 2. In this sense, they represent a "closure" of the rational numbers. However, there are operations with respect to which the real numbers lack closure themselves. We all know that the square root of a negative number, such as $\sqrt{-1}$, is not a real number. Similarly, some real polynomials can be factored as $x^2 - 1 = (x - 1)(x + 1)$, where the factors specify the zeros of the polynomial. However, $x^2 + 1$ cannot be factorized into a product of two real factors. Somehow, it does not feel "right" that similar polynomials behave so strikingly differently.

INFO Complex numbers were introduced and used long before they were understood conceptually. Their first appearances can be traced to early studies of geometric objects in the ancient world. Complex numbers became an element of mainstream mathematics in the early seventeenth century when mathematicians worked on the solution theory of algebraic equations.

imaginary numbers The term **imaginary numbers** was coined by

René Descartes 1596–1650
French philosopher and mathematician. Descartes is considered the founding father of modern western philosophy and made important contributions to mathematics. The *Cartesian* coordinate system is named after him and he introduced analytical geometry, the bridge between algebra and geometry.

Descartes in 1637, who wrote "sometimes only imaginary, that is one can imagine as many as I said in each equation, but sometimes there exists no quantity that matches that which we imagine." The confused wording of this sentence suggests that Descartes was ill at ease with objects that were apparently quite useful, but hard to conceptualize within seventeenth-century mathematics. It took almost three hundred more years before the modern theory of number fields was invented and a sound conceptual framework for complex numbers came to existence. In the meantime, numerous mathematicians – Euler, Gauss, Abel, Jacobi, Cauchy, Riemann and various others more – contributed to the applied theory of complex numbers. Interested readers are encouraged to study these developments and sense the difficulties mathematicians had in working with a concept which was irresistibly interesting, yet hard to grasp.

The complex numbers are an extension of the real numbers large enough to accommodate all algebraic operations commonly associated with "numbers". In the following we sketch this extension in the contemporary language of mathematical fields. We start by giving $\sqrt{-1}$ a name,

$$\sqrt{-1} \equiv i, \tag{L8}$$

imaginary unit where i is called the **imaginary unit**. If we accept the existence of this object as a valid number, the problem of taking the square root of negative reals disappears. For positive r we define,

$$r > 0 : \qquad \sqrt{-r} \equiv \sqrt{-1}\sqrt{r} = i\sqrt{r}.$$

complex numbers By squaring (L8) we also know that $i^2 = -1$. Now let us define the **complex numbers** as the set

$$\mathbb{C} \equiv \{z = x + iy \,|\, x, y \in \mathbb{R}\}. \tag{L9}$$

We call $x \equiv \mathrm{Re}(z)$ the **real part** and $y \equiv \mathrm{Im}(z)$ the **imaginary part** of the complex number z. If one of these vanishes the notation is simplified by writing $0 + iy \equiv iy$ or $x + i0 \equiv x$. Thus, real numbers are complex numbers with vanishing imaginary part, implying the embedding $\mathbb{R} \subset \mathbb{C}$.

Next, define the *addition and multiplication* of complex numbers, as

$$z + z' = (x + iy) + (x' + iy') \equiv (x + x') + i(y + y'), \tag{L10}$$

$$zz' = (x + iy)(x' + iy') \equiv (xx' + ixy' + iyx' + i^2yy') = (xx' - yy') + i(xy' + yx').$$

These definitions are such that i behaves as an "ordinary" number, except for the identification $i^2 = -1$. Addition and multiplication are closed in $\mathbb{C}$, i.e. both the sum and product of two complex numbers again produce a complex number. This means that we are on the way towards constructing a number *field*.

Indeed, it is straightforward to show that $(\mathbb{C}, +)$ forms an additive group (do it!). A little more work is required to show that multiplication defines a group structure, too. We first need to know how to construct the inverse of a given complex number, $z = x + iy \in \mathbb{C}$. To **complex conjugate** this end, the **complex conjugate**, $\bar{z}$,[10] of z is defined as the complex number obtained by inverting the imaginary part of z,

$$\boxed{\bar{z} \equiv x - iy.}$$

Equation (L10) then yields

$$z\bar{z} = x^2 + y^2, \tag{L11}$$

meaning that $z\bar{z}$ is real. If z is nonzero, $x^2 + y^2 \neq 0$ and the result can be used to construct the inverse of z. We know that $z\bar{z}/(z\bar{z}) = 1$, which means that the inverse of z is given by

$$z^{-1} = \frac{\bar{z}}{z\bar{z}} \overset{\text{(L11)}}{=} \frac{x - iy}{x^2 + y^2} \in \mathbb{C}. \tag{L12}$$

This expression is "explicit" in the sense that for any $z = x + iy$ the inverse is obtained as a rational function of x and y.

When encountered for the first time, these definitions may feel alien. However, complex numbers are as easy to handle as real ones. Just keep the rule $i^2 = -1$ in mind and otherwise compute products as usual, for example, ($\rightarrow$ L1.3.1-4)

$$(2 + 3i)(1 + 2i) = -4 + 7i, \qquad i(4 + 6i) = -6 + 4i.$$

These relations illustrate that one may compute with complex numbers much as with real numbers. For example, a power z^n with integer n is defined as the n-fold product $z^n = zz \ldots z$.

[10] The complex conjugate is equivalently denoted by an asterisk, $z^* \equiv \bar{z} \equiv x - iy$.

INFO Adding and multiplying complex numbers is so straightforward that a word of caution is in order: more general functional operations, such as taking non-integer powers, can be much more subtle.

As an example, consider the paradoxical calculation $1 = \sqrt{1} = \sqrt{(-1)(-1)} \overset{?}{=} \sqrt{-1}\sqrt{-1} = ii = -1$. What goes wrong here (in the third step) is that an identity holding for positive real numbers, $\sqrt{xy} = \sqrt{x}\sqrt{y}$, has been naively, and incorrectly, generalized to the square root function for general complex numbers. The proper generalizations of some functional operations involving powers will be addressed in later parts of the text. For example, Section C5.2 introduces e^z, the exponential function with complex argument; and Chapter C9 concludes with a discussion of the complex square root function, $z^{1/2}$. For now, we just remark that the identity $x^{ab} = (x^a)^b$ holds for fractional exponents, $a, b \in \mathbb{Q}$, only if the base is real and positive, $x \in \mathbb{R}^+$.

complex plane   It is often useful to represent the complex numbers as points in a two-dimensional **complex plane**. The complex plane plays a role analogous to the one-dimensional line representing the reals. In it, a complex number $z = x+iy$ is represented by a point with coordinates (x, y), such that its real and imaginary parts define the abscissa and ordinate, x and y, respectively. The horizontal axis represents the real numbers, the vertical axis the purely imaginary ones, and generic complex numbers populate the plane. Note that z can also be written in the form

$$z = |z|(\cos\phi + i\sin\phi), \tag{L13}$$

polar representation   which defines the **polar representation** of complex numbers. Here ϕ is the angle between the real axis and a line connecting the points $(0, 0)$ and (x, y) in the complex plane, and $|z| \equiv \sqrt{z\bar{z}} = \sqrt{x^2 + y^2}$ is the length of this line. Angle and length are called the **argument**, $\arg(z) = \phi$, and **modulus**, $\mod z = |z|$, of z, respectively. The argument of z is only defined modulo 2π, i.e. an argument $-\pi$ is equivalent to $+\pi$, where $\arg(z) \in [0, 2\pi)^{11}$ is the conventional choice for its range of values. The complex conjugate, $\bar{z} = x - iy$, is represented by $(x, -y)$, the reflection of (x, y) at the x-axis. This implies $|\bar{z}| = |z|$ and $\arg(\bar{z}) = -\arg(z) \mod 2\pi$.

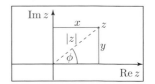

EXERCISE Show that the product of two complex numbers, $z_j = |z_j|(\cos\phi_j + i\sin\phi_j)$, with $j = 1, 2$, can be written as $z_1 z_2 = |z_1||z_2|(\cos(\phi_1 + \phi_2) + i\sin(\phi_1 + \phi_2))$. Illustrate this result with a sketch showing z_1, z_2 and $z_1 z_2$ in the complex plane. ($\rightarrow$ L1.3.5-6)

Complex numbers are *powerful tools in physics and mathematics*: algebraic operations which are not globally defined on the real numbers – such as square roots, logarithms, etc. – do exist as complex functions. Polynomials of degree n always have n complex zeros and can be decomposed into n factors, for example, $z^2 - 1 = (z + 1)(z - 1)$ or $z^2 + 1 = (z+i)(z-i)$. Throughout this text we will encounter numerous applications where these and other features of the complex numbers are of importance. Generally speaking,

---

11 Referring for a more detailed discussion to the info box below, we use notation in which $[a, b)$ is an interval with right boundary point excluded, $b \notin [a, b)$. However, the left boundary point is included, $a \in [a, b)$. The type of the brackets indicates which situation is realized. For example, $(0, 1)$ contains neither 0 nor 1.

one may say that complex numbers are more frequently used in the mathematics of physics than real numbers!

norm of real number   **INFO** To any two real or complex numbers one may assign a real value specifying the "distance" between them. In the case of real numbers, $x, y \in \mathbb{R}$, this is the **norm** $|x - y| \in \mathbb{R}$, i.e. the **absolute value** of their difference. For two complex numbers, $z, w \in \mathbb{C}$, the **(complex) norm** $|z - w|$ is the modulus of their difference. Norm functions are important in many ways. Specifically, they are used to distinguish between different types of real intervals and domains in the complex plane.

bounded set   For example, we call $U \subset \mathbb{F} = \mathbb{R}$ or $\mathbb{C}$ a **bounded set** of $\mathbb{F}$ if there exists a positive real number r such that $\forall u, v \in U, |u - v| < r$. The extent of U is then finite in the sense that no two elements have a distance exceeding r. A subset U is called **open** if any point $u \in U$ is enclosed within a region **open set**   of nonzero extent which itself is fully included in U. More formally, this means that there must exist a positive $\epsilon \in \mathbb{R}$ such that $\forall v \in \mathbb{F}$ with $|v - u| < \epsilon$, $v \in U$. Heuristically, one may imagine open sets as sets whose boundaries are excluded from the set itself, i.e. as sets with "soft" boundaries.

A real interval is an **open interval** if its boundaries are excluded; this is indicated by round brack- **closed**   ets, $(0, 1) \equiv \{u \in \mathbb{R} \mid 0 < u < 1\}$. If the endpoints are included one obtains a **closed interval**, **interval**   generally denoted by square brackets, $[0, 1] \equiv \{u \in \mathbb{R} \mid 0 \leq u \leq 1\}$. (Closed intervals are not open because their end points do not contain neighborhoods fully contained in them.) The exclusion of just one endpoint, as in $[0, 1) \equiv \{u \in \mathbb{R} \mid 0 \leq u < 1\}$, defines a **semi-open interval**.

closure   More generally, the **closure**, cl(U), of a set is defined as the union of the set and all of its limit points. A limit point of a set is a point for which every neighborhood contains at least one point belonging to the set. For example, the number 1 is not contained in $[0, 1)$ but it is still one of its limit points. This can be seen by inspection of the sequence $1 - \frac{1}{n}$. Every neighborhood of 1, no matter how small, contains elements of this sequence with sufficiently large n. This shows that cl($[0, 1)$) = $[0, 1]$. As an example of an open set in the *complex* numbers consider the open disk $D \equiv \{z \in \mathbb{C} \mid |z| < 1\} \subset \mathbb{C}$. The circle $|z| = 1$ is not included in D but it is contained in the closure cl(D) = $\{z \in \mathbb{C} \mid |z| \leq 1\}$.

More rigorous definitions of these terms are usually discussed in introductory mathematics courses. The more advanced discipline of mathematical *topology* addresses the extension of the concepts of openness, compactness, closure, etc. to sets more general than the number fields. While this is a very interesting subject, and not entirely irrelevant to physics, it is beyond the scope of the present text.

L1.4 Summary and outlook

In this chapter we have introduced various funda-mental structures of mathematics, notably sets, maps, groups, and eventually fields. We gave several exam-ples indicating that all these structures are tailored to specific tasks, both in mathematics and physics. For example, the set of lattice translations of a crystal – evidently a set of physical importance – realizes a group, and this classification is a powerful aid in the understanding of crystal structures.

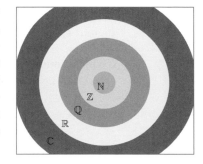

hierarchy of numbers   At the end of this chapter we have arrived at a *hier-archy of numbers*, $\mathbb{N} \subset \mathbb{Z} \subset \mathbb{Q} \subset \mathbb{R} \subset \mathbb{C}$. Each new member of the hierarchy realizes a new level of structure and admits operations which its predecessor cannot accommodate, $\mathbb{Q}$ closes under division whereas $\mathbb{Z}$ does not, etc. We have seen how mathematics provides the proper structures to describe the algebraic features of all number sets: beginning with $\mathbb{Z}$ the number sets form groups, and beginning with $\mathbb{Q}$ they are fields. That this understanding is much more than a formality is seen from the historical fact that for hundreds of years the complex numbers remained somewhat of a mystery. The situation changed only after the concept of the number field had been introduced.

Given the supreme potency of the complex numbers one may wonder if the "less powerful" numbers can be abandoned altogether. The answer is no, they remain universally useful. Generally speaking it is good practice to solve problems in terms of number sets just large enough to achieve what needs to be done. For example, we do not use real numbers to count the number of balls in a box, we use integers, etc.

We are now ready to advance to the next hierarchical level, vector spaces. Whereas numbers have a "norm" specifying their magnitude, vectors are objects of a given magnitude *and* direction. As we will see, this added feature makes them indispensable tools in the mathematical description of physics.

Vector spaces

Many objects of physical interest can be described in terms of a single number. Examples include the temperature of a body, its mass or volume, the energy required to move a body, or the number of gas molecules in a container. Quantities of this type are called **scalars**. However, scalars do not suffice to describe even many daily-life situations. For example, if a person who got lost asks for guidance the answer generally includes a direction and a distance. The two pieces of information can be combined into a "*vector*" whose length and geometric orientation encode distance and direction respectively. Vectorial quantities play an important role in physics, and in this chapter we introduce their mathematics from a perspective broad enough to include types of vectors that cannot be visualized in easy geometric ways. Such "non-visual" realizations are ubiquitous in physics – for example, they are key to the mathematical description of quantum mechanics and the theory of relativity – and the beauty of the overarching mathematical framework is that they can all be understood in a unified manner.

However, before turning to the general level, let us begin by introducing a concrete realization of a set of vectors. This example will anticipate the key mathematical structures of vectors and motivate the general definition of Section L2.2.

L2.1 The standard vector space $\mathbb{R}^n$

In this section, we will define $\mathbb{R}^n$ as an important class of vector spaces. (Sets of vectors that are complete in a sense to be defined a little further down are called vector "spaces".) The spaces $\mathbb{R}^n$ can be looked at from two different perspectives: first they are vector spaces in their own right, second they provide a "language" in which all other vector spaces can be described. This bridging functionality makes them important from both a fundamental and an applied perspective. However, before turning to a mathematical formulation of these statements, let us demonstrate the appearance of vectors and their relation to $\mathbb{R}^n$ on a daily-life example.

A motivating example

Consider the layout of a kitchen as shown in Fig. L4. How can the information contained in the plan be described quantitatively? The first step must be the definition of a unit of length, such as centimeters or inches. Second, a system of coordinates has to be specified. The latter is defined by two axes along which distances are to be measured. In a rectangular room, directions parallel to the walls would be a natural choice, e.g. the axes labeled 1 and

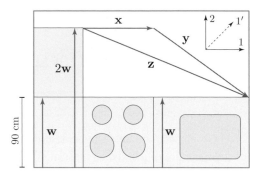

Fig. L4   The layout of a kitchen described in terms of various vectors.

2 in the figure. However, the choice is arbitrary and the axes labeled $1'$ and 2 would define an alternative and equally valid coordinate system as well.

Given a coordinate system, a *vector* describing the position of two points relative to each other is specified through two numbers fixing the separation between the points in the coordinate directions. These numbers define the components of the vector in the chosen system of coordinates. For example, the vector labeled $\mathbf{x}$ in the figure describes points shifted relative to each other by 90 cm in the 1-direction and 0 cm in the 2-direction. For brevity, we write $\mathbf{x} = \begin{pmatrix} 90 \\ 0 \end{pmatrix}$. Likewise, $\mathbf{w} = \begin{pmatrix} 0 \\ 90 \end{pmatrix}$, $\mathbf{y} = \begin{pmatrix} 120 \\ -90 \end{pmatrix}$, etc. A vector may be graphically represented by an arrow connecting its two defining points. The length of the arrow measures the distance between the points and its direction their relative orientation. Note that the *same* arrow is obtained for any two points that have the same relative distance and orientation, irrespective of their actual location. For example, the arrow denoted by $\mathbf{w}$ in the figure describes the separation of any two points shifted relative to each other by 90 cm in the 2-direction. We should think of a vector as an object that can be shifted (but not rotated or stretched) to any desired point of origin.

concatenation   Two vectors can be **concatenated** to define a new vector. For example, the vector $\mathbf{z}$ in the figure is obtained by concatenation of $\mathbf{x}$ and $\mathbf{y}$ and denoted by $\mathbf{z} = \mathbf{x} + \mathbf{y}$. The two components of $\mathbf{z}$ are given by the sum of the components of $\mathbf{x}$ and $\mathbf{y}$, respectively, $\mathbf{z} = \begin{pmatrix} 90 \\ 0 \end{pmatrix} + \begin{pmatrix} 120 \\ -90 \end{pmatrix} = \begin{pmatrix} 210 \\ -90 \end{pmatrix}$. (Exercise: draw the vector $\mathbf{w}+\mathbf{x}$ and compute its components.) Similarly, a vector can be multiplied by a real number $a \in \mathbb{R}$ to change its length. For example, $2\mathbf{w} = \begin{pmatrix} 0 \\ 180 \end{pmatrix}$ is a vector with doubled components and thus corresponds to a vector of doubled length, as indicated in the figure. If $a < 0$, the direction of the vector is inverted, for example $-\mathbf{w}$ has the same length as $\mathbf{w}$ but points downwards.

EXERCISE   Suppose we had decided to use the axes labeled $1'$ and 2 in Fig. L4 as coordinate axes. Assume that the angle between $1'$ and 2 is 45 deg. The component representations of the vectors $\mathbf{x}$, $\mathbf{y}$, $\mathbf{w}$, etc., change accordingly. Specifically, which of the following three representations of the vector $\mathbf{x}$ is correct?

$$\text{(a)} \quad \mathbf{x} = 90 \begin{pmatrix} \sqrt{2} \\ 0 \end{pmatrix}, \qquad \text{(b)} \quad \mathbf{x} = \begin{pmatrix} 45 \\ 0 \end{pmatrix}, \qquad \text{(c)} \quad \mathbf{x} = 90 \begin{pmatrix} \sqrt{2} \\ -1 \end{pmatrix}. \qquad \text{(L14)}$$

A systematic way to find the answer is to represent $\mathbf{x} = \mathbf{x}_{1'} + \mathbf{x}_2$ as the sum of two vectors where $\mathbf{x}_{1'}$ and $\mathbf{x}_2$ point in the direction of $1'$ and 2, respectively. We then need to find out how long $\mathbf{x}_{1'}$ and $\mathbf{x}_2$ need to be if $\mathbf{x}$ has length 90. Compute the $(1', 2)$ coordinate representation of the other vectors shown in the figure.

The space of all two-component[12] vectors $\mathbf{x} = \binom{x^1}{x^2}$ is called $\mathbb{R}^2$ (spoken "r-two"). Our discussion above shows that there are different ways to look at vectors and their representation through elements of $\mathbb{R}^2$. (i) We can think of vectors as objects geometrically defined as classes of arrows in the plane. Arrows are unique up to translation. (ii) Once a system of coordinates has been chosen, each of these arrows is uniquely described through a two-component element of $\mathbb{R}^2$. However, keep in mind that the description changes if different coordinates are chosen. (iii) Elements of $\mathbb{R}^2$ are vectors in their own right in that the geometrically defined vector operations concatenation and stretching correspond to equivalent operations in $\mathbb{R}^2$. The example suggests that $\mathbb{R}^2$ is a "reference" or standard vector space in terms of which vectors defined in different (geometric) ways can be described. However, the concrete numerical "language" in which $\mathbb{R}^2$ represents a geometric vector depends on a choice of coordinates. The situation is not so different from human languages, which describe identical objects in different ways. However, before formulating the connection between component vectors and generic vectors (which have not been defined yet) in generality, let us extend the definition of $\mathbb{R}^2$ to objects containing an arbitrary number of components.

Definition of $\mathbb{R}^n$

standard vector space The definition of $\mathbb{R}^2$ affords an obvious generalization to vectors with an arbitrary number of components: we define the so-called **standard vector space** $\mathbb{R}^n$ as the set of all multicomponent objects,

$$\mathbb{R}^n = \left\{ \mathbf{x} = \begin{pmatrix} x^1 \\ x^2 \\ \vdots \\ x^n \end{pmatrix} \middle| x^1, x^2, \ldots, x^n \in \mathbb{R} \right\}. \tag{L15}$$

component vector The elements $\mathbf{x}$ of $\mathbb{R}^n$ are n-**component vectors**. In the introductory parts of this text, vectors will generally be denoted by boldface symbols, $\mathbf{x}, \mathbf{y}$, etc.[13] The components of a vector $\mathbf{x}$ are referred to by x^i, although the alternative notation $(\mathbf{x})^i$ will be occasionally used as well. For example $(\mathbf{y})^1 = y^1 = 120$ in the kitchen example above. To save space

[12] The superscripts on x^1 and x^2 are indices (not powers of x!) distinguishing the first from the second component. The reason why we use superscripts rather than subscripts will be explained on p. 24.
[13] Since the boldface convention is inconvenient in handwriting, a variety of *alternative notations for vectors* exist: physicists and engineers often write $\vec{v}$. However, the repeated drawing of arrows costs time and more time-efficient alternative notations include $\vec{v}$, $\bar{v}$, or $\underline{v}$. Mathematicians often prefer a totally "naked" notation, v. This is OK as long as it is made clear that $v \in V$ is a vector and *not* a number. We will use this notation in later chapters of the text when vectors belonging to different spaces are handled at the same time. However, whichever notation is chosen, consistency and clearly stated definitions are imperative. The convention $\underline{v}$ may be a good compromise between efficiency and explicitness.

transpose   we often use the in-line notation $\mathbf{x} = (x^1, \ldots, x^n)^{\mathrm{T}}$, where "T" is spoken "**transpose**".[14] Finally, the number n is called the **dimension** of $\mathbb{R}^n$.

Much as a group is more than a simple set of elements (it is a set plus rules of composition), $\mathbb{R}^n$ is more than just a set of multicomponent objects: vectors can be added to each other and they can be multiplied by real numbers. As illustrated in our introductory discussion, the sum $\mathbf{z} = \mathbf{x} + \mathbf{y}$ of two vectors is the vector with components $z^i = x^i + y^i$. For example,

$$\begin{pmatrix} 1.5 \\ 2 \\ 0 \end{pmatrix} = \begin{pmatrix} 0.5 \\ -3 \\ 1 \end{pmatrix} + \begin{pmatrix} 1 \\ 5 \\ -1 \end{pmatrix}.$$

Likewise, the multiplication of a vector by a number is defined component-wise, i.e. the vector $a\mathbf{x}$ has components ax^i, for example

$$2 \begin{pmatrix} 1.5 \\ 2 \\ 0 \end{pmatrix} = \begin{pmatrix} 3 \\ 4 \\ 0 \end{pmatrix}.$$

Notice, however, that elements of $\mathbb{R}^n$ cannot be multiplied with each other,[15] nor divided by each other.

The vector space $\mathbb{R}^n$ is just one example of many other vector spaces encountered in physics and mathematics. In the next two sections we define vector spaces in general terms and introduce a number of important spaces to be discussed in more depth later in the text.

L2.2   General definition of vector spaces

Above, we introduced two different perspectives of vectors. The first was geometrical ("arrows"), the second algebraic in that it emphasized the operations that can be performed with vectors – addition and multiplication by numbers. In this section we upgrade the algebraic description to a definition of vector spaces in general. The algebraic approach is motivated by its generality and the fact that the vectors relevant to physics often do not have a visual geometric interpretation. The situation resembles that with groups which, likewise, were defined by the operations defined for them. That approach, too, was motivated by the observation that identical algebraic properties describe a multitude of very different realizations of groups.

Vector space definition

Vectors are objects that can be added to each other and multiplied by elements of a number field $\mathbb{F}$. (So far, we discussed the case $\mathbb{F} = \mathbb{R}$.) The corresponding formal definition reads as follows.

---

[14] At this stage $\mathbf{x} = (x^1, \ldots, x^n)^{\mathrm{T}}$ is just a space-saving alternative to the column notation (L15). The actual mathematical meaning of transposition is discussed later in Section L5.

[15] However, different types of vector "multiplications" will be introduced below.

vector space   An $\mathbb{F}$-**vector space** is a triple, $(V, +, \cdot)$, consisting of a set, V, a **vector addition** rule,

$$+ \; : \; V \times V \to V, \qquad (\mathbf{v}, \mathbf{w}) \mapsto \mathbf{v} + \mathbf{w}, \tag{L16}$$

and a rule for **multiplication by scalars**,

$$\cdot \; : \; \mathbb{F} \times V \to V, \qquad (a, \mathbf{v}) \mapsto a \cdot \mathbf{v} \equiv a\mathbf{v}, \tag{L17}$$

vector space
axioms
such that the following **vector space axioms** hold. (i) The addition of vectors, $(V, +)$, defines an abelian group. The neutral element of addition, $\mathbf{0}$, is called the null vector; the inverse element of a vector is called the negative vector, $-\mathbf{v}$. (ii) Scalar multiplication satisfies the following rules, $\forall a, b \in \mathbb{F}, \mathbf{v}, \mathbf{w} \in V$:

(a) $(a + b)\mathbf{v} = a\mathbf{v} + b\mathbf{v}$ (distributivity of scalar ...),

(b) $a(\mathbf{v} + \mathbf{w}) = a\mathbf{v} + a\mathbf{w}$ (... and vector addition),

(c) $(ab)\mathbf{v} = a(b\mathbf{v})$ (associativity of scalar multiplication),

(d) $1\mathbf{v} = \mathbf{v}$ (neutral element of $\mathbb{F}$ leaves vectors invariant).

$$\tag{L18}$$

Comments:

▷ The first part of the definition, (L16), formalizes the addition of vectors. In the case of $\mathbb{R}^n$ the null vector is given by $\mathbf{0} = (0, \ldots, 0)^\mathsf{T}$. We may think of it as an arrow shrunk to a point (and hence not pointing anywhere). Addition of this object to another vector does not do anything. The negative vector, $-\mathbf{v}$, can be imagined as a vector pointing in the direction opposite to $\mathbf{v}$, such that $\mathbf{v} + (-\mathbf{v}) = \mathbf{0}$. Equivalently, one may think of $-\mathbf{v}$ as $(-1)\mathbf{v}$ (multiplication of $\mathbf{v}$ by $-1 \in \mathbb{F}$). Axiom (a) above then states that $0\mathbf{v} = (1 - 1)\mathbf{v} = \mathbf{v} - \mathbf{v} = \mathbf{0}$.

▷ Relations (a) to (d) appear to be so obvious that they hardly seem worth mentioning. However, they are required to ensure that the "algebraic" properties of a vector space match the geometric understanding of directed objects ("arrows"). Without these specifications the definition would not be sharp enough to exclude "weird spaces" outside the useful category of vector spaces.

▷ The definition does not make reference to vector components, nor to the "dimension" of vectors. We conclude that these must be secondary characteristics deducible from the general definition.

linear
combination
▷ Given $a, b, c \in \mathbb{F}$ and $\mathbf{u}, \mathbf{v}, \mathbf{w} \in V$, the combination $a\mathbf{u} + b\mathbf{v}$ also lies in V. The same is true for $a\mathbf{u} + b\mathbf{v} + c\mathbf{w}$, etc. Expressions of this sort are called **linear combinations** of vectors.

▷ As mentioned previously, sets fulfilling the criteria above are generally called "spaces", a space of functions, the space of matrices, etc.

null space   ▷ The smallest of all spaces is the **null space**, $\{\mathbf{0}\}$. It contains only one vector, the null vector $\mathbf{0}$. The null space satisfies all vector space axioms. For example, $a\mathbf{0} = \mathbf{0} \in \{\mathbf{0}\}$, etc.

The definition above introduces the concept of a vector in its most general form, including realizations where geometric visualizations are not natural. In practice, the question whether a given set is a vector space is always answered by checking the defining criteria above. Geometric visualizations can, but need not, be involved. In some cases they can even be counterproductive.

Covariant notation

We conclude this section with some remarks on notation. Below, we will frequently consider sums $\mathbf{v}_1 a^1 + \mathbf{v}_2 a^2 + \cdots$ of vectors $\mathbf{v}_1, \mathbf{v}_2, \ldots$ with coefficients $a^1, a^2, \ldots$. Notice that we write the coefficients, a^i, with superscript^i indices while the vectors, $\mathbf{v}_i$, carry subscript$_i$ indices. Superscript and subscript indices are called **contravariant** and **covariant indices**, respectively. Notation adopting this index positioning convention is **covariant notation** called **covariant notation**[16] and it will be used throughout this text (for a specific description of covariant notation, see p. 32). At this stage, this may seem to be a purely technical convention. However, as we progress we will see that the distinction between co- and contravariant objects becomes more and more important both from a physical and a mathematical perspective (see info section below). Anticipating this development, we use covariant notation from the very beginning. However, it should not go unnoted that this approach is not standard and that most texts on linear algebra prefer an all-indices-downstairs notation.

INFO Vectors are members of a more general category of objects known as tensors (the topic of Chapter L10). For example, L5 *matrices*, which are perhaps familiar from high school and which will be introduced in Chapter L5, also belong to this family of objects.

It is common practice in physics to treat every object that carries a single index (such as $\mathbf{x} \leftrightarrow \{x^i\}$) as if it were a vector. However, many of the "vectors" routinely encountered in physics are actually **tensors** not vectors, but objects of different structure known as **tensors**. Important examples of such tensors-in-disguise include electric and magnetic fields, mechanical forces and currents, and more. In cases where the non-vectorial nature of such quantities becomes too apparent to ignore, they are assigned special names, such as "pseudo-vector" or "axial vector". However, physicists do not easily let go of the vector association as such. Depending on the research field one is working in, this practice can be either harmless or a potent source of confusion. The latter is the case in a growing number of disciplines including the theory of relativity, particle physics, topological condensed matter theory, quantum information theory, and others. It is probably fair to say that the only reason why the physics literature sticks to its all-is-vector culture is social inertia. The indiscriminate identification of single-component objects with vectors does not "simplify" anything. On the contrary, it obscures connections that become clear within a more differentiated approach. On the other hand, a fully reformed approach which, for example, would describe a magnetic field as an alternating tensor of second degree rather than as a vector, might be too radical. Students trained in this way would not be able to communicate with colleagues speaking a more traditional language, so this is not a viable solution.

In this text, we aim to strike a middle ground. Tensor calculus and the ensuing interpretation of physical objects are explained in the advanced Chapters L10 and V4 to V6 later in this text. However, from the beginning we will pay careful attention to the consistent positioning of co- and contravariant indices. This is done because *covariant notation is an important aid in discriminating between objects that are fundamentally vectors and others that are not*. Occasionally, we will run into trouble and realize that the use of covariant notation leads to inconsistent index positions. This is the way by which the notation signals that an object truly different from a vector has been encountered. Depending on the context, we will fix the situation right away or, on a few occasions, refer to a section of Chapter L10 where the origin of the problem is explained.

---

[16] In covariant notation, powers of scalars, x^2, and contravariant components, x^2, are denoted by the same notation. Which of the two is meant generally follows from the context. In cases with potential for confusion use brackets, e.g. $(x^1)^2 + (x^2)^2$ for the sum of squares of two covariant vector components.

L2.3 Vector spaces: examples

In the following we introduce a number of examples which all play an important role as vectors spaces in physics.

The standard vector spaces

We have already introduced $\mathbb{R}^n$ as the standard vector space defined over the real numbers, $\mathbb{F} = \mathbb{R}$. The alternative choice $\mathbb{F} = \mathbb{C}$ defines the **complex standard vector space** $\mathbb{C}^n$. This is the set of all vectors, $\mathbf{z} = (z^1, \ldots, z^n)^{\mathrm{T}}$, with components $z^i \in \mathbb{C}$. At first sight $\mathbb{C}^n$ may seem to be "more complicated" than $\mathbb{R}^n$ (inasmuch as complex numbers carry more structure than real numbers). However, we will see that the opposite is true and in many instances we will prefer to work with $\mathbb{C}^n$.

EXERCISE Convince yourself that the standard vector spaces $\mathbb{R}^n$, $\mathbb{C}^n$ fulfill the vector space axioms.

Affine and Euclidean spaces

Consider an infinite d-dimensional space, for example an infinite two-dimensional plane or the three-dimensional space we live in. The mathematical abstractions of these objects are called *affine spaces*, A. Elements $P \in A$ are called **points**.

points

Affine spaces are almost, but not quite, vector spaces. To understand the difference, notice that a vector space contains the neutral element of addition, $\mathbf{0}$, as its distinguished origin, or null vector. By contrast, affine spaces are the mathematical formalization of idealized infinite space and therefore do not contain a "special" point. To establish the connection between an affine space, A, and a vector space, V, of the same dimension one needs to pick an arbitrary reference point, $O \in A$, and identify it with the origin, $\mathbf{0} \in V$. For example, if the focus is on describing our solar system (which lies in three-dimensional affine space) it would be natural to choose the center of the Sun as a reference point. Each point $P \in A$ can then be identified with a vector, $\mathbf{v}$, representing the arrow from O to P, as illustrated in Fig. L5. If another point, $Q \in A$, is represented by the vector $\mathbf{w}$ then the linear combination $\mathbf{u} = \mathbf{w} - \mathbf{v}$ represents the arrow from P to Q.

Note that the points P and Q are independent of the choice of reference point, but the vectors representing them are not. If a different reference point, O', is chosen then P and Q

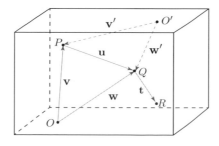

Fig. L5   On the definition of affine space.

are described by different vectors, $\mathbf{v}'$ and $\mathbf{w}'$, respectively. However, the vector representing the arrow from P to Q remains the same, $\mathbf{w}' - \mathbf{v}' = \mathbf{w} - \mathbf{v}$. As an example, take $P =$ (center of Earth) and $Q =$ (center of Venus). The vectors representing P and Q depend on whether the centers of the Sun or of Jupiter are chosen as reference points. However, the vector connecting the center of Earth to that of Venus is independent of the choice of reference point.

affine space

The preceding description of affine spaces is made precise as follows: consider a set of points, $A = \{P, Q, \ldots\}$ subject to the following three conditions. (i) There exists a vector space V such that to any ordered pair of points, $(P, Q) \in A \times A$, a vector $\mathbf{u} \in V$ may be uniquely assigned. We call $\mathbf{u}$ the *difference vector* from P to Q. (ii) For any point $P \in A$ and any vector $\mathbf{u} \in V$ there exists a unique point $Q \in A$ such that $\mathbf{u}$ is the difference vector from P to Q. (iii) For any three points P, Q and $R \in A$, with difference vectors $\mathbf{u}$ from P to Q and $\mathbf{t}$ from Q to R, respectively, the difference vector from P to R is given by $\mathbf{u} + \mathbf{t}$. If these conditions are met, A is called an **affine space**. Once a point $O \in A$ has been chosen as reference point, A becomes identifiable with V and there is a bijection between points $P \in A$ and the difference vectors $\mathbf{v} \in V$ connecting O and P. This identification is sometimes written as $V = (A, O)$, and the correspondence between P and $\mathbf{v}$ as $P = O + \mathbf{v}$. For example, in this language criterion (iii) above assumes the form $P + \mathbf{u} + \mathbf{t} = Q + \mathbf{t} = R$.

Euclidean space

In the particular case where $V = \mathbb{R}^d$ we call $A \equiv \mathbb{E}^d$ d-dimensional **Euclidean space**. This denotation hints at the fact that Euclidean space possesses structures that a generic affine space need not have: to vectors of $\mathbb{E}^d$ lengths and angles and other elements of Euclidean (!) geometry, to be introduced in Chapter L3, may be assigned. Both generic affine spaces and Euclidean spaces play important roles in physics. Their description in terms of vector spaces is so natural that the distinction between affine and vector spaces is easily forgotten. However, occasionally it has to be remembered to avoid confusion!

Function spaces

Let $f : I \to \mathbb{R}$, $t \mapsto f(t)$ be a real function defined over a finite interval, I (see Fig. L6). The set containing all these functions is called $L^2(I)$.[17] Two functions, f, g may be added to each other to obtain a new function, $f + g$, in the same set. That new function is defined as the superposition of f and g, $(f + g)(t) \equiv f(t) + g(t)$. Likewise, the product of a function with a number, $a \in \mathbb{R}$, defines another function, af, via $(af)(t) \equiv af(t)$. This shows that $L^2(I)$ is a vector space, and that one may think of the functions contained in it as vectors. Give yourself some time to let this message sink in! ($\to$ L2.3.3)

To make the vectorial interpretation of functions more concrete, consider storing the function $f(t)$ on a computer, with $I = [0, \tau]$. This may be done by discretizing the interval into a large number, N, of small intervals of width τ/N, each centered on a point t_i, $i = 1, \ldots, N$ (see Fig. L6). One then samples N representative readouts of the function, $f^i = f(t_i)$. These values define an N-dimensional vector, $\mathbf{f} \equiv (f^1, \ldots, f^N)$, which may be saved as a discrete approximation of the function. The number of components, N,

[17] The notation $L^2(I)$ for the set of functions defined on I is standard in mathematics. Its definition requires one additional condition, namely "square integrability" to be discussed in Chapter L9. However, for the moment, this additional condition is not of relevance.

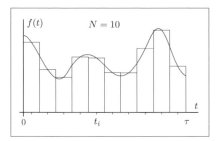
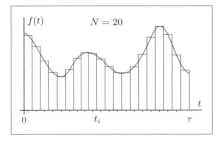

Fig. L6   The discretization of a function in terms of N discrete values yields an N-component vector. The larger N is, the more closely the discretized function approximates the continuous one.

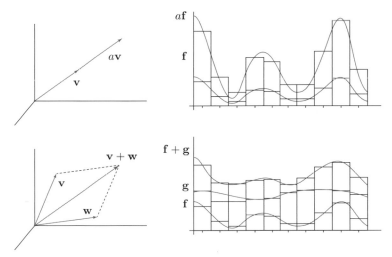

Fig. L7   Visualization of vector multiplication and addition, respectively, for two-dimensional vectors in $\mathbb{E}^2$ (left), and discretized functions (right).

may be increased to make the approximation of the "continuous" function f as accurate as desired.[18]

Given two functions, f, g, with discrete representations, $\mathbf{f}, \mathbf{g}$, a discretized function, $\mathbf{f}+\mathbf{g}$, can be defined by addition of the individual function values: $(\mathbf{f} + \mathbf{g})^i = (\mathbf{f})^i + (\mathbf{g})^i$ (see Fig. L7, bottom right). Similarly, the discrete function $a\mathbf{f}$, $a \in \mathbb{R}$, may be defined by component-wise multiplication, $(a\mathbf{f})^i = a(\mathbf{f})^i$ (Fig. L7, top right). This construction shows that one may work with discretized functions just as with N-component vectors and that the set of N-step discretized functions is identical to $\mathbb{R}^N$.

Heuristically, $L^2(I)$ may be interpreted as the $N \to \infty$ limit of the discretization spaces $\mathbb{R}^N$, and functions are "infinitely-high"-dimensional vectors with components $f(t) \leftrightarrow f^i = f(t_i)$. This view of functions is very important. It makes the connection between calculus and linear algebra tangible and plays an important role in various fields of physics.

[18] For example, if $f(t)$ represents an audio signal as a function of time, sampling rates with $N = 44,100$ for a time interval of 1 second correspond to the resolution of standard CD recordings.

EXAMPLE In the following we introduce a few more examples which may help in building familiarity with the concept of vector spaces:

▷ *Number fields as vector spaces*: for $n = 1$, the vector space $\mathbb{R}^n$ reduces to the real numbers, $\mathbb{R}^1 = \mathbb{R}$. (The set of vectors with just one real component is trivially equivalent to the real numbers.) Likewise, $\mathbb{Q}^1$ is a $\mathbb{Q}$-vector space ($\to$ L2.3.1), and $\mathbb{C}^1$ is a $\mathbb{C}$-vector space. However, one may also think of $\mathbb{C}^1 \cong \mathbb{C}$ as an $\mathbb{R}$-vector space ($\to$ L2.3.2): any complex number, z, can be multiplied by a real number, a, to yield another complex number az, and complex numbers can be added to each other. One may decompose $z = x + iy$ into real and imaginary parts to uniquely describe it by a *pair* of real numbers, $z \leftrightarrow (x, y)^{\mathrm{T}}$. This shows that $\mathbb{C} \cong \mathbb{R}^2$ can be identified with $\mathbb{R}^2$. The identification of $\mathbb{C}$ with the real vector space $\mathbb{R}^2$ is often useful in applications. For example, physical problems defined in two-dimensional space are sometimes described in a "complex notation" in which each point is represented by a complex number. This is done because complex numbers are often more convenient to work with than two-component vectors.

polynomials ▷ Let $P_2 \equiv \{a_2 x^2 + a_1 x + a_0 | a_{0,1,2} \in \mathbb{R}\}$ denote the set of all **polynomials** in the variable x of degree 2.[19] For two polynomials, $p(x) \equiv a_2 x^2 + a_1 x + a_0$ and $q(x) \equiv b_2 x^2 + b_1 x + b_0$, the sum, $(p+q)(x) = (a_2 + b_2)x^2 + (a_1 + b_1)x + (a_0 + b_0)$, is again a polynomial of degree two, and so is the product with a real number, $ap(x) = (aa_2)x^2 + (aa_1)x + (aa_0)$. This shows that P_2 is a vector space. Since these polynomials are uniquely identified by three coefficients, $p \leftrightarrow (a_2, a_1, a_0)$, we have a bijection $P_2 \cong \mathbb{R}^3$. Exercise: think of generalizations to polynomials of arbitrary degree ($\to$ L2.3.4), or to polynomials in more than one variable. For example, the polynomials of degree 2 in two variables, x and y, have the form $a_{22}x^2y^2 + a_{21}x^2y + a_{12}xy^2 + a_{20}x^2 + a_{02}y^2 + a_{11}xy + a_{10}x + a_{01}y + a_{00}$, with real coefficients a_{ij}. How many components are required to uniquely describe the polynomials in two variables of maximal degree five?

▷ For some exotic examples of vector spaces with more contrived addition and multiplication rules, see problems L2.3.5-6.

L2.4 Basis and dimension

A common property of all vectors discussed above was that they could be represented through a list of components, $\mathbf{v} \leftrightarrow (v^1, \ldots, v^n)^{\mathrm{T}}$. However, we also observed that the component representation was not unique. For example, the vector $\mathbf{x}$ in Fig. L4 has a representation $\mathbf{x} = (90, 0)^{\mathrm{T}}$ if cm are used as a unit of length and the coordinate axes are oriented as indicated. It would change to $\mathbf{x} \simeq (35.4, 0)^{\mathrm{T}}$ if inches were used, or to $\mathbf{x} \simeq (63.3, 63.3)^{\mathrm{T}}$ if the coordinate axes were rotated by 45 deg. However, irrespective of the chosen representation, *two* numbers are needed to describe it. For any vector space, the number of components required to specify a vector is a characteristic property of that space,[20] called its *dimension*. Notice that the dimension is not mentioned in the fundamental definition of vector spaces given in Section L2.2. This shows that it must be an attribute following from the vector space axioms. In the following we discuss how this happens.

Given a vector space V and a set S containing m of its vectors,

$$S \equiv \{\mathbf{v}_1, \ldots, \mathbf{v}_m\}, \quad \mathbf{v}_i \in V, \tag{L19}$$

linear span   the **linear span** (or **linear hull**) of S is defined as the set of all linear combinations of the elements of S:

---

[19] The fact that we are discussing polynomials here implies that x^2 means x squared, cf. footnote 16.

[20] Readers to whom this sounds trivial may try to prove this statement. It is not as easy as one might think.

$$\text{span}(S) \equiv \{\mathbf{v}_1 a^1 + \mathbf{v}_2 a^2 + \cdots + \mathbf{v}_m a^m \,|\, a^1, \ldots, a^m \in \mathbb{F}\}. \tag{L20}$$

For $\mathbf{u}, \mathbf{w} \in \text{span}(S)$, the linear combination $a\mathbf{u} + b\mathbf{w}$, $a, b \in \mathbb{F}$ again lies in span(S), and this shows that span(S) is a vector space in itself.

subspace   We call a vector space, $W \subset V$, embedded in V a **subspace** of V. This includes the extremes $W = \{\mathbf{0}\}$ of the zero-dimensional null space containing just the null vector, and the full space, $W = V$. At any rate, the linear hull, span(S), is a subspace of V, non-empty if S contains at least one non-vanishing vector. An interesting question to be addressed in the next sections is how large a set of vectors has to be to include the full vector space in its span, span(S) = V.

Linear independence

Suppose $S = \{\mathbf{v}_1, \ldots, \mathbf{v}_n\}$ has the property that one of its elements, say $\mathbf{v}_1$, can be represented as a linear combination of the others,

$$\mathbf{v}_1 = \mathbf{v}_2 b^2 + \cdots + \mathbf{v}_m b^m. \tag{L21}$$

linearly dependent vectors   In this case, the vectors defining S are called **linearly dependent vectors**, and S is called a linearly dependent set of vectors. Such sets contain redundancy in that some of their elements may be removed without diminishing the span. For example, with $S' \equiv S \backslash \{\mathbf{v}_1\}$ we have (why?)

$$\text{span}(S) = \text{span}(S'). \tag{L22}$$

The upper row of Fig. L8 shows linearly dependent sets containing three and four vectors of $\mathbb{E}^3$, respectively. Conversely, if none of the elements of S can be obtained by linear combination of the others, $\mathbf{v}_1$ to $\mathbf{v}_m$ are called **linearly independent** vectors.

For later reference we note that there is an alternative way to test linear independence: the vectors in S are linearly dependent if they can be linearly combined to form a non-trivial representation of the zero vector, $\mathbf{0}$, i.e. if there exist non-vanishing coefficients,

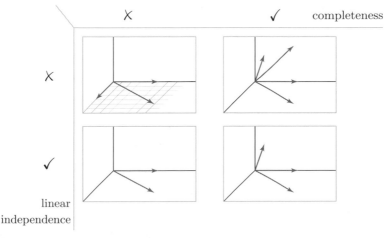

Fig. L8   The concepts of linear independence and completeness, illustrated for the example $\mathbb{E}^3$.

$$\{a^1,\ldots,a^m\} \neq \{0,\ldots,0\}, \qquad \mathbf{v}_1 a^1 + \mathbf{v}_2 a^2 + \cdots + \mathbf{v}_m a^m = \mathbf{0} . \tag{L23}$$

Why are the two conditions equivalent? If (L21) holds, then we have a representation $\mathbf{0} = -\mathbf{v}_1 + \mathbf{v}_2 b^2 + \cdots + \mathbf{v}_m b^m$, as in (L23). Conversely, if (L23) holds we may pick a non-vanishing coefficient, say a^1, to obtain the linear combination of $\mathbf{v}_1 = -\frac{1}{a^1}(\mathbf{v}_2 a^2 + \cdots + \mathbf{v}_n a^n)$ through the other vectors. This demonstrates linear dependence in the sense of Eq. (L21).

EXAMPLE Consider the vectors

$$\mathbf{v}_1 = \begin{pmatrix} 1 \\ 0 \end{pmatrix}, \qquad \mathbf{v}_2 = \begin{pmatrix} 1 \\ 2 \end{pmatrix}, \qquad \mathbf{v}_3 = \begin{pmatrix} -1 \\ -1 \end{pmatrix} . \tag{L24}$$

The set $S \equiv \{\mathbf{v}_1, \mathbf{v}_2, \mathbf{v}_3\}$ is linearly dependent because $\mathbf{v}_1 = -(\mathbf{v}_2 + 2\mathbf{v}_3)$. However, the set $S' = \{\mathbf{v}_1, \mathbf{v}_2\}$ is linearly independent, because $\mathbf{v}_1 a^1 + \mathbf{v}_2 a^2 = (a^1 + a^2, 2a^2)^{\mathrm{T}}$, which equals $\mathbf{0}$ only if $a^1 = a^2 = 0$. Similarly, $S'' = \{\mathbf{v}_1, \mathbf{v}_3\}$ is linearly independent, as is $S''' = \{\mathbf{v}_2, \mathbf{v}_3\}$. ($\rightarrow$ L2.4.1-2)

It is often useful to eliminate redundancy by working with linearly independent sets. This is done by removing redundant vectors of S until one arrives at a linearly independent set of reduced cardinality. That reduced set need not be unique, as shown by the example of S', S'' and S''' above. However, its span does not depend on which of the linearly dependent vectors are removed and S', S'' and S''' all span the same vector space.

Completeness

There is another important feature which the sets like (L19) may or may not have: a set
completeness $S = \{\mathbf{v}_1, \ldots, \mathbf{v}_n\}$ containing n vectors is called **complete** if

$$\mathrm{span}(S) = V. \tag{L25}$$

In this case, every vector $\mathbf{v} \in V$ can be written as a linear combination of the vectors $\mathbf{v}_i$. For example, the set $S = \{\mathbf{v}_1, \mathbf{v}_2, \mathbf{v}_3\}$ of Eq. (L24) is complete in $\mathbb{R}^2$ because any vector $\mathbf{v} = (a, b)^{\mathrm{T}}$ can be represented as $\mathbf{v} = \mathbf{v}_1 a + \mathbf{v}_2 b + \mathbf{v}_3 b$. The reduced sets S', S'' and S''', too, are all complete in $\mathbb{R}^2$. Examples of sets complete in the Euclidean space $\mathbb{E}^3$ are shown in the second column of Fig. L8.

Basis

basis   A set S that is both complete *and* linearly independent is called a **basis** of the vector space V. These properties guarantee (i) that each element $\mathbf{v} \in V$ can be expressed as a linear combination of basis elements (completeness),

$$\mathbf{v} = \mathbf{v}_1 a^1 + \mathbf{v}_2 a^2 + \cdots + \mathbf{v}_n a^n, \tag{L26}$$

and (ii) that this linear combination is unique (linear independence). To understand how uniqueness follows from linear independence, suppose that $\mathbf{v}$ could also be represented in a different way, say $\mathbf{v} = \mathbf{v}_1 b^1 + \mathbf{v}_2 b^2 + \cdots + \mathbf{v}_n b^n$. Subtracting the second representation from the first, we obtain $\mathbf{0} = \mathbf{v} - \mathbf{v} = \mathbf{v}_1(a^1 - b^1) + \mathbf{v}_2(a^2 - b^2) + \cdots + \mathbf{v}_n(a^n - b^n)$. This is a representation of the null vector and so the assumed linear independence of the basis vectors requires that all coefficients must be zero, $b^j = a^j$. This in turn means that

expansion

the two representations of **v** have to be identical. We call the representation of a vector in a given basis its **expansion**, and the corresponding coefficients the *expansion coefficients* or *components* with respect to that basis.

Each vector space has a basis. For given realizations of vector spaces this statement is usually straightforward to verify by the constructive specification of a basis. However, the general proof is not straightforward and will not be given here.[21] We also note that for a given space there is infinite freedom in the choice of a basis. The expansion coefficients of vectors depend on this choice and therefore change under changes of basis. If vectors are regarded as invariant "objects" one may think of their components as descriptions in a "language" tied to a basis. Tools for the computation of vector expansion coefficients and their transformation under changes of bases will be introduced in Section L5.6.

EXERCISE Show that the pair $\{\mathbf{v}_1, \mathbf{v}_2\}$ of vectors defined in Eq. (L24) defines a basis of $\mathbb{R}^2$. Draw the two basis vectors and an arbitrary other vector **v** of your choice. Compute its two expansion coefficients algebraically, and represent **v** graphically as a linear combination of $\mathbf{v}_1$ and $\mathbf{v}_2$. Repeat the exercise, to show that $\{\mathbf{v}_2, \mathbf{v}_3\}$, and $\{\mathbf{v}_1, \mathbf{v}_3\}$ are bases, too. Explain why the pair $\{\mathbf{e}_1, \mathbf{e}_2\}$ defined by $\mathbf{e}_1 = \mathbf{v}_1 = (1,0)^{\mathrm{T}}$ and $\mathbf{e}_2 = (0,1)^{\mathrm{T}}$ is a basis more convenient to work with than the others.

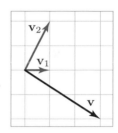

dimension

If the number n of elements of a basis is finite, it is unique. Any other basis then has the same number of elements, and n is called the **dimension** of the space. As an exercise, assume the existence of two bases of different cardinality n, m and show that the assumptions of linear independence and completeness lead to a contradiction.

INFO Many problems in *physics* are described in vector spaces whose dimensionality is different from the three dimensions of ambient space. For example, crystalline structures are often effectively two-dimensional. Einstein's theory of relativity adds "time" to three-dimensional space, and is thus formulated in four-dimensional space-time. Functions describing physical phenomena can be discretized as N-dimensional vectors (with $N \gg 1$). Quantum mechanics is formulated in vector spaces whose dimension is determined by the number of particles under consideration. These and many more examples motivate the study of vector spaces of arbitrary dimensionality.

EXERCISE The microscopic structure of *graphite* is defined by stacked two-dimensional sheets of carbon.[22] The carbon atoms of each sheet form a regular hexagonal lattice, as shown in the figure. Choosing the position of an arbitrary carbon atom as a point of origin, the position of any other atom in the plane is described by a two-dimensional vector. Convince yourself that each of these vectors may be represented by a linear

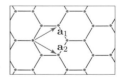

infinite-dimensional vector space

[21] Whereas the proof is relatively elementary for spaces with bases of finite cardinality, in the opposite case of **infinite-dimensional vector spaces** the situation is more involved. The function spaces $L^2(I)$ introduced above are examples of this type. Fortunately, the majority of vector spaces relevant to physics are finite dimensional, or can be made finite without significant loss of physical information. For example, we have discussed above how $L^2(I)$ can be approximated to any desired precision by an N-dimensional vector space $\mathbb{R}^N$.

[22] As of 2005 it has become possible to isolate individual atomic layers of graphite. The ensuing two-dimensional crystalline material is known as *graphene*. For its discovery, A. Geim and K. Novoselov were awarded the 2010 Nobel prize in physics.

combination of two suitably chosen "basis" vectors, e.g. the vectors denoted by $\mathbf{a}_1$ and $\mathbf{a}_2$ in the figure. Work out the linear combinations representing the positions of a few atoms of your choice. Try to derive a general formula specifying the position of all atoms as linear combinations of the basis vectors.

Einstein summation convention

Linear combinations such as $\mathbf{v}_1 a^1 + \mathbf{v}_2 a^2 + \cdots + \mathbf{v}_n a^n$ appear very frequently in the following and it is worthwhile to discuss a few notation conventions. First, note that the summation involves a contravariant (superscript) and a covariant (subscript) index.

▷ An index summation always runs over a pair of co- and contravariant indices, e.g. $\mathbf{w} = \mathbf{v}_1 a^1 + \mathbf{v}_2 a^2 + \mathbf{v}_3 a^3$.

▷ An unsummed ("free") index always appears at the same position on both sides of an equation, e.g. $\mathbf{w}_i = \mathbf{v}_1 A^1{}_i + \mathbf{v}_2 A^2{}_i + \mathbf{v}_3 A^3{}_i$.

These structures are hallmarks of the *covariant notation* and they provide useful consistency checks. For example, equations of the form $w^i(\ldots) = v^i(\ldots)$ have consistent index positions but $w(\ldots) = v_i(\ldots)$ and $w^i(\ldots) = v_i(\ldots)$ do not. Inconsistencies of this type are generally due to some computational mistake, or they indicate some deeper form of trouble (for an example of the latter type, see the discussion of Section L3).

Expressions of the architecture $A_1 B^1 + A_2 B^2 + \cdots + A_n B^n$ appear so frequently that various abbreviating notations have been introduced:

$$A_1 B^1 + A_2 B^2 + \cdots + A_n B^n \equiv \sum_{i=1}^{n} A_i B^i \equiv \sum_i A_i B^i \equiv A_i B^i. \qquad \text{(L27)}$$

In the third representation the upper and lower limits of the sum are implicit. In the last one we have introduced the **Einstein summation convention**, according to which indices occurring pairwise on one side of an equation are to be summed over. Such index pairs are called "pairwise repeated indices" or "dummy indices" and their summation is called a **contraction** of indices. The Einstein summation convention assumes that the summation range is specified by the context. For example, the Einstein representation of the argument formulated after Eq. (L26) reads: if $\mathbf{v} = \mathbf{v}_i a^i = \mathbf{v}_i b^i$ then $\mathbf{0} = \mathbf{v} - \mathbf{v} = \mathbf{v}_i (a^i - b^i)$, implying $a^i = b^i$. Since dummy indices are summed over, they can be relabeled at will, e.g. $A_i B^i = A_j B^j$.

Einstein summation convention

contraction of indices

In much of the rest of the text, the Einstein convention is applied. However, to ease the transition we use it in parallel with more expansive representations for a while. ($\to$ L2.4.3-4)

Vector space bases: examples

The concept of a basis is very important to the description of vector spaces. So much so that the choice of a suitable basis usually comes first in the work with a new vector space. Some spaces have a "canonical" basis[23] and some do not. In the following we revisit the examples of Section L2.3 to illustrate this point.

canonical

[23] The attribute "**canonical**" stands for "natural" or "standard" but does not have a mathematically precise definition.

▷ The natural basis, $\{e_j | j = 1, \ldots, n\}$, of the *standard vector spaces* $\mathbb{R}^n$ and $\mathbb{C}^n$ contains the basis vectors

$$e_j = (0, \ldots, 1, \ldots, 0)^{\mathrm{T}}, \tag{L28}$$

where the 1 stands at position number j. (Verify the linear independence and completeness of this set.) The components of the standard basis vectors may be alternatively specified as $(e_j)^i = \delta^i_{\ j}$, where the **Kronecker delta** $\delta^i_{\ j}$ is defined by[24]

$$\delta^i_{\ j} = \begin{cases} 1 & \text{for } i = j, \\ 0 & \text{for } i \neq j. \end{cases} \tag{L29}$$

In the basis (L28), the expansion of a general vector $\mathbf{x} = (x^1, \ldots, x^n)^{\mathrm{T}}$ assumes the form

$$\mathbf{x} = e_1 x^1 + e_2 x^2 + \cdots + e_n x^n = e_j x^j.$$

This shows that the components of an $\mathbb{R}^n$-vector and its expansion coefficients in the standard basis *coincide*. This is the defining feature of that basis.

▷ By contrast, the *Euclidean spaces*,[25] $\mathbb{E}^d$, do not favor particular directions over others and therefore do not possess a "canonical" basis. However, in many cases the identity of a suitable basis is determined by the context. For example, the kitchen layout of Fig. L4 favors a basis of vectors v_i, $i = 1, 2$ parallel to the walls of the room. The representation of generic vectors as $\mathbf{x} = 90 v_1$, or $2\mathbf{w} = 180 v_2$, then defines component representations as $\mathbf{x} = (90, 0)^{\mathrm{T}}$, $2\mathbf{w} = (0, 180)^{\mathrm{T}}$.

▷ The choice of a basis is particularly important in the case of *function spaces* such as $L^2(I)$. For definiteness, consider the N-dimensional space obtained by discretizing functions $f : [0, 1] \to \mathbb{R}$, $t \mapsto f(t)$ into N-component vectors, $\mathbf{f} = (f^1, \ldots, f^N)^{\mathrm{T}} \in \mathbb{R}^N$, where $f^i = f(t_i)$. In this case it is preferable to work with basis vectors $\delta_j = N e_j$, i.e. the standard basis vectors $\{e_j\}$ of $\mathbb{R}^N$ scaled by a factor of N. The basis vector δ_j may be viewed as a "discrete function", $\delta_j : \{1, \ldots, N\} \to \mathbb{R}$, $i \mapsto (\delta_j)^i$, vanishing for all values of i, except for $i = j$, where it equals N. We can think of δ_j as a discretized version of a box-shaped function $\delta_j : [0, 1] \to \mathbb{R}$, $t \mapsto \delta_j(t)$, which equals zero everywhere except in an interval of width $1/N$ centered on time t_j, in which it takes the constant value N. A general function vector can now be expanded as

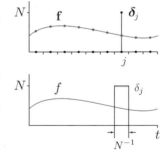

$$\mathbf{f} = \frac{1}{N}(\delta_1 f^1 + \delta_2 f^2 + \cdots + \delta_N f^N) = \frac{1}{N} \delta_j f^j. \tag{L30}$$

This shows how a discretized function can be expanded in terms of a "standard basis". However, it is less obvious what becomes of this strategy in the limit $N \to \infty$. We will address this question later in the text, see Section C6.1 and Chapter L9.

---

[24] For later reference we note that the Kronecker delta appears in different contexts, and with differently positioned indices, $\delta^i_{\ j}, \delta_i^{\ j}, \delta_{ij}, \delta^{ij}$. All of these are defined in the same way, as unity if $i = j$, and zero otherwise.

[25] Unless stated otherwise, we assume that a point of origin has been chosen so that $\mathbb{E}^d$ can be identified with a vector space.

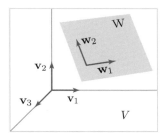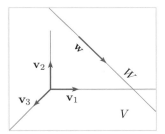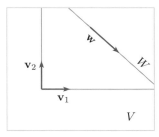

Fig. L9   Left: a two-dimensional subspace (plane) in three-dimensional space. Center: a one-dimensional subspace (line) in three-dimensional space. Right: a one-dimensional subspace in two-dimensional space.

EXERCISE Returning to the examples discussed on p. 28, consider $\mathbb{R}$ as a vector space, and show that the number $\{1\}$ (or any other set containing just one non-vanishing number) is a basis. Show that $\{1, i\}$ defines a basis of $\mathbb{C}$ if it is interpreted as a two-dimensional $\mathbb{R}$-vector space. Why is $\{1, 2\}$ not a basis? Show that the set $\{1, x, x^2\}$, containing three polynomials, forms a basis of the space of polynomials P_2. What would be a basis of the space of polynomials in two variables x and y up to degree 2?

Subspaces

subspace   If the span W of a set of vectors in V is not complete, then $W \subsetneq V$ and W is called a true **subspace** of V. For example, if $\mathbf{w}_1$ and $\mathbf{w}_2$ are linearly independent vectors in $\mathbb{R}^3$, then $W = \mathrm{span}(\{\mathbf{w}_1, \mathbf{w}_2\})$ is a two-dimensional subspace of three-dimensional space.

Subspaces of dimension one and two are called **lines** and **planes**, respectively. Examples include planes in three-dimensional space ($m = 2, n = 3$), lines in three-dimensional space ($m = 1, n = 3$), or lines in two-dimensional space ($m = 1, n = 2$), as illustrated in Fig. L9. Subspaces of higher dimension can no longer be visualized. For example, the space of polynomials of degree 2 is a three-dimensional subspace of the infinite-dimensional space $L^2(I)$.

As with vectors, subspaces are defined only up to parallel translation. For example, a parallel translation of the plane shown in Fig. L9, left, would still represent the same plane.

L2.5 Vector space isomorphism

REMARK In this section, connections between vectors of a general n-dimensional real space, V, and component vectors in $\mathbb{R}^n$ will be addressed. To distinguish the former from the latter, a caret notation, $\hat{\mathbf{v}} \in V$, for general vectors is used. The component vector representing $\hat{\mathbf{v}}$ in $\mathbb{R}^n$ is denoted by the same symbol without caret, $\mathbf{v}$. Nothing in this section relies on the choice $\mathbb{F} = \mathbb{R}$; to adapt it to complex vector spaces, simply replace $\mathbb{R}^n$ by $\mathbb{C}^n$ everywhere.

Once a basis $\{\hat{\mathbf{v}}_j\}$ has been chosen for an n-dimensional vector space V, every vector $\hat{\mathbf{v}}$ can be expanded as $\hat{\mathbf{v}} = \hat{\mathbf{v}}_1 v^1 + \hat{\mathbf{v}}_2 v^2 + \cdots + \hat{\mathbf{v}}_n v^n$. This expansion assigns to $\hat{\mathbf{v}}$ an n-tuple[26] of

n-tuple   [26] In mathematics an n-**tuple** is an ordered list of n objects. For example $(1, 4, 2, 6)$ is a 4-tuple. It is ordered in the sense that it must be distinguished from the differently ordered list $(4, 1, 2, 6)$.

real numbers, v^1 to v^n, which together can be viewed as an element of $\mathbb{R}^n$. In other words, the basis defines a map

$$\phi_{\hat{\mathbf{v}}} : V \to \mathbb{R}^n, \qquad \hat{\mathbf{v}} = \hat{\mathbf{v}}_j v^j \mapsto \phi_{\hat{\mathbf{v}}}(\hat{\mathbf{v}}) \equiv \begin{pmatrix} v^1 \\ \vdots \\ v^j \\ \vdots \\ v^n \end{pmatrix}, \tag{L31}$$

where the subscript in $\phi_{\hat{\mathbf{v}}}$ indicates that the map is specific to the basis $\{\hat{\mathbf{v}}_1, \ldots, \hat{\mathbf{v}}_n\}$. Under this map, the basis vectors $\hat{\mathbf{v}}_j$ themselves are assigned to the standard basis vectors of $\mathbb{R}^n$, $\phi_{\hat{\mathbf{v}}}(\hat{\mathbf{v}}_j) = \mathbf{e}_j$.

We saw above that for a given basis the assignment (vector) $\leftrightarrow$ (components) is unique. Every vector has a unique component representation and every set of components corresponds to a unique vector. This is another way of saying that the map $\phi_{\hat{\mathbf{v}}}$ is bijective. However, it is more than that: the sum of two vectors, $\hat{\mathbf{v}} + \hat{\mathbf{w}}$, is represented as $\hat{\mathbf{v}} + \hat{\mathbf{w}} = (\hat{\mathbf{v}}_j v^j) + (\hat{\mathbf{v}}_j w^j) = \hat{\mathbf{v}}_j(v^j + w^j)$ so that its components are given by the sum, $v^j + w^j$, of the components of $\hat{\mathbf{v}}$ and $\hat{\mathbf{w}}$, respectively. The same fact may be expressed as $\phi_{\hat{\mathbf{v}}}(\hat{\mathbf{v}} + \hat{\mathbf{w}}) = \phi_{\hat{\mathbf{v}}}(\hat{\mathbf{v}}) + \phi_{\hat{\mathbf{v}}}(\hat{\mathbf{w}})$. Notice that the two "+" signs in this equation are defined in different vector spaces: the plus on the left acts in V, that on the right in $\mathbb{R}^n$ (see Fig. L10). An analogous statement holds for scalar multiplication, $\phi_{\hat{\mathbf{v}}}(a\hat{\mathbf{v}}) = a\phi_{\hat{\mathbf{v}}}(\hat{\mathbf{v}})$.

In the language of Section L1.2, $\phi_{\hat{\mathbf{v}}}$ defines a bijective *homomorphism* between the spaces $(V, +)$ and $(\mathbb{R}^n, +)$, i.e. it is a vector space **isomorphism**. We have argued that the existence of an isomorphism between two spaces means that they are "practically identical", $V \cong \mathbb{R}^n$. Since this link to $\mathbb{R}^n$ can be established for any n-dimensional vector space V, the former is justly called the "standard" vector space. However, one always has to keep in mind that

isomorphism

> the isomorphism $V \cong \mathbb{R}^n$ is not canonical.

For a different basis of V, $\{\hat{\mathbf{w}}_1, \ldots, \hat{\mathbf{w}}_n\}$, a different isomorphism, $\phi_{\hat{\mathbf{w}}}$, and a different component representation are obtained.

EXAMPLE Fig. L11 illustrates the *isomorphism between two-dimensional Euclidean space and* $\mathbb{R}^2$. In Euclidean space, vector addition and scalar multiplication are geometric operations – the concatenation of vector-arrows and their stretching by scalar factors. These operations are compatible with

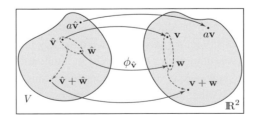

Fig. L10   On the isomorphism between a general two-dimensional real vector space V and $\mathbb{R}^2$.

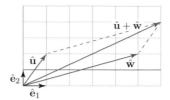 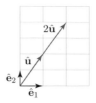

Fig. L11   Figure illustrating the isomorphism between two-dimensional Euclidean space and $\mathbb{R}^2$. Geometrically defined vector addition/multiplication is compatible with the algebraic operation on $\mathbb{R}^2$-component representations.

the algebraic addition and scalar multiplication of the corresponding $\mathbb{R}^2$ component representations, irrespective of what basis is chosen. For example, in a basis defined by the horizontal and vertical vectors $\mathbf{e}_1$ and $\mathbf{e}_2$, the geometric vectors $\hat{\mathbf{v}}$ and $\hat{\mathbf{w}}$ have the representations $\mathbf{v} = \left(\begin{smallmatrix}1\\2\end{smallmatrix}\right)$ and $\mathbf{w} = \left(\begin{smallmatrix}5\\2\end{smallmatrix}\right)$. Left: the geometric sum $\hat{\mathbf{v}} + \hat{\mathbf{w}}$ has components $\mathbf{v} + \mathbf{w} = \left(\begin{smallmatrix}6\\4\end{smallmatrix}\right)$. Right: the geometrically stretched vector $2\hat{\mathbf{v}}$ has components $2\mathbf{v} = \left(\begin{smallmatrix}2\\4\end{smallmatrix}\right)$. In either case, the same results are obtained by addition and multiplication in $\mathbb{R}^2$.

Thanks to the correspondence $V \cong \mathbb{R}^n$, vector calculations can be performed either in V or in $\mathbb{R}^n$. In the latter case one first applies the map $\phi_{\hat{\mathbf{v}}}$ to assign components to vectors, $\hat{\mathbf{v}}, \hat{\mathbf{w}}, \ldots$, does computations with the component representations, $\mathbf{v}, \mathbf{w}, \ldots$, and finally uses the inverse of the map $\phi_{\hat{\mathbf{v}}}$ to recover V-vectors from $\mathbb{R}^n$-vectors. The correspondence between vectors and their components is so tight that the symbol $\phi_{\hat{\mathbf{v}}}$ is often omitted and a notation $\hat{\mathbf{v}} = (v^1, \ldots, v^n)^\mathrm{T}$ is used. Although this is illegitimate (because it equates a vector in $\hat{\mathbf{v}} \in V$ with a vector $\mathbf{v} = (v^1, \ldots, v^n)^\mathrm{T} \in \mathbb{R}^n$), the notation is ubiquitous and one just has to accept its presence. However, in this text we avoid it and keep using the caret to distinguish between vectors and their component representation.

L2.6 Summary and outlook

In this chapter we introduced the important concept of vector spaces. Starting from a geometric motivation, we emphasized a view in which vectors were characterized by the operations defined on them – addition and scalar multiplication – and not so much by concrete realizations. This more general approach was motivated by the frequent occurrence of vectors without visual representation in physics. It also allows one to understand very different realizations of vectors in unified terms.

We saw that the general definition of vector spaces led to various secondary definitions, including that of the dimension of a vector space, linear dependence, completeness and that of a basis. Vector space bases were the key to the description of vectors through component representations, or elements of the standard space $\mathbb{R}^n$ (or $\mathbb{C}^n$). The existence of this component language is very important and means that the mathematics of $\mathbb{R}^n$ is a template describing all other vector spaces at once.

So far we have not done much with vector spaces other than defining them. Building on this foundation there are two directions to move forward. The first is the definition of additional structures required to perform actual geometric operations with vectors, the measurement of lengths and angles, etc. The second will be the discussion of *maps* preserving the fundamental structures of vector spaces. We will discuss these two extensions in turn, beginning with the "geometrization" of vector spaces.

Euclidean geometry

REMARK In this chapter the concepts of vector space *scalar products* are introduced. Scalar products play an important role in most areas of physics and we will discuss them in due generality. At the same time, they are needed very early in the physics curriculum and students have to get acquainted with scalar products as quickly as possible. The operation required in introductory lecture courses in physics is called the "standard scalar product" of $\mathbb{R}^n$. We have taken this as an incentive to include a fast track to the reading of this chapter. The first two sections, L3.1 and L3.2, introduce the standard scalar product and the essential geometric structures following from it. These sections contain the knowledge required in introductory courses. At the same time, they serve as an introduction to Sections L3.3 and L3.4, where general scalar products are introduced. These sections should be read as soon as possible, before venturing into the advanced chapters of this text. However, in the case of an acute shortage of time, they can be skipped at first reading.

In most of this chapter, real vector spaces are considered. Complex Euclidean geometry is the subject of Section L3.4.

In school, vectors are usually introduced as arrows of specified length and direction. The reason why these attributes have not been mentioned so far is that they are not included in the general definition of vector spaces. There are vector spaces for which the concept of length is not meaningful. Often, however, it is, and in these cases additional mathematical structure on top of the vector space axioms is required. Specifically, we need to introduce the *Euclidean scalar product* as a product operation defined for pairs of vectors.

Euclid
Often referred to as the father of geometry. In his influential work *Elements* Euclid formulated the principles of Euclidean geometry; it served as the main textbook on this subject for more than 2000 years. Little is known about the date of his birth and death, or about his personal life. He was active around 300 BC.

INFO When introducing mathematical concepts it is generally good practice to progress from minimal to more complex structures. We already discussed this principle in connection with the ascent sets–groups–fields. The situation with vector spaces is similar. Some physical theories require basic vector space structures such as completeness or linear independence and nothing more. For example, in classical mechanics the coordinate vector, $\mathbf{q}$, of a point particle and its momentum, $\mathbf{p}$, are often combined into a composite vector $\mathbf{x} = \binom{\mathbf{q}}{\mathbf{p}}$. To these vectors no "length" can be meaningfully assigned. (What would be the "length" of a vector having coordinates and momenta as components?) In such contexts a scalar product is not only superfluous but even obscures the physical contents.

L3.1 Scalar product of $\mathbb{R}^n$

As a warmup to the discussion of scalar products in general vector spaces, we consider the standard space $\mathbb{R}^n$. The goal is to equip $\mathbb{R}^n$ with an instrument to determine the length of vectors and angles between them.

Definition

standard scalar product

A scalar product of an $\mathbb{F}$-vector space is a function that takes two vectors as arguments to produce an ($\mathbb{F}$-valued) number. The **standard scalar product** of $\mathbb{R}^n$ is an example of this type of operation. It is defined as[27]

$$\langle\,,\,\rangle : \mathbb{R}^n \times \mathbb{R}^n \to \mathbb{R}, \qquad (\mathbf{v}, \mathbf{w}) \mapsto \langle \mathbf{v}, \mathbf{w} \rangle \equiv v^1 w^1 + v^2 w^2 + \cdots + v^n w^n. \qquad \text{(L32)}$$

The scalar product assigns to pairs of vectors, $(\mathbf{v}, \mathbf{w})$, the real number $\langle \mathbf{v}, \mathbf{w} \rangle$. Note the occurrence of a pair of contravariant indices in Eq. (L32). This is in violation of the rules of covariant notation (see p. 32) and signals that the definition does not seem to be complete yet. The missing element will be identified in Section L3.3; however, for the moment we continue to work with the definition as it is. In any case, we will focus on general properties of the scalar product throughout this section and the explicit summation formula in (L32) will not be used.

Key properties of the scalar product Eq. (L32) include

(i) *symmetry*: $\qquad\qquad\qquad\qquad \langle \mathbf{v}, \mathbf{w} \rangle = \langle \mathbf{w}, \mathbf{v} \rangle,$

(ii) *linearity in each argument*: $\quad \langle a\mathbf{v}, \mathbf{w} \rangle = a \langle \mathbf{v}, \mathbf{w} \rangle$ and

$$\qquad\qquad\qquad\qquad\qquad\qquad \langle \mathbf{u} + \mathbf{v}, \mathbf{w} \rangle = \langle \mathbf{u}, \mathbf{w} \rangle + \langle \mathbf{v}, \mathbf{w} \rangle, \text{ and}$$

(iii) *positive definiteness*: $\qquad \langle \mathbf{v}, \mathbf{v} \rangle > 0$ for all $\mathbf{v} \neq \mathbf{0}.$ $\qquad\qquad\qquad$ (L33)

The geometric structures defined by a scalar product rely solely on these three properties. This will become apparent in our discussion below, which makes repeated reference to (i)–(iii) but never to the specific formula (L32).

norm

Given a scalar product, the **norm of a vector** is defined as

$$\|\mathbf{v}\| = \sqrt{\langle \mathbf{v}, \mathbf{v} \rangle}. \qquad \text{(L34)}$$

The norm deduced from the standard scalar product of $\mathbb{R}^n$ defines the **geometric length** of a vector.[28] To see this, consider

length

$\mathbb{R}^2$, where $\left\|\binom{a}{b}\right\|! = \sqrt{a^2 + b^2}$ is the Pytha-

Pythagoras ca. 570–495 BC
Greek philosopher and mathematician. Best known for his theorem relating the length of the three sides of a right-angled triangle, Pythagoras had great influence on the philosophical and religious teaching of the late 6th century BC.

gorean length familiar from school. It is left as an exercise to show that the $\mathbb{R}^3$-norm assigns to a vector its geometric length, i.e. the length one would measure with a ruler.

---

[27] Other ways of writing the scalar product of $\mathbb{R}^n$ include $\mathbf{v}^\mathrm{T}\mathbf{w}$ or $\mathbf{v} \cdot \mathbf{w}$. Referring to the latter notation the scalar product is sometimes called the **dot product**. However, all these conventions are specific to $\mathbb{R}^n$ and we prefer to use the more general $\langle \mathbf{v}, \mathbf{w} \rangle$.

[28] The more general denotation "norm" is also used in vector spaces where (L34) does not have an interpretation as "length".

A scalar product is the starting point for the definition of various features describing the geometry of vectors – angles, length, parallelity and orthogonality, etc. These concepts are the subject of "Euclidean geometry" and will be discussed next.

Euclidean geometry

Euclidean geometry **Euclidean geometry** addresses geometric concepts based on the the scalar product (L32). However, it is important to note that these elements of geometry are not limited to the low-dimensional spaces, $\mathbb{R}^2, \mathbb{R}^3$, for which straightforward visualizations exist. We have seen above that general vector spaces of arbitrary dimension n can be identified with $\mathbb{R}^n$ once a basis has been chosen. For example, polynomials of degree 5 are elements of a six-dimensional vector space, and hence can be identified with elements of $\mathbb{R}^6$. Euclidean geometry then makes it possible to define an "angle" between different polynomials. In view of such generalizations one should not be too strongly attached to the visual representation of geometry and also keep its algebraic formulation (in formulas) in mind.

Cauchy–Schwarz inequality The most fundamental relation of Euclidean geometry is the **Cauchy–Schwarz inequality**: $\forall \mathbf{v}, \mathbf{w} \in V$,

$$|\langle \mathbf{v}, \mathbf{w} \rangle| \leq \|\mathbf{v}\| \|\mathbf{w}\|. \tag{L35}$$

The proof of this inequality illustrates how nontrivial results may be derived from the general properties of the scalar product. For $\mathbf{w} = \mathbf{0}$ Eq. (L35) holds trivially. For $\mathbf{w} \neq \mathbf{0}$, define the number $a \equiv \langle \mathbf{v}, \mathbf{w} \rangle / \|\mathbf{w}\|^2$, and consider the vector $\mathbf{v} - a\mathbf{w}$. Since its norm is greater than or equal to zero we have $0 \leq \langle \mathbf{v} - a\mathbf{w}, \mathbf{v} - a\mathbf{w} \rangle = \|\mathbf{v}\|^2 - 2a\langle \mathbf{v}, \mathbf{w} \rangle + a^2 \|\mathbf{w}\|^2 = \|\mathbf{v}\|^2 - (\langle \mathbf{v}, \mathbf{w} \rangle)^2 / \|\mathbf{w}\|^2$, where in the last step the definition of a was used. Multiply this inequality by $\|\mathbf{w}\|^2$, rearrange terms, and take the square root to arrive at Eq. (L35). For **colinearity** **colinear vectors**, i.e. for vectors "pointing in parallel directions", $\mathbf{v} = b\mathbf{w}$ with $b \in \mathbb{R}$, the inequality becomes an equality.[29] However, if $\mathbf{u}$ and $\mathbf{v}$ are not colinear their scalar product is strictly smaller than the product of their norms, and the inequality (L35) quantifies the degree of "misalignment".

angle This interpretation motivates the definition of the **angle**, $\angle(\mathbf{v}, \mathbf{w})$, between two vectors as

$$\angle(\mathbf{v}, \mathbf{w}) \equiv \arccos\left(\frac{\langle \mathbf{v}, \mathbf{w} \rangle}{\|\mathbf{v}\| \|\mathbf{w}\|} \right), \tag{L36}$$

or the equivalent representation

$$\langle \mathbf{v}, \mathbf{w} \rangle = \cos(\angle(\mathbf{v}, \mathbf{w})) \, \|\mathbf{v}\| \|\mathbf{w}\|. \tag{L37}$$

Before elucidating in what sense this is an "angle" we note that the definition makes mathematical sense: from Eq. (L35) we know that $\langle \mathbf{v}, \mathbf{w} \rangle / \|\mathbf{v}\| \|\mathbf{w}\| \in [-1, 1]$, and so the inverse of the cosine function, arccos, can be applied to produce a value between 0 and π. These two values represent the extreme limits of complete alignment and anti-alignment, $\langle \mathbf{v}, \mathbf{w} \rangle = \pm \|\mathbf{v}\| \|\mathbf{w}\|$, respectively.

---

[29] If $\mathbf{v} = b\mathbf{w}$, then $\|\mathbf{v}\| = |b| \|\mathbf{w}\|$ and $\langle \mathbf{v}, \mathbf{w} \rangle = |b| \langle \mathbf{w}, \mathbf{w} \rangle = |b| \|\mathbf{w}\|^2 = \|\mathbf{v}\| \|\mathbf{w}\|$.

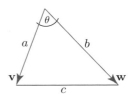

On this basis, let us now give equation (L36) a geometric interpretation. Consider the triangle in $\mathbb{E}^2$ shown in the figure. Its sides are
defined by the vectors $\mathbf{v}$, $\mathbf{w}$ and $\mathbf{v} - \mathbf{w}$, with side lengths $a \equiv \|\mathbf{v}\|$,
$b \equiv \|\mathbf{w}\|$ and $c \equiv \|\mathbf{v} - \mathbf{w}\|$. The geometric angle enclosed by $\mathbf{v}$ and
$\mathbf{w}$ is θ. *If* we identify this geometric angle with the one derived
from the Cauchy–Schwarz equation, $\theta \equiv \angle(\mathbf{v}, \mathbf{w})$, Eq. (L37) can be written as $\langle \mathbf{v}, \mathbf{w} \rangle = ab\cos(\theta)$. Now consider the vector identity $\langle (\mathbf{v} - \mathbf{w}), (\mathbf{v} - \mathbf{w}) \rangle = \langle \mathbf{v}, \mathbf{v} \rangle + \langle \mathbf{w}, \mathbf{w} \rangle - 2 \langle \mathbf{v}, \mathbf{w} \rangle$.
With the above identifications, it assumes the form $c^2 = a^2 + b^2 - 2ab\cos(\theta)$, which is the
familiar "law of cosines"[30]

$$a^2 + b^2 - c^2 = 2ab\cos(\theta). \tag{L38}$$

In other words, the identification of the Cauchy–Schwarz angle of Eq. (L36) with the
geometric angle follows from basic geometric considerations in the space $\mathbb{E}^2$. However,
the definition (L36) holds more generally and can be used to quantify the misalignment between two vectors even in contexts where these vectors do not have a visual
interpretation.

INFO Scalar products play an important role in most areas of physics. As an example, consider
mechanical work. If a constant force, $\mathbf{F}$, is applied to move an object along a straight line to induce
a certain displacement, $\mathbf{s}$, the force performs work, W. Both force and displacement are vectorial
quantities and the work done by the force is given by

$$W = \mathbf{F} \cdot \mathbf{s}. \tag{L39}$$

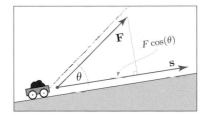

This equation can be read as the *definition* of force. The
norm $F \equiv \|\mathbf{F}\|$ quantifies its magnitude, and the direction
of $\mathbf{F}$ is the direction in which the force acts. Similarly, $s \equiv \|\mathbf{s}\|$ is the length of a displacement $\mathbf{s}$, whose direction may
differ from that of $\mathbf{F}$ by an arbitrary angle, $\theta = \angle(\mathbf{F}, \mathbf{s})$ (see
figure). Only the component of force parallel to $\mathbf{s}$, $F\cos\theta$,
effectively performs work, and the total amount of work is
proportional to the length of the displacement. This leads to
$W = Fs\cos(\theta)$, which can be equivalently expressed as (L39). Notice that in experiments, Eq. (L39)
really *is* applied to define forces. For example, the Coulomb force acting on charged particles can be
determined by displacing a test particle of a given charge in definite ways and measuring the required
work. If this is done for sufficiently many linearly independent displacements (how many?) a force
vector is determined.

L3.2 Normalization and orthogonality

Scalar products can be used to define vectors of definite length. **Unit normalized** vectors,
$\|\hat{\mathbf{w}}\| = 1$, are sometimes denoted by a caret, $\hat{\mathbf{w}}$.[31] For a given vector $\mathbf{w}$, a colinear unit
vector is obtained by division by its norm, or **normalization**,

---

[30] The law of cosines, which holds for arbitrary triangles in $\mathbb{E}^2$, is usually taught in school. It may be proven by
subdividing the triangle into two right-angled triangles and applying the Pythagorean theorem to each.

[31] Exceptions to this convention include unit vectors denoted by $\mathbf{e}$, such as $\mathbf{e}_j$ of Eq. (L28), for which the caret
is omitted. To be on the safe side, it is good practice to always define unit vectors explicitly. Also, the current
definition of $\hat{\mathbf{w}} \in \mathbb{R}^n$ must be distinguished from that of (not necessarily normalized) elements $\hat{\mathbf{w}} \in V$ of
generic vector spaces, see Section L2.5.

$$\hat{\mathbf{w}} \equiv \frac{\mathbf{w}}{\|\mathbf{w}\|}. \tag{L40}$$

orthogonality   Two vectors $\mathbf{v}$ and $\mathbf{w}$ are called **orthogonal** if $\langle \mathbf{v}, \mathbf{w} \rangle = 0$ and this is indicated by $\mathbf{v} \perp \mathbf{w}$. If two vectors are **parallel** to each other (in the sense that the Cauchy–Schwarz inequality becomes an equality) we write $\mathbf{v} \parallel \mathbf{w}$. For given $\mathbf{w}$, any vector **projection** $\mathbf{v}$ can be decomposed as $\mathbf{v} = \mathbf{v}_\perp + \mathbf{v}_\parallel$, where the **projection**, $\mathbf{v}_\parallel$ (spoken "v-parallel"), and the **orthogonal complement**, $\mathbf{v}_\perp$ (spoken "v-perpendicular"), are parallel and orthogonal to $\mathbf{w}$, respectively (see figure).

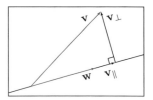

An explicit formula for this decomposition can be obtained from the ansatz[32] $\mathbf{v}_\parallel = \hat{\mathbf{w}}\, a$, where a is a coefficient. The value of a can now be determined by requiring that $\mathbf{v}_\perp = \mathbf{v} - \mathbf{v}_\parallel$ be orthogonal to $\hat{\mathbf{w}}$, i.e. that $0 = \langle \hat{\mathbf{w}}, \mathbf{v}_\perp \rangle = \langle \hat{\mathbf{w}}, \mathbf{v} \rangle - \langle \hat{\mathbf{w}}, \hat{\mathbf{w}} \rangle\, a$. From $\langle \hat{\mathbf{w}}, \hat{\mathbf{w}} \rangle = 1$ follows $a = \langle \hat{\mathbf{w}}, \mathbf{v} \rangle$. Projection and orthogonal complement are therefore given by

$$\boxed{\begin{aligned} \mathbf{v}_\parallel &= \hat{\mathbf{w}}\,\langle \hat{\mathbf{w}}, \mathbf{v} \rangle, \\ \mathbf{v}_\perp &= \mathbf{v} - \hat{\mathbf{w}}\langle \hat{\mathbf{w}}, \mathbf{v} \rangle. \end{aligned}} \tag{L41}$$

Note that the projection can also be written as $\mathbf{v}_\parallel = \cos(\angle(\mathbf{v}, \mathbf{w}))\,\|\mathbf{v}\|\,\hat{\mathbf{w}}$. This follows from Eq. (L37) and is consistent with an elementary geometric construction. ($\rightarrow$ L3.2.1-2)

parallelogram   The above relations can be applied to derive a useful formula for the area of the **parallelogram**, $A(\mathbf{v}, \mathbf{w})$, spanned by two vectors $\mathbf{v}, \mathbf{w} \in \mathbb{R}^n$. From elementary geometry in $\mathbb{E}^2$ we know that $A(\mathbf{v}, \mathbf{w}) = \|\mathbf{v}_\perp\|\|\mathbf{w}\|$: the area equals the length of one edge of the parallelogram, $\|\mathbf{w}\|$, multiplied by its height relative to this edge, $\|\mathbf{v}_\perp\|$ (see figure). This can be equivalently represented as

$$A(\mathbf{v}, \mathbf{w}) \overset{(\text{L41})}{=} \|\mathbf{v} - \hat{\mathbf{w}}\langle \hat{\mathbf{w}}, \mathbf{v} \rangle\|\|\mathbf{w}\| = \left[\|\mathbf{v}\|^2 \|\mathbf{w}\|^2 - \langle \mathbf{w}, \mathbf{v} \rangle^2 \right]^{1/2} \tag{L42}$$

$$\overset{(\text{L37})}{=} \|\mathbf{v}\|\|\mathbf{w}\| \left[1 - \cos^2(\angle(\mathbf{v}, \mathbf{w})) \right]^{1/2} = \|\mathbf{v}\|\|\mathbf{w}\| \sin(\angle(\mathbf{v}, \mathbf{w})). \tag{L43}$$

These expressions will be repeatedly used in later discussions.

The decomposition of vectors into perpendicular and parallel components has many applications. More generally, scalar products can be used to decompose vectors into contributions pointing in arbitrary directions:

---

ansatz   [32] The German word **ansatz** roughly means "well-motivated starting point". It is used for mathematical expressions which generally contain undetermined elements to be fixed at later stages of a calculation. The present ansatz contains the as yet undetermined parameter, a.

EXAMPLE Let us illustrate the utility of vector decomposi-
statics   tions with an example from **statics**. A mass is suspended
by a rigid rod and a rope. We aim to assess the sta-
bility of the construction, which means that we need to
know the forces acting on rope and rod, respectively. We
first decompose the gravitational force, **F**, exerted by the
body as

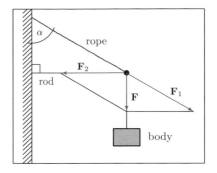

$$\mathbf{F} = \mathbf{F}_1 + \mathbf{F}_2 \qquad (L44)$$

into contributions $\mathbf{F}_1$ and $\mathbf{F}_2$ acting in the directions parallel to the rope and rod, respectively. Expressing each
force as unit vector times norm, $\mathbf{F} = \hat{\mathbf{F}}F$, $\mathbf{F}_1 = \hat{\mathbf{F}}_1 F^1$,
and $\mathbf{F}_2 = \hat{\mathbf{F}}_2 F^2$, the goal is to find the two unknowns F^1 and F^2 in terms of the known force, F, and
the angle α. To this end, we write Eq. (L44) as $\hat{\mathbf{F}}F = \hat{\mathbf{F}}_1 F^1 + \hat{\mathbf{F}}_2 F^2$ and take scalar products with $\hat{\mathbf{F}}$
and $\hat{\mathbf{F}}_2$ to obtain

$$\langle \hat{\mathbf{F}}, \hat{\mathbf{F}} \rangle F = \langle \hat{\mathbf{F}}, \hat{\mathbf{F}}_1 \rangle F^1 + \langle \hat{\mathbf{F}}, \hat{\mathbf{F}}_2 \rangle F^2,$$

$$\langle \hat{\mathbf{F}}_2, \hat{\mathbf{F}} \rangle F = \langle \hat{\mathbf{F}}_2, \hat{\mathbf{F}}_1 \rangle F^1 + \langle \hat{\mathbf{F}}_2, \hat{\mathbf{F}}_2 \rangle F^2.$$

We know that $\langle \hat{\mathbf{F}}, \hat{\mathbf{F}} \rangle = \langle \hat{\mathbf{F}}_2, \hat{\mathbf{F}}_2 \rangle = 1$ and deduce by elementary geometry that $\langle \hat{\mathbf{F}}, \hat{\mathbf{F}}_1 \rangle = \cos(\alpha)$,
$\langle \hat{\mathbf{F}}, \hat{\mathbf{F}}_2 \rangle = 0$, and $\langle \hat{\mathbf{F}}_2, \hat{\mathbf{F}}_1 \rangle = \cos(\frac{\pi}{2} + \alpha) = -\sin(\alpha)$. This leads to the solutions

$$F^1 = \frac{F}{\cos \alpha}, \qquad F^2 = F \tan \alpha.$$

Can you explain intuitively why F^1 grows indefinitely in the limit $\alpha \to \frac{\pi}{2}$?

Orthonormal bases

In Section L2.4 we argued that the fixation of a basis plays an important role in the work
with vector spaces. One of the advantages of having a scalar product is that it can be
used to define "orthonormal" bases, $\{\mathbf{e}_j\}$, which offer a maximal degree of simplicity and
orthonormal   convenience. We call a basis an **orthonormal basis** if its elements are (i) unit normalized,
basis $\|\mathbf{e}_j\| = 1$, and (ii) pairwise orthogonal, $\mathbf{e}_i \perp \mathbf{e}_j$ for $i \neq j$. The first property means that
the expansion coefficients, v^i, of general vectors $\mathbf{v} = \mathbf{e}_i v^i$ in the different "directions",
i, of the basis are measured relative to the same unit length. The second that there is no
redundancy in the description and that the coefficient v^i measures the extent of the vector
in the direction orthogonal to all $j \neq i$. The combination of these two features defines the
qualifier "ortho-normal" and leads to the condition (see footnote 11 of Chapter L2)

$$\langle \mathbf{e}_i, \mathbf{e}_j \rangle = \delta_{ij}. \qquad (L45)$$

The simplest example of an orthonormal basis is the standard basis of $\mathbb{R}^n$ introduced in
Section L2.4, $(\mathbf{e}_j)^i = \delta^i_j$. However, it is not the only one. For example, it is straightfor-
ward to verify that $\mathbf{e}_1 \equiv \frac{1}{\sqrt{2}}(1, 1)^{\mathrm{T}}$, $\mathbf{e}_2 \equiv \frac{1}{\sqrt{2}}(-1, 1)^{\mathrm{T}}$ defines an alternative orthonormal basis
(sketch the basis vectors) of $\mathbb{R}^2$. We will generally use the notation $\mathbf{e}_j$ for orthonormal basis
vectors of $\mathbb{R}^n$.[33]

---

[33] In the rare case that two such bases need to be used simultaneously, we will distinguish them by primes, $\{\mathbf{e}_j\}$
and $\{\mathbf{e}'_j\}$.

Orthonormal bases simplify the work with vector spaces. By way of example, consider the expansion of a general vector, $\mathbf{v} = \mathbf{e}_i v^i$. Obtaining the expansion coefficients, v^i, for a general basis can be difficult, as we have seen in a number of examples above. However, if $\{\mathbf{e}_j\}$ is an orthonormal basis, all one needs to do is compute the scalar product $\langle \mathbf{e}_i, \mathbf{v} \rangle = \langle \mathbf{e}_i, \mathbf{e}_j \rangle v^j = \delta_{ij} v^j = v^i$. The expansion coefficients are thus obtained via scalar products, $v^i = \langle \mathbf{e}_i, \mathbf{v} \rangle$, which can be computed with little effort.

At the same time, the notation $v^i = \langle \mathbf{e}_i, \mathbf{v} \rangle$ is notationally awkward in that the same index has different positions on the left and right, contravariant vs. covariant. This inconsistency is one of the occasions where the *covariant notation* signals that "something is missing". Going back one step, we note that the equation was obtained from $\langle \mathbf{e}_i, \mathbf{v} \rangle = \delta_{ij} v^j$. This expression *is* notationally consistent, due to the extra symbol δ_{ij}. We will learn in the next section that for scalar products more general than Eq. (L32), the Kronecker-δ_{ij} gets replaced by a nontrivial quantity, g_{ij}. For the time being, we note that notational consistency can be restored by keeping track of the Kronecker-δ. The excess baggage thus incurred can be conveniently absorbed in the notation by the definitions

$$v_i \equiv \delta_{ij} v^j, \qquad \mathbf{e}^i \equiv \delta^{ij} \mathbf{e}_j, \tag{L46}$$

covariant notation for scalar product i.e. we "raise" or "lower" indices via contraction with a Kronecker-δ. This enables us to formulate *covariant* expressions for the components of a vector, $v_i = \langle \mathbf{e}_i, \mathbf{v} \rangle$ or $v^i = \langle \mathbf{e}^i, \mathbf{v} \rangle$, and for its expansion in an orthonormal basis:

$$\mathbf{v} = \mathbf{e}_i \langle \mathbf{e}^i, \mathbf{v} \rangle. \tag{L47}$$

Of course the numerical values of the quantities related via Eq. (L46) are identical, e.g. $v^3 = v_3 = 5$, or $\mathbf{e}_1 = \mathbf{e}^1 = (1, 1)^{\mathsf{T}}$. This is another way of saying that

> in vector spaces with standard scalar product the co- or contravariant positioning of indices is computationally irrelevant, $v_i = v^i$, $\mathbf{e}^j = \mathbf{e}_j$.

However, the flip side of this statement is that the index position does become important in more general situations, to be addressed in Section L3.3.

We finally note that the index lowering operation can be applied to Eq. (L32) to formulate the *standard scalar product in covariant notation* as

$$\langle \mathbf{v}, \mathbf{w} \rangle = v^i w_i = v_i w^i. \tag{L48}$$

The rationale behind this notation is explained in Section L3.3.

EXAMPLE As an example for constructions using an orthonormal basis, let us represent the geometric area, $A(\mathbf{v}, \mathbf{w})$, spanned by two vectors $\mathbf{v}, \mathbf{w} \in \mathbb{R}^n$ in terms of their components in the standard basis, $\{\mathbf{e}_j\}$. To this end, we insert the expansions $\mathbf{v} = \mathbf{e}_i v^i$, $\mathbf{w} = \mathbf{e}_j w^j$ into Eq. (L42), and obtain:

$$A^2(\mathbf{v}, \mathbf{w}) = \|\mathbf{v}\|^2 \|\mathbf{w}\|^2 - \langle \mathbf{v}, \mathbf{w} \rangle^2 = \sum_{ij} \left[(v^i)^2 (w^j)^2 - (v^i w^i)(v^j w^j) \right]$$

$$= \sum_{i<j} \left[(v^i)^2 (w^j)^2 + (v^j)^2 (w^i)^2 - 2(v^i w^i)(v^j w^j) \right] = \sum_{i<j} \left[v^i w^j - v^j w^i \right]^2. \quad \text{(L49)}$$

In the double sum of the first line terms with $i = j$ cancel. We split the remaining contribution into sums $\sum_{i<j}$ and $\sum_{j<i}$, and in the latter sum relabel indices as $i \leftrightarrow j$ to arrive at the second line.

For two-dimensional vectors, $n = 2$, this reduces to

$$A(\mathbf{v}, \mathbf{w}) = |v^1 w^2 - v^2 w^1|. \quad \text{(L50)}$$

The general result (L49) will be needed in Chapter C4.2 where we discuss area integrals in general coordinates. It is left as an exercise to verify that for three-dimensional vectors, $n = 3$, it reproduces the cross product formula (L91), $A(\mathbf{v}, \mathbf{w}) = \|\mathbf{v} \times \mathbf{w}\|$, to be derived by different means in Section L4.2.

Orthonormalization

We will see that orthonormal bases have striking advantages which generally simplify calculations. Unless there are strong reasons in favor of a different choice, one really should aim to work with orthonormal bases.[34] The standard basis Eq. (L28) of $\mathbb{R}^n$ is orthonormal by design. However, we have demonstrated the existence of other orthonormal bases and this raises the question of how they can be found, or if a general basis can be systematically changed into an orthonormal one.

Gram–Schmidt orthonormalization. The answer to these questions is affirmative and is provided by a procedure known as **Gram–Schmidt orthonormalization**. Gram–Schmidt orthonormalization is an iterative algorithm for constructing from the vectors of a general basis $\{\mathbf{v}_j\}$ an orthonormal basis $\{\mathbf{e}_j\}$ (different in general from the standard basis of Eq. (L28), despite our use of the same symbol!). Its first step is the choice of one of the basis vectors, say $\mathbf{v}_1$. (This choice is arbitrary, but the orthonormal basis obtained in the end depends on it.) This vector is normalized to yield the first vector of the new basis, $\mathbf{e}_1 \equiv \mathbf{v}_1 / \|\mathbf{v}_1\|$. Next define $\mathbf{v}_{2,\perp} \equiv \mathbf{v}_2 - \mathbf{e}_1 \langle \mathbf{e}^1, \mathbf{v}_2 \rangle$ as the orthogonal complement of $\mathbf{v}_2$ with respect to $\mathbf{e}_1$, obtained by subtracting the component of $\mathbf{v}_2$ parallel to $\mathbf{e}_1$ (see Eq. (L41)). Here we used the identification $\mathbf{e}^1 \equiv \mathbf{e}_1$ or $e^i = \delta^{ij} e_j$ to obtain a consistent covariant notation. The vector $\mathbf{v}_{2,\perp}$ is non-vanishing. (Why? Remember linear independence!) Moreover, it is perpendicular to $\mathbf{e}^1$ by construction,

$$\langle \mathbf{e}^1, \mathbf{v}_{2,\perp} \rangle = \langle \mathbf{e}^1, \mathbf{v}_2 - \mathbf{e}_1 \langle \mathbf{e}^1, \mathbf{v}_2 \rangle \rangle = \langle \mathbf{e}^1, \mathbf{v}_2 \rangle - \langle \mathbf{e}^1, \mathbf{v}_2 \rangle = 0.$$

Normalization of $\mathbf{v}_{2,\perp}$ yields $\mathbf{e}_2 \equiv \mathbf{v}_{2,\perp} / |\mathbf{v}_{2,\perp}|$. We continue in this manner to define $\mathbf{v}_{3,\perp} \equiv \mathbf{v}_3 - \mathbf{e}_1 \langle \mathbf{e}^1, \mathbf{v}_3 \rangle - \mathbf{e}_2 \langle \mathbf{e}^2, \mathbf{v}_3 \rangle$, then normalize it, and so on, until we arrive at $\mathbf{e}_n$:

$$
\begin{aligned}
\mathbf{v}_1, && \mathbf{e}_1 &\equiv \mathbf{v}_{1,\perp} / \|\mathbf{v}_{1,\perp}\|, \\
\mathbf{v}_{2,\perp} &\equiv \mathbf{v}_2 - \mathbf{e}_1 \langle \mathbf{e}^1, \mathbf{v}_2 \rangle, & \mathbf{e}_2 &\equiv \mathbf{v}_{2,\perp} / \|\mathbf{v}_{2,\perp}\|, \\
\vdots && \vdots && \vdots && \vdots \\
\mathbf{v}_{i,\perp} &\equiv \mathbf{v}_i - \sum_{j=1}^{i-1} \mathbf{e}_j \langle \mathbf{e}^j, \mathbf{v}_i \rangle, & \mathbf{e}_i &\equiv \mathbf{v}_{i,\perp} / \|\mathbf{v}_{i,\perp}\|, \\
\vdots && \vdots && \vdots && \vdots \\
\mathbf{v}_{n,\perp} &\equiv \mathbf{v}_n - \sum_{j=1}^{n-1} \mathbf{e}_j \langle \mathbf{e}^j, \mathbf{v}_n \rangle, & \mathbf{e}_n &\equiv \mathbf{v}_{n,\perp} / \|\mathbf{v}_{n,\perp}\|.
\end{aligned}
\quad \text{(L51)}
$$

Gram–Schmidt orthonormalization

[34] Geometric contexts where non-orthogonal bases are to be favored will be discussed in Section V6.

The result of this procedure is an orthonormal basis, $\{\mathbf{e}_1, \ldots, \mathbf{e}_n\}$. We repeat that the basis resulting from a Gram–Schmidt procedure is non-canonical in that it depends on the order in which the vectors are orthonormalized (see the example below). It is also worth noting that the algorithm may be applied to sets $U = \{\mathbf{v}_1, \ldots, \mathbf{v}_m\}$, $m < n$, of vectors containing fewer vectors than the dimensionality of $\mathbb{R}^n$. In this case, an orthonormal basis of the m-dimensional subspace span(U) is obtained.

EXERCISE Apply the *Gram–Schmidt orthonormalization* algorithm to the $\mathbb{R}^3$-basis

$$\mathbf{v}_1 = \begin{pmatrix} 0 \\ 0 \\ 2 \end{pmatrix}, \quad \mathbf{v}_2 = \begin{pmatrix} 2 \\ 2 \\ 3 \end{pmatrix}, \quad \mathbf{v}_3 = \begin{pmatrix} -4 \\ 0 \\ 4 \end{pmatrix}. \tag{L52}$$

Before doing the calculation think a little and try to anticipate the geometric orientation of the orthonormal basis. Show how different bases are produced depending on whether you start the procedure with $\mathbf{v}_1, \mathbf{v}_2$ or $\mathbf{v}_3$. ($\rightarrow$ L3.3.5-6)

L3.3   Inner product spaces

REMARK In this section, as in Section L2.5, a caret symbol is used to distinguish elements of general vector spaces, $\hat{\mathbf{v}} \in V$, from their component representations, $\mathbf{v} \in \mathbb{R}^n$, relative to a basis. Scalar products in V are denoted by $\langle \hat{\mathbf{v}}, \hat{\mathbf{w}} \rangle_V$ and those in $\mathbb{R}^n$ by $\langle \mathbf{v}, \mathbf{w} \rangle_{\mathbb{R}^n}$. Beware: this definition of $\hat{\mathbf{v}}$ is unrelated to the unit-vector notation of the previous sections.

While our previous discussion focused on the standard scalar product Eq. (L32) of $\mathbb{R}^n$, the algebraic form of that formula was not essential. Rather, all results were obtained from the three fundamental features (i)–(iii) listed in Eq. (L33). In this section we invert the logic of the argument and *define* scalar products for general vector spaces as vector-pairing operations obeying the criteria (i)–(iii). Equation (L32) then has the status of one of many possible realizations of scalar products in $\mathbb{R}^n$. Generalized scalar products can be abstract and need not have straightforward geometric interpretations. At the same time, they are powerful computational aids and often carry physical significance. For example, the vector spaces relevant to the formulation of quantum theory all have scalar products. These product operations are directly related to physical observables, but only in few cases admit straightforward geometric visualizations.

Scalar product: general definition

A **scalar product** of an $\mathbb{R}$-vector space V is a map

$$\langle \, , \, \rangle : V \times V \rightarrow \mathbb{R}, \qquad (\hat{\mathbf{v}}, \hat{\mathbf{w}}) \mapsto \langle \hat{\mathbf{v}}, \hat{\mathbf{w}} \rangle, \tag{L53}$$

with the following properties ($\hat{\mathbf{u}}, \hat{\mathbf{v}}, \hat{\mathbf{w}} \in V, a \in \mathbb{R}$):

(i)   *symmetry*: $\langle \hat{\mathbf{v}}, \hat{\mathbf{w}} \rangle = \langle \hat{\mathbf{w}}, \hat{\mathbf{v}} \rangle$,

(ii) *linearity*: $\langle a\hat{\mathbf{v}}, \hat{\mathbf{w}} \rangle = a\langle \hat{\mathbf{v}}, \hat{\mathbf{w}} \rangle$,

$\langle \hat{\mathbf{u}} + \hat{\mathbf{v}}, \hat{\mathbf{w}} \rangle = \langle \hat{\mathbf{u}}, \hat{\mathbf{w}} \rangle + \langle \hat{\mathbf{v}}, \hat{\mathbf{w}} \rangle$,

(iii) *positive definiteness*: $\hat{\mathbf{v}} \neq \mathbf{0} : \langle \hat{\mathbf{v}}, \hat{\mathbf{v}} \rangle > 0$. \tag{L54}

inner
product
The general scalar product is often called the **inner product** while denotations such as "scalar product" or "dot product" are reserved for the standard scalar product of $\mathbb{R}^n$. We will adopt this convention but not be strict about it. A vector space equipped with an inner product is called an **inner product space**, a **normed vector space**, or a *Euclidean vector space*. In cases where the presence of an inner product is to be emphasized one uses the pair notation $(V, \langle\, ,\, \rangle)$. Finally, the inner product is sometimes denoted by the symbol g. In this case, the space (V, g) is defined through the map $g : V \times V \rightarrow \mathbb{R}, (\mathbf{v}, \mathbf{w}) \mapsto g(\mathbf{v}, \mathbf{w})$.

Euclidean
space
INFO In the literature, the term **Euclidean space** is used in three different ways:

 ▷ a general vector space V equipped with a scalar product, $(V, \langle\, ,\, \rangle)$;
 ▷ the standard space $\mathbb{R}^n$ with its standard scalar product Eq. (L32);
 ▷ the affine space $\mathbb{E}^n$ discussed on p. 25.

While this may be an overloaded usage of a single denotation, it is backed by a rationale which will become evident as we go along. To anticipate, once a proper (orthonormal) basis of a general inner product space, $(V, \langle\, ,\, \rangle)$, has been chosen, the geometry of the latter becomes equivalent to that of $(\mathbb{R}^n, \langle\, ,\, \rangle)$ with its standard scalar product. At the same time, geometric tools (i.e. a ruler) may be applied to define orthonormal basis vectors in the vector spaces $V = (A, O)$ associated with the affine space, A, of our intuition. In this case, every affine point may be identified with a component vector, thus the affine spaces A and $\mathbb{E}^d$ become equivalent, as do the associated vector spaces V and $\mathbb{R}^d$. As discussed in the previous section, this construction is such that geometrically defined angles and norms coincide with the angles and norms obtained from the standard scalar product of $\mathbb{R}^n$. These connections are so tight that the same terminology "Euclidean space" is used in all cases.

EXAMPLE

 ▷ In the *Euclidean spaces* $\mathbb{E}^2$ and $\mathbb{E}^3$, lengths and angles between vectors can be determined geometrically. One may then *define* the scalar product of two vectors $\mathbf{v}$ and $\mathbf{w}$ through Eq. (L37), i.e. the product of their length and the enclosed angle. As discussed in the previous section, this geometrically constructed scalar product obeys all criteria required by the general definition.
 ▷ In the *function space* $L^2(I)$ (see p. 26), consider two functions $f, g : I \rightarrow \mathbb{R}$ and define the map $\langle\, ,\, \rangle : I \times I \rightarrow \mathbb{R}$ by

$$\langle f, g \rangle = \int_I dx\, f(x)g(x), \tag{L55}$$

where the shorthand $\int_I \equiv \int_a^b$ for $I = [a, b]$ was used. This operation defines a scalar product. ($\rightarrow$ L3.3.1)

 ▷ For an example of an unconventional scalar product defined on $\mathbb{R}^2$, see problem L3.3.2.

EXERCISE Identities derived from scalar products often have an intuitive interpretation. For example,
triangle
inequality
in the Euclidean space $\mathbb{E}^3$ the **triangle inequality**,

$$\|\hat{\mathbf{v}}\| + \|\hat{\mathbf{w}}\| \geq \|\hat{\mathbf{v}} - \hat{\mathbf{w}}\|, \tag{L56}$$

states that the sum of the lengths of two sides of a triangle exceeds the length of the third side. However, one should also learn to understand such relations in more general ways, not tied to a geometric picture. At present, try to prove the triangle identity from the general definition of the scalar product. *Hint:* Use the Cauchy–Schwarz inequality to show that $(\|\hat{\mathbf{v}}\| + \|\hat{\mathbf{w}}\|)^2 \geq \|\hat{\mathbf{v}} - \hat{\mathbf{w}}\|^2$, which implies (L56). Discuss the interpretation of the triangle inequality in the case of the function space scalar product (L55).

Metric tensor

metric tensor   Consider an inner product space $(V, \langle\ ,\ \rangle_V)$ where the subscript on $\langle\ ,\ \rangle_V$ emphasizes that the product is specific to V. Given a basis $\{\hat{\mathbf{v}}_i\}$, we define the **metric tensor**[35] $g \equiv \{g_{ij}\}$,

$$g_{ij} \equiv \langle\hat{\mathbf{v}}_i, \hat{\mathbf{v}}_j\rangle_V, \tag{L57}$$

as an $n \times n$ array containing all inner products of basis vectors. The symmetry of the product implies the relation $g_{ij} = g_{ji}$, so that g is specified by $n(n-1)/2$ real numbers. This data fully specifies the inner product in the chosen basis because the product of two generic vectors, $\hat{\mathbf{x}} = \hat{\mathbf{v}}_i x^i$ and $\hat{\mathbf{y}} = \hat{\mathbf{v}}_j y^j$, can be obtained from the metric tensor as

$$\langle\hat{\mathbf{x}}, \hat{\mathbf{y}}\rangle_V = \langle\hat{\mathbf{v}}_i x^i, \hat{\mathbf{v}}_j y^j\rangle_V = x^i \langle\hat{\mathbf{v}}_i, \hat{\mathbf{v}}_j\rangle_V y^j = x^i g_{ij} y^j.$$

At the same time, the components x^i, y^j define the representations $\mathbf{x}, \mathbf{y}$ of the vectors $\hat{\mathbf{x}}, \hat{\mathbf{y}}$ in the standard vector space $\mathbb{R}^n$. We take this as an incentive to define a *generalized scalar product of* $\mathbb{R}^n$ as

$$\langle\mathbf{x}, \mathbf{y}\rangle_{\mathbb{R}^n} \equiv x^i g_{ij} y^j, \tag{L58}$$

where $\mathbf{x} = (x^1, \ldots, x^n)^{\mathrm{T}}$ and $\mathbf{y} = (y^1, \ldots, y^n)^{\mathrm{T}}$ as usual. With this definition we obtain

$$\langle\hat{\mathbf{x}}, \hat{\mathbf{y}}\rangle_V = \langle\mathbf{x}, \mathbf{y}\rangle_{\mathbb{R}^n} = x^i g_{ij} y^j, \tag{L59}$$

i.e. the equality of the inner product in V with that of the representing vectors in $\mathbb{R}^n$. Keep in mind, however, that both the component representations of the vectors and the form of the metric tensor are specific to a basis. Also note that if the basis $\{\hat{\mathbf{v}}_i\}$ is orthonormal, $\langle\hat{\mathbf{v}}_i, \hat{\mathbf{v}}_j\rangle_V = \delta_{ij}$, the metric tensor is trivial, $g_{ij} = \delta_{ij}$ (and vice versa), and the $\mathbb{R}^n$ scalar product reduces to the standard scalar product of Eq. (L32).

isometry   INFO   Given two vector spaces, $(V, \langle\ ,\ \rangle_V)$ and $(W, \langle\ ,\ \rangle_W)$, a map $F : V \to W$ is called an **isometry** if $\forall \hat{\mathbf{x}}, \hat{\mathbf{y}} \in V, \langle\hat{\mathbf{x}}, \hat{\mathbf{y}}\rangle_V = \langle F(\hat{\mathbf{x}}), F(\hat{\mathbf{y}})\rangle_W$, i.e. if the V-scalar product of the two arguments of the map is equal to the W-scalar product of their images. The definition of the scalar product (L58) is such that the component representation, $\phi_{\mathbf{v}} : V \to \mathbb{R}^n, \hat{\mathbf{x}} \to \mathbf{x}$, becomes an isometry of the vector spaces $(V, \langle\ ,\ \rangle_V)$ and $(\mathbb{R}^n, \langle\ ,\ \rangle_{\mathbb{R}^n})$. Whenever possible one should aim to work with isometries to benefit from the fact that they leave the scalar product, and hence also lengths, angles, etc., invariant.

It is customary to abbreviate the notation by introducing components with covariant indices as[36]

$$x_j \equiv x^i g_{ij}. \tag{L60}$$

index lowering   This parallels an operation performed in the previous section for the case $g_{ij} = \delta_{ij}$, i.e. $x_j = x^i \delta_{ij}$. This **index lowering** convention is introduced to compactly represent the inner

---

[35] For the general definition of tensors, see Chapter L10.
[36] Since the metric tensor is symmetric, $g_{ij} = g_{ji}$, Equation (L60) can equivalently be written as $x_j \equiv g_{ji} x^i$.

product as $x^i g_{ij} y^j = x_j y^j$: this hides the metric in a changed index position, and the notation remains consistently covariant. Inner products are now represented as

$$\langle \hat{\mathbf{x}}, \hat{\mathbf{y}} \rangle_V = \langle \mathbf{x}, \mathbf{y} \rangle_{\mathbb{R}^n} = x_j y^j. \tag{L61}$$

When using this notation the positioning of indices (upstairs vs. downstairs) is of course crucially important. Only if the metric is trivial, $g_{ij} = \delta_{ij}$, do we have an equality $x_j = x^j$.

inverse metric tensor   For later reference, we note that it is often convenient to introduce an **inverse metric tensor**, $\{g^{ij}\}$, through the relation

$$g_{kj} g^{ji} = \delta_k{}^i, \tag{L62}$$

index raising   where i and k are arbitrary, the repeated index j is summed over, and $\delta_k{}^j$ is defined in footnote 24. The inverse metric defines an **index raising** relation analogous to Eq. (L60):

$$x^i = x_j g^{ji}. \tag{L63}$$

The index lowering and raising relations (L60) and (L63) are consistent with each other in the sense that $x^i = x_j g^{ji} = (x^k g_{kj}) g^{ji} = x^k \delta_k{}^j = x^i$. It also implies a *symmetry in covariant summations*,

$$x_j y^j = x^j y_j. \tag{L64}$$

This useful relation follows directly from the definition of the raising and lowering operations.

EXAMPLE If the metric tensor has elements

$$g_{11} = g_{22} = 1, \qquad g_{12} = g_{21} = \frac{1}{\sqrt{2}}, \tag{L65}$$

it is straightforward to verify that its inverse is given by $g^{11} = g^{22} = 2$ and $g^{12} = g^{21} = -\sqrt{2}$.[37]

To *summarize*, a generic inner product, $\langle \,, \rangle_V$, of a vector space V motivates the definition of the metric tensor, Eq. (L57). Its components g_{ij} define a non-standard inner product (L58) of $\mathbb{R}^n$. The correspondence $(V, \langle \,, \rangle_V) \cong (\mathbb{R}^n, \langle \,, \rangle_{\mathbb{R}^n})$ is an isometry for which the inner products of vectors and of their component representations are equal and given by (L59).

EXERCISE Define a generalized inner product of $\mathbb{R}^n$ through Eq. (L58) with *a priori* unspecified coefficients g_{ij}. What conditions have to be imposed on the coefficients g_{ij} to satisfy the criteria (L54) defining an inner product? Show that these conditions hold if $g_{ij} = \langle \hat{\mathbf{v}}_i, \hat{\mathbf{v}}_j \rangle_V$ is defined through an inner product of a vector space V.

expansion of vectors   An inner product is a powerful aid in the *computation of the expansion coefficients* x^i of general vectors, $\hat{\mathbf{x}} = \hat{\mathbf{v}}_i x^i$, in a basis $\{\hat{\mathbf{v}}_i\}$. Let us assume the presence of an inner product with metric tensor, g_{ij}, and inverse, g^{ij}. We may then define a set of **contravariant basis vectors** $\{\hat{\mathbf{v}}^i\}$ by an index raising relation analogous to Eq. (L63):

[37] In many applications, "off-diagonal" elements of the metric tensor vanish, $g_{ij} = 0$ for $i \neq j$. In this simple case the inverse metric tensor is likewise diagonal, $g^{ij} = 0$ for $i \neq j$, and the diagonal elements are obtained as $g^{ii} = (g_{ii})^{-1}$. Methods for obtaining the inverse of generic metric tensors are discussed in Section L5.4.

$$\hat{\mathbf{v}}^i \equiv g^{ij}\hat{\mathbf{v}}_j. \tag{L66}$$

The quick calculation $\langle \hat{\mathbf{v}}^i, \hat{\mathbf{v}}_k \rangle_V = \langle g^{ij}\hat{\mathbf{v}}_j, \hat{\mathbf{v}}_k \rangle_V = g^{ij}\langle \hat{\mathbf{v}}_j, \hat{\mathbf{v}}_k \rangle_V = g^{ij}g_{jk} = \delta^i{}_k$ shows that the two sets of vectors $\{\hat{\mathbf{v}}^i\}$ and $\{\hat{\mathbf{v}}_k\}$ satisfy the *orthonormality relation*

$$\langle \hat{\mathbf{v}}^i, \hat{\mathbf{v}}_k \rangle_V = \delta^i{}_k. \tag{L67}$$

This relation greatly facilitates the computation of expansion coefficients. To obtain x^i, simply take the inner product between $\mathbf{v}^i$ and $\mathbf{x}$, $\langle \hat{\mathbf{v}}^i, \hat{\mathbf{x}} \rangle_V = \langle \hat{\mathbf{v}}^i, \hat{\mathbf{v}}_k x^k \rangle_V = \langle \hat{\mathbf{v}}^i, \hat{\mathbf{v}}_k \rangle_V x^k = \delta^i{}_k x^k = x^i$. This leads to the result

$$\boxed{\hat{\mathbf{x}} = \hat{\mathbf{v}}_i \langle \hat{\mathbf{v}}^i, \hat{\mathbf{x}} \rangle_V.} \tag{L68}$$

The expansion coefficients of a generic vector $\hat{\mathbf{x}}$ can thus be found using a four-step program: (1) compute the metric tensor, g_{ij}, and (2) its inverse, g^{ij}. Then (3) construct contravariant basis vectors $\hat{\mathbf{v}}^i = g^{ij}\hat{\mathbf{v}}_j$ from covariant ones, and (4) compute the components of $\hat{\mathbf{x}}$ using the inner product $x^i = \langle \hat{\mathbf{v}}^j, \hat{\mathbf{x}} \rangle_V$. Steps (1)–(3) of this program are generally easy and need to be performed only once. The final step, (4), too, is much easier than the computation of expansion coefficients without employing an inner product. ($\rightarrow$ L3.3.7-8)

EXAMPLE Consider the vector space $V = \mathbb{E}^2$ equipped with its geometrically defined inner product (L37). As basis we use the vectors $\{\hat{\mathbf{v}}_1, \hat{\mathbf{v}}_2\}$ indicated in the figure. In coordinates corresponding to the grid they have a component representation $\hat{\mathbf{v}}_1 = (1,0)^T$ and $\hat{\mathbf{v}}_2 = \frac{1}{\sqrt{2}}(1,1)^T$, respectively. It is straightforward to verify (do it!) that the metric tensor defined by this basis is given by Eq. (L65). What is the component representation of the vector $\mathbf{w}$ in the given basis? Compute its norm from the component representation and check that the result agrees with the geometrically computed length.

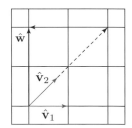

Orthonormal bases of inner product spaces

Given an inner product space, $(V, \langle \ , \ \rangle_V)$, it is generally convenient to work with bases for which the metric tensor $\{g_{ij}\}$ assumes a simple form. The metric becomes simplest for bases, $\{\hat{\mathbf{e}}_i\}$, whose vectors have unit norm, $||\hat{\mathbf{e}}_i|| = \sqrt{\langle \hat{\mathbf{e}}_i, \hat{\mathbf{e}}_i \rangle_V} = 1$, and are mutually orthogonal, $\langle \hat{\mathbf{e}}_i, \hat{\mathbf{e}}_j \rangle_V = 0, i \neq j$. These two criteria define the metric tensor as

$$g_{ij} = \langle \hat{\mathbf{e}}_i, \hat{\mathbf{e}}_j \rangle_V \equiv \eta_{ij} = \delta_{ij}. \tag{L69}$$

orthonormal basis A basis obeying these properties is called an **orthonormal basis**. It is customary to reserve the symbol η_{ij} for the metric tensor in orthonormal bases. For the positive definite inner products to be discussed, $\eta_{ij} = \delta_{ij}$ is just the Kronecker-δ. However, there exist generalizations to indefinite products (see the info section below) for which η_{ij} has more structure.

Orthonormal bases simplify all operations involving inner products. For example, the inner product between vectors $\hat{\mathbf{x}} = \hat{\mathbf{e}}_i x^i$ and $\hat{\mathbf{y}} = \hat{\mathbf{e}}_j y^j$ expanded in an orthonormal basis assumes the form (see Eq. (L59))

$$\langle \hat{\mathbf{x}}, \hat{\mathbf{y}} \rangle_V = \langle \hat{\mathbf{e}}_i x^i, \hat{\mathbf{e}}_j y^j \rangle_V = x^i \langle \hat{\mathbf{e}}_i, \hat{\mathbf{e}}_j \rangle_V y^j = x^i \delta_{ij} y^j = \langle \mathbf{x}, \mathbf{y} \rangle_{\mathbb{R}^n} , \qquad \text{(L70)}$$

where in the last equality we encounter the *standard inner product* (L32) of the component representation in $\mathbb{R}^n$, $x^i \delta_{ij} y^j = x^j y^j$. This leads to the conclusion that

> the inner product of vectors represented in an orthonormal basis equals the standard $\mathbb{R}^n$-inner product of their components.

The construction above suggests a reinterpretation of the standard inner product formula (L32). Written as it was in Eq. (L32), the formula contained a summation over two contravariant indices and did not conform to the conventions of covariant notation. The present discussion suggests rewriting (L32) as $x^i y^i = x^i \delta_{ij} y^j$. While at first sight the inclusion of the Kronecker-δ may look artificial, we now understand that it represents a particularly simple metric, $g_{ij} = \eta_{ij} = \delta_{ij}$. The misalignment of indices simply meant that a trivial metric tensor, $g = \eta$, was missing. The inclusion of that structure leads to the covariantly consistent formula, anticipated on formal grounds in Eq. (L48).

expansion of vectors in orthonormal basis   Another advantage of orthonormal bases is that they ease the computation of *vector expansion coefficients*. The evaluation of the general expansion formula (L68) simplifies because $\hat{\mathbf{e}}^i = \delta^{ij} \hat{\mathbf{e}}_j = \hat{\mathbf{e}}_i$, so that the expansion assumes the simple form

$$\hat{\mathbf{x}} = \hat{\mathbf{e}}_i \langle \hat{\mathbf{e}}^i, \hat{\mathbf{x}} \rangle_V = \mathbf{e}_i x^i, \qquad \text{(L71)}$$

where $x^i = \langle \hat{\mathbf{e}}^i, \hat{\mathbf{x}} \rangle_V = \langle \hat{\mathbf{e}}_i, \hat{\mathbf{x}} \rangle_V$ is straightforwardly obtained by taking the inner product of $\hat{\mathbf{x}}$ with the corresponding basis vector. ($\rightarrow$ L3.3.3-4)

Given that orthonormal bases are convenient to work with, one may ask how to find them. A similar question was raised in Section L3.2 in connection with bases of $\mathbb{R}^n$. There we showed that a general $\mathbb{R}^n$-basis could be transformed into an orthonormal one with the help of the *Gram–Schmidt orthonormalization* algorithm. The very same algorithm can be applied to general bases $\{\hat{\mathbf{v}}_i\}$ of an inner product space. All that needs to be done compared to Section L3.2 is put a caret on vectors $\mathbf{v}_i \rightarrow \hat{\mathbf{v}}_i$. The procedure then produces an orthonormal basis $\{\hat{\mathbf{e}}_i\}$ from a general one.

To *summarize*, there are two options for working with a generic inner product, $\langle \, , \, \rangle_V$.

▷ For a generic basis $\{\hat{\mathbf{v}}_i\}$, a metric tensor, g_{ij}, needs to be introduced. The $\mathbb{R}^n$-inner product for the component representations of vectors is then a non-standard inner product in $\mathbb{R}^n$.

▷ If one works with an orthonormal basis $\{\hat{\mathbf{e}}_i\}$, the metric tensor is trivial, $\eta_{ij} = \delta_{ij}$. Co- and contravariant components are equal, $x_i = x^i$, $\hat{\mathbf{e}}^i = \hat{\mathbf{e}}_i$, and the corresponding inner product of components is the standard inner product of $\mathbb{R}^n$.

Obviously, the second approach is simpler and one will usually aim to work with orthonormal bases of inner products. However, sometimes there are compelling reasons to single out a non-orthonormal basis. The covariant representation of the metric introduced above is then the best way to describe the situation.

Positive indefinite metric

The condition of positive definiteness in the definition of inner products is sometimes abandoned. This leads to the definition of inner products which are **positive semidefinite** ($\langle \hat{\mathbf{v}}, \hat{\mathbf{v}} \rangle = 0$ for $\hat{\mathbf{v}} \neq \mathbf{0}$ is permitted) or **indefinite** ($\langle \hat{\mathbf{v}}, \hat{\mathbf{v}} \rangle < 0$ is permitted). Positive indefinite inner products play an important role in various areas of physics, notably in *special and general relativity*. In spaces with indefinite inner product, nontrivial vectors of vanishing "norm" exist, $\langle \hat{\mathbf{v}}, \hat{\mathbf{v}} \rangle = 0$. One may then ask whether there may exist vectors $\hat{\mathbf{v}} \neq \mathbf{0}$ such that $\langle \hat{\mathbf{v}}, \hat{\mathbf{w}} \rangle = 0$ for all other $\hat{\mathbf{w}} \in V$. The existence of such vectors is not excluded by the definition and inner products permitting them are called **degenerate**. For example, the trivial map $\langle \hat{\mathbf{v}}, \hat{\mathbf{w}} \rangle \equiv 0$ defines a positive semidefinite degenerate inner product. However, in physics such products do not play a role, and non-degeneracy is generally required.

Most of the general structures discussed above carry over to the case of indefinite inner products. In a given basis, they are represented by a metric tensor, g_{ij}, and they are conveniently described in covariant notation as above. However, some elements of the theory need to be modified. For example, the Gram–Schmidt algorithm relies on the positive definiteness of an inner product. For semi- and indefinite inner products, vectors of ill-defined norm arise (the square root of a vector with $\langle \hat{\mathbf{v}}, \hat{\mathbf{v}} \rangle < 0$ does not yield a real number), and this invalidates essential steps of the normalization procedure. However, the procedure may be adapted (in a manner not discussed here) to obtain a basis $\{\mathbf{e}_j\}$ in which the inner product is represented by a metric tensor $g_{ij} \equiv \eta_{ij}$, with diagonal elements $\eta_{ii} = 1$ for $i = 1, \dots, p$ and $\eta_{ii} = -1$ for $i = p+1, \dots, n$, with vanishing off-diagonal elements, $\eta_{i \neq j} = 0$. Here η is the standard symbol for the representation of the metric in this form. The number of positive and negative elements, $(p, n - p)$, defines the **signature** of the metric. No matter what diagonal basis is chosen, the basis vectors can always be ordered such that the signature remains the same. The η-symbol introduced in the previous section in Eq. (L69) describes a positive definite metric with signature $(n, 0)$.

indefinite inner product

signature

INFO An indefinite metric of great physical significance appears in the theory of relativity. The four-dimensional **Minkowski metric** has signature $(1, 3)$, with $\eta_{00} = 1$ and $\eta_{ii} = -1$, $i = 1, 2, 3$ in a diagonal basis, $\{\mathbf{e}_0, \mathbf{e}_1, \mathbf{e}_2, \mathbf{e}_3\}$.[38] Physically, Minkowski vector spaces are understood as **space-time**, where $\mathbf{e}_0$ represents a "time-like" direction, and $\mathbf{e}_i$ are "space-like" directions. For further information on the Minkowski metric, consult the info section on p. 113.

Minkowski metric

space-time

Note that the Minkowski metric admits vectors of vanishing norm ($\mathbf{e}_0 + \mathbf{e}_1$ is an example). However, it is non-degenerate. To see this, suppose that $\mathbf{v} = \mathbf{e}_\mu v^\mu$ (summed over $\mu = 0, 1, 2, 3$) has vanishing inner product with all other vectors. Then $\langle \mathbf{e}_\eta, \mathbf{v} \rangle = 0$ certainly holds for the basis vectors $\mathbf{e}_\eta$, thus $v^0 = \langle \mathbf{e}_0, \mathbf{v} \rangle = 0$ and $v^i = -\langle \mathbf{e}_i, \mathbf{v} \rangle = 0$. Hence all the expansion coefficients of the vector $\mathbf{v}$ vanish, so that it must be the null vector.

L3.4 Complex inner product

REMARK This section can be skipped at first reading. It is a prerequisite for Chapters L8 and C6.

---

[38] In the literature one often finds an alternative convention, where $(3, 1)$ is used as signature for the Minkowski metric, with $\eta_{00} = -1, \eta_{ii} = 1$. This global sign change has no physical consequences.

So far our focus has been on real inner product spaces. However, many spaces of relevance to physics are complex and in this case too, inner products provide powerful additional structure. For example, the vector spaces of quantum mechanics are all complex and their inner products generally describe operations of physical significance.

complex inner product   A **complex inner product** of a $\mathbb{C}$-vector space, V, is a map

$$\langle \, , \rangle : V \times V \to \mathbb{C}, \qquad (\hat{\mathbf{v}}, \hat{\mathbf{w}}) \mapsto \langle \hat{\mathbf{v}}, \hat{\mathbf{w}} \rangle, \tag{L72}$$

with the following properties:

(i)   *symmetry*: $\langle \hat{\mathbf{v}}, \hat{\mathbf{w}} \rangle = \overline{\langle \hat{\mathbf{w}}, \hat{\mathbf{v}} \rangle}$,

(ii) *complex linearity*: $\langle a\hat{\mathbf{v}}, \hat{\mathbf{w}} \rangle = \bar{a}\langle \mathbf{v}, \mathbf{w} \rangle$,

$\langle \hat{\mathbf{v}}, a\hat{\mathbf{w}} \rangle = a\langle \hat{\mathbf{v}}, \hat{\mathbf{w}} \rangle$,

$\langle \hat{\mathbf{u}} + \hat{\mathbf{v}}, \hat{\mathbf{w}} \rangle = \langle \hat{\mathbf{u}}, \hat{\mathbf{w}} \rangle + \langle \hat{\mathbf{v}}, \hat{\mathbf{w}} \rangle$,

(iii) *positive definiteness*: $\hat{\mathbf{v}} \neq \mathbf{0} : \langle \hat{\mathbf{v}}, \hat{\mathbf{v}} \rangle > 0$. $\tag{L73}$

These properties differ from their real analogues, Eq. (L54), in that complex conjugation appears in the symmetry relation, and in the linearity condition for the first (but not the second!) argument vector. Also note from (i) that $\langle \hat{\mathbf{v}}, \hat{\mathbf{v}} \rangle$, needed to compute the norm of complex vectors, is real. This is an important feature which would be lost were it not for the complex conjugation in the symmetry relation. A complex vector space, V, equipped **unitary** with an inner product, $(V, \langle \, , \rangle)$, is called a **unitary vector space**.
vector space

A generic inner product of a vector space with basis $\{\hat{\mathbf{v}}_i\}$ is described by a metric tensor $g_{ij} \equiv \langle \hat{\mathbf{v}}_i, \hat{\mathbf{v}}_j \rangle$ defined in analogy to the real metric, Eq. (L58). However, the complex symmetry relation now implies that $g_{ij} = \bar{g}_{ji}$. As in the real case, the inner product between general vectors, $\hat{\mathbf{u}} = \hat{\mathbf{v}}_i u^i$ and $\hat{\mathbf{w}} = \hat{\mathbf{v}}_i w^i$, is obtained as

$$\boxed{\langle \hat{\mathbf{u}}, \hat{\mathbf{w}} \rangle_V = \overline{u^i} g_{ij} w^j \equiv \langle \mathbf{u}, \mathbf{w} \rangle_{\mathbb{C}^n},} \tag{L74}$$

where the complex conjugation of the left vector components is required to satisfy the complex linearity property (ii) of Eq. (L73). On the right, $\mathbf{u} = (u^1, \dots, u^n)^T$ and $\mathbf{w} = (w^1, \dots, w^n)^T$ are the $\mathbb{C}^n$-vectors representing $\hat{\mathbf{u}}$ and $\hat{\mathbf{w}}$ in the basis, and we have introduced the $\mathbb{C}^n$-inner product representing the V-inner product.

As in real vector spaces, it is often convenient to work with orthonormal bases, $g_{ij} = \delta_{ij}$. In this case, the $\mathbb{C}^n$-product becomes the **standard complex scalar product**, $\langle \mathbf{u}, \mathbf{v} \rangle = \sum_i \bar{u}^i v^i$. The standard product is often written as $\mathbf{u}^\dagger \mathbf{v}$, where the component vector $\mathbf{u}^\dagger$ is defined as $\mathbf{u}^\dagger \equiv \bar{\mathbf{u}}^T = (\bar{u}^1, \dots, \bar{u}^n)$. This is the complex analogue of the real standard scalar product notation $\mathbf{x}^T \mathbf{y}$. For example, with $\mathbf{u} = (1, i)^T$ and $\mathbf{v} = (4 + 2i, 1)^T$, $\mathbf{u}^\dagger = (1, -i)$ and $\mathbf{u}^\dagger \mathbf{v} = 4 + i$.

L3.5 Summary and outlook

In this chapter we introduced scalar or inner products as an additional structure on top of the general vector space axioms. Inner products connect the abstract concept of vectors to (Euclidean) geometry. They are instrumental to the definition of length, angles, areas and other elements of our geometric intuition. At the same time, we stressed that one must not be attached too strongly to visual representations of the geometric structures constructed from inner products. For example, in inner product function spaces it is perfectly valid to define an angle between two functions. This angle has all the properties otherwise known from "geometric angles", but cannot be visualized.

We also saw that inner products are powerful computational aids. They generally ease algebraic computations of vector components and can be used to define orthonormal bases with pleasant algebraic properties. Finally, we observed that the connection between a general vector space and the standard vector space of the same dimension can be extended to accommodate an inner product: a general inner product defines a metric tensor which in turn defines an inner product of the standard space. These definitions are engineered in such a way that geometric structures are equivalently described in both spaces. We saw that the covariant notation used in this text conveniently describes this correspondence in such a way that the bookkeeping of elements of the metric tensor becomes nearly automatic.

Inner products will play an important role throughout the rest of the text. However, readers should really keep in mind that they are added structure on top of the fundamental vector space axioms. For example, in the final Chapter V7 of the text we will discuss the mathematical structures underlying electrodynamics. Electrodynamics is a field closely related to relativity theory, which in turn crucially relies on the properties of the (Minkowski) metric. In this context, it is very important to understand which properties of electrodynamics depend on a metric (i.e. the possibility to measure lengths) and which do not. The covariant formalism makes it easy to keep these dependencies in sight and this is one of its principal advantages over alternative formulations.

Vector product

In this chapter we focus on the Euclidean space $\mathbb{E}^3 \simeq$ $\mathbb{R}^3$. This space is special not only because it is the space of our daily experience but also because it admits the definition of a product operation between vectors that is very different from the scalar product discussed above. This *vector product* assigns to two vectors $\mathbf{v}, \mathbf{w} \in \mathbb{R}^3$ another "vector" which is usually denoted $\mathbf{v} \times \mathbf{w}$. (The quotation marks hint at the fact that $\mathbf{v} \times \mathbf{w}$ actually is not a true vector, a point to be discussed below.) In the following, we introduce the vector product from two different perspectives. The first is geometric and the second emphasizes the algebraic features of the vector product.

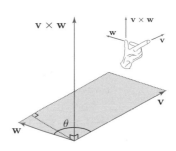

L4.1 Geometric formulation

vector product The **vector product** or **cross product** is a map, $\times : \mathbb{E}^3 \times \mathbb{E}^3 \to V, (\mathbf{v}, \mathbf{w}) \mapsto \mathbf{v} \times \mathbf{w}$, assigning to two $\mathbb{E}^3$-vectors, $\mathbf{v}$ and $\mathbf{w}$, an element, $\mathbf{v} \times \mathbf{w}$, of another three-dimensional vector space, V. The mathematical identity of V is discussed in precise terms in Section L10.9. For the moment we note that being three-dimensional, $V \cong \mathbb{E}^3$ is isomorphic (in one-to-one correspondence) to $\mathbb{E}^3$ and each of its elements can be described as a three-component object. For this reason, the physics literature often does not distinguish between V and $\mathbb{E}^3$ and considers the cross product as a map,

$$\times : \mathbb{E}^3 \times \mathbb{E}^3 \to \mathbb{E}^3, \qquad (\mathbf{v}, \mathbf{w}) \mapsto \mathbf{v} \times \mathbf{w}, \tag{L75}$$

between vectors in $\mathbb{E}^3$. The image, $\mathbf{v} \times \mathbf{w}$, of the cross product is implicitly defined by the following geometric properties.

1 *Perpendicularity:* by definition, $\mathbf{v} \times \mathbf{w}$ points in a direction perpendicular to the two-dimensional plane[39] spanned by $\mathbf{v}$ and $\mathbf{w}$ (unless $\mathbf{v} \parallel \mathbf{w}$, in which case $\mathbf{v} \times \mathbf{w} \equiv \mathbf{0}$). In other words, $\mathbf{v} \times \mathbf{w}$ is *perpendicular to both $\mathbf{v}$ and $\mathbf{w}$*.

---

[39] Here it is essential that we are operating in $\mathbb{E}^3$. In $\mathbb{E}^2$ a direction perpendicular to a two-dimensional plane does not exist and in $\mathbb{E}^{n>3}$ a plane does not uniquely identify a perpendicular direction.

right-hand rule

2   *Orientation:* perpendicularity to a plane still leaves two possible directions, "upwards" or "downwards". The orientation of $\mathbf{v} \times \mathbf{w}$ is defined according to a **right-hand rule**: if index and middle finger of a right hand point in the direction of $\mathbf{v}$ and $\mathbf{w}$, respectively, then its thumb indicates the direction of $\mathbf{v} \times \mathbf{w}$ (see the figure above).

3   *Norm:* by definition, $\|\mathbf{v} \times \mathbf{w}\|$ is equal to the geometric area of the parallelogram spanned by $\mathbf{v}$ and $\mathbf{w}$. According to Eq. (L43), this area is given by

$$\|\mathbf{v} \times \mathbf{w}\| = \|\mathbf{v}\|\|\mathbf{w}\| \sin\theta, \tag{L76}$$

where $\theta = \angle(\mathbf{v}, \mathbf{w}) \in [0, \pi]$ is the angle between $\mathbf{v}$ and $\mathbf{w}$, as defined in Eq. (L37).

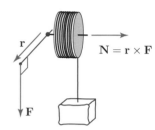

INFO The vector product plays an important role in *physics*. Generally speaking, vector products appear whenever the two physical concepts "vector" and "rotation" meet.

Let us illustrate this point with an example from mechanics. Consider a weight lifted by a lever. The influence of the lever on the weight depends on three factors: (i) the magnitude and (ii) direction of the lifting force, $\mathbf{F}$, and (iii) the position, $\mathbf{r}$, of the point at which it acts relative to the closest point on the axis of rotation. The applied force is maximally efficient if it acts in a direction perpendicular

torque

to $\mathbf{r}$. All these factors are combined in the definition of **torque**,

$$\mathbf{N} = \mathbf{r} \times \mathbf{F}. \tag{L77}$$

The torque is defined to be perpendicular to both $\mathbf{r}$ and $\mathbf{F}$ and this defines an imaginary axis around which it tries to induce rotational motion. A torque acts *efficiently* if it is aligned with an axis around which mechanical motion is actually possible, such as the cylindrical axis of the structure shown in the figure. Also notice that torque is always defined relative to a point of origin, at present the point on the axis relative to which $\mathbf{r}$ is computed. This point has deliberately been chosen such that $\mathbf{N}$ is colinear with the cylindrical axis. For different choices of the origin, different values of $\mathbf{r}$ and of $\mathbf{N}$ would result.

Taking the norm, $\|\mathbf{N}\| = \|\mathbf{r}\|\|\mathbf{F}\|\|\sin\angle(\mathbf{r}, \mathbf{F})|$, we see that the torque is largest if $\mathbf{r} \perp \mathbf{F}$. The norm

law of levers

also expresses the **law of levers**, according to which the effect of the force depends on the product of its magnitude, $\|\mathbf{F}\|$, and the distance of application relative to the rotation axis, $\|\mathbf{r}\|$.

From its geometric construction it follows that the vector product is

antisymmetric: $\mathbf{v} \times \mathbf{w} = -\mathbf{w} \times \mathbf{v}$,   (L78a)

distributive: $\mathbf{u} \times (\mathbf{v} + \mathbf{w}) = \mathbf{u} \times \mathbf{v} + \mathbf{u} \times \mathbf{w}$,   (L78b)

in general not associative: $\mathbf{u} \times (\mathbf{v} \times \mathbf{w}) \neq (\mathbf{u} \times \mathbf{v}) \times \mathbf{w}$.   (L78c)

The lack of associativity can be shown by constructing counter-examples (consider, for example, the vector product of three orthonormal basis vectors). The verification of distributivity on the basis of the geometric definition of the vector product is tricky – consider it a challenging exercise – and will not be discussed here. However, distributivity will follow as a trivial consequence of the alternative algebraic definition of the vector product to be discussed in the next section.

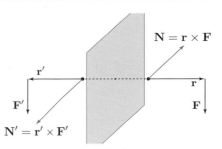

vector product is not a vector!

INFO We mentioned above that *the vector product,* $\mathbf{v} \times \mathbf{w}$, *is not an element of the space* $\mathbb{E}^3$ *in which* $\mathbf{v}$ *and* $\mathbf{w}$ *are defined*. Although $\mathbf{v} \times \mathbf{w}$ does live in a three-dimensional vector space, V, this space is different from the $\mathbb{E}^3$ of the argument vectors, and this difference shows in various ways. As a physically motivated example, let us consider the torque, $\mathbf{N} = \mathbf{r} \times \mathbf{F}$, of two vectors assumed to be perpendicular for simplicity, $\mathbf{r} \perp \mathbf{F}$, and study its properties under reflection[40] at a plane. Let us denote the image of a vector $\mathbf{v}$ under this reflection by $\mathbf{v}'$. For reflection at the shaded plane in the figure we thus have $\mathbf{r}' = -\mathbf{r}$ and $\mathbf{F}' = \mathbf{F}$. The torque $\mathbf{N} = \mathbf{r} \times \mathbf{F}$ is parallel to the plane. If it were an element of $\mathbb{E}^3$, its reflection would thus be equal to $\mathbf{N}$. However, the torque is actually an element of V, and as such its reflection is *defined* to be the torque computed from the reflected argument vectors, $\mathbf{N}' \equiv \mathbf{r}' \times \mathbf{F}' = -\mathbf{r} \times \mathbf{F}$. This vector points in the direction *opposite* to $\mathbf{N}$, showing that under planar reflections the cross product transforms differently from an $\mathbb{E}^3$-vector. In view of this oddity, the vector product of two vectors is sometimes called a **pseudo-vector** or **axial vector**. However, the mathematically clean view, presented in Section L10.9, is that the cross product lives in the three-dimensional vector space of "covariant tensors of first degree",

axial vector

above denoted by V. In three dimensions, these tensors are described by three components, and hence resemble vectors. Much of the physics literature ignores the distinction between covariant tensors of first degree and vectors, and treats them as vectors all the same. However, the above construction shows that both physically and mathematically there is a price to be paid for this indiscriminating approach. Cross products are different from vectors, and if they are nevertheless treated as vectors one has to live with strange transformation behavior. Confusion can be avoided by keeping this point in mind.

INFO We know from daily experience that rotating bodies resist changes of their axis of rotation. For example it takes a strong force to change the rotation of a wheel in motion and this is what keeps bicycles from falling. The same principle maintains the rotational axis of the planets of the solar system in their motion around the Sun.

angular momentum

The quantity that is "conserved" in free rotational motion is called **angular momentum**. For any body with mass m and velocity $\mathbf{v}$, its angular momentum relative to a point O, is defined as (see figure)

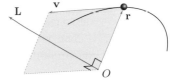

$$\mathbf{L} \equiv m\mathbf{r} \times \mathbf{v}, \tag{L79}$$

where $\mathbf{r}$ is the vector connecting O and the body. The angular momentum vector $\mathbf{L}$ is perpendicular to the plane spanned by $\mathbf{r}$ and the direction of instantaneous motion specified by $\mathbf{v}$. In the particular case of a body travelling along a planar orbit, e.g. Earth orbiting around the Sun, the angular momentum relative to a point in the orbital plane stands perpendicular to that plane.

The *conservation law* expressing the "stability" of rotational motion reads

$$\frac{d\mathbf{L}}{dt} = \mathbf{N}, \tag{L80}$$

reflection   [40] Formally, a **reflection at a plane** is a map, $\mathbb{E}^3 \to \mathbb{E}^3, \mathbf{v} \equiv \mathbf{v}_{\parallel} + \mathbf{v}_{\perp} \mapsto \mathbf{v}' \equiv \mathbf{v}_{\parallel} - \mathbf{v}_{\perp}$, where $\mathbf{v} = \mathbf{v}_{\parallel} + \mathbf{v}_{\perp}$ is a decomposition of the argument into components parallel and perpendicular to the plane, respectively. The reflection inverts the sign of the perpendicular component.

where $\mathbf{N} = \mathbf{r} \times \mathbf{F}$ is the torque acting on the body relative to the reference point used to define angular momentum. In the absence of torque, angular momentum is conserved. Notice that the absence of torque does not necessitate the absence of forces. For example, Earth experiences a gravitational force, $\mathbf{F}$, from the Sun. However, that force is radial, $\mathbf{F} \parallel \mathbf{r}$, i.e. directed along the line connecting the centers of Earth and Sun. This means that it does not create a torque, and so the rotational motion of our planet is (approximately) conserved. For an extended discussion of angular momentum, consult a lecture course in classical mechanics.

L4.2 Algebraic formulation

REMARK The non-vectorial nature of the vector product shows not only in its geometric features (see the preceding info block) but also algebraically: relations involving the vector product generally lead to index configurations violating the conventions of covariant notation (see p. 32). The proper algebraic treatment of the vector product requires the concept of "tensors" and will be discussed in Section L10.9. In that section all indices will be consistently in place, and much as in Section L3.3 this happens via the appearance of the metric tensor, g_{ij}, or δ_{ij} if an orthonormal basis is used. However, these structures cannot yet be meaningfully introduced here, and therefore we will end up with a number of index inconsistencies, similar to those in Eq. (L32). Since all definitions in this section assume an orthonormal basis, $g_{ij} = \delta_{ij}$, these mismatches are algebraically inconsequential.

The geometric definition of the vector product presented in Section L4.1 is intuitive but cumbersome to work with. We aim to introduce a more efficient formulation and to this end consider an orthonormal basis, $\{\mathbf{e}_1, \mathbf{e}_2, \mathbf{e}_3\}$. The vector product makes reference to the right-hand orientation, and so it will be natural to label the basis vectors such that $\mathbf{e}_1, \mathbf{e}_2, \mathbf{e}_3$ point in the direction of the index finger, middle finger, and thumb of a right hand, respectively. A basis obeying this criterion is called **positively oriented**. Any basis can be converted into a positively oriented one by a relabeling of basis vectors.[41]

orientation

The positive orientation of a basis defines a cyclic ordering of basis vectors as shown in the figure. The geometrically defined vector product of any two consecutive vectors in this sequence yields the third, e.g. $\mathbf{e}_2 \times \mathbf{e}_3 = \mathbf{e}_1$. The product in "reverse order" comes with a minus sign, e.g. $\mathbf{e}_2 \times \mathbf{e}_1 = -\mathbf{e}_3$. It is convenient to introduce notation representing these relations in compact form: we call a triple of three unequal indices, (i, j, k), **cyclically ordered** if they are ordered in the sequence 123, 231 or 312, as indicated in the figure, and anti-cyclically ordered if the sequence of ordering is reversed, 213, 321 or 132. The cross product of basis vectors of a right-handed orthonormal basis satisfies $\mathbf{e}_i \times \mathbf{e}_j = \pm\mathbf{e}_k$, where the upper or lower sign applies to cyclically or anti-cyclically ordered indices (i, j, k), respectively.

cyclically ordered

Note that the three indices (i, j, k) are cyclically or anti-cyclically ordered if they are obtained from the ordered sequence $(1, 2, 3)$ by an even or odd permutation, respectively (see the discussion of Eq. (L5) on p. 12). For example, the cyclic permutation $(3, 1, 2)$ is

---

[41] For example, if $\{\mathbf{e}_1, \mathbf{e}_2, \mathbf{e}_3\}$ is positively oriented, then $\{\mathbf{e}_2, \mathbf{e}_1, \mathbf{e}_3\}$ is negatively oriented.

even, the anti-cyclic $(2,1,3)$ odd. The **Levi–Civita symbol**, introduced on p. 12 to do the bookkeeping of the positive/negative signum of even/odd permutations, thus assumes the value

$$\epsilon_{ijk} \equiv \begin{cases} 1, & (i,j,k) \text{ cyclic}, \\ -1, & (i,j,k) \text{ anti-cyclic}, \\ 0, & \text{otherwise}. \end{cases} \tag{L81}$$

vector product and Levi–Civita symbol

The third variant applies to configurations with coinciding indices, e.g. $\epsilon_{112} = 0$. Using the Levi–Civita symbol, the product of all possible combinations of basis vectors is represented as

$$\mathbf{e}_i \times \mathbf{e}_j = \epsilon_{ijk}\, \mathbf{e}_k. \tag{L82}$$

Note that the equation conveniently includes the vanishing of the cross product of identical vectors, $\mathbf{e}_i \times \mathbf{e}_i = 0$. A reformulation of Eq. (L82) respecting the rules of covariance[42] is

$$\boxed{(\mathbf{e}_i \times \mathbf{e}_j) \cdot \mathbf{e}_k = \epsilon_{ijk}.} \tag{L83}$$

Although the Levi–Civita formulation of the cross product Eq. (L82) may not look very intuitive, it is the key to fault-proof computations with cross products. We first note that the vector product between generic vectors, $\mathbf{u} = \mathbf{e}_i u^i$ and $\mathbf{w} = \mathbf{e}_j w^j$, can now be represented as

$$\mathbf{u} \times \mathbf{w} = (\mathbf{e}_i u^i) \times (\mathbf{e}_j w^j) = u^i w^j (\mathbf{e}_i \times \mathbf{e}_j) \overset{(L82)}{=} u^i w^j \epsilon_{ijk}\, \mathbf{e}_k. \tag{L84}$$

This relation shows that the kth component of $\mathbf{u} \times \mathbf{w}$ is given by[42]

$$\boxed{(\mathbf{u} \times \mathbf{w})^k = u^i w^j \epsilon_{ijk}.} \tag{L85}$$

Formulated in column-vector notation, this reads ($\rightarrow$ L4.2.1-2)

$$\begin{pmatrix} u^1 \\ u^2 \\ u^3 \end{pmatrix} \times \begin{pmatrix} w^1 \\ w^2 \\ w^3 \end{pmatrix} = \begin{pmatrix} u^2 w^3 - u^3 w^2 \\ u^3 w^1 - u^1 w^3 \\ u^1 w^2 - u^2 w^1 \end{pmatrix}. \tag{L86}$$

This relation is often taught in high school. However, it leads to cumbersome expressions in calculations with more than two vectors, and we suggest not using it in such cases.

contraction identity

The Levi–Civita symbol obeys the **contraction identity** ($\rightarrow$ L4.2.3-4)

$$\boxed{\epsilon_{ijk}\epsilon_{mnk} = \delta_{im}\delta_{jn} - \delta_{in}\delta_{jm},} \tag{L87}$$

---

[42] Equations (L82) and (L85) show how vector product relations violate the rules of covariant notation: in the former, the index k sits downstairs twice on the right; in the latter, it sits upstairs on the left, downstairs on the right. As mentioned on p. 58, the reason is that we are working in an orthonormal basis whose metric tensor, $g_{ij} = \delta_{ij}$, is not tracked by the notation. Covariant versions of these formulas are discussed in Section L6.4.

where, as usual with pairs of indices, k is summed over. (Verify this identity, ideally without using the hint given in the footnote.[43]) This identity is instrumental to the simplification of expressions involving more than one cross product. It does so by converting products of cross products into combinations of scalar products. ($\rightarrow$ L4.3.1-2)

bac-cab
formula

EXERCISE As an example for the application of Eq. (L87) verify the "**bac-cab formula**",

$$\mathbf{a} \times (\mathbf{b} \times \mathbf{c}) = \mathbf{b}(\mathbf{a} \cdot \mathbf{c}) - \mathbf{c}(\mathbf{a} \cdot \mathbf{b}), \tag{L88}$$

named as such for the characteristic order of vectors on the right. ($\rightarrow$ L4.3.1)

EXAMPLE Let us illustrate the usage of the Levi–Civita symbol by checking that Eq. (L85) conforms with the geometric definition of the vector product. The orthogonality $\mathbf{u} \perp (\mathbf{u} \times \mathbf{w})$ is verified by taking the scalar product:

$$\mathbf{u} \cdot (\mathbf{u} \times \mathbf{w}) \overset{(L85)}{=} u^k u^i w^j \epsilon_{ijk} = -u^k u^i w^j \epsilon_{kji} = -u^i u^k w^j \epsilon_{ijk}. \tag{L89}$$

In the second equality we used the antisymmetry of the Levi–Civita symbol and in the third relabeled the summation indices $i \leftrightarrow k$ (the dummy index in a summation can always be relabeled without changing the result). Comparing the second and fourth terms in (L89), we see that $\mathbf{u} \cdot (\mathbf{u} \times \mathbf{w})$ equals its negative, and therefore must vanish. Similarly one shows that $\mathbf{w} \perp (\mathbf{u} \times \mathbf{w})$. We thus confirm that the vector product computed by (L85) is perpendicular to the plane spanned by $\mathbf{u}$ and $\mathbf{w}$. Its orientation (upward or downward) relative to this plane follows the right-hand rule as described by (L82). A little more work is needed to verify Eq. (L76), according to which the norm of $\mathbf{u} \times \mathbf{w}$ is equal to the area of the parallelogram spanned by $\mathbf{u}$ and $\mathbf{w}$. Using the contraction identity, we find:

$$\|\mathbf{u} \times \mathbf{w}\|^2 \overset{(L34)}{=} (\mathbf{u} \times \mathbf{w}) \cdot (\mathbf{u} \times \mathbf{w}) \overset{(L85)}{=} (u^i w^j \epsilon_{ijk})(u^m w^n \epsilon_{mnk})$$

$$\overset{(L87)}{=} u^i w^j u^m w^n (\delta_{im}\delta_{jn} - \delta_{in}\delta_{jm}) = u^i w^j u^i w^j - u^i w^j u^j w^i$$

$$= (\mathbf{u} \cdot \mathbf{u})(\mathbf{w} \cdot \mathbf{w}) - (\mathbf{u} \cdot \mathbf{w})^2. \tag{L90}$$

Taking the square root, we indeed obtain the area $A(\mathbf{u}, \mathbf{w})$ of the stated parallelogram:

$$\|\mathbf{u} \times \mathbf{w}\| = [(\mathbf{u} \cdot \mathbf{u})(\mathbf{w} \cdot \mathbf{w}) - (\mathbf{u} \cdot \mathbf{w})^2]^{1/2} \overset{(L43)}{=} \|\mathbf{u}\|\|\mathbf{w}\|\sin(\angle(\mathbf{u}, \mathbf{w})) = A(\mathbf{u}, \mathbf{w}). \tag{L91}$$

L4.3 Further properties of the vector product

vector
product
identities

The algebraic definition of the vector product leads to a number of secondary relations:

Grassmann identity:	$\mathbf{u} \times (\mathbf{v} \times \mathbf{w}) = \mathbf{v}(\mathbf{u} \cdot \mathbf{w}) - \mathbf{w}(\mathbf{u} \cdot \mathbf{v}),$	(L92)
Jacobi identity:	$\mathbf{u} \times (\mathbf{v} \times \mathbf{w}) + \mathbf{v} \times (\mathbf{w} \times \mathbf{u}) + \mathbf{w} \times (\mathbf{u} \times \mathbf{v}) = \mathbf{0},$	(L93)
Lagrange identity:	$(\mathbf{v} \times \mathbf{w}) \cdot (\mathbf{t} \times \mathbf{u}) = (\mathbf{v} \cdot \mathbf{t})(\mathbf{w} \cdot \mathbf{u}) - (\mathbf{v} \cdot \mathbf{u})(\mathbf{w} \cdot \mathbf{t}),$	(L94)
	$(\mathbf{v} \times \mathbf{w})^2 = \|\mathbf{v}\|^2\|\mathbf{w}\|^2 - (\mathbf{v} \cdot \mathbf{w})^2.$	(L95)

---

[43] Without loss of generality, assume $(i, j) = (1, 2)$. The sum over k in Eq. (L87) then yields nonzero only for $k = 3$. For this value of k, $\epsilon_{mnk} = 1$ if $(m, n) = (1, 2) = (i, j)$ while $\epsilon_{mnk} = -1$ if $(m, n) = (2, 1) = (j, i)$. This agrees with the value produced by the combination of Kronecker δs on the r.h.s. of Eq. (L87).

All of these have geometric interpretations which, however, are not entirely obvious (try to find them as an exercise). Their algebraic proofs, utilizing the contraction identity (L87), are more straightforward and likewise left as an exercise. ($\rightarrow$ L4.3.1-2)

triple product

It is sometimes useful to combine the cross and scalar products into a third-order product operation known as the **triple product**,

$$\mathbb{E}^3 \times \mathbb{E}^3 \times \mathbb{E}^3 \to \mathbb{R}, \quad (\mathbf{u}, \mathbf{v}, \mathbf{w}) \mapsto (\mathbf{u} \times \mathbf{v}) \cdot \mathbf{w} \stackrel{(L82)}{=} \epsilon_{ijk} u^i v^j w^k. \qquad (L96)$$

The triple product specifies the volume of the parallelepiped[44] spanned by its three argument vectors:

$$\mathrm{Vol}(\mathbf{u}, \mathbf{v}, \mathbf{w}) = |(\mathbf{u} \times \mathbf{v}) \cdot \mathbf{w}|. \qquad (L97)$$

INFO To *understand* Eq. (L97), notice that the volume of a parallelepiped is given by the area of one of its faces times its height in the direction perpendicular to that face. (This statement generalizes the area formula for parallelograms, area = (base line) × (height), to three dimensions.) For example, the volume of the parallelepiped shown in the figure can be computed as the product of the shaded area, A, spanned by $\mathbf{u}$ and $\mathbf{v}$, and the length, s, of the projection of $\mathbf{w}$ onto a line

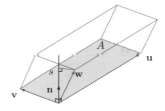

perpendicular to that area. This volume, As, can conveniently be produced by a combination of scalar and vector products: $\mathbf{u} \times \mathbf{v} = A\mathbf{n}$, where $\mathbf{n}$ is a unit vector perpendicular to the base area, and $\mathbf{n} \cdot \mathbf{w} = s$. Thus $|(\mathbf{u} \times \mathbf{v}) \cdot \mathbf{w}| = As$, as stated.[45] Finally, the volume does not depend on which of the three different faces is chosen for the construction. This freedom is reflected in the *cyclic invariance* of the triple product,

$$(\mathbf{u} \times \mathbf{v}) \cdot \mathbf{w} = (\mathbf{v} \times \mathbf{w}) \cdot \mathbf{u} = (\mathbf{w} \times \mathbf{u}) \cdot \mathbf{v}, \qquad (L98)$$

which follows from the cyclic invariance of the Levi–Civita symbol in Eq. (L96).

The triple product can be used to diagnose whether three vectors are linearly independent. If, and only if, they are linearly independent, they span a parallelepiped with nonzero volume, and $(\mathbf{u} \times \mathbf{v}) \cdot \mathbf{w} \neq 0$. The non-vanishing of the triple product is therefore a test for linear independence ($\rightarrow$ L4.3.3-4).

L4.4 Summary and outlook

In this short chapter we introduced the vector product as an operation producing "vectors" from a product operation taking two vectors as arguments. We emphasized that the resulting object is not actually a vector, although it shares many properties with vectors. The differences from conventional vectors showed up in anomalous transformation behavior under reflections, and technically in problems with the covariant formulation. In physics courses these anomalies are often swept under the rug, which is OK as long as one is

---

parallelepiped

[44] A **parallelepiped** is the three-dimensional generalization of a parallelogram.

[45] The absolute value is needed due to the antisymmetry of the vector product, $(\mathbf{u} \times \mathbf{v}) \cdot \mathbf{w} = -(\mathbf{v} \times \mathbf{u}) \cdot \mathbf{w}$. The scalar triple product can thus be either positive or negative, depending on whether its argument vectors satisfy a right-hand rule or not. However, in both cases $|(\mathbf{u} \times \mathbf{v}) \cdot \mathbf{w}|$ gives the volume of the parallelepiped.

prepared to encounter occasional oddities. At the same time, we noted that the vector product *does* admit a mathematically clean formulation, to be discussed later in Section L10.9. The reason why we have discussed it at this early stage in the text, and in its traditional formulation as a (pseudo-)vector, is that it is required in introductory physics courses on mechanics and electricity. We illustrated this point on various examples discussing the role of the vector product in the description of rotational motion.

After this digression we now return to the main track and the discussion of *maps* between vector spaces.

L5   Linear maps

REMARK Much of the following discussion applies to both real and complex vector spaces. To avoid excessive notation we will focus on the complex case throughout. At all stages the imaginary part of complex numbers may be set to zero (i.e. replacing $a \in \mathbb{C}$ by $a \in \mathbb{R}$) to obtain the corresponding theory of real matrices. In the few cases where the real and the complex theory differ, both variants will be discussed in turn.

At this point, we understand the structure of vector spaces. However, beyond this descriptive level not much has really been "done" with them. This will now change when we consider maps between vector spaces. Of particular interest is a class of maps called "linear". Linear maps are to vector spaces what homomorphisms are to groups. They are compatible with the algebraic operations addition and scalar multiplication. For example, the rotation (a linear operation) of the sum of two vectors gives the same result as the addition of the individually rotated vectors. In this sense, rotations are compatible with the linear structure of vector spaces. Both in physics and mathematics, linear maps are very important. Many maps of physical significance – rotations, reflections, and virtually all maps appearing in the mathematical description of quantum mechanics – are linear. At the same time, linear maps provide simple approximations to more complicated functions: much like a smooth one-dimensional function looks approximately linear if one zooms in strongly enough, functions between higher-dimensional spaces, too, admit local approximations through linear maps.

Once a basis has been chosen, linear maps are described by mathematical objects called matrices. Matrices are to linear maps what component vectors are to general vectors. They are the objects with which calculations with linear maps are performed. In this chapter, the general theory of linear maps and matrices is introduced. This will be followed by three more chapters in which practical aspects of working with matrices stand in the foreground.

L5.1   Linear maps

linear map   A map $F : V \rightarrow W$ between two vector spaces (this includes the case $V = W$ of maps acting within one vector space) is called a **linear map** if

$$F(a\mathbf{v} + b\mathbf{w}) = aF(\mathbf{v}) + bF(\mathbf{w}),$$

(L99)

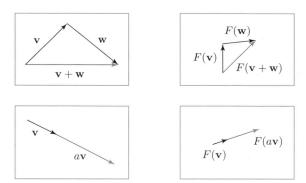

Fig. L12   Illustration of linearity. Top panels: the map F rotates vectors by 45 deg and shrinks them by a factor of 2. Application of the map to the sum of two vectors, $F(\mathbf{v} + \mathbf{w})$, gives the same result as the addition of the individually mapped vectors, $F(\mathbf{v}) + F(\mathbf{w})$. Bottom panels: the same with regard to scalar multiplication: $F(a\mathbf{v}) = aF(\mathbf{v})$, sketched for $a = 3$.

for all $a, b \in \mathbb{R}$ and $\mathbf{v}, \mathbf{w} \in V$.[46] The definition requires that the same result is obtained no matter whether vectors are first added in V and then mapped into W, or first mapped and then added in W (as illustrated in Fig. L12). For example, the isomorphisms $\phi_{\mathbf{v}} : V \to \mathbb{R}^n$ of section L2.5, mapping V-vectors onto $\mathbb{R}^n$-component vectors, satisfy this condition and are examples of linear maps.

It is customary to denote linear maps by capitalized early latin letters, i.e. $A, B, \ldots$ instead of $F, G, \ldots$. The brackets enclosing the argument vector are usually dropped and one writes $A\mathbf{v}$ instead of $A(\mathbf{v})$.

EXAMPLE A **photo** is a map of three-dimensional objects onto a two-dimensional image. This defines a linear map, $\mathbb{E}^3 \to \mathbb{E}^2$: doubling the object size makes the image twice as large, and the displacement of the object (formally, the addition of a fixed vector to the vectors defining it) leads to a proportionally displaced image. Notice, however, that it is not in general possible to reconstruct the original object from its image: the photographic "map" is not invertible. By contrast, photos taken of two-dimensional objects do allow for reconstruction. This anticipates a point to be discussed in more detail below: invertible linear maps can exist only between vector spaces of equal dimensionality.

INFO The importance of linear maps is also reflected in the physics curriculum, where linear algebra is routinely taught in the first or second terms. It was not always like this. When the "modern" theory of quantum mechanics was formulated in the third decade of the twentieth century, linear maps and their description through matrices were unfamiliar to the majority of physicists. They were certainly unknown to Werner Heisenberg when he formulated the foundations of the operator approach to *quantum mechanics*. It was Max Born, together with his collaborator Pascual Jordan, who realized that Heisenberg's theory could be formulated in the language of linear maps, and this observation was published in the joint paper M. Born, W. Heisenberg, P. Jordan, Zur Quantenmechanik II. *Zeitschrift für Physik* **35**, 557 (1926). The authors of that paper considered matrices so important that they suggested the name "Matrizenmechanik" (matrix mechanics) for their formulation of quantum mechanics. Although this denotation did not stick, the first formulation of quantum mechanics in terms of linear maps marked the beginning of quantum theory as it is taught to date. Since then linear algebra has been an indispensable tool in physics.

---

[46] As an example of a *non*linear map consider $F : V \to V, \mathbf{v} \mapsto \mathbf{v}\|\mathbf{v}\|$. Why is this map not linear?

L5.2 Matrices

In this section we consider linear maps between standard vector spaces, $V = \mathbb{C}^n$, $W = \mathbb{C}^m$. We will see how in this particular case linear maps are represented by matrices. Later, in Section L5.5, it will turn out that this discussion already covered most of the mathematics of linear maps between general spaces. The logic parallels that of the introductory Chapter L2, where we first discussed the standard spaces to later realize that their mathematics is representative of that of general spaces.

The emergence of matrices

Consider the simplest of all complex vector spaces, $\mathbb{C}^1 \cong \mathbb{C}$. A general linear map multiplies "vectors",[47] $x \in \mathbb{C}$, by a fixed complex number, $x \mapsto Ax$, $A \in \mathbb{C}$. It is indeed straightforward to verify (do it!) that this operation satisfies the linearity criterion. At the same time, the linearity criterion excludes the appearance of powers different from one ($x \mapsto Bx^\alpha$, $\alpha \neq 1$), and of additive constants ($x \mapsto Ax + C$) (why?). So $x \mapsto Ax$ really is the most general linear map of the one-dimensional vector space. ($\rightarrow$ L5.1.1-2)

Now let us generalize this construction to maps of the form $A : \mathbb{C}^2 \to \mathbb{C}$. Here, $\mathbf{x} = (x^1, x^2)^\mathrm{T}$ has two components and $A\mathbf{x} \in \mathbb{C}$ is a number depending on these. Again, it is not difficult to verify that the most general linear map is of first order in the arguments, $A\mathbf{x} = A_1 x^1 + A_2 x^2$, with complex coefficients A_1, A_2. For a map $A : \mathbb{C}^2 \to \mathbb{C}^2$ the image $A\mathbf{x}$ has two components which must both be linear functions of x^1 and x^2. The image vector can therefore be parametrized as

$$A \begin{pmatrix} x^1 \\ x^2 \end{pmatrix} = \begin{pmatrix} A^1{}_1 x^1 + A^1{}_2 x^2 \\ A^2{}_1 x^1 + A^2{}_2 x^2 \end{pmatrix}, \tag{L100}$$

through four complex numbers $\{A^i{}_j\}$.

EXERCISE Everything said so far likewise applies to real vector spaces. As an example, consider the map $A : \mathbb{R}^2 \to \mathbb{R}^2$ described by

$$A \begin{pmatrix} x^1 \\ x^2 \end{pmatrix} = \begin{pmatrix} \cos(\theta)\, x^1 - \sin(\theta)\, x^2 \\ \sin(\theta)\, x^1 + \cos(\theta)\, x^2 \end{pmatrix}, \tag{L101}$$

and let it act on some simple vectors such as $\mathbf{x} = (1, 0)^\mathrm{T}$, or $\mathbf{x} = (1, 1)^\mathrm{T}$. Convince yourself that A describes the *rotation* of vectors by an angle θ. Formulate geometric arguments (no formulas) for why rotations of space are linear.

The generalization to maps between vector spaces of arbitrary dimension should now be obvious. A linear map $A : \mathbb{C}^n \mapsto \mathbb{C}^m$ is specified by

$$\mathbf{x} = \begin{pmatrix} x^1 \\ x^2 \\ \vdots \\ x^m \end{pmatrix} \longmapsto A\mathbf{x} = \begin{pmatrix} A^1{}_1 x^1 + A^1{}_2 x^2 + \cdots + A^1{}_n x^n \\ A^2{}_1 x^1 + A^2{}_2 x^2 + \cdots + A^2{}_n x^n \\ \vdots \\ A^m{}_1 x^1 + A^m{}_2 x^2 + \cdots + A^m{}_n x^n \end{pmatrix} = \begin{pmatrix} A^1{}_j x^j \\ A^2{}_j x^j \\ \vdots \\ A^m{}_j x^j \end{pmatrix}, \tag{L102}$$

---

[47] For one-component vectors, we abandon the boldface convention and write $x = \mathbf{x} = (x)$ for simplicity.

so that the ith component of the vector $A\mathbf{x}$ is given by

$$\boxed{i = 1, \ldots m: \qquad (A\mathbf{x})^i = A^i{}_1 x^1 + A^i{}_2 x^2 + \cdots + A^i{}_n x^n = A^i{}_j x^j.}$$ (L103)

In other words,

$$\boxed{\text{a linear map } A : \mathbb{C}^n \to \mathbb{C}^m \text{ is fully specified by } m \times n \text{ complex numbers } \{A^i{}_j\}.}$$

This means that linear maps are simpler than generic maps. For example, just a single number is required to specify a linear function in the case $n = m = 1$. By contrast, the description of a generic function, $F : \mathbb{C} \to \mathbb{C}, x \mapsto F(x)$, requires the specification of infinitely many function values.

In the following, arrays of sums as in Eq. (L102) will appear so often that it pays to

matrix   switch to a more *efficient notation*: we define the rectangular **matrix**

$$A = \begin{pmatrix} A^1{}_1 & \cdots & A^1{}_j & \cdots & A^1{}_n \\ \vdots & & \vdots & & \vdots \\ A^i{}_1 & \cdots & A^i{}_j & \cdots & A^i{}_n \\ \vdots & & \vdots & & \vdots \\ A^m{}_1 & \cdots & A^m{}_j & \cdots & A^m{}_n \end{pmatrix}.$$ (L104)

This **matrix representation** fully specifies the map, A, and is customarily denoted by the same symbol, A. Occasionally, a more marked distinction needs to be drawn between a map, $\hat{A}$, and its matrix representation, A. In such cases the former will carry a caret.

matrix   The entries, $A^i{}_j$, of a matrix A are called its **components** or **elements**. The full matrix is
elements   often denoted as $A = \{A^i{}_j\}$, where the index range is left implicit. In the representation of matrix elements as $A^i{}_j$, the left index i categorically labels **rows** and the right index j **columns**. This convention applies also when non-covariant notation, A_{ij}, is used. Finally, notice that the limiting case of an $m \times 1$ matrix with m rows and just one column is an m-component *vector* with components $A^i{}_1$. Conversely, a $1 \times n$ matrix with only one row is an n-component "vector" with components $A^1{}_j$ and components arranged in a horizontal order. The term vector is put in quotes because the row-object carries its indices in a covariant position. This signals that an n-component object different from a genuine vector is at hand.

Matrix–vector multiplication

The action of this matrix on a vector $\mathbf{x}$ to yield the vector $A\mathbf{x}$ of Eq. (L102), expressed more commonly as the *multiplication of a vector by a matrix*, is now defined through the relation:

$$
\begin{pmatrix} A^1{}_j x^j \\ \vdots \\ A^i{}_j x^j \\ \vdots \\ A^m{}_j x^j \end{pmatrix} \equiv \begin{pmatrix} A^1{}_1 & \cdots & A^1{}_j & \cdots & A^1{}_n \\ \vdots & & \vdots & & \vdots \\ A^i{}_1 & \cdots & A^i{}_j & \cdots & A^i{}_n \\ \vdots & & \vdots & & \vdots \\ A^m{}_1 & \cdots & A^m{}_j & \cdots & A^m{}_n \end{pmatrix} \begin{pmatrix} x^1 \\ \vdots \\ x^j \\ \vdots \\ x^n \end{pmatrix}. \tag{L105}
$$

The ith component of $A\mathbf{x}$ is computed by moving stepwise from left to right along the ith row of the matrix A, and at the same time from top to bottom along the single column of the vector representing $\mathbf{x}$. At each step the corresponding matrix element $A^i{}_j$ and vector element x^j are multiplied and the results added up, $A^i{}_1 x^1 + \cdots + A^i{}_n x^n$, to obtain $(A\mathbf{x})^i = A^i{}_j x^j$, as given in Eq. (L103).

EXAMPLE We illustrate the operation of multiplying a vector by a matrix with some examples:

$$
\begin{pmatrix} 1 & 4 \\ 5 & 3 \end{pmatrix} \cdot \begin{pmatrix} 2 \\ 1 \end{pmatrix} = \begin{pmatrix} 1 \cdot 2 + 4 \cdot 1 \\ 5 \cdot 2 + 3 \cdot 1 \end{pmatrix} = \begin{pmatrix} 6 \\ 13 \end{pmatrix},
$$

$$
\begin{pmatrix} 2 & 5 \\ 3 & 3 \\ 6 & 1 \end{pmatrix} \cdot \begin{pmatrix} 3 \\ 2 \end{pmatrix} = \begin{pmatrix} 2 \cdot 3 + 5 \cdot 2 \\ 3 \cdot 3 + 3 \cdot 2 \\ 6 \cdot 3 + 1 \cdot 2 \end{pmatrix} = \begin{pmatrix} 16 \\ 15 \\ 20 \end{pmatrix},
$$

$$
\begin{pmatrix} 1 & 3 & 2 \end{pmatrix} \cdot \begin{pmatrix} 4 \\ -1 \\ 2 \end{pmatrix} = 1 \cdot 4 + 3 \cdot (-1) + 2 \cdot 2 = 5,
$$

$$
2 \cdot 2 = 4.
$$

For an $m \times n$ matrix $A = \{A^i{}_j\}$, the jth column defines a vector,

$$
\mathbf{A}_j = (A^1{}_j, \ldots, A^m{}_j)^{\mathrm{T}} = \begin{pmatrix} A^1{}_j \\ \vdots \\ A^m{}_j \end{pmatrix}, \tag{L106}
$$

with components $(\mathbf{A}_j)^i = A^i{}_j$. For example, the first column of the 2×2 matrix $A = \begin{pmatrix} a & b \\ c & d \end{pmatrix}$ defines $\mathbf{A}_1 = \begin{pmatrix} a \\ c \end{pmatrix}$. A general matrix can be written as an n-tuple,

$$
A = (\mathbf{A}_1, \ldots, \mathbf{A}_n), \tag{L107}
$$

formed by its n column vectors. Likewise, the ith row of A can be identified with the transpose of a vector, $\mathbf{A}^i = (A^i{}_1, \ldots, A^i{}_n)$. For example, the second row of the 2×2 matrix corresponds to $\mathbf{A}^2 = (c, d)$. We can think of a general matrix as a stack of m of these objects,

$$
A = \begin{pmatrix} \mathbf{A}^1 \\ \vdots \\ \mathbf{A}^m \end{pmatrix}. \tag{L108}
$$

Matrix action on basis vectors   The *action of A on the jth standard basis vector*, $\mathbf{e}_j = (0, \ldots, 1, \ldots, 0)^{\mathrm{T}}$ (with the 1 at position j), is given by

$$Ae_j = \begin{pmatrix} A^1{}_1 & \cdots & A^1{}_j & \cdots & \cdots & A^1{}_n \\ \vdots & & \vdots & & & \vdots \\ & & & & & \\ \vdots & & \vdots & & & \vdots \\ A^i{}_1 & \cdots & A^i{}_j & \cdots & \cdots & A^i{}_n \\ \vdots & & \vdots & & & \vdots \\ & & & & & \\ A^m{}_1 & \cdots & A^m{}_j & \cdots & \cdots & A^m{}_n \end{pmatrix} \cdot \begin{pmatrix} 0 \\ \vdots \\ 1 \\ \vdots \\ \\ 0 \end{pmatrix} = \begin{pmatrix} A^1{}_j \\ \vdots \\ \\ \vdots \\ A^i{}_j \\ \vdots \\ \\ A^m{}_j \end{pmatrix}, \tag{L109}$$

i.e. by the jth column of the matrix A. In the column vector notation introduced above this can be written as

$$Ae_j = \mathbf{A}_j. \tag{L110}$$

Linear maps are often *defined* by their action on the standard basis vectors, $\mathbf{e}_j \mapsto \mathbf{v}_j$. In this definition, n vectors $\mathbf{v}_j$ are specified as the image vectors of the standard basis vectors. Equation (L110) implies that the matrix representing the map can then be represented as an array containing the n image vectors, $\mathbf{v}_j = \mathbf{A}_j$, as columns, $A = (\mathbf{v}_1, \mathbf{v}_2, \ldots, \mathbf{v}_n)$.

EXAMPLE If $A : \mathbb{C}^2 \to \mathbb{C}^3$ is a map whose action on the standard basis of $\mathbb{C}^2$ is defined by the first two expressions in Eq. (L111), then its matrix representation A is given by the third:

$$\begin{pmatrix} 1 \\ 0 \end{pmatrix} \overset{A}{\mapsto} \begin{pmatrix} 3 \\ 2 \\ 1 \end{pmatrix}, \quad \begin{pmatrix} 0 \\ 1 \end{pmatrix} \overset{A}{\mapsto} \begin{pmatrix} -1 \\ 0 \\ 1 \end{pmatrix}, \quad A = \begin{pmatrix} 3 & -1 \\ 2 & 0 \\ 1 & 1 \end{pmatrix}. \tag{L111}$$

INFO The set of all complex matrices containing m rows and n columns is denoted by $\mathrm{mat}(m, n, \mathbb{C})$, and by $\mathrm{mat}(n, \mathbb{C})$ for square matrices. Two matrices $A, B \in \mathrm{mat}(m, n, \mathbb{C})$ may be added component-wise to obtain a matrix $A + B$ in the same set with components $(A + B)^i{}_j \equiv A^i{}_j + B^i{}_j$. Similarly, a matrix, A, may be multiplied by a scalar, $a \in \mathbb{C}$, to obtain a matrix aA with matrix elements $(aA)^i{}_j \equiv aA^i{}_j$. With these definitions, $\mathrm{mat}(m, n, \mathbb{C})$ becomes a complex vector space. (Show that all the vector space axioms are satisfied.) Since each matrix is specified by $m \cdot n$ components, we know that the complex dimension of the *vector space of complex matrices* is mn.

The transpose and the adjoint of a matrix

We conclude our introduction of matrices by defining two operations that will become increasingly important as we go along. Given an $m \times n$ matrix A we may define an $n \times m$ matrix A^{T} (spoken A-transpose) by exchanging rows and columns. This is illustrated in the following examples:

$$A = \begin{pmatrix} 2 & 3 & 1 \\ 2 & 4 & 7 \end{pmatrix} \longrightarrow A^{\mathrm{T}} = \begin{pmatrix} 2 & 2 \\ 3 & 4 \\ 1 & 7 \end{pmatrix},$$

$$A = \begin{pmatrix} 4 & 2 \\ 1 & 5 \end{pmatrix} \longrightarrow A^{\mathrm{T}} = \begin{pmatrix} 4 & 1 \\ 2 & 5 \end{pmatrix},$$

$$A = (1 \ 3 \ 0) \longrightarrow A^{\mathrm{T}} = \begin{pmatrix} 1 \\ 3 \\ 0 \end{pmatrix}. \tag{L112}$$

The third example shows how the transpose of a row vector, $(1, 3, 0)^{\mathrm{T}}$, yields a column vector. Previously we had used this notation as a shorthand for column vectors, e.g.
transpose $(0, 1)^{\mathrm{T}} = \binom{0}{1}$. For a matrix with elements $A^i{}_j$, the elements of the **transpose matrix** are given by

$$(A^{\mathrm{T}})_j{}^i = A^i{}_j, \tag{L113}$$

where on both sides of the equation the left index labels rows, the right index columns. For example, $(A^{\mathrm{T}})_2{}^1 = 3$ in the first example given above. According to Eq. (L113), the transpose of a matrix is obtained by sliding its indices horizontally, $A^i{}_j \to A_{j \leftarrow}{}^{\to i}$, thus interchanging row and column indices. The prescription works in both directions, so that for a matrix with elements $B_j{}^i$, $(B^{\mathrm{T}})^i{}_j = B_j{}^i$ via the reverse slide $B_{j \to}{}^{\leftarrow i}$. In covariant notation, covariant indices remain covariant (a subscript index remains in a subscript position), and the same with contravariant indices. The rationale behind this structure will be discussed later in Section L10.3. In non-covariant notation, transposition amounts to an index interchange, $(A^{\mathrm{T}})_{ij} = A_{ji}$.

adjoint   For a *complex vector space*, we define the **adjoint matrix**, $A^\dagger$ (spoken "A-adjoint" or
dagger, †   "A- **dagger**"), through transposition followed by complex conjugation:

$$A^\dagger = \overline{A^{\mathrm{T}}}, \qquad (A^\dagger)_j{}^i = \overline{A^i{}_j}. \tag{L114}$$

For example,

$$A = \begin{pmatrix} 1 + 2\mathrm{i} & 5 \\ 4 - \mathrm{i} & 3\mathrm{i} \end{pmatrix} \longrightarrow A^\dagger = \begin{pmatrix} 1 - 2\mathrm{i} & 4 + \mathrm{i} \\ 5 & -3\mathrm{i} \end{pmatrix}. \tag{L115}$$

L5.3 Matrix multiplication

The power of matrix calculus becomes apparent when we consider the *composition* of linear maps. For example, consider the consecutive rotation, stretching and again rotation of a figure on a computer screen. All three operations are linear maps and the joint operation is the composition of these maps.

The composition, $C \equiv B \circ A : \mathbb{C}^n \to \mathbb{C}^l$, of two linear maps, $A : \mathbb{C}^n \to \mathbb{C}^m$ and $B : \mathbb{C}^m \to \mathbb{C}^l$, is again linear and hence described by a matrix. The $l \times n$ matrix $\{C^k{}_j\}$ representing map C can be found from the $l \times m$ and $m \times n$ matrices $\{B^k{}_i\}$ and $\{A^i{}_j\}$ of B and A, respectively. To this end, observe that A maps a $\mathbb{C}^n$-vector with components x^j onto a $\mathbb{C}^m$-vector with components $A^i{}_j x^j$. This image vector, in turn, is mapped by B onto a $\mathbb{C}^l$-vector with components $B^k{}_i \left(A^i{}_j x^j \right) = \left(B^k{}_i A^i{}_j \right) x^j$. The matrix of the composite map, $C = B \circ A$, thus has matrix elements

$$\boxed{C^k{}_j = B^k{}_i A^i{}_j.} \tag{L116}$$

**matrix multi-
plication**
The product matrix, C, representing the composite map, is denoted as $C = B \cdot A$, and even more frequently just $C = BA$. Writing the matrix product in an array representation as

$$
\begin{pmatrix}
C^1_1 & \cdots & C^1_j & \cdots & C^1_n \\
\vdots & & \vdots & & \vdots \\
C^k_1 & \cdots & C^k_j & \cdots & C^k_n \\
\vdots & & \vdots & & \vdots \\
C^l_1 & \cdots & C^l_j & \cdots & C^l_n
\end{pmatrix}
\equiv
\begin{pmatrix}
B^1_i A^i_1 & \cdots & B^1_i A^i_j & \cdots & B^1_i A^i_n \\
\vdots & & \vdots & & \vdots \\
B^k_i A^i_1 & \cdots & B^k_i A^i_j & \cdots & B^k_i A^i_n \\
\vdots & & \vdots & & \vdots \\
B^l_i A^i_1 & \cdots & B^l_i A^i_j & \cdots & B^l_i A^i_n
\end{pmatrix}
$$

$$
\equiv
\begin{pmatrix}
B^1_1 & \cdots & B^1_i & \cdots & B^1_m \\
\vdots & & \vdots & & \vdots \\
B^k_1 & \cdots & B^k_i & \cdots & B^k_m \\
\vdots & & \vdots & & \vdots \\
B^l_1 & \cdots & B^l_i & \cdots & B^l_m
\end{pmatrix}
\cdot
\begin{pmatrix}
A^1_1 & \cdots & A^1_j & \cdots & A^1_n \\
\vdots & & \vdots & & \vdots \\
A^i_1 & \cdots & A^i_j & \cdots & A^i_n \\
\vdots & & \vdots & & \vdots \\
A^m_1 & \cdots & A^m_j & \cdots & A^m_n
\end{pmatrix}, \quad \text{(L117)}
$$

the element C^k_j is obtained by moving from left to right along the kth row of B and at the same time from top to bottom along the jth column of A, multiplying the corresponding matrix elements and adding up: $C^k_j = B^k_1 A^1_j + \cdots + B^k_m A^m_j = B^k_i A^i_j$, in agreement with Eq. (L116). The take-home message here is:

> The composition of linear maps is described by the product of their representing matrices, where the product operation is defined by Eq. (L116).

EXAMPLE Here are a few examples of matrix products: ($\rightarrow$ L5.3.1-6)

$$
\begin{pmatrix} 1 & 4 \\ 5 & 3 \end{pmatrix} \cdot \begin{pmatrix} 2 & 1 \\ 1 & 3 \end{pmatrix} = \begin{pmatrix} 1\cdot 2 + 4\cdot 1 & 1\cdot 1 + 4\cdot 3 \\ 5\cdot 2 + 3\cdot 1 & 5\cdot 1 + 3\cdot 3 \end{pmatrix} = \begin{pmatrix} 6 & 13 \\ 13 & 14 \end{pmatrix},
$$

$$
\begin{pmatrix} 1 & 2 \\ 3 & 3 \\ 2 & 4 \end{pmatrix} \cdot \begin{pmatrix} 3 & 1 \\ 2 & 1 \end{pmatrix} = \begin{pmatrix} 1\cdot 3 + 2\cdot 2 & 1\cdot 1 + 2\cdot 1 \\ 3\cdot 3 + 3\cdot 2 & 3\cdot 1 + 3\cdot 1 \\ 2\cdot 3 + 4\cdot 2 & 2\cdot 1 + 4\cdot 1 \end{pmatrix} = \begin{pmatrix} 7 & 3 \\ 15 & 6 \\ 14 & 6 \end{pmatrix},
$$

$$
(2) \cdot (2) = (4).
$$

The *composition of more than two maps* is defined by straightforward extension of the definition above. For example, if three maps are applied in succession as A, then B, then C, the composite map is obtained as $C \circ (B \circ A) = (C \circ B) \circ A = C \circ B \circ A$, where the associativity of the composition means that the brackets are superfluous. In the language of matrices this means $C \cdot B \cdot A = C \cdot (B \cdot A) = (C \cdot B) \cdot A$, where all "$\cdot$" operations are matrix products. When working with products containing more than two matrices it sometimes pays to think about the "most economic" way to compute the product; matrix multiplication is time consuming and some orders of taking products are more efficient than others. In general, the proliferation of terms to be added/multiplied makes higher-order multiplications cumbersome. Sometimes, however, matrix multiplication leads to simple results, as is illustrated by the following example.

EXERCISE Verify that the product of the matrices A, B and A' given below has the "diagonal" form indicated on the right:

$$A = \frac{1}{\sqrt{2}} \begin{pmatrix} 1 & 1 \\ -1 & 1 \end{pmatrix}, \quad B = \begin{pmatrix} 0 & 1 \\ 1 & 0 \end{pmatrix}, \quad A' = \frac{1}{\sqrt{2}} \begin{pmatrix} 1 & -1 \\ 1 & 1 \end{pmatrix}, \quad ABA' = \begin{pmatrix} 1 & 0 \\ 0 & -1 \end{pmatrix}.$$

In Section L7 we will discuss why the product ABA' assumes a simple form.

Properties of matrix multiplication

Matrix multiplication is one of the most important operations of linear algebra. The matrix product is $(a, a' \in \mathbb{C})$

▷ *associative*: $C \cdot (B \cdot A) = (C \cdot B) \cdot A = C \cdot B \cdot A,$ (L118)

▷ *distributive*: $C \cdot (aB + a'B') = aC \cdot B + a'C \cdot B',$ (L119)

▷ in general *not commutative*: $A \cdot B \neq B \cdot A,$ (L120)

▷ factor-reversing under *transposition*: $(A \cdot B)^{\mathrm{T}} = B^{\mathrm{T}} \cdot A^{\mathrm{T}}.$ (L121)

associativity The **associativity** of matrix multiplication reflects the associativity of the composition of linear maps. It also follows from the definition (L117) of matrix multiplication: $C^l{}_k(B^k{}_i A^i{}_j) = (C^l{}_k B^k{}_i)A^i{}_j$. **Distributivity** is a trivial consequence of the definition. The **noncommutativity** **lack of commutativity** is very important. It means that linear maps carried out in different orders generally lead to different results, as illustrated in the exercise below. Finally, Eq. (L121) for the transpose of products follows from $((A \cdot B)^{\mathrm{T}})_j{}^i = (A \cdot B)^i{}_j = A^i{}_k B^k{}_j = (A^{\mathrm{T}})_k{}^i (B^{\mathrm{T}})_j{}^k = (B^{\mathrm{T}})_j{}^k (A^{\mathrm{T}})_k{}^i = (B^{\mathrm{T}} \cdot A^{\mathrm{T}})_j{}^i$. (The fourth equality follows from the fact that individual matrix elements of A and B, which are numbers, do commute, although the matrices themselves in general do not commute.)

algebra **INFO** A (complex) **algebra** is a $\mathbb{C}$-vector space V with a product operation

$$V \times V \to V, \qquad (u, v) \mapsto u \cdot v,$$

subject to the following conditions $(u, v, w \in V, \ c \in \mathbb{C})$:

▷ $(u + v) \cdot w = u \cdot w + v \cdot w,$

▷ $u \cdot (v + w) = u \cdot v + u \cdot w,$

▷ $c(v \cdot w) = (cv) \cdot w = v \cdot (cw).$

Our discussion above shows that the space of $n \times n$ matrices forms an algebra, $(\mathrm{mat}(n, \mathbb{C}), \cdot)$. Its elements are matrices, $A, B, \ldots$, and its product operation is matrix multiplication, $A \cdot B = C$. Due to the associativity of this operation, the matrix algebra is called an associative algebra. *Real algebras* over $\mathbb{R}$-vector spaces are defined analogously.

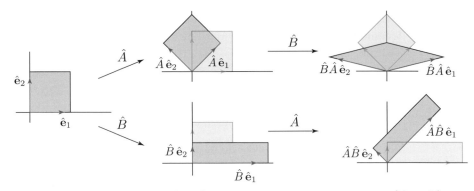

Fig. L13   Illustration in $\mathbb{E}^2$ of the action of the maps $\hat{A}$ and $\hat{B}$ defined in Eq. (L122), and of their compositions $\hat{A}\hat{B}$ and $\hat{B}\hat{A}$.

EXERCISE Consider two maps in $\mathbb{R}^2$, A and B, with matrix representations

$$A = \frac{1}{\sqrt{2}} \begin{pmatrix} 1 & -1 \\ 1 & 1 \end{pmatrix}, \qquad B = \begin{pmatrix} 2 & 0 \\ 0 & \frac{1}{2} \end{pmatrix}. \tag{L122}$$

Compute the action of these matrices on the standard vectors $(1,0)^{\mathrm{T}}$ and $(0,1)^{\mathrm{T}}$. Convince yourself that A describes a counter-clockwise rotation by an angle $\pi/4$, and B stretching and shrinking by a factor two in the 1- and 2-directions, respectively (see Fig. L13). Now compute the matrix products

$$BA = \begin{pmatrix} 2 & 0 \\ 0 & \frac{1}{2} \end{pmatrix} \frac{1}{\sqrt{2}} \begin{pmatrix} 1 & -1 \\ 1 & 1 \end{pmatrix} = \frac{1}{\sqrt{2}} \begin{pmatrix} 2 & -2 \\ \frac{1}{2} & \frac{1}{2} \end{pmatrix},$$

$$AB = \frac{1}{\sqrt{2}} \begin{pmatrix} 1 & -1 \\ 1 & 1 \end{pmatrix} \begin{pmatrix} 2 & 0 \\ 0 & \frac{1}{2} \end{pmatrix} = \frac{1}{\sqrt{2}} \begin{pmatrix} 2 & -\frac{1}{2} \\ 2 & \frac{1}{2} \end{pmatrix},$$

describing rotating then stretching, or stretching then rotating, respectively. Let these composites act on the standard basis vectors to explore how they are affected.

L5.4   The inverse of a matrix

In Section L1.1 we learned that if a map is bijective then an inverse map exists. Specifically, for an invertible linear map, $A : \mathbb{C}^n \to \mathbb{C}^m$, there is an inverse map, $A^{-1} : \mathbb{C}^m \to \mathbb{C}^n$, such that $A^{-1}A$ is the identity map on $\mathbb{C}^n$ (and AA^{-1} the identity on $\mathbb{C}^m$). These statements raise a number of questions: do invertible maps exist between spaces of arbitrary dimension n and m? How can we know if a map possesses an inverse? If it does, how can it be obtained?

The answer to the first question is that

> invertible maps $A : \mathbb{C}^n \to \mathbb{C}^m$ exist only
> between spaces of equal dimension, $n = m$.

This statement is intuitively understandable. If $A : \mathbb{C}^n \to \mathbb{C}^m$, and $m > n$, then the target space is "too big" to be surjectively covered. Conversely, if $m < n$ then it is "too small" for an injective assignment. Consider, then, maps between spaces of equal dimension, $n = m$. According to Section L1.1, bijectivity requires both injectivity and surjectivity. Whereas these are generally independent features, the situation with linear maps is simpler.

> For linear maps between spaces of equal dimension, $A : \mathbb{C}^n \to \mathbb{C}^n$, the conditions of injectivity and surjectivity are equivalent.

criteria for invertibility of linear maps

According to this statement (which is proven in the next section), invertibility may be verified by testing either one of the two criteria. In practice, however, one does not usually verify injectivity or surjectivity "by hand". Instead, one of the following two useful *criteria for the invertibility of a map* $A : \mathbb{C}^n \to \mathbb{C}^n$ is checked:

▷ consider any basis, $\{\mathbf{v}_j\}$, and verify whether the set of image vectors, $\{A\mathbf{v}_j\}$, is again a basis;

▷ check that no non-zero vector is mapped to zero: $\forall \mathbf{v} \neq \mathbf{0} : A\mathbf{v} \neq \mathbf{0}$.

If either one of these conditions is met, A is invertible (for a proof, see the next subsection).

kernel
We finally remark that the **kernel** of a linear map $A : \mathbb{C}^n \to \mathbb{C}^m$ is defined as the set of vectors which are annihilated by the map, $\mathrm{Ker}(A) \equiv \{\mathbf{v} \in \mathbb{C}^n | A\mathbf{v} = \mathbf{0}\} \subset A$. The kernel of A is a subspace of $\mathbb{C}^n$ (why?). The condition above states that the kernel of an invertible matrix is the null space, $\mathrm{Ker}(A) = \{\mathbf{0}\}$. Likewise, the image of A, $\mathrm{Im}(A) = A(\mathbb{C}^n)$, is a subspace of $\mathbb{C}^m$ (why?), and its dimension, $\dim(\mathrm{Im}(A))$, is called the **rank** of the map. The **rank** invertibility criteria for a map between spaces of equal dimension, n, require that its image span the full space $\mathbb{C}^n$.

> An invertible matrix $A : \mathbb{C}^n \to \mathbb{C}^n$ has maximal rank n.

The dimension formula

REMARK In this subsection we verify the various statements made in the previous one. No reference to specific properties of the standard vector spaces will be made and we therefore consider linear maps, $A : V \to W$, between general spaces of dimensions n and m, respectively. The section can be skipped on first reading if time is short.

Above, we introduced the kernel, $\mathrm{Ker}(A)$, and the image, $\mathrm{Im}(A)$, of a map $A : \mathbb{C}^n \to \mathbb{C}^m$ as subspaces of $\mathbb{C}^n$ and $\mathbb{C}^m$, respectively. In the same way, the image of a general map is the space spanned by its image vectors in W, and the kernel the space of annihilated vectors in V. All that we seem to know *a priori* about the dimensions of these spaces is that they are smaller than or equal to n and m, respectively. However, for linear maps the kernel and image dimensions are in fact related by a stronger relation, known as the **dimension** **dimension formula**:

$$\dim(\mathrm{Ker}(A)) + \dim(\mathrm{Im}(A)) = \dim(V). \qquad \text{(L123)}$$

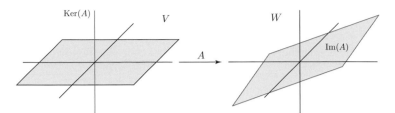

Fig. L14   Schematic of the one-dimensional kernel and the two-dimensional image of a linear map, $A : V \to W$, between two
three-dimensional vector spaces. For a discussion, see the info section below.

Figure L14 illustrates the situation for a real map with $n = 3$ and $\dim(\text{Ker}(A)) = 1$,
$\dim(\text{Im}(A)) = 2$. The dimension of the kernel and of the image add to three, as stated by
the formula. Notice that no reference to the dimension of the target space is made.

proof of   **INFO** The *proof of* Eq. (L123) is straightforward. Let $k \equiv \dim(\text{Ker}(A)) \leq n \equiv \dim(V)$. Now construct
dimension   a basis $\{\mathbf{v}_j\}$ of V such that $\{\mathbf{v}_1, \ldots, \mathbf{v}_k\}$ span $\text{Ker}(A)$. (For a trivial kernel, $\text{Ker}(A) = \{\mathbf{0}\}$, this set is
formula   empty.) The image of A is then spanned by the $n - k$ vectors $\{\mathbf{w}_j = A\mathbf{v}_{k+j} | j = 1, \ldots, n - k\}$, i.e.
$\text{span}\{\mathbf{w}_1, \ldots, \mathbf{w}_{n-k}\} = \text{Im}(A)$. It remains to be shown that these vectors are linearly independent;
if they are, $\text{span}\{\mathbf{w}_1, \ldots, \mathbf{w}_{n-k}\} = \text{Im}(A)$ is an $n - k$ dimensional subspace of W, which implies
$\dim(\text{Im}(A)) + \dim(\text{Ker}(A)) = (n - k) + k = n$ as claimed by the formula. To show that the vectors
$\mathbf{w}_j$ are linearly independent, assume the opposite. In this case there would exist a nontrivial linear
combination yielding zero, $\mathbf{0} = \sum_j a^j \mathbf{w}_j = \sum_j a^j A\mathbf{v}_{k+j} = A\left(\sum_j a^j \mathbf{v}_{k+j}\right)$. This, however, is a
contradiction, because the linear combination $\sum_j a^j \mathbf{v}_{k+j}$ does not lie in $\text{Ker}(A)$ and hence it cannot
map to the null vector.

Equation (L123) has a number of important consequences. For example, it implies that
invertible maps can exist only between spaces of equal dimension, $n = m$. This follows from Eq. (L123) because an invertible map must be surjective and injective and this
requires $\dim(\text{Im}(A)) = m$ and $\dim(\text{Ker}(A)) = 0$, respectively. Our relation thus assumes
the form $n = m$. The formula also implies that if $n = m$, then injectivity and surjectivity
are equivalent. To understand why, assume surjectivity, $\dim(\text{Im}(A)) = n$. Equation (L123)
then states $\dim(\text{Ker}(A)) = 0$, which means injectivity. The reverse conclusion, injectivity
implying surjectivity, is shown in the same way.

Finally, notice that all properties relying on the dimension formula which were discussed
above for $\mathbb{C}^n$ – bijectivity, invertibility criteria, rank of a map, etc. – carry over to the case
of general maps. One just needs to replace $\mathbb{C}^n = V$ and $\mathbb{C}^m = W$ in the discussion of the
previous section.

Matrix inversion

A matrix, A, representing a map $A : \mathbb{C}^n \to \mathbb{C}^n$ has as many rows as columns and is
therefore called a **square matrix**. Assume that A is invertible and an inverse map A^{-1}
inverse   exists. The matrix, A^{-1}, representing the inverse map is called the **inverse matrix** of the
matrix   matrix A. Its defining property, $A^{-1}A = AA^{-1} = \mathbb{1}$, is a matrix equation where $\mathbb{1}$ denotes
unit matrix   the **unit matrix**,

$$\mathbb{1} = \begin{pmatrix} 1 & 0 & \cdots & \\ 0 & 1 & & \\ \vdots & & \ddots & 0 \\ & & 0 & 1 \end{pmatrix}, \tag{L124}$$

with matrix elements $\mathbb{1}^i_{\ j} = \delta^i_{\ j}$. The unit matrix leaves vectors invariant, $\mathbb{1}\mathbf{v} = \mathbf{v}$, and represents the unit map in $\mathbb{C}^n$.

INFO Recall the concept of groups introduced in Section L1.2. The composition of two invertible matrices is invertible, invertible matrices have an inverse, and the identity matrix is the neutral element. This means that the set of invertible, complex $n \times n$ matrices forms a group, the **general linear group**, GL($n, \mathbb{C}$). The restriction to invertible, real $n \times n$ matrices defines GL($n, \mathbb{R}$).

general linear group

The explicit form of the matrix equation defining the inverse reads as

$$A^{-1}A = \begin{pmatrix} (A^{-1})^1_{\ 1} & (A^{-1})^1_{\ 2} & \cdots & \\ (A^{-1})^2_{\ 1} & (A^{-1})^2_{\ 2} & & \\ \vdots & & \ddots & \\ & & & (A^{-1})^n_{\ n} \end{pmatrix} \begin{pmatrix} A^1_{\ 1} & A^1_{\ 2} & \cdots & \\ A^2_{\ 1} & A^2_{\ 2} & & \\ \vdots & & \ddots & \\ & & & A^n_{\ n} \end{pmatrix}$$

$$= \begin{pmatrix} 1 & 0 & \cdots & \\ 0 & 1 & & \\ \vdots & & \ddots & \\ & & & 1 \end{pmatrix},$$

or in a more compact representation,

$$\boxed{i, j = 1, \ldots, n: \qquad (A^{-1})^i_{\ k} A^k_{\ j} = \delta^i_{\ j}.} \tag{L125}$$

EXAMPLE For $n = 2$, a straightforward check shows that the matrix inverse is obtained as

$$A = \begin{pmatrix} a & b \\ c & d \end{pmatrix} \qquad \Rightarrow \qquad A^{-1} = \frac{1}{ad - bc} \begin{pmatrix} d & -b \\ -c & a \end{pmatrix}. \tag{L126}$$

Notice that no inverse exists if $ad = bc$. The reason is that in this case A does not satisfy the invertibility criteria formulated in the previous section: the non-vanishing vector $(d, -c)^{\mathrm{T}}$ is annihilated by A, and the image vectors $A\mathbf{e}_1$ and $A\mathbf{e}_2$ are linearly dependent (check!).

Unfortunately, there is no quick and painless way of *computing matrix inverses* for general n. For $n \geq 4$ matrix inversion is often done on a computer. In fact, the optimization of matrix inversion algorithms is a field of active research in computer science, which shows that the problem is of high relevance, including in applied science.

In low dimensions, $n = 2, 3, 4, \ldots$, a matrix can be constructively inverted as follows: start from the column vector representation Eq. (L107), $A^{-1} = (\mathbf{a}_1, \mathbf{a}_2, \ldots, \mathbf{a}_n)$.

Substituting this into the matrix equation $AA^{-1} = \mathbb{1}$ we observe that the jth column vector, $\mathbf{a}_j$, is determined by the equation

$$A\mathbf{a}_j = \mathbf{e}_j. \tag{L127}$$

For each j, this defines a system of n linear equations, $A^i{}_k(\mathbf{a}_j)^k = \delta^i{}_j$, one for each component i. Altogether $n \times n$ scalar linear equations, or n vector equations, need to be solved. The n solution vectors, $\mathbf{a}_j$, then define A^{-1}. An efficient scheme for solving linear systems of equations is discussed in the next subsection.

Solving systems of linear equations

In linear algebra one often needs to solve problems of the form

$$A\mathbf{x} = \mathbf{b}, \tag{L128}$$

system of linear equations

where A is a given $m \times n$ matrix, $\mathbf{b} \in \mathbb{C}^m$ a given vector, and $\mathbf{x} \in \mathbb{C}^n$ is sought. Written in components, this assumes the form of a **system of linear equations**,

$$A^i{}_j x^j = b^i, \qquad i = 1, \ldots, m. \tag{L129}$$

For $\mathbf{b} = \mathbf{0}$ the system is called *homogenous*, otherwise *inhomogeneous*. A homogeneous system determines $\mathbf{x}$ only up to a multiplicative constant: if $\mathbf{x}$ satisfies $A\mathbf{x} = \mathbf{0}$ then any $(c\mathbf{x})$, $c \in \mathbb{C}$ does too.

There are various ways to approach problems of this type. The most straightforward one is to proceed by iteration: pick any of the equations $A^i{}_j x^j = b^i$ and solve for x^1 in terms of the unknowns x^2 to x^n as $x^1 = -\frac{1}{A^i{}_1}\left(b^i - \sum_{j=2}^n A^i{}_j x^j\right)$. (The choice to begin with x^1 is arbitrary. It might be more convenient to start with some other component, ideally one for which the r.h.s. contains the largest number of nonzero matrix elements and hence the smallest number of variables x^j.) Substitute this result into the remaining equations and the problem has been reduced to one of $m-1$ equations for the $n-1$ variables $x^2, \ldots, x^n$. This procedure must now be repeated. *If* the system possesses solution(s) they can be obtained by iteratively solving all m equations in this manner.

Let us illustrate the program on the example of the homogeneous system $A\mathbf{x} = \mathbf{0}$ defined by the matrix

$$A = \begin{pmatrix} 6 & -1 & 5 \\ 2 & 0 & 2 \\ -8 & 1 & -7 \end{pmatrix}. \tag{L130}$$

Expanded into a system of three equations, this reads

$$6x^1 - x^2 + 5x^3 = 0,$$
$$2x^1 + 2x^3 = 0,$$
$$-8x^1 + x^2 - 7x^3 = 0. \tag{L131}$$

Start with the second equation to obtain $x_3 = -x_1$. When this result is inserted into the first and third equations, they simplify to

$$x^1 - x^2 = 0,$$
$$-x^1 + x^2 = 0.$$

The first of these equations now implies $x^1 = x^2$. The fact that this is compatible with the second equation signals that the system is solvable. Defining $x^3 = c$, we obtain $x^1 = x^2 = -c$ and hence the set of solutions

$$\mathbf{x} = c \begin{pmatrix} -1 \\ -1 \\ 1 \end{pmatrix},$$

parametrized by a free running variable, c.

In general, depending on the number of equations, m, and on the number of variables, n, the iteration through a system of linear equations will lead to one of three possible outcomes: ($\rightarrow$ L5.4.1-2)

uniquely solvable systems   ▷ If there are *as many equations as variables*, $m = n$, the system may or may not have a solution. It does not if one of the equations states a contradiction (such as $0 \cdot x^n = b^m$, where b^m is non-vanishing). If no contradiction is encountered and if the system is homogeneous, $\mathbf{b} = \mathbf{0}$, the final equation specifies x^n only up to a multiplicative constant, see the example above. If the system is inhomogeneous its solution is unique, and the final equation uniquely specifies $x^n = x^n(A, b)$ in terms of the given coefficients A^i_j, and b^i. One may now iterate backwards by expressing x^{n-1} through x^n, then x^{n-2} through x^{n-1} and x^n, until all $x^j(A, b)$ are specified.

under-determined systems   ▷ If there are *fewer equations than variables*, $m < n$, the procedure ends at a point where all m equations have been processed but $n - m$ variables, $x^{m+1,...,n}$, are still unspecified. These variables then have the status of free parameters. There exists an infinite family of solutions parametrized by the set of variables $\{x^{m+1}, \dots, x^n\}$. Systems of this type are called *under-determined*. For an under-determined system one may express m variables, $x^j(A, b, x^{m+1,...,n}), j = 1, \dots, m$, as functions of the given coefficients and the $n - m$ free parameters.

over-determined systems   ▷ If there are *more independent equations than variables*, $m > n$, we run out of variables before the last $m - n$ equations have been processed. Such systems are *over-determined* and in general have no solutions.[48]

It is good practice to always substitute the result $\mathbf{x}$ back into the defining equations to check that no mistakes have been made. Then, confirm by matrix–vector multiplication that the solution indeed satisfies $A\mathbf{x} = \mathbf{b}$.

---

[48] After all variables have been eliminated, the remaining $m - n$ equations assume a purely numerical form, such as $3 = 4$ or $3 = 3$. An equation like $3 = 4$ signals a contradiction and means that the system is insoluble. Equations like $3 = 3$ mean that the system contained redundancies and that not all of its equations were truly independent. In such cases the system has effectively fewer equations, and may be soluble after all.

EXAMPLE Let us illustrate the algorithm for computing matrix inverses with the example:

$$A = \begin{pmatrix} 1 & 1 & -1 \\ 1 & 0 & -\frac{1}{2} \\ -1 & -1 & \frac{3}{2} \end{pmatrix}. \tag{L132}$$

Following the general algorithm, we need to solve three systems of equations, $A\mathbf{a}_j = \mathbf{e}_j$, for $j = 1, 2, 3$. Each of these is processed according to the solution scheme discussed above. In this way we find

$$\mathbf{a}_1 = \begin{pmatrix} 1 \\ 2 \\ 2 \end{pmatrix}, \qquad \mathbf{a}_2 = \begin{pmatrix} 1 \\ -1 \\ 0 \end{pmatrix}, \qquad \mathbf{a}_3 = \begin{pmatrix} 1 \\ 1 \\ 2 \end{pmatrix},$$

and this is then combined into the matrix

$$A^{-1} = \begin{pmatrix} 1 & 1 & 1 \\ 2 & -1 & 1 \\ 2 & 0 & 2 \end{pmatrix}. \tag{L133}$$

Check by matrix multiplication that $AA^{-1} = \mathbb{1}$ indeed holds.

We finally note that the solution of linear equations and the inversion of matrices can be streamlined using algorithms such as **Gaussian elimination**. ($\rightarrow$ L5.4.1-4)

L5.5 General linear maps and matrices

REMARK Throughout this section, vectors $\hat{\mathbf{v}} \in V$ of general vector spaces will carry a caret to distinguish them from their column vector representations, $\mathbf{v} \in \mathbb{C}^n$. General linear maps, $\hat{A} : V \rightarrow W$, are distinguished by the same symbol from their matrix representations, $A : \mathbb{C}^n \rightarrow \mathbb{C}^m$.

While our discussion so far was restricted to the standard vector spaces $\mathbb{C}^n$, we now consider linear maps between generic spaces, V. To motivate this generalization, consider the example of Euclidean space $\mathbb{E}^3$. Pick a vector $\hat{\mathbf{x}}$ and consider rotating space around the direction of $\hat{\mathbf{x}}$ by an angle θ. In this way a map, A, obeying the linearity criteria is defined (why?). This raises questions like how the action of A can be described in terms of formulas or how this linear map can be included as a building block in more complicated operations. For example, we might want to describe a succession of two rotations around different rotation axes, $\hat{\mathbf{x}}$ and $\hat{\mathbf{x}}'$. Such operations are no longer easily visualized and we need an efficient formalism for their description.

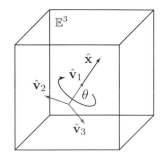

To this end, let $\hat{A} : V \rightarrow W$ be an arbitrary linear map between vector spaces. In both V and W, we pick (not necessarily orthonormal) bases, $\{\hat{\mathbf{v}}_j\}$ and $\{\hat{\mathbf{w}}_i\}$. The discussion includes linear maps operating within one vector space (such as the rotation above), $V = W$, in which case identical bases may be chosen, $\hat{\mathbf{w}}_j = \hat{\mathbf{v}}_j$. We may now apply the map $\hat{A}$ to the basis vectors $\hat{\mathbf{v}}_j$ and expand their image vectors, $\hat{\mathbf{u}}_j \equiv A\hat{\mathbf{v}}_j \in W$, in the $\{\hat{\mathbf{w}}_i\}$ basis as

$$j = 1, \ldots, n: \qquad \hat{\mathbf{u}}_j \equiv \hat{A}\hat{\mathbf{v}}_j = \hat{\mathbf{w}}_i A^i{}_j, \qquad \text{(L134)}$$

where the coefficients $A^i{}_j$ specify the action of the map. Note that for the purposes of the present discussion it is convenient to write the coefficients describing the map *behind* the vectors. The ordering $\hat{\mathbf{w}}_i A^i{}_j$ puts the summation indices i next to each other and looks more natural than the (identical) expression $A^i{}_j \hat{\mathbf{w}}_i$.

EXAMPLE In the example of rotations in $\mathbb{E}^3$ above, it would be convenient to choose a basis containing a unit vector $\hat{\mathbf{v}}_1$ pointing in the direction of the rotation axis. This vector can be complemented by two mutually orthogonal unit vectors, $\hat{\mathbf{v}}_2 \perp \hat{\mathbf{v}}_3$, in the plane perpendicular to $\hat{\mathbf{v}}_1$ to yield a basis $\{\hat{\mathbf{v}}_1, \hat{\mathbf{v}}_2, \hat{\mathbf{v}}_3\}$. It is then not difficult to see that the rotation acts as

$$\hat{A}\hat{\mathbf{v}}_1 = \hat{\mathbf{v}}_1,$$
$$\hat{A}\hat{\mathbf{v}}_2 = \hat{\mathbf{v}}_2 \cos\theta - \hat{\mathbf{v}}_3 \sin\theta,$$
$$\hat{A}\hat{\mathbf{v}}_3 = \hat{\mathbf{v}}_2 \sin\theta - \hat{\mathbf{v}}_3 \cos\theta.$$

Given the basis $\{\hat{\mathbf{v}}_j\}$, any V-vector, $\hat{\mathbf{x}}$, may be represented by a vector $\mathbf{x} \equiv \phi_{\hat{\mathbf{v}}}(\hat{\mathbf{x}})$ in $\mathbb{C}^n$ (see Section L2.5). Specifically, the basis vectors $\hat{\mathbf{v}}_j$ get mapped onto the standard basis vectors $\phi_{\hat{\mathbf{v}}}(\hat{\mathbf{v}}_j) = \mathbf{e}_j = (0, \ldots, 1, \ldots, 0)^T \in \mathbb{C}^n$. In the same way, vectors $\hat{\mathbf{y}} \in W$ are represented by vectors $\mathbf{y} \equiv \phi_{\hat{\mathbf{w}}}(\hat{\mathbf{y}}) \in \mathbb{C}^m$, and basis vectors $\hat{\mathbf{w}}_i$ by standard basis vectors $\mathbf{f}_i = (0, \ldots, 1, \ldots, 0)^T \in \mathbb{C}^m$. Equation (L134) states that a basis vector with component representation $\mathbf{e}_j$ is mapped onto one with component representation $\mathbf{u}_j \equiv \mathbf{f}_i A^i{}_j = (A^1{}_j, A^2{}_j, \ldots, A^m{}_j)^T$. We observe that the map $\hat{A}$ defines an assignment of $\mathbb{C}^n$-standard basis vectors to $\mathbb{C}^m$-component vectors. As discussed in Section L5.2, this defines an $m \times n$ matrix, $A = (\mathbf{u}_1, \ldots, \mathbf{u}_n)$, containing the image component vectors as columns, and the numbers $A^i{}_j$ (see Eq. (L104)) as elements. This matrix acts on basis vectors as $A\mathbf{e}_j = \mathbf{u}_j$. Much as $\mathbf{e}_j$ represents the vector $\hat{\mathbf{v}}_j$, $A : \mathbf{e}_j \to \mathbf{u}_j$ is the matrix representing the map $\hat{A} : \hat{\mathbf{v}}_j \mapsto \hat{\mathbf{u}}_j$. The situation is summarized in the diagram above. Always remember that the matrix representation, A, is specific to a choice of basis.

matrix representing a map

$$V \xrightarrow{\hat{A}} W$$
$$\hat{\mathbf{v}}_j \longmapsto \hat{\mathbf{u}}_j = \hat{\mathbf{w}}_i A^i{}_j$$
$$\phi_{\hat{\mathbf{v}}} \downarrow \qquad\qquad \downarrow \phi_{\hat{\mathbf{w}}}$$
$$\mathbf{e}_j \longmapsto \mathbf{u}_j = \mathbf{f}_i A^i{}_j$$
$$\mathbb{C}^n \xrightarrow{A} \mathbb{C}^m$$

EXERCISE The construction of matrix representations of general rotations in two and three dimensions is described in problems L5.5.1-2. Use that approach to find the matrix representing the rotation map discussed on p. 78.

The discussion above shows how a generic linear map, $\hat{A}$, may be described by a matrix, A. Matrix representations are powerful aids in computations. The *standard procedure in working with linear maps* is as follows.

▷ The first step is the choice of a basis tailored to the action of the linear map. For example, basis vectors pointing in the direction of rotation axes are natural choices. For a reflection at a plane, vectors within that plane would be convenient basis vectors, and they might be complemented by vectors perpendicular to the plane, etc.

▷ Next, one constructs the matrix representing the map, as discussed above, and the column-vector representation of V-vectors, $\mathbf{x} \mapsto x$.

▷ Concrete calculations are then usually performed by matrix algebra. For example, the composition of two linear maps is described by a matrix product, etc.

▷ At the end of the computation, the map $\mathbf{x} \mapsto \hat{\mathbf{x}}$ may be applied to switch back to V-vectors.

EXERCISE The *real-valued polynomials* of degree n, $p(x) = \sum_{j=1}^{n} c_n x^n$, form a vector space, P_n (see the discussion on p. 28). Discuss why differentiation, $\mathrm{d}_x : P_n \to P_{n-1}$, $p(x) \mapsto \mathrm{d}_x p(x)$, is a linear map between P_n and P_{n-1}. What is the form of the matrix representing this map in a basis formed by the polynomials $\{1, x, \ldots, x^n\}$? Write down the 4×5 matrix describing this map for the particular case $n = 4$.

L5.6 Matrices describing coordinate changes

REMARK Throughout this section, vectors in general spaces will be denoted $\hat{\mathbf{x}}$ with a caret. Their $\mathbb{C}^n$-component representation relative to a basis $\{\hat{\mathbf{v}}_j\}$ will be denoted $\mathbf{x}$, and the representation relative to a basis $\{\hat{\mathbf{v}}'_j\}$ by $\mathbf{x}'$. The formulas discussed in this section are easiest to read if expansion coefficients are written to the right of vectors, $\hat{\mathbf{x}} = \hat{\mathbf{v}}_j x^j$. However, this notation convention is not mandatory.

Above, we discussed how a choice of basis $\{\hat{\mathbf{v}}_j\}$ of a vector space V defines an isomorphism, $\phi_{\hat{\mathbf{v}}} : V \to \mathbb{C}^n$, sending vectors, $\hat{\mathbf{x}} = \hat{\mathbf{v}}_j x^j \in V$, to their component representations, $\mathbf{x} = (x^1, \ldots, x^n)^{\mathrm{T}}$, and linear maps, $\hat{A} : V \to V$, to their matrix representations, A. If we now choose a different basis, $\{\hat{\mathbf{v}}'_j\}$, the component representation of vectors changes, $\mathbf{x} \mapsto \mathbf{x}'$, and so does the matrix representation, $A \mapsto A'$, but the vectors, $\hat{\mathbf{x}}$, and linear maps, $\hat{A}$, themselves remain invariant. In a sense, a basis change is a change of the "language" by which the invariant objects $\hat{\mathbf{x}}$ and $\hat{A}$ are described in $\mathbb{C}^n$. Such changes are important operations and in this section we will learn how to describe them efficiently.

EXAMPLE Let us revisit the kitchen example of p. 20 to exemplify how the change of coefficients caused by a change of bases can be computed in elementary terms. For example, on p. 20 we asked how the components of a vector with representation $\hat{\mathbf{x}} = (90, 0)^{\mathrm{T}}$ change if we switch from coordinates measured along the walls to a system in which one of the coordinate directions is rotated by 45 deg (see figure, where the vector $\hat{\mathbf{x}}$ has been shortened for better visibility). Such situations can be conveniently described as basis changes. The phrase "coordinates along the walls" means that the vector $\hat{\mathbf{x}}$ in the figure is expanded in two unit length basis vectors, $\hat{\mathbf{e}}_1$ and $\hat{\mathbf{e}}_2$, parallel to the walls, $\hat{\mathbf{x}} = \hat{\mathbf{e}}_1 90 + \hat{\mathbf{e}}_2 0$. This

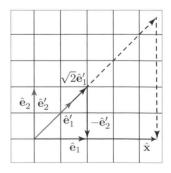

defines a component vector as $\mathbf{x} = (90, 0)^{\mathrm{T}}$. Suppose now the first coordinate direction is changed to lie along the diagonal dashed line, while the second is kept fixed. This defines a basis in which $\hat{\mathbf{e}}_1$ is replaced by $\hat{\mathbf{e}}_1' = \frac{1}{\sqrt{2}}(\hat{\mathbf{e}}_1 + \hat{\mathbf{e}}_2)$ and $\hat{\mathbf{e}}_2' = \hat{\mathbf{e}}_2$. To find the coefficients x'^i of a vector $\hat{\mathbf{x}} = \hat{\mathbf{e}}_1' x'^1 + \hat{\mathbf{e}}_2' x'^2$ expanded in the new basis, we first represent the old basis vectors as linear combinations of the new ones: $\hat{\mathbf{e}}_1 = \hat{\mathbf{e}}_1' \sqrt{2} - \hat{\mathbf{e}}_2'$ and $\hat{\mathbf{e}}_2 = \hat{\mathbf{e}}_2'$ (see figure). Now substitute this result into the old basis vector expansion to obtain $\hat{\mathbf{x}} = \hat{\mathbf{e}}_1 90 + \hat{\mathbf{e}}_2 0 = (\hat{\mathbf{e}}_1' \sqrt{2} - \hat{\mathbf{e}}_2')90 + \hat{\mathbf{e}}_2' 0 = \hat{\mathbf{e}}_1' \sqrt{2}\, 90 - \hat{\mathbf{e}}_2' 90$. We thus obtain the new component representation $\mathbf{x}' = 90(\sqrt{2}, -1)^{\mathrm{T}}$. The geometric interpretation of this expansion is shown in the figure.

The discussion above shows that basis changes are not really complicated operations. All that needs to be done is solving linear equations containing expansion coefficients in one basis as linear functions of coefficients in another basis. At the same time, it is evident that such calculations can become cumbersome in higher dimensions where lots of coefficients are involved. In the next subsection, we will discuss how the required operations can be streamlined to maximal efficiency.

$$\hat{\mathbf{v}}_j x^j = \hat{\mathbf{x}} = \hat{\mathbf{v}}_{j'}' x^{j'}$$

However, before that, it is worth understanding the *change of representation* under a change of basis conceptually. The situation is summarized in the diagram, where the maps $\phi_{\hat{\mathbf{v}}}$ and $\phi_{\hat{\mathbf{v}}'}$ assign to vectors $\hat{\mathbf{x}}$ the component representations $\mathbf{x}$ and $\mathbf{x}'$, respectively. These maps are vector space isomorphisms (invertible linear maps), which means that the composite map,

$$T = \phi_{\hat{\mathbf{v}}'} \circ \phi_{\hat{\mathbf{v}}}^{-1} : \mathbb{C}^n \to \mathbb{C}^n, \qquad \mathbf{x} \mapsto \phi_{\hat{\mathbf{v}}'}(\phi_{\hat{\mathbf{v}}}^{-1}(\mathbf{x})) = \mathbf{x}', \tag{L135}$$

is an isomorphism, too. The linear map $T : \mathbb{C}^n \to \mathbb{C}^n$, $\mathbf{x} \mapsto T\mathbf{x} = \mathbf{x}'$, describes how the coordinates of the vector $\hat{\mathbf{x}}$ change upon a change of basis. Being a linear map $\mathbb{C}^n \to \mathbb{C}^n$, we can think of T as a square matrix. We next learn how to identify this *transformation matrix* in practical terms.

Transformation matrix

REMARK In this subsection, we use primed indices like $x'^{i'}$ to label the components of vector representations $\mathbf{x}'$ in a basis, $\{\hat{\mathbf{v}}_{i'}'\}$. The notation does not look nice, but helps to remember to which basis an index relates. After some familiarity with basis transformations has been gained we will stop using it. For brevity, the bases $\{\hat{\mathbf{v}}_j\}$ and $\{\hat{\mathbf{v}}_{i'}'\}$ will be referred to as the "old" and the "new" basis throughout this section.

In the old and the new basis representation, respectively, the expansion of a V-vector $\hat{\mathbf{x}}$ assumes the form

$$\hat{\mathbf{x}} = \hat{\mathbf{v}}_j x^j,$$

$$\hat{\mathbf{x}} = \hat{\mathbf{v}}_{i'}' x'^{i'},$$

and this defines the representation vectors $\mathbf{x} = (x^1, \ldots, x^n)^{\mathrm{T}}$ and $\mathbf{x}' = (x'^1, \ldots, x'^n)^{\mathrm{T}}$. Assume that the expansion of the old basis vectors in the new ones is given by

$$\hat{\mathbf{v}}_j = \hat{\mathbf{v}}'_{i'} \, T^{i'}{}_j. \tag{L136}$$

Substituting this expansion into the first of the equations above, $\hat{\mathbf{x}} = \hat{\mathbf{v}}'_{i'} T^{i'}{}_j x^j$, and comparing with the second equation, we obtain the identification

$$x'^{i'} = T^{i'}{}_j x^j. \tag{L137}$$

transformation   Equation (L105) states that the basis change is described by the **transformation matrix**
matrix

$$T = \begin{pmatrix} T^1{}_1 & \cdots & T^1{}_j & \cdots & T^1{}_n \\ \vdots & & \vdots & & \vdots \\ T^{i'}{}_1 & \cdots & T^{i'}{}_j & \cdots & T^{i'}{}_n \\ \vdots & & \vdots & & \vdots \\ T^n{}_1 & \cdots & T^n{}_j & \cdots & T^n{}_n \end{pmatrix}, \tag{L138}$$

as $\mathbf{x}' = T\mathbf{x}$. From the construction of this matrix it follows that it affords different interpretations.

▷ The jth column, $\mathbf{T}_j$, contains the expansion coefficients, $T^{i'}{}_j$, of the jth old basis vector $\hat{\mathbf{v}}_j$ in the new basis $\{\hat{\mathbf{v}}'_{i'}\}$ (Eq. (L136)).

▷ The basis vector $\hat{\mathbf{v}}_j$ had the representation $\mathbf{e}_j$ in the old basis. In the new basis it is represented by $\mathbf{T}_j = T\mathbf{e}_j$, the jth column of $\mathbf{T}_j$.

▷ The vectors $\mathbf{x}'$ and $\mathbf{x}$ representing $\hat{\mathbf{x}}$ in the new and old basis, respectively, are related to each other by matrix multiplication, $\mathbf{x}' = T\mathbf{x}$ (Eq. (L137)).

The transformation matrix T describes how the form of a $\mathbb{C}^n$-vector changes from the old to the new representation. However, it also contains information on the *inverse transformation*, i.e. the question of how a vector assuming a known form in the new representation looked in the old one. To understand how, notice that a change from the old basis to the new one and then back to the old one amounts to the identity operation. We just discussed how old-to-new is described by a matrix T. The change new-to-old must undo the effect of this transformation, and is therefore represented by the *inverse of the transformation matrix*, T^{-1}. A transformation from the old to the new representation and then back then amounts to the identity operation, $T^{-1} \cdot T = \mathbb{1}$.

EXAMPLE Consider two $\mathbb{E}_2$-bases, $\{\hat{\mathbf{v}}_1, \hat{\mathbf{v}}_2\}$ and $\{\hat{\mathbf{v}}'_1, \hat{\mathbf{v}}'_2\}$, shown in the figure. The old basis vectors are expressed in the new basis as

$$\hat{\mathbf{v}}_1 = \tfrac{3}{4}\hat{\mathbf{v}}'_1 + \tfrac{1}{3}\hat{\mathbf{v}}'_2,$$
$$\hat{\mathbf{v}}_2 = -\tfrac{1}{8}\hat{\mathbf{v}}'_1 + \tfrac{1}{2}\hat{\mathbf{v}}'_2, \tag{L139}$$

and this defines the matrix

$$T = \begin{pmatrix} \tfrac{3}{4} & -\tfrac{1}{8} \\ \tfrac{1}{3} & \tfrac{1}{2} \end{pmatrix}. \tag{L140}$$

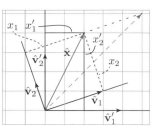

(For example, from the first term in the expression $\hat{v}_2 = \hat{v}'_1 T^1{}_2 + \hat{v}'_2 T^2{}_2$, we obtain $T^1{}_2 = -\frac{1}{8}$, etc.) Now consider a vector $\hat{\mathbf{x}} = \hat{v}_j x^j$ with components $\mathbf{x} = (1, 2)^T$ in the old basis. According to the discussion above, its representation in the new basis, $\hat{\mathbf{x}} = \hat{v}'_{i'} x'^{i'}$, is obtained as

$$\mathbf{x} \mapsto \mathbf{x}' = T\mathbf{x} = \begin{pmatrix} \frac{3}{4} & -\frac{1}{8} \\ \frac{1}{3} & \frac{1}{2} \end{pmatrix} \begin{pmatrix} 1 \\ 2 \end{pmatrix} = \begin{pmatrix} \frac{1}{2} \\ \frac{4}{3} \end{pmatrix}.$$

inverse
transforma-
tion

We conclude that $\hat{\mathbf{x}} = 1\hat{v}_1 + 2\hat{v}_2 = \frac{1}{2}\hat{v}'_1 + \frac{4}{3}\hat{v}'_2$, which can be confirmed by inspection of the figure. The inverse transformation can be obtained using Eq. (L126),

$$T^{-1} = \frac{12}{5} \begin{pmatrix} \frac{1}{2} & \frac{1}{8} \\ -\frac{1}{3} & \frac{3}{4} \end{pmatrix}. \tag{L141}$$

(Verify that $T^{-1}T = \mathbb{1}$.) For example, in the new basis the old basis vector $\hat{v}_1$ has the representation $\hat{v}_i \overset{(\text{L139})}{=} (\frac{3}{4}, \frac{1}{3})^T$. Under the inverse transformation this changes to

$$T^{-1} \begin{pmatrix} \frac{3}{4} \\ \frac{1}{3} \end{pmatrix} - \begin{pmatrix} 1 \\ 0 \end{pmatrix},$$

confirming that in the old basis $\hat{v}_1$ is described by a standard basis vector. ($\to$ L5.6.1-2 (a-d)).

Change of matrix representation

We now understand how basis changes define transformation matrices, T, describing the change of component vectors as $\mathbf{x}' = T\mathbf{x}$. Next we address the related question of how matrix representations of linear maps, $A : V \to V, \hat{\mathbf{x}} \mapsto \hat{\mathbf{y}} = \hat{A}\hat{\mathbf{x}}$, transform under a change of basis.

The defining property of the matrices A and A' representing $\hat{A}$ in the old and the new basis, respectively, is that $\mathbf{y} - A\mathbf{x}$ and $\mathbf{y}' = A'\mathbf{x}'$. Substitution of $\mathbf{y}' = T\mathbf{y}$ and $\mathbf{x}' = T\mathbf{x}$ into the second relation yields $T\mathbf{y} = A'T\mathbf{x}$. We multiply this vector relation by T^{-1} to obtain $\mathbf{y} = T^{-1}A'T\mathbf{x} = A\mathbf{x}$. Since this equality holds for arbitrary $\mathbf{x}$, we have the identification

$$A = T^{-1}A'T. \tag{L142}$$

This relation can be inverted by multiplying from the left and right by T and T^{-1}, respectively, to obtain $TAT^{-1} = T(T^{-1}A'T)T^{-1} = A'$, or

$$\boxed{A' = TAT^{-1}.} \tag{L143}$$

This formula states that the application of A' to a vector in the *new* representation amounts to a succession of three steps:

1. Apply the matrix T^{-1} to pass from the vector's new representation to the old one.

2. Apply the old form of the matrix, A.

3. Apply T to transform the result back to the new representation.

similarity
transforma-
tions

Read: the matrix representation in the new basis is obtained from the old representation by applying the inverse transformation matrices. The transformations (L142) and (L143) are sometimes called **similarity transformations**. The name indicates that matrices related by a similarity transformation describe the same linear map, albeit in a different representation.

EXAMPLE Let us illustrate this transformation procedure with the example of the vectors defined in Eq. (L139). Consider a map stretching all vectors in the $\hat{\mathbf{v}}_1$-direction by a factor of 2. For example, the vector $\hat{\mathbf{x}}$ in the figure on p. 82 maps to the long-dashed one. In the $\hat{\mathbf{v}}$-basis, this transformation assumes a simple form: $\hat{\mathbf{v}}_1 \mapsto 2\hat{\mathbf{v}}_1$ and $\hat{\mathbf{v}}_2 \mapsto \hat{\mathbf{v}}_2$, and is described by

$$A = \begin{pmatrix} 2 & 0 \\ 0 & 1 \end{pmatrix}. \tag{L144}$$

The application of the transformation matrix (L140) and its inverse (L141) yields the representation of the map in the $\hat{\mathbf{v}}$-basis:

$$A' = TAT^{-1} = \begin{pmatrix} \frac{3}{4} & -\frac{1}{8} \\ \frac{1}{3} & \frac{1}{2} \end{pmatrix} \begin{pmatrix} 2 & 0 \\ 0 & 1 \end{pmatrix} \frac{12}{5} \begin{pmatrix} \frac{1}{2} & \frac{1}{8} \\ -\frac{1}{3} & \frac{3}{4} \end{pmatrix} = \frac{1}{10} \begin{pmatrix} 19 & \frac{9}{4} \\ 4 & 11 \end{pmatrix}. \tag{L145}$$

It is instructive to check that this more complicated representation correctly describes the stretching operation. Application of the matrix on the right to the vector $\mathbf{x}' = \left(\frac{1}{2}, \frac{4}{3} \right)^{\mathrm{T}}$ representing $\hat{\mathbf{x}}$ in the $\hat{\mathbf{v}}'$-basis produces the image vector $A'\mathbf{x}' = \left(\frac{5}{4}, \frac{5}{3} \right)^{\mathrm{T}}$. This means that $\hat{\mathbf{x}}$ is mapped to $\frac{5}{4}\hat{\mathbf{v}}'_1 + \frac{5}{3}\hat{\mathbf{v}}'_2$. Inspect the figure to verify that this expansion is consistent with the graphical representation of the transformation. ($\rightarrow$ L5.6.1-4)

Matrix trace

The discussion above shows how matrix representations of linear maps change under changes of basis. In view of this "volatility" of representations, it is natural to ask whether there are at least some characteristics of matrices which remain invariant under a change of basis. Such features indeed exist and the simplest of them (others will be discussed later) is the **trace of a matrix** A. The trace is defined as the sum of all matrix elements on the diagonal:

trace

$$\boxed{\mathrm{tr}(A) \equiv \sum_i A^i{}_i.} \tag{L146}$$

The most important property of the trace is its **cyclic invariance**: for two matrices A, B,

$$\boxed{\mathrm{tr}(AB) = \mathrm{tr}(BA).}$$

This identity is verified as $\mathrm{tr}(AB) = (AB)^i{}_i = A^i{}_j B^j{}_i = B^j{}_i A^i{}_j = (BA)^j{}_j = \mathrm{tr}(BA)$. The generalization of this identity to products of several matrices reads as

$$\mathrm{tr}(AB \dots CDE) = \mathrm{tr}(EAB \dots CD),$$

and motivates the denotation "cyclic" invariance.

Table L1 Formulas for matrix representations of linear maps and their changes under transformations between bases $\{\hat{\mathbf{v}}_j\}$ and $\{\hat{\mathbf{v}}'_{i'}\}$ of V. The space W is spanned by $\{\hat{\mathbf{w}}_i\}$.

general linear map, $\hat{A} : V \to W$	$\hat{A}(\hat{\mathbf{v}}_j) = \hat{\mathbf{w}}_i A^i{}_j$
matrix representation, A	$A = \{A^i{}_j\}$
image of standard basis vector $\mathbf{e}_j$ under A	jth column vector of A, $\mathbf{A}_j$
matrix representation of $\hat{A} : \hat{\mathbf{x}} \mapsto \hat{\mathbf{y}}$	$\mathbf{y} = A\mathbf{x}$
basis transformation within V	$\hat{\mathbf{v}}_j = \hat{\mathbf{v}}'_{i'} T^{i'}{}_j$
matrix representation, T	$T = \{T^{i'}{}_j\}$
jth old standard basis vector in new representation	jth column vector of T, $\mathbf{T}_j$
relation between old and new representations of $\hat{\mathbf{x}}$	$\mathbf{x}' = T\mathbf{x}$
inverse transformation	$\hat{\mathbf{v}}'_{i'} = \hat{\mathbf{v}}_j (T^{-1})^j{}_{i'}$
i'th new standard basis vector in old representation	i'th column vector of T^{-1}, $\mathbf{T}^{-1}_{i'}$
matrix representation of $\hat{A}$ in new basis	$A' = TAT^{-1}$

The invariance of the trace under changes of bases is a direct consequence of this relation: with $A' = TAT^{-1}$ we have

$$\text{tr}(A') = \text{tr}(TAT^{-1}) = \text{tr}(T^{-1}TA) = \text{tr}(A).$$

For example, the matrix A of Eq. (L144) has the trace $\text{tr}(A) = \frac{2}{3} + 1 = \frac{5}{3}$, equal to the trace of its transform, $\text{tr}(A') = \frac{7}{10} + \frac{29}{30} = \frac{5}{3}$.

We finally notice that the *matrix trace* is invariant under transposition,

$$\text{tr}(A) = \text{tr}(A^{\text{T}}),$$

where $\text{tr}(A^{\text{T}}) \equiv \sum_i (A^{\text{T}})_i{}^i$. This follows trivially from $A^i{}_i = (A^{\text{T}})_i{}^i$.

L5.7 Summary and outlook

In this chapter we introduced matrix representations as the language describing linear maps in standard vector spaces. We discussed how the action of linear maps is described by matrix–vector multiplication, how the composition of linear maps becomes the multiplication of matrices, and how the inverse of a linear map defines the inverse of a matrix. We also observed that, depending on the choice of basis, matrix representations can be simple or complicated (cf. Eq. (L144) vs. Eq. (L145)). It is favorable to work with simple representations and in the following chapters we will show how they can be found. All these subsequent discussions make extensive reference to results derived in this chapter. For convenience, Table L1 provides a reference chart summarizing the essential results in one table.

Determinants

At the end of the previous chapter we mentioned that the trace of a square matrix – the sum of its diagonal elements – does not change under changes of basis. There exists one more basis-invariant scalar quantity characterizing a square matrix, A: the determinant, $\det(A)$. The important role played by the determinant is somewhat difficult to describe before it has been defined and applied. However, let us mention in advance that the determinant provides a powerful test for the invertibility of a matrix (it is invertible if and only if the determinant is non-vanishing), and plays a key role in obtaining the simplest possible matrix representation of a linear map. In this chapter we will define the determinant, and discuss its characteristic properties. In later chapters it will then be applied in various different contexts.

L6.1 Definition and geometric interpretation

In this section we introduce the determinant through one of several equivalent definitions. We then discuss the determinant of low-dimensional matrices from a geometric perspective to provide some early intuition for the new concept.

Definition of the determinant

determinant   The **determinant** is a function

$$\det : \mathrm{mat}(n, \mathbb{C}) \to \mathbb{C}, \qquad A \mapsto \det(A),$$

producing numbers from square matrices. It is sometimes denoted by $\det(A) \equiv |A|$, e.g. $\det \left(\begin{smallmatrix} 1 & 2 \\ 0 & 3 \end{smallmatrix} \right) \equiv \left| \begin{smallmatrix} 1 & 2 \\ 0 & 3 \end{smallmatrix} \right|$.

The importance of this function shows in that there are several equivalent definitions for it, some requiring methodology

Gottfried Wilhelm Leibniz 1646–1716
A German mathematician and philosopher. Leibniz developed infinitesimal calculus independently of Newton. Being fascinated with automated computation, he invented various types of mechanical calculators, and refined the binary number system. In the humanities Leibniz is known for his "philosophical optimism", e.g. the view that our Universe is the best a god could possibly have created.

which has not yet been introduced. Much of our discussion in this part of the text is based **Leibniz rule** on the **Leibniz rule**. This formula states that the determinant of an $n \times n$ matrix, $A = \{A^i{}_j\}$, is given by

$$\det(A) = \sum_{P \in S_n} \mathrm{sgn}(P) A^{P1}{}_1 A^{P2}{}_2 \ldots A^{Pn}{}_n. \tag{L147}$$

Here P denotes the permutation $P^{1,2,\ldots,n}_{P1,P2,\ldots,Pn}$ replacing the ordered sequence $(1, 2, \ldots, n)$ by $(P1, P2, \ldots, Pn)$ (see Section L1.2), and the summation extends over all $n!$ permutations in the permutation group S_n. In the three cases of lowest dimension, $n = 1, 2, 3$, this formula assumes the following forms:

$$n = 1: \qquad \det(A) = A^1{}_1,$$

$$n = 2: \qquad \det(A) = A^1{}_1 A^2{}_2 - A^2{}_1 A^1{}_2, \tag{L148}$$

$$n = 3: \qquad \det(A) = A^1{}_1 A^2{}_2 A^3{}_3 - A^2{}_1 A^1{}_2 A^3{}_3 - A^3{}_1 A^2{}_2 A^1{}_3$$
$$- A^1{}_1 A^3{}_2 A^2{}_3 + A^2{}_1 A^3{}_2 A^1{}_3 + A^3{}_1 A^1{}_2 A^2{}_3. \tag{L149}$$

For example,

$$\det \begin{pmatrix} 1 & 3 \\ 2 & 5 \end{pmatrix} = 1 \cdot 5 - 2 \cdot 3 = -1,$$

$$\det \begin{pmatrix} 2 & -3 & 1 \\ 1 & 4 & 3 \\ 0 & 2 & 2 \end{pmatrix} = 2 \cdot 4 \cdot 2 - 1 \cdot (-3) \cdot 2 - 0 \cdot 4 \cdot 1$$

$$- 2 \cdot 2 \cdot 3 + 1 \cdot 2 \cdot 1 + 0 \cdot (-3) \cdot 3 = 12. \tag{L150}$$

For larger n the number of terms grows quickly and the manual computation of determinants becomes cumbersome. However, several efficient alternatives to brute force computations of determinants will be introduced in the next section and various later chapters.

Geometric interpretation of determinants

The determinants of real matrices afford an important *geometric interpretation*: the absolute value of the determinant of an $n \times n$ matrix A, $|\det(A)|$, equals the *volume* of the n-dimensional parallelepiped spanned by the column vectors of A. The general proof of this identity requires integration theory and will not be discussed here. However, for two- and three-dimensional matrices the statement can be checked by straightforward computation.

determinants compute volumes For a two-dimensional matrix $A = (\mathbf{A}_1, \mathbf{A}_2)$ with column vectors $\mathbf{A}_1$ and $\mathbf{A}_2$, Eq. (L148) gives

$$n = 2: \quad |\det(A)| = |A^1{}_1 A^2{}_2 - A^2{}_1 A^1{}_2| \stackrel{(L50)}{=} A(\mathbf{A}_1, \mathbf{A}_2), \tag{L151}$$

where $A(\mathbf{v}, \mathbf{w})$ is the area of the parallelogram spanned by $\mathbf{v}$ and $\mathbf{w}$ (see Eq. (L50)). Similarly, for a three-dimensional matrix $A = (\mathbf{A}_1, \mathbf{A}_2, \mathbf{A}_3)$, Eq. (L149) gives

$$n = 3: \quad |\det(A)| = |\epsilon_{ijk} A^i{}_1 A^j{}_2 A^k{}_3| \stackrel{(L96)}{=} |(\mathbf{A}_1 \times \mathbf{A}_2) \cdot \mathbf{A}_3| \stackrel{(L97)}{=} V(\mathbf{A}_1, \mathbf{A}_2, \mathbf{A}_3), \tag{L152}$$

where the Levi–Civita symbol generates the sign factors occurring in Eq. (L149), and $V(\mathbf{u}, \mathbf{v}, \mathbf{w})$ is the volume of the parallelepiped spanned by $\mathbf{u}, \mathbf{v}$ and $\mathbf{w}$.

L6.2 Computing determinants

Besides the Leibniz rule, various other methods for computing determinants exist. Depending on the structure of the matrices one is dealing with, these alternatives can be much more efficient than the original definition. In this section we introduce a few approaches to the computation of determinants frequently occurring in practice.

Laplace rule

Laplace rule   The **Laplace rule** expresses the determinant of an $n \times n$ matrix through n or fewer determinants of submatrices of dimension $(n-1)\times(n-1)$. It states that $\det(A)$ can be computed via either of the representations

$$\det(A) = \sum_i A^i{}_j (-)^{i+j} M^i{}_j = \sum_j A^i{}_j (-)^{i+j} M^i{}_j. \tag{L153}$$

Here $(-)^{i+j} \equiv (-1)^{i+j}$ and the **minor**, $M^i{}_j$, is defined as the determinant of the $(n-1)\times(n-1)$ matrix obtained by crossing out row i and column j of A.[49] In the first representation, one chooses an arbitrary column j of A and sums over all rows, i. It can be shown that the result does not depend on which j is chosen. Ideally one will pick a column containing as many vanishing matrix elements as possible. Since the corresponding minors are multiplied by $A^i{}_j = 0$, they need not be calculated and work can be avoided. Likewise, in the second representation an arbitrary row i is fixed and a summation over columns j performed.

The Laplace rule reduces the computation of one n-dimensional determinant to that of at most n determinants of reduced dimension, $n - 1$. The procedure can be iterated and each of the minors again computed by the Laplace rule. This recursive approach can greatly simplify the computation of determinants, in particular in the case of sparse matrices containing lots of zeros.

EXAMPLE The Laplace rule expansion of the determinant of Eq. (L150) along column $j = 1$ yields

$$\begin{vmatrix} 2 & -3 & 1 \\ 1 & 4 & 3 \\ 0 & 2 & 2 \end{vmatrix} = 2 \underbrace{\begin{vmatrix} 4 & 3 \\ 2 & 2 \end{vmatrix}}_{M^1{}_1} - 1 \underbrace{\begin{vmatrix} -3 & 1 \\ 2 & 2 \end{vmatrix}}_{M^2{}_1} + 0 \underbrace{\begin{vmatrix} -3 & 1 \\ 4 & 3 \end{vmatrix}}_{M^3{}_1} = 2(4 \cdot 2 - 2 \cdot 3) - 1(-3 \cdot 2 - 2 \cdot 1) = 12.$$

We recover the answer of Eq. (L150), but with less effort, since fewer terms are involved. ($\to$ L6.2.1-2)

We will not show the equivalence of the Laplace and Leibniz rules. However, it is instructive to verify that when applied to a general 3×3 matrix, Laplace's rule reproduces all the terms of the Leibniz expansion (L149). The general proof is left as a challenging exercise in combinatorics. (Attempt it using Eq. (L159) once you have reached that formula below.)

---

[49] The product $(-)^{i+j} M^i{}_j$ is called the **cofactor** of the matrix element $A^i{}_j$.

Determinants of special types of matrices

triangular matrix

There are a few *types of matrices for which the calculation of determinants can be simplified*. Consider, for example, the case of **triangular matrices**, for which $A^i_{\ j} = 0$ for either all $i > j$ (upper triangular matrix), or all $i < j$ (lower triangular matrix). These matrices are called "triangular" because all matrix elements to the lower left of the diagonal, or upper right of the diagonal, respectively, vanish by the above condition. The *determinant of a triangular matrix* is simply given by $\det(A) = \prod_i A^i_{\ i}$.[50] Show how this is a straightforward consequence of the Leibniz rule. (*Hint*: Think which permutations obey the condition $Pi \geq i$ for all i.) A **diagonal matrix**, D, is a matrix which has non-vanishing

diagonal matrix

elements only on the diagonal, $D^i_{\ j} = \delta^i_{\ j}\lambda_j$. Diagonal matrices are often defined by the notation $D = \mathrm{diag}(\lambda_1, \dots, \lambda_n)$. They are a particularly simple form of triangular matrix hence the determinant of a diagonal matrix is given by the product of its diagonal matrix elements, $\det(D) = \prod_i \lambda_i$.

block matrix

Any square matrix, $X \in \mathrm{mat}(n, \mathbb{C})$, can be represented as a **matrix block structure**

$$X = \begin{pmatrix} A & B \\ C & D \end{pmatrix}, \tag{L154}$$

where $A \in \mathrm{mat}(r, \mathbb{C})$ and $D \in \mathrm{mat}(s, \mathbb{C})$ are square matrices, $r + s = n$, and the complementary blocks are rectangular, $B \in \mathrm{mat}(r, s, \mathbb{C})$ and $C \in \mathrm{mat}(s, r, \mathbb{C})$.[51] It can be shown that the determinant of a matrix in block representation is given by

$$\det \begin{pmatrix} A & B \\ C & D \end{pmatrix} = \det(A - BD^{-1}C)\det(D). \tag{L155}$$

This formula reduces the determinant of a block matrix to the product of determinants of the $r \times r$ matrix $(A - BD^{-1}C)$ and the $s \times s$ matrix D. Whether or not this representation simplifies the calculation of the determinant depends on the structure of the blocks A, B, C, D. If one of the off-diagonal blocks vanishes, $B = 0$ or $C = 0$, the formula collapses to

$$\det \begin{pmatrix} A & 0 \\ C & D \end{pmatrix} = \det \begin{pmatrix} A & B \\ 0 & D \end{pmatrix} = \det(A)\det(D), \tag{L156}$$

a great simplification compared to the computation of an $(r + s)$-dimensional determinant. A take-home message: when computing determinants, always look out for block structures.

EXERCISE Verify Eq. (L155) for the case of 2×2 matrices with one-dimensional blocks, or 3×3 matrices with blocks of dimension two and one.

INFO The block representation becomes significant in cases where a matrix contains *blocks of distinct physical or mathematical significance*. For example, consider an atom containing r electrons and s nucleons (the protons and neutrons forming its nucleus). Let $X^i_{\ j}$ be the strength of the magnetic interaction between these particles. The interaction strength between the electrons, described by $X^i_{\ j}$, $i, j \leq r$, i.e. by the matrix block A, will be qualitatively different from that between

---

[50] The symbol $\prod$ is the product analogue of the $\sum$-symbol, $\prod_{i=1}^n x_i = x_1 \cdot x_2 \cdot \cdots \cdot x_n$.
[51] More generally, one can define block structures where A and D need not be square.

the nucleons (block D), or the electron–nucleon interaction B, C. The magnetic interaction matrix therefore naturally carries a block structure.

Alternative representations of the Leibniz rule

Determinants appear in various mathematical contexts and it is important to recognize them as such even if they are not represented exactly as in (L147). For later reference, we here review a number of frequently occurring *alternative determinant representations* building on the Leibniz rule.

All these reformulations are obtained by different representations of the permutations in Eq. (L147). For example, that relation sums over products of matrix elements in which column indices appear in ascending order. Alternatively, we may reorder each product to bring the row indices into ascending order.[52] The sum then reads $\sum_{P \in S_n} \text{sgn}(P) A^1{}_{P^{-1}1} \cdots A^n{}_{P^{-1}n}$, where $P^{-1}j$ is the inverse permutation acting on j. We recall that $\text{sgn}(P) = \text{sgn}(P^{-1})$, and note that the sum over all permutations in S_n equals the sum over all inverses of permutations, $\sum_P F(P^{-1}) = \sum_P F(P)$, where F is an arbitrary function. The Leibniz formula (L147) may therefore be rewritten as

$$\det(A) = \sum_{P \in S_n} \text{sgn}(P) A^1{}_{P1} A^2{}_{P2} \cdots A^n{}_{Pn}, \tag{L157}$$

Leibniz formula based on row indices with ascending row indices. Also note that the ascending index configuration $(1, 2, \ldots, n)$ appearing in these expressions is not "special". It is the identity permutation, and can be replaced by any other permutation, $Q(1, 2, \ldots, n) \equiv (i_1, \ldots, i_n)$. This rearrangement leaves the formula unchanged, up to a multiplicative factor $\text{sgn}(Q)$. To see this, note that

$$\sum_{P \in S_n} \text{sgn}(P) A^{P1}{}_{Q1} \cdots A^{Pn}{}_{Qn} = \sum_{P \in S_n} \text{sgn}(P) A^{Q^{-1}P1}{}_1 \cdots A^{Q^{-1}Pn}{}_n$$

$$= \sum_{P \in S_n} \text{sgn}(QQ^{-1}P) A^{Q^{-1}P1}{}_1 \cdots A^{Q^{-1}Pn}{}_n$$

$$= \text{sgn}(Q) \sum_{P \in S_n} \text{sgn}(Q^{-1}P) A^{Q^{-1}P1}{}_1 \cdots A^{Q^{-1}Pn}{}_n$$

$$= \text{sgn}(Q) \sum_{P \in S_n} \text{sgn}(P) A^{P1}{}_1 \cdots A^{Pn}{}_n,$$

where various of the identities described above were used. The final sum equals $\text{sgn}(Q)$ times $\det(A)$, and so we have obtained another Leibniz rule clone,

$$\det(A) \, \text{sgn}(Q) = \sum_{P \in S_n} \text{sgn}(P) A^{P1}{}_{Q1} A^{P2}{}_{Q2} \cdots A^{Pn}{}_{Qn}. \tag{L158}$$

We finally note that the Leibniz formula can be compactly represented using the Levi–Civita symbol, $\epsilon_{i_1, \ldots, i_n} = \text{sgn}(P^{1,2,\ldots,n}_{i_1,\ldots,i_n}) = \epsilon^{i_1,\ldots,i_n}$, defined in Eq. (L7),

---

[52] For example, $A^2{}_1 A^1{}_2 A^3{}_3$ can be reordered as $A^1{}_2 A^2{}_1 A^3{}_3$, or more generally, $\prod_i A^{Pi}{}_i = \prod_j A^j{}_{P^{-1}j}$. The permutation of column indices in the second expression is the inverse of the permutation of row indices in the first.

$$\det(A) \overset{(L147)}{=} \epsilon_{i_1,i_2,\ldots,i_n} A^{i_1}{}_1 A^{i_2}{}_2 \ldots A^{i_n}{}_n \overset{(L157)}{=} \epsilon^{i_1,i_2,\ldots,i_n} A^1{}_{i_1} A^2{}_{i_2} \ldots A^n{}_{i_n}.$$

Leibniz formula via Levi–Civita symbol

With $\mathrm{sgn}(Q^{1,2,\ldots,n}_{j_1,\ldots,j_n}) = \epsilon_{j_1,\ldots,j_n}$, Eq. (L158) assumes the form

$$\boxed{\det(A)\, \epsilon_{j_1,\ldots,j_n} = \epsilon_{i_1,i_2,\ldots,i_n} A^{i_1}{}_{j_1} A^{i_2}{}_{j_2} \ldots A^{i_n}{}_{j_n}.} \qquad (L159)$$

This is a convenient and versatile representation of the determinant which will be frequently used in later parts of the text.

L6.3   Properties of determinants

Much like the trace, the determinant defines a "fingerprint" of a linear map which does not change under changes of basis. In this section we define various useful properties of this function. Although they all follow from the definition Eq. (L147) the general proof of some identities is not entirely obvious. In such cases, we refer to lecture courses in linear algebra, or to the example on p. 93 below where the situation is illustrated for two-dimensional matrices, $n = 2$.

determinants of diagonal matrices

1.   We already noted that for a *diagonal matrix* $D = \mathrm{diag}(\lambda_1,\ldots,\lambda_n)$, the determinant is given by the product of the diagonal elements,

$$\det(D) = \prod_{i=1}^{n} \lambda_i. \qquad (L160)$$

The determinant of a diagonal matrix containing zeros on the diagonal vanishes.

2.   The determinant is *invariant under transposition*,

$$\boxed{\det(A^{\mathrm{T}}) = \det(A).} \qquad (L161)$$

To prove this, use the definition of the transpose in the Leibniz formula to obtain
$$\det(A^{\mathrm{T}}) \overset{(L147)}{=} \sum_P \mathrm{sgn}(P) \prod_i (A^{\mathrm{T}})_{Pi}{}^{i} = \sum_P \mathrm{sgn}(P) \prod_i A^i{}_{Pi} \overset{(L157)}{=} \det(A).$$
Under complex conjugation the determinant changes as

$$\overline{\det(A)} = \det(\overline{A}), \qquad (L162)$$

invariance under transposition

where $\overline{A} = \{\overline{A^i{}_j}\}$ is the complex conjugate of the matrix. Combining this with Eq. (L161), we obtain

$$\det(A^{\dagger}) = \overline{\det(A)}, \qquad (L163)$$

where the adjoint is defined in Eq. (L114).

antisymmetry under row or column exchange

3.   The determinant is *antisymmetric* under the pairwise *exchange* of *columns* or *rows*,

$$\det(\ldots, \mathbf{A}_i, \ldots, \mathbf{A}_j, \ldots) = -\det(\ldots, \mathbf{A}_j, \ldots, \mathbf{A}_i, \ldots), \qquad (L164)$$

$$\det \begin{pmatrix} \vdots \\ \mathbf{A}^i \\ \vdots \\ \mathbf{A}^j \\ \vdots \end{pmatrix} = -\det \begin{pmatrix} \vdots \\ \mathbf{A}^j \\ \vdots \\ \mathbf{A}^i \\ \vdots \end{pmatrix}, \tag{L165}$$

where the ellipses represent unchanged columns or rows. The sign change implies that the determinant of matrices containing identical rows or columns equals its own negative, and therefore vanishes, e.g. $\det(\ldots, \mathbf{A}_i, \ldots, \mathbf{A}_i, \ldots) = 0$.

multilinearity 4. **Multilinearity**: the determinant is *linear in each column*, $(r, r' \in \mathbb{C})$

$$\det(\ldots, r\mathbf{A}_j + r'\mathbf{A}'_j, \ldots) = r\det(\ldots, \mathbf{A}_j, \ldots) + r'\det(\ldots, \mathbf{A}'_j, \ldots), \tag{L166}$$

where $\mathbf{A}_j$ and $\mathbf{A}'_j$ are column vectors in the jth slot. Similarly, it is *linear in each row*,

$$\det \begin{pmatrix} \vdots \\ r\mathbf{A}^i + r'\mathbf{A}'^i \\ \vdots \end{pmatrix} = r\det \begin{pmatrix} \vdots \\ \mathbf{A}^i \\ \vdots \end{pmatrix} + r'\det \begin{pmatrix} \vdots \\ \mathbf{A}'^i \\ \vdots \end{pmatrix}. \tag{L167}$$

Notice the general "symmetry" between relations affecting rows and columns, respectively. Convince yourself that this follows from the invariance of the determinant under transposition, Eq. (L161).

5. For $r \in \mathbb{C}$, the determinant obeys the relation

$$\det(rA) = r^n \det(A), \tag{L168}$$

where $(rA)_{ij} \equiv rA_{ij}$. This follows by repeated application of Eq. (L166), with $r' = 0$.

6. The determinant of a matrix vanishes if and only if it contains *linearly dependent rows or columns*. If the columns are linearly dependent, e.g. if $\mathbf{A}_1 = \sum_{i=2}^{n} c^i \mathbf{A}_i$, then Eq. (L166) may be applied to reduce $\det(A)$ to a sum of determinants of the form $\det(\mathbf{A}_i, \ldots, \mathbf{A}_i, \ldots)$. The antisymmetry relation (L164) implies that each determinant in the sum vanishes individually. Now recall from Section L5.4 that a square matrix has linearly dependent rows and columns if and only if it is non-invertible. Combined with our preceding conclusion, this implies that if A is non-invertible, then $\det(A) = 0$. In point 9 below we will justify the converse statement: if $\det(A) = 0$, then A is non-invertible, which in turn implies linearly dependent rows and columns.

determinant 7. The *determinant vanishes if and only if a matrix is non-invertible*, i.e. if it annihilates
test of   a non-vanishing vector. This criterion follows from the arguments given in point 6. It
invertibility provides a powerful test for the invertibility of a matrix: compute its determinant and if a non-vanishing result is obtained, invertibility is guaranteed.

determinant
of product of
matrices

8. The *determinant of a product of matrices* equals the product of determinants,

$$\det(AB) = \det(A)\det(B).\tag{L169}$$

This important formula is shown by multiple application of Eq. (L159) and reordering:

$$\det AB \overset{(L159)}{=} \epsilon_{i_1,\dots,i_n}(AB)^{i_1}{}_1\dots(AB)^{i_n}{}_1 \overset{(L116)}{=} \epsilon_{i_1,\dots,i_n}A^{i_1}{}_{j_1}B^{j_1}{}_1\dots A^{i_n}{}_{j_n}B^{j_n}{}_n$$

$$= \epsilon_{i_1,\dots,i_n}A^{i_1}{}_{j_1}\dots A^{i_n}{}_{j_n}B^{j_1}{}_1\dots B^{j_n}{}_n \overset{(L159)}{=} \det A\,\epsilon_{j_1,\dots,j_n}B^{j_1}{}_1\dots B^{j_n}{}_n$$

$$\overset{(L159)}{=} \det A\,\det B.\tag{L170}$$

9. Equation (L169) implies an important formula for the *inverse of determinants*. From Eq. (L160) we know that the unit matrix has unit determinant, $\det(\mathbb{1}) = 1$. Now use $1 = \det(\mathbb{1}) = \det(AA^{-1}) = \det(A)\det(A^{-1})$ to obtain

$$\det(A^{-1}) = \frac{1}{\det(A)}.\tag{L171}$$

It follows that if $\det(A) = 0$, then $\det(A^{-1})$ does not exist, implying that A is not invertible.

determinant
invariant
under basis
change

10. An important consequence of Eqs. (L169) and (L171) is that

> the determinant of a matrix is invariant under a change of basis.

Indeed, if a basis change transforms a matrix A to $A' = TAT^{-1}$, then

$$\det(A') = \det(TAT^{-1}) \overset{(L169)}{=} \det(T)\det(A)\det(T^{-1}) \overset{(L171)}{=} \det(A).\tag{L172}$$

EXAMPLE It is instructive to verify the *properties of the determinant* for the simple case of a 2×2 matrix

$$A = \begin{pmatrix} a & b \\ c & d \end{pmatrix},\tag{L173}$$

with determinant

$$\det(A) = ad - bc.\tag{L174}$$

1. For a diagonal matrix, $b = c = 0$, Eq. (L174) indeed reduces to $\det A = ad$, the product of diagonal elements.

2. Invariance under transposition follows from the invariance of (L174) under exchange $c \leftrightarrow b$.

3. Column and row antisymmetry is illustrated by $\det \begin{pmatrix} b & a \\ d & c \end{pmatrix} = \det \begin{pmatrix} c & d \\ a & b \end{pmatrix} = bc - ad = -\det(A)$.

4. Column-linearity, (L166), is illustrated by $\det \begin{pmatrix} (a+a') & b \\ (c+c') & d \end{pmatrix} = (a + a')d - b(c + c') = (ad - bc) + (a'd - bc') = \det \begin{pmatrix} a & b \\ c & d \end{pmatrix} + \det \begin{pmatrix} a' & b \\ c' & d \end{pmatrix}$. Row linearity is shown in the same way.

5. To verify Eq. (L168) consider $\det(rA) = \det \begin{pmatrix} ra & rb \\ rc & rd \end{pmatrix} = (ra)(rd) - (rb)(rc) = r^2(ad - bc) = r^2\det(A)$.

6. If A contains linearly dependent columns, $\left(\begin{smallmatrix} a \\ c \end{smallmatrix}\right) = \lambda \left(\begin{smallmatrix} b \\ d \end{smallmatrix}\right)$, we have $\frac{a}{b} = \lambda = \frac{c}{d}$ and $ad - bc = 0$, showing that $\det(A)$ vanishes.

7. If the determinant vanishes, the matrix elements are related by $ad = bc$. Verify that the same condition is required to obtain a nontrivial solution of the equation $A\mathbf{v} = 0$.

8. Equation (L169) is verified by defining two 2×2 matrices, A and B, computing the product AB, and comparing its determinant with the product of the determinants of the individual matrices.

9. The inverse of the matrix (L173) is given by Eq. (L126). Taking its determinant we indeed obtain $\det(A^{-1}) = \frac{1}{(ad-bc)^2}(da - bc) = \frac{1}{ad-bc} = 1/\det(A)$.

INFO For the purpose of illustration, let us discuss the *proof of the exchange identity* (L165). The proofs of other identities are of comparable complexity and they are all based on group properties of permutations. For example, using the notation of Section L1.2, let $P_{[12]}$ be the pair permutation exchanging the numbers 1 and 2 from a set of n numbers, $P_{[12]}\{3, 4, 2, 1\} = \{3, 4, 1, 2\}$. For an arbitrary permutation, P, the composition $P' = P \circ P_{[1,2]}$ is again a permutation – the group property. The composite permutation P' acts as $P'2 = P1$, $P'1 = P2$, and $P'l = Pl$ for $l > 2$. We also know that $\mathrm{sgn}(P') = -\mathrm{sgn}(P)$, because P and P' differ by one pair permutation (if P is even P' is odd, and vice versa). Now, consider two matrices A and A' differing by an exchange of the first and the second columns:

$$A = (\mathbf{A}_1, \mathbf{A}_2, \mathbf{A}^3, \dots, \mathbf{A}^n), \qquad A' = (\mathbf{A}_2, \mathbf{A}_1, \mathbf{A}^3, \dots, \mathbf{A}^n). \tag{L175}$$

Thus $A'^i{}_1 = A^i{}_2$ and $A'^i{}_2 = A^i{}_1$, while $A'^i{}_j = A^i{}_j$ for all columns other than 1 and 2. Now let us apply the Leibniz rule to the computation of the respective determinants:

$$\det(A') = \sum_{P \in S_n} \mathrm{sgn}(P) A'^{P1}{}_1 A'^{P2}{}_2 A'^{P3}{}_3 \dots A'^{Pn}{}_n = \sum_{P \in S_n} \mathrm{sgn}(P) A^{P1}{}_2 A^{P2}{}_1 A^{P3}{}_3 \dots A^{Pn}{}_n$$

$$= -\sum_{P \in S_n} \mathrm{sgn}(P') A^{P2}{}_2 A^{P1}{}_1 A^{P3}{}_3 \dots A^{Pn}{}_n = -\sum_{P' \in S_n} \mathrm{sgn}(P') A^{P'1}{}_1 A^{P'2}{}_2 A^{P'3}{}_3 \dots A^{P'n}{}_n$$

$$= -\det(A).$$

In the second equality we used the exchange relation (L175) between A' and A, in the third the relation between P' and P, and in the fourth that if P runs over all permutations in S_n so does P', $\sum_{P \in S_n} = \sum_{P' \in S_n}$. The final equality is just the definition of the determinant where the summation "variable" P' replaces P.

A more concise proof of the exchange identity can be given using Eq. (L159) and exploiting the antisymmetry of the Levi–Civita symbol under the interchange of two indices. Try it!

L6.4 Some applications

In this section, we discuss some applications of the determinant in the context of linear algebra and calculus. All of these are relevant to the physics curriculum and will be referred to at later stages of the text. However, if time is short, this section can be skipped at first reading.

Special linear group

Above, we have seen that invertible matrices have non-vanishing determinants. The set of all these matrices defines the **general linear group** $\mathrm{GL}(n, \mathbb{C})$. This group contains the subset $\mathrm{SL}(n, \mathbb{C}) \subset \mathrm{GL}(n, \mathbb{C})$, defined by a *unit* determinant,

general linear group

$$SL(n, \mathbb{C}) = \{A \in GL(n, \mathbb{C}) | \det(A) = 1\}. \tag{L176}$$

special linear group   This set is called the **special linear group**. It is a group (with matrix multiplication as composition) because it contains the multiplicative unit element, $\mathbb{1} \in SL(n, \mathbb{C})$, and the unit determinant is preserved under multiplication: if $\det(A) = \det(B) = 1$, then $\det(AB) = \det(A)\det(B) = 1$. The special linear group $SL(n, \mathbb{C})$ is one of several subgroups of $GL(n, \mathbb{C})$ carrying physical and mathematical significance. We will return to this point in Chapter L8 after a few more matrix structures have been introduced.

Determinant of metric tensor

REMARK We assume here a positive definite metric. Generalizations to positive indefinite cases are addressed in Section L10.9.

Consider a basis $\{\mathbf{v}_1, \ldots, \mathbf{v}_n\}$ of $\mathbb{R}^n$. It defines two closely related structures. First, the transformation from the standard basis to the basis $\{\mathbf{v}_i\} = \{\mathbf{e}_k J^k{}_i\}$ is described by the matrix containing these vectors as columns, $J \equiv (\mathbf{v}_1, \ldots, \mathbf{v}_n)$. With later applications to calculus in mind, we denote it by the symbol J, rather than by T as in Section L5.6. Second, the geometry of the basis vectors is described by the metric tensor, with elements $g_{ij} \equiv \langle \mathbf{v}_i, \mathbf{v}_j \rangle$.

To see the connection between these two important objects, compute g_{ij} explicitly; the result can be expressed as the product of J and its transpose:

$$g_{ij} = \langle \mathbf{v}_i, \mathbf{v}_j \rangle = \langle \mathbf{e}_k, \mathbf{e}_l \rangle J^k{}_i J^l{}_j \overset{(L113)}{=} (J^T)_i{}^k \delta_{kl} J^l{}_j = (J^T)_i{}^k J^k{}_j = (J^T J)_{ij}.$$

determinant of metric tensor   In applications, one is often interested in the determinant of the matrix J.[53] An important spin-off of the above formula is that it connects $\det(J)$ with the **determinant of the metric tensor**. Although the latter differs from a conventional matrix in that both indices of g_{ij} are covariant, its determinant, $g \equiv \det(g)$,[54] is defined by a Leibniz formula analogous to (L159) for standard matrices,

$$g \equiv \det g \equiv \epsilon^{i_1, i_2, \ldots, i_d} g_{i_1 1} g_{i_2 2} \cdots g_{i_d d}, \tag{L177}$$

where the Levi–Civita symbol $\epsilon^{i_1, i_2, \ldots, i_n} \equiv \epsilon_{i_1, i_2, \ldots, i_n}$ does the bookkeeping of the sign of the permutations.[55]

The matrix relation $g = J^T J$ implies $\det(g) = \det(J^T J) \overset{(L169)}{=} \det(J^T)\det(J) \overset{(L161)}{=} \det(J)^2$,[56] or

$$\boxed{|\det(\mathbf{v}_1, \ldots, \mathbf{v}_n)| = \sqrt{\det(g)} = \sqrt{g}.} \tag{L178}$$

---

[53] For example, determinants of transformation matrices routinely appear in higher-dimensional integration theory. In that context, the basis vectors $\mathbf{v}_j$ are derived from a generalized coordinate system and the multidimensional volume element spanned by them is computed using $\det(J)$ (see Section C4.2).

[54] Following standard conventions, the symbol g is used for all three: the abstract metric, its representation as a tensor, and the determinant of that tensor. Which particular g is meant generally follows from the context.

[55] In some texts, a covariant Levi–Civita *tensor*, $\tilde{\epsilon}$ is defined. It differs from the combinatorial Levi–Civita symbol in that $\tilde{\epsilon}^{i_1, i_2, \ldots, i_n} \neq \tilde{\epsilon}_{i_1, i_2, \ldots, i_n}$.

[56] Note that $\det(g)$ is positive (since equal to $\det(J)^2$), i.e. the metric is positive definite. The reason is that related by a basis transformation to the positive definite trivial metric, δ_{ij}.

This formula will be used repeatedly later in the text. For later reference, we also note that

orientation   the **orientation of a basis** is determined by the sign of $\det(\mathbf{v}_1, \ldots, \mathbf{v}_n)$:

$$\det(\mathbf{v}_1, \ldots, \mathbf{v}_n) \begin{cases} > 0, & \text{right-handed,} \\ < 0, & \text{left-handed.} \end{cases} \tag{L179}$$

For example, flipping $\mathbf{v}_i \rightarrow -\mathbf{v}_i$ for one basis vector turns a right-handed basis into a left-handed one.

Covariant formulation of vector product

In Chapter L4 we discussed the **vector product**. Difficulties with the covariant notation indicated that the vector product of two vectors is not a genuine vector. In the following, we show how the metric tensor and its determinant can be applied to make the formula for the cross product covariant. The actual meaning of the repair construction is explained later in Section L10.9.

Among the formulas discussed in Chapter L4 one *was* covariant, namely that for the triple product, $(\mathbf{u} \times \mathbf{v}) \cdot \mathbf{w} \overset{(L96)}{=} \epsilon_{ijk} u^i v^j w^k \overset{(L159)}{=} \det(\mathbf{u}, \mathbf{v}, \mathbf{w})$. Although this formula was derived for vectors expanded in a Cartesian basis, $\mathbf{u} = \mathbf{e}_i u^i$, its covariance suggests promoting it

triple   to a general definition of the **triple product**
product

$$\boxed{(\mathbf{u} \times \mathbf{v}) \cdot \mathbf{w} \equiv \det(\mathbf{u}, \mathbf{v}, \mathbf{w}).} \tag{L180}$$

Now consider a general right-handed basis, $\{\mathbf{v}_1, \mathbf{v}_2, \mathbf{v}_3\}$, with metric tensor $g_{ij} = g(\mathbf{v}_i, \mathbf{v}_j)$. The triple product of the three basis vectors can be expressed as

$$(\mathbf{v}_i \times \mathbf{v}_j) \cdot \mathbf{v}_m = \det(\mathbf{v}_i, \mathbf{v}_j, \mathbf{v}_m) \overset{(L159)}{=} \det(\mathbf{v}_1, \mathbf{v}_2, \mathbf{v}_3) \epsilon_{ijm} \overset{(L178)}{=} \sqrt{g} \, \epsilon_{ijm}. \tag{L181}$$

Since the basis vectors are linearly independent, Eq. (L181) implies

$$\mathbf{v}_i \times \mathbf{v}_j = \sqrt{g} \, \epsilon_{ijl} \, g^{lk} \, \mathbf{v}_k. \tag{L182}$$

This formula is a covariant generalization of Eq. (L82). For Cartesian coordinates, with $g^{lk} = \delta^{lk}$, it reduces to that earlier formula, $\mathbf{e}_i \times \mathbf{e}_j = \epsilon_{ijl} \mathbf{e}^l$, with $\mathbf{e}^l = \delta^{lk} \mathbf{e}_k = \mathbf{e}_l$. Applied to the evaluation of a general cross product, it leads to a covariant version of Eq. (L84),

$$\boxed{\mathbf{u} \times \mathbf{w} = \sqrt{g} \, u^i w^j \, \epsilon_{ijl} \, g^{lk} \mathbf{v}_k,} \tag{L183}$$

covariant   constituting a **covariant definition of the vector product**. The factor $\sqrt{g} \, \epsilon_{ijl} \, g^{lk}$ attached
vector   to the basis vector $\mathbf{v}_k$ on the right indicates that the vector product is different from a
product   conventional vector. We return to the discussion of this point in Section L10.9.

L6.5 Summary and outlook

In this chapter we introduced the determinant as a complex-valued function defined on the set of square matrices. We defined various representations of the determinant which all had in common that they were constructed from sums over permutations of products of matrix elements. The heavy reference to permutations in these definitions implied a large set of symmetries. Among these, the most important were (i) the multiplicativity of the determinant under matrix multiplication, Eq. (L169), (ii) its invariance under changes of basis, (L172), and (iii) the fact that a non-vanishing determinant implies the invertibility of its matrix. We also saw that the determinant in real vector spaces affords a geometric interpretation as the volume spanned by the rows or columns defining the argument matrices.

Later in the text, we will see that its geometric and algebraic properties make the determinant an important tool in various mathematical and physical contexts. A first example is discussed in the next chapter, where we explore how the determinant is applied to the simplification of matrix representations.

Matrix diagonalization

We have seen that matrices change under basis transformations. For example, the matrices A and A' in Eqs. (L144) and (L145) describe the same linear map. One matrix looks simple, the other complicated. This situation routinely arises in linear algebra. One has often described a linear map in some basis. Since matrix operations – multiplications, determinant calculations, etc. – are computationally costly, one would like to find a basis in which the map assumes "its simplest possible matrix representation". This objective immediately raises a number of questions: what is the simplest form of a matrix one can possibly hope to obtain? Is it unique? And do there exist algorithms for finding such representations? These questions will be addressed in the present chapter.

L7.1 Eigenvectors and eigenvalues

The simplest possible action of a linear map on a vector is multiplication by a scalar,

$$\boxed{\hat{A}\hat{\mathbf{v}} = \lambda\hat{\mathbf{v}}.}$$

(L184)

eigenvectors   Nonzero vectors with this property are called **eigenvectors** of the map, and λ are the corresponding **eigenvalues**.[57] One may ask if eigenvectors can be found for any linear map, or if it is possible to construct bases consisting entirely of eigenvectors. We will address these questions in the next sections and eventually define a constructive algorithm for the simplification of matrix representations.

Bases of eigenvectors and eigenvalues

Consider an n-dimensional vector space V, and suppose that one has managed to find n linearly independent eigenvectors, $\hat{\mathbf{v}}_j$, of a linear map $\hat{A}$ acting in this space, $\hat{A}\hat{\mathbf{v}}_j = \lambda_j\hat{\mathbf{v}}_j$, where $\lambda_j \in \mathbb{C}$ is the jth eigenvalue. One may then choose these eigenvectors as a basis, $\{\hat{\mathbf{v}}_j\}$, of V. In this basis, the eigenvectors are represented by the standard basis vectors, $\{\mathbf{e}_j\}$, **eigenbasis**   of $\mathbb{C}^n$. The matrix, say D, representing the map in this **eigenbasis** acts on the jth standard basis vector as $D\mathbf{e}_j = \lambda_j\mathbf{e}_j$. This means that D assumes the form of a *diagonal matrix*,

---

[57] The word "eigen", loaned from German, means "own".

$$D = \begin{pmatrix} \lambda_1 & & & \\ & \lambda_2 & & \\ & & \ddots & \\ & & & \lambda_n \end{pmatrix},$$

or $D = \mathrm{diag}(\lambda_1, \lambda_2, \ldots, \lambda_n)$ in short. This diagonal form is the simplest possible representation of the map $\hat{A}$ one can hope to achieve. For example, the vectors $\hat{\mathbf{v}}_j$ in the figure on p. 82 are eigenvectors of a map $\hat{A}$ stretching $\hat{\mathbf{v}}_1$ by a factor of 2 and leaving $\hat{\mathbf{v}}_2$ unchanged. In this basis, the matrix representing the map assumes the diagonal form Eq. (L144).

If a basis different from the eigenbasis $\{\hat{\mathbf{v}}_j\}$ is used, the matrix representation changes to

$$D \overset{(\mathrm{L143})}{\longmapsto} A = TDT^{-1},$$

which in general will not be diagonal anymore. (In the example on p. 82, $\{\hat{\mathbf{v}}_j'\}$ was one such basis, and (L145) the non-diagonal matrix.) Notice, however, that not all signatures of the diagonal representation are lost: the *eigenvalues*, λ_j, were defined in the "abstract space", V. This means that the set of eigenvalues $\{\lambda_j\}$ is a basis-invariant feature of the linear map $\hat{A}$.

What do the above observations imply for the *simplification of matrix representations*? In general, one starts from a situation in which a linear map $\hat{A}$ has been represented in *some* basis, $\{\hat{\mathbf{w}}_j\}$, of V through a matrix A. The discussion above indicates that its representation can be simplified by transforming to a basis of eigenvectors, $\{\hat{\mathbf{v}}_j\}$. As a first step one should therefore aim to identify as many linearly independent eigenvectors as possible. In principle, one may search for these vectors in V by analyzing the eigenvector equation $\hat{A}\hat{\mathbf{v}}_j = \lambda_j\hat{\mathbf{v}}_j$. (For example, if $\hat{A}$ is a rotation, vectors lying along the rotation axes do not change and therefore are eigenvectors with unit eigenvalue.) However, as we will see, the more systematic approach is to work with the matrix representation, A, of $\hat{A}$, in the given basis $\{\hat{\mathbf{w}}_j\}$. Let $\mathbf{v}_j \in \mathbb{C}^n$ represent the eigenvector $\hat{\mathbf{v}}_j$ in this basis. The eigenvector equation now assumes the form of a matrix equation,

$$\boxed{A\mathbf{v}_j = \lambda_j\mathbf{v}_j\,.} \tag{L185}$$

If n linearly independent solutions of these equations can be found, $\{\mathbf{v}_j\}$ defines a basis of eigenvectors in $\mathbb{C}^n$. In this basis, the linear map $\hat{A}$ is represented by a diagonal matrix D, as discussed above. Equivalently, the change from the given representation, A, to the diagonal one, D, is described through the transformation formula Eq. (L142),[58]

$$D = T^{-1}AT, \tag{L186}$$

where $T \equiv (\mathbf{v}_1, \ldots, \mathbf{v}_n)$ is the transformation matrix mapping the standard basis to the eigenbasis of $\mathbb{C}^n$, $T\mathbf{e}_j = \mathbf{v}_j$. To verify Eq. (L186) independently from the general discussion of Section L5.6 on matrix transformations, note that $D\mathbf{e}_j = \lambda_j\mathbf{e}_j$ acts on the standard basis vector $\mathbf{e}_j$ by scalar multiplication. On the other hand, $T\mathbf{e}_j = \mathbf{v}_j$ first sends $\mathbf{e}_j$ to the eigenvector representation in the $\{\hat{\mathbf{w}}_j\}$-basis, then $A\mathbf{v}_j = \lambda_j\mathbf{v}_j$ acts according to the eigenvector equation, and finally, $T^{-1}\lambda_j\mathbf{v}_j = \lambda_j\mathbf{e}_j$ transforms back to the standard vector. Thus, $(T^{-1}AT)\mathbf{e}_j = \lambda_j\mathbf{e}_j = D\mathbf{e}_j$ as stated.

---

[58] Comparing to Section L5.6, associate $\{\mathbf{v}_j\}$ and $\{\mathbf{w}_j\}$ here with the "odd" and "new" bases there, and D, A here with A, A' there.

INFO The problem of finding simple (diagonal) representations of matrices is of great *practical importance* not just in physics but also in engineering, computer science, biology, and other fields. Matrices play an important role as computational tools in all these disciplines. However, the generic representation especially of large matrices does not tell much about their action. They simply look like huge arrays of numbers which can be hard to interpret. The situation becomes much more transparent if a diagonal representation has been found. This not only drastically simplifies calculations, but also makes the interpretation of the map evident. The importance of the problem shows in that physicists, mathematicians and computer scientists are investing a lot of effort into improving algorithms for the constructive "diagonalization" of high-dimensional linear maps. On modern computers, the diagonalization of matrices of dimension 10 000 has become a routine operation that can be accomplished within a few minutes. However, the numerical cost grows with the third power of the dimension and the diagonalization of much larger matrices is prohibitive.

L7.2 Characteristic polynomial

The discussion above shows that the key to the diagonalization of matrices lies in their eigenvectors and eigenvalues. How can these be found? An eigenvector, $\mathbf{v}$, obeys the equation $A\mathbf{v} = \lambda\mathbf{v}$ or $(A - \lambda\mathbb{1})\mathbf{v} = 0$, where λ is its as yet unknown eigenvalue. If this equation is to have a non-vanishing solution, then the matrix $A - \lambda\mathbb{1}$ must have a vanishing determinant, $Z(\lambda) \equiv \det(A - \lambda\mathbb{1}) = 0$ (because it annihilates a non-vanishing vector, $\mathbf{v}$). This condition is a necessary and sufficient criterion for λ to be an eigenvalue.

> Every eigenvalue λ of a matrix A satisfies the condition $\det(A - \lambda\mathbb{1}) = 0$.

EXAMPLE For the matrix A of Eq. (L130) a calculation of $Z(\lambda)$, best performed via the Laplace rule (L153), yields a polynomial of degree three,

$$Z(\lambda) = \det(A - \lambda\mathbb{1}) = \begin{vmatrix} 6-\lambda & -1 & 5 \\ 2 & -\lambda & 2 \\ -8 & 1 & -7-\lambda \end{vmatrix} = -\lambda^3 - \lambda^2 + 2\lambda = -\lambda(\lambda-1)(\lambda+2). \quad \text{(L187)}$$

Each of its zeros, $\lambda_1 = 0$, $\lambda_2 = 1$, $\lambda_3 = -2$, is an eigenvalue of A.

For an n-dimensional matrix, the evaluation of the determinant $\det(A - \lambda\mathbb{1})$ by the Leibniz formula (L147) reads

$$Z(\lambda) = \sum_{P \in S_n} \text{sgn}(P)(A - \lambda\mathbb{1})^{P1}{}_1 (A - \lambda\mathbb{1})^{P2}{}_2 \dots (A - \lambda\mathbb{1})^{Pn}{}_n. \quad \text{(L188)}$$

The sum over products of n factors, each containing λ to zeroth or first order, yields a polynomial of degree n in λ, which may be expressed as

$$Z(\lambda) = \sum_{m=0}^{n} c_m \lambda^m, \quad \text{(L189)}$$

with coefficients $c_m \in \mathbb{C}$ depending on the matrix elements $A^i{}_j$.

characteristic polynomial

The polynomial $Z(\lambda)$ is called the **characteristic polynomial** of A. It is a "characteristic" feature of the matrix A in that it does not change under transformations of basis. This follows from the fact that for $A' = TAT^{-1}$, we have $\det(A' - \lambda\mathbb{1}) = \det(TAT^{-1} - \lambda T\mathbb{1}T^{-1}) = \det(T(A - \lambda\mathbb{1})T^{-1}) = \det(A - \lambda\mathbb{1})$, where $T\mathbb{1}T^{-1} = \mathbb{1}$ and the invariance of the determinant Eq. (L172) were used. The invariance of the characteristic polynomial is expected, because a quantity determining eigenvalues should not depend on the choice of basis.

coefficients of characteristic polynomial

There is not so much that can be said in general about the *coefficients of the characteristic polynomial*. For $\lambda = 0$ we obtain $Z(0) = \det(A)$ by definition of the characteristic polynomial, and $Z(0) = c_0$ according to Eq. (L189). This yields the identification $c_0 = \det(A)$. The two highest possible powers, λ^n and λ^{n-1}, are obtained from the contribution of the unit permutation, $Pj = j$, to the sum, $\prod_j (A^j_{\ j} - \lambda) = (-\lambda)^n + (-\lambda)^{n-1}\sum_{j=1}^n A^j_{\ j} + \cdots$. Here, the ellipses denote terms of order λ^{n-2} and less. We thus conclude $c_n = (-)^n$ and $c_{n-1} = (-)^{n-1}\sum_{j=1}^n A^j_{\ j} = (-)^{n-1}\mathrm{tr}(A)$. All other coefficients, $c_{n-2}, \ldots, c_1$, have a more complicated structure. Summarizing, the characteristic polynomial has the structure

$$Z(\lambda) = (-\lambda)^n + (-\lambda)^{n-1}\mathrm{tr}(A) + \cdots + \det(A). \tag{L190}$$

Once a value λ has been found for which the characteristic polynomial vanishes, $Z(\lambda) = 0$, the corresponding eigenvector, $\mathbf{v}$, is obtained by solving the system of linear equations $(A - \lambda\mathbb{1})^i_{\ j}v^j = 0$ for the coefficients v^j. This can be done by the methods detailed on p. 76. Before discussing how the program can be applied in general, let us consider the simple case of a 2×2 matrix.

EXAMPLE Consider the 2×2 matrix

$$A = \begin{pmatrix} 1 & -1 \\ -\frac{1}{2} & \frac{3}{2} \end{pmatrix}. \tag{L191}$$

This matrix acts on the unit vectors $\mathbf{e}_1$ and $\mathbf{e}_2$ as shown in Fig. L15. It simultaneously stretches and rotates the vectors, which leads to a distortion of the plane, as indicated in the figure. Now let us identify the eigenvectors, $\mathbf{v}_1$ and $\mathbf{v}_2$, of the matrix. Following the above procedure, the first step is to set up the characteristic polynomial

$$Z(\lambda) = \det(A - \lambda\mathbb{1}) = \det \begin{pmatrix} 1 - \lambda & -1 \\ -\frac{1}{2} & \frac{3}{2} - \lambda \end{pmatrix} = (1-\lambda)(\tfrac{3}{2} - \lambda) - \tfrac{1}{2} = \lambda^2 - \tfrac{5}{2}\lambda + 1.$$

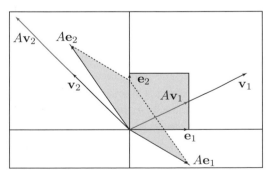

Fig. L15   Action of the matrix (L191) in the two-dimensional plane.

The equation $Z(\lambda) = 0$ is quadratic and its two solutions are given by $\lambda_1 = \frac{1}{2}$ and $\lambda_2 = 2$. We may now find the corresponding eigenvectors by solution of

$$(A - \lambda_1 \mathbb{1})\mathbf{v}_1 = \begin{pmatrix} \frac{1}{2} & -1 \\ -\frac{1}{2} & 1 \end{pmatrix}\begin{pmatrix} v_1^1 \\ v_1^2 \end{pmatrix} = \begin{pmatrix} 0 \\ 0 \end{pmatrix} \quad \Rightarrow \quad \mathbf{v}_1 = c_1 \begin{pmatrix} 2 \\ 1 \end{pmatrix},$$

$$(A - \lambda_2 \mathbb{1})\mathbf{v}_2 = \begin{pmatrix} -1 & -1 \\ -\frac{1}{2} & -\frac{1}{2} \end{pmatrix}\begin{pmatrix} v_2^1 \\ v_2^2 \end{pmatrix} = \begin{pmatrix} 0 \\ 0 \end{pmatrix} \quad \Rightarrow \quad \mathbf{v}_2 = c_2 \begin{pmatrix} 1 \\ -1 \end{pmatrix},$$

where $c_{1,2}$ are arbitrary constants which may be set to unity, $c_{1,2} = 1$. We may now verify the eigenvector property by checking $A\mathbf{v}_j = \lambda_j \mathbf{v}_j$. It is easy to make mistakes when computing eigenvectors, so a check should be a routine element of the program. In the directions specified by $\mathbf{v}_1$ and $\mathbf{v}_2$, the matrix A acts by multiplication by the factors $\frac{1}{2}$ and 2, respectively. The two eigenvectors are linearly independent, and the matrix transforming A into a diagonal representation is given by (see the general discussion of Section L7.1)

$$T = (\mathbf{v}_1, \mathbf{v}_2) = \begin{pmatrix} 2 & -1 \\ 1 & 1 \end{pmatrix}, \qquad T^{-1} = \frac{1}{3}\begin{pmatrix} 1 & 1 \\ -1 & 2 \end{pmatrix}.$$

It is straightforward to verify that $T^{-1}AT = \mathrm{diag}(\frac{1}{2}, 2)$ assumes a diagonal form.

EXERCISE Consider the matrices shown in the example of p. 71 and discuss how they can be interpreted as elements of a diagonalization procedure.

EXERCISE Confirm that the matrix A of Eq. (L130) has eigenvalues given by $\lambda_0 = 0$, $\lambda_1 = 1$ and $\lambda_2 = -2$.

L7.3 Matrix diagonalization

We now have everything in place to discuss the general procedures for the simplification of matrix representations. In this section, we will explore under which conditions a matrix can be transformed into a diagonal representation, and discuss what options for simplification remain if this is not the case.

Eigenvalue spectrum

The previous section has shown that the zeros of the characteristic polynomial, $Z(\lambda)$, are key to the simplification of matrix representations via diagonalization. This raises the question whether characteristic polynomials categorically *have* zeros. The answer depends on whether the vector space is real or complex – this is one of the few cases where the distinction turns out to be crucially important. For example, the characteristic polynomial of the real matrix

$$A = \begin{pmatrix} 0 & 1 \\ -1 & 0 \end{pmatrix} \tag{L192}$$

is given by $Z(\lambda) = \lambda^2 + 1$ and does not have *real* zeros. This means that no real eigenvalues can be found and that the matrix is not diagonalizable by real matrices.

The situation in the complex vector space $\mathbb{C}^n$ is different. According to the **fundamental theorem of algebra** (whose proof is a subject of "algebra", not "linear algebra"), every polynomial of degree n has an equal number of complex zeros, $\lambda_j \in \mathbb{C}$, and can hence be factorized as

$$Z(\lambda) = \prod_{j=1}^{n} (\lambda_j - \lambda), \qquad \lambda_j \in \mathbb{C}. \tag{L193}$$

The set of n complex eigenvalues of an n-dimensional matrix is called its **eigenvalue spectrum**. For example, the characteristic polynomial of the matrix A in Eq. (L192) can be factored as $Z(\lambda) = \lambda^2 + 1 = (\lambda - i)(\lambda + i)$ and its spectrum contains the two elements $\lambda_1 = i$ and $\lambda_2 = -i$. Considered as an element of mat$(2, \mathbb{C})$ the matrix A therefore is diagonalizable. An important corollary of Eq. (L193) is that

> the determinant of a matrix equals the product of its eigenvalues,
> $$\det(A) = \prod_j \lambda_j.$$

This follows from the observation that, by Eq. (L190), $Z(0) = \det(A)$, while Eq. (L193) states that this equals $Z(0) = \prod_j \lambda_j$. On a similar note, the general equation (L190) also states that the coefficient of the second highest degree, λ^{n-1}, is given by $(-)^{n-1}\text{tr}(A)$. From Eq. (L193), the same coefficient follows as $(-)^{n-1}\sum_j \lambda_j$. This leads to the conclusion that

> the trace of a matrix equals the sum of its eigenvalues, $\text{tr}(A) = \sum_j \lambda_j.$

These two results provide consistency checks for any computation of the eigenvalue spectrum.

EXERCISE Show that the matrix A of Eq. (L192) has eigenvectors $\mathbf{v}_1 = (-i, 1)^T$ and $\mathbf{v}_2 = (i, 1)^T$. Verify that

$$T = (\mathbf{v}_1, \mathbf{v}_2) = \begin{pmatrix} -i & i \\ 1 & 1 \end{pmatrix}, \qquad T^{-1} = \frac{1}{2} \begin{pmatrix} i & 1 \\ -i & 1 \end{pmatrix}$$

transforms the matrix to the diagonal form $T^{-1}AT = \text{diag}(i, -i)$. ($\rightarrow$ L7.3.1-4)

We conclude that

> if a real matrix cannot be diagonalized within the matrix space mat$(n, \mathbb{R})$, it may still be diagonalizable by complex matrices in mat$(n, \mathbb{C})$.

Matrices with non-degenerate spectra

The existence of n eigenvalues is a necessary but not sufficient condition for the diagonalizability of a matrix. Complications may arise if eigenvectors coincide, $\lambda_i = \lambda_j$.
degenerate Such eigenvalues are called **degenerate** and the number, r, of eigenvalues λ_j of a given
eigenvalues value is called their **degree of degeneracy**. For example, the unit matrix has n degenerate
eigenvalues equal to 1.

   The absence of degeneracies safeguards the diagonalizability of matrices. This follows
from the following fact (proven in the info section below):

> The eigenvectors, $\mathbf{v}_j$, corresponding to different eigenvalues, λ_j, are linearly
> independent.

Specifically, if a spectrum is non-degenerate, all n eigenvalues are different. According
to the statement above the n eigenvectors then are all linearly independent and form an
eigenbasis. In Section L7.1, we have seen that this implies diagonalizability through the
transformation matrix $T = (\mathbf{v}_1, \ldots, \mathbf{v}_n)$, with $D = T^{-1}AT$. We thus conclude that

> Matrices with non-degenerate spectra are diagonalizable.

INFO The *linear independence of eigenvectors with different eigenvalues* is best shown by induction.
Take two eigenvectors, $\mathbf{v}_1$ and $\mathbf{v}_2$, and assume that a nontrivial linear combination, $\mathbf{0} = c_1\mathbf{v}_1 + c_2\mathbf{v}_2$,
exists, with $c_1, c_2 \neq 0$. Subtract the two equations

$$\mathbf{0} = A\mathbf{0} = A(c_1\mathbf{v}_1 + c_2\mathbf{v}_2) = c_1\lambda_1\mathbf{v}_1 + c_2\lambda_2\mathbf{v}_2,$$

$$\mathbf{0} = \lambda_2\mathbf{0} = c_1\lambda_2\mathbf{v}_1 + c_2\lambda_2\mathbf{v}_2$$

from each other to obtain $c_1(\lambda_1 - \lambda_2)\mathbf{v}_1 = \mathbf{0}$. This is a contradiction because $c_1, \lambda_1 - \lambda_2$ and $\mathbf{v}_1$ are
all non-vanishing. The two vectors $\mathbf{v}_1, \mathbf{v}_2$ therefore cannot be linearly dependent. Now assume that
the first i eigenvectors are linearly independent, for some $i \in \{1, \ldots, n-1\}$, and assume that a linear
combination exists for which $\mathbf{0} = \sum_{j=1}^{i+1} c_j\mathbf{v}_j$. Arguing as before, we then find that

$$\mathbf{0} = A\mathbf{0} = A\left(\sum_{j=1}^{i+1} c_j\mathbf{v}_j\right) = \sum_{j=1}^{i+1} c_j\lambda_j\mathbf{v}_j,$$

$$\mathbf{0} = \lambda_{i+1}\mathbf{0} = \sum_{j=1}^{i+1} c_j\lambda_{i+1}\mathbf{v}_j.$$

Subtraction yields $0 = \sum_{j=1}^{i} c_j(\lambda_j - \lambda_{i+1})\mathbf{v}_j$, in contradiction to the starting assumption. We are
therefore led to the conclusion that the eigenvectors $\mathbf{v}_j$ are all linearly independent.

Matrices with degenerate spectra

The situation gets a little more involved if degenerate eigenvalues occur. We first note
that the eigenvectors corresponding to an eigenvalue, λ, span a subspace of $\mathbb{C}^n$ called

eigenspace the **eigenspace** of that eigenvalue. This is because for any two eigenvectors $\mathbf{v}, \mathbf{w}$ with $A\mathbf{v} = \lambda\mathbf{v}$ and $A\mathbf{w} = \lambda\mathbf{w}$, the linear combination $c\mathbf{v} + d\mathbf{w}$, $c, d \in \mathbb{C}$, is again an eigenvector with the same eigenvalue. For non-degenerate eigenvalues, λ_j, the eigenspaces are all one-dimensional and are spanned by the corresponding eigenvectors, $\mathbf{v}_j$. Eigenvalues of degeneracy r can have eigenspaces whose dimensionality, s, ranges anywhere between 1 and r. (For an extreme example, consider the unit matrix, which has $r = s = n$ degenerate eigenvalues equal to 1, where the standard basis vectors $\mathbf{e}_j$ span the n-dimensional eigenspace $\mathbb{C}^n$.) If the eigenspaces of all r-fold degenerate eigenvalues have maximal dimension, $s = r$, the total number of linearly independent eigenvectors equals n, the dimension of the vector space. This defines an eigenbasis diagonalizing the matrix. ($\rightarrow$ L7.3.5-6)

However, if r-fold degenerate eigenvectors with eigenspaces of dimensionality $s < r$ occur, diagonalizability is lost. To illustrate the phenomenon with a simple example, consider the matrix $A = \left(\begin{smallmatrix} 1 & b \\ 0 & 1 \end{smallmatrix}\right)$. Its characteristic polynomial, $Z(\lambda) = (\lambda - 1)^2$, has the two-fold degenerate zero $\lambda = 1$. However, for $b \neq 0$ the corresponding eigenvector equation, $(A - 1 \cdot \mathbb{1})\mathbf{v} = \left(\begin{smallmatrix} 0 & b \\ 0 & 0 \end{smallmatrix}\right)\mathbf{v}$, has only one solution, $\mathbf{v} = (0, c)^\mathsf{T}$, where c is a normalization constant. The eigenspace for $\lambda = 1$ is one-dimensional and the matrix cannot be diagonalized.

INFO In physical applications, non-diagonalizable matrices with degenerate eigenvalues do not occur very often. Still it is good to know what the simplest possible representation of a matrix with degenerate eigenvalues looks like. An example is shown in the schematic below:

$$\begin{pmatrix} \ddots & & & & & & \\ & \lambda & 1 & & & & \\ & 0 & \lambda & & & & \\ & & & \lambda & & & \\ & & & & \mu & 1 & 0 \\ & & & & 0 & \mu & 1 \\ & & & & 0 & 0 & \mu \\ & & & & & & \ddots \end{pmatrix}, \tag{L194}$$

where λ and μ are both three-fold degenerate eigenvalues, having eigenspaces of dimension 2 and 1, respectively. In general, a matrix with a set of eigenvalues $\{\lambda_a\}$, with degeneracy r_a and eigenspace dimensionality s_a for eigenvalue λ_a, can be reduced to one containing **Jordan blocks**, each having $r_a - s_a + 1$ copies of eigenvalue λ_a on the main diagonal and $r_a - s_a$ copies of unity on the next diagonal. Blocks with $s_a > 1$ are complemented by $s_a - 1$ copies of the eigenvalue on the main diagonal. Equation (L194) illustrates the situation for an eigenvalue λ with $r = 3$ and $s = 2$, and an eigenvalue μ with $r = 3$ and $s = 1$. Matrices represented in this way are said to be in **Jordan form**.

For a general discussion of algorithms for transforming matrices into a Jordan form we refer to specialized textbooks on linear algebra. However, the general idea may be illustrated using a simple example. Consider the matrix

Jordan blocks

$$A = \begin{pmatrix} 0 & 1 \\ -1 & -2 \end{pmatrix}. \tag{L195}$$

Its characteristic polynomial is $Z(\lambda) = (1 + \lambda)^2$, with two degenerate zeros, $\lambda = -1$. The corresponding eigenvector equation, $\mathbf{0} = \left(\begin{smallmatrix} 1 & 1 \\ -1 & -1 \end{smallmatrix}\right)\mathbf{v}$, yields (always up to normalization) only one solution, $\mathbf{v}_1 = \left(\begin{smallmatrix} 1 \\ -1 \end{smallmatrix}\right)$. We aim to transform A into its Jordan representation, $A' = \left(\begin{smallmatrix} -1 & b \\ 0 & -1 \end{smallmatrix}\right)$, where

b is a parameter to be determined. To this end we consider an ansatz for the transformation matrix of the form $T = (\mathbf{v}_1, \mathbf{w})$, where $\mathbf{v}_1$ is A's eigenvector and $\mathbf{w}$ is a complementing vector of the new basis, which likewise remains to be determined. The equation fixing $\mathbf{w}$ and b is then given by $T^{-1}AT = A'$. It is a straightforward exercise to compute T^{-1} and to write out the matrix equation above in terms of four equations for the coefficients of $T^{-1}AT$. A solution of these equations is given by $b = 1$, $\mathbf{w} = \left(\begin{smallmatrix}1\\0\end{smallmatrix}\right)$. Substituting this into the defining equation, we obtain

$$\begin{pmatrix}0 & -1\\1 & 1\end{pmatrix}\begin{pmatrix}0 & 1\\-1 & -2\end{pmatrix}\begin{pmatrix}1 & 1\\-1 & 0\end{pmatrix} = \begin{pmatrix}-1 & 1\\0 & -1\end{pmatrix},$$

as can be checked by direct matrix multiplication. The matrix on the r.h.s. is the Jordan representation of A. The computation of general Jordan representations generalizes the above program to the solution of systems of linear equations of higher order.

Matrix diagonalization algorithm

We are now in a position to summarize the algorithm for diagonalizing matrices $A \in \text{mat}(n, \mathbb{C})$: ($\rightarrow$ L7.3.1-6)

1. Compute the characteristic polynomial $Z(\lambda)$ and find its zeros, λ_j.

2. If the eigenvalues are all different the matrix is diagonalizable. In this case, find the eigenvectors by solving the linear system $(A - \lambda_j \mathbb{1})\mathbf{v}_j = 0$, for $j = 1, \ldots, n$.

3. The matrix $T = (\mathbf{v}_1, \ldots, \mathbf{v}_n)$ describes the transformation into a diagonal form, $T^{-1}AT = D \equiv \text{diag}(\lambda_1, \ldots, \lambda_n)$.

4. Check that no mistakes have been made and explicitly verify that $A = TDT^{-1}$ holds.

degenerate eigenvalues In the case of *eigenvalues with degeneracy r*, the first step is to find as many eigenvectors as possible. If r linearly independent eigenvectors $\mathbf{v}_1, \ldots, \mathbf{v}_r$ can be found, we include them as part of our transformation matrix T and A remains diagonalizable. In the exceptional case where only $s < r$ eigenvectors can be found, A is not diagonalizable. It then contains a Jordan block of size $r - s + 1$, which must be computed by procedures similar to those exemplified on p. 105.

real matrices The characteristic polynomials of *real matrices* often have fewer real zeros than their rank. In such cases they are not diagonalizable by real transformation matrices but may still be complex-diagonalizable, as discussed in the exercise of Section L7.3.

One final remark: the discussion above covered all possible scenarios and this may somewhat over-emphasize the role played by non-diagonalizable matrices. In fact, most matrices encountered in *physical applications* are diagonalizable and this includes the real diagonalizability of real matrices. The point is that the matrices relevant to mechanics, electrodynamics, quantum mechanics and other physical sub-fields usually obey conditions which grant diagonalizability from the outset. For example, we will show in Chapter L8 that real matrices which equal their own transpose, $A = A^T$, are categorically diagonalizable, as are matrices which obey the condition $A^T = A^{-1}$. The characteristic polynomials of such matrices factorize and even degenerate eigenvalues are not harmful to diagonalizability. Similar statements apply to complex matrices obeying the conditions $A = A^{\dagger}$

or $A^\dagger = A^{-1}$. Matrices relevant to mechanics, electrodynamics or quantum mechanics generally satisfy one of these equations.

L7.4 Functions of matrices

REMARK This section can be skipped on first reading. It requires familiarity with Taylor series, Chapter C5.

Given a square matrix, $A \in \text{mat}(n, \mathbb{C})$, the product $AA \in \text{mat}(n, \mathbb{C})$ is again a square matrix. This observation suggests generalizing the scalar function $f : \mathbb{C} \to \mathbb{C}, z \mapsto z^2 = zz$ to a function defined on the set of n-dimensional matrices, $f : \text{mat}(n, \mathbb{C}) \to \text{mat}(n, \mathbb{C}), A \mapsto AA$. We keep the notation simple and denote this function again by f.

This idea can be extended to arbitrary functions $f : \mathbb{C} \to \mathbb{C}, z \mapsto f(z)$ possessing a Taylor expansion around $z = 0$. Given the Taylor series representation,

$$f(z) = \sum_{m=0}^{\infty} \frac{f^{(m)}(0)}{m!} z^m,$$

a function $f : \text{mat}(n, \mathbb{C}) \to \text{mat}(n, \mathbb{C}), A \mapsto f(A)$ is defined by

$$f(A) = \sum_{m=0}^{\infty} \frac{f^{(m)}(0)}{m!} A^m, \qquad \text{(L196)}$$

where

$$A^m \equiv \underbrace{AA \ldots A}_{m \text{ times}}, \qquad \text{and} \quad A^0 \equiv \mathbb{1}.$$

exponential matrix function For example, the *exponential function of a matrix* is now defined as $\exp(A) = \sum_{m=0}^{\infty} \frac{A^m}{m!}$, etc. In many ways, matrix functions behave as ordinary functions. For example, derivatives $f'(A)$ are obtained by computation of the ordinary derivatives $f'(z)$ and substitution of A for z. (This is possible because the derivative of a series is again a series.)

computational rules However, one must be careful not to apply relations which rely on the commutativity of numbers. For example, the relation $\exp(z+z') = \exp(z)\exp(z')$ does not extend to matrices, $\exp(A + B) \neq \exp(A)\exp(B)$ in general. The origin of the inequality can be understood by separate Taylor expansion of the two sides of the (in)equality in A and B up to second order. For the l.h.s. we have $1+(A+B)+\frac{1}{2}(A+B)^2+\cdots = 1+(A+B)+\frac{1}{2}(A^2+AB+BA+B^2)+\cdots$, while the r.h.s. yields $(1+A+\frac{1}{2}A^2+\cdots)(1+B+\frac{1}{2}B^2+\cdots) = 1+(A+B)+\frac{1}{2}(A^2+2AB+B^2)$. The two expressions are different, unless the matrices A and B commute, $AB = BA$. The rule of thumb is that functions of a single matrix, $f(A)$, behave like ordinary functions (for example, $\exp(A + A) = \exp(A)\exp(A)$), because the commutativity issue does not arise. However, when functions of different matrices appear, one has to be careful.

The function of a matrix becomes even easier to evaluate if the matrix is known in diagonal form, $A = TDT^{-1}$. In this case, $A^m = (TDT^{-1})(TDT^{-1})\ldots(TDT^{-1}) = TD^mT^{-1}$, where D^m is a diagonal matrix containing the mth power, λ_j^m, of A's eigenvalues on its diagonal. The matrix function can now be represented as

$$f(A) = T\left(\sum_{m=0}^{\infty}\frac{f^{(m)}(0)}{m!}D^m\right)T^{-1} = Tf(D)T^{-1}, \tag{L197}$$

where $f(D) = \mathrm{diag}(f(\lambda_1),\ldots,f(\lambda_n))$ is a diagonal matrix containing the complex functions $f(\lambda_j)$ as its diagonal elements. We will return to the discussion of matrix functions in Section L8.3 below. ($\rightarrow$ L7.4.1-2)

L7.5 Summary and outlook

In this chapter we have developed the methodology for diagonalizing matrices. We showed that the procedure involves three logical steps, the computation of the characteristic polynomial, obtaining the eigenvalues of its zeros, and solving the ensuing linear equations for the eigenvectors. While this program is straightforward as a matter of principle, its three steps are all computationally costly, which makes matrix diagonalization a topic of serious study in (numerical) mathematics. We also developed principal criteria for the diagonalizability of matrices and saw that once again the complex numbers outperform the reals: while the spectra of n-dimensional matrices generically contain n complex eigenvalues, the eigenvalues of real matrices are not necessarily real. This implies that real matrices are often not diagonalizable by real transformations. In the complex case, diagonalizability can be jeopardized only by the appearance of degenerate eigenvalues; however, we reasoned that this situation does not frequently occur in physics applications.

What safeguards the diagonalizability of matrices relevant to physics is a connection between two important concepts of linear algebra so far introduced separately: matrix algebra and inner products. In the next two chapters we will show that this linkage defines families of matrices possessing rather strong mathematical properties, of which categoric diagonalizability is only one.

Unitarity and Hermiticity

REMARK In this chapter, a few differences between real and complex vector spaces need to be addressed. Although we will avoid repetition and discuss both cases in parallel, some features of the real and complex maps addressed in this chapter are denoted by different vocabulary. In such cases we will use a bracket notation, mentioning the complex keywords followed by their (real versions). If vocabulary does not play a crucial role, we follow the same policy as in earlier chapters and use wording appropriate to the more general, complex case.

In both physics and mathematics one is often interested in maps that are linear and at the same time preserve the scalar product between vectors. For example, the rotation of vectors is a linear map which does not change lengths, nor angles between vectors. It therefore falls into the above category. Other elementary examples include the reflection of vectors at planes, or points. However, the importance of inner-product-preserving maps extends beyond applications with an easy geometric visualization: later in the text, we will discuss how quantum mechanics is described by vector spaces of functions. All these spaces are inner product spaces, and inner-product-preserving linear maps play a distinguished physical role. This chapter introduces the general concept of such maps and discusses their most important mathematical properties. We will see that various "nice" features – invertibility, diagonalizability, etc. – are automatically granted for maps of this type. This makes their handling easier than that of generic linear maps.

L8.1 Unitarity and orthogonality

The defining feature of an inner-product-preserving map is that

$$\forall \hat{\mathbf{u}}, \hat{\mathbf{w}} \in V, \qquad \langle \hat{A}\hat{\mathbf{u}}, \hat{A}\hat{\mathbf{w}} \rangle = \langle \hat{\mathbf{u}}, \hat{\mathbf{w}} \rangle. \tag{L198}$$

unitary maps Depending on whether V is complex or real, maps obeying this criterion are called **unitary** or **orthogonal maps**, respectively. As usual, in such situations the real orthogonal maps are included in the complex unitary maps as a subset.

Unitary maps have a trivial kernel. The reason is that for every non-vanishing vector $\hat{\mathbf{u}}$ we have $\|\hat{A}\hat{\mathbf{u}}\|^2 = \langle \hat{A}\hat{\mathbf{u}}, \hat{A}\hat{\mathbf{u}} \rangle = \langle \hat{\mathbf{u}}, \hat{\mathbf{u}} \rangle = \|\hat{\mathbf{u}}\|^2 \neq 0$, i.e. the image $\hat{A}\hat{\mathbf{u}}$ cannot be the null vector. From our discussion in Section L5.4, we conclude that *unitary and orthogonal maps are invertible.*

Given two unitary maps, $\hat{A}, \hat{B}$, the product $\hat{A}\hat{B}$ is again unitary:

$$\langle \hat{A}\hat{B}\hat{\mathbf{u}}, \hat{A}\hat{B}\hat{\mathbf{w}}\rangle = \langle \hat{A}(\hat{B}\hat{\mathbf{u}}), \hat{A}(\hat{B}\hat{\mathbf{w}})\rangle = \langle \hat{B}\hat{\mathbf{u}}, \hat{B}\hat{\mathbf{w}}\rangle = \langle \hat{\mathbf{u}}, \hat{\mathbf{w}}\rangle.$$

Similarly, the inverse of a unitary map $\hat{A}$ is also unitary:

$$\langle \hat{A}^{-1}\hat{\mathbf{u}}, \hat{A}^{-1}\hat{\mathbf{w}}\rangle = \langle \hat{A}(\hat{A}^{-1}\hat{\mathbf{u}}), \hat{A}(\hat{A}^{-1}\hat{\mathbf{w}})\rangle = \langle \hat{\mathbf{u}}, \hat{\mathbf{w}}\rangle,$$

where the unitarity of $\hat{A}$ was used. Moreover, the unit map is trivially unitary. We conclude that the set of unitary maps defines a group embedded in the larger group of invertible maps. This group is called the **unitary group**, U(n), and its subgroup defined by real orthogonal matrices the **orthogonal group**, O(n).

unitary group

We finally note that linear maps in the unitary group are frequently denoted by Latin letters starting with $\hat{U}, \hat{V}, \ldots \in$ U(n) and elements of the orthogonal group as $\hat{O}, \hat{P}, \hat{Q}, \ldots \in$ O(n).

Unitary and orthogonal matrices

Let us now explore what unitarity of $\hat{U}$ implies for the associated matrices U. Given that we work in an inner product space, it is natural to consider an *orthonormal basis*, $\langle \hat{\mathbf{e}}_i, \hat{\mathbf{e}}_j\rangle = \delta_{ij}$. The inner product of two vectors is then given by Eq. (L74), $\langle \hat{\mathbf{u}}, \hat{\mathbf{w}}\rangle = \overline{u^i}\delta_{ij}w^j$. Representing the coefficients of $\hat{U}\hat{\mathbf{w}}$ as $(U\mathbf{w})^l = U^l_{\ j}w^j$[59] and those of $\hat{U}\hat{\mathbf{u}}$ as $(U\mathbf{u})^m = U^m_{\ k}u^k$, and using $\overline{U^m_{\ k}u^k} = \overline{U^m_{\ k}}\,\overline{u^k}$, the unitarity condition (L198) becomes

$$\overline{U^m_{\ k}}\,\overline{u^k}\,\delta_{mj}U^j_{\ l}w^l = \overline{u^k}\delta_{kl}w^l. \tag{L199}$$

This condition must hold for arbitrary $\hat{\mathbf{u}}$ and $\hat{\mathbf{w}}$, implying the matrix condition $\overline{U^m_{\ k}}\delta_{mj}U^j_{\ l} = \delta_{kl}$. Recalling the definition (L114) of the *adjoint matrix*, $(U^\dagger)_k^{\ m} = \overline{U^m_{\ k}}$, this may be rewritten as

$$(U^\dagger)_k^{\ m}\delta_{mj}U^j_{\ l} = \delta_{kl}. \tag{L200}$$

To *simplify the notation*, we define

$$(U^\dagger)^i_{\ j} \equiv \delta^{ik}(U^\dagger)_k^{\ m}\delta_{mj}. \tag{L201}$$

Conceptually, this definition changes covariant indices to contravariant ones and vice versa. This is done by application of the index raising and lowering relations (L63) and (L60), with standard metric $g_{ij} = \delta_{ij}$.[60] In the info section on p. 112 below, the definition of unitarity is extended to the case of non-standard inner products, $g_{ij} \neq \delta_{ij}$. For such inner products, the index-positioning becomes essential, and $(U^\dagger)^i_{\ j} \neq (U^\dagger)_i^{\ j}$.

[59] Recall that $\mathbf{w}$ and U refer to the $\mathbb{C}^n$-component representation of the vector $\hat{\mathbf{w}}$ and the matrix representation of the linear map $\hat{U}$ in a given basis, respectively.

[60] For the standard metric, the index raising operation does not affect the concrete values of matrix elements, $(U^\dagger)^i_{\ j} = (U^\dagger)_i^{\ j}$. For example, for $U = \begin{pmatrix} a & b \\ c & d \end{pmatrix}$, we have $U^\dagger = \overline{U}^\mathsf{T} = \begin{pmatrix} \bar{a} & \bar{c} \\ \bar{b} & \bar{d} \end{pmatrix}$, with $(U^\dagger)^1_{\ 2} = (U^\dagger)_1^{\ 2} = \bar{c}$, etc.

Multiplying the unitarity condition (L200) by δ^{ik} and using Eq. (L201), it assumes the form

$$(U^\dagger)^i{}_j U^j{}_l = \delta^i{}_l.$$ (L202)

unitary matrix Matrices obeying this condition are called **unitary matrices**. In a similar manner, the orthogonality condition of a real map, $\hat{O}$, implies the matrix relation $(O^T)_k{}^m \delta_{mj} O^j{}_l = \delta_{kl}$. This is the same as Eq. (L200) above, only the complex conjugation is absent. Indices can be raised/lowered as $(O^T)^i{}_j \equiv \delta^{ik}(O^T)_k{}^m \delta_{mj}$ to obtain the condition

$$(O^T)^i{}_j O^j{}_l = \delta^i{}_l.$$ (L203)

orthogonal matrix Matrices obeying this condition are called **orthogonal matrices**. The essential statement made by Eqs. (L202) and (L203) is that

> the adjoint, $U^\dagger$, of a unitary matrix U equals its inverse. The same holds true for the transpose, O^T, of an orthogonal matrix O. Conversely, a matrix whose inverse is given by its adjoint (transpose) is unitary (orthogonal).

In other words, the inverse of unitary and orthogonal matrices can be obtained without any elaborate calculation. Consider, for example, the matrix $U = \frac{1}{\sqrt{2}} \begin{pmatrix} 1 & i \\ i & 1 \end{pmatrix}$. It is straightforward to verify that it is unitary: $U^\dagger = \frac{1}{\sqrt{2}} \begin{pmatrix} 1 & -i \\ -i & 1 \end{pmatrix}$ obeys the condition (L202) and the matrices are inverse to each other, $U^\dagger U = \mathbb{1}$.

In an index-free notation the defining equations read

$$\begin{aligned} \hat{U} \text{ unitary} &\quad\Leftrightarrow\quad U^\dagger U = \mathbb{1}, \\ \hat{O} \text{ orthogonal} &\quad\Leftrightarrow\quad O^T O = \mathbb{1}. \end{aligned}$$ (L204)

They imply an economic way to *test for the unitarity (orthogonality) of a matrix*: build the adjoint (transpose), $U^\dagger$ (O^T), and check whether $UU^\dagger = \mathbb{1}$ ($OO^T = \mathbb{1}$). While unitarity and orthogonality cannot usually be "seen" with the naked eye (is $U = \frac{1}{\sqrt{2}} \begin{pmatrix} i & 1 \\ 1 & i \end{pmatrix}$ unitary?) this operation can be performed with relatively little effort. ($\to$ L8.1.1-2)

For completeness, we mention that the *non-covariant index representation* (all indices downstairs) of the unitarity/orthogonality conditions reads as

$$U^\dagger_{ij} U_{jl} = \delta_{il}, \qquad O^T_{ij} O_{jl} = \delta_{il},$$ (L205)

where $U^\dagger_{ij} = \overline{U_{ji}}$ and $O^T_{ij} = O_{ji}$.

The groups of unitary and orthogonal matrices

We have seen that the abstract unitary and orthogonal maps form subgroups, U(n) and O(n), of the group of invertible linear maps, respectively. Likewise, the sets of unitary and orthogonal matrices define subgroups of the group of invertible matrices, GL(n, $\mathbb{C}$) and GL(n, $\mathbb{R}$).

unitary
matrix group

These matrix groups are denoted by the same symbols, U(n) and O(n), as their abstract siblings, respectively.[61] They are called the **group of unitary matrices** and the **group of orthogonal matrices**, respectively, and are defined as

$$U(n) = \{U \in GL(n, \mathbb{C}) \mid U^\dagger = U^{-1}\},$$

$$O(n) = \{O \in GL(n, \mathbb{R}) \mid O^T = O^{-1}\}. \tag{L206}$$

Their group property follows from the fact that they are matrix representations of the groups U(n) and O(n) introduced in the previous section (think about this point). However, it is a good exercise to check the group criteria explicitly.

INFO As mentioned in the beginning of the chapter, orthogonal and unitary maps play an important *role in physics*. The reason is that many physically important linear transformations preserve the norm of vectors. As an example, consider a two-dimensional real vector space. The matrix

$$R(\theta) = \begin{pmatrix} \cos\theta & -\sin\theta \\ \sin\theta & \cos\theta \end{pmatrix} \tag{L207}$$

describes a rotation of vectors by the angle θ. This can be seen by inspection of the image vectors obtained by application of $R(\theta)$ to the horizontal vector $\mathbf{e}_x = (1, 0)^T$. The inverse of this matrix describes a rotation by the same amount in the opposite direction and is given by $R^{-1}(\theta) = R(-\theta)$. Using $\cos(-\theta) = \cos\theta$ and $\sin(-\theta) = -\sin\theta$ this gives

$$R^{-1}(\theta) = \begin{pmatrix} \cos\theta & \sin\theta \\ -\sin\theta & \cos\theta \end{pmatrix} = R^T(\theta). \tag{L208}$$

(Multiply the two matrices to confirm that $RR^{-1} = \mathbb{1}$.) The reflection of vectors at the origin, $\mathbf{x} \mapsto -\mathbf{x}$, another orthogonal map, is represented by the negative of the unit matrix, $-\mathbb{1}$. For example, a 90 deg rotation ($\theta = \pi/2$) followed by a reflection and another 90 deg rotation is not expected to have any effect. Show this by verifying that $R(\pi/2)(-\mathbb{1})R(\pi/2) = \mathbb{1}$.

The most important applications of *unitary maps* in physics are found in *quantum mechanics*. A brief teaser introduction to the description of quantum phenomena in terms of unitary maps on p. 123 can be studied after some more material has been introduced.

unitarity and
non-
orthonormal
bases

INFO In applications, we often work with *non-orthonormal bases* $\{\hat{\mathbf{v}}_i\}$, for which the inner product $\langle \hat{\mathbf{v}}_i, \hat{\mathbf{v}}_j \rangle = g_{ij}$ defines the elements of a metric tensor. In Section L3.3 we saw that the $\mathbb{C}^n$-inner product compatible with this situation is given by $\langle \mathbf{u}, \mathbf{w} \rangle = \overline{u^i} g_{ij} w^j$. Representing the coefficients of $\hat{U}\hat{\mathbf{u}}$ as $(U\mathbf{u})^m = U^m{}_k u^k$, the unitarity condition (L198) becomes $\overline{U^m{}_k u^k} g_{mj} U^j{}_l w^l = \overline{v^k} g_{kl} w^l$. Comparison with Eq. (L199) shows that the presence of a metric tensor amounts to replacing δ_{ij} by g_{ij}. Accordingly, the condition (L200) for the adjoint of the transformation now reads

$$(U^\dagger)_k{}^m g_{mj} U^j{}_l = g_{kl}. \tag{L209}$$

The difference from the orthonormal relation (L200) is the generalization $\delta_{ij} \to g_{ij}$. The fact that (L200) required a Kronecker-δ on its left for covariant consistency was an early hint to the presence of a more general structure. This level of "alertness" should be considered a strength of the covariant

---

[61] In the abstract context, the symbol U(n) denotes the set of linear maps of an n-dimensional complex vector space obeying the criterion (L198). In the matrix context, the same symbol denotes the set of matrices obeying the condition (L204). Once a basis has been chosen, each element of the abstract group $U(n)$ corresponds to one of the matrix group $U(n)$, and this assignment is compatible with the rules of group composition. The groups are therefore "almost identical", and it is justified to denote them by the same name.

notation. For example, the generalization to non-orthonormal bases in the non-covariant formulation of unitarity, Eq. (L205) is not as obvious.

INFO Consider an **indefinite metric** with **signature** $(p, n-p)$. In an orthonormal basis it is represented by a diagonal tensor $\eta = \text{diag}(1, \ldots, 1, -1, \ldots, -1)$, containing p elements 1 and $n-p$ elements -1 on its diagonal. Complex matrices preserving this inner product obey the relation $(U^\dagger)_l{}^i \eta_{ij} U^j{}_m = \eta_{lm}$. This is the defining relation of the **pseudo-unitary group**, $U(p, n-p)$. The analogous relation for the real case, $(O^T)_l{}^i \eta_{ij} O^j{}_m = \eta_{lm}$, defines the **pseudo-orthogonal group** $O(p, n-p)$. As an example, consider $n = 2$ and $p = 1$. The condition

pseudo-unitary group

$$\Lambda^T \begin{pmatrix} 1 & \\ & -1 \end{pmatrix} \Lambda = \begin{pmatrix} 1 & \\ & -1 \end{pmatrix} \tag{L210}$$

is satisfied by matrices of the form $\Lambda = \begin{pmatrix} \cosh\alpha & \sinh\alpha \\ \sinh\alpha & \cosh\alpha \end{pmatrix}$.

Lorentz group

A group of great physical significance is the **Lorentz group**, $O(1,3)$, preserving the form of a signature-$(1,3)$ metric in four-dimensional space. Its elements are real matrices satisfying

$$\Lambda^T \eta \Lambda = \eta \tag{L211}$$

for the Minkowski metric introduced on p. 52. Elements, $\Lambda \in O(1,3)$, of the Lorentz group are called the **Lorentz transformations** and play an important role in the theory of *special relativity*.

Lorentz transformation

In this theory, space-time points are represented by vectors $x = (x^0, x^1, x^2, x^3)^T$, where the non-boldface notation is standard. The coordinates $x^{1,2,3}$ describe three-dimensional space, and $x^0 = ct$ is time, t, scaled by the speed of light, c, to give all four components the same dimensions of length. Following standard conventions, we use Greek indices $\mu, \nu = 0, 1, 2, 3$ to label the components, x^μ, of a space-time point x, and Latin indices $i, j = 1, 2, 3$ for its spatial components, x^i.

Hendrik Antoon Lorentz 1853–1928
Dutch physicist and recipient of the 1902 Nobel prize (with Peter Zeeman) for the explanation of the Zeeman effect. Lorentz identified the coordinate transformation equations stabilizing the Minkowski metric, and in this way laid the mathematical foundations of special relativity. The importance of his work reflects in Einstein's quotation (1953): "For me personally he meant more than all the others I have met on my life's journey."

inertial coordinate systems

Coordinate systems $\{x^\mu\}$ and $\{x'^\mu\}$ related to each other by a Lorentz transformation, $x'^\mu = \Lambda^\mu{}_\nu x^\nu$, are called **inertial** relative to each other. This invariance condition is the mathematical formulation of a fundamental axiom of special relativity: *in all inertial coordinate systems, light moves with constant velocity, c.* To see why, assume that the tip of a light ray is tracked in the unprimed system. It moves with velocity c, and so we have the condition that at any t, $(ct)^2 = x^i x_i$, or $0 = x^\mu \eta_{\mu\nu} x^\nu$. In the primed system the same event is described by coordinates, x'^μ, which again must obey the relation $(ct')^2 = x'^i x'_i$ or $x'^\mu \eta_{\mu\nu} x'^\nu = 0$. The Lorentz transformation $x'^\mu = \Lambda^\mu{}_\nu x^\nu$ is designed to respect this condition and hence correctly incorporates the constancy-of-light axiom of relativity.

Examples of Lorentz transformations include *time reversal*, $\Lambda^0{}_0 = -1$, $\Lambda^i{}_j = \delta^i{}_j$, with all other elements zero, i.e. the reflection of the time coordinate; *space reflection*, $\Lambda^0{}_0 = 1$, $\Lambda^i{}_j = -1$; and *rotations of space*, $\Lambda^0{}_0 = 1$, $\Lambda^i{}_j = A^i{}_j$, where A is a rotation (i.e. orthogonal) matrix. The most interesting Lorentz transformations are the *Lorentz boosts*, which mix space and time coordinates. By way of example, we consider a transformation that does not affect the coordinates x^2 and x^3. It is then straightforward to verify that the condition (L211) requires Λ to have the form

$$\Lambda = \begin{pmatrix} \cosh\lambda & \sinh\lambda & & \\ \sinh\lambda & \cosh\lambda & & \\ & & 1 & \\ & & & 1 \end{pmatrix}, \qquad (L212)$$

where λ is a free parameter. The Lorentz boost describes the transformation between two coordinate systems moving relative to each other in the x^1 direction with relative velocity v. The dependence of the Lorentz parameter, λ, on the boost velocity v follows from comparing the coordinate dependence of specific events in the two coordinate systems. For example, at time t of the unprimed system, the origin of the boosted system ($x'^1 = 0$) is at $x^1 = vt$. Since x'^1 depends on ct and x^1 through the Lorentz transformation $x'^\mu = \Lambda^\mu{}_\nu x^\nu$, we obtain $0 = x'^1 = \sinh\lambda x^0 + \cosh\lambda x^1 = \sinh\lambda ct + \cosh\lambda vt$. This leads to the identification of the boost parameter λ as $\tanh\lambda = -v/c$. For velocities $|v| \ll c$ much smaller than the speed of light, we have $\lambda \simeq -v/c$, hence $\cosh\lambda \simeq 1$ and $\sinh\lambda \simeq -v/c$. The transformed coordinates are then given by $x'^1 \simeq x^1 + \lambda x^0 = x^1 - vt$ and $x'^0 \simeq x^0 + \lambda x^1$, which is equivalent to $t' = t + \mathcal{O}(x^1 v/c^2)$. This shows how for small velocities the Lorentz transformation leaves time approximately invariant, $t' \simeq t$, and just describes the change of coordinates $x'^1 = x^1 - vt$ between two moving coordinate frames. (These effectively non-relativistic transformations are called **Galilei transformations**.) However, for relativistic velocities, $|v| \sim c$, the Lorentz transformation mixes space and time coordinates inseparably and leads to the principal effects of special relativity, the dilation of time, length contraction, the non-invariance of mass, etc. For a physical discussion of these effects we refer to courses in special relativity.

Galilei trans-
formations

Eigenvalues

Unitary matrices are invertible and therefore possess non-vanishing eigenvalues. However, there is an even stronger statement, constraining the form of their eigenvalues: if $U\mathbf{v} = \lambda\mathbf{v}$, then $\langle \mathbf{v}, \mathbf{v} \rangle = \langle U\mathbf{v}, U\mathbf{v} \rangle = \langle \lambda\mathbf{v}, \lambda\mathbf{v} \rangle = |\lambda|^2 \langle \mathbf{v}, \mathbf{v} \rangle$ implies that $|\lambda|^2 = 1$:

eigenvalues
of unitary
matrices

> the n eigenvalues, λ_j, of a unitary matrix are complex unit-modular numbers, $\lambda_j = e^{i\phi_j}$, with real $\phi_j \in [0, 2\pi)$.

The same argument applied to an orthogonal matrix, O, shows that its eigenvalues must have unit modulus, too, $|\lambda| = 1$. However, these eigenvalues need not be real, i.e. there is no guarantee that the characteristic polynomial of an orthogonal matrix has real zeros. For example, the orthogonal matrix $\begin{pmatrix} 0 & 1 \\ 1 & 0 \end{pmatrix}$ has the two imaginary unit-modular eigenvalues $\pm i$. If the eigenvalues of an orthogonal matrix are real, they must equal ± 1, the only two real numbers with modulus one.

Special unitary and special orthogonal matrices

The determinant of a matrix equals the product of its eigenvalues. We have seen that for unitary matrices the latter are unit-modular numbers. Since the product of unit-modular numbers is again unit-modular (why?), we know that[62]

---

[62] Notice that even if an orthogonal matrix has no real eigenvalues its determinant is real by construction. At the same time, the determinant is the product of n (possibly complex) unit modular eigenvalues. Combining these two facts we conclude that the product must be real and unit-modular, i.e. it must equal ± 1.

> The *determinant of a unitary matrix* is a complex number of *unit modulus*, $\det(U) = e^{i\phi}$, where $\phi \in [0, 2\pi)$ is real. Likewise, the *determinant of an orthogonal matrix* is a real number of unit modulus, $\det(O) = \pm 1$.

It is instructive to prove this statement without reference to the eigenvalues. To this end, take a unitary matrix, U, define $z = \det(U)$, and compute

$$|z|^2 = \det(U)\overline{\det(U)} = \det(U)\det(U^\dagger) = \det(UU^\dagger) = \det(\mathbb{1}) = 1,$$

where Eq. (L163) has been used. The same construction applied to an orthogonal matrix shows that $\det(O) = \pm 1$.

Unitary (orthogonal) matrices with determinants possessing the special value $\det(U) = 1$ ($\det O = 1$) are called *special unitary (orthogonal) matrices*. The unit-determinant property is preserved under matrix multiplication, if $\det(U) = \det(V) = 1$ then $\det(UV) = 1$, and under matrix inversion, if $\det(U) = 1$ then $\det(U^{-1}) = 1$. We also know that the unit matrix has determinant 1. This shows that the set of special unitary (orthogonal) matrices forms a sub*group* of the set of unitary (orthogonal) matrices. This group is known as the **special unitary (orthogonal) group**, SU(n) (SO(n)),

special unitary group

$$SU(n) \equiv \{U \in U(n) \mid U^\dagger = U^{-1}, \det(U) = 1\},$$
$$SO(n) \equiv \{O \in O(n) \mid O^T = O^{-1}, \det(O) = 1\}. \tag{L213}$$

We note that the special unitary group can also be understood as a subgroup of SL($n, \mathbb{C}$), the group of unit-determinant (but not necessarily unitary) complex matrices. Similarly, the special orthogonal group can be understood as a subgroup of the group SL($n, \mathbb{R}$) of real matrices with unit determinant.

special unitary groups in physics

INFO Special unitary matrices play an important *role in physics*, notably in quantum mechanics and particle physics. For example, the quantum mechanics of spin (which is the quantum general-ization of classical angular momentum) is mathematically described in terms of the group SU(2). ($\rightarrow$ L5.3.3-4) The groups SU(2) and SU(3) played a decisive role in the 1960s when their mathemat-ical structure was linked to the properties of known elementary particles and the *standard model* of matter emerged.

The groups SO(2) and SO(3) feature in *classical mechanics* where they describe the mathematics of rotations in two- and three-dimensional space, respectively. We have argued above that rotations are mathematically represented by orthogonal matrices. In fact, they are *special* orthogonal matrices. This is because any rotation by a certain angle can be continuously "deformed" to a unit operation by reducing the angle. Consider, for example, the rotation matrix $R(\theta)$ defined in Eq. (L207). It can be deformed to the unit matrix by a continuous reduction of θ to zero. The unit matrix has unit determi-nant, and the same must hold for any *continuous* deformation of it. A sudden "jump" to a determinant -1 would be in conflict with continuity. By contrast, matrices describing reflections, for example the matrix $R = \begin{pmatrix} 1 & \\ & -1 \end{pmatrix}$ describing a reflection at the x-axis (why?), can have determinant -1.

Summarizing, the set of complex (real) matrix groups encountered so far contains mat($n, \mathbb{C}$) $\supset$ GL($n, \mathbb{C}$) $\supset$ U(n), SL($n, \mathbb{C}$) $\supset$ SU(n) (and analogously for the real case). There are a few more groups of relevance to the physics curriculum; however, the ones above occur most frequently. The hierarchical relation between them is illustrated in Fig. L16.

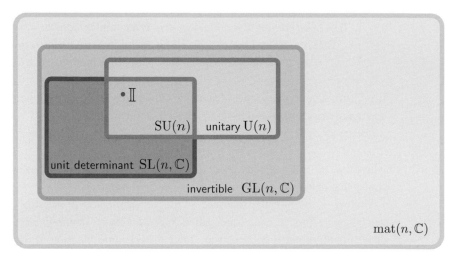

Fig. L16   The most imortant matrix subgroups of $\mathrm{mat}(n, \mathbb{C})$. The "smallest" group, $\mathrm{SU}(n) = \mathrm{SL}(n, \mathbb{C}) \cap \mathrm{U}(n)$, is the intersection of the groups of unit determinant, $\mathrm{SL}(n, \mathbb{C})$, and the unitary group, $\mathrm{U}(n)$. For the real case, replace $\mathbb{C} \to \mathbb{R}$ and $\mathrm{U} \to \mathrm{O}$.

INFO For any unitary matrix U with determinant $\det(U) = \mathrm{e}^{\mathrm{i}\phi}$, a matrix of unit determinant may be defined as $U' = \mathrm{e}^{-\mathrm{i}\phi/n} \, U$. This follows from $\det(U') = \det(\mathrm{e}^{-\mathrm{i}\phi/n}U) = (\mathrm{e}^{-\mathrm{i}\phi/n})^n \det(U) = \mathrm{e}^{-\mathrm{i}\phi}\mathrm{e}^{\mathrm{i}\phi} = 1$, where Eq. (L168) has been used. For example, the unitary matrix $U = \frac{1}{\sqrt{2}}\left(\begin{smallmatrix} \mathrm{i} & \mathrm{i} \\ \mathrm{i} & \mathrm{i} \end{smallmatrix}\right)$ has determinant $\mathrm{e}^{\mathrm{i}\pi} \overset{(C93)}{=} -1$. Multiply it by $\mathrm{e}^{-\mathrm{i}\pi/2} \overset{(C89)}{=} -\mathrm{i}$ to obtain the special unitary matrix $U' = \frac{1}{\sqrt{2}}\left(\begin{smallmatrix} 1 & -\mathrm{i} \\ -\mathrm{i} & 1 \end{smallmatrix}\right)$. Since matrices differing by a multiplicative factor are "almost equivalent", the manipulation above is often used to pass from a unitary matrix to its slightly simpler unit-determinant version. For orthogonal matrices, this prescription does not work since $\det(O)^{-1/n} = (-1)^{-1/n}$ is not a real number, so multiplication by it takes one outside the set of real matrices.

Unitary and orthogonal basis changes

In Section L5.6 we considered a basis transformation from a basis $\{\hat{\mathbf{v}}_j\}$ to a new basis $\{\hat{\mathbf{v}}'_i\}$, and learned that finding the matrix representing the inverse transformation, T^{-1}, generally requires inverting the transformation matrix T. Much less work is required if we work with orthonormal bases, i.e. if both the old and new bases, $\{\hat{\mathbf{e}}_j\}$ and $\{\hat{\mathbf{e}}'_i\}$, are orthonormal. In this case, the transformation matrix $\hat{\mathbf{e}}_j = \hat{\mathbf{e}}'_i T^i{}_j$ preserves the inner product, $\langle \hat{\mathbf{e}}_i, \hat{\mathbf{e}}_j \rangle = \langle \hat{\mathbf{e}}'_i, \hat{\mathbf{e}}'_j \rangle = \delta_{ij}$. This means that *the transformation matrix is unitary (orthogonal)* and that its inverse is obtained "for free" just by building $T^\dagger$ (T^{T}).

EXERCISE Apply elementary trigonometry to compute the matrix describing the transformation between the basis vectors shown in the figure. Verify its orthogonality by building the transpose and checking that Eq. (L203) holds.

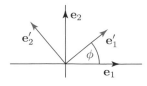

Writing $T \equiv U$ to emphasize the unitarity of the transform, these statements can be made more concrete by inspection of the inverse relation, $\hat{\mathbf{e}}'_i = \hat{\mathbf{e}}_j (U^{-1})^j{}_i$. Multiplication of Eq. (L202) from the right by U^{-1} gives $(U^{-1})^i{}_j = (U^\dagger)^i{}_j$, and this yields the relations

$$\hat{\mathbf{e}}_j = \hat{\mathbf{e}}'_i \, U^i{}_j, \qquad \hat{\mathbf{e}}'_j = \hat{\mathbf{e}}_i \, (U^\dagger)^i{}_j. \tag{L214}$$

For the orthonormality-preserving transformations $T \equiv O$ of a *real vector space*, they are to be replaced by

$$\hat{\mathbf{e}}_j = \hat{\mathbf{e}}'_i \, O^i{}_j, \qquad \hat{\mathbf{e}}'_j = \hat{\mathbf{e}}_i \, (O^\mathrm{T})^i{}_j. \tag{L215}$$

L8.2 Hermiticity and symmetry

Besides the unitary (orthogonal) maps there exists a second family of linear maps defined in relation to the inner product. They are called Hermitian (symmetric) maps and they, too, are of great importance to physics, notably to quantum mechanics. Given that unitarity and Hermiticity are both defined in relation to an inner product, it should not be surprising that these two families of maps are intimately connected. We will discuss their relation after the mathematical properties of Hermitian and symmetric maps have been discussed. Following the same strategy as above we will first address these structures on the general level of maps between vector spaces, and then investigate the resulting matrix structures. Finally, the importance of Hermitian matrices in quantum mechanics will be discussed within the framework of a case study.

Hermitian and symmetric maps and matrices

Hermitian map A linear map, $\hat{A} : V \to V$, of a complex (real) inner product space is called a **Hermitian (symmetric) map** if

$$\boxed{\forall \hat{\mathbf{u}}, \hat{\mathbf{w}} \in V, \qquad \langle \hat{A}\hat{\mathbf{u}}, \hat{\mathbf{w}} \rangle = \langle \hat{\mathbf{u}}, \hat{A}\hat{\mathbf{w}} \rangle.} \tag{L216}$$

Unlike with unitary (orthogonal) maps the relation above does not define a group property: if $\hat{A}, \hat{B}$ are Hermitian (symmetric) we know that $\langle \hat{A}\hat{B}\hat{\mathbf{u}}, \hat{\mathbf{w}} \rangle = \langle \hat{B}\hat{\mathbf{u}}, \hat{A}\hat{\mathbf{w}} \rangle = \langle \hat{\mathbf{u}}, \hat{B}\hat{A}\hat{\mathbf{w}} \rangle$. However, this does not equal $\langle \hat{\mathbf{u}}, \hat{A}\hat{B}\hat{\mathbf{w}} \rangle$, unless $\hat{A}\hat{B} = \hat{B}\hat{A}$. The composition of two Hermitian (symmetric) maps, $\hat{A}\hat{B}$, therefore is not Hermitian (symmetric) in general. However, the absence of a group structure notwithstanding, the matrices representing Hermitian (symmetric) maps possess strong mathematical structure, to be discussed next.

The availability of an inner product suggests representing Hermitian (symmetric) linear maps in an orthonormal basis $\{\hat{\mathbf{e}}_j\}$. Writing the components of $\hat{A}\hat{\mathbf{u}}$ as $(A\mathbf{u})^m = A^m{}_k u^k$, the Hermiticity condition $\langle \hat{A}\hat{\mathbf{u}}, \hat{\mathbf{w}} \rangle = \langle \hat{\mathbf{u}}, \hat{A}\hat{\mathbf{w}} \rangle$ then takes the form

$$\overline{A^m{}_k} \, \overline{u^k} \delta_{mj} w^j = \overline{u^k} \delta_{kl} A^l{}_j w^j \, .$$

This must hold for arbitrary $\hat{\mathbf{u}}, \hat{\mathbf{w}}$, which requires $\overline{A^m}_k \delta_{mj} = \delta_{kl} A^l_j$. Recalling the definition (L114) of the adjoint matrix, $(A^\dagger)_k{}^m = \overline{A^m}_k$, and multiplying with δ^{ik} we obtain

$$\delta^{ik}(A^\dagger)_k{}^m \delta_{mj} \equiv (A^\dagger)^i_j = A^i_j. \tag{L217}$$

The same construction carried out for a symmetric matrix acting on a real vector space shows that $(A^T)^i_j = A^i_j$. We have thus found that

> the matrices, A, representing Hermitian (symmetric) linear maps in an orthonormal basis are equal to their adjoint (transpose), $A^i_j = (A^\dagger)^i_j$
> $(A^i_j = (A^T)^i_j)$.

EXAMPLE The matrices

$$A \equiv \begin{pmatrix} 1 & 0 & -i \\ 0 & 1 & 0 \\ i & 0 & 1 \end{pmatrix} \qquad B \equiv \begin{pmatrix} 3 & -4 \\ -4 & -3 \end{pmatrix} \tag{L218}$$

are Hermitian and symmetric, respectively.

Eigenvalues and determinant

All n eigenvalues, λ, of a Hermitian (symmetric) matrix are *real*. To see this, let $\mathbf{v}$ be λ's eigenvector and compute $\lambda\langle\mathbf{v},\mathbf{v}\rangle = \langle\mathbf{v},\lambda\mathbf{v}\rangle = \langle\mathbf{v},A\mathbf{v}\rangle = \langle A\mathbf{v},\mathbf{v}\rangle = \langle\lambda\mathbf{v},\mathbf{v}\rangle = \overline{\lambda}\langle\mathbf{v},\mathbf{v}\rangle$. Since $\langle\mathbf{v},\mathbf{v}\rangle \neq 0$, the first and the last entry in this chain of equalities require $\lambda = \overline{\lambda}$. The result also implies that unlike a generic real matrix, a symmetric matrix has n real eigenvalues. We know that its characteristic polynomial has n zeros (which for a generic matrix may be complex). However, the argument above shows that these solutions must be real. To summarize,

> an n-dimensional Hermitian (symmetric) matrix has n real eigenvalues.

As a corollary we observe that

> the determinant of a Hermitian matrix is real.

positive definite Hermitian matrices

This is because it is the product of its n real eigenvalues. (Of course, if a symmetric matrix is real, then its determinant is real, too.) For later reference, we note that a Hermitian matrix with positive eigenvalues is called **positive definite**, or positive *semi*definite if zero eigenvalues occur. A matrix with only negative eigenvalues is called negative definite, and in all other cases **indefinite**.

EXERCISE Show that the eigenvalues of the Hermitian matrix A of Eq. (L218) are given by $\{1, 0, 2\}$ and those of the symmetric matrix B by $\{5, -5\}$.

Diagonalization

Hermitian (symmetric) matrices have the important property that they can always be diagonalized. The transformation matrices effecting the diagonalization are unitary (orthogonal). These statements are proven in the info section below. However, not only can they be diagonalized as a matter of principle, the diagonalization procedure is also much simpler than that for generic matrices. The key simplification lies in the fact that

> eigenvectors $\mathbf{v}_1$ and $\mathbf{v}_2$ corresponding to different eigenvalues, $\lambda_1 \neq \lambda_2$, of a Hermitian matrix are orthogonal, $\langle \mathbf{v}_1, \mathbf{v}_2 \rangle = 0$.

To show this, consider two different eigenvalues $\lambda_1 \neq \lambda_2$. Then $\lambda_1 \langle \mathbf{v}_1, \mathbf{v}_2 \rangle = \langle \lambda_1 \mathbf{v}_1, \mathbf{v}_2 \rangle = \langle A\mathbf{v}_1, \mathbf{v}_2 \rangle = \langle \mathbf{v}_1, A\mathbf{v}_2 \rangle = \lambda_2 \langle \mathbf{v}_1, \mathbf{v}_2 \rangle$ implies $0 = (\lambda_1 - \lambda_2)\langle \mathbf{v}_1, \mathbf{v}_2 \rangle$. Since $\lambda_1 - \lambda_2 \neq 0$, this equality requires $\langle \mathbf{v}_1, \mathbf{v}_2 \rangle = 0$. This observation suggests starting the diagonalization by computing as many eigenvectors $\mathbf{v}_j$ as there are different eigenvalues λ_j. Choosing these vectors to be normalized, we know that they form an orthonormal set, $\langle \mathbf{v}_i, \mathbf{v}_j \rangle = \delta_{ij}$. If all eigenvalues are different they form a basis and the matrix $T = (\mathbf{v}_1, \ldots, \mathbf{v}_n)$ transforms A into diagonal form, $D = T^{-1}AT$. ($\rightarrow$ L8.2.1-6)

The procedure becomes a little more complicated if degenerate eigenvalues λ of degeneracy r are present. The diagonalizability of A means that r linearly independent eigenvectors with eigenvalue λ exist. Complications like those with the Jordan matrices discussed on p. 105 do not arise. Instead, it is always possible to find r linearly independent solutions of the eigenvalue equation $(A - \lambda \mathbb{1})\mathbf{v} = 0$. In a second step, these solutions may be orthonormalized by the Gram–Schmidt procedure detailed in Section L3.2. The full set of eigenvectors then transforms A to diagonal form.

diagonaliz-
ability of
Hermitian
matrices

INFO   A proof of principle for the **diagonalizability of Hermitian matrices**, A, goes as follows. Pick one of the eigenvalues, λ_1, with (normalized) eigenvector $\mathbf{v}_1$. Next define $V_1 \subset V$ to be the subspace of V containing all vectors orthogonal to $\mathbf{v}_1$. The key observation now is that A acts *within* V_1, i.e. for $\mathbf{w} \in V_1$, $A\mathbf{w} \in V_1$ is orthogonal to $\mathbf{v}_1$, too. To see this, compute $\langle \mathbf{v}_1, A\mathbf{w} \rangle = \langle A\mathbf{v}_1, \mathbf{w} \rangle = \langle \lambda_1 \mathbf{v}_1, \mathbf{w} \rangle = \lambda_1 \langle \mathbf{v}_1, \mathbf{w} \rangle = 0$, where in the last step the assumed orthogonality of $\mathbf{w}$ and $\mathbf{v}_1$ was used.

The procedure is now iterated by picking a second normalized eigenvector $\mathbf{v}_2 \in V_1$ (whether it has the same eigenvalue as $\mathbf{v}_1$ or not does not matter) and determining a subspace, $V_2 \subset V_1$, of vectors in V_1 perpendicular to $\mathbf{v}_2$. (Elements of this subspace are automatically perpendicular to $\mathbf{v}_1$ as well, because we are working inside V_1.) In each step, the dimension of the spaces $V_1, V_2, \ldots$ reduces by one. The procedure is continued until a one-dimensional vector space, $V_{n-1} \subset \cdots \subset V_1 \subset V$, spanned by a single normalized basis vector, $\mathbf{v}_n$, is reached.

The algorithm above yields a basis of orthonormal eigenvectors, $\langle \mathbf{v}_i, \mathbf{v}_j \rangle = \delta_{ij}$, which means that the transformation matrix, $T \equiv (\mathbf{v}_1, \mathbf{v}_2, \ldots, \mathbf{v}_n)$, is unitary (orthogonal), see Section L8.1. As a drawback, in each step a subspace of vectors perpendicular to a set of given vectors needs to be determined. This is generally cumbersome and for this reason Hermitian matrices are usually diagonalized differently, as discussed in the main text above.

EXERCISE Apply the above procedure to show that the matrices of Eq. (L218) are diagonalized by

$$A = TDT^{\mathrm{T}}, \qquad D = \mathrm{diag}(1,0,2), \qquad T = \begin{pmatrix} 0 & \frac{1}{\sqrt{2}} & \frac{1}{\sqrt{2}} \\ 1 & 0 & 0 \\ 0 & \frac{-i}{\sqrt{2}} & \frac{i}{\sqrt{2}} \end{pmatrix},$$

$$B = TDT^{\dagger}, \qquad D = \mathrm{diag}(5,-5), \qquad T = \frac{1}{\sqrt{5}}\begin{pmatrix} 2 & 1 \\ -1 & 2 \end{pmatrix}.$$

We conclude this section by *summarizing* the most essential properties of Hermitian and symmetric matrices:

▷ Hermitian and symmetric matrices are diagonalizable;

▷ their eigenvalues, $\{\lambda_j\}$, are real, and

▷ an orthonormal basis of eigenvectors can always be found;

▷ the transformation matrices, T, to an orthonormal basis of eigenvectors are unitary, $T^{-1} = T^{\dagger}$ (symmetric, $T^{-1} = T^{\mathrm{T}}$), which means that

▷ Hermitian and symmetric matrices are representable as $A = TDT^{\dagger}$, and $A = TDT^{\mathrm{T}}$, respectively, with $D = \mathrm{diag}(\lambda_1, \ldots, \lambda_n)$.

L8.3   Relation between Hermitian and unitary matrices

REMARK This section discusses connections between Hermitian and unitary matrices and can be skipped at first reading. It requires familiarity with the concept of functions of matrices introduced in Section L7.4.

Both Hermitian and unitary matrices are defined with reference to an inner product, which suggests that they must be connected in some way. To understand the relation between these two sets of matrices, let A be a Hermitian matrix and consider its exponential,

exponential representation of unitary matrices

$$U \equiv \exp(iA), \tag{L219}$$

where the exponential function is defined in Eq. (L196). We claim that U is unitary. To see this, compute the Hermitian adjoint,

$$U^{\dagger} = \left(\sum_m \frac{(iA)^m}{m!}\right)^{\dagger} = \sum_m \frac{((iA)^m)^{\dagger}}{m!} = \sum_m \frac{(-iA)^m)}{m!} = \exp(-iA) = U^{-1}.$$

Here, we used $(X^m)^{\dagger} = (X^{\dagger})^m$, and $(iA)^{\dagger} = -iA$. In the last equality we noted that $\exp(-iA)\exp(iA) = \mathbb{1}$, i.e. $U^{-1} = \exp(-iA)$. We have thus found that

the exponential of i·(a Hermitian matrix) is unitary.

EXAMPLE Consider the Hermitian matrix $A = \theta \begin{pmatrix} 0 & i \\ -i & 0 \end{pmatrix}$, where θ is a real parameter. Multiplication by i yields the matrix $iA = \theta J$, with $J = \begin{pmatrix} 0 & -1 \\ 1 & 0 \end{pmatrix}$. We observe that $J^2 = -\mathbb{1}$, hence $J^{2m} = (-)^m \mathbb{1}$ and $J^{2m+1} = (-)^m J$. If we split the Taylor series for $\exp(iA)$ into two parts, involving even and odd powers of (iA), we thus obtain

$$\exp(iA) = \sum_{m=0}^{\infty} \left(\frac{1}{2m!}(iA)^{2m} + \frac{1}{(2m+1)!}(iA)^{2m+1} \right) = \sum_{m=0}^{\infty} \left(\frac{(-)^m \theta^{2m}}{2m!}\mathbb{1} + \frac{(-)^m \theta^{2m+1}}{(2m+1)!}J \right)$$

$$= \cos(\theta)\mathbb{1} + \sin(\theta)J = \begin{pmatrix} \cos(\theta) & -\sin(\theta) \\ \sin(\theta) & \cos(\theta) \end{pmatrix}.$$

It is straightforward to verify that the resulting matrix is unitary.

In fact, an even stronger statement can be made: a Hermitian matrix, A, of dimension n is fixed by n^2 real parameters. To understand this counting, note that the relation $A^\dagger = A$, or $\overline{A^j}_i = A^i_j$, requires all n diagonal elements, A^i_i, to be real. The $n(n-1)/2$ elements A^i_j, $i > j$, defining the upper right triangle of the matrix can be chosen as arbitrary complex numbers. The elements of the lower left triangle are then fixed through the above Hermiticity condition. Noting that a complex number contains two real parameters, we conclude that $n + 2\frac{n(n-1)}{2} = n^2$ free real parameters need to be specified to define a Hermitian matrix. For example, a general two-dimensional Hermitian matrix is of the form $A = \begin{pmatrix} a & b+ic \\ b-ic & d \end{pmatrix}$, and thus described by $4 = 2^2$ real parameters a, b, c, d.

The unitary matrices of dimension n, too, are parametrized by n^2 real parameters. This follows from the fact that the relation $U^\dagger U = \mathbb{1}$, or $(U^\dagger)^i_k U^k_i = \delta^i_j$, can be understood as a set of n^2 *real* equations[63] constraining the n^2 complex or $2n^2$ real parameters describing an arbitrary complex matrix. Each equation effectively fixes one free parameter, so that the set of unitary matrices can be parametrized in terms of $2n^2 - n^2 = n^2$ real parameters. For example, a two-dimensional unitary matrix, $U = \begin{pmatrix} r & s \\ t & u \end{pmatrix}$, is constrained by the condition $U^\dagger U = \begin{pmatrix} \bar{r} & \bar{t} \\ \bar{s} & \bar{u} \end{pmatrix} \begin{pmatrix} r & s \\ t & u \end{pmatrix} = \begin{pmatrix} 1 & 0 \\ 0 & 1 \end{pmatrix}$. Forming the matrix product, we see that this implies two real equations, $|r|^2 + |t|^2 = |s|^2 + |u| = 1$, and a complex one, $\bar{r}s + \bar{t}u = 0$. (The fourth component relation is the complex conjugate of the third relation and does not introduce further constraints.) Since a complex relation implies two separate real equations for the real and imaginary parts, we have a total of four real equations for the eight real parameters entering the complex numbers r, t, s, u. This leaves $4 = 2^2$ free real parameters determining a two-dimensional unitary matrix.

The above considerations lead to the key conclusion that unitary and Hermitian matrices of the same dimension contain equally many free real parameters. We have also seen that for a Hermitian matrix A, the matrix $U = \exp(iA)$ is unitary. This suggests that *every* unitary matrix can be expressed as the exponential of i times a Hermitian matrix. It is

---

[63] The counting follows from the observation that for $i > j$, $(U^\dagger)^i_k U^k_j = 0$ is an equation fixing the *complex* number $(U^\dagger)^i_k U^k_j$. This gives a total of twice as many, $2n(n-1)/2 = n(n-1)$, real equations. The equations for $i < j$ are obtained by complex conjugation of those for $i > j$ (why?) and must not be counted separately. For $i = j$, $(U^\dagger)^i_k U^k_i = \sum_k |U^k_i|^2 = 1$ are n real conditions, so that we have a total of $n + n(n-1) = n^2$ real equations.

nontrivial to show that this is indeed the case and that (L219) represents a proper *exponential parametrization of the group of unitary matrices*. This representation plays a rather important role in physics. For example, in quantum mechanics (see Section L8.4 below), physical observables are represented by Hermitian matrices, the evolution of observables in time is described by unitary matrices, and the exponential representation establishes the correspondence between these two descriptions. Another important consequence of the exponential representation is that a unitary matrix can be factorized as

$$U = e^{iA} = e^{iA/N}e^{iA/N} \dots e^{iA/N}, \tag{L220}$$

i.e. as a product of N factors $\exp(iA/N)$. (Explain on the basis of the results of Section L7.4 why this relation holds.) If $N \gg 1$ is very large, the matrices A/N (containing the matrix elements of A divided by N) are close to zero and an expansion $\exp(iA/N) \simeq \mathbb{1} + iA/N$ is permissible. The decomposition (L220) then represents a possibly complicated unitary matrix as a product of a large number of relatively simple (close to the unit matrix) factors, described by "small" anti-Hermitian matrices iA/N.[64] Hermitian matrices representing unitary matrices in this way are often called *generators of unitary matrices*. The discussion above can thus be summarized by saying that

> the group of unitary matrices, U, is generated by Hermitian matrices, A, through the exponential representation, $U = \exp(iA)$.

EXERCISE As an example, consider the rotation matrix $U = \begin{pmatrix} \cos\theta & -\sin\theta \\ \sin\theta & \cos\theta \end{pmatrix}$ as in the example above. For sufficiently large N, the matrix is factorized by the matrices $\mathbb{1} + iA/N$, with generators $A = \theta \begin{pmatrix} & i \\ -i & \end{pmatrix}$. Convince yourself that the application of this matrix to a two-dimensional vector, $\mathbf{u}$, generates an "infinitesimal rotation" of $\mathbf{u}$ by an angle θ/N. Specifically, check that the norm of the transformed vector equals that of $\mathbf{u}$, up to corrections of $\mathcal{O}(N^{-2})$ neglected in the first-order expansion in iA/N. In the limit $N \to \infty$, the decomposition (L220) thus describes a finite-angle rotation as a product of an infinitely large number of "infinitesimal" rotations. Representations of this type play an important role in many physical applications. ($\to$ L8.3.1-2)

diagonaliza-
bility of
unitary
matrices

The correspondence between the unitary and Hermitian matrices also shows why *unitary matrices are diagonalizable*, an assertion made in Section L8.1 but not proven so far. The existence of a parametrization $U = \exp(iA)$, and the diagonalizability of Hermitian matrices $A = TDT^{-1}$, imply the representation $U = \exp(iTDT^{-1}) = T\exp(iD)T^{-1}$, where Eq. (L197) was used. This shows that U is diagonalizable and that the diagonal matrix of eigenvalues, $\exp(iD)$, contains the exponentials, $\exp(i\lambda_j)$, of the real eigenvalues, λ_j, of A. These are unit-modular complex numbers, and we see how the reality of the eigenvalues of Hermitian matrices and the unit-modularity of the eigenvalues of unitary matrices imply each other.

A similar relation holds between the set of *anti*symmetric real matrices, $A^{\mathsf{T}} = -A$, and the orthogonal matrices $O^{\mathsf{T}} = O^{-1}$. It is a good exercise (try it!) to show that $\exp(A) = O$ is

anti-
Hermitian
matrix

[64] A matrix X is an **anti-Hermitian matrix** if $X^{\dagger} = -X$. The matrices iA are anti-Hermitian because $(iA)^{\dagger} = -iA^{\dagger} = -iA$.

orthogonal and that the sets of *antisymmetric matrices and orthogonal matrices* contain the same number of parameters, $n(n-1)/2$. However, in this case, $\exp(A)$ does not cover the full group of orthogonal matrices. (Only the subgroup $SO(n) \subset O(n)$ of unit-determinant orthogonal matrices is obtained.) This correspondence and its applications in physics are discussed in advanced lecture courses.

L8.4 Case study: linear algebra in quantum mechanics

REMARK Two formulas towards the end of this section require familiarity with Chapter C5 on Taylor series. Nevertheless, most of this section should be accessible prior to reading that chapter.

We mentioned above that unitary and Hermitian maps are of key relevance to the mathematical description of quantum mechanics. Although this text cannot introduce quantum mechanics as such, let us conclude this chapter with an outlook to the application of linear algebra in this field. We begin by reviewing various key axioms of quantum mechanics, emphasizing their formulation through concepts of linear algebra. They are "axioms" in that they cannot be proven from more fundamental principles (much like Newton's laws of classical mechanics cannot be proven). They are believed to be valid because of their success in predicting and explaining experimental observations.

▷ The physical state of a system is described by a unit-normalized vector, $\hat{\psi} \in V$, $\langle\hat{\psi},\hat{\psi}\rangle = 1$, defined in a complex inner product space. The dimension of that space, N, may be infinite, in which case a number of extra conditions need to be imposed. Vector spaces equipped with these properties are called **Hilbert spaces** (see Chapter L9 for more information). Following standard conventions we denote Hilbert space vectors, $\hat{\psi}$, in a non-boldface notation. Describing **quantum mechanical states** of the system, the vectors $\hat{\psi}$ are called **state vectors** or just *states*.[65] It is usually advantageous to work with orthonormal bases $\{\hat{e}_j\}$, and decompositions of general states as $\hat{\psi} = \hat{e}_j\psi_j$. In non-relativistic quantum mechanics, the metric is generally trivial and it is customary to use non-covariant subscript notation for state components, $\psi_j \equiv (\hat{\psi})_j$. The collection of components, $\psi = (\psi_1,\ldots,\psi_N)^T \in \mathbb{C}^N$, is called the **wave function**[66] of the state $\hat{\psi}$.

As a concrete example, consider a one-dimensional ring comprising N equally spaced atoms, as in Fig. L17. Such geometries can be realized in molecular systems, for example the benzene molecule forms a ring with $N = 6$ sites. Our "system" in this context is an *electron*, a quantum particle free to move along the ring by hopping from one atom to the next. (The chemistry of benzene is strongly influenced by quasi-free bond electrons doing just this.)

Hilbert space

quantum state

wave function

Dirac notation

[65] In the **Dirac notation** routinely used in quantum mechanics, states are denoted as $|\psi\rangle \equiv \hat{\psi}$, basis vectors as $|j\rangle \equiv \hat{e}_j$, and inner products as $\langle\psi|\psi'\rangle \equiv \langle\hat{\psi},\hat{\psi}'\rangle$.

[66] Strictly speaking, the terminology "wave function" should be reserved for the limit of infinite-dimensional Hilbert spaces, $N \to \infty$, where ψ_i is replaced by a function, $\psi(x)$, of a continuous variable. However, it is often used in finite-dimensional contexts just the same.

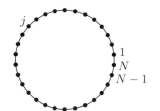 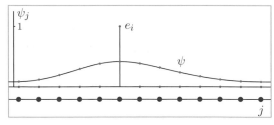

Fig. L17 Cartoon of a one-dimensional "quantum ring" of atoms. Individual atoms are labeled by j. The components ψ_j of a quantum mechanical state describing an electron on this ring are a measure of the "probability amplitude" of finding it at site j, and $|\psi_j|^2$ is the corresponding probability.

> In this context the jth basis state, $\hat{e}_j$, with components $e_{j,i} \equiv (\hat{e}_j)_i = \delta_{ji}$, describes a physical state in which the electron is found with certainty (probability one) at site j. A general state, $\hat{\psi} = \hat{e}_j \psi_j$, describes a "superposition", i.e. a physical state in which the probability of finding the electron is delocalized over different sites. The compo-

probability amplitude nents ψ_j are a measure for the "**probability amplitude**" that the electron is located at atom j, and the real number $|\psi_j|^2$ gives the actual probability of finding it at j in a position measurement. The unit normalization, $1 = \langle \hat{\psi}, \hat{\psi} \rangle = \sum_j |\psi_j|^2$, means that these probabilities add to unity: the electron will be found with certainty *somewhere* on the ring. This probabilistic interpretation of the wave function is axiomatic and defines a cornerstone of quantum mechanics.

> **Physical observables** are quantities which can be measured. Position, momentum, angular momentum and many others are observables in this sense. According to another axiom, each observable is represented by a Hermitian linear map, $\hat{A}$. In the present context these maps are called **(Hilbert space) operators**, and their eigenvectors, $\hat{\psi}_n$,

eigenstates   are called **eigenstates**. The full set of real eigenvalues, $\{\lambda_n\}$, of a Hermitian operator
measurement is called its **spectrum**.[67] Each single **measurement of the observable** must yield an eigenvalue, λ_n, of $\hat{A}$ as result (another axiom).

In a given basis, the action $\hat{\psi} \mapsto \hat{A}\hat{\psi}$ of an operator $\hat{A}$ on a Hilbert space state is represented by the multiplication of a matrix A onto a component vector ψ in $\mathbb{C}^N$, and the eigenvalue equation, $\lambda_n \hat{\psi}_n = \hat{A}\hat{\psi}_n$, has the matrix representation $\lambda_n \psi_{n,j} = A_{ji}\psi_{n,i}$. For example, the *position operator*, $\hat{X}$, describing the position of the electron on the ring, is represented by a diagonal matrix,

$$
X = \begin{pmatrix} 1 & & & & & \\ & 2 & & & & \\ & & 3 & & & \\ & & & \ddots & & \\ & & & & N-1 & \\ & & & & & N \end{pmatrix},
$$

---

[67] The denotation "spectrum" is physically motivated and reflects the fact that the eigenvalues of operators carrying physical significance are often determined by spectroscopic methods.

with elements $X_{ji} = j\delta_{ji}$. The eigenvalue equation, $\lambda_n \psi_{n,j} = X_{ji}\psi_{n,i} = j\psi_{n,j}$, is solved by the standard basis vectors of $\mathbb{C}^N$, with components $\psi_{n,j} = e_{n,j} = \delta_{jn}$ and eigenvalues $\lambda_n = n$. Therefore, the eigenstates of $\hat{X}$ are just the basis states, $\hat{\psi}_n = \hat{e}_n$. If the system is described by $\hat{e}_n$, a measurement of its position yields n with certainty.

In general, a position measurement may find the electron at one of the N possible sites. If it is observed at n, the position operator eigenvalue, $\lambda_n = n$, has been measured. However, *before* the measurement it cannot be known with certainty where the electron will be found – a hallmark of the probabilistic nature of quantum mechanics. At best one can make statements about the average result of many repeated measurements. For example, an electron with probability amplitude $\psi_1 = \psi_2 = \frac{1}{\sqrt{2}}$, $\psi_{j>2} = 0$, is at site 1 or 2 with equal probability, $|\psi_1|^2 = |\psi_2|^2 = \frac{1}{2}$. In this case, a position measurement is expected to yield the results 1 or 2 with probability $\frac{1}{2}$, and the *expected* value of the measurement will be $1 \cdot \frac{1}{2} + 2 \cdot \frac{1}{2} = \frac{3}{2}$. Notice that the expected value may take fractional values even if individual measurements yield integer results.

expectation
value

An axiom of quantum mechanics condenses all this into the mathematical statement that the **expectation value**, $\langle \hat{A} \rangle$, of the measurement of an observable, $\hat{A}$, on a system in a state $\hat{\psi}$ is given by the inner product,

$$\langle \hat{A} \rangle = \langle \hat{\psi}, \hat{A}\hat{\psi} \rangle.$$

The Hermiticity of $\hat{A}$ guarantees that this value is real, $\overline{\langle \hat{\psi}, \hat{A}\hat{\psi} \rangle} = \langle \hat{A}\hat{\psi}, \hat{\psi} \rangle = \langle \hat{\psi}, \hat{A}\hat{\psi} \rangle$, as required for a measurable quantity. The meaning of the formula is easiest to understand in a basis in which $\hat{A}$ is represented by a diagonal matrix. For example, the expectation value of the position operator $\hat{X}$ is obtained as

$$\langle \hat{X} \rangle = \langle \hat{\psi}, \hat{X}\hat{\psi} \rangle = \sum_j \overline{\psi_j}(X\psi)_j = \sum_j j |\psi_j|^2.$$

This formula expresses the fact that the result j is found with probability $|\psi_j|^2$ and that the expected value is the sum over all these contributions. For the state mentioned above, application of this formula indeed yields $\langle \hat{X} \rangle = 1 \left|\frac{1}{\sqrt{2}}\right|^2 + 2\left|\frac{1}{\sqrt{2}}\right|^2 = 1.5$.

▷ Unlike with classical physics, where measurements can be non-invasive and purely observational, *a quantum measurement on a state $\hat{\psi}$ generally causes a state change*. This is asserted by the **measurement postulate** of quantum mechanics: if the measurement yields a particular eigenvalue λ_n, the system will be in the eigenstate $\hat{\psi}_n$ immediately after the measurement. For example, if an electron has been measured at position j, then after the measurement it is in the state $\hat{e}_j$. This reflects the fact that right after it has been measured it is known to be at the measurement site with certainty.

▷ In classical physics, the instantaneous position, $\mathbf{x}$, of a particle does not contain the full information about its motion. In addition, one needs to know its velocity, $\mathbf{v}$, or momentum, $\mathbf{p} = m\mathbf{v}$, where m is the particle mass. The pair $(\mathbf{x}, \mathbf{p})$ fully specifies the state of the particle in the sense that knowledge of $(\mathbf{x}, \mathbf{p})(0)$ at an initial time $t = 0$ is sufficient information to solve Newton's equations and to predict the future motion $(\mathbf{x}, \mathbf{p})(t)$.

Turning back to the quantum ring, let us illustrate how these structures change in the quantum world. In lecture courses on quantum mechanics, it is shown that the Hermitian operator $\hat{P}$ of the observable "momentum" is defined by the action $(P\psi)_j = \frac{1}{2i}(\psi_{j+1} - \psi_{j-1})$.[68] The corresponding Hermitian (check!) matrix reads

$$P = \frac{1}{2i} \begin{pmatrix} 0 & 1 & & & & & -1 \\ -1 & 0 & 1 & & & & \\ & -1 & 0 & 1 & & & \\ & & & \ddots & & & \\ & & & & -1 & 0 & 1 \\ 1 & & & & & -1 & 0 \end{pmatrix},$$

where all the empty positions are filled with zeros.[69] We now ask what possible values a *quantum measurement of momentum* can yield. The measurement postulate formulated above states that these must be eigenvalues of the matrix P. These eigenvalues are determined by the matrix eigenvalue equation $\lambda\psi_j = (P\psi)_j$ for the components of an λ-eigenvector, $\hat{\psi} = \hat{e}_j\psi_j$. The solution of the equation can be found from an ansatz, $\psi_j = c\exp(zj)$, where c and z are complex parameters. For any j different from 1 and N the substitution of this expression into the eigenvalue equation yields

$$\lambda\psi_j = (P\psi)_j = \frac{1}{2i}(\psi_{j+1} - \psi_{j-1}) = \frac{c}{2i}\left(e^{z(j+1)} - e^{z(j-1)}\right) = \frac{1}{2i}\left(e^z - e^{-z}\right)\psi_j.$$

This shows that the eigenvalue has the form $\lambda = \frac{1}{2i}(e^z - e^{-z})$. A constraint for the parameter z follows from setting $j = N$ in the eigenvalue equation,[69]

$$\lambda\psi_N = (P\psi)_N = \frac{1}{2i}(\psi_1 - \psi_{N-1}) = \frac{c}{2i}\left(e^z - e^{z(N-1)}\right) = \frac{1}{2i}\left(e^z e^{-zN} - e^{-z}\right)\psi_N.$$

This requires $\exp(-zN) = 1$, and for values of z satisfying this condition the equation is solved for all j. The eigenvalue condition is resolved by any of the N different choices (see Eq. (C92)) $z = i2\pi l/N$, where $l = 1, \ldots, N$. Each integer l corresponds to one of N different eigenvectors $\hat{\psi}_l$, with components $\psi_{l,j}$ and eigenvalues λ_l,

$$\psi_{l,j} = \frac{1}{\sqrt{N}}e^{i\frac{2\pi l}{N}j}, \qquad \lambda_l = \sin(2\pi l/N), \qquad l = 1, \ldots, N, \qquad \text{(L221)}$$

where we have chosen the second free parameter, $c = 1/\sqrt{N}$, to obtain unit normalization, $\sum_j |\psi_{l,j}|^2 = 1$, and used $\frac{1}{2i}(e^z - e^{-z}) \overset{(C91)}{=} \sin(2\pi l/N)$. The N linearly independent (why?) eigenstates $\hat{\psi}_l = \hat{e}_j\psi_{l,j}$ define a basis, the *eigenbasis of the momentum operator* $\hat{P}$.

The measurement axiom states that a measurement of the momentum of a quantum particle on a ring must yield one of the *discrete* values $\lambda_l = \sin(2\pi l/N)$. Unlike in

---

momentum operator

[68] For a heuristic argument why the momentum operator $\hat{P}$ might have something to do with motion, notice that it maps the localized state $\hat{e}_j$ to $\frac{1}{2i}(\hat{e}_{j+1} - \hat{e}_{j-1})$, i.e. it displaces an electron located at site j to the two neighboring sites $j \pm 1$.

[69] Due to the closure of the ring, sites 1 and N are adjacent to each other. The corner elements 1 and -1 in the momentum matrix ensure that these sites are connected by the momentum operator as $(P\psi)_N = \frac{1}{2i}(\psi_1 - \psi_{N-1})$ and $(P\psi)_1 = \frac{1}{2i}(\psi_2 - \psi_N)$.

classical physics, the momentum of a quantum particle is *quantized*. Neighboring eigen-values differ by $|\sin(2\pi(l+1)/N) - \sin(2\pi l/N)| \simeq |\cos(2\pi l/N)|2\pi/N = \mathcal{O}(N^{-1})$, where a first-order Taylor expansion (see Eq. (C82)) was applied and we used the fact that the cosine is of order unity. This shows that the eigenvalue spacing is inversely pro-portional to the system size, N. For large systems, the effects of quantization become less pronounced.

The diagonalization procedure also shows that the *eigenstates of the momentum oper-ator* extend over the whole ring, with components of uniform magnitude, $|\psi_{l,j}|^2 = N^{-1}$. Suppose one had measured the position of the electron and obtained an eigenvalue n of the position operator $\hat{X}$ as an answer. Immediately after the measurement the state of the electron is then described by the position eigenstate $\hat{e}_n$, whose wave function, $e_{n,j} = \delta_{n,j}$, is concentrated at site n. But this is very different from any of the delocalized momen-tum eigenstates! Conversely, suppose one had measured the momentum and obtained an eigenvalue $\lambda_l = \sin(2\pi l/N)$. After the measurement the system is described by the eigenstate $\hat{\psi}_l$. Its wave function, $\psi_{l,j}$, is spread out over the entire ring, in stark contrast to the position eigenfunctions. These observations show that the observables *position and momentum cannot be simultaneously determined with certainty*. The more accurately one is determined, the more undetermined the other becomes. The limits in accuracy to which both observables can be determined simultaneously are given by the **Heisenberg uncertainty relation**, which we do not discuss here. In lecture courses on quantum mechanics it is shown how the "incompatibility" of simultaneous measurements of observables is at the root of many quantum phenomena.

Heisenberg uncertainty relation

The case study above demonstrates the importance of Hermitian operators in quantum mechanics. For a discussion of the equally important role played by unitary operators we refer to specialized lecture courses.

L8.5 Summary and outlook

In this chapter we introduced the important classes of unitary and Hermitian linear trans-formations. We saw that these transformations possess strong mathematical features and are easier to handle than generic linear maps: unitary maps are categorically diagonaliz-able and invertible, and their inverses are easily obtained by taking Hermitian adjoints. The eigenvalues are confined to have unit modulus and the same holds for the matrix determi-nants. Similarly, Hermitian maps have real eigenvalues and determinants, and are likewise diagonalizable. We showed that Hermitian and unitary maps are closely related, which explains why they possess equally high levels of mathematical structure. The connection is so close that the Hermitian maps provide "parametrizations" of the group of unitary maps via the exponential matrix function.

We also saw that Hermitian and unitary maps are of high relevance to the description of physical phenomena. Physical operations such as rotations or reflections preserve the norm of vectors and are therefore unitary. The largest class of applications is found in quantum mechanics. We provided a brief introduction to the subject and discussed how the axioms of quantum mechanics are effectively formulated through the mathematics of Hermitian maps and their eigenvectors. These connections, and the equally important role played by unitary maps, are addressed at a much deeper level in lecture courses on quantum mechanics.

As a spin-off, our discussion showed how quantum mechanics encodes information on physical states in wave functions. We considered a case where the wave function reduces to a vector in a finite-dimensional $\mathbb{C}^N$. However, in many applications the restriction to finite N is unphysical, or just inconvenient, and the wave function really is a function of continuous variables. One is then led to do linear algebra in function spaces, which is the subject of the next chapter.

L9   Linear algebra in function spaces

REMARK This chapter looks at the mathematics of functions (i.e. mathematics commonly addressed within the framework of *calculus*) from the perspective of linear algebra. It should be read at a relatively late stage and requires familiarity with major parts of Part C. Specifically, we will make reference to Section C6.1 on the δ-function and to Section C6.2 on Fourier series. Some familiarity with linear differential equations, discussed in Chapter C7, is also required. Throughout this chapter we will indicate the correspondence between functions and vectors using the notation $f \leftrightarrow \mathbf{v}$, where the l.h.s. contains an object belonging to a function space, and the r.h.s. the analogous object from a finite-dimensional vector space. Occasionally we will consider spaces with nontrivial metrics and familiarity with Section L3.3 and covariant notation is required to understand these parts of the chapter.

Earlier in Part L (Section L2.3) we introduced function spaces as an example of vector spaces. However, so far we have not discussed any of the central concepts of linear algebra – changes of basis, linear maps, etc. – in this context. This extension will be the subject of the present chapter. It provides important foundations for the mathematical understanding of various physical disciplines and notably of quantum mechanics. The mathematical framework of quantum mechanics is essentially a synthesis of analysis and linear algebra, and the most efficient way to comprehend it is to regard functions as vectors to which all operations of linear algebra may be applied. In this chapter we will discuss how this works in practice.

Function spaces differ from the conventional vector spaces discussed so far in two respects. The first is different notation. For example, the "components" specifying a function f are denoted $f(x)$, whereas v^i is used for those of a vector $\mathbf{v}$. As with any change of notation it may take some time to get used to this, but after a while the linear-algebraic way of handling functions will begin to feel natural. The second difference is more substantial: function spaces are infinite-dimensional. For example, we need infinitely many "components" $f(x)$ to fully describe a function f, and indices labeling function bases run over infinite index sets, etc. Infinite-dimensionality may also lead to existence problems. For example, linear maps of function spaces can be thought of as infinitely large matrices. Traces and determinants of such matrices then assume the form of infinite products and sums whose convergence must be checked. All this indicates that the mathematically rigorous treatment of infinite-dimensional vector spaces requires substantial extensions of the framework of finite-dimensional linear algebra, and this is the subject of **functional analysis**. While a mathematically rigorous introduction to functional analysis is beyond the scope of this text, we will point out convergence issues where they occur and suggest

functional analysis

pragmatic ways of handling them. This approach should be sufficient for the majority of situations encountered in physics.

Throughout, we will consider function spaces, $X \equiv \{f : I \to \mathbb{C}\}$, containing functions mapping a *bounded* domain of definition, $I \subset \mathbb{R}$, into the complex numbers. (Generalizations to bounded domains in $\mathbb{R}^d$ are straightforward, we restrict ourselves to $d = 1$ merely for notational simplicity.) The restriction to bounded domains of definition simplifies the discussion of some of the convergence issues mentioned above. Some of the modifications required for the treatment of unbounded domains will be listed at the end of the chapter.

L9.1 Bases of function space

As with finite-dimensional vector spaces, the mathematics of function spaces heavily relies on the choice of suitable bases. We start with the introduction of the infinite-dimensional analogue of the standard basis. This will be followed by the discussion of more general choices, tailored to the requirements of specific applications.

The standard basis of a function space

In Section L2.3 we established a correspondence between functions, $f \in X$, and vectors, $\mathbf{v} \in \mathbb{C}^n$,

$$
\begin{aligned}
f &\longleftrightarrow \mathbf{v}, \\
f(x) &\longleftrightarrow v^i, \\
x &\longleftrightarrow i.
\end{aligned}
\tag{L222}
$$

This means that if a function f is considered as a vector, then the function values $f(x)$ are its "components", and $x \in I$ plays the role of the index $i = \{1, \dots, n\}$. We also pointed out in Section L3.3 the existence of a natural inner product,

$$
\langle f, g \rangle = \int_I dx \overline{f(x)} g(x) \qquad \longleftrightarrow \qquad \langle \mathbf{u}, \mathbf{v} \rangle = \sum_i \overline{u^i} v^i.
\tag{L223}
$$

square integrable The space of all **square integrable** functions, $\langle f, f \rangle = \int_I dx |f(x)|^2 < \infty$, is denoted $L^2(I)$. In some applications, the standard inner product is generalized as

$$
\langle f, g \rangle \equiv \int_I dx \overline{f(x)} w(x) g(x) \qquad \longleftrightarrow \qquad \langle \mathbf{u}, \mathbf{v} \rangle = \sum_{ij} \overline{u^i} g_{ij} v^j \equiv \sum_j \overline{u_j} v^j,
\tag{L224}
$$

Hilbert space where $w(x)$ is a positive "weight function".[70] A function space equipped with an inner product is called a **Hilbert space**.[71]

---

[70] One might consider even more general inner products, i.e. $\langle f, g \rangle \equiv \int_I dx \, dy \overline{f(x)} w(x, y) g(y)$, where the weight function, $w(x, y) = \overline{w(y, x)}$, assumes the role of the metric, $g_{ij} = \overline{g_{ji}}$ (check that this defines an inner product if w has suitable positivity properties). However, such generalizations do not often occur in practice and we will not discuss them.

[71] Hilbert spaces are generalizations of the Euclidean spaces discussed in Chapter L3. They are inner product spaces equipped with an extra condition ensuring that the norm, (L34), of a vector exists. Denoting the

When we speak of $f(x)$ as the components of f, implicit reference to a basis is made. Indeed, the function f can be "expanded" as

$$f = \int dy\, \delta_y f(y) \qquad \longleftrightarrow \qquad \mathbf{v} = \sum_j \mathbf{e}_j\, v^j, \qquad (L225)$$

where the δ-functions, $\delta_y \leftrightarrow \mathbf{e}_j$, now assume the role of *basis functions* and $\int dy \leftrightarrow \sum_j$ is a "sum" over all of them.

To better understand the analogy $\delta_y \leftrightarrow \mathbf{e}_j$, recall that the distinguishing property of the finite-dimensional standard basis $\{\mathbf{e}_j\}$ is its orthonormality, $\langle \mathbf{e}_i, \mathbf{e}_j \rangle \equiv g_{ij} = \delta_{ij}$. It is a basis in which the metric assumes the form of a unit matrix and contra- and covariant indices are equivalent, $v_j = v^i g_{ij} = v^j$ and $\mathbf{e}^i = g^{ij}\mathbf{e}_j = \mathbf{e}_i$. The component representation of these basis

David Hilbert 1862–1943
One of the most influential mathematicians of his time. Hilbert is considered one of the last "universal" mathematicians, capable of overseeing the field as a whole. He made important contributions not only to many areas of mathematics but also to physics, notably to the development of general relativity and to the mathematical foundations of quantum mechanics.

vectors, $(\mathbf{e}_j)^i = \delta^i{}_j$, has zeros everywhere except for a one at position j. The δ-functions mimic these properties in the infinite-dimensional case. This was pointed out early in the text, on p. 33, where we argued that the basis function multiplying the coefficient $f(y)$ in an expansion such as (L225) must be focused on the point y with infinite precision. The functions δ_y have the required properties that $\delta_y(x)$ vanishes for $x \neq y$ in such a way that its "infinitely narrow" support is compensated by the infinite amplitude at $x = y$, $\delta_y(y) = \infty$, i.e.

$$\delta_y(x) = \delta(x - y) \qquad \longleftrightarrow \qquad (\mathbf{e}_j)^i = \delta^i{}_j, \qquad (L226)$$

assumes the role of the Kronecker-δ valued components of a standard basis vector. The components, $f(x)$, of a function expanded in the standard basis can then be expressed as

$$f(x) = \int dy\, \underbrace{\delta_y(x)}_{\delta(x-y)} f(y) \qquad \longleftrightarrow \qquad v^i = \sum_j \underbrace{(\mathbf{e}_j)^i}_{\delta^i{}_j}\, v^j. \qquad (L227)$$

Much as the standard basis of $\mathbb{R}^n$ is orthonormal, $\langle \mathbf{e}_i, \mathbf{e}_j \rangle = \delta_{ij}$, the δ-function basis $\{\delta_j\}$ satisfies an *orthonormality relation*, too: $\langle \delta_x, \delta_y \rangle = \int_I dz\, \delta_x(z)\delta_y(z) = \int_I dz\, \delta(z - x)\delta(z - y) = \delta(x - y)$, so we have the correspondence

$$\langle \delta_x, \delta_y \rangle = \delta(x - y) \qquad \longleftrightarrow \qquad \langle \mathbf{e}_i, \mathbf{e}_j \rangle = \delta_{ij}. \qquad (L228)$$

components of a vector f by f_k, this amounts to existence conditions on sums such as $\langle f, f \rangle = \sum_k |f_k|^2 < \infty$. In the case of finite-dimensional Euclidean spaces this condition is trivially fulfilled. While the detailed discussion of the Hilbert condition for infinite-dimensional spaces is beyond the scope of the present text, we note that the condition $\langle f, f \rangle = \int_I dx |f(x)|^2 < \infty$ states the finiteness of a norm of f and hence defines a Hilbert space.

Table L2   Summary of the linear algebraic interpretation of basis changes in function space, in the context of Fourier transformations discussed in the next subsection. Einstein summation over the repeated indices α or k is used. For completeness, the table makes reference to a general metric, $g = \{g_{ij}\}$, and an index lowering and raising convention, $v_j \equiv v^i g_{ij}$, $\mathbf{e}^i = g^{ij}\mathbf{e}_j$, is used (see Eqs. (L60), (L66)). Similarly, $v_\beta = v^\alpha g_{\alpha\beta}$, $\mathbf{w}^\alpha = g^{\alpha\beta}\mathbf{w}_\beta$. Our discussion in the main text assumes orthonormal bases, $g_{ij} = \delta_{ij}$ and $g_{\alpha\beta} = \delta_{\alpha\beta}$, so that $v_j = v^j$ and $\mathbf{w}^\alpha = \mathbf{w}_\alpha$, etc. Readers not yet familiar with these index conventions may regard all indices as subscripts.

	vector space		function space	
	invariant	components	invariant	components
elements	$\mathbf{v}$	$v^i = \langle \mathbf{e}^i, \mathbf{v}\rangle$	f	$f(x) = \langle \delta_x, f\rangle$
inner product	$\langle \mathbf{u}, \mathbf{v}\rangle$	$\overline{u^i}g_{ij}v^j \equiv \overline{u_j}v^j$	$\langle f, g\rangle$	$\int dx\, \overline{f(x)}g(x)$
standard basis	$\mathbf{e}_j$	$(e_j)^i = \delta^i{}_j$	δ_y	$\delta_y(x) = \delta(x-y)$
alternative basis	$\mathbf{w}_\alpha$	$(\mathbf{w}_\alpha)^i = \langle \mathbf{e}^i, \mathbf{w}_\alpha\rangle$	ψ_k	$\psi_k(x) = \langle \delta_x, \psi_k\rangle = \frac{1}{\sqrt{L}}e^{ikx}$
orthonormality	$\langle \mathbf{w}_\alpha, \mathbf{w}_\beta\rangle \equiv g_{\alpha\beta} = \delta_{\alpha\beta}$	$\overline{(\mathbf{w}_\alpha)_i}(\mathbf{w}_\beta)^i = \delta_{\alpha\beta}$	$\langle \psi_k, \psi_p\rangle = \delta_{kp}$	$\frac{1}{L}\int dx\, e^{i(p-k)x} = \delta_{kp}$
expansion	$\mathbf{v} = \mathbf{w}_\alpha v^\alpha$	$v^i = (\mathbf{w}_\alpha)^i v^\alpha$	$f = \psi_k \tilde{f}_k$	$f(x) = \frac{1}{\sqrt{L}}\sum_k e^{ikx}\tilde{f}_k$
coefficients	$v^\alpha = \langle \mathbf{w}^\alpha, \mathbf{v}\rangle$	$v^\alpha = \overline{(\mathbf{w}^\alpha)_i}v^i$	$\tilde{f}_k = \langle \psi_k, f\rangle$	$\tilde{f}_k = \frac{1}{\sqrt{L}}\int dx\, e^{-ikx}f(x)$
completeness	$\langle \mathbf{e}^i, \mathbf{e}_j\rangle = \langle \mathbf{e}^i, \mathbf{w}_\alpha\rangle\langle \mathbf{w}^\alpha, \mathbf{e}_j\rangle$	$\delta^i{}_j = (\mathbf{w}_\alpha)^i \overline{(\mathbf{w}^\alpha)_j}$	$\delta(x-y) = \langle \delta_x, \psi_k\rangle\langle \psi_k, \delta_y\rangle$	$\delta(x-y) = \frac{1}{L}\sum_k e^{ik(x-y)}$

The components of a vector can be obtained by taking the inner product with a basis vector, $\langle \mathbf{e}^i, \mathbf{v}\rangle = \sum_j \langle \mathbf{e}^i, \mathbf{e}_j\rangle v^j = \sum_j \delta^i{}_j v^j = v^i$. Likewise, the "components" of f can be obtained as $\langle \delta_x, f\rangle = \int dy\, \delta_x(y)f(y) = f(x)$, i.e.

$$f(x) = \langle \delta_x, f\rangle \qquad\longleftrightarrow\qquad v^i = \langle \mathbf{e}^i, \mathbf{v}\rangle. \tag{L229}$$

We note that for function spaces covariant notation of indices is not defined. If a metric enters the stage, it needs to be written in explicit form and cannot be "hidden" in raised or lowered indices.

Non-standard bases of function space

The relations above would be of little more than pedagogical value if no interesting function bases different from the standard δ-basis were available. We have already met one important example of a non-standard basis, the basis of Fourier functions discussed in Section C6.2: Fourier series representations of functions can be seen as a change of basis, as summarized in Table L2. To understand this correspondence in detail, we consider the space $X = \{f \in L^2(I)\}$ of square integrable functions defined on an interval, $I = (x_0, x_0+L)$, of length L. Now consider the set of functions $\{\psi_k \in X | k \in \frac{2\pi}{L}\mathbb{Z}\}$, where

$$\psi_k(x) \equiv \frac{1}{\sqrt{L}}\exp(ikx). \tag{L230}$$

Fourier basis functions   Apart from the normalization factor $L^{-1/2}$, these *Fourier basis functions* coincide with the Fourier modes $\exp(ikx)$ introduced in Section C6.2. As we are going to show next, the set $\{\psi_k\}$ defines an *orthonormal basis* of $L^2(I)$ different from the standard basis $\{\delta_y\}$. To explore these connections we again refer to the analogous situation in a finite-dimensional vector space, with an orthonormal system of basis vectors, $\{\mathbf{w}_\alpha | \alpha = 1, \ldots, N\}$, different from the standard basis $\{\mathbf{e}_j | j = 1, \ldots, N\}$.[72]

The function values

$$\psi_k(x) \qquad \longleftrightarrow \qquad (\mathbf{w}_\alpha)^i \qquad\qquad\qquad (\text{L231})$$

are the components of the new basis vectors written in terms of the old basis. In the language of Section L5.6 they define the entries of an "infinite-dimensional" transformation matrix $(T^{-1})_{x,k}$. Thanks to the orthonormalization of the standard basis, $\{\delta_y\} \leftrightarrow \{\mathbf{e}_j\}$, we may understand the function *values*, $\psi_k(x)$ (i.e. the analogue of vector components) as inner products taken between the Fourier functions, ψ_k (the analogue of vectors), and the basis functions, δ_y, of the standard basis (see Eq. (L229))

$$\psi_k(x) = \langle \delta_x, \psi_k \rangle \qquad \longleftrightarrow \qquad (\mathbf{w}_\alpha)^i = \langle \mathbf{e}^i, \mathbf{w}_\alpha \rangle. \qquad\qquad (\text{L232})$$

It is straightforward to check the *orthonormalization of the Fourier basis*:

$$\langle \psi_k, \psi_p \rangle = \int_I \mathrm{d}x\, \overline{\psi_k(x)} \psi_p(x) = \frac{1}{L} \int_0^L \mathrm{d}x\, e^{i(-k+p)x} \stackrel{(\text{C123})}{=} \delta_{kp}$$

$$\longleftrightarrow$$

$$\langle \mathbf{w}_\alpha, \mathbf{w}_\beta \rangle = \sum_i \overline{(\mathbf{w}_\alpha)_i} (\mathbf{w}_\beta)^i = \delta_{\alpha\beta}. \qquad\qquad (\text{L233})$$

The second line of Eq. (L233) states the orthonormality of the $\{\mathbf{w}_\alpha\}$ basis; the first line shows that the Fourier modes $\{\psi_k\}$ satisfy an analogous orthogonality relation.

Completeness relations

Equation (L233) shows that the functions $\{\psi_k\}$ are orthonormal and hence linearly independent. However, we do not yet know whether they represent a *complete* set. Unlike an n-dimensional vector space where n mutually orthogonal vectors automatically form a basis, $L^2(I)$ is infinite-dimensional. But ∞ is not a well-defined number and no counting scheme can determine whether the *infinitely many* functions $\{\psi_k\}$ suffice to span it. (Maybe twice as many functions, $2 \times \infty = \infty$, would be needed for that task?) Unlike with finite-dimensional vector spaces, completeness needs to be established in different ways.

In the next section we will see that for most function bases of practical interest, completeness is granted and need not be checked manually. Occasionally, however, this needs to be done and here we show how. A set of functions $\{\psi_k\}$ is complete if every function can be expanded as $f = \sum_k \psi_k c_k$, where c_k are expansion coefficients. Taking the inner product $\langle \psi_k, f \rangle$ and using the orthonormality relation (L233) we obtain the identification $c_k = \langle \psi_k, f \rangle$, so completeness requires the existence of expansions

$$f = \sum_k \psi_k \langle \psi_k, f \rangle \qquad \longleftrightarrow \qquad \mathbf{v} = \sum_\alpha \mathbf{w}_\alpha \langle \mathbf{w}^\alpha, \mathbf{v} \rangle. \qquad (\text{L234})$$

---

[72] We use different basis indices, j and α, respectively, to foster comparison to the functions $\{\delta_y\}$ and $\{\psi_k\}$ which, likewise, are labeled by different indices x and k.

An equivalent condition is that every element of a function basis, for example those of the standard basis $\{\delta_y\} \leftrightarrow \{\mathbf{e}_j\}$, is expandable:

$$\forall y \in I: \quad \delta_y = \sum_k \psi_k \langle \psi_k, \delta_y \rangle \qquad \longleftrightarrow \qquad \forall j = 1, \ldots, N: \quad \mathbf{e}_j = \sum_\alpha \mathbf{w}_\alpha \langle \mathbf{w}^\alpha, \mathbf{e}_j \rangle.$$

(L235)

Since a generic function can be expanded in the standard basis, Eq. (L235) suffices to guarantee expandability in elements of the $\{\psi_k\}$ basis. Taking inner products $\langle \delta_y, \ \rangle$ of this relation with generic standard basis vectors we obtain the equivalent set of relations

$$\delta(x - y) \overset{(L228)}{=} \langle \delta_x, \delta_y \rangle \overset{(L235)}{=} \sum_k \langle \delta_x, \psi_k \rangle \langle \psi_k, \delta_y \rangle \overset{(L232)}{=} \sum_k \psi_k(x)\overline{\psi_k(y)}$$

$$\longleftrightarrow$$

$$\delta^i_j \overset{(L228)}{=} \langle \mathbf{e}^i, \mathbf{e}_j \rangle \overset{(L235)}{=} \sum_\alpha \langle \mathbf{e}^i, \mathbf{w}_\alpha \rangle \langle \mathbf{w}^\alpha, \mathbf{e}_j \rangle \overset{(L232)}{=} \sum_\alpha (\mathbf{w}_\alpha)^i \overline{(\mathbf{w}^\alpha)_j}.$$

(L236)

These relations are easy to conceptualize: in the finite-dimensional case (corresponding to Eq. (L232)), we are considering a basis change, $\mathbf{e}_j = \mathbf{w}_\alpha T^\alpha_j$, between two orthonormal bases, described by a *unitary* transformation, $(T^\dagger)^i_\alpha T^\alpha_j = \delta^i_j$. The second line of Eq. (L236) expresses this relation in terms of inner products, using $T^\alpha_j = \langle \mathbf{w}^\alpha, \mathbf{e}_j \rangle = \overline{(\mathbf{w}^\alpha)_j}$ and $(T^\dagger)^i_\alpha = (T^{-1})^i_\alpha = \langle \mathbf{e}^i, \mathbf{w}_\alpha \rangle = (\mathbf{w}_\alpha)^i$. In an analogous manner, the left expression in Eq. (L235) identifies $\langle \psi_k, \delta_y \rangle = \overline{\psi_k(y)} = T_{k,y}$ (i.e. $\psi_k(x) = (T^{-1})_{x,k}$) as the elements of an infinite-dimensional unitary "matrix", describing the change from the standard basis $\{\delta_y\}$ to the basis $\{\psi_k\}$. Equations like

$$\boxed{\delta(x - y) = \sum_k \psi_k(x)\,\overline{\psi_k(y)}}$$

(L237)

are called **completeness relations**. For the specific case of the Fourier functions (L230), we have checked the completeness explicitly, see Eq. (C124). The orthonormality relation (L233) and the completeness relation prove that

> Fourier series expansion amounts to a change of basis in function space.

Below, we will introduce a few more examples of function bases and demonstrate how their completeness follows in different ways from general criteria. However, before that we need to adapt another key concept of linear algebra to function spaces.

L9.2 Linear operators and eigenfunctions

The natural maps between vector spaces are the linear maps compatible with the vector space operations addition and scalar multiplication. Infinite-dimensional limits of such maps are called linear operators. In this section, we introduce these objects and the operations of function space linear algebra relating to them. All this will be not much more than a straightforward adaption of structures familiar from the finite-dimensional case.

Linear operators

linear operator

Linear maps $\hat{A} : X \to X, f \mapsto \hat{A}f$ which send functions, $f \in X$, to new functions, $\hat{A}f$, are called **linear operators**. Linearity means that $\hat{A}(cf + dg) = c\hat{A}f + d\hat{A}g, f, g \in X$, $c, d \in \mathbb{C}$.

EXAMPLE Consider the space $X \subset L^2([0, 1])$ of complex-valued functions on the unit interval, subject to the periodicity condition $f(0) = f(1)$. Linear operators $\hat{A} : X \to X$ respecting the periodicity condition are easily constructed. For example, consider the function $h \in X$ with $h(x) = \cos(2\pi x)$. Multiplication by h defines the linear operator $\hat{A}_h : X \to X, f \mapsto hf$, where $(hf)(x) = h(x)f(x)$. Like h and f, the function $\hat{A}_h f = hf$ is periodic, so that $\hat{A}_h$ acts *within* the space X. Multiplication by h satisfies the linearity criterion, $\hat{A}_h(cf + dg) = \hat{A}_h f + \hat{A}_h g$, and hence defines a linear operator on the space X.

differential operator

The linear operators playing the most important role in applications involve derivative operations and are called **differential operators**. As an example consider the operator $-\mathrm{i}\,\mathrm{d}_x$ acting on functions by differentiation, e.g. $-\mathrm{i}\,\mathrm{d}_x \cos(2\pi x) = 2\pi\,\mathrm{i}\,\sin(2\pi x)$, where the factor of $-\mathrm{i}$ has been introduced for later convenience. This map, too, satisfies the periodicity condition (why?) and linearity, and so defines a linear operator in X. Later in the chapter we will see that this operator plays an important role in the description of periodic functions, and we will use it as a role model to illustrate various generic features of differential operators. Sums and products of linear operators are again linear operators. For example $(-\mathrm{i}\,\mathrm{d}x)^2 = -\mathrm{d}_x^2$ and $-\mathrm{d}_x^2 + \cos(2\pi x)$, too, act linearly in X.

Eigenfunctions

In previous sections we have seen that the essential information on a linear map, $A : \mathbb{C}^n \to \mathbb{C}^n$, is contained in its eigenvectors, satisfying $A\mathbf{v}_j = \lambda_j\mathbf{v}_j$, where λ_j is the corresponding eigenvalue. If a basis of eigenvectors could be found, the linear map A assumed the simple form of a diagonal matrix. General vectors could be expanded in eigenvectors and the action of the A was essentially under control.

The situation with linear operators in function spaces is similar. For an operator $\hat{A} : X \to X$, a function $\tilde{f}_k$ satisfying the relation

$$\hat{A}\tilde{f}_k = \lambda_k \tilde{f}_k \tag{L238}$$

eigenfunction of linear operator

is called an **eigenfunction** with eigenvalue λ_k. Here, the index k plays a role similar to the discrete index j. The difference is that this counting index will generally run over an infinite set, reflecting the infinite-dimensionality of the space. Typical examples include $k \in \mathbb{Z}$, or double-indices such as $(k_1, k_2) \in \mathbb{Z} \times \mathbb{Z}$. If I is unbounded, dense sets of eigenvalue indices may occur, see Section L9.4 at the end of this chapter.

It will often be convenient to normalize eigenfunctions. As with vectors this is done by computing the square of the norm $\mathcal{N} \equiv \langle \tilde{f}_k, \tilde{f}_k \rangle = \int_I \mathrm{d}x\,|\tilde{f}_k|^2$, and defining $f_k \equiv \frac{1}{\sqrt{\mathcal{N}}}\tilde{f}_k$. If I is non-compact, this normalization may not be straightforward and we refer to Section L9.4 for a discussion of this case.

EXAMPLE As an example consider the space of periodic functions on the unit interval discussed in Section L9.1. The eigenfunction equation for the linear operator $(-i)d_x$ reads

$$-id_x f(x) = \lambda f(x), \tag{L239}$$

where we temporarily omitted the counting index k. It is easily verified by substitution that this equation is solved by the function $ae^{i\lambda x}$, where $a, \lambda \in \mathbb{C}$, and λ features as the eigenvalue. This result raises two questions: the first is whether there are other eigenfunctions with the same eigenvalue. The answer follows from the theory of differential equations. Equation (L239) is an ordinary first-order linear differential equation. In Section C7.7 we show that up to normalization the solution to these equations is unique, where the freedom of normalization is reflected by the arbitrary prefactor a.[73] The second question is whether the eigenfunctions actually belong to the function space X. They do if the periodicity condition $f(0) = f(1)$, or $a = ae^{\lambda}$, is obeyed. This is satisfied if and only if $\lambda = 2\pi i k$, $k \in \mathbb{Z}$. The proper eigenvalues of our operator in X are therefore given by $\lambda = \lambda_k \equiv 2\pi i k$, and $a \exp(2\pi i k x)$ are the corresponding eigenfunctions. Finally, we verify that the norm of these functions is given by $|a|$, so for $a = 1$ (or some other unit-modular constant) we have unit normalization.

To summarize our results, we have found that the linear operator $-id_x$ possesses the set of eigenfunctions

$$f_k(x) \equiv e^{2\pi i k x}, \qquad k \in \mathbb{Z}. \tag{L240}$$

These functions are just the *Fourier modes* on the unit interval. We have identified the Fourier basis as the basis of eigenfunctions of the linear operator $-i\,d_x : X \to X$.

The discussion of the preceding example contains a few general *guiding principles for the identification of eigenfunctions*.

▷ Technically, the "eigenequations" $\hat{A}f = \lambda f$ associated with a linear differential operator are linear differential equations. Start by identifying a complete set of linearly independent solutions. The cardinality of that set depends both on the order of the highest derivative operator contained in $\hat{A}$ and on the dimensionality of I. If $\dim(I) = 1$ as in the example above, the order of $\hat{A}$ determines the number of linearly independent solutions. For higher-dimensional I the situation can become more complicated (think of the Fourier expansion of higher-dimensional functions defined in Eq. (C141), for example).

▷ Next, check that the general solutions actually lie in the function space X, i.e. that they satisfy the defining properties of elements of X. It may happen that some solutions have to be discarded (such as those with $\lambda \notin 2\pi i \mathbb{Z}$ discussed above). The appropriately restricted set then defines the set of eigenfunctions.

▷ Finally, it may be convenient to normalize the functions, as discussed above.

---

[73] The situation would be different had we considered the operator $-d_x^2 = ((-i)d_x)^2$. Its eigenequation, $-d_x^2 f = \lambda f$, is a second-order differential equation with an eigenspace spanned by *two* linearly independent solution functions, $e^{\pm\lambda x}$. The general solution is then given by all linear combinations $c_+ e^{+\lambda x} + c_- e^{-\lambda x}$ with constants $c_\pm$. More generally, an nth-order differential operator has n linearly independent solutions and an n-dimensional eigenspace.

L9.3   Self-adjoint linear operators

In Section L8.2 we discussed the specific properties of Hermitian linear maps. We learned that a Hermitian matrix can always be diagonalized. The diagonalizability property means that the set of eigenvectors of a Hermitian matrix is complete. Further, the set of eigenvectors could be chosen so as to form an orthonormal basis. All these niceties carry over to the case of linear operators.

Definition

self adjoint linear operator

Assume our function space X is equipped with an inner product, (L223). For concreteness, we may consider the space X of periodic functions on the interval $[0, L]$ with complex standard inner product, $\langle f, g \rangle = \int_0^L dx \overline{f(x)} g(x)$. An operator $\hat{A} : X \to X$ is called **self-adjoint** if

$$\forall f, g \in X : \langle \hat{A} f, g \rangle = \langle f, \hat{A} g \rangle \quad \longleftrightarrow \quad \forall \mathbf{u}, \mathbf{v} \in \mathbb{C}^n : \langle A\mathbf{u}, \mathbf{v} \rangle = \langle \mathbf{u}, A\mathbf{v} \rangle. \quad (L241)$$

We observe that a self-adjoint linear operator is the analogue of a Hermitian matrix. In fact, it is common practice (in physics) to use the terminology of "Hermitian operators", and we will do so in the following. For example, the operator $(-i)d_x$ considered above enjoys the Hermiticity property:

$$\langle -id_x f, g \rangle = \int_0^L dx \overline{(-i)d_x f(x)} g(x) = i \int_0^L dx\, d_x \overline{f(x)} g(x) = \int_0^L dx \overline{f(x)}\,(-i)d_x g(x)$$

$$= \langle f, -id_x g \rangle,$$

where we integrated by parts, noting that no boundary terms arise due to the assumed periodicity of the integrand.[74]

EXERCISE   Recapitulate the arguments of Section L8.2 to verify that they carry over to the case of function spaces.

As in the case of finite-dimensional vector spaces, the Hermiticity of a linear operator makes strong statements about its eigenvalues and eigenfunctions: all eigenvalues λ_k are real; eigenfunctions f_k with different eigenvalues are mutually orthogonal, $\langle f_k, f_{k'} \rangle = 0$, if $\lambda_k \neq \lambda_{k'}$; and the full system of eigenfunctions is complete. The Fourier eigenfunctions of $-id_x$ are an example of this.

completeness of Hermitian operator

Importantly, the knowledge that an operator is Hermitian and that we have found all of its eigenfunctions is sufficient to establish the completeness of that set of functions. In most applications the equations determining eigenfunctions assume the form of linear differential equations. The number of independent solutions of these equations is determined by the order of the differential operator, the dimensionality of the space in which the problem is realized, and the definition of boundary conditions. Once we know that all solutions are

---

[74]   At this point it becomes evident why we included a factor $(-i)$ in the definition of the differential operator: it serves to make the operator Hermitian.

under control, completeness is granted by the Hermiticity of the operator, and no explicit verification of completeness relations is necessary. This explains the statement above that explicit verifications of completeness are often not needed.

INFO Hermitian differential operators play an important role in both mathematics and physics. For example, the *axioms of quantum mechanics* state that to each physical "observable" (position, momentum, angular momentum, etc.), there corresponds one Hermitian operator $\hat{A}$ (see case study in Section L8.4). The expected value of a measurement of the observable for a system described by a physical "state vector" $\hat{\psi}$ is defined by $\langle \hat{\psi}, \hat{A}\hat{\psi} \rangle$. Depending on the context, $\hat{\psi}$ may belong to a finite-dimensional vector space or to a function space; in the latter case $\psi(x) = \langle \delta_x, \hat{\psi} \rangle$ is called a wave function. A great deal can be learned about a physical observable from the eigenfunctions of its corresponding operator. If a specific operator plays an especially important role in a problem, then its eigenfunctions define the natural basis to work with. In the case of infinite-dimensional operators, the eigenequations are determined by differential equations. For example, the two differential operators discussed in the examples below (Eqs. (L242) and (L247)) are relevant to the quantum mechanical description of the hydrogen atom.

Example: Legendre polynomials

Consider the the real-valued functions on the interval $[-1, 1]$, $X \equiv \{f : [-1, 1] \rightarrow \mathbb{R}\}$ (no boundary conditions specified) with inner product $\langle f, g \rangle = \int_{-1}^{1} dx\, f(x)g(x)$. On this space, we define the second-order differential operator

$$\hat{A} = \frac{d}{dx}(1 - x^2)\frac{d}{dx}. \tag{L242}$$

It is straightforward to check that this operator is *symmetric* relative to the above inner product, $\langle \hat{A}f, g \rangle = \langle f, \hat{A}g \rangle$.

EXERCISE Integrate by parts to verify the symmetry of the differential operator (L242). Show why no boundary terms arise.

Legendre polynomials

The eigenequation of $\hat{A}$, $\hat{A}P_l = \lambda_l P_l$, is called the *Legendre differential equation* and the eigenfunctions P_l are known as **Legendre polynomials**. (We will see shortly why they are polynomials in x.) Finding a complete set of solutions of the Legendre equation is a nontrivial task, often discussed in lecture courses on ordinary differential equations, theoretical

Adrien Marie Legendre 1752–1833
French mathematician known for numerous contributions, notably the identification of Legendre polynomials and the Legendre transform. The watercolor caricature is the only existing portrait of Legendre. (Images published prior to 2005 mistakenly showed the portrait of a French politician of the same name.)

electrodynamics, or quantum mechanics. Referring to the info section below for a quick sketch of the solution strategy, we here just state the result: solutions of the Legendre differential equation are found for eigenvalues $\lambda_l = -l(l + 1)$, where $l = 0, 1, 2, \ldots$ is a positive integer. The corresponding Legendre polynomials can be represented in various ways, among them the so-called **Rodrigues formula**,

Rodrigues formula

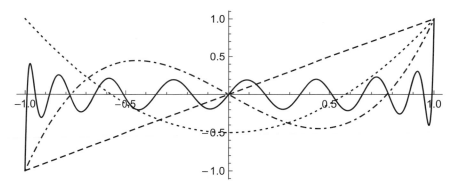

The Legendre polynomials P_1 (dashed), P_2 (dotted), P_3 (dash-dotted), P_{17} (solid).

$$P_l(x) = \frac{1}{2^l l!} \frac{d^l}{dx^l}(x^2 - 1)^l. \qquad (L243)$$

A list of the first four polynomials reads,

$$P_0(x) = 1, \quad P_1(x) = x, \quad P_2(x) = \tfrac{1}{2}(3x^2 - 1), \quad P_3(x) = \tfrac{1}{2}(5x^3 - 3x),$$
$$P_4(x) = \tfrac{1}{8}(35x^4 - 30x^2 + 3).$$

It is customary to normalize the Legendre polynomials as $P_l(1) = 1$, i.e. to fix their value at $x = 1$. For a visual representation of a few Legendre polynomials, see Fig. L18.

INFO Let us sketch the derivation of the result (L243).[75] The fact that the differential operator $\hat{A}$ contains a polynomial $(1 - x^2)$ suggests that polynomial solutions to the eigenequation might exist. (This is a weak argument, but better than none.) Indeed, we may check by direct substitution that $P_0 \equiv 1$ is a solution with eigenvalue $\lambda_0 = 0$, and $P_1 \equiv x$ one with eigenvalue $\lambda_1 = -2$. Encouraged by these findings, we may speculate that other solutions of the equation $\hat{A}P(x) = \lambda P(x)$ can be represented as a series,

$$P(x) = \sum_{j=0}^{\infty} a_j x^{\alpha+j},$$

where the parameter $0 \leq \alpha < 1$ has been introduced to allow for fractional, yet positive powers of x. Acting on this ansatz with the Legendre differential operator (do it!) we obtain the function

$$\hat{A}P(x) = \sum_{j=0}^{\infty} \Big[a_j(\alpha+j)(\alpha+j-1)x^{\alpha+j-2} - a_j(\alpha+j)(\alpha+j+1)x^{\alpha+j} \Big].$$

It will be convenient to combine the two contributions to this series into one. This is done by rearranging terms as

$$\sum_{j=0}^{\infty} a_j(\alpha+j)(\alpha+j-1)x^{\alpha+j-2} = X(x) + \sum_{l=2}^{\infty} a_l(\alpha+l)(\alpha+l-1)x^{\alpha+l-2}$$

$$= X(x) + \sum_{j=0}^{\infty} a_{j+2}(\alpha+j+2)(\alpha+j+1)x^{\alpha+j},$$

---

[75] A complete treatment would need to address convergence issues whose discussion is beyond the scope of this text.

where the first term, $X(x) \equiv a_0\alpha(\alpha-1)x^{\alpha-2} + a_1(\alpha+1)\alpha x^{\alpha-1}$, contains the first two summands, $j = 0, 1$, of the series, and we relabeled, $j = l-2$, in the second line. Combining terms we have

$$\hat{A}P(x) = X(x) + \sum_{j=0}^{\infty}\left[a_{j+2}(\alpha+j+2)(\alpha+j+1) - a_j(\alpha+j)(\alpha+j+1)\right]x^{\alpha+j}.$$

This series must equal the series $\lambda P(x)$. Representing this condition in the form of a series, we get

$$0 \stackrel{!}{=} \hat{A}P(x) - \lambda P(x)$$

$$= X(x) + \sum_{j=0}^{\infty}\left[a_{j+2}(\alpha+j+2)(\alpha+j+1) - a_j\big((\alpha+j)(\alpha+j+1)+\lambda\big)\right]x^{\alpha+j}. \tag{L244}$$

The global vanishing of the left-hand side requires that the coefficients of each power $x^{\alpha+k}$ are individually zero (why?). We first notice that terms of $\mathcal{O}(x^{\alpha-2}, x^{\alpha-1})$ are contained only in the contribution $X(x)$. These two must both be zero, which leads to the condition $\alpha = 0$ if either a_0 or a_1 are different from zero. To eliminate terms containing higher powers, $x^{\alpha+j}$, we need

$$a_{j+2} = \frac{(\alpha+j)(\alpha+j+1) + \lambda}{(\alpha+j+2)(\alpha+j+1)}a_j.$$

This relation recursively fixes coefficients as $a_0 \to a_2 \to a_4 \to \ldots$ and $a_1 \to a_3 \to a_5 \to \ldots$. To avoid a solution vanishing everywhere, at least one of its "anchors", a_0 or a_1, must be non-vanishing, which in turn requires that the fractional power $\alpha = 0$ vanishes. Turning to the termination of the series, we seek solutions of *polynomial* form, i.e. we require that $a_j = 0$ after a finite number of terms. Inspection of the recursion relation shows that this condition requires the eigenvalue to assume the form $\lambda \equiv -l(l+1)$, where $l \in \mathbb{N}_+$ is a positive integer. Assume that l is an even/odd integer. We then observe that the series of even/odd coefficients, anchored by a_0 or a_1, terminates, while the odd/even series remains infinite. (These are the two linearly independent solutions of the Legendre differential equation at given l, there won't be other solutions.) Closer inspection shows that the non-terminating series has convergence issues and that we should discard it, by taking $a_1 = 0$ or $a_0 = 0$, respectively. It is easy to compute the first few good solutions by hand and as a result one obtains the list (L244). A less easy exercise (try it!) is to verify that the recurrence relations defining the Legendre polynomials are generated by the Rodrigues formula (L243).

EXERCISE Prove the *orthogonality of the Legendre polynomials*,

$$\int_{-1}^{1} dx\, P_l(x)P_{l'}(x) = \tfrac{2}{2l+1}\delta_{ll'}, \tag{L245}$$

for $l = 2$ and $l' = 3$. The general proof is not easy, unless we use our linear algebraic background knowledge: use $\hat{A}P_l = l(l+1)P_l$ and the symmetry of $\hat{A}$, $\int(\hat{A}P_l)P_{l'} = \int P_l(\hat{A}P_{l'})$ to show that the integral vanishes unless $l = l'$. Finally, the Legendre polynomials can also be generated by a **Gram–Schmidt orthonormalization** procedure, as in Section L3.2. Starting from $P_0(x) = 1$, at each step one orthogonalizes x^l relative to all previously generated polynomials $P_{l'<l}(x)$, while normalizing according to Eq. (L245). Try this for the first few polynomials.

In the next section we introduce the framework in which the Legendre differential operator appears in applications.

Example: spherical harmonics

REMARK Requires Sections V2.3 and C4.4.

In physics and mathematics we often work with problems defined on spheres. Using spherical coordinates, (θ, ϕ), we saw in Section C4.4 (see example on p. 254) that the natural surface element defined by a coordinate increment of $(d\theta, d\phi)$ is given by $dS \equiv \sin\theta \, d\theta \, d\phi$. This suggests consideration of the inner product

$$\langle f, g \rangle = \int_0^\pi d\theta \int_0^{2\pi} d\phi \, \sin\theta \overline{f(\theta, \phi)} g(\theta, \phi), \tag{L246}$$

defined on the space $X = L^2(S^2)$ of square-integrable functions on the coordinate domain $S^2 = \{(\theta, \phi) \,|\, \theta \in (0, \pi), \phi \in (0, 2\pi)\}$, subject to the periodicity condition $f(\theta, 0) = f(\theta, 2\pi)$.

Laplacian on unit sphere   In electrodynamics and quantum mechanics one is often interested in the **Laplacian on the unit sphere** $\Delta : X \to X$. This operator is obtained from the spherical coordinate representation of the three-dimensional Laplace operator, Eq. (V100), by fixing the radius, $r = 1$, and dropping the r-derivative,

$$\Delta = \frac{1}{\sin\theta} \partial_\theta \sin\theta \, \partial_\theta + \frac{1}{\sin^2\theta} \partial_\phi^2. \tag{L247}$$

An instructive little calculation (do it!) shows that the Laplace operator is Hermitian relative to the inner product Eq. (L246).[76]

spherical eigenfunctions   Let us now proceed to identify the *eigenfunctions*, f_λ, of the **Laplace operator**, i.e. the solutions of the second-order partial differential equation $\Delta f_\lambda(\theta, \phi) = \lambda f_\lambda(\theta, \phi)$. Progress with this equation is made by transforming it to two simpler ones. We start by multiplying the equation with $\sin^2\theta$ and rearranging terms,

$$\left[\left(\sin\theta \partial_\theta \sin\theta \, \partial_\theta - \lambda \sin^2\theta \right) + \partial_\phi^2 \right] f(\theta, \phi) = 0,$$

where the subscript on f_λ was omitted for notational clarity. The point now is that the differential operator in this equation is the sum of two terms, one depending only on θ, the other only on ϕ. Multivariate operators which separate into additive single-variate contributions are called **separable**. They have the nice property that their eigenfunctions are products of single-variate functions. In the present ansatz, this is established by substitution of a product ansatz, $f(\theta, \phi) = g(\theta)h(\phi)$, into the equation and multiplying from the left by $g^{-1}(\theta)h^{-1}(\phi)$:

separable differential operator

$$g^{-1}(\theta) \left(\sin\theta \partial_\theta \sin\theta \, \partial_\theta - \lambda \sin^2\theta \right) g(\theta) = -h^{-1}(\phi)\partial_\phi^2 h(\phi).$$

The right side of this equation does not depend on θ, which means that the left side must be independent of θ too. Conversely, the left side must be independent of ϕ. Since the left side

---

[76] Conceptually, its Hermiticity is inherited from that of the Laplacian, $\Delta = \partial_x^2 + \partial_y^2 + \partial_z^2$, relative to the standard three-dimensional scalar product, $\langle f, g \rangle = \int dx dy dz \overline{f(x, y, z)} \, g(x, y, z)$. The restriction of this operator to the subset of spherical functions must be Hermitian too.

equals the right side, they are independent of both θ and ϕ, and hence constant. Denoting this constant by m^2, we obtain two separate ordinary differential equations,

$$\left(\sin\theta d_\theta \sin\theta\, d_\theta - \lambda \sin^2\theta \right) g(\theta) = m^2 g(\theta),$$

$$d_\phi^2 h(\phi) = -m^2 h(\phi).$$

The second of these is solved by the Fourier modes $h = h_m$, where $h_m(\phi) \equiv \exp(i\phi m)$ and the periodicity condition, $h_m(0) = h_m(2\pi)$, requires integer values m. In the first equation, we apply a variable substitution, $x \equiv \cos\theta \in [-1,1]$. Defining $g(\theta(x)) \equiv P(x)$ and using $(\sin\theta)^{-1}d_\theta = d_x$ and $\sin^2\theta = 1 - x^2$, the equation assumes the form

$$\left(d_x(1 - x^2)d_x - \lambda - m^2(1 - x^2)^{-1} \right) P(x) = 0.$$

For $m = 0$ this is just the Legendre differential equation discussed in the previous section. In this case, we have the solutions $P(x) = P_l(x)$, with corresponding eigenvalues $\lambda = -l(l+1)$. The generalization of the solutions to arbitrary m are known as **Legendre functions**, P_l^m. It can be shown that these functions exist for $m \in \{-l, \ldots, l\}$ and that their eigenvalues are given by $-l(l+1)$, independent of m. For positive m they are defined by

$$P_l^m = (-)^m (1 - x^2)^{m/2} \frac{d^m}{dx^m} P_l(x),$$

while the solutions for negative m are $P_l^{m<0} = \frac{(l-m)!}{(l+m)!} P_l^{-m}$. Summarizing, we have found that the eigenfunctions of the spherical Laplace operator are given by the **spherical harmonics**

$$\boxed{Y_l^m(\theta,\phi) \equiv \sqrt{\frac{(2l+1)}{4\pi}\frac{(l-m)!}{(l+m)!}}\, e^{im\phi} P_l^m(\cos\theta),}$$

where the prefactor ensures unit normalization $\left\langle Y_l^m, Y_j^n \right\rangle = \delta_{lj}\delta^{mn}$. (The proof of this is not straightforward.) We note that some texts define $Y_l^m(\theta,\phi)$ with an additional phase factor, $(-)^m$.

EXERCISE Convince yourself of the generality of the above separability argument: partial differential equations $\hat{A}f = 0$, defined by a separable differential operator of n variables, $\hat{A}(x_1, \partial_1, \ldots, x_n, \partial_n) = \sum_i \hat{A}_i(x_i, \partial_i)$, are solved by products, $f(x_1, \ldots, x_n) = \prod_i f_i(x_i)$, where the f_i are solutions of the ordinary differential equations $\hat{A}_i f_i = cf_i$, with a common constant c.

The first few spherical harmonics are given by

$$l = 0: \qquad Y_0^0(\theta,\phi) = \frac{1}{\sqrt{4\pi}},$$

$$l = 1: \qquad Y_1^0(\theta,\phi) = \sqrt{\frac{3}{4\pi}}\cos\theta,$$

$$Y_1^{\pm 1}(\theta,\phi) = \sqrt{\frac{3}{8\pi}}\sin\theta\, e^{\pm i\phi},$$

Legendre functions *(margin note)*

spherical harmonics *(margin note)*

examples of spherical harmonics *(margin note)*

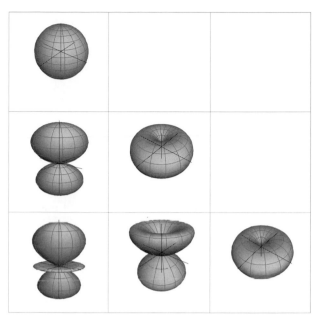

Plot of the first few spherical harmonics, Y_0^0 (first row), Y_1^0, $Y_1 \pm 1$ (second row), $Y_2^0, Y_2^{\pm 1}, Y_2^{\pm 2}$ (third row). The surfaces are represented in the **polar representation**, in which the spherical coordinate points $(r, \theta, \phi) \equiv (|Y_l^m|^2(\theta, \phi), \theta, \phi)$ are plotted as a function of the angles (θ, ϕ), such that the radial coordinate indicates the value of $|Y_l^m|^2$.

$$l = 2: \quad Y_2^0(\theta, \phi) = \sqrt{\tfrac{5}{16\pi}}(3\cos^2\theta - 1),$$

$$Y_2^{\pm 1}(\theta, \phi) = \mp\sqrt{\tfrac{15}{8\pi}}\cos\theta\,\sin\theta\,e^{\pm i\phi},$$

$$Y_2^{\pm 2}(\theta, \phi) = \mp\sqrt{\tfrac{15}{32\pi}}\sin^2\theta\,e^{\pm 2i\phi}. \tag{L248}$$

We notice a trend of increasingly complex dependence on the angular arguments. Fig. L19 shows a graphic representation of these functions. The resemblance of these figures to the atomic orbitals familiar from chemistry classes is not accidental. The spherical harmonics play a central role in the solution of the Schrödinger equations for atoms and their angular dependence reflects the shell structure of atoms and molecules. These connections are typically discussed in lecture courses on quantum mechanics.

The spherical harmonics are the complete set of eigenfunctions of a Hermitian differential operator acting on functions $f \in L^2(S^2)$ on the sphere. From linear algebra we know that these functions define a *function basis on the sphere*, hence one may expand functions in spherical harmonics as

$$f(\theta, \phi) = \sum_{l=0}^{\infty} \sum_{m=-l}^{l} a_l^m\, Y_l^m(\theta, \phi),$$

$$a_l^m = \langle Y_l^m, f \rangle = \int_0^\pi \sin\theta\,d\theta \int_0^{2\pi} d\phi\,\overline{Y_l^m(\theta, \phi)}\,f(\theta, \phi).$$

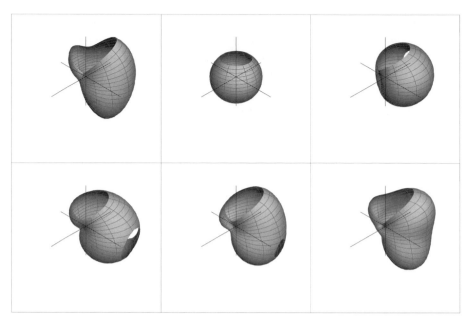

Fig. L20   First panel: a randomly generated positive function f on the sphere plotted in the polar representation, $(f(\theta, \phi), \theta, \phi)$. Remaining panels: expansion of this function in terms of spherical harmonics up to level $l = 4$. For visual clarity the plots of all functions are limited to an angular window $\theta \in [0.6, \pi - 0.6]$.

The orthonormality of the spherical harmonics, $\langle Y_l^m, Y_j^n \rangle = \delta_{lj} \delta^{mn}$, implies that the expansion coefficients are obtained by the straightforward computation of scalar products, as indicated on the right.

Functions of one variable can often be approximated in terms of only a few Fourier harmonics. Similarly, low-order spherical harmonics expansions are generally sufficient to obtain good descriptions of functions on the sphere (provided the latter do not exhibit rapid variations). For an illustration of this point, consider the randomly generated function $f(\theta, \phi)$ shown in the top left panel of Fig. L20. The remaining panels show approximations of this function by linear combinations of the first few spherical harmonics, terminating at $l = 0$ (second panel) up to $l = 4$ (last panel). The $l = 4$ approximation already does a rather good job at describing the function through the 24 expansion coefficients determining an $l = 4$ expansion. What the example illustrates is that the spherical harmonics are the "Fourier modes of the sphere" and that they play an equally useful role in the description of functions with angular variation.

L9.4 Function spaces with unbounded support

We conclude this chapter with some remarks on function spaces with *unbounded* support, for example the space $X = L^2(\mathbb{R})$ of square integrable functions on the real axis. Most of the concepts developed in previous sections carry over to this case. For example, as discussed in Section C6.3, Fourier series become Fourier transforms when functions defined on the entire real axis are considered. In doing so, the general structure of the concept remains unchanged, but sums become integrals. The importance one attributes to such differences depend a lot on one's individual perspective.

Rigorous perspective: When we pass to spaces of functions with unbounded support, many complications arise. For example, the eigenfunctions $\psi_k(x) \equiv \exp(ikx)$ of the differential operator $(-i)d_x$ lie *outside X*, which follows from the fact that $|\psi_k|^2 = 1$ is not integrable over the whole real line. This lack of integrability implies that scalar products between the eigenfunctions cannot be taken, unless properly "regularized" (see below). Relatedly, the Fourier index k now becomes a continuous variable, and we need to specify how sums over eigenfunctions turn into integrals. Other elements of linear algebra whose generalization to infinite dimensions is not trivial include traces, determinants, and matrices describing basis changes. However, being physicists, we may ask how severe these complications are from the point of view of a pragmatic perspective.

Pragmatic perspective: In Section C6.3 we have seen that the non-integrability of the Fourier modes ψ_k could be dealt with by introducing convergence factors (see Eq. (C139)). Alternatively, it is often legitimate, and convenient, to consider functions on a large but finite domain of definition, $(-L/2, L/2)$, and send $L \to \infty$ in the end. (In practice this means making L larger than any other length scale of the physics problem at hand.) As long as L remains finite, the concepts discussed in previous sections remain applicable. Provided nothing dangerous happens as the limit $L \to \infty$ is taken, the case of an unbounded integration domain is then effectively under control. Pragmatic strategies of this sort usually work in physics. However, there are important exceptions to the rule. For example, the quantum mechanics of particles moving at relativistic velocities is described by Dirac (differential) operators. Such operators show poor convergence behavior and their 'regularization' is a delicate subject. These issues are discussed in lecture courses on quantum field theory at a late stage of the physics curriculum.

L9.5   Summary and outlook

In this chapter we extended the concepts of linear algebra to function spaces. This led to a synthesis of linear algebra and calculus which is of key importance to various areas of physics, and in particular to quantum mechanics. Whereas the structures of this framework are of (linear) algebraic nature, concrete operations are done by calculus. For example, the structure of a Fourier series reflects the completeness of Fourier modes in a function space, while the inner products required to obtain the expansion coefficients are computed as integrals. This job division between linear algebra and calculus is powerful and can be a great aid in rationalizing mathematical structures which otherwise may remain obscure. For example, the completeness relations essential to Fourier analysis are not easy to understand before they are interpreted as "matrix relations" connecting function spaces.

As mentioned above, these concepts take center stage in quantum mechanics. How strongly they are emphasized in teaching depends on the details of the curriculum, and on the preferences of the lecturer. However, experience shows that the principles of linear algebra generally facilitate the understanding of calculus in quantum mechanics. Learning to look at functions from this perspective is well-invested time which will certainly pay dividends at later stages of the physics curriculum.

L10 Multilinear algebra

REMARK In this chapter, vectors, $\mathbf{v}$, matrices, A, and other objects of linear algebra will be discussed from a unified perspective. It no longer makes sense to distinguish vectors by a boldface notation and we will denote them as $v \in V$ throughout. Unless mentioned otherwise we will work with real vector spaces. Throughout this chapter $a, a', a_1, \ldots \in \mathbb{R}$ are scalar coefficients. If they appear, the condition $a \in \mathbb{R}$ is implied.

This chapter not only elucidates fundamental structures of linear algebra but also introduces various new concepts. What makes its reading somewhat dry is that these cannot really be convincingly motivated nor applied within the restricted framework of this chapter. However, they will take center stage in Chapters V4 to V6, on differential geometry, which deals with the synthesis of linear algebra, calculus and geometry. In view of this application gap, this chapter may be skipped at first reading and studied at a later stage as an (essential) preparation for the discussion of differential geometry. Alternatively, one may read its first sections, up to and including Section L10.3, where the meaning of several operations which so far have been purely formal – the matrix transpose, the raising and lowering of indices, etc. – is explained. The final sections can then be read at a later stage, before entering Chapter V4.

Linear algebra is the mathematical discipline of objects that satisfy certain linearity criteria. So far, we have seen two representatives of these, vectors and matrices. However, there are other classes of "linear objects" and the collective term for all of them – including vectors and matrices – is *tensors*. Much as vectors can be generalized to vector *fields*, one may define tensor fields. Tensors and tensor fields play an important role in various fields of physics including general relativity, hydrodynamics, quantum information and others.

In this chapter we introduce tensor calculus as a unifying approach to linear algebra. The extension to tensor fields, along with an introductory discussion of physical applications, is the subject of Chapters V4 to V7.

L10.1 Direct sum and direct product of vector spaces

Starting from an n-dimensional vector space, V, multilinear algebra builds vector spaces of richer structure by hierarchical constructions. This is achieved starting from two basic constructors, the direct sum and the direct product of vector spaces.

Direct sum

direct sum of vector spaes

Consider two real vector spaces, U and V, of dimension m and n, respectively. Their **direct sum**, $U \oplus V$, is defined as the set of ordered pairs of elements of U and V,

$$U \oplus V \equiv \{(u,v) \mid u \in U, v \in V\}. \tag{L249}$$

For these pairs, addition and scalar multiplication are defined as

$$(u,v) + (u',v') = (u + u', v + v'),$$

$$a(u,v) = (au, av),$$

making $U \oplus V$ a real vector space. Given bases $\{e_i\}$ and $\{f_j\}$ of U and V, respectively, a general element of $U \oplus V$ is expanded as $\sum_i (e_i, 0) u^i + \sum_j (0, f_j) v^j$. This shows that the $m + n$ vectors $(e_i, 0)$ and $(0, f_j)$ form a basis of $U \oplus V$ and that $\dim(U \oplus V) = m + n$. Any element of $U \oplus V$ can be decomposed as $(u,v) = (u,0) + (0,v)$ into a contribution of U and V. In this way, the sum $U \oplus V$ contains U and V as natural subspaces. For example, $\mathbb{R}^3 = \mathbb{R}^2 \oplus \mathbb{R}$ is the direct sum of $\mathbb{R}^2$ and $\mathbb{R}$, see the figure. A component representation of vectors in $U \oplus V$ is obtained by concatenating the component vectors of U and V. In the example of $\mathbb{R}^3 = \mathbb{R}^2 \oplus \mathbb{R}$, two-dimensional vectors, $(a,b)^\mathrm{T} \in \mathbb{R}^2$, and one-dimensional vectors, $c \in \mathbb{R}$, are concatenated to obtain a component representation, $(a,b,c)^\mathrm{T}$, of $\mathbb{R}^3$. Likewise, the basis vectors $e_1 = (1,0)^\mathrm{T}$ and $e_2 = (0,1)^\mathrm{T}$ of $\mathbb{R}^2$ and that, $f_1 = 1$, of $\mathbb{R}$ yield the three basis vectors of $\mathbb{R}^3$, $(e_1, 0) = (1,0,0)^\mathrm{T}$, $(e_2, 0) = (0,1,0)^\mathrm{T}$ and $(0, f_1) = (0,0,1)^\mathrm{T}$.

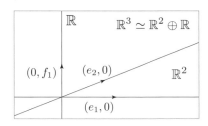

The construction can be iterated to yield direct sums of higher order. For example, given three vector spaces, U, V, W, with bases $\{e_i\}, \{f_j\}, \{g_k\}$, the direct sum $U \oplus V \oplus W$ is obtained as $(U \oplus V) \oplus W = U \oplus (V \oplus W) \equiv U \oplus V \oplus W$ (why is this construction associative?), with basis vectors $(e_i, 0, 0), (0, f_j, 0)$ and $(0, 0, g_k)$. In this way, the standard vector space $\mathbb{R}^n$ may be considered as the direct sum of n copies of $\mathbb{R}$.

Tensor product

tensor product of vector spaces

Besides the direct sum, there exists a second option to build a vector space from two constituent spaces U and V, the **direct product** or **tensor product**, $U \otimes V$ (Latin: *tendo* – I stretch, span). This space is defined as the set of formal linear combinations, $a_{11} u_1 \otimes v_1 + a_{12} u_1 \otimes v_2 + a_{21} u_2 \otimes v_1 + a_{22} u_2 \otimes v_2 + \cdots$, of pairs $u \otimes v$, $u \in U, v \in V$. Within this set the identifications

$$(u + u') \otimes v = u \otimes v + u' \otimes v,$$

$$u \otimes (v + v') = u \otimes v + u \otimes v',$$

$$(au) \otimes v = u \otimes (av) \equiv a(u \otimes v)$$

are declared. These rules define addition and scalar multiplication in $U \otimes V$. Given bases e_i and f_j of U and V, they lead to the expansion $u \otimes v = \sum_{ij} u^i v^j e_i \otimes f_j$, for example $(e_1 + 3e_2) \otimes (5f_4 + 2f_5) = 5e_1 \otimes f_4 + 2e_1 \otimes f_5 + 15e_2 \otimes f_4 + 6e_2 \otimes f_5$. This decomposition may be applied to each term in the sum, so that general elements of $U \otimes V$ afford a representation as $\sum_{ij} a^{ij} e_i \otimes f_j$:

$$U \otimes V = \left\{ \sum_{ij} a^{ij} e_i \otimes f_j \middle| a^{ij} \in \mathbb{R} \right\}. \tag{L250}$$

We conclude that $\{e_i \otimes f_j\}$ defines a basis of $U \otimes V$ and that the dimension of the latter is mn. However, unlike with the direct sum, there is no natural component representation of $U \otimes V$. For example, for $U = V = \mathbb{R}^3$, elements of the nine-dimensional space $U \otimes V$ may be represented as nine-component vectors, however, the connection between this representation and the component vectors of the spaces U and V is not particularly transparent. There is also no intuitive graphical representation of $U \otimes V$, even for low-dimensional U and V; it is generally preferable to work with linear combinations of basis vectors, as in Eq. (L250). Finally, notice – an important but easily forgotten fact – that not every element of $U \otimes V$ can be represented as a product $u \otimes v$. For example, $e_1 \otimes f_2 + e_2 \otimes f_1$ cannot.

As with the direct sum, the tensor product can be iterated to build tensor products of higher order: given three vector spaces, U, V, W, with bases $\{e_i\}, \{f_j\}, \{g_k\}$, the tensor product $U \otimes V \otimes W$ is obtained as $(U \otimes V) \otimes W = U \otimes (V \otimes W) \equiv U \otimes V \otimes W$ (why is the product operation associative?). A basis is provided by the tensor products of basis vectors, $\{e_i \otimes f_j \otimes g_k\}$. The extension to products of higher order is obvious. ($\rightarrow$ L10.1.1-2)

The product $U \otimes V$ is a vector space and one may consider *linear maps* acting in it. A general linear transformation is represented by a matrix $\{A^{ij}{}_{kl}\}$ with $(mn)^2$ components and transforms a vector with components a^{kl} to $A^{ij}{}_{kl} a^{kl}$. Within the set of general linear transformations there exist simpler subclasses frequently occurring in practice. An important example is transformations separately acting on the two factors. For $A : U \rightarrow U$ and $B : V \rightarrow V$, these are denoted as $A \otimes B$, and their index structure is given by $(A \otimes B)^{ij}{}_{kl} = A^i{}_k B^j{}_l$. Various examples of such maps will be encountered below.

INFO Tensor spaces play a very important role in *physical applications*, notably in quantum mechanics. To appreciate why, consider the example discussed on p. 123. There, we argued that the state of a quantum particle moving on an N-site ring is described by a vector, $\hat{\psi} \in V \equiv \mathbb{C}^N$. Now suppose that the particle is a physical electron. The electron is an elementary particle with a property called

spin   **spin**. Heuristically, we can think of spin as a compass needle pointing in one of only two possible directions, say up and down. According to the principles of quantum mechanics the two alternatives correspond to the basis states of a two-dimensional vector space, for example spin up $\leftrightarrow \hat{s}_1 \equiv (1, 0)^T$ and spin down $\leftrightarrow \hat{s}_2 \equiv (0, 1)^T$. A general spin configuration of the particle is then described by a two-dimensional vector, $\hat{\chi} \in U \equiv \mathbb{C}^2$. For example, $\hat{\chi} = \frac{1}{\sqrt{3}} \hat{s}_1 + \frac{\sqrt{2}}{\sqrt{3}} \hat{s}_2$ describes a state where "spin up" is realized with probability $(1/\sqrt{3})^2 = 1/3$ and "spin down" with probability $2/3$.

Now consider the situation where the electron, carrying spin, is free to move on the ring. The joint information is contained in states, $\hat{\Psi} \in U \otimes V = \mathbb{C}^2 \otimes \mathbb{C}^N$, living in the tensor product[77] of the spaces U and V of spin and position, respectively. For example, a spin-up electron at site j is

---

[77] Tensor products of complex spaces are defined in analogy to the real case. The details are discussed in lecture courses in quantum mechanics and are not essential for the present discussion.

in state $\hat{s}_1 \otimes \hat{e}_j$. However, the quantum particle may also be in a superposition state, for example an equal-probability superposition, $\frac{1}{\sqrt{2}}\hat{s}_1 \otimes \hat{e}_j + \frac{1}{\sqrt{2}}\hat{s}_2 \otimes \hat{e}_{j'}$, of spin up at j and spin down at j'. A general configuration is described as

$$\sum_i \sum_j c^{ij} \hat{s}_i \otimes \hat{e}_j, \tag{L251}$$

with complex coefficients c^{ij}. The probability of a combined position/spin measurement is given by $|c^{ij}|^2$, where $\sum_{ij} |c^{ij}|^2 = 1$.

According to the general principles discussed on p. 123, physical observables are represented by Hermitian linear maps acting on the tensor product space. This includes transformations separately acting on the spin and position sectors, $A \otimes \mathbb{1}$ and $\mathbb{1} \otimes B$, where A and B are 2×2 and $n \times n$ Hermitian matrices, but also more complicated operations $c_1 A_1 \otimes B_1 + c_2 A_2 \otimes B_2$, where the two factors transform in correlated ways.

The example illustrates how the mathematical structure of tensor products is tailored to the description of composite quantum systems. For further discussion of this point we refer to lecture courses in quantum mechanics.

L10.2 Dual space

dual space Each vector space V has a partner space called its **dual space**, V^*. The dual space is defined as the set of all linear maps w of V into the real numbers,

$$w : V \to \mathbb{R}, \qquad v \mapsto w(v) \equiv wv, \tag{L252}$$

where linearity implies the properties $w(u + v) = wu + wv$ and $w(av) = a\,wv$, and the standard notation $wv \equiv w(v)$ is used. To heuristically understand the "duality" attribute, consider $v = e_j v^j$ expanded in a basis. Any linear map can then be represented as $wv = w_j v^j$ and is described by a set of coefficients w_i. One may consider these as components of a vector, only that they carry covariant, rather than contravariant, indices. The set of all these objects defines the dual space, V^*, which clearly has the same dimension as V. In the following we make the interpretation of V^* as twin vector space of V concrete.

First note that vector addition and multiplication by scalars, making V^* a *vector space*, are naturally defined as $(w + w')(v) = wv + w'v$ and $(aw)(v) = a\,wv$. The elements of V^*

dual vector are called **dual vectors**.[78]

dual basis To understand the "duality" $V \leftrightarrow V^*$, let $\{e_j\}$ be a basis of V and define its **basis**, $\{e^i\}$, (note the upper index) by the condition that

$$\forall j : \quad e^i(e_j) = e^i e_j \equiv \delta^i{}_j. \tag{L253}$$

This fixes the action of e^i on a general vector, $v = e_j v^j$, as $e^i v = e^i(e_j v^j) = e^i e_j v^j = v^i$. The dual vector e^i thus maps v onto its ith component,

$$e^i v = v^i. \tag{L254}$$

---

[78] Alternative denotations include **covectors** or **one-forms**.

A dual vector can be expanded as $w = w_i\, e^i$, and its *co*variant components are given by

$$w_j = we_j. \tag{L255}$$

The action of w on v then yields

$$wv = (w_i e^i)(e_j v^j) = w_i v^i. \tag{L256}$$

Notice the apparent symmetry between vectors and dual vectors. Indeed, V^* is a vector space and one may ask what *its* dual is. Equation (L253) suggests an interpretation of the vector $e_j \in V$ as a map $e_j : V^* \to \mathbb{R}$, $e^i \mapsto e^i e_j = \delta^i{}_j$. In spite of the unusual notation in which the symbol denoting the map, e_j, appears to the right of its argument, e^i, this is a valid definition. It shows that vectors $v \in V$ afford an interpretation of linear maps of V^* into the reals, and that

$$(V^*)^* = V,$$

i.e. the dual of the dual space of V^*, is V itself. The action of general vectors $v \in V$ on dual vectors $w \in V^*$ is again given by Eq. (L256), and this equation can be read both as w-acts-on-v or v-acts-on-w.

This shows in what sense V and V^* are dual to each other. However, in spite of the intimate connection between the spaces, there exists no canonical bijection mapping elements $v \in V$ to elements $w \in V^*$ or vice versa. The construction above, which assigned dual basis vectors e^i to basis vectors e_i, does define a map $V \to V^*$. However, this map requires the prior fixation of a basis, and hence is not canonical. (In a different basis, $\{f_i\}$, the prior basis vector e_i would be mapped onto a dual vector different from e^i; think about this point.) On p. 153 we will explain that a canonical identification $V \leftrightarrow V^*$ does exist if V is an inner product space.

Finally, it is sometimes useful to think about the connection between vector spaces and their duals in a *component representation*, where a basis is fixed and vectors, $v \leftrightarrow (v^1, \ldots, v^n)^T$, are identified as column vectors or $n \times 1$ matrices. Equation (L253) then defines a representation, $w \leftrightarrow (w_1, \ldots, w_n)$, of dual vectors through row vectors (note the absence of the transposition symbol), or $1 \times n$ matrices. In this picture, the pairing $wv = w_i v^i$ assumes the form of a multiplication of a $1 \times n$ matrix with an $n \times 1$ matrix.($\to$ L10.2.1-4)

EXAMPLE Consider $\mathbb{R}^3$ with its standard scalar product. Fix a vector u and define a linear map as $uv \equiv \langle u, v \rangle = u^i v^i$. This map computes the scalar product of the argument vectors with u and satisfies the criteria required for dual vectors. More generally, one may consider an inner product space, (V, g), pick some $u \in V$ and define $uv \equiv \langle u, v \rangle = u^i g_{ij} v^j$. In this way, a linear map is defined for each $u \in V$. In the subsection after next we will see that this observation defines an important bridge between vector spaces and their dual spaces.

Co- and contravariant transformation

Suppose we choose a different basis, $e_j \mapsto e'_j \equiv e_i (T^{-1})^i{}_j$. The corresponding vector components transform as $v^i \mapsto v'^i = T^i{}_j v^j$, so that

$$v = e'_i v'^i = (e_j (T^{-1})^j{}_i)(T^i{}_k v^k) = e_j v^j$$

remains invariant. The dual vectors e'^i associated with the new basis vectors e'_j are defined as $e'^i(e'_j) = \delta^i{}_j$. This condition implies that the expansion of the new dual basis vectors in the old basis reads as $e'^i = T^i{}_j e^j$ (show this!). The components of a dual vector must then transform as $w_j \mapsto w'_j = w_i(T^{-1})^i{}_j$, so as to leave $w = w_i e^i = w'_i e'^i$ invariant.

Observe that the co- or contravariance of an index fixes the *transformation behavior* of the object it refers to. Regardless of whether they represent vectors or coefficients, contravariant objects transform with T and covariant ones with T^{-1}, $(x^i = e^i, v^i)$

general co- and con- travariant transforma- tion

$$ x^i \mapsto x'^i = T^i{}_j x^j, \qquad x_j \mapsto x'_j = x_i(T^{-1})^i{}_j. \tag{L257}$$

This transformation behavior ensures the invariance of all three types of index contractions, $v = e_i v^i$, $w = w_i e^i$, and $wu = w_i v^i$. Although the placement of transformation matrix elements to the right of covariant objects is most natural, one may change the order by writing $x_j \mapsto (T^{-1T})_j{}^i x_i$.

EXAMPLE In $\mathbb{R}^2$ consider the basis $e'_1 = \binom{1}{1}$, $e'_2 = \binom{2}{3}$. The transformation from the standard basis is defined as $e'_j = e_i(T^{-1})^i{}_j$, with $T^{-1} = \binom{1\ 2}{1\ 3}$. By the principles discussed above, the primed dual basis is obtained as $e'^i = T^i{}_j e^j$. Inverting T^{-1} we find $T = \begin{pmatrix} 3 & -2 \\ -1 & 1 \end{pmatrix}$, hence $e'^1 = (3, -2)$ and $e'^2 = (-1, 1)$. Check that these dual vectors satisfy the dual basis criterion (L253). ($\rightarrow$ L10.2.5-6)

INFO In physics there is a tendency to indiscriminately regard objects carrying single indices – forces, velocities, current densities, etc. – as "vectors". However, many of these objects afford a more natural interpretation as dual vectors. Consider the *example of mechanical force*. The force, F, acting on a particle is determined by measuring the work required to move the particle along small displacements. Work, W, is a scalar and displacements, $s \in \mathbb{R}^3$, are three-dimensional vectors.

force as a dual vector

Force is a function *defined* through the relation $F(s) = W$. Since the work required to go along two consecutive small segments, $s + s'$, is additive, $F(s + s') = F(s) + F(s') = W + W'$, this function is linear. In other words, F is a dual vector, $F \in (\mathbb{R}^3)^*$. In a basis, the assignment of force to work reads $W = F_i s^i$, where F_i are the covariant components defining the force dual vector through work measurement.

angular momentum

Another physics example of a dual vector is **angular momentum**, L. Let us describe the rotational motion of a body around a center by a vector ω, where $|\omega|$ quantifies the frequency of the rotation, and $\omega/|\omega|$ the direction of the rotation axis. Then the kinetic energy (a number) stored in the rotational motion can be expressed as $\frac{1}{2}L_i \omega^i$, where L_i are the components of angular momentum. (Consider the motion of a point particle on a circle to convince yourself that this is so.) This shows that L is a map of vectors to numbers, i.e. a dual vector. Other examples of dual vectors in physics include the *electric and magnetic fields*, E and H, and *mechanical momentum*, p. In all these cases, the dual vector identification follows from physical rather than mathematical reasoning. The concept of dual vectors is as intuitive as that of vectors and the all-is-vector culture of physics can be considered an artifact of traditional teaching.

If we accept that forces are more naturally described by dual vectors, the question presents itself of how one may switch between the dual and the direct representation. As mentioned above, the canonical passage between a dual F and a standard vector representing F requires additional structure in the form of an inner product – see the next section. Depending on the context, the necessity to introduce that excess baggage may be harmless, or seriously obscure the natural interpretation of physical quantities.

Metric provides a canonical connection between space and dual space

The fact that the physics community mostly does not distinguish between vectors and dual vectors testifies to the intimate connection between these objects. A canonical identification between them indeed exists for vector spaces with inner products. (In physics texts not distinguishing between vectors and their duals – the vast majority of the literature – an inner product is implicitly assumed and used for the identification.) Formally, this identification

canonical isomorphism is a linear map

$$J : V \to V^*, \qquad v \mapsto J(v),$$

where the dual vector $J(v)$ is defined through the condition

$$J : V \to V^*, \quad v \mapsto J(v),$$
$$\forall u \in V : J(v)u = g(v, u). \tag{L258}$$

Here the second line defines $J(v)$, and g is the inner product of V. This condition defines the value of the dual vector $J(v)$ on $u \in V$ as $g(v, u)$. The map J is a bijection (which we verify below via explicit construction of its inverse), and it is linear in the sense that $J(av + a'v') = aJ(v) + a'J(v')$. It thus defines a vector space *isomorphism*.

The covariant components, $J(v)_i$, in a basis $\{e^i\}$ dual to $\{e_j\}$[79] are obtained as $J(v)_j = $

index lowering $J(v)e_j = g(v, e_j) = v^i g_{ik}(e_j)^k = v^i g_{ij} = v_j$, where in the last equality the index lowering convention (L60) was used. The result,

$$J(v)_j = v^i g_{ij} = v_j, \tag{L259}$$

identifies the previously formal index-lowering operation as a map from vectors to dual vectors:

> The canonical isomorphism j maps vectors, v, with contravariant components, v^i, to dual vectors, $J(v)$, with covariant components, $v_j = v^i g_{ij}$.

Other index-changing operations, too, can now be understood in a more meaningful way. First note that the *inverse of J* is defined through the condition $wu = J(J^{-1}(w))(u)$, where $w \in V^*$ and $u \in V$. The left-hand side equals $w_j u^j$ and the right-hand side $g(J^{-1}(w), u) = J^{-1}(w)^l g_{lj} u^j$. This implies the condition $w_j = J^{-1}(w)^l g_{lj}$, which can be inverted via the

index raising inverse of the metric tensor, g^{ji}, to yield

$$J^{-1}(w)^i = w_j g^{ji} = w^i. \tag{L260}$$

Thus, the inverse isomorphism J^{-1} maps dual vectors, w, with covariant components, w_i, to vectors, $J^{-1}w$, with contravariant components, $w^i = w_j g^{ij}$.

> Index lowering or raising is equivalent to passing from a vector space to its dual vector space or back, in a component language.

[79] Note that the dual basis vector e^i, defined by the condition $e^i e_j = \delta^i_j$, differs from the assignment $J(e_i)$. Rather, $J(e_i) = J(e_i)_l e^l = \delta_i{}^k g_{kl} e^l = g_{il} e^l$. This equals e_i only in the case of an orthonormal basis, $g_{ij} = \delta_{ij}$.

We finally note that the isomorphism J defines a *metric of dual space*. The latter is defined as $g^* : V^* \times V^* \to \mathbb{R}, (w, w') \mapsto g^*(w, w')$ via the condition

$$g^*(w, w') = g(J^{-1}(w), J^{-1}(w')). \tag{L261}$$

Using a basis representation, and defining $(g^*)^{ij} = g^*(e^i, e^j)$, a quick check[80] shows that

$$(g^*)^{ij} = g^{ij}.$$

Thus, the contravariant components of the canonical metric of dual space equal the components of the inverse of the metric tensor. ($\to$ L10.2.7-8)

INFO In *physics*, the connection between a vector space and its dual space is implicitly used when, e.g., the work along a line segment is calculated as the scalar product between the vectors representing the segment and the force. However, in the previous section we argued that force actually is a *dual* vector, with covariant components F_i. The work done along a segment with contravariant components s^i is then given by $W = F_i s^i$. Physics describes force by a vector with components F^i and the work by the scalar product, $W = \langle F, s \rangle = g(F, s) = F^j g_{ji} s^j$. Comparison of the two descriptions shows that $F_i = g_{ij} F^j$. One may reason that the dual vector approach is more natural in that it (i) introduces force via a measurement protocol (see the previous section), and (ii) does not require a scalar product for the computation of work. On the other hand, the usage of a metric required by the all-is-vector approach is mostly harmless. Exceptions include cases where the metric itself plays a key role (such as in the theory of gravity), or cases where it obscures physically important structures (such as in the understanding of topological structures).

L10.3   Tensors

Vectors and dual vectors are the basic elements from which all objects of linear algebra are built. The key to these constructions is the *tensor product* of vector spaces. Of particular interest are tensor products of real vector spaces and their duals. In this section we introduce these product spaces, discuss their properties, and consider a number of examples.

Definition of tensors

Introducing the notation $\otimes^q V \equiv V \otimes \cdots \otimes V$ for the product of q identical spaces we define

$$T^q_{\ p}(V) \equiv (\otimes^q V) \otimes (\otimes^p V^*). \tag{L262}$$

tensor   This is the product of the spaces $\otimes^q V$ and $\otimes^p V^*$, which in turn are products of the basic spaces, V and V^*. Elements $t \in T^q_{\ p}(V)$ are called **tensors** of contravariant degree q and covariant degree p. If a basis $\{e_j\}$ has been chosen, elements $t \in T^q_{\ p}$ can be represented as

---

[80] The left-hand side of Eq. (L261) yields $w_i (g^*)^{ij} w'_j$ and the right-hand side $g(e_k w^k, e_l w'^l) = w^k g_{kl} w'^l = w_i g^{ik} g_{kl} g^{lj} w'_j = w_i g^{ij} w'_j$.

$$t = t^{j_1, \cdots, j_q}{}_{i_1, \cdots, i_p} \, e_{j_1} \otimes \cdots \otimes e_{j_q} \otimes e^{i_1} \otimes \cdots \otimes e^{i_p}, \qquad \text{(L263)}$$

where $t^{j_1, \cdots, j_q}{}_{i_1, \cdots, i_p}$ are the *coefficients of the tensor* (think about this point).

It is useful to interpret *tensors as multilinear maps*: much like a dual vector and a vector are maps of the vector space V and the dual space V^* into the reals, respectively, a general tensor $t \in T^q{}_p$ defines a multilinear[81] map assigning to q dual vectors and p vectors a number,

$$t: \qquad \left(\otimes^q V^*\right) \otimes \left(\otimes^p V\right) \longrightarrow \mathbb{R},$$
$$(w^1, \ldots, w^q, v_1, \ldots, v_p) \longmapsto t(w^1, \ldots, w^q, v_1, \ldots, v_p). \qquad \text{(L264)}$$

The image, $t(w^1, \ldots, w^q, v_1, \ldots, v_p) \in \mathbb{R}$, is obtained by application of the q vectorial factors e_{j_k} in t to the corresponding dual-vector arguments, w^k, and the p dual-vectorial factors e^{i_l} to the corresponding vector arguments v_l,

$$t(w^1, \ldots, w^q; v_1, \ldots, v_p) = t^{j_1, \cdots, j_q}{}_{i_1, \ldots, i_p} \, (e_{j_1} w^1) \ldots (e_{j_q} w^q) \, (e^{i_1} v_1) \ldots (e^{i_p} v_p).$$

For example, a tensor $t = t^{ij}{}_k e_i \otimes e_j \otimes e^k \in T^2{}_1(V)$ acts on its three arguments as $t(w, w', v) = t^{ij}{}_k (e_i w) \otimes (e_j w') \otimes (e^k v) = t^{ij}{}_k w_i w'_j v^k$. The components of a tensor in a given basis are obtained from its action on the basis vectors and the dual basis vectors,

$$t^{j_1, \cdots, j_q}{}_{i_1, \ldots, i_p} = t(e^{j_1}, \ldots, e^{j_q}, e_{i_1}, \ldots, e_{i_p}). \qquad \text{(L265)}$$

Under a *transformation of bases*,[82] $e_j \mapsto e'_j = e_i(T^{-1})^i{}_j$, the coefficients of a tensor transform co- and contravariantly according to their degree. For example, for $t \in T^1{}_2(V)$,

$$t^i{}_{jk} \mapsto t'^i{}_{jk} = T^i{}_{i'} \, t^{i'}{}_{j'k'} \, (T^{-1})^{j'}{}_j (T^{-1})^{k'}{}_k.$$

In an index-free notation, this transformation map is given by $T \otimes T^{-1} \otimes T^{-1}$.

We finally note that a tensor can be applied to an incomplete set of arguments to produce a tensor of lowered rank. For example the application of $t = t^{ij}{}_k e_i \otimes e_j \otimes e^k \in T^2{}_1(V)$ to $(w, ., .)$ ($w \in V^*$, second and third argument left empty) yields $t(w, ., .) = (t^{ij}{}_k w_i) e_j \otimes e^k \in T^1{}_1$, a tensor of lower rank with components $t^{ij}{}_k w_i$. The pairwise summation of indices appearing in these expressions is called an **index contraction** of tensors. For further discussion of such operations, see Section L10.7. ($\rightarrow$ L10.3.1-4)

index contraction

Examples of tensor classes

The tensors of lowest degree, $T^0{}_0 = \mathbb{R}$, are just numbers. Tensor spaces of low order, $p, q = 0, 1, 2$, contain familiar objects of linear algebra, as summarized in the following.

multilinear map
[81] A **multilinear map** is separately linear in each of its arguments, i.e. $t(\ldots av + a'v', \ldots) = at(\ldots v, \ldots) + a't(\ldots v', \ldots)$, $a, a' \in \mathbb{R}$. The mathematics of such maps is the subject of multilinear algebra.
[82] Do not confuse matrix elements of the transformation matrix, $T^i{}_j$, with the denotation of the tensor space, $T^q{}_p = T^q{}_p(V)$.

Tensors of degree $(1, 0)$ *and* $(0, 1)$. The tensors of lowest nontrivial degree are the *vectors*, $T^1_{\ 0} = V$, and *dual vectors*, $T^0_{\ 1} = V^*$, respectively. Equation (L263) shows how these are the building blocks of more complex tensorial structures.

Tensors of degree $(1, 1)$. Tensors of first contra- and covariant degree have the form $A = e_i \otimes e^j A^i_{\ j} \in T^1_{\ 1}(V)$. Up to now, we have considered coefficients of the form $A^i_{\ j}$ as the components of *matrices*. The interpretation of A as a matrix (or linear map $V \to V$) becomes apparent upon application of the tensor to an incomplete set of arguments, $(., v)$, containing a vector and an empty dual vector slot. This yields $A(., v) = e_i A^i_{\ j} v^j \in V$, which is a vector whose components $A^i_{\ j} v^j$ are obtained by application of the matrix $A^i_{\ j}$ to the argument v. In this way, the tensor is identified as a linear map, $V \to V$, $v \mapsto A(., v)$. However, the tensor formulation affords more varied interpretations of elements $A \in T^1_{\ 1}(V)$. For example, we can think of A as a map, $A(w, v) = A^i_{\ j} w_i v^j$, assigning a number to a pair comprising a dual vector and vector. In index-free notation, this reads as wAv, where w is a row vector with covariant components w_i. Or, we let A act on $(w, .)$ with the vectorial argument left open, to obtain the dual vector $A(w, .) = A^i_{\ j} w_i e^j$, which amounts to the action of the matrix A to its left as wA. Depending on the context, all these different views have advantages and illustrate the flexibility of the tensor formulation of linear algebra.

Finally, note that under a basis transformation, the coefficients of the tensor transform as

$$A^i_{\ j} \mapsto A'^i_{\ j} = T^i_{\ i'} A^{i'}_{\ j'} (T^{-1})^{j'}_{\ j}. \tag{L266}$$

In an index-free notation this reads as $A \mapsto A' = TAT^{-1}$, in which we recognize the familiar transformation behavior of matrices.

transpose   INFO   For every tensor $A \in V \otimes V^*$, a **transpose** is defined as $A^T \in V^* \otimes V$, with interchanged order of V and its dual V^*. This object is defined by the condition that $A^T(v, w) = A(w, v)$ for arbitrary arguments. Expanding it as $A^T = (A^T)_j^{\ i} e^j \otimes e_i$, the application to a pair of arguments yields the number $A^T(v, w) = v^j (A^T)_j^{\ i} w_i$. The component representation of the defining condition thus reads $v^j (A^T)_j^{\ i} w_i = w_i A^i_{\ j} v^j$, implying $(A^T)_j^{\ i} = A^i_{\ j}$. In Eq. (L113) this relation served as a formal definition of a transposed matrix.

Tensors of degree $(0, 2)$ *or* $(2, 0)$. Tensors of second covariant degree, $t \in T^0_{\ 2}(V)$, are called **bilinear form**   **bilinear forms**. They define bilinear maps, $t : V \otimes V \to \mathbb{R}$, $(u, v) \mapsto t(u, v)$. A prominent example is the *metric*, g, of a vector space introduced in Section L3.3. There, we defined a general inner product of a vector space, $g(u, v) = u^i g_{ij} v^j$, through a set of coefficients g_{ij}. In tensor language, this is equivalent to the definition of a second-degree covariant tensor, $g = g_{ij} e^i \otimes e^j$. In physics, the coefficients $\{g_{ij}\}$ are often considered as a matrix. This can be a source of confusion, because under a change of basis, $e_j \mapsto e'_j = e_i (T^{-1})^i_{\ j}$, the metric $g_{ij} \mapsto g'_{ij} = g_{i'j'} (T^{-1})^{i'}_{\ i} (T^{-1})^{j'}_{\ j}$ transforms differently from a matrix (see (L266), or the previous subsection).

inertia tensor   INFO   Another example of a bilinear form is the **inertia tensor**, I, of a rigid body. Describing the rotational motion of the body by a rotation vector ω, the kinetic energy stored in the rotation is given by $T = \frac{1}{2} I(\omega, \omega) = \frac{1}{2} \omega^i I_{ij} \omega^j$. Likewise, the components of the angular momentum are obtained as $L_i = I_{ij} \omega^j$. ($\to$ L8.2.5-6)

L10.4 Alternating forms

We next introduce a subclass of tensors which plays an important role in applications and deserves special attention: consider the space, $T^0{}_p$, of multilinear maps of $\otimes^p V$ into the reals. Now define $\Lambda^p(V) \subset T^0{}_p$ as[83]

$$\Lambda^p(V) \equiv \{\phi : \otimes^p V \to \mathbb{R} \mid \phi \text{ multilinear \& alternating}\}, \qquad (\text{L267})$$

where "alternating" means antisymmetry under exchange of any of the vector arguments: **alternating form** $\phi(..., u, ..., v, ...) = -\phi(..., v, ..., u, ...)$. The elements of $\Lambda^p(V)$ are called **(alternating) forms of degree** p, or *p-forms* for short.[84]

EXAMPLE An example of a two-form is $\phi = e^1 \otimes e^2 - e^2 \otimes e^1$. Applied to two vectors u and v it yields the antisymmetric combination $\phi(u, v) = u^1 v^2 - u^2 v^1$. The *triple product* discussed in Chapter L4 is a three-form in $\Lambda^3(\mathbb{R}^3)$: it maps three vectors onto a number and is antisymmetric under exchange of its arguments. The matrix *determinant* can be interpreted as an n-form in $\Lambda^n(\mathbb{R}^n)$: take an $n \times n$ matrix, $A = (v_1, ..., v_n)$, and consider it as a rack of n column vectors, $v_j \in \mathbb{R}^n$. We can then write $\det(A) = \det(v_1, ..., v_n)$, where the latter representation is the image of the multilinear form det, evaluated on n argument vectors v_j. The determinant is linear in each entry and antisymmetric under argument exchange, and this makes it an n-form. Expanded in a tensor basis the determinant assumes the form

$$\det = \sum_{P \in S_n} \text{sgn}(P) e^{P1} \otimes e^{P2} ... e^{Pn}, \qquad (\text{L268})$$

where S_n is the permutation group of n objects and the sum runs over all permutations P.

The expansion of a general p-form in basis one-forms reads as

$$\phi = \sum_{i_1 < i_2 < \cdots < i_p} \phi_{i_1,...,i_p} \sum_{P \in S_p} \text{sgn}(P) e^{iP1} \otimes e^{iP2} ... e^{iPp}. \qquad (\text{L269})$$

Here, P is an element of the group of permutations of p objects, S_p. By definition, the coefficients of the expansion obey the antisymmetrization condition $\phi_{Pi_1,...,Pi_p} = \text{sgn}(P)\phi_{i_1,...,i_p}$. For this reason, the sum over indices in (L269) may be limited to ordered index configurations $i_1 < i_2 < \cdots < i_p$. For example, the $p = 2$ forms in an ($n = 3$)-dimensional space afford the expansion

$$\phi = \phi_{12}(e^1 \otimes e^2 - e^2 \otimes e^1) + \phi_{23}(e^2 \otimes e^3 - e^3 \otimes e^2) + \phi_{31}(e^3 \otimes e^1 - e^1 \otimes e^3),$$

where the antisymmetry $\phi_{ij} = -\phi_{ji}$ is used. However, it is sometimes convenient to avoid the ordering condition and instead sum over all index combinations as

---

[83] The notation where $\Lambda^p(V) \subset T^0{}_p$ carries the degree-index p upstairs is not ideal but standard and we will use it here too.

[84] When using the shorthand nomenclature p-form, be aware that the general bilinear forms discussed earlier need not be alternating. For example, the metric $g(u, v) = g(v, u)$ is a symmetric bilinear form and therefore not a two-form.

$$\phi = \frac{1}{p!} \sum_{i_1,i_2,\dots,i_p} \phi_{i_1,\dots,i_p} \sum_{P \in S_p} \text{sgn}(P) e^{iP1} \otimes e^{iP2} \dots e^{iPp}. \tag{L270}$$

The terms with coinciding indices vanish, and the redundant summation over unordered index pairs is compensated by the prefactor $1/p!$ (think about this point!).[85]

top-form   If $p = n$, there is only one ordered configuration, $(i_1, \dots, i_n) = (1, \dots, n)$. Forms in the space $\Lambda^n(\mathbb{R}^n)$, called **top-forms**, are thus determined by a single number, $\phi_{1,\dots,n}$. Specifically, the application of the n-form with $\phi_{1,\dots,n} = 1$ to a set of n vectors, $v_1, \dots, v_n$, from $\mathbb{R}^n$ yields

$$\phi(v_1, \dots, v_n) = \sum_{P \in S_n} \text{sgn}(P) e^{P1} \otimes \dots e^{Pn}(v_1, \dots, v_n) = \sum_{P \in S_n} \text{sgn}(P)(v_1)^{P1} \dots (v_n)^{Pn}. \tag{L271}$$

One may interpret the n-tuple $A \equiv (v_1, \dots, v_n)$ as a matrix with elements $A^i_{\ j} = (v_j)^i$, to identify the value $\phi(v_1, v_2, \dots, v_n) \equiv \phi(A) = \det(A)$ as the determinant of A (see Eq. (L147)). In two and three dimensions, the top-forms $\det(v_1, v_2)$ and $\det(v_1, v_2, v_3)$ yield the area of the parallelogram or the volume of the parallelepiped defined by the argument vectors, respectively (see Section L6.1). This interpretation generalizes to arbitrary dimensions: in $\mathbb{R}^n$, the top-form (L271) defines the geometric volume spanned by its argument vectors, provided the basis $\{e_i\}$ is orthonormal, $g(e_i, e_j) = \delta_{ij}$. We will return to this point in Section L10.9.

We conclude this section with a summary of the most important *mathematical properties of alternating forms*, all following from the general discussion above.

▷ The sum $\phi + \phi'$ of two alternating forms, $\phi, \phi' \in \Lambda^p(V)$, is again an alternating form, i.e. $\Lambda^p(V)$ is a vector space.

▷ We define $\Lambda^0(V) = \mathbb{R}$. $\Lambda^1(V) = V^*$, the space of one-forms, is the dual vector space.

▷ For $\dim V = n$, $\Lambda^{p>n}(V) = \{0\}$ is the null space, since an n-dimensional vector space does not support alternating p-forms if $p > n$. This is best seen by considering the action of forms $\phi(e_{j_i}, \dots, e_{j_p})$ on basis vectors. If $p > n$, identical basis vectors will appear repeatedly in the argument (because we have only n different ones). However, $\phi(\dots, e_j, \dots, e_j, \dots) = -\phi(\dots, e_j, \dots, e_j, \dots) = 0$ by antisymmetry.

▷ For $\dim V = n$, the dimension of the vector space of alternating p forms is $\dim(\Lambda^p(V)) = \binom{n}{p}$. This follows since the number of ordered p-tuples, $1 \leq i_1 < i_2 < \dots < i_p \leq n$, is $\binom{n}{p}$, and the sum in (L269) extends over as many basis forms.

($\rightarrow$ L10.4.1-2)

L10.5 Visualization of alternating forms

Much as vectors can be visualized as arrows, alternating forms in low-dimensional spaces, $\dim(V) = 1, 2, 3$, too, afford pictorial representations. These visualizations are not so much

[85] For example, for $p = 2$ we have a 2!-fold redundancy, $\sum_{ij} \phi_{ij}(e^i \otimes e^j - e^j \otimes e^i) = \phi_{12}(e^1 \otimes e^2 - e^2 \otimes e^1) + \phi_{21}(e^2 \otimes e^1 - e^1 \otimes e^2) = 2\phi_{12}(e^1 \otimes e^2 - e^2 \otimes e^1)$, since both factors in the product are antisymmetric.

instruments for computations as aids for the intuitive understanding of forms. Alternating p-forms in n-dimensional space are represented through periodically repeated patterns of $(n - p)$-dimensional linear structures, such as lines, planes, volumes, etc. These patterns are designed to be "paired" with p vectors to yield the value of the form on the p argument vectors.

For example, a *one-form in two-dimensional space*, ϕ, is represented by a system of parallel lines of specified slope and inter-line spacing (see Fig. L21(b)). These lines define a pattern of strips (one is shown shaded) tiling the plane. The strips are given a sense of orientation by naming one side "upside", the other "downside". In the figure, this choice is indicated by an orientation arrow, labeled o, pointing in the chosen upward direction. The value of the form on a vector, $\phi(v)$, is now defined graphically as follows: The modulus, $|\phi(v)|$, equals the (generally fractional) number of strips pierced by v (in panel(b) this would be about 2.5). Note that this number is invariant under parallel translation of both the arrow and the strip pattern, as it should be. The sign of $\phi(v)$ is defined to be positive/negative if v is aligned upward/downward relative to o. It is straightforward to verify (do it!) that these rules are compatible with the linearity criteria defining differential forms. The algebraic coefficients, ϕ^i, defining the expansion of the form in a given dual basis, $\phi = \phi_1 e^1 + \phi_2 e^2$, are obtained by graphical evaluation of the form on basis vectors, $\phi^i = \phi(e_i)$.

In a similar manner, a *two-form in two-dimensional space*, ω, is defined by a lattice of tiles (one is shown shaded in Fig. L21(c)) of arbitrary shape, but with a specified number of tiles per unit area (this number can be fractional). An orientation is chosen by distinguishing between anti-clockwise (mathematically positive) and clockwise (negative) orientated forms. The modulus of the form acting on two vectors, $|\omega(u, v)|$, is obtained by counting

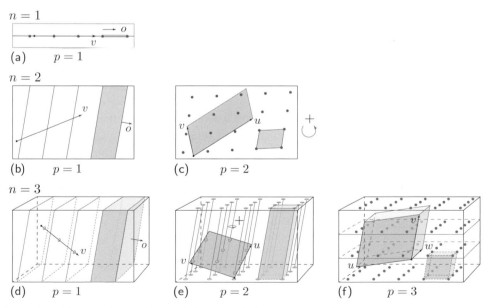

Fig. L21   Visualization of alternating forms in (a) one-, (b, c) two- and (d–f) three-dimensional space. Discussion, see text.

the (generally fractional) number of tiles covered by the parallelogram spanned by u and v (in panel (c), this number would equal ca. 4.5). Thus, $|\omega(u, v)|$ equals the area of the parallelogram spanned by u and v, in units of the tile area. For an anti-clockwise orientated form, $\text{sgn}(\omega(u, v))$ yields a positive sign when the orientation of v relative to u is anti-clockwise (left panel), and a negative sign otherwise (right panel). For a clockwise oriented form the assignment is opposite. These rules are consistent with the linearity and antisymmetry criteria of differential two-forms. The algebraic coefficient, ω_{12}, defining the expansion of the form in a dual basis, $\omega = \omega_{12}e^1 \wedge e^2$, is obtained by graphical evaluation of $\omega(e_1, e_2)$ on basis vectors.

EXERCISE Discuss how a *one-form in one-dimensional space* is defined through a periodic pattern of points on the real line, plus a sense of orientation, see Fig. L21(a). How is the value of the form on a one-dimensional vector computed, and how is its expansion in a basis obtained?

A *one-form in three-dimensional space*, ϕ, is defined in conceptual analogy to the one-form in two-dimensional space, only that the strips are replaced by slabs defined by a system of equi-spaced parallel planes (see Fig. L21(d), where one slab is shown shaded). As in the two-dimensional case its value on a vector, $\phi(v)$, is obtained by determining the number of slabs pierced by v, where the sign follows from the alignment of v relative to that of an orienting direction, o. A *two-form in three-dimensional space*, ϕ, is a lattice of parallel lines of specified direction, density of lines per unit area and orientation (upwards or downwards) along the lines, see Fig. L21(e). The lines define a pattern of parallel rods (one shown shaded), and absolute values, $|\phi(u, v)|$, are obtained by counting the number of parallelogram-shaped rod cross sections intersected by the parallelogram defined by u and v (about 5.5 in the figure). As with the two-forms in two-dimensional space, an orientation is provided by assigning a sense of revolution to the lines. Finally a *three-form in three-dimensional space*, ϕ, is a lattice of points with specified density of points per unit volume, see Fig. L21(f). The points define a pattern of parallelepiped-shaped cells, and $|\phi(u, v, w)|$ is obtained by counting how many of these are contained in the parallelepiped spanned by u, v and w, thus giving its volume in units of the cell volume.

In general, a *p-form in n-dimensional space* is a pattern of identical $(n - p)$-dimensional linear structures.[86] The value of the form is obtained by determining how many of these are covered by the generalized parallelepiped spanned by p vectors. For top-forms, $n - p = 0$, the subunits have finite extent in all n dimensions, which explains why top-forms measure the geometric volume of n-dimensional parallelepipeds.

EXERCISE Think more about the pictures in Fig. L21 and make sure you are comfortable with the rules of assignment, the senses of orientation, the number of coefficients required to uniquely specify a form, etc. Explain how the fractional counting of lines or grid-areas implied by the procedures above can be approximated through integer counting by reducing the separation between the points,

---

[86] An $(n-p)$-dimensional structure has infinite extent in $n-p$ dimensions. For example, strips in two-dimensional space or rods in three-dimensional space are both one-dimensional $(n - p = 2 - 1 = 3 - 2 = 1)$.

lines or planes used to represent the various forms, respectively. Think how a graphical procedure in terms of an infinitely dense pattern of lines or planes should be designed.

L10.6 Wedge product

Alternating forms can be multiplied by an operation called the wedge product to yield alternating forms of higher degree. The wedge product of two one-forms, is a two-form defined as $(\phi \wedge \psi)(v, w) = \phi(v)\psi(w) - \phi(w)\psi(v)$. It acts antisymmetrically, $(\phi \wedge \psi)(v, w) = -(\phi \wedge \psi)(w, v)$, on two arguments and thus defines a two-form. More generally, the **wedge product** or **(exterior product)** of a p-form and a q-form is a $(p + q)$-form, defined as

wedge product

$$\wedge : \Lambda^p(V) \otimes \Lambda^q(V) \to \Lambda^{p+q}(V), \qquad (\phi, \psi) \mapsto \phi \wedge \psi,$$

$$(\phi \wedge \psi)(v_1, \ldots, v_{p+q}) \equiv \frac{1}{p!q!} \sum_{P \in S_{p+q}} \operatorname{sgn} P \, \phi(v_{P(1)}, \ldots, v_{P(p)}) \psi(v_{P(p+1)}, \ldots, v_{P(p+q)}).$$

(L272)

Here, S_{p+q} is the permutation group of $p + q$ objects and the sum runs over all permutations P. The wedge product of two forms thus feeds the $p + q$ arguments in all possible combinations to the factor forms, where $\operatorname{sgn}(P)$ guarantees an overall antisymmetric expression, and the prefactors $1/(p!q!)$ account for redundant argument permutations yielding identical results (why?). For $p = 0$, ϕ is a number and we define $\phi \wedge \psi(v) = \phi \psi(v)$, analogously for $q = 0$. Important *properties of the wedge product* include ($\phi \in \Lambda^p(V), \psi \in \Lambda^q(V), \lambda \in \Lambda^r(V)$):

▷ **bilinearity**, $(\phi_1 + \phi_2) \wedge \psi = \phi_1 \wedge \psi + \phi_2 \wedge \psi$ and $(a\phi) \wedge \psi = a(\phi \wedge \psi)$.
▷ **associativity**, $\phi \wedge (\psi \wedge \lambda) = (\phi \wedge \psi) \wedge \lambda \equiv \phi \wedge \psi \wedge \lambda$.
▷ **graded commutativity**, $\phi \wedge \psi = (-)^{pq} \psi \wedge \phi$.

The change of the degree of forms by the wedge product is motivation to define a space containing all spaces of fixed degree p, with $0 \le p \le n$, as subspaces via the direct sum

$$\Lambda(V) \equiv \bigoplus_{p=0}^{n} \Lambda^p(V). \tag{L273}$$

This vector space has dimension $\dim(\Lambda(V)) = \sum_{p=0}^{n} \dim(\Lambda^p(V)) = \sum_{p=0}^{n} \binom{n}{p} = 2^n$, where in the last equality we used the binomial formula. The most important feature of $\Lambda(V)$ is that it is a vector space endowed with a product operation, thus forming an *algebra* (see the definition on p. 71). The product, $\wedge$, of the algebra $(\Lambda(V), \wedge)$ is associative and antisymmetric, which is the defining property of a **Grassmann algebra**.

Grassmann algebra

A natural basis of $\Lambda^p(V)$ is given by the set of forms,

$$\bigwedge_{a=1}^{p} e^{i_a} \equiv e^{i_1} \wedge \cdots \wedge e^{i_p}, \qquad 1 \le i_1 < \cdots < i_p \le n. \tag{L274}$$

To see this, notice (i) that these forms are alternating and therefore elements of $\Lambda^p(V)$, (ii) that they are linearly independent, and (iii) that there are $\binom{n}{p}$ of them. For example, for $n = 3$ and $p = 2$, we have the $3 = \binom{3}{2}$ linearly independent forms, $e^1 \wedge e^2, e^2 \wedge e^3, e^3 \wedge e^1$.

The three criteria (i)–(iii) guarantee the basis property. Any p-form can be represented in the above basis as

$$\phi = \sum_{i_1 < \cdots < i_p} \phi_{i_1,\ldots,i_p} \bigwedge_{a=1}^{p} e^{i_a}. \tag{L275}$$

The coefficients $\phi_{i_1,\ldots,i_p} \in \mathbb{R}$ are given by $\phi_{i_1,\ldots,i_p} = \phi(e_{i_1},\ldots,e_{i_p})$ and therefore are antisymmetric under exchange of index arguments (for example $\phi_{123} = -\phi_{213}$). Alternatively ϕ may be represented by an unrestricted sum with a compensating factor of $1/p!$ (see (L270)):[87]

$$\phi = \frac{1}{p!} \sum_{i_1,\ldots,i_p} \phi_{i_1,\ldots,i_p} \bigwedge_{a=1}^{p} e^{i_a}. \tag{L276}$$

To illustrate Eq. (L275), we note that the zero-, ..., three-forms in $\mathbb{R}^3$ can be represented as

$$p = 0: \quad \phi = \phi_1,$$
$$p = 1: \quad \phi = \phi_1 e^1 + \phi_2 e^2 + \phi_3 e^3,$$
$$p = 2: \quad \phi = \phi_{12} e^1 \wedge e^2 + \phi_{23} e^2 \wedge e^3 + \phi_{31} e^3 \wedge e^1,$$
$$p = 3: \quad \phi = \phi_{123} e^1 \wedge e^2 \wedge e^3. \tag{L277}$$

Notice that there are $8 = 2^3$ independent coefficients in all, and that the one- and two-forms are described by three coefficients each. The formulas above illustrate the importance of the wedge product: it allows us to build forms of arbitrary complexity from the much simpler one-forms. ($\rightarrow$ L10.6.1-4)

L10.7 Inner derivative

One can think of a p-form as a machine with p slots into which vectors are fed as arguments. It is sometimes useful to feed a p-form only one vector, v, to produce a form of degree lower by one, $p - 1$. The corresponding map, denoted by $i_v : \Lambda^p(V) \rightarrow \Lambda^{p-1}(V)$, is called the **inner derivative** and defined by its action on an arbitrary $(p - 1)$-tuple, $(v_1,\ldots,v_{p-1})$, as

inner
derivative

$$(i_v\phi)(v_1,\ldots,v_{p-1}) \equiv \phi(v, v_1,\ldots,v_{p-1}). \tag{L278}$$

Here v, indicated as a subscript on the left, acts as an additional argument to be supplied to the p-form ϕ. The components of $i_v\phi$ are obtained by contraction of one of the components of ϕ with those of v (as follows from Eq. (L263)),

$$(i_v\phi)_{i_1,\ldots,i_{p-1}} = v^i \phi_{i,i_1,\ldots,i_{p-1}}.$$

[87] For example, for $p = 2$ we have $\frac{1}{2!} \sum_{ij} \phi_{ij} e^i \wedge e^j = \frac{1}{2}(\phi_{12} e^1 \wedge e^2 + \phi_{21} e^2 \wedge e^1) = \phi_{12} e^1 \wedge e^2$, see footnote 85.

Notice that in Eq. (L278) the seemingly ad hoc choice to feed v into the first argument slot of ϕ is inconsequential: due to the antisymmetry of ϕ, we have, e.g., $v^i \phi_{i,i_1,\ldots,i_{p-1}} = -v^i \phi_{i_1,i,\ldots,i_{p-1}}$, i.e. the contracted index can be permuted at the expense of, at most, a minus sign.

The definition of the inner derivative implies the following properties.

▷ i_v is a linear map, $i_v(\phi + \phi') = i_v\phi + i_v\phi'$.

▷ It is also linear in its "parametric argument", $i_{v+w} = i_v + i_w$.

graded product rule ▷ i_v obeys the **graded product rule**:

$$i_v(\phi \wedge \psi) = (i_v\phi) \wedge \psi + (-)^p \phi \wedge (i_v\psi), \qquad \phi \in \Lambda^p(V),\ \psi \in \Lambda^q(V). \qquad \text{(L279)}$$

This resembles the product rule of differentiation, modified by an extra sign $(-1)^p$, and motivates the terminology "inner derivative".

▷ The inner derivative is antisymmetric in the sense that $i_u \circ i_v = -i_v \circ i_u$. In particular, $(i_v)^2 = 0$.

EXERCISE Gain familiarity with the *Leibniz rule* by computing the components of the inner derivative of a simple form, for example $i_v(e^1 \wedge e^2)$.

Let us illustrate the inner derivative on the *example* of the three-form $\phi = e^1 \wedge e^2 \wedge e^3$ in $\mathbb{R}^3$. A quick calculation shows that ($\to$ L10.7.1-2)

$$i_u(\phi) = u^1 e^2 \wedge e^3 + u^2 e^3 \wedge e^1 + u^3 e^1 \wedge e^2,$$
$$i_v i_u(\phi) = (u^2 v^3 - u^3 v^2)e^1 + (u^3 v^1 - u^1 v^3)e^2 + (u^1 v^2 - u^2 v^1)e^3,$$
$$i_w i_v i_u(\phi) = \det(u, v, w).$$

INFO The second of these lines contains an interesting message. The components of the one-form $i_v i_u(\phi)$ are those of the *vector product* $u \times v$. This is in line with our earlier observation that the vector product is not a real vector – it does not transform as a vector under linear transformation, but, as we now understand, as a one-form. Recalling Eq. (L277), we note that in $d = 3$ there are three different objects characterized by three components: vectors $v = e_1 v^1 + e_2 v^2 + e_3 v^3$, one-forms $\phi = \phi_1 e^1 + \phi_2 e^2 + \phi_3 e^3$, and two-forms $\omega = \omega_{12} e^1 \wedge e^2 + \omega_{23} e^2 \wedge e^3 + \omega_{31} e^3 \wedge e^1$. In physics, they are all indiscriminately treated as vectors. Only if the non-vectorial transformation behavior of the forms-in-disguise becomes too apparent is terminology like "pseudo-vector" used instead. While it would be better to accept forms as what they are, old habits are hard to change and it may take a while before forms become part of the physics mainstream culture. We will return to the discussion of this point in Section V6.3 after the form-analogue of vector fields has been introduced.

L10.8 Pullback

Given a linear map, $F : V \to U$, between two vector spaces, a form, $\phi \in \Lambda^p(U)$, defined on U may be "pulled back" by F to become a form, $F^*(\phi) \in \Lambda^p(V)$, defined on V. The **pullback** operation F^*, mapping forms on U to forms on V, is defined as

$$F^* : \Lambda^p(U) \to \Lambda^p(V), \qquad \phi \mapsto F^*\phi,$$
$$(F^*\phi)(v_1, \ldots, v_p) \equiv \phi(Fv_1, \ldots, Fv_p). \tag{L280}$$

The idea behind this definition is illustrated in the figure on the example of the pullback of one-forms between two-dimensional vector spaces: the action of $F^*\phi$ on vectors $v \in V$ is defined by the action of ϕ on their image vectors, $Fv \in U$, as $(F^*\phi)(v) \equiv \phi(Fv)$. (The figure uses the visualization discussed in Section L10.5, where the value of a one-form on two-dimensional vectors is obtained by counting line-crossings.) In the present context, it is useful to think of $F : V \to U, v \mapsto Fv$

pushforward   as a map describing a **pushforward**[88] of vectors from V to U. The pullback pulls one-forms in the reverse direction, $F^* : \Lambda^1(U) \to \Lambda^1(V), \phi \to F^*\phi$. Eq. (L280) defines the extension of the pushforward–pullback duality to forms of higher degree. Notice that the pullback operation imposes no conditions on F (besides linearity), nor on the dimensionality of V and U.

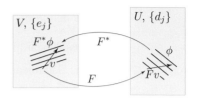

Let $\{e_j\}$ and $\{d_i\}$ be bases of V and U, respectively. The map F is then represented by a component matrix, $\{F^i_j\}$, as $Fe_j = d_i F^i_j$. In the same way, the pullback operation on the dual basis vector, $d^i \in \Lambda^1(U)$, is described as $F^*d^i = (F^*)^i_j e^j$. The matrix $(F^*)^i_j$ representing the map is identified from the definition of pullback by evaluating this equation on a basis vector of V: on the left we obtain $(F^*d^i)e_j \overset{(\text{L280})}{=} d^i(Fe_j) = d^i(d_k F^k_j) = F^i_j$, and on the right just $(F^*)^i_j$. This shows that $(F^*)^i_j = F^i_j$, or

$$F^*d^i = F^i_j e^j. \tag{L281}$$

Notice how this formula parallels Eq. (L257) for a contravariant basis change. In the particular case where $U = V$ and F *is* a change of bases, the pullback formula indeed describes the associated change of the dual basis vectors (think about this point).

The action of F^* on forms of higher degree follows from the following general properties of the pullback operation:

▷ F^* is linear, $F^*(\phi + \psi) = F^*\phi + F^*\psi$.

▷ F^* acts on all factors of wedge products, $F^*(\phi \wedge \psi) = (F^*\phi) \wedge (F^*\psi)$.

▷   Pullback reverses the order of compositions, $(F \circ G)^* = G^* \circ F^*$.

All three rules are immediate consequences of the definition, and the second may be applied to compute the pullback of general forms, as given by Eq. (L275): iterated application of the second identity to the expansion $\phi = \frac{1}{p!} \sum \phi_{i_1,\ldots,i_p} \bigwedge_{a=1}^{p} d^{i_a}$ of a form $\phi \in \Lambda^p(U)$ leads to $F^*\phi = \frac{1}{p!} \sum \phi_{i_1,\ldots,i_p} \bigwedge_{a=1}^{p} (F^*d^{i_a})$. The pullbacks of the individual

---

[88]   The operation is commonly denoted as "pushforward", although "push-forward" or "push forward" would seem to be more natural ways of writing.

one-forms are obtained from Eq. (L281) and we obtain ($\rightarrow$ L10.8.1-2)

$$F^*\phi = \frac{1}{p!} \sum_{i_1,\dots,i_p} \phi_{i_1,\dots,i_p} \bigwedge_{a=1}^{p} F^{i_a}{}_{j_a} d^{j_a}.$$ (L282)

Later on, in Chapters V5 and V6, we will understand that this operation is implicitly used in many routine operations of physics calculus, notably in the manipulation of integrals.

L10.9 Metric structures

We have seen that a metric provides a canonical passage between a vector space and its dual. In this final section, we introduce various other elements of multilinear algebra owing their existence to a metric. As with many other concepts introduced in this chapter, they will "come to life" only later when we combine linear algebra and calculus in Part V. At the same time, it may be interesting to see in its own right how much structure is added to a tensor space when an inner product enters the stage.

Canonical volume form

In Eq. (L181) we defined the volume spanned by three basis vectors, e_i, e_j, e_k, in $\mathbb{R}^3$ through the modulus of the triple product $(e_i \times e_j) \cdot e_k = \sqrt{g}\,\epsilon_{ijk}$, where $g \equiv \det(g)$ denotes the determinant, (L177), of the metric tensor, assumed positive definite. The volume spanned by three general vectors is then defined by the modulus of $(u \times v) \cdot w = \sqrt{g}\,\epsilon_{ijk} u^i v^j w^k$. The structure of this expression suggests the definition of the three-from $\omega = \sqrt{g}\,e^1 \wedge e^2 \wedge e^3$. This form is defined such that $\omega(u, v, w)$ equals $(u \times v) \cdot w$.

Unlike with the triple product, no reference to structures specific to three dimensions is made in this expression. This suggests the definition of the **canonical volume form**,

volume form

$$\omega \equiv \sqrt{|g|}e^1 \wedge \cdots \wedge e^n,$$ (L283)

of an n-dimensional vector space with metric g (not necessarily positive definite). It defines the volume spanned by n vectors as

$$\mathrm{Vol}(v_1, \dots, v_n) = |\omega(v_1, \dots, v_n)| = \sqrt{|g|}\,|\epsilon_{i_1,\dots,i_n}(v_1)^{i_1}\dots(v_n)^{i_n}|$$
$$= \sqrt{|g|}\,|\det(v_1, \dots, v_n)|.$$ (L284)

This definition is designed to ensure invariance under basis transformations. Let us verify this. Under $e_j \mapsto e'_j = e_i(T^{-1})^i{}_j$, the wedge product of one-forms is mapped to

$$\bigwedge_{i=1}^{n} e'^i = \bigwedge_{i=1}^{n}(T^i{}_{j_i} e^{j_i}) = T^1{}_{j_1}\dots T^n{}_{j_n}\,\epsilon^{j_1,\dots,j_n}\bigwedge_{j=1}^{n} e^j = \det(T)\bigwedge_{j=1}^{n} e^j.$$

The second step brought the forms into ascending order, using an ϵ-symbol to track signs.

Moreover the metric tensor changes as $g_{ij} \mapsto g'_{ij} = g_{i'j'}(T^{-1})^{i'}{}_{i}(T^{-1})^{j'}{}_{j}$, implying $\det(g') = \det(g)\det(T)^{-2}$, or $\sqrt{|g'|} = \sqrt{|g|}\,|\det(T)|^{-1}$. Under orientation-preserving transformations with $\det(T) > 0$ (see Eq. (L179)), we thus have $\sqrt{|g'|}\bigwedge_{i=1}^{n} e'^{i} = \sqrt{|g|}\bigwedge_{j=1}^{n} e^{j}$. The volume form preserves its form under coordinate transformations, and hence is canonical.

EXERCISE Consider the volume form in the two-dimensional case, $n = 2$, and show that it yields the geometric area spanned by two vectors.

The volume form is key to the **covariant definition of the vector product**, stated in Eq. (L183) in coordinates. Assume a *positive definite metric*. The vector product $u \times v$ is defined as the vector whose scalar product with any other vector, w, i.e. $(u \times v) \cdot w \equiv g(u \times v, w)$, equals the volume, $\omega(u, v, w)$, spanned by the three vectors if they form a right-handed system. For a left-handed system, the sign changes. This identity can be written as $g(u \times v, w) \stackrel{\text{(L258)}}{=} (J(u \times v))w \equiv \omega(u, v, w) = (i_v i_u \omega)w$, where the definition (L278) of the inner derivative was used. Comparison yields the one-form $J(u \times v) = i_v i_u \omega$, and its dual is the desired vector,

$$u \times v = J^{-1} i_v i_u \omega. \tag{L285}$$

Using $i_v i_u \omega = \sqrt{g}\,\epsilon_{ijk} u^i v^j e^k$, we find $u \times v = \sqrt{g}\,u^i v^j \epsilon_{ijl} g^{lk} e_k$, in agreement with the earlier definition, Eq. (L183). The fact that (L285) makes reference to an orientation (via the definition of ω) indicates that, like Eq. (L183), it does not define a vector of genuine contravariant transformation under orientation changing maps (like space reflection).

Hodge star

The spaces of p-forms and $(n-p)$-forms, $\Lambda^p(V)$ and $\Lambda^{n-p}(V)$, are of the same dimension, $\binom{n}{p}$. This suggests that there might be a linear bijection, $\Lambda^p(V) \to \Lambda^{n-p}(V)$, assigning p-forms to $(n-p)$-forms. For a given basis, it is straightforward to construct bijections assigning to the basis p-forms of $\Lambda^p(V)$ basis $(n-p)$-forms of $\Lambda^{n-p}(V)$. For example, for $n = 3$ one may consider $e^1 \leftrightarrow e^2 \wedge e^3, e^2 \leftrightarrow e^3 \wedge e^1, e^3 \leftrightarrow e^1 \wedge e^2$, and $1 \leftrightarrow e^1 \wedge e^2 \wedge e^3$, where the constant 1 defines a zero-form. However, as with maps between vectors and dual vectors, the reference of this map to a basis makes it not canonical.

In the following we apply a metric to define a canonical bijection, $* : \Lambda^p(V) \to$

Hodge star $\Lambda^{n-p}(V)$, known as the **Hodge star**. This map is designed such that for orthonormal bases it reduces to the one given above. Considering $n = 3$, the assignment above can be written as $e^i \mapsto \frac{1}{2}\epsilon_{ijk} e^j \wedge e^k$. The non-canonical nature of this map shows in that it changes form under a change of bases, $e_j \mapsto e'_j = e_k(T^{-1})^k{}_j$. At the same time, the index structure of the expression does not look nice: in violation of covariance, the index i appears upstairs on the left and downstairs on the right. These deficiencies are both repaired via the definition

$$e^i \mapsto *e^i = \tfrac{1}{2}\sqrt{|g|}\,g^{il}\epsilon_{ljk}\, e^j \wedge e^k.$$

Here, the index raising operation via the inverse metric, g^{ij}, repairs the index positioning, and the factor $\sqrt{|g|}$ ensures invariance under orientation-preserving transformations (as shown below). The definition is linear, thus a general one-form is mapped to a two-form as

$$\phi = \phi_i e^i \mapsto *\phi = \tfrac{1}{2}\sqrt{|g|}\,\phi_i g^{il}\epsilon_{ljk}\, e^j \wedge e^k. \tag{L286}$$

The generalization of this expression to arbitrary p and n defines the *Hodge star in coordinates* as a map with the following structure:

$$* : \Lambda^p(V) \to \Lambda^{(n-p)}(V), \qquad \phi \mapsto *\phi,$$

$$*\phi = *\left(\frac{1}{p!}\phi_{i_1,\ldots,i_p}e^{i_1}\wedge\cdots\wedge e^{i_p}\right) = \frac{\sqrt{|g|}}{p!(n-p)!}\phi^{j_1,\ldots,j_p}\epsilon_{j_1,\ldots,j_p j_{p+1},\ldots,j_n}e^{j_{p+1}}\wedge\cdots\wedge e^{j_n},$$

$$\text{(L287)}$$

or

$$(*\phi)_{j_{p+1},\ldots,j_n} = \frac{1}{p!}\sqrt{|g|}\phi_{i_1,\ldots,i_p}g^{i_1 j_1}\ldots g^{i_p j_p}\epsilon_{j_1,\ldots,j_p j_{p+1},\ldots,j_n}. \qquad \text{(L288)}$$

For a one-form in three-dimensional space this definition reduces to the prototype Eq. (L286).

A second application of the star maps this $(n-p)$-form back to a form of degree $n - (n-p) = p$. Indeed, it is straightforward to verify that the Hodge star operation is self-involutory up to a sign factor,

$$**\phi = \text{sgn}(g)(-)^{p(n-p)}\phi. \qquad \text{(L289)}$$

EXERCISE Consider a three-dimensional vector space with a diagonal metric of signature $(3,0)$ or $(0,3)$, in an orthonormal basis, $g_{ij} = \eta_{ij} = \pm\delta_{ij}$. Verify that

$$1 \mapsto *1 = e^1\wedge e^2\wedge e^3,$$

$$\lambda = \lambda_i e^i \mapsto *\lambda = \pm\lambda_1(e^2\wedge e^3) \pm \lambda_2(e^3\wedge e^1) \pm \lambda_3(e^1\wedge e^2),$$

$$\phi = \tfrac{1}{2!}\phi_{jk}e^j\wedge e^k \mapsto *\phi = \phi_{23}\,e^1 + \phi_{31}\,e^2 + \phi_{12}\,e^3,$$

$$\omega = \tfrac{1}{3!}\omega_{ijk}e^i\wedge e^j\wedge e^k \mapsto *\omega = \pm\omega_{123},$$

and $**\psi = \text{sgn}(g)\,\psi$ in all cases. Prove (L289) for $n = 4$, or better still for general n, and track the origin of the factor $\text{sgn}(g)$. It features, for example, in applications involving the Minkowski metric with signature $(1,3)$, where $\text{sgn}(g) = -1$. ($\to$ L10.9.1-2)

The direct proof of the form invariance of the Hodge star operation is straightforward but also rather technical. ($\to$ L10.9.3-4) In the info section below we take the different approach by first defining a general Hodge star operation in a coordinate-invariant manner and then translating back to its coordinate representation, Eq. (L288).

INFO We here define the Hodge star without reference to coordinates, and from there establish contact with the coordinate definition. A first observation is that, starting from the inner product of 1-forms, $g^{ij} = \langle e^i, e^j \rangle$, an **inner product of p-forms** may be defined as

$$\langle e^{i_1}\wedge\cdots\wedge e^{i_p}, e^{j_1}\wedge\cdots\wedge e^{j_p}\rangle \equiv \det(\{\langle e^{i_a}, e^{j_b}\rangle\}) = \det(\{g^{i_a j_b}\}) = \epsilon_{a_1,\ldots,a_p}g^{i_{a_1} j_1}\ldots g^{i_{a_p} j_p}.$$

For example, $\det(e^3\wedge e^1, e^5\wedge e^7) = \det\begin{pmatrix} g^{35} & g^{37} \\ g^{15} & g^{17} \end{pmatrix}$. Since every p-form may be represented by a linear combination of basis forms, $e^{i_1}\wedge\cdots\wedge e^{i_p}$, this formula defines an inner product on $\Lambda^p(V)$:

$$\langle \phi, \psi \rangle = \frac{1}{(p!)^2} \phi_{i_1,\dots,i_p} \psi_{j_1,\dots,j_p} \langle e^{i_1} \wedge \cdots \wedge e^{i_p}, e^{j_1} \wedge \cdots \wedge e^{j_p} \rangle$$

$$= \frac{1}{(p!)^2} \phi_{i_1,\dots,i_p} \epsilon_{a_1,\dots,a_p} g^{i_a_1 j_1} \dots g^{i_a_p j_p} \psi_{j_1,\dots,j_p} = \frac{1}{(p!)^2} \phi_{i_1,\dots,i_p} \epsilon_{a_1,\dots,a_p} \psi^{i_{a_1},\dots,i_{a_p}}$$

$$= \frac{1}{p!} \phi_{i_1,\dots,i_p} \psi^{i_1,\dots,i_p}, \tag{L290}$$

where we raised indices and used the antisymmetry of form components. The representation

$$\boxed{\langle \phi, \psi \rangle = \frac{1}{p!} \phi_{i_1,\dots,i_p} \psi^{i_1,\dots,i_p}} \tag{L291}$$

invariant
definition

uses indices but is manifestly basis invariant because all indices are covariantly contracted. The **coordinate invariant definition of the Hodge star** now reads as

$$* : \Lambda^p(V) \to \Lambda^{n-p}(V), \qquad \phi \mapsto *\phi,$$

$$\forall \psi \in \Lambda^p(V) : \quad \psi \wedge (*\phi) = \langle \psi, \phi \rangle \, \omega. \tag{L292}$$

In words: wedging an arbitrary test p-form, ψ, with the to-be-defined $*\phi$ must yield a top-dimensional form equal to the canonical volume form, ω, of Eq. (L283), times the scalar product of the test form with ϕ. To make this implicit definition concrete, we evaluate it on basis forms, $\psi = e^{i_1} \wedge \cdots \wedge e^{i_p}$. The two sides of the definition then yield

$$e^{i_1} \wedge \cdots \wedge e^{i_p} \wedge (*\phi) = \frac{1}{(n-p)!} (*\phi)_{j_{p+1},\dots,j_n} e^{i_1} \wedge \cdots \wedge e^{i_p} \wedge e^{j_{p+1}} \wedge \cdots \wedge e^{j_n}$$

$$= \frac{1}{(n-p)!} (*\phi)_{j_{p+1},\dots,j_n} \epsilon^{i_1,\dots,i_p j_{p+1},\dots,j_n} e^1 \wedge \cdots \wedge e^n$$

$$\overset{(L283)}{=} \frac{1}{(n-p)!} \frac{1}{\sqrt{|g|}} \epsilon^{i_1,\dots,i_p j_{p+1},\dots,j_n} (*\phi)_{j_{p+1},\dots,i_n} \, \omega,$$

$$\langle e^{i_1} \wedge \cdots \wedge e^{i_p}, \phi \rangle \omega = \frac{1}{p!} \phi_{j_1,\dots,j_p} \langle e^{i_1} \wedge \cdots \wedge e^{i_p}, e^{j_1} \wedge \cdots \wedge e^{j_p} \rangle \omega$$

$$\overset{(L290)}{=} \frac{1}{p!} \epsilon_{a_1,\dots,a_p} g^{i_{a_1} j_1} \dots g^{i_{a_p} j_p} \phi_{j_1,\dots,j_p} \, \omega = \frac{1}{p!} \epsilon_{a_1,\dots,a_p} \phi^{i_{a_1},\dots,i_{a_p}} \, \omega = \phi^{i_1,\dots,i_p} \, \omega.$$

Comparing these results, we arrive at the equation

$$\frac{1}{(n-p)!} \epsilon^{i_1,\dots,i_p j_{p+1},\dots,j_n} (*\phi)_{j_{p+1},\dots,j_n} = \sqrt{|g|} \phi^{i_1,\dots,i_p}.$$

For a fixed $(n-p)$-tuple, $(k_{p+1},\dots,k_n)$, multiply this equation with $\epsilon_{i_1,\dots,i_p,k_{p+1},\dots,k_n}$ and sum over i-indices. Using the relations $\epsilon_{i_1,\dots,i_p,k_{p+1},\dots,k_n} \epsilon^{i_1,\dots,i_p j_{p+1},\dots,j_n} = p! \epsilon^{j_{p+1},\dots,j_n}_{k_{p+1},\dots,k_n}$ and

coordinate
represen-
tation

$\epsilon^{j_{p+1},\dots,j_n}_{k_{p+1},\dots,k_n} (*\phi)_{j_{p+1},\dots,j_n} = (n-p)! (*\phi)_{k_{p+1},\dots,k_n}$, this leads to the **coordinate representation of the Hodge star operation** stated in Eq. (L288),

$$(*\phi)_{j_{p+1},\dots,j_n} = \frac{1}{p!} \sqrt{|g|} \phi^{i_1,\dots,i_p} \epsilon_{i_1,\dots,i_p j_{p+1},\dots,j_n}. \tag{L293}$$

In words: the components of the form $*\phi$ are obtained by (i) raising the p indices of the components of the original form, (ii) contracting the result with the first p indices of the n-index ϵ-symbol, and (iii) multiplying by $\sqrt{|g|}/p!$. Using this representation for the components of $*\phi$, we arrive at Eq. (L288).

L10.10 Summary and outlook

In this chapter we introduced the calculus of tensor products as an organizing framework broad enough to accommodate all elements of linear algebra. The ensuing multilinear structures emerged as products of vector spaces and their duals, where a canonical connection between these different factors was established via the metric isomorphism. Important low-order constructions in this hierarchy included vectors and dual vectors, matrices and their transposes, and bilinear forms. Previously formal operations like the raising and lowering of indices or taking matrix transposes could now be understood as natural operations involving these spaces. In the second part of the chapter, we introduced alternating forms as a particularly important class of covariant tensors. We showed that the combination of form-algebra with an inner product defines structures such as canonical volume forms, or the Hodge star operation. However, in this chapter we caught only a first glimpse of the application potential of these objects. Their true potential will become apparent in Chapters V4 to V6, when the connection between linear algebra, calculus, geometry and physics is developed.

Problems: Linear Algebra

The problems come in odd–even-numbered pairs, labeled ε for "example" and ρ for "practice". Each example problem prepares the reader for tackling the subsequent practice problem. The solutions to the odd-numbered example problems are given in Chapter SL. A password-protected manual containing solutions of all even-numbered practice problems will be made available to instructors.

P.L1   Mathematics before numbers (p. 3)

P.L1.1   Sets and maps (p. 3)

εL1.1.1 Composition of maps (p. 7)

Let $\mathbb{N}_0$ denote the set of all natural numbers including zero, and $\mathbb{Z}$ the set of all integers. Consider the following two maps:

$$A : \mathbb{Z} \to \mathbb{Z}, \qquad n \mapsto A(n) = n + 1,$$
$$B : \mathbb{Z} \to \mathbb{N}_0, \qquad n \mapsto B(n) = |n| \equiv n \cdot \text{sign}(n).$$

(a)   Find the composite map $C = B \circ A$, i.e. specify its domain, image and action on n.

(b)   Which of the above maps A, B and C are surjective? Injective? Bijective?

ρL1.1.2 Composition of maps (p. 7)

(a)   Consider the set $S = \{-2, -1, 0, 1, 2\}$. Find its image, $T = A(S)$, under the map $n \mapsto A(n) = n^2$. Is the map $A : S \to T$ surjective? Injective? Bijective?

(b)   Find the image, $U = B(T)$, of the set T from part (a) under the map $n \mapsto B(n) = \sqrt{n}$.

(c)   Find the composite map $C = B \circ A$.

(d)   Which of the above maps A, B and C are surjective? Injective? Bijective?

P.L1.2   Groups (p. 7)

εL1.2.1 The group $\mathbb{Z}_2$ (p. 8)

(a)   Show that $\mathbb{Z}_2 \equiv (\{0, 1\}, +)$, where the addition operation $+$ is defined by the adjacent composition table, is an abelian group.

$+$	0	1
0	0	1
1	1	0

(b) Construct a group isomorphic to $\mathbb{Z}_2$, using two integers as group elements and standard multiplication of integers as group operation. Set up the corresponding composition table.

ₚL1.2.2 The groups of addition modulo 5 and rotations by multiples of 72 deg (p. 8)

(a) Consider the set $\mathbb{Z}_5 = \{0, 1, 2, 3, 4\}$, endowed with the group operation

$$\boldsymbol{+} : \mathbb{Z}_5 \times \mathbb{Z}_5 \to \mathbb{Z}_5, \qquad (p, p') \mapsto p \boldsymbol{+} p' \equiv (p + p') \bmod 5.$$

Set up the composition table for the group $(\mathbb{Z}_5, \boldsymbol{+})$. Which element is the neutral element? For a given $n \in \mathbb{Z}$, which element is the inverse of n?

(b) Let $r(\phi)$ denote a rotation by ϕ degrees about a fixed axis, with $r(\phi + 360) = r(\phi)$. Consider the set of rotations by multiples of 72 deg,

$$\mathcal{R}_{72} = \{r(0), r(72), r(144), r(216), r(288)\},$$

and the group $(\mathcal{R}_{72}, \cdot)$, where the group operation $\cdot$ involves two rotations in succession:

$$\cdot : \mathcal{R}_{72} \times \mathcal{R}_{72} \to \mathcal{R}_{72}, \qquad (r(\phi), r(\phi')) \mapsto r(\phi) \cdot r(\phi') \equiv r(\phi + \phi').$$

Set up the multiplication table for this group. Which element is the neutral element? Which element is the inverse of $r(\phi)$?

(c) Explain why the groups $(\mathbb{Z}_5, \boldsymbol{+})$ and $(\mathcal{R}_{72}, \cdot)$ are isomorphic.

(d) Let $(\mathbb{Z}_n, \boldsymbol{+})$ denote the group of integer addition modulo n of the elements of the set $\mathbb{Z}_n = \{0, 1, \ldots, n - 1\}$. Which group of discrete rotations is isomorphic to this group?

ₑL1.2.3 Group of discrete translations in one dimension (p. 8)

In this problem we show that discrete translations on an infinite, one-dimensional lattice form a group. Let us denote the lattice constant, i.e. the fixed distance between neighboring lattice points, by $\lambda \subset \mathbb{R}^+$, a positive, real number. The lattice $\mathbb{G}$ consists of the set of all integer multiples of λ, i.e. $\mathbb{G} \equiv \lambda\mathbb{Z} \equiv \{x \in \mathbb{R} | \exists n \in \mathbb{Z} : x = \lambda \cdot n\}$, where $\cdot$ is the usual multiplication rule in $\mathbb{R}$. Note that for any given $x \in \mathbb{G}$, n is uniquely determined. On this lattice we define "translation" by the group operation

$$T : \quad \mathbb{G} \times \mathbb{G} \to \mathbb{G}, \quad (x, y) \mapsto T(x, y) \equiv x + y,$$

where $+$ denotes the usual addition of real numbers. Since this operation is symmetric, it can be visualized in two equivalent ways: $T(x, y)$ describes (i) a "shift" or a "translation" of lattice point x by the distance y, or (ii) a translation of lattice point y by the distance x. [Figure (a), where $\lambda = \frac{1}{3}$, shows both visualizations of $T(\frac{2}{3}, \frac{4}{3})$.]

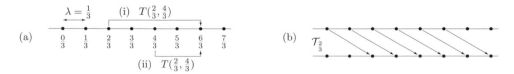

(a) Show that $(\mathbb{G}, T)$ forms an abelian group.

(b)   For a given $y \in \mathbb{G}$ we now define, in accordance with visualization (i), a "translation" of the lattice by y, i.e. each lattice point x is "shifted" by y:

$$\mathcal{T}_y : \quad \mathbb{G} \to \mathbb{G}, \quad x \mapsto \mathcal{T}_y(x) \equiv T(x, y).$$

[Figure (b), where $\lambda = \frac{1}{3}$, shows $\mathcal{T}_{\frac{2}{3}}$.] Now consider the set of all such translations, $\mathbb{T} \equiv \{\mathcal{T}_y, y \in \mathbb{G}\}$. Show that $(\mathbb{T}, +)$ forms an abelian group, where $+$ is defined as

$$+ : \quad \mathbb{T} \times \mathbb{T} \to \mathbb{T}, \quad (\mathcal{T}_x, \mathcal{T}_y) \mapsto \mathcal{T}_x + \mathcal{T}_y \equiv \mathcal{T}_{T(x,y)}.$$

Remark: The set $\mathbb{T}$ underlying this group consists of maps (namely translations), illustrating that the set underlying a group need not be "simple".

ₚL1.2.4 Group of discrete translations on a ring (p. 8)

In this problem we show that discrete translations on a finite, one-dimensional lattice with periodic boundary conditions form a group. Consider a ring with radius $0 < R \in \mathbb{R}$ and lattice constant $\lambda = 2\pi R/N$ with $N \in \mathbb{N}$, thus $\mathbb{G} \equiv \lambda(\mathbb{Z} \bmod N) \equiv \{x \in \mathbb{R} | \exists n \in \{0, 1, \dots, N-1\} : x = \lambda \cdot n\}$, where $\cdot$ is the usual multiplication rule in $\mathbb{R}$. Note that for any given $x \in \mathbb{G}$, n is uniquely determined. The ring forms a "periodic" structure: when counting its sites, 0λ and $N\lambda$ describe the same lattice site, the same is true for 1λ and $(1 + N)\lambda$, for 2λ and $(2 + N)\lambda$, etc. On this lattice we define a group operation, corresponding to a "translation", using addition modulo N:

$$T : \quad \mathbb{G} \times \mathbb{G} \to \mathbb{G}, \quad (x, y) = (\lambda \cdot n_x, \lambda \cdot n_y) \mapsto T(x, y) \equiv \lambda \cdot ((n_x + n_y) \bmod N).$$

Here $+$ is the usual addition of integers, and $n \bmod N$ (spoken as "n mod N") is defined as the integer remainder after division of n by N (e.g. $9 \bmod 8 = 1$). [For $N = 8$, figure (a) shows two visualizations of the translation $T(4\lambda, 5\lambda)$: as a "shift" of the lattice site 4λ by the distance 5λ along the ring, or of the site 5λ by the distance 4λ.]

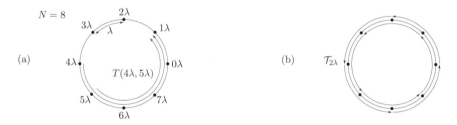

(a)   Show that $(\mathbb{G}, T)$ forms an abelian group.

(b)   For a given $y \in \mathbb{G}$ we now define a "translation" of the lattice by y,

$$\mathcal{T}_y : \quad \mathbb{G} \to \mathbb{G}, \quad x \mapsto \mathcal{T}_y(x) \equiv T(x, y),$$

i.e. each site x is "shifted" by y along the ring. [For $N = 8$, figure (b) shows the translation $\mathcal{T}_{2\lambda}$.] Now consider the set of all such translations, $\mathbb{T} \equiv \{\mathcal{T}_y, y \in \mathbb{G}\}$. Show that $(\mathbb{T}, +)$ forms an abelian group, where the group operation $+$ is defined as

$$+ : \quad \mathbb{T} \times \mathbb{T} \to \mathbb{T}, \quad (\mathcal{T}_x, \mathcal{T}_y) \mapsto \mathcal{T}_x + \mathcal{T}_y \equiv \mathcal{T}_{T(x,y)}.$$

εL1.2.5 The permutation group S_3 (p. 11)

A map which reorders n labeled objects is called a **permutation** of these objects. For example, $1234 \overset{[4312]}{\longmapsto} 4312$ is a permutation of the four numbers in the string 1234, where we use [4312] as shorthand for the map $1 \mapsto 4$, $2 \mapsto 3$, $3 \mapsto 1$ and $4 \mapsto 2$. Similarly, if the same permutation is applied to the string 2314, it yields $2314 \overset{[4312]}{\longmapsto} 3142$. (In general, $[P(1)...P(n)]$ denotes the map $j \mapsto P(j)$ which replaces j by $P(j)$, for $j = 1, ..., n$.) Two permutations performed in succession again yield a permutation. For example, acting on 1234 with $P = [4312]$ followed by $P' = [2413]$ yields $1234 \overset{[4312]}{\longmapsto} 4312 \overset{[2413]}{\longmapsto} 3124$, thus the resulting permutation is $P' \circ P = [3124]$.

The set of all possible permutations of n numbers, denoted by S_n, contains $n!$ elements. Viewing $P' \circ P$ (perform first P, then P') as a group operation,

$$\circ : S_n \times S_n \to S_n, \qquad (P', P) \mapsto P' \circ P,$$

we obtain a group, $(S_n, \circ)$, the **permutation group**, usually denoted simply by S_n.

(a) Complete the adjacent composition table for S_3, in which the entries $P' \circ P$ are arranged such that those with fixed P' sit in the same row, those with fixed P in the same column.

$P' \circ P$	[123]	[231]	[312]	[213]	[321]	[132]
[123]	[123]	[231]	[312]	[213]	[321]	[132]
[231]		[312]	[123]	[321]	[132]	[213]
[312]			[231]	[132]	[213]	[321]
[213]				[312]	[231]	
[321]					[312]	
[132]						

(b) Which element is the neutral element of S_3? How can we see from the multiplication table that every element has a unique inverse?

(c) Is S_3 an abelian group? Justify your answer.

ρL1.2.6 Decomposing permutations into sequences of pair permutations (p. 11)

Consider the permutation group S_n defined in the previous problem. Any permutation can be decomposed into a sequence of **pair permutations**, i.e. permuations which exchange just two objects, leaving the others unchanged. Examples:

$123 \overset{[321]}{\longmapsto} 321 \overset{[132]}{\longmapsto} 231$ $\qquad\qquad\Rightarrow\quad [231] = [132] \circ [321].$

$1234 \overset{[2134]}{\longmapsto} 2134 \overset{[3214]}{\longmapsto} 2314$ $\qquad\Rightarrow\quad [2314] = [3214] \circ [2134],$

$1234 \overset{[3214]}{\longmapsto} 3214 \overset{[1324]}{\longmapsto} 2314$ $\qquad\Rightarrow\quad [2314] = [1324] \circ [3214],$

$1234 \overset{[4231]}{\longmapsto} 4231 \overset{[1432]}{\longmapsto} 2431 \overset{[1243]}{\longmapsto} 2341 \overset{[4231]}{\longmapsto} 2314 \Rightarrow [2314] = [4231] \circ [1243] \circ [1432] \circ [4231].$

The last three lines illustrate that a given permutation can be pair-decomposed in several ways, and that these may or may not involve different numbers of pair exchanges. However, one may convince oneself (try it!) that all pair decompositions of a given permutation have the same **parity**, i.e. the number of exchanges is either always **even** or always **odd**.

To find a "minimal" (shortest possible) pair decomposition of a given permutation, say [2413], we may start from the naturally ordered string 1234 and rearrange it to its desired form, 2413, one pair permutation at a time, bringing the 2 to the first slot, then the 4 to the second slot, etc. This yields $1234 \overset{[2134]}{\longmapsto} 2134 \overset{[4231]}{\longmapsto} 2431 \overset{[3214]}{\longmapsto} 2413$, hence $[2413] = [3214] \circ [4231] \circ [2134]$.

Find a minimal pair decomposition and the parity of each of the following permutations:

(a) [132],   (b) [231],   (c) [3412],   (d) [3421],   (e) [15234],   (f) [31542].

P.L1.3   Fields (p. 12)

ₜL1.3.1 Complex numbers – elementary computations (p. 15)

Consider the complex numbers $z_1 = 12 + 5i$, $z_2 = -3 + 2i$ and $z_3 = a - ib$, with $a, b \in \mathbb{R}$. Compute (a) $\bar{z}_1$, (b) $z_1 + z_2$, (c) $z_1 + \bar{z}_3$, (d) $z_1 z_2$, (e) $\bar{z}_1 z_3$ and (f) z_1 / z_2. (Present each answer in the form $x + iy$.) Also compute (g) $|z_1|$, (h) $|z_1 + z_2|$ and (i) $|az_2 + 3z_3|$.

ₚL1.3.2 Complex numbers – elementary computations (p. 15)

Consider the complex numbers $z_1 = 3 + ai$ and $z_2 = b - 2i$ with $a, b \in \mathbb{R}$. Compute (a) $\bar{z}_1$, (b) $z_1 - z_2$, (c) $z_1 \bar{z}_2$ and (d) $\bar{z}_1 / z_2$. (Present each answer in the form $x + iy$.) Also, compute (e) $|z_1|$ and (f) $|bz_1 - 3z_2|$. [Check your results for $a = 2$, $b = 3$: (a) $3 - 2i$, (b) $4i$, (c) $5 + 12i$, (d) 1, (e) $\sqrt{13}$, (f) 12.]

ₜL1.3.3 Algebraic manipulations with complex numbers (p. 15)

For $z = x + iy \in \mathbb{C}$, bring each of the following expressions into standard form, i.e. write them as (real part) + i(imaginary part):

(a) $z + \bar{z}$,   (b) $z - \bar{z}$,   (c) $z \cdot \bar{z}$,   (d) $\dfrac{z}{\bar{z}}$,

(e) $\dfrac{1}{z} + \dfrac{1}{\bar{z}}$,   (f) $\dfrac{1}{z} - \dfrac{1}{\bar{z}}$,   (g) $z^2 + z$,   (h) z^3.

[Check your results for $x = 2$, $y = 1$: (a) 4, (b) i2, (c) 5, (d) $\frac{3}{5} + i\frac{4}{5}$, (e) $\frac{4}{5}$, (f) $-i\frac{2}{5}$, (g) $5 + i5$, (h) $2 + i11$.]

ₚL1.3.4 Algebraic manipulations with complex numbers (p. 15)

For $z = x + iy \in \mathbb{C}$, bring each of the following expressions into standard form:

(a) $(z + i)^2$,   (b) $\dfrac{z}{z + 1}$,   (c) $\dfrac{\bar{z}}{z - i}$.

[Check your results for $x = 1$, $y = 2$: (a) $-8 + i6$, (b) $\frac{3}{4} + i\frac{1}{4}$, (c) $-\frac{1}{2} - i\frac{3}{2}$.]

ₜL1.3.5 Multiplying complex numbers – geometrical interpretation (p. 16)

(a)   Consider the polar representation, $z_j = (\rho_j \cos \phi_j, \rho_j \sin \phi_j)$, of two complex numbers, z_1 and z_2 with $\phi_j \in [0, 2\phi)$. Show that multiplying them, $z_3 = z_1 z_2$, yields the relations $\rho_3 = \rho_1 \rho_2$ and $\phi_3 = (\phi_1 + \phi_2) \mathrm{mod}(2\pi)$. [The $\mathrm{mod}(2\pi)$ is needed since we restricted polar angles to lie in the interval $[0, 2\pi)$.] To this end, the following trigonometric identities are useful:

$$\cos \phi_1 \cos \phi_2 - \sin \phi_1 \sin \phi_2 = \cos(\phi_1 + \phi_2),$$
$$\sin \phi_1 \cos \phi_2 + \cos \phi_1 \sin \phi_2 = \sin(\phi_1 + \phi_2).$$

(b) For $z_1 = \sqrt{3} + i$, $z_2 = -2 + 2\sqrt{3}i$, compute the product $z_3 = z_1 z_2$, as well as $z_4 = 1/z_1$ and $z_5 = \bar{z}_1$. Find the polar representation (with $\phi \in [0, 2\pi)$) of all five complex numbers and sketch them in the complex plane (in one diagram). Is your result for z_3 consistent with (a)?

‣L1.3.6 Multiplying complex numbers – geometrical interpretation (p. 16)

For $z_1 = \frac{1}{\sqrt{8}} + \frac{1}{\sqrt{8}}i$, $z_2 = \sqrt{3} - i$, compute the product $z_3 = z_1 z_2$, as well as $z_4 = 1/z_1$ and $z_5 = \bar{z}_1$. Find the polar representation (with $\phi \in [0, 2\pi)$) of all five complex numbers and sketch them in the complex plane (in one diagram).

‣L1.3.7 Field axioms for $\mathbb{F}_4$ (p. 12)

For a field, the requirement that the addition and multiplication rules be distributive imposes powerful constraints on both composition rules. The present problem illustrates this fact for a discrete field involving just four elements.

Equip the set $\mathbb{F}_4 = \{0, 1, a, b\}$ with "multiplication" and "addition" rules chosen such that $(\mathbb{F}_4, +, \cdot)$ is a field, with 0 and 1 as neutral elements of addition and multiplication, respectively. To this end, complete the given composition tables in such a way that the properties of a field are fulfilled.

$\cdot$	0	1	a	b
0				
1		1	a	b
a		a		
b		b		

$+$	0	1	a	b
0	0	1	a	b
1	1			
a	a			
b	b			

Hint: Start with the multiplication table!

P.L2   Vector spaces (p. 19)

P.L2.3   Vector spaces: examples (p. 25)

‣L2.3.1 Vector space axioms: rational numbers (p. 28)

(a) Show that the set $\mathbb{Q}^2 = \{\binom{x^1}{x^2} \mid x^1, x^2 \in \mathbb{Q}\}$, consisting of all pairs of rational numbers, forms a $\mathbb{Q}$-vector space over the field of rational numbers.

(b) Is it possible to construct a vector space from the set of all pairs of integers, $\mathbb{Z}^2 = \{\binom{x^1}{x^2} \mid x^1, x^2 \in \mathbb{Z}\}$? Justify your answer!

‣L2.3.2 Vector space axioms: complex numbers (p. 28)

Show that the complex numbers $\mathbb{C}$ form an $\mathbb{R}$-vector space over the field of real numbers.

‣L2.3.3 Vector space of real functions (p. 26)

Let $F \equiv \{f : \mathbb{R} \to \mathbb{R}, x \mapsto f(x)\}$ be the set of real functions. Show that $(F, +, \cdot)$ is an $\mathbb{R}$-vector space, where the addition of functions, and their multiplication by scalars, are defined as follows:

$$+ : F \times F \to F \qquad (f, g) \mapsto f + g, \qquad \text{with} \qquad f + g : x \mapsto [f + g](x) \equiv f(x) + g(x), \qquad (1)$$

$$\bullet : \mathbb{R} \times F \to F \qquad (\lambda, f) \mapsto \lambda \bullet f, \qquad \text{with} \qquad \lambda \bullet f : x \mapsto \ [\lambda \bullet f](x) \equiv \lambda f(x). \tag{2}$$

Remark regarding notation: It is important to distinguish the "name" of a function, f, from the "function value", $f(x)$, which it returns when evaluated at the argument x. The sum of the functions f and g is a function named $f + g$. Equation (1) states that its function value at x, denoted by $[f + g](x)$, is by definition equal to $f(x) + g(x)$, the sum of the function values of f and g at x. (For emphasis, in this problem we use square brackets to indicate the function name; elsewhere we'll use round brackets.) The product of the number c and the function f yields a function named $c \bullet f$. Equation (2) states that its function value at x, denoted by $[c \bullet f](x)$, is by definition equal to $cf(x)$, the product of c with the function value of f at x.

ₚL2.3.4 Vector space of polynomials of degree n (p. 28)

The vector space of all real functions is infinite-dimensional. However, if only functions of a prescribed form are considered, the corresponding vector space can be finite-dimensional. As an example, it is shown in this problem that the set of all polynomials of degree n form a vector space of dimension $n + 1$, isomorphic to $\mathbb{R}^{n+1}$.

[*Remark on the notation*: In the context of the present problem on polynomials, x^k means "x to the power of k", and a_k is "the coefficient of x^k". This is in contrast to the notation that we have adopted elsewhere when discussing vectors, where x^k stands for the kth component of the vector $\mathbf{x} = \sum_k \mathbf{v}_k x^k$ with respect to a basis of vectors $\{\mathbf{v}_k\}$. Every notational convention has exceptions!]

Let $p_\mathbf{a}$ denote a polynomial in the variable $x \in \mathbb{R}$ of degree n:

$$p_\mathbf{a} : \mathbb{R} \to \mathbb{R}, \qquad x \mapsto p_\mathbf{a}(x) \equiv a_0 x^0 + a_1 x^1 + \cdots + a_n x^n. \tag{1}$$

$p_\mathbf{a}$ is uniquely specified by its $n + 1$ real coefficients $a_0, a_1, \ldots, a_n$, which for notational brevity we arrange into an $(n + 1)$-tuplet, $\mathbf{a} = (a_0, a_1, \ldots, a_n)^\mathrm{T} \in \mathbb{R}^{n+1}$. Let $P_n = \{p_\mathbf{a} | \mathbf{a} \in \mathbb{R}^{n+1}\}$ denote the set of all such polynomials of degree n. The natural definitions for adding such polynomials, or multiplying them by a scalar $c \in \mathbb{R}$, are:

$$p_\mathbf{a} + p_\mathbf{b} : \mathbb{R} \to \mathbb{R}, \qquad x \mapsto p_\mathbf{a}(x) + p_\mathbf{b}(x), \tag{2}$$

$$c \bullet p_\mathbf{a} : \mathbb{R} \to \mathbb{R}, \qquad x \mapsto c\, p_\mathbf{a}(x), \tag{3}$$

where on the right side the usual addition and multiplication in $\mathbb{R}$ is used.

(a) Show that the above addition and scalar multiplication rules imply the following composition rules in P_n:

addition of polynomials: $+ : \quad P_n \times P_n \to P_n, \qquad (p_\mathbf{a}, p_\mathbf{b}) \mapsto p_\mathbf{a} + p_\mathbf{b} \equiv p_{\mathbf{a}+\mathbf{b}},$

multiplication by a scalar: $\bullet : \quad \mathbb{R} \times P_n \to P_n, \qquad (c, p_\mathbf{x}) \mapsto c \bullet p_\mathbf{a} \equiv p_{c\mathbf{a}},$

where $\mathbf{a} + \mathbf{b}$ and $c\mathbf{a}$ denote the usual addition and scalar multiplication in $\mathbb{R}^{n+1}$.

(b) Show that $(P_n, +, \bullet)$ is an $\mathbb{R}$-vector space, and that it is isomorphic to $\mathbb{R}^{n+1}$.

(c) Find a set $n + 1$ of polynomials, $\{p_{\mathbf{a}_0}, \ldots, p_{\mathbf{a}_n}\} \subset P_n$, forming a basis for this vector space.

ₑL2.3.5 Vector space with unusual composition rule (p. 28)

The axioms that define a vector space can be satisfied in many different ways. These may involve unconventional definitions of vector addition and scalar multiplication. The purpose of the present problem is to illustrate this point.

For any $a \in \mathbb{R}$, let $V_a \equiv \{v_x\}$ be a set whose elements v_x, labeled by real numbers $x \in \mathbb{R}$, satisfy the following composition rules:

addition: $+ : \quad V_a \times V_a \to V_a, \quad (v_x, v_y) \mapsto v_x + v_y \equiv v_{x+y+a},$

multiplication by a scalar: $\cdot : \quad \mathbb{R} \times V_a \to V_a, \quad (\lambda, v_x) \mapsto \lambda \cdot v_x \equiv v_{\lambda x + a(\lambda - 1)}.$

The a and x labels, being real numbers, satisfy the usual addition and scalar multiplication rules of $\mathbb{R}$; e.g. in V_2 we have: $v_3 + v_4 = v_{3+4+2} = v_9$ and $3 \cdot v_4 = v_{3 \cdot 4 + 2(3-1)} = v_{16}$. Show that the triple $(V_a, +, \cdot)$ represents an $\mathbb{R}$-vector space, with v_{-a} and 1 being the neutral elements for addition and scalar multiplication, respectively, and v_{-x-2a} the additive inverse of v_x.

ₚL2.3.₆ Vector space with unusual composition rule (p. 28)

For any $\mathbf{a} \in \mathbb{R}^2$, let $V_\mathbf{a} \equiv \{v_\mathbf{x}\}$ be a set whose elements $v_\mathbf{x}$, labeled by vectors $\mathbf{x} \in \mathbb{R}^2$, satisfy the following composition rules:

addition: $+ : \quad V_\mathbf{a} \times V_\mathbf{a} \to V_\mathbf{a}, \quad (v_\mathbf{x}, v_\mathbf{y}) \mapsto v_\mathbf{x} + v_\mathbf{y} \equiv v_{\mathbf{x}+\mathbf{y}-\mathbf{a}},$

multiplication by a scalar: $\cdot : \quad \mathbb{R} \times V_\mathbf{a} \to V_\mathbf{a}, \quad (\lambda, v_\mathbf{x}) \mapsto \lambda \cdot v_\mathbf{x} \equiv v_{\lambda \mathbf{x} + f(\mathbf{a}, \lambda)}.$

Here $f(\mathbf{a}, \lambda)$ is a function of $\mathbf{a}$ and λ, whose form will be determined below.

(a)   Show that $V_\mathbf{a}$, endowed with the composition rule $+$, forms an abelian group, and specify the neutral element of addition and the additive inverse of $v_\mathbf{x}$.

(b)   Find the specific form of f that ensures that the triple $(V_\mathbf{a}, +, \cdot)$ forms an $\mathbb{R}$-vector space.

(c)   Would a similar construction work for $\mathbf{a}, \mathbf{x} \in \mathbb{R}^n$ (with n a positive integer) instead of $\mathbb{R}^2$?

P.L2.4   Basis and dimension (p. 28)

ₑL2.4.₁ Linear Independence (p. 30)

(a)   Are the vectors $v_1 = (0, 1, 2)^\mathsf{T}$, $v_2 = (1, -1, 1)^\mathsf{T}$ and $v_3 = (2, -1, 4)^\mathsf{T}$ linearly independent?

(b)   Depending on whether your answer is yes or no, find a vector v_2' such that v_1, v_2' and v_3 are linearly dependent or independent, respectively, and show explicitly that they have this property.

ₚL2.4.₂ Linear independence (p. 30)

(a)   Are the vectors $v_1 = (1, 2, 3)^\mathsf{T}$, $v_2 = (2, 4, 6)^\mathsf{T}$ and $v_3 = (-1, -1, 0)^\mathsf{T}$ linearly independent?

(b)   Depending on whether your answer is yes or no, find a vector, v_2' such that v_1, v_2' and v_3 are linearly dependent or independent, respectively, and show explicitly that they have this property.

ₑL2.4.₃ Einstein summation convention (p. 32)

Let $a_1, a_2, b_1, b_2 \in \mathbb{R}$. Which of the following statements, formulated using the Einstein summation convention, are true and which are false? Justify your answers!

(a) $a_i b^i \overset{?}{=} b^j a_j$,

(b) $a_i \delta^i_{\ j} b^j \overset{?}{=} a_k b^k$,

(c) $a_i b^j a_j b^k \overset{?}{=} a_k b^l a_l b^i$,

(d) $a_1 a_i b^1 b^i + b^2 a_j a_2 b^j \overset{?}{=} (a_i b^i)^2$.

$_p$L2.4.4 Einstein summation convention (p. 32)

Let $a_1 = 1$, $a_2 = 2$, $b^1 = -1$, $b^2 = x$. Evaluate the following expressions, formulated using the Einstein summation convention, as functions of x:

(a) $a_i b^i$,   (b) $a_i a_j b^i b^j$,   (c) $a_1 a_j b^2 b^j$.

[Check your results for $x = 3$: (a) 5, (b) 25, (c) 15.]

P.L3   Euclidean geometry (p. 38)

P.L3.2   Normalization and orthogonality (p. 41)

$_E$L3.2.1 Angle, orthogonal decomposition (p. 42)

(a)   Find the angle between the vectors $\mathbf{a} = (3, 4)^T$ and $\mathbf{b} = (7, 1)^T$.

(b)   Consider the vectors $\mathbf{c} = (3, 1)^T$ and $\mathbf{d} = (-1, 2)^T$. Decompose $\mathbf{c} = \mathbf{c}_\| + \mathbf{c}_\perp$ into components parallel and perpendicular to $\mathbf{d}$, respectively. Sketch all four vectors.
[Check your results: $\|\mathbf{c}_\|\| = \frac{1}{\sqrt{5}}$, $\|\mathbf{c}_\perp\| = \frac{7}{\sqrt{5}}$.]

$_p$L3.2.2 Angle, orthogonal decomposition (p. 42)

(a)   Find the angle between the vectors $\mathbf{a} = (2, 0, \sqrt{2})^T$ and $\mathbf{b} = (\sqrt{2}, 1, 1)^T$.

In the figure, the points P, Q and R have coordinate vectors $\mathbf{p} = (-1, -1)^T$, $\mathbf{q} = (2, 1)^T$ and $\mathbf{r} = (-1, -1 + 13a)^T$, with a a positive real number. The line RS is perpendicular to the line PQ.

(b)   Find the coordinate vector $\mathbf{s}$ of S, expressed as a function of a.
       Hint: Let $\mathbf{c}$ denote the vector from P to Q, and $\mathbf{d}$ the vector from P to R, then decompose $\mathbf{d} = \mathbf{d}_\| + \mathbf{d}_\perp$ into components parallel and perpendicular to $\mathbf{c}$.

(c)   Find the distance $\overline{RS}$ from R to S and the distance $\overline{PS}$ from P to S.

[Check your results for $a = 1$: (b) $\mathbf{s} = (5, 3)^T$, (c) $\overline{RS}^2 + \overline{PS}^2 = 169$.]

P.L3.3   Inner product spaces (p. 46)

$_E$L3.3.1 Inner product for vector space of continuous functions (p. 47)

This problem illustrates a particularly important example of an inner product: in the space of continuous functions, an inner product can be defined via integration.

   Let V be the vector space of *continuous* real functions defined on an interval $I \in \mathbb{R}$, $f : I \to \mathbb{R}$, with the usual composition rules of vector addition and scalar multiplication:

$$\forall f, g \in V : \qquad f + g : I \to \mathbb{R}, \qquad x \mapsto (f + g)(x) \equiv f(x) + g(x),$$
$$\forall f \in V, \lambda \in \mathbb{R} : \qquad \lambda \cdot f : I \to \mathbb{R}, \qquad x \mapsto (\lambda \cdot f)(x) \equiv \lambda \, (f(x)) .$$

(a)   Show that the following map defines an inner product on V:

$$\langle \cdot, \cdot \rangle : V \times V \to \mathbb{R}, \qquad (f, g) \mapsto \langle f, g \rangle \equiv \int_I dx f(x) g(x) .$$

(b)   Now consider $I = [-1, 1]$. Compute $\langle f_1, f_2 \rangle$ for $f_1(x) \equiv \sin\left(\frac{x}{\pi}\right)$ and $f_2(x) \equiv \cos\left(\frac{x}{\pi}\right)$.

P L3.3.2 Unconventional inner product (p. 47)

The defining properties of an inner product on $\mathbb{R}^n$ are of course satisfied not only by the "standard" definition, $\langle \mathbf{x}, \mathbf{x} \rangle = \sum_{i=1}^n (x^i)^2$; there are infinitely many other bilinear forms that do so, too. The present problem illustrates this with a simple example. Show that the following map defines an inner product on the vector space $\mathbb{R}^2$:

$$\langle \cdot, \cdot \rangle : \mathbb{R}^2 \times \mathbb{R}^2 \to \mathbb{R}, \qquad (\mathbf{x}, \mathbf{y}) \mapsto x_1 y_1 + x_1 y_2 + x_2 y_1 + 3 x_2 y_2 .$$

E L3.3.3 Projection onto an orthonormal basis (p. 51)

(a)   Show that the vectors $\mathbf{e}_1' = \frac{1}{\sqrt{2}} (1, 1)^T$, $\mathbf{e}_{2'} = \frac{1}{2}(1, -1)^T$ form an orthonormal basis for $\mathbb{R}^2$.

(b)   Express the vector $\mathbf{w} = (-2, 3)^T$ in the form $\mathbf{w} = \mathbf{e}_1' w^1 + \mathbf{e}_2' w^2$, by computing its components w^i with respect to the basis $\{\mathbf{e}_i'\}$ through projection onto the basis vectors. [Check your results: $\sum_{i=1}^2 w^i = -2\sqrt{2}$.]

P L3.3.4 Projection onto an orthonormal basis (p. 51)

(a)   Show that the vectors $\mathbf{e}_1' = \frac{1}{9}(4, -1, 8)^T$, $\mathbf{e}_2' = \frac{1}{9}(-7, 4, 4)^T$ and $\mathbf{e}_3' = \frac{1}{9}(-4, -8, 1)^T$ form an orthonormal basis in $\mathbb{R}^3$.

(b)   Let $\mathbf{w} = \mathbf{e}_i' w^i$ be the decomposition of $\mathbf{w} = (1, 2, 3)^T$ in this basis. Find the components w^i. [Check your results: $\sum_{i=1}^3 w^i = \frac{22}{9}$.]

E L3.3.5 Gram–Schmidt orthonormalization (p. 46)

Apply the Gram–Schmidt procedure to the following set of linearly independent vectors $\{\mathbf{v}_1, \mathbf{v}_2, \mathbf{v}_3\}$ to construct an orthonormal set $\{\mathbf{e}_1', \mathbf{e}_2', \mathbf{e}_3'\}$ with the same span and with $\mathbf{e}_1' || \mathbf{v}_1$.

$$\mathbf{v}_1 = (1, -2, 1)^T, \qquad\qquad \mathbf{v}_2 = (1, 1, 1)^T, \qquad\qquad \mathbf{v}_3 = (0, 1, 2)^T .$$

P L3.3.6 Gram–Schmidt orthonormalization (p. 46)

Apply the Gram–Schmidt procedure to each of the following sets of linearly independent vectors $\{\mathbf{v}_1, \mathbf{v}_2, \mathbf{v}_3\}$ to construct an orthonormal set $\{\mathbf{e}_1', \mathbf{e}_2', \mathbf{e}_3'\}$ with the same span and with $\mathbf{e}_1' || \mathbf{v}_1$.

(a) $\quad \mathbf{v}_1 = (-2, 0, 2)^T, \qquad\qquad \mathbf{v}_2 = (2, 1, 0)^T, \qquad\qquad \mathbf{v}_3 = (3, 6, 5)^T .$

(b) $\quad \mathbf{v}_1 = (1, 1, 0, 0)^T, \qquad\qquad \mathbf{v}_2 = (0, 0, 1, 1)^T, \qquad\qquad \mathbf{v}_3 = (0, 1, 1, 0)^T .$

εL3.3.7 Non-orthonormal basis vectors and metric (p. 50)

Consider the vectors $\hat{\mathbf{v}}_1 = \binom{2}{0}$ and $\hat{\mathbf{v}}_2 = \binom{1}{1}$, written as column vectors in the standard basis of $\mathbb{R}^2$. (In this problem we use the notation of Section L3.3: vectors in the inner product space $\mathbb{R}^2$ carry a caret, e.g. $\hat{\mathbf{x}}$, and their components w.r.t. a given basis do not, e.g. $\mathbf{x}$.)

(a) Write the standard basis vector $\hat{\mathbf{e}}_1 = \binom{1}{0}$ as a linear combination of $\hat{\mathbf{v}}_1$ and $\hat{\mathbf{v}}_2$. Ditto for $\hat{\mathbf{e}}_2 = \binom{0}{1}$. Do $\{\hat{\mathbf{v}}_1, \hat{\mathbf{v}}_2\}$ form a basis for $\mathbb{R}^2$?

(b) Let $\hat{\mathbf{x}} = \hat{\mathbf{v}}_1 x^1 + \hat{\mathbf{v}}_2 x^2$ and $\hat{\mathbf{y}} = \hat{\mathbf{v}}_1 y^1 + \hat{\mathbf{v}}_2 y^2$ be two vectors in $\mathbb{R}^2$, whose components w.r.t. $\hat{\mathbf{v}}_1$ and $\hat{\mathbf{v}}_2$ are given by $\mathbf{x} = (x^1, x^2)^T = (3, -4)^T$ and $\mathbf{y} = (y^1, y^2)^T = (-1, 3)^T$ respectively. Express $\hat{\mathbf{x}}$ and $\hat{\mathbf{y}}$ as column vectors in the standard basis of $\mathbb{R}^2$ and compute their scalar product $\langle \hat{\mathbf{x}}, \hat{\mathbf{y}} \rangle_{\mathbb{R}^2}$.

(c) If the scalar product $\langle \hat{\mathbf{x}}, \hat{\mathbf{y}} \rangle_{\mathbb{R}^2}$ is expressed through the components x^i of $\hat{\mathbf{x}}$ and y^i of $\hat{\mathbf{y}}$ w.r.t. the non-orthogonal basis $\{\hat{\mathbf{v}}_1, \hat{\mathbf{v}}_2\}$, then it takes the form of an inner product with a metric: $\langle \hat{\mathbf{x}}, \hat{\mathbf{y}} \rangle_{\mathbb{R}^2} = \langle \mathbf{x}, \mathbf{y} \rangle_g = x^i g_{ij} y^j$, with $g_{ij} = \langle \hat{\mathbf{v}}_i, \hat{\mathbf{v}}_j \rangle_{\mathbb{R}^2}$. Compute the components of the metric explicitly (specifically: find g_{11}, g_{12}, g_{21} and g_{22}).

(d) The inner product from (c) can be written as $\langle \hat{\mathbf{x}}, \hat{\mathbf{y}} \rangle_{\mathbb{R}^2} = (x^i g_{ij}) y^j = x_j y^j$, with $x_j = x^i g_{ij}$, thus "hiding" the metric by absorbing it into the definition of covariant components (with subscript indices). Compute $\langle \hat{\mathbf{x}}, \hat{\mathbf{y}} \rangle_{\mathbb{R}^2}$ in this manner, by first finding x_1 and x_2. [Check: is the result consistent with that from (b)?]

ₚL3.3.8 Non-orthonormal basis vectors and metric (p. 50)

Consider the vectors $\hat{\mathbf{v}}_1 = (2, 1, 2)^T$, $\hat{\mathbf{v}}_2 = (1, 0, 1)^T$ and $\hat{\mathbf{v}}_3 = (1, 1, 0)^T$, written as column vectors in the standard basis of $\mathbb{R}^3$. (We use the same notational conventions as for problem L3.3.7.)

(a) Write the standard basis vector $\hat{\mathbf{e}}_1 = (1, 0, 0)^T$ as a linear combination of $\hat{\mathbf{v}}_1$, $\hat{\mathbf{v}}_2$ and $\hat{\mathbf{v}}_3$. Ditto for $\hat{\mathbf{e}}_2 = (0, 1, 0)^T$ and $\hat{\mathbf{e}}_3 = (0, 0, 1)^T$. Do $\hat{\mathbf{v}}_1$, $\hat{\mathbf{v}}_2$ and $\hat{\mathbf{v}}_3$ form a basis for $\mathbb{R}^3$?

(b) Let $\hat{\mathbf{x}} = \hat{\mathbf{v}}_1 x^1 + \hat{\mathbf{v}}_2 x^2 + \hat{\mathbf{v}}_3 x^3$ and $\hat{\mathbf{y}} = \hat{\mathbf{v}}_1 y^1 + \hat{\mathbf{v}}_2 y^2 + \hat{\mathbf{v}}_3 y^3$ be two vectors in $\mathbb{R}^3$, whose components w.r.t. $\hat{\mathbf{v}}_1$, $\hat{\mathbf{v}}_2$ and $\hat{\mathbf{v}}_3$ are given by $\mathbf{x} = (x^1, x^2, x^3) = (2, -5, 3)^T$ and $\mathbf{y} = (y^1, y^2, y^3) = (4, -1, -2)^T$, respectively. Express $\hat{\mathbf{x}}$ and $\hat{\mathbf{y}}$ as column vectors in the standard basis of $\mathbb{R}^3$ and compute their scalar product $\langle \hat{\mathbf{x}}, \hat{\mathbf{y}} \rangle_{\mathbb{R}^3}$.

(c) Find the components of the metric $g_{ij} = \langle \hat{\mathbf{v}}_i, \hat{\mathbf{v}}_j \rangle_{\mathbb{R}^3}$ explicitly.

(d) Now calculate the scalar product of $\hat{\mathbf{x}}$ and $\hat{\mathbf{y}}$ using the formula $\langle \hat{\mathbf{x}}, \hat{\mathbf{y}} \rangle_{\mathbb{R}^3} = \langle \mathbf{x}, \mathbf{y} \rangle_g = x^i g_{ij} y^j = x_j y^j$, with $x_j = x^i g_{ij}$, and carry out the sum over i and j explicitly. [Check: is the result consistent with that from (b)?]

P.L4 Vector product (p. 55)

P.L4.2 Algebraic formulation (p. 58)

εL4.2.1 Elementary computations with vectors (p. 59)

Given the vectors $\mathbf{a} = (4, 3, 1)^T$ and $\mathbf{b} = (1, -1, 1)^T$:

(a) calculate $\|\mathbf{b}\|$, $\mathbf{a} - \mathbf{b}$, $\mathbf{a} \cdot \mathbf{b}$ and $\mathbf{a} \times \mathbf{b}$;

(b)   decompose $\mathbf{a} \equiv \mathbf{a}_\| + \mathbf{a}_\perp$ into two vectors parallel and perpendicular to $\mathbf{b}$, respectively;

(c)   calculate $\mathbf{a}_\| \cdot \mathbf{b}$, $\mathbf{a}_\perp \cdot \mathbf{b}$, $\mathbf{a}_\| \times \mathbf{b}$ and $\mathbf{a}_\perp \times \mathbf{b}$. Do these results match your expectations?

[Check your results: (a) $\mathbf{a} \cdot \mathbf{b} + \sum_i (\mathbf{a} \times \mathbf{b})^i = -4$, (b) $\sum_i (\mathbf{a}_\|)^i = \frac{2}{3}$, $\sum_i (\mathbf{a}_\perp)^i = 7\frac{1}{3}$.]

ₚL4.2.2 Elementary computations with vectors (p. 59)

Given the vectors $\mathbf{a} = (2, 1, 5)^T$ and $\mathbf{b} = (-4, 3, 0)^T$:

(a)   calculate $\|\mathbf{b}\|$, $\mathbf{a} - \mathbf{b}$, $\mathbf{a} \cdot \mathbf{b}$ and $\mathbf{a} \times \mathbf{b}$;

(b)   decompose $\mathbf{a} \equiv \mathbf{a}_\| + \mathbf{a}_\perp$ into two vectors parallel and perpendicular to $\mathbf{b}$, respectively;

(c)   calculate $\mathbf{a}_\| \cdot \mathbf{b}$, $\mathbf{a}_\perp \cdot \mathbf{b}$, $\mathbf{a}_\| \times \mathbf{b}$ and $\mathbf{a}_\perp \times \mathbf{b}$. Do these results match your expectations?

[Check your results: (a) $\mathbf{a} \cdot \mathbf{b} + \sum_i (\mathbf{a} \times \mathbf{b})^i = -30$, (b) $\sum_i (\mathbf{a}_\|)^i = \frac{1}{5}$, $\sum_i (\mathbf{a}_\perp)^i = 7\frac{4}{5}$.]

ₑL4.2.3 Levi–Civita tensor (p. 59)

(a)   Is the statement $a^i b^j \epsilon_{ij2} \overset{?}{=} -a^k \epsilon_{k2l} b^l$ true or false? Justify your answer.

Express the following k-sums over products of two Levi–Civita tensors in terms of Kronecker delta functions. Check your answers by also writing out the k-sums explicitly and evaluating each term separately.

(b) $\epsilon_{1ik}\epsilon_{kj1}$,   (c) $\epsilon_{1ik}\epsilon_{kj2}$.

ₚL4.2.4 Levi–Civita tensor (p. 59)

(a)   Is the statement $a^i a^j \epsilon_{ij3} \overset{?}{=} b^m b^n \epsilon_{mn2}$ true or false? Justify your answer.

Express the following k-sums over products of two Levi–Civita tensors in terms of Kronecker delta functions.

(b) $\epsilon_{1ik}\epsilon_{23k}$,   (c) $\epsilon_{2jk}\epsilon_{ki2}$,   (d) $\epsilon_{1ik}\epsilon_{k3j}$.

P.L4.3   Further properties of the vector product (p. 60)

ₑL4.3.1 Grassmann identity (BAC-CAB) and Jacobi identity (p. 61)

(a)   Prove the Grassmann (or BAC-CAB) identity for arbitrary vectors $\mathbf{a}, \mathbf{b}, \mathbf{c} \in \mathbb{R}^3$:

$$\mathbf{a} \times (\mathbf{b} \times \mathbf{c}) = \mathbf{b}(\mathbf{a} \cdot \mathbf{c}) - \mathbf{c}(\mathbf{a} \cdot \mathbf{b}).$$

Hint: Expand the three vectors in an orthonormal basis, e.g. $\mathbf{a} = \mathbf{e}_i a^i$, and use the identity $\epsilon_{ijk}\epsilon_{mnk} = \delta_{im}\delta_{jn} - \delta_{in}\delta_{jm}$ for the Levi–Civita tensor. If you prefer, you may equally well write all indices downstairs, e.g. $\mathbf{a} = \mathbf{e}_i a_i$, since in an orthonormal basis $a_i = a^i$.

(b)   Use the Grassmann identity to derive the Jacobi identity:

$$\mathbf{a} \times (\mathbf{b} \times \mathbf{c}) + \mathbf{b} \times (\mathbf{c} \times \mathbf{a}) + \mathbf{c} \times (\mathbf{a} \times \mathbf{b}) = \mathbf{0}.$$

(c)   Check both identities explicitly for $\mathbf{a} = (1, 1, 2)^T$, $\mathbf{b} = (3, 2, 0)^T$ and $\mathbf{c} = (2, 1, 1)^T$ by separately computing all terms they contain.

ₚL4.3.2 Lagrange identity (p. 61)

(a)   Prove the Lagrange identity for arbitrary vectors $\mathbf{a}, \mathbf{b}, \mathbf{c}, \mathbf{d} \in \mathbb{R}^3$:

$$(\mathbf{a} \times \mathbf{b}) \cdot (\mathbf{c} \times \mathbf{d}) = (\mathbf{a} \cdot \mathbf{c})(\mathbf{b} \cdot \mathbf{d}) - (\mathbf{a} \cdot \mathbf{d})(\mathbf{b} \cdot \mathbf{c}).$$

Hint: Work in an orthonormal basis and use the properties of the Levi–Civita tensor.

(b)   Use (a) to compute $\|\mathbf{a} \times \mathbf{b}\|$ and express the result in terms of $\|\mathbf{a}\|$, $\|\mathbf{b}\|$ and the angle ϕ between $\mathbf{a}$ and $\mathbf{b}$.

(c)   Check the Lagrange identity explicitly for the vectors $\mathbf{a} = (2, 1, 0)^T$, $\mathbf{b} = (3, -1, 2)^T$, $\mathbf{c} = (3, 0, 2)^T$, $\mathbf{d} = (1, 3, -2)^T$ by separately computing all its terms.

ₑL4.3.3 Scalar triple product (p. 61)

This problem illustrates an important relation between the scalar triple product and the question of whether three vectors in $\mathbb{R}^3$ are linearly independent or not.

(a)   Compute the scalar triple product $S(y) = \mathbf{v}_1 \cdot (\mathbf{v}_2 \times \mathbf{v}_3)$, of $\mathbf{v}_1 = (1, 0, 2)^T$, $\mathbf{v}_2 = (3, 2, 1)^T$ and $\mathbf{v}_3 = (-1, -2, y)^T$ as a function of the variable y. [Check your result: $S(1) = -4$.]

(b)   By solving the vector equation $\mathbf{v}_i a^i = \mathbf{0}$, find that value of y for which $\mathbf{v}_1$, $\mathbf{v}_2$ and $\mathbf{v}_3$ are *not* linearly independent.

(c)   What is the value of $S(y)$ for the value of y found in (b)? Interpret your result!

ₚL4.3.4 Scalar triple product (p. 61)

Compute the volume, $V(\phi)$, of the parallelepiped spanned by three unit vectors $\mathbf{v}_1$, $\mathbf{v}_2$ and $\mathbf{v}_3$, each pair of which encloses a mutual angle of ϕ (with $0 \leq \phi \leq \frac{2}{3}\pi$; why is this restriction needed?).

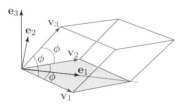

Check your results: (i) What do you expect for $V(\frac{\pi}{2})$ and $V(\frac{2}{3}\pi)$? (ii): $V(\frac{\pi}{3}) = \frac{1}{\sqrt{2}}$.

Hint: Choose the orientation of the parallelepiped such that $\mathbf{v}_1$ and $\mathbf{v}_2$ both lie in the plane spanned by $\mathbf{e}_1$ and $\mathbf{e}_2$, and that $\mathbf{e}_1$ bisects the angle between $\mathbf{v}_1$ and $\mathbf{v}_2$ (see figure).

P.L5   Linear maps (p. 63)

P.L5.1   Linear maps (p. 63)

ₑL5.1.1 Checking linearity (p. 65)

Consider the map $F : \mathbb{R} \rightarrow \mathbb{R}$, $F(v) = 2v + B$. For which values of the constant B is the map linear? Justify your answer by evaluating both $F(v + w)$ and $F(v) + F(w)$.

ₚL5.1.2 Checking linearity (p. 65)

Consider the map $F : \mathbb{R} \to \mathbb{R}$, $F(v) = 2v^{\alpha}$. For which values of the constant α is the map linear? Justify your answer by evaluating both $F(v + w)$ and $F(v) + F(w)$.

P.L5.3 Matrix multiplication (p. 69)

ₑL5.3.1 Matrix multiplication (p. 70)

Compute all possible products of pairs of the following matrices, including their squares, where possible:

$$P = \begin{pmatrix} 4 & -3 & 1 \\ 2 & 2 & -4 \end{pmatrix}, \quad Q = \begin{pmatrix} 3 & 0 & 1 \\ 1 & 2 & 5 \\ 1 & -6 & -1 \end{pmatrix}, \quad R = \begin{pmatrix} 3 & 0 \\ 1 & 2 \\ 1 & -6 \end{pmatrix}.$$

[Check your results: The sum of all elements of the first column of the following matrix products is: $\sum_i (PQ)^i{}_1 = 14$, $\sum_i (PR)^i{}_1 = 14$, $\sum_i (QR)^i{}_1 = 16$, $\sum_i (RP)^i{}_1 = 12$, $\sum_i (QQ)^i{}_1 = 16$.]

ₚL5.3.2 Matrix multiplication (p. 70)

Compute all possible products of pairs of the following matrices, including their squares, where possible:

$$P = \begin{pmatrix} 2 & 0 & 3 \\ -5 & 2 & 7 \\ 3 & -3 & 7 \\ 2 & 4 & 0 \end{pmatrix}, \quad Q = \begin{pmatrix} -3 & 1 \\ -1 & 0 \\ 2 & 1 \end{pmatrix}, \quad R = \begin{pmatrix} 6 & -1 & 4 \\ 4 & 4 & -4 \\ -4 & -4 & 6 \end{pmatrix}.$$

[Check your results: The sum of all elements of the first column of the following matrix products is: $\sum_i (PQ)^i{}_1 = 25$, $\sum_i (PR)^i{}_1 = -44$, $\sum_i (RQ)^i{}_1 = -5$, $\sum_i (RR)^i{}_1 = 8$.]

ₑL5.3.3 Spin-$\frac{1}{2}$ matrices (p. 70)

The **"spin"** of a quantum mechanical particle is a type of internal angular momentum. The description of quantum mechanical spin requires three matrices, S_x, S_y and S_z, whose commutators satisfy the **SU(2) algebra**. The **commutator** of two matrices is defined as $[A, B] \equiv AB - BA$. The SU(2) algebra is defined by the relations $[S_i, S_j] = i\epsilon_{ijk}S_k$, where ϵ_{ijk} is the antisymmetric Levi–Civita symbol (with $\epsilon_{xyz} = 1$, $\epsilon_{yxz} = -1$, etc.). The description of quantum mechanical particles with spin s, where $s \in \frac{1}{2}\mathbb{Z}$, utilizes a representation of the SU(2) algebra in terms of matrices of dimension $(2s + 1) \times (2s + 1)$. They have the property that the matrix $\mathbf{S}^2 \equiv S_x^2 + S_y^2 + S_z^2$ equals $s(s + 1)\mathbb{1}$.

The following matrices are used to describe quantum mechanical particles with spin $s = \frac{1}{2}$:

$$S_x = \frac{1}{2}\begin{pmatrix} 0 & 1 \\ 1 & 0 \end{pmatrix}, \quad S_y = \frac{1}{2}\begin{pmatrix} 0 & -i \\ i & 0 \end{pmatrix}, \quad S_z = \frac{1}{2}\begin{pmatrix} 1 & 0 \\ 0 & -1 \end{pmatrix}.$$

(a) Compute $\mathbf{S}^2$. Is the result consistent with the expected form $s(s + 1)\mathbb{1}$?

(b) Verify that S_x, S_y and S_z satisfy the SU(2) algebra $[S_i, S_j] = i\epsilon_{ijk}S_k$.

⊳L5.3.4 Spin-1 matrices (p. 70)

The following matrices are used to describe quantum mechanical particles with spin $s = 1$:

$$S_x = \frac{1}{\sqrt{2}} \begin{pmatrix} 0 & 1 & 0 \\ 1 & 0 & 1 \\ 0 & 1 & 0 \end{pmatrix}, \qquad S_y = \frac{1}{\sqrt{2}} \begin{pmatrix} 0 & -i & 0 \\ i & 0 & -i \\ 0 & i & 0 \end{pmatrix}, \qquad S_z = \begin{pmatrix} 1 & 0 & 0 \\ 0 & 0 & 0 \\ 0 & 0 & -1 \end{pmatrix}.$$

(a)   Compute $\mathbf{S}^2 \equiv S_x^2 + S_y^2 + S_z^2$. Is the result consistent with the expected form $s(s+1)\mathbb{1}$?

(b)   Verify that S_x, S_y and S_z satisfy the SU(2) algebra $[S_i, S_j] = i\epsilon_{ijk}S_k$.

εL5.3.5 Matrix multiplication (p. 70)

Let A and B be $N \times N$ matrices with matrix elements $A^i{}_j = a_j \delta^i{}_m$ and $B^i{}_j = b_i \delta^i{}_j$, for a fixed choice of $m \in \{1, 2, \ldots, N\}$. *Remark:* Since the indices i and j are specified on the left, they are *not* summed over on the right even though in the expression for $B^i{}_j$ the index i appears twice on the right.

(a)   For $N = 3$ and $m = 2$, write these matrices explicitly in the usual matrix representation and calculate the matrix product AB explicitly.

(b)   Calculate the product AB for arbitrary $N \in \mathbb{N}$ and $1 \leq m \leq N$. [Check your result: The sum of the diagonal elements yields: $\sum_{i=1}^{N}(AB)^i{}_i = a_m b_m$.]

⊳L5.3.6 Matrix multiplication (p. 70)

Let A and B be $N \times N$ matrices with matrix elements $A^i{}_j = a_i \delta^i{}_{N+1-j}$ and $B^i{}_j = b_i \delta^i{}_j$. *Remark:* Since the indices i and j are specified on the left, they are *not* summed over on the right even though the index i appears twice in $B^i{}_j$ on the right.

(a)   For $N = 3$ and $m = 2$, write these matrices explicitly in the usual matrix representation and calculate the matrix product AB explicitly.

(b)   Calculate the product AB for arbitrary $N \in \mathbb{N}$ and $1 \leq m \leq N$. [Check your result: If N is odd, the sum of the diagonal elements yields: $\sum_{i=1}^{N}(AB)^i{}_i = a_{(N+1)/2}\, b_{(N+1)/2}.$]

P.L5.4   The inverse of a matrix (p. 72)

εL5.4.1 Gaussian elimination and matrix inversion (p. 78)

Gaussian elimination is a convenient bookkeeping scheme for solving a linear system of equations of the form $A\mathbf{x} = \mathbf{b}$. For example, consider the system

$$A^1{}_1\, x^1 + A^1{}_2\, x^2 + A^1{}_3\, x^3 = b^1 ,$$
$$A^2{}_1\, x^1 + A^2{}_2\, x^2 + A^2{}_3\, x^3 = b^2 ,$$
$$A^3{}_1\, x^1 + A^3{}_2\, x^2 + A^3{}_3\, x^3 = b^3 .$$

It can be solved by a sequence of steps, each of which involves taking a linear combination of rows, chosen such that the system is brought into the form

$$1\, x^1 + 0\, x^2 + 0\, x^3 = c^1 ,$$
$$0\, x^1 + 1\, x^2 + 0\, x^3 = c^2 ,$$
$$0\, x^1 + 0\, x^2 + 1\, x^3 = c^3 .$$

The solution can then be read off from the right-hand side, $(x^1, x^2, x^3)^{\mathrm{T}} = (c^1, c^2, c^3)^{\mathrm{T}}$.

During these manipulations, time and ink can be saved by refraining from writing down the x^is over and over again. Instead, it suffices to represent the linear system by an **augmented matrix**, containing the coefficients in array form, with a vertical line instead of the equal signs. This augmented matrix is to be manipulated in a sequence of steps, each of which involves taking a linear combination of rows, chosen such that the left side is brought into the form of the unit matrix. The right column then contains the desired solution for $(x^1, x^2, x^3)^T$.

$$
\begin{array}{ccc|c}
x^1 & x^2 & x^3 & \\
A^1_{\ 1} & A^1_{\ 2} & A^1_{\ 3} & b^1 \\
A^2_{\ 1} & A^2_{\ 2} & A^2_{\ 3} & b^2 \\
A^3_{\ 1} & A^3_{\ 2} & A^3_{\ 3} & b^3
\end{array}
\quad \longrightarrow \quad
\begin{array}{ccc|c}
x^1 & x^2 & x^3 & \\
1 & 0 & 0 & c^1 \\
0 & 1 & 0 & c^2 \\
0 & 0 & 1 & c^3
\end{array}
$$

Gaussian elimination is also useful for matrix inversion. The inverse of A has the form $A^{-1} = (\mathbf{a}_1, \ldots, \mathbf{a}_n)$, where the jth column is the solution of the linear system $A\mathbf{a}_j = \mathbf{e}_j$. The computation of all n vectors $\mathbf{a}_j$ can be done simultaneously by setting up an augmented matrix with n columns on the right, each containing an $\mathbf{e}_j$. After manipulating the augmented matrix such that the left side is the unit matrix, the columns on the right contain the desired vectors $\mathbf{a}_j$.

$$
\begin{array}{ccc|ccc}
A^1_{\ 1} & A^1_{\ 2} & A^1_{\ 3} & 1 & 0 & 0 \\
A^2_{\ 1} & A^2_{\ 2} & A^2_{\ 3} & 0 & 1 & 0 \\
A^3_{\ 1} & A^3_{\ 2} & A^3_{\ 3} & 0 & 0 & 1
\end{array}
\quad \longrightarrow \quad
\begin{array}{ccc|ccc}
1 & 0 & 0 & a^1_{\ 1} & a^1_{\ 2} & a^1_{\ 3} \\
0 & 1 & 0 & a^2_{\ 1} & a^2_{\ 2} & a^2_{\ 3} \\
0 & 0 & 1 & a^3_{\ 1} & a^3_{\ 2} & a^3_{\ 3}
\end{array}
$$

(a) Solve the following system of linear equations using Gaussian elimination.

$$
\begin{array}{rcrcrcr}
3x^1 & + & 2x^2 & - & x^3 & = & 1, \\
2x^1 & - & 2x^2 & + & 4x^3 & = & -2, \\
-x^1 & + & \frac{1}{2}x^2 & - & x^3 & = & 0.
\end{array}
$$

[Check your result: The norm of $\mathbf{x}$ is $\|\mathbf{x}\| = 3$.]

(b) How does the solution change when the last equation is removed?

(c) What happens if the last equation is replaced by $-x^1 + \frac{2}{7}x^2 - x^3 = 0$?

(d) The system of equations given in (a) can also be expressed in the form $A\mathbf{x} = \mathbf{b}$. Calculate the inverse A^{-1} of the 3×3 matrix A using Gaussian elimination. Verify your answer to (a) using $\mathbf{x} = A^{-1}\mathbf{b}$.

PL5.4.2 Gaussian elimination and matrix inversion (p. 78)

Consider the linear system of equations $A\mathbf{x} = \mathbf{b}$, with

$$
A = \begin{pmatrix}
8 - 3a & 2 - 6a & 2 \\
2 - 6a & 5 & -4 + 6a \\
2 & -4 + 6a & 5 + 3a
\end{pmatrix}. \tag{1}
$$

(a) For $a = \frac{1}{3}$, use Gaussian elimination to compute the inverse matrix A^{-1}. (*Remark:* It is advisable to avoid the occurrence of fractions until the left side has been brought into row echelon form.) Use the result to find the solution $\mathbf{x}$ for $\mathbf{b} = (4, 1, 1)^T$. [Check your result: The norm of $\mathbf{x}$ is $\|\mathbf{x}\| = \sqrt{117}/18$.]

(b) For which values of a can the matrix A *not* be inverted?

(c) If A can be inverted, the system of equations $A\mathbf{x} = \mathbf{b}$ has a unique solution for every $\mathbf{b}$, namely $\mathbf{x} = A^{-1}\mathbf{b}$. If A cannot be inverted, then either the solution is not unique, or no solution exists at all – it depends on $\mathbf{b}$ which of these two cases arises. Decide this for $\mathbf{b} = (4, 1, 1)^{\mathrm{T}}$ and the values for a found in (b), and determine $\mathbf{x}$, if possible.

ᴇL5.4.3 Matrix inversion (p. 78)

Let M_n be an $n \times n$ matrix with matrix elements $(M_n)^i_{\;j} = \delta^i_{\;j} m + \delta^1_{\;j}$, with $i,j = 1, \ldots, n$.

(a) Find the inverse matrices M_2^{-1} and M_3^{-1}. Verify in both cases that $M_n^{-1} M_n = \mathbb{1}$.

(b) Use the results from (a) to formulate an ansatz for the form of the inverse matrix M_n^{-1} for arbitrary n. Check your ansatz by calculating $M_n^{-1} M_n$.

(c) Give a compact formula for the matrix elements $(M_n^{-1})^i_{\;j}$. Check its validity by showing that $\sum_l (M_n^{-1})^i_{\;l} (M_n)^l_{\;j} = \delta^i_{\;j}$ holds, by explicitly performing the sum on l.

ᴘL5.4.4 Matrix inversion (p. 78)

Let M_n be an $n \times n$ matrix with matrix elements $(M_n)^i_{\;j} = m \, \delta^i_{\;j} + \delta^{i+1}_{\;j}$ with $i,j = 1, \ldots, n$.

(a) Find the inverse matrices M_2^{-1} and M_3^{-1}. Verify in both cases that $M_n^{-1} M_n = \mathbb{1}$.

(b) Use the results from (a) to formulate a guess at the form of the inverse matrix M_n^{-1} for arbitrary n. Check your guess by calculating $M_n^{-1} M_n$.

(c) Give a compact formula for the matrix elements $(M_n^{-1})^i_{\;j}$. Check its validity by showing that $\sum_l (M_n^{-1})^i_{\;l} (M_n)^l_{\;j} = \delta^i_{\;j}$ holds, by explicitly performing the sum over l.

P.L5.5 General linear maps and matrices (p. 78)

ᴇL5.5.1 Two-dimensional rotation matrices (p. 79)

A rotation in two dimensions is a linear map, $R : \mathbb{R}^2 \to \mathbb{R}^2$, that rotates every vector by a given angle about the origin without changing its length.

(a) Let the (2×2)-dimensional rotation matrix R_θ describing a rotation by the angle θ be defined by $\mathbf{e}_j \overset{R_\theta}{\longmapsto} \mathbf{e}'_j = \mathbf{e}_i (R_\theta)^i_{\;j}$. Find R_θ by proceeding as follows. Make a sketch that illustrates the effect, $\mathbf{e}_j \overset{R_\theta}{\longrightarrow} \mathbf{e}'_j$, of the rotation on the two basis vectors $\mathbf{e}_j$ ($j = 1, 2$) (e.g. for $\theta = \frac{\pi}{6}$). The image vectors $\mathbf{e}'_j$ of the basis vectors $\mathbf{e}_j$ yield the columns of the matrix R_θ.

(b) Write down the matrix R_{θ_i} for the angles $\theta_1 = 0, \theta_2 = \pi/4, \theta_3 = \pi/2$ and $\theta_4 = \pi$. Compute the action of R_{θ_i} ($i = 1, 2, 3, 4$) on $\mathbf{a} = (1, 0)^{\mathrm{T}}$ and $\mathbf{b} = (0, 1)^{\mathrm{T}}$, and make a sketch to visualize the results.

(c) The composition of two rotations is again a rotation. Show that $R_\theta R_\phi = R_{\theta+\phi}$. *Hint:* Utilize the following "addition theorems":

$$\cos(\theta + \phi) = \cos\theta\cos\phi - \sin\theta\sin\phi\,,$$
$$\sin(\theta + \phi) = \sin\theta\cos\phi + \cos\theta\sin\phi\,.$$

Remark: A geometric proof of these theorems (not requested here) follows from the figure by inspecting the three right-angled triangles with diagonals of length 1, $\cos\phi$ and $\sin\phi$.

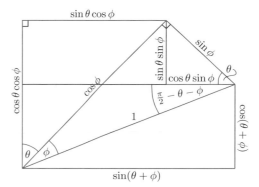

(d) Show that the rotation of an arbitrary vector $\mathbf{r} = (x, y)^{\mathrm{T}}$ by the angle θ does not change its length, i.e. that $R_\theta \mathbf{r}$ has the same length as $\mathbf{r}$.

P L5.5.2 Three-dimensional rotation matrices (p 79)

Rotations in three dimensions are represented by (3×3)-dimensional matrices. Let $R_\theta(\mathbf{n})$ be the rotation matrix that describes a rotation by the angle θ about an axis whose direction is given by the unit vector $\mathbf{n}$. Its elements are defined via $\mathbf{e}_j \overset{R_\theta(\mathbf{n})}{\longmapsto} = \mathbf{e}_l(R_\theta(\mathbf{n}))^l{}_j$.

(a) Find the three rotation matrices $R_\theta(\mathbf{e}_i)$ for rotations about the three Cartesian coordinate axes $\mathbf{e}_1$, $\mathbf{e}_2$ and $\mathbf{e}_3$, by proceeding as follows. Use three sketches, one each for $i = 1, 2, 3$, illustrating the effect, $\mathbf{e}_j \overset{R_\theta(\mathbf{e}_i)}{\longmapsto} \mathbf{e}'_j$, of a rotation about the i axis on all three basis vectors $\mathbf{e}_j$ ($j = 1, 2, 3$) (e.g. for $\theta = \frac{\pi}{6}$). The image vectors $\mathbf{e}'_j$ of the basis vectors $\mathbf{e}_j$ yield the columns of $R_\theta(\mathbf{e}_i)$.

(b) It can be shown ($\to$ L8.3.2) that for a general direction, $\mathbf{n} = (n_1, n_2, n_3)^{\mathrm{T}}$, of the axis of rotation, the matrix elements have the following form:

$$(R_\theta(\mathbf{n}))^i{}_j = \delta_{ij}\cos\theta + n_i n_j(1 - \cos\theta) - \epsilon_{ijk}\, n_k \sin\theta \qquad (\epsilon_{ijk} = \text{Levi–Civita symbol}).$$

Use this formula to find the three rotation matrices $R_\theta(\mathbf{e}_i)$ ($i = 1, 2, 3$) explictly. Are your results consistent with those from (a)?

(c) Write down the following rotation matrices explicitly, and compute and sketch their effect on the vector $\mathbf{v} = (1, 0, 1)^{\mathrm{T}}$:

(i) $A = R_\pi(\mathbf{e}_3)$,   (ii) $B = R_{\frac{\pi}{2}}(\frac{1}{\sqrt{2}}(\mathbf{e}_3 - \mathbf{e}_1))$.

(d) Rotation matrices form a group. Use A and B from (c) to illustrate that this group is not commutative (in contrast to the two-dimensional case!).

(e) Show that a general rotation matrix R satisfies the relation $\mathrm{Tr}(R) = 1 + 2\cos\theta$, where the "trace" of a matrix R is defined by $\mathrm{Tr}(R) = \sum_i (R)^i{}_i$.

(f) The product of two rotation matrices is again a rotation matrix. Consider the product $C = AB$ of the two matrices from (c), and find the corresponding unit vector $\mathbf{n}$ and rotation angle θ. *Hint:* These are uniquely defined only up to an arbitrary sign, since $R_\theta(\mathbf{n})$ and $R_{-\theta}(-\mathbf{n})$ describe the same rotation. (To be concrete, fix this sign by choosing the component n_2 positive.) $|\theta|$ and $|n_i|$ are fixed by the trace and the diagonal elements of the rotation matrix, respectively; their relative sign is fixed by the off-diagonal elements. [Check your result: $n_2 = 1/\sqrt{3}$.]

P.L5.6 Matrices describing coordinate changes (p. 80)

ₑL5.6.1 Basis transformations and linear maps in $\mathbb{E}^2$ (p. 84)

Remark on notation: For this problem we denote vectors in Euclidean space $\mathbb{E}^2$ using hats (e.g. $\hat{\mathbf{v}}_j$, $\hat{\mathbf{x}}, \hat{\mathbf{y}} \in \mathbb{E}^2$). Their components with respect to a given basis are vectors in $\mathbb{R}^2$ and are written without hats (e.g. $\mathbf{x}, \mathbf{y} \in \mathbb{R}^2$).

Consider two bases for the Euclidean vector space $\mathbb{E}^2$, one old, $\{\hat{\mathbf{v}}_j\}$, and one new, $\{\hat{\mathbf{v}}_i'\}$, with

$$\hat{\mathbf{v}}_1 = \tfrac{3}{4}\hat{\mathbf{v}}_1' + \tfrac{1}{3}\hat{\mathbf{v}}_2', \quad \hat{\mathbf{v}}_2 = -\tfrac{1}{8}\hat{\mathbf{v}}_1' + \tfrac{1}{2}\hat{\mathbf{v}}_2'.$$

(a) The relation $\hat{\mathbf{v}}_j = \hat{\mathbf{v}}_i' T^i{}_j$ expresses the old basis in terms of the new basis. Find the transformation matrix $T = \{T^i{}_j\}$. [Check your result: $\sum_j T^1{}_j = \tfrac{5}{8}$.]

(b) Find the matrix T^{-1}, and use the inverse transformation $\hat{\mathbf{v}}_i' = \hat{\mathbf{v}}_j (T^{-1})^j{}_i$ to express the new basis in terms of the old basis. [Check your result: $\hat{\mathbf{v}}_1' - 4\hat{\mathbf{v}}_2' = -8\hat{\mathbf{v}}_2$.]

(c) Let $\hat{\mathbf{x}}$ be a vector with components $\mathbf{x} = (1, 2)^{\mathrm{T}}$ in the old basis. Find its components $\mathbf{x}'$ in the new basis. [Check your result: $\sum_i x'^i = \tfrac{11}{6}$.]

(d) Let $\hat{\mathbf{y}}$ by a vector with components $\mathbf{y}' = (\tfrac{3}{4}, \tfrac{1}{3})^{\mathrm{T}}$ in the new basis. Find its components $\mathbf{y}$ in the old basis. [Check your result: $\sum_j y^j = 1$.]

(e) Let $\hat{A}$ be the linear map defined by $\hat{\mathbf{v}}_1' \overset{\hat{A}}{\mapsto} 2\hat{\mathbf{v}}_1'$ and $\hat{\mathbf{v}}_2' \overset{\hat{A}}{\mapsto} \hat{\mathbf{v}}_2'$. First find the matrix representation A' of this map in the new basis, then use a basis transformation to find its matrix representation A in the old basis. [Check your result: $(A)^2{}_1 = -\tfrac{3}{5}$.]

(f) Let $\hat{\mathbf{z}}$ be the image vector onto which the vector $\hat{\mathbf{x}}$ is mapped by $\hat{A}$, i.e. $\hat{\mathbf{x}} \overset{\hat{A}}{\mapsto} \hat{\mathbf{z}}$. Find its components $\mathbf{z}'$ with respect to the new basis by using A', and its components $\mathbf{z}$ with respect to the old basis by using A. Are your results for $\mathbf{z}'$ and $\mathbf{z}$ consistent? [Check your result: $\mathbf{z}' = (1, \tfrac{4}{3})^{\mathrm{T}}$.]

(g) Now make the choice $\hat{\mathbf{v}}_1 = 3\hat{\mathbf{e}}_1 + \hat{\mathbf{e}}_2$ and $\hat{\mathbf{v}}_2 = -\tfrac{1}{2}\hat{\mathbf{e}}_1 + \tfrac{3}{2}\hat{\mathbf{e}}_2$ for the old basis, where $\hat{\mathbf{e}}_1 = (1, 0)^{\mathrm{T}}$ and $\hat{\mathbf{e}}_2 = (0, 1)^{\mathrm{T}}$ are the standard Cartesian basis vectors of $\mathbb{E}^2$. What are the components of $\hat{\mathbf{v}}_1', \hat{\mathbf{v}}_2', \hat{\mathbf{x}}$ and $\hat{\mathbf{z}}$ in the standard basis $\mathbb{E}^2$? [Check your results: $\|\hat{\mathbf{v}}_1'\| = 4$, $\|\hat{\mathbf{v}}_2'\| = 3$, $\|\hat{\mathbf{x}}\| = 2\sqrt{5}$, $\|\hat{\mathbf{z}}\| = 4\sqrt{2}$.]

(h) Make a sketch (with $\hat{\mathbf{e}}_1$ and $\hat{\mathbf{e}}_2$ as unit vectors in the horizontal and vertical directions respectively), showing the old and new basis vectors, as well as the vectors $\hat{\mathbf{x}}$ and $\hat{\mathbf{z}}$. Are the coordinates of these vectors, discussed in (c) and (f), consistent with your sketch?

ₚL5.6.2 Basis transformations and linear maps in $\mathbb{E}^2$ (p. 83)

Remark on notation: For this problem we denote vectors in Euclidean space $\mathbb{E}^2$ using hats (e.g. $\hat{\mathbf{v}}_j$, $\hat{\mathbf{x}}, \hat{\mathbf{y}} \in \mathbb{E}^2$). Their components with respect to a given basis are vectors in $\mathbb{R}^2$ and are written without hats (e.g. $\mathbf{x}, \mathbf{y} \in \mathbb{R}^2$).

Consider two bases for the Euclidean vector space $\mathbb{E}^2$, one old, $\{\hat{\mathbf{v}}_j\}$, and one new, $\{\hat{\mathbf{v}}_i'\}$, with

$$\hat{\mathbf{v}}_1 = \tfrac{1}{5}\hat{\mathbf{v}}_1' + \tfrac{3}{5}\hat{\mathbf{v}}_2', \quad \hat{\mathbf{v}}_2 = -\tfrac{6}{5}\hat{\mathbf{v}}_1' + \tfrac{2}{5}\hat{\mathbf{v}}_2'.$$

(a) The relation $\hat{\mathbf{v}}_j = \hat{\mathbf{v}}_i' T^i{}_j$ expresses the old basis in terms of the new basis. Find the transformation matrix $T = \{T^i{}_j\}$. [Check your result: $\sum_j T^2{}_j = 1$.]

(b) Find the matrix T^{-1}, and use the inverse transformation $\hat{\mathbf{v}}'_i = \hat{\mathbf{v}}_j (T^{-1})^j{}_i$ to express the new basis in terms of the old basis. [Check your result: $\hat{\mathbf{v}}'_1 + 3\hat{\mathbf{v}}'_2 = 5\hat{\mathbf{v}}_1$.]

(c) Let $\hat{\mathbf{x}}$ be a vector with components $\mathbf{x} = (2, -\frac{1}{2})^{\mathrm{T}}$ in the old basis. Find its components $\mathbf{x}'$ in the new basis. [Check your result: $\sum_i x'^i = 2$.]

(d) Let $\hat{\mathbf{y}}$ by a vector with components $\mathbf{y}' = (-3, 1)^{\mathrm{T}}$ in the new basis. Find its components $\mathbf{y}$ in the old basis. [Check your result: $\sum_j y^j = \frac{5}{2}$.]

(e) Let $\hat{A}$ be the linear map defined by $\hat{\mathbf{v}}_1 \overset{\hat{A}}{\mapsto} \frac{1}{3}(\hat{\mathbf{v}}_1 - 2\hat{\mathbf{v}}_2)$ and $\hat{\mathbf{v}}_2 \overset{\hat{A}}{\mapsto} -\frac{1}{3}(4\hat{\mathbf{v}}_1 + \hat{\mathbf{v}}_2)$. First find the matrix representation A of this map in the old basis, then use a basis transformation to find its matrix representation A' in the new basis. [Check your result: $(A')^2{}_1 = \frac{2}{3}$.]

(f) Let $\hat{\mathbf{z}}$ be the image vector onto which the vector $\hat{\mathbf{x}}$ is mapped by $\hat{A}$, i.e. $\hat{\mathbf{x}} \overset{\hat{A}}{\mapsto} \hat{\mathbf{z}}$. Find its components $\mathbf{z}$ with respect to the old basis by using A, and its components $\mathbf{z}'$ with respect to the new basis by using A'. Are your results for $\mathbf{z}$ and $\mathbf{z}'$ consistent? [Check your result: $\mathbf{z}' = \frac{1}{3}(5, 1)^{\mathrm{T}}$.]

(g) Now make the choice $\hat{\mathbf{v}}_1 = \hat{\mathbf{e}}_1 + \hat{\mathbf{e}}_2$ and $\hat{\mathbf{v}}_2 = 2\hat{\mathbf{e}}_1 - \hat{\mathbf{e}}_2$ for the old basis, where $\hat{\mathbf{e}}_1 = (1, 0)^{\mathrm{T}}$ and $\hat{\mathbf{e}}_2 = (0, 1)^{\mathrm{T}}$ are the standard Cartesian basis vectors of $\mathbb{E}^2$. What are the components of $\hat{\mathbf{v}}'_1$, $\hat{\mathbf{v}}'_2$, $\hat{\mathbf{x}}$ and $\hat{\mathbf{z}}$ in the standard basis $\mathbb{E}^2$? [Check your results: $\|\hat{\mathbf{v}}'_1\| = \frac{\sqrt{41}}{4}$, $\|\hat{\mathbf{v}}'_2\| = \frac{\sqrt{89}}{4}$, $\|\hat{\mathbf{x}}\| = \|\hat{\mathbf{z}}\| = \frac{\sqrt{29}}{2}$.]

(h) Make a sketch (with $\hat{\mathbf{e}}_1$ and $\hat{\mathbf{e}}_2$ as unit vectors in the horizontal and vertical directions, respectively), showing the old and new basis vectors, as well as the vectors $\hat{\mathbf{x}}$ and $\hat{\mathbf{z}}$. Are the coordinates of these vectors, discussed in (c) and (f), consistent with your sketch?

P.L5.6.3 Basis transformations (p. 84)

Consider the following three linear transformations in $\mathbb{R}^3$, with standard basis $\{\mathbf{e}_1, \mathbf{e}_2, \mathbf{e}_3\}$.

A : Rotation about the third axis by the angle $\theta_3 = \frac{\pi}{4}$, in the right-hand positive direction. *Hint:* Use the compact notation $\cos\theta_3 = \sin\theta_3 = s$.

B : Dilation (stretching) of the first axis by the factor $s_1 = 3$;

C : Rotation about the second axis by the angle $\theta_2 = \frac{\pi}{2}$, in the right-hand positive direction.

Hint: To understand what "right-hand positive" means, imagine wrapping your right hand around the axis of rotation, with your thumb pointing in the positive direction. Your other fingers will be curled in the direction of "positive rotation".

(a) Find the matrix representations (with respect to the standard basis) of A, B and C.

(b) What is the image, $\mathbf{y} = B\mathbf{x}$, of the vector $\mathbf{x} = (1, 1, 1)^{\mathrm{T}}$ under the dilation B?

(c) What is the image, $\mathbf{z} = D\mathbf{x}$, of $\mathbf{x}$ under the composition of all three maps, $D = C \cdot B \cdot A$? [Check your result: $z^2 = \sqrt{2}$.]

(d) Now consider a new basis, $\{\mathbf{e}'_i\}$, defined by a rotation of the standard basis by A, i.e. $\mathbf{e}_j \overset{A}{\mapsto} \mathbf{e}'_j$. Draw the new and old basis vectors in the same figure. Find the transformation matrix $T = \{T^i{}_j\}$, and specify the matrix elements of the transformation between the old and the rotated bases using $\mathbf{e}_j = \mathbf{e}'_i T^i{}_j$.

(e) In the $\{\mathbf{e}'_i\}$ basis, let the vectors $\mathbf{x}$ and $\mathbf{y}$ considered above be represented by $\mathbf{x} = \mathbf{e}'_i x'^i$ and $\mathbf{y} = \mathbf{e}'_i y'^i$. Find the corresponding components $\mathbf{x}' = (x'^1, x'^2, x'^3)^{\mathrm{T}}$ and $\mathbf{y}' = (y'^1, y'^2, y'^3)^{\mathrm{T}}$. [Check your results: $x'^1 = \sqrt{2}, y'^3 = 1$.]

(f)   Let B' denote the dilation B in the rotated basis. Find B' by the appropriate transformation of the matrix B, and use the result to calculate the image $\mathbf{y}'$ of $\mathbf{x}'$ under B'. [Does the result match that from (e)?]

pL5.6.4   Basis transformations and linear maps (p. 84)

Consider the following three linear transformations in $\mathbb{R}^3$, with standard basis $\{\mathbf{e}_1, \mathbf{e}_2, \mathbf{e}_3\}$.

A :   Rotation around the first axis by the angle $\theta_1 = -\frac{\pi}{3}$ in the right-handed sense, i.e. a left-handed rotation. *Hint:* Use the compact notation $\cos\theta_1 = c$, $\sin\theta_1 = s$.

B :   Dilation of the first and second axes by the factors $s_1 = 2$ and $s_2 = 4$ respectively.

C :   A reflection in the (2,3)-plane.

(a)   Find the matrix representations (using the standard basis) of A, B, C. Which of these transformations commute with each other (i.e. for which pairs of matrices does $M_1 M_2 = M_2 M_1$)?

(b)   What is the image, $\mathbf{y} = CA\mathbf{x}$, of the vector $\mathbf{x} = (1, 1, 1)^{\mathrm{T}}$ under the transformation CA?

(c)   Find the vector $\mathbf{z}$ whose image under the composition of all three transformations, $D = C \cdot B \cdot A$, gives $\mathbf{y}$ [*Hint:* $D^{-1} = A^{-1} B^{-1} C^{-1}$.] [Check your result: $z^3 = \frac{1}{16}(7 - 3\sqrt{3})$.]

(d)   Now consider a new basis, $\{\mathbf{e}'_i\}$, defined by a rotation and reflection CA of the standard basis, $\mathbf{e}'_i \overset{CA}{\mapsto} \mathbf{e}_i$. [*Caution:* In the example problem the order was reversed!] Sketch the old and new bases in the same picture. [*Note:* The new basis vectors are a left-handed system! Why?] Find the transformation matrix T, and specify the matrix elements of the transformation between the old and the new basis, with $\mathbf{e}_j = \mathbf{e}'_i T^i{}_j$.

(e)   In the $\{\mathbf{e}'_i\}$-basis, let the vectors $\mathbf{z}$ and $\mathbf{y}$ considered above be represented by $\mathbf{z} = \mathbf{e}'_i z'^i$ and $\mathbf{y} = \mathbf{e}'_i y'^i$. Find the corresponding components $\mathbf{z}' = (z'^1, z'^2, z'^3)^{\mathrm{T}}$ and $\mathbf{y}' = (y'^1, y'^2, y'^3)^{\mathrm{T}}$. [Check your results: $z'^3 = \frac{1}{2}(1 - \sqrt{3})$, $y'^2 = \frac{1}{2}(-1 + \sqrt{3})$.]

(f)   Let D' denote the representation of D in the new basis. Find D' by an appropriate transformation of the matrix D, and use the result to find the image $\mathbf{y}'$ of $\mathbf{z}'$ under D'. [Does the result match the one from (e)?].

P.L6   Determinants (p. 86)

P.L6.2   Computing determinants (p. 88)

eL6.2.1   Computing determinants (p. 88)

Compute the determinants of the following matrices by expanding them along an arbitrary row or column. *Hint:* The more zeros the row or column contains, the easier the calculation.

$$A = \begin{pmatrix} 2 & 1 \\ 5 & -3 \end{pmatrix}, \quad B = \begin{pmatrix} 3 & 2 & 1 \\ 4 & -3 & 1 \\ 2 & -1 & 1 \end{pmatrix}, \quad C = \begin{pmatrix} a & a & a & 0 \\ a & 0 & 0 & b \\ 0 & 0 & b & b \\ a & b & b & 0 \end{pmatrix}.$$

[Check your result: If $a = 1$, $b = 2$, then $\det(C) = -4$.]

pL6.2.2 Computing determinants (p. 88)

(a)   Compute the determinant of the matrix $D = \begin{pmatrix} 1 & c & 0 \\ d & 2 & 3 \\ 2 & 2 & e \end{pmatrix}$.

[Check your result: If $c = 1$, $d = 3$, $e = 2$, then $\det(C) = -2$.]
(i) Which values must c and d have to ensure that $\det(D) = 0$ for all values of e?
(ii) Which values must d and e have to ensure that $\det(D) = 0$ for all values of c?
Could you have found the results of (i), (ii) *without* explicitly calculating $\det(D)$?

Now consider the two matrices $A = \begin{pmatrix} 2 & -1 & -3 & 1 \\ 0 & 1 & 5 & 5 \end{pmatrix}$ and $B = \begin{pmatrix} 2 & 1 \\ 6 & 6 \\ -2 & 8 \\ -2 & -2 \end{pmatrix}$.

(b)   Compute the product AB, as well as its determinant $\det(AB)$ and inverse $(AB)^{-1}$.

(c)   Compute the product BA, as well as its determinant $\det(BA)$ and inverse $(BA)^{-1}$.
Is it possible to calculate the determinant and the inverse of A and B?

P.L7   Matrix diagonalization (p. 98)

P.L7.3   Matrix diagonalization (p. 102)

eL7.3.1 Matrix diagonalization (p. 103)

For each of the following matrices, find the eigenvalues λ_j and a set of eigenvectors $\mathbf{v}_j$. Also find a similarity transformation, T, and its inverse, T^{-1}, for which $T^{-1}AT$ is diagonal.

(a) $A = \begin{pmatrix} -1 & 6 \\ -2 & 6 \end{pmatrix}$,   (b) $A = \begin{pmatrix} -i & 0 \\ 2 & i \end{pmatrix}$,   (c) $A = \begin{pmatrix} 1 & 0 & -1 \\ 0 & 2i & 0 \\ 1 & 0 & 1 \end{pmatrix}$.

[Consistency checks: Do the eigenvalues satisfy $\sum_j \lambda_j = \text{tr}(A)$ and $\prod_j \lambda_j = \det(A)$? Does $T^{-1}AT$ yield a matrix, $D = \text{diag}\{\lambda_j\}$, containing the eigenvalues on the diagonal, or conversely, does TDT^{-1} reproduce A? Which of the latter two checks do you find more efficient?]

pL7.3.2 Matrix diagonalization (p. 103)

For each of the following matrices, find the eigenvalues λ_j and a set of eigenvectors $\mathbf{v}_j$. For definiteness, choose the first element of each eigenvector equal to unity, $v^1_j = 1$. Find a similarity transformation, T, and its inverse, T^{-1}, for which $T^{-1}AT$ is diagonal.

(a) $A = \begin{pmatrix} 4 & -6 \\ 3 & -5 \end{pmatrix}$,   (b) $A = \begin{pmatrix} 2-i & 1+i \\ 2+2i & -1+2i \end{pmatrix}$,   (c) $A = \begin{pmatrix} -1 & 1 & 0 \\ 1 & 1 & 1 \\ 3 & -1 & 2 \end{pmatrix}$.

[Consistency checks: Do the sum and the product of all eigenvalues yield $\text{tr}(A)$ and $\det(A)$, respectively? Let D be the diagonal matrix containing all eigenvalues; does TDT^{-1} yield A?]

εL7.3.3 Diagonalizing a matrix that depends on a variable (p. 103)

Consider the matrix $A = \begin{pmatrix} x & 1 & 0 \\ 1 & 2 & 1 \\ 3-x & -1 & 3 \end{pmatrix}$, which depends on the variable $x \in \mathbb{R}$. Find the eigenvalues λ_j and eigenvectors $\mathbf{v}_j \in \mathbb{R}^3$ of A as functions of x, with $j = 1, 2, 3$.
Hints: One of the eigenvalues is $\lambda = x$. (Of course the other results, too, can depend on x.) Avoid fully multiplying out the characteristic polynomial; try instead to directly bring it to a completely factorized form! [Check your results: For $x = 4$, two of the (unnormalized) eigenvectors are given by $(1, -2, -1)^{\mathsf{T}}$ and $(1, -1, -2)^{\mathsf{T}}$.]

ρL7.3.4 Diagonalizing a matrix depending on two variables: qubit (p. 103)

A qubit (for "quantum bit" = quantum version of a classical bit) is a manipulable two-level quantum system. The simplest version of a qubit is described by the matrix $H = \begin{pmatrix} B & \bar{\Delta} \\ \Delta & -B \end{pmatrix}$, with $B \in \mathbb{R}$ and $\Delta \in \mathbb{C}$.

(a) Calculate the eigenvalues E_j (choose $E_1 < E_2$) and normalized eigenvectors $\mathbf{v}_1$ and $\mathbf{v}_2$ of H as a function of B, Δ and $X \equiv [B^2 + |\Delta|^2]^{1/2}$.

(b) Show that the eigenvectors can be brought to the form $\mathbf{v}_1 = \frac{1}{\sqrt{2}} \begin{pmatrix} -\sqrt{1-Y} \\ e^{i\phi}\sqrt{1+Y} \end{pmatrix}$ and $\mathbf{v}_2 = \frac{1}{\sqrt{2}} \begin{pmatrix} \sqrt{1+Y} \\ e^{i\phi}\sqrt{1-Y} \end{pmatrix}$, where $e^{i\phi}$ is the phase factor of $\Delta \equiv |\Delta|e^{i\phi}$. How does Y depend on B and X? On three diagrams arranged below each other, each showing two curves, sketch first E_1 and E_2, second $|v^1{}_1|^2$ and $|v^2{}_1|^2$, the squares of the absolute values of the components of the eigenvector $\mathbf{v}_1$, and third $|v^1{}_2|^2$ and $|v^2{}_2|^2$, the squares of the absolute values of the components of the eigenvector $\mathbf{v}_2$, all as functions of $B/|\Delta| \in \{-\infty, \infty\}$ for fixed $|\Delta|$.

Background information: The first sketch shows the so called "avoided crossing", a typical trait of a quantum bit. The second and third sketches show that the eigenvectors "exchange their roles" if B/Δ goes from $-\infty$ to $+\infty$. Both these properties have been detected in many experiments.

εL7.3.5 Degenerate eigenvalue problem (p. 105)

Consider the the matrix $A = \begin{pmatrix} 2 & -1 & 2 \\ -1 & 2 & -2 \\ 2 & -2 & 5 \end{pmatrix}$.
Find its eigenvalues λ_j, a set of *orthonormal* eigenvectors $\mathbf{v}_j$, and a similarity transformation T, as well as its inverse, T^{-1}, such that $T^{-1}AT$ is diagonal. *Hint:* One eigenvalue is $\lambda_1 = 1$.
[Consistency checks: Do the sum and the product of all eigenvalues yield tr(A) and det(A), respectively? Let D be the diagonal matrix containing all eigenvalues; does TDT^{-1} yield A?]

ρL7.3.6 Degenerate eigenvalue problem (p. 105)

For each of the following matrices, find the eigenvalues λ_j, a set of *orthonormal* eigenvectors $\mathbf{v}_j$, and a similarity transformation, T, and its inverse, T^{-1}, for which $T^{-1}AT$ is diagonal.

(a) $A = \begin{pmatrix} 15 & 6 & -3 \\ 6 & 6 & 6 \\ -3 & 6 & 15 \end{pmatrix}$,   (b) $A = \begin{pmatrix} -1 & 0 & 0 & 2i \\ 0 & 7 & 2 & 0 \\ 0 & 2 & 4 & 0 \\ -2i & 0 & 0 & 2 \end{pmatrix}$.

Hint: Both these matrices have a pair of degenerate eigenvalues. Call these $\lambda_2 = \lambda_3$. One of the corresponding eigenvectors is $\mathbf{v}_3 = \frac{1}{\sqrt{3}}(1, 1, 1)^{\mathsf{T}}$ for (a) and $\mathbf{v}_3 = \frac{1}{\sqrt{5}}(0, 1, -2, 0)^{\mathsf{T}}$ for (b).

[Consistency checks: Do the sum and the product of all eigenvalues yield tr(A) and det(A), respectively? Let D be the diagonal matrix containing all eigenvalues; does TDT^{-1} yield A?]

P.L7.4   Functions of matrices (p. 107)

ᴇL7.4.1   Functions of matrices (p. 108)

The purpose of this problem is to gain familiarity with the concept of a "function of a matrix". Let f be an analytic function, with Taylor series $f(x) = \sum_{l=0}^{\infty} c_l x^l$, and $A \in \text{mat}(\mathbb{R}, n, n)$ a square matrix, then $f(A)$ is defined as $f(A) = \sum_{l=0}^{\infty} c_l A^l$, with $A^0 = \mathbb{1}$.

(a)   A matrix A is called "nilpotent" if an $l \in \mathbb{N}$ exists such that $A^l = 0$. Then the Taylor series of $f(A)$ ends after l terms. Example with $n = 2$: Compute e^A for $A = \begin{pmatrix} 0 & a \\ 0 & 0 \end{pmatrix}$.

(b)   If $A^2 \propto \mathbb{1}$, then $A^{2m} \propto \mathbb{1}$ and $A^{2m+1} \propto A$, and the Taylor series for $f(A)$ has the form $f_0 \mathbb{1} + f_1 A$. Example with $n = 2$: Compute e^A explicitly for $A = \theta \tilde{\sigma}$, with $\tilde{\sigma} = \begin{pmatrix} 0 & -1 \\ 1 & 0 \end{pmatrix}$.

   [Check your result: If $\theta = -\frac{\pi}{6}$, then $e^A = \frac{1}{2} \begin{pmatrix} \sqrt{3} & 1 \\ -1 & \sqrt{3} \end{pmatrix}$.]

(c)   If A is diagonalizable, then $f(A)$ can be expressed in terms of its eigenvalues. Let T be the similarity transformation that diagonalizes A, with diagonal matrix $D = T^{-1}AT$ and diagonal elements $D = \text{diag}(\lambda_1, \lambda_2, \dots, \lambda_n)$. Show that the following relations then hold:

$$ f(A) = Tf(D)T^{-1} = T \begin{pmatrix} f(\lambda_1) & 0 & \cdots & 0 \\ 0 & f(\lambda_2) & \ddots & \vdots \\ \vdots & \ddots & \ddots & 0 \\ 0 & \cdots & 0 & f(\lambda_n) \end{pmatrix} T^{-1}. $$

   Remark: Both equalities are to be established independently of each other.

(d)   Now compute the matrix function e^A from (b) using diagonalization, as in (c).

ᴘL7.4.2   Functions of matrices (p. 108)

Express each of the following matrix functions explicitly in terms of a matrix:

(a) e^A, with $A = \begin{pmatrix} 0 & a & 0 \\ 0 & 0 & b \\ 0 & 0 & 0 \end{pmatrix}$.

(b) e^B, with $B = b\sigma_1$ and $\sigma_1 = \begin{pmatrix} 0 & 1 \\ 1 & 0 \end{pmatrix}$, using the Taylor series of the exponential function.

   [Check your result: If $b = \ln 2$, then $e^B = \frac{1}{4} \begin{pmatrix} 5 & 3 \\ 3 & 5 \end{pmatrix}$.]

(c)   The same function as in (b), now by diagonalizing B.

(d) e^C, with $C = i\theta \, \Omega$, where $\Omega = n_j S_j$, while $\mathbf{n} = (n_1, n_2, n_3)^{\mathsf{T}}$ is a unit vector ($\|\mathbf{n}\| = 1$) and S_j are the spin-$\frac{1}{2}$ matrices: $S_1 = \frac{1}{2} \begin{pmatrix} 0 & 1 \\ 1 & 0 \end{pmatrix}$, $S_2 = \frac{1}{2} \begin{pmatrix} 0 & -i \\ i & 0 \end{pmatrix}$, $S_3 = \frac{1}{2} \begin{pmatrix} 1 & 0 \\ 0 & -1 \end{pmatrix}$.

   Hint: Start by computing Ω^2 (for this, the property $S_i S_j + S_j S_i = \frac{1}{2}\delta_{ij}\mathbb{1}$ of the spin-$\frac{1}{2}$ matrices is useful), and then use the Taylor series of the exponential function.

   $\left[$Check your result: If $\theta = -\frac{\pi}{2}$ and $n_1 = -n_2 = n_3 = \frac{1}{\sqrt{3}}$, then $e^C = \frac{1}{\sqrt{6}} \begin{pmatrix} \sqrt{3} - i & 1 - i \\ -1 - i & \sqrt{3} + i \end{pmatrix}.\right]$

   Remark: The exponential form e^C is a representation of SU(2) transformations, the group of all special unitary transformations in $\mathbb{C}^2$. Its elements are characterized by three continuous real parameters (here θ, n_1 and n_2, with $n_3 = \sqrt{1 - n_1^2 - n_2^2}$). The S_j matrices are "generators" of these transformations; they satisfy the SU(2) algebra, i.e. their commutators yield $[S_i, S_j] = i\epsilon_{ijk}S_k$. (See Section L8.3 for further discussion.)

P.L8　Unitarity and hermiticity (p. 109)

P.L8.1　Unitarity and orthogonality (p. 109)

ɛL8.1.1　Orthogonal and unitary matrices (p. 111)

(a)　Is the matrix A given below an orthogonal matrix? Is B unitary?

$$A = \begin{pmatrix} \sin\theta & \cos\theta \\ -\cos\theta & \sin\theta \end{pmatrix}, \qquad B = \frac{1}{1-i}\begin{pmatrix} 2 & 1+i & 0 \\ 1+i & -1 & 1 \\ 0 & 2 & i \end{pmatrix}.$$

(b)　Let $\mathbf{x} = (1,2)^T$. Calculate $\mathbf{a} = A\mathbf{x}$ explicitly, as well as the norm of $\mathbf{x}$ and $\mathbf{a}$. Does the action of A on $\mathbf{x}$ conserve its norm?

(c)　Let $\mathbf{y} = (1,2,i)^T$. Calculate $\mathbf{b} = B\mathbf{y}$ explicitly, and also the norm of $\mathbf{y}$ and $\mathbf{b}$. Does the action of B on $\mathbf{y}$ conserve its norm?

ₚL8.1.2　Orthogonal and unitary matrices (p. 111)

(a)　Determine whether the following matrices are orthogonal or unitary:

$$A = \begin{pmatrix} 0 & 3 & 0 \\ 2 & 0 & 1 \\ -1 & 0 & 2 \end{pmatrix}, \quad B = \frac{1}{3}\begin{pmatrix} 1 & 2 & -2 \\ -2 & 2 & 1 \\ 2 & 1 & 2 \end{pmatrix}, \quad C = \frac{1}{\sqrt{2}}\begin{pmatrix} i & 1 \\ -1 & -i \end{pmatrix}.$$

(b)　Let $\mathbf{x} = (1,2,-1)^T$. Calculate $\mathbf{a} = A\mathbf{x}$ and $\mathbf{b} = B\mathbf{x}$ explicitly. Also, calculate the norm of $\mathbf{x}$, $\mathbf{a}$ and $\mathbf{b}$. Which of these norms should be equal? Why?

(c)　Let $\mathbf{y} = (1,i)^T$. Calculate $\mathbf{c} = C\mathbf{y}$ explicitly, and also determine the norm of $\mathbf{y}$ and $\mathbf{c}$. Should the norms be equal? Why?

P.L8.2　Hermiticity and symmetry (p. 117)

ɛL8.2.1　Diagonalizing symmetric or Hermitian matrices (p. 119)

For each of the following matrices, find the eigenvalues λ_j and a set of eigenvectors $\mathbf{v}_j$. Also find a similarity transformation, T, and its inverse, T^{-1}, for which $T^{-1}AT$ is diagonal.

(a) $A = \begin{pmatrix} 3 & -4 \\ -4 & -3 \end{pmatrix}$, $\qquad$ (b) $A = \begin{pmatrix} 1 & i \\ -i & 1 \end{pmatrix}$, $\qquad$ (c) $A = \begin{pmatrix} 1 & 0 & -i \\ 0 & 1 & 0 \\ i & 0 & 1 \end{pmatrix}$.

Hint: Each of these matrices is either symmetric or Hermitian. Therefore T can be chosen to be either orthogonal or unitary respectively, which facilitates computing its inverse using $T^{-1} = T^T$.

or $T^{-1} = T^{\dagger}$. To achieve this, the columns of T, containing the eigenvectors $\mathbf{v}_j$, must form an orthonormal system w.r.t. to the real or complex scalar product, respectively. It is therefore advisable to normalize all eigenvectors as $\|\mathbf{v}_j\| = 1$. Moreover, recall that non-degenerate eigenvectors of symmetric or Hermitian matrices are guaranteed to be orthogonal.

[Consistency checks: Do the sum and the product of all eigenvalues yield tr(A) and det(A), respectively? Let D be the diagonal matrix containing all eigenvalues; does TDT^{-1} yield A?]

ₚL8.2.2 Diagonalizing symmetric or Hermitian matrices (p. 119)

For each of the following matrices, find the eigenvalues λ_j and a set of eigenvectors $\mathbf{v}_j$. Also find a similarity transformation, T, and its inverse, T^{-1}, for which $T^{-1}AT$ is diagonal.

(a) $A = \dfrac{1}{10}\begin{pmatrix} -19 & 3 \\ 3 & -11 \end{pmatrix}$,   (b) $A = \begin{pmatrix} 0 & 1 & 0 \\ 1 & -1 & 1 \\ 0 & 1 & 0 \end{pmatrix}$,   (c) $A = \begin{pmatrix} 1 & i & 0 \\ -i & 2 & -i \\ 0 & i & 1 \end{pmatrix}$.

[Consistency checks: Do the sum and the product of all eigenvalues yield tr(A) and det(A), respectively? Let D be the diagonal matrix containing all eigenvalues; does TDT^{-1} yield A?]

ₑL8.2.3 Spin-$\frac{1}{2}$ matrices: eigenvalues and eigenvectors (p. 119)

The following matrices are used to describe quantum mechanical particles with spin $\frac{1}{2}$:

$$ S_x = \tfrac{1}{2}\begin{pmatrix} 0 & 1 \\ 1 & 0 \end{pmatrix}, \qquad S_y = \tfrac{1}{2}\begin{pmatrix} 0 & -i \\ i & 0 \end{pmatrix}, \qquad S_z = \tfrac{1}{2}\begin{pmatrix} 1 & 0 \\ 0 & -1 \end{pmatrix}. $$

For each matrix S_j ($j = x, y, z$), compute its two eigenvalues $\lambda_{j,a}$ and normalized eigenvectors $\mathbf{v}_{j,a}$ ($a = 1, 2$). Choose the phase of the eigenvector normalization factor in such a way that the 1-component, $v^1_{j,a}$ (or, if it vanishes, the 2-component), is positive and real.

[Check your results: All three matrices have the same eigenvalues, and $\sum_{a=1}^{2} \lambda_{j,a} = 0$.]

ₚL8.2.4 Spin-1 matrices: eigenvalues and eigenvectors (p. 119)

The following matrices are used to describe quantum mechanical particles with spin 1:

$$ S_x = \frac{1}{\sqrt{2}}\begin{pmatrix} 0 & 1 & 0 \\ 1 & 0 & 1 \\ 0 & 1 & 0 \end{pmatrix}, \qquad S_y = \frac{1}{\sqrt{2}}\begin{pmatrix} 0 & -i & 0 \\ i & 0 & -i \\ 0 & i & 0 \end{pmatrix}, \qquad S_z = \begin{pmatrix} 1 & 0 & 0 \\ 0 & 0 & 0 \\ 0 & 0 & -1 \end{pmatrix}. $$

For each matrix S_j ($j = x, y, z$), compute its three eigenvalues $\lambda_{j,a}$ and normalized eigenvectors $\mathbf{v}_{j,a}$ ($a = 1, 2, 3$). Choose the phase of the eigenvector normalization factor in such a way that the 1-component, $v^1_{j,a}$ (or, if it vanishes, the 2- or 3-component), is positive and real.

[Check your results: All three matrices have the same eigenvalues, and $\sum_{a=1}^{3} \lambda_{j,a} = 0$.]

ₑL8.2.5 Inertia tensor (p. 156)

The inertia tensor of a rigid body composed of point masses is defined as

$$ \widetilde{I}_{ij} = \sum_a m_a \widetilde{I}_{ij}(\mathbf{r}_a, \mathbf{r}_a), \quad \text{with} \quad \widetilde{I}_{ij}(\mathbf{r}, \mathbf{r}') \equiv \delta_{ij}\mathbf{r} \cdot \mathbf{r}' - (\mathbf{e}_i \cdot \mathbf{r})(\mathbf{e}_j \cdot \mathbf{r}'), $$

where m_a and $\mathbf{r}_a = (r^1{}_a, r^2{}_a, r^3{}_a)^{\mathrm{T}}$ are, respectively, the mass and position of point mass a. The eigenvalues of the inertia tensor are known as the rigid body's *moments of inertia*.

Consider a rigid body consisting of three point masses, $m_1 = 4$, $m_2 = M$ and $m_3 = 1$, at positions $\mathbf{r}_1 = (1, 0, 0)^{\mathrm{T}}$, $\mathbf{r}_2 = (0, 1, 2)^{\mathrm{T}}$ and $\mathbf{r}_3 = (0, 4, 1)^{\mathrm{T}}$, respectively. Determine its inertia tensor $\widetilde{I}$ and moments of inertia as functions of M. (Eigenvectors are not required.) [Check your results: If $M = 5$, then $\lambda_1 = 42$, $\lambda_2 = 39$, $\lambda_3 = 11$.]

Remark: For the purposes of diagonalizing the inertia tensor, it may be viewed as a matrix. (Its properties as a "tensor" differ from those of a matrix w.r.t. coordinate transformations, as discussed in Section L10.3, but coordinate transformations do not concern us here.)

ₚL8.2.6 Inertia tensor (p. 156)

Consider a rigid body consisting of two point masses, $m_1 = \frac{2}{3}$ and $m_2 = 3$, at positions $\mathbf{r}_1 = (2, 2, -1)^{\mathrm{T}}$ and $\mathbf{r}_2 = \frac{1}{3}(2, -1, 2)^{\mathrm{T}}$, respectively.

(a) Show that its inertia tensor has the following form: $\widetilde{I} = \begin{pmatrix} 5 & -2 & 0 \\ -2 & 6 & 2 \\ 0 & 2 & 7 \end{pmatrix}$.

(b) Find the moments of inertia (eigenvalues). (*Hint:* One eigenvalue is $\lambda = 3$.)

(c) Construct matrices T and T^{-1} that diagonalize the inertia tensor.

P.L8.3   Relation between Hermitian and unitary matrices (p. 120)

ₑL8.3.1 Exponential representation of two-dimensional rotation matrix (p. 122)

The matrix $R_\theta = \begin{pmatrix} \cos\theta & -\sin\theta \\ \sin\theta & \cos\theta \end{pmatrix}$ describes a rotation by the angle θ in $\mathbb{R}^2$. Use the following "infinite product decomposition" to find an exponential representation of this matrix.

(a) A rotation by the angle θ can be represented as a sequence of m rotations, each by the angle θ/m: $R_\theta = [R_{(\theta/m)}]^m$. For $m \to \infty$ we have $\theta/m \to 0$, thus the matrix $R_{(\theta/m)}$ can be written as $R_{(\theta/m)} = \mathbb{1} + (\theta/m)\tilde{\sigma} + \mathcal{O}\big((\theta/m)^2\big)$. Find the matrix $\tilde{\sigma}$.

(b) Now use the identity $\lim_{m\to\infty}[1 + x/m]^m = e^x$ to show that $R_\theta = e^{\theta\tilde{\sigma}}$.
Remark: Justification for this identity. We have $e^x = [e^{x/m}]^m = [1 + x/m + \mathcal{O}\big((x/m)^2\big)]^m$. In the limit $m \to \infty$ the terms of order $\mathcal{O}\big((x/m)^2\big)$ can be neglected.

[Check your result: Does the Taylor series for $e^{\theta\tilde{\sigma}}$ reproduce the matrix for R_θ given above?]

Remark: The procedure illustrated here, by which an infinite sequence of identical, infinitesimal transformations is exponentiated, is a cornerstone of the theory of "Lie groups", whose elements are associated with continuous parameters (here the angle θ). In that context the Hermitian matrix $i\tilde{\sigma}$ is called the "generator" of the rotation.

ₚL8.3.2 Exponential representation three-dimensional rotation matrix (p. 122)

In $\mathbb{R}^3$, a rotation by an angle θ, about an axis whose direction is given by the unit vector $\mathbf{n} = (n_1, n_2, n_3)$, is represented by a 3×3 matrix that has the following matrix elements: ($\to$ L5.5.2)

$$(R_\theta(\mathbf{n}))_{ij} = \delta_{ij}\cos\theta + n_i n_j(1 - \cos\theta) - \epsilon_{ijk} n_k \sin\theta \qquad (\epsilon_{ijk} = \text{Levi–Civita symbol}). \qquad (1)$$

The goal of the following steps is to supply a justification for Eq. (1).

(a) Consider first the three matrices $R_\theta(\mathbf{e}_i)$ for rotations by the angle θ about the three coordinate axes $\mathbf{e}_i$, with $i = 1, 2, 3$. Elementary geometrical considerations yield:

$$R_\theta(\mathbf{e}_1) = \begin{pmatrix} 1 & 0 & 0 \\ 0 & \cos\theta & -\sin\theta \\ 0 & \sin\theta & \cos\theta \end{pmatrix}, \quad R_\theta(\mathbf{e}_2) = \begin{pmatrix} \cos\theta & 0 & \sin\theta \\ 0 & 1 & 0 \\ -\sin\theta & 0 & \cos\theta \end{pmatrix}, \quad R_\theta(\mathbf{e}_3) = \begin{pmatrix} \cos\theta & -\sin\theta & 0 \\ \sin\theta & \cos\theta & 0 \\ 0 & 0 & 1 \end{pmatrix}.$$

For each of these matrices, use an infinite product decomposition of the form $R_\theta(\mathbf{n}) = \lim_{m \to \infty} [R_{\theta/m}(\mathbf{n})]^m$ to obtain an exponential representation of the form $R_\theta(\mathbf{e}_i) = e^{\theta \tau_i}$. Find the three 3×3 matrices τ_1, τ_2 and τ_3. [Check your results: The τ_i commutators yield $[\tau_i, \tau_j] = \epsilon_{ijk} \tau_k$. This is the so-called SO(3) algebra, which underlies the representation theory of three-dimensional rotations. Moreover, $\tau_1^2 + \tau_2^2 + \tau_3^2 = -2\mathbb{1}$.]

(b) Now consider a rotation by the angle θ about an arbitrary axis $\mathbf{n}$. To find an exponential representation for it using an infinite product decomposition, we need an approximation for $R_{\theta/m}(\mathbf{n})$ up to first order in the small angle θ/m. It has the following form:

$$R_{\theta/m}(\mathbf{n}) = R_{n_1\theta/m}(\mathbf{e}_1) R_{n_2\theta/m}(\mathbf{e}_2) R_{n_3\theta/m}(\mathbf{e}_3) + \mathcal{O}\big((\theta/m)^2\big). \tag{2}$$

Intuitive justification: if the rotation angle θ/m is sufficiently small, the rotation can be performed in three substeps, each about a different direction $\mathbf{e}_i$, by the "partial" angle $n_i\theta/m$. The prefactors n_i ensure that for $\mathbf{n} = \mathbf{e}_i$ (rotation about a coordinate axis i) only *one* of the three factors in (2) is different from $\mathbb{1}$, namely the one that yields $R_{\theta/m}(\mathbf{e}_i)$; for example, for $\mathbf{n} = \mathbf{e}_2 = (0, 1, 0)^{\mathrm{T}}$: $R_{0\theta/m}(\mathbf{e}_1) R_{1 n_2\theta/m}(\mathbf{e}_2) R_{0\theta/m}(\mathbf{e}_3) = R_{n_2\theta/m}(\mathbf{e}_2)$.

Show that such a product decomposition of $R_\theta(\mathbf{n})$ yields the following exponential representation:

$$R_\theta(\mathbf{n}) = e^{\theta\Omega}, \quad \Omega = n_i \tau_i = \begin{pmatrix} 0 & -n_3 & n_2 \\ n_3 & 0 & -n_1 \\ -n_2 & n_1 & 0 \end{pmatrix}, \quad (\Omega)_{ij} = -\epsilon_{ijk} n_k. \tag{3}$$

(c) Show that Ω, the "generator" of the rotation, has the following properties:

$$(\Omega^2)_{ij} = n_i n_j - \delta_{ij}, \quad \Omega^l = -\Omega^{l-2} \quad \text{for } 3 \le l \in \mathbb{N}. \quad \text{[Cayley–Hamilton theorem]} \tag{4}$$

Hint: First compute Ω^2 and Ω^3, then the form of $\Omega^{l>3}$ will be obvious.

(d) Show that the Taylor expansion of $R_\theta(\mathbf{n}) = e^{\theta\Omega}$ yields the following expression:

$$R_\theta(\mathbf{n}) = \mathbb{1} + \Omega \sin\theta + \Omega^2(1 - \cos\theta), \tag{5}$$

and that its matrix elements correspond to Eq. (1).

P.L10 Multilinear algebra (p. 147)

P.L10.1 Direct sum and direct product of vector spaces (p. 147)

L10.1.1 Direct sum and direct product (p. 149)

Consider the two vectors $u = e_i u^i = (2, 1)^{\mathrm{T}}$ and $v = e_i v^i = (3, -1)^{\mathrm{T}}$, where $\{e_i\}$ denotes the canonical basis of $\mathbb{R}^2$.

(a) Compute the direct sums $u \oplus v$ and $v \oplus u$ in $\mathbb{R}^2 \oplus \mathbb{R}^2 = \mathbb{R}^4$.

(b)   Compute the direct products $u \otimes v$ and $v \otimes u$ as linear combinations of the basis states $e_i \otimes e_j$ of $\mathbb{R}^2 \otimes \mathbb{R}^2$.

[Check your results: (a) $u \oplus v - v \oplus u = (-1, 2, 1, -2)^{\mathrm{T}}$; (b) $u \otimes v - v \otimes u = -5e_1 \otimes e_2 + 5e_2 \otimes e_1$.]

⊳L10.1.2 Direct sum and direct product (p. 149)

Consider the two vectors $u = e_i u^i = (-1, 2)^{\mathrm{T}}$ and $v = e_i v^i = (1, -3)^{\mathrm{T}}$ in $\mathbb{R}^2$, and a number $a \in \mathbb{R}$.

(a)   Compute the direct sum, $au \oplus 3v + 2v \oplus au$, in $\mathbb{R}^2 \oplus \mathbb{R}^2 = \mathbb{R}^4$.

(b)   Compute the direct product, $au \otimes 3v - v \otimes 2u$, as a linear combination of the basis states $e_i \otimes e_j$ of $\mathbb{R}^2 \otimes \mathbb{R}^2$.

[Check your results: If $a = 1$ then (a): $(1, -4, 2, -7)^{\mathrm{T}}$, (b): $-e_1 \otimes e_1 + 5e_1 \otimes e_2 - 6e_2 \otimes e_2$.]

P.L10.2   Dual space (p. 150)

⊳L10.2.1 Dual vector in $\mathbb{R}^{2*}$: height of slanted plane (p. 151)

Consider a slanted plane containing the origin, $S = \{(x^1, x^2, x^3) \mid h_1 x^1 + h_2 x^2 - x^3 = 0\} \subset \mathbb{R}^3$, for fixed $h_1, h_2 \in \mathbb{R}$. Its "height" relative to the $x^1 x^2$-plane, $h \equiv x^3 = h_1 x^1 + h_2 x^2$, viewed as a function of the position, $x = (x^1, x^2)^{\mathrm{T}}$, in that plane, constitutes a linear map, $h : \mathbb{R}^2 \to \mathbb{R}, x \mapsto h(x) = h_j x^j$, specified by the dual vector $h = (h_1, h_2) \in \mathbb{R}^{2*}$.

(a)   Let the height of the plane above the points $x_1 = \binom{2}{1}$ and $x_2 = \binom{5}{3}$ be $h(x_1) = 3$ and $h(x_2) = a$, respectively. Find the dual vector h.

(b)   What is the height of the plane above the point $x_3 = \binom{7}{3}$?

[Check your results: If $a = 5$, then (a) gives $h = (4, -5)$, (b) gives $h(x_3) = 13$.]

⊳L10.2.2 Dual vector in $\mathbb{R}^{3*}$: temperature in room (p. 151)

Suppose the temperature in a room varies as a function of position according to the linear map $T : \mathbb{R}^3 \to \mathbb{R}, x \mapsto T(x) = T_j x^j$, specified by the dual vector $T = (T_1, T_2, T_3) \in \mathbb{R}^{3*}$.

(a)   Let the temperature at the points $x_1 = (1, 1, 0)^{\mathrm{T}}, x_2 = (0, 1, 1)^{\mathrm{T}}, x_3 = (0, 0, 1)^{\mathrm{T}}$ be $T(x_1) = 2$, $T(x_2) = 1 + a, T(x_3) = a$ (with $0 < a \in \mathbb{R}$). Find the dual vector T.

(b)   What is the temperature at the point $x_4 = (3, 2, 1)^{\mathrm{T}}$?

[Check your results: If $a = 2$, then (a): $T = (1, 1, 2)$, (b): $T(x_4) = 7$.]

⊳L10.2.3 Dual vectors in $\mathbb{R}^{2*}$ (p. 151)

Let $w : \mathbb{R}^2 \to \mathbb{R}$ be a linear map, which sends the vectors $x_1 = \binom{a}{1}$ and $x_2 = \binom{a}{2}$ (with $a \neq 0$) from $\mathbb{R}^2$ onto the numbers $w(x_1) = 1$ and $w(x_2) = -1$.

(a)   Find its image, $w(e_j)$, for the standard basis vectors, $e_1 = \binom{1}{0}$ and $e_2 = \binom{0}{1}$.

(b)   Find its image, $w(y)$, for the vector $y = \binom{2}{b}$.

[Check your results: If $a = 3, b = 2$, then $w(e_1) = 1$, $w(y) = -2$.]

(c) More generally , let $w(x_j)$ $(j = 1, 2)$ be the images of two general, linearly independent vectors, $x_j = e_i(x_j)^i = e_i X^i_j \in \mathbb{R}^2$, where X is the matrix whose jth column contains the components of x_j. Show that the image, $w(y)$, of another vector, $y = e_j y^j \in \mathbb{R}^2$, under the map w is given by $w(y) = w(x_i)(X^{-1})^i_j y^j$. [Check that this formula correctly reproduces the results obtained in (a) and (b).]

ₚL10.2.4 Dual vectors in $\mathbb{R}^{4*}$ (p. 151)

Let $w : \mathbb{R}^4 \to \mathbb{R}, x \mapsto w(x)$ be the linear map which sends the vectors

$$x_1 = \begin{pmatrix} 1 \\ 0 \\ 0 \\ 1 \end{pmatrix}, \quad x_2 = \begin{pmatrix} -1 \\ 1 \\ 0 \\ 0 \end{pmatrix}, \quad x_3 = \begin{pmatrix} 0 \\ -1 \\ 1 \\ 0 \end{pmatrix}, \quad x_4 = \begin{pmatrix} 0 \\ 0 \\ -1 \\ 1 \end{pmatrix},$$

onto the numbers $w(x_1) = 2$, $w(x_2) = 1$, $w(x_3) = 0$, $w(x_4) = 1$. Find the image, $w(y)$, of the vector $y = (a, 2, a, 2)^{\mathrm{T}}$ under this map. [Check your result: If $a = -1$ then $w(y) = 5$.]

ₑL10.2.5 Basis transformation for vectors and dual vectors (p. 152)

Let $e'_1 = \binom{2}{3}$ and $e'_2 = \binom{3}{5}$ be a basis for $\mathbb{R}^2$. Find the corresponding dual basis, $\{e'^1, e'^2\}$, of $\mathbb{R}^{2*}$. [Check your result: $e'^1 + e'^2 = (2, -1)$.]

ₚL10.2.6 Basis transformation for vectors and dual vectors (p. 152)

Let $e'^1 = (1, 1, 0)$, $e'^2 = (0, 1, 1)$ and $e'^1 = (1, 0, 1)$ be a dual basis for $\mathbb{R}^{3*}$. Find the corresponding basis $\{e'_j\}$ for $\mathbb{R}^3$. [Check your result: $\sum_j e'_j = (1, 1, 1)^{\mathrm{T}}$.]

ₑL10.2.7 Canonical map between vectors and dual vectors via metric (p. 154)

Let V be a real two-dimensional vector space with non-canonical basis $\{e_j\}$ and metric $g(e_i, e_j) = g_{ij} = \binom{2\ 3}{3\ 5}$. Let $J : V \to V^*$ denote the canonical map induced by the metric.

(a) Consider the vector $u_1 = e_1 + a e_2$ (with $a \in \mathbb{R}$). Find its dual, $J(u) \in V^*$, as a linear combination of the dual basis vectors $\{e^i\}$.

(b) Consider the dual vector $w = e^1 - 2a e^2 \in V^*$ (with $a \in \mathbb{R}$). Find its dual, $J^{-1}(w_1) \in V$, as a linear combination of the basis vectors $\{e_j\}$.

[Check your results: If $a = -1$ then (a) $J(u) = -e^1 - 2e^2$, (b) $J^{-1}(w) = -e_1 + e_2$.]

(c) For the vector $v = 2e_1 - e_2$, compute $J(u)v$. Does the result agree with $g(u, v)$?

ₚL10.2.8 Canonical map between vectors and dual vectors via metric: hexagonal lattice (p. 154)

For a hexagonal lattice with bond length 1, a natural choice of basis, $\{v_{\alpha=\pm}\}$, is $v_\pm = \frac{1}{2}(e_1 \pm \sqrt{3}e_2) = \frac{1}{2}\binom{1}{\pm\sqrt{3}}$, where $\{e_i\}$ is the standard basis of $\mathbb{R}^2$.

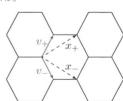

(a) Compute the metric $g_{\alpha\beta} = \langle v_\alpha, v_\beta \rangle$ $(\alpha, \beta \in \{+, -\})$, where $\langle e_i, e_j \rangle = \delta_{ij}$ is the standard scalar product in $\mathbb{R}^2$. Also compute the inverse metric, $g^{\alpha\beta}$. [Check your result: Use $g_{\alpha\beta}$ to verify that the angle between v_+ and v_- is $\frac{2\pi}{3}$.]

(b)   Find the dual basis, $\{v^\alpha\}$, defined by the condition $v^\alpha v_\beta = \delta^\alpha{}_\beta$. [Check your result: $v^+ - v^- = (0, \frac{2}{\sqrt{3}})$.]

(c)   Compute the metric of the dual space, $(g^*)^{\alpha\beta} = \langle v^\alpha, v^\beta \rangle$, where $\langle e^i, e^j \rangle = \delta^{ij}$ is the standard scalar product in $\mathbb{R}^{2*}$. Does it agree with $g^{\alpha\beta}$? [Check your result: Use $(g^*)^{\alpha\beta}$ to verify that the angle between v^+ and v^- is $\frac{\pi}{3}$.]

(d)   Consider the vectors $x_+ = 2v_+ + v_-$ and $x_- = v_+ + 2v_-$. Find the corresponding dual vectors, $x^\pm \equiv J(x_\pm)$, where $J : \mathbb{R}^2 \to \mathbb{R}^{2*}$ is the canonical isomorphism induced by the metric. [Check your result: $x^+ - x^- = 3v^-$.]

(e)   Compute the scalar products $g(x_+, x_-)$ and $g^*(x^+, x^-)$. [Check: They should agree!]

Remark: In physics applications, the lattice spanned by integer multipliers of the dual basis vector is called the "reciprocal lattice". It plays a central role in understanding, e.g, the diffraction pattern of X-rays scattering off the lattice.

P.L10.3 Tensors (p. 154)

εL10.3.1 Linear transformations in tensor spaces (p. 155)

Let $\{e_i\}$ be a basis for a real vector space V, and $A : V \to V$, $e_i \mapsto Ae_i \equiv e_k A^k{}_i$ a linear transformation on V. Let the action of A on the tensor space $T^2{}_0 = V \otimes V$ be defined as $A : T^2{}_0 \to T^2{}_0(V)$, $e_i \otimes e_j \mapsto (Ae_i) \otimes (Ae_j) = e_k \otimes e_l A^k{}_i A^l{}_j$. Now consider the case $V = \mathbb{R}^2$, and $A = \frac{1}{\sqrt{2}}\begin{pmatrix} 1 & 1 \\ 1 & -1 \end{pmatrix}$. Compute the images, $t' = At$, of the following tensors $t \in T^2{}_0(\mathbb{R}^2)$:

(a) $t_+ = \frac{1}{\sqrt{2}}(e_1 \otimes e_1 + e_2 \otimes e_2)$,   (b) $t_- = \frac{1}{\sqrt{2}}(e_1 \otimes e_2 - e_2 \otimes e_1)$.

[Check your results: For case (a) and (b), we have $At'_\pm = \pm t_\pm$.]

(c)   Consider $V = \mathbb{R}^n$, a general linear transformation A on V, and a general tensor, $t = e_i \otimes e_j t^{ij}$ in the tensor space $T^2{}_0(\mathbb{R}^2)$. Denote its image under A by $t' = At = e_k e_l t'^{kl}$. Show that $t'^{kl} = A^k{}_i A^l{}_j t^{ij}$.

(d)   A tensor t is "invariant under A" if its coefficients remain unchanged, i.e. $t'^{ij} = t^{ij}$. Consider the special case that $t^{ij} = \delta^{ij}$, so that $t = \sum_i e_i \otimes e_i$. (In quantum mechanics such a tensor is called "maximally entangled".) Which class of linear transformations leaves such a tensor invariant?

ρL10.3.2 Linear transformations in tensor spaces (p. 155)

Consider the linear transformation $A : \mathbb{R}^3 \to \mathbb{R}^3$, $e_i \mapsto (Ae_i) = e_k A^k{}_i$, with $A = \begin{pmatrix} 0 & a & 0 \\ -a & 0 & 1 \\ 0 & -1 & 0 \end{pmatrix}$, and the tensor $t = e_1 \otimes e_2 + 3e_2 \otimes e_3 \in T^2{}_0(\mathbb{R}^3)$. Find its image, $t' = At$, expanded in the basis $\{e_i \otimes e_j\}$, in two ways:

(a)   Compute each vector Ae_i explicitly and then evaluate $t' = (Ae_1) \otimes (Ae_2) + 3(Ae_2) \otimes (Ae_3)$.

(b)   Identify the components t^{ij} of $t = e_i \otimes e_j t^{ij}$, and compute $t' = e_k \otimes e_l t'^{kl}$ using $t'^{kl} = A^k{}_i A^l{}_j t^{ij}$.

[Check your result: If $a = 2$, then $t'^{kl} = \begin{pmatrix} 0 & 6 & 0 \\ -4 & 0 & 2 \\ 0 & -3 & 0 \end{pmatrix}$.]

For the above example, the direct method of (a) may seem simpler than the component transformation of (b). However, that is only because both t' and A had a fairly simple structure. For the following example, the method of (b) is definitely advisable.

(c)   For $t = e_i \otimes e_j t^{ij}$ with $t^{ij} = \begin{pmatrix} 1 & 2 & 1 \\ 2 & a & 2 \\ 1 & 2 & 1 \end{pmatrix}$ and $A = \begin{pmatrix} 3 & -2 & 3 \\ 2 & -1 & 2 \\ 3 & -1 & 3 \end{pmatrix}$, compute $t' = At$.

[Check your result: If $a = 2$, then $t'^{kl} = \begin{pmatrix} -4 & 0 & 4 \\ 0 & 2 & 6 \\ 4 & 6 & 14 \end{pmatrix}$.]

$_{\varepsilon}$L10.3.3 Tensors in $T^1{}_2(V)$ (p. 155)

Let $u = 3e_1 + ae_2$ and $v = be_1 - e_2$ be elements of $\mathbb{R}^2$, with canonical basis $\{e_j\}$, and $w = 5e^1 - 2ce^2$ an element of $\mathbb{R}^{2*}$, with dual basis $\{e^i\}$. Evaluate the action of the tensor $t = e_2 \otimes e^1 \otimes e^2 \in T^1{}_2(\mathbb{R}^2)$ on its arguments, for

(a) $t(w; u, v)$,   (b) $t(w; ., u)$,   (c) $t(.; v, u)$.

[Check your results: If $a = 1$, $b = 2$, $c = 3$, then (a): 18, (b): $-6e^1$, (c): $2e_2$.]

$_{P}$L10.3.4 Tensors in $T^2{}_2(V)$ (p. 155)

Let $u = 2e_1 - ae_2 + ae_3$ and $v = e_1 + be_2 - 3e_3$ be elements of $\mathbb{R}^3$, with canonical basis $\{e_j\}$, and $w = 4e^1 + 2ce^2 - 5e^3$ an element of $\mathbb{R}^{3*}$, with dual basis $\{e^i\}$. Evaluate the action of the tensor $t = 2e_1 \otimes e_2 \otimes e^1 \otimes e^3 - e_2 \otimes e_1 \otimes e^3 \otimes e^2 \in T^2{}_2(\mathbb{R}^3)$ on its arguments, for:

(a) $t(w, w; u, v)$,   (b) $t(., w; v, u)$,   (c) $t(w, .; u, .)$.

[Check your results: If $a = 2$, $b = 1$, $c = -1$, then (a): 112, (b): $-8e_1 - 24e_2$, (c): $16e_2 \otimes e^3 + 4e_1 \otimes e^2$.]

P.L10.4 Alternating forms (p. 157)

$_{\varepsilon}$L10.4.1 Three-form in $\mathbb{R}^3$ (p. 158)

Consider the three-form $\omega = \sum_{P \in S_3} \operatorname{sgn}(P) e^{P1} \otimes e^{P2} \otimes e^{P3} \in \Lambda^3(\mathbb{R}^3)$, and three vectors, u, v, $w \in \mathbb{R}^3$, with components $\{u^i\}$, $\{v^i\}$ and $\{w^i\}$.

(a)   Write out all six terms in the sum for ω explicitly.

(b)   Evaluate $\phi \equiv \omega(., u, w)$ and $\omega(u, v, w)$ in terms of the components of their argument vectors. To which familiar constructions do the resulting expressions correspond?

(c)   Evaluate ϕ and $\omega(u, v, w)$ for $u = (1, 1, a)^T$, $v = (1, a, 1)^T$, $w = (a, 1, 1)^T$. [Check your results: If $a = 2$ then $\phi = e^1 + e^2 - 6 e^3$ and $\omega(u, v, w) = -4$.]

$_{P}$L10.4.2 Two-form in $\mathbb{R}^3$ (p. 158)

Consider the two-form $\phi = \phi^i \epsilon_{ijk} e^j \otimes e^k \in \Lambda^2(\mathbb{R}^3)$, and two vectors, u, $v \in \mathbb{R}^3$, with components $\{u^i\}$, $\{v^i\}$.

(a)   Write out all six terms in the sum for ϕ explicitly.

(b)   Evaluate $\phi(u, v)$ in terms of the components, $\{u^i\}$ and $\{v^i\}$, of its argument vectors. What familiar construction does the result correspond to?

(c)   Evaluate $\phi(u, v)$ for $(\phi_1, \phi_2, \phi_3) = (1, 4, 3)$, $u = e^1 + 2e^2$, $v = ae^2 + 3e^3$.
[Check your result: If $a = 3$ then $\phi(u, v) = 3$.]

P.L10.6 Wedge product (p. 161)

ᴇL10.6.1 Alternating forms in $\Lambda^2(\mathbb{R}^3)$ (p. 162)

Evaluate the action of each of the following two-forms from $\Lambda^2(\mathbb{R}^3)$ on the vectors $u = 3e_1 - ae_2$ and $v = 2e_1 + be_3$:

(a) $(e^1 \wedge e^3)(u, v)$,   (b) $(e^1 \wedge e^3)(v, u)$,   (c) $(e^3 \wedge e^2)(u, u)$,   (d) $(e^2 \wedge e^1)(u, v)$.

[Check your results: If $a = 1$, $b = 2$ then (a): 6, (d): -2.]

(e)   Expand the three-form $e^1 \wedge e^2 \wedge e^3 \in \Lambda^3(\mathbb{R}^3)$ as a linear combination of the forms $e^i \otimes e^j \otimes e^k \in T^0_{\;3}(\mathbb{R}^3)$.

ᴘL10.6.2 Alternating forms in $\Lambda(\mathbb{R}^4)$ (p. 162)

Evaluate the action of the following two- and three-forms, elements of the Grassmann algebra $\Lambda(\mathbb{R}^4)$, on the vectors $u = ae_2 - e_3$, $v = be_1 + 2e_3$, $w = ce_3 - 5e_4$.

(a) $(e^1 \wedge e^3)(u, v)$,   (b) $(e^1 \wedge e^2 \wedge e^4)(u, v, w)$.

[Check your results: If $a = 3$, $b = c = 1$, then (a): 1, (b): 15.]

ᴇL10.6.3 Wedge products in the Grassmann algebra $\Lambda(\mathbb{R}^4)$ (p. 162)

Any element of the Grassmann algebra $\Lambda(\mathbb{R}^4)$ can be brought to the canonical form $\phi = \sum_{i_1 < ... < i_4} \phi_{i_1,...,i_p} e^{i_1} \wedge ... \wedge e^{i_4}$. Let $\phi_A = e^3$, $\phi_B = e^4$, $\phi_C = e^3 \wedge e^1 + e^2 \wedge e^4$, $\phi_D = e^1 \wedge e^2 \wedge e^3$, be four elements of this algebra, and bring the following wedge products into canonical form:

(a) $\phi_B \wedge \phi_A$,   (b) $\phi_A \wedge \phi_C$,   (c) $\phi_A \wedge \phi_D$,   (d) $\phi_B \wedge \phi_D$.

ᴘL10.6.4 Wedge products in the Grassmann algebra $\Lambda(\mathbb{R}^n)$ (p. 162)

Consider $\Lambda(\mathbb{R}^n)$, the algebra of the alternating forms defined on $\mathbb{R}^n$. Let $\phi_A = \sum_{i=1}^{n} e^i$, $\phi_B = \sum_{i<j} e^i \wedge e^j$ and $\phi_C = \sum_{i<j<k} e^i \wedge e^j \wedge e^k$. Which of the following forms is nonzero?

(a) $\phi_A \wedge \phi_A$,   (b) $\phi_B \wedge \phi_B$,   (c) $\phi_C \wedge \phi_C$.

(d)   Let $\phi_D \in \Lambda^p(\mathbb{R}^n)$ be a p-form with $p \leq n/2$. For which values of p is $\phi_D \wedge \phi_D = 0$?

P.L10.7 Inner derivative (p. 162)

ᴇL10.7.1 Inner derivative of two-form (p. 163)

Consider the two-form $\phi = e^1 \wedge e^2 + e^2 \wedge e^3 + e^3 \wedge e^1 \in \Lambda^2(\mathbb{R}^3)$. Compute its inner derivative, $i_u\phi$, for the vector $u = e_1 - ae_2$. [Check your result: For $a = 2$, $i_u\phi = 2e^1 + e^2 - 3e^3$.]

ᴘL10.7.2 Inner derivative of three-form (p. 163)

Consider the form $\omega = e^1 \wedge e^2 \wedge e^3 \in \Lambda^3(\mathbb{R}^3)$. Compute the inner derivatives

(a) $i_u\phi$,   (b) $i_{u,v}\phi$,

for the vectors $u = e_1 - ae_2$ and $v = 2e_2 + be_3$.
[Check your results: If $a = 1, b = 1$ then (a): $e^2 \wedge e^3 - e^3 \wedge e^1$, (b): $-e^1 - e^2 + 4e^3$.]

P.L10.8 Pullback (p. 163)

ᴇL10.8.1 Pullback from $\mathbb{R}^2$ to $\mathbb{R}^2$ (p. 165)

Consider the linear map $F : \mathbb{R}^2 \to \mathbb{R}^2$, $Fe_j = e_i F^i_{\ j}$, with $F = \begin{pmatrix} 2 & a \\ a & 1 \end{pmatrix}$. Compute the pullback of the following forms:

(a) $\phi = e^1 - e^2$,   (b) $\omega = e^1 \wedge e^2$.

[Check your results: If $a = 3$ then (a): $-e^1 + 2e^2$, (b): $-7e^1 \wedge e^2$.]

ᴘL10.8.2 Pullback from $\mathbb{R}^3$ to $\mathbb{R}^2$ (p. 165)

Consider the linear map $F : \mathbb{R}^2 \to \mathbb{R}^3$, $Ff_j = e_i F^i_{\ j}$, where $\{f_j\}$ and $\{e_i\}$ are bases of $\mathbb{R}^2$ and $\mathbb{R}^3$, respectively, with $F = \begin{pmatrix} a & a \\ 1 & 2 \\ 3 & a \end{pmatrix}$. Compute the pullback of the following forms:

(a) $\phi = e^1 \wedge e^3 + e^2 \wedge e^1$,   (b) $\omega = e^1 \wedge e^2 \wedge e^3$.

[Check your results: If $a = 1$, then (a): $-3f^1 \wedge f^2$, (b): 0.]

P.L10.9 Metric structures (p. 165)

ᴇL10.9.1 Hodge duals on a three-dimensional vector space (p. 167)

(a)   Consider a three-dimensional vector space with a general metric, g. Write down the general formulas for the action of the Hodge star on each of the following forms:

$$1, \qquad \lambda = \lambda_i e^i, \qquad \phi = \tfrac{1}{2!}\phi_{ji} e^j \wedge e^k, \qquad \omega = e^1 \wedge e^2 \wedge e^3.$$

(b)   Now consider the standard metric $g_{ij} = \delta_{ij}$. How do the expressions from (a) simplify? [Check your results: If $\lambda = e^2$ and $\phi = e^3 \wedge e^2$, then $*\lambda = e^3 \wedge e^1$ and $*\phi = -e^1$.]

(c)   For the metric of (b), verify that $**\lambda = \lambda$ and $**\phi = \phi$.

(d)   Show that for a general metric, g, we have $**\lambda = \mathrm{sgn}(g)\lambda$ and $**\phi = \mathrm{sgn}(g)\,\phi$.

(e)   Consider $(g_{ij}) = \mathrm{diag}(-1, -1, -1)$, a diagonal metric with signature $(0, 3)$. How do the results from (b) change? [Check your results: Two do not change, two pick up a sign.]

pL10.9.2 Hodge dual of three-form in $\mathbb{R}^4$ for signature $(1, 3)$ metric (p. 167)

Consider the form $\lambda = \frac{1}{3!}\lambda_{\mu\nu\sigma}e^\mu \wedge e^\nu \wedge e^\sigma \in \Lambda^3(\mathbb{R}^4)$, with a general metric. Compute

(a) $*\lambda = \frac{1}{3!}(*\lambda)_\tau e^\tau$ and   (b) $**\lambda = \frac{1}{3!}\lambda_{\alpha\beta\gamma}e^\alpha \wedge e^\beta \wedge e^\gamma$

in terms of the components $\lambda_{\mu\nu\sigma}$. Show that $**\lambda = -\text{sgn}(g)\lambda$.

(c) Show that for $(g_{\mu\nu}) = \text{diag}(1, -1, -1, -1)$, a diagonal metric with signature $(1, 3)$, the results from (a) simplify to

$$(*\lambda)_0 = \lambda_{123}, \quad (*\lambda)_1 = \lambda_{023}, \quad (*\lambda)_2 = \lambda_{031}, \quad (*\lambda)_3 = \lambda_{012}.$$

εL10.9.3 Coordinate invariance of the Hodge star operation: $p = 1, n = 3$ (p. 167)

Prove that the Hodge star action on a one-form in a three-dimensional vector space,

$$e^i \mapsto *e^i = \frac{1}{2}\sqrt{|g|}g^{il}\epsilon_{ljk}e^j \wedge e^k, \tag{1}$$

is coordinate invariant by proceeding as follows. Consider a transformation to a new basis, $e_j \mapsto e'_j = e_i(T^{-1})^i{}_j$. (i) Start from the expression on the left, $*e^i$, and express the argument of the Hodge star operation in terms of the primed basis vectors, using $e^i = (T^{-1})^i{}_k e'^k$. (ii) Assume the yet-to-be-established invariance of the Hodge star operation and apply $*$ to e'^j according to the primed version of (1), involving the inverse, $(g')^{-1}$, of the transformed metric, $g' = gT^{-1}T^{-1}$. (iii) Finally, transform back to the unprimed basis. If the result coincides with the application of $*$ according to the unprimed version of Eq. (1), the invariance property is established.

pL10.9.4 Coordinate invariance of the Hodge star operation: general p and n (p. 167)

Use a strategy analogous to that of problem L10.9.3 to prove the coordinate invariance of the general Hodge star operation on a p-form in an n-dimensional vector space,

$$*\bigwedge_{a=1}^{p} e^{i_a} = \frac{\sqrt{|g|}}{(d-p)!}\prod_{a=1}^{p}g^{i_a j_a}\,\epsilon_{j_1,\dots,j_n}\bigwedge_{b=p+1}^{n} e^{j_b}.$$

C

CALCULUS

Part C of this book introduces the elements of calculus[1] required in the first years of the physics curriculum. We start with a recapitulation of one-dimensional differentiation and integration. Although this may be material familiar to many readers, we will provide interpretations of differentiation and integration not normally emphasized in school. We then turn to higher dimensions and discuss how differentiation can be applied to understand the behavior of functions depending on several parameters. Next we discuss the integration of multidimensional functions and functions defined on higher-dimensional geometric domains, such as spheres. The generalized concepts of differentiation and integration are the basis for the increasingly advanced elements of calculus discussed in later parts of the chapter, including Taylor series, Fourier calculus, differential equations, functional calculus, and the calculus of functions depending on complex variables.

---

[1] Although the connotation of "**analysis**" is a shade more rigorous than "**calculus**", the two terms are almost synonymous with each other.

Introductory remarks

The mathematics of physics is all about differentiating and integrating. The reasons for this are deeply rooted in the foundations of our science. To understand why, consider the situation before the Age of Enlightenment. At that time scientific knowledge was accumulated empirically, for example through the tabulation of the motion of celestial bodies. Although people were aware that a complete tabulation

Sir Isaac Newton
1642–1727
English physicist, mathematician and natural philosopher who had enormous influence on the development of modern science. He laid the foundations of classical mechanics as it is taught to this day. Newton shares credit with Leibniz for the development of modern differential and integral calculus. Newton's way of doing scientific research set new standards which greatly promoted the progress of science in the centuries to follow.

of all planets and stars is out of the question, no alternative method was known. The situation changed when it became understood that a more rewarding approach was to monitor small incremental *changes* in the motion of celestial bodies. For example, interesting quantities to study were the changes, $\mathbf{v}(t + \delta) - \mathbf{v}(t)$, accumulated in a body's velocity during small increments of time, δ. For sufficiently small δ such a change is approximately proportional to δ, and it made sense to shift the focus of attention to the study of the rate-change, or *derivative*, of the velocity, $\mathbf{v}'(t) \equiv \lim_{\delta \to 0} \delta^{-1}(\mathbf{v}(t + \delta) - \mathbf{v}(t))$. The great step forward came with the observation that these incremental changes are *universal* and can be described through relatively simple physical laws equally applicable to all bodies. This realization, which found its quantitative expression in Newton's famous laws of mechanical motion, defined the starting point of modern physics. From then on the laws of nature were often encoded in "differential relations" describing rate changes of physical quantities. From such laws the actual behavior of a physical object, for example, the full time-dependent profile, $\mathbf{v}(t)$, of a planet's velocity, can be reconstructed through the twin sisters of differentiation and integration.

In the next chapter we introduce the concept of differentiation on the important example of one-dimensional functions familiar from high school. However, we will do so in a manner that differs from the school approach in that it affords straightforward generalizations to cases involving more complex functions.

C1   Differentiation of one-dimensional functions

In school, differentiation is often introduced as a tool to describe the slope of a function defined on a one-dimensional interval. However, differentiation is a concept much more general than that: much like the surface of Earth looks flat when viewed locally, even very complicated functions assume a simple (linear) appearance if we "zoom in" closely and look at them from close up. For example, what has been said above amounts to the statement that for short time differences, δ, velocity is a function linear in δ, $\mathbf{v}(t + \delta) \simeq \mathbf{v}(t) + \delta\,\mathbf{v}'(t)$. More generally, the overarching objective of differentiation is to describe functions locally in terms of simple linear approximations. In this chapter we introduce this idea on the example of one-dimensional functions familiar from high school. This will set the stage for the generalization to more complicated functions discussed in later chapters.

C1.1   Definition of differentiability

We start by recapitulating the definition of differentiability as it is usually taught in school. Heuristically, a function $f : \mathbb{R} \to \mathbb{R}, x \mapsto f(x)$ is differentiable at x if in the vicinity of x it may be approximated by a well-defined tangent, and if that tangent does not have infinite slope, see Fig. C1.

The construction of a tangent effectively monitors the changes of the function in the limit of small increments of its arguments. As a first step towards a more rigorous definition, we need to discuss what is meant by the term "**limit**". Intuitively, f approaches the limit $f(x) = c$ if deviations from the value c become arbitrarily small for arguments y sufficiently close to x.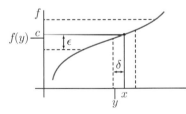
We then write $\lim_{y \to x} f(y) = c$, or $\lim_{\delta \to 0} f(x + \delta) = f(x)$, and say that "the limit exists". An equivalent formulation is to say that f converges to $f(x) = c$ in the limit $y \to x$.

limit   INFO   There are various ways to turn the intuitive formulation into a rigorous *definition of a limit*. Referring for in-depth discussions to lecture courses in mathematics, we mention the Weierstrass–Jordan criterion (also known as the ϵ–δ **criterion**), which says that f converges to $f(x) = c$ if for any $\epsilon > 0$ there exists a $\delta > 0$ such that for all arguments y which are δ-close to x, $|y - x| < \delta$, the function values $f(y)$ are ϵ-close to $f(x)$, $|f(y) - f(x)| < \epsilon$ (see figure).

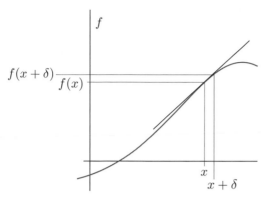
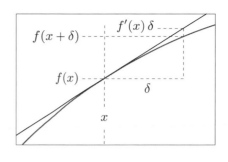

Fig. C1   Differentiation of a function. Discussion, see text.

differentiability   We now define a function $f : \mathbb{R} \rightarrow \mathbb{R}$, $x \mapsto f(x)$ to be **differentiable** if the **difference quotient**, $\frac{1}{\delta}[f(x + \delta) - f(x)]$, has a well-defined limit when $\delta \rightarrow 0$. In this case, it probes the local slope of f at x (see Fig. C1) and the limit,

$$\frac{\mathrm{d}f(x)}{\mathrm{d}x} \equiv \lim_{\delta \to 0} \frac{1}{\delta}[f(x + \delta) - f(x)], \qquad (\text{C1})$$

derivative   is called the **derivative** of f at x. The limiting form of the difference quotient on the r.h.s. is sometimes called the **differential quotient**. Alternative denotations of the derivative include

$$f'(x) = \frac{\mathrm{d}f(x)}{\mathrm{d}x} = \frac{\mathrm{d}f(y)}{\mathrm{d}y}\bigg|_{y=x} = \frac{\mathrm{d}}{\mathrm{d}x}f(x) = \mathrm{d}_x f(x) = f_x(x).$$

They are all defined in the same way by the r.h.s. of Eq. (C1).

   Before discussing the properties of the differential quotient defined by Eq. (C1), it is worthwhile understanding the conditions under which the limit $\delta \rightarrow 0$ exists. A first, if not sufficient, existence condition is the *continuity* of f at x.

Continuity and differentiability of functions

The existence of the differential quotient in Eq. (C1) requires that f have no "jumps" at x, which in mathematical terminology is called the absence of discontinuities. For example, the function shown in the left panel of Fig. C2 has a unit jump at zero, implying the

continuity   divergence of the limit of $\delta^{-1}(f(0 + \delta) - f(0)) = \delta^{-1}$. **Continuity** is a necessary (but not sufficient, see discussion below) prerequisite for differentiability. Using the above terminology of limits, a function is continuous at x if $\lim_{y \to x} f(y) = f(x)$, i.e. if it converges to $f(x)$ for arguments approaching x.

   To understand that *continuity does not imply differentiability* consider the second panel of Fig. C2. It shows a function which *is* continuous, but not differentiable at $x = 0$. The reason is the presence of an "edge", i.e. $f(x) = +x$ for any $x > 0$ and $f(x) = -x$ for $x < 0$. For *positive* values of δ we then have $\delta^{-1}[f(\delta) - f(0)] = \delta^{-1}[+\delta - 0] = 1$, while for negative values $\delta^{-1}[f(\delta) - f(0)] = \delta^{-1}[-\delta - 0] = -1$. This means that the limit is

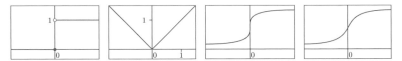

Fig. C2   Left: a function that is not continuous at $x = 0$. The solid dot indicates that $f(0) = 0$. For all strictly positive values $x > 0, f(x) = 1$. Middle panels: functions that are continuous but not differentiable at $x = 0$. Right: a smooth function, differentiable throughout its domain of definition.

not unambiguously defined, and therefore the differentiability criterion fails. Another thing that may go wrong is that "infinitely strong" slopes appear (third panel.) For example, the function $f(x) = 3x^{1/3}$ has the derivative (see below for a summary of differentiation rules) $f'(x) = x^{-2/3}$ and this does not exist at $x = 0$, where the function crosses the y-axis with "infinite slope". Finally, functions such as that shown in the fourth panel have well-defined tangents everywhere and therefore are differentiable.

    In this text, we often require global differentiability and for this reason we generally con-

open interval sider functions defined on **open intervals**, $I \equiv (a, b)$ (see the discussion of openness on p. 17). Openness is required to safeguard the existence of the differential quotient throughout the entire interval: by definition, an interval I is open if any $x \in U \subset I$ lies in a neighborhood, $U = \{y | |y - x| \leq \epsilon\} \subset I$, entirely contained in I. The differential quotient can then be computed within U. By contrast, the differential quotient cannot be computed at the boundary points of a closed interval, $[a, b]$, because for any $\delta > 0, b + \delta$ is outside $[a, b]$ and $f(b + \delta)$ is not defined there. If a function defined on an open interval needs to be evaluated *at* an endpoint, its value is *defined* via a limit, e.g. $f(b) \equiv \lim_{x \to b} f(x)$.

Interpretation of the derivative

In school it is pointed out that the derivative, $f'(x)$, determines the slope of f at x. This view applies to the situation with one-dimensional functions but is too narrow to capture the meaning of derivatives in more general contexts. A more versatile interpretation is as follows: before taking the limit $\delta \to 0$, consider a fixed but very small value of δ. The right-hand side of (C1) will then be a very good approximation to the derivative, i.e. $\frac{1}{\delta}[f(x + \delta) - f(x)] \simeq f'(x)$. Now rewrite this equation as

$$f(x + \delta) \simeq f(x) + f'(x)\,\delta. \tag{C2}$$

This tells us that in the immediate neighborhood of x, the function f can be approximated by a function[2] which is linear in δ, namely $f(x) + f'(x)\,\delta$ (see Fig. C1). We may set $x + \delta \equiv y$ to formulate the linear approximation as

$$f(y) \simeq f(x) + f'(x)\,(y - x).$$

However, it has to be understood that this equation holds only for arguments y very close to the fixed value x where the derivative is taken. Summarizing,

---

[2] A function $g(x)$ is called linear in x if it is of the form $x \mapsto g(x) = ax + b, \, a, b \in \mathbb{R}$. This should be distinguished from the slightly more restrictive definition $x \mapsto ax$ of linear *maps* used in Part L.

> derivatives provide local approximations to functions by linear functions.

We will soon see that this interpretation carries over to more general contexts, including situations where the notion of "slope" is not defined. By contrast, the approximation-by-linear-functions view is generally valid and provides the key to understanding even the most involved derivative operations.

INFO The linear approximation (C2) is often applied to actually compute derivatives. To illustrate this principle, consider the function $f(x) = x^3$. Then $f(x+\delta) = (x+\delta)^3 = f(x) + 3x^2\delta + 3x\delta^2 + \delta^3$. Now, δ is assumed to be very small, hence δ^2 is even smaller, and δ^3 smaller still. For example, for $\delta = 10^{-2}$ we have $\delta^2 = 10^{-4}$ and $\delta^3 = 10^{-6}$. This illustrates that for δ approaching zero, terms beyond linear order become negligible compared to the linear ones. It is standard to represent this smallness as

$$f(x+\delta) = f(x) + 3x^2\delta + \mathcal{O}(\delta^2),$$

where the notation $\mathcal{O}(\delta^2)$ (spoken "order-δ^2") indicates that terms of order δ^2 and higher are neglected.[3] Rearranging terms we have $\delta^{-1}(f(x+\delta) - f(x)) = 3x^2 + \delta^{-1} \times \mathcal{O}(\delta^2)$. In the limit $\delta \to 0$ the second term on the right-hand side vanishes and comparison with (C2) leads to the identification $d_x x^3 = 3x^2$. Applied to a positive integer power, $f(x) = x^n$, the same argument gives $f(x+\delta) = f(x) + nx^{n-1}\delta + \mathcal{O}(\delta^2)$, and $\frac{d}{dx}x^n = nx^{n-1}$. ($\to$ C1.1.1-2)

Notice that we did not take the limit $\delta \to 0$ in the expansions above – that would have yielded a trivial equation, $f(x) = f(x)$. Instead, we took δ to be nonzero but "arbitrarily small". Variables, δ, assuming values smaller than any other in a specific mathematical context are said to be **infinitesimally** infinitesimally small. The attribute **infinitesimal** usually implies that a limit $\delta \to 0$ will eventually **small** be taken. Still, it can be advantageous to keep the variable temporarily finite and use its smallness as an aid in computations (ignoring terms of $\mathcal{O}(\delta^2)$, etc.).

As an instructive example, we note that a geometric construction may be applied to show that the derivative of the sine function is given by $d_\phi \sin(\phi) = \cos(\phi)$. To this end, apply geometric reasoning to verify that an infinitesimal increment of the argument of the sine function changes its value from $\sin(\phi)$ to $\sin(\phi+\delta) \simeq \sin\phi + \delta\cos\phi$. Use a similar argument to show that $d_\phi \cos(\phi) = -\sin(\phi)$.

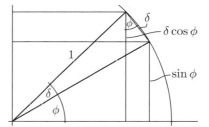

Derivatives of higher order and smoothness

higher-order **Derivatives of higher order** are defined by the iteration of ordinary derivatives. For **derivative** example, the second derivative of a function is defined as

$$\frac{d^2 f(x)}{dx^2} = \frac{d}{dx}\left(\frac{df(x)}{dx}\right). \tag{C3}$$

---

[3] The mathematically precise definition of the symbol $\mathcal{O}$ is as follows: given two functions, $g(x), h(x)$, we write $g(x) = \mathcal{O}(h(x))$ in the limit $x \to 0$ if there exists a constant, c, such that $|g(x)| < c|h(x)|$ for sufficiently small x. For example, $x^2 + 3x^3$ is $\mathcal{O}(x^2)$ because $|x^2 + 3x^3|/|x^2| < c$, for $c > 1$ and x sufficiently small. The notation always makes reference to a limit which, however, need not be 0. For example, $1/(x^2 + x^3)$ is $\mathcal{O}(x^{-2})$ in the limit $x \to \infty$.

One sometimes says that a derivative is taken by "applying the derivative operator $\frac{d}{dx}$ to a function". The mathematical formulation of this statement reads

$$\frac{d^n f(x)}{dx^n} \equiv \frac{d^n}{dx^n} f(x) \equiv \underbrace{\frac{d}{dx}\left(\frac{d}{dx}\cdots\left(\frac{d}{dx}f(x)\right)\right)}_{n \text{ factors}}. \tag{C4}$$

For example,

$$\frac{d}{dx}(x^2 \sin(x)) = 2x \sin(x) + x^2 \cos(x),$$

$$\frac{d^2}{dx^2}(x^2 \sin(x)) = (2 - x^2)\sin(x) + 4x \cos(x),$$

$$\frac{d^3}{dx^3}(x^2 \sin(x)) = (-6x)\sin(x) + (6 - x^2)\cos(x). \tag{C5}$$

smooth function

Functions that can be differentiated infinitely many times are called **smooth functions**. Examples of such functions include polynomials and trigonometric functions. By contrast, the function defined by

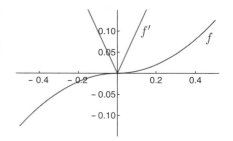

$$f(x) = \frac{1}{2}\begin{cases} +x^2, & x \geq 0, \\ -x^2, & x < 0, \end{cases}$$

is differentiable at $x = 0$, but not smooth. Indeed, $f'(x) = |x|$ and this cannot be differentiated at zero, i.e. the function above is differentiable, but not two-fold differentiable. Although it looks smooth the function is not smooth in the mathematical sense. We finally note that higher-order derivatives are sometimes represented in terms of the alternative *notation*

$$f^{(n)}(x) \equiv d_x^n f(x) \equiv \frac{d^n}{dx^n} f(x), \tag{C6}$$

which indicates the order of differentiation as a superscript. It is imperative to put the latter in parentheses, $f^{(n)}$, to avoid confusion with the nth power, f^n, of the function f.

C1.2 Differentiation rules

We summarize here the most important rules of differentiation. These identities may be familiar from high school and they are routinely proven in introductory courses in mathematics. In the following, $f, g : \mathbb{R} \to \mathbb{R}$ are smooth functions, and $a \in \mathbb{R}$.

product rule ▷ **Product rule (Leibniz rule)**

$$\boxed{\frac{d(fg)}{dx} = \frac{df(x)}{dx}g(x) + f(x)\frac{dg(x)}{dx}.} \tag{C7}$$

chain rule   ▷   **Chain rule**

$$\frac{df\big(g(x)\big)}{dx} = \frac{df(y)}{dy}\bigg|_{y=g(x)} \frac{dg(x)}{dx}. \tag{C8}$$

The essence of the chain rule is that the rate of change of the function $f(g(x))$ is determined by that of the function $f(y)$ at $y = g(x)$, multiplied by that of $g(x)$ with x. It is worth taking a moment to understand this statement in intuitive terms.

In particular, $\frac{df(ax)}{dx} = a\frac{df(y)}{dy}\big|_{y=ax}$ and

$$\frac{d}{dx}\frac{1}{g(x)} = -\frac{1}{\big(g(x)\big)^2}\frac{dg(x)}{dx},$$

where the latter identity follows from the choice $f(y) = 1/y$, and $d_y(1/y) = -1/y^2$.

inverse   ▷   **Derivative of inverse functions.** For a bijective function, f, the **inverse function**, f^{-1},[4]
function   is defined through $f(f^{-1}(x)) = x$. Differentiating both sides in x and using the chain rule on the left, one finds the formula

$$\frac{df^{-1}(x)}{dx} = \frac{1}{\frac{df(y)}{dy}\big|_{y=f^{-1}(x)}}, \tag{C9}$$

expressing the derivative of the inverse through that of the function itself. For example,

$$\frac{d}{dx}\ln(x) = \frac{1}{\exp'(y)\big|_{y=\ln(x)}} = \frac{1}{\exp(y)\big|_{y=\ln(x)}} = \frac{1}{x}.$$

INFO Differentiation formulas such as the chain rule, or the derivative of inverse functions, can always be derived by application of the basic rule (C2). Let us illustrate this idea on a heuristic *proof of the chain rule,* Eq. (C8). Using the abbreviation $g(x) = y$, Eq. (C2) may be applied to linearize first g, then f, and obtain

$$f(g(x+\delta)) \simeq f(g(x) + g'(x)\delta) = f(y + g'(x)\delta) \simeq f(y) + f'(y)g'(x)\delta. \tag{C10}$$

In the second equality we noted that for infinitesimally small δ the product $g'(x)\delta$ is likewise small. The function f may therefore be linearized in it as indicated. Rearranging terms and dividing by δ we obtain $\delta^{-1}\big[f(g(x+\delta)) - f(g(x))\big] \simeq f'(y)g'(x)$. Remembering the definition of the derivative, Eq. (C1), we arrive at the chain rule. Use similar reasoning to derive the (simpler!) product rule.

($\rightarrow$ C1.2.1-2)

C1.3 Derivatives of selected functions

For reference we list the derivatives of some functions frequently occurring in practice.

---

[4] The inverse function of a function f is usually denoted by f^{-1} and this must not be confused with the inverse of the function *value,* $(f(y))^{-1} = 1/f(y)$. (For example, $f(y) = y^2$ has the inverse function $f^{-1}(x) = \sqrt{x}$, different from $(f(y))^{-1} = 1/y^2$.) Which quantity is meant should generally be evident from the context.

▷ *Power functions*:

$$\frac{\mathrm{d}}{\mathrm{d}x}\mathrm{d}x^{\alpha} = \alpha x^{\alpha-1}. \tag{C11}$$

The formula also applies to fractional powers, e.g. for $\alpha = 1/3$ we have $\frac{\mathrm{d}x^{1/3}}{\mathrm{d}x} = \frac{1}{3}x^{-2/3}$.

▷ *Exponential function and logarithm*:

$$\exp'(x) = \exp(x), \qquad \ln'(x) = \frac{1}{x}. \tag{C12}$$

▷ *Trigonometric functions*: ($\rightarrow$ C1.3.1)

$$\sin'(x) = \cos(x), \qquad \cos'(x) = -\sin(x), \qquad \tan'(x) = \frac{1}{\big(\cos(x)\big)^2}. \tag{C13}$$

▷ *Hyperbolic functions*:[5] ($\rightarrow$ C1.3.2)

$$\sinh'(x) = \cosh(x), \quad \cosh'(x) = \sinh(x), \quad \tanh'(x) = \frac{1}{(\cosh(x))^2}. \tag{C15}$$

▷ *Inverse trigonometric and hyperbolic functions* are generally labeled by "arc", as in $\arcsin \equiv \sin^{-1}$. They are differentiated using (C9):[6] ($\rightarrow$ C1.3.3-4)

$$\arcsin'(x) = \frac{\pm 1}{\sqrt{1-x^2}}, \qquad \arccos'(x) = \frac{\mp 1}{\sqrt{1-x^2}}, \qquad \arctan'(x) = \frac{1}{1+x^2}, \tag{C16}$$

$$\mathrm{arcsinh}'(x) = \frac{1}{\sqrt{1+x^2}}, \quad \mathrm{arccosh}'(x) = \frac{\pm 1}{\sqrt{x^2-1}}, \quad \mathrm{arctanh}'(x) = \frac{1}{1-x^2}. \tag{C17}$$

Derivatives of more complicated functions can be computed with the help of the product and chain rules. ($\rightarrow$ C1.3.5-8) For example,

$$\frac{\mathrm{d}}{\mathrm{d}x}x^2\exp(3x) = 2x\exp(3x) + 3x^2\exp(3x), \qquad \frac{\mathrm{d}}{\mathrm{d}x}\ln(x^2+5) = \frac{1}{x^2+5}2x.$$

C1.4 Summary and outlook

In this introductory chapter we have reviewed the basic idea of differentiation on the example of one-dimensional real-valued functions. We discussed why differentiation is so important to physics and emphasized its interpretation as a linear approximation of smooth functions. In this way of thinking it is frequently useful to keep the infinitesimal parameter

---

hyperbolic functions

[5] The **hyperbolic sine, cosine, tangent functions** are defined as:

$$\sinh(x) = \tfrac{1}{2}(e^x - e^{-x}), \qquad \cosh(x) = \tfrac{1}{2}(e^x + e^{-x}), \qquad \tanh(x) = \frac{\sinh x}{\cosh x}. \tag{C14}$$

[6] Two possible signs appear in cases where the functions f are non-monotonic. Intervals with definite slope of f then define distinct "branches" of f^{-1}. In each of these, the sign of $(f^{-1})'$ is determined by that of f'. For example, $\cos'(x)$ is positive for $x \in (-\pi, 0)$ and negative for $x \in (0, \pi)$, hence $\mathrm{arccosh}'(x) \gtrless 0$ in the corresponding branches of arccosh. ($\rightarrow$ C1.3.3-4)

δ entering the construction of the differential quotient (C1) finite and to work with effectively linearized representations of functions, as in (C2). The utility of such representations became evident in a number of cases, including the explicit computation of derivatives via the manipulation of difference quotients with finite δ. We also discussed various more technical aspects of differentiation including continuity requirements, differentiation rules, and the derivatives of various important classes of functions. Although these concepts have been discussed within the framework of one-dimensional functions, they are of general relevance and will play an important role in our subsequent discussion of multidimensional functions, starting in Chapter C3.

However, before generalizing to the multidimensional case, we first introduce integration as the twin operation of differentiation. Following the same logic as above, we begin with the case of one-dimensional functions, once more staying at a level familiar to many readers from high school. This will allow us to keep the intimate connection between integration and differentiation in sight when we advance to higher dimensions.

Integration of one-dimensional functions

Integration is as important to physics as differentiation. Whereas incremental changes in physical quantities are monitored by differentiation, integration is applied to *sum* over small increments. For example, once the incremental change in the coordinates of a stellar body has been understood, an integration procedure needs to be applied to sum over increments and obtain the change of its position over finite time spans. This simple analogy already indicates that, quite generally, differentiation and integration are mutually inverse operations.

In the following, we adopt a strategy similar to that of the previous chapter and introduce the concept of integration on the example of one-dimensional real functions. The technical aspects of this operation will be familiar to many readers from high school. However, we stress an interpretation of integration which is not usually emphasized in school and which extends to integrals over more complex functions, to be discussed in later chapters.

C2.1 The concept of integration

In school, integration is introduced as an operation to determine the area under a function. However, only a small minority of the integrals encountered in physics can be interpreted in this way. A more general view is to think of **integrals as sums** integrals as generalized *sums*. Let us introduce this interpretation on a simple example: suppose we are given a two-dimensional painted surface, S, and want to determine its geometric area, A. A practical approach to solving this task would first choose a reference shape of known area A_0, a square say. One might then count the number, $N(A_0)$, of squares fitting into the area (see figure). An estimate of A would then be given by

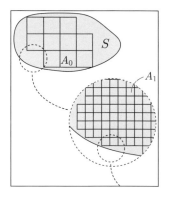

$$A \simeq \sum_{\ell=1}^{N(A_0)} A_0 = N(A_0)A_0,$$

where the index ℓ enumerates the squares. Of course, this estimate generally contains an error because parts of the area remain uncovered. However, the accuracy may be refined by turning to squares of smaller area, A_1, and counting the number, $N_1 > N_0$, of squares required to cover S in this refined way. This leaves less uncovered excess area and leads to the improved estimate

$$A \simeq \sum_{\ell=1}^{N(A_1)} A_1 = N(A_1)A_1.$$

In principle, the procedure may be iterated down to "infinitely many" squares of infinitesimally small area and in this limit the true value of A will be recovered. The limiting operation is called an "integral". All integrals have in common that they can be interpreted as limits of sums conceptually similar to that considered above.

It is straightforward to generalize the above procedure to more complicated settings. For example, consider a surface, S, coated with an inhomogeneous distribution of a massive substance (e.g. paint), as in the figure where darker/lighter areas represent regions of stronger/weaker coverage. We describe the system through two Cartesian coordinates, (x, y), and a *mass distribution function*, $\rho(x, y)$, defined in such a way that $\rho(x, y)\delta_x\delta_y$ equals the mass of the substance present in a small rectangle of area $\delta_x\delta_y$ at the coordinate point (x, y).

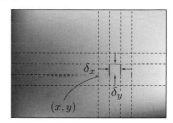

An estimate for the total mass, M, of the substance may be obtained by discretizing the total area, A, of the surface into a system of $N(\delta_x\delta_y) \propto A/\delta_x\delta_y$ infinitesimal rectangles at points (x_ℓ, y_ℓ), where the index ℓ enumerates the rectangles. Summing over the respective weights we obtain the estimate $M \simeq \delta_x\delta_y \sum_{\ell=1}^{N(\delta_x\delta_y)} \rho(x_\ell, y_\ell)$. We may now proceed to finer and finer discretizations to generate a sequence[7] of increasingly accurate estimates, which in the limit of infinitely small discretization areas, $\delta_x, \delta_y \to 0$, approaches the true value of M:

$$M = \lim_{\delta_x, \delta_y \to 0} \delta_x\delta_y \sum_{\ell=1}^{N(\delta_x\delta_y)} \rho(x_\ell, y_\ell) \equiv \int_S dxdy\, \rho(x, y).$$

the symbol $\int$

Here, the symbolic notation appearing on the r.h.s. of the equation is *defined* by the expression on its left: the *integral* symbol $\int$ stands for an infinitely refined sum carried out over the area S, indicated as a subscript. That the summation is over a set of two-dimensional "surface elements" $\delta_x\delta_y$, built with reference to coordinates (x, y), is indicated by the symbol $dxdy$. However, we repeat that all this notation is "implicit" in the sense that the actual definition of the integral is given by the sequence of sums on the l.h.s. of the equation. Each of these sums can be computed in concrete ways, either manually, or on a computer,

sequence

[7] An (infinite) **sequence**, $(a_n)_{n\in\mathbb{N}} = (a_0, a_1, \dots)$ is an infinite and sequentially ordered collection of objects. For example, $a_n = 1/n$ defines the sequence $(1, 1/2, 1/3, \dots)$. The sequence converges to a **limit**, $\lim_{n\to\infty} a_n \equiv a$, if for increasing n the values a_n converge to the value a. For example, the sequence $a_n = 1/n$ converges to 0. **Convergence** means that for any $\epsilon > 0$ there exists a threshold $n_\epsilon \in \mathbb{N}$ such that for $n > n_\epsilon$, $|a_n - a| < \epsilon$.

and at any desired level of accuracy. The important and general statement conveyed by this discussion is that[8]

> almost any integral encountered in physics can be represented as the limit of a sequence of finite sums, each computable by "conventional techniques".

The sequences of sums representing integrals are generally called **Riemann sums**. All Riemann sums have the structure

$$\text{Riemann sum} = \lim_{\delta \to 0} \delta \sum_{\ell=1}^{N(\delta)} X_\ell, \quad \text{(C18)}$$

where X_ℓ is the quantity to be summed and the index ℓ enumerates subdivisions of a summation domain that has been divided

Bernhard Riemann
1826–1866
A German mathematician who made break-through contributions to analysis, number theory and differential geometry. Riemann gave the concept of integration a precise meaning. He also introduced various foundations of modern geometry, including the concept of Riemannian manifolds fundamental to the later formulation of general relativity.

into $N(\delta) \propto \delta^{-1}$ compartments. This proportionality ensures that the smallness of δ is balanced by the increase in the number of summation steps. In the following we will discuss various concrete examples of Riemann summation procedures.

C2.2 One-dimensional integration

In this section we apply the program outlined above to one-dimensional functions $f\colon \mathbb{R} \to \mathbb{R}, \, y \mapsto f(y)$. The quantities to be summed now are the values $f(y)\delta$ obtained by multiplying function values with small increments, δ, in the argument variable y. We observe that $f(y)$ plays a role analogous to that of the mass distribution discussed in the previous section. In the present one-dimensional context the result of the summation procedure will be the geometric area between the function graph and the horizontal axis (see the figure below). We next discuss how the summation procedure is made quantitative.

One-dimensional Riemann sums

In the one-dimensional case, the integration domain (i.e. the analogue of the area S in our example above) is a real interval, say $[0, x]$. Proceeding in analogy to the previous discussion, the domain is partitioned into $N(\delta_0) \equiv x/\delta_0$ intervals of small width δ_0. For finite δ_0 we will call these increment intervals **bins**. They define a partitioning of the area between the function graph and the horizontal axis into strips of width δ. Let $f_\ell \equiv f(y_\ell)$, with $\ell = 1, \dots, N(\delta_0)$, be the value of f at a point y_ℓ somewhere in the ℓth bin. The area of the corresponding strip is approximately given by $\delta_0 f_\ell$, and summation leads to the estimate

---

[8] We write "almost any" because there are rare cases of integrals which cannot be computed along the lines of our construction above. For further comments on this point, see Section C2.2.

$$F(x) \simeq \delta_0 \sum_{\ell=1}^{N(\delta_0)} f_\ell$$

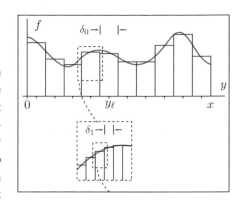

for the total area. Note that for finite δ_0 the value of this estimate depends on the arbitrary choice of the readout point y_ℓ within the ℓth bin – left edge, right edge, center? – and therefore contains arbitrariness. However, as is indicated by the figure, the dependence of the individual strip areas on the positioning of y_ℓ diminishes upon passing to bins of higher resolution. This point is discussed in more detail and generality on p. 448.

integral The limiting case of an infinitely refined sum is called the **integral** of the function:

$$F(x) = \lim_{\delta \to 0} \delta \sum_{\ell=1}^{x/\delta} f_\ell \equiv \int_0^x \mathrm{d}y \, f(y). \tag{C19}$$

The interpretation in terms of sums also shows how *integration and differentiation are "inverse" operations*. To understand this point, let us ask how $F(x)$ varies as a function of x. An approximate answer can be found by considering its Riemann sum at a small but fixed value of δ:

$$F(x + \delta) = \delta \sum_{\ell=1}^{\frac{x}{\delta}+1} f_\ell = \delta \sum_{\ell=1}^{\frac{x}{\delta}} f_\ell + \delta f_{\frac{x}{\delta}+1} = F(x) + \delta f(x),$$

where in the last step we conveniently put the readout position to the left of the bin, $f_{\frac{x}{\delta}+1} = f(x)$. (Why is the arbitrariness of this choice inessential?) We divide by δ and take the limit $\delta \to 0$ to obtain

$$\frac{\mathrm{d}F(x)}{\mathrm{d}x} \equiv \lim_{\delta \to 0} \frac{1}{\delta} \big[F(x + \delta) - F(x) \big] = f(x).$$

This shows that the rate at which an integral, $F(x) = \int_0^x \mathrm{d}y \, f(y)$, changes under variation of the integration boundary, x, is given by the value of the integrand at the boundary, $f(x)$.

fundamental theorem of calculus The reciprocity between integration and differentiation is summarized by the **fundamental theorem of calculus**:

$$F(x) = \int_0^x \mathrm{d}y \, f(y) \quad \Rightarrow \quad \frac{\mathrm{d}F(x)}{\mathrm{d}x} = f(x). \tag{C20}$$

Later in the text, we will see that similar relations hold for more general classes of integrals and derivatives. They all follow from the interpretation of integrals as sums and of derivatives as measures of small increments.

Definite and indefinite integrals

primitive   Any function, $F(x)$, whose derivative equals $f(x)$, $\frac{d}{dx}F(x) = f(x)$, is called a **primitive**
function   **function** or **anti-derivative** of f. The terminology anti-derivative emphasizes that passing
anti-   from f to F is the opposite of passing from f to f'. We write "a" instead of "the" primitive
derivative   function because for any constant C, the function $F(x) + C$ is an equally valid primitive
function, $\frac{d}{dx}(F(x) + C) = f(x)$.

Even in the higher-dimensional integration theory to be discussed in later sections, actual
calculations come down to successions of one-dimensional integrals. Equation (C20) indicates that primitive functions are the key to the computation of these integrals and this
explains their general importance. The connection between integrals and the primitive
function is underpinned by the notation

$$\int dx f(x) \equiv F(x) + C, \tag{C21}$$

indefinite   where the symbol on the l.h.s. is called an **indefinite integral**. The indefinite integral
integral   is just another denotation for the whole class of primitive functions with unspecified
integration   **integration constant**, C. For example, $\int dx\, x^4 = \frac{1}{5}x^5 + C$, since $\frac{d}{dx}\left(\frac{1}{5}x^5 + C\right) = x^4$,
constant   irrespective of the value of C. In integral tables the additive constant is often omitted,
although its presence is implicitly assumed.

Knowing a primitive function, the value of a **definite integral**, i.e. an integral over a
definite interval $[a, b]$, is obtained as

$$\int_a^b dy f(y) = F(b) - F(a) \equiv F(x)\Big|_a^b \equiv \Big[F(x)\Big]_a^b. \tag{C22}$$

This relation follows from Eq. (C20) and the observation that $\int_a^b dy f = \int_0^b dy f - \int_0^a dy f$,
i.e. the summed area from a to b equals that from 0 to b minus that from 0 to a. Note that in
the difference on the r.h.s. the integration constant drops out and there is no arbitrariness in
the definite integral. The general consistency of the additivity of integrals with Eq. (C22) is
seen from relations such as $\int_a^b dy f(y) = \int_a^c dy f(y) + \int_c^b dy f(y)$, which is compatible with
$F(b) - F(a) = [F(b) - F(c)] + [F(c) - F(a)]$.

EXAMPLE $\int_0^{\pi/2} dy \cos(y) = \Big[\sin(x)\Big]_0^{\pi/2} = \sin(\pi/2) - \sin(0) = 1.$ ($\rightarrow$ C2.2.1-2)

When doing an integral over a definite interval, $[a, b]$, the first step is usually to "compute
the indefinite integral" and find a primitive function of the integrand. However, unlike with
the derivative of functions, not every integral can be solved in closed form – sometimes it is
just not possible to find a suitable primitive function. Nevertheless, there exists a huge body
of solution strategies (analytical, approximate or numerical) and satisfactory solutions to
most integration problems can be found. Some general rules and hints in this regard are
summarized in Section C2.4.

We finally note that the *integral over an open interval*, (a, b), gives the same result as that over its closure, $[a, b]$. The reason is that the estimate of the bin width entering the Riemann sum construction does not depend on the presence or absence of the isolated endpoints, a, b, in which the two intervals differ. (In fact, the notation $\int_a^b dx f(x)$ does not even distinguish between the two cases.) As mentioned previously, we will mostly work with open intervals in this text. However, where integration is concerned the difference between "open" and "closed" is conveniently irrelevant.

EXERCISE In the above formulas for integrals, $\int_a^b dy f(y)$, an order of integration boundaries, $a < b$ is assumed. Sometimes, integrals in which the order is reversed are encountered, usually at intermediate steps of calculations. Such integrals are then defined as

$$\int_b^a dy f(y) = -\int_a^b dy f(y). \qquad (C23)$$

In this way, consistency with Eq. (C22) is established.

Integrability

Not all integrals are well-defined. Much as a function can vary too rapidly to be differentiable, it can diverge too strongly to be "summable". More precisely, an integral over a specified interval is said to "exist" if and only if the Riemann sum (C19) converges to a finite value in the limit $\delta \to 0$. If it does not, we say that the integral "does not exist"
Riemann integrable or that the integrand is not **Riemann integrable**.[9] For example, consider the function $f(y) = 1/y$, which has a **singularity** (i.e. a point of divergence)

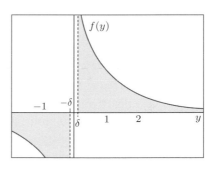

at $y = 0$. The integral $\int_1^2 dy \, y^{-1} = \ln(y)|_1^2 = \ln(2)$ exists, but $\int_0^2 dy \, y^{-1} = \ln(y)|_0^2$ does not because the primitive function, $F(y) = \ln(y)$, diverges at zero (see figure). Note that the divergence of an integrand at a singularity does not necessarily imply the non-existence of its integral. For example $y^{-1/2}$ has a singularity at $y = 0$, however, the integral
integrable singularity $\int_0^1 dy \, y^{-1/2} = 2y^{1/2}|_0^1 = 2$ does exist. This is an example of an **integrable singularity**. More generally, any power, $y^{-\alpha}, \alpha < 1$ has the primitive function $-\frac{1}{\alpha-1} y^{-(\alpha-1)}$, which is well behaved at $y = 0$. However, $y^{-\alpha}, \alpha \geq 1$ are examples of *non-integrable singularities*.

Finally, there are integrals that trick one into believing that they are Riemann-doable, although they are not. As an example, consider the integral $\int_{-3}^2 dy \, y^{-1}$. A naive evaluation through the primitive function $F(y) = \ln(y)$ yields the result $\ln(2) - \ln(-3)$. This looks like the difference of two finite numbers. However, the appearance of the (ill-defined)

---

[9] There exist more general integration schemes – the relevant keyword is **Lebesgue integrability** – often discussed in advanced lecture courses of calculus. However, in view of the rarity of functions which are Lebesgue-but not Riemann-integrable we do not address this generalization here.

logarithm of a negative real number makes the result questionable. Indeed, the integrand contains a *non-integrable singularity* at $y = 0$, whereas the application of Eq. (C19) requires integrability throughout the entire domain of integration.

One may make sense of an integral with an *isolated singularity* at y_0 by considering the expression

$$P \int_a^b dy f(y) \equiv \lim_{\delta \to 0} \left(\int_a^{y_0-\delta} dy + \int_{y_0+\delta}^b dy \right) f(y). \tag{C24}$$

For finite δ the singularity is avoided by removal of a small region around it. If the limit **principal** $\delta \to 0$ of an infinitesimally small cutout region exists, $P \int_a^b dy f(y)$ is called the **principal** **value** **value integral** of the function around the singularity. This expression must not be identified **integral** with the integral of the function $f(y)$, which does not exist if the limits $\lim_{\delta \to 0} \int_a^{y_0-\delta} dy f$ and $\lim_{\delta \to 0} \int_{y_0+\delta}^b dy f$ do not exist separately. (If they do, then $P \int dy f = \int dy f$ by construction.) Principal value integrals can be finite if the diverging contributions to an integral from the left and the right of a singularity almost cancel each other. For example, for $a < 0 < b$, the principal value integral (see the figure above for an illustration)

$$P \int_a^b \frac{dy}{y} = \lim_{\delta \to 0} \left[\int_a^{-\delta} \frac{dy}{y} + \int_\delta^b \frac{dy}{y} \right] = \lim_{\delta \to 0} \left[-\int_\delta^{-a} \frac{dy'}{y'} + \int_\delta^b \frac{dy}{y} \right]$$

$$= \lim_{\delta \to 0} \left[-\ln(-a) + \ln \delta + \ln b - \ln \delta \right] = \ln\left(\frac{b}{|a|} \right),$$

is finite. In the second equality we used a substitution[10] $y' = -y$ and in the third noted that the $\pm \ln \delta$ contributions from the "inner boundaries" at $\mp \delta$ cancel. We will discuss applications of principal value integrals in Chapter C9.

As a corollary we note that a *criterion for the integrability of a function*, f, is the integrability of its *modulus*, $|f|$. The integral $\int dy |f(y)| \geq |\int dy f(y)|$ is an upper bound for the modulus of an integral (why?) and if it exists, the integral of f exists with certainty. In the integral of the modulus, potential singularities are all counted with equal sign and a spurious cancellation mechanism as discussed above will not go undetected.

C2.3 Integration rules

The fundamental theorem of calculus, Eq. (C20), is called "fundamental" for a reason: from it, other integration identities can be derived with little effort. In the following, we discuss two important secondary identities, the rule of integration by parts, and that of substitution of variables.

---

[10] Readers not familiar with variable substitutions in integrals from high school will find the concept explained in the next section.

Integration by parts

Consider the function $F(x) = u(x)v(x)$, where u and v are differentiable functions. The product rule of differentiation, Eq. (C7), then states that $F' = u'\,v + u\,v'$, where we omitted the arguments for clarity. This means that

$$F(b) - F(a) = \int_a^b dx\, F'(x) = \int_a^b dx\big[u'(x)\,v(x) + u(x)\,v'(x)\big].$$

integration by parts   Rearranging terms we obtain the formula for **integration by parts**,

$$\boxed{\int_a^b dx\, u(x)\, v'(x) = \big[u(x)\,v(x)\big]_a^b - \int_a^b dx\, u'(x)\, v(x).}$$
(C25)

This rule is often formulated without explicit reference to boundaries:

$$\int dx\, u(x)\, v'(x) = u(x)\,v(x) - \int dx\, u'(x)\, v(x).$$
(C26)

This relation is useful in cases where the integral on the right is easier to do than that on the left.

EXAMPLE Consider the integral $\int dx\, x\, e^x$. With $u(x) = x$ and $v'(x) = \exp(x)$ we have $u'(x) = 1$ and $v(x) = \exp(x)$. Integration by parts then yields

$$\int dx\, x\, e^x = x\, e^x - \int dx\, e^x = e^x(x - 1).$$

As a check, we note that differentiating the result indeed reproduces the integrand, $x e^x$. ($\to$ C2.3.1-2)

Substitution of variables

Much as Eq. (C25) follows from the product rule of differentiation, an integration formula for changes of variables follows from the chain rule (C8). Consider a function $f(y) \equiv d_y F(y)$. Let $y(x)$ be a *monotonically increasing* differentiable function of the variable x. Then $b > a$ implies $y(b) > y(a)$, and application of the fundamental theorem yields $F(y(b)) - F(y(a)) = \int_{y(a)}^{y(b)} dy\, d_y F(y) = \int_{y(a)}^{y(b)} dy\, f(y)$. On the other hand, we may consider $F(y(x))$ as a function of x. Applying the fundamental theorem once more, but this time with reference to the variable x, we obtain $F(y(b)) - F(y(a)) = \int_a^b dx\, d_x F(y(x)) = \int_a^b dx\, d_y F\big|_{y(x)} d_x y(x) = \int_a^b dx\, f(y(x)) d_x y(x)$, where the chain rule was used. Equating the

substitution of variables   two results yields the rule of **substitution of variables**,

$$\boxed{\int_a^b dx\, \frac{dy(x)}{dx} f(y(x)) = \int_{y(a)}^{y(b)} dy\, f(y).}$$
(C27)

EXAMPLE Consider the integral $I(z) = \int_0^z dx\, x\, e^{-x^2}$. Define $y(x) = x^2$ and write the integral as $I(z) = \frac{1}{2}\int_0^z dx\, \frac{dy(x)}{dx} e^{-y(x)} = \frac{1}{2}\int_0^{z^2} dy\, e^{-y} = -\frac{1}{2}[e^{-y}]_0^{z^2} = -\frac{1}{2}[e^{-z^2} - 1]$. As a check, differentiate the result to verify that the integrand is reproduced. ($\to$ C2.3.3-4)

For a monotonically *decreasing* function y the same construction yields

$$\int_a^b dx\, \frac{dy(x)}{dx} f(y(x)) = -\int_{y(b)}^{y(a)} dy\, f(y),$$

where now $y(b) < y(a)$. Since the derivative of a decreasing function is negative, we may absorb the minus sign by writing $-d_x y = |d_x y|$. Both variants may therefore be subsumed in a single equation, known as the *indefinite version of the rule of substitution of variables*,

$$\int dx\, \left|\frac{dy(x)}{dx}\right| f(y(x)) = \int dy\, f(y). \tag{C28}$$

INFO Consider the substitution rule (C27). Formulas describing the change of variables in integrals generally contain derivative factors such as $\frac{dy}{dx}$ above. The following *dirty trick* is a mnemonic for remembering the placement of such factors: suppose dx and dy were ordinary "variables" and $\frac{dy}{dx}$ an ordinary ratio. The structure "$dx\frac{dy}{dx} = dy$" would then be an ordinary formula for fractions. The mathematically precise formulation of this mnemonic is discussed in Chapter V5 (see Eq. (V148)).

EXAMPLE Consider the integral $I = \int dx\frac{1}{\sqrt{1-x^2}}$. Recalling that $\frac{d}{dx}\arcsin x = \frac{1}{\sqrt{1-x^2}}$ (Eq. (C16)), we have $I = \arcsin x$. Equivalently, we may use the substitution $x = \sin y$, which is helpful since $\sqrt{1-x^2} = \cos y$. Recalling the trick $dx = dy\frac{dx}{dy} = dy\cos y$, the integral is computed as $I = \int dy\cos y\frac{1}{\cos y} = \int dy = y = \arcsin x$.

This strategy is exemplary for a more general procedure. The derivatives of the inverse trigonometric functions in Eqs. (C16) are key to the computation of integrals involving $\frac{1}{\sqrt{1-x^2}}$ or $\frac{1}{1+x^2}$ by **trigonometric substitutions** $x = \sin y$, $\cos y$ or $\tan y$. Similarly, by Eq. (C17), **hyperbolic substitutions** $x = \sinh y$, $\cosh y$ or $\tanh y$ may be applied to integrals with $\frac{1}{\sqrt{1+x^2}}$, $\frac{1}{\sqrt{x^2-1}}$ or $\frac{1}{1-x^2}$. ($\to$ C2.3.5-8)

trigonometric substitution of variables

INFO Above we provided a formal proof of the rule of substitution of variables by application of the fundamental theorem. However, variable substitutions appear very frequently, not just in one-dimensional contexts, and it is well to understand the meaning of Eq. (C27) intuitively. To this end, let us return to the description of integration as sums over increasingly fine discretization "grids". The point to notice now is that these grids need not be evenly spaced. The freedom to choose discretizations of varying width is the *principle behind all variable substitution rules* of integration.

non-uniform discretization of functions

Consider a function $f : [\tilde{a}, \tilde{b}] \to \mathbb{R}$, $y \mapsto f(y)$ for which regions of rapid variation alternate with ones where changes are slow (see the figure). In this case, it might make sense to introduce a system of bins of varying width: rapid changes would call for a finer discretization through a large number of narrow bins, while fewer and wider bins would suffice to describe regions of modest variation. (On a computer such flexible sampling leads to higher efficiency and saves memory without sacrificing accuracy.)

A variant of the Riemann sum over N bins, $[y_\ell, y_{\ell+1}]$, of varying width, $y_{\ell+1} - y_\ell$, reads

$$\int_{\tilde{a}}^{\tilde{b}} dy f(y) = \lim_{N \to \infty} \sum_{\ell=0}^{N-1} [y_{\ell+1} - y_\ell] f(y_\ell). \qquad (C29)$$

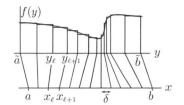

To compute the sum (C29) in concrete terms we need to specify the points y_ℓ. To this end, we introduce an interval $[a, b]$ and a *monotonically increasing* function

$$y : [a, b] \to [\tilde{a}, \tilde{b}], \qquad x \mapsto y(x), \qquad (C30)$$

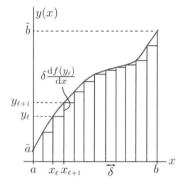

where $y(a) = \tilde{a}$, $y(b) = \tilde{b}$. This function is defined such that for a *uniform* discretization of $[a, b]$ into N points $x_\ell = a + \ell\delta$, with $\delta = (b - a)/N$ and $\ell = 0, \ldots, N - 1$, the values $y_\ell \equiv y(x_\ell)$ define the points of the desired y-discretization. For example (see figure), a region of rapid variation of $y(x)$ leads to widely spaced points y_ℓ, and hence wide bins $y_{\ell+1} - y_\ell$.

Using $y_{\ell+1} - y_\ell = y(x_\ell + \delta) - y(x_\ell) \simeq \delta \frac{dy(x_\ell)}{dx}$, we now represent the Riemann sum as

$$\int_{\tilde{a}}^{\tilde{b}} dy f(y) = \lim_{N \to \infty} \sum_{\ell=0}^{N-1} \big[y(x_\ell + \delta) - y(x)\big] f(y(x_\ell))$$

$$\overset{(C2)}{\simeq} \lim_{N \to \infty} \sum_{\ell=0}^{N-1} \delta \frac{dy(x_\ell)}{dx} f(y(x_\ell)) = \int_a^b dx \frac{dy}{dx} f(y(x)). \qquad (C31)$$

Here, the factor $\frac{dy}{dx}$ describes the way in which the uniform x-grid gets distorted to generate the non-uniform y-grid. For example, regions where $\frac{dy}{dx}$ is large contribute to the x-integral with increased weight because they correspond to wide grid spacings in the original y-representation. Recalling that $\tilde{a} = y(a)$ and $\tilde{b} = y(b)$ we recognize the rule of substitution of variables, Eq. (C27) above.

Later in the text, we will meet various other identities describing the change of variables in integrals. However, all these formulas rely on constructions similar to that discussed above. It may be a good idea to spend a little time and let the geometric interpretation sink in, both in the discrete and the continuum representation.

C2.4 Practical remarks on one-dimensional integration

Although there exists no general recipe to compute the primitive function for arbitrary f, the majority of integrals encountered in the physics curriculum involve standard functions – polynomials, exponentials, logarithms, trigonometric functions, etc. With time and practice the integrals of these functions will become familiar. A number of important examples of such "basic" integrals are implicit in the derivatives listed in Section C1.3, we just need to read the equations from right to left. For example,

$$(\ln(x))' = \frac{1}{x} \qquad \Leftrightarrow \qquad \int dx \frac{1}{x} = \ln(x).$$

How do we approach integrals if the solution is not immediately obvious? The following list contains a number of useful procedures and guiding principles:

▷ It often helps to start from an *educated guess* for the primitive $F(x)$. Sometimes one just needs to play around a little to improve an initially not-quite-correct guess and arrive at a function satisfying $\frac{d}{dx}F(x) = f(x)$.

▷ If the integrand contains functions whose derivative looks more inviting than the function itself, *try to integrate by parts.* ($\to$ C2.3.1-2) For example,

$$\int dx\, x \ln(x) = \int dx\, \frac{1}{2}\frac{dx^2}{dx}\ln(x) \overset{(C26)}{=} \frac{1}{2}x^2 \ln(x) - \frac{1}{2}\int dx\, x^2 \frac{d\ln(x)}{dx}$$

$$= \frac{1}{2}x^2 \ln(x) - \frac{1}{2}\int dx\, x^2 \frac{1}{x} = \frac{x^2}{2}\left(\ln(x) - \frac{1}{2}\right).$$

▷ If an integral contains terms more complicated than the elementary functions listed in Section C1.3, *try substitutions.* ($\to$ C2.3.3-8) An expression containing $dx\frac{1}{x}$ might call for the substitution $y = \ln(x)$, which results in $dy = dx\frac{1}{x}$. For example,

$$\int dx\, \frac{1}{x}\frac{1}{a + \ln(x)} = \int dy\, \frac{1}{a + y} = \ln(a + y) = \ln(a + \ln(x)).$$

Similarly, the combination $dx\, x$ suggests the substitution $y = x^2$, with $dx\, x = \frac{1}{2}dy$.

▷ There are families of functions whose integrals look complicated but are known to be doable. An important example are the **rational functions**, i.e. functions $f(x) = P(x)/Q(x)$ which can be written as a ratio of two polynomials. These can be integrated using a technique called **partial fraction decomposition**. ($\to$ C2.4.1-2) Other examples of integrable families include *rational functions of trigonometric functions* (ratios of polynomials in the functions $\sin(x), \cos(x)$ and $\tan(x)$), and *polynomials in exponential functions*. For the corresponding integration strategies we refer to textbooks on calculus. Try to memorize the families of functions mentioned above to be able to recognize their integrals as doable when you meet them.

integral of rational functions

▷ *Computer algebra* packages such as Mathematica® or Maple® can be powerful aids for solving even very complex integration problems. However, we recommend not to use these packages excessively: integrals encountered in physics often have a structure that "reflects" the underlying physics, and if one lets a computer do the job one loses touch with this structure. On the same note, "manual" struggling with an integral is usually rewarded with added insight into the problem. It is therefore good practice to seriously try to solve integrals by hand before turning to a computer.

integration by computer algebra

▷ As a compromise between the manual and the fully automated solution of integrals one may use **integral tables**. The primary reference in this context is I.S. Gradshteyn and I.M. Ryzhik, *Table of Integrals, Series, and Products*, Academic Press, 7th edition, 2007. This book tabulates thousands of integrals.

▷ No matter how the primitive function has been obtained, *always* check it by differentiation.

▷ *Many integrals are not expressible through elementary functions*. For example, the **Gaussian function**, $\exp(-x^2)$, does not have an elementary primitive. In cases where an "important" function cannot be integrated to elementary functions, its primitive *defines* what is called a **special function**. For example, the integral of the Gaussian function defines the **error function**

$$\int_0^y dx\, e^{-x^2} \equiv \frac{\sqrt{\pi}}{2} \text{erf}(y).$$

special function

▷ For some types of *definite integrals* there exist methods which avoid the need to find the primitive function, and some of these will be discussed in Section C9.4 on complex calculus. Such shortcuts are helpful in cases where the indefinite integrals cannot be expressed in elementary terms. For example, the **Gaussian integral**,

Gaussian integral

$$\int_{-\infty}^{\infty} dx\, e^{-x^2} = \sqrt{\pi}, \tag{C32}$$

can be computed ($\rightarrow$ C2.4.3-4) without reference to its indefinite integral, the error function. Other integrals in the same league include the exponential integrals, $\int_0^\infty dx\, x^n\, e^{-x}$ ($\rightarrow$ C2.4.5), and general Gaussian integrals, $\int_0^\infty dx\, x^{2n}\, e^{-x^2}$ ($\rightarrow$ C2.4.6).

numerical integration

▷ Any (Riemann integrable) function can be integrated *numerically* on a computer. In this case, a computer is employed to evaluate the Riemann discretization sums. The accuracy of the result can be increased by reducing the discretization step size, and/or turning to non-uniform discretizations (see info section on p. 224) adjusted to the profile of the integrand. For discussion of discretization grids tailored to obtain rapid convergence, etc., we refer to textbooks on numerical integration.

INFO We conclude with a remark on integrals of complex-valued functions, $f : \mathbb{R} \rightarrow \mathbb{C}$, $x \mapsto f(x) \equiv u(x) + iv(x)$. These are defined as separate integrals for the real and imaginary parts, $\int dx f(x) \equiv \int dx u(x) + i \int dx v(x)$. Often, however, it is best not to forcibly decompose a complex-valued function but to treat the imaginary unit, i, as an ordinary number. As an example, consider the integral $\int_0^1 dx(x+i)^2 = \frac{1}{3}(x+i)^3 \big|_0^1 = \frac{1}{3}(1+i)^3 - \frac{1}{3}i^3 = -\frac{2}{3} + i$. Check that the same result is obtained if $(x + i)^3$ is decomposed into real and imaginary parts, and the two contributions are integrated separately.

Integrals over functions of a complex variable, $f : \mathbb{C} \rightarrow \mathbb{C}$, are discussed in Chapter C9.

C2.5 Summary and outlook

In this chapter we introduced the idea of integration as a refined way of summation on the example of one-dimensional functions. Many aspects of this discussion, notably the "reciprocity" of integration and differentiation, carry over to the generalized integrals addressed in later chapters. For example, the substitution rule has various higher-dimensional generalizations. Although these may look a little more complicated than the one-dimensional one, the construction principles always reflect the discussion of the info section on p. 224. We also discussed various integration techniques specific to one-dimensional functions. These, too, continue to play an important role in more general contexts: higher-dimensional integrals are usually broken down to successions of one-dimensional ones, which then need to be processed by the methods reviewed above.

We have now reached a good basis for generalizing the concepts of differentiation and integration to functions defined in higher-dimensional spaces, and this is the subject to which we turn next.

C3

 Partial differentiation

Consider a function depending on more than one variable, such as the water depth, $D(\mathbf{r})$, beneath a boat at position $\mathbf{r} = (x, y)$ on a lake, or the air pressure, $P(T, V)$, in a container of volume V at temperature T. One may ask how these quantities change if only *one* of the variables is varied: how does the water depth vary if the boat moves in the x-direction at fixed y? Or how does the pressure in the container change upon increasing the temperature at fixed volume? The present chapter introduces *partial derivatives* as the mathematical tools to address such questions.

C3.1 Partial derivative

partial derivative

Consider a function $f : \mathbb{R}^d \rightarrow \mathbb{R}$, $\mathbf{x} \mapsto f(\mathbf{x}) = f(x^1, \dots, x^d)$ depending on d variables $x^1, \dots, x^d$. The **partial derivative** of f with respect to x^i probes how $f(\mathbf{x})$ changes if only the single variable x^i is varied. It is defined as the ordinary derivative of f w.r.t. to x^i taken at fixed values of the other variables:

$$\frac{\partial f(\mathbf{x})}{\partial x^i} \equiv \lim_{\delta \to 0} \frac{1}{\delta} \left[f(x^1, \dots, x^i + \delta, \dots, x^d) - f(x^1, \dots, x^i, \dots, x^d) \right]. \qquad \text{(C33)}$$

The symbol ∂ indicates that this is a *partial* derivative of a multidimensional function, in contrast to the ordinary derivative (written as d) of a one-dimensional function. Other frequently used notations include[11]

$$\frac{\partial f(\mathbf{x})}{\partial x^i} \equiv \partial_{x^i} f(\mathbf{x}) \equiv \partial_i f(\mathbf{x}).$$

EXAMPLE The examples below are partial derivatives written in different notations:

$$\partial_1 \left[(x^1)^2 x^2 + x^3 \right] = 2x^1 x^2,$$
$$\partial_{x^2} \left[(x^1)^2 x^2 \cos(x^2) + x^3 \right] = (x^1)^2 \left[\cos(x^2) - x^2 \sin(x^2) \right],$$
$$\partial_y \left[(x^2 + y^3) \sin(x + y^2) \right] = 3y^2 \sin(x + y^2) + (x^2 + y^3) 2y \cos(x + y^2).$$

---

[11] In *covariant notation* where component indices are written as superscripts (x^i), the symbol ∂_i carries a subscript. The rationale behind this convention will become clear in Chapter V3 (see Eq. (V69)). However, an easy way to memorize it is to note that $\partial / \partial x^i$ is an object carrying a superscript symbol in the "*denominator*". Much as with a double fraction ($1/(1/5) = 5$) this corresponds to a symbol with inverted index position in the "numerator", $\partial / \partial x^i = \partial_i$.

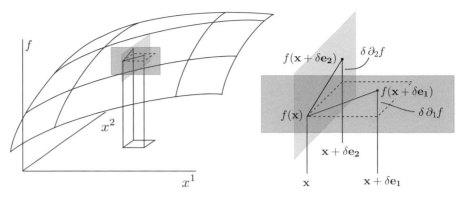

Fig. C3   Partial derivatives illustrated for a function $f : \mathbb{R}^2 \to \mathbb{R}$.

It is sometimes useful to write Eq. (C33) in a vectorial notation where the variables x^i define a vector as $\mathbf{x} = \sum_i^d \mathbf{e}_i x^i$. We then have

$$\frac{\partial f(\mathbf{x})}{\partial x^i} = \lim_{\delta \to 0} \frac{1}{\delta} \big(f(\mathbf{x} + \delta\,\mathbf{e}_i) - f(\mathbf{x})\big). \tag{C34}$$

Figure C3 illustrates the interpretation of the partial derivative on a two-dimensional example. The shaded planes indicate how one variable is kept constant in the process. The variation of the other variable yields the partial derivative as an ordinary derivative taken in the direction of the corresponding coordinate axis.

Since partial derivatives are ordinary derivatives taken w.r.t. one out of d variables, they are as easy to compute as one-dimensional derivatives ($\to$ C3.1.1-2). All the differentiation rules discussed in Section C1.2 directly carry over to partial differentiation. For example, the product rule reads:

$$\partial_i \big(f(\mathbf{x})\, g(\mathbf{x})\big) = \big(\partial_i f(\mathbf{x})\big)\, g(\mathbf{x}) + f(\mathbf{x})\, \big(\partial_i g(\mathbf{x})\big).$$

C3.2 Multiple partial derivatives

Just as with multiple ordinary derivatives (see Eq. (C4)), *multiple partial derivatives* are obtained by repeatedly taking single derivatives. For example, the symbols $\partial^2_{x^i,x^j}$ or just $\partial^2_{i,j}$ indicate a double partial derivative in which one first differentiates in the variable x^j, and then the result in x^i. If $i = j$, this is an ordinary second-order derivative in x^i and generally abbreviated as $\partial^2_{x^i,x^i} \equiv \partial^2_{x^i} \equiv \partial^2_i$. For example, with $x^1 \equiv x$ and $x^2 \equiv y$,

$$\partial^2_x \big(x^3 y^2\big) = \partial_x \big(3x^2 y^2\big) = 6xy^2, \qquad \partial^2_y \big(x^3 y^2\big) = \partial_y \big(2x^3 y\big) = 2x^3.$$

Mixed derivatives in different variables generally are to be taken in the order specified by the notation:

$$\partial^2_{i,j} f(\mathbf{x}) \equiv \partial^2_{x^i,x^j} f(\mathbf{x}) \equiv \partial_{x^i} \partial_{x^j} f(\mathbf{x}) \equiv \partial_{x^i} \big(\partial_{x^j} f(\mathbf{x})\big).$$

Schwarz's theorem

However, for smooth functions **Schwarz's theorem** states that the order in which partial derivatives are taken does not matter:[12]

$$\partial^2_{i,j} f(\mathbf{x}) = \partial_{x^i} \partial_{x^j} f(\mathbf{x}) = \partial_{x^j} \partial_{x^i} f(\mathbf{x}) = \partial^2_{j,i} f(\mathbf{x}) \qquad (f \text{ "smooth"}). \tag{C35}$$

For example ($\rightarrow$ C3.2.1-2),

$$\partial^2_{x,y} \cos(x e^y) = \partial_x\left(- \sin(x e^y) \, x \, e^y\right) = -\cos(x e^y) \, x \, e^{2y} - \sin(x e^y) \, e^y,$$
$$\partial^2_{y,x} \cos(x e^y) = \partial_y\left(- \sin(x e^y) \, e^y\right) = -\cos(x e^y) \, x \, e^{2y} - \sin(x e^y) \, e^y.$$

In physics, multiple partial derivatives appear frequently and changes in the order of derivatives are applied to simplify calculations or even prove statements. However, it is important to remember that such operations rely on the smoothness condition and that there exist (a few) treacherous functions which look smooth but are not (in the sense of the definition on p. 212). In such cases, the exchange of derivatives may be invalid.

EXAMPLE Consider the function

$$f(x, y) = \begin{cases} \frac{xy(x^2 - y^2)}{x^2 + y^2}, & (x, y) \neq (0, 0), \\ 0, & (x, y) = (0, 0). \end{cases}$$

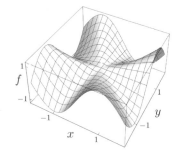

non-commuting partial derivatives

The function looks smooth and is partially differentiable everywhere. However, at $(x, y) = (0, 0)$ the partial derivatives do not commute: $\partial_x \partial_y f\big|_{(0,0)} \neq \partial_y \partial_x f\big|_{(0,0)}$ (check this). This signifies that the smoothness conditions required by Schwarz's theorem are not given. In mathematics, it is good practice to check the required criteria *before* a derivative is carried out. Physicists tend to be more cavalier and assume the commutativity of derivatives. This approach becomes dangerous in the (admittedly very rare) cases where functions look smooth, but are not in a mathematical sense. Premature differentiation may then lead to errors, which, however, are generally easy to track.

C3.3 Chain rule for functions of several variables

In physics one frequently encounters situations in which a multivariate function, $f(\mathbf{g}) = f(g^1, \ldots, g^d)$, depends on a parameter variable, x, indirectly via the dependence $g^i = g^i(x)$ of its arguments on x. For example, the pressure, $P(V, T)$, of a gas in a piston depends on the available volume, V, and temperature, T. This dependence may become time-dependent, $P(t) = P(V(t), T(t))$, if temperature, $T(t)$, and pressure, $P(t)$, vary in time. In such cases, it is natural to ask how the composite function $f(\mathbf{g}(x))$ varies with x. The answer to this question is provided by a generalization of the chain rule to be introduced in this section.

---

[12] Actually, Schwarz's theorem does not require smoothness but the weaker condition that all second-order partial derivatives, $\partial^2_{i,j} f$, be *continuous* at $\mathbf{x}$.

An auxiliary relation

We first ask how a function $f(\mathbf{y}) = f(y^1, \dots, y^d)$ changes under the simultaneous variation of *all* its arguments, $\mathbf{y} \to \mathbf{y} + \delta\mathbf{z}$, where $\mathbf{z} \in \mathbb{R}^d$ is arbitrary and δ is infinitesimal. Before answering this question in general, let us consider a function depending on just two arguments, $d = 2$. In this case, the rate of change is described by the difference quotient, $\frac{1}{\delta}\left[f(y^1 + \delta z^1, y^2 + \delta z^2) - f(y^1, y^2)\right]$. We aim to reduce this expression to one containing the more familiar difference quotients of ordinary derivatives in single variables. This can be achieved by the insertion of $0 = -f(y^1 + \delta z^1, y^2) + f(y^1 + \delta z^1, y^2)$. In this way, the difference quotient can be written as the sum of two such quotients,

$$
\frac{1}{\delta}\left[f(y^1 + \delta z^1, y^2 + \delta z^2) - f(y^1, y^2)\right]
$$
$$
= \frac{1}{\delta}\left[f(y^1 + \delta z^1, y^2 + \delta z^2) - f(y^1 + \delta z^1, y^2)\right] + \frac{1}{\delta}\left[f(y^1 + \delta z^1, y^2) - f(y^1, y^2)\right]
$$
$$
\stackrel{(C33)}{\simeq} z^2 \frac{\partial f(y^1 + \delta z^1, y^2)}{\partial y^2} + z^1 \frac{\partial f(y^1, y^2)}{\partial y^1} \xrightarrow{\delta \to 0} z^2 \frac{\partial f(y^1, y^2)}{\partial y^2} + z^1 \frac{\partial f(y^1, y^2)}{\partial y^1}. \tag{C36}
$$

In the second line, the second difference quotient probes the function's increment when its first argument y^1 changes to $y^1 + \delta z^1$, at fixed y^2. Similarly, the first quotient probes the function's increment when its second argument y^2 changes to $y^2 + \delta z^2$, at fixed $y^1 + \delta z^1$. Figure C4 visualizes this decomposition of the full increment into two separate contributions, indicated by thick vertical lines.

In the first approximate equality of the third line, the two increments are expressed by the corresponding partial derivatives, the first taken at $(y^1 + \delta z^1, y^2)$, the second at (y^1, y^2). Finally, f was assumed to be smooth, and so $\partial_{y^2} f$ is likewise smooth. In particular, it is continuous. This implies $\lim_{\delta \to 0} \partial_{y^2} f(y^1 + \delta z^1, y^2) = \partial_{y^2} f(y^1, y^2)$, i.e. in the limit of infinitesimal δ the slight shift in the evaluation point of the partial derivative does not matter, and this point is made in the final equality. Similar lines of reasoning will be applied in several other cases below. Before reading on, make sure that you understand the logic of the construction above well.

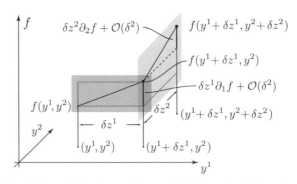

Fig. C4   The qualitative picture behind relation (C36), the specialization of Eq. (C38) to $d = 2$ dimensions. Discussion, see text.

The construction immediately generalizes to functions depending on more than two arguments and the result then reads

$$\lim_{\delta \to 0} \frac{1}{\delta}\left[f(\mathbf{y} + \delta\mathbf{z}) - f(\mathbf{y})\right] = \sum_{j=1}^{d} \frac{\partial f(\mathbf{y})}{\partial y^j} z^j. \tag{C37}$$

This identity states that the net change of the function is obtained by computing its partial derivatives, ∂_{y^j}, in the directions of the individual variables, weighting each with the component of the increment vector, z^j, and adding up.

A version of this formula describing the "linearization" of f in small yet not necessarily infinitesimal variations of δ reads

$$f(\mathbf{y} + \delta\mathbf{z}) - f(\mathbf{y}) \simeq \sum_{j=1}^{d} \frac{\partial f(\mathbf{y})}{\partial y^j} \delta z^j. \tag{C38}$$

Notice how this relation embodies the essence of differentiation: the local structure of a function, i.e. the difference of function values between nearby points on the l.h.s., is approximately described by a function that is linear in the argument displacements, $\propto \delta z^j$. The linearization on the r.h.s. is the higher-dimensional analogue of the straight line of Figure C1 and the corresponding Eq. (C2).

Chain rule

Let us now turn back to the setting mentioned in the beginning of the section and consider the composite function (see Fig. C5)

$$f \circ \mathbf{g} : \mathbb{R} \to \mathbb{R}, \quad x \mapsto f\big(\mathbf{g}(x)\big) = f\big(g^1(x), \dots, g^d(x)\big) \equiv f(x),$$

where $\mathbf{g} : \mathbb{R} \to \mathbb{R}^d$, $x \mapsto \mathbf{g}(x) = \big(g^1(x), \dots, g^d(x)\big)^{\mathrm{T}}$ defines the dependence of the d arguments of f on a single parameter x. Notice that the dependence $x \mapsto f(x)$ defines an ordinary real-valued function of a single variable and so it must be possible to compute the derivative $\mathrm{d}_x f(x)$. We compute this derivative by explicit linearization of the functions involved in the process:

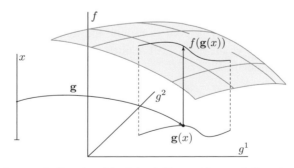

Fig. C5   Geometric description relevant to the discussion of the chain rule illustrated for $n = 2$. Discussion, see text.

$$\frac{df(\mathbf{g}(x))}{dx} = \lim_{\delta \to 0} \frac{1}{\delta}\left[f(\mathbf{g}(x+\delta)) - f(\mathbf{g}(x))\right] = \lim_{\delta \to 0} \frac{1}{\delta}\left[f(\mathbf{g}(x) + \delta\, d_x\mathbf{g}(x)) - f(\mathbf{g}(x))\right],$$

where in the last step we used $g^j(x+\delta) = g^j(x) + \delta\frac{g^j(x)}{dx}$, and introduced the shorthand notation $d_x\mathbf{g} \equiv \left(\frac{dg^1}{dx}, \ldots, \frac{dg^d}{dx}\right)^{\mathrm{T}}$. We may now apply Eq. (C37) with the identifications $\mathbf{y} = \mathbf{g}(x)$ and $\mathbf{z} = d_x\mathbf{g}(x)$, to obtain

$$\boxed{\frac{df(\mathbf{g}(x))}{dx} = \sum_{j=1}^{d} \left.\frac{\partial f(\mathbf{y})}{\partial y^j}\right|_{\mathbf{y}=\mathbf{g}(x)} \frac{\partial g^j(x)}{dx} = \sum_{j=1}^{d} \frac{\partial f(\mathbf{g}(x))}{\partial g^j} \frac{\partial g^j(x)}{dx},} \qquad \text{(C39)}$$

where the right-most expression defines a shorthand for the middle one. This is one of various versions of a *chain rule for a function of several variables*. The rationale underlying this formula is similar to that of the ordinary chain rule, Eq. (C8):

The change of a function $f(\mathbf{g}(x))$ under variations of the argument x multiplicatively depends on both the change of $f(\mathbf{g})$ with g^j and the change of $g^j(x)$ with x. The total rate changes in the different variables, $\partial_{g^j}f\partial_x g^j$, need to be added to obtain the full variation, as in Eq. (C39).

EXAMPLE Chain rules appear frequently in *physical applications*. Consider, for example, a mobile particle in a volume with nontrivial temperature profile, $T(\mathbf{x})$. The trajectory of the particle is described by a curve $\mathbf{r}(t)$ and the instantaneous ambient temperature "felt" by the particle at time t is $T(\mathbf{r}(t))$. The rate of change in temperature with time is described by the derivative $\frac{dT(\mathbf{r}(t))}{dt}$, for which Eq. (C39) yields

$$\frac{dT(\mathbf{r}(t))}{dt} = \sum_{j=1}^{3} \frac{\partial_j T(\mathbf{r}(t))}{\partial r^j} \frac{dr^j(t)}{dt}.$$

Generalized chain rules

The chain rule has two extensions which straightforwardly follow from Eq. (C39). The *first generalization* replaces the scalar function f by a vectorial function,

$$\mathbf{f}: \mathbb{R}^d \to \mathbb{R}^m, \qquad \mathbf{y} \mapsto \mathbf{f}(\mathbf{y}) = \left(f^1(\mathbf{y}), \ldots, f^m(\mathbf{y})\right).$$

This function may be composed with the function $\mathbf{g}(x)$ to yield $\mathbf{f} \circ \mathbf{g} : \mathbb{R} \to \mathbb{R}^m, x \mapsto \mathbf{f}(\mathbf{g}(x))$. The chain rule (C39) applies to each component $f^i(\mathbf{g}(x))$ separately. Using the vectorial notation

$$\frac{d\mathbf{f}}{dx} \equiv \left(\frac{df^1}{dx}, \ldots, \frac{df^m}{dx}\right)^{\mathrm{T}}, \qquad \frac{\partial\mathbf{f}}{\partial y_j} \equiv \left(\frac{\partial f^1}{\partial y_j}, \ldots, \frac{\partial f^m}{\partial y_j}\right)^{\mathrm{T}}, \qquad \text{(C40)}$$

etc., we may write the generalized chain rule as

$$\boxed{\frac{d\mathbf{f}(\mathbf{g}(x))}{dx} = \sum_{j=1}^{d} \frac{\partial\mathbf{f}(\mathbf{g})}{\partial g^j} \frac{dg^j(x)}{dx}.} \qquad \text{(C41)}$$

To formulate the *second generalization* we introduce a function $\mathbf{g}(\mathbf{x})$ of $n > 1$ variables x^k,

$$\mathbf{g} : \mathbb{R}^n \to \mathbb{R}^d, \qquad \mathbf{x} \mapsto \mathbf{g}(\mathbf{x}) = \left(g^1(x^1, \ldots, x^n), \ldots g^d(x^1, \ldots, x^n)\right)^{\mathrm{T}},$$

and compose it with $\mathbf{f}$ to yield

$$\mathbf{f} \circ \mathbf{g} : \mathbb{R}^n \to \mathbb{R}^m, \qquad \mathbf{x} \mapsto \mathbf{f}(\mathbf{g}(\mathbf{x})) = \mathbf{f}\left(g^1(x^1, \ldots, x^n), \ldots, g^d(x^1, \ldots, x^n)\right).$$

We may now ask how the component f^i changes if one variable x^k is varied while all others are kept fixed. By definition, this amounts to taking the partial derivative $\partial_{x^k} f^i$. Remembering that this is just an ordinary derivative in x^k taken at fixed $x^{l \neq k}$, Eq. (C41) **general chain** may be applied to obtain **the most general form of the chain rule** ($\to$ C3.3.1-2) **rule**

$$\boxed{\frac{\partial \mathbf{f}(\mathbf{g}(\mathbf{x}))}{\partial x^k} = \sum_{j=1}^{d} \frac{\partial \mathbf{f}(\mathbf{g}(\mathbf{x}))}{\partial g^j} \frac{\partial g^j(\mathbf{x})}{\partial x^k}.} \tag{C42}$$

INFO As an *example application*, consider a jet engine whose output power $W(T, P)$ depends on **chain rule in** both the temperature, T, and the pressure, P, in the combustion chamber. These two quantities **thermody-** in turn depend on the fuel injection rate, κ, and the chamber volume, V. The task is to opti- **namics** mize the function $W(T(\kappa, V), P(\kappa, V))$ with respect to κ and V. To this end, one needs to know the partial derivatives $\partial_\kappa W$ and $\partial_V W$ (here $m = 1$, $n = 2$, $d = 2$). Application of Eq. (C42) yields

$$\partial_\kappa W = \partial_T W \partial_\kappa T + \partial_P W \partial_\kappa P, \qquad \partial_V W = \partial_T W \partial_V T + \partial_P W \partial_V P,$$

where we have used shorthand notations for the partial derivatives, $\partial_T = \frac{\partial}{\partial T}$, etc. An optimization procedure would now seek points where these derivatives vanish, i.e. configurations where the adjustable parameters are such that the engine output is at an extremum.

EXERCISE Consider the two functions

$$\mathbf{f}(y^1, y^2) = \begin{pmatrix} y^1 \cos(y^2) \\ y^1 \sin(y^2) \end{pmatrix}, \qquad \mathbf{g}(x^1, x^2) = \begin{pmatrix} ((x^1)^2 + (x^2)^2)^{1/2} \\ \arctan(x^2/x^1) \end{pmatrix}.$$

Show that $\dfrac{\partial f^i(g^1(x^1, x^2), g^2(x^1, x^2))}{\partial x^j} = \delta^i{}_j$. How would you interpret this result?

C3.4 Extrema of functions

REMARK Requires Chapter L8.

extremum A function of a single variable, $f(x)$, has an **extremum** at x_0 if its derivative vanishes there, $f'(x_0) = 0$. Whether this extremum is a maximum or a minimum depends on whether the second derivative, $f''(x)$, is negative or positive.

In the same way, an extremum, $\mathbf{x}_0$, of a multidimensional func-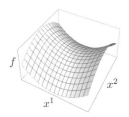
tion, $f(\mathbf{x})$, is defined by the vanishing of all partial derivatives,
$\partial_{x^i} f(\mathbf{x}_0) = 0$. An important difference from the one-dimensional
case is that there now exist d different directions in which the argument can deviate from $\mathbf{x}_0$. In general, deviations in some directions
will cause the function to increase, in others to decrease. In such

saddle point   cases, the extremum is called a **saddle point** of f. For example,
the figure shows a saddle point at $\mathbf{x}_0 = (0,0)$ for the function $f(\mathbf{x}) = (x^1)^2 - (x^2)^2$.

As in the one-dimensional case, the nature of extrema in multivariate functions is
Hesse matrix   revealed through their second derivatives. The key diagnostic tool is the **Hesse matrix**,
or just *Hessian*, H, at $\mathbf{x}_0$, defined as

$$H_{ij} = \frac{\partial^2 f(\mathbf{x}_0)}{\partial x^i \partial x^j}. \tag{C43}$$

Note that the Hesse matrix is symmetric, $H = H^T$. It can therefore be diagonalized by
an orthogonal transformation (Chapter L8), and one may inspect its real eigenvalues.
Without proof, we mention that the function has a *maximum* at $\mathbf{x}_0$ if the Hesse matrix
is **negative definite**, meaning that all eigenvalues are negative. This generalizes the one-
dimensional criterion of a maximum having negative second derivative. Conversely, a
minimum requires a positive definite Hesse matrix. The more generic case of a saddle point
is realized for indefinite Hesse matrices. For example, the function $f(\mathbf{x}) = (x^1)^2 - (x^2)^2$
shown in the figure has an indefinite diagonal Hesse matrix, $H = \text{diag}(2, -2)$, indicative of
a saddle point. For completeness, we mention that the indefiniteness test is sufficient, but
not necessary for the existence of a saddle point. Anomalies may occur for functions with
vanishing second derivatives (the analogue of the one-dimensional function x^3 which has
vanishing first and second derivative at zero). For example, the function $(x^1)^4 - (x^2)^6$ has
a saddle point at $(0,0)$, but the Hesse matrix vanishes there.

C3.5 Summary and outlook

In this chapter we introduced partial differentiation as a means to probe the variation
of multivariate functions. Partial derivatives monitor the rate at which such functions
change if just one of their arguments is varied, and all others are kept fixed. All rules
of ordinary differentiation are equally applicable to partial derivatives. The same goes
for the interpretation of derivatives as effective linearizations of functions. The application of this idea to functions with indirect variable dependences led to higher-dimensional
variants of the chain rule, the most general one being Eq. (C42). These rules are
required to describe the change of functions depending on multiple, mutually correlated
variables.

Partial derivatives are the workhorses used to break even very complex derivatives down
to manageable "ordinary derivatives" in individual scalar variables. They are easy to get

used to, not least because they appear on a daily basis in the work of any physicist. Much as ordinary derivatives are "dual" to integrals over single variables, partial derivatives are dual to successive integrations over several variables. In the next chapter, we introduce this first extension of one-dimensional integrals, which will then become the basis of the more general multidimensional integrals discussed in later parts of the text.

C4 — Multidimensional integration

REMARK Familiarity with Sections V2.1 to V2.3 on curvilinear coordinates is required. The later sections of the chapter also require Chapter L6 on matrix determinants, and Section V2.5 on Jacobians.

In physics, one often needs to integrate ("sum") over the values of functions defined in higher-dimensional spaces. A cartoon of the general situation has been discussed in Section C2.1, where we asked how the total mass carried by a surface coated with a substance of a given "mass density" can be obtained. More generally, integration problems arise when the many incremental changes accumulated by a function in a given context (differentiation) need to be resummed (integration) to obtain the change of the function at large. In one-dimensional contexts, this task is achieved by the high-school variant of integration, which effectively samples the area enclosed by the graph of a function. Building on the understanding of this procedure, we discuss here the extension of integration to higher dimensions.

Higher-dimensional integration theory is a subject of considerable depth and needs to be introduced with an appropriate level of care. At the same time, multidimensional integration techniques are required early on in the physics curriculum. We here resolve this dilemma by including a fast track to integration in this chapter. It provides a quick introduction to the integrals generally required by first- and second-term experimental physics lecture courses – integration over functions defined in two-dimensional and three-dimensional space, and on two-dimensional surfaces. These integrals are under control after reading Section C4.1 and the first subsection in each of C4.2, C4.3 and C4.4. However, these sections by themselves do not introduce integration at the level of depth required in later stages of the curriculum; the chapter should therefore be revisited and read in full at the earliest possible convenience.

C4.1 Cartesian area and volume integrals

Integrals over higher-dimensional structures can always be reduced to successions of one-dimensional integrals. This reduction is best introduced on the example of "cuboids" – rectangles in two-dimensional space, boxes in three-dimensional space, etc. Once the principles are understood, the extension to the more complex integrals discussed in later parts of the chapter will be straightforward.

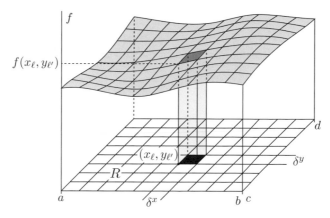

Fig. C6   On the concept of two-dimensional integration over a function.

Integration over rectangles

Consider a function of two variables, defined on a rectangle, $R = [a, b] \times [c, d]$:[13]

$$f : [a, b] \times [c, d] \subset \mathbb{R}^2 \to \mathbb{R}, \qquad (x, y)^{\mathrm{T}} \mapsto f(x, y). \tag{C44}$$

In physical applications, f will usually represent some kind of "density". For example, it might be a mass density in the sense that $f(x, y)\delta^x \delta^y$ represents the mass of a substance contained in a small rectangle with area $\delta^x \delta^y$ at the point $(x, y)^{\mathrm{T}}$. In this case, the integral would compute the total mass contained in the full rectangle R. In the visualization in Fig. C6, the mass contained in such a small rectangle is represented by the volume of the column above that rectangle, and the total mass by the volume under the floating surface defined by $f(x, y)$.

Following the discussion of Section C2.1, we tile the rectangle R by a set of infinitesimal rectangular cells and then sum the contributions of all cells. The summation procedure is set up by dividing the interval $[a, b]$ into N_x bins of infinitesimal width $\delta^x = (b - a)/N_x$, and similarly for the interval $[c, d]$, with $\delta^y = (d - c)/N_y$. Next, the function values are read out as $f(x_\ell, y_{\ell'})$, where x_ℓ and $y_{\ell'}$ lie in the ℓth x-bin and ℓ'th y-bin, respectively. The exact positioning of these coordinates within the bins is not essential (see the analogous discussion in Section C2.2). For example, one possible choice is $x_\ell = \ell \delta^x$ with $\ell = 0, \ldots, N_x{-}1$, and $y_{\ell'} = \ell' \delta^y$ with $\ell' = 0, \ldots, N_y{-}1$. One may now sum over column volumes, $f(x_\ell, y_{\ell'})\delta^x \delta^y$, and in the limit $\delta^x, \delta^y \to 0$ obtain the **two-dimensional integral** as

two-dimensional integral

$$\boxed{\int_R \mathrm{d}x\mathrm{d}y\, f(x, y) \equiv \lim_{\delta^x, \delta^y \to 0} \delta^x \delta^y \sum_\ell \sum_{\ell'} f(x_\ell, y_{\ell'}).} \tag{C45}$$

---

[13] Referring to the definition of Cartesian products of sets, Eq. (L1), the rectangle is defined as the set of points $[a, b] \times [c, d] \equiv \{(x, y) \,|\, x \in [a, b], y \in [c, d]\}$.

This construction not only defines the integral but also contains the recipe for its *practical computation*. In the limit $\delta^y \to 0$, at fixed δ^x and fixed first coordinate x_ℓ, the integral converges to a one-dimensional integral of the function $f(x_\ell, y)$ over y,

$$\lim_{\delta^y \to 0} \delta^y \sum_{\ell'} f(x_\ell, y_{\ell'}) = \int_c^d dy\, f(x_\ell, y) \equiv I(x_\ell),$$

whose value $I(x_\ell)$ depends on the value of x_ℓ. The insertion of $I(x_\ell)$ into the remaining sum, followed by a limit $\delta^x \to 0$, leads to another one-dimensional integral, now over x: $\lim_{\delta^x \to 0} \delta^x \sum_\ell I(x_\ell) = \int_a^b dx\, I(x)$. We conclude that the area integral is given by

$$\int_R dx\, dy\, f(x, y) = \int_a^b dx \int_c^d dy\, f(x, y) = \int_c^d dy \int_a^b dx\, f(x, y), \qquad (\text{C46})$$

where $\int dy\, f(x, y)$ means "integrate $f(x, y)$ over the second argument, y, at a fixed value of the first argument, x". The second equality holds since the construction above could have been formulated in the reverse order – first integrate over x, then over y.

EXAMPLE As an example, consider the function $f : [0, 2] \times [0, 1] \to \mathbb{R}, (x, y) \mapsto f(x, y) = xy + y^2$. It can be integrated in either order to obtain identical results:

$$\int_0^2 dx \int_0^1 dy\, f(x, y) = \int_0^2 dx \left[\frac{1}{2} y^2 x + \frac{1}{3} y^3\right]_0^1 = \int_0^2 dx \left(\frac{1}{2} x + \frac{1}{3}\right) = \left[\frac{1}{4} x^2 + \frac{1}{3} x\right]_0^2 = \frac{5}{3},$$

$$\int_0^1 dy \int_0^2 dx\, f(x, y) = \int_0^1 dy \left[\frac{1}{2} x^2 y + xy^2\right]_0^2 = \int_0^1 dy \left(2y + 2y^2\right) = \left[y^2 + \frac{2}{3} y^3\right]_0^1 = \frac{5}{3}.$$

Fubini's theorem
The fact that the order of integration does not matter is known as **Fubini's theorem** ($\to$ C4.1.1-2). Generally speaking, integrals are defined as Riemann sums over the cells tiling the integration domains. Due to the commutativity of addition the order in which one sums over these is arbitrary. This statement holds for all types of integrals to be discussed in subsequent chapters.

INFO Apart from rare exceptions, Fubini's theorem holds if the double integral over a function, performed in either order, exists. More precisely, the *condition granting Fubini interchangeability* is that $\int_R dx dy |f(x, y)|$, i.e. the integral over the *modulus* of the function, must exist. To appreciate the relevance of the modulus, consider the function $f(x, y) = (x^2 - y^2)/(x^2 + y^2)^2$. It is straightforward to verify that

$$\int_0^1 dx \int_0^1 dy\, \frac{x^2 - y^2}{(x^2 + y^2)^2} = \frac{\pi}{4}.$$

However, the integral done in reverse order yields the negative value, $-\pi/4$ (see C4.1.3(b)). To understand what is happening here, notice that for x, y approaching zero while $x > y$ the integrand contains a strong positive divergence. For $x < y$ the divergence is negative. The integrals over the respective regions, $1 \geq x > y \geq 0$ and $0 \leq x < y \leq 1$, do not exist. Likewise, the integral of the modulus $|f(x, y)|$ over the full square, $0 \leq x, y \leq 1$, does not exist either (because in this case the two singularities add). However, doing the double integral over the function itself, we obtain a result of $\infty - \infty$ type, where the two ∞s come from $x > y$ and $x < y$, respectively. The naive evaluation of the integral tricks one into believing that the difference of the two infinities is finite, either $\pi/4$

or $-\pi/4$, depending on the order in which the x- and y-integrals are performed. However, the sign discrepancy is a manifestation of the fact that the difference of two ∞s is actually not well defined. While the double integrals make formal sense, they do not represent a well-defined *area* integral.

The general message is that before doing an integral one should check that the integral over the modulus of the integrated function exists (see the discussion on p. 221). If not, one is generally working with an ill-defined expression. ($\rightarrow$ C4.1.3-4)

cuboid The tiling construction described above can readily be generalized to *integrals over higher-dimensional* **cuboids**. For example, consider a function $f(x,y,z)$ on $C = [a,b] \times [c,d] \times [e,f] \subset \mathbb{R}^3$. The separate discretization along each dimension divides C into a large number of small cubicles. In the limit, the Riemann sum over all these leads to the triple integral

$$\int_C dx\,dy\,dz\,f(x,y,z) = \int_a^b dx \int_c^d dy \int_e^f dz\,f(x,y,z), \tag{C47}$$

where the order of integrations is again arbitrary. The extension to cuboids of higher dimension should be obvious.

Integration over domains with spatially varying boundaries

Many functions of practical interest are defined on non-rectangular domains. Integrals over such functions can often be computed by straightforward adaption of the above strategy: the integration domain is tiled by infinitesimal rectangular cells (or boxes in three-dimensional settings). However, the number of cells in one direction may now depend on the cell index in other directions. For a two-dimensional example, consider the circular disk, D, shown in the figure. In this case, the number of cells in the y-direction is largest close to the center of the x-axis at $x_\ell \simeq 0$. As a consequence, the lower and upper summation thresholds for $y_{\ell'}$ now depend on x_ℓ. Let us denote them by $c_-(x_\ell)$ and $c_+(x_\ell)$, respectively, and the lower and upper thresholds for x_ℓ by a_- and a_+. The discrete approximation of the integral then assumes the form

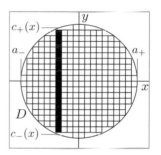

$$\delta^x \delta^y \sum_{a_- \leq x_\ell < a_+} \sum_{c_-(x_\ell) \leq y_{\ell'} < c_+(x_\ell)} f(x_\ell, y_{\ell'}).$$

We take the limit $\delta^x, \delta^y \rightarrow 0$ to obtain the integral representation

$$\int_D dx\,dy\,f(x,y) \equiv \int_{a_-}^{a_+} dx \int_{c_-(x)}^{c_+(x)} dy\,f(x,y).$$

Here, the integration boundaries of the "inner" y-integral, $I(x) = \int_{c_-(x)}^{c_+(x)} dy\,f(x,y)$, depend on the integration variable of the "outer" x-integral. However, this is no cause for concern – one simply integrates over y to find $I(x)$, and subsequently over x to obtain $\int_{a_-}^{a_+} dx\,I(x)$. ($\rightarrow$ C4.1.5-10).

EXAMPLE Consider a *disk of radius R* and let us determine its area, A, by integration. In this case, the boundaries $a_\pm = \pm R$ are set by the disk radius, and $c_\pm(x) = \pm\sqrt{R^2 - x^2}$. This gives

$$A = \int_{-R}^{R} dx \int_{-\sqrt{R^2-x^2}}^{\sqrt{R^2-x^2}} dy\, 1 = 2\int_{-R}^{R} dx\, \sqrt{R^2 - x^2} = \left[x\sqrt{R^2 - x^2} + R^2 \arctan\left(\frac{x}{\sqrt{R^2 - x^2}} \right) \right]_{-R}^{R}.$$

(The integral can be done using the substitution $x = R\cos u$. Verify the last equality by differentiating the result of the integration.) For $x = \pm R$ the first term on the right vanishes while the argument of the arctan assumes the value $\pm\infty$. Since $\arctan(\pm\infty) = \pm\frac{\pi}{2}$ we arrive at the expected result $A = \pi R^2$, i.e. the familiar area enclosed by a circle of radius R. Note that the computation is somewhat unwieldy; there should be easier ways to obtain the area of the surface of a disk, and we will introduce them in the next section.

Fubini's theorem on the interchangeability of integration orders extends to non-cuboidal integration domains. For example, the above construction for the disk could have been organized in such a way that the integration over x is performed first and that over y second. The freedom to choose the order of integration becomes relevant when one order is more convenient than the other. As an example, let us apply two-dimensional integration to compute the area enclosed between the curve $y = \cos(x)$ and the x- and y-axes (see figure). Integrating first over y, then x, turns out to be easier than the reverse order:

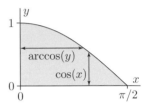

$$A = \int_0^{\pi/2} dx \int_0^{\cos(x)} dy = \int_0^{\pi/2} dx\, \cos(x) = \left[\sin(x) \right]_0^{\pi/2} = 1,$$

$$A = \int_0^1 dy \int_0^{\arccos(y)} dx = \int_0^1 dy\, \arccos(y) = \left[-\sqrt{1 - y^2} - y\arccos(y) \right]_0^1 = 1.$$

C4.2 Curvilinear area integrals

The integration procedure described in the preceding section uses Cartesian coordinates. However, these coordinates are not ideal for the description of integration domains possessing rotational or other symmetries. This is illustrated by the above example, where the integration over a disk in Cartesian coordinates led to cumbersome expressions. In this section we introduce more powerful techniques and learn how to integrate over two-dimensional structures in arbitrary coordinates.

Describing areas by curvilinear coordinates

REMARK Knowledge of Sections V2.1 to V2.3 on curvilinear coordinates is required for this section.

Let us turn back to the example of *integration over a circular disk, D*. Again, we start by introducing a discretization grid; however, this time it will be defined such that the symmetries of the integration domain are taken into account. To this end, consider the representation of D in terms of the polar coordinates introduced in Section V2.1,

$$\mathbf{r}: U \equiv (0, R) \times (0, 2\pi) \to D, \qquad \mathbf{y} \equiv (\rho, \phi)^{\mathrm{T}} \mapsto \mathbf{r}(\rho, \phi) \equiv (\rho \cos \phi, \rho \sin \phi)^{\mathrm{T}}. \quad \text{(C48)}$$

Observe that the circular domain of integration, D, is now parametrized by the rectangular coordinate domain, $U = (0, R) \times (0, 2\pi)$.

INFO As always with curvilinear coordinate descriptions, we take the coordinate domain to be *open*. This is done to ensure global differentiability of the map, see the discussion at the end of Section V2.1 (p. 409). The *openness of the coordinate intervals* often implies that an integration domain can be almost, but not *fully*, covered by a single coordinate map. For example, the image of the map above does not cover the boundary of the disk, where $R = 1$, nor the intersection of the disk with the positive real axis, where $\phi = 0$. However, these excluded regions are of dimension lower than two and their exclusion does not affect a two-dimensional integral over a continuous function. The heuristic picture behind this statement is that an "infinitely thin" line does not contribute to the summation over areas. A more formal justification of this statement will be given in the context of Eq. (C51) below. In the following, when saying that a d-dimensional integration domain, M, is *covered* by a system of coordinates, we mean that the coordinates parametrize all of M, except perhaps for subsets of lower dimension. For completeness, we mention that situations where the full coverage of a domain by coordinates is essential are addressed in Chapter V4.

We now introduce a set of points, $\{(\rho_\ell, \phi_{\ell'})^{\mathrm{T}}\}$, with $\rho_\ell = \ell \delta^\rho$, $\phi_{\ell'} = \ell' \delta^\phi$, $0 \le \ell \le R/\delta^\rho$, $0 \le \ell' \le 2\pi/\delta^\phi$, defining the corners of a system of rectangular cells of area $\delta^\rho \delta^\phi$ covering U. The coordinate map $\mathbf{r}(\rho, \phi)$ sends this rectangular grid onto a "distorted grid" of image points, $\mathbf{r}(\rho_\ell, \phi_{\ell'})$, whose corners define a set of area elements, $\delta S_{\ell\ell'}$, in D. These have the shape of "distorted rectangles" tiling D in a spider-web pattern, as illustrated in Fig. C7. The covering generated in this fashion reflects the rotational symmetry of the disk – a key advantage relative to the Cartesian grid of p. 241.

Geometrically distorted area elements

The strategy just described is not limited to polar coordinates. Integration over non-rectangular domains often starts with a coverage generated by curvilinear coordinates. All subsequent steps are of general nature and it therefore makes sense to introduce them for a generic two-dimensional coordinate system, $\mathbf{r}: U \to M, \mathbf{y} \mapsto \mathbf{r}(\mathbf{y})$, parametrizing some two-dimensional integration domain M. In the end of the section we will turn back to polar coordinates, $\mathbf{y} = (\rho, \phi)^{\mathrm{T}}$ and $M = D$, and do specific integrals over the disk.

Let us denote the points of a tiling grid in U by $\mathbf{y}_{\ell\ell'} \equiv (y_\ell^1, y_{\ell'}^2)^{\mathrm{T}} \equiv (\ell \delta^1, \ell' \delta^2)^{\mathrm{T}}$ and let $\mathbf{r}(\mathbf{y}_{\ell\ell'})$ define the induced grid in M. The integral of a function, $f: M \to \mathbb{R}, \mathbf{r} \mapsto f(\mathbf{r})$, over M is then defined as the Riemann sum over the area elements,

$$\int_M \mathrm{d}S f(\mathbf{r}) \equiv \lim_{\delta^1, \delta^2 \to 0} \sum_{\ell\ell'} |\delta S_{\ell\ell'}| f(\mathbf{y}_{\ell\ell'}), \qquad \text{(C49)}$$

where the notation $f(\mathbf{y}) \equiv f(\mathbf{r}(\mathbf{y}))$ is used and $|\delta S_{\ell\ell'}|$ is the geometric area of the area element $\delta S_{\ell\ell'}$. If f represents the density of a quantity such as mass, then the summand $|\delta S_{\ell\ell'}| f(\mathbf{y}_{\ell\ell'})$ gives the amount of this quantity associated with that area element.

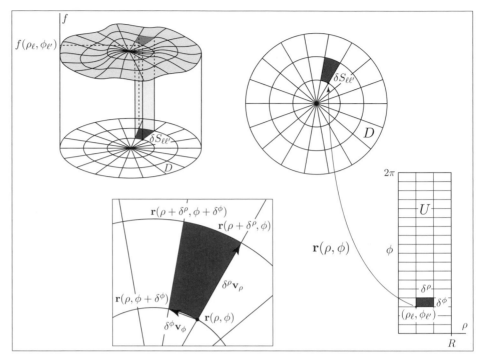

Fig. C7   Integration in two dimensions using polar coordinates. A rectangular coordinate domain U (bottom right) is used to parametrize the disk D (top right). This leads to area elements $\delta S_{\ell\ell'}$ shaped like distorted rectangles (bottom left). The integration of a function $f(\rho, \phi)$ over the disk amounts to the summation over these shapes, weighted with the product of the base areas $|\delta S_{\ell\ell'}|$ and the heights $f(\rho_\ell, \phi_{\ell'})$ (top left). The arrows shown in the bottom left panel are defined in Eq. (C50).

concept of   Equation (C49) remains formal as long as the dependence of the area elements $|\delta S_{\ell\ell'}|$ on
area   the coordinate points $\mathbf{y}_{\ell\ell'}$ has not been specified. To this end, we temporarily suppress the
elements   indices ℓ, ℓ' and note that an element δS labeled by $\mathbf{y}$ is defined by the four corner points
$\mathbf{r}(y^1, y^2)$, $\mathbf{r}(y^1 + \delta^1, y^2)$, $\mathbf{r}(y^1 + \delta^1, y^2 + \delta^2)$ and $\mathbf{r}(y^1, y^2 + \delta^2)$. These points are connected
by corresponding coordinate lines (see Fig. C7). What simplifies the computation of the
enclosed area is the proximity of the corner points to each other: in the limit of infinitesimally small δ^1 and δ^2, the curvature of the coordinate lines between the points becomes
negligibly small and the shape of δS approaches that of a *parallelogram* spanned by the
two vectors

$$\mathbf{r}(y^1 + \delta^1, y^2) - \mathbf{r}(y^1, y^2) \simeq \delta^1 \, \partial_{y^1} \mathbf{r}(\mathbf{y}) = \delta^1 \, \mathbf{v}_1(\mathbf{y}),$$
$$\mathbf{r}(y^1, y^2 + \delta^2) - \mathbf{r}(y^1, y^2) \simeq \delta^2 \, \partial_{y^2} \mathbf{r}(\mathbf{y}) = \delta^2 \, \mathbf{v}_2(\mathbf{y}). \tag{C50}$$

In the last equalities of each line we noted that (see Section V2.2) the tangent vectors to the
coordinate lines, $\partial_{y^i} \mathbf{r}$, equal the basis vectors, $\mathbf{v}_i$, of the coordinate basis of the $\mathbf{y}$-coordinate
system. The vectors spanning δS are thus given by the scaled basis vectors $\delta^1 \mathbf{v}_1$ and $\delta^2 \mathbf{v}_2$,
and $|\delta S|$ is the area of the corresponding parallelogram.

There are three different ways to describe the geometric area of this particular
parallelogram. All have advantages and we will discuss them in turn. The first approach

is based on elementary geometry and suffices for a first introduction to the subject. The other two formulations are more general and distinctly more powerful. They are introduced in the next subsection, where integration in two dimensions is discussed from a general perspective.

Area element from geometric construction

The first approach describes the *area element by geometric construction*. The area of the parallelogram spanned by two vectors $\mathbf{v}_1$ and $\mathbf{v}_2$ enclosing an angle $\angle(\mathbf{v}_1, \mathbf{v}_2)$ is given by $A(\mathbf{v}_1, \mathbf{v}_2) = \|\mathbf{v}_1\| \|\mathbf{v}_2\| \sin(\angle(\mathbf{v}_1, \mathbf{v}_2))$. Using Eq. (L76), this may be rewritten as $\|\mathbf{v}_1 \times \mathbf{v}_2\|$. (The latter notation implicitly assumes that the vectors $\mathbf{v}_1$ and $\mathbf{v}_2$ span a two-dimensional plane in three-dimensional space,

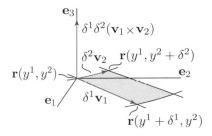

see the figure. Their cross product, $\mathbf{v}_1 \times \mathbf{v}_2$, then points in the 3-direction perpendicular to the plane and its norm gives the required parallelogram area.) In this notation, the area of δS is expressed as

$$|\delta S| \simeq \delta^1 \delta^2 A(\mathbf{v}_1, \mathbf{v}_2) = \delta^1 \delta^2 \|\mathbf{v}_1 \times \mathbf{v}_2\| = \delta^1 \delta^2 \|\partial_{y^1} \mathbf{r}(\mathbf{y}) \times \partial_{y^2} \mathbf{r}(\mathbf{y})\|. \tag{C51}$$

It remains to substitute this expression into Eq. (C49) and perform the summation over indices ℓ, ℓ'. In the limit of an infinitely fine discretization, each sum $\delta \sum_\ell \to \int dy$ becomes an integral over a coordinate interval. The Riemann sum thus assumes the form of a double integral,

$$\int_M dS f(\mathbf{r}) = \int_U dy^1 dy^2 \|\partial_{y^1} \mathbf{r}(\mathbf{y}) \times \partial_{y^2} \mathbf{r}(\mathbf{y})\| f(\mathbf{y}). \tag{C52}$$

Note that the final integral extends over a rectangular coordinate domain and hence falls into the category of integrals discussed in the previous chapter. The geometric distortion of the coordinate lines in the image domain, M, enters through the factor $\|\partial_{y^1} \mathbf{r} \times \partial_{y^2} \mathbf{r}\|$. This factor mediates between the *rectangular* shape of the coordinate cells in U (convenient for integration) and the *distorted* shape of the image cells in M (convenient for tiling a general integration area). The formal expression

$$dS = dy^1 dy^2 \|\partial_{y^1} \mathbf{r}(\mathbf{y}) \times \partial_{y^2} \mathbf{r}(\mathbf{y})\| \tag{C53}$$

area element by geometric construction is sometimes called the **area element** or the **integration measure** of two-dimensional integration. The latter terminology refers to the right-hand side of the defining equation (C52) as a "measure" of geometric areas in the integration domain.[14] In the following, we will refer to both dS and its finite analogue δS as "area elements".

The result (C52) also shows why the *assumed openness of the coordinate domain* does not matter. For a rectangular open domain $U = (a, b) \times (c, d)$, the double integral becomes $\int_U dy^1 dy^2 = \int_a^b dy^1 \int_c^d dy^2$. However, as discussed in Section C2.2, integrals over open

---

[14] The mathematically precise definition of measures is a subject of "measure theory". However, we do not enter this discussion here.

and closed intervals yield the same values, i.e. the same expression would be obtained for the integration over the product of intervals $[a, b] \times [c, d]$ parametrizing a closed coordinate domain.

Integration in polar coordinates

Let us now return to *polar coordinates* and evaluate the expressions above in that concrete context. Equation (C51) applied to the coordinate basis vectors (V25) of the polar coordinate system, $\mathbf{v}_\rho = \partial_\rho \mathbf{r} = \mathbf{e}_\rho$ and $\mathbf{v}_\phi = \partial_\phi \mathbf{r} = \mathbf{e}_\phi \rho$, yields

$$\|\partial_{y^1} \mathbf{r}(\mathbf{y}) \times \partial_{y^2} \mathbf{r}(\mathbf{y})\| = \|\mathbf{v}_\rho \times \mathbf{v}_\phi\| = \rho, \qquad (C54)$$

area element in polar coordinates

and the **polar area element**

$$\boxed{dS = \rho \, d\rho \, d\phi.} \qquad (C55)$$

The proportionality to ρ means that the area element *increases* in the radial direction. The geometric reason is that the extension of δS in the ϕ-direction, given by $\rho \, \delta^\phi$, increases linearly with the radial coordinate (see Fig. C7). Substituting this result into Eq. (C52), we obtain

$$\int_D dS f(\mathbf{r}) = \int_0^R d\rho \int_0^{2\pi} d\phi \, \rho f(\rho, \phi) \qquad (C56)$$

as a formula for the *integration in polar coordinates* over the disk D.

EXAMPLE Returning to the example on p. 242, the *geometric area of a circular disk* of radius R is now simply obtained by integrating the constant function $f(\mathbf{r}) = 1$ over the disk D:

$$A = \int_0^R \rho \, d\rho \int_0^{2\pi} d\phi \, 1 = \int_0^R \rho \, d\rho \, 2\pi = \pi R^2.$$

This computation is simpler and more elegant than its Cartesian counterpart of p. 242. Since the integration domains of ϕ and ρ are independent, the integrals over these variables factorize. This is the essential advantage of polar coordinates over Cartesian coordinates and it is owing to the fact that the former are adjusted to the rotational symmetry of the disk.

integration in polar coordinates

Polar representations are particularly well-suited for integrating functions which are rotationally symmetric and hence depend on only the radial coordinate, $f(\rho, \phi) = f(\rho)$. Consider, for example, a surface carrying a mass density increasing quadratically with the distance from the origin, $\rho_m(\rho) = \kappa \rho^2$, where κ is a constant. The total mass carried by the surface is obtained by integration:

$$M = \int_D dS \, \rho_m(\rho) = \int_0^R \rho \, d\rho \int_0^{2\pi} d\phi \, \kappa \rho^2 = \int_0^R \rho \, d\rho \, (2\pi \kappa \rho^2) = \tfrac{1}{2}\pi \kappa R^4.$$

The analogous calculation in Cartesian coordinates would be significantly harder.

The result (C56) may be straightforwardly generalized to *integration over domains without rotational symmetry* ($\rightarrow$ C4.2.1-2). For example, the integral of a function $f(\mathbf{r})$ over a quarter of a disk, parametrized as $\{\mathbf{r}(\rho, \phi) \mid \rho \in (0, R), \phi \in (0, \pi/2)\}$, is given by

$$\int_0^R \rho \, d\rho \int_0^{\pi/2} d\phi \, f(\rho, \phi).$$

EXAMPLE As a less trivial example, consider the heart-shaped area shown in the figure, defined such that for any given angle, $\phi \in (-\pi, \pi)$, the distance from the origin to the boundary of the heart is given by $\rho_b(\phi) = (1 - |\phi|/\pi)$. This means that its area is given by

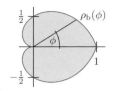

$$A = \int_{-\pi}^{\pi} d\phi \int_0^{\rho_b(\phi)} \rho \, d\rho \, 1 = \int_{-\pi}^{\pi} d\phi \, \tfrac{1}{2} \rho_b^2(\phi) = \tfrac{1}{2} \int_{-\pi}^{\pi} d\phi \left[1 - |\phi|/\pi\right]^2 = \tfrac{1}{3}\pi.$$

Jacobian and metric representations of area element

REMARK Requires Section V2.5 on Jacobians, and Chapter L6 on matrix determinants.

Above, we applied geometric reasoning to obtain the area of the surface element δS. We introduce here two more approaches to the same problem which will lead to alternative representations of the area integral. Depending on the context, application of either of these methods can be favorable. An important feature of the procedures introduced in this section is that they afford transparent generalizations to integrals in arbitrary dimensions.

The second method expresses the *area element as a matrix determinant*. To this end assume the presence of a Cartesian basis, $\{\mathbf{e}_a\}$, in the integration domain. Adopting the notation of Section V2.2, the coordinate image points can then be expanded as $\mathbf{r}(\mathbf{y}) = \mathbf{e}_a x^a(\mathbf{y})$, where the Cartesian expansion coefficients $x^a(\mathbf{y})$ are functions of the coordinates $\mathbf{y}$. The partial derivatives of $\mathbf{r}(\mathbf{y})$ in the coordinates y^j yield the expansion of the coordinate basis vectors as $\mathbf{v}_j(\mathbf{y}) = \mathbf{e}_a v^a{}_j(\mathbf{y})$, with components $J^a{}_j = \frac{\partial x^a}{\partial y^j}$, see Eq. (V19). The advantage of this Cartesian representation is that the area spanned by the coordinate basis vectors can be expressed through the determinant formula Eq. (L151) (which assumes an expansion in a Cartesian basis): the area spanned by $\mathbf{v}_1$ and $\mathbf{v}_2$ is given by $A(\mathbf{v}_1, \mathbf{v}_2) = |\det J|$, where the 2×2 matrix $J = (\mathbf{v}_1, \mathbf{v}_2)$ contains the components $J^a{}_j = (\mathbf{v}_j)^a$ as columns. In the present context, where $J^a{}_j = \frac{\partial x^a}{\partial y^j}$, this is often expressed through the suggestive notation

$$J \equiv \frac{\partial \mathbf{x}}{\partial \mathbf{y}} \equiv \frac{\partial(x^1, x^2)}{\partial(y^1, y^2)} \equiv \begin{pmatrix} \frac{\partial x^1}{\partial y^1} & \frac{\partial x^1}{\partial y^2} \\ \frac{\partial x^2}{\partial y^1} & \frac{\partial x^2}{\partial y^2} \end{pmatrix}, \tag{C57}$$

where $\mathbf{x}_j = (x^1, x^2)^{\mathsf{T}}$ is the component vector representing $\mathbf{r}$ in the Cartesian basis $\{\mathbf{e}_a\}$. (Keep in mind that all these quantities are functions of the generalized coordinates, $\mathbf{x} = \mathbf{x}(\mathbf{y})$, etc.)

Jacobi matrix The matrix J is called the **Jacobi matrix** (or just **Jacobian**) of the map $\mathbf{r} : U \to M$, $\mathbf{y} \mapsto \mathbf{r}(\mathbf{y})$. (Confusingly, the determinant $|\det(J)| = |\det(\frac{\partial \mathbf{x}}{\partial \mathbf{y}})|$ is likewise called the *Jacobian* of the map. In cases where unambiguous phrasing is required we will refer to it as the **Jacobi determinant**.) Comparison with Eq. (C51) shows that the **area element via Jacobi determinant** is given by

$$|\delta S| = \delta^1 \delta^2 |\det(J)| = \delta^1 \delta^2 \left| \det\left(\frac{\partial(x^1, x^2)}{\partial(y^1, y^2)}\right) \right| = \delta^1 \delta^2 \left| \frac{\partial x^1}{\partial y^1} \frac{\partial x^2}{\partial y^2} - \frac{\partial x^1}{\partial y^2} \frac{\partial x^2}{\partial y^1} \right|. \tag{C58}$$

A straightforward check shows that for polar coordinates, Eq. (C54) is indeed reproduced.

    The third approach expresses the *area element via the metric tensor*. Here, the starting point is Eq. (L42) for the parallelogram area,

$$A(\mathbf{v}_1, \mathbf{v}_2) = |\langle \mathbf{v}_1, \mathbf{v}_1 \rangle \langle \mathbf{v}_2, \mathbf{v}_2 \rangle - \langle \mathbf{v}_1, \mathbf{v}_2 \rangle^2|^{1/2} = |g_{11} g_{22} - g_{12} g_{21}|^{1/2} = |\det(g(\mathbf{y}))|^{1/2},$$
$$\text{(C59)}$$

where in the second step we noted that the scalar products $\langle \mathbf{v}_i, \mathbf{v}_j \rangle \overset{(V21)}{=} g_{ij}(\mathbf{y})$ define the metric tensor, and $\det(g)$ is the determinant of the matrix $\{g_{ij}\}$. We thus obtain

$$|\delta S| = \delta^1 \delta^2 |\det(g(\mathbf{y}))|^{1/2}. \tag{C60}$$

This formula expresses the area element through the metric tensor, $g(\mathbf{y})$, defined by the coordinate basis vectors. For example, in polar coordinates, Eq. (V22) yields $\sqrt{\det(g(\rho, \phi))} = \sqrt{g_{\rho\rho} g_{\phi\phi}} = \rho$, so that we again arrive at Eq. (C54).

    Generally speaking, the metric determinant is the "strongest" of the three representations discussed above. Unlike the Jacobi determinant, it does not make reference to Cartesian representations of the vectors $\mathbf{v}_1$ and $\mathbf{v}_2$. We will also see in Section C4.4 that it generalizes to integrals for which no Jacobian determinant exists. On the other hand, there are situations where the Jacobian formula, or the elementary geometric procedure discussed first above, are more convenient; it is certainly good to know all three.

Two-dimensional area integrals – summary

Summarizing, we now have three representations for the area spanned by $\mathbf{v}_1$ and $\mathbf{v}_2$, and
**area element
summarized**
this implies three alternative representations for the integral of a function, Eq. (C52):

$$\int_M dS f(\mathbf{r}) = \int_U dy^1 dy^2 \begin{Bmatrix} \|\partial_{y^1} \mathbf{r}(\mathbf{y}) \times \partial_{y^2} \mathbf{r}(\mathbf{y})\| \\ \left| \det\left(\frac{\partial(x^1, x^2)}{\partial(y^1, y^2)} \right) \right| \\ |\det(g(\mathbf{y}))|^{1/2} \end{Bmatrix} f(\mathbf{r}(\mathbf{y})). \tag{C61}$$

Each of these expresses the curvilinear integration of f over M in terms of

▷ an integral over the underlying *coordinate domain, U*,

▷ the *function evaluated in curvilinear coordinates, $f(\mathbf{r}(\mathbf{y}))$*, and

▷ any of the *rescaling factors*, $\|\partial_1 \mathbf{r} \times \partial_2 \mathbf{r}\|$, $\left|\frac{\partial \mathbf{x}}{\partial \mathbf{y}}\right|$, $|\det(g(\mathbf{y}))|^{1/2}$, which all characterize the geometric area in the integration domain corresponding to an infinitesimal area $\delta^1 \delta^2$ in the coordinate domain. This area element generally varies as a function of $\mathbf{y}$.

EXAMPLE As an instructive application of the second line of Eq. (C61), consider the integral

$$I \equiv \int_{\mathbb{R}^2} dx dy f\big((x/a)^2 + (y/b)^2\big).$$

The integrand depends on the Cartesian coordinates $\mathbf{x} \equiv (x, y)^{\mathrm{T}}$ only via the combined variable $\mu^2 \equiv (x/a)^2 + (y/b)^2$. This suggests a coordinate transformation, $\mathbf{x}(\mathbf{y}) = (x(\mathbf{y}), y(\mathbf{y}))^{\mathrm{T}} = (a\mu\cos\phi, b\mu\sin\phi)^{\mathrm{T}}$, to **generalized polar coordinates**, $\mathbf{y} \equiv (\mu, \phi)^{\mathrm{T}}$. Its Jacobi matrix is[15]

$$\frac{\partial \mathbf{x}}{\partial \mathbf{y}} = \frac{\partial(x, y)}{\partial(\mu, \phi)} = \begin{pmatrix} \frac{\partial x}{\partial \mu} & \frac{\partial x}{\partial \phi} \\ \frac{\partial y}{\partial \mu} & \frac{\partial y}{\partial \phi} \end{pmatrix} = \begin{pmatrix} a\cos\phi & -a\mu\sin\phi \\ b\sin\phi & b\mu\cos\phi \end{pmatrix},$$

with Jacobi determinant $\left|\det\left(\frac{\partial \mathbf{x}}{\partial \mathbf{y}}\right)\right| = \mu ab$. We may now pass to a (μ, ϕ) integration as

$$I \overset{(\text{C61})}{=} \int_0^\infty d\mu\,(\mu ab) \int_0^{2\pi} d\phi\, f(\mu^2),$$

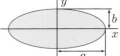

where the integration boundaries are chosen such that $M = \mathbb{R}^2$ is covered. This integral is easier to compute than the original expression. For example, consider the function $f(\mu^2) = 1$ for $\mu^2 \le 1$ and 0 otherwise, so that the integrand equals one on the ellipsoidal area shown in the figure, and vanishes elsewhere. The integral I should then yield the area, πab, of an ellipse with semi-axes a and b. Doing the integral, we indeed obtain $I = \int_0^1 d\mu(\mu ab) \int_0^{2\pi} = \pi ab$.

Jacobian vs. metric determinant

The considerations discussed above imply the *equality of the Jacobian and metric expressions* for the area element,

$$\left| \det\left(\frac{\partial \mathbf{x}}{\partial \mathbf{y}}\right) \right| = |\det(g(\mathbf{y}))|^{1/2}. \tag{C62}$$

We have established this equality in two dimensions by geometric reasoning. However, the final formula does not make visible reference to the two-dimensionality of the vectors and one may suspect that it is of more general validity. That this is indeed the case is shown in Section L6.4, see Eq. (L178).

C4.3   Curvilinear volume integrals

The concepts developed above are straightforwardly generalized to higher dimensions. Of particular importance to applications are integrals over three-dimensional space, or volume integrals. For example, the mass of a three-dimensional structure is obtained by integrating a mass density function over its volume. This section explains how to do integrals of this type.

Geometric representation of the volume element

Three-dimensional volumes, $V \subset \mathbb{R}^3$, such as balls, cylinders or the generic structure shown in Fig. C8, can be described by a three-dimensional extension of the curvilinear

---

[15] Notice that for generalized polar coordinates the coordinate basis vectors, $\mathbf{v}_\mu = \partial_\mu \mathbf{r}$ and $\mathbf{v}_\theta = \partial_\theta \mathbf{r}$, are not orthogonal. This coordinate system thus has non-orthogonal coordinate lines.

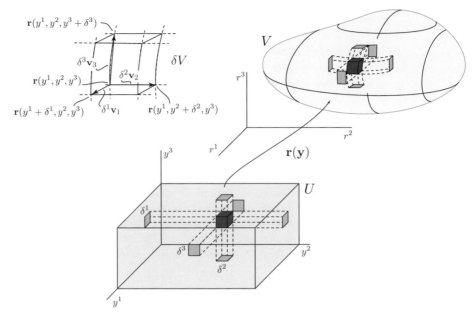

Fig. C8   On the definition of three-dimensional volume integrals.

coordinates discussed in the previous section. We define coordinates, $\mathbf{y} \equiv (y^1, y^2, y^3)^{\mathrm{T}}$, on a domain, U, and a smooth map, $\mathbf{r} : U \to V$, $\mathbf{y} \mapsto \mathbf{r}(\mathbf{y})$, parametrizing the integration domain in these coordinates. For example, the *unit radius ball*, $B \equiv \{\mathbf{r} \in \mathbb{R}^3 \mid \|\mathbf{r}\| \leq 1\}$, is conveniently described in spherical coordinates, Eq. (V36a), through a map $\mathbf{r} : U \to B$, $\mathbf{y} = (r, \theta, \phi)^{\mathrm{T}} \mapsto \mathbf{r}(r, \theta, \phi)$.

Once a system of good coordinates has been established, volume integrals may be constructed in analogy to the one- and two-dimensional integrals discussed above (see Fig. C8). Let us assume that the coordinate domain is given by the Cartesian product of three intervals, $U = (a^1, b^1) \times (a^2, b^2) \times (a^3, b^3)$ as in $U = (0, 1) \times (0, \pi) \times (0, 2\pi)$ for the unit radius ball described in spherical coordinates (r, θ, ϕ). The domain U is partitioned into a large number of boxes with corner points at $\mathbf{y}_{\boldsymbol{\ell}} \equiv (y^1_{\ell_1}, y^2_{\ell_2}, y^3_{\ell_3})^{\mathrm{T}}$, $y^i_{\ell_i} \equiv \ell_i \delta^i$ ($i = 1, 2, 3$, no summation) and volume $\delta^1 \delta^2 \delta^3$. The indices enumerating these points run in the ranges $0 \leq \ell_i \leq (b^i - a^i)/\delta^i$, and $\boldsymbol{\ell}$ is a shorthand for $\boldsymbol{\ell} = (\ell_1, \ell_2, \ell_3)^{\mathrm{T}}$. Under the coordinate map $\mathbf{r}(\mathbf{y})$ these boxes get sent onto distorted volume elements, $\delta V_{\boldsymbol{\ell}}$, in V, bounded by the coordinate lines running through the corners $\mathbf{r}(\mathbf{y}_{\boldsymbol{\ell}})$ (see Fig. C8).

By construction, the system of volume elements, $\{\delta V_{\boldsymbol{\ell}}\}$, covers the target volume. The volume integral of a function $f : V \to \mathbb{R}$, $\mathbf{r} \mapsto f(\mathbf{r})$ may thus be defined as the sum

$$\int_V \mathrm{d}V f \equiv \lim_{\delta^i \to 0} \sum_{\boldsymbol{\ell}} |\delta V_{\boldsymbol{\ell}}| f(\mathbf{y}_{\boldsymbol{\ell}}), \qquad (C63)$$

where $f(\mathbf{y}) \equiv f(\mathbf{r}(\mathbf{y}))$ and $|\delta V_{\boldsymbol{\ell}}|$ is the geometric volume of $\delta V_{\boldsymbol{\ell}}$. This formula is the three-dimensional analogue of the two-dimensional Eq. (C49).

concept of volume elements *(margin note)*

Next we need a formula for the volume elements. Proceeding in analogy to the two-dimensional case, and suppressing the box index ℓ for brevity, we note that for small δ^i, δV can be approximated by a parallelepiped spanned by the vectors $\delta^i \, \partial_{y_i} \mathbf{r}(\mathbf{y}) = \delta^i \, \mathbf{v}_i$ $(i = 1, 2, 3$, no summation), see Fig. C8. Its volume can be computed by a *geometric construction* similar to that applied in the two-dimensional case: according to Eq. (L97), the volume of the parallelepiped spanned by the vectors $\mathbf{v}_1$, $\mathbf{v}_2$ and $\mathbf{v}_3$ is given by the triple product $|(\mathbf{v}_1 \times \mathbf{v}_2) \cdot \mathbf{v}_3|$. The volume of δV thus equals

$$|\delta V| = \delta^1 \delta^2 \delta^3 |(\mathbf{v}_1 \times \mathbf{v}_2) \cdot \mathbf{v}_3| = \delta^1 \delta^2 \delta^3 \left|\left(\partial_{y^1} \mathbf{r}(\mathbf{y}) \times \partial_{y^2} \mathbf{r}(\mathbf{y})\right) \cdot \partial_{y^3} \mathbf{r}(\mathbf{y})\right|.$$

This expression is the three-dimensional analogue of Eq. (C51) for the two-dimensional area element $|\delta S|$. Substituting it into the Riemann sum (C63) and taking the limit we obtain

$$\boxed{\int_V dV f(\mathbf{r}) = \int_U dy^1 dy^2 dy^3 \left|(\partial_{y^1} \mathbf{r} \times \partial_{y^2} \mathbf{r}) \cdot \partial_{y^3} \mathbf{r}\right| f(\mathbf{y})} \qquad \text{(C64)}$$

for the three-dimensional volume integral. The combination

$$dV \equiv dy^1 dy^2 dy^3 |(\partial_{y^1} \mathbf{r} \times \partial_{y^2} \mathbf{r}) \cdot \partial_{y^3} \mathbf{r}| \qquad \text{(C65)}$$

volume element by geometric construction

is called the **volume element** or **integration measure** of the integral.

volume element in spherical coordinates

EXAMPLE The coordinate basis vectors in *spherical coordinates* are given by (see Eq. (V37)) $\mathbf{v}_r = \mathbf{e}_r$, $\mathbf{v}_\theta = \mathbf{e}_\theta \, r$, $\mathbf{v}_\phi = \mathbf{e}_\phi \, r \sin\theta$. The orthonormality of the *local* spherical basis vectors, $\mathbf{e}_i$, implies $|(\mathbf{e}_r \times \mathbf{e}_\theta) \cdot \mathbf{e}_\phi| = 1$. By Eq. (C65), the **volume element in spherical coordinates** is thus given by

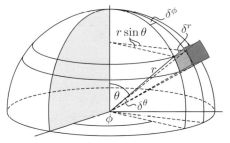

$$dV = r^2 \, dr \, \sin\theta \, d\theta \, d\phi. \qquad \text{(C66)}$$

Here, the factor $r^2 \sin\theta$ reflects the fact that the dimensions of the distorted box δV in θ- and ϕ-directions are given by $r \delta^\theta$ and $r \sin\theta \, \delta^\phi$, respectively. The factor $\sin\theta$ is best understood by exploring how the volume element shrinks upon approaching the north and south pole of the sphere, respectively (think about this point).

For example, the integral of a function over a ball, B, of radius R has the form

$$\int_B dV f(\mathbf{r}) = \int_0^R r^2 dr \int_0^\pi \sin\theta d\theta \int_0^{2\pi} d\phi \, f(r, \theta, \phi).$$

For $f = 1$ this integral yields the result $\frac{4}{3}\pi R^3$, the well-known formula for the volume of the ball. ($\to$ C4.3.1-6)

volume element in cylindrical coordinates

EXERCISE Verify that the **volume element in cylindrical coordinates** is given by

$$dV = \rho \, d\rho \, d\phi \, dz. \qquad \text{(C67)}$$

Jacobian and metric representations of volume element

Above, we applied geometric reasoning to express the volume $|\delta V|$ of the element δV as a triple product. Proceeding in analogy to Section C4.2 we now introduce alternative representations of the same quantity, as a Jacobian and a metric determinant, respectively.

As in our previous discussion, the Jacobian representation of the volume element is based on a Cartesian expansion of the coordinate basis vectors, $\mathbf{v}_j = \mathbf{e}_a J^a{}_j$. Recalling Eq. (L152), the volume of the parallelepiped spanned by these vectors may be represented as $V(\mathbf{v}_1, \mathbf{v}_2, \mathbf{v}_3) = |\det(J)|$, where the 3×3 matrix $J = (\mathbf{v}_1, \mathbf{v}_2, \mathbf{v}_3)$ contains the Cartesian component representations, $(\mathbf{v}_j)^a = J^a{}_j = \frac{\partial x^a}{\partial y^j}$, of these vectors as columns. As in the two-

volume element via Jacobi matrix and metric tensor

dimensional case, this motivates the definition of the **Jacobi matrix** as $J = \frac{\partial \mathbf{x}}{\partial \mathbf{y}} \equiv \frac{\partial(x^1, x^2, x^3)}{\partial(y^1, y^2, y^3)}$, so that $|\delta V| = \delta^1 \delta^2 \delta^3 \det(J)$.

The third representation is based on Eq. (C62), which holds in arbitrary dimensions (see Section L6.4): $|\det(J)| = |\det(g)|^{1/2}$, where $g_{ij}(\mathbf{y}) = \langle \mathbf{v}_i, \mathbf{v}_j \rangle$ is the *metric tensor* in the coordinate basis. We have thus obtained two more representations for the volume element,

$$|\delta V| = \delta^1 \delta^2 \delta^3 \left| \det\left(\frac{\partial(x^1, x^2, x^3)}{\partial(y^1, y^2, y^3)} \right) \right| = \delta^1 \delta^2 \delta^3 \left| \det(g(\mathbf{y})) \right|^{1/2}.$$

In conceptual analogy to Eq. (C61), a volume integral can now be represented by any of the three formulas

$$\int_V dV f(\mathbf{r}) = \int_U dy^1 dy^2 dy^3 \left\{ \begin{array}{l} \left| (\partial_{y^1} \mathbf{r} \times \partial_{y^2} \mathbf{r}) \cdot \partial_{y^3} \mathbf{r} \right| \\[2mm] \left| \det\left(\frac{\partial(x^1, x^2, x^3)}{\partial(y^1, y^2, y^3)} \right) \right| \\[2mm] \left| \det(g(\mathbf{y})) \right|^{1/2} \end{array} \right\} f(\mathbf{y}). \qquad \text{(C68)}$$

For later reference we note that the determinants of the metric tensor in *cylindrical and spherical coordinates* are given by (see Eqs. (V37) and (V32))

$$\text{cylindrical:} \quad \sqrt{\det(g(\rho, \phi, z))} = \sqrt{g_{\rho\rho} g_{\phi\phi} g_{zz}} = \rho,$$

$$\text{spherical:} \quad \sqrt{\det(g(r, \theta, \phi))} = \sqrt{g_{rr} g_{\theta\theta} g_{\phi\phi}} = r^2 \sin\theta. \qquad \text{(C69)}$$

These formulas lead back to Eqs. (C67) and (C66), as they should. As an instructive exercise, re-derive them from the Jacobian perspective.

C4.4 Curvilinear integration in arbitrary dimensions

REMARK Requires Chapter L6 on matrix determinants.

In this section we consider integrals over generic d-dimensional objects embedded in n-dimensional space, with $d \leq n$. Once more, the construction of these integrals is based on a suitable integration "measure". The definition of these measures in turn relies on the metric, and the ensuing integrals will be generalized variants of the third representations in Eqs. (C61) and (C68), respectively.

Consider a smooth subset of n-dimensional space, $M \subset \mathbb{R}^n$. By "smooth" we mean that M is the image of a smooth map, $\mathbf{r} : U \to M$, $\mathbf{y} \mapsto \mathbf{r}(\mathbf{y})$, where $U \subset \mathbb{R}^d$ is a d-dimensional coordinate domain.[16] In this case, d coordinates, $\mathbf{y} = (y^1, \dots, y^d)^{\mathrm{T}}$, are required to parametrize M and we call it a "d-dimensional" structure. For example, a sphere of unit radius is a $(d=2)$-dimensional object embedded in $(n=3)$-dimensional $\mathbb{R}^3$ which can be parametrized by two spherical coordinates (θ, ϕ). Without loss of generality, we assume $U = (a^1, b^1) \times \cdots \times (a^d, b^d)$ to be a d-dimensional cuboid.

Consider U discretized by a d-dimensional lattice of coordinate points, as discussed on p. 250, only that the index i now runs from 1 to d. The assignment $\mathbf{y} \mapsto \mathbf{r}(\mathbf{y})$ maps this lattice onto a distorted lattice of image points in M. These define the corners of generalized d-dimensional volume elements covering M. Each element δV can be approximated by a d-dimensional parallelepiped spanned by the d vectors $\delta^i \partial_{y^i} \mathbf{r}(\mathbf{y}) = \delta^i \mathbf{v}_i \in \mathbb{R}^n$ ($i = 1, \dots, d$, no summation).

Next we need formulas for the volume, $|\delta V|$, of these generalized parallelepipeds. As a warm-up to the discussion of general n and d, let us discuss an instructive example.

Example: integration over a two-dimensional surface in three-dimensional space

In practice, one often needs to integrate over a $(d = 2)$-dimensional surface, M, embedded in $(n = 3)$-dimensional space, $\mathbb{R}^3$, see Fig. C9. The "volume elements" then actually are *surface* elements, δS, embedded in $\mathbb{R}^3$. They are spanned by the pair of three-dimensional vectors $\delta^1 \partial_{y^1} \mathbf{r} = \delta^1 \mathbf{v}_1$ and $\delta^2 \partial_{y^2} \mathbf{r} = \delta^2 \mathbf{v}_2$. We know two expressions for the geometric area of such parallelograms: the norm of their vector product, Eq. (C51); and Eq. (L42) which in the present context assumes the form of Eq. (C59). The area of the surface element δS is therefore given as

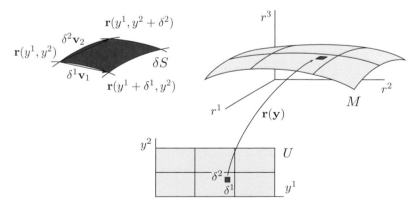

Fig. C9   Integral over a two-dimensional surface in three-dimensional space. Discussion, see text.

[16] Again, we accept the exclusion of regions of dimensions $< d$ in M from the image $\mathbf{r}(U)$.

$$|\delta S| = \delta^1 \delta^2 \|\partial_{y^1}\mathbf{r} \times \partial_{y^2}\mathbf{r}\| = \delta^1 \delta^2 |\det(g(\mathbf{y}))|^{1/2}. \tag{C70}$$

general two-dimensional surface integral This in turn means that the integral over a two-dimensional surface M embedded in $\mathbb{R}^3$ is defined as

$$\int_M \mathrm{d}S\, f(\mathbf{r}) = \int_U \mathrm{d}y^1 \mathrm{d}y^2 \left\{ \begin{array}{c} \|\partial_{y^1}\mathbf{r} \times \partial_{y^2}\mathbf{r}\| \\[2mm] |\det(g(\mathbf{y}))|^{1/2} \end{array} \right\} f(\mathbf{r}(\mathbf{y})). \tag{C71}$$

The first of these two representations is often encountered in introductory texts. It utilizes the vector product, and is therefore limited to the present situation of dimensions, $d = 2, n = 3$. The second representation, however, holds for a $(d = 2)$-dimensional surface embedded in a space of arbitrary dimension $n \geq 2$. This can be traced to the fact that the formula for the area element, Eq. (C59), is valid for any n. Also notice that the surface integral does not afford a representation in terms of a Jacobian $\det(\partial \mathbf{x}/\partial \mathbf{y})$, the reason being that Jacobians can be defined only for $d = n$. These observations suggest that integral formulas based on the metric determinant may be the "most general" representations. This impression will be corroborated by the discussion of the general case below.

EXAMPLE Let us use Eq. (C71) to compute the area of a *two-dimensional sphere in three-dimensional space*. We apply (V36a) with $r = R$ to parametrize a sphere of radius R by spherical coordinates, $\mathbf{y} = (y^1, y^2)^\mathsf{T} = (\theta, \phi)^\mathsf{T}$. From Eq. (V37) we find $\partial_{y^1}\mathbf{r} = \partial_\theta \mathbf{r} = \mathbf{v}_\theta = \mathbf{e}_\theta R$ and $\partial_{y^2}\mathbf{r} = \partial_\phi \mathbf{r} = \mathbf{v}_\phi = \mathbf{e}_\phi R \sin\theta$ for the coordinate basis vectors, and $g_{\theta\theta} = R^2$, $g_{\phi\phi} = \sin^2(\theta)R^2$, $g_{\theta\phi} = g_{\phi\theta} = 0$ for the elements of the metric tensor. Both $\|\mathbf{v}_\theta \times \mathbf{v}_\phi\|$ and $|\det(g)|^{1/2}$ yield $R^2 \sin(\theta)$. We thus obtain the area as ($\rightarrow$ C4.4.1)

$$A = \int_0^\pi \mathrm{d}\theta \int_0^{2\pi} \mathrm{d}\phi\, R^2 \sin(\theta) = 4\pi R^2. \tag{C72}$$

EXERCISE As another example consider the surface shown in the figure. It is rotationally symmetric in the xy-plane, and the Cartesian height coordinate is given by $z = \frac{1}{3}(a^3 - (x^2 + y^2)^{3/2})$ for $x^2 + y^2 < a^2$. We aim to compute the geometric area of this surface. To this end, we introduce polar coordinates, $\mathbf{y} = (\rho, \phi)^\mathsf{T}$, in the xy-plane and obtain the parametrization $\mathbf{r}(\mathbf{y}) = (x, y, z)^\mathsf{T}(\mathbf{y}) = (\rho \cos\phi, \rho \sin\phi, \frac{1}{3}(a^3 - \rho^3))^\mathsf{T}$. Show that the metric determinant reads as

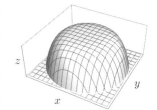

$$[\det(g(\mathbf{y}))]^{1/2} = \rho(1 + \rho^4)^{1/2}.$$

Use this result to confirm that the area is given by $A = \frac{\pi}{2}(a^2\sqrt{1 + a^4} + \mathrm{arcsinh}(a^2))$. Discuss the results in the limits $a \gg 1$ and $a \ll 1$, respectively. ($\rightarrow$ C4.4.2-6)

Integration over objects of arbitrary dimension: metric tensor

We now turn to the generic case $d \leq n$. We need the volume of a d-dimensional parallelepiped, $\delta V \subset \mathbb{R}^n$, spanned by the vectors $\delta^i \mathbf{v}_i$, with $i = 1, \ldots, d$. Referring to

Section V6.2 for a general argument, we state that this volume is given by the metric determinant,

$$|\delta V| = (\delta^1 \ldots \delta^d) \, |\det(g(\mathbf{y}))|^{1/2}, \tag{C73}$$

volume integral in arbitrary dimensions

where $g_{ij} = \langle \mathbf{v}_i, \mathbf{v}_j \rangle$ is computed using the standard scalar product of the embedding space $\mathbb{R}^n$. The integral over M is thus defined as

$$\int_M dV f(\mathbf{r}) = \int_U dy^1 \ldots dy^d \, \left| \det(g(\mathbf{y})) \right|^{1/2} f(\mathbf{r}(\mathbf{y})). \tag{C74}$$

All multidimensional integration formulas descend from this powerful result. It holds for arbitrary $d \le n$ and encompasses all the special cases discussed so far. To recapitulate, the application of this formula requires

▷ a parametrization of the integration domain, M, by a coordinate map, $\mathbf{r}(\mathbf{y})$,

▷ computation of the partial derivative vectors, $\mathbf{v}_i = \partial_{y^i} \mathbf{r}$, the elements of the metric tensor, $g_{ij}(\mathbf{y}) = \langle \mathbf{v}_i, \mathbf{v}_j \rangle$, and its determinant $\det(g(\mathbf{y}))$, and finally

▷ the computation of the integral over the coordinate domain U.

Integration over d-dimensional volumes in d-dimensional space: Jacobian

In the special case $d = n$ there exists an alternative representation of the volume element in terms of a Jacobian. Although this case, too, is covered by Eq. (C74), the Jacobian **volume element from d-dimensional Jacobi matrices** formulation is widely used and we discuss it for completeness. The **Jacobi matrix** of the coordinate map $\mathbf{y} \mapsto \mathbf{r}(\mathbf{y})$ generalizes Eq. (C57): it is defined as the matrix, $J(\mathbf{v}_1, \ldots, \mathbf{v}_d)$, whose columns contain the Cartesian coordinates, $\{J^a{}_j\} = \{\frac{\partial x^a}{\partial y^j}\}$, of the vectors, $\mathbf{v}_j = \partial_{y^j} \mathbf{r}(\mathbf{y})$:

$$\frac{\partial \mathbf{r}}{\partial \mathbf{y}} \equiv \frac{\partial(x^1, \ldots, x^d)}{\partial(y^1, \ldots, y^d)} \equiv \begin{pmatrix} \frac{\partial x^1}{\partial y^1} & \frac{\partial x^1}{\partial y^2} & \cdots & \frac{\partial x^1}{\partial y^d} \\ \frac{\partial x^2}{\partial y^1} & \frac{\partial x^2}{\partial y^2} & \cdots & \frac{\partial x^2}{\partial y^d} \\ \vdots & \vdots & \ddots & \vdots \\ \frac{\partial x^d}{\partial y^1} & \frac{\partial x^d}{\partial y^2} & \cdots & \frac{\partial x^d}{\partial y^d} \end{pmatrix}. \tag{C75}$$

From Eq. (C62) we know that $\left| \det\left(\frac{\partial \mathbf{x}}{\partial \mathbf{y}} \right) \right| = |\det(g(\mathbf{y}))|^{1/2}$, and this implies that Eq. (C74) has the equivalent representation

$$\int_M dV f(\mathbf{r}) = \int_U dy^1 \ldots dy^d \, \left| \det\left(\frac{\partial(x^1, \ldots, x^d)}{\partial(y^1, \ldots, y^d)} \right) \right| f(\mathbf{r}(\mathbf{y})). \tag{C76}$$

Notice the structural similarity of this formula to the one-dimensional substitution rule (C28). In the next section we discuss how Eqs. (C28) and (C76) can be understood as special cases of a general formula describing variable changes in integrals of arbitrary dimensionality.

C4.5   Changes of variables in higher-dimensional integration

Equation (C76) affords an interesting interpretation as a generalization of the one-dimensional substitution rule (C28). To understand this, consider an integral over a d-dimensional volume, M, in d-dimensional space. Assume that $\mathbf{r}(\mathbf{x}) = \mathbf{e}_a x^a$ has been parametrized by a Cartesian coordinate system. In this case, the basic integration formulas of Section C4.1 may be applied to represent the integral as

$$\int_M \mathrm{d}V f(\mathbf{r}) = \int_M \mathrm{d}x^1 \dots \mathrm{d}x^d f(\mathbf{x}),$$

where the boundaries of the x^a-integrals must be chosen so as to obtain a full coverage of M. Alternatively, we may introduce a map, $\mathbf{x} : U \to M$, $\mathbf{y} \mapsto \mathbf{x}(\mathbf{y})$, to cover M by a different system of coordinates, $\mathbf{y}$, and represent the integral through Eq. (C76). The equality of the two representations leads to the formula

variable change in arbitrary dimensions

$$\boxed{\int_M \mathrm{d}x^1 \dots \mathrm{d}x^d f(\mathbf{x}) = \int_U \mathrm{d}y^1 \dots \mathrm{d}y^d \left| \det\left(\frac{\partial \mathbf{x}}{\partial \mathbf{y}}\right) \right| f(\mathbf{x}(\mathbf{y})).} \qquad \text{(C77)}$$

This formula is valid independent of the geometric context in which it has been derived. In particular, it does not rely on an interpretation of $\mathbf{x}$ as a *Cartesian* coordinate vector. It describes, rather, a *change of integration variables*, $\mathbf{x} \to \mathbf{x}(\mathbf{y})$, in general d-dimensional integrals and extends the one-dimensional formula (C28) to higher dimensions. The notation $\mathbf{x} \to \mathbf{x}(\mathbf{y})$ is a shorthand for saying: "a re-parametrization of variables, $\mathbf{y} \mapsto \mathbf{x}(\mathbf{y})$, is applied to convert an integral over $\mathbf{x}$ to an integral over $\mathbf{y}$". The appearance of the Jacobian in this formula may be remembered from the mnemonic $\prod_a \mathrm{d}x^a \leftrightarrow \prod_j \mathrm{d}y^j \det\left(\frac{\partial \mathbf{x}}{\partial \mathbf{y}}\right)$, which has a status similar to that of the trick mentioned after Eq. (C28).

Equation (C77) motivated from a different perspective

To obtain a better understanding of the generality of formula (C77), consider another change of variables (see figure), $\mathbf{y} : T \to U$, $\mathbf{z} \mapsto \mathbf{y}(\mathbf{z})$. One now has two options to express Eq. (C77) as an integral over $\mathbf{z}$. The first is to parametrize $\mathbf{x}$ through $\mathbf{z}$ via the composite map, $\mathbf{x} \circ \mathbf{y} : T \to M$, $\mathbf{z} \mapsto \mathbf{x}(\mathbf{y}(\mathbf{z})) \equiv \mathbf{x}(\mathbf{z})$. Application of Eq. (C77) to $\mathbf{x} \to \mathbf{x}(\mathbf{z})$ then yields

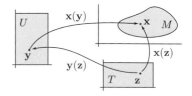

$$\int_M \mathrm{d}x^1 \dots \mathrm{d}x^d f(\mathbf{x}) = \int_T \mathrm{d}z^1 \dots \mathrm{d}z^d \left| \det\left(\frac{\partial \mathbf{x}}{\partial \mathbf{z}}\right) \right| f(\mathbf{x}(\mathbf{z})).$$

The second is to apply the variable change $\mathbf{y} \to \mathbf{y}(\mathbf{z})$ to the integral on the r.h.s. of Eq. (C77):

$$\int_U dy^1 \ldots dy^d \left| \det\left(\frac{\partial \mathbf{x}}{\partial \mathbf{y}}\right) \right| f(\mathbf{x}(\mathbf{y}))$$

$$= \int_T dz^1 \ldots dz^d \left| \det\left(\frac{\partial \mathbf{y}}{\partial \mathbf{z}}\right) \right| \left| \det\left(\frac{\partial \mathbf{x}}{\partial \mathbf{y}}\right) \right|_{\mathbf{y}(\mathbf{z})} f(\mathbf{x}(\mathbf{y}(\mathbf{z}))),$$

where the notation emphasizes that in the integral on the right, all functions have to be expressed through the **z**-coordinates.

Since the preceding two equations represent the same integral, we conclude that the Jacobian determinants occuring therein must satisfy the relation

$$\left| \det\left(\frac{\partial \mathbf{x}}{\partial \mathbf{z}}\right) \right| = \left| \det\left(\frac{\partial \mathbf{x}}{\partial \mathbf{y}}\right) \right|_{\mathbf{y}(\mathbf{z})} \left| \det\left(\frac{\partial \mathbf{y}}{\partial \mathbf{z}}\right) \right|. \qquad (C78)$$

Indeed, the validity of Eq. (C78) follows from an important property of the Jacobi matrix: application of the chain rule (C42) (with the identification $f^i = x^i$, $g^j = y^j$, $x^k = z^k$) gives

$$\frac{\partial x^i(\mathbf{y}(\mathbf{z}))}{\partial z^k} = \left.\frac{\partial x^i(\mathbf{y})}{\partial y^j}\right|_{\mathbf{y}(\mathbf{z})} \frac{\partial y^j(\mathbf{z})}{\partial z^k}.$$

This formula has the suggestive shorthand notation $\frac{\partial x^i}{\partial z^k} = \frac{\partial x^i}{\partial y^j}\frac{\partial y^j}{\partial z^k}$, or just

$$\boxed{\frac{\partial \mathbf{x}}{\partial \mathbf{z}} = \frac{\partial \mathbf{x}}{\partial \mathbf{y}}\frac{\partial \mathbf{y}}{\partial \mathbf{z}}.} \qquad (C79)$$

Equation (C79) states that Jacobi matrices are multiplicative: the Jacobian of the transformation $\mathbf{x} \to \mathbf{x}(\mathbf{z})$ equals the product of those of $\mathbf{x} \to \mathbf{x}(\mathbf{y})$ and $\mathbf{y} \to \mathbf{y}(\mathbf{z})$. The matrix product identity for determinants, $\det(AB) = \det(A)\det(B)$, then directly implies Eq. (C78).

To summarize, Eq. (C77) is the generalization of the one-dimensional substitution rule (C28) to changes of variables in generic multidimensional integrals. Judiciously applied transformations of variables often lead to drastic simplifications in the calculation of such integrals. ($\to$ C4.5.1-6).

C4.6 Summary and outlook

In this chapter we introduced the higher-dimensional integration techniques required in the early physics curriculum. Building on the general understanding of integration as generalized (Riemann) summation, we began with a straightforward construction of integrals over cuboidal domains, line segments, rectangles, boxes, etc. We then moved on to the important subject of integration over more general structures, where the usage of symmetry-adapted coordinates became vital. In all cases integration turned out to be an algorithm of three consecutive steps. (i) Covering the integration domain by suitable coordinates, preferably defined on a cuboidal coordinate domain. (ii) Determining the geometric distortion factors by which the cuboidal line, surface, volume elements of the coordinate domain differ from the distorted line, surface, volume elements defined by the coordinate map in the integration domain. This step really is at the heart of the matter of all integration and we provided three alternative solutions, each tailored to different situations. Finally, (iii) computing the integral over the coordinate domain, weighted with a function of interest and said distortion factor.

We discussed various types of integrals distinguished by the dimension of the integration domain (the number of coordinates required to parametrize it) and the dimensionality of the space in which the domain is embedded. This led to a perhaps somewhat overwhelming variety of integrals, which, however, are all relevant in practice. In Section V5.4 we will introduce a more geometric perspective of integration and demonstrate that the various integrals introduced above are not as different as they might seem. However, for the time being we leave the subject of integration and turn back to the "local" analysis of functions by advanced techniques of differentiation.

Taylor series

Depending on the type of information they encode, mathematical functions may be simple or complicated. Sometimes they are defined "implicitly",[17] or they may be the results of measurements in which case no analytic representation exists. While the description of a function in full generality may be a difficult task, it is often sufficient to understand its behavior in the vicinity of a specific point of interest.

EXAMPLE The binding potentials stabilizing a chemical molecule such as O_2 are complicated functions $V(r)$ of the inter-atomic distances, r. However, at temperatures far below those where the molecule disintegrates, the inter-atomic separations are close to an equilibrium value, $r = a$. Much of the observable physics of the molecule can then be understood from the profile of $V(r)$ for values of r close to a.

In this chapter we introduce methodology capable of describing the "local" structure of functions even if the global structure is not known. In the next chapter we then take a complementary point of view and focus on the global profile of functions.

C5.1 Taylor expansion

In Chapter C1 we discussed how the derivative of a function $f(y)$ yields a local approximation through a linear function. This is made explicit in Eq. (C2), $f(y + \delta) \simeq f(y) + \delta f'(y)$, where the $\simeq$ sign indicates that the quality of the approximation depends on the range over which it is applied. The reason is that even for small δ, $f(y + \delta)$ generally is not linear in δ but may depend on arbitrary powers, $\delta, \delta^2, \delta^3, \ldots$. However, for δ very small, say $\delta = 10^{-9}$, these terms rapidly decrease as $10^{-9}, 10^{-18}, 10^{-27}$, and this explains why for very small δ a linear approximation may be good enough. For larger δ, however, we should

idea of Taylor expansion consider an "expansion" of the form

$$f(y + \delta) = c_0 + c_1\delta + c_2\delta^2 + c_3\delta^3 + \cdots, \tag{C80}$$

where $c_0 = f(y)$, $c_1 = f'(y)$, and $c_{i \geq 2}$ are coefficients that need to be determined by methods presented below. However, let us first discuss the conceptual meaning of the above representation.

---

[17] For example, the function might be expressed as an integral $\int^x dy\, g(y)$ for which no closed representation is known.

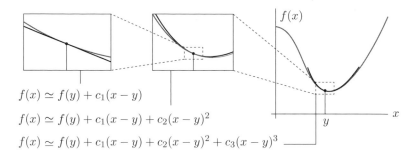

$$f(x) \simeq f(y) + c_1(x - y)$$

$$f(x) \simeq f(y) + c_1(x - y) + c_2(x - y)^2$$

$$f(x) \simeq f(y) + c_1(x - y) + c_2(x - y)^2 + c_3(x - y)^3$$

Fig. C10   Schematic on the interpretation of a Taylor series expansion. Discussion, see text.

Local expansions of functions

Defining $x = y + \delta$, Eq. (C80) represents the function at s as

$$f(x) = c_0 + c_1(x - y) + c_2(x - y)^2 + c_3(x - y)^3 + \cdots = \sum_{n=0}^{\infty} c_n(x - y)^n . \qquad \text{(C81)}$$

This equation represents $f(x)$ in the vicinity of a fixed reference point, y, as a power series in the deviations $(x - y)$. Series representations of this type are generally called *Taylor series*. If only a finite number of terms are kept, an approximation of f by a polynomial of finite order in $(x - y)$ is obtained. For increasing $|x - y|$ an increasing number of terms is required for an accurate representation of f. The sit-

Brook Taylor 1685–1731
A British mathematician best known for introducing the concept of Taylor series to mathematics. The series appeared as part of his work on generalizing infinitesimal calculus to a calculus of finite differences (the precise description of how a function changes upon finite changes of the argument). The importance of this line of thinking remained unrecognized until four decades after Taylor's death when Lagrange understood its powers.

uation is illustrated in Fig. C10, where on the smallest scales (left panel), the function looks nearly linear and can be approximated by a linear function. At larger scales (middle panel), the curvature of f becomes noticeable and a parabolic representation by a quadratic polynomial becomes appropriate. On yet larger scales (right panel), a cubic polynomial representation is required, etc.

Definition of Taylor series

The values of the expansion coefficients are easily determined. To see how, differentiate $f(x)$ n times and then set $x = y$. On the left-hand side this yields $\left[d_x^n f(x)\right]_{x=y} = d_y^n f(y)$, by definition of the derivative. We assume that this derivative can be computed analytically (or perhaps numerically if the function is the result of a measurement). Turning to the right-hand side, we note that the only non-vanishing contribution to $\left[d_x^n \sum_l c_l(x - y)^l\right]_{x=y}$ comes from the term with $l = n$: for $l < n$, we have $d_x^n(x - y)^l = 0$, and for $l > n$, $d_x^n(x - y)^l =$

const $\cdot (x - y)^{l-n}$, which vanishes at $x = y$. The term of nth order yields[18] $d_x^n (x - y)^n c_n = 1 \cdot 2 \cdots (n - 1) \cdot n c_n \equiv n! c_n$. Here, $n!$ is the **factorial** of n, defined as $n! \equiv 1 \cdot 2 \cdots (n - 1) \cdot n$ for positive integers and $0! \equiv 1$. Equating the two expressions for the n-fold derivative at $x = y$, we identify $c_n = \frac{1}{n!} d_y^n f(y)$, so that the expansion (C81) assumes the form

$$f(x) = \sum_{n=0}^{\infty} \frac{1}{n!} \frac{d^n f(y)}{dy^n} (x - y)^n. \tag{C82}$$

definition of Taylor series

This is the **Taylor series expansion** of the function f around y. The expression on the right is somewhat formal in that it implicitly assumes the convergence of the series. In general, this feature depends on the function f, and on the magnitude of the parameter $|x-y|$. The set of values for which the series converges is called its **radius of convergence** around the point y. It needs to be established individually for each function, and each value of y. (For the rationale behind the denotation "radius" of convergence, see the discussion of complex Taylor series in Section C5.2.)

radius of convergence

> **INFO** A series, $S_n \equiv \sum_{l=0}^{n} a_l$, converges for $n \to \infty$ if the sequence of numbers $\{S_0, S_1, S_2, \dots\}$ tends to a limiting value. There exist several *tests for the convergence of a series*. As a useful example, and without proof, we mention the **ratio test**, which states that convergence is guaranteed if $\lim_{n\to\infty} |a_{n+1}/a_n| = r$ converges to a number $r < 1$. (If $r > 1$ the series diverges, and for $r = 1$ the situation is inconclusive.) For later reference, we note that the ratio test also works for series with complex coefficients, $a_n \in \mathbb{C}$.

Examples of Taylor series

Consider the function $f(x) = \exp(x)$. The derivatives of this function are obtained as $\exp^{(n)}(y) \equiv \frac{d^n \exp(y)}{dy^n} = \exp(y)$, so that the series assumes the form

$$\exp(x) = \exp(y) \sum_{n=0}^{\infty} \frac{(x - y)^n}{n!}.$$

(Fig. C11 visualizes these expressions for $y = 1$ up to expansions of sixth order.) Setting $y = 0$, we obtain the famous **exponential series**,

exponential series

$$\exp(x) = \sum_{n=0}^{\infty} \frac{x^n}{n!}. \tag{C83}$$

This series converges for all x. Subjecting it to the ratio test (see info section above), we obtain $(x^{n+1}/(n + 1)!)/(x^n/n!) = x/(n + 1)$ for the ratio of successive summands. For any x this converges to $r = 0 < 1$ in the limit $n \to \infty$, which proves the convergence.

sine and cosine series

As another important example, consider the **Taylor expansions of the sine and cosine functions** around zero. The identities $\sin(0) = 0$, $\cos(0) = 1$, $d_y \sin(y) = \cos(y)$ and

---

[18] For example, $d_x^3 (x - y)^3 = 3 d_x^2 (x - y)^2 = 2 \cdot 3 d_x^1 (x - y)^1 = 1 \cdot 2 \cdot 3$.

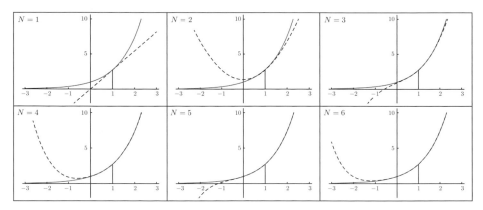

The function $\exp(x)$ (solid curves) and six approximate representations (dashed curves) around $y = 1$, obtained by truncating the Taylor series (C83) at $n = N$, with $N = 1, \ldots, 6$.

$d_y \cos(y) = -\sin(y)$ imply the vanishing of all even derivatives of the sine function at zero, $\sin^{(2n)}(0) = 0$, and alternating behavior of the odd derivatives, $\sin^{(2n+1)}(0) = (-1)^n$. Likewise, $\cos^{(2n)}(0) = (-1)^n$, $\cos^{(2n+1)}(0) = 0$. We thus obtain

$$\sin(x) = \sum_{n=0}^{\infty} \frac{(-1)^n}{(2n+1)!} x^{2n+1}, \qquad \cos(x) = \sum_{n=0}^{\infty} \frac{(-1)^n}{(2n)!} x^{2n}. \qquad \text{(C84)}$$

Again, these series have coefficients of $\mathcal{O}(x^n/n!)$ and therefore infinite radius of convergence.

If a Taylor series does exist, it need not necessarily converge for all values of x. Consider, for example, the function $f(x) = \frac{1}{1-x}$. Differentiating this function at $x = 0$, we obtain $f^{(n)}(0) = n!$, so that the Taylor series takes the form

$$\frac{1}{1-x} = \sum_{n=0}^{\infty} x^n. \qquad \text{(C85)}$$

geometric series   This is known as the **geometric series**, one of the most important series in mathematics. The convergence of the right-hand side is limited to values $|x| < 1$. (Apply the ratio criterion to discuss why.) This reflects the fact that for $x \nearrow 1$,[19] we hit the divergence of the left-hand side.

As another example, consider the *logarithm*, $f(x) = \ln(1 - x)$. Its series expansion is closely related to that of the geometric series, since $f'(x) = -(1 - x)^{-1}$. Indeed, it is **logarithmic series**   straightforward to verify that the **logarithmic series** assumes the form

---

[19] The symbol $x \nearrow a$ indicates that x approaches the value a from below. Similarly, $x \searrow a$ stands for approach from above.

$$\ln(1 - x) = -\sum_{n=0}^{\infty} \frac{1}{n} x^n. \qquad (C86)$$

As for the geometric series, the radius of convergence is $|x| < 1$.

C5.2 Complex Taylor series

In previous parts of the text we have often seen that "complex mathematics", defined for complex numbers, can be more flexible and powerful than real mathematics. Taylor series are a prime example of this principle. In this section we discuss the straightforward extension from real to complex Taylor series and study the consequences.

Series representation of complex functions

Functions of complex variables are often defined by straightforward extension of real definitions. For example, the real power function x^n, with integer exponent n, has the complex extension z^n. In this way, the series discussed in the previous section can all be promoted to complex series, $\sum_{n=0}^{\infty} a_n z^n$. In such cases, the convergence properties in the now complex domain of definition must be re-examined, a point to which we return at the end of this section.

Complex series representations are often used to *define* complex extensions of functions which need not be polynomials themselves. For example, the complex extension of the exponential Taylor series (C83) defines a complex exponential function as

$$\exp(z) \equiv \sum_{n=0}^{\infty} \frac{z^n}{n!}. \qquad (C87)$$

By the complex version of the ratio test mentioned above, the series converges for arbitrary z and hence defines $\exp(z)$ in the entire complex plane. The complex sine and cosine functions, and many others, are defined analogously.

Complex functions and their series representations have many powerful features whose discussion will be the subject of Chapter C9. The point we want to emphasize at present is that complex series representations can disclose relations between functions which are difficult to obtain otherwise. As an example, let us explore what happens if a variable substitution, $z \rightarrow iz$, is applied to Eq. (C87):

$$\exp(iz) = \sum_{m=0}^{\infty} \frac{(iz)^m}{m!} = \sum_{n=0}^{\infty} \frac{(-1)^n}{(2n)!} z^{2n} + i \sum_{n=0}^{\infty} \frac{(-1)^n}{(2n+1)!} z^{2n+1},$$

where in the second equality we split the summation into even ($m = 2n$) and odd ($m = 2n + 1$) powers of x, and used $i^{2n} = (i^2)^n = (-1)^n$. Comparison with (C84) shows that the two contributions coincide with the cos- and the sin-series, respectively. All three series have infinite radius of convergence and therefore can be considered as equivalent to the functions they represent. This identification leads to the famous **Euler formula**:

Euler formula

$$\exp(\mathrm{i}z) = \cos(z) + \mathrm{i}\sin(z). \qquad \text{(C88)}$$

Leonhard Euler 1707–1783
A Swiss mathematician and physicist. Euler played a pioneering role in the development of modern analysis, but also contributed to number theory, graph theory and applied mathematics. In physics he worked on problems of mechanics, fluid dynamics, astronomy and others. Euler is generally considered one of the greatest mathematicians of all time. A famous contemporary mathematician exclaimed: "Read Euler, read Euler, he is the master of us all!"

Remarkably, this simple relation between three elementary functions is not straightforward to prove by means other than series expansion. At the same time, the Euler formula is immensely useful in applications. It is often evaluated on real arguments, $z = x$, in which case it assumes the form

$$\mathrm{e}^{\mathrm{i}x} = \cos(x) + \mathrm{i}\sin(x). \qquad \text{(C89)}$$

To mention one application of this formula, in the "polar representation" of complex numbers, Eq. (L13), $z = |z|(\cos\phi + \mathrm{i}\sin\phi)$ can now be written as

$$z = |z|\mathrm{e}^{\mathrm{i}\phi}. \qquad \text{(C90)}$$

In this way, a complex number is expressed as the product of its modulus, $|z|$, and a unit-modular *phase factor* $\mathrm{e}^{\mathrm{i}\phi}$, 2π-periodic in the argument ϕ. This representation plays an important role in numerous applications in physics and engineering.

INFO It is worthwhile to list a few important *corollaries of the Euler formula*: since $\cos(-z) = \cos(z)$ and $\sin(-z) = -\sin(z)$, Eq. (C88) immediately leads to

$$\cos(z) = \tfrac{1}{2}\left(\mathrm{e}^{\mathrm{i}z} + \mathrm{e}^{-\mathrm{i}z}\right), \qquad \sin(z) = \tfrac{1}{2\mathrm{i}}\left(\mathrm{e}^{\mathrm{i}z} - \mathrm{e}^{-\mathrm{i}z}\right). \qquad \text{(C91)}$$

At the particular values $z = 2\pi n$, $n \in \mathbb{Z}$, the Euler relation reduces to the identity

$$\mathrm{e}^{\mathrm{i}2\pi n} = 1, \qquad \text{(C92)}$$

which we will see plays an important role within the framework of Fourier calculus, Section C6.2. Similarly, inserting $z = \pi$ into Eq. (C88) leads to

$$\mathrm{e}^{\mathrm{i}\pi} = -1, \qquad \text{(C93)}$$

a magic formula or $\mathrm{e}^{\mathrm{i}\pi} + 1 = 0$, a relation famously connecting five of the most important numbers in mathematics.
The Euler formula can also be used to prove the trigonometric addition theorems: ($\to$ C5.2.1-2)

$$\cos(a + b) = \cos a \cos b - \sin a \sin b, \qquad \sin(a + b) = \cos a \sin b + \sin a \cos b, \qquad \forall a, b \in \mathbb{C}.$$

EXERCISE The functions $\sinh(z)$ and $\cosh(z)$, defined by complex extension of Eqs. (C14), satisfy the following identities (implied directly by their definition, and Eqs. (C88)):

$$\exp(z) = \cosh(z) + \sinh(z),$$
$$\sinh(z) = \mathrm{i}\sin(-\mathrm{i}z), \qquad \cosh(z) = \cos(-\mathrm{i}z). \qquad \text{(C94)}$$

Find the Taylor series expansions around $z = 0$ of all terms in these relations and verify that they are mutually consistent.

Complex differentiability and series representations

Above, we defined complex functions through series representations. However, it is often more natural to proceed in the opposite direction. Let us start from a function $f : \mathbb{C} \to \mathbb{C}, z \mapsto f(z)$. We aim to construct a series representation for it, and much as in the real case, this requires a notion of differentiability. The *complex derivative*, $f'(z)$, is defined in analogy to the derivative of a real function, Eq. (C1),

$$f'(z) \equiv \frac{df(z)}{dz} \equiv \lim_{\delta \to 0} \frac{1}{\delta}[f(z + \delta) - f(z)], \qquad (C95)$$

where δ is an infinitesimal complex number. The limit on the right-hand side exists if a unique value is obtained independent of the way in which δ is sent to zero.[20] In this

complex differentiability case, the function is called "**complex differentiable**". As in the real case, higher-order derivatives are defined by repeated differentiation, $d_z^2 f(z) = d_z f'(z)$, etc. For example, the function $(1 - z)^{-1}$ is complex differentiable around $z = 0$ and its first two derivatives are given by $d_z(1 - z)^{-1} = (1 - z)^{-2}$ and $d_z^2(1 - z)^{-1} = 2(1 - z)^{-3}$, respectively.

complex Taylor series If a function is infinitely often differentiable at z_0, its **complex Taylor series** is defined as

$$f(z) = \sum_{n=0}^{\infty} \frac{1}{n!} \frac{d^n f(z_0)}{dz_0^n} (z - z_0)^n. \qquad (C96)$$

As in the real case, the equality of the two sides of the equation is shown by n-fold differentiation at $z = z_0$. For example, the complex generalization of the geometric series is

$$\frac{1}{1 - z} = \sum_{n=0}^{\infty} z^n. \qquad (C97)$$

Comparison to the real case (C85) shows that the complex series is obtained by complex generalization $x \to z$ in the latter. Indeed it is good practice to

> always think of Taylor series as complex series. A real series is then understood as the restriction to real arguments, $z = x + iy \to x$.

EXAMPLE As an example illustrating how *complex Taylor series are superior to real ones*, consider the function

$$f(x) = e^{-1/x^2}. \qquad (C98)$$

issues with real Taylor series The function f is infinitely often differentiable at $x = 0$ and all its derivatives vanish, $f^{(n)}(0) = 0$.[21] The Taylor series expansion thus predicts $f(x) = \sum_n 0 \cdot (x^n/n!) = 0$. However, this is incorrect, since

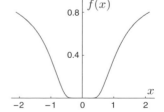

---

[20] Unlike in the real case, where an infinitesimal parameter can be sent to zero only in two ways, from the positive or negative direction, the complex δ can be sent to zero along arbitrary "paths" in the complex plane. For further discussion of this point, see Chapter C9.

[21] To see this, note that all derivatives are proportional to e^{-1/x^2}, times potentially diverging power laws. However, the exponential vanishes so rapidly for $x \to 0$ that it suppresses all other contributions to yield zero.

f is different from the zero function. This frustrating ambiguity – how can one tell in advance whether the series gets it right or not? – disappears if f is interpreted as the real restriction of the *complex* function $f(z) = \exp(-1/z^2)$. At $z = 0$, this function lacks complex differentiability. To see why, explore the limiting behavior of $\exp(-1/z^2)$ for $z = \delta$ and $z = i\delta$, respectively. One obtains two different limiting values, $e^{-1/0} = 0$ or $e^{+1/0} = \infty$, in violation of the differentiability criterion, which requires the existence of a unique limiting value. A series representation of $\exp(-1/z^2)$ around $z = 0$ therefore does not exist, and the same is true for the restriction of this function, $\exp(-1/x^2)$, to the real axis.

Although the above example illustrates that a rigorous discussion of Taylor series convergence is best carried out for functions of a complex variable, pathologies such as the one above are rare. In most cases, the convergence of real Taylor series can be addressed within the framework of real calculus.

radius of convergence   Generally, the **radius of convergence** of a Taylor series around a point z_0 is the maximum value, R, such that the series converges for all z inside the circle $|z - z_0| < R$. As an example, consider the function $\frac{1}{1-z}$. Its Taylor series around a point z_0, obtained via Eq. (C96), is given by $\sum_{n=0}^{\infty} \left(\frac{z-z_0}{1-z_0}\right)^n$ (check!). This series converges if the modulus of each term is smaller than unity, $|z - z_0| < |1 - z_0|$, which reveals the radius of convergence as $|1 - z_0|$. Setting imaginary parts to zero, we conclude that the real series of the function $\frac{1}{1-x}$ around a real number x_0 has a radius of convergence given by $|x - x_0| < |1 - x_0|$.

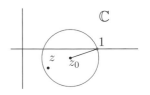

C5.3 Finite-order expansions

Taylor expansions of finite order are often applied to approximate a function $f(x)$ in the neighborhood of a point y by a polynomial of Nth order:

$$f_N(x) \equiv \sum_{n=0}^{N} \frac{f^{(n)}(y)}{n!}(x - y)^n. \tag{C99}$$

The advantage of this representation, called an *expansion of order* $\mathcal{O}(x-y)^N$, is that the information on the local behavior of even very complicated functions can be encoded in $N+1$ *numbers*, the derivatives $f^{(n)}(y)$. However, the example shown in Fig. C11 shows that the accuracy of the approximation generally depends on the order, N, the range, $|x - y|$, and on the local profile of f. It should be evident that a rapidly varying function is less easy to approximate than a slowly varying one. ($\rightarrow$ C5.3.1-2)

The accuracy of a finite-order Taylor expansion is measured by its difference from the actual function, $|f(x) - f_N(x)|$. Rigorous bounds for approximation errors are derived in lecture courses on mathematics and we here restrict ourselves to stating one principal result: let I be an interval in which the function's higher derivatives are bounded as $|f^{(n)}(x)| < \alpha C^n$, where α, C are constant. The approximation error is then bounded as

$$|f(x) - f_N(x)| < \alpha \frac{(C|x - y|)^{N+1}}{(N + 1)!}.$$

For large N, the factorial in the denominator grows more rapidly than the power in the numerator and the error converges to zero. We also observe that the magnitudes of the function derivatives, which are measures of the rapidity of its variation, enter the estimate.

INFO If functions violate the above convergence criterion it is often possible to identify a specific value of N for which an optimal approximation is obtained. In such cases, the truncation of the series both at values smaller *and* larger than the optimal N produces larger errors. The systematic discussion of these so-called **asymptotic expansions** is beyond the scope of this text.

asymptotic
expansion

C5.4 Solving equations by Taylor expansion

Taylor expansions can be applied to find approximate solutions of equations which are too complicated to be solved in closed form. To introduce the idea let us consider the example of an equation that *can* be solved exactly:

$$0 = y^2 - 2\epsilon y - 1. \tag{C100}$$

Considered as an equation for y, this quadratic equation has the exact solution

$$y(\epsilon) = \epsilon \pm (1 + \epsilon^2)^{1/2}, \tag{C101}$$

where the notation indicates that the solution for y depends on ϵ. If ϵ is small, this can be approximated by the first few terms of the corresponding Taylor series (check!),

$$y(\epsilon) = \pm 1 + \epsilon \pm \tfrac{1}{2}\epsilon^2 + \mathcal{O}(\epsilon^3). \tag{C102}$$

Let us now discuss how to find this approximation assuming that the exact solution $y(\epsilon)$ was unknown.

For $\epsilon = 0$, the equation is trivially solved by $y(0) = \pm 1$. One expects that for $|\epsilon| \ll 1$ two distinct solutions for $y(\epsilon)$ emanate from these limiting values. An approximate solution may thus be represented as a series

$$y(\epsilon) = y_0 + y_1\epsilon + \tfrac{1}{2!}y_2\epsilon^2 + \mathcal{O}(\epsilon^3), \tag{C103}$$

whose coefficients y_i need to be determined such that the quadratic equation (C100) is satisfied order by order in an expansion in ϵ. We already know that $y_0 = \pm 1$. To determine y_1 and y_2 the series is substituted into the quadratic equation. Retaining only terms up to $\mathcal{O}(\epsilon^2)$ (i.e. neglecting terms of order ϵ^3 and higher arising from the substitution), one obtains

$$0 = \left[y_0^2 + y_1^2\epsilon^2 + 2y_0y_1\epsilon + y_0y_2\epsilon^2 \right] - 2\epsilon(y_0 + y_1\epsilon) - 1 + \mathcal{O}(\epsilon^3),$$

or, sorting according to powers of ϵ,

$$0 = (y_0^2 - 1) + (2y_0y_1 - 2y_0)\epsilon + (y_1^2 + y_0y_2 - 2y_1)\epsilon^2 + \mathcal{O}(\epsilon^3).$$

The equation is satisfied if the polynomial on the right-hand side vanishes identically. This requires the individual vanishing of the coefficients in brackets, and hence the equations

$$0 = y_0^2 - 1 \, ,$$
$$0 = 2y_0 y_1 - 2y_0 \, , \tag{C104}$$
$$0 = y_1^2 + y_0 y_2 - 2y_1 \, .$$

Notice that the coefficients y_n appear successively: the first equation contains y_0, the second y_0 and y_1, the third all coefficients up to y_2, etc. Also observe that each time a new coefficient $y_{n \geq 1}$ appears, it enters the corresponding equation *linearly*. These two properties do not depend on the particular form of our equation, but are general features of the solution strategy of solving an equation using a series ansatz (think about this point). They allow us to solve the equations (C104) iteratively, determining first y_1 through y_0, then y_2 through y_0 and y_1, etc. This yields $y_0 = \pm 1$, $y_1 = 1$ and $y_2 = \pm 1$, so that the ansatz (C103) indeed reproduces the expansion (C102) up to the desired order.

Let us conclude with a few *general remarks on the procedure*. First, we built the approach on the *a priori* assumption that the solution can be expanded in ϵ. If this assumption turns out to be illegitimate, the equations will signal it by a breakdown of the hierarchical construction. For example, the simple equation $\left(y(\epsilon) \right)^2 = \epsilon$ cannot be expanded in ϵ. This can be understood by inspection of its exact solution, $y(\epsilon) = \pm\sqrt{\epsilon}$, which cannot be Taylor expanded around $\epsilon = 0$ (why?). Attempting a series ansatz as above, one readily finds that no solvable hierarchy of equations ensues.[22]

For equations whose solutions can be expanded, the procedure above generally works. To phrase the three-step algorithm in general terms, consider an equation $F(y, \epsilon) = 0$ and assume that for $\epsilon = 0$ the solution of the equation, $F(y, 0) = 0$, is known as $y = y_0$. An approximate solution for small $\epsilon \neq 0$ is then found as follows.

▷ Start by substituting the power series ansatz $y(\epsilon) = \sum_{n=0} \frac{1}{n!} y_n \epsilon^n$ into the equation.

▷ Expand the resulting expression, $F\left(\sum_n \frac{1}{n!} y_n \epsilon^n, \epsilon \right) = 0$, in powers of ϵ^n to obtain another power series in ϵ, of the form $\sum_n F_n(y_0, \ldots, y_n) \epsilon^n = 0$. Here each coefficient F_n is linear in y_n and in general can depend on all $y_{i \leq n}$.

▷ This power series must vanish for all ϵ, hence each of its coefficients must vanish identically, $F_n = 0$. Solving these equations iteratively yields the coefficients y_n. ($\rightarrow$ C5.4.1-4)

perturbation theory | Iterative solution strategies of this type are called **perturbative solutions**. The procedure is "perturbative" in the sense that for small ϵ the solution is only weakly deformed or "perturbed" from its $\epsilon = 0$ value, y_0.

INFO Perturbative solutions of algebraic equations, and of the differential equations to be discussed later, play a very important role in physics: physical problems generally present themselves in the

---

[22] Explicitly: inserting the ansatz $y(\epsilon) = y_0 + y_1 \epsilon + \mathcal{O}(\epsilon^2)$ into $y^2 = \epsilon$ yields $y_0^2 + 2y_0 y_1 \epsilon + (y_1 \epsilon)^2 + \mathcal{O}(\epsilon^2) = \epsilon$. Equating coefficients with the same power of ϵ yields $y_0 = 0$ and $y_0 y_1 = 1$, which would imply $y_1 = \infty$.

form of equations. These equations often contain "small" parameters, ϵ, and simplify if they are expanded in these. For example, the electric current, $I(V)$, flowing through a metal depends on the applied voltage, V. For $V = 0$ no current flows, $I(0) = 0$. Since externally applied fields are generally tiny compared to the internal fields in solids, V may be considered a "small perturbation". This suggests an expansion of the current to linear order, $I = 0 + gV$, where the unknown coefficient, g, defines the *linear conductance* of the system. The computation of g for realistic solids can still be complicated, but it is much easier than the (generally impossible) computation of the full function $I(V)$ for general V. Perturbative expansions are so useful that a plethora of different variants of *perturbation theory* have been developed. They play an important role in almost all sub-disciplines of physics, and are prominently discussed in lecture courses in theoretical physics.

C5.5 Higher-dimensional Taylor series

REMARK Requires Section V3.2. The physical understanding of the example contained in this section requires basic familiarity with electrostatics.

general higher-dimensional Taylor series

The concept of Taylor expansion can straightforwardly be generalized to multi-dimensional functions, $f : \mathbb{R}^d \to \mathbb{R}$. The expansion of $f(\mathbf{x})$ in the vicinity of a fixed argument, $\mathbf{y}$, reads

$$f(\mathbf{x}) = \sum_{n=0}^{\infty} \frac{1}{n!}((\mathbf{x} - \mathbf{y}) \cdot \nabla)^n f(\mathbf{y}), \qquad \text{(C105)}$$

where $\mathbf{y} \cdot \nabla \equiv \sum_{i=1}^{d} y^i \partial_{y^i}$. Here, the derivatives in ∇ are defined to act exclusively on $f(\mathbf{y})$, but not on the factors $(\mathbf{x} - \mathbf{y})$. For example, the expansion up to second order reads as

$$f(\mathbf{x}) \simeq f(\mathbf{y}) + \sum_{i=1}^{d}(x - y)^i \partial_{y^i} f(\mathbf{y}) + \frac{1}{2}\sum_{i,j=1}^{d}(x - y)^i (x - y)^j \partial^2_{y^i,y^j} f(\mathbf{y}). \qquad \text{(C106)}$$

Equation (C105) is proven in the same way as the one-dimensional Eq. (C82). An n-fold partial derivative, $\partial^n_{x^{i_1},...,x^{i_n}}|_{\mathbf{x}=\mathbf{y}}$, applied to the left-hand side of the equation, yields $\partial^n_{y^{i_1},...,y^{i_n}} f(\mathbf{y})$. On the right, the derivatives act on the factors $(x - y)^i$ multiplying the derivative operators ∂_{y^i}. For $n = 2$, inspection of Eq. (C106) shows that the two sides of the equation coincide. Some more bookkeeping is required to do the calculation for general n, however the principal procedure remains the same. ($\to$ C5.5.1-2)

Unlike with the infinitely extended Taylor series discussed in the previous sections, multidimensional expansions are usually truncated after the first few orders. The reason is that high-order multidimensional expansions are not of much practical usefulness and at the same time tedious to compute.

EXAMPLE The electric potential created by a charged particle at $\mathbf{r}_0$ is given by $\varphi(\mathbf{r}) = \frac{q}{|\mathbf{r}-\mathbf{r}_0|}$,[23] where q is the charge. At the point $\mathbf{r}$ this potential creates an electric field

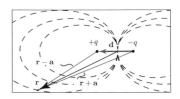

$$\mathbf{E} \equiv -\nabla\varphi(\mathbf{r}) = q\frac{\mathbf{r}-\mathbf{r}_0}{|\mathbf{r}-\mathbf{r}_0|^3}.$$

electric dipole   Consider now an **electric dipole**, i.e. a system of two opposite electric charges, $\pm q$, sitting at positions $\pm\mathbf{a}$ relative to the origin of a coordinate system. When the electric potentials generated by these charges, $\varphi_\pm(\mathbf{r}) = \pm q/|\mathbf{r}\mp\mathbf{a}| \simeq \pm q/|\mathbf{r}|$, are observed at remote points, $r = \|\mathbf{r}\| \gg a$, they largely cancel out. The residual contribution is captured by a first-order Taylor expansion of the function $(1/|\mathbf{r}-\mathbf{a}|)$ in the increment $\mathbf{a}$ around $\mathbf{a} = 0$. Using $\partial_i(1/|\mathbf{r}|) = -x^i/|\mathbf{r}|^3$, we find

$$\varphi_\pm(\mathbf{r}) = \frac{q}{|\mathbf{r}\mp\mathbf{a}|} = \pm\frac{q}{|\mathbf{r}|} + \frac{q\mathbf{a}\cdot\mathbf{r}}{|\mathbf{r}|^3} + \mathcal{O}(\mathbf{a}^2).$$

The potential created by the two charges, $\varphi(\mathbf{r}) = \varphi_+(\mathbf{r}) + \varphi_-(\mathbf{r})$, is thus given by

$$\varphi(\mathbf{r}) \simeq \frac{2q\mathbf{a}\cdot\mathbf{r}}{|\mathbf{r}|^3} \equiv \frac{\mathbf{d}\cdot\mathbf{e}_r}{r^2}.$$

Here $\mathbf{e}_r \equiv \mathbf{r}/r$ is a unit vector in the $\mathbf{r}$-direction, and the *dipole moment*, $\mathbf{d} \equiv q(2\mathbf{a})$, is a vector connecting the positions of the two opposite charges, multiplied by their magnitude. Contours of the dipole potential (lines along which the potential remains constant) are indicated by dashed lines in the figure above. Dipole fields appear in many different contexts. For example, biological membranes often consist of layers of molecules stacked in such a way that the membrane does not carry a net charge but does create a dipole potential.

Verify that the *dipole electric field*, $\mathbf{E} = -\nabla\varphi$, is given by $\mathbf{E}(\mathbf{r}) \simeq \big(3(\mathbf{d}\cdot\mathbf{e}_r)\mathbf{e}_r - \mathbf{d}\big)/r^3$, and discuss the spatial profile of this field.

C5.6 Summary and outlook

In this chapter we introduced Taylor series as a powerful diagnostic tool for differentiable functions. We saw that even low-order Taylor expansions yield accurate descriptions of the local profile of a function, with accuracy bounds that can be quantitatively specified. Infinite-order expansions can reveal relations between functions which are difficult to obtain by other means. We saw that the convergence issues arising in connection with such infinite series are best addressed within the framework of a complex extension of Taylor series. In fact, we reasoned that it is good practice to think even of real Taylor series as restrictions of complex series. Existence and convergence issues which otherwise may remain mysterious can be clarified in this way.

Even if Taylor series can sometimes be applied to establish global connections between functions, the tool as such is geared towards a description of local features. In the next chapter, we adopt a complementary perspective and introduce a series representation of functions tailored to the description of global structures.

---

[23] Here, and in various other examples from electromagnetism in this text, we express physical quantities in **CGS units** (centimetre–gram–second units). CGS units are an alternative to **SI units** (International System of units) widely used in physics and engineering. Referring for a detailed comparison of units to lecture courses in experimental physics, we note that the choice of CGS units leads to slightly more compact formulas but otherwise is inessential to our discussion.

CGS units

Fourier calculus

REMARK In this chapter we will occasionally draw connections to linear algebra. They mostly appear in info sections and are meant as optional pedagogical assets. A deepened perspective of connections between calculus and linear algebra is developed in Chapter L9.

In the previous chapter we introduced Taylor expansions as a means to describe the local profile of a function through a finite number of real or complex expansion coefficients. We now turn to the discussion of a different diagnostic tool which shifts the focus from local to global structures. The general idea is indicated in the figure, where the dashed line is a 30th-order Taylor expansion

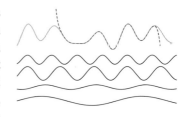

of the topmost function around its local minimum. We observe that the expansion does an excellent job in an extended region but fails drastically when the distance from the expansion center becomes too large. However, the function itself is just the sum of the curves shown below it, four sine functions, $r_n \sin(nt + \phi_n)$, $n = 1, 2, 3, 4$, of increasing oscillation period. Therefore the full profile of the function is encoded in just eight real numbers, the four weights, r_n, and phase shifts, ϕ_n. Apparently, these parameters provide a more efficient "encoding" of the function than using a large number of Taylor coefficients.

idea of Fourier calculus

The general idea of *Fourier analysis* is to represent functions as sums or series over "simple" functions. These building blocks are generally functions with periodic modulation, sin, cos, or exponential functions of purely imaginary arguments. Fourier calculus is applied both to engineer and to analyze functions. For example in an analogue synthesizer, complex

Jean Baptiste Fourier
1768–1830
A French mathematician and physicist best known for the invention of Fourier calculus. Fourier applied his new concept to the study of physical phenomena such as heat conduction or the physics of vibrations. He is also considered the discoverer of the greenhouse effect.

sounds are built by superposition of monochromatic oscillations. Conversely, compression algorithms analyze acoustic signals and determine the weights, r_n, and phases, ϕ_n, required for their accurate reproduction from minimal sets of stored data.

In this chapter we introduce the methodology of Fourier calculus and illustrate its application on various examples. However, before turning to the core of this discussion, we need to digress and introduce a particular "function" which will become a very important tool in the development of the theory.

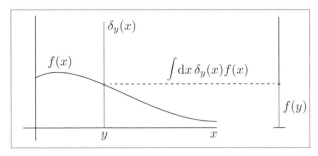

On the definition of the δ-function. For the "function" $\delta_y(x)$ to have the required property Eq. (C107), it must vanish for all values of x except $x = y$, where it has to be "infinitely large".

C6.1 The δ-function

qualitative discussion The δ-**function**, $\delta_y(x)$, is a strange-looking object resembling an infinitely sharp needle. It vanishes for all $x \neq y$, but is infinitely large at $x = y$. Clearly this "function" cannot be a function in the sense of any conventional definition. Instead, δ_y is defined via an implicit condition: it is a map $\delta_y : \mathbb{R} \to \mathbb{R}, x \mapsto \delta_y(x)$, obeying the condition that for any continuous function $f : \mathbb{R} \to \mathbb{R}$,

$$\int_{\mathbb{R}} dx\, \delta_y(x) f(x) = f(y). \tag{C107}$$

The condition states that for any function f the integral of f multiplied by δ_y "projects out" the function value $f(y)$. For example, $\int_{\mathbb{R}} dx\, \delta_2(x) x^3 = 2^3 = 8$. A moment's thought[24] shows that the function $\delta_y(x)$ must vanish for all values of x except for $x = y$, see Fig. C12. On the other hand, for the particular "test function" $f = 1$ we obtain the normalization condition

$$\int_{\mathbb{R}} dx\, \delta_y(x) = 1. \tag{C108a}$$

We are thus dealing with a "function" that vanishes everywhere except at one point, $x = y$. If $\delta_y(y)$ were finite the integral would vanish, and so we have the conditions:

$$\delta_y(x) = \begin{cases} 0, & x \neq y, \\ \infty, & x = y. \end{cases} \tag{C108b}$$

This extreme behavior implies that the δ-"function" cannot be a function in an ordinary sense.

However, much as $0 = \lim_{\epsilon \to 0} \epsilon$ can be thought of as a limiting value, one may try to construct a family of functions, δ_y^ϵ, such that for any finite ϵ, δ_y^ϵ is well-defined and the extreme behavior of δ_y recovered in the limit $\lim_{\epsilon \to 0} \delta_y^\epsilon = \delta_y$. In the next sections we introduce concrete realizations of such constructions.

[24] If $\delta_y(x)$ were non-vanishing for $x \neq y$, it would be possible to devise functions $f(x)$ such that the integral $\int dx\, \delta_y(x) f(x)$ yields a value different from $f(y)$. Think about this point.

Construction of the δ-function

Gaussian representation of δ-function

It is not difficult to define families of functions, δ_y^ϵ, satisfying the required convergence criterion. As an example, consider the family of **Gaussian functions**,

$$\delta_y^\epsilon(x) \equiv \frac{1}{\epsilon\sqrt{\pi}}e^{-(x-y)^2/\epsilon^2}.$$

For all values of the real parameter $\epsilon > 0$, these functions satisfy the normalization condition $\int_\mathbb{R} dx\, \delta_y^\epsilon(x) = 1$. The region within which δ_y^ϵ is not negligibly small is centered around y and of width ϵ. With decreasing ϵ it shrinks to a point: $\lim_{\epsilon\to 0}\delta_y^\epsilon(x) = 0$ for $x \neq y$. On the other hand, the function values at y diverge, $\lim_{\epsilon\to 0}\delta_y^\epsilon(y) = \infty$. The figure illustrates this behavior for the three values $\epsilon = 0.2, 0.07, 0.02$, respectively.

In the limit $\epsilon \to 0$ the defining criteria Eqs. (C108) are satisfied and in this sense we have $\lim_{\epsilon\to 0}\delta_y^\epsilon = \delta_y$. Also notice that $\delta_y^\epsilon(x)$ depends on the parameter y and the argument x only through the difference $x - y$, and is symmetric in this difference. In particular, we have $\delta_y^\epsilon(x) = \delta_0^\epsilon(x-y) \equiv \delta^\epsilon(x-y)$, where $\delta_0^\epsilon \equiv \delta^\epsilon$ is centered around 0. This leads to the alternative representation

$$\delta^\epsilon(x-y) \equiv \frac{1}{\epsilon\sqrt{\pi}}e^{-(x-y)^2/\epsilon^2}. \tag{C109}$$

In a similar manner, we write $\delta_y(x) \equiv \delta_x(y) \equiv \delta(x-y)$, where $\delta(x-y) \equiv \lim_{\epsilon\to 0}\delta^\epsilon(x-y)$. To summarize, we *constructively define the δ-function* as the limit,

$$\delta(x) \equiv \lim_{\epsilon\to 0}\delta^\epsilon(x), \tag{C110}$$

of a one-parameter family of functions defined by two conditions, unit normalization, $\int dx\, \delta^\epsilon(x) = 1$, and vanishing support[25] except at $x = 0$ in the limit $\epsilon \to 0$. Above, we realized δ^ϵ through Gaussian functions; however, there exist many other convenient representations – any family of functions describing symmetric peaks of unit weight can be used.

EXERCISE Use the auxiliary identity $\int_\mathbb{R} dx\frac{1}{1+x^2} = \pi$ to show that the family of functions

$$\delta^\epsilon(x) = \frac{1}{\pi}\frac{\epsilon}{\epsilon^2 + x^2} \tag{C111}$$

converges to the δ-function. Equation (C111) defines the *Lorentzian representation of the δ-function.* ($\to$ C6.1.3-4).

Before turning to the discussion of important mathematical properties of the δ-function, some more remarks on the general definition are in order.

support of a function

---

[25] The **support**, $\mathrm{supp}(f) \equiv \{x \in \mathbb{R}\,|\,f(x) \neq 0\}$, of a function $f : \mathbb{R} \to \mathbb{R}$ is the subset of arguments on which f is non-vanishing.

First, it is possible to *restrict the definition of the δ-function to finite intervals*. Intuitively, a needle-sharp function does not notice whether it is defined on an unbounded or bounded domain. Somewhat more rigorously, any point y contained in an open interval $I = (a, b)$ is surrounded by a neighborhood of finite width Δ likewise contained in I. For $\epsilon \ll \Delta$,

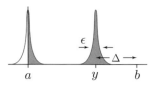

the support of δ_y^ϵ, too, is almost entirely contained in I, and we have the asymptotically exact normalization $\int_I dx\, \delta_y^\epsilon(x) \simeq 1$. This implies $\int_I dx\, \delta_y(x) f(x) = f(y)$, as before.[26]

Second, we note that an *abstract formulation of the defining relation* (C107), making no reference to the function arguments,

$$f = \int dy\, \delta_y\, f(y), \qquad \text{(C112)}$$

is sometimes useful. Eq. (C112) represents the function f as a sum (integral) over functions δ_y with complex weights $f(y)$. Its equivalence to the previous (C107) is seen by inserting arguments, $f \to f(x)$ and $\delta_y \to \delta_y(x)$, exchanging variables, $x \leftrightarrow y$, and using $\delta_x(y) = \delta_y(x)$.

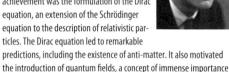

Paul Adrien Maurice Dirac
1902–1984
An English physicist and one of the founding fathers of quantum mechanics and quantum field theory. His most striking single achievement was the formulation of the Dirac equation, an extension of the Schrödinger equation to the description of relativistic particles. The Dirac equation led to remarkable predictions, including the existence of anti-matter. It also motivated the introduction of quantum fields, a concept of immense importance in modern physics.

Third, in calculations, *the δ-function always appears under an integral* operation. In exceptional cases (for an example, see the next section), where an integral cannot be evaluated by the definition $\int dx\, \delta(y-x) f(x) = f(y)$, a family of representing functions, δ^ϵ, needs to be engaged. One then evaluates the integrals of regular functions, $\int dx \delta^\epsilon(x-y) f(x)$, and takes the limit $\epsilon \to 0$ on the resulting expressions.

δ-functions simplify calculations

Finally, notice that *δ-functions should be considered as friends* which generally simplify calculations. Even if f is a very complicated function, the integral $\int dx\, \delta(y-x) f(x) = f(y)$ reduces to the evaluation of a single function value – we say the δ-function "collapses the integral" to a number – a very welcome feature.

INFO The concept of δ-functions was introduced by P.A.M. Dirac as a tool for the description of point charges in physics. A point charge at y is an object whose charge density is zero everywhere except at the point y, where it diverges. Its charge profile is described by δ_y as in (C108b). It took several decades to formulate Dirac's description in precise mathematical terms. The result was an extension of the concept of functions, known as *distributions*. For any nonzero ϵ, δ^ϵ is a regular function, but the limit $\delta = \lim_{\epsilon \to 0} \delta^\epsilon$ is not. To understand the limit one may note that δ_y extracts the number $f(y)$ from a function, $f : I \to \mathbb{R}$, via the integral $f(y) = \int dx\, \delta_y(x) f(x)$. This suggests an interpretation of δ_y as a map from the space of functions to the real numbers. This map acts on functions as $\delta_y[f] = f(y)$, where we followed the convention of using square brackets, $[f]$, to enclose the argument of a map acting on functions. For technical reasons, one has to restrict the set of arguments to **test functions** which are smooth and have compact support. Maps assigning numbers

distributions

to test functions are called **distributions**, and δ_y belongs to this class of objects. For this reason,

---

[26] For a closed integration interval, $I = [a, b]$, an exceptional situation arises if y coincides with one of the boundaries, $y = a$ or b. In this case only one half of the symmetric peak δ_y^ϵ lies within the integration interval (see figure above). This implies $\lim_{\epsilon \to 0} \int_a^b dx\, \delta_a^\epsilon(x) f(x) = \int_a^b \delta(x-a) f(x) = \frac{1}{2} f(a)$, and similarly for $y = b$.

mathematicians prefer the denotation δ-*distribution*. However, we here follow physics parlance and use the more sloppy terminology "δ-function".

Properties of the δ-function

For later reference, we summarize here various useful *relations obeyed by the δ-function*. They all follow directly from its definition (C107).

▷ For any smooth function, we have the *defining property*

$$\int dx\, \delta(x-y) f(x) = f(y), \qquad\qquad (\text{C113})$$

which implies the unit normalization $\int dx\, \delta(x) = 1$.

▷ Integrals of the δ-function against functions that are not smooth need to be processed with the help of a representing family, $\{\delta^\epsilon\}$. As an example, consider the function $f(x) = a$ for $x < 0$ and $f(x) = b$ for $x \geq 0$. If $b \neq a$ it has a discontinuous step of height $b - a$ at $x = 0$. Substitution of any representation of δ_0^ϵ leads to $\int dx\, \delta_0^\epsilon(x) f(x) = a \int_{-\infty}^0 dx\, \delta_0^\epsilon(x) + b \int_0^\infty dx\, \delta_0^\epsilon(x) = \frac{1}{2}(a + b)$, where the symmetry $\delta_0^\epsilon(x) = \delta_0^\epsilon(-x)$ and footnote 26 was used. This leads to the identification $\int dx f(x) \delta_0(x) = \frac{1}{2}(a + b)$, hence the δ-function extracts the "average" value of f at the step.

▷ In applications, one often encounters δ-functions whose arguments differ from the integration variable. As an example, consider the integral $\int dx\, \delta(cx) f(x)$, where c is a constant. Defining the new variable $u = cx$ and applying the substitution rule Eq. (C28), this becomes $\int du |c|^{-1} \delta(u) f(u/c) = |c|^{-1} f(0/c) = |c|^{-1} f(0)$, where Eq. (C113) was applied to the u-integral. The result,

$$\int dx\, \delta(cx) f(x) = \frac{1}{|c|} f(0), \qquad\qquad (\text{C114})$$

is equivalent to the *scaling relation*

$$\delta(cx) = \frac{1}{|c|} \delta(x). \qquad\qquad (\text{C115})$$

This shows that the δ-function is "inversely proportional" to factors appearing in its arguments. For example, $\int_{\mathbb{R}} dx\, \delta(2x - \pi) \sin x = \int_{\mathbb{R}} dx\, \delta(2(x - \frac{1}{2}\pi)) \sin x = \frac{1}{2} \sin \frac{1}{2}\pi = \frac{1}{2}$.

▷ Eq. (C115) implies symmetry of the δ-function under changes of sign, $\delta(x) = \delta(-x)$.

▷ Another frequently occurring expression reads $\int dx\, \delta(g(x)) f(x)$, where g is a function. Since $\delta(g(x)) = 0$ for $g(x) \neq 0$ this integral receives contributions only from an (infinitesimal) neighborhood of the zeros of g. Assume that x_0 is such a zero, assume differentiability of g there, and expand $g(x) = g(x_0) + g'(x_0)(x - x_0) + \cdots = g'(x_0)(x - x_0) + \cdots$. We substitute this expression into the integral, note that for $x \to x_0$ the higher-order terms in the expansion can be neglected, and apply Eq. (C114) to obtain

variable change in δ-function integrals

$$\int dx\, \delta(g(x)) f(x) = \frac{f(x_0)}{\left| \dfrac{dg(x_0)}{dx} \right|}.$$

In cases where g has more than one zero, at the points $\{x_i\}, i = 0, 1, \ldots$, the contributions of all these to the integral need to be added together. In this way we arrive at the most general representation for the *change of variables under a δ-function*:

$$\int dx\, \delta(g(x))f(x) = \sum_i \frac{f(x_i)}{\left|\frac{dg(x_i)}{dx}\right|}. \tag{C116}$$

Remember: the defining property (C113) only applies to combinations $\int dx\, \delta(x)(\ldots)$ where the integration variable itself features as an argument of the δ-function. If other expressions appear as arguments, Eq. (C116) needs to be applied. Finally, note that Eq. (C116) implies that integrals for which $g'(x_i) = 0$, such as $\int dx\, \delta(x^2)$, are ill-defined.

EXAMPLE Consider the integral $I = \int_{-\infty}^{\infty} dx\, \delta(x^2 + 3x - 10) \cdot (2x + 1)$. The function $g(x) = x^2 + 3x - 10$ has zeros at $x_0 = 2$ and $x_1 = -5$, and at these points its derivative, $g'(x) = 2x + 3$, gives $g'(x_0) = 7$ and $g'(x_1) = -7$. Equation (C116) thus yields

$$I = \frac{f(x_0)}{|g'(x_0)|} + \frac{f(x_1)}{|g'(x_1)|} = \frac{1}{|7|}[2 \cdot 2 + 1] + \frac{1}{|-7|}[2 \cdot (-5) + 1] = -\frac{4}{7}.$$

▷ One sometimes encounters *derivatives* acting on the δ-function, $\delta'(x)$. Although the δ-function is not differentiable in any conventional sense, a representing family, δ^ϵ, can be engaged to make sense of this expression:

$$\int dx\, \delta'(x)f(x) = \lim_{\epsilon \to 0} \int dx\, \delta^{(\epsilon)'}(x)f(x) = -\lim_{\epsilon \to 0} \int dx\, \delta^\epsilon(x)f'(x) = -f'(y).$$

Here we have integrated by parts, and the narrow support of the δ-function implies the absence of boundary terms. This yields the identification

$$\int dx\, \delta_y'(x)f(x) = -\int dx\, \delta_y(x)f'(x) = -f'(y). \tag{C117}$$

▷ Consider the integral $\int_{-\infty}^{x} dy\, \delta(y)f(y)$. It yields 0 for $x < 0$, $f(0)$ for $x > 0$, and $\frac{1}{2}f(0)$ for $x = 0$, since then the integration boundary coincides with the center of the δ-function (see footnote 26). The frequent occurrence of this expression motivates the definition of the **Heaviside step function**,[27]

Heaviside step function

$$\Theta(x) = \begin{cases} 1 & \text{for } x > 0, \\ \frac{1}{2} & \text{for } x = 0, \\ 0 & \text{for } x < 0. \end{cases} \tag{C118}$$

[27] In the literature, one often finds a simplified version of the Heaviside function, $\Theta(x) = 1$ for $x \geq 0$ and 0 for $x < 0$, which is not perfectly symmetric relative to the jump point, $x = 0$.

The above results may now be written as

$$\int_{-\infty}^{x} dy\, \delta(y)f(y) = \Theta(x)f(0).$$

For $f = 1$ we have

$$\int_{-\infty}^{x} dy\, \delta(y) = \Theta(x), \tag{C119}$$

which is an alternative definition of the Heaviside function. Differentiation of this relation yields

$$\boxed{\Theta'(x) = \delta(x).} \tag{C120}$$

This formula states that the Θ-function is constant almost everywhere except at the jump-point $x = 0$, where its derivative is singular. Of course, neither the Θ-function nor the δ-function are truly differentiable. All the expressions above have to be understood as limits of appropriately defined δ^{ϵ}-sequences. ($\rightarrow$ C6.1.1-4)

C6.2 Fourier series

REMARK In this section various remarks on the conceptual relation between Fourier calculus and linear algebra will be made. While these connections are quite illuminating, they assume familiarity with general inner products of vector spaces (Section L3.3). Readers not yet acquainted with this topic may ignore all comments concerning linear algebra in this section.

We now turn to the principal theme of this chapter, the representation of functions as sums of periodic functions. For a concrete motivation, suppose we are confronted with recorded data such as that in Fig. C13. (The plot shows a recording of about 1 sec of spoken human language.) How can the "essence" of the measured signal be efficiently described? The data clearly contains high frequency fluctuations modulated by structures of lower frequency. How can one disentangle and understand these different contributions? Fourier calculus tackles such problems by representing arbitrary signals as superpositions of simple periodic functions. The information on the parent function is then contained in the coefficients with

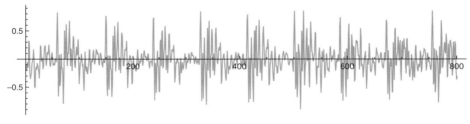

Fig. C13   Recorded sound intensity of about one second of spoken human language. The data consists of 800 measured values, connected by lines for better visibility.

which the elementary functions enter the superposition. As we will see, the knowledge of these coefficients contains the key to the solution of the problem outlined above and of many others.

The idea of Fourier calculus

As stated above, the idea of Fourier calculus is to represent a given function f as a sum over many simple functions. In most cases these are the **harmonic functions** $\exp(ikx)$, $\cos(kx)$ and $\sin(kx)$.[28] Here we focus on the expansion in terms of exp-functions, which play the most important role in physical applications. The straightforward modification to cos or sin series is the subject of problems C6.2.3-4.

Consider a complex-valued function, f, defined on an open interval of width L, $I \equiv (x_0, x_0 + L)$, where the starting point x_0 is arbitrary. We aim to represent f as a

Fourier series

$$f(x) = \frac{1}{L} \sum_k \exp(ikx)\tilde{f}_k, \tag{C121}$$

Fourier series

Fourier modes

containing **Fourier modes**, $\exp(ikx)$, weighted by complex-valued **Fourier coefficients**, $\tilde{f}_k$. The sum extends over all values

$$k = \frac{2\pi n}{L}, \qquad n \in \mathbb{Z}. \tag{C122}$$

INFO The rationale behind this expansion is best understood from the perspective of linear algebra (!). In Section L2.3 we introduced a view considering functions, f, as vectors. In this interpretation, $f(x)$ corresponds to a vector component, and this being so, there must be an analogue of the standard basis, $\{\mathbf{e}_j\}$, i.e. a basis whose vectors have components vanishing for all but one entry, $(\mathbf{e}_j)^i = \delta^i{}_j$. For the vector space of functions the role of the basis vectors $\mathbf{e}_j$ is taken by the δ-functions, δ_y, whose function values, $\delta_y(x)$, likewise vanish for all values of x except one, $x = y$. The relation Eq. (C112) is then interpreted as the expansion of the vector f in the "standard basis" $\{\delta_y\}$, with complex coefficients $f(y)$. This point of view is discussed in more detail in Chapter L9. For the moment we just note that much as in finite-dimensional linear algebra, function spaces, too, can have different useful bases. At present, our objective is to establish the Fourier modes $\exp(ikx)$ as an alternative to the standard basis functions, $\delta_y(x)$. Equation (C121) anticipates that this program will be successful and formulates the expansion of a general function in this basis, with coefficients $\tilde{f}_k$. In essence, Fourier calculus is the basis change from the standard, δ-function basis to a basis of harmonic functions. This view will prove useful throughout and make the conceptual meaning of key formulas transparent.

---

[28] A function $\psi(x)$ is harmonic if it obeys the condition $\mathrm{d}_x^2 \psi(x) = c\,\psi(x)$. Harmonic functions reproduce themselves, up to a multiplicative constant, c, under two-fold differentiation. There exists a generalized definition (and a corresponding generalization of Fourier calculus) in which the two-fold derivative is replaced by a more complicated Laplace operator (see Section V3.5, Eq. (V98)). However, this extension is beyond the scope of this text.

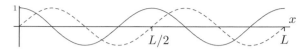

Fig. C14   Real part, $\cos(kx)$ (solid), and imaginary part, $(\sin(kx))$ (dashed), of the Fourier mode $\exp(ikx)$, for $k = 2(2\pi/L)$.

Properties of Fourier modes

The Fourier basis functions are characterized through three crucially important mathematical properties, *spatial periodicity*, *orthonormality* and *completeness*. First, and in striking contrast to the standard basis functions, δ_y, the Fourier modes are non-vanishing throughout the entire interval, I. At the interval boundaries, x_0 and $x_0 + L$, the Fourier modes assume identical values, $e^{ikx_0} = e^{ik(x_0+L)}$, a feature referred to as **periodic boundary conditions**. This property is intended and useful,[29] and follows from choosing the k-values as $k = 2\pi n/L$, so that $e^{ikL} = e^{i2\pi n} \overset{(C92)}{=} 1$. In the physics literature, the discretely spaced or "quantized" values k are often called **Fourier momenta**.[30] Within the interval I, the functions $\exp(ikx)$ are oscillatory (see Fig. C14), where the oscillation period, $\lambda \equiv 2\pi/k = L/n$, decreases with n. Fourier modes often appear in the context of wave-like phenomena where λ defines the **wave length** of a mode and k is its **wave number**.

periodic boundary conditions

Fourier momenta

wave length

Second, the Fourier modes are mutually orthogonal to each other:

$$\boxed{\frac{1}{L}\int_I dx\, e^{i(k-k')x} = \delta_{kk'}.}$$ (C123)

The relation is proven by noting that for $k \neq k'$,

$$\frac{1}{L}\int_I dx\, e^{i(k-k')x} = \frac{1}{iL(k-k')}\, e^{i(k-k')x}\Big|_{x_0}^{L+x_0} = 0,$$

due to the periodic boundary conditions. However, for $k = k'$, $\exp(i(k-k')x) = 1$, and the integral trivially yields unity.

orthonormality relation

INFO Equation (C123) really is an **orthonormality relation** in the sense of linear algebra. Referring to Chapter L9 for an in-depth discussion, we note that $\langle f, g \rangle \equiv \int_I dx\, \overline{f(x)}g(x)$ is an inner product on the set of functions on I. Defining **Fourier basis functions**, $\psi_k(x) \equiv \frac{1}{\sqrt{L}}\exp(ikx)$, Eq. (C123) states that these form an orthonormal set, $\langle \psi_{k'}, \psi_k \rangle = \delta_{kk'}$.

completeness relation

Third, the Fourier modes satisfy the **completeness relation** defined on the open interval $x \in (-L, L)$,

$$\frac{1}{L}\sum_k \exp(ikx) = \delta(x).$$ (C124)

---

[29] In many applications, Fourier analysis is applied to functions defined on a *ring* of width L. These functions are naturally periodic, and hence must be expanded in functions which are periodic themselves.

[30] Readers who have studied Section L8.4 may note that these values define the quantized eigenvalues of the quantum mechanical momentum operator on a ring, and that the Fourier modes are the corresponding eigenfunctions – hence the denotation.

(In applications of this relation, the argument x features as a difference, $x = x_1 - x_2$, of two points $x_1, x_2 \in I$. This means that x is bounded by $\pm L$, and that $x = 0$ naturally occurs as an argument.) This relation is called a completeness relation because it is the key to proving that any function (vector) can be expanded in Fourier modes and that the latter form a basis of function space. The linear algebraic meaning of the structure of this relation is explained in the second info section below.

INFO   Referring for a rigorous treatment to problem C6.1.6, we here restrict ourselves to a *semi-quantitative discussion* of Eq. (C124). The sum vanishes for $x \neq 0$ because in this case the phase factors $\exp(ikx)$ are distributed quasi-randomly around the unit circle (see figure,[31] and recall Eq. (C90)). Upon summation, these values tend to cancel each other, and in the limit of an infinite sum zero is obtained. For $x = 0$, $\exp(ikx) = 1$, and the sum diverges. This is the behavior expected of a δ-function. The proper unit normalization, (C108a), is confirmed by computing the integral,

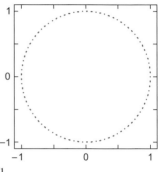

$$\int_{-L/2}^{L/2} dx \frac{1}{L} \sum_k e^{ikx} \stackrel{(C123)}{=} \sum_k \delta_{k0} = 1.$$

INFO   Equation (C124) has an analogue in linear algebra. A transformation $\mathbf{e}_j = \mathbf{w}_\alpha T^\alpha{}_j$ between the standard basis $\{\mathbf{e}_j\}$ and a general orthonormal basis $\{\mathbf{w}_a\}$ of $\mathbb{C}^n$ is unitary, $(T^\dagger)^i{}_\alpha T^\alpha{}_j = \delta^i{}_j$ (see Eq. (L202)). With the associations $\{\mathbf{e}_j\} \leftrightarrow \{\delta_y\}$, $\{\mathbf{w}_a\} \leftrightarrow \{\psi_k\}$ and $i \leftrightarrow x$, the matrix elements $(T^\dagger)^i{}_\alpha = (T^{-1})^i{}_\alpha = \langle \mathbf{e}^i, \mathbf{w}_\alpha \rangle$ correspond to $(T^{-1})_{x,k} \equiv \langle \delta_x, \psi_k \rangle = \psi_k(x) = \frac{1}{\sqrt{L}} \exp(ikx)$, and the unitarity condition to the completeness relation (C124). For a detailed discussion, see Section L9.1.

For later reference, we note that Eq. (C124) affords extension to the full real axis. The L-periodicity of the Fourier modes, $\exp(ikx) = \exp(ik(x + nL))$, in the integers n means that for arbitrary x,

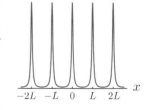

$$\boxed{\frac{1}{L} \sum_k \exp(ikx) = \sum_{m=-\infty}^{\infty} \delta(x - mL).}$$ (C125)

The right-hand side is a periodic succession of δ-peaks. ($\rightarrow$ C6.1.6)

Fourier series construction

Let us now address the question under what criteria functions can be represented as sums over Fourier modes. We first note that for many functions of interest, an expansion as in (C121) fails for a lack of convergence. The Fourier modes have unit modulus, $|e^{ikx}| = 1$, and so the series converges only if the coefficients $\tilde{f}_k$ decay sufficiently fast. While this

---

[31] The figure depicts $\exp(ix2\pi n/L)$ as points in the complex plane, for $n = 0, \ldots, 500$ and $x/L = (\sqrt{2} + \sqrt{3})/100$.

condition severely limits the class of expandable functions, an efficient way to improve the situation is to redefine the series as

$$f(x) = \frac{1}{L} \lim_{\epsilon \to 0} \sum_k e^{ikx - \epsilon|k|} \tilde{f}_k, \tag{C126}$$

**convergence-
generating
factor**

where here and in the following the notation $\epsilon \to 0$ means that ϵ is sent to zero coming from positive values. For any finite ϵ the series now converges, unless the coefficients $\tilde{f}_k$ increase exponentially in k. The factor $e^{-\epsilon|k|}$ is called a **convergence-generating factor**, or just *convergence factor*. It is customary to not write this factor explicitly even if its presence is required to render a series convergent. However, if seemingly ill-defined series such as $\sum_k e^{ikx}$ are encountered here or in the literature, one may assume that the presence of a convergence factor is implicit.

Series whose convergence is safeguarded in this manner can represent most functions occurring in practice, including aperiodic functions or even singular functions. (In view of our discussion above, this freedom may be interpreted as a consequence of the function-basis property of the Fourier modes.) To understand this point more concretely, *assume* that a function f has a Fourier representation. In this case, the orthogonality of the Fourier modes, (C123), can be used to identify its Fourier coefficients $\tilde{f}_k$ as follows: multiply $f(x)$ by $\exp(-ikx)$ and integrate over I:

$$\int_I dx\, e^{-ikx} f(x) = \int_I dx\, e^{-ikx} \left[\frac{1}{L} \sum_{k'} e^{ik'x} \tilde{f}_{k'} \right]$$

$$= \sum_{k'} \tilde{f}_{k'} \frac{1}{L} \int_I dx\, e^{-i(k-k')x} \overset{(C123)}{=} \sum_{k'} \tilde{f}_{k'} \delta_{kk'} = \tilde{f}_k.$$

We thus have the identification

$$\boxed{\tilde{f}_k = \int_I dx\, e^{-ikx} f(x).} \tag{C127}$$

INFO The structure of Eq. (C127) affords a natural linear algebraic interpretation. The series (C121) is an expansion, $f = \sum_k \psi_k (\frac{1}{\sqrt{L}} \tilde{f}_k)$, of f in the Fourier basis, $\{\psi_k\}$. Much as the coefficients of a vector, $\mathbf{v} = \mathbf{w}_\alpha v^\alpha$, in an orthonormal basis $\{\mathbf{w}_\alpha\}$ are obtained by taking inner products, $v^\alpha = \langle \mathbf{w}^\alpha, \mathbf{v} \rangle$ (see Eq. (L68)), the Fourier coefficients are obtained as $\tilde{f}_k = \sqrt{L} \langle \psi_k, f \rangle$.

Notice that the coefficients $\tilde{f}_k$ are generally complex, even if the function $f(x)$ is real. However, in the latter case, $\overline{f(x)} = f(x)$, we have the symmetry relation

$$\overline{\tilde{f}_k} = \int_I dx\, e^{+ikx} f(x) = \tilde{f}_{-k}. \tag{C128}$$

EXAMPLE Consider the "sawtooth" function defined on the interval $I = [0, 1]$ (see Fig. C15)

$$f(x) = \begin{cases} -x, & x \in \left(0, \frac{1}{2}\right), \\ 1 - x, & x \in \left(\frac{1}{2}, 1\right). \end{cases} \tag{C129}$$

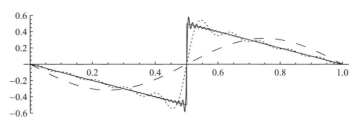

Fig. C15   Fourier representation of the function (C129) in terms of a finite series with maximum Fourier index $n_{\max} = 1$ (long dashed), 10 (dashed) and 80 (solid).

Here $L = 1$. The Fourier coefficients for non-vanishing k are obtained from Eq. (C127) as

$$k \neq 0 : \quad \tilde{f}_k = \int_0^{1/2} dx\, e^{-ikx}(-x) + \int_{1/2}^1 dx\, e^{-ikx}(1-x) = \frac{e^{-ik/2}}{ik}, \qquad (C130)$$

where partial integration was used to integrate $x e^{-ikx}$. For $k = 0$ we have $\tilde{f}_0 = \int_0^1 dx\, f(x) = 0$.

The above procedure yields the coefficients $\tilde{f}_k$ *provided* the Fourier expansion exists. To explore under which conditions this is the case, we substitute Eq. (C127) for $\tilde{f}_k$ into (C121) for $f(y)$, assume that the order of the summation over k and the integration over x can be exchanged, and obtain

$$f(y) = \frac{1}{L} \sum_k e^{iky} \int_I dx\, e^{-ikx} f(x) = \int_I dx \left(\frac{1}{L} \sum_k e^{ik(y-x)} \right) f(x)$$

$$\stackrel{(C124)}{=} \int_I dx\, \delta(x - y) f(x). \qquad (C131)$$

In the crucial third equality we used the completeness relation (C124), which identifies the expression in brackets as a δ-function. The final integral yields $f(y)$, so the ansatz of f as a sum over Fourier modes faithfully reproduces f as required. However, in deriving this relation, we tacitly assumed that the Fourier coefficients, $\tilde{f}_k$, are finite and that the sum over them exists. For functions obeying rather mild *Dirichlet conditions* (see second info section below) this is the case, and Fourier expandability is granted.

INFO   The expansion $\mathbf{v} = \mathbf{w}_\alpha \langle \mathbf{w}^\alpha, \mathbf{v} \rangle$ of a vector $\mathbf{v}$ in an orthonormal basis $\{\mathbf{w}_\alpha\}$ implies $v^i = w_\alpha{}^i (\overline{w^\alpha}_j v^j)$ for its components in the standard basis. Holding for arbitrary v^i, this requires the *finite-dimensional completeness relation* $w_\alpha{}^i \overline{w^\alpha}_j = \delta^i{}_j$. The Fourier completeness identity, $\frac{1}{L} \sum_k \exp(ik(x_1 - x_2)) = \delta(x_1 - x_2)$, is a similar relation in function space, with $\alpha \leftrightarrow k$, $i \leftrightarrow x$, $w_\alpha{}^i \leftrightarrow \psi_k(x) = L^{-1/2} \exp(ikx)$.

**Dirichlet
conditions**   **INFO**   The three **Dirichlet conditions** sufficient for the Fourier expandability of a function read:

▷   the integral of the modulus of the function must exist: $\int_I |f(x)| dx < \infty$;

▷   the number of local extrema of f in I must be finite;

▷ f must contain only finitely many discontinuities in I.[32]

---

[32]   At the point of a discontinuity, $\lim_{\delta \to 0} f(y \pm \delta) = f_\pm$, the Fourier series representation yields the average value, $(f_+ + f_-)/2$. This is a consequence of Eq. (C131), and the fact that $\int dx f(x) \delta(x - y) = (f_+ + f_-)/2$, see footnote 26.

An example of a function failing this test is given by

$$f(x) = \begin{cases} 0, & x \in (-\pi, 0], \\ \sin(1/x), & x \in (0, \pi). \end{cases}$$

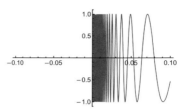

The infinitely many extrema accumulating in the vicinity of $x = 0$ spoil its Fourier-expandability.

EXAMPLE The Fourier coefficients Eq. (C130) of the "sawtooth function", Eq. (C129), define the representation

$$f(x) = \sum_{k \neq 0} e^{ikx} \frac{e^{-ik/2}}{ik} = \frac{1}{\pi} \sum_{n \in \mathbb{Z} \setminus \{0\}} \frac{e^{i2\pi nx}}{2i} \frac{e^{-i\pi n}}{n} = \frac{1}{\pi} \sum_{n > 0} \sin(2\pi nx) \frac{(-1)^n}{n}. \qquad (C132)$$

Figure C15 illustrates how this series models the function through sin-functions. The long-dashed, dashed and solid curves are obtained by truncation of the sum at the index $n_{max} = 1, 10, 80$, respectively. Notice the efficiency of these approximations in regions where the function is smooth. However, problems arise in the neighborhood of sharp corners where a large number of terms is required to obtain satisfactory agreement.

INFO Notice that even for an elaborate representation of $f(x)$ in terms of 80 Fourier contributions the series visualized in Fig. C15 "overshoots" near the "corners" of the function. It can be shown that the excess peak does not diminish if more terms are included: its height remains at a level of $\mathcal{O}(10\%)$ of the function value, only the width of the excess region shrinks.

This phenomenon is known as *ringing*. Ringing is notorious in audio and video compression algorithms such as MP3 or AAC, or JPEG, which all rely on Fourier signal encoding. The term "ringing" alludes to the fact that in compressed acoustic data, such overshooting becomes audible as a sharp "ringing" noise, accompanying the reproduction of dynamical sound sources (such as drums). The JPEG compressed reproduction of a star shown in the figure illustrates how the same effect spoils the accurate reproduction of sharp edges in visual data.

The example above illustrates an important *general feature of Fourier representations*: the series (C121) encodes the information carried by the function $f(x)$. For large k the exponential functions oscillate rapidly, i.e. functions with large k carry the information about the "fine structure" of $f(x)$. Conversely, the information on large-scale structures is contained in slowly oscillating contributions. The appearance of k and x in the product $k \cdot x$ in the Fourier modes shows that scales of characteristic length Δx are described by Fourier modes with index $k \sim \Delta x^{-1}$. This fact is known as **Fourier reciprocity**,

Fourier reciprocity

> Fourier modes of large/small values of k describe structures at small/large scales, $k \sim x^{-1}$.

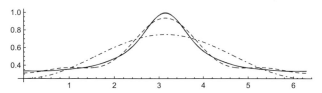

Fig. C16   Fourier representations of the function Eq. (C133): the dashed, dash-dotted and solid black lines are approximations by 1, 3 and 6 Fourier modes. The last is already almost indistinguishable from the actual function.

EXAMPLE For functions devoid of sharp singularities, the inclusion of just a few Fourier modes can suffice to obtain excellent approximations. Consider, for example, the function

$$f(x) = \text{Re}\, \frac{1}{2 + e^{ix}} = \frac{2 + \cos x}{5 + 4\cos x} \tag{C133}$$

on the interval $x \in (0, 2\pi)$. The Fourier coefficients of this function are best computed by geometric series expansion (see Eq. (C85)),

$$f(x) = \frac{1}{2}\text{Re}\, \frac{1}{1 + \frac{1}{2}e^{ix}} = \frac{1}{2}\text{Re} \sum_{n=0}^{\infty} \left(-\frac{1}{2}e^{ix}\right)^n.$$

Substituting this expression into Eq. (C127) and using the orthogonality relation (C123), we obtain $\tilde{f}_0 = \pi$ and $\tilde{f}_{k\neq 0} = \frac{\pi}{2}\left(-\frac{1}{2}\right)^{|k|}$. Figure C16 illustrates how the approximation of the function by only a few Fourier modes yields excellent results.

INFO Although the essential architecture is always the same, *details of the definition of Fourier series* vary depending on the application field and/or the scientific community. For example, in some fields it is customary to rescale all Fourier coefficients as $\tilde{f}_k \to \sqrt{L}\tilde{f}_k$, such that the prefactor multiplying Eq. (C121) changes to $1/\sqrt{L}$. The right-hand side of Eq. (C127) then likewise carries a factor of $1/\sqrt{L}$. In texts using Fourier calculus it is therefore common practice to open with a remark such as: "In this text, we will define Fourier series as $f(x) = \ldots$".

In physics and engineering the Fourier modes are defined as $\exp(ikx)$ if x is a space-like argument. However, just as often Fourier calculus is applied to functions depending on a time-like variable, t. (For example, Fig. C13 shows a time-like signal.) In such cases, the Fourier modes are more frequently defined to include a minus sign, $\exp(-i\omega t)$, and the variable ω assumes the role of k. The Fourier representation then reads as

$$f(t) = \frac{1}{T} \sum_{\omega} e^{-i\omega t}\tilde{f}_{\omega}, \qquad \tilde{f}_{\omega} = \int_I dt\, e^{+i\omega t}f(t). \tag{C134}$$

Here, $f(t)$ is defined on a time interval I of duration T and $\omega = 2\pi n/T$ is called the *frequency* of the Fourier mode.

Fourier analysis of periodic function

INFO Fourier series are frequently applied to the analysis of functions periodically repeating themselves after an interval of length, L, as in Fig. C17. A continuous **periodic function** satisfies the periodicity condition $f(x) = f(x + L)$. As a consequence, any interval, I, of length L contains the full information required to reconstruct the function $f(y)$ for arbitrary arguments: represent y as $y = x + nL$, with $x \in I$ and an integer n, to obtain $f(y) = f(x + nL) = f(x)$.

A periodic function can be represented by the Fourier series (C121), where the Fourier coefficients, Eq. (C127), are obtained by integration over an arbitrary choice of the reference interval I. The L-periodicity of the Fourier modes, $\exp(ik(x + L)) = \exp(ikx)$, then guarantees that the Fourier

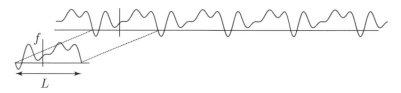

L

The full "information" about a periodic function is stored in an arbitrary single period.

series correctly reproduces the periodicity properties of f. For example, the Fourier series (C132) of the single sawtooth function Eq. (C129) defines a periodic extension to a sequence of sawtooths if its domain of definition is extended to the full real axis. ($\rightarrow$ C6.2.1-2) If the periodic function is an even or odd function of its argument, $f(-x) = \pm f(x)$, the corresponding Fourier series can be

cosine, sine   expressed as a sum over cosine or sine functions, known as a **cosine** or **sine series**, respectively.
series   ($\rightarrow$ C6.2.3-4)

C6.3 Fourier transform

The concept of Fourier representations can straightforwardly be adapted to functions defined on the entire real axis. Formally, this extension is achieved by sending the width of the support interval I to infinity. For example, starting from $I \equiv (-\frac{L}{2}, \frac{L}{2})$ one may consider the limit $L \rightarrow \infty$, in which f is defined on the entire real axis, $f : \mathbb{R} \rightarrow \mathbb{C}$.

Definition of the Fourier transform

For $L \rightarrow \infty$ the spacing, $\delta k = \frac{2\pi}{L}$, between successive Fourier series wave numbers $k = \frac{2\pi}{L}n$ tends to zero. The Fourier sum then assumes the form of a Riemann sum, which in the limit becomes an integral over a continuous variable k:

$$\frac{1}{L}\sum_{k}(\ldots) = \frac{1}{2\pi}\delta k \sum_{n}(\ldots) \rightarrow \frac{1}{2\pi}\int_{-\infty}^{\infty} dk\,(\ldots). \qquad (C135)$$

In the same limit, the Fourier coefficients become functions of a continuous variable, $\tilde{f}_k \rightarrow \tilde{f}(k)$. Introducing the shorthand notation $\int \frac{dk}{2\pi} = \frac{1}{2\pi}\int dk$, the discrete relation

completeness   Eq. (C125) now becomes a continuum **completeness relation**,[33]
relation

$$\boxed{\int \frac{dk}{2\pi}\,e^{ikx} = \delta(x),} \qquad (C136)$$

where $\pm\infty$-integration boundaries are omitted for notational brevity. For later reference we note an equivalent relation which is "dual" to Eq. (C136) in the sense that the variables x and k are exchanged,

[33] Also notice that in the limit $L \rightarrow \infty$ the periodicity interval, L, of the sequence of δ-functions on the right-hand side of Eq. (C125) approaches infinity, so that only a single peak, $\delta(x)$, remains.

$$\int dx\, e^{-ikx} = 2\pi\,\delta(k). \tag{C137}$$

Due to the completeness of Fourier modes, a function, $f(x)$, may be reconstructed from its full set of Fourier coefficients, $\tilde{f}(k)$. This fact is expressed by the continuum versions of **Fourier transform** Eqs. (C127) and (C121),

$$\tilde{f}(k) = \int_{-\infty}^{\infty} dx\, e^{-ikx} f(x), \tag{C138a}$$

$$f(x) = \int_{-\infty}^{\infty} \frac{dk}{2\pi}\, e^{ikx} \tilde{f}(k). \tag{C138b}$$

The first equation defines the **Fourier transform**[34] of the function, the second is the **inverse Fourier transform**. The functions f and $\tilde{f}$ define a **Fourier transform pair**. In cases where a specific function is to be Fourier transformed and the use of tildes becomes cumbersome, the alternative notation $\mathcal{F}f(x) = \tilde{f}(k)$, assigning to a function of x its Fourier-transformed partner function of k, and the inverse $\mathcal{F}^{-1}\tilde{f}(k) = f(x)$, is used. For example, $\mathcal{F}\exp(-|x|) = \frac{2}{1+k^2}$ and $\mathcal{F}^{-1}\frac{2}{1+k^2} = \exp(-|x|)$ (see below).

To once more see how completeness is key to the Fourier relations, substitute Eq. (C138a) into Eq. (C138b), exchange the order of integrations and obtain

$$f(y) = \int \frac{dk}{2\pi} e^{iky} \int dx\, e^{-ikx} f(x) = \int dx \left(\int \frac{dk}{2\pi} e^{ik(y-x)} \right) f(x)$$

$$\overset{\text{(C136)}}{=} \int dx\,\delta(y-x) f(x) = f(y).$$

Due to the completeness relation (C136), the inverse transform indeed recovers $f(y)$ from $\tilde{f}(k)$, as required.

EXAMPLE The Fourier transform of $f(x) = e^{-\gamma|x|}$, $\gamma > 0$, is given by

$$\tilde{f}(k) \overset{\text{(C138a)}}{=} \int_{-\infty}^{\infty} dx\, e^{-ikx} e^{-\gamma|x|} = \int_0^{\infty} dx \left[e^{-x(ik+\gamma)} + e^{-x(-ik+\gamma)} \right] = \frac{1}{ik+\gamma} + \frac{1}{-ik+\gamma} = \frac{2\gamma}{k^2+\gamma^2},$$

a Lorentzian peak. Notice the Fourier reciprocity: the scales on which $f(x)$ and $\tilde{f}(k)$ exhibit strong variation, $1/\gamma$ and γ, respectively, are inversely proportional to each other. The inverse Fourier transform,

$$f(x) \overset{\text{(C138b)}}{=} \int_{-\infty}^{\infty} \frac{dk}{2\pi} e^{ikx} \frac{2\gamma}{k^2+\gamma^2},$$

can not be computed by elementary means. However, it can be done using the powerful contour integration methods introduced in Chapter C9 (see p. 352), which readily yield $e^{-\gamma|x|}$, as expected.

INFO The above constructions relied on an exchange of integrals $\int dk \int dx(\dots) \rightarrow \int dx \int dk(\dots)$. Fubini's theorem, discussed in Section C4.1, states that this operation is OK provided the involved integrals individually exist. While generic criteria for the *existence of Fourier integrals* are difficult to

[34] In general, an *integral transform* of a function, f, is a set of data, $\{\tilde{f}\}$, from which the function can be fully recovered.

state, the rule of thumb is that functions that can be integrated in the presence of a convergence factor can be transformed. A convergence factor may be introduced by modification of the exponential functions in the Fourier integrals,

$$
\int \frac{dk}{2\pi} \exp(ikx) \rightarrow \int \frac{dk}{2\pi} \exp(ikx - \epsilon|k|),
$$

$$
\int dx \exp(-ikx) \rightarrow \int dx \exp(-ikx - \epsilon|x|), \tag{C139}
$$

existence of Fourier integrals

where ϵ is positive and sent to zero after all integrals have been computed. As before, the convergence factors are usually not indicated explicitly. Due to their presence, the Fourier integrals of functions growing no faster than power laws all exist. Modified versions of the Fourier transform, such as the Laplace transform defined on p. 291, can be applied to handle functions of even stronger divergence.

Convergence factors may also be engaged to avoid the explicit appearance of δ-functions. For example, in the presence of a convergence generator, the completeness integral (C136) yields a Lorentzian representation of the δ-function,

$$
\int \frac{dk}{2\pi} e^{ikx - \epsilon|k|} = \frac{1}{\pi} \frac{\epsilon}{x^2 + \epsilon^2}. \tag{C140}
$$

This turns into the "true" δ-function, $\delta(x)$, only in the limit $\epsilon \rightarrow 0$, reproducing Eq. (C136).

Sometimes, it is more convenient to work with alternative realizations of convergence factors. For example, the *Gaussian convergence factor*, $\exp(ikx) \rightarrow \exp(ikx - \epsilon k^2)$, generates even more powerful convergence at large values of k, and can lead to k-integrals which are easier to compute. (Which realization of a convergence factor is most convenient must be decided on a case-by-case basis.) As an exercise, verify that $\lim_{\epsilon \rightarrow 0} \int \frac{dk}{2\pi} e^{ikx - \epsilon k^2}$ produces a Gaussian representation of the δ-function. ($\rightarrow$ C6.3.3)

Generalized definitions of Fourier integrals

In practice, one often needs to Fourier transform higher-dimensional functions, $f : \mathbb{R}^d \rightarrow \mathbb{C}$, $\mathbf{x} \mapsto f(\mathbf{x})$. This is achieved by individual Fourier transformation in each of the variables contained in $\mathbf{x} = (x^1, \ldots, x^d)$. Defining

$$
\tilde{f}(k_1, \ldots, k_d) \equiv \int dx^1 \, e^{-ix^1 k_1} \ldots \int dx^d \, e^{-ix^d k_d} f(x^1, \ldots, x^d),
$$

Fourier transform of multi-dimensional functions

and combining the products of exponentials to a single exponential, we obtain the **Fourier transform of multidimensional functions**,

$$
\boxed{
\begin{aligned}
\tilde{f}(\mathbf{k}) &= \int dx^1 \ldots dx^d \, e^{-i\mathbf{k} \cdot \mathbf{x}} f(\mathbf{x}), \\
f(\mathbf{x}) &= \int \frac{dk_1}{2\pi} \ldots \frac{dk_d}{2\pi} \, e^{i\mathbf{k} \cdot \mathbf{x}} \tilde{f}(\mathbf{k}),
\end{aligned}
} \tag{C141}
$$

where $\mathbf{k} \cdot \mathbf{x} \equiv \sum_j k_j x^j$, $\mathbf{k} = (k_1, \ldots, k_d)^{\mathrm{T}}$,[35] and the second line contains the inverse transform. Corresponding d-dimensional δ-functions are defined as products, $\delta(\mathbf{x}) \equiv \delta(x^1) \ldots \delta(x^d)$.

[35] Here, we use that $(k_1, \ldots, k_d)^{\mathrm{T}}$ is a covariant object whose components should be labeled by subscripts (see V. I. Arnold, *Mathematical Methods of Classical Mechanics*, Springer Verlag 1978). The covariance of k_j and the contravariance of x^i make the argument in the exponent, $k_j x^j$, a coordinate invariant scalar. For example, if the arguments $x^j \rightarrow cx^j$ are scaled by a number, $k_j \rightarrow k_j c^{-1}$ transforms by the inverse of that number (Fourier reciprocity!) but $k_j x^j$ remains invariant.

The **Fourier transform of functions** *f(t)* **depending on time-like arguments** is defined as

$$
\tilde{f}(\omega) = \int_{-\infty}^{\infty} dt\, e^{i\omega t} f(t),
$$

$$
f(t) = \int_{-\infty}^{\infty} \frac{d\omega}{2\pi} e^{-i\omega t} \tilde{f}(\omega),
\tag{C142}
$$

where ω is a frequency-like variable. As with the definition of the Fourier series (C134), the signs in the exponents are exchanged relative to the "space-like" Fourier transform, (C138). The rationale behind this convention becomes evident once Fourier transforms of functions $f(x, t)$ depending on space- and time-like arguments are considered. For further discussion of this point we refer to lecture courses on classical electrodynamics.

Properties of the Fourier transform

REMARK In this section, the Fourier transforms of composite expressions such as $xf(x)$ will be considered. For notational economy, we use the $\mathcal{F}(xf(x))(k)$ notation throughout.

Functions and their Fourier transforms are related to each other by a multitude of useful relations, summarized here for later reference and application. With few exceptions, all identities are immediate consequences of the definition.

▷ The *integral* of $f(x)$ is proportional to $(\mathcal{F}f)(0)$, and the integral of $(\mathcal{F}f)(k)$ to $f(0)$:

$$
(\mathcal{F}f)(0) = \int dx f(x), \qquad f(0) = \int \frac{dk}{2\pi} (\mathcal{F}f)(k).
\tag{C143}
$$

▷ For an even or odd function, $f(-x) = \pm f(x)$, the Fourier transform is likewise even or odd, $(\mathcal{F}f)(-k) = \pm(\mathcal{F}f)(k)$.[36] ($\to$ C6.3.1-2)

▷ Under *complex conjugation* the Fourier transform behaves as

$$
\overline{(\mathcal{F}f)(k)} = (\mathcal{F}\bar{f})(-k),
\tag{C144}
$$

where $\mathcal{F}\bar{f}$ is the Fourier transform of the complex conjugate function $\bar{f}$. For real-valued functions the relation simplifies to $\overline{(\mathcal{F}f)(k)} = (\mathcal{F}f)(-k)$. For real functions which are even/odd, this becomes $\overline{(\mathcal{F}f)(k)} = \pm(\mathcal{F}f)(k)$, i.e. the Fourier transforms of such functions are purely real/imaginary.

▷ The *Fourier transform of the exponential function* is a δ-function, and vice versa:

$$
\mathcal{F}e^{iqx} = 2\pi\,\delta(k - q), \qquad \mathcal{F}\delta_y(x) = e^{-iky}.
\tag{C145}
$$

Important special cases are $q = 0$ and $y = 0$, respectively, i.e.

$$
\mathcal{F}1 = 2\pi\,\delta(k), \qquad \mathcal{F}\delta(x) = 1.
\tag{C146}
$$

---

[36] This follows from $(\mathcal{F}f)(-k) = \int dx\, e^{+ikx} f(x) \overset{x \to -x}{=} \int dx\, e^{-ikx} f(-x) = \pm \int dx\, e^{-ikx} f(x) = \pm(\mathcal{F}f)(k).$

▷ *The Fourier transform converts derivatives into multiplicative factors.* To see what is meant by this statement, consider the derivative of a function, $d_x f(x)$. When $f(x)$ is represented as a Fourier integral (C138b), its x-dependence is contained solely in the Fourier modes e^{ikx}, and the action of d_x produces a factor ik under the k-integral. The **derivatives to multiplicative factors** Fourier transform, $(\mathcal{F}d_x f)(k)$, is thus given by

$$(\mathcal{F}d_x f)(k) = ik(\mathcal{F}f)(k), \tag{C147}$$

where $(\mathcal{F}f)(k)$ is the transform of $f(x)$. A more formal check of this statement reads as

$$(\mathcal{F}d_x f)(k) = \int dx\, e^{-ikx} d_x f(x) = -\int dx\, (d_x e^{-ikx}) f(x) = ik \int dx\, e^{-ikx} f(x)$$
$$= ik(\mathcal{F}f)(k),$$

where we integrated by parts and boundary terms at infinity are suppressed by the implicit convergence factors. In the same way one verifies the inverse property: a Fourier transform given by $d_k(\mathcal{F}f)(k)$ has inverse Fourier transform $-ixf(x)$, or

$$d_k(\mathcal{F}f)(k) = -i(\mathcal{F}xf)(k). \tag{C148}$$

The key observation here is that

> derivatives simplify under Fourier transformation – they get converted to
> multiplicative factors.

For future reference, we note the sign factors in these relations depend on the conventions used in the definition of the Fourier transform. For example, the temporal Fourier transform of Eq. (C142) leads to

$$(\mathcal{F}d_t f)(\omega) = -i\omega(\mathcal{F}f)(\omega), \qquad d_\omega(\mathcal{F}f)(\omega) = +i(\mathcal{F}tf)(\omega). \tag{C149}$$

In problems involving many derivatives it is often convenient to pass to a Fourier representation, work for a while there, and only later transform back to the original representation. We will discuss such strategies in the next chapter when we solve differential equations.

convolution   ▷ The **convolution** of two functions is defined as

$$(f * g)(x) \equiv \int dy f(x - y)g(y). \tag{C150}$$

The definition is general, and likewise applies to functions depending on the Fourier variable, k:

$$(\mathcal{F}f * \mathcal{F}g)(k) \equiv \int dp(\mathcal{F}f)(k - p)(\mathcal{F}g)(p).$$

Convolution is a commutative operation, $f * g = g * f$, as follows from a change of integration variables, $x' = x - y$. The **Fourier convolution theorem** states that the convolution of two functions, $(f * g)(x)$, has a Fourier transform given by the product of their respective Fourier transforms:

$$(\mathcal{F}(f * g))(k) = (\mathcal{F}f)(k)\,(\mathcal{F}g)(k)\,. \tag{C151}$$

Conversely, the Fourier transform of a product, $f(x)g(x)$, equals the convolution of the Fourier transforms,

$$(\mathcal{F}(fg))(k) = \frac{1}{2\pi}(\mathcal{F}f * \mathcal{F}g)(k). \tag{C152}$$

These relations are proven by direct calculation. For example, Eq. (C151) follows via

$$(\mathcal{F}(f * g))(k) = \int dx\,e^{-ikx} \int dy f(x - y)g(y)$$

$$= \int dx\,e^{-ikx} \int dy \int \frac{dp}{2\pi} e^{ip(x-y)}(\mathcal{F}f)(p) \int \frac{dq}{2\pi}\,e^{iqy}(\mathcal{F}g)(q)$$

$$= \int \frac{dp}{2\pi}\frac{dq}{2\pi} \underbrace{\int dx\,e^{i(p-k)x}}_{2\pi\,\delta(p-k)} \underbrace{\int dy\,e^{i(q-p)y}}_{2\pi\,\delta(q-p)}(\mathcal{F}f)(p)(\mathcal{F}g)(q) = (\mathcal{F}f)(k)\,(\mathcal{F}g)(k).$$

The converse relation, Eq. (C152), is shown similarly. Observe that

> Fourier transformation turns a convolution (complicated) into a product (simple), and vice versa.

Convolutions appear in many different contexts. For example, in applied mathematics and engineering they are often applied to smooth ragged signal structures (see the next subsection). ($\rightarrow$ C6.3.3-4)

▷ The Fourier transform *preserves the inner product*,

$$\int dx \overline{f(x)}g(x) = \int \frac{dk}{2\pi}\, \overline{(\mathcal{F}f)(k)}\,(\mathcal{F}g)(k), \tag{C153}$$

Parseval's theorem

a result known as **Parseval's theorem**. When applied to the case $f = g$, Eq. (C153) reduces to the **Plancherel theorem**:

$$\int dx\,|f(x)|^2 = \frac{1}{2\pi} \int dk\,|(\mathcal{F}f)(k)|^2. \tag{C154}$$

INFO Equation (C153) can be verified directly from the definition of the Fourier transform and the orthogonality of the Fourier modes (exercise). Conceptually, it reflects the fact that the Fourier modes define an orthonormal basis. Much as the expansion $\mathbf{v} = \mathbf{e}_j v^j = \mathbf{w}_\alpha v^\alpha$ of a vector in a finite-dimensional orthonormal basis, $\{\mathbf{w}_\alpha\}$, preserves the inner product, $\overline{u_i}v^i = \overline{u_\alpha}v^\alpha$, the same is true for the infinite-dimensional basis of Fourier modes.

Table C1 Key relations satisfied by Fourier series coefficients.

function		series coefficients
$\int_I dx f(x)$	$=$	$\tilde{f}_0$
$f(0)$	$=$	$\frac{1}{L}\sum_k \tilde{f}_k$
$f(x) = \pm f(-x)$	$\Leftrightarrow$	$\tilde{f}_k = \pm \tilde{f}_{-k}$
$-$		$\overline{\tilde{f}_k} = \overline{\tilde{f}_{-k}}$
$d_x f(x)$	$\Leftrightarrow$	$ik\tilde{f}_k$
$(f * g)(x)$	$\Leftrightarrow$	$\tilde{f}_k \tilde{g}_k$
$f(x)\,g(x)$	$\Leftrightarrow$	$(\tilde{f} * \tilde{g})_k$
$\int_I dx \overline{f(x)} g(x)$	$=$	$\frac{1}{L}\sum_k \overline{\tilde{f}_k}\, \tilde{g}_k$
$\int_I dx \, |f(x)|^2$	$=$	$\frac{1}{L}\sum_k |\tilde{f}_k|^2$

INFO Most of the relations discussed in this section also hold for discrete *Fourier series*. The convolution of two periodic functions with period L, on a finite interval, I, of length L, is defined as

$$(f * g)(x) = \int_I dy f(x - y) g(y), \tag{C155}$$

and the convolution of discrete Fourier coefficients as

$$(\tilde{f} * \tilde{g})_k = \frac{1}{L}\sum_p \tilde{f}_{k-p} \tilde{g}_p. \tag{C156}$$

For convenience, the corresponding Fourier series relations are summarized in Table C1. Their verification is left as an exercise. ($\rightarrow$ C6.3.5-6)

INFO We finally note that there exist variants of the Fourier transform tailored to the solution of **Laplace** specific tasks. A prominent example is the **Laplace transform**, **transform**

$$(\mathcal{L}f)(s) \equiv \int_0^\infty dt f(t) e^{-st}. \tag{C157}$$

Here, $f(t)$ is a "time-dependent" function, and the complex parameter s is a "frequency-like" variable. (Imaginary parts in frequency-like variables can be interpreted as finite damping rates, see the discussion in Section C7.3). The required existence of the integral is an implicit part of the definition, i.e. the Laplace transform $(\mathcal{L}f)(s)$ is defined only if the integral on the right-hand side exists. For example, for the δ-function, $\delta(t)$, the Laplace transform exists for all s and yields[37] $(\mathcal{L}\delta)(s) = 1$, while that of the exponential function, e^{-t}, exists for all arguments $\text{Re}(s) > -1$, where it yields $(s + 1)^{-1}$.

The Laplace transform shares many essential properties with the Fourier transform. For example, it is manifestly linear, $\mathcal{L}(f + g) = \mathcal{L}f + \mathcal{L}g$, and satisfies relations such as (verify) $(\mathcal{L}(tf))(s) = -d_s(\mathcal{L}f)(s)$, and $(\mathcal{L}d_t f)(s) = s(\mathcal{L}f)(s) - f(0)$. An important difference is that the inverse transform, $f(t)$, of a Laplace represented function $(\mathcal{L}f)(s)$ is not as easily obtained as in the Fourier case. Rather, the inverse transformation relies on complex function integral techniques (as introduced in Chapter C9). For further discussion of the Laplace transform and applications where it is applied as a problem solver, we refer to the literature.

---

[37] Here the integration domain for Eq. (C157) is taken as $(-\epsilon, \infty)$, with $\epsilon \searrow 0$, so that it encloses the *full* δ-peak.

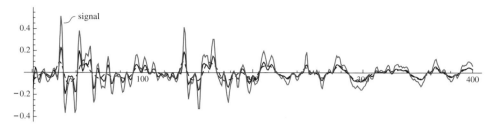

Fig. C18   Decomposition of a sound signal of 0.1 sec duration in terms of 500 (dashed) or 1000 (solid) Fourier modes. The latter compresses the raw data (5000 data points) by a factor 5 and already gives a decent approximation to the full signal.

Fourier transform applications

Fourier transformation is a powerful tool in science and engineering. Prominent areas of applications include the following.

▷ The transformation of derivatives into multiplicative factors (see Eq. (C147)) makes the Fourier transform an aid in the solution of *differential equations*. We will return to this point in the next chapter.

▷ Fourier transformations are used to *analyze and manipulate measurement data*. This point is illustrated by an example discussed below, and in the case study of Section C6.4.

▷ Fourier transformation plays an important role in the *compression of acoustic or visual data*. The basic idea is to decompose data into Fourier modes and to discard modes with wave numbers or frequencies exceeding a certain threshold (see Fig. C18). Actual compression algorithms are refined implementations of this approach.

▷ Fourier transformation plays a key role in *imaging* algorithms (see the info section below).

In the following, we illustrate the utility of the Fourier transform on the example of *noise reduction*. Data recorded in experimental physics and engineering inevitably contains *random noise*, and Fourier transformation techniques are frequently applied to extract the underlying signals. To understand the principle, consider a function $f(x) = f_s(x) + f_n(x)$, where $f_s(x)$ represents a "signal" varying slowly on a scale ℓ, while $f_n(x)$ is a "noise" func-

smoothing of functions   tion fluctuating rapidly on scales $\sim \Delta \ll \ell$. The noise contribution can be reduced by convolution of f with a suitably chosen "smearing function". Consider, for example, the *box function*

$$g(x) = \begin{cases} 1/\epsilon, & |x| < \epsilon/2, \\ 0, & \text{otherwise,} \end{cases} \tag{C158}$$

normalized to have unit weight, $\int \mathrm{d}x\, g(x) = 1$, and take its width, ϵ, to be larger than the noise length scale but smaller than the signal length scale, $\Delta < \epsilon < \ell$. Then the convolution

$$f(x) \mapsto (f * g)(x) = \int \mathrm{d}y f(x-y) g(y) = \frac{1}{\epsilon} \int_{-\epsilon/2}^{\epsilon/2} \mathrm{d}y f(x-y)$$

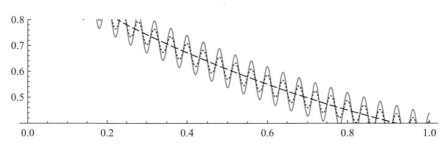

Fig. C19   Plot of the function $f = f_s + f_n$ defined by Eq. (C159) (solid line), for $\ell = 1, \eta = 0.07, K = 50\pi$. The dotted and dashed lines show its convolution with the box function (C158), for $\epsilon = 0.01$ and 0.1, respectively.

effectively "averages" f over an interval of width ϵ around x and hence smooths the function. If $\epsilon \ll \ell$, it damps out the noise but leaves the signal essentially unaffected.

The effect of convolving the signal with a smearing function becomes even more transparent in the Fourier domain. To illustrate the principle, consider the toy example

$$f_s(x) = \Theta(x)e^{-x/\ell}, \qquad f_n(x) = \eta\cos(Kx), \tag{C159}$$

shown in Fig. C19. Here, f_s represents a "signal" decaying exponentially on the scale ℓ, while f_n simulates "noise" with a single wave number, $|k| = K$, fluctuating at the scale $\Delta = 2\pi/K \ll \ell$. Their Fourier transforms[38] and that of the smearing function are (check!)

$$\tilde{f}_s(k) = \frac{1}{ik + 1/\ell}, \qquad \tilde{f}_n(k) = \pi\eta\big[\delta(k + K) + (k - K)\big], \qquad \tilde{g}(k) = \frac{2}{\epsilon k}\sin(\epsilon k/2).$$

Notice that $\tilde{g}(0) = 1$, which reflects the normalization $\int dx g(x) = 1$. Since the smeared signal is a convolution, its Fourier transform becomes a simple product, $(\mathcal{F}(f * g))(k) = \tilde{f}(k)\tilde{g}(k)$, with

$$\tilde{f}_s(k)\,\tilde{g}(k) \overset{k\sim 1/\ell}{\simeq} f_s(k), \qquad \tilde{f}_n(k)\,\tilde{g}(k) \simeq \mathcal{O}(1/\epsilon K) \ll 1.$$

The first relation reflects that k-values relevant to variations of the signal have small wave numbers, $|k| \lesssim 1/\ell \ll 1/\epsilon$, for which $\tilde{g}(k) \simeq \tilde{g}(0) = 1$. Multiplication with $\tilde{g}(k)$ has little effect and does not alter the signal. By contrast, noise fluctuations have large wave numbers, $|k| = K$, for which $\tilde{g}(K) \sim \frac{1}{K\epsilon} \ll 1$. The factor $\tilde{g}(k)$ therefore strongly suppresses the noise, so that the convolution $(f * g)(x) \simeq f_s(x)$ essentially recovers the signal, as desired.

EXERCISE Consider the *Gaussian smearing function*, $g(x) = \frac{1}{\sqrt{\pi}\epsilon}e^{-x^2/\epsilon^2}$. Show that its Fourier transform is again a Gaussian, $\tilde{g}(k) = e^{-(k\epsilon/2)^2}$, and that its convolution with the signal and noise functions of Eq. (C159) yields

$$(f_s * g)(x) \simeq e^{-x/\ell + (\epsilon/\ell)^2/4} \quad \text{(for } x \gg \epsilon\text{)}, \qquad (f_n * g)(x) = e^{-(K\epsilon/2)^2}\cos(Kx).$$

Discuss in what sense convolution with g has little/strong effect on signal/noise.
($\rightarrow$ C6.3.3-4)

---

[38] The Fourier transforms of experimentally recorded data would have to be computed by numerical methods, see below.

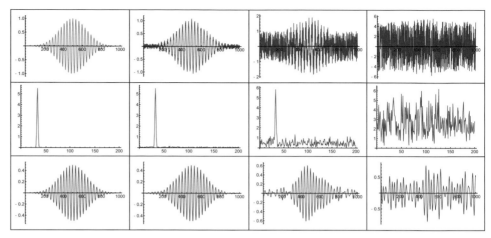

Fig. C20   Top row: synthetically generated data (left), masked by noise of increasing strength (toward right). Middle row: numerically computed Fourier spectra. Bottom row: signal reconstructed from the low frequency components of the Fourier spectrum. For discussion, see text.

The above discussion may make one wonder why we bothered to evoke the Fourier picture where it seems that a straightforward convolution in x-space already did the job. The reason is that an efficient approach to data cleansing will often *design* suitable convolution functions based on the structure of the Fourier data. The principle is illustrated in Fig. C20 on data synthetically generated on a computer. An artificial "signal" (top left) is masked by noise of strength 10%, 100% and 500% relative to the signal amplitude, respectively (top row). The middle row shows the (absolute value of the) numerically computed Fourier transforms of the data. Note that for noise strengths below the maximum one, a peak at low frequencies is clearly distinguishable from an erratic background. The signal can now be approximately recovered by removing all Fourier data except for the one around the non-stochastic signal. Specifically, the signals in the bottom row were obtained from the restriction of the Fourier data to a frequency window, $[25, 45]$, containing the peak. Note that this is equivalent to multiplication with a convolution function selecting a frequency *window*, and not just frequencies below a certain cutoff-frequency as in the previous discussion. Also notice that the choice of optimal frequencies requires knowledge of the Fourier transform and could not have been guessed on the basis of the raw data.

With this choice of frequency window, even noise equal in strength to the signal (third column) can be effectively removed and a fairly reasonable reproduction of the signal be obtained. For noise strength several times larger than the signal (right), no clearly visible structure in the Fourier data is left. However, even then, application of the "frequency frequency" reveals the presence of non-stochastic oscillations in the signal.

noise reduction by Fourier transform

tomographic imaging

INFO **Tomographic imaging** is an important tool for visualization in medicine and technology. Conceptually, it relies on a variant of the Fourier transform known as the Radon transform. To understand the idea, consider a thin slice of some substance, for example a section of a human skull, shown in Fig. C21. We assume the object to be described by a yet unknown density profile, $f(\mathbf{r})$, where $\mathbf{r} = (x, y)^{\mathrm{T}}$ are Cartesian coordinates. The strategy is to obtain information on $f(\mathbf{r})$ by exposing the

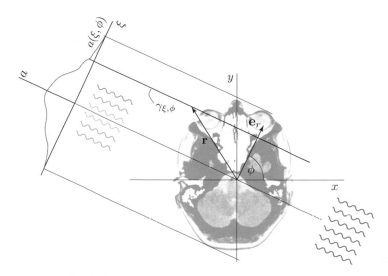

Fig. C21   The idea behind tomographic imaging. Discussion, see text.

object to spatially directed electromagnetic radiation. For each angle ϕ of the incident radiation, the signal transmitted by the substance is recorded at a detector behind it. The objective is to recover f from the knowledge of this data.

We start by defining an angle ϕ such that the polar unit vector, $\mathbf{e}_r = (\cos \phi, \sin \phi)^{\mathrm{T}}$, is perpendicular to the direction of incidence, and thereby parallel to the radiation wave fronts. The radiation transmitted by the substance defines an absorption function, $a(\xi, \phi)$, where ξ is the distance from the central axis of the radiation beam. Let us assume that all radiation arriving at a point specified by the coordinates (ϕ, ξ) has reached it along the straight-line path $\gamma_{\xi,\phi}$ indicated in the figure. All points, $\mathbf{r} \in \gamma_{\xi,\phi}$, on this path have the *same* projection onto the direction of $\mathbf{e}_r$, $\mathbf{r} \cdot \mathbf{e}_r = x \cos(\phi) + y \sin(\phi) = \xi$. We also assume that the radiation loss incurred locally along the path is proportional to the local density of tissue, $f(\mathbf{r})$. The total loss along the path can then be expressed as an integral along the trajectory,

$$a(\xi, \phi) = c \int \mathrm{d}x \, \mathrm{d}y \, f(\mathbf{r}) \delta(x \cos \phi + y \sin \phi - \xi), \qquad (\text{C160})$$

where the δ-function restricts the integration variables to the line $\gamma_{\xi,\phi}$, and c is a constant of proportionality.

Radon   Equation (C160) defines the **Radon transform** of the function f. In a tomographic scan it is
transform   obtained by recording the absorption profiles, $a(\xi, \phi_j)$, for a discrete set of incidence angles, ϕ_j. (Changes in the angular direction are responsible for the infamous clicking noise audible during computer tomography.) The desired density profile, $f(\mathbf{r})$, is reconstructed from the measured signal, $a(\xi, \phi)$, in two steps, both involving Fourier transforms. One first transforms $a(\xi, \phi)$ with respect to the variable ξ at a fixed angle ϕ,

$$\tilde{a}(k, \phi) \equiv \int \mathrm{d}\xi \, \mathrm{e}^{-\mathrm{i}\xi k} a(\xi, \phi) \overset{(\text{C160})}{=} c \int \mathrm{d}x \, \mathrm{d}y \, \mathrm{e}^{-\mathrm{i}(k \cos \phi \, x + k \sin \phi \, y)} f(\mathbf{r})$$

$$\equiv c \int \mathrm{d}x \, \mathrm{d}y \, \mathrm{e}^{-\mathrm{i}(k_x x + k_y y)} f(\mathbf{r}) \overset{(\text{C141})}{=} c \tilde{f}(\mathbf{k}). \qquad (\text{C161})$$

In the second equality we substituted Eq. (C160) and performed the ξ-integral over the δ-function, which projects ξ to the value $\xi = x \cos \phi + y \sin \phi$. In the third equality we identified k and ϕ as polar coordinates of a two-dimensional vector, $\mathbf{k} = (k_x, k_y)^{\mathrm{T}} \equiv (k \cos \phi, k \sin \phi)^{\mathrm{T}}$. This vector can be viewed as a Fourier vector "conjugate" to $\mathbf{r}$, because the final equality shows that $\tilde{a}(k,\phi) = c\tilde{f}(\mathbf{k})$ is just proportional to the Fourier transform, $\tilde{f}(\mathbf{k}) = \tilde{f}(k_x, k_y)$, of the density $f(\mathbf{r})$.

In a second step the inverse Fourier transform is applied to obtain $f(\mathbf{r})$ from $\tilde{f}(\mathbf{k})$,

$$cf(\mathbf{r}) \stackrel{(C141)}{=} \int \frac{dk_x dk_y}{(2\pi)^2} e^{i(k_x x + k_y y)} c\tilde{f}(\mathbf{k}) \stackrel{(C161)}{=} \int_0^\infty \frac{dk\, k}{(2\pi)^2} \int_0^{2\pi} d\phi\, \tilde{a}(k,\phi) e^{ik(\cos\phi x + \sin\phi y)},$$

where in the second equality we changed from Cartesian coordinates to polar coordinates, $\int dk_x\, dk_y = \int k\, dk \int d\phi$.

To summarize, the two-dimensional density profile, $f(\mathbf{r})$, can be be obtained from the absorption signal, $a(\xi,\phi)$, by first computing its one-dimensional Fourier transform $\tilde{a}(k,\phi)$, and from there performing a double inverse transform to obtain $f(\mathbf{r})$. In spite of various oversimplifications, this defines the algorithm underlying tomographic image analysis.

C6.4 Case study: frequency comb for high-precision measurements

REMARK The solutions to the questions formulated below may be found in Section S.C6.4.

frequency comb The most precise method we have today of measuring optical frequencies employs an optical **frequency comb**. John L. Hall (University of Colorado, USA) and Theodor W. Hänsch (Ludwig–Maximilians–Universität Munich) shared the 2005 Nobel prize in physics for the development of this technology.

An optical cavity is used to generate a precisely periodic series of light pulses, whose Fourier spectrum has the form of a frequency comb. This comb has more than a million "teeth"; they span more than an octave and their frequencies are known to a precision of better than one part in 10^{15}. To measure the frequency of a beam of light in the optical spectrum, it is superimposed with the frequency comb. Whichever tooth is closest in frequency to the frequency to be measured will combine with the latter to form a "beat" in the radio frequency range, allowing the unknown frequency to be determined to better than one part in 10^{15}.

In the following, we describe the mathematical principles underlying the frequency comb technique. The main idea is simple: the Fourier transform of a periodic signal yields a periodic δ-function – the frequency comb. We are interested in finding the positions, heights and widths of the comb's "teeth", and how these change when the input signal is not perfectly periodic, as is bound to be the case in the lab.

Let $p(t)$ be a periodic function with period τ. It has a Fourier series representation, $p(t) = \frac{1}{\tau} \sum_{m \in \mathbb{Z}} e^{-i\omega_m t} \tilde{p}_m$, whose discrete Fourier frequencies, $\omega_m = m\omega_r$, are multiples of $\omega_r = 2\pi/\tau$. The same function also has a Fourier integral representation, $p(t) = \int_{-\infty}^{\infty} \frac{d\omega}{2\pi} e^{-i\omega t} \tilde{p}(\omega)$.

Q1: Show that the Fourier transform $\tilde{p}(\omega)$ is given by a sum of discrete δ functions – the "frequency comb" – whose teeth coincide with the discrete Fourier frequencies ω_m, weighted with the discrete Fourier coefficients $\tilde{p}_m$:

$$\tilde{p}(\omega) = \omega_r \sum_m \tilde{p}_m \, \delta(\omega - \omega_m). \tag{C162}$$

We henceforth consider a periodic function of the form $p(t) \equiv \sum_{n \in \mathbb{Z}} f(t - n\tau)$, consisting of a series of shifted versions of the same "seed function", $f(t)$. For example, if the seed function describes a single peak, the periodic function describes a train of such peaks. Let $f(t) = \int_{-\infty}^{\infty} \frac{d\omega}{2\pi} e^{-i\omega t} \tilde{f}(\omega)$ denote the Fourier integral representation of the seed function.

Q2: Show that the coefficients $\tilde{p}_m$ of the Fourier series representation of $p(t)$ are determined by the Fourier transform of the seed function via $\tilde{p}_m = \tilde{f}(\omega_m)$. *Hint:* Insert the Fourier integral representation of the seed function $f(t)$ into the definition of $p(t)$. Bring the resulting expression for $p(t)$ into the form of a Fourier series by using the **Poisson summation formula**,

Poisson summation formula

$$\sum_n \tilde{F}(2\pi n) = \sum_m F(m), \tag{C163}$$

which for an arbitrary Fourier-transformable function F relates infinite sums over function values of the Fourier transform pair, F and $\tilde{F}$ (see problems C6.3.7-8). Read off the Fourier coefficients $\tilde{p}_m$ from the Fourier series thus obtained for $p(t)$.

To be concrete, let us now consider a seed function having the form of a Gaussian peak, $f_G(t) = \frac{1}{\sqrt{2\pi T^2}} e^{-t^2/(2T^2)}$. Let $p_G(t)$ denote a periodic train of such peaks. Moreover, we take the Gaussian peak width, T, to be much smaller than the period of the periodic train, $T \ll \tau$.

Q3: According to Eq. (C162), the Fourier transform, $\tilde{p}_G(\omega)$, of the Gaussian train constitutes a frequency comb. Find a formula for $\tilde{p}_G(\omega)$ and identify the positions and weights of the comb's peaks. Show that the peak weights are governed by a Gaussian envelope whose width is inversely proportional to that of the Gaussian seed function – Fourier reciprocity!

An optical frequency comb generator emits a beam of light whose electric field (which we present in a simplified manner) has the form $E(t) = e^{-i\omega_c t} p(t)$. Here $e^{-i\omega_c t}$ represents the oscillation of a "carrier signal", with carrier frequency ω_c. It is modulated by a pulse train $p(t)$, a precisely periodic series of short pulses (with pulse duration/period $= T/\tau \ll 10^{-6}$). The pulse train and the carrier signal are typically "incommensurate": the carrier frequency ω_c typically lies *between* two teeth (rather than *on* a tooth) of the frequency comb generated by the pulse train. It thus has the generic form $\omega_c = N\omega_r + \omega_{\text{off}}$, with an "offset frequency" $\omega_{\text{off}} \in (0, \omega_r)$. Due to this incommensurability, the combined signal $E(t)$ is not periodic: the phase of the carrier signal relative to the pulse train shifts or "slips" from one pulse to the next by an amount $\Delta\phi = \omega_{\text{off}}\tau < 2\pi$.

offset frequency

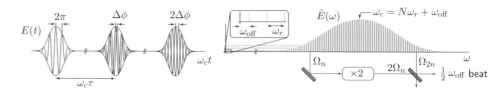

Q4: Show that the Fourier spectrum of $E(t)$ forms a frequency comb, whose "center" has been shifted from 0 to ω_c, and whose teeth are shifted relative to the Fourier frequencies ω_n by the offset frequency ω_{off}:

$$\tilde{E}(\omega) = \omega_r \sum_n \tilde{f}(\omega_{n-N})\delta(\omega - \omega_n - \omega_{\mathrm{off}}). \tag{C164}$$

Remark: Precise frequency measurements using a frequency comb require accurate knowledge of the teeth positions, $\Omega_n = n\omega_r + \omega_{\mathrm{off}}$, and hence also of ω_r and ω_{off}. The frequency $\omega_r = 2\pi/\tau$ is typically very stable, but ω_{off} undergoes slow, irregular fluctuations as a function of time. Trying to measure frequencies with such a comb would be like trying to measure distances with a slightly shaking ruler. The key insight, due to Hänsch, that made it possible to "control this shaking" is that ω_{off} can be measured accurately if the teeth span at least a full octave. In this case, a tooth near the lower end of the comb, with frequency Ω_n, has the property that twice its frequency again lies within the range of the comb. By **stabilizing the offset frequency** doubling Ω_n (a standard procedure in optics) and superimposing the resulting signal with the comb, one thus obtains beats between $2\Omega_n$ and a tooth of frequency Ω_{2n}. The beat frequency is $\frac{1}{2}(2\Omega_n - \Omega_{2n}) = \frac{1}{2}[2(n\omega_r + \omega_{\mathrm{off}}) - (2n\omega_r + \omega_{\mathrm{off}})] = \frac{1}{2}\omega_{\mathrm{off}}$. By observing the beat signal one may thus monitor ω_{off} and stabilize its value via a feedback loop. This ultimately allows a frequency comb to be stabilized to a precision of one part in 10^{15}.

So far, we have assumed the pulse train $p(t)$ to be strictly periodic. Deviations from perfect periodicity cause the teeth of the frequency comb to be **broadened of comb's teeth** broadened. To illustrate this, let us consider a "truncated" pulse train existing only for a finite amount of time, of the form $p_\gamma(t) = \sum_{n\in\mathbb{Z}} f(t - n\tau)e^{-|n|\tau\gamma}$, with $\tau\gamma \ll 1$. The factor $e^{-|n|\tau\gamma}$ suppresses the contributions for which $|n| \gtrsim 1/(\tau\gamma) \gg 1$, thereby "truncating" the series.

Q5: Compute the Fourier transform $\tilde{p}_\gamma(\omega)$ of the truncated pulse train. What shape do the individual teeth have? Show that their width is inversely proportional to the "duration" of the truncated pulse chain – Fourier reciprocity again! *Hint:* To compute $\tilde{p}_\gamma(\omega) = \int_{-\infty}^{\infty} dt\, e^{i\omega t} p_\gamma(t)$, express $p_\gamma(t)$ in terms of the Fourier representation of $f(t)$. Use the substitution $t' = t - n\tau$ in the time integral to arrive at an expression of the form $\tilde{p}_\gamma(\omega) = S(\omega)\tilde{f}(\omega)$, where $S(\omega)$ is given by a geometric series. Evaluate this series and analyze the shape of the peak at $\omega \simeq m\omega_r$ in the limit $\gamma\tau \ll 1$.

To summarize, we have encountered the following general relationships, which, remarkably, all come into play in the frequency comb measuring technique. (1) A periodic function $p(t)$ has a discrete Fourier *series* representation, with discrete Fourier frequencies ω_n. Therefore its Fourier *integral* representation, $\tilde{p}(\omega)$, must consist of a series of δ-functions at these discrete frequencies – forming a frequency comb. (2) For a periodic function of the form $p(t) = \sum_n f(t - n\tau)$, where $f(t)$ is some seed function, the envelope of the frequency comb corresponds to the Fourier transform of the seed function, $\tilde{p}_m = \tilde{f}(\omega_m)$. (3) Fourier reciprocity applies: if the seed function $f(t)$ describes a peak, then the narrower this peak, the broader the peak described by its Fourier transform, $\tilde{f}(\omega_m)$, and hence the broader the envelope of the frequency comb. (4) When the periodic function $p(t)$ is multiplied by a periodic carrier signal whose frequency is incommensurate with that of the comb, then the comb is shifted by an offset frequency. (5) If $p(t)$ is truncated to lie within some bounded time interval, then the teeth of the frequency comb are broadened – Fourier reciprocity again.

C6.5 Summary and outlook

In this chapter we introduced the powerful machinery of Fourier transformation. We emphasized a view in which it is understood as a change of basis in function space. In this interpretation, the δ-function frequently featuring in Fourier analysis assumes the role of a "standard basis vector"; $f(x)$ are the "components" of a function, f, in this basis; Fourier transformation represents f as a sum over different basis vectors, the Fourier modes $\exp(ikx)$, and $\tilde{f}(k)$ are the corresponding coefficients. Most formulas of Fourier calculus afford a straightforward linear algebraic interpretation reflecting this view (see Section L9.1).

We saw that a change to the Fourier basis can be useful for various reasons: Fourier representations can simplify mathematical operations which are complicated in the original operation. (This parallels the situation in linear algebra, where a map may look simple or complicated depending on the chosen basis.) They can be applied to encode the information on functions in a relatively small set of Fourier coefficient data, to analyze data, or to modify it. The applications mentioned as examples above illustrated how several of these aspects may be combined into powerful computational tools. For example, in noise reduction one first analyzes a signal, then compresses its Fourier data, to finally synthesize a Fourier inverse transformed function which contains the original signal but not the excessive noise data. In the next chapter we will show how a different type of "Fourier engineering" can be used to solve differential equations by transforming them to a Fourier basis.

Differential equations

A differential equation (DEQ) is an equation containing both a function and its derivatives. A solution is a function for which the conditions defined by the DEQ are satisfied. This means that if the solution and its derivatives are substituted into the DEQ, an equality results. For example,

$$\frac{df(x)}{dx} = cf(x), \qquad c \in \mathbb{R} \tag{C165}$$

is a differential equation – an equation involving both f and f'. It is solved by all functions of the form $f(x) = \mu \exp(cx), \mu \in \mathbb{R}$. Each value of μ defines a different solution. We observe that the solution of a differential equation need not be unique. The set of **general** all solutions of a DEQ is called its **general solution**. Additional conditions need to be **solution** imposed to fix a unique specific solution. For example, one might require that the solution of Eq. (C165) obey a **boundary or initial condition** such as $f(0) = 1$. This would fix a **particular** **particular solution** with $\mu = 1$. **solution**

Before turning to a more substantial discussion of the mathematics, let us indicate why *differential equations are important to physics*. Physics aims to make quantitative predictions for observable phenomena. For example, an objective of celestial mechanics is the prediction of the positions of planets at specified times. Such predictions are obtained on the basis of fundamental physical laws which generally are formulated in differential form: they state how a physical quantity X changes if a physical quantity Y acts over a small span of time or space. For example, Newton's second law, $md_t\mathbf{v} = \mathbf{F}$, can be written as $\mathbf{v}(t + \delta) - \mathbf{v}(t) \simeq \delta\,\mathbf{F}(t)/m$. In this form it states how the velocity, $\mathbf{v}$, of a body of mass m changes if a force, $\mathbf{F}$, is applied over a small time δ. Similar equations encode the laws of electrodynamics, quantum mechanics, relativity and other fields. Predictions about physical processes extending over finite intervals of time are obtained by solving the differential equations, where the uniqueness of the solution requires the specification of boundary data. For example, Newton's equations for the motion of a planet have a unique solution specified by the planet's position and velocity at a given initial time. This illustrates the tight connection between making physical predictions and the solution of differential equations, summarized in general terms in Fig. C22.

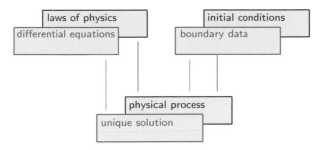

Fig. C22   The role of differential equations in physics. Discussion, see text.

C7.1 Typology of differential equations

There exist many *different types of differential equations*: a DEQ can involve the first derivative of a function, f', or higher-order derivatives, $f^{(n)}$; it may be an equation for a one-dimensional function, $f(x)$, or for a higher-dimensional function, $f(\mathbf{x})$; it may be an equation for more than one function, $f_i, i = 1, \ldots$; it may depend linearly on the function (as in Eq. (C165)), or it may depend on f in complicated ways, etc. Different types of differential equations call for distinct solution strategies and in many cases the solutions are not known in closed form. For all these reasons, the theory of differential equations is diverse and difficult to survey. It is therefore all the more important to know the most important criteria for distinguishing between different types of differential equations and their mathematical complexity.[39]

ordinary vs. partial differential equation   ▷ There exist two major families, ordinary and partial DEQs. **Ordinary differential equations** (ODEs) contain derivatives, d_x, in only one variable x. **Partial differential equations** (PDEs) involve several variables, $x^1, x^2, \ldots$ and their derivatives, $\partial_{x^1}, \partial_{x^2}, \ldots$

For example,

$$d_x f(x) = -\Gamma f(x), \qquad\qquad\qquad \text{ordinary,}$$

$$(\partial_x^2 - \partial_t^2) f(x, t) = 0, \qquad\qquad\qquad \text{partial,}$$

are an ordinary and a partial differential equation describing the decay of a quantity and the propagation of a wave front, respectively.

nth-order differential equation   ▷ A differential equation of *nth **order**** contains derivatives of nth and lower order. The majority of differential equations relevant to physics are of order 2 or less.

For example,

$$(d_t^2 + \omega^2) x(t) = 0, \qquad\qquad\qquad \text{2nd order,} \qquad\qquad \text{(C166)}$$

is a second-order equation describing the motion of a pendulum.

▷ A *system of differential equations* is a set of $n > 1$ coupled differential equations.

---

[39] In all equations the differentiated function is unknown, and other functions occurring in the DEQ are part of its definition and given.

For example, the system of $n = 2$ equations

$$d_t x(t) = v(t),$$
$$m d_t v(t) = f(x(t)), \tag{C167}$$

describes the motion of a particle with coordinate x and velocity v under the influence of a force f through Newton's laws.

▷ A **linear differential equation** contains the function in question only to linear order. **Nonlinear differential equations** depend on the solution in more complicated ways. For example,

$$\partial_x^2 \phi(x) = -\rho(x) \qquad \text{linear, ordinary, 2nd order}$$
$$d_t^2 x(t) = -\omega^2 \sin x(t) \qquad \text{nonlinear, ordinary, 2nd order,}$$

are the linear Poisson equation for the potential created by a static charge distribution, and the nonlinear equation of a mathematical pendulum, respectively.

In the following, we discuss the above classes separately and introduce different types of solution strategies. The focus will be on ordinary DEQs, which are much easier to solve than partial DEQs. A few comments on the latter are included at the end of the chapter.

C7.2 Separable differential equations

We begin our discussion with a specific class of differential equations known as separable differential equations. Separable equations afford straightforward solutions and often appear as building blocks in more complicated structures. It therefore makes sense to discuss them first.

separable differential equation   A **separable differential equation** has the form

$$d_t f(t) = h(f(t)) \, g(t), \tag{C168}$$

where g and h are given functions. The equation is "separable" in that the right-hand side splits into factors containing the unknown function, f, and the variable, t, respectively, as arguments. This structure motivates a solution strategy called **separation of variables**: rearrange terms such that all f and t dependence appears on separate sides of the equation,

separation of variables

$$\frac{df}{dt} \frac{1}{h(f(t))} = g(t).$$

Knowing that $d_t f$ describes small increments of the function f under variations of the variable t, we may sum (integrate) over increments to describe the change of f at large:

$$\int dt \frac{df}{dt} \frac{1}{h(f(t))} = \int dt \, g(t). \tag{C169}$$

The structure of the left-hand side suggests temporarily interpreting $f(t)$ as an independent "integration variable" (hence the denotation separation of variables), and to apply the substitution rule (C28)[40] to pass to an integration over f,

$$\int df \frac{1}{h(f)} = \int dt\, g(t). \tag{C170}$$

quadrature At this stage, we have reduced the solution of the equation to the computation of integrals. (In the parlance of the field, this is known as a solution up to **quadrature**, where "quadrature" is historical terminology for integration.) Doing the integrals, we obtain some function, $X(f)$, on the left and some function, $G(t)$, on the right. In a final step, one needs to solve the algebraic equation $X(f) = G(t)$ for f to obtain solutions $f(t)$ as functions of t.

EXAMPLE As an example, consider

$$d_t f(t) = \big(f(t)\big)^2 e^t. \tag{C171}$$

Here, $h(f) = f^2$ and $g(t) = e^t$. We transform the equation to

$$\int \frac{df}{f^2} = \int dt\, e^t,$$

which leads to the algebraic equation $-f^{-1} = e^t + c$. The presence of an integration constant (why is the equation determined up to one and not two independent constants?) reflects that the general differential equation has a family of solutions, depending on one free parameter. Solving for f yields $f(t) = -\frac{1}{e^t+c}$. These functions can be differentiated in t to check that they solve Eq. (C171).

Above, we considered the general differential equation and the example illustrated how this yields a family of general solutions containing an undetermined parameter, c. This constant can be fixed by requiring one additional condition, such as $f(t_0) = f_0$, or $f'(t_0) = C$. In cases where t is a time-like variable, this is called an **initial condition**; if the independent variable is space-like, the denotation **boundary condition** is alternatively used. In either case, one algebraic equation is required to uniquely fix the solution to the differential equation.

uniqueness via initial condition **INFO** Referring for a more detailed discussion to the info section on p. 305, we note that the reason why *an initial condition uniquely specifies a solution* is best understood by approximating the derivative in Eq. (C168) as a difference quotient, $d_t f \simeq \delta^{-1}(f(t+\delta) - f(t))$, and reorganizing the equation as $f(t+\delta) = f(t) + \delta h(f(t))g(t)$. Applied to the time argument $t = t_0$, $f(t_0 + \delta) = f(t_0) + \delta h(f_0)g(t_0)$, this shows how the function at a slightly later time step, $f(t_0 + \delta)$, is obtained from the known initial data, $f(t_0)$ and $g(t_0)$. The procedure may now be iterated to obtain $f(t_0 + 2\delta)$ from $f(t_0 + \delta)$, $g(t_0 + \delta)$, etc. In the limit of infinitely many discretization steps, the actual function $f(t)$ is obtained. This is the solution strategy often applied in the numerical solution of differential equations. It works, provided the functions appearing on the right-hand side of the differential equation do not vary too violently (see info section on p. 305 below). Finally, notice that the argument did not rely on the separability of the equation. Appropriately modified versions of it can be applied to demonstrate the uniqueness of the solution of more general ordinary and partial differential equations.

---

[40] We assume here positive $d_t f = h(f) > 0$. For negative h a sign change must be applied. Note that regions of positive and negative h are separated by a zero, $h = 0$, where division by h is not permitted. In such cases, the different sectors need to be treated separately.

In the concrete case of the separable equation (C169), the specification of initial data, $f(t_0) = f_0$, means that the integral appearing in the solution program is naturally written as a definite integral over a variable, s, ranging from the initial argument, t_0, to t,

$$\int_{t_0}^{t} ds \frac{df(s)}{ds} \frac{1}{h(f(s))} = \int_{t_0}^{t} ds\, g(s).$$

We again pass to f as an integration variable, keeping in mind that $f(t_0) = f_0$,

$$\int_{f_0}^{f(t)} df \frac{1}{h(f)} = \int_{t_0}^{t} ds\, g(s). \tag{C172}$$

Performing the integrals, we obtain an algebraic equation for $f(t)$, this time without undetermined elements. As a final step, this equation has to be solved for f to obtain a specific function $f(t)$ satisfying the required initial condition. ($\rightarrow$ C7.2.1-6)

EXAMPLE Consider the preceding example of Eq. (C171), subject to the condition $f(0) = 3$. Doing the definite integrals in Eq. (C172) we obtain $-(f(t))^{-1} + \frac{1}{3} = e^t - 1$, or $f(t) = (\frac{4}{3} - e^t)^{-1}$. Alternatively, this result could have been obtained by requiring that the general solution found above satisfy the initial condition, thus fixing the integration constant as $c = -(e^{t_0} + f_0^{-1}) = -\frac{4}{3}$.

INFO When confronting a new equation it is often a good idea to dispense with mathematical rigor and play a little bit to develop ideas for solution strategies. In this context, a useful trick is to write differential quotients as $d_t f = \frac{df}{dt}$ and work with them as if they were ordinary fractions. Applied to the separable differential equation (C168), this would motivate rewriting it as $\frac{df}{h(f)} = g(t)dt$. This reflects the fact that small increments, df, are proportional to small increments, dt. The summation (integration) over increments then immediately yields $\int df\, h(f)^{-1} = \int dt\, g(t)$, reproducing Eq. (C170).

logistic
differential
equation

EXAMPLE Consider the **logistic differential equation**,

$$d_t f = cf(\Gamma - f), \tag{C173}$$

where c and Γ are positive constants, and an initial condition $f(0) = f_0$ is imposed. This equation is often employed to model *population growth*. The rationale is that a small population of individuals will initially grow at a constant rate. Denoting the population size by f and the growth rate by $c\Gamma$, this is described by the linear equation $d_t f = c\Gamma f$. This equation predicts exponential growth, $f(t) = f_0 \exp(c\Gamma)$. The logistic equation takes into account that at some point the host medium will reach its load capacity and the exponential growth must come to an end. The DEQ describes this by assuming that the growth rate, $d_t f$, diminishes as f approaches a limiting value Γ from below. Applying the procedure of Eq. (C172) outlined above, we compute

$$\int_{f_0}^{f(t)} \frac{df}{cf(\Gamma - f)} = \frac{1}{c\Gamma} \ln\left(\frac{f}{\Gamma - f}\right)\Big|_{f_0}^{f(t)} = \frac{1}{c\Gamma} \ln\left(\frac{f(t)}{f_0} \frac{\Gamma - f_0}{\Gamma - f(t)}\right) = \int_{0}^{t} ds = t.$$

Solution of this equation for $f(t)$ leads to

$$f(t) = \frac{\Gamma f_0}{(\Gamma - f_0)e^{-c\Gamma t} + f_0}.$$

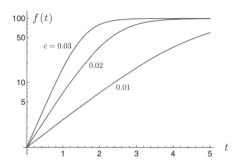

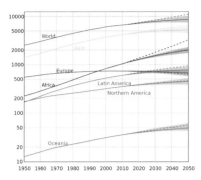

Fig. C23   Left: logistic growth for system parameters $\Gamma = 100, f_0 = 1$, and $c = 0.01, 0.02, 0.03$, where the largest value of c corresponds to the steepest growth. Notice that the vertical axis is in logarithmic units. Right: demographic predictions of world population growth.

A plot of f for three different growth rates, c, is shown in the left panel of Fig. C23. The right panel shows predictions for the population growth on various continents. Can we interpret these as logistic growth profiles?

INFO The above solution strategy relied on the specific structure of the separable differential equation. However, even if more general equations,

$$d_t f = g(f(t), t),\qquad\qquad (C174)$$

may call for different computational approaches, it is good to know that they can be solved as a matter of principle, provided the given function, $g(f, t)$, does not fluctuate too wildly. If initial data, $f(t_0) = f_0$, is provided, the solutions are unique. In the sections below, we will often refer to the *unique solutions of differential equations*. As background information, the general argument demonstrating their existence is given below.

For definiteness, consider the equation Eq. (C174) on an interval I containing the initial argument $t_0 \equiv 0$. We first impose a condition on g which will later enable us to construct a solution by a recursive algorithm: a function $g : I \to \mathbb{R}$ is called **Lipschitz continuous** if

$$\exists K \in \mathbb{R}^+ : \quad \forall x, y \in I : \quad |g(x) - g(y)| \le K|x - y|.$$

The figure illustrates this statement graphically.

If a function g is Lipschitz continuous, with *Lipschitz constant* K, then its graph, $\{(x, g(x))|x \in I\}$, always lies outside the (shaded) "cone" formed by two lines of slope K and $-K$ intersecting it at any given point $(y, g(y))$, i.e. (the modulus of) its slope is always less than K. The function g in the figure satisfies this criterion, whereas $\tilde{g}$ does not. In essence, the condition states that the derivative of g must never become infinitely large. For example, the function $\sqrt{|x|}$ is not Lipschitz continuous at $x = 0$, where its slope, $\lim_{x \searrow 0} d_x \sqrt{x} = \frac{1}{2\sqrt{x}}$, diverges.

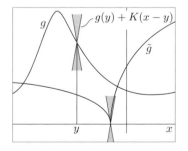

Lipschitz continuity is required for the existence and uniqueness of solutions of DEQs. The idea of the existence proof is that the differential equation (C174) can be approximated by the discrete equation (see previous info section on p. 303),

$$f(t + \delta) \simeq f(t) + \delta\, g(f(t)),$$

where for notational simplicity we suppressed the second argument of the function $g(f, t) \to g(f)$. This approximation works provided g does not change too rapidly as a function of its argument.

(Left margin: **Lipschitz continuity**)

Under this condition the readout argument on the right, $\tilde{t} \in [t, t + \delta]$, may be chosen arbitrarily and in the limit $\delta \to 0$ the discrete expression converges to the differential equation. The degree of continuity required to grant the existence of the limit is expressed by the Lipschitz condition. To see this, we define $\tilde{t} = t + \epsilon$ and write $g(f(\tilde{t})) \simeq g(f(t) + \epsilon f'(t)) \equiv g(f(t)) + X$. The Lipschitz continuity of g, with Lipschitz constant K, ensures that $|X| \leq K\epsilon |f'(t)| \leq K|f'(t)|\delta$. The error X introduced by the ambiguity in choosing $\tilde{t}$ therefore is of $\mathcal{O}(\delta)$, so we have

$$f(t + \delta) = f(t) + \delta g(f(t)) + \mathcal{O}(\delta^2).$$

We conclude that $f(t + \delta)$ may be obtained from $f(t)$ up to an error vanishing in the limit $\delta \to 0$, provided g is Lipschitz continuous. The full solution may thus be constructed iteratively, starting from $t = 0$:

$$f(0) = f_0,$$
$$f(\delta) = f_0 + \delta\, g(f_0),$$
$$f(2\delta) = f(\delta) + \delta\, g(f(\delta)) = f_0 + \delta\, g(f_0) + \delta\, g(f(\delta)),$$
$$f(3\delta) = f(2\delta) + \delta g(f(2\delta)) = f_0 + \delta\, g(f_0) + \delta\, g(f(\delta)) + \delta\, g(f(2\delta)),$$
$$\vdots \qquad \vdots$$
$$f(N\delta) = f_0 + \delta \sum_{\ell=0}^{N-1} g(f(\ell\delta)).$$

Notice that in the limit $\delta \to 0$, $N \to \infty$, taken at fixed $t = N\delta$, this expression converges to the **integral equation**[41] $f(t) = f_0 + \int_0^t ds\, g(f(s))$. Differentiation of the left- and right-hand sides makes the equivalence of the integral equation to the DEQ (C174) explicit. However, the main point of the discussion is that if g is Lipschitz continuous, a unique solution can be constructed by iteration. The mathematically precise formulation of these statements is made by the **Picard–Lindelöf theorem** discussed in mathematics courses on differential equations.

integral equation [margin note]

C7.3 Linear first-order differential equations

After this initial study, we now turn to a systematic discussion of differential equations, starting with those of least complexity. The simplest differential equations are linear in the sought function and of first order in derivatives. An equation of this type can always be represented as (why?)

$$\boxed{d_t f(t) = g(t)f(t) + h(t),} \qquad (C175)$$

where g and h are given. For unique solubility, the equation needs to be supplied with an initial condition, $f(t_0) = f_0$, where we set $t_0 = 0$ for simplicity, throughout. Equations of this form play an important role in the theory of electric circuits and in signal processing, where t is a time-like variable.

homogeneous differential equation [margin note]

We first consider the **homogeneous equation** defined by the absence of the f-independent term, $h = 0$. This equation is separable in the sense of the previous discussion and can be solved accordingly. Integration of the equation $\int_{f_0}^{f(t)} \frac{df}{f} = \int_0^t dr\, g(r)$ immediately leads to

---

[41] An *integral equation* relates a function to an integral containing the same function in its integrand.

$$f(t) = f_0 \exp(\Phi(t)), \qquad \Phi(t) \equiv \int_0^t dr\, g(r). \qquad\qquad \text{(C176)}$$

That this is a solution is easily verified by differentiating, $f'(t) = f_0 \exp(\Phi(t))\, d_t \Phi(t) = f(t)g(t)$. For $t = 0$, the integral vanishes, $\Phi(0) = 0$, and we obtain $f(0) = f_0$, as required.

We next turn to the **inhomogeneous differential equation** (C175) with non-vanishing h. The strategy for solving the general problem was introduced by Euler and is called **variation of constants**. Euler's proposal was to "vary the constant" f_0 of the homogeneous solution, Eq. (C176), by making an ansatz in which it is replaced by a function, $c(t)$,

variation of constants

$$f(t) = c(t)\, e^{\Phi(t)}. \qquad\qquad \text{(C177)}$$

The question now is how to find a function c such that the DEQ (C175) is solved. To this end, we substitute the ansatz (C177) into that equation and bring all f-dependence to the left-hand side. This leads to

$$h(t) \overset{\text{(C175)}}{=} (d_t - g(t))\left(c(t)e^{\Phi(t)}\right) = (c'(t) + c\Phi'(t) - cg(t))e^{\Phi(t)} = c'(t)e^{\Phi(t)},$$

where we used $\Phi'(t) \overset{\text{(C176)}}{=} g(t)$. Therefore, the function c is determined by the condition $c' = e^{-\Phi}h$. Integrating this equation over time, we obtain $c(t) = \int_0^t ds\, h(s)e^{-\Phi(s)} + c(0)$. The substitution of this result into (C177) yields

$$f(t) = c(0)e^{\Phi(t)} + e^{\Phi(t)} \int_0^t ds\, h(s)\, e^{-\Phi(s)}.$$

At $t = 0$ the integral vanishes, and $\Phi(0) = 0$, and so the constant $c(0) = f(0) = f_0$ is fixed by the boundary condition. This leads to the final result

$$\boxed{f(t) = f_0\, e^{\int_0^t ds\, g(s)} + \int_0^t ds\, h(s)\, e^{\int_s^t dr\, g(r)}.} \qquad\qquad \text{(C178)}$$

Verify that this function solves Eq. (C175). ($\rightarrow$ C7.3.1-2)

EXAMPLE Linear differential equations play an important role in applications. As an example, consider an RC-circuit containing a resistor of resistance R, a capacitor of capacitance C, and a time-dependent voltage source, $V(t)$, in series.

We aim to compute the time-dependent current flow, $I(t)$, through the circuit. According to Kirchhoff's voltage law, the sum of the voltage drops across each of the three circuit elements equals zero. The voltage drops by $-V(t)$ at the source, by RI at the resistor (Ohm's law) and by Q/C at the capacitor, where $Q = Q(t)$ is the time-dependent charge on the capacitor. We thus have $Q(t)/C + RI(t) = V(t)$. We also know that the rate of change of the capacitor's charge is equal to the current, $d_t Q(t) = I(t)$. The function $Q(t)$ therefore satisfies the inhomogeneous linear differential equation

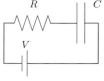

$$d_t Q + \frac{1}{\tau} Q = \frac{V(t)}{R},$$

RC-time

where the constant $\tau = RC$ defines the circuit's **RC-time**. Comparing to Eq. (C175) we have the identification $g(t) = -1/\tau$ and $h(t) = V(t)/R$. The integration over the constant function g yields $\Phi(t) = \int_0^t ds\, g = -t/\tau$ and so the general solution (C178) assumes the form

$$Q(t) = Q_0 e^{-t/\tau} + \frac{1}{R} \int_0^t ds\, V(s)\, e^{-(t-s)/\tau}.$$

Differentiation in t yields

$$I(t) = -\frac{Q_0}{\tau} e^{-t/\tau} - \frac{1}{R\tau} \int_0^t ds\, V(s)\, e^{-(t-s)/\tau} + \frac{V(t)}{R},$$

where the first two terms come from the differentiation of the exponential functions and the third from the differentiation in the upper limit of the integration boundary. The exponential factors in the solution show that the RC-time sets the rate at which the circuit responds to changes in the external voltage. Also note that for constant voltage, $V = $ const., the current vanishes for time scales $t \gg \tau$ (check this!). The physical reason is that the presence of the capacitor forbids a stationary (i.e. time-independent) current flow through the circuit.

C7.4 Systems of linear first-order differential equations

A **system of linear first-order differential equations** is a set of first-order DEQs for n unknown functions, $\{f^1, \ldots, f^n\}$, linear in all f^i. In its most general form it is given by

$$d_t f^1(t) = A^1{}_1(t) f^1(t) + A^1{}_2(t) f^2(t) + \cdots + A^1{}_n(t) f^n(t) + g^1(t),$$

$$\vdots \qquad \vdots$$

$$d_t f^n(t) = A^n{}_1(t) f^1(t) + A^n{}_2(t) f^2(t) + \cdots + A^n{}_n(t) f^n(t) + g^n(t),$$

where the coefficients $A^i{}_j(t)$ and $g^i(t)$ may be functions of t. Defining a matrix $A = \{A^i{}_j\}$ and combining the functions f^i into a vector, $\mathbf{f} = (f^1, \ldots, f^n)^{\mathrm{T}}$, and likewise, $\mathbf{g} = (g^1, \ldots, g^n)^{\mathrm{T}}$, this assumes the compact form

$$\boxed{d_t \mathbf{f}(t) = A(t)\mathbf{f}(t) + \mathbf{g}(t).} \tag{C179}$$

A unique solution of this equation is determined if n initial conditions, $f^i(0) = f_0^i$, or $\mathbf{f}(0) = \mathbf{f}_0$, are provided. Equivalently, one may say that the general solution of the DEQ contains n undetermined constants. This statement follows from a straightforward adaption of the argument given in the info section of p. 303: representing the derivative as an approximate difference quotient,

$$\mathbf{f}(t + \delta) = \mathbf{f}(t) + \delta \left(A(t)\mathbf{f}(t) + \mathbf{g}(t) \right),$$

we note that the solution function $\mathbf{f}(t + \delta)$ at an incremented argument is obtained from $\mathbf{f}(t)$ and the known data $A(t)$, $\mathbf{g}(t)$. Starting from $\mathbf{f}(t_0)$, a solution may thus be constructed by iteration.

Linear differential equations are much easier to solve than nonlinear ones. Much of this simplicity is owed to a **superposition principle**: if $\mathbf{f}_1$ and $\mathbf{f}_2$ solve the equation with inhomogeneity $\mathbf{g}_1$ and $\mathbf{g}_2$, respectively, then $\mathbf{f}_1 + \mathbf{f}_2$ is a solution for the inhomogeneity $\mathbf{g}_1 + \mathbf{g}_2$. (Check!) This feature is at the root of many physical phenomena in quantum mechanics,

superposition principle

electrodynamics and other disciplines described by linear differential equations. Mathematically, it is key to the construction of solutions with complicated inhomogeneities from those with simpler ones. This strategy will be discussed in Section C7.5.

In spite of the great simplification provided by the superposition principle, general linear differential equations can be hard to solve. The problem becomes easier in cases where **system with constant coefficients** $A(t) = A$ is constant in time. This defines the **system of first-order linear differential equations with constant coefficients**,

$$d_t\mathbf{f}(t) = A\mathbf{f}(t) + \mathbf{g}(t). \tag{C180}$$

The problem becomes simpler still if $\mathbf{g} = 0$ and the DEQ becomes a *homogeneous system*,

$$\boxed{d_t\mathbf{f}(t) = A\mathbf{f}(t).} \tag{C181}$$

We next discuss how the system (C181) can be solved in closed form.

EXERCISE Consider the case $n = 1$, i.e. $d_t f = af + g$. Solve this DEQ with the initial condition $f(0) = f_0$.

Solution of the homogeneous linear equation Eq. (C181)

The appearance of a matrix, A, in the system (C181) suggests considering its eigenvectors, $\mathbf{v}_j$ and eigenvalues, λ_j. At the same time, the structure of the equation, $d_t \times$ (solution) $=$ (constant) $\times$ (solution), indicates an exponential solution. Combining these thoughts, one may try an ansatz (no summation over j)

$$\mathbf{f}(t) = \mathbf{v}_j e^{\lambda_j t}. \tag{C182}$$

That this solves the equation with initial condition $\mathbf{f}(0) = \mathbf{v}_j$ is readily checked by substitution, $d_t\mathbf{f} = d_t\mathbf{v}_j e^{\lambda_j t} = \lambda_j\mathbf{v}_j e^{\lambda_j t} = A\mathbf{v}_j e^{\lambda_j t} = A\mathbf{f}$. Assuming that A is diagonalizable, i.e. that the set $\{\mathbf{v}_j\}$ defines a basis of $\mathbb{C}^n$, we may generalize this result to a full solution of the problem.

1. Expand an arbitrary initial condition in terms of the eigenvectors,

$$\mathbf{f}_0 \equiv \sum_j \mathbf{v}_j c^j, \tag{C183}$$

2. to obtain the solution (check!)

$$\boxed{\mathbf{f}(t) = \sum_j \mathbf{v}_j c^j e^{\lambda_j t}.} \tag{C184}$$

Observe how the structure of the solution relies on the superposition principle: linear combinations of individual solutions of the form (C182) are again solutions. This principle is used here to construct a solution with arbitrary initial condition by linear combination. ($\rightarrow$ C7.4.1-2)

INFO (Requires Section L7.4 on functions of matrices.) Systems of linear equations play an important role in physics, for example in classical (see the next section) and quantum mechanics. In these fields it is often preferable to work with *"invariant" solutions of the system* (C181), i.e. solutions not making explicit reference to the eigenvectors of the matrix A. The structure of the equation suggests that there might be a solution of the form $\exp(At)$. Indeed, we may apply the methods of Section L7.4 to define

$$\mathbf{f}(t) = e^{At}\,\mathbf{f}(0), \tag{C185}$$

where the exponential $\exp(At) \equiv \sum_m \frac{1}{m!}(At)^m$ acts as a matrix on the vector of initial conditions, $\mathbf{f}(0)$. That this is a solution can be checked by direct computation:

$$d_t\mathbf{f} = d_t e^{At}\mathbf{f}(0) = d_t \sum_{m=0}^{\infty} \frac{t^m}{m!} A^m \mathbf{f}(0)$$

$$= \sum_{m=1}^{\infty} \frac{t^{m-1}}{(m-1)!} A^{m-1+1}\mathbf{f}(0) = A \sum_{m'=0}^{\infty} \frac{t^{m'}}{m'!} A^{m'}\mathbf{f}(0) = A\,e^{At}\mathbf{f}(0) = A\mathbf{f}(t),$$

where in the fourth equality we renamed the summation index as $m' = m - 1$. Equation (C185) does not assume the matrix A to be diagonalizable. However, if it is, $A = TDT^{-1}$, we may use (L197) to reformulate the result as

$$\mathbf{f}(t) = (T\,e^{Dt}\,T^{-1})\mathbf{f}(0) = T\,e^{Dt}(T^{-1}\mathbf{f}(0)). \tag{C186}$$

This equation establishes the relationship with the previous solution (C184). To see this, recall (see Section L7.1) that $T = (\mathbf{v}_1, \dots, \mathbf{v}_n)$ contains the eigenvectors of A as columns, $T\mathbf{e}_j = \mathbf{v}_j$. We may then write the expansion (C183) of the initial condition in eigenvectors as $\mathbf{f}(0) = \sum_j (T\mathbf{e}_j)c^j$, which shows that $T^{-1}\mathbf{f}(0) = \sum_j \mathbf{e}_j c^j \equiv \mathbf{c}$. Substitution of this representation into Eq. (C186) reproduces (C184),

$$\mathbf{f}(t) = T e^{Dt} \sum_j \mathbf{e}_j c^j = T \sum_j \mathbf{e}_j c^j\, e^{\lambda_j t} = \sum_j \mathbf{v}_j c^j\, e^{\lambda_j t},$$

where in the second equality we used that the diagonal matrix, D, acts as $D\mathbf{e}_j = \lambda_j\,\mathbf{e}_j$.

Behavior of the homogeneous solution

The behavior of the solution (C184) crucially depends on the eigenvalues λ_j. In the exceptional case that all eigenvalues are imaginary, $\lambda_j \in i\mathbb{R}$, the factors $\exp(\lambda_j t) = \exp(\pm i|\lambda_j|t)$ are purely oscillatory, with frequencies $2\pi/|\lambda_j|$, and the vector $\mathbf{f}(t)$ is a superposition of n such contributions. More generally, consider the case of complex eigenvalues, and denote the eigenvalue with the largest real part by λ_1. This eigenvalue dominates the solution at large times, because $|\exp(\lambda_1 t)| = \exp((\operatorname{Re}\lambda_1)t)$ is exponentially larger than all other $|\exp(\lambda_j t)| = \exp((\operatorname{Re}\lambda_j)t)$, $j > 1$. (This is true also if $\operatorname{Re}\lambda_1$ is negative or zero.) For sufficiently large times we may thus approximate the solution by

$$\mathbf{f}(t) \sim c\,e^{\lambda t}\mathbf{v},$$

where the subscript on λ_1 has been omitted for brevity. We now have to distinguish between several different types of long-time solutions.

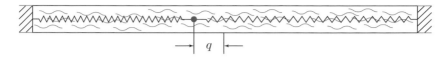

Fig. C24   A massive particle acted upon by two springs, all in a tank filled by some viscous liquid.

> ▷ For $\lambda > 0$ *real and positive*, the solution grows exponentially.

> ▷ For $\lambda < 0$ *real and negative*, it shrinks to zero.

> ▷ For $\lambda = i\omega$ *purely imaginary*, the solution is oscillatory in time, with frequency ω.

> ▷ For $\lambda = c + i\omega$ *complex with positive real part, $c > 0$*, the norm of the solution vector grows exponentially and the vector itself performs oscillatory motion.

> ▷ For $\lambda = c + i\omega$ *complex with negative real part, $c < 0$*, the vector oscillates and shrinks.

In the next section we illustrate these different types of solutions with a concrete example.

Application: damped oscillator

Phenomena described by linear differential equations include oscillations, damping by frictional motion, or instabilities of mechanical systems. As an example, we consider *a particle of mass m attached to two springs*, see Fig. C24. At the particle coordinate $q = 0$ the combined force $F(q)$ exerted by the springs vanishes, $F(0) = 0$. We aim to describe what happens at small deviations away from $q = 0$. The force acting to restore equilibrium then is approximately linear in q, $F_{\text{springs}} = -m\Omega^2 q$, where Ω is a constant. To make the
friction force   problem more realistic we assume the presence of a **friction force**, $F_{\text{friction}} = -2m\gamma\,d_t q$, where $\gamma > 0$ is a friction constant and the proportionality of the force to minus the velocity means that friction slows the motion.[42] The differential equations describing the motion of the particle then read

$$d_t q = v,$$
$$d_t v = -2\gamma v - \Omega^2 q, \tag{C187}$$

where the first equation defines the velocity in terms of the particle coordinate and the second is Newton's law, $F_{\text{total}} = m d_t v$. These equations define the dynamics of the **damped**
harmonic   **harmonic oscillator**. They need to be solved with an initial condition, $(q(0), p(0))^{\text{T}} =$
oscillator $(q_0, v_0)^{\text{T}}$. In matrix notation the problem takes the form

$$d_t \begin{pmatrix} q \\ v \end{pmatrix} = \begin{pmatrix} 0 & 1 \\ -\Omega^2 & -2\gamma \end{pmatrix} \begin{pmatrix} q \\ v \end{pmatrix}, \tag{C188}$$

where the 2×2 matrix now assumes the role of the matrix A of the previous section. Its eigenvalues, obtained by solving a quadratic equation, are

$$\lambda_\pm = -\gamma \pm i\sqrt{\Omega^2 - \gamma^2}, \tag{C189}$$

---

[42]   This force is caused by molecules of the liquid colliding with the particle, thus impeding its motion.

and the corresponding eigenvectors are given by $\mathbf{v}_\pm = (1, \lambda_\pm)^{\mathrm{T}}$. Recalling Eq. (C184), the *general solution of the homogeneous oscillator equation* then assumes the form

$$\mathbf{y}(t) \equiv \begin{pmatrix} q(t) \\ v(t) \end{pmatrix} = c_+ \begin{pmatrix} 1 \\ \lambda_+ \end{pmatrix} e^{\lambda_+ t} + c_- \begin{pmatrix} 1 \\ \lambda_- \end{pmatrix} e^{\lambda_- t}, \tag{C190}$$

where $c_\pm$ are constants determined by the initial conditions. For definiteness, let us consider a situation where at time $t = 0$ the particle is released at coordinate $q(0) = q_0$ and zero velocity, $v(0) = 0$. We then have $\mathbf{y}_0 = (q_0, 0)^{\mathrm{T}}$ and the expansion of this starting vector in the eigenvectors of the problem yields $\mathbf{y}_0 = \frac{q_0}{\lambda_- - \lambda_+}(\lambda_- \mathbf{v}_+ - \lambda_+ \mathbf{v}_-)$. Substitution of this representation into the general solution yields the specific solution

$$\mathbf{y}(t) = \frac{q_0}{\lambda_- - \lambda_+} \left(\lambda_- \mathbf{v}_+ e^{\lambda_+ t} - \lambda_+ \mathbf{v}_- e^{\lambda_- t} \right). \tag{C191}$$

Let us discuss the behavior of this solution in a number of physically distinct cases.

▷ In the *frictionless case*, $\gamma = 0$, the eigenvalues $\lambda_\pm = \pm i\Omega$ are purely imaginary, and

$$\mathbf{y}(t) \stackrel{(\mathrm{C191})}{=} q_0 \begin{pmatrix} \cos(\Omega t) \\ -\Omega \sin(\Omega t) \end{pmatrix},$$

i.e. the particle performs *oscillatory motion* at a frequency set by Ω. The trajectory $\mathbf{y}(t)$ is shown in the bottom left panel of Fig. C25, and the corresponding q-coordinate by the dashed line in the upper panel.

underdamped harmonic oscillator ▷ In the **underdamped regime** of weak to intermediate friction, $\gamma < \Omega$, the eigenvalues are complex with a negative real part, $\lambda_\pm = -\gamma \pm i\Omega_r$, with $\Omega_r = \sqrt{\Omega^2 - \gamma^2}$:

$$\mathbf{y}(t) \stackrel{(\mathrm{C191})}{=} q_0 \, e^{-\gamma t} \begin{pmatrix} \cos(\Omega_r t) + \frac{\gamma}{\Omega_r} \sin(\Omega_r t) \\ -\Omega_r \left(1 + \frac{\gamma^2}{\Omega_r^2}\right) \sin(\Omega_r t) \end{pmatrix}.$$

This describes *damped oscillations* with a reduced frequency Ω_r and damping rate γ (Fig. C25, bottom middle, and solid line in upper panel).

overdamped harmonic oscillator ▷ In the **overdamped regime** of strong friction, $\gamma > \Omega$, the eigenvalues are purely real, $\lambda_\pm = -\gamma \pm \gamma_r$, with $\gamma_r = \sqrt{\gamma^2 - \Omega^2}$:

$$\mathbf{y}(t) \stackrel{(\mathrm{C191})}{=} q_0 \, e^{-\gamma t} \begin{pmatrix} \cosh(\gamma_r t) + \frac{\gamma}{\gamma_r} \sinh(\gamma_r t) \\ \gamma_r \left(1 - \frac{\gamma^2}{\gamma_r^2}\right) \sinh(\gamma_r t) \end{pmatrix}.$$

The damping is so strong that the particle no longer performs oscillatory motion. Instead, it slowly "creeps" from its point of origin back to the equilibrium position at $q = 0$ (Fig. C25, bottom right, and dotted line in upper panel). For very strong damping, $\gamma \gg \Omega$, one may Taylor-expand $\gamma_r \simeq \gamma - \Omega^2/(2\gamma)$, implying $\lambda_+ \simeq -\Omega^2/(2\gamma)$ and $\lambda_- \simeq -2\gamma$, so that $|\lambda_+| \ll |\lambda_-|$. The large negative value of λ_- causes contributions proportional to $\exp(\lambda_- t)$ to damp out rapidly, while the longer-lived terms proportional to $\exp(\lambda_+ t)$ decay as $\exp\left[-\frac{\Omega^2 t}{2\gamma}\right]$.

▷ In the special case of **critical damping**, $\lambda = \Omega$, the eigenvalues are degenerate, $\lambda_+ = \lambda_-$, and the matrix in Eq. (C188) is not diagonalizable. ($\to$ C7.4.3-4)

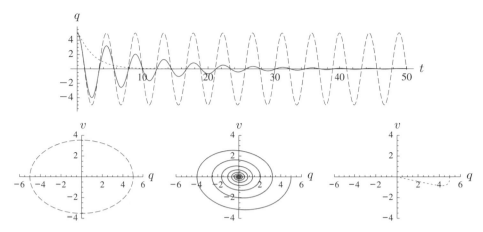

Fig. C25   Three distinct types of behaviors of a harmonic oscillator. Upper panel: coordinate $q(t)$ of the particle shown in Fig. C24 in the case of no damping (dashed), underdamping (solid), and overdamping (dotted). Bottom: the trajectories $\mathbf{y}(t) = (q(t), v(t))^{\mathsf{T}}$ of the particle.

INFO   The two equations Eq. (C187) describing the damped oscillator are equivalent to a single second-order equation, obtained by substitution of the first into the second:

$$d_t^2 q + 2\gamma\, d_t q + \Omega^2 q = 0. \tag{C192}$$

harmonic oscillator differential equation

This **harmonic oscillator equation** is commonly employed as the starting point for discussions of damped motion in physics textbooks. Above, we traded this single second-order equation for a system of two first-order DEQs. This is an example of the general fact, to be addressed in Section C7.6, that higher-order DEQs can be converted to systems of first-order DEQs, which often are more convenient to analyze. ($\rightarrow$ C7.4.5-6)

The inhomogeneous system of linear first-order DEQs with constant coefficients, Eq. (C180), can be solved by a generalized version of the variation of constants procedure discussed in Section C7.3 ($\rightarrow$ C7.4.7-8). However, instead of elaborating that approach here, we discuss inhomogeneities from a more general perspective in the next section, where a powerful solution strategy exploiting the linearity of differential equations will be introduced.

C7.5 Linear higher-order differential equations

A *general ordinary linear differential equation* can be written as

$$\hat{L}(t)f(t) = g(t),$$

$$\hat{L}(t) = h^{(0)}(t) + h^{(1)}(t)\frac{d}{dt} + h^{(2)}(t)\frac{d^2}{dt^2} + \cdots, \tag{C193}$$

where the "coefficients" $h^{(n)}(t)$ are functions, and the ellipsis represents terms containing higher-order derivatives. A few comments on this equation follow.

linear
differential
operator

▷ The formal expression $\hat{L}(t)$ is called a **linear differential operator**. It is an "operator" in the sense that

$$\hat{L}(t)f(t) = h^{(0)}(t)f(t) + h^{(1)}(t)f'(t) + h^{(2)}(t)f''(t) + \cdots \qquad \text{(C194)}$$

operates on a function to produce a new function. The operator is linear because $\hat{L}(c_1f_1 + c_2f_2) = c_1\hat{L}f_1 + c_2\hat{L}f_2$, where we omitted the argument t and $c_{1,2}$ are constants.

▷ In practice linear differential *equations of higher than second order* occur rarely. Most equations of relevance to physics are of first or second order.

▷ The unique solution of an nth-order linear differential equation requires the specification of n *initial conditions*. For example, the solution of a second-order equation can be made unique by specifying the function's value at two times, $f(t_i) = f_i$, $i = 1, 2$, or the function value at one time and that of its derivative at another, $f(t_0) = f_0, f'(t_1) = d_1$, etc.

▷ Linear differential equations play a key *role in physics*. Electrodynamics and quantum mechanics are linear theories in the sense that their fundamental laws – the Maxwell equations and the Schrödinger equation, respectively – assume the form of linear differential equations.[43] The fundamental equations of other theories can often be approximated by linear differential equations in physically relevant limits. For example, "linearized" versions of Einstein's equations of general relativity describe gravitational waves and other phenomena caused by moderately weak sources of gravitation.

Even complicated-looking linear differential equations can often be solved. The key point is the *superposition principle* following from their linearity (see the discussion of the linear first-order DEQ in Section C7.4): if f_i are solutions of Eq. (C193), with inhomogeneities g_i, the superposition with constant coefficients, $\sum_i c_i f_i$, is a solution to a similar linear equation with inhomogeneity $\sum_i c_i g_i$. (Check!) The superposition principle suggests representing complicated inhomogeneities as sums of simpler ones, and first trying to solve the equation for these. This solution strategy goes by the name "Green function method" and plays an important role in both physics and mathematics.

INFO There are important *physical consequences of the superposition principle*. As an example, consider Maxwell's equations (to be discussed in chapter V7), whose solutions are electromagnetic fields generated in response to charges and currents, assuming the role of inhomogeneities. If two such sources generate two electromagnetic fields, then the field generated by the combined sources will be the sum or *superposition* of the individual fields. This physical superposition principle is responsible for phenomena like wave interference, the formation of superimposed wave patterns resulting from the addition of individual waves.

Green function methods

The simplest inhomogeneities possible for a differential equation are δ-functions. The function $g(t) = \delta(t - u)$ vanishes everywhere except at $t = u$, and in this sense

---

[43] Both the Maxwell equations and the Schrödinger equation are partial linear equations, i.e. they contain derivatives with respect to multiple variables. However, most of our discussion specific to the linearity of DEQs carries over to these cases.

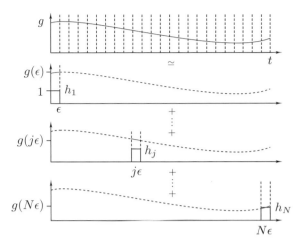

Fig. C26   A function, g, can be approximately represented as a weighted sum over a set of "box functions", h_j, having unit height, small width ϵ and support near the discrete coordinates $u_j = j\epsilon$. In the limit of a large number of discretization steps, the sum $g(t) \simeq \sum_j h_j(t)g(u_j)$ is a good approximation of the continuous function $g(t)$. For $\epsilon \to 0$ the scaled functions $\epsilon^{-1}h_j$ turn into functions $\delta(t - u_j)$, and the scaled sum $\epsilon \sum_j \to \int du$ becomes an integral. This illustrates how the formal expression (C196) can be understood as a "sum" over functions $\delta(t - u)$ with "weights" $g(u)$.

has minimal mathematical structure. The solution of a linear DEQ (C193) with a δ-inhomogeneity,

$$\hat{L}G(t, u) = \delta(t - u), \tag{C195}$$

Green function   is called a **Green function** for historical reasons. The Green function depends on the position of the singularity of the δ-inhomogeneity (i.e. for each u we have a different function) and its second argument keeps track of this dependence.

Assume we had managed to compute the Green function for all values of u. The superposition principle may then be evoked to construct a general solution to the DEQ (C193) for an arbitrary inhomogeneity g. To see this, notice that any function $g(t)$ may be expressed as a superposition of δ-functions,

> **George Green 1793–1841**
> He owned and worked a Nottingham windmill. His only schooling consisted of four terms in 1801/1802, and where he learned his mathematical skills remains a mystery. Green published only ten mathematical works, the first and most important at his own expense in 1828, "An essay on the application of mathematical analysis to the theories of electricity and magnetism." He left his mill, became an undergraduate at Cambridge in 1833 at the age of 40, then a Fellow of Gonville and Caius College in 1839.

$$g(t) \overset{(C113)}{=} \int du\, \delta(t - u)g(u). \tag{C196}$$

In this formula, the function g is represented by a "sum" (integral) over all values of u. The constant, t-independent coefficients in this sum are the values $g(u)$, and the functions entering the superposition are the δ-functions, $\delta(t - u)$.

EXERCISE Formally, Eq. (C196) is proven by doing the integral over u and using the defining property of the δ-function. However, in the present context, it is more useful to think of it in the spirit of the discrete representation shown in Fig. C26. Before reading the caption of that figure, try to understand in which sense it depicts a discrete representation of the continuum Eq. (C196).

The representation of the inhomogeneity, g, as a "sum" over δ-functions with "coefficients" $g(u)$ implies, via the superposition principle, that the solution of the original DEQ (C193), too, assumes the form of a sum,

$$f(t) = \int du\, G(t,u)\, g(u),$$ (C197)

with the same coefficients, $g(u)$. That this is a solution can be verified by direct computation: the differential operator, $\hat{L}(t)$, is linear and acts on functions of t (not u!), hence

$$\hat{L}(t)f(t) = \int du\,(\hat{L}(t)G(t,u))\,g(u) \stackrel{\text{(C195)}}{=} \int du\,\delta(t-u)g(u) = g(t).$$

As mentioned above, this solution is not unique unless *boundary conditions* are provided. Specifically, for a differential operator of nth degree, the fixation of n conditions selects a uniquely defined specific solution.[44]

solution of DEQ by Green functions

Unless boundary data is provided, any solution, f_h, of the homogeneous differential equation, $\hat{L}(t)f_h(t) = 0$, may be added to a solution, f, of the inhomogeneous equation, to obtain another solution, $f(t) + f_h(t)$: $\hat{L}(f + f_h) = \hat{L}f + \hat{L}f_h = g + 0 = g$. Conversely, if f_1 and f_2 are solutions of the inhomogeneous equation,[45] their difference is a solution of the homogeneous equation: $\hat{L}(f_1 - f_2) = g - g = 0$. This means that the *general solution of the inhomogeneous equation* can be written as $f + f_h$, where f is a fixed but arbitrary solution of the inhomogeneous equation, and f_h is the general solution of the homogeneous problem. Finally, the specification of boundary conditions selects a specific solution.

Of course, the solution (C197) remains formal as long as the Green function $G(t,u)$ is not known. In the theory of inhomogeneous linear differential equations, finding the Green function is more or less equivalent to solving the entire problem. Accordingly, sophisticated analytical tools revolving around Green functions have been developed, and their discussion is a standard topic of theoretical physics courses. In the next section, we will illustrate the computation of Green functions of certain classes of linear DEQs on a simple yet important example.

Application: driven damped oscillator

REMARK Requires familiarity with Chapter C6 on Fourier calculus.

Consider the damped oscillator problem described by Eq. (C187), but now including an external driving force, $\xi(t)$, see Fig. C27. The driving adds to the balance of forces, hence

---

[44] To understand why n conditions are required, we note (see the discussion of Section C7.6 below) that an nth-order linear differential equation can be transformed to a system of n first-order equations. The solubility of such systems in dependence on n boundary conditions follows from the remarks made in Section C7.4. We here favor the terminology "boundary" conditions over "initial" conditions because the additional data need not be given at an initial (time) argument. For example, the solution of a second-order differential equation can be fixed by specification of the value of the solution at two different time arguments, $f(t_1)$ and $f(t_2)$.

[45] Equation (C197) does not specify a unique solution because prior to the fixation of boundary conditions the Green function is not uniquely defined either. The addition of an arbitrary solution of the homogeneous equation to the Green function then defines another valid Green function.

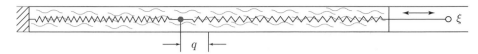

Fig. C27   An oscillating particle subject to friction and external driving. Discussion, see text.

the second of Eqs. (C187) generalizes to $d_t v = -2\gamma - \Omega^2 q + \xi$. We substitute the first equation, $d_t q = v$ into the second to obtain the second-order linear equation

$$\left(d_t^2 + 2\gamma d_t + \Omega^2\right) q(t) = \xi(t). \tag{C198}$$

Comparison with Eq. (C193) shows that the differential operator governing this equation is given by $\hat{L} = d_t^2 + 2\gamma d_t + \Omega^2$. Its coefficient functions, $h^{(2)} = 1$, $h^{(1)} = 2\gamma$, $h^{(0)} = \Omega^2$, are time-independent constants, and this facilitates the solution of the problem.

INFO Equation (C198) finds many *applications in the natural sciences.* Depending on the context, it describes mechanical, electrical, chemical or biological systems in which a quantity of interest (q) is subject to effective forces restoring equilibrium (Ω^2), friction (γ) and external influence (ξ). In the mechanical context, this situation is realized for the majority of realistic systems performing oscillatory motion. An example of an electrical system described by these equations is a resonator where the role of q is played by a time-dependent voltage, oscillatory motion is caused by the interplay of a capacitor and a coil, friction by a resistor, and the external forcing is due to an external voltage. For the discussion of this and other applications we refer to specialized courses.

Green function of harmonic oscillator

The Green function of the oscillator problem obeys the equation

$$\left(d_t^2 + 2\gamma d_t + \Omega^2\right) G(t-u) = \delta(t-u). \tag{C199}$$

Heuristically, it is a solution of the problem in the presence of a δ-function force acting at time $t = u$. Due to the absence of time-dependent coefficients we anticipate the existence of a solution, $G(t, u) = G(t - u)$, depending only on the difference between the time t at which the solution is evaluated and the time u at which the force acts (think about this point). From this function, the desired solution of Eq. (C198) is obtained as

$$q(t) \overset{(C197)}{=} \int du\, G(t-u)\,\xi(u). \tag{C200}$$

The key to the computation of the Green function lies in the *Fourier transform* identity (C149), $(\mathcal{F}d_t f)(\omega) = -i\omega(\mathcal{F}f)(\omega)$. Repeated application of this identity leads to $(\mathcal{F}d_t^m f)(\omega) = (-i\omega)^m(\mathcal{F}f)(\omega)$, i.e. under the Fourier transform derivatives d_t change to algebraic factors $-i\omega$. To make use of this feature, we substitute t by $t + u$ in equation (C199) for the Green function to rewrite it as $(d_t^2 + 2\gamma d_t + \Omega^2)G(t) = \delta(t)$. In a second step we Fourier transform both the left- and right-hand sides to obtain

$$\left(-\omega^2 - i2\gamma\omega + \Omega^2\right) \tilde{G}(\omega) = 1, \tag{C201}$$

where we noted that the Fourier transform of a δ-function equals unity, Eq. (C146). Note that Eq. (C201) is much simpler than its counterpart, Eq. (C199), since the Fourier transformation has converted the differential operator $\hat{L}$ into an algebraic factor. We divide by it to obtain the Fourier transform of the Green function,

$$\tilde{G}(\omega) = \frac{1}{-\omega^2 - i2\gamma\omega + \Omega^2}. \tag{C202}$$

This completes the most important step in solving the inhomogeneous DEQ. All that remains is the computation of integrals. Depending on the type of the driving force, however, these integrals can be nontrivial and one of three different strategies may be favorable.

▷ One can first compute the *inverse Fourier transform of the Green function* as

$$G(t) = \int \frac{d\omega}{2\pi} e^{-i\omega t} \tilde{G}(\omega). \tag{C203}$$

The solution $q(t)$ is then obtained by substituting the result for $G(t)$ into Eq. (C200). For completeness, we discuss the behavior of the function $G(t)$ in the info section below.

▷ Alternatively, one may observe that the solution of the problem has the form of a *convolution*, $q(t) \overset{(C200)}{=} \int du\, G(t-u)\xi(u) = (G * \xi)(t)$, of the Green function and the driving force. Equation (C151) then implies that $\tilde{q}(\omega) = \tilde{G}(\omega)\tilde{\xi}(\omega)$: the Fourier transform of the solution is obtained as the product of the Fourier-transformed Green function, (C202), and the Fourier transform, $\tilde{\xi}(\omega)$, of the driving force (which needs to be computed from the given $\xi(t)$). In a final step, one computes $q(t)$ from $\tilde{q}(\omega)$ by inverse Fourier transformation.

▷ For driving forces with simple time dependence it may be preferable to compute the result by *direct substitution* of the Fourier representation Eq. (C203) into the convolution integral Eq. (C200). We give an example of this strategy in the second info section below.

INFO Let us discuss the *temporal behavior of the harmonic oscillator Green function*. The inverse Fourier integral, Eq. (C203), leading from $\tilde{G}(\omega)$ to $G(t)$, actually is hard to compute, unless one employs the complex contour integration techniques introduced in Chapter C9. As shown in the example on p. 352, the result in the underdamped regime, $\gamma < \Omega$, is given by

$$G(t) = \Theta(t)\, e^{-\gamma t} \frac{1}{\Omega_r} \sin(\Omega_r t) + G_0(t), \qquad \Omega_r \equiv (\Omega^2 - \gamma^2)^{1/2}, \tag{C204}$$

where $\Theta(t)$ is the Heaviside step function (C118), and $G_0(t)$ is an arbitrary solution of the homogeneous equation. Its general form is given by the $q(t)$ component of Eq. (C190), which in the present notation reads

$$G_0(t) = e^{-\gamma t}(c_+ e^{i\Omega_r t} + c_- e^{-i\Omega_r t}).$$

The constants $c_\pm$ entering here via the homogeneous solution serve as the two undetermined constants needed for the general solution. Notice that, regardless of the choice of these constants, G_0 decays exponentially in time. A very natural choice of boundary conditions would be $G(t \to \pm\infty) = 0$, implying the absence of oscillatory motion at both negative infinity, prior to the action of the

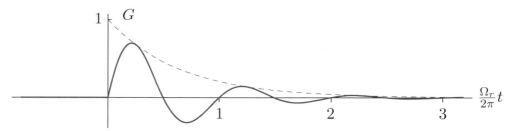

Fig. C28   Green function of the harmonic oscillator for the underdamped configuration, $\gamma = \Omega/4$.

δ-function, and positive infinity, when the damped oscillation has fully relaxed. This requires the choice $c_\pm = 0$, and consequently $G_0 = 0$. In this case, the Green function is given by the first term in Eq. (C204), depicted in Fig. C28. It affords an intuitive interpretation: prior to the action of the δ-function force at $t = u$ the oscillator is at rest, $G(t - u)\big|_{t<u} = 0$. At $t = u$ it receives a kick from the δ-force, $\delta_u(t)$, and subsequently performs oscillatory motion, attenuated by the damping rate γ. ($\rightarrow$ C7.5.1-2)

INFO As an example of the third solution strategy mentioned above, let us consider a *harmonic oscillator subject to periodic driving*, $\xi(t) = \xi_0 \cos(\omega_d t)$. We use the Euler formula, Eq. (C91), substitute Eq. (C203) into Eq. (C200) and obtain

$$q(t) = \xi_0 \int du \int \frac{d\omega}{2\pi} \frac{e^{-i\omega(t-u)} \frac{1}{2}\left[e^{-i\omega_d u} + e^{+i\omega_d u}\right]}{-\omega^2 - i2\gamma\omega + \Omega^2}$$

$$= \xi_0 \int \frac{d\omega}{2\pi} \frac{e^{-i\omega t} \frac{1}{2}\left[\delta(\omega - \omega_d) + \delta(\omega + \omega_d)\right]}{-\omega^2 - i2\gamma\omega + \Omega^2} = \mathrm{Re}\left[\frac{\xi_0\, e^{-i\omega_d t}}{-\omega_d^2 - i2\gamma\omega_d + \Omega^2}\right]$$

$$= \frac{\xi_0}{(\Omega^2 - \omega_d^2)^2 + 4\gamma^2\omega_d^2}\left[(\Omega^2 - \omega_d^2)\cos(\omega_d t) + 2\gamma\omega_d \sin(\omega_d t)\right].$$

The second equality follows from the time/frequency incarnation of Eq. (C137), which reads $\int du \exp(i(\omega \mp \omega_d)u) = 2\pi \delta(\omega \mp \omega_d)$, and the third from the subsequent ω-integration over the δ-functions.

For an in-depth discussion of this result we refer to lecture courses in mechanics. Notice, however, that the time-dependence of the forced oscillations is periodic (sin, cos) in the driving frequency, and that the amplitude becomes largest when the driving frequency equals the natural frequency of the oscillator, $\omega_d = \Omega$. This observation is at the root of all *resonance phenomena*. In the limit of small damping, $\gamma \rightarrow 0$, the response may actually diverge at the resonance frequency. Resonance phenomena may occur in the presence of even moderate driving (whose strength is governed by the prefactor ξ_0) if only the damping is weak enough. As an example, we mention the evacuation of TechnoMart, a 37-story high rise in Seoul, in 2011, which became necessary due to a resonance building up when a group of only 17 aerobic enthusiasts performed a rhythmic exercise.

Linear algebraic interpretation of the Green function

REMARK Requires Chapter L9.

Although the solution based on Green functions follows a well-motivated logic, it may look somewhat opaque to first-time readers. However, as we are going to show here, the formalism becomes rather transparent when interpreted from the perspective of *linear algebra*. To this end, let us consider the functions f and g in (C193) as infinite-dimensional limits of finite-dimensional vectors, $\mathbf{f}$ and $\mathbf{g}$, serving as their discrete representations (in the spirit of Fig. L6 on p. 27). The linear operator, $\hat{L}$, then acts as a finite-dimensional linear map, L, and the linear DEQ assumes the form of a matrix equation, $L\mathbf{f} = \mathbf{g}$. When written in this form, it is evident how to solve the equation: multiply from the left by the *inverse* of the linear map (assuming that it exists), to obtain $\mathbf{f} = L^{-1}\mathbf{g}$, or $f^i = (L^{-1})^i{}_j g^j$. In the infinite-dimensional limit, vector components become function values, $f^i \to f(t)$, and sums become integrals. The equation for the solution will therefore assume the form

$$f(t) = \int du\,(\hat{L}^{-1})(t, u)g(u).$$

Comparison with Eq. (C197) shows that

$$G(t, u) = (\hat{L}^{-1})(t, u). \qquad (C205)$$

> The Green function is the inverse of the operator defining a linear differential equation.

To connect this general view with the concrete formulas used to compute the Green function above, recall that the inverse, L, of a matrix, G, is defined through the relation $L^i{}_j G^j{}_k = \delta^i{}_k$. In the "continuum limit", $\delta^i{}_k \to \delta(t - u)$ becomes a δ-function, and so the equation should assume the form $\int ds\,\hat{L}(t, s)G(s, u) = \delta(t - u)$. This looks almost, but not quite, like Eq. (C195), the difference being that the latter does not contain an integral over the running variable, s. The reason for this is that $\hat{L}$ is not a totally generic linear operator in function space, but one that is effectively diagonal. Much as the application of a diagonal matrix, $D^i{}_j = d^i \delta^i{}_j$, to a generic matrix, $(DA)^i{}_j = d^i A^i{}_j$, does not contain an index summation, $(\hat{L}f)(t) = \hat{L}(t)f(t)$ does not contain an integral over a running variable. In this sense, $\hat{L}$ is "almost diagonal", which explains why the defining equation for its continuum inverse assumes the form $\hat{L}(t)G(t, u) = \delta(t, u)$. To summarize: we have identified the Green function as the inverse of the almost diagonal differential operator. Many formulas involving Green functions afford easy interpretations emphasizing this matrix perspective.

There remains one unexplained subtlety, though. In our discussion above we talked much about "*homogeneous solutions*" of the equation, $\hat{L}f = 0$. In the finite-dimensional context this becomes $L\mathbf{f} = 0$, i.e. the existence of a non-vanishing homogeneous solution would be equivalent to the existence of a zero eigenvalue, in conflict with the assumed invertibility of the operator. There is, however, no reason for concern: *prior* to fixing the boundary conditions the operator $\hat{L}$ is indeed not invertible. In our discussion above this showed up through the non-uniqueness of the solution for the Green function. But after fixing boundary conditions, invertibility is granted and G is uniquely specified.[46]

---

[46] In this argument we tacitly assume that the assumed boundary conditions define a function *space*. For example, the conditions $f(\pm\infty) = 0$ satisfy this criterion: the linear combination of two functions vanishing at infinity again vanishes at infinity. However, $f(t_1) = f(t_2) = 1$ is an example of boundary conditions not defining a function space. For the rigorous algebraic interpretation of this situation we refer to specialized courses on differential equations.

differential
operators as
matrices

INFO Let us try to understand in what sense a differential operator corresponds to an *almost diagonal matrix*. A derivative acts on a function as $d_t f(t) = \lim_{\delta \to 0} \delta^{-1}(f(t + \delta) - f(t))$. In a representation discretized as $t = i\delta$, this corresponds to the "discrete derivative" $(df)^i \equiv f^{i+1} - f^i$. The matrix representing d contains -1 on the main diagonal and 1 on its neighboring diagonal: $d^i_{\ j} = \delta^{i+1}_{\ \ j} - \delta^i_{\ j}$. Upon multiplying this matrix by a function, $(df)^i = d^i_{\ j} f^j = f^{i+1} - f^i$, the j-index summation collapses to just two terms. Likewise, the continuum representation, $(df)(t) = \int du\, d(t, u) f(u) = d_t f(t)$, does not contain an integral over an intermediate variable. If we multiply matrices to represent higher derivatives, $d^n \leftrightarrow d^n_t$, the diagonals "shift", for example, $(d^2 f)_i = f_{i+2} - 2f_{i+1} + f_i$, however the fact remains that in the limit of very large dimensions, the matrices look almost diagonal. The multiplication of derivatives by time-dependent functions, $h(t)d_t \leftrightarrow h_i(\delta^{i+1}_{\ \ j} - \delta^i_{\ j})$, does not change this structure, either. This explains the absence of integrals in the product $\hat{L}G$.

Note, however, that the inverse of an almost diagonal matrix need not be almost diagonal at all. This is exemplified by the fact that G, the inverse of the almost diagonal $\hat{L}$, has non vanishing "matrix elements", $G(t, u)$, even for large separations $|t - u|$. As an instructive exercise, try to compute the inverse of $d^i_{\ j}$ for low matrix dimensions (or perhaps even general matrix dimension) to explore this point.

C7.6   General higher-order differential equations

We next consider general DEQs, of higher order in the number of derivatives and possibly nonlinear. For example, Newton's equation, $\ddot{q} = \frac{1}{m} F(q)$, is an equation relating second derivatives of the coordinate function q to a force, $F(q)$, which may depend nonlinearly on q. The generalization to a generic **nth-order differential equation** is often expressed in the form

nth-order
differential
equation

$$H\big(f^{(n)}(t), f^{(n-1)}(t), f^{(n-2)}(t), \ldots, f^{(0)}(t), t\big) = 0, \qquad (C206)$$

with an arbitrary function $H(y^{n+1}, \ldots, y^1, t)$. Newton's equation fits into this scheme as $\ddot{q} - \frac{1}{m} F(q) = 0$, i.e. $H(y^3, y^1, t) = y^3 - \frac{1}{m} F(y^1)$. The unique solution of an nth-order differential equation requires the specification of n boundary conditions. These may be provided via the specification of function values at n different points, $f(t_i) = f_i$, or the specification of function values and derivatives at a single point, $f^{(l)}(t_0) = f_l$ for $l = 0$ to $n - 1$, or combinations thereof.

Reformulation as a system of first-order equations

It is often convenient to transform the nth-order equation into an equivalent system of n first-order equations. To this end, we define a set of n functions, $x^1(t) \equiv f(t)$ and $x^l(t) \equiv f^{(l-1)}(t) = d_t x^{l-1}(t)$ for $l = 2, \ldots, n$. The differential equation (C206) then becomes equivalent to the *system of n first-order differential equations*,

$$d_t x^1 = x^2,$$
$$d_t x^2 = x^3,$$
$$\vdots \quad \vdots$$
$$d_t x^{n-1} = x^n,$$
$$H(d_t x^n, x^n, x^{n-1}, \ldots, x^1, t) = 0, \qquad (C207)$$

for the vector of functions, $\mathbf{x} = (x^1, \ldots, x^n)^{\mathrm{T}}$. Once $\mathbf{x}(t)$ is found, the function of interest, $f = x^1$, is given by the first component of the solution vector. Although the system of n first-order equations amounts to just a rewriting of the original problem, this change of representation is often advantageous. Specifically, we will see in the next section that systems of first-order differential equations can be handled using powerful geometric methods. The reformulation as a system of first-order equations also demonstrates that an n-component vector containing initial data, $x^l(t_0) = f^{(l-1)}(t_0)$ for $l = 1$ to n, defines a unique solution.

Newton equation

EXAMPLE Returning to the one-dimensional *Newton equation*,

$$\mathrm{d}_t^2 q = \frac{1}{m} F(q),$$

we define the vector $\mathbf{x} \equiv (q, p)^{\mathrm{T}}$ comprising the particle's coordinate, $x^1 = q$, and its momentum, $x^2 = p = mv = m\mathrm{d}_t q.$[47] Newton's equation can now be equivalently expressed as

$$\mathrm{d}_t q = \frac{1}{m} p,$$
$$\mathrm{d}_t p = F(q),$$

which corresponds to Eq. (C187) from Section C7.4. In an analogous manner, Newton's equation in d-dimensional space, $\mathrm{d}_t^2 \mathbf{q} = \frac{1}{m} \mathbf{F}(\mathbf{q})$, assumes the form

$$\mathrm{d}_t \mathbf{q} = \frac{1}{m} \mathbf{p},$$
$$\mathrm{d}_t \mathbf{p} = \mathbf{F}(\mathbf{q}),$$

which is a first-order equation for the $2d$-component vector $\mathbf{x} \equiv \begin{pmatrix} \mathbf{q} \\ \mathbf{p} \end{pmatrix}$. The $2d$-dimensional space hosting the vectors $\mathbf{x}$ is called the **phase space** of a mechanical system. It plays an important role in the modern theory of mechanical systems.

phase space

Systems of first-order equations

systems of differential equations

We have seen above how an nth-order differential equation can be transformed to a particular system of first-order equations, (C207). The general *system of coupled first-order differential equations* is defined by[48]

$$\mathrm{d}_t x^1 = g^1(x^1, x^2, \ldots, x^n, t),$$
$$\mathrm{d}_t x^2 = g^2(x^1, x^2, \ldots, x^n, t),$$

$$\vdots \qquad \vdots$$

$$\mathrm{d}_t x^n = g^n(x^1, x^2, \ldots, x^n, t), \qquad\qquad (C208)$$

---

[47] Slightly deviating from the general scheme, a constant, m, is included in the definition of $x^2 = m\mathrm{d}_t x^1$.

[48] Equation (C207) fits into this scheme by defining $g^i(x^1, \ldots, x^n, t) = x^{i+1}$, $i = 1, \ldots, n-1$, solving the algebraic equation, $H(\mathrm{d}_t x^n, x^n, \ldots, x^1, t) = 0$, for $\mathrm{d}_t x^n$, and expressing the result in the form $\mathrm{d}_t x^n \equiv g^n(x^1, \ldots, x^n, t)$.

where $x^i = x^i(t)$ are the sought solutions. We introduce a compact vector notation, $\mathbf{x} \equiv (x^1, \ldots, x^n)^T$, $\mathbf{g} = (g^1, \ldots, g^n)^T$, to represent the system as

$$\boxed{d_t\mathbf{x}(t) = \mathbf{g}(\mathbf{x}(t), t).}$$
(C209)

This notation suggests an interpretation of $\mathbf{x}(t)$ as a curve. The curve is defined by the condition that at every instant of time, t, its velocity, $d_t\mathbf{x}(t)$, equals the given vector function $\mathbf{g}(\mathbf{x}(t))$. If the function $\mathbf{g}$ does not carry explicit time dependence, $\mathbf{g}(\mathbf{x}, t) = \mathbf{g}(\mathbf{x})$, the system is called **autonomous**.

autonomous differential equation

Systems of first-order differential equations play an important role not only in physics but also in biology, chemistry, engineering and the social sciences. They are used to describe the time evolution of multicomponent quantities (the coordinates and momenta of a mechanical system, the concentrations of chemical compounds, the population numbers describing a multi-species habitat, the stock market value of a system of companies, etc.) in response to "forces" driving that evolution (generalized forces, chemical reactions, environmental changes, economic market forces, etc.). The connection between cause and effect is then represented in terms of a system of differential equations where the coupling between the equations expresses the interaction between the agents of the systems.

EXAMPLE Let us illustrate the application of coupled first-order DEQs with a toy model for ecological interdependence, the **Lotka–Volterra system**.[49] The LV system describes a population of prey (rabbits) and predators (foxes), containing r rabbits and f foxes, $\mathbf{x} = (r, f)^T$. The number of rabbits is assumed to increase at a constant rate α, describing their proliferation, and to diminish at a rate βf, due to the presence of the f foxes. Conversely, the population of foxes is controlled by a mortality rate, γ, and a proliferation rate, δr, proportional to the available food resources, i.e. the rabbit population. Expressed as a system of two differential equations, this model assumes the form

Lotka– Volterra (LV) system

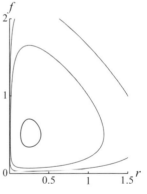

$$d_t r = (\alpha - \beta f)r,$$
$$d_t f = (-\gamma + \delta r)f.$$
(C210)

The analytic solution of this system of equations is possible but complicated. A plot of numerically computed solutions shows a periodic pattern in the population balance: an abundance of rabbits causes a flourishing of the fox population, which leads to a decimation of the rabbits. This in turn suppresses the fox population, and the cycle starts again. The precise form of the cycle in the rf-plane depends on the numerical values chosen for the rates $\alpha, \beta, \gamma, \delta$.

fixed point

Observe that the system possesses a **fixed point**, where the populations remain stationary: for $f = f^* \equiv \alpha/\beta$ and $r = r^* \equiv \gamma/\delta$, the right-hand side of the system vanishes, which means that $d_t r = d_t f = 0$. Consequently the populations remain stationary. Fixed points are important

[49] A.J. Lotka, *Elements of Physical Biology*, Williams and Wilkins, (1925); V. Volterra, Mem. Acad. Lincei Roma 2, 31 (1926).

characteristics of systems of DEQs in general. Methods of finding them and exploring what happens in their vicinity will be introduced in the next section.

Although the LV system is based on oversimplifying assumptions, it describes important aspects of population fluctuations. More complicated models of ecological systems are often constructed by generalization of LV-type differential equations. Generally speaking, finite systems of equations can never faithfully describe "reality". The goal of modeling nature or society in terms of systems of differential equations is to reduce real-world processes down to a manageable level of complexity which is still sufficiently "realistic" to have predictive power.

Closed-form solutions of systems of DEQs can be found only in exceptional cases, and this motivates the development of *qualitative methods* for their description. One frequently employs a language in which t is considered as a time-like variable, and the system (C209) interpreted as a **dynamical system**. For a given $\mathbf{x}$, the curve $\mathbf{x}(t)$ solving the system with initial condition $\mathbf{x}(0) = \mathbf{x}$ is called a **trajectory of the system**. The full information on all trajectories is carried by the **flow** of the DEQ.

dynamical system

flow

Mathematically, the flow is a map,

$$\boldsymbol{\Phi} : I \times M \to M, \qquad (t, \mathbf{x}) \mapsto \mathbf{x}(t) \equiv \boldsymbol{\Phi}_t(\mathbf{x}), \quad \text{(C211)}$$

where I is a time interval and M, the domain of definition of the functions $\mathbf{x}$, is often called the **phase space** of the system.[50] The flow map obeys the "initial condition" $\boldsymbol{\Phi}_0(\mathbf{x}) = \mathbf{x}$. For finite t, the flow, $\boldsymbol{\Phi}_t(\mathbf{x}) = \mathbf{x}(t)$, is defined by the "trajectory" $\mathbf{x}(t)$ passing through $\mathbf{x}$. As such the flow obeys the composition rule

phase space

$$\boldsymbol{\Phi}_{t+s}(\mathbf{x}) = \mathbf{x}(t + s) = \boldsymbol{\Phi}_t(\mathbf{x}(s)) = \boldsymbol{\Phi}_t(\boldsymbol{\Phi}_s(\mathbf{x})),$$

i.e. the trajectory point $\mathbf{x}(t + s)$ can be understood as the endpoint of a trajectory of duration t starting at $\mathbf{x}(s) = \boldsymbol{\Phi}_s(\mathbf{x})$. This composition rule can be used to extend the definition of flow to negative times: $\mathbf{x} = \boldsymbol{\Phi}_0(\mathbf{x}) = \boldsymbol{\Phi}_{(-t)+t}(\mathbf{x}) = \boldsymbol{\Phi}_{-t}(\boldsymbol{\Phi}_t(\mathbf{x}))$. Also notice that

$$d_t \boldsymbol{\Phi}_t(\mathbf{x}) = d_t \mathbf{x}(t) = \mathbf{g}(\mathbf{x}(t)) = \mathbf{g}(\boldsymbol{\Phi}_t(\mathbf{x})),$$

i.e. considered as a function of t, the flow is a solution of the DEQ. Plotting the *flow lines*, $\boldsymbol{\Phi}_t(\mathbf{x})$, for various different initial points $\mathbf{x}$ gives the trajectories starting at these points ($\to$ C7.6.1-2). In the exceptional case of a "*stationary flow*", i.e. a time-independent solution, $\boldsymbol{\Phi}_t(\mathbf{x}^*) = \mathbf{x}^*$, we call the point $\mathbf{x}^*$ a stationary point, or a **fixed point** of the system. The fixed-point property, $d_t \mathbf{x}^* = \mathbf{0}$, implies the condition $\mathbf{g}(\mathbf{x}^*) = \mathbf{0}$. Finding the fixed points is, thus, equivalent to finding the zeros of $\mathbf{g}$, and this is usually the first step in the analysis of a system of DEQs. In a second step, one then analyzes the behavior of the system in the vicinity of its stationary points.

fixed point

[50] The phase space defined by the time-dependent coordinates and momenta, $(\mathbf{q}, \mathbf{p})^{\mathrm{T}}$, of a mechanical system is an important example of this notion. In fact, physicists tend to reserve the term "phase space" for this particular realization.

Deterministic chaos: introduction in a nutshell

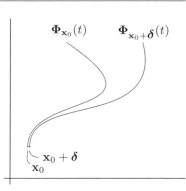

deterministic chaos

Before turning to the discussion of near-fixed-point dynamics, let us stay for a moment at the global level. Above, we argued that DEQs possess unique solutions provided the defining equations are sufficiently well-behaved. Often, however, the formal existence criterion has only limited practical value. The reason is that the flows of many systems of DEQs exhibit the phenomenon of **deterministic chaos**. A defining feature of chaotic flows is their exponential sensitivity to initial conditions: consider two initial configurations, $\mathbf{x}_0$ and $\mathbf{x}_0 + \boldsymbol{\delta}$, differing only by an infinitesimal amount $\|\boldsymbol{\delta}\|$. If the flow is chaotic, the trajectories $\boldsymbol{\Phi}_{\mathbf{x}_0}(t)$ and $\boldsymbol{\Phi}_{\mathbf{x}_0+\boldsymbol{\delta}}(t)$ starting from these two configurations will deviate strongly from each other no matter how small their initial difference $\|\boldsymbol{\delta}\|$. The deviation grows exponentially in time, i.e. there exists a **Lyapunov exponent**, λ, such that

Lyapunov exponent

$$\|\boldsymbol{\Phi}_{\mathbf{x}+\boldsymbol{\delta}}(t) - \boldsymbol{\Phi}_{\mathbf{x}}(t)\| \sim \|\boldsymbol{\delta}\| \exp(\lambda t). \tag{C212}$$

This means that for times $t \gg \lambda^{-1}$ even tiny changes in the initial conditions take a drastic effect on the course of the trajectory. The mathematician and meteorologist Edward Lorenz coined a metaphor for this phenomenon by saying that the flap of a butterfly's wings in Brazil could set off a tornado in Texas.

EXAMPLE *The phenomenon of chaos* occurs even in very simple systems. For example, the $n = 3$ system,

$$\frac{\mathrm{d}x}{\mathrm{d}t} = \sigma(y - x), \qquad \frac{\mathrm{d}y}{\mathrm{d}t} = x(\rho - z) - y, \qquad \frac{\mathrm{d}z}{\mathrm{d}t} = xy - \beta z, \tag{C213}$$

with constants σ, ρ, β was proposed by Lorenz[51] as a model of atmospheric convection phenomena. In spite of its relatively simple form – three functions g^i of quadratic order in the variables (x, y, z) – it cannot be solved analytically. However, a numerical solution reveals its sensitivity to variations in the boundary conditions, as shown in Fig. C29.

  The structure of a typical trajectory in the three-dimensional space of variables is illustrated in Fig. C29(a). Panels (b) to (e) show the curve $\boldsymbol{\Phi}_{\mathbf{x}_0}(t)$ plotted until times $t = 0.1, 0.3, 1.5,$ and 5. These values correspond to different stages of the dynamics, also visible in Fig. C29(a): An initial sweep, (b), from the starting point to a center region during time $t \lesssim 0.1$ is followed by a spiraling motion, (c), up to a time $t \simeq 1.5$. At larger times, (d,e), the trajectory traces out a two-winged structure known as the **Lorenz attractor**. This is an example of a **strange attractor**, a region in space which binds or attracts trajectories to it. The attribute "strange" is well deserved. Trajectories captured by the Lorenz attractor perform perpetual motion about it and their flow traces an infinitely filigree pattern having a "fractal geometry". This defines an object that looks almost, but not quite, like a surface in space. In mathematics, these structures are characterized in terms of a **fractal dimension** ($d \simeq 2.05$ for the Lorenz attractor). The fascinating physics and mathematics of chaotic dynamics is explored in fields including *chaos theory*, *nonlinear dynamics* and *turbulence* and for further discussion we refer to texts introducing these disciplines.

Lorenz attractor

fractal dimension

[51] E.N. Lorenz, Deterministic non-periodic flow. *Journal of the Atmospheric Sciences* **20**, 130 (1963).

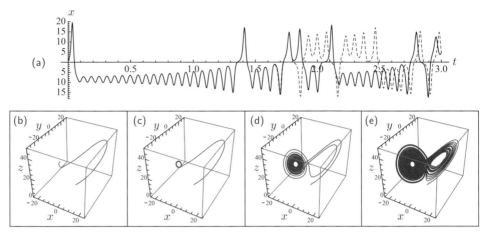

Fig. C29   (a) A single component, x, of two trajectories (solid, dashed) of the Lorenz system (C213). Both trajectories are computed for the parameters $\sigma = 10, \rho = 28, \beta = 8/3$, but are started using slightly different initial conditions, $\mathbf{x}_0 = (x_0, y_0, z_0)^\mathrm{T} = (1, 1, 1)^\mathrm{T}$ and $(1.01, 1, 1)^\mathrm{T}$, respectively. For short times the trajectories are visually indistinguishable, but the uncontrolled growth of the relative deviation becomes apparent at times $t \sim 2$ (arbitrary units). Each of panels (b) to (e) illustrates a different stage of the dynamics of the first trajectory, as discussed in the text.

C7.7   Linearizing differential equations

Consider a set of differential equations, $\mathrm{d}_t \mathbf{x} = \mathbf{f}(\mathbf{x})$, for a solution vector, $\mathbf{x} = (x^1, \ldots, x^n)^\mathrm{T}$, where $\mathbf{f}$ is the vector of generalized forces describing the evolution. For concreteness, think of $\mathbf{x}$ as the vector of coordinates specifying a complex mechanical system, or that of populations in a biological context. Let us assume that the system is initially at rest and remains so. In mathematical terms, this means that it is initialized at a stationary point, $\mathbf{x}^*$, of the differential equation, where $\mathrm{d}_t \mathbf{x}^* = \mathbf{f}(\mathbf{x}^*) = \mathbf{0}$. A perturbation of the system will cause a deviation away from the stationary point, $\mathbf{x}^* \to \mathbf{x}^* + \mathbf{y}$. Since $\mathbf{x}^* + \mathbf{y}$ no longer is a stationary point, $\mathbf{y}(t)$ now becomes a function of time. For example, it may perform a damped oscillatory motion describing the vibrational relaxation, $\mathbf{y}(t) \to \mathbf{0}$, of an elastic mechanical system. However, the perturbation may also cause more dramatic effects. For example, the small perturbation of a spherical body initially at rest on the top a hill may cause accelerated motion *away* from the initial configuration.

Even if the full equation, $\mathrm{d}_t \mathbf{x} = \mathbf{f}(\mathbf{x})$, describing the system is complicated, motion in the vicinity of fixed points can generally be described analytically. To understand how, we substitute $\mathbf{x} = \mathbf{x}^* + \mathbf{y}$ into the equation and obtain

$$\mathrm{d}_t(\mathbf{x}^* + \mathbf{y}) = \mathrm{d}_t \mathbf{y} = \mathbf{f}(\mathbf{x}^* + \mathbf{y}),$$

since $\mathbf{x}^*$ is time-independent. For sufficiently small $\mathbf{y}$ we may Taylor expand the r.h.s. to first order in the small increment $\mathbf{y}$. Equation (C106) then tells us that

$$f^i(\mathbf{x}^* + \mathbf{y}) \simeq f^i(\mathbf{x}^*) + \mathbf{y} \cdot \nabla f^i(\mathbf{x}^*) = \sum_j \frac{\partial f^i(\mathbf{x}^*)}{\partial x^j} y^j,$$

where the fixed-point condition $f^i(\mathbf{x}^*) = 0$ was used again. From these equations we obtain a **system of linear differential equations with constant coefficients**,

$$\mathrm{d}_t y^i = \sum_{j=1}^{n} A^i{}_j y^j, \qquad A^i{}_j = \frac{\partial f^i(\mathbf{x}^*)}{\partial x^j},$$

for the deviations y^i. The constancy (i.e. independence of time) of the coefficient matrix A follows from its definition as a fixed point property. Formulated in vector/matrix notation, the equation assumes the compact form

$$\boxed{\mathrm{d}_t \mathbf{y} = A\,\mathbf{y}.} \tag{C214}$$

We may now relate back to our discussion of Section C7.4 to predict which types of dynamical behaviors can arise in the vicinity of fixed points. The structure of the general solution (C184) implies that the behavior depends on the structure of A's eigenvalues, λ_j. These eigenvalues are generally complex and depending on their value one may observe:

- ▷ *oscillatory motion* around the fixed point: all eigenvalues purely imaginary, $\mathrm{Re}(\lambda_j) = 0$;
- ▷ *damped oscillatory motion*: eigenvalues have finite negative real part, $\mathrm{Re}(\lambda_j) < 0$;
- ▷ *attenuated motion back to the fixed point*: eigenvalues real and negative, $\mathrm{Re}(\lambda_j) < 0$, $\mathrm{Im}(\lambda_j) = 0$;
- ▷ *instability*: there exist eigenvalues with positive real part, $\mathrm{Re}(\lambda_j) > 0$, for at least one j.

In the last case, a deviation in the direction $\mathbf{v}_j$ will grow exponentially, $\sim \mathbf{v}_j e^{\lambda_j t}$, hence the fixed point is *unstable* and the system will flow away from it. In such cases, the condition that the deviation $\mathbf{y}$ is small holds only for short time scales and different solution methods must be applied to describe the dynamics at longer time scales. (→ C7.7.1-4)

C7.8 Partial differential equations

Partial differential equations are differential equations involving derivatives w.r.t. several variables. A simple example is the **wave equation in one dimension**,

wave equation

$$(v^2\partial_x^2 - \partial_t^2)u(x,t) = 0, \tag{C215}$$

where x and t are a spatial and temporal coordinate, respectively, v is a constant and $u(x,t)$ is a function representing the medium undergoing wave-like motion (the pressure of a gas, the height of a water wave, etc.).

From a physical perspective, the two most important facts about differential equations are as follows.

- ▷ They are of profound importance to all disciplines of physics. This follows from our reasoning on p. 300: the laws of physics are naturally expressed in terms of differential equations and most involve more than one variable.
- ▷ Their solution theory is much more complex than that of ordinary differential equations.

Partial differential equations take center stage in the physics curriculum, examples include the Hamilton equations (mechanics), the Maxwell equations (electrodynamics), the Schrödinger equation (quantum mechanics) and the Einstein equations (general relativity). Due to the mathematical complexity of these equations, lecture courses generally devote much effort to discussing the appropriate solution schemes.

The *typology of partial DEQs*, too, is much richer than that of ordinary DEQs. Again, we have to discriminate between linear and nonlinear equations, or equations of different order. On top of that, however, further criteria exist for classifying different types of partial DEQs whose discussion is beyond the scope of this text. Also, the questions of existence and uniqueness of solutions becomes more complicated. For example, it is straightforward to verify that for arbitrary one-dimensional functions f and g, the functions $u_1(x, t) \equiv f(x - vt)$ and $u_2(x, t) \equiv g(x + vt)$ both solve equation (C215). This illustrates that it is not enough to specify an "initial condition": if $f = g$, the two solutions obey the same initial condition, $u_1(x, 0) = u_2(x, 0) = f(x)$, but for finite times they are clearly distinct. In the case of the wave equation, different information is required to fix a solution.[52]

How do we know that all solutions of a partial DEQ have been found? How much additional information is required to uniquely specify a unique solution and in what form can this information be provided? These are questions of considerable depth which are addressed in mathematics lecture courses on partial DEQs and, from a more applied perspective, in all lecture courses on theoretical physics.

C7.9 Summary and outlook

Differential equations describe how a function changes under the influence of other functions. In cases where the sought function is a physical quantity, and the influencing functions represent physical mechanisms affecting it, the equation defines a physical law. This shows that differential equations provide the natural language for the abstraction of physical observation.

We saw that even ordinary differential equations, with derivatives in only one variable, show a remarkably rich typology. The different types of equations call for different solution schemes, and determine their level of complexity. For example, we saw that linear equations obey a superposition principle which makes them drastically simpler than generic equations. At the same time, the superposition principle is reflected in the presence of physical phenomena (wave interference), and in the absence of others (like chaos, which is an essentially nonlinear phenomenon.)

Given the remarkable complexity of even ordinary differential equations, the statement that partial differential equations, representing the majority of physical laws, are even more complex may sound intimidating. However, in practice the situation is often not quite as severe: in many cases of physical interest, physical symmetries permit the reduction

---

[52] Unique solutions of the wave equation are fixed by initial conditions in the form $u(x, 0) = g(x)$ and $\partial_t u(x, t) = h(x)$, or boundary data such as $u(0, t) = u(L, t) = 0$. For a discussion of these statements we refer to lecture courses in mechanics and electrodynamics.

of partial differential equations to ordinary equations, or at least to systems of coupled ordinary equations. These can then be approached by the techniques introduced in this chapter. For example, the Schrödinger equation of the hydrogen atom is a partial differential equation in three variables. However, due to the rotational symmetry of the system, this equation can be traded for three decoupled, ordinary differential equations in the variables (r, θ, ϕ) of a spherical coordinate system. The solution of these equations is routinely discussed in lecture courses of quantum mechanics, and the solution functions determine the atomic orbitals, key to the structure of the periodic table of elements (which is another powerful illustration that the solution of differential equations can lead to a wealth of insight). In cases where no symmetries are present, the solution of partial equations generally requires numerical methods. The construction of powerful numerical solution schemes has become a science in its own right, but not one to be discussed here.

C8   Functional calculus

REMARK Knowledge of Chapters V1 and V2 is required. Reference to Section V3.2 on differentials of functions is made but not essentially required.

In standard calculus, one deals with functions $F(\mathbf{v})$ having vectors, $\mathbf{v} \in \mathbb{R}^d$, as arguments.
functional   Functional calculus generalizes the setting to functions $F[f]$ taking functions as arguments. Since "function of a function" does not sound nice, such maps are called **functionals**. It is customary to indicate the argument of a functional in square brackets. To understand why functionals have a lot in common with ordinary functions, recall that their arguments can always be discretized as $f(t) \rightarrow \{f^\ell | \ell = 1,\ldots,N\}$, so that they may be interpreted as $N \rightarrow \infty$ limits of N-dimensional vectors (Fig. L6). This indicates that one may work with functionals much as with ordinary functions. All standard operations of calculus have generalizations to functionals.

EXAMPLE (a) Consider the set of functions $f : (0,1) \rightarrow \mathbb{R}$ mapping the interval $(0,1)$ into the reals. For $0 < a < 1$, $\delta_a[f] \equiv f(a)$ defines a functional reading out the value of the argument function at a. In this way, the function f is mapped to the number $f(a)$. (b) The functional $\mathrm{Av}[f] \equiv \int_0^1 f(x)\mathrm{d}x$ yields the average value of f over the domain of definition. (c) A map assigning to each curve, γ, its geometric length, $L[\gamma]$, is another example of a functional, called the length functional.

C8.1 Definitions

To be concrete, we will focus on functionals taking maps,

$$\mathbf{r} : I \rightarrow \mathbb{R}^n, \quad t \mapsto \mathbf{r}(t), \tag{C216}$$

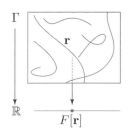

as arguments, where I is an open interval in the real numbers. In physical applications, the functions $\mathbf{r}$ describe the trajectories of objects as functions of a time-like variable, t. While functionals defined on such trajectories appear early in the physics curriculum and deserve particular attention, much of the discussion below immediately carries over to other functionals.

A functional is a smooth map,

$$F : \Gamma \rightarrow \mathbb{R}, \quad \mathbf{r} \mapsto F[\mathbf{r}], \tag{C217}$$

where Γ denotes the set of all argument functions (C216). Sometimes, boundary conditions fixing the value of the argument trajectories at the initial and final time of the interval, $I = (t_0, t_1)$, are imposed. The argument domain, Γ, then is restricted to a set of trajectories obeying $\mathbf{r}(t_0) = \mathbf{r}_0$ and $\mathbf{r}(t_1) = \mathbf{r}_1$, i.e. all elements of Γ have the same initial point $\mathbf{r}_0$ and final point $\mathbf{r}_1$.[53] In either case, F assigns real numbers to trajectories in the set.

local functional

In many applications, the value taken by a functional on a curve is determined by local properties such as the local position, $\mathbf{r}(t)$, of the trajectory and its velocity, $\dot{\mathbf{r}}(t)$. The value of a **local functional** on a curve can be represented as a time integral, $F[\mathbf{r}] = \int_I dt L(\mathbf{r}(\mathbf{y}(t)), \dot{\mathbf{r}}(\mathbf{y}(t), \dot{\mathbf{y}}(t)))$, over a function, L, depending on the instantaneous position, $\mathbf{r}(\mathbf{y})$, and velocity, $\dot{\mathbf{r}}(\mathbf{y}, \dot{\mathbf{y}})$.[54] Here, the notation indicates that the concrete evaluation of position and velocity requires a coordinate representation, $\mathbf{r}(\mathbf{y})$, where $\mathbf{y} = (y^1, \ldots, y^d)^T$ is a vector of generalized coordinates.[55] The time derivatives, $\dot{\mathbf{r}} = \dot{\mathbf{r}}(\mathbf{y}, \dot{\mathbf{y}})$, then become functions of coordinate derivatives, and sometimes the coordinates themselves. For example, in a Cartesian representation, a two-dimensional vector, $\mathbf{r}(\mathbf{x}) = \mathbf{e}_1 x^1 + \mathbf{e}_2 x^2$, is parametrized by $\mathbf{x} = (x^1, x^2)^T$, and its time derivative, $\dot{\mathbf{r}}(\dot{\mathbf{x}}) = \mathbf{e}_1 \dot{x}^1 + \mathbf{e}_2 \dot{x}^2$, by $\dot{\mathbf{x}} = (\dot{x}^1, \dot{x}^2)^T$. However, if $\mathbf{r}(\mathbf{y}) = \mathbf{c}_\rho \rho$ is represented in polar coordinates, $\mathbf{y} = (\rho, \phi)^1$, the velocity assumes the form $\dot{\mathbf{r}}(\mathbf{y}, \dot{\mathbf{y}}) \overset{(V29)}{=} \mathbf{e}_\rho \dot{\rho} + \mathbf{e}_\phi \rho\dot{\phi}$ and depends on both $\dot{\mathbf{y}}$ and $\mathbf{y}$.

The shorthand notation for a local functional reads

$$S[\mathbf{r}] = \int_I dt\, L(\mathbf{y}, \dot{\mathbf{y}}), \qquad (C218)$$

where $L(\mathbf{y}, \dot{\mathbf{y}})$ is a function of coordinates and their derivatives. Functionals of this structure are often called **action function-**

action

als, S, and the function L is called their

Lagrangian

Lagrangian function, or just **Lagrangian**. Often a more implicit notation,

Joseph-Louis Lagrange
1736–1813
A mathematician who excelled in all fields of analysis, number theory and celestial mechanics. In 1788 he published *Mécanique Analytique*, which formulated Newtonian mechanics in the then modern language of differential equations.

$$S[\mathbf{r}] = \int_I dt\, L(\mathbf{r}, \dot{\mathbf{r}}),$$

is used, where it is understood that the concrete evaluation of the integral requires a coordinate representation. Action functionals play an important role in almost all areas of physics.

length functional

EXAMPLE The *length of a curve*, $\mathbf{r} : I \to \mathbb{R}^n$, $t \mapsto \mathbf{r}(t)$, is given by the local **length functional**,

$$L[\mathbf{r}] \equiv \int_I dt\, L(\mathbf{y}, \dot{\mathbf{y}}), \qquad L(\mathbf{y}, \dot{\mathbf{y}}) = \|\dot{\mathbf{r}}(\mathbf{y}, \dot{\mathbf{y}})\|, \qquad (C219)$$

(see Eq. (V8)). If $n = 2$ and Cartesian coordinates are used, $L(\mathbf{x}, \dot{\mathbf{x}}) = ((\dot{x}^1)^2 + (\dot{x}^2)^2)^{1/2}$, whereas $L(\mathbf{y}, \dot{\mathbf{y}}) = (\dot{\rho}^2 + \rho^2\dot{\phi}^2)^{1/2}$ for polar coordinates.

---

[53] The end-point values of a smooth function on an open interval, $I = (t_0, t_1)$, are defined via a limit, $\mathbf{r}(t_{0,1}) \equiv \lim_{t \to t_{0,1}} \mathbf{r}(t)$.

[54] The generalization to more complicated functionals depending on the local acceleration, $L(\mathbf{r}, \dot{\mathbf{r}}, \ddot{\mathbf{r}})$, or the time argument, $L(\mathbf{r}, \dot{\mathbf{r}}, t)$, is straightforward but does not add anything essential to our discussion.

[55] The coordinate dimension, d, can be smaller than n, if $\mathbf{r}$ is confined to lower-dimensional regions in space. For example, if $\mathbf{r}$ lies on a sphere embedded in $\mathbb{R}^{n=3}$, it is described by $d = 2$ spherical coordinates, $\mathbf{y} = (\theta, \phi)^T$.

EXAMPLE Consider a particle of mass m subject to a potential $V(\mathbf{r})$. For example, $V(\mathbf{r}) = \frac{1}{2}m\Omega^2\mathbf{r}^2$ for the quadratic potential of a multidimensional harmonic oscillator. The kinetic energy of the particle is given
by $T(\dot{\mathbf{r}}) = \frac{1}{2}m\dot{\mathbf{r}}^2$. One may then define the *action functional*,

$$S[\mathbf{r}] \equiv \int_I dt\, L(\mathbf{r}, \dot{\mathbf{r}}) = \int_I dt\, \left(\tfrac{1}{2}m\dot{\mathbf{r}}^2 - V(\mathbf{r})\right), \tag{C220}$$

Lagrangian of classical mechanics

where $L(\mathbf{r}, \dot{\mathbf{r}}) = T(\dot{\mathbf{r}}) - V(\mathbf{r}) = \frac{1}{2}m\dot{\mathbf{r}}^2 - V(\mathbf{r})$ is known as the **Lagrange function of classical mechanics**. We will return to the discussion of this functional later in the chapter.

non-local functional

An example of a **non-local functional** is one which assigns to closed curves ($\mathbf{r}(t_0) = \mathbf{r}(t_1)$) the number of knots contained in them. This value can not be obtained by a single integration over time.

C8.2 Functional derivative

Much like ordinary functions, functionals are characterized by their *local extrema*. An extremum of a functional is an argument function such that infinitesimal deviations away from it leave the functional unchanged. This will be a maximum or a minimum only in exceptional cases. Since there are infinitely many ways to deform a function, it is much more likely that some finite deviations away from an extremal function lead to an increase of the functional and others to a decrease. In such cases the extremal curve defines the functional analogue of a saddle point (see Section C3.4).

As with ordinary functions, the condition for a function to be extremal is the vanishing of suitably defined derivatives. Consider an argument function, $\mathbf{r}(t)$, and an infinitesimal deformation thereof, $\mathbf{r}(t) + \delta\,\mathbf{u}(t)$, where $\mathbf{u}(t)$ is another function, and δ an infinitesimally small parameter. (If boundary conditions, $\mathbf{r}(t_{0,1}) = \mathbf{r}_{0,1}$, are imposed, the variation of

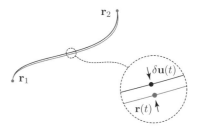

differentiable functional

the argument at these points has to vanish, $\mathbf{u}(t_{0,1}) = \mathbf{0}$.) The functional is called **differentiable** at $\mathbf{r}$ if the differences $F[\mathbf{r} + \delta\,\mathbf{u}] - F[\mathbf{r}]$ exist for all deformations $\mathbf{u}$ and vanish linearly in δ. In this case,

$$dF_{\mathbf{r}}[\mathbf{u}] = \lim_{\delta \to 0} \frac{1}{\delta}\left(F[\mathbf{r} + \delta\,\mathbf{u}] - F[\mathbf{r}]\right) \tag{C221}$$

differential

is called the **differential** of the functional F at $\mathbf{r}$. The differential plays a role analogous to that of the differential of a function introduced in Section V3.2. As with the ordinary differential, $dF_{\mathbf{r}}[c_1\mathbf{u}_1 + c_2\mathbf{u}_2] = c_1 dF_{\mathbf{r}}[\mathbf{u}_1] + c_2 F_{\mathbf{r}}[\mathbf{u}_2]$ $(c_{1,2} \in \mathbb{R})$ is linear in the increment arguments (why?) but may depend nonlinearly on the trajectory $\mathbf{r}$. A function, $\mathbf{r}$, is called

extremal function

an **extremal function** of F if $dF_{\mathbf{r}} = 0$.

EXAMPLE Consider the length functional $L[\mathbf{r}]$ of Eq. (C219), restricted to curves beginning and ending at fixed initial and final points, $\mathbf{r}_0$ and $\mathbf{r}_1$. Assuming a Cartesian coordinate representation, $\mathbf{r} = \mathbf{e}_i x^i$, and $\mathbf{u} = \mathbf{e}_i u^i$, the norm of the velocity is obtained as $\|\dot{\mathbf{r}}\| = \|\dot{\mathbf{x}}\| = (\dot{x}_i \dot{x}^i)^{1/2}$,[56] and analogously for the shifted curve $\|\dot{\mathbf{r}} + \delta\dot{\mathbf{u}}\|$. The variation of the functional can then be computed by a first-order expansion of the integrand,

$$dL_{\mathbf{r}}[\mathbf{u}] \overset{(C221)}{=} \lim_{\delta \to 0} \int_{t_0}^{t_1} dt \, \frac{1}{\delta} \big[\|\dot{\mathbf{r}} + \delta\dot{\mathbf{u}}\| - \|\dot{\mathbf{r}}\| \big] = \int_{t_0}^{t_1} dt \, \frac{\dot{x}_i \dot{u}^i}{\|\dot{\mathbf{x}}\|} = -\int_{t_0}^{t_1} dt \left[\frac{d}{dt} \frac{\dot{x}_i}{\|\dot{\mathbf{x}}\|} \right] u^i, \quad (C222)$$

where in the last step we integrated by parts, and the boundary terms vanish due to the absence of variation at the outer points, $\mathbf{u}(t_0) = \mathbf{u}(t_1) = \mathbf{0}$. The differential vanishes if the integrand on the right equals zero for all variational curves. This in turn requires the vanishing of the functions multiplying $u^i(t)$ for all intermediate times,

$$\frac{d}{dt} \left(\frac{\dot{x}_i}{\|\dot{\mathbf{x}}\|} \right) = 0. \quad (C223)$$

Intuitively, we know that the connections of extremal (shortest) length between two points are straight lines. Indeed, it is easy to check that the *stationarity condition* (C223) is satisfied by any straight connection, $\mathbf{r}(t) = \mathbf{r}_0 + (\mathbf{r}_1 - \mathbf{r}_0)f(t)$, where $f : (t_0, t_1) \to (0, 1)$ can be any function increasing monotonically from 0 to 1.

C8.3 Euler–Lagrange equations

Calculating extrema of functionals by explicit manipulation of the corresponding integrals is always an option, but can be tedious in practice. However, we have seen in the example above that for local functionals, the procedure involves a succession of three operations: variation of the ordinary function, L, an integration by parts, and from there the identification of a differential equation determining the extremum.

Derivation of the Euler–Lagrange equations

In the present section, we formulate this program in generality. Consider a general action functional,

$$S[\mathbf{y}] = \int_{t_0}^{t_1} dt \, L(\mathbf{y}(t), \dot{\mathbf{y}}(t), t), \quad (C224)$$

defined on a set of functions, $\mathbf{y} : I \to \mathbb{R}^d$, $t \mapsto \mathbf{y}(t)$, connecting two specified points, $\mathbf{y}_0 = \mathbf{y}(t_0)$ and $\mathbf{y}_1 = \mathbf{y}(t_1)$, over the course of a time interval $I = (t_0, t_1)$. (Think of these functions as the coordinate representations of trajectories $\mathbf{r}(\mathbf{y}(t))$.) The differential of this functional can be computed in a similar manner as in the example of the length functional, see Eq. (C222). We substitute an infinitesimally deformed function, $\mathbf{y}(t) + \delta\mathbf{u}(t)$, into the Lagrangian and expand to first order to obtain

---

[56] Recall that in Cartesian coordinates the placement of indices is a matter of cosmetics, $\dot{x}_i \dot{x}^i = \dot{x}^i \dot{x}^i$. However, the notation $\dot{\mathbf{r}}^2 = \dot{y}_i \dot{y}^i = \dot{y}^i g_{ij} \dot{y}^j$ makes sense even in curvilinear coordinates where the appearance of a metric can be absorbed in covariant indices, see the example of the sphere on p. 336.

$$dS_{\mathbf{y}}[\mathbf{u}] \overset{(C221)}{=} \int_{t_0}^{t_1} dt \lim_{\delta \to 0} \frac{1}{\delta} \Big[L(\mathbf{y} + \delta\mathbf{u}, \dot{\mathbf{y}} + \delta\dot{\mathbf{u}}, t) - L(\mathbf{y}, \dot{\mathbf{y}}, t) \Big]$$

$$= \int_{t_0}^{t_1} dt \left[(\partial_{y^i} L) u^i + (\partial_{\dot{y}^i} L) \dot{u}^i \right],$$

where $\partial_{\dot{y}^i} L$ is sloppy but standard notation for the partial derivative of L in its second argument, $\partial_{\dot{y}^i} L = \partial_{z^i} L(\mathbf{y}, \mathbf{z})|_{\mathbf{z}=\dot{\mathbf{y}}}$. Integrating by parts, and noting that boundary terms vanish due to the absence of variation at the outer points, $\mathbf{u}(t_0) = \mathbf{u}(t_1) = \mathbf{0}$, we obtain

$$dS_{\mathbf{y}}[\mathbf{u}] = \int_{t_0}^{t_1} dt \left[\partial_{y^i} L - d_t(\partial_{\dot{y}^i} L) \right] u^i . \tag{C225}$$

The functional is extremal, $dS_{\mathbf{y}}[\mathbf{u}] = 0$, for all $\mathbf{u}$, if the expression in bracket vanishes, i.e. if L obeys the **Euler–Lagrange equation**,

Euler–Lagrange equations

$$\boxed{\frac{d}{dt} \frac{\partial L}{\partial \dot{y}^i} - \frac{\partial L}{\partial y^i} = 0, \quad i = 1, \dots, d.} \tag{C226}$$

EXAMPLE Consider the *action functional of classical mechanics*, Eq. (C220), in a d-dimensional Cartesian representation, $\mathbf{r}(\mathbf{x})$, for which $\dot{\mathbf{r}}^2 = \dot{\mathbf{x}}^2 = \dot{x}_i \dot{x}^i$,

$$S[\mathbf{x}] = \int_I dt \left(\tfrac{1}{2} m \dot{\mathbf{x}}^2 - V(\mathbf{x}) \right).$$

Its Euler–Lagrange equations are obtained as

$$\frac{d}{dt} \frac{\partial L}{\partial \dot{x}^i} - \frac{\partial L}{\partial x^i} = m d_t \dot{x}_i + \partial_{x^i} V = 0, \qquad i = 1, \dots, d.$$

Recalling that $-\partial_{x^i} V = F_i$ defines the components of mechanical force, F_i, and raising the index, $x_i = x^i$, and $F_i = F^i$ in Cartesian coordinates, we recognize the *Newton equations*,

$$m \ddot{x}^i = F^i,$$

or $m\ddot{\mathbf{x}} = \mathbf{F}$ in vector notation. This shows that the trajectories, $\mathbf{x}(t)$, solving the Newton equations are extremal functions of the action functional of classical mechanics. This correspondence is known as the **action principle**. We return to its interpretation in the next section.

action principle

INFO Sometimes, a Lagrangian $L(\mathbf{y}, \dot{\mathbf{y}}, t)$ does not depend on all coordinates y^i of a problem. As an example, consider the Lagrangian of the previous example for a potential $V(\mathbf{x}) = V(x^1, x^2)$, independent of the coordinate x^3. In this case, the Euler–Lagrange equation of that coordinate reduces to $d_t(m\dot{x}^3) = 0$: the momentum $m\dot{x}^3$ in the 3-direction does not change in time. This is an example of a **conservation law**, where independence of the Lagrangian on a coordinate implies the time-independence, or conservation, of a physical observable. More generally, if L does not depend on a coordinate y^j, this implies

conservation law

$$\frac{d}{dt} \frac{\partial L}{\partial \dot{y}^j} = 0. \tag{C227}$$

canoncial momentum

The quantity $\partial_{\dot{y}^j} L \equiv p_j$ is called the **canonical momentum** associated with the variable y^j, and independence of the Lagrangian on that variable implies a conservation law, $d_t p_j = 0$.

If a Lagrangian does not depend explicitly on time, $L(\mathbf{y}, \dot{\mathbf{y}}, t) = L(\mathbf{y}, \dot{\mathbf{y}})$, i.e. $\partial_t L = 0$, there is another conservation law: define the *Hamilton function* or **Hamiltonian** as

$$H(\mathbf{y}, \dot{\mathbf{y}}) \equiv \left(\partial_{\dot{y}^i} L(\mathbf{y}, \dot{\mathbf{y}})\right)\dot{y}^i - L(\mathbf{y}, \dot{\mathbf{y}}). \tag{C228}$$

If $\partial_t L = 0$, this function is constant, $H(\mathbf{y}(t), \dot{\mathbf{y}}(t)) = \text{const.}$, along any extremal trajectoy, i.e. any $\mathbf{y}(t)$ solving the Euler–Lagrange equations. Indeed, in this case its total time derivate vanishes,

$$d_t H = \sum_i \left[(\partial_{\dot{y}^i} L)\ddot{y}^i + (d_t(\partial_{\dot{y}^i} L))\dot{y}^i - (\partial_{y^i} L)\dot{y}^i - (\partial_{\dot{y}^i} L)\ddot{y}^i\right] = 0,$$

where the second and third terms cancel due to the Euler–Lagrange equations (C226). The equation $H(\mathbf{y}, \dot{\mathbf{y}}) = \text{const.}$ is a first-order differential equation in the variables $y^i(t)$. It should be understood as a necessary condition the solutions $\mathbf{y}(t)$ have to obey. Since the equation $H = \text{const.}$ is generally simpler than the full Euler–Lagrange equations, its solution often is the first step in the analysis of an extremal problem with time-independent Lagrangian. ($\rightarrow$ C8.3.1-2)

Consider again the Lagrangian of classical mechanics, Eq. (C220). If the potential is time-independent, then $\partial_t L = 0$, hence the Hamiltonian is conserved. In a Cartesian representation, $(\partial_{\dot{x}^i} L)\dot{x}^i = m \sum_i (\dot{x}^i)^2 = m\dot{\mathbf{x}}^2$, thus $H(\mathbf{x}, \dot{\mathbf{x}}) = \frac{1}{2}m\dot{\mathbf{x}}^2 + V(\mathbf{x})$. This is simply the total energy of a point particle, comprising the sum of its kinetic and potential energy. We conclude that in mechanics, time-independent potentials lead to conservation of the total energy along a trajectory.

Coordinate invariance of the Euler–Lagrange equations

An extremal function, $\mathbf{r} : I \rightarrow \mathbb{R}^n$, is extremal no matter what coordinate representation is used for $\mathbf{r}(t)$. If $\mathbf{r}(t)$ is parametrized in two different coordinate representations, $\mathbf{y}(t)$ and $\mathbf{z}(t)$, the Euler–Lagrange equations must hold for both sets of coordinates, y^i and z^i,

$$\left(d_t \partial_{\dot{y}^i} - \partial_{y^i}\right) L(\mathbf{y}, \dot{\mathbf{y}}) = 0, \tag{C229a}$$

$$\left(d_t \partial_{\dot{z}^i} - \partial_{z^i}\right) M(\mathbf{z}, \dot{\mathbf{z}}) = 0, \tag{C229b}$$

where $M(\mathbf{z}, \dot{\mathbf{z}}) = L(\mathbf{y}(\mathbf{z}), \dot{\mathbf{y}}(\mathbf{z}, \dot{\mathbf{z}}))$ is the $\mathbf{y}$-Lagrange function expressed in $\mathbf{z}$-coordinates and velocities. The important point here is that in either representation the equations involve a differential operator of the same form, $(d_t \partial_{\dot{\alpha}^i} - \partial_{\alpha^i})$, $\alpha = y, z, x, \ldots$, acting on the respective coordinate representation of the Lagrange function.

It is instructive to show this invariance by explicit calculation. First note that $\dot{y}^i = \frac{\partial y^i}{\partial z^j}\dot{z}^j$. In Euler–Lagrange calculus, $\dot{y}^i$ and $\dot{z}^j$ have the status of independent variables. Differentiation of this relation therefore leads to $\frac{\partial \dot{y}^i}{\partial \dot{z}^j} = \frac{\partial y^i}{\partial z^j}$. Second, note that the partial derivatives of M and L are related by

$$\frac{\partial M}{\partial \dot{z}^i} = \frac{\partial L}{\partial \dot{y}^j}\frac{\partial \dot{y}^j}{\partial \dot{z}^i} = \frac{\partial L}{\partial \dot{y}^j}\frac{\partial y^j}{\partial z^i}, \qquad \frac{\partial M}{\partial z^i} = \frac{\partial L}{\partial y^j}\frac{\partial y^j}{\partial z^i} + \frac{\partial L}{\partial \dot{y}^j}\frac{\partial \dot{y}^j}{\partial z^i}.$$

Inserting these relations into the left-hand side of Eq. (C229b), we find

$$\frac{d}{dt}\frac{\partial M}{\partial \dot{z}^i} - \frac{\partial M}{\partial z^i} = \frac{d}{dt}\left(\frac{\partial L}{\partial \dot{y}^j}\frac{\partial y^j}{\partial z^i}\right) - \left(\frac{\partial L}{\partial y^j}\frac{\partial y^j}{\partial z^i} + \frac{\partial L}{\partial \dot{y}^j}\frac{\partial \dot{y}^j}{\partial z^i}\right)$$

$$= \left[\frac{d}{dt}\frac{\partial L}{\partial \dot{y}^j} - \frac{\partial L}{\partial y^j}\right]\frac{\partial y^j}{\partial z^i} + \frac{\partial L}{\partial \dot{y}^j}\left[\frac{d}{dt}\frac{\partial y^j}{\partial z^i} - \frac{\partial \dot{y}^j}{\partial z^i}\right] = 0.$$

The first bracket in the final equation vanishes due to Eq. (C229a), and the second because $d_t \partial_{\dot{z}^i} y^j - \partial_{z^i} \dot{y}^j = \partial_{z^i}(\dot{y}^j - \dot{y}^j) = 0$.

INFO Newton's equations assume the form $m\ddot{x}^i = F^i(\mathbf{x})$ only in Cartesian coordinates. For example, a change to polar coordinates, $(x^1, x^2) = \rho(\cos\phi, \sin\phi)$, leads to differential equations different in structure from the two-dimensional Cartesian Newton equations. (Compute them by substitution of the polar representation into the Cartesian equations.) At the same time, the *physical symmetries* of a problem often favor a coordinate change. For example, a two-dimensional potential with rotational symmetry is naturally represented by a function $V(\rho)$ depending on the radial coordinate of a polar system.

conservation laws   In such situations, Euler–Lagrange calculus is a convenient alternative to the Newtonian formalism. The statement that trajectories of mechanical motion, $\mathbf{r}(t)$, extremize an action functional with Lagrangian function $L = T - V$ is independent of coordinates. In a problem with rotational symmetry, one is therefore free to evaluate the extremal principle in a polar representation, $L(\rho, \dot{\rho}, \dot{\phi}) = \frac{1}{2}m(\dot{\rho}^2 + \rho^2\dot{\phi}^2) - V(\rho)$, where Eq. (V29) was used to compute $\dot{\mathbf{r}}^2$. Note that the Lagrangian does not depend on the variable ϕ, reflecting rotation symmetry. The Euler–Lagrange equations, (C226), for $y^1 = \rho$ and $y^2 = \phi$ now assume the form

$$d_t(m\dot{\rho}) = -\partial_\rho V(\rho) + m\rho\dot{\phi}^2,$$

$$d_t(m\rho^2\dot{\phi}) = 0.$$

The second equation is solved as $m\rho^2\dot{\phi} \equiv l = \text{const.}$, which shows the conservation of angular momentum, l. Substituting $\dot{\phi} = l/(m\rho^2)$ into the first equation transforms it to an ordinary differential equation for ρ,

$$m\ddot{\rho} = -\partial_\rho V(\rho) + \frac{l^2}{m\rho^3}.$$

This equation describes the radial motion of a body in a rotationally symmetric potential and is discussed in lecture courses on mechanics. Its derivation illustrates the fact that the presence of a symmetry in a problem generally reduces the number of independent differential equations by one. Lagrangian calculus is optimally suited to expose this reduction and is the method of choice in the solution of problems with symmetries.

EXERCISE Consider a unit sphere parametrized by spherical coordinates, $\mathbf{y} = (\theta, \phi)^{\mathrm{T}}$. What are the trajectories of shortest length connecting two points on it? The answer is found by variation of the length functional, $L[\mathbf{r}]$, of Eq. (C219). The velocity of a trajectory is given by $\dot{\mathbf{r}} = \mathbf{e}_\theta\dot{\theta} + \mathbf{e}_\phi\sin\theta\dot{\phi}$ (Eq. (V39)), hence $\dot{\mathbf{r}}^2 = \dot{y}^i g_{ij}\dot{y}^j = \dot{y}_i\dot{y}^i$ with the metric $g = \mathrm{diag}(1, \sin^2\theta)$, and the Lagrangian function reads $L(\mathbf{y}, \dot{\mathbf{y}}) = \|\dot{\mathbf{r}}\| = (\dot{\theta}^2 + \sin^2\theta\dot{\phi}^2)^{1/2}$. Verify that the Euler–Lagrange equations (C226), evaluated for $y^1 = \theta$ and $y^2 = \phi$, take the form

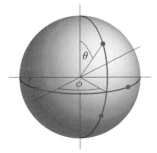

$$d_t\left(\frac{\dot{\theta}}{(\dot{\theta}^2 + \sin^2\theta\dot{\phi}^2)^{1/2}}\right) - \frac{\sin\theta\cos\theta\dot{\phi}^2}{(\dot{\theta}^2 + \sin^2\theta\dot{\phi}^2)^{1/2}} = 0, \qquad d_t\left(\frac{\sin^2\theta\dot{\phi}}{(\dot{\theta}^2 + \sin^2\theta\dot{\phi}^2)^{1/2}}\right) = 0.$$

Explain why trajectories in north–south direction, having $\dot{\phi} = 0$, are solutions. Show that trajectories along the equator, $\theta = \pi/2$, are solutions, too. Why are trajectories with constant $\theta \neq \pi/2$ and varying ϕ not minimal-length trajectories? General solutions of these equations lie on great circles, i.e.

geodesics

intersections of the sphere with planes through the origin. ($\rightarrow$ C8.3.3-4) Think how this observation is relevant to air travel.

In general, trajectories of minimal length on smooth surfaces are called **geodesics**. The general study of geodesics is an important problem in general relativity where the role of the surface is taken by curved space-time, and geodesics define the trajectories of free-moving bodies. The general equation defining geodesics is obtained by variation of the length functional (C219) under conditions where the norm $\|\dot{\mathbf{r}}\|$ must be computed using a metric describing the curvature of the underlying surface. While this construction is beyond the scope of this text, the problem above led to the geodesic equations on a sphere, where the metric is given by (V37).

C8.4 Summary and outlook

In this chapter we introduced the concepts of functional calculus required in the first or second year of the physics curriculum. We defined functionals as functions of functions and discussed methods for the identification of their extrema. We reasoned that a good way to understand these definitions is to think of functions as limits of infinitely high-dimensional vectors, $f(x) \leftrightarrow f^i$, and of functionals as limits of functions mapping such vectors into the reals. The extremal conditions then become equivalent to the vanishing of all partial derivatives, or the vanishing of a high-dimensional generalization of the function differential. The practical evaluation of these conditions is particularly easy for local functionals, which at the same time are the most important functionals in physical applications. In their case, the extremal condition is equivalent to the solution of a class of differential equations known as Euler–Lagrange equations.

We reasoned that the form of the Euler–Lagrange equation is preserved under coordinate changes, which makes functional calculus a powerful tool in cases where the right choice of coordinates is crucial. Specifically, for the local action functional of mechanics, whose Lagrangian is the difference of kinetic and potential energy of a point particle, we demonstrated that the Cartesian incarnation of the Euler–Lagrange equation equals the Newton equations. However, unlike the Newton equations, the variational equations remain form-invariant under a coordinate change and we demonstrated how this invariance can be put to advantage in the solution of problems with symmetries.

Given the identification of the Newton differential equations as variational equations of an action functional, one may wonder whether other differential equations fundamental to physics, the Schrödinger equation, the Maxwell equations, the Einstein equations, etc., might afford similar interpretations. The answer is affirmative, and has led to the definition of the *action principles* of quantum mechanics, electrodynamics, relativity and various other disciplines. In each case, the fundamental equations can be obtained as extrema of a corresponding action functional. These functionals are the gateway into the modern formulation of classical and quantum field theory. Their mathematics builds on the principles introduced above and exhibits a high level of universality and common structures. For this reason, functional calculus is ideally suited to explore connections between seemingly different areas of physics and has become an indispensable tool of modern theoretical physics.

Calculus of complex functions

Although the complex number field $\mathbb{C}$ may appear to be more complicated than the real one, we had frequent occasion to observe that the "complex" descriptions of problems can be simpler than real ones – think of Fourier calculus, or the diagonalization of matrices as examples. However, the applications discussed so far are no more than a prelude to the full potential of complex number representations. The latter will be unlocked in the present chapter where we introduce the complex version of *calculus*, complex differentiation, integration, etc. The ensuing framework will be much more powerful, and in many ways simpler than real calculus. For example, the application of complex integration theorems to even nominally real integrals often leads to results difficult to obtain otherwise.

Readers not yet familiar with complex calculus may find these remarks perplexing: in Part L, we reasoned that the complex numbers $z = x + iy$ can be parametrized through two real coordinates x, y and therefore define a two-dimensional real vector space, $\mathbb{C} \simeq \mathbb{R}^2$. This suggests that complex calculus might be a variant of two-dimensional real calculus. However, what this argument misses is that complex numbers can be multiplied with each other. This means that we are dealing with an upgraded variant of $\mathbb{R}^2$, turned into a two-dimensional real *algebra* (see p. 71) by the presence of a product operation. It is this added feature which gives complex calculus its strength.

C9.1 Holomorphic functions

Complex calculus addresses the properties of differentiable complex functions, $f : U \rightarrow \mathbb{C}, z \mapsto f(z)$, where $U \subset \mathbb{C}$ is an open subset of $\mathbb{C}$. The concept of complex differentiability was introduced in Section C5.2. To repeat, f is differentiable at z if the limit

$$f'(z) \equiv \frac{df(z)}{dz} \equiv \lim_{\delta \to 0} \frac{1}{\delta}(f(z + \delta) - f(z)) \tag{C230}$$

exists. Here, the existence criterion requires that the limiting value be independent of the particular direction in the complex plane in which the complex parameter δ is sent to zero. If f is differentiable for all $z \in U$ it is called **holomorphic** or **analytic** in U. Examples of functions holomorphic in all of $\mathbb{C}$ include the monomials $z^l, l \in \mathbb{N}$, convergent power series, or the functions $\exp(z), \sin(z), \cos(z)$ (by virtue of their power series representations). In these cases, holomorphy is established by direct computation of the limit, for example, $\delta^{-1}((z + \delta)^2 - z^2) = 2z + \delta \to 2z$.

holomorphic function

Recall that the product rule and the chain rule of differentiation are direct consequences of the limit definition. In the complex case, this leads to

$$\text{product rule:} \quad \frac{d}{dz}(fg)(z) = f'(z)g(z) + f(z)g'(z),$$

$$\text{chain rule:} \quad \frac{d}{dz}(f(g(z))) = f'(g(z))g'(z), \quad \text{(C231)}$$

where the existence of all derivatives is assumed.

Cauchy–Riemann differential equations

Let $f(z)$ be a holomorphic function of $z = x + iy$ and decompose it into real and imaginary parts, $f(x+iy) = u(x,y)+iv(x,y)$. Now compute the complex derivative of f in two different ways, for $\delta = \delta_x$ purely real or $\delta = i\delta_y$ imaginary, respectively:

$$f'(z) = \lim_{\delta_x \to 0} \frac{1}{\delta_x}\left[f(x+iy+\delta_x) - f(x+iy)\right] = \partial_x f(x+iy) = \partial_x u(x,y) + i\partial_x v(x,y),$$

$$f'(z) = \lim_{\delta_y \to 0} \frac{1}{i\delta_y}\left[f(z+iy+i\delta_y) - f(z)\right] = -i\partial_y f(x+iy) = -i\partial_y u(x,y) + \partial_y v(x,y).$$

Since f is holomorphic, the two expressions must be equal, $\partial_x u + i\partial_x v = -i\partial_y u + \partial_y v$. Noting that this equality must hold for real and imaginary part separately, we obtain the **Cauchy–Riemann differential equations**,

Cauchy–Riemann equations

$$\boxed{\partial_x u = \partial_y v, \quad \partial_y u = -\partial_x v.} \quad \text{(C232)}$$

Augustin-Louis Cauchy
1789–1857
A French mathematician generally considered as one of the fathers of modern analysis and in particular of complex analysis. However, Cauchy also contributed to many other areas of mathematics and physics including algebra, number theory, wave mechanics, and elasticity.

These two first-order partial differential equations express the condition of complex differentiability in the language of real functions.

EXAMPLE Consider the function $f(z) = u(z) + iv(z) = (1 + iz)^2$. Substituting $z = x + iy$ we obtain $u(x,y) = 1-2y-x^2+y^2$ and $v(x,y) = 2x-2xy$. These functions indeed satisfy the Cauchy–Riemann equations: $\partial_x u = -2x = \partial_y v$ and $\partial_y u = -2 + 2y = -\partial_x v$. ($\to$ C9.1.1-2)

complex vs. two-dimensional real differentiability

INFO It is instructive to compare the conditions of *complex and two-dimensional real differentiability*. Identifying $\mathbb{C} \simeq \mathbb{R}^2$ with a two-dimensional real space, $f(x,y) = u(x,y) + iv(x,y)$ can be interpreted as a particular vector-valued function, $(x,y) \mapsto (u(x,y), v(x,y))$. Such functions are differentiable in the real sense if the partial derivatives $\partial_x u$, $\partial_y u$ and $\partial_x v$, $\partial_y v$ exist. The Cauchy–Riemann equations define a more stringent condition. Retracing their derivation, we note that they originate in the fact that $f(x,y)$ depends on its arguments only in the specific combination $x + iy$. The complex representation of a function depending on $x = (z+\bar{z})/2$ and $y = -i(z-\bar{z})/2$ in arbitrary combinations would be one depending on both z and $\bar{z}$. For example, the function $z\bar{z} = x^2 + y^2$ fails the Cauchy–Riemann test and is not holomorphic.

Analyticity

One can show that a function is analytic in U if and only if it can be expanded in a *complex Taylor series* around each $z_0 \in U$, i.e. if there exists a representation (see Section C5.2)

$$f(z) = \sum_{n=0}^{\infty} a_n(z - z_0)^n, \qquad a_n = \frac{f^{(n)}(z_0)}{n!}. \tag{C233}$$

radius of convergence The maximal value ρ such that the series converges for all $z \in U$ lying within ρ of z_0, $|z - z_0| \le \rho$, defines its **radius of convergence** about z_0. The disk defined by the radius of convergence is fully contained in U, as indicated in the figure. Here are a few examples of complex functions and their analyticity properties:

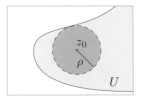

 ▷ $\exp(z), \sin(z), \cos(z)$ are power series expandable around any $z \in \mathbb{C}$ and therefore globally analytic;
 ▷ $\bar{z} = x - iy$ – is not analytic, because it violates Eqs. (C232);
 ▷ $|z|$ – is also not analytic, for the same reason;
 ▷ $\frac{1}{z-w}$ – is analytic in $\mathbb{C}\backslash\{w\}$.

singularity A point $z_0 \in \mathbb{C}$ where a function f is not analytic is called a **singularity**. Notice that a singularity need not imply diverging behavior. For example, $z = 0$ is a singularity of the function $|z|$.

Geometric interpretation of holomorphy

Let us revisit the $\mathbb{R}^2$-interpretation of complex functions to give the concept of holomorphy a geometric meaning. As mentioned above, the function $f(z) = u(x, y) + iv(x, y)$ defines a map, $\mathbf{y} \equiv (x, y)^{\mathrm{T}} \mapsto \mathbf{r}(\mathbf{y}) \equiv (u(x, y), v(x, y))^{\mathrm{T}}$, between the real and imaginary parts of z and those of f. Under this map, curves in the xy-plane obtained by keeping one of the coordinates constant (Fig. C30 left) are mapped to curves in the uv-plane, with tangent

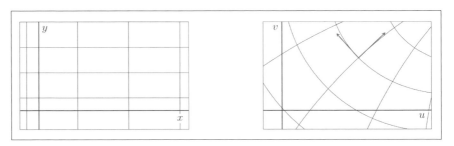

Fig. C30 A holomorphic map, $f(x + iy) = u(x, y) + iv(x, y)$, sends a perpendicular grid of coordinate lines in the xy-plane to a distorted but angle-preserving grid in the uv-plane.

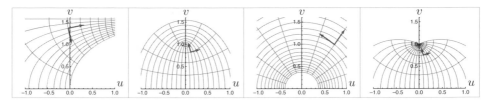

Fig. C31   Image curves of the holomorphic functions (left to right) $\log(z)$, $\sinh(z)$, $\exp(z)$ and $\tanh(z)$, plotted for arguments $z = x + iy$, with $x \in [-1, 1]$ and $y \in [0, 2\pi]$, discretized into an argument grid with 20×20 lines. The arrows are the tangents at arguments $(x_0, y_0) = (0.6, \pi/5)$.

vectors $\partial_x \mathbf{r}$ and $\partial_y \mathbf{r}$ (right). If f is holomorphic, the Cauchy–Riemann equations ensure that these vectors are orthogonal to each other,

$$\partial_x \mathbf{r} \cdot \partial_y \mathbf{r} = \partial_x u \, \partial_y u + \partial_x v \, \partial_y v \overset{(C232)}{=} \partial_x u \, \partial_y u - \partial_y u \, \partial_x u = 0.$$

conformal map

We conclude that the images of curves in the xy-plane crossing at $90\deg$ likewise cross at $90\deg$ in the uv-plane. This is a signature of an angle-preserving, or **conformal** map, and it is a direct consequence of holomorphy. Figure C31 illustrates this feature for various conformal maps.

C9.2 Complex integration

REMARK Requires Section V3.6.

In this section, we define integrals over complex functions. Due to the isomorphy $\mathbb{C} \simeq \mathbb{R}^2$ these integrals have a lot in common with integrals in two-dimensional real space. However, they have much stronger mathematical properties than these – due to the Cauchy–Riemann equations, as we will see – and can often be computed without any explicit calculation of principal functions and similar objects. This makes complex integration a powerful tool even for the computation of real integrals which cannot be easily computed by different methods.

Definition of complex integrals

Consider a *curve in the complex plane*, $\gamma : I \to \mathbb{C}$, $t \mapsto z(t) = x(t) + iy(t)$. If $\mathbb{C} \simeq \mathbb{R}^2$ were identified with two-dimensional real space, the canonical objects to integrate along the curve would be two-dimensional vector fields. In the following, we will show that their role is now taken by holomorphic functions.

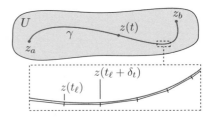

To this end, consider a function $f(z)$ holomorphic in a domain U containing γ. Proceeding as in the construction of the line integral, we discretize "time" into segments of size $\delta_t = t_{\ell+1} - t_\ell$, thus partitioning the curve into segments, $z(t_{\ell+1}) - z(t_\ell)$. Now weight each segment with $f(z_{t_\ell})$ and form a Riemann form a sum to obtain

$$\sum_l f(z(t_\ell))(z(t_{\ell+1}) - z(t_\ell)) \simeq \delta_t \sum_l f(z_{t_\ell})\tfrac{d}{dt}z(t_\ell) \xrightarrow{\delta_l \to 0} \int_I dt \, \tfrac{d}{dt}z(t)f(z(t)).$$

complex line integral This construction leads to the definition of the **complex line integral** or **contour integral** of a holomorphic function $f(z)$ along a curve γ as

$$\boxed{\int_\gamma dz f(z) \equiv \int_I dt \, \frac{dz(t)}{dt} f(z(t)).} \tag{C234}$$

As always with line integrals, $\int_\gamma dz f$ is symbolic notation for the expression on the right.

Suppose the curve $\gamma : (t_a, t_b) \to \mathbb{C}$ starts at $z_a = z(t_a)$ and ends at $z_b = z(t_b)$. If f can be expressed as the derivative, $f = F'$, of a holomorphic function F, the line integral can be computed as $\int_{t_a}^{t_b} dt \frac{dz(t)}{dt}\frac{dF(z(t))}{dz} = \int_{t_a}^{t_b} dt \frac{dF(z(t))}{dt} = F(z(t_b)) - F(z(t_a))$, a result compactly expressed as

$$\int_{z_a}^{z_b} dz f(z) = F(z_b) - F(z_a). \tag{C235}$$

This generalizes the fundamental theorem of calculus to contour integrals of functions of complex arguments. Equation (C235) states that the integral of a function $f = F'$ is independent of the shape of the curve connecting z_a and z_b. To understand this remarkable fact at an elementary level, note that the Riemann sum for $\int_{t_a}^{t_b} dt \frac{dF(z(t))}{dt}$ has the form $\sum_\ell [F(z(t_{\ell+1})) - F(z(t_\ell))]$. This sums up the changes in $F(z)$ across each discretization interval along the curve and hence yields its total change from start to end, irrespective of the curve's shape. As we will explain later, the path independence originates in the structure of the Cauchy–Riemann equations. We will see how the condition $f = F'$ relates to the holomorphy of f itself, and how holomorphy makes complex line integrals much easier to handle than two-dimensional real integrals.

EXAMPLE Consider the contour integral $\int_\gamma dz\, f(z)$ of the function $f(z) = z$ along (i) a straight line, $\gamma_1 : z(t) = (1+i)t$, connecting 0 with $1+i$, and (ii) a parabola, $\gamma_2 : z(t) = t + it^2$, between the same points, with $t \in (0, 1)$:

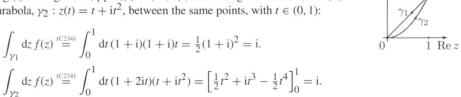

$$\int_{\gamma_1} dz\, f(z) \overset{(C234)}{=} \int_0^1 dt\,(1+i)(1+i)t = \tfrac{1}{2}(1+i)^2 = i.$$

$$\int_{\gamma_2} dz\, f(z) \overset{(C234)}{=} \int_0^1 dt\,(1+2it)(t+it^2) = \left[\tfrac{1}{2}t^2 + it^3 - \tfrac{1}{2}t^4\right]_0^1 = i.$$

Integration along the two paths yields the same result, illustrating the path independence of the contour integral. The reason is that $f(z) = z = \tfrac{1}{2}d_z z^2$ is the derivative of a holomorphic function.

EXAMPLE Consider the integral of $(z - z_0)^n$ $(n \in \mathbb{Z})$ along an arc S_ϕ, of radius R, central point z_0, and opening angle ϕ, connecting the points $z_a = z_0 + R$ and $z_b = z_0 + R e^{i\phi}$. With the parametrization $z(t) = z_0 + R e^{it}$, $t \in (0, \phi)$, this becomes

$$\int_{S_\phi} dz\, (z - z_0)^n = \int_0^\phi dt\, \frac{dz(t)}{dt} (z(t) - z_0)^n = \int_0^\phi dt\, (iR e^{it})(R e^{it})^n$$

$$= \begin{cases} \frac{R^{n+1}}{(n+1)} \left(e^{i(n+1)\phi} - 1 \right), & n \neq -1, \\ i\phi, & n = -1. \end{cases}$$

For $\phi = 2\pi$ the contour is a closed circle of radius R. Since $\exp(i2\pi(n+1)) = 1$, the integral along $S_{2\pi}$ vanishes for any $n \neq -1$. This includes negative powers $n \leq -2$, even though in this case $(z - z_0)^n$ is singular at z_0. Only for $n = -1$ is a finite result, $\int_{S_{2\pi}} dz\, (z - z_0)^{-1} = 2\pi i$, obtained. If the circle is traversed in a clockwise direction the answer changes sign (check this!). To summarize,

$$\oint_{S_\pm} dz\, (z - z_0)^n = \pm 2\pi i\, \delta_{n,-1}, \tag{C236}$$

where $S_\pm$ denotes a circle of constant radius centered on z_0, traversed in a **positively oriented** (counter-clockwise) direction or a negatively oriented (clockwise) direction, respectively.

To appreciate the significance of this result, let $f(z)$ be a function holomorphic inside the integration contour. The function then affords an expansion $f(z) = \sum_{n \geq 0} a^n (z - z_0)^n$ around the central point. By Eq. (C236), each term in the series vanishes under integration, which means that the function itself has a vanishing contour integral. We conclude that the line integrals along circular contours over holomorphic functions vanish. In the next section we will extend this result and generalize it to arbitrary integration contours.

INFO The result (C236) implies a useful representation of the coefficients, a_n, of the complex series, Eq. (C233) of holomorphic functions, $f(z) = \sum_{n \geq 0} a_n (z - z_0)^n$: let S be a circle inside the domain of **coefficients** holomorphy of the function. For fixed n consider the counter-clockwise contour integral of $f(z)/(z - z_0)^{n+1}$ along that circle,
of complex
Taylor series
from
integrals

$$\oint_S dz\, \frac{f(z)}{(z - z_0)^{n+1}} = \sum_{m \geq 0} a_m \oint_S dz\, (z - z_0)^{m-n-1} = 2\pi i a_n,$$

where in the last step Eq. (C236) was used. We thus obtain the representation

$$a_n = \frac{1}{2\pi i} \oint_S dz\, \frac{f(z)}{(z - z_0)^{n+1}}, \tag{C237}$$

which is useful in various contexts and will be applied in the proof of Liouville's theorem below.

Cauchy's theorem

Let U be a simply connected domain in the complex plane. We recall (see Section V3.4) that a domain is simply connected if it is connected (any two points in U can be connected by a curve in U) and any

**Cauchy's
theorem**

closed curve in U can be shrunk to a point (see figure, where the left domain is simply connected, the right not). Let f be a function holomorphic in U and γ a closed curve in U. **Cauchy's theorem** then states that

$$\oint_\gamma \mathrm{d}z f(z) = 0, \qquad\qquad\qquad\text{(C238)}$$

i.e. the line integral of a holomorphic function along a closed curve is zero. Before proving the theorem, we note an equally important consequence.

> If a function is holomorphic in a simply connected domain U, its contour integral along a curve in U depends only on the endpoints of the curve and not on its shape.

The reason is that two different curves, γ_1 and γ_2, with the same terminal points, z_a and z_b, may be concatenated to form a closed curve,[57] $\gamma = \gamma_1 \cup (-\gamma_2)$, where $-\gamma_2$ denotes the reversed version of γ_2. This leads to $\int_{\gamma_1} \mathrm{d}z f - \int_{\gamma_2} \mathrm{d}z f = \oint_\gamma \mathrm{d}z f = 0$, where the last equality is due to Cauchy. This path independence underlies the complex version of the fundamental theorem of calculus, Eq. (C235), encountered above.

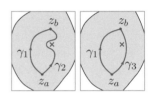

Thanks to this corollary, integration contours connecting two specified points may be arbitrarily deformed, as long as the deformation does not pass through regions where f is not holomorphic. For example, in the figure, where the cross represents a point of non-holomorphy, the curve γ_2 is a valid deformation of γ_1, but γ_3 is not.

To *prove Cauchy's theorem* we rephrase the definition of the contour integral in the language of real vector fields. Upon insertion of $f(x+iy) = u(x,y)+iv(x,y)$ into Eq. (C234), the real and imaginary parts of $\int_\gamma \mathrm{d}z f(z)$ take the form $\int_I \mathrm{d}t\,(\dot{x}(t)u(t) - \dot{y}(t)v(t))$ and $\int_I \mathrm{d}t\,(\dot{x}(t)v(t) + \dot{y}(t)u(t))$, respectively. Both these expressions have the structure of two-dimensional line integrals (see (V12)) of real vector fields, $(u, -v)^{\mathrm{T}}$ and $(v, u)^{\mathrm{T}}$, respectively, integrated along a closed curve, γ, with parametrization $\mathbf{r}(t) = (x(t), y(t))^{\mathrm{T}}$ in the two-dimensional plane. Stokes's theorem (V106) may now be applied to convert these line integrals to area integrals over the surface, S, enclosed by γ.[58] This leads to the expressions

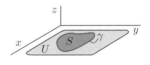

$$\mathrm{Re}\oint_\gamma \mathrm{d}z f = \oint_\gamma \mathrm{d}\mathbf{r}\cdot(u,-v,0)^{\mathrm{T}} = \int_S \mathrm{d}S(\nabla\times(u,-v,0)^{\mathrm{T}})_z = \int_S \mathrm{d}S\big(\partial_x(-v)-\partial_y u\big) = 0,$$

$$\mathrm{Im}\oint_\gamma \mathrm{d}z f = \oint_\gamma \mathrm{d}\mathbf{r}\cdot(v,u,0)^{\mathrm{T}} = \int_S \mathrm{d}S(\nabla\times(v,u,0)^{\mathrm{T}})_z = \int_S \mathrm{d}S\big(\partial_x u-\partial_y v\big) = 0,$$

---

[57] Recall the very similar reasoning employed in Section V3.4 to show the path independence of integrals of gradient fields.

[58] Stokes's theorem applies to three-dimensional vector fields. Here, the two-dimensional xy-plane may be considered as the $z = 0$ plane of a three-dimensional space to make it applicable.

where the final equalities follow from the Cauchy–Riemann equations (C232). This establishes Cauchy's theorem. ($\to$ C9.2.1-2)

Cauchy's theorem is a powerful ally in both complex *and* real integration theory. Integrals over real functions can often be processed by interpreting the integral as a complex one, followed by application of the theorem. In the following example we apply this strategy to the computation of a real integral over a rational function.

EXAMPLE Consider the integral $I \equiv \int_{-\infty}^{\infty} dx\, r(x)$ over the rational function $r(x) = \frac{x^2-1}{(x^2+1)^2}$. Integrals of this type can be computed by elementary yet somewhat laborious variable substitutions. (Attempt the above integral in that way!) Complex integration is an elegant and less work-intensive alternative to this approach. First note that $r(x) = \mathrm{Re}(x+i)^{-2}$, and represent the integral over the real axis as a contour integral of the function $f(z) \equiv (z+i)^{-2}$ in the complex plane: write

$$I = \mathrm{Re}\int_{-\infty}^{\infty} dx\, \frac{1}{(x+i)^2} = \mathrm{Re}\lim_{R\to\infty}\int_{\Gamma_0} dz\, f(z),$$

where $\Gamma_0 = (-R,R)$ is a segment of the real axis. Next we employ an often-used trick and *close the contour* by joining the endpoints of Γ_0 with a radius-R semicircle, Γ_+, in the upper complex plane. In the limit $R\to\infty$, the added contribution, $\int_{\Gamma_+} dz f(z)$, equals zero, because for large $|z|=R$ the integrand decays as $f(z)\sim R^{-2}$, whereas the radius of the circle grows only as $\sim R$. (Verify this asymptotic vanishing via an explicit parametrization, $z(\phi)=Re^{i\phi}$, of the semicircular line integral.) The combined contour, $\Gamma_0\cup\Gamma_+ = $ ⟳, is closed and confined to the upper half-plane, where $f(z)$ is holomorphic. (The only singularity of f lies at $z=-i$, indicated by the cross in the figure.) Cauchy's theorem then states that $\int dz f(z)=0$ for any R. On the other hand, in the limit $R\to\infty$, only $\lim_{R\to\infty}\Gamma_0 = \mathbb{R}$ contributes to the integral, and so we have demonstrated that $I=0$.

The contour-closing trick described in the above example is quite powerful and it is worth formulating general criteria for its applicability. To this end, consider a semicircle in the upper or lower half-plane, Γ_+, parametrized as $z(\phi)=Re^{i\phi}$, $\phi\in(0,\pm\pi)$. Then, the line integral $\int_{\Gamma_\pm} dz f(z) = iR\int d\phi\, e^{i\phi}f(Re^{i\phi})$ vanishes in the limit $R\to\infty$ provided that the function has the property

$$\lim_{|z|\to\infty} zf(z) = 0. \tag{C239}$$

In that case, a semicircle at infinity can be "added" to the contour without changing the value of the contour integral.

C9.3 Singularities

Most functions of interest are not holomorphic throughout all of $\mathbb{C}$. For example, **Liouville's theorem** states that every function f that is both bounded (i.e. $|f(z)| < M$ for some $M>0$) and holomorphic in all of $\mathbb{C}$, must necessarily be constant. This means that interesting bounded functions (of which there are many) must contain points or regions where they are not holomorphic.

Liouville's theorem

INFO   The *proof of Liouville's theorem* beautifully illustrates the power of series expansions in complex calculus: let f be holomorphic in all of $\mathbb{C}$ *and* bounded, $|f(z)| < M$ for some $M > 0$. Holomorphicity guarantees a series representation, $f(z) = \sum_{n=0}^{\infty} a_n z^n$, with infinite radius of convergence, where 0 has been chosen as the center of expansion for convenience. Consider a circle of radius R centered on this point and represent the coefficients via the contour integral (C237). Choosing the parametrization $z(\phi) = Re^{i\phi}$, this leads to the estimate

$$|a_n| \leq \frac{1}{2\pi} \oint dz \left| \frac{f(z)}{z^{n+1}} \right| \leq \frac{1}{2\pi} (2\pi R) \frac{M}{R^{n+1}} = MR^{-n}, \tag{C240}$$

where the factor $2\pi R$ in the numerator comes from the length of the integration contour. This means that $|a_n| \leq MR^{-n}$ for arbitrarily large R, which in turn implies $a_n = 0$, unless $n = 0$. The series thus collapses to $f(z) = a_0$, showing that f is a constant.

Classification of singularities

As we will see below, complex singularities are interesting objects which can even be potent allies in the integration of functions. Compared to real functions, there exists a richer spectrum of singularities, where "singularity" need not mean divergent behavior. Figure C32 provides a schematic of the different types, ordered according to their degree of severity. In the following, we define these cases, and then address their role in complex integration.

isolated singularity   Complex singularities may be isolated or extended. A function has an **isolated singularity** at $z_0 \in U$ if it is holomorphic on $U \backslash \{z_0\}$, where U is an open neighborhood of z_0. For example, the singularity of the function z^{-1} at $z = 0$ is isolated because the function can be series expanded around any point different from zero. By contrast, the square **extended singularity** root function, $z^{1/2}$, has an **extended singularity** along the negative real axis related to its notorious sign ambiguity, e.g. $\sqrt{4} = \pm 2$. We will return to the discussion of this point in Section C9.6.

removable singularity   Turning to the isolated singularities, the least singular of these is a **removable singularity**. A removable singularity is an isolated point where $f(z)$ is not defined. It is removable in the sense that a function value at the critical point may be defined to obtain a holomorphic function. The canonical example in this context is the function $\text{sinc}(z) \equiv \sin(z)/z$. This function has a problem at $z_0 = 0$ where it shows a 0/0 ambiguity. However, recalling that the sine function behaves as $\sin z = z + \mathcal{O}(z^3)$ for small arguments, one may define $\text{sinc}(0) \equiv 1$ and remove the singularity. Patched in this way, the function becomes globally holomorphic and its Taylor series representation around $z = 0$ follows from that of the sine function.

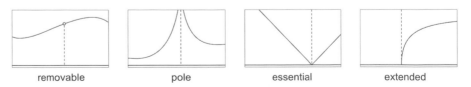

removable   pole   essential   extended

Fig. C32   A one-dimensional cartoon of function singularities in ascending order of severity.

pole   A **pole** is an example of a more serious singularity. The function $f(z)$ has a pole at z_0 if $1/f(z)$ has a zero at z_0. In this case, there exists a neighborhood U of z_0 such that f is analytic on $U\backslash\{z_0\}$ but not at z_0, and a holomorphic function, $g : U \to \mathbb{C}$, with non-vanishing $g(z_0)$, such that for all $z \in U\backslash\{z_0\}$,

$$f(z) = \frac{g(z)}{(z - z_0)^n}, \qquad (C241)$$

with n a positive integer. The smallest n for which such a representation exists is called the *order of the pole*; poles of order 1 are called *simple* poles. A pole is isolated in the sense that it has a neighborhood which does not contain any other poles. A few examples:

▷ $f_1(z) = \frac{1}{(z-i)^2}$ has a pole of order 2 at $z_0 = i$.

▷ $f_2(z) = \frac{1}{(z+1)(z-1)^2}$ has a pole of order 1 at $z_1 = -1$ and a pole of order 2 at $z_2 = 1$.

▷ $f_3(z) = \frac{1}{e^{iz}+1}$ has a pole of order 1 at $z_0 = \pi$. To see this, expand the exponential function in powers of $\delta = z - \pi$ as $e^{i(\pi+\delta)} = -1[1 + i\delta - \delta^2/2 + \mathcal{O}(\delta^3)]$. This leads to $f_3(z) \simeq \frac{1}{\delta(-i+\delta/2)}$, which has the form $\frac{g(z)}{z-\pi}$, with $g(z) \simeq \frac{1}{-i+(z-\pi)/2}$ near $z \simeq \pi$.

essential singularity   An isolated singularity which is neither removable nor a pole is called an **essential singularity**. For example, the function $|z|$ has an essential singularity at $z = 0$ (but notice the absence of any divergences!).

Laurent series

The nomenclature distinguishing between poles and essential singularities suggests that poles are somehow considered "non-essential". But why is this? The answer lies in the analyticity of the function g in (C241), which in turn implies the existence of a Taylor series expansion, $g(z) = \sum_{m\geq 0} b_m(z - z_0)^m$. If we substitute this into the pole expression (C241), we obtain the series representation

$$f(z) = \sum_{m=-n}^{\infty} a_m(z - z_0)^m, \qquad (C242)$$

Laurent series   where the coefficients $a_m = b_{m+n}$ are determined by the expansion of g. Series of this type, starting at a negative exponent $-n$, are called **Laurent series**. The singularity of the function f is now encoded in a finite number of simple functions $(z - z_0)^{m<0}$, and in this sense is non-essential.

meromorphic functions   Functions that are holomorphic except for finitely many points where they have poles are called **meromorphic functions**. The terminology (inspired by the Greek word *meros*="part") suggests that they stand halfway between the holomorphic and the truly singular functions. The coefficient $a_{-1} \equiv \mathrm{Res}(f, z_0)$ of the Laurent expansion at a pole is **residue**   called the **residue** of f at z_0. In the next section will see that the residue plays a central role in the complex integration of meromorphic functions.

EXAMPLE The second function in the list above, $f_2(z) = \frac{1}{(z+1)(z-1)^2}$, is meromorphic in $\mathbb{C}$. In the vicinity of its second-order pole at $z_2 = 1$, the first factor may be Taylor-expanded as

$$\frac{1}{z+1} = \frac{1}{2[1 + \frac{1}{2}(z-1)]} = \sum_{m=0}^{\infty} \frac{1}{2} \left(-\frac{1}{2}\right)^m (z-1)^m.$$

The Laurent series representation of $f_2(z)$ near $z_2 = 1$ therefore has the form

$$\frac{1}{(z+1)(z-1)^2} \stackrel{z \cong 1}{=} \sum_{m=-2}^{\infty} \frac{1}{2} \left(-\frac{1}{2}\right)^{m+2} (z-1)^m, \tag{C243}$$

and the coefficient of $(z-1)^{-1}$ yields the corresponding residue, $\text{Res}(f_2, 1) = -\frac{1}{4}$.

INFO As we will see in the next section, residues play a central role in the computation of various types of integrals, both real and complex. It is therefore important to know how to *compute the residues of meromorphic functions*. One approach is to represent the function f in the form (C241), followed by a Taylor expansion of the function g. In practice, g is often easy to identify by inspection and this method works reasonably well. Alternatively, one may compute the derivative

$$\text{Res}(f, z_0) = \frac{1}{(n-1)!} \left[\partial_z^{n-1} \left((z - z_0)^n f(z) \right) \right]_{z = z_0}, \tag{C244}$$

where n is the order of the pole of f at z_0. To understand this formula consider the Laurent series of f and notice that

$$\partial_z^{n-1} \left((z - z_0)^n f(z) \right) = \partial_z^{n-1} \sum_{m=-n}^{\infty} a_m (z - z_0)^{n+m} = \sum_{m=-1}^{\infty} \frac{(n+m)!}{m!} a_m (z - z_0)^{m+1},$$

where in the second step we used that $(z - z_0)^{n+m}$ vanishes under the derivative if $n + m$ is smaller than $n - 1$. If we now set $z = z_0$, all terms in the series except the lowest, $m = -1$, vanish. So we are left with $a_{-1}(n-1)!$, which shows that Eq. (C244) picks out a_{-1}, the residue of the function at z_0.

If f has a pole of order 1 and is known in the representation $f(z) = \frac{g(z)}{z - z_0}$, the residue formula (C244) reduces to $\text{Res}(f, z_0) = g(z_0)$.

EXAMPLE The residues of the function f_2 from p. 347, with poles at $z_1 = -1$ and $z_2 = 1$, are:

$$\text{Res}(f_2, -1) = \left[(z+1) f(z) \right]_{z=-1} = \frac{1}{4}.$$

$$\text{Res}(f_2, 1) = \left[\partial_z \left((z-1)^2 f(z) \right) \right]_{z=1} = \left[\partial_z \frac{1}{z+1} \right]_{z=1} = -\frac{1}{4}.$$

The second line reproduces the residue obtained above via the series expansion (C243). ($\rightarrow$ C9.3.1-2)

C9.4 Residue theorem

Suppose we want to integrate a meromorphic function f around a path γ encircling one or more of its poles (see Fig. C33(a)) in the mathematically positive direction. Is there anything that can be said in general about the outcome of the integration? Not much,

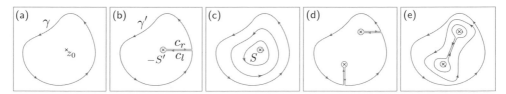

Fig. C33   (a) Integration of a meromorphic function along a contour enclosing a pole. (b) Composite contour excluding the pole. (c) The
original contour can be deformed (without changing the value of the integral) to one encircling the pole along a small circle.
This integral in turn yields $2\pi\mathrm{i}$ times the corresponding residue of the function at the pole. (d, e) Analogous constructions
can be used for an arbitrary number of poles inside the contour.

one might suspect – should the result of the integration not depend on the choice of the
integration curve?

However, it is one of the marvels of complex calculus that the value of such integrals is
independent of the shape of the contour and determined solely by the pole structure of f.
To see this, consider the composite contour $\gamma_c = \gamma' \cup c_l \cup (-S') \cup c_r$ shown in Fig. C33(b),
where γ' is the curve γ cut at a single point, c_l and c_r run side by side from this point to
the neighborhood of the singularity and back, and $-S'$ is a small cut circle surrounding the
singularity in the mathematically negative, clockwise direction (hence the minus sign in the
notation). The composite contour γ_c was purposefully chosen such that f is holomorphic
in its interior and so Cauchy's theorem tells us that

$$\oint_{\gamma_c} \mathrm{d}z f = \int_{\gamma'} \mathrm{d}z f + \int_{c_r} \mathrm{d}z f + \int_{-S'} \mathrm{d}z f + \int_{c_l} \mathrm{d}z f = 0.$$

The integrals along $c_{l,r}$ mutually cancel out because they run along geometrically identical stretches traversed in opposite directions, $\int_{c_r} + \int_{c_l} = \int_{c_r} + \int_{-c_r} = 0$. Moreover, the
integral over γ' equals that over γ because removing a single point from an integration
domain does not change the value of an integral. Finally, $\int_{-S'} \mathrm{d}z f = -\oint_S \mathrm{d}f$, where S is
the positively traversed closed circle. Combining these observations, we arrive at the result

$$\oint_{\gamma} \mathrm{d}z f = \oint_S \mathrm{d}z f. \tag{C245}$$

This formula states that the original contour integral equals one tightly encircling the singularity. Intuitively, the original contour may be deformed into S as if it were a rubber
band draped around a peg (the pole) – since the domain enclosed between γ and S does
not contain singularities, Cauchy's theorem implies that the value of the integral remains
unchanged during this deformation.

If there are several poles the construction may be generalized as (Fig. C33(d,e))

$$\oint_{\gamma} \mathrm{d}z f = \sum_i \oint_{S_i} \mathrm{d}z f, \tag{C246}$$

where the sum is over all poles lying within γ, and each S_i is a small circle around such a
pole.

The residue theorem

In the previous section, we have seen that integration over meromorphic functions can be reduced to an integration over circular contours around their poles. Now assume that the function in question has been expanded in a Laurent series (C242) around one of these singularities. Equation (C236) then tells us that the counter-clockwise integration along the corresponding circle yields $2\pi i$ times the coefficient of order $m = -1$, i.e. the residue of the function at the singularity:

$$\oint_{S_i} dz f(z) = 2\pi i \operatorname{Res}(f, z_i). \tag{C247}$$

residue theorem Combining this formula with Eq. (C246), we arrive at the **residue theorem**,

$$\oint_{\gamma} dz f(z) = 2\pi i \sum_i \operatorname{Res}(f, z_i). \tag{C248}$$

(If the integration is clockwise, $\oint$, the answer changes sign.) The integral of a meromorphic function along a closed contour traversed in the mathematically positive or negative direction equals $\pm 2\pi i$ times the sum over the residues of all the poles enclosed by the contour. ($\rightarrow$ C9.4.1-2)

Examples

EXAMPLE Like Cauchy's theorem, the theorem of residues may be applied to the computation of real integrals. As an example, consider the integral

$$I \equiv \int_{-\infty}^{\infty} dx \frac{1}{x^2 + a^2}.$$

We proceed as in the example of p. 345 and interpret I as an integral of the complex function $f(z) = \frac{1}{z^2 + a^2}$ over the real axis. Since $\lim_{|z| \to \infty} z f(z) = 0$, Eq. (C239) legitimizes the closure of the integration contour by an infinite semicircle in either the upper or lower half-plane. This closed contour integral may be dealt with by the residue theorem. Assuming that $a > 0$, the factorization $f(z) = \frac{1}{(z-ia)(z+ia)}$ shows that f has two simple poles, sitting at $z_\pm = \pm ia$ in the upper and lower half-plane, with residues $\pm \frac{1}{2ia}$, respectively, see Fig. C34(a). A contour closed in the upper

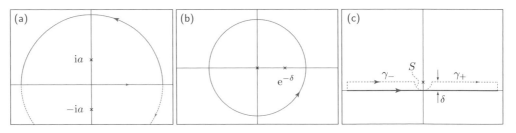

Fig. C34   Three examples of contour integration via the theorem of residues. Discussion, see text.

half-plane encircles the upper pole in the counter-clockwise direction and the residue theorem gives $I = 2\pi i \operatorname{Res}(f, z_+) = \frac{\pi}{a}$. Alternatively, a closure in the lower half-plane encircling the lower pole clockwise leads to the same result, $I = -2\pi i \operatorname{Res}(f, z_-) = \frac{\pi}{a}$. ($\to$ C9.4.3-6)

EXAMPLE Occasionally, one encounters integrals over complex functions which do not yet have the canonical form, Eq. (C234), of a complex contour integral. As an example, consider the integral (which finds applications in quantum mechanics)

$$I \equiv \int_0^{2\pi} d\phi \, \frac{1}{1 - e^\delta e^{i\phi}},$$

where $\delta > 0$. Defining $z = e^{i\phi}$ and noting that $\frac{dz}{d\phi} = iz$, we can express this integral as

$$I = \oint_S dz \, \frac{1}{iz(1 - e^\delta z)},$$

where the integration contour is the unit circle around the origin (see Fig. C34(b)). There are two first-order poles, at $z_1 = 0$ and $z_2 = e^{-\delta}$, with residues i and $-i$, respectively. Both lie within the contour and the residue theorem yields $I = 0$.

An alternative way to obtain this result is to expand the ϕ-representation of the integral in a power series in $e^{i\phi}$ and to show that each term in the expansion vanishes (try it!). As an exercise you may explore what happens if $\delta < 0$. Does the integral still vanish? Compute its value by an explicit expansion and by the theorem of residues.

EXAMPLE As another important example, consider the integral

$$I \equiv \int_{-\infty}^{\infty} dx \, \frac{f(x)}{x - i\delta}, \tag{C249}$$

where $\delta > 0$ is infinitesimal, $f(z)$ is analytic in a strip enclosing the entire real axis and assumed to decay for $|x| \to \infty$ to make the integral existent. Now consider deforming the integration contour by shifting it upwards from the real axis to obtain the dotted contour shown in Fig. C34(c). It consists of two infinitesimally short vertical pieces at $\operatorname{Re} z = \pm\infty$, making negligible contributions to the integral, two horizontal segments, $\gamma_- = \{z(x) = x + i\delta | x \in (-\infty, -\delta)\}$ and $\gamma_+ = \{z(x) = x + | x \in (\delta, \infty)\}$, and an infinitesimal, positively oriented semicircle connecting them, of radius δ and centered on the point $i\delta$, $S = \{z(\phi) = i\delta + \delta e^{i\phi} | \phi \in (\pi, 2\pi)\}$. Since no singularities of the integrand are crossed during this deformation, the integral remains unchanged (Cauchy's theorem):

$$I = \int_{\mathbb{R}} dz \, \frac{f(z)}{z - i\delta} = \int_{\gamma_- \cup S \cup \gamma_+} dz \, \frac{f(z)}{z - i\delta}.$$

Computing the three contributions separately we obtain

$$\int_{\gamma_-} dz \, \frac{f(z)}{z - i\delta} = \int_{-\infty}^{-\delta} dx \frac{f(x + i\delta)}{x} \simeq \int_{-\infty}^{-\delta} dx \frac{f(x)}{x},$$

$$\int_{\gamma_+} dz \, \frac{f(z)}{z - i\delta} = \int_{\delta}^{\infty} dx \frac{f(x + i\delta)}{x} \simeq \int_{\delta}^{\infty} dx \frac{f(x)}{x},$$

$$\int_S dz \, \frac{f(z)}{z - i\delta} = \int_{\pi}^{2\pi} d\phi \, (i\delta \, e^{i\phi}) \frac{f(i\delta + \delta e^{i\phi})}{\delta e^{i\phi}} \simeq i\pi f(0),$$

where the last equalities are based on the assumed continuity of f, i.e. the assumption that f does not vary noticeably over scales δ. We combine these results to obtain

$$\int_{-\infty}^{\infty} dx \frac{f(x)}{x - i\delta} = P \int_{-\infty}^{\infty} dx \frac{f(x)}{x} + i\pi f(0),$$

where the first term is a principal value integral, as defined in Eq. (C24). The result above is often

**Dirac
identity**

abbreviated as the **Dirac identity**,

$$\frac{1}{x - i\delta} = P \frac{1}{x} + i\pi \delta(x). \tag{C250}$$

Here it is understood that this formula makes sense only under an integral, and in the limit $\delta \to 0$. For example, with $f(x) = \frac{1}{1+x^2}$, we have

$$\int_{\mathbb{R}} dx \frac{1}{1 + x^2} \frac{1}{x - i\delta} = P \int_{-\infty}^{\infty} dx \frac{1}{x(1 + x^2)} + i\pi \int_{-\infty}^{\infty} dx \frac{\delta(x)}{1 + x^2} = i\pi,$$

where the principal value integral vanishes since its integrand is odd under $x \leftrightarrow (-x)$.

EXERCISE Compute the imaginary part of $(x - i\delta)^{-1}$ and convince yourself that in the limit $\delta \to 0$, it yields a representation of a δ-function (times π). This is an alternative way to understand the appearance of $i\pi \delta(x)$ in Eq. (C250).

EXAMPLE As a concrete and final example, we consider the inverse Fourier transform,

$$G(t) = \int_{-\infty}^{\infty} \frac{d\omega}{2\pi} e^{-i\omega t} \tilde{G}(\omega) = \int_{\mathbb{R}} dz f(z), \qquad f(z) = \frac{1}{2\pi} e^{-izt} \tilde{G}(z),$$

**Green
function of
harmonic
oscillator**

of the *Green function of the damped harmonic oscillator*, with $\tilde{G}(z) = \frac{-1}{z^2 + 2iz\gamma - \Omega^2}$ (Eq. (C202)). Provided that $\lim_{|z| \to \infty} z f(z) = 0$, it can be computed by closing the contour along an infinite semi-circle in the upper or lower half-plane. For $t = 0$ this is guaranteed by $\tilde{G}(z) \sim z^{-2}$. For $t \neq 0$, however, the exponential dependence of the factor e^{-izt} requires a more careful analysis. Exponential decay, rather than divergence, of this factor requires $\mathrm{Re}(-izt) < 0$. This in turn implies that the choice of contour is dictated by the sign of t: for $t > 0$ we need $\mathrm{Im}(z) < 0$ and must close the contour in the lower half-plane, for $t < 0$ in the upper half-plane (see Fig. C35).

The poles of $f(z)$ are determined by the zeros of the quadratic denominator, $z_\pm = -i\gamma \pm \sqrt{\Omega^2 - \gamma^2}$. Since the friction coefficient γ of the harmonic oscillator equation is positive, they lie in the lower half-plane. This means that the upper half-plane contour relevant to $t < 0$ does not enclose any singularities and $G(t < 0) = 0$ by Cauchy's theorem. For $G(t \geq 0)$, the lower half-plane contour does enclose the poles. In the *underdamped case*, $\gamma < \Omega$, there are two simple poles, $z_\pm = -i\gamma \pm \Omega_r$

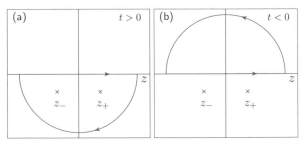

Fig. C35   Computation of the Green function of the underdamped harmonic oscillator as a residue integral (discussion, see text).

(depicted in Fig. C35), with residues $\text{Res}(f, z_\pm) = \mp \frac{e^{-(\gamma \pm i\Omega_r)t}}{2\pi(2\Omega_r)}$, where $\Omega_r = (\Omega^2 - \gamma^2)^{1/2}$. Summing over the residues according to the residue theorem (with minus sign for clockwise contours) one readily finds

$$G(t) = \tfrac{1}{\Omega_r} \sin(\Omega_r t) e^{-\gamma t}, \qquad (t \geq 0). \tag{C251}$$

Combining the results in a single formula, we arrive at the first term in Eq. (C204). The Green functions for the critically damped ($\gamma = \Omega$) and overdamped ($\gamma > \Omega$) regimes can be computed in a similar manner. ($\to$ C9.4.7-8)

C9.5 Essential singularities

We finally turn to the discussion of complex functions with singularities that can not be dealt with via generalized series expansions. The square root $\sqrt{z}$ is the simplest representative of this class. This function appears so frequently that it is easy to forget how strange it really is. To begin with, it is no one-to-one function at all: $\sqrt{4} = \pm 2$ has two solutions, i.e. $\sqrt{z}$ is multivalued. It is customary to restrict oneself to one "branch" of the function, e.g. $\sqrt{4} \equiv 2$, but the fact remains that there is an ambiguity. Also recall that the non-existence of $\sqrt{-1}$ in the reals was motivation to introduce the complex numbers in the first place. However, this extension does not settle the ambiguity problem, which in fact looks even worse than in the real case! To understand what is going on, let us parametrize complex numbers as $z = re^{i\phi}, r \geq 0$ and define $z^{1/2} = r^{1/2}e^{i\phi/2}$, where $r^{1/2} \geq 0$ is the positive branch of the real square root. This function is a valid "square root", because $(z^{1/2})^2 = (r^{1/2})^2(e^{i\phi/2})^2 = z$.

ambiguity of square root function

Now consider what happens with the real part, $\text{Re}(z^{1/2}) = r^{1/2}\cos(\phi/2)$, of this function as we move once around the unit circle, $r = 1$, starting and ending with the "almost real" arguments $z = 1 + i\delta$ and $z = 1 - i\delta$, respectively. During this circular motion, ϕ smoothly changes from δ to $2\pi - \delta$, and the square root function from $\sqrt{1 + i\delta} = \cos(\delta/2) \overset{\delta \to 0}{=} 1$ to $\sqrt{1 - i\delta} = \cos(\pi - \delta/2) \overset{\delta \to 0}{=} -1$. At first sight, this looks like an acceptable result: the complex square root function incorporates the two branches of the real square root in such a way that near the real axis the small imaginary part of the argument in $\sqrt{1 \pm i\delta} \simeq \pm 1$ signals which branch to pick. The price to be paid for that "switch functionality" is a singular jump of height $1 - (-1) = 2$ as the positive real axis at $r = 1$ is crossed in the imaginary direction, see the top left panel of Fig. C36 for a visualization.

However, on closer inspection it becomes apparent that the construction recipe above remains ambiguous. First, even the complex square root function has two branches: $\pm z^{1/2} = \pm r^{1/2}e^{i\phi/2}$ both square to z and each is a legitimate option. (The sign-inverted branch is shown in the top center of Fig. C36). So, the question of which branch to pick is still with us. However, what is really unacceptable is that the construction is "parametrization dependent" and therefore does not make for a valid definition: we might have decided to parametrize the complex plane by a differently chosen angle, $\varphi \in (-\pi, \pi)$, as shown in

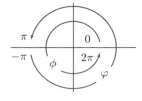

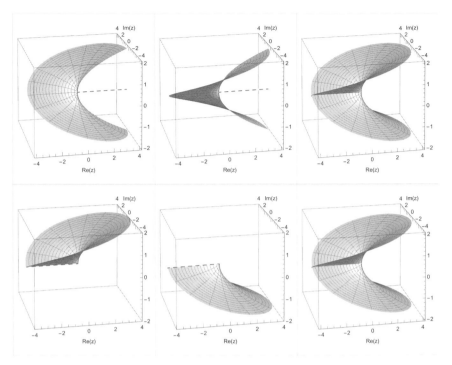

The two branches of the multi-valued function $\sqrt{z}$. Top row: The left and middle panels show the real part of the branches $\pm z^{1/2} = \pm r^{1/2}e^{i\phi}$, the right panel their combination, $\{z^{1/2}, -z^{1/2}\}$, with $\phi \in (0, 2\pi)$. Bottom row: Same as top row, but for branches $\pm z^{1/2} = \pm r^{1/2}e^{i\varphi}$ defined with a different angular coordinate, $\varphi = (-\pi, \pi)$. Further discussion, see text.

the figure. If we now define $z = re^{i\varphi}$, the two branches $\sqrt{z} = \pm r^{1/2}e^{i\varphi/2}$ are very different from the ones before (compare the bottom left and center panels of Fig. C36 with those in the top row).

All this indicates that $z^{1/2}$ cannot be an ordinary function. As a first step towards a better understanding of $z^{1/2}$, let us combine the two branches $\pm z^{1/2}$ into a set, $S \equiv \{(z^{1/2}, -z^{1/2})|z \in \mathbb{C}\}$. This set affords several interpretations. We can think of it as the set of solutions, w, of the complex equation $w^2 = z$, i.e. $w = \pm z^{1/2}$. Alternatively, it may be interpreted as the image of a "bi-valued" function, i.e. a function taking two values at each z. However, the most constructive view is geometrical: S defines a two-sheeted "surface", where the two vertically superimposed sheets represent the choices $\pm z^{1/2}$. A visualization of the real part of this surface[59] is shown in the right panels of Fig. C36, where both the sheets are represented in one plot. The most apparent features of this construction are: (i) the ensuing double-sheeted surfaces are globally smooth – they do not contain jumps of any kind; and (ii) the surfaces obtained for the two different parametrizations above are identical. This latter observation is important and indicates that the construction $S = \{(z^{1/2}, -z^{1/2})|z \in \mathbb{C}\}$ might contain the key to a good understanding of the

---

[59] Calling S a "surface" is metaphoric inasmuch as its elements are complex numbers. Surfaces in the traditional sense are obtained when real and imaginary parts of S are considered separately. However, this separation is not really natural in the present context.

square root: no matter which specific "coordinate representation" for $z^{1/2}$ is chosen, be it $z^{1/2} = \pm r^{1/2}\mathrm{e}^{\mathrm{i}\phi}$, or $\pm r^{1/2}\mathrm{e}^{\mathrm{i}\varphi}$, or yet another one, the union of the $+$ and $-$ branches always leads to the same surface, S.

C9.6 Riemann surfaces

Riemann surface

The construction above suggests a new way to think about a whole of class of multi-valued inverse functions. Consider a set of functions $f_1, \dots, f_n : U \to \mathbb{C}$, all solving the equation $g(f(z)) = z$ defining the inverse of a function $g(w)$. For example, for $g(w) = w^2$, the equation $(f(z))^2 = z$ has two solutions, $f_{1,2}(z) = \pm z^{1/2}$. We now define a set, $\bigcup_i \{f_i(z) | z \in U\}$, containing the combined images of these functions. This set is the n-sheeted **Riemann surface** of inverse function of $g(z)$. Figure C37 shows the real and imaginary parts of the

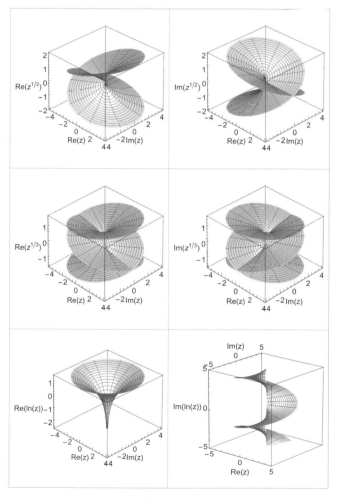

Fig. C37   Riemann surfaces of the three functions $z^{1/2}$ (top), $z^{1/3}$ (center), and $\ln(z)$ (bottom). The plots show the real (left) and imaginary (right) parts of the functions over the complex plane. Further discussion, see text.

Riemann surfaces of the two-sheeted function $z^{1/2}$, the three-sheeted function $z^{1/3}$ and the infinite-sheeted function $\ln(z)$.

Riemann
sheet
branch cuts

The individual **Riemann sheets**, $f_i(z)$, of the surface are glued together at lines which are called **branch cuts** and which emanate at **branch points**. Branch points are essential singularities – they are the end points of cut lines and therefore fundamentally distinct from pole singularities – whose positions in the complex plane follow from the definition of the function. However, both the choice of Riemann sheets

and that of the branch cuts are not canonical. For example, in our first/second parametrization above, the cut line is along the positive/negative real axis (indicated by dashed straight lines in Fig. C36) and we noted the differences in the Riemann sheets corresponding to the two choices. Either way, the square root function has one branch point at $z = 0$ and the other at $|z| = \infty$, and the choice of the line connecting them is arbitrary – Fig. C36 shows two possibilities.

For an example of a *function with two branch points* at finite values of z, consider $f(z) = \sqrt{1 - z^2} = \sqrt{(1 - z)(1 + z)}$. This function has branch points at $z = \pm 1$, corresponding to the essential singularities of the two square root factors. Depending on how one chooses the branch cuts of those, $f(z)$ has either a finite cut, $(-1, 1)$, along the real axis connecting the two branch points, or two disconnected segments, $(-\infty, -1) \cup (1, \infty)$, connecting ± 1 with $\pm\infty$.

Each Riemann sheet $f_i(z)$ has a discontinuity at the cut lines. Approaching the line from opposing directions, different values for f_i are obtained.[60] For example, a common choice for the branch cut of the function $\ln(z)$ is the negative real axis, $\mathbb{R}^-$. Each sheet of the function then has a discontinuity or jump at the branch cut, $\lim_{\delta \to 0} \ln(-r + i\delta) = \lim_{\delta \to 0} \ln(-r - i\delta) + 2\pi i$, where the function values on either side of the jump yield the same result when inserted into the inverse function, $\lim_{\delta \to 0} e^{\ln(-r+i\delta)} = \lim_{\delta \to 0} e^{\ln(-r-i\delta)+2\pi i}$. However as Fig. C37 shows, the Riemann surface of ln itself is a smooth object. We may interpret the singularity by saying that upon crossing the cut we smoothly pass from one Riemann sheet, f_i, to the next one, f_{i+1}, where $\lim_{\delta \to 0} f_i(-r+i\delta) = \lim_{\delta \to 0} f_{i+1}(-r - i\delta)$.

Riemann surfaces demystify the spurious ambiguities otherwise observed with functions such as $z^q, q \in \mathbb{Q}$. They establish a beautiful connection between the theory of complex functions and the geometry of two-dimensional surfaces. In fact, Riemann surfaces are two-dimensional real manifolds in the sense of our discussion of Section V4.1. However, a comprehensive discussion of their rich geometry is beyond the scope of this text and interested readers are referred to courses in complex analysis.

INFO Non-meromorphic functions with cut singularities are often realized as primitive functions of functions with poles. As an example, consider the function z^{-1} with its pole at $z = 0$. Away from that singularity it is holomorphic, and $\ln(z) + C$, with arbitrary complex C, are its primitive functions. The integral of z^{-1} along any path inside a domain of analyticity of z^{-1} then yields $\int_{z_a}^{z_b} dz\, z^{-1} \stackrel{(C235)}{=} \ln(z_b) - \ln(z_a)$, independent of the integration constant. However, now consider a

---

[60] In Fig. C36, the second parametrization of the square root function appears continuous at the cut line. However, this is because only the real part is plotted. The imaginary part is discontinuous (check this).

situation where the initial point is placed at some finite distance R from the origin, for example, $z_a = R$, and the final point dragged along the circle of radius R, $z_b = Re^{i\phi}$. In this case, the integral between the points yields $\ln(Re^{i\phi}) - \ln(R) = i\phi$. Once the circle is closed, a seemingly paradoxical situation arises. On the one hand, the initial and final points are now identical, and one would expect a vanishing integral. On the other hand, by the continuity of the above construction, one would expect $i2\pi$. The paradox is resolved by noting that upon closure of the integration contour, Cauchy's theorem is no longer applicable, because now the pole of z^{-1} is enclosed. In this case, the integration does yield $2\pi i$, as given by Eq. (C236). The fact that one can move from $z_0 = R$ back to this point either encircling the origin, or omitting it, is accounted for by the multi-valuedness of the logarithm. This function has an infinite set of branches, $\ln(z) + 2\pi i n$ (all inverse to the exponential function, $\exp(\ln(z) + 2\pi i n) = z$). Each time the cut line emanating from zero is crossed by a closed integration contour, a sheet-change takes place, and the result of the integral changes by $2\pi i$.

C9.7 Summary and outlook

In this chapter we extended the framework of calculus to the realm of complex numbers. In spite of the nominal equivalence $\mathbb{C} \simeq \mathbb{R}^2$, we saw that the condition of complex differentiability is much more stringent than that of real two-dimensional differentiability. As with many other properties of complex numbers, this had to do with the fact that there exists a multiplication inside $\mathbb{C}$, which gives this set a richer structure than $\mathbb{R}^2$. The Cauchy–Riemann equations were the formal way to describe the complex differentiability conditions, and they were in turn responsible for almost all the powerful structures of complex differentiation and integration discussed in the chapter. In hindsight, one may observe that the Cauchy–Riemann equations (C232) for a function depending on $z = x + iy$ via a real and imaginary part, $u(x, y) + iv(x, y)$, are reminiscent of the vanishing of the curl of a vector field with components u and v in the xy-plane. Cauchy's theorem, or the path independence of complex line integrals, are consequences of the fact that due to this vanishing of the "Cauchy–Riemann-curl", line integrals over holomorphic functions have a lot in common with integrals over gradient fields. At the same time, the existence of a rich spectrum of complex singularities, notably poles and essential singularities, lead to geometric structures not present in this form in two-dimensional real calculus. This became apparent in the discussion of Riemann surfaces, geometric (!) objects which finally demystified the strange behavior of the square root function mentioned at the very beginning of this text.

The beauty of complex functions manifests itself in the computational power of complex calculus and the fact that, e.g., their mathematics is mandatory in any engineering curriculum. Beyond these plain computational aspects, the study of two-dimensional models describable by complex theories has become an active research field of theoretical physics in its own regard. Two-dimensional worlds have physical properties distinct from those of any other dimension, a fact directly traceable to the particular structure of complex differentiability. The ensuing phenomena are studied in disciplines such as two-dimensional statistical mechanics or conformal field theory and they are exemplary for the direct linkage of seemingly abstract mathematical structures with amazing physical effects.

Problems: Calculus

The problems come in odd–even-numbered pairs, labeled ε for "example" and ρ for "practice". Each example problem prepares the reader for tackling the subsequent practice problem. The solutions to the odd-numbered example problems are given in Chapter SC. A password-protected manual containing solutions of all even-numbered practice problems will be made available to instructors.

P.C1 Differentiation of one-dimensional functions (p. 208)

P.C1.1 Definition of differentiability (p. 208)

ε**C1.1.1 Differentiation of polynomials** (p. 211)

Compute the first and second derivatives of the following polynomials. [Check your results against those in square brackets, where $[a; b, c]$ stands for $f'(a) = b, f''(a) = c$.]

(a) $f(x) = 3x^3 + 2x - 1$, $\qquad$ [2; 38, 36]; $\qquad$ (b) $f(x) = x^4 - 2x^2 + 2$, $\qquad$ [2; 24, 44].

ρ**C1.1.2 Differentiation of polynomials** (p. 211)

Compute the first and second derivatives of the following polynomials. [Check your results against those in square brackets, where $[a; b, c]$ stands for $f'(a) = b, f''(a) = c$.]

(a) $f(x) = 4x^5 - x^3 + 2$, $\qquad$ $[\frac{1}{2}; \frac{1}{2}, 7]$; $\qquad$ (b) $f(x) = x^3 - 2x^2 - x + 9$, $\qquad$ [3; 14, 14].

P.C1.2 Differentiation rules (p. 212)

ε**C1.2.1 Product rule and chain rule** (p. 213)

Compute the first derivative of the following functions.
[Check your results against those in square brackets, where $[a, b]$ stands for $f'(a) = b$.]

(a) $f(x) = x \sin x$, $\qquad$ $\left[\frac{\pi}{4}, \frac{1}{\sqrt{2}}\left(1 + \frac{\pi}{4}\right)\right]$; $\qquad$ (b) $f(x) = \cos[\pi(x^2 + x)]$, $\qquad$ $\left[\frac{3\pi}{4}, -1\right]$;

(c) $f(x) = \frac{1}{7 - x^2}$, $\qquad$ $\left[3, \frac{3}{2}\right]$; $\qquad$ (d) $f(x) = \frac{x-1}{x+1}$, $\qquad$ $\left[3, \frac{1}{8}\right]$.

ρ**C1.2.2 Product rule and chain rule** (p. 213)

Compute the first derivative of the following functions.
[Check your results against those in square brackets, where $[a, b]$ stands for $f'(a) = b$.]

(a) $f(x) = (x + \frac{1}{\pi}) \sin[\pi (x + \frac{1}{4})]$, $[0, \sqrt{2}]$;   (b) $f(x) = -x^2 \cos(\pi x)$, $[\frac{1}{3}, -\frac{1}{3} + \frac{\pi}{6\sqrt{3}}]$;

(c) $f(x) = \cos[\pi \sin(x)]$, $[\frac{\pi}{6}, -\frac{\sqrt{3}}{2}\pi]$;   (d) $f(x) = -\cos^4(\frac{3}{\pi}x^2 - x)$, $[\frac{\pi}{2}, 2]$;

(e) $f(x) = \frac{1}{x^3 - 2x^2}$, $[3, -\frac{5}{27}]$;   (f) $f(x) = \frac{x^2 - 2}{x^2 + 1}$, $[2, \frac{12}{25}]$.

P.C1.3 Derivatives of selected functions (p. 213)

ℰC1.3.1 Differentiation of trigonometric functions (p. 214)

Show that the trigonometric functions

$$\tan x = \frac{\sin x}{\cos x}, \qquad \csc x = \frac{1}{\sin x}, \qquad \sec x = \frac{1}{\cos x}, \qquad \cot x = \frac{\cos x}{\sin x} = \frac{1}{\tan x}$$

satisfy the following identities:

(a) $\frac{d}{dx} \tan x = 1 + \tan^2 x = \sec^2 x$,   (b) $\frac{d}{dx} \cot x = -1 - \cot^2 x = -\csc^2 x$.

ℙC1.3.2 Differentiation of hyperbolic functions (p. 214)

Show that the hyperbolic functions

$$\sinh x = \tfrac{1}{2}(e^x - e^{-x}), \qquad \cosh x = \tfrac{1}{2}(e^x + e^{-x}), \qquad \tanh x = \frac{\sinh x}{\cosh x},$$

$$\operatorname{csch} x = \frac{1}{\sinh x}, \qquad \operatorname{sech} x = \frac{1}{\cosh x}, \qquad \coth x = \frac{\cosh x}{\sinh x} = \frac{1}{\tanh x}$$

satisfy the following identities:

(a) $\cosh^2 x - \sinh^2 x = 1$,

(b) $\frac{d}{dx} \sinh x = \cosh x$,   (c) $\frac{d}{dx} \cosh x = \sinh x$,

(d) $\frac{d}{dx} \tanh x = 1 - \tanh^2 x = \operatorname{sech}^2 x$,   (e) $\frac{d}{dx} \coth x = 1 - \coth^2 x = -\operatorname{csch}^2 x$.

ℰC1.3.3 Differentiation of inverse trigonometric functions (p. 214)

Compute the following derivatives of inverse trigonometric functions, f^{-1}, to verify Eq. (C16). For each case, make a qualitative sketch showing $f(x)$ and $f^{-1}(x)$. If f is non-monotonic, consider domains with positive or negative slope separately. [Check your results: $[a, b]$ stands for $(f^{-1})'(a) = b$.]

(a) $\frac{d}{dx} \arcsin x$, $[\frac{1}{3}, \frac{3}{\sqrt{8}}]$;   (b) $\frac{d}{dx} \arccos x$, $[\frac{1}{2}, \frac{2}{\sqrt{3}}]$;   (c) $\frac{d}{dx} \arctan x$, $[1, \frac{1}{2}]$.

Hint: The identity $\sin^2 x + \cos^2 x = 1$ is useful for (a) and (b), $\sec^2 x = 1 + \tan^2 x$ for (c).

ℙC1.3.4 Differentiation of inverse hyperbolic functions (p. 214)

Compute the following derivatives of inverse hyperbolic functions, f^{-1}, to verify Eq. (C17). For each case, make a qualitative sketch showing $f(x)$ and $f^{-1}(x)$. If f is non-monotonic, consider domains with positive or negative slope separately. [Check your results: $[a, b]$ stands for $(f^{-1})'(a) = b$.]

(a) $\frac{d}{dx} \operatorname{arcsinh} x$, $[2, \frac{1}{\sqrt{5}}]$;   (b) $\frac{d}{dx} \operatorname{arccosh} x$, $[2, \frac{1}{\sqrt{3}}]$;   (c) $\frac{d}{dx} \operatorname{arctanh} x$, $[\frac{1}{2}, \frac{4}{3}]$.

Hint: The identity $\cosh^2 x = 1 + \sinh^2 x$ is useful for (a) and (b), $\operatorname{sech}^2 x = 1 - \tanh^2 x$ for (c).

$_\mathrm{E}$C1.3.5 Differentiation of powers, exponentials, logarithms (p. 214)

Compute the first derivative of the following functions.
[Check your results against those in square brackets, where $[a, b]$ stands for $f'(a) = b$.]

(a) $f(x) = -\dfrac{1}{\sqrt{2x}}$, $\left[2, \frac{1}{8}\right]$; (b) $f(x) = \dfrac{x^{1/2}}{(x+1)^{1/2}}$, $\left[3, \dfrac{1}{16\sqrt{3}}\right]$;

(c) $f(x) = e^x(2x - 3)$, $\left[1, e\right]$; (d) $f(x) = 3^x$, $\left[-1, \dfrac{\ln 3}{3}\right]$;

(e) $f(x) = x \ln x$, $\left[1, 1\right]$; (f) $f(x) = x \ln(9x^2)$, $\left[\frac{1}{3}, 2\right]$.

$_\mathrm{P}$C1.3.6 Differentiation of powers, exponentials, logarithms (p. 214)

Compute the first derivative of the following functions.
[Check your results against those in square brackets, where $[a, b]$ stands for $f'(a) = b$.]

(a) $f(x) = \sqrt[3]{x^2}$, $\left[8, \frac{1}{3}\right]$; (b) $f(x) = \dfrac{x}{(x^2+1)^{1/2}}$, $\left[1, \dfrac{1}{\sqrt{8}}\right]$;

(c) $f(x) = -e^{(1-x^2)}$, $\left[1, 2\right]$; (d) $f(x) = 2^{x^2}$, $[1, 4\ln 2]$;

(e) $f(x) = 2\dfrac{\sqrt{\ln x}}{x}$, $\left[e, -\dfrac{1}{e^2}\right]$; (f) $f(x) = \ln\sqrt{x^2+1}$, $\left[1, \frac{1}{2}\right]$.

$_\mathrm{E}$C1.3.7 L'Hôpital's rule (p. 214)

Consider the following question: what is the limiting value of the ratio, $\lim_{x\to x_0}\dfrac{f(x)}{g(x)}$, if the functions f and g both vanish at the point x_0? The naive answer, $\dfrac{f(x_0)}{g(x_0)} \overset{?}{=} \dfrac{0}{0}$, is ill-defined. However, if both functions have a finite slope at x_0, we may use a linear approximation (see Eq. (C2)) for both, $f(x_0 + \delta) \simeq 0 + \delta f'(x_0)$ and $g(x_0 + \delta) \simeq 0 + \delta g'(x_0)$, to obtain $\lim_{x\to x_0}\dfrac{f(x)}{g(x)} = \dfrac{f'(x_0)}{g'(x_0)}$. This result is a special case of **L'Hôpital's rule**.

The general formulation of L'Hôpital's rule is: if either $\lim_{x\to x_0} f(x) = \lim_{x\to x_0} g(x) = 0$ or $\lim_{x\to x_0} |f(x)| = \lim_{x\to x_0} |g(x)| = \infty$, and the limit $\lim_{x\to x_0}\dfrac{f'(x)}{g'(x)}$ exists, then

$$\lim_{x\to x_0} \frac{f(x)}{g(x)} = \lim_{x\to x_0} \frac{f'(x)}{g'(x)}. \tag{1}$$

The proof of this general statement is nontrivial, but is a standard topic in calculus textbooks.

Use L'Hôpital's rule to evaluate the following limits as functions of the real number a: [Check your results against those in square brackets, where $[a, b]$ means that the limit $L(a) = b$.]

(a) $\lim_{x\to 1} \dfrac{x^2 + (a-1)x - a}{x^2 + 2x - 3}$, $[3, 1]$; (b) $\lim_{x\to 0} \dfrac{\sin(ax)}{x + ax^2}$, $[2, 2]$.

If not only f and g but also f' and g' all vanish at x_0, the limit on the r.h.s. of L'Hôpital's rule may be evaluated by applying the rule a second time (or $n + 1$ times, if the derivatives up to $f^{(n)}$ and $g^{(n)}$ all vanish at x_0). Use this procedure to evaluate the following limits:

(c) $\lim_{x\to 0} \dfrac{1 - \cos(ax)}{\sin^2 x}$, $[4, 8]$; (d) $\lim_{x\to 0} \dfrac{x^3}{\sin(ax) - ax}$, $\left[2, -\frac{3}{4}\right]$.

(e) Use L'Hôpital's rule to show that $\lim_{x\to 0}(x \ln x) = 0$ (with $x > 0$). This result implies that for $x \to 0$, "x decreases more quickly than $\ln(x)$ diverges", i.e. "linear beats log".

ₚC1.3.8 L'Hôpital's rule (p. 214)

Use L'Hôpital's rule (possibly multiple times) to evaluate the following limits as functions of the real number a. [Check your results: $[a, b]$ means that the limit $L(a) = b$.]

(a) $\displaystyle\lim_{x \to a} \frac{x^2 + (2-a)x - 2a}{x^2 - (a+1)x + a}$, $[2, 4]$;   (b) $\displaystyle\lim_{x \to 0} \frac{\sinh(x)}{\tanh(ax)}$, $\left[2, \frac{1}{2}\right]$;

(c) $\displaystyle\lim_{x \to 0} \frac{e^{x^2} - 1}{(e^{ax} - 1)^2}$, $\left[2, \frac{1}{2}\right]$;   (d) $\displaystyle\lim_{x \to 0} \frac{\cosh(ax) + \cos(ax) - 2}{x^4}$, $\left[2, \frac{4}{3}\right]$.

(e) Use L'Hôpital's rule to show that for $\alpha \in \mathbb{R}$ and $0 < \beta \in \mathbb{R}$ we have

$$\lim_{x \to 0} (x^\beta \ln^\alpha x) = 0 \quad \text{(with } x > 0\text{)},$$

i.e. "any positive power law beats any power of log".

P.C2 Integration of one-dimensional functions (p. 216)

P.C2.2 One-dimensional integration (p. 218)

ₑC2.2.1 Elementary integrals (p. 220)

Compute the following integrals. $\left[\text{Check your results: (a) } I(2) = \frac{15}{2}; \text{ (b) } I(\ln 2) = \frac{7}{3}.\right]$

(a) $I(x) = \displaystyle\int_1^x dy(2y^3 - 2y + 3)$,   (b) $I(x) = \displaystyle\int_0^x dy\, e^{3y}$.

ₚC2.2.2 Elementary integrals (p. 220)

Compute the following integrals. $\left[\text{Check your results: (a) } I(6) = \ln 2; \text{ (b) } I(\ln 9) = \frac{4}{3}.\right]$

(a) $= \displaystyle\int_0^x dy\, \frac{1}{2y+4}$,   (b) $I(x) = \displaystyle\int_0^x dy\, \sinh\left(\tfrac{1}{2}y\right)$.

P.C2.3 Integration rules (p. 222)

ₑC2.3.1 Integration by parts (p. 223)

Integrals of the form $I(z) = \int_{z_0}^z dx\, u(x)v'(x)$ can be written as $I(z) = [u(x)v(x)]_{z_0}^z - \int_{z_0}^z dx\, u'(x)v(x)$ using integration by parts. This is useful if $u'v$ can be integrated – either directly, or after further integrations by parts [see (b)], or after other manipulations [see (e), (f)]. When doing such a calculation, it is advisable to clearly indicate the factors u, v', v and u'. Always check that the derivative $I'(z) = dI/dz$ of the result reproduces the integrand! If a single integration by parts suffices to calculate $I(z)$, its derivative exhibits the cancellation pattern $I' = u'v + uv' - u'v = uv'$ [see (a), (c), (d)]; otherwise, more involved cancellations occur [see (b), (e), (f)].

Integrate the following integrals by parts. [Check your results against those in square brackets, where $[a, b]$ stands for $I(a) = b$.]

(a) $I(z) = \displaystyle\int_0^z dx\, x\, e^{2x}$, $\left[\tfrac{1}{2}, \tfrac{1}{4}\right]$;   (b) $I(z) = \displaystyle\int_0^z dx\, x^2\, e^{2x}$, $\left[\tfrac{1}{2}, \tfrac{e}{8} - \tfrac{1}{4}\right]$;

(c) $I(z) = \int_0^z dx \ln x,$ $\qquad [1, -1];$ $\qquad$ (d) $I(z) = \int_0^z dx \ln x \frac{1}{\sqrt{x}},$ $\qquad [1, -4];$

(e) $I(z) = \int_0^z dx \sin^2 x,$ $\qquad [\pi, \frac{\pi}{2}];$ $\qquad$ (f) $I(z) = \int_0^z dx \sin^4 x,$ $\qquad [\pi, \frac{3\pi}{8}].$

pC2.3.2 **Integration by parts** (p. 223)

Integrate the following integrals by parts. [Check your results against those in square brackets, where [a, b] stands for I(a) = b.]

(a) $I(z) = \int_0^z dx \, x \sin(2x),$ $\quad [\frac{\pi}{2}, \frac{\pi}{4}];$ $\qquad$ (b) $I(z) = \int_0^z dx \, x^2 \cos(2x),$ $\quad [\frac{\pi}{2}, -\frac{\pi}{4}];$

(c) $I(z) = \int_0^z dx \, (\ln x) \, x,$ $\quad [1, -\frac{1}{4}];$ $\qquad$ (d) $I(z) \overset{[n > -1]}{=} \int_0^z dx \, (\ln x) \, x^n,$ $\quad [1, \frac{-1}{(n+1)^2}];$

(e) $I(z) = \int_0^z dx \, \cos^2 x,$ $\quad [\pi, \frac{\pi}{2}];$ $\qquad$ (f) $I(z) = \int_0^z dx \, \cos^4 x,$ $\quad [\pi, \frac{3}{8}\pi].$

fC2.3.3 **Integration by substitution** (p. 224)

Integrals of the form $I(z) = \int_{z_0}^z dx \, y'(x) f(y(x))$ can be written as $I(z) = \int_{y(z_0)}^{y(z)} dy f(y)$ by using the substitution $y = y(x)$, $dy = y'(x)dx$. When doing such integrals, it is advisable to explicitly write down $y(x)$ and dy, to ensure that you correctly identify the prefactor of $f(y)$. Always check that the derivative $I'(z) = dI/dz$ of the result reproduces the integrand! You'll notice that the factor $y'(z)$ emerges via the chain rule for differentiating composite functions.

Calculate the following integrals by substitution. [Check your results against those in square brackets, where [a, b] stands for I(a)=b.]

(a) $I(z) = \int_0^z dx \, x \cos(x^2 + \pi),$ $\quad [\sqrt{\frac{\pi}{2}}, -\frac{1}{2}];$ $\qquad$ (b) $I(z) = \int_0^z dx \sin^3 x \cos x,$ $\quad [\frac{\pi}{4}, \frac{1}{16}];$

(c) $I(z) = \int_0^z dx \sin^3 x,$ $\quad [\frac{\pi}{3}, \frac{5}{24}];$ $\qquad$ (d) $I(z) = \int_0^z dx \cosh^3 x,$ $\quad [\ln 2, \frac{57}{64}];$

(e) $I(z) = \int_0^z dx \frac{\sqrt{1 + \ln(x+1)}}{x+1},$ $\quad [e^3 - 1, \frac{14}{3}];$ $\qquad$ (f) $I(z) = \int_0^z dx \, x^3 e^{-x^4},$ $\quad [\sqrt[4]{\ln 2}, \frac{1}{8}].$

pC2.3.4 **Integration by substitution** (p. 224)

Calculate the following integrals by substitution. [Check your results versus those in square brackets, where [a, b] stands for I(a) = b.]

(a) $I(z) = \int_0^z dx \, x^2 \sqrt{x^3 + 1},$ $\quad [2, \frac{52}{9}];$ $\qquad$ (b) $I(z) = \int_0^z dx \sin x \, e^{\cos x},$ $\quad [\frac{\pi}{3}, e - \sqrt{e}];$

(c) $I(z) = \int_0^z dx \cos^3 x,$ $\quad [\frac{\pi}{4}, \frac{5}{6\sqrt{2}}];$ $\qquad$ (d) $I(z) = \int_0^z dx \sinh^3 x,$ $\quad [\ln 3, \frac{44}{81}];$

(e) $I(z) = \int_0^z dx \frac{\sin \sqrt{\pi x}}{\sqrt{x}},$ $\quad [\frac{\pi}{9}, \frac{1}{\sqrt{\pi}}];$ $\qquad$ (f) $I(z) = \int_0^z dx \sqrt{x} \, e^{\sqrt{x^3}},$ $\quad [(\ln 4)^{2/3}, 2].$

fC2.3.5 $\sqrt{1 - x^2}$ **integrals by trigonometric substitution** (p. 224)

For integrals involving $\sqrt{1 - x^2}$, the substitution $x = \sin y$ may help, since it gives $\sqrt{1 - x^2} = \cos y$. Use it to compute the following integrals $I(z)$; check your answers by calculating $\frac{dI(z)}{dz}$.

[Check your results: (a) $I(\frac{1}{\sqrt{2}}) = \frac{\pi}{4}$; (b) for $a = \frac{1}{2}$, $I(\sqrt{2}) = \frac{\pi}{4} + \frac{1}{2}$.]

(a) $I(z) = \int_0^z dx \, \frac{1}{\sqrt{1-x^2}}$ $(|z| < 1)$,   (b) $I(z) = \int_0^z dx \, \sqrt{1 - a^2 x^2}$ $(|az| < 1)$.

Hint: For (b), use integration by parts for the $\cos^2 y$ integral emerging after the substitution.

P.C2.3.6 $\sqrt{1 + x^2}$ integrals by hyberbolic substitution (p. 224)

For integrals involving $\sqrt{1 + x^2}$, the substitution $x = \sinh y$ may help, since it gives $\sqrt{1 + x^2} = \cosh y$. Use it to compute the following integrals $I(z)$; check your answers by calculating $\frac{dI(z)}{dz}$.
$\left[\text{Check your results: (a) } I(\frac{3}{4}) = \ln 2; \text{ (b) for } a = \frac{1}{2}, I(\frac{3}{2}) = \ln 2 + \frac{15}{16}.\right]$

(a) $I(z) = \int_0^z dx \, \frac{1}{\sqrt{1 + x^2}}$;   (b) $I(z) = \int_0^z dx \, \sqrt{1 + a^2 x^2}$.

P.C2.3.7 $1/(1 - x^2)$ integrals by hyperbolic substitution (p. 224)

For integrals involving $1/(1-x^2)$, the substitution $x = \tanh y$ may help, since it gives $1-x^2 = \text{sech}^2 y$. Use it to compute the following integrals $I(z)$; check your answers by calculating $\frac{dI(z)}{dz}$.
$\left[\text{Check your results: (a) } I(\frac{3}{5}) = \ln 2; \text{ (b) for } a = 3, I(\frac{1}{5}) = \frac{1}{6} \ln 2 + \frac{5}{32}.\right]$

(a) $I(z) = \int_0^z dx \, \frac{1}{1 - x^2}$ $(|z| < 1)$,   (b) $I(z) = \int_0^z dx \, \frac{1}{(1 - a^2 x^2)^2}$ $(|az| < 1)$.

Hint: For (b), use integration by parts for the $\cosh^2 y$ integral emerging after the substitution.

P.C2.3.8 $1/(1 + x^2)$ integrals by trigonometric substitution (p. 224)

For integrals involving $1/(1+x^2)$, the substitution $x = \tan y$ may help, since it gives $1 + x^2 = \sec^2 y$. Use it to compute the following integrals $I(z)$; check your answers by calculating $\frac{dI(z)}{dz}$.
$\left[\text{Check your results: (a) } I(1) = \frac{\pi}{4}; \text{ (b) for } a = \frac{1}{2}, I(2) = \frac{\pi}{4} + \frac{1}{2}.\right]$

(a) $I(z) = \int_0^z dx \, \frac{1}{1 + x^2}$,   (b) $I(z) = \int_0^z dx \, \frac{1}{(1 + a^2 x^2)^2}$.

P.C2.4   Practical remarks on one-dimensional integration (p. 225)

P.C2.4.1   Partial fraction decomposition (p. 226)

A function f is called a **rational function** if it can be expressed as a ratio $f(x) = P(x)/Q(x)$ of two polynomials, P and Q. Integrals of rational functions can be computed using a **partial fraction decomposition**, a procedure that expresses f as the sum of a polynomial (possibly with degree 0) and several ratios of polynomials with simpler denominators. To achieve this, the denominator Q is factorized into a product of polynomials, q_j, of lower degree, $Q(x) = \prod_j q_j(x)$, and the function f is written as $f(x) = \sum_j p_j(x)/q_j(x)$. The form of the polynomials p_j in the numerators is fixed uniquely by the form of the polynomials P and q_j. (Since a partial fraction decomposition starts with a common denominator and ends with a sum of rational functions, it is in a sense the inverse of the procedure of adding rational functions by finding a common denominator.) If a complete factorization of Q is used, this yields a decomposition of the integral $\int dx f(x)$ into a sum of integrals that can be solved by elementary means. Here we illustrate the method using some simple examples; for a systematic treatment, consult textbooks on calculus.

Use partial fraction decomposition to compute the following integrals, for $z \in (0, 2)$:

(a) $I(z) = \int_0^z dx \, \dfrac{3}{(x+1)(x-2)}$,

(b) $I(z) = \int_0^z dx \, \dfrac{3x}{(x+1)^2(x-2)}$.

[Check your results: (a) $I(3) = -\ln 8$, (b) $I(3) = -\ln 4 + \frac{3}{4}$.]

ₚC2.4.2 **Partial fraction decomposition** (p. 226)

Use partial fraction decomposition to compute the following integrals, for $z \in (0, 1)$:

(a) $I(z) = \int_0^z dx \dfrac{x+2}{x^3 - 3x^2 - x + 3}$,

(b) $I(z) = \int_0^z dx \dfrac{4x-1}{(x+2)(x-1)^2}$.

$\left[\text{Check your results: (a) } I(\tfrac{1}{2}) = \tfrac{5}{8}\ln 5 - \tfrac{1}{2}\ln 3, \text{ (b) } I(\tfrac{1}{2}) = 1 - \ln\left(\tfrac{5}{2}\right).\right]$

The final four problems of this section should best be done after reading Section C4.2, which is needed for the Gaussian integral of Problem C2.4.3.

ₑC2.4.3 **Elementary Gaussian integrals** (p. 227)

This problem requires Section C4.2 on two-dimensional integration with polar coordinates.

(a) Show that the two-dimensional Gaussian integral $I = \int_{-\infty}^{\infty}\int_{-\infty}^{\infty} dxdy \, e^{-(x^2+y^2)}$ has the value $I = \pi$. *Hint:* Use polar coordinates; the radial integral can be solved by substitution.

(b) Now calculate the one-dimensional Gaussian integral

$$ I_0(a) = \int_{-\infty}^{\infty} dx \, e^{-ax^2} \quad (a \in \mathbb{R}, \, a > 0). $$

Hint: $I = \left[I_0(1)\right]^2$. Explain why! [Check your result: $I_0(\pi) = 1$.]

(c) Compute the one-dimensional Gaussian integral with a linear term in the exponent:

$$ I_1(a, b) = \int_{-\infty}^{\infty} dx \, e^{-ax^2+bx} \quad (a, b \in \mathbb{R}, a > 0). $$

Hint: Write the exponent in the form $-ax^2 + bx = -a(x - C)^2 + D$ (called **completing the square**), then substitute $y = x - C$ and use the result from (b).
[Check your result: $I_1(1, 2) = \sqrt{\pi}\mathrm{e}$.]

ₚC2.4.4 **Gaussian integrals with linear term in exponent** (p. 227)

Compute the following Gaussian integrals:

(a) $I_1(c) = \int_{-\infty}^{\infty} dx \, e^{-3(x+c)x}$,

(b) $I_2(c) = \int_{-\infty}^{\infty} dx \, e^{-\frac{1}{2}(x^2+3x+\frac{c}{4})}$,

(c) $I_3(c) = \int_{-\infty}^{\infty} dx \, e^{-2(x+3)(x-c)}$.

$\left[\text{Check your results: } I_1(2) = \sqrt{\tfrac{\pi}{3}}\mathrm{e}^3, I_2(1) = \sqrt{2\pi}\,\mathrm{e}, I_3(-3) = \sqrt{\tfrac{\pi}{2}}.\right]$

ᴇC2.4.5 Definite exponential integrals (p. 227)

Calculate the integral $I_n(a) = \int_0^\infty dx\, x^n e^{-ax}$ (with $a \in \mathbb{R}$, $a > 0$, $n \in \mathbb{N}$) using two different methods: (a) repeated partial integration, and (b) repeated differentiation.

(a) Calculate I_0, I_1 and I_2 by using partial integration where necessary. Then use partial integration to show that

$$I_n(a) = \frac{n}{a} I_{n-1}(a)$$

for all $n \geq 1$. Use this relation iteratively to determine $I_n(a)$ as a function of a and n. [Check your result: $I_3(2) = \frac{3}{8}$.]

(b) Show that taking n derivatives of $I_0(a)$ with respect to a yields

$$I_n(a) = (-1)^n \frac{d^n I_0(a)}{da^n}.$$

Then calculate these derivatives for a few small values of n. From the emerging pattern, deduce the general formula for $I_n(a)$.

ᴘC2.4.6 General Gaussian integrals (p. 227)

Determine the value of the x^{2n} Gaussian integral, $I_n(a) = \int_{-\infty}^\infty dx\, x^{2n} e^{-ax^2}$ (with $a \in \mathbb{R}$, $a > 0$, $n \in \mathbb{N}$), using two different methods: (a) repeated partial integration, and (b) repeated differentiation.

(a) Starting from the Gaussian integral $I_0(a) = \sqrt{\frac{\pi}{a}}$, compute the integrals I_1 and I_2 by using partial integration where necessary. Then use partial integration to show that

$$I_n(a) = \frac{2n - 1}{2a} I_{n-1}(a)$$

holds for all $n \geq 1$. Use this relation iteratively to determine $I_n(a)$ as a function of a and n. [Check your result: $I_3(3) = \sqrt{\frac{\pi}{3}} \frac{5}{72}$.]

(b) Show that taking n derivatives of $I_0(a)$ with respect to a yields

$$I_n(a) = (-1)^n \frac{d^n I_0(a)}{da^n}.$$

Then calculate these derivatives for a few small values of n. From the emerging pattern, deduce the general formula for $I_n(a)$.

P.C3   Partial differentiation (p. 229)

P.C3.1   Partial derivative (p. 229)

ᴇC3.1.1 Partial derivatives (p. 230)

Compute the partial derivatives $\partial_x f(x, y)$ and $\partial_y f(x, y)$ of the following functions.
[Check your results against those in square brackets.]

(a) $f(x, y) = x^2 y^3 - 2xy$, $[\partial_x f(2, 1) = 2, \quad \partial_y f(1, 2) = 10]$;

(b) $f(x, y) = \sin[xe^{2y}]$, $\left[\partial_x f(0, \frac{1}{2}) = e, \quad \partial_y f(\pi, 0) = -2\pi\right]$.

pC3.1.2 Partial derivatives (p. 230)

Compute the partial derivatives $\partial_x f(x, y)$ and $\partial_y f(x, y)$ of the following functions.
[Check your results against those in square brackets.]

(a) $f(x, y) = \dfrac{x^2}{y^3} + \dfrac{4y}{x}$,　　　　　　　　　$\left[\partial_x f(2, 1) = 3, \quad \partial_y f(3, 3) = 1\right]$;

(b) $f(x, y) = \ln\left(x^2 \sin(y)\right)$,　　　　　　　　$\left[\partial_x f(2, 1) = 1, \quad \partial_y f(1, \frac{\pi}{4}) = 1\right]$;

(c) $f(x, y) = e^{-x^2 \cos(y)}$,　　　　　　　　$\left[\partial_x f(1, \pi) = 2e, \quad \partial_y f(1, \frac{\pi}{2}) = 1\right]$;

(d) $f(x, y) = \sinh\left(\frac{x}{y}\right)$,　　　　　　　　$\left[\partial_x f(\ln 2, 1) = \frac{5}{4}, \quad \partial_y f(\ln 2, 1) = -\frac{5}{4} \ln 2\right]$.

P.C3.2　Multiple partial derivatives (p. 230)

eC3.2.1 Partial derivatives of first and second order (p. 231)

Consider the function $f : \mathbb{R}^2\backslash(0, 0)^\mathsf{T} \to \mathbb{R}$, $\mathbf{r} = (x, y)^\mathsf{T} \mapsto f(\mathbf{r}) = \frac{x}{r} + 1$, with $r = \sqrt{x^2 + y^2}$.
Calculate all possible partial derivatives of first and second order.

pC3.2.2 Partial derivatives of first and second order (p. 231)

Consider the function $f : \mathbb{R}^3 \to \mathbb{R}$, $\mathbf{r} = (x, y, z)^\mathsf{T} \mapsto f(\mathbf{r}) = z^2 e^{xy}$. Calculate all possible partial derivatives of first and second order.

P.C3.3　Chain rule for functions of several variables (p. 231)

eC3.3.1 Chain rule for functions of two variables (p. 235)

This problem aims to illustrate the inner life of the chain rule for a function of several variables.
Consider the function $f : \mathbb{R}^2 \to \mathbb{R}$, $\mathbf{y} = (y^1, y^2)^\mathsf{T} \mapsto f(\mathbf{y}) = \|\mathbf{y}\|^2$ and the vector field
$\mathbf{g} : \mathbb{R}_+^2 \to \mathbb{R}^2$, $\mathbf{x} = (x^1, x^2)^\mathsf{T} \mapsto \mathbf{g}(\mathbf{x}) = (\ln x^2, 3\ln x^1)^\mathsf{T}$, then $f(\mathbf{g}(\mathbf{x}))$ gives the norm of $\mathbf{g}$ as a
function of $\mathbf{x}$. Find the partial derivatives $\partial_{x^1} f(\mathbf{g}(\mathbf{x}))$ and $\partial_{x^2} f(\mathbf{g}(\mathbf{x}))$ as functions of x^1 and x^2 in two
ways,

(a)　by first computing $f(\mathbf{x}) = f(\mathbf{g}(\mathbf{x}))$ as function of $\mathbf{x}$ and then taking partial derivatives;

(b)　by using the chain rule $\partial_{x^k} f(\mathbf{g}(\mathbf{x})) = \sum_j \partial_{g^j} f(\mathbf{g}(\mathbf{x})) \, \partial_{x^k} g^j(\mathbf{x})$.

Why do both routes yield the same answer? Identify the similarities in both computations!
[Check your results: If $x^1 = 9$, $x^2 = 2$, then $\partial_{x^1} f = 4 \ln 3$, $\partial_{x^2} f = \ln 2$.]

pC3.3.2 Chain rule for functions of two variables (p. 235)

Consider the function $f : \mathbb{R}^2 \to \mathbb{R}$, $\mathbf{y} = (y^1, y^2)^\mathsf{T} \mapsto f(\mathbf{y}) = \mathbf{y} \cdot \mathbf{a}$, where $\mathbf{a} = (a^1, a^2)^\mathsf{T} \in \mathbb{R}^2$, and
the vector field $\mathbf{g} : \mathbb{R}^2 \to \mathbb{R}^2$, $\mathbf{x} = (x^1, x^2)^\mathsf{T} \mapsto \mathbf{g}(\mathbf{x}) = \mathbf{x}(\mathbf{x} \cdot \mathbf{b})$, where $\mathbf{b} = (b^1, b^2)^\mathsf{T} \in \mathbb{R}^2$. Compute
the partial derivatives $\partial_{x^k} f(\mathbf{g}(\mathbf{x}))$ (with $k = 1, 2$) as functions of $\mathbf{x}$,

(a)　by first computing $f(\mathbf{g}(\mathbf{x}))$ explicitly and then taking partial derivatives;

(b)　by using the chain rule $\partial_{x^k} f(\mathbf{g}(\mathbf{x})) = \sum_j \partial_{g^j} f(\mathbf{g}(\mathbf{x})) \partial_{x^k} g^j(\mathbf{x})$.

[Check your result: If $\mathbf{a} = (0, 1)^\mathsf{T}$, $\mathbf{b} = (1, 0)^\mathsf{T}$, then $\partial_{x^1} f(\mathbf{g}(\mathbf{x})) = x^2$, $\partial_{x^2} f(\mathbf{g}(\mathbf{x})) = x^1$.]
Hint: If compact notation is used, such as $\mathbf{a} \cdot \mathbf{x} = a_l x^l$ and $\partial_{x^k} x^l = \delta^l_k$, the computations are quite
short.

P.C4 Multidimensional integration (p. 238)

P.C4.1 Cartesian area and volume integrals (p. 238)

ᴇC4.1.1 Fubini's theorem (p. 240)

Verify Fubini's theorem for the following integrals of the function $f(x, y) = x\sqrt{x^2 + y}$.
[Check your result: $I(1) = \frac{2}{15}(2^{5/2} - 2)$.]

(a) $I(a) = \int_0^a dx \int_0^1 dy\, f(x, y),$ (b) $I(a) = \int_0^1 dy \int_0^a dx\, f(x, y).$

ᴘC4.1.2 Fubini's theorem (p. 240)

Verify Fubini's theorem for the following integrals of the function $f(x, y) = xy^2 \sin(x^2 + y^3)$.
[Check your result: $I(\sqrt{\pi/2}) = \frac{1}{3}$.]

(a) $I(a) = \int_0^a dx \int_0^{\pi^{1/3}} dy\, f(x, y),$ (b) $I(a) = \int_0^{\pi^{1/3}} dy \int_0^a dx\, f(x, y).$

ᴇC4.1.3 Violation of Fubini's theorem (p. 241)

Fubini's theorem holds only if the integrand is sufficiently well-behaved that the integral of its *modulus* over the integration domain exists. Here we explore a counterexample.

(a) Integrate the function $f(x, y) = \frac{x^2 - y^2}{(x^2 + y^2)^2}$ over the rectangle $R_a = \{a \leq x \leq 1, 0 \leq y \leq 1\}$, with $0 < a \in \mathbb{R}$, using two different orders of integration:

$$I_A(a) = \int_a^1 dx \int_0^1 dy\, f(x, y), \qquad\qquad I_B(a) = \int_0^1 dy \int_a^1 dx\, f(x, y).$$

Verify that $I_A(a) = I_B(a)$. [Check your results: $I_{A,B}(\sqrt{3}) = -\frac{\pi}{12}$.]

Hint: First show that $f(x, y) = \dfrac{\partial}{\partial y}\dfrac{y}{x^2 + y^2} = -\dfrac{\partial}{\partial x}\dfrac{x}{x^2 + y^2}.$

Set $a = 0$ for the remainder of this problem.

(b) Show that $I_A(0) = -I_B(0)$ if these integrals are recomputed, setting $a = 0$ from the outset. Which of the two, $I_A(0)$ or $I_B(0)$, agrees with the $a \to 0$ limit from part (a)?

(c) Show that the integral $I_C = \int_{R_0} dxdy\, |f(x, y)|$ does not exist. To this end, split the integration domain $R_{a=0}$ into two parts, $R_0 = R_0^+ \cup R_0^-$, chosen such that $f \geq 0$ on R_0^+ and $f \leq 0$ on R_0^- (see figure). Then $I_C = \int_{R_0^+ \cup R_0^-} dxdy\, |f(x, y)| = I_0^+ - I_0^-$, with $I_0^{\pm} = \int_{R_0^{\pm}} dx\, dy\, f(x, y)$.
Compute the contributions $I_0^{\pm}$ separately and show that $I_0^+ = -I_0^- = \infty$.

As seen in (a) and (b), Fubini's theorem applies for $a > 0$, but not for $a = 0$, because then the integral over the *modulus* of the function does not exist, $I_C = I_0^+ - I_0^- = \infty$, as seen in (c). This happens because for $a = 0$ the integration domain touches a point where f diverges – the origin: as

$(x,y)^{\mathrm{T}}$ approaches $(0,0)^{\mathrm{T}}$, the integrand tends to $+\infty$ for $x > y$ or $-\infty$ for $x < y$. According to (c), the integrals over the positive or negative "branches" of f diverge, $I_0^{\pm} = \pm\infty$. Hence the integral $I_0 = \int_{R_0} dxdy f(x,y)$ is *not defined*: it yields $\infty - \infty$ contributions, and the extent to which these cancel depends on the integration order, as seen in (b).

One may make sense of the integral I_0 by **regularizing** it, i.e. by modifying the integration domain to avoid the singularity. For example, consider the domain $R_\delta = R_0 \backslash S_\delta$, obtained from R_0 by removing an infinitesimal square adjacent to the origin, $S_\delta = \{0 \le x \le \delta, 0 \le y \le \delta\}$.

(d)   Compute the integral $I_\delta = \int_{R_\delta} dxdy f(x,y)$ using the method of (c), splitting the integration domain as $R_\delta = R_\delta^+ \cup R_\delta^-$ (see figure). Discuss the limit $I_{\delta \to 0}$. Why is it well-defined?

P C4.1.4 Violation of Fubini's theorem (p. 241)

(a)   Compute the integral of the function $f(x,y) = \frac{xy(x^2-y^2)}{(x^2+y^2)^3}$ over the rectangle $R_a = \{a \le x \le 1, 0 \le y \le 1\}$, with $0 < a \in \mathbb{R}$, using two different orders of integration:

$$I_A(a) = \int_a^1 dx \int_0^1 dy f(x,y), \qquad\qquad I_B(a) = \int_0^1 dy \int_a^1 dx f(x,y).$$

Verify that $I_A(a) = I_B(a)$. [Check your results: $I_{A,B}(\frac{1}{3}) = \frac{1}{10}$.]

Hint: First show that $f(x,y) = \dfrac{\partial}{\partial y}\dfrac{xy^2}{2(x^2+y^2)^2} = -\dfrac{\partial}{\partial x}\dfrac{x^2 y}{2(x^2+y^2)^2}$.

(b)   Show that $I_A(0) = -I_B(0)$ if these integrals are recomputed with $a = 0$ from the outset.

(c)   Compute $I_C = \int_{R_0} dxdy\,|f(x,y)|$ and explain why Fubini's theorem is violated in (b).

(d)   Compute the regularized integral $I_\delta = \int_{R_\delta} dxdy f(x,y)$, where the integration domain $R_\delta = R_0 \backslash S_\delta$ is obtained from $R_{a=0}$ by removing an infinitesimal square adjacent to the origin, $S_\delta = \{0 \le x \le \delta, 0 \le y \le \delta\}$. Discuss the limit $I_{\delta \to 0}$. Why is it well-defined?

E C4.1.5 Two-dimensional integration (Cartesian coordinates) (p. 241)

Calculate the surface integral $I(a) = \int_{G_a} dx\,dy f(x,y)$ of the function $f(x,y) = xy$, over the area $G = \{(x,y) \subset \mathbb{R}^2; 0 \le y \le 1; 1 \le x \le a - y\}$, with $2 \le a \in \mathbb{R}$.
[Check your result: $I(2) = \frac{5}{24}$.]

P C4.1.6 Two-dimensional integration (Cartesian coordinates) (p. 241)

Calculate the surface integral $I(a) = \int_G dx\,dy f(x,y)$ of the function $f(x,y) = y^2 + x^2$ over the surface $G = \{(x,y) \in \mathbb{R}^2; 0 \le x \le 1; 0 \le y \le e^{ax}\}$, with $a \in \mathbb{R}$. *Hint:* For $\int dx\, x^2 e^{ax}$, use partial integration twice! [Check your result: $I(1) = e + (e^3 - 19)/9$.]

E C4.1.7 Area enclosed by curves (Cartesian coordinates) (p. 241)

Consider the curve $\gamma_1 : \mathbb{R} \to \mathbb{R}^2, t \mapsto (t, b(1 - t/a))^{\mathrm{T}}$ and the closed curve $\gamma_2 : (0, 2\pi) \subset \mathbb{R} \to \mathbb{R}^2, t \mapsto (a\cos t, b\sin t)^{\mathrm{T}}$ in Cartesian coordinates, with $0 < a, b \in \mathbb{R}$.

(a)   Sketch the curves γ_1 and γ_2.

(b)   Compute the area $S(a,b)$ enclosed by γ_2. [Check your result: $S(1,1) = \pi$.]

(c) γ_1 divides the area enclosed by γ_2 into two parts. Find the area $A(a, b)$ of the smaller part by computing an area integral. Check your result using elementary geometrical considerations.

P(4.1.8) **Area enclosed by curves (Cartesian coordinates)** (p. 241)

Consider the curves $\gamma_1 : \mathbb{R} \to \mathbb{R}^2, t \mapsto ((t - 2a)^2 + 2a^2, t)^T$ and $\gamma_2 : \mathbb{R} \to \mathbb{R}^2, t \mapsto (2(t - a)^2, t)^T$ in Cartesian coordinates, with $0 < a \in \mathbb{R}$.

(a) Sketch the curves γ_1 and γ_2.

(b) Compute the finite area $S(a)$ enclosed between these curves. [Check your result: $S(\frac{1}{2}) = \frac{4}{3}$.]

E(4.1.9) **Area integral for volume of a pyramid (Cartesian coordinates)** (p. 241)

Consider the pyramid bounded by the xy-plane, the yz-plane, the xz-plane and the plane $E = \{(x, y, z) \in \mathbb{R}^3, z = c(1 - x/a - y/b)\}$, with $0 < a, b, c \in \mathbb{R}$.

(a) Make a qualitative sketch of the pyramid. Find its volume $V(a, b, c)$ using geometric arguments. [Check your result: $V(1, 1, 1) = \frac{1}{6}$.]

(b) Compute $V(a, b, c)$ by integrating the height $h(x, y)$ of the pyramid over its base area in the xy-plane.

P(4.1.10) **Area integral for volume of ellipsoidal tent (Cartesian coordinates)** (p. 241)

A tent has a flat, ellipsoidal base, given by the equation $(x/a)^2 + (y/b)^2 \leq 1$. The shape of the tent's roof is given by the height function $h(x, y) = c[1 - (x/a)^2 - (y/b)^2]$.

(a) Give a qualitative sketch of the shape of the tent, for $a = 2, b = 1$ and $c = 2$.

(b) Calculate the volume V of the tent via a surface integral of the height function. Use Cartesian coordinates. [Check your result: If $a = b = c = 1$, then $V = \pi/2$.]
   Hint: Show by a suitable trigonometric substitution that $\int_0^1 dx\, (1 - x^2)^{3/2} = \frac{3}{16}\pi$.

P.C4.2 Curvilinear area integrals (p. 242)

E(4.2.1) **Area of an ellipse (generalized polar coordinates)** (p. 246)

(a) Let $f : \mathbb{R}^2 \to \mathbb{R}$ be a function of the coordinates x and y that depends on only the combined variable $(x/a)^2 + (y/b)^2$. Show that a two-dimensional area integral of f over $\mathbb{R}^2$ can be written as

$$I = \int_{\mathbb{R}^2} dxdy f\left((x/a)^2 + (y/b)^2\right) = 2\pi ab \int_0^\infty d\mu\, \mu f(\mu^2),$$

by transforming from Cartesian coordinates to generalized polar coordinates, defined as follows:

$$x = \mu a \cos\phi, \qquad\qquad y = \mu b \sin\phi,$$
$$\mu^2 = (x/a)^2 + (y/b)^2, \qquad \phi = \arctan(ay/bx).$$

Hint: For $a = b = 1$, they correspond to polar coordinates. For $a \neq b$, the local basis is *not* orthogonal!

(b) Using a suitable function f, calculate the area of an ellipse with semi-axes a and b, with a and b defined by $(x/a)^2 + (y/b)^2 \leq 1$.

P.C4.2.2 Area integral for volume (generalized polar coordinates) (p. 246)

In the following, use generalized polar coordinates in two dimensions, defined as $x = \mu a \cos\phi$, $y = \mu b \sin\phi$, with $a, b \in \mathbb{R}$, $a > b > 0$. Calculate the volume $V(a, b, c)$ of the following objects T, E and C, as a function of the length parameters a, b and c.

(a) T is a tent with an elliptical base with semi-axes a and b. The height of its roof is described by the height function $h_T(x, y) = c\left[1 - (x/a)^2 - (y/b)^2\right]$.

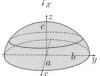

(b) E is an ellipsoid with semi-axes a, b and c, defined by $(x/a)^2 + (y/b)^2 + (z/c)^2 \leq 1$.

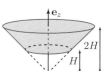

(c) C is a cone with height c and an elliptical base with semi-axes a and b. All cross sections parallel to the base are elliptical, too. *Hint:* Augment the generalized polar coordinates by another coordinate, z (in analogy to passing from polar to cylindrical coordinates).

[Check your answers: if $a = 1/\pi$, $b = 2$, $c = 3$, then (a) $V_T = 3$, (b) $V_E = 8$, (c) $V_C = 2$.]

P.C4.3   Curvilinear volume integrals (p. 249)

E.C4.3.1 Volume and moment of inertia (cylindrical coordinates) (p. 251)

The moment of inertia of a rigid body with respect to a given axis of rotation is defined as $I = \int_V dV \, \rho_0(\mathbf{r}) d_\perp^2(\mathbf{r})$, where $\rho_0(\mathbf{r})$ is the density at the point $\mathbf{r}$, and $d_\perp(\mathbf{r})$ the perpendicular distance from $\mathbf{r}$ to the rotation axis.

Let $F = \{\mathbf{r} \in \mathbb{R}^3 \,|\, H \leq z \leq 2H, \sqrt{x^2 + y^2} \leq az\}$ be a homogeneous conical frustum (cone with tip removed) centered on the z-axis. Calculate, using cylindrical coordinates,

(a) its volume, $V_F(a)$, and

(b) its moment of inertia, $I_F(a)$, with respect to the z-axis, as functions of the dimensionless, positive scale factor a, the length parameter H and the mass M of the frustum. [Check your results: $V_F(3) = 21\pi H^3$, $I_F(1) = \frac{93\pi}{70}MH^2$.]

P.C4.3.2 Volume and moment of inertia (cylindrical coordinates) (p. 251)

Consider the homogeneous rigid bodies C, P and B specified below, each with density ρ_0. For each body, use cylindrical coordinates to compute its volume, $V(a)$, and moment of inertia, $I(a) = \rho_0 \int_V dV \, d_\perp^2$, with respect to the axis of symmetry, as functions of the dimensionless, positive scale factor a, the length parameter R and the mass of the body, M.

(a) C is a hollow cylinder with inner radius R, outer radius aR and height $2R$. [Check your results: $V_C(2) = 6\pi R^3$, $I_C(2) = \frac{15}{6}MR^2$.]

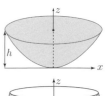

(b) P is a paraboloid with height $h = aR$ and curvature $1/R$, defined by $P = \{\mathbf{r} \in \mathbb{R}^3 \,|\, 0 \leq z \leq h, (x^2 + y^2)/R \leq z\}$. [Check your results: $V_P(2) = 2\pi R^3$, $I_P(2) = \frac{2}{3}MR^2$.]

(c) B is the bowl obtained by taking a sphere, $S = \{\mathbf{r} \in \mathbb{R}^3 \,|\, x^2 + y^2 + (z - aR)^2 \leq a^2 R^2\}$, with radius aR, centered on the point $P: (0, 0, aR)^T$, and cutting a cone from it, $C = \{\mathbf{r} \in \mathbb{R}^3 \,|\, (x^2 + y^2) \leq (a - 1)z^2, a \geq 1\}$, which is symmetric about the z axis, with apex at the origin. [Check your results: $V_B\left(\frac{4}{3}\right) = \frac{16}{9}\pi R^3$, $I_B\left(\frac{4}{3}\right) = \frac{14}{15}MR^2$. What do you get for $a = 1$? Why?]

Hint: First, for a given z, find the radial integration boundaries, $\rho_1(z) \leq \rho \leq \rho_2(z)$, then the z integration boundaries, $0 \leq z \leq z_m$. What do you find for z_m, the maximal value of z?

ɛC4.3.3 Volume of a buoy (spherical coordinates) (p. 251)

Consider a buoy, with its tip at the origin, bounded from above by a sphere centered on the origin, with $x^2 + y^2 + z^2 \leq R^2$, and from below by a cone with tip at the origin, with $z \geq a\sqrt{(x^2 + y^2)}$.

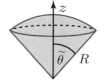

(a)   Show that the half angle at the tip of the cone is given by $\widetilde{\theta} = \arctan(1/a)$.

(b)   Use spherical coordinates to calculate the volume $V(R, a)$ of the buoy as a function of R and a. [Check your results: $V(2, \sqrt{3}) = (16\pi/3)(1 - \sqrt{3}/2)$.]

ᵖC4.3.4 Volume integral over quarter-sphere (spherical coordinates) (p. 251)

Use spherical coordinates to calculate the volume integral $F(R) = \int_Q dV f(\mathbf{r})$ of the function $f(\mathbf{r}) = xy$ on the quarter-sphere Q, defined by $x^2 + y^2 + z^2 \leq R^2$ and $x, y \geq 0$. Sketch Q. [Check your result: $F(2) = \frac{64}{15}$.]

ɛC4.3.5 Wave functions of two-dimensional harmonic oscillator (polar coordinates) (p. 251)

The quantum mechanical treatment of a two-dimensional harmonic oscillator leads to so-called "wave functions",

$$\Psi_{nm} : \mathbb{R}^2 \to \mathbb{C}, \mathbf{r} \mapsto \Psi_{nm}(\mathbf{r}), \quad \text{with} \quad n \in \mathbb{N}_0, \quad m \in \mathbb{Z}, \quad m = -n, -n+2, \ldots, n-2, n,$$

which have a factorized form when written in terms of polar coordinates, $\Psi_{nm}(\mathbf{r}) = R_{n|m|}(\rho)Z_m(\phi)$, with $Z_m(\phi) = \frac{1}{\sqrt{2\pi}}e^{im\phi}$. The wave functions satisfy the following "orthogonality relation":

$$O_{nn'}^{mm'} \equiv \int_{\mathbb{R}^2} dS\, \overline{\Psi}_{nm}(\mathbf{r})\Psi_{n'm'}(\mathbf{r}) = \delta_{nn'}\delta_{mm'} .$$

Verify these for $n = 0$, 1 and 2, where the radial wave functions have the form:

$$R_{00}(\rho) = \sqrt{2}e^{-\rho^2/2}, \quad R_{11}(\rho) = \sqrt{2}\rho e^{-\rho^2/2}, \quad R_{22}(\rho) = \rho^2 e^{-\rho^2/2}, \quad R_{20}(\rho) = \sqrt{2}[\rho^2 - 1]e^{-\rho^2/2}.$$

Proceed as follows. Due to the product form of the wave function Ψ, each area integral separates into two factors that can be calculated separately, $O_{nn'}^{mm'} = P_{nn'}^{|m||m'|}\widetilde{P}^{mm'}$, where P is a radial integral and $\widetilde{P}$ an angular integral.

(a)   Find general expressions for P and $\widetilde{P}$ as integrals over R- or Z-functions, respectively.

(b)   Compute the angular integral $\widetilde{P}^{mm'}$ for arbitrary values of m and m'.

(c)   Now compute those radial integrals that arise in combination with $\widetilde{P} \neq 0$, namely $P_{00}^{00}, P_{11}^{11}, P_{22}^{22}, P_{22}^{00}$ and P_{20}^{00}.

Hint: The Euler identity, $e^{i2\pi k} = 1$ if $k \in \mathbb{Z}$, is useful for evaluating the angular integral, and $\int_0^\infty dx\, x^n e^{-x} = n!$ for the radial integrals. ($\rightarrow$ C2.4.5)

Background information: The functions $\Psi_{nm}(\mathbf{r})$ are the "eigenfunctions" of a quantum mechanical particle in a two-dimensional harmonic potential, $V(\mathbf{r}) \propto \mathbf{r}^2$, where n and m are "quantum numbers" that specify a particular "eigenstate". A particle in this state is found with probability $|\Psi_{nm}(\mathbf{r})|^2 dS$ within the area element dS at position $\mathbf{r}$. The total probability of being found anywhere in $\mathbb{R}^2$ equals 1, hence the normalization integral yields $O_{nn}^{mm} = 1$ for *every* eigenfunction $\Psi_{nm}(\mathbf{r})$. The fact that the area integral of two eigenfunctions vanishes if their quantum numbers are not equal reflects the fact that the eigenfunctions form an orthonormal basis in the space of square-integrable complex functions on $\mathbb{R}^2$.

ₚC4.3.6 Wave functions of the hydrogen atom (spherical coordinates) (p. 251)

Show that the volume integral, $P_{nlm} = \int_{\mathbb{R}^3} dV |\Psi_{nlm}(\mathbf{r})|^2$, for the following functions $\Psi_{nlm}(\mathbf{r}) = R_{nl}(r)Y_l^m(\theta,\phi)$, with spherical coordinates $\mathbf{r} = \mathbf{r}(r,\theta,\phi)$, yields $P_{nlm} = 1$:

(a) $\Psi_{210}(\mathbf{r}) = R_{21}(r)Y_1^0(\theta,\phi)$, $R_{21}(r) = \dfrac{re^{-r/2}}{\sqrt{24}}$, $Y_1^0(\theta,\phi) = \left(\dfrac{3}{4\pi}\right)^{1/2}\cos\theta$;

(b) $\Psi_{320}(\mathbf{r}) = R_{32}(r)Y_2^0(\theta,\phi)$, $R_{32}(r) = \dfrac{4\,r^2 e^{-r/3}}{81\sqrt{30}}$, $Y_2^0(\theta,\phi) = \left(\dfrac{5}{16\pi}\right)^{1/2}(3\cos^2\theta - 1)$.

(c)   Show that the "overlap integral" $O = \int_{\mathbb{R}^3} dV\, \overline{\Psi}_{320}(\mathbf{r})\Psi_{210}(\mathbf{r})$ yields zero.

Hint: $I_n = \int_0^\infty dx\, x^n\, e^{-x} = n!$ ($\rightarrow$ C2.4.5).

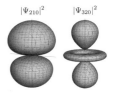

Background information: The $\Psi_{nlm}(\mathbf{r})$ are quantum mechanical "eigenfunctions" of the hydrogen atom, where n, l and m are "quantum numbers" which specify the quantum state of the system. A particle in this state is found with probability $|\Psi_{nm}(\mathbf{r})|^2 dV$ within the volume element dV at position $\mathbf{r}$. The total probability for being found anywhere in $\mathbb{R}^3$ equals 1, hence $P_{nlm} = 1$ holds for *every* eigenfunction $\Psi_{nm}(\mathbf{r})$.

The figures each show a surface on which $|\Psi_{nlm}|^2$ has a constant value. The eigenfunctions form an orthonormal basis in the space of square-integrable complex functions on $\mathbb{R}^3$, hence the volume integral of two eigenfunctions vanishes if their quantum numbers are not equal.

P.C4.4 Curvilinear integration in arbitrary dimensions (p. 252)

ₑC4.4.1 Surface integral: area of a sphere (p. 254)

Consider a sphere S with radius R. Compute its area, A_S, using (a) Cartesian coordinates, and (b) spherical coordinates, by proceeding as follows.

(a)   Choose Cartesian coordinates, with the origin at the center of the sphere. Its area is twice that of the half-sphere S_+ lying above the xy-plane. S_+ can be parametrized as

$$\mathbf{r} : D \rightarrow S_+, \qquad (x,y)^T \mapsto \mathbf{r}(x,y) = (x, y, \sqrt{R^2 - x^2 - y^2})^T,$$

where $D = \{(x,y)^T \in \mathbb{R}^2 | x^2 + y^2 < R^2\}$ is a disk of radius R. Use this parametrization to compute the area of the sphere as $A_S = 2\int_D dxdy\, \|\partial_x\mathbf{r} \times \partial_y\mathbf{r}\|$.

(b)   Now choose spherical coordinates and parametrize the sphere as

$$\mathbf{r} : U \to S, \qquad (\theta, \phi)^{\mathrm{T}} \mapsto \mathbf{r}(\theta, \phi) = R(\sin\theta\cos\phi, \sin\theta\sin\phi, \cos\theta)^{\mathrm{T}},$$

with $U = (0, \pi) \times (0, 2\pi)$. Compute its area, $A_S = \int_U d\theta\, d\phi\, \|\partial_\theta \mathbf{r} \times \partial_\phi \mathbf{r}\|$.

P.C4.4.2 Surface integral: area of slanted face of rectangular pyramid (p. 254)

Consider the pyramid shown in the sketch. Find a parametrization of its slanted face, F_{slant}, of the form

$$\mathbf{r} : U \subset \mathbb{R}^2 \to F_{\text{slant}} \subset \mathbb{R}^3, \quad (x, y)^{\mathrm{T}} \mapsto \mathbf{r}(x, y),$$

i.e. specify the domain U and the Cartesian vector $\mathbf{r}(x, y)$. Then compute the area of the slanted face as $A_{\text{slant}} = \int_U dx\, dy\, \|\partial_x \mathbf{r} \times \partial_y \mathbf{r}\|$.

$\left[\text{Check your result: If } a = 2, \text{ then } A_{\text{slant}} = \frac{\sqrt{53}}{12}. \right]$

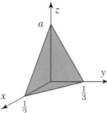

E.C4.4.3 Volume and surface integral: parabolic solid of revolution (p. 254)

Consider a parabolic solid of revolution, P, bounded from above by the plane $z = z_{\max}$, and from below by the surface of revolution obtained by rotating the parabola $z(x) = x^2$ about the z-axis.

(a)   Calculate the volume, V, of the body P.

(b)   Calculate the surface area, A, of the curved part of the surface of P.

$\left[\text{Check your results: For } z_{\max} = \frac{3}{4} \text{ we have } V = \frac{9\pi}{32} \text{ and } A = \frac{7\pi}{6}. \right]$

P.C4.4.4 Surface integral: hyperbolic solid of revolution (Gabriel's horn) (p. 254)

Consider the solid body, K, generated by rotating the function $\rho(z) = 1/z$, with $1 \leq z \leq a$, about the z-axis. This shape is known as Gabriel's horn or Torticelli's trumpet.

(a)   Compute the volume, $V(a)$, of the body K.
      [Check your result: $V(2) = \frac{\pi}{2}$.]

(b)   Write down the integral for the surface area of this solid, $A(a)$, and calculate its derivative, $A'(a) = \frac{d}{da}A(a)$. [Check your result: $A'(1) = 2\sqrt{2}\pi$.]

(c)   Find a lower bound for the value of the integral $A(a)$ by using the inequality $\sqrt{z^{-4} + 1} \geq 1$.

(d)   How large are the volume and (the lower bound for) the area in the limit as $a \to \infty$?

E.C4.4.5 Surface area of a circular cone (p. 254)

Consider a circular cone, C, of radius R and height h. Compute the area, $A_C(R, h)$, of its (slanted) conical surface S_C as a function of R and h. [Check your result: $A_C(3, 4) = 15\pi$.]

P.C4.4.6 Surface area of an elliptical cone (p. 254)

Consider an elliptical cone, C, with semi-axes a and b and height h. Use generalized polar coordinates to show that the area, A_C, of its (slanted) conical surface S_C is given by an integral of the form

$$A_C = \int_{S_C} dS = P \int_0^{2\pi} d\phi \, \sqrt{1 + Q \sin^2 \phi} \,,$$

and find $P(a,b,h)$ and $Q(a,b,h)$ as functions of a, b and h. *Remark:* This integral belongs to the class

elliptical
integrals

of so-called **elliptical integrals**, which cannot be solved in closed form.
[Check your results: If $a = 3$, $b = 2$ and $h = 4$, then $P = 5$ and $Q = \frac{4}{5}$.]

P.C4.5   Changes of variables in higher-dimensional integration (p. 256)

ᴇC4.5.1 Variable transformation for two-dimensional integral (p. 257)

(a)   Consider the transformation of variables $x = \frac{1}{2}(X+Y)$, $y = \frac{1}{2}(X-Y)$. Invert it to find $X(x,y)$ and
$Y(x,y)$. Compute the Jacobian matrices $J = \frac{\partial(x,y)}{\partial(X,Y)}$ and $J^{-1} = \frac{\partial(X,Y)}{\partial(x,y)}$, and their determinants.
[Check your results: Verify that $JJ^{-1} = \mathbb{1}$ and $(\det J)(\det J^{-1}) = 1$.]

Use the transformation from (a) to compute the following integrals as $\int dXdY$ integrals:

(b) $I_1 = \int_S dxdy$, integrated over the square $S = \{0 \le x \le 1, 0 \le y \le 1\}$.

(c) $I_2(n) = \int_T dxdy \, |x - y|^n$, integrated over the triangle $T = \{0 \le x \le 1, 0 \le y \le 1 - x\}$.
[Check your result: $I_2(1) = \frac{1}{6}$.]

ᴘC4.5.2 Variable transformation for two-dimensional integral (p. 257)

(a)   Consider the transformation of variables $x = \frac{3}{5}X + \frac{3}{5}Y$ and $y = \frac{3}{5}X - \frac{2}{5}Y$. Invert it to find
$X(x,y)$ and $Y(x,y)$. Compute the Jacobian matrices $J = \frac{\partial(x,y)}{\partial(X,Y)}$ and $J^{-1} = \frac{\partial(X,Y)}{\partial(x,y)}$, and their
determinants. [Check your results: Verify that $JJ^{-1} = \mathbb{1}$ and $(\det J)(\det J^{-1}) = 1$.]

(b)   Compute the integral $I(a) = \int_{T_a} dxdy \, \cos\left[\pi\left(\frac{2}{3}x + y\right)^3\right](x - y)$ over
the trapezoid T_a enclosed by the lines $x = 0$, $y = 1 - \frac{2}{3}x$, $y = 0$ and
$y = a - \frac{2}{3}x$, with $a \in (0, 1)$. *Hint:* Express $I(a)$ as an $\int dXdY$ integral
using the transformation from (a).
[Check your result: $I(2^{-1/3}) = -\frac{1}{8\pi}$.]

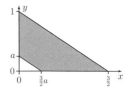

ᴇC4.5.3 Three-dimensional Gaussian integral via linear transformation (p. 257)

Calculate the following three-dimensional Gaussian integral (with $a, b, c > 0$, $a, b, c \in \mathbb{R}$):

$$I = \int_{\mathbb{R}^3} dx \, dy \, dz \; e^{-\left[a^2(x+y)^2 + b^2(z-y)^2 + c^2(x-z)^2\right]}.$$

Hint: Use the substitution $u = a(x + y)$, $v = b(z - y)$, $w = c(x - z)$ and calculate the Jacobian
determinant, using $J = \left|\frac{\partial(u,v,w)}{\partial(x,y,z)}\right|^{-1}$. You may use $\int_{-\infty}^{\infty} dx \, e^{-x^2} = \sqrt{\pi}$. [Check your result: If
$a = b = c = \sqrt{\pi}$, then $I = \frac{1}{2}$.]

ᴘC4.5.4 Three-dimensional Lorentzian integral via linear transformation (p. 257)

Calculate the following triple Lorentz integral (with $a, b, c, d > 0$, $a, b, c, d \in \mathbb{R}$):

$$I = \int_{\mathbb{R}^3} dx \, dy \, dz \; \frac{1}{[(xd + y)^2 + a^2]} \cdot \frac{1}{[(y + z - x)^2 + b^2]} \cdot \frac{1}{[(y - z)^2 + c^2]}.$$

Hint: Use the change of variables $u = (xd + y)/a$, $v = (y + z - x)/b$, $w = (y - z)/c$ and compute the Jacobian determinant using $J = \left| \frac{\partial(u,v,w)}{\partial(x,y,z)} \right|^{-1}$. You may use $\int_{-\infty}^{\infty} dx (x^2 + 1)^{-1} = \pi$. [Check your result: If $a = b = c = \pi$, $d = 2$, then $I = \frac{1}{5}$.]

ₑC4.5.5 Multidimensional Gaussian integrals (p. 257)

[This problem and the next presume knowledge about matrix diagonalization (Section L7.3) and the simplifications arising for symmetric matrices (Section L8.2).]

Multiple Gaussian integrals are integrals of the form

$$I = \int_{\mathbb{R}^n} dx^1 \ldots dx^n \, e^{-\mathbf{x}^T A \mathbf{x}},$$

where $\mathbf{x} = (x^1, ..., x^n)^T$ and the matrix A is symmetric and positive definite (i.e. all eigenvalues of A are > 0). The characteristic property of this class of integrals is that the exponent is a "quadratic form", i.e. a *quadratic* function of all integration variables. In general this function contains mixed terms, but these can be removed by a basis transformation: Let T be the similarity transformation that diagonalizes A, so that $D = T^{-1}AT$ is diagonal, with eigenvalues $\lambda_1, ..., \lambda_n$. Since A is symmetric, T can be chosen orthogonal, with $T^{-1} = T^T$ and $\det T = 1$. Now define $\tilde{\mathbf{x}} = (\tilde{x}^1, ..., \tilde{x}^n)^T$ by $\tilde{\mathbf{x}} \equiv T^T \mathbf{x}$, then we have

$$\mathbf{x}^T A \mathbf{x} = \mathbf{x}^T T D T^T \mathbf{x} = \tilde{\mathbf{x}}^T D \tilde{\mathbf{x}} = \sum_i \lambda_i (\tilde{x}^i)^2. \tag{1}$$

When expressed through the new variables $\tilde{\mathbf{x}}$, the exponent thus no longer contains any mixed terms, so that the Gaussian integral can be solved by the variable substitution $\mathbf{x} = T\tilde{\mathbf{x}}$:

$$I = \int_{\mathbb{R}^n} dx^1 \ldots dx^n \, e^{-\mathbf{x}^T A \mathbf{x}} = \int_{\mathbb{R}^n} d\tilde{x}^1 \ldots d\tilde{x}^n \, J \, e^{-\sum_i^n \lambda_n (\tilde{x}^i)^2} = \sqrt{\frac{\pi}{\lambda_1}} \cdots \sqrt{\frac{\pi}{\lambda_n}} = \boxed{\sqrt{\frac{\pi^n}{\det A}}}.$$

We have exploited two facts here: (i) Since $\partial x^i / \partial \tilde{x}^j = T^i_{\ j}$, the Jacobian determinant of the variable substitution equals the determinant of T and thus is equal to 1:

$$J = \left| \frac{\partial(x^1, ..., x^n)}{\partial(\tilde{x}^1, ..., \tilde{x}^n)} \right| = \left| \det \begin{pmatrix} \frac{\partial x^1}{\partial \tilde{x}^1} & \cdots & \frac{\partial x^1}{\partial \tilde{x}^n} \\ \vdots & & \vdots \\ \frac{\partial x^n}{\partial \tilde{x}^1} & \cdots & \frac{\partial x^n}{\partial \tilde{x}^n} \end{pmatrix} \right| = \left| \det \begin{pmatrix} T^1_{\ 1} & \cdots & T^1_{\ n} \\ \vdots & & \vdots \\ T^n_{\ 1} & \cdots & T^n_{\ n} \end{pmatrix} \right| = |\det T| = 1.$$

(ii) The product of the eigenvalues of a matrix equals its determinant, $\prod_i^n \lambda_i = \det A$.

Now use the above strategy to compute the following integral ($a > 0$):

$$I(a) = \int_{\mathbb{R}^2} dx \, dy \, e^{-[(a+3)x^2 + 2(a-3)xy + (a+3)y^2]}.$$

Execute all steps of the above argumentation explicitly:

(a) Bring the exponent into the form $-\mathbf{x}^T A \mathbf{x}$, with $\mathbf{x} = (x, y)^T$ and A symmetric. Identify and diagonalize the matrix A. In particular, explicitly write out equation (1) for the present case.

(b) Find T. Calculate the Jacobian determinant explicitly.

(c) What is the value of the Gaussian integral? $\left[\text{Check your result}: I(1) = \frac{\pi}{2\sqrt{3}}.\right]$

pC4.5.6 Three-dimensional Gaussian integrals (p. 257)

Compute the following three-dimensional Gaussian integral ($a > 0$):

$$I(a) = \int_{\mathbb{R}^3} dx\, dy\, dz\; e^{-\left[(a+2)x^2 + (a+2)y^2 + (a+2)z^2 + 2(a-1)xy + 2(a-1)yz + 2(a-1)xz\right]}$$

(a) Bring the exponent into the form $-\mathbf{x}^{\mathrm{T}}A\mathbf{x}$, with $\mathbf{x} = (x, y, z)^{\mathrm{T}}$ and A symmetric.

(b) Diagonalize the matrix A. You do not need to compute the corresponding similarity transformation explicitly.

(c) Compute $I(a)$ by expressing it as a product of three one-dimensional Gaussian integrals. $\left[\text{Check your result}: I(3) = \frac{1}{9}\sqrt{\pi^3}.\right]$

P.C5 Taylor series (p. 259)

P.C5.2 Complex Taylor series (p. 263)

eC5.2.1 Addition theorems for sine and cosine (p. 264)

Prove the addition theorems for sine and cosine, for any $a, b \in \mathbb{C}$:

(a) $\cos(a + b) = \cos a \cos b - \sin a \sin b$,   (b) $\sin(a + b) = \cos a \sin b + \sin a \cos b$.

Hint: Use the Euler formula on both sides of $e^{i(a+b)} = e^{ia}e^{ib}$.

pC5.2.2 Powers of sine and cosine (p. 264)

Use the Euler–de Moivre identity to prove the following identities, for any $a \in \mathbb{C}$:

(a) $\cos^2 a = \frac{1}{2} + \frac{1}{2}\cos(2a)$, $\sin^2 a = \frac{1}{2} - \frac{1}{2}\cos(2a)$;

(b) $\cos^3 a = \frac{3}{4}\cos a + \frac{1}{4}\cos(3a)$, $\sin^3 a = \frac{3}{4}\sin a - \frac{1}{4}\sin(3a)$.

P.C5.3 Finite-order expansion (p. 266)

eC5.3.1 Taylor series (p. 266)

Taylor expand the following functions. You may choose either to calculate the coefficients of the Taylor series by taking the corresponding derivatives, or to use the known Taylor expansions of $\sin(x)$, $\cos(x)$, $\frac{1}{1-x}$ and $\ln(1 + x)$.

(a) $f(x) = \frac{1}{1-\sin(x)}$ around $x = 0$, up to and including fourth order.

(b) $g(x) = \sin(\ln(x))$ around $x = 1$, up to and including second order.

(c) $h(x) = e^{\cos x}$ around $x = 0$, up to and including second order.

$\left[\text{Check your results: The highest} - \text{order term requested in each case is}: \text{(a)}\ \frac{2}{3}x^4,\ \text{(b)} -\frac{1}{2}(x-1)^2,\right.$
$\left.\text{(c)} -\frac{1}{2}ex^2.\right]$

P$C5.3.2 Taylor series (p. 266)

Taylor expand the following functions. You may choose either to calculate the coefficients of the Taylor series by taking the corresponding derivatives, or to use the known Taylor expansions of $\sin(x)$, $\cos(x)$, $\frac{1}{1-x}$ and $\ln(1+x)$.

(a) $f(x) = \frac{\cos(x)}{1-x}$ around $x = 0$. Keep all terms up to and including third order.

(b) $g(x) = e^{\cos(x^2 + x)}$ about $x = 0$, up to and including third order.

(c) $h(x) = e^{-x} \ln(x)$ around $x = 1$, up to and including third order.

[Check your results: The highest-order term requested in each case is: (a) $\frac{1}{2}x^3$, (b) $-e x^3$, (c) $\frac{4}{3}e^{-1}(x - 1)^3$.]

P.C5.4 Solving equations by Taylor expansion (p. 267)

E$C5.4.1 Series expansion for iteratively solving an equation (p. 268)

Solve the equation $e^{y-1} = 1 - \epsilon y$ for y, to second order in the small parameter ϵ, using the ansatz $y(\epsilon) = y_0 + y_1\epsilon + \frac{1}{2!}y_2\epsilon^2 + \mathcal{O}(\epsilon^3)$. Use both of the following approaches.

(a) Method 1: **expansion of equation**. Insert the ansatz for $y(\epsilon)$ into the given equation, Taylor expand each term to order $\mathcal{O}(\epsilon^2)$, and collect terms having the same power of ϵ to obtain an equation of the form $0 = \sum_n F_n\epsilon^n$. The coefficient of each ϵ^n must vanish, yielding a hierarchy of equations, $F_n = 0$. Starting from $n = 0$, solve these successively for the y_n, using knowledge of the previously determined $y_{i<n}$ at each step. [Check your results: $y_2 = 1$.]

(b) Method 2: **repeated differentiation**. Method 1 can be viewed from the following perspective: the given equation is written in the form $0 = \mathcal{F}(y(\epsilon), \epsilon) \equiv F(\epsilon)$, and the r.h.s. is brought into the form $\sum_n F_n\epsilon^n$. The latter process can be streamlined by realizing that $F_n = \frac{1}{n!}d_\epsilon^n F(\epsilon)|_{\epsilon=0}$. Hence, the nth equation in the hierarchy, $F_n = 0$, can be set up by simply differentiating the given equation n times and then setting ϵ to zero, $0 = d_\epsilon^n F(\epsilon)|_{\epsilon=0}$. Use this approach to find a hierarchy of equations for y_0, y_1 and y_2.

Hint: Since $F(\epsilon)$ depends on ϵ both directly and via $y(\epsilon)$, the chain rule must be used when computing derivates, e.g. $d_\epsilon F(\epsilon) = \partial_y \mathcal{F}(y, \epsilon)y' + \partial_\epsilon \mathcal{F}(y, \epsilon)$.

Remark: Method 2 has the advantage that it systematically proceeds order by order: information from $\mathcal{O}(\epsilon^n)$ is generated at just the right time, namely when it is needed in step n for computing y_n. As a result, this method is often more convenient than method 1, particularly if the dependence of $\mathcal{F}(y, \epsilon)$ on y is nontrivial.

P$C5.4.2 Series expansion for iteratively solving an equation (p. 268)

Solve the equation $\ln\left[(x + 1)^2\right] + e^y = 1 - y$ for y, to second order in the small parameter x, using the ansatz $y(x) = y_0 + y_1 x + \frac{1}{2!}y_2 x^2 + \mathcal{O}(x^3)$. Use both the methods described in problem C5.4.1:

(a) Method 1: expansion of equation; and (b) method 2: repeated differentiation.

Which one do you find more convenient? [Check your results: $y_2 = \frac{1}{2}$.]

εC5.4.3 Series expansion of inverse function (p. 268)

This problem illustrates how the series expansion of an inverse function can be computed by expansion of the equation defining the inverse function.

The inverse, $g(x)$, of the function $f(x)$ fulfills the defining equation $f(g(x)) = x$. To find the series expansion of the inverse function around some point x_0, we may use the ansatz $g(x_0 + x) \equiv y(x) \equiv \sum_{n=0}^{\infty} \frac{1}{n!} y_n x^n$, and determine the coefficients $y_n \equiv y^{(n)}(0)$ by iteratively solving the equation $f(y(x)) = x_0 + x$ for $y(x)$. In this manner, calculate the series expansion of the following functions around $x = 0$, up to and including second order in x:

(a) $\ln(1 + x)$,   (b) 2^x.

[Check your results: (a) $y_2 = -1$, (b) $y_2 = \ln^2(2)$.]

ₚC5.4.4 Series expansion of inverse function (p. 268)

Find the series expansion of $\arcsin(x)$ around $x = 0$, up to and including order three, using both of the following methods:

(a) Find the expansion of $\arcsin(x) \equiv y(x)$ by iteratively solving the equation $\sin[y(x)] = x$.

(b) Since the sine function is odd, so is its inverse, hence it can be represented by the ansatz $\arcsin(x) = c_1 x^1 + \frac{1}{3!} c_3 x^3 + \mathcal{O}(x^5)$. Determine the coefficients c_1 and c_3 by expanding the equation $\arcsin(\sin(y)) = y$ in powers of y, using the known series expansion for $\sin(y)$. [Check your results: $c_3 = 1$.]

P.C5.5   Higher-dimensional Taylor series (p. 269)

εC5.5.1 Taylor expansion in two dimensions (p. 269)

Find the Taylor expansion of the function $g(x, y) = e^x \cos(x + 2y)$ in x and y, around the point $(x, y) = (0, 0)$. Calculate explicitly all terms up to and including second order,

(a) by multiplying out the series expansions for the exponential and cosine functions;

(b) by using the formula for the Taylor series of a function of two variables.

[Check your results: The mixed second-order term in each case is: (a) $-2xy$, (b) $-2xy$.]

ₚC5.5.2 Taylor expansion in two dimensions (p. 269)

For the following functions, calculate the Taylor expansion in x and y around the point $(x, y) = (0, 0)$, up to and including second order:

(a) $f(x, y) = e^{-(x+y)^2}$,   (b) $g(x, y) = \dfrac{1 + x}{\sqrt{1 + xy}}$.

[Check your results: The mixed second-order term in each case is: (a) $-2xy$, (b) $-\frac{1}{2}xy$.]

P.C6   Fourier calculus (p. 271)

P.C6.1   The δ-function (p. 272)

ᴇC6.1.1 Integrals with δ-function (p. 277)

Calculate the following integrals (with $a \in \mathbb{R}$):

(a) $I_1(a) = \displaystyle\int_{-\infty}^{\infty} dx\, \delta(x - \pi) \sin(ax);$

(b) $I_2(a) = \displaystyle\int_{\mathbb{R}^3} dx^1 dx^2 dx^3\, \delta(\mathbf{x} - \mathbf{y}) \|\mathbf{x}\|^2,$ with $\mathbf{y} = (a, 1, 2)^{\mathsf{T}};$

(c) $I_3(a) = \displaystyle\int_0^a dx\, \delta(x - \pi) \frac{1}{a + \cos^2(x/2)};$

(d) $I_4(a) = \displaystyle\int_0^3 dx\, \delta(x^2 - 6x + 8)\sqrt{e^{ax}};$

(e) $I_5(a) = \displaystyle\int_{\mathbb{R}^2} dx^1 dx^2\, \delta(\mathbf{x} - a\mathbf{y})\, \mathbf{x} \cdot \mathbf{y},$ with $\mathbf{y} = (1, 3)^{\mathsf{T}}.$ *Remark:* $\delta(\mathbf{x}) = \delta(x^1)\delta(x^2).$

[Check your results: $I_1(\tfrac{1}{2}) = 1,\ I_2(1) = 6,\ I_3(\pi) = \frac{1}{2\pi},\ I_4(\ln 2) = 1,\ I_5(1) = 10.$]

ᴩC6.1.2 Integrals with δ-function (p. 277)

Calculate the following integrals (with $a \in \mathbb{R}, n \in \mathbb{N}$):

(a) $I_1(a) = \displaystyle\int_1^4 dx\, \delta(x - 2)(a^x + 3);$

(b) $I_2(a) = \displaystyle\int_{\mathbb{R}^2} dx^1 dx^2\, \delta(\mathbf{x} - \mathbf{y})(x^1 + x^2)^2\, e^{3 - x^1},$ with $\mathbf{y} = (3, a)^{\mathsf{T}};$

(c) $I_3(a) = \displaystyle\int_{-1}^1 dx\, \sqrt{2 + 2x}\, \delta(ax - 2),$ with $a \neq 0;$

(d) $I_4(a) = \displaystyle\int_{-\infty}^{\infty} dx\, \delta(3^{-x} - 9)(1 - x^a);$

(e) $I_5(n) = \displaystyle\int_{-\pi/2}^{9\pi/2} dx\, \cos(nx)\, \delta(\sin x);$

(f) $I_6(a) = \displaystyle\int_{\mathbb{R}^2} dx^1 dx^2\, \delta(\mathbf{x} - \mathbf{y}) e^{\|\mathbf{x}\|^2},$ with $\mathbf{y} = (a, -a)^{\mathsf{T}}.$

$\left[\text{Check your results: } I_1(3) = 12,\, I_2(-5) = 4,\, I_3(2) = \tfrac{1}{2},\, I_4(3) = \frac{1}{\ln 3},\, I_5(7) = 1,\, I_6\left(\frac{1}{\sqrt{2}}\right) = e.\right]$

ᴇC6.1.3 Lorentz representation of the Dirac delta function (p. 277)

Explain why in the limit $\epsilon \to 0^+$, the Lorentz peak function $\delta^\epsilon(x)$ given below is a representation of the Dirac delta function $\delta(x)$. To this end, compute (i) the height, (ii) the width x_{w} (defined by $\delta^\epsilon(x_{\mathrm{w}}) = \tfrac{1}{2}\delta^\epsilon(0),\ x_{\mathrm{w}} > 0$) and (iii) the area of the peak. How do these quantities behave for $\epsilon \to 0^+$? Furthermore, calculate the functions (iv) $\Theta^\epsilon(x) = \int_{-\infty}^x dx'\, \delta^\epsilon(x')$ and (v) $\delta'^\epsilon(x) = \frac{d}{dx}\delta^\epsilon(x)$. Sketch

Θ^ϵ, $\epsilon\delta^\epsilon$ and $\epsilon^2\delta'^\epsilon$ as functions of x/ϵ in three separate sketches (one beneath the other, with aligned y-axes and the same scaling for the x/ϵ-axes).

Lorentz peak: $\delta^\epsilon(x) = \dfrac{\epsilon/\pi}{x^2 + \epsilon^2}$.

Hint: When calculating the peak weight, use the substitution $x = \epsilon \tan y$.

Remark: Lorentzian functions are common in physics. Example: the spectrum energy of a discrete quantum state, which is weakly coupled to the environment, has the form of a Lorentzian function, the width of which is determined by the strength of the coupling to the environment. As the coupling strength approaches zero, we obtain a δ peak.

pC6.1.4 Representations of the Dirac delta function (p. 277)

Explain why in the limit $\epsilon \to 0^+$, each of the three peak-shaped functions $\delta^\epsilon(x)$ given below is a representation of the Dirac delta function $\delta(x)$. To this end, compute (i) the height, (ii) the width x_w (defined by $\delta^\epsilon(x_w) = \frac{1}{2}\delta^\epsilon(0)$, $x_w > 0$) and (iii) the area of each peak. How do these quantities behave for $\epsilon \to 0^+$? Furthermore, calculate the functions (iv) $\Theta^\epsilon(x) = \int_{-\infty}^{x} dx'\,\delta^\epsilon(x')$ and (v) $\delta'^\epsilon(x) = \frac{d}{dx}\delta^\epsilon(x)$. For each peak shape, sketch Θ^ϵ, $\epsilon\delta^\epsilon$ and $\epsilon^2\delta'^\epsilon$ as functions of x/ϵ in three separate sketches (one beneath the other, with aligned y-axes and the same scaling for the x/ϵ-axes).

(a) Gaussian peak: $\delta^\epsilon(x) = \dfrac{1}{\epsilon\sqrt{\pi}} e^{-(x/\epsilon)^2}$.

Hint: The function $\Theta^\epsilon(x)$ cannot be calculated in terms of elementary functions; instead write it in terms of the "error function", $\mathrm{Erf}(z) = \frac{2}{\sqrt{\pi}}\int_0^z dy\,e^{-y^2}$, with $\mathrm{Erf}(\infty) = 1$.

Remark: Gaussians appear very often in physics. Example: a quantum mechanical harmonic oscillator with spring constant k and potential energy $\frac{1}{2}kx^2$ has a Gaussian wave function for its ground state, with width $\sim 1/\sqrt{k}$.

(b) Derivative of the Fermi function: $\delta^\epsilon(x) = \dfrac{1}{4\epsilon}\dfrac{1}{\cosh^2[x/(2\epsilon)]}$.

Hint: When calculating the peak weight, use the substitution $y = \tanh[x/(2\epsilon)]$.

Remark: In condensed matter physics and nuclear physics the function $\delta^\epsilon(x)$ plays an important role: it arises as the derivative of the so-called **Fermi function**, $f(E) = \frac{1}{e^{E/k_B T}+1} = \Theta^{k_B T}(-E)$, with $-\frac{d}{dE}f(E) = \delta^{k_B T}(E)$, where $f(E)$ is the occupation probability of a fermionic single-particle state with energy E as a function of the system's temperature T (k_B is the so-called Boltzmann constant). In the limit of zero temperature, $T \to 0$, the derivative of the Fermi function reduces to a Dirac δ-function.

(c) Second derivative of the absolute value function: $\delta^\epsilon(x) = \dfrac{1}{2}\dfrac{\epsilon^2}{(x^2 + \epsilon^2)^{3/2}}$.

Hint: When calculating the peak weight, use the substitution $x = \epsilon \tan y$.

Remark: This peak form can be written as $\delta^\epsilon(x) = \frac{d^2}{dx^2}\frac{1}{2}|x|_\epsilon$, where $|x|_\epsilon = (\epsilon^2 + x^2)^{1/2}$ represents a "smeared" version of the absolute value function, with $\lim_{\epsilon \to 0}|x|_\epsilon = |x|$. [Using $\epsilon \neq 0$ "smears out" the sharp "kink" in $|x|$ at $x = 0$.] The first and second derivatives of $\frac{1}{2}|x|_\epsilon$ yield "smeared" versions of the step function $\Theta(x)$ and the Dirac delta function $\delta(x)$, respectively. To illustrate this, include a sketch of the function $\frac{1}{\epsilon}|x|_\epsilon$ above your sketch of Θ^ϵ.

εC6.1.5 Series representation of the coth function (p. 280)

Compute the following series for $y \in \mathbb{R}^+$, by expressing each as a geometric series in $\omega \equiv e^{-y}$.

(a) $\displaystyle\sum_{n=0}^{\infty} e^{-y(n+1/2)}$,

(b) $\displaystyle\sum_{n=0}^{\infty} (-1)^n e^{-y(n+1/2)}$,

(c) $\displaystyle\sum_{n\in\mathbb{Z}} e^{-y|n|}$.

ρC6.1.6 Series representation of the periodic δ-function (p. 280)

Show that the function $\delta^\epsilon(x)$, defined by

$$\delta^\epsilon(x) = \frac{1}{L} \sum_k e^{ikx - \epsilon|k|}, \quad k = 2\pi n/L, \quad n \in \mathbb{Z}, \quad x, \epsilon, L \in \mathbb{R}, \quad 0 < \epsilon \ll L, \tag{1}$$

has the following properties:

(a) $\delta^\epsilon(x) = \delta^\epsilon(x+L)$. $\tag{2}$

(b) $\displaystyle\int_{-L/2}^{L/2} dx\, \delta^\epsilon(x) = 1$. *Hint:* Treat $k = 0$ and $k \neq 0$ separately in $\sum_k$. $\tag{3}$

(c) $\delta^\epsilon(x) = \dfrac{1}{2L}\left[\dfrac{1+w}{1-w} + \dfrac{1+\overline{w}}{1-\overline{w}}\right] = \dfrac{1}{L}\dfrac{1 - e^{-4\pi\epsilon/L}}{1 + e^{-4\pi\epsilon/L} - 2e^{-2\pi\epsilon/L}\cos(2\pi x/L)}$, $\tag{4}$

where $w = e^{2\pi(ix-\epsilon)/L}$ and $\overline{w} = e^{2\pi(-ix-\epsilon)/L}$.
Hint: Write out the sum in Eq. (1) as a geometric series in powers of w and $\overline{w}$.

(d) $\displaystyle\lim_{\epsilon\to 0} \delta^\epsilon(x) = 0$ for $x \neq mL$, with $m \in \mathbb{Z}$. *Hint:* Start from Eq. (4).

(e) $\delta^\epsilon(x) \simeq \dfrac{\epsilon/\pi}{\epsilon^2 + x^2}$ for $|x|/L \ll 1$ and $\epsilon/L \ll 1$.

Hint: Taylor expand the numerator in Eq. (4) up to first order in $\tilde{\epsilon} = 2\pi\epsilon/L$, and the denominator up to second order in $\tilde{\epsilon}$ and $\tilde{x} = 2\pi x/L$.

(f) Sketch the function $\delta^\epsilon(x)$ qualitatively for $\epsilon/L \ll 1$ and $x \in [-\frac{7}{2}L, \frac{7}{2}L]$.

(g) Deduce that in the limit of $\epsilon \to 0$, $\delta^\epsilon(x)$ represents a periodic δ function, with

$$\delta^0(x) = \frac{1}{L}\sum_k e^{ikx} = \sum_{m\in\mathbb{Z}} \delta(x - mL).$$

P.C6.2 Fourier series (p. 277)

εC6.2.1 Fourier series of the sawtooth function (p. 285)

Let $f(x)$ be a sawtooth function, defined by $f(x) = x$ for $-\pi < x < \pi$, $f(\pm\pi) = 0$ and $f(x + 2\pi) = f(x)$. Calculate the Fourier coefficients $\tilde{f}_n$ in the representation $f(x) = \frac{1}{L}\sum_n e^{ik_n x}\tilde{f}_n$. How should k_n and L be chosen? Sketch the function $f(x)$, as well as the sum of the $n = 1$ and $n = -1$ terms of the Fourier series (i.e. the first term of the corresponding sine series). [Check your result: $\tilde{f}_6 = \frac{1}{3}i\pi$.]

pC6.2.2 Fourier series (p. 285)

Determine the Fourier series for the following periodic functions, i.e. calculate the Fourier coefficients $\tilde{f}_n$ in the representation $f(x) = \frac{1}{L}\sum_n e^{ik_n x}\tilde{f}_n$. How should k_n and L be chosen in each case? Sketch the functions first.

$$(a)\, f(x) = |\sin x|, \qquad (b)\, f(x) = \begin{cases} 4x & \text{for} \quad -\pi \le x < 0, \\ 2x & \text{for} \quad 0 \le x < \pi, \end{cases} \quad \text{and} \ f(x+2\pi) = f(x).$$

[Check your results: (a) $\tilde{f}_3 = -\frac{2}{35}$, (b) $\tilde{f}_3 = \frac{2}{9}(2 - 9i\pi)$.]

εC6.2.3 Cosine series (p. 278)

For the function $f : I \to \mathbb{C}$, $x \mapsto f(x)$, with $I = [-L/2, L/2]$, consider the Fourier series representation $f(x) = \frac{1}{L}\sum_k e^{ikx}\tilde{f}_k$, with $k = \frac{2\pi n}{L}$ and $n \in \mathbb{Z}$.

(a) Show that the Fourier coefficients are given by $\tilde{f}_k = \int_{-L/2}^{L/2} dx\, e^{-ikx}f(x)$.

(b) Now let f be an even function, i.e. $f(x) = f(-x)$. Show that then the Fourier coefficients are given by $\tilde{f}_k = 2\int_0^{L/2} dx\, \cos(kx)f(x)$, and furthermore, that $f(x)$ can be represented by a cosine series of the form $f(x) = \frac{1}{2}a_0 + \sum_{k>0} a_k \cos(kx)$, with $k = \frac{2\pi n}{L}$ and $n \in \mathbb{N}_0$. Find a_k, expressed through $\tilde{f}_k$.

(c) Now consider the following function: $f(x) = 1$ for $|x| < L/4$, $f(x) = -1$ for $L/4 < |x| < L/2$. Sketch it, and compute the coefficients $\tilde{f}_k$ and a_k of the corresponding Fourier and cosine series. $\left[\text{Check your result : if } k = \frac{2\pi}{L}, \text{ then } a_k = \frac{4}{\pi} \text{ and } \tilde{f}_k = \frac{2L}{\pi}.\right]$

pC6.2.4 Sine series (p. 278)

For the function $f : I \to \mathbb{C}$, $x \mapsto f(x)$, with $I = [-L/2, L/2]$, consider the Fourier series representation $f(x) = \frac{1}{L}\sum_k e^{ikx}\tilde{f}_k$, with $k = \frac{2\pi n}{L}$ and $n \in \mathbb{Z}$, with Fourier coefficients $\tilde{f}_k = \int_{-L/2}^{L/2} dx\, e^{-ikx}f(x)$.

(a) Let f be an odd function, i.e. $f(x) = -f(-x)$. Show that then the Fourier coefficients are given by $\tilde{f}_k = -2i\int_0^{L/2} dx\, \sin(kx)f(x)$, and furthermore, that $f(x)$ can be represented by a sine series of the form $f(x) = \sum_{k>0} b_k \sin(kx)$ with $k = \frac{2\pi n}{L}$ and $n \in \mathbb{N}_0$. What does b_k look like when expressed through $\tilde{f}_k$?

(b) Now consider the following function: $f(x) = 1$ for $0 < x < L/2$, $f(x) = -1$ for $-L/2 < x < 0$. Sketch it, and compute the coefficients $\tilde{f}_k$ and b_k of the corresponding Fourier and sine series. $\left[\text{Check your result : if } k = \frac{2\pi}{L}, \text{ then } b_k = \frac{4}{\pi} \text{ and } \tilde{f}_k = \frac{2L}{i\pi}.\right]$

P.C6.3 Fourier transform (p. 285)

εC6.3.1 Properties of Fourier transformations (p. 288)

Demonstrate the following properties of the Fourier transformation, where a is an arbitrary real constant.

(a) The Fourier transform of $f(x - a)$ is $e^{-ika}\tilde{f}(k)$.

(b) The Fourier transform of $f(ax)$ is $\tilde{f}(k/a)/|a|$, where $a \neq 0$.

ₚC6.3.2 Properties of Fourier transformations (p. 288)

Prove that the following properties of the Fourier transform hold in two dimensions, where $\mathbf{a} \in \mathbb{R}^2$, $\alpha \in \mathbb{R} \setminus \{0\}$ and R is a rotation matrix.

(a)   The Fourier transform of $f(\mathbf{x} - \mathbf{a})$ is $e^{-i\mathbf{k}\cdot\mathbf{a}} \tilde{f}(\mathbf{k})$.

(b)   The Fourier transform of $f(\alpha\mathbf{x})$ is $\frac{1}{|\alpha|^2} \tilde{f}(\mathbf{k}/\alpha)$.

(c)   The Fourier transform of $f(R\mathbf{x})$ is $\tilde{f}(R\mathbf{k})$.

ₑC6.3.3 Fourier transformation of a Gauss peak (p. 290)

Show that the Fourier transform of a normalized Gaussian distribution with width σ, $g^{[\sigma]}(x) = \frac{1}{\sqrt{2\pi}\sigma}e^{-x^2/2\sigma^2}$, with $\int_{-\infty}^{\infty} dx\, g^{[\sigma]}(x) = 1$, is given by $\tilde{g}_k^{[\sigma]} = e^{-\sigma^2 k^2/2}$. *Hint*: The Fourier integral can be calculated by completing the square in the exponent.

ₚC6.3.4 Convolution of Gauss peaks (p. 290)

The purpose of this exercise is to illustrate the following statement. "The fine structure of a function (e.g. noise in a test signal) can be smoothed out via convolution with a peaked function of suitable width."

   A normalized Gaussian function with width σ has the form $g^{[\sigma]}(x) = \frac{1}{\sqrt{2\pi}\sigma}e^{-x^2/2\sigma^2}$. Show that the convolution of two normalized Gaussians with widths σ_1 and σ_2 is again a normalized Gaussian, with width $\sigma = \sqrt{\sigma_1^2 + \sigma_2^2}$, i.e. show that $\left(g^{[\sigma_1]} * g^{[\sigma_2]}\right)(x) = g^{[\sigma]}(x)$. Do this via two different methods, (a) and (b):

(a)   Calculate the convolution integral by completing the square in the exponent.

(b)   Use the convolution theorem, $\left(\widetilde{g^{[\sigma_1]} * g^{[\sigma_2]}}\right)(k) = \tilde{g}^{[\sigma_1]}(k)\tilde{g}^{[\sigma_2]}(k)$, and the known form of the Fourier transform of a Gaussian, $\tilde{g}^{[\sigma_j]}(k)$.

(c)   Draw two qualitative sketches, the first of $g^{[\sigma_1]}(x)$, $g^{[\sigma_2]}(x)$ and $g^{[\sigma]}(x)$, the second of their respective Fourier spectra $\tilde{g}^{[\sigma_1]}(k)$, $\tilde{g}^{[\sigma_2]}(k)$ and $\left(\widetilde{g^{[\sigma_1]} * g^{[\sigma_2]}}\right)(k)$. Explain, with reference to the sketch, why the convolution of a function (here $g^{[\sigma_1]}$) with a peaked function (here $g^{[\sigma_2]}$) leads to a broadened version of the first function.

Let $f^{[\sigma_1]}(x) = \sum_{n=-5}^{5} g_n^{[\sigma_1]}(x)$, with $g_n^{[\sigma_1]}(x) = g^{[\sigma_1]}(x - nL)$, be a "comb" of 11 identical, normalized Gaussian peaks of width σ_1, with peak-to-peak distance L, and let $F^{[\sigma_2]}(x) = \left(f * g^{[\sigma_2]}\right)(x)$ be the convolution of this comb with a normalized Gaussian peak of width σ_2.

(d)   Find a formula for $F^{[\sigma_2]}(x)$, expressed as a sum over the normalized Gaussians. What is the width of each of these peaks?

(e)   The sketch shows $F^{[\sigma_2]}(x)$ for $\sigma_1/L = \frac{1}{4}$ and four values of σ_2/L: $\frac{1}{100}$, $\frac{1}{4}$, $\frac{1}{2}$ and $\frac{3}{4}$. Explain the observed behavior based on your formula from part (d). Why does the fine structure vanish in $F^{[\sigma_2]}(x)$ for $\sigma_2 \gtrsim \frac{1}{2}L$?

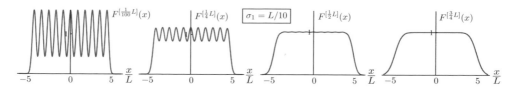

(f) Regarding the introductory statement about using convolutions to smoothen noisy functions: explain in general how the width of the peaked function should be chosen to smooth out the noise.

ₑC6.3.5 Parseval's identity and convolution (p. 291)

Let $f(x)$ be a sawtooth function, defined by $f(x) = x$ for $-\pi < x < \pi$, $f(\pm\pi) = 0$ and $f(x + 2\pi) = f(x)$. In the Fourier representation $f(x) = \frac{1}{2\pi} \sum_{n\in\mathbb{Z}} e^{\mathrm{i}nx}\tilde{f}_n$, its Fourier coefficients are $\tilde{f}_0 = 0$, $\tilde{f}_{n\neq0} = 2\pi\mathrm{i}(-1)^n/n$ ($\rightarrow$ C6.2.1). Let $g(x) = \sin x$.

(a) Using this concrete example, check that Parseval's identity holds, by computing both the integral $\int_{-\pi}^{\pi} dx \overline{f(x)}g(x)$ and the sum $(1/2\pi)\sum_n \overline{\tilde{f}_n}\tilde{g}_n$ explicitly.

(b) Prove the famous identity $\sum_{n=1}^{\infty} \frac{1}{n^2} = \frac{\pi^2}{6}$, by computing the integral $\int_{-\pi}^{\pi} dx f^2(x)$ in two ways: first, by direct integration, and second, by expressing it as a sum over Fourier modes using Parseval's identity.

(c) Calculate the convolution $(f * g)(x)$ both by directly computing the convolution integral and by using the convolution theorem and a summation of Fourier coefficients.

ₚC6.3.6 Performing an infinite series using the convolution theorem (p. 291)

This problem illustrates how a complicated sum may be calculated explicitly using the convolution theorem.

Consider the periodic function $f_\gamma(t) = f_\gamma(0)e^{\gamma t}$ for $t \in [0, \tau)$ and $f(t + \tau) = f(t)$, with $f_\gamma(0) = 1/(e^{\gamma\tau} - 1)$. Take both γ and τ to be positive numbers, so that $f_{\pm\gamma}(0) \geq 0$.

(a) Consider a Fourier series representation of $f_\gamma(t)$ of the following form:

$$f_\gamma(t) = \frac{1}{\tau}\sum_{\omega_n} e^{-\mathrm{i}\omega_n t}\tilde{f}_{\gamma,n}, \qquad \tilde{f}_{\gamma,n} = \int_0^\tau dt e^{\mathrm{i}\omega_n t}f_\gamma(t), \quad \text{with} \quad \omega_n = 2\pi n/\tau, \quad n \in \mathbb{Z}.$$

Show that the Fourier coefficients are given by $\tilde{f}_{\gamma,n} = 1/(\mathrm{i}\omega_n + \gamma)$.

(b) Use this result and the convolution theorem to express the following series as a convolution of f_γ and $f_{-\gamma}$:

$$S(t) = \sum_{n=-\infty}^{\infty} \frac{e^{-\mathrm{i}\omega_n t}}{\omega_n^2 + \gamma^2} = -\tau \int_0^\tau dt' f_\gamma(t - t')f_{-\gamma}(t'). \tag{1}$$

(c) Sketch the functions $f_\gamma(t - t')$ and $f_{-\gamma}(t')$ occurring in the convolution theorem as functions of t', for $t' \in [-\tau, 2\tau]$. Assume $0 \leq t \leq \tau$ and show that the convolution integral (1) is given by the following expression:

$$S(t) = \frac{\tau\left[\sinh\left(\gamma\left(t - \tau\right)\right) - \sinh\left(\gamma t\right)\right]}{2\gamma\left[1 - \cosh\left(\gamma\tau\right)\right]}.$$

Hint: The integral $\int_0^\tau dt'$ involves an interval of t' values for which $t - t'$ lies outside of $[0, \tau)$. It is therefore advisable to split the integral into two parts, with $\int_0^t dt'$ and $\int_t^\tau dt'$.

εC6.3.7 Poisson summation formula (p. 297)

Poisson summation formula

(a) Let $f : \mathbb{R} \to \mathbb{C}$ be a function admitting a Fourier-integral representation of the form $f(x) = \int_{-\infty}^{\infty} \frac{dk}{2\pi} e^{ikx} \tilde{f}(k)$. Show that it satisfies the Poisson summation formula,

$$\sum_{m \in \mathbb{Z}} f(m) = \sum_{n \in \mathbb{Z}} \tilde{f}(2\pi n),$$

which states that the sum of function values, $f(m)$, over all integers equals the sum over all Fourier coefficients, $\tilde{f}(2\pi n)$.

Hint: Multiply the completeness relation for discrete Fourier modes, $\frac{1}{L} \sum_{n \in \mathbb{Z}} e^{-i2\pi ny/L} = \sum_{m \in \mathbb{Z}} \delta(y - Lm)$, by $f(y/L)$, then integrate over $x = y/L$.

(b) Use the Poisson summation formula for the function $f(x) = e^{-a|x|}$ (with $0 < a \in \mathbb{R}$) to prove the following identity:

$$\sum_{n \in \mathbb{Z}} \frac{2a}{(2\pi n)^2 + a^2} = \coth(a/2).$$

ᵖC6.3.8 Poisson resummation formula for Gaussians (p. 297)

Poisson resummation formula

Use the Poisson summation formula ($\to$ C6.3.7) for $f(x) = e^{-(ax^2 + bx + c)}$ (with $0 < a \in \mathbb{R}$) to prove the **Poisson resummation formula** for infinite sums over discrete Gaussians:

$$\sum_{m \in \mathbb{Z}} e^{-(am^2 + bm + c)} = \sqrt{\frac{\pi}{a}} e^{\left(\frac{b^2}{4a} - c\right)} \sum_{n \in \mathbb{Z}} e^{-\frac{1}{a}\left(\pi^2 n^2 + i\pi nb\right)}.$$

Note that this is an example of Fourier reciprocity: the widths of the discrete Gaussian functions on the left and right are proportional to $1/a$ and a/π^2 respectively.

P.C7   Differential equations (p. 300)

P.C7.2   Separable differential equations (p. 302)

εC7.2.1 Separation of variables (p. 304)

A first-order differential equation is called **autonomous** if it has the form $\dot{x} = f(x)$, i.e. the right-hand side is time independent (non-autonomous equations have $\dot{x} = f(x,t)$). (Here we use the shorthand $\dot{x} \equiv d_t x$.) Such an equation can be solved by separation of variables.

(a) Consider the autonomous differential equation $\dot{x} = x^2$ for the function $x(t)$. Solve it by separation of variables for two different initial conditions: (i) $x(0) = 1$ and (ii) $x(2) = -1$. [Check your results: (i) $x(-2) = \frac{1}{3}$, and (ii) $x(2) = -1$.]

(b) Sketch your solutions qualitatively. Convince yourself that your sketches for the function $x(t)$ and its derivative $\dot{x}(t)$ satisfy the relation specified by the differential equation.

ₚC7.2.2 Separation of variables (p. 304)

(a) Consider the differential equation $y' = -x^2/y^3$ for the function $y(x)$. Solve it by separation of variables, for two different initial conditions: (i) $y(0) = 1$, and (ii) $y(0) = -1$.
$$\left[\text{Check your result}: \ (i)\ y(-1) = \left(\tfrac{7}{3}\right)^{1/4}, \ (ii)\ y(-1) = -\left(\tfrac{7}{3}\right)^{1/4}. \right]$$

(b) Sketch your solutions qualitatively. Convince yourself that your sketches for the function $y(x)$ and its derivative $y'(x)$ satisfy the relation specified by the differential equation.

ₑC7.2.3 Separation of variables: barometric formula (p. 304)

The standard barometric formula for atmospheric pressure, $p(x)$, as a function of the height, x, is given by: $\frac{dp(x)}{dx} = -\alpha \frac{p(x)}{T(x)}$. Solve this equation with initial value $p(x_0) = p_0$ for the case of a linear temperature gradient, $T(x) = T_0 - b(x - x_0)$.
[Check your result: If $\alpha, b, T_0, x_0, p_0 = 1$, then $p(1) = 1$.]

ₚC7.2.4 Separation of variables: bacterial culture with toxin (p. 304)

A bacterial culture is exposed to the effects of a toxin. The death rate induced by the toxin is proportional to the number, $n(t)$, of bacteria still alive in the culture at a time t and the amount of toxin, $T(t)$, remaining in the system, which is given by $\tau n(t)T(t)$, where τ is a positive constant. On the other hand, the natural growth rate of the bacteria in the culture is exponential, i.e. it grows with a rate $\gamma n(t)$, with $\gamma > 0$. In total, the number of bacteria in the culture is given by the differential equation

$$\dot{n} = \gamma n - \tau n T(t), \quad \text{for } t \geq 0.$$

(a) Find the general solution to this linear DEQ, with initial condition $n(0) = n_0$.

(b) Assume now that the toxin is injected into the system at a constant rate $T(t) = at$, where $a > 0$. Use a qualitative analysis of the differential equation (i.e. without solving it explicitly), to show that the bacterial population grows up to a time $t = \gamma/(a\tau)$, and decreases thereafter. Furthermore, show that as $t \to \infty$, $n(t) \to 0$, i.e. the bacterial culture is practically wiped out.

(c) Now find the explicit solution, $n(t)$, to the differential equation and sketch $n(t)$ qualitatively as a function of t. Convince yourself that the sketch fulfills the relation between $n(t)$, $\dot{n}(t)$ and t that is specified by the differential equation. [Check your result: If $\tau = 1$, $a = 1$, $n_0 = 1$ and $\gamma = \sqrt{\ln 2}$, then $n(\sqrt{\ln 2}) = \sqrt{2}$.]

(d) Find the time t_h at which the number of bacteria in the culture drops to half the initial value. [Check your result: If $\tau = 4$, $a = 2/\ln 2$ and $\gamma = 3$, then $t_h = \ln 2$.]

ₑC7.2.5 Substitution and separation of variables (p. 304)

Often differential equations can be solved by convenient substitution.

(a) Consider the differential equation $y' = f(y/x)$ for the function $y(x)$. Show that the substitution $y = ux$ can be used to convert it into a separable differential equation for the function $u(x)$, which can be solved using separation of variables.

(b) Use this method to solve the equation $xy' = 2y + x$ with the initial condition $y(1) = 0$. [Check your result: $y(2) = 2$.]

ₚC7.2.6 Substitution and separation of variables (p. 304)

Consider differential equations of the type

$$y'(x) = f(ax + by(x) + c). \tag{1}$$

(a) Substitute $u(x) = ax + by(x) + c$ and find a differential equation for $u(x)$.

(b) Find an implicit expression for the solution $u(x)$ of the new differential equation using an integral that contains the function f. *Hint:* Separation of variables!

(c) Use the substitution strategy of (a,b) to solve the differential equation $y'(x) = e^{x+3y(x)+5}$, with initial condition $y(0) = 1$.
$$\left[\text{Check your result} : y(\ln(e^{-8} + 3) - 2\ln 2) = \tfrac{1}{3}\left(2\ln 2 - \ln(e^{-8} + 3) - 5\right).\right]$$

(d) Check: Solve the differential equation given in (c) directly (without substitution) using separation of variables. Is the result in agreement with the result from (c)?

(e) Solve the differential equation $y'(x) = [a(x + y) + c]^2$ with initial condition $y(x_0) = y_0$ using the substitution given in (a).
[Check your result: If $x_0 = y_0 = 0$ and $a = c = 1$, then $y(0) = 0$.]

P.C7.3 Linear first-order differential equations (p. 306)

ᴇC7.3.1 Inhomogeneous linear differential equation: variation of constant (p. 307)

Solve the inhomogeneous differential equation $\dot{x} + 2x = t$ with $x(0) = 0$, as follows:

(a) Determine the general solution of the homogeneous equation.

(b) Then find a special (particular) solution to the inhomogeneous problem by means of variation of constants. $\left[\text{Check your result} : x(-\ln 2) = \tfrac{3}{4} - \tfrac{1}{2}\ln 2.\right]$

ᴘC7.3.2 Inhomogeneous linear differential equation, variation of constants (p. 307)

The function $x(t)$ satisfies the inhomogeneous differential equation

$$\dot{x}(t) + tx(t) = e^{-\frac{t^2}{2}}, \qquad \text{with initial condition} \quad x(0) = x_0. \tag{1}$$

(a) Find the solution, $x_h(t)$, of the corresponding homogeneous equation with $x_h(0) = x_0$.

(b) Find the particular solution, $x_p(t)$, of the inhomogeneous equation (1), with $x_p(0) = 0$ using variation of constants, $x_p(t) = c(t)x_h(t)$. What is the general solution?
[Check your result: If $x_0 = 0$, then $x(1) = e^{-1/2}$.]

(c) For a differential equation of the form $\dot{x}(t) + a(t)x(t) = b(t)$ (ordinary, first-order, linear and inhomogeneous), the sum of the homogeneous and inhomogeneous solutions has the form:

$$x(t) = x_h(t) + x_p(t) = x_h(t) + c(t)x_h(t) = (1 + c(t))x_h(t) = \tilde{c}(t)x_h(t).$$

The initial condition $x(0) = x_0$ can therefore also be satisfied by imposing on $x_h(t)$ and $\tilde{c}(t)$ the initial conditions $x_h(0) = 1$ and $\tilde{c}(0) = x_0$. Use this approach to construct a solution to the differential equation (1) of the form $x(t) = \tilde{c}(t)x_h(t)$. Does the result agree with the result as obtained in (b)? This example illustrates the general fact that the same initial condition can be implemented in more than one way.

P.C7.4 Systems of linear first-order differential equations (p. 308)

ᴇC7.4.1 Linear homogeneous differential equation with constant coefficients (p. 309)

Use an exponential ansatz to solve the following differential equation:

$$\dot{\mathbf{x}}(t) = A \cdot \mathbf{x}(t), \qquad A = \frac{1}{5}\begin{pmatrix} 3 & -4 \\ -4 & -3 \end{pmatrix}, \qquad \mathbf{x}(0) = (2, 1)^{\mathrm{T}}.$$

ᴘC7.4.2 Linear homogeneous differential equation with constant coefficients (p. 309)

Use an exponential ansatz to solve the following differential equation:

$$\dot{\mathbf{x}}(t) = A \cdot \mathbf{x}(t), \qquad A = \frac{1}{2}\begin{pmatrix} 3 & -1 \\ -1 & 3 \end{pmatrix}, \qquad \mathbf{x}(0) = (1, 3)^{\mathrm{T}}.$$

ᴇC7.4.3 System of linear differential equations with non-diagonizable matrix (p. 312)

We consider a procedure to solve the differential equation

$$\dot{\mathbf{x}} = A \cdot \mathbf{x} \tag{1}$$

for the case of a matrix $A \in \mathrm{Mat}(n, \mathbb{R})$ that has $n - 1$ distinct eigenvalues λ_j and associated eigenvectors $\mathbf{v}_j$, with $j = 1, \ldots, n - 1$, where the eigenvalue λ_{n-1} is a two-fold zero of the characteristic polynomial but has only *one* eigenvector. Such a matrix is not diagonalizable. However, it can be brought into the so-called Jordan normal form:

$$T^{-1}AT = J, \qquad J = \begin{pmatrix} \lambda_1 & 0 & \cdots & \cdots & 0 \\ 0 & \lambda_2 & 0 & \cdots & 0 \\ 0 & 0 & \ddots & \cdots & 0 \\ \vdots & \cdots & \cdots & \lambda_{n-1} & 1 \\ 0 & \cdots & \cdots & 0 & \lambda_{n-1} \end{pmatrix}, \qquad T = (\mathbf{v}_1, \ldots, \mathbf{v}_{n-1}, \mathbf{v}_n). \tag{2}$$

Using $A = TJT^{-1}$, as well as $\mathbf{v}_j = T\mathbf{e}_j$ and $J\mathbf{e}_j = \lambda_j\mathbf{e}_j + \delta_{jn}\mathbf{e}_{j-1}$, one finds that this is equivalent to

$$A \cdot \mathbf{v}_j = \lambda_j\mathbf{v}_j + \mathbf{v}_{j-1}\delta_{jn}, \quad \forall j = 1, \ldots, n. \tag{3}$$

For $j = 1, \ldots, n - 1$ this corresponds to the usual eigenvalue equation, and $\mathbf{v}_j$ to the usual eigenvectors. $\mathbf{v}_n$, however, is not an eigenvector, but is rather determined by the following equation:

$$(A - \mathbb{1}\lambda_n)\mathbf{v}_n = \mathbf{v}_{n-1}. \tag{4}$$

Since $(A - \mathbb{1}\lambda_n)$ is not invertible, this equation does not uniquely fix the vector $\mathbf{v}_n$. Different choices of $\mathbf{v}_n$ lead (via (2)) to different similarity transformation matrices T, but they all yield the same form for the Jordan matrix J.

The λ_j and $\mathbf{v}_j$ thus obtained can be used to find a solution for the DEQ (1), using an exponential ansatz together with "variation of the constants":

$$\mathbf{x}(t) = \sum_{j=1}^{n} \mathbf{v}_j e^{\lambda_j t} c^j(t), \quad \text{with} \quad \lambda_n \equiv \lambda_{n-1}. \tag{5}$$

The coefficients $c^j(t)$ can be determined by inserting this ansatz into (1):

$$0 = \left(\frac{\mathrm{d}}{\mathrm{d}t} - A\right)\mathbf{x}(t) = \sum_{j=1}^{n} \mathbf{v}_j e^{\lambda_j t}\left[\lambda_j c^j(t) + \dot{c}^j(t) - \lambda_j c^j(t)\right] - \mathbf{v}_{n-1}e^{\lambda_n t}c^n(t). \tag{6}$$

Comparing coefficients of $\mathbf{v}_j$ we obtain:

$\mathbf{v}_{j \neq n-1}$: $\dot{c}^j(t) = 0$ $\Rightarrow$ $\boxed{c^j(t) = c^j(0) = \text{const.}}$,   (7)

$\mathbf{v}_{n-1}$: $\dot{c}^{n-1}(t) = c^n(t)$ $\Rightarrow$ $\boxed{c^{n-1}(t) = c^{n-1}(0) + t\, c^n(0)}$.   (8)

The values of $c^j(0)$ are fixed by the initial conditions $\mathbf{x}(0)$:

$$\mathbf{x}(0) = \sum_j \mathbf{v}_j c^j(0) = T\mathbf{c}(0) \quad \Rightarrow \quad \mathbf{c}(0) = T^{-1}\mathbf{x}(0).$$ (9)

Now use this method to find the solution of the DEQ

$$\dot{\mathbf{x}} = A\mathbf{x}, \quad \text{with} \quad A = \frac{1}{3}\begin{pmatrix} 7 & 2 & 0 \\ 0 & 4 & -1 \\ 2 & 0 & 4 \end{pmatrix} \quad \text{and} \quad \mathbf{x}(0) = \begin{pmatrix} 1 \\ 1 \\ 1 \end{pmatrix}.$$ (10)

(a) Show that the characteristic polynomial for A has a simple zero, say λ_1, and a two-fold zero, say $\lambda_2 = \lambda_3$. $\left[\text{Check : do your results satisfy } \sum_j \lambda_j = \text{Tr}(A) \text{ and } \prod_j \lambda_j = \det(A)?\right]$

(b) Show that the eigenspaces associated with λ_1 and λ_2 are both one-dimensional (which implies that A is not diagonalizable), and find the corresponding normalized eigenvectors $\mathbf{v}_1$ and $\mathbf{v}_2$.

(c) Use Eq. (4) to find a third, normalized vector $\mathbf{v}_3$, having the property that A is brought into a Jordan normal form using $T = (\mathbf{v}_1, \mathbf{v}_2, \mathbf{v}_3)$. While doing so, exploit the freedom of choice that is available for $\mathbf{v}_3$ to choose the latter orthonormal to $\mathbf{v}_1$ and $\mathbf{v}_2$. [*Remark:* For the present example orthonormality is achievable (and useful, since then $T^{-1} = T^{\text{T}}$ holds), but this is generally not the case.]

(d) Now use an ansatz of the form (5) to find the solution $\mathbf{x}(t)$ to the DEQ (10). [Check your result: $\mathbf{x}(\ln 2) = (2, 4, 0)^{\text{T}} + \frac{4}{3}(1 + \ln 2)(2, -1, 2)^{\text{T}}$.]

(e) Check your result explicitly by verifying that it satisfies the DEQ.

$_\text{P}$C7.4.4 System of linear differential equations with non-diagonizable matrix: critically damped harmonic oscillator (p. 312)

Consider a critically damped harmonic oscillator, described by the second-order DEQ

$$\ddot{x} + 2\gamma\dot{x} + \gamma^2 x = 0.$$ (1)

By introducing the variables $\mathbf{x} \equiv (x, v)^{\text{T}}$, with $v \equiv \dot{x}$ and $\dot{v} = \ddot{x} = -\gamma^2 x - 2\gamma v$, this equation can be transcribed into a system of two first-order DEQs:

$$\begin{pmatrix} \dot{x} \\ \dot{v} \end{pmatrix} = \begin{pmatrix} 0 & 1 \\ -\gamma^2 & -2\gamma \end{pmatrix} \begin{pmatrix} x \\ v \end{pmatrix}.$$ (2)

To solve the matrix equation (2), $\dot{\mathbf{x}} = A\mathbf{x}$, we may try the ansatz $\mathbf{x}(t) = \mathbf{v}e^{\gamma t}$, leading to the eigenvalue problem $\lambda\mathbf{v} = A\mathbf{v}$. For the damped harmonic oscillator, this eigenvalue problem turns out to have degenerate eigenvalues. To deal with this complication, proceed as follows.

(a) Find the degenerate eigenvalue, λ, its eigenvector, $\mathbf{v}$, and the corresponding solution, $\mathbf{x}(t)$, of Eq. (2). Verify that its first component, $x(t)$, is a solution of (1). We will call this solution $x_1(t)$ henceforth.

(b) Find a second solution, $x_2(t)$, of Eq. (1) via variation of constants, by inserting the ansatz $x_2(t) = c(t)x_1(t)$ into Eq. (1). Find a differential equation for $c(t)$ and solve this equation.

(c)   Using a linear combination of $x_1(t)$ and $x_2(t)$, find the solution $x(t)$ satisfying $x(0) = 1$, $\dot{x}(1) = 1$.
$\left[\text{Check your result : if } \gamma = 2, \text{ then} x(\ln 2) = \frac{1}{4}\left(1 - \ln 2(2 + e^2)\right).\right]$

(d)   The critically damped harmonic oscillator can be thought of as the limit $\lambda \to \Omega$ of both the overdamped and underdamped harmonic oscillator. Their general solution has the form $x(t) = c_+ e^{\gamma_+ t} + c_- e^{\gamma_- t}$, where $\gamma_\pm = -\gamma \pm \sqrt{\gamma^2 - \Omega^2}$ in the overdamped case and $\gamma_\pm = -\gamma \pm i\sqrt{\Omega^2 - \gamma^2}$ in the underdamped case. For both cases, show that a Taylor expansion of the general solution for small values of ϵt, with $\epsilon \equiv \sqrt{|\gamma^2 - \Omega^2|}$, yields expressions which can be written as linear combinations of the solutions to the critically damped harmonic oscillator found in (a) and (b).

ₑC7.4.5 Coupled oscillations of two point masses (p. 313)

Consider a system of two point masses, with masses m_1 and m_2, which are connected to two fixed walls and to each other by means of three springs (spring constants K_1, K_{12} and K_2) (see sketch). The equations of motion for both masses are

$$m_1\ddot{x}^1 = -K_1 x^1 - K_{12}(x^1 - x^2),$$
$$m_2\ddot{x}^2 = -K_2 x^2 - K_{12}(x^2 - x^1).$$

(a)   Bring the system of equations into the form $\ddot{\mathbf{x}}(t) = -A \cdot \mathbf{x}(t)$, with $\mathbf{x} = (x^1, x^2)^{\mathsf{T}}$. What is the form of matrix A?
[Check your result: $\det A = [K_1 K_2 + (K_1 + K_2)K_{12}]/(m_1 m_2)$.]

(b)   Using the ansatz $\mathbf{x}(t) = \mathbf{v}\cos(\omega t)$, this system of differential equations can be converted to an algebraic eigenvalue problem. Find the form of this eigenvalue problem.

(c)   Set $m_1 = m_2$, $K_2 = m_1 \Omega^2$, $K_1 = 4K_2$ and $K_{12} = 2K_2$ (note that Ω has the dimension of frequency). Find the eigenvalues, λ_j, and the eigenvectors, $\mathbf{v}_j$, of the matrix $\frac{1}{\Omega^2}A$, and therefore the corresponding **eigenfrequencies**, ω_j, and **eigenmodes**, $\mathbf{x}_j(t)$, of the coupled masses (with $\mathbf{x}_j(0) = \mathbf{v}_j$). [Check your result: $\lambda_1 + \lambda_2 = 9$.]

(d)   Make a sketch of both eigenmodes $\mathbf{x}_j(t)$ which shows both the $j = 1$ and 2 cases on the same set of axes. Comment on the physical behavior that you observe!

ₚC7.4.6 Coupled oscillations of three point masses (p. 313)

Consider a system consisting of three masses, m_1, m_2 and m_3, coupled through two identical springs, each with spring constant k (see sketch). The equations of motion for the three masses read:

$$m_1\ddot{x}^1 = -k(x^1 - x^2),$$
$$m_2\ddot{x}^2 = -k\left([x^2 - x^1] - [x^3 - x^2]\right),$$
$$m_3\ddot{x}^3 = -k(x^3 - x^2),$$

(a)   Bring this system of equations into the form $\ddot{\mathbf{x}}(t) = -A \cdot \mathbf{x}(t)$, with $\mathbf{x} = (x^1, x^2, x^3)^{\mathsf{T}}$. What is the matrix A? [Check your result: $\det(A) = 0$.]

(b)   By making the ansatz $\mathbf{x}(t) = \mathbf{v}\cos(\omega t)$, this system of equations can be reduced to an algebraic eigenvalue problem. Find this eigenvalue equation.

(c)   From now on, set $m_1 = m_3 = m$, $m_2 = \frac{2}{3}m$, and $k = m\Omega^2$. (Ω has the dimension of a frequency.) Find the eigenvalues, λ_j, and normalized eigenvectors, $\mathbf{v}_j$, of the matrix $\frac{1}{\Omega^2}A$, and thus the corresponding eigenfrequencies, ω_j, and eigenmodes, $\mathbf{x}_j(t)$, of the coupled masses (with $\mathbf{x}_j(0) = \mathbf{v}_j$). [Check your result: $\lambda_1 + \lambda_2 + \lambda_3 = 5$.]

(d) Sketch the three eigenmodes $\mathbf{x}_j(t)$ as functions of time: for each $j = 1$, 2 and 3, make a separate sketch that displays the three components, $x_j^1(t)$, $x_j^2(t)$ and $x_j^3(t)$, on the same axis. Comment on the physical behavior that you observe!

EC7.4.7 Inhomogeneous linear differential equation of second order: driven overdamped harmonic oscillator (p. 313)

Consider the following driven, overdamped harmonic oscillator with $\gamma > \Omega$:

Differential equation: $\ddot{x} + 2\gamma\dot{x} + \Omega^2 x = f_A(t),$ (1)

Initial value: $x(0) = 0, \quad \dot{x}(0) = 1,$ (2)

Driving function: $f_A(t) = \begin{cases} f_A & \text{for} \quad t \geq 0, \\ 0 & \text{for} \quad t < 0. \end{cases}$

For $t > 0$, find a solution to this equation of the form $x(t) = x_h(t) + x_p(t)$, where the homogeneous solution, $x_h(t)$, solves the homogeneous DEQ, with initial values (2), and the particular solution, $x_p(t)$, solves the inhomogeneous DEQ, with initial values $x_p(0) = \dot{x}_p(0) = 0$. Proceed as follows.

(a) Rewrite as a matrix equation: write the DEQ (1) in the matrix form

$$\dot{\mathbf{x}} = A \cdot \mathbf{x} + \mathbf{b}(t), \quad \text{with} \quad \mathbf{x} \equiv (x, \dot{x})^{\mathsf{T}} \equiv (x^1, x^2)^{\mathsf{T}}. \tag{3}$$

Find the matrix A, the driving force vector $\mathbf{b}(t)$ and the initial value $\mathbf{x}_0 = \mathbf{x}(0)$.

(b) Homogeneous solution: find the solution, $\mathbf{x}_h(t)$, of the homogeneous DEQ $(3)|_{\mathbf{b}(t)=0}$ having the initial value $\mathbf{x}_h(0) = \mathbf{x}_0$. Use the ansatz $\mathbf{x}_h(t) = \sum_j c_h^j \mathbf{x}_j(t)$, with $\mathbf{x}_j(t) = \mathbf{v}_j e^{\lambda_j t}$, where λ_j and $\mathbf{v}_j$ ($j = 1, 2$) are the eigenvalues and the eigenvectors of A. What does the corresponding solution, $x_h(t) = x_h^1(t)$, of the homogeneous differential equation $(1)|_{f_A(t)=0}$ look like?
$$\left[\text{Check your result: If } \gamma = \sqrt{2}\ln 2 \text{ and } \Omega = \ln 2, \text{ then } x_h(1) = \tfrac{3}{4}\tfrac{2^{-\sqrt{2}}}{\ln 2}. \right]$$

(c) Particular solution: using the ansatz $\mathbf{x}_p(t) = \sum_j c_p^j(t)\mathbf{x}_j(t)$ (variation of constants), find the particular solution for the inhomogeneous differential equation (3) having the initial value $\mathbf{x}_p(0) = \mathbf{0}$. What is the corresponding solution, $x_p(t) = x_p^1(t)$, of the inhomogeneous DEQ (1)? $\left[\text{Check your result : if } \gamma = 3\ln 2, \Omega = \sqrt{5}\ln 2 \text{ and } f_A = 1, \text{ then} x_p(1) = \tfrac{49}{640}\tfrac{1}{(\ln 2)^2}. \right]$

(d) Qualitative discussion: the desired solution of the inhomogeneous DEQ (1) is given by $x(t) = x_h(t) + x_p(t)$. Sketch your result for this function qualitatively for the case $f_A < 0$, and explain the behavior as $t \to 0$ and $t \to \infty$.

PC7.4.8 Inhomogeneous linear differential equation of third order (p. 313)

Consider the following third-order inhomogeneous linear differential equation:

Differential equation: $\dddot{x} - 6\ddot{x} + 11\dot{x} - 6x = f_A(t),$ (1)

Initial value: $x(0) = 1, \quad \dot{x}(0) = 0, \quad \ddot{x}(0) = a, \quad$ with $\quad a \in \mathbb{R}.$ (2)

Driving: $f_A(t) = \begin{cases} e^{-bt} & \text{for} \quad t \geq 0, \\ 0 & \text{for} \quad t < 0, \end{cases}$ with $\quad 0 < b \in \mathbb{R}.$ (3)

For $t > 0$, find a general solution to this equation of the form $x(t) = x_h(t) + x_p(t)$, where $x_h(t)$ and $x_p(t)$ are the homogeneous and particular solutions to the homogeneous and inhomogeneous differential equation that have the initial values (2) or $x_p(0) = \dot{x}_p(0) = \ddot{x}_p(0) = 0$ respectively. Proceed as follows:

(a)   Write the differential equation (1) in the matrix form

$$\dot{\mathbf{x}} = A \cdot \mathbf{x} + \mathbf{b}(t), \quad \text{with } \mathbf{x} \equiv (x, \dot{x}, \ddot{x})^{\mathrm{T}} \equiv (x^1, x^2, x^3)^{\mathrm{T}}, \quad \mathbf{x}_0 = (x(0), \dot{x}(0), \ddot{x}(0))^{\mathrm{T}}. \tag{4}$$

(b)   Find the homogeneous solution $\mathbf{x}_h(t)$ of $(4)|_{\mathbf{b}(t)=0}$ with $\mathbf{x}_h(0) = \mathbf{x}_0$; then $x_h(t) = x^1{}_h(t)$.

(c)   Find the inhomogeneous solution $\mathbf{x}_p(t)$ of (4), with $\mathbf{x}_p(0) = \mathbf{0}$; then $x_p(t) = x^1{}_p(t)$.

Hint: The eigenvalues $\lambda_1, \lambda_2, \lambda_3$ of A are integers, with $\lambda_1 = 1$.

P.C7.5   Linear higher-order differential equations (p. 308)

ᴇC7.5.1 Green function of $(d_t + a)$ (p. 319)

Let $\hat{L}(t) = (d_t + a)$ be a first-order differential operator, and a be a positive, real constant. The corresponding Green function is defined by the differential equation

$$\hat{L}(t)G(t) = \delta(t). \tag{1}$$

(a)   Show that the ansatz

$$G(t) = \Theta(t)x_h(t), \quad \text{with } \Theta(t) = \begin{cases} 1 & \text{for } t > 0, \\ 0 & \text{for } t < 0, \end{cases}$$

satisfies the defining equation (1), provided that $x_h(t)$ is a solution to the homogeneous equation $\hat{L}(t)x_h(t) = 0$ with initial condition $x_h(0) = 1$. [*Hint:* The initial condition guarantees that $\delta(t)x_h(t) = \delta(t)$.]

(b)   Determine $G(t)$ explicitly by solving the homogeneous equation for $x_h(t)$. $\left[\text{Check your result:} \right.$ $G(\frac{1}{a}\ln 2) = \frac{1}{2}.\Big]$

(c)   Calculate the Fourier integral $\tilde{G}(\omega) = \int_{-\infty}^{\infty} dt\, e^{i\omega t}G(t)$. $\Big[\text{Check your result : for } a = 1,$ $|\tilde{G}(a)| = \frac{1}{\sqrt{2}}.\Big]$

(d)   Consistency check: Alternatively, determine $\tilde{G}(\omega)$ via a Fourier transformation of the defining equation (1). Is the result in agreement with the result from part (c) of the exercise?

(e)   Find a solution to the inhomogeneous differential equation, $(d_t + a)x(t) = e^{2at}$, by convolving the function $G(t)$ with the inhomogeneity. Verify the obtained solution explicitly by inserting it into the differential equation.

ᴘC7.5.2 Green function of critically damped harmonic oscillator (p. 319)

A driven, critically damped harmonic oscillator with frequency $\Omega > 0$ and damping rate $\gamma = \Omega$ satisfies the equation $\hat{L}(t)\,q(t) = g(t)$, with $\hat{L}(t) = (d_t^2 + 2\Omega d_t + \Omega^2)$. The corresponding Green function is defined by the differential equation

$$\hat{L}(t)\,G(t) = \delta(t). \tag{1}$$

(a)   Show that the ansatz

$$G(t) = \Theta(t)q_h(t), \quad \text{with} \quad \Theta(t) = \begin{cases} 1 & \text{for } t > 0, \\ 0 & \text{for } t < 0, \end{cases}$$

satisfies the defining equation (1) if $q_h(t)$ is a solution of the homogeneous equation $\hat{L}(t)\,q_h(t) = 0$, with initial values $q_h(0) = 0$ and $d_t q_h(0) = 1$. [*Hint:* The initial values ensure that $\delta(t)q_h(t) = \delta(t)q_h(0) = 0$ and $\delta(t)d_t q_h(t) = \delta(t)d_t q_h(0) = \delta(t)$.]

(b) Determine $G(t)$ explicitly by solving the homogeneous equation for $q_h(t)$, using the ansatz $q_h(t) = (c_1 + c_2 t)e^{-\Omega t}$ (see Problem C7.4.4). [Check your result: If $\Omega = 1$, then $G(1) = 1/e$.]

(c) Compute the Fourier integral $\tilde{G}(\omega) = \int_{-\infty}^{\infty} dt\, e^{i\omega t} G(t)$. [Check your result: If $\Omega = 1$, then $|\tilde{G}(\Omega)| = \frac{1}{2}$.]

(d) Consistency check: Find $\tilde{G}(\omega)$ in an alternative way by Fourier transforming the defining equation (1). Does the result agree with that of part (c)?

(e) Find a solution of the inhomogeneous differential equation, $\hat{L}(t)\, q(t) = g_0 \sin(\omega_0 t)$, by convolving $G(t)$ with the inhomogeneity. Check explicitly that your result satisfies this equation. [*Hint:* It is advisable to represent the sine function by $\text{Im}\left[e^{i\omega_0 t}\right]$ and use $e^{i\omega_0 t}$ as inhomogeneity, and to take the imaginary part only at the very end of the calculation.]

P.C7.6 General higher-order differential equations (p. 313)

ᴇC7.6.1 **Field lines in two dimensions** (p. 324)

The behavior of a vector field, $\mathbf{u}(\mathbf{r})$, is often indicated graphically by sketching its **field lines**. A field line is a curve such that the tangent vector at any point along the curve points in the direction of the field at that point. If $\mathbf{r}(t)$ is a parametrization of a field line, its shape is thus determined by the requirement $\dot{\mathbf{r}}(t) \| \mathbf{u}(\mathbf{r}(t))$. This can be used to set up a differential equation whose solution describes the shape of the field lines.

To illustrate the procedure, let us consider a two-dimensional vector field in two dimensions, $\mathbf{u} : \mathbb{R}^2 \to \mathbb{R}^2, \mathbf{r} = (x, y)^\mathsf{T} \mapsto \mathbf{u}(\mathbf{r}) = (u_x(\mathbf{r}), u_y(\mathbf{r}))^\mathsf{T}$. Let $\mathbf{r}(t) = (x(t), y(t))^\mathsf{T}$ be a parametrization of a field line, then the components of its tangent vector satisfy the equation

$$\frac{\dot{y}(t)}{\dot{x}(t)} = \frac{u_y(\mathbf{r}(t))}{u_x(\mathbf{r}(t))}.$$

Alternatively, we can parametrize the field line as $(x, y(x))^\mathsf{T}$, viewing y as function of x. If $x(t)$ changes as a function of time, so does $y(t) = y(x(t))$, in a manner satisfying the relation $\dot{y}(t) = \frac{dy(x(t))}{dx}\dot{x}(t)$, or $\frac{dy(x(t))}{dx} = \frac{\dot{y}(t)}{\dot{x}(t)}$. Combining this with the above equation, we obtain

$$\frac{dy(x)}{dx} = \frac{u_x(\mathbf{r})}{u_y(\mathbf{r})}.$$

This is a differential equation for $y(x)$, whose solution describes the shape of the field lines. Different choices for the initial conditions of the DEQ yield different field lines.

Consider the vector field $\mathbf{u}(\mathbf{r}) = (-ay, x)^\mathsf{T}$, with $a > 0$. Set up and solve a differential equation for its field lines, $y(x)$. Sketch some representative lines for the case $a = \frac{1}{2}$. $\Big[$Check your results: For $a = \frac{1}{2}$, a field line passing through the point $(x, y)^\mathsf{T} = (3, 0)^\mathsf{T}$ also passes through $(1, 4)^\mathsf{T}.\Big]$

ᴘC7.6.2 **Field lines of electric quadrupole field in two dimensions** (p. 324)

Consider the field $\mathbf{E} = F(x, -3z)^\mathsf{T}$ in the xz-plane, generated by an electric quadrupole. The constant F governs the field strength. The shape of the field lines can be described by expressing z as a function of x. Find a formula for $z(x)$ by solving a suitable differential equation. Sketch some field lines in each quadrant of the xz-plane to illustrate their shape. [Check your results: A field line passing through the point $(x, z)^\mathsf{T} = (2, 1)^\mathsf{T}$ also passes through $(1, 8)^\mathsf{T}.]$

P.C7.7 Linearizing differential equations (p. 326)

ᴇC7.7.1 Fixed points of a differential equation in one dimension (p. 327)

Consider the autonomous differential equation $\dot{x} = f_\lambda(x) = (x^2 - \lambda)^2 - \lambda^2$ for the real function $x(t)$, with $\lambda \in \mathbb{R}$.

(a)   Find the fixed points of this differential equation as a function of λ for (i) $\lambda \leq 0$, and (ii) $\lambda > 0$. [Check your results: For $\lambda = 2$, the fixed points lie at 0, 2 and -2.]

(b)   Make two separate sketches of $f(x)$ as a function of x for the following fixed values of λ: (i) $\lambda = -1$ and (ii) $\lambda = +1$, and mark on your sketches the fixed points found in (a).

(c)   Determine the stability of each of these fixed points via a graphical analysis of the function, and show the flow of $x(t)$ in the neighborhood of these fixed points on the sketch from (b).

ᴘC7.7.2 Fixed points of a differential equation in one dimension (p. 327)

Consider the differential equation $\dot{x} = f(x) = \tanh[5(x-3)]\tanh[5(x+1)]\sin(\pi x)$ for the real-valued function $x(t)$.

(a)   Find the fixed points of this differential equation. [*Hint:* There are infinitely many!]

(b)

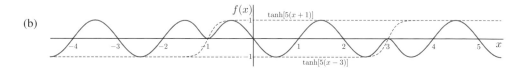

Redraw the above sketch of $f(x)$ as a function of x for $x \in [-4, 5]$, and mark on it the fixed points that you found in (a).

(c)   From an analysis of your sketch, determine the stability of each of these fixed points, and show the flow of $x(t)$ near the fixed points in the sketch from (b).

ᴇC7.7.3 Stability analysis in two dimensions (p. 327)

The function $\mathbf{x} : \mathbb{R} \to \mathbb{R}^2$, $t \mapsto \mathbf{x}(t)$ satisfies the following differential equation, with $0 < c \in \mathbb{R}$:

$$\dot{\mathbf{x}} = \begin{pmatrix} \dot{x} \\ \dot{y} \end{pmatrix} = \mathbf{f}(\mathbf{x}) = \begin{pmatrix} 2x^2 - xy \\ c(1-x) \end{pmatrix}.$$

(a)   Find the fixed point, $\mathbf{x}^*$, of the differential equation.

(b)   For a small displacement, $\boldsymbol{\eta} = \mathbf{x} - \mathbf{x}^*$, from the fixed point, linearize the differential equation and bring it into the form $\dot{\boldsymbol{\eta}} = A\boldsymbol{\eta}$. Find the matrix A.

(c)   Check that the matrix elements of A are given by $A^i{}_j = \left(\frac{\partial f^i}{\partial x^j} \right)|_{\mathbf{x}=\mathbf{x}^*}$.

(d)   Find the eigenvalues and eigenvectors of A.

(e)   Analyze the stability of the fixed point: for displacements relative to the fixed point, in which directions do these displacements grow or shrink the fastest? On which timescales?

[Check your results: If $c = 3$, then (a) $\|\mathbf{x}^*\| = \sqrt{5}$, (b) $\det(A) = -3$, (d) eigenvalues: $\lambda_+ = 3$, $\lambda_- = -1$; eigenvectors: $\mathbf{v}_+ = (3, -3)^\mathsf{T}$ and $\mathbf{v}_- = (1, 3)^\mathsf{T}$.]

ρC7.7.4 **Stability analysis in three dimensions** (p. 327)

Consider the following autonomous differential equation:

$$\dot{\mathbf{x}} = \begin{pmatrix} \dot{x} \\ \dot{y} \\ \dot{z} \end{pmatrix} = \begin{pmatrix} x^{10} - y^{24} \\ 1 - x \\ -3z - 3 \end{pmatrix}.$$

(a) Find the fixed points of this equation. [Check your result: For all fixed points, $\|\mathbf{x}^*\| = \sqrt{3}$.]

(b) Show that the fixed points are in general unstable, but that they are stable to deviations in certain directions. Determine the linear approximation to this equation for small deviations about the fixed point, and calculate the eigenvalues and eigenvectors of the corresponding matrix, A. [Check your results: For all fixed points, $|\det(A)| = 72$. Some of the eigenvalues for these fixed points are 6, 4, 12, −2.]

(c) Identify the stable directions, and the respective characteristic timescale for each deviation from the fixed point to decay to zero.

P.C8   Functional calculus (p. 330)

P.C8.3   Euler–Lagrange equations (p. 333)

ɛC8.3.1 **Snell's law** (p. 335)

Let $\mathbf{r} : I \to \mathbb{R}^2$, $x \mapsto \mathbf{r}(x) = (x, y(x))^\mathsf{T}$ describe a curve in $\mathbb{R}^2$ in Cartesian coordinates, with the x-coordinate as curve parameter.

(a) Show that the length functional, $\int_I dx \|d_x \mathbf{r}(x)\|$, can be expressed as a functional of $y(x)$, $L[y] = \int_I dx L(y(x), y'(x))$, with Lagrangian $L(y, y') = \sqrt{1 + y'^2}$, where $y'(x) = d_x y(x)$.

(b) Set up the Euler–Lagrange equation for the curve extremizing the length functional. Show that its solutions are straight lines.

A light beam, traversing a medium, travels along a straight line only if the medium is homogeneous. In an inhomogeneous medium, the speed of light, $v(\mathbf{r}) = c/n(\mathbf{r})$, depends on position, where c is its speed in vacuum and $n(\mathbf{r})$ the refraction index. According to **Fermat's principle**, the light beam then follows the path which minimizes the traversal time.

(c) Consider a beam traveling within the xy-plane along a path $\mathbf{r} : (a, b) \to \mathbb{R}^2$, $x \mapsto \mathbf{r}(x) = (x, y(x))^\mathsf{T}$. Let $t(s(x))$ be the time taken to traverse the path up to an arc length $s(x)$, given by the length functional $s(x) = \int_a^x d\bar{x} \|d_{\bar{x}} \mathbf{x}(\bar{x})\|$. Show that the traversal time for a given path is computed by the functional

$$t[\mathbf{r}] = \frac{1}{c} \int_a^b dx\, n(x, y(x)) \sqrt{1 + y'^2(x)}.$$

Hint: Use the chain rule to find the change, δ_t, in traversal time associated with a small increment, δ_x, in the curve parameter. Then set up an integral that sums up these changes.

(d) Suppose the beam traverses a boundary, lying within the yz-plane, between two media with different refraction indices, n_A for $x < 0$ and n_B for $x > 0$. **Snell's law of refraction** states that the angles of incidence, α, and of refraction, β, measured relative to the normal to the boundary plane (see figure), satisfy

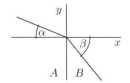

$$\frac{\sin \beta}{\sin \alpha} = \frac{n_A}{n_B}.$$

Derive this law from Fermat's principle.

ₚC8.3.2 Fermat's principle (p. 335)

Consider a light beam traveling with speed $v(\mathbf{r}) = c/n(\mathbf{r})$ within the xy-plane in an inhomogeneous medium, where c is its speed in vacuum and the refraction index has the form $n(x, y) = n_0/y$. According to **Fermat's principle**, the beam follows the path, $\mathbf{r} : x \mapsto \mathbf{r}(x) = (x, y(x))^T$, which minimizes the traversal time, given by the functional ($\rightarrow$ C8.3.1)

$$t[\mathbf{r}] = \int dx\, L(y(x), y'(x)), \qquad L(y, y') = \frac{1}{c}\frac{n_0}{y}\sqrt{1 + y'^2}.$$

(a) The x-independence of L implies the existence of a conserved quantity, $H(y, y') = (\partial_{y'} L)y' - L = h$. Use this relation to derive a differential equation for extremal paths of the form

$$y' = \pm\sqrt{r^2/y^2 - 1},$$

and determine the value of the constant r.

(b) Solve this equation by separation of variables. Show that the solutions describe circles.

ₑC8.3.3 Length functional in two dimensions, in polar coordinates (p. 337)

This problem illustrates the use of polar coordinates for the length functional in $\mathbb{R}^2$.

Let curves in $\mathbb{R}^2$ be parametrized as $\mathbf{r} : I \rightarrow \mathbb{R}^2$, $\rho \mapsto \mathbf{r}(\rho) = (\rho \cos \phi(\rho), \rho \sin \phi(\rho))^T$, where ρ serves as curve parameter, while the function $\phi(\rho)$ determines the curve shape.

(a) Show that the length functional, $L[\phi] = \int_I d\rho\, \|d_\rho \mathbf{r}(\rho)\|$, takes the form

$$L[\phi] = \int_I d\rho\, L(\phi(\rho), \phi'(\rho), \rho), \quad \text{with} \quad \phi' = \frac{d\phi}{d\rho}, \qquad L(\phi, \phi', \rho) = \sqrt{1 + \rho^2 \phi'^2}.$$

(b) Set up the Euler–Lagrange equation minimizing this functional. Integrate it, and solve the resulting equation for ϕ'. Show that this leads to

$$\phi'(\rho) = \frac{\rho_0}{\rho\sqrt{\rho^2 - \rho_0^2}}, \tag{1}$$

a differential equation for $\phi(\rho)$, where ρ_0 is an integration constant.

(c) The solutions of this differential equation are straight lines. To show this, note that along a straight line, ϕ and ρ are related by (see figure)

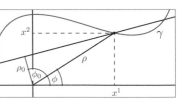

$$\cos(\phi - \phi_0) = \rho_0/\rho. \tag{2}$$

Show that this relation satisfies Eq. (1).

Now use an alternative curve parametrization, where both polar coordinates, $\mathbf{y} = (\rho, \phi)^T$, are functions of time, $\mathbf{r} : I \to \mathbb{R}^2, t \mapsto \mathbf{r}(\mathbf{y}(t)) = (\rho(t) \cos \phi(t), \rho(t) \sin \phi(t))^T$. Then there are two Euler–Lagrange equations, hence the analysis is more involved, but nevertheless instructive.

(d) Show that the length functional, $L[\mathbf{y}] = \int_I dt \| d_t \mathbf{r}(t) \|$, now takes the form

$$L[\mathbf{y}] = \int_I dt L(\mathbf{y}(t), \dot{\mathbf{y}}(t)), \quad \text{with} \quad L(\mathbf{y}, \dot{\mathbf{y}}) = \left(\dot{\rho}^2 + \rho^2 \dot{\phi}^2 \right)^{1/2}.$$

(e) Set up the Euler–Lagrange equations extremizing the length functional, $d_t(\partial_{\dot{y}^i} L) = \partial_{y^i} L$, for the variables $y^1 = \rho$ and $y^2 = \phi$. Show that both these equations are satisfied if the straight-line relation (2) holds. *Hint:* Start by showing that along curves satisfying relation (2), the Lagrangian reduces to $L = \rho^2 \dot{\phi} / \rho_0$.

pC8.3.4 Geodesics on the unit sphere (p. 337)

The purpose of this problem is to find the geodesics, i.e. trajectories of shortest length connecting two points, on the unit sphere, S^2, by extremizing the length functional on S^2.

Let the unit sphere, embedded in $\mathbb{R}^3$, be parametrized by spherical coordinates,

$$\mathbf{r} : (0, \pi) \times (0, 2\pi) \to S^2 \subset \mathbb{R}^3, \quad \mathbf{y} = (\theta, \phi) \mapsto \mathbf{r}(\mathbf{y}) = (\cos \phi \sin \theta, \sin \phi \sin \theta, \cos \theta)^T.$$

Let curves on S^2 be parametrized as $\mathbf{y} : (0, \pi) \to S^2, \; \theta \mapsto \mathbf{y}(\theta) = (\theta, \phi(\theta))^T$, using θ as curve parameter. (This parametrization is not applicable for curves of constant θ; for those, ϕ has to be used as curve parameter.)

(a) Show that the length functional on S^2, $L[\mathbf{r}] = \int d\theta \, \| d_\theta \mathbf{r}(\mathbf{y}(\theta)) \|$, takes the form

$$L[\mathbf{r}] = \int d\theta \, L(\phi(\theta), \phi'(\theta), \theta) \quad \text{with} \quad \phi' = \tfrac{d\phi(\theta)}{d\theta}, \quad L(\phi, \phi', \theta) = \sqrt{1 + \sin^2 \theta \, \phi'^2}.$$

(b) Set up the Euler–Lagrange equation minimizing this functional. Show that any meridian (a great circle connecting the north and south poles) solves this equation.

It is well known that any great circle is a geodesic on S^2. Verify this fact as follows.

(c) Show that the Euler–Lagrange equation leads to the differential equation

$$\phi' = \frac{d}{\sin \theta \sqrt{\sin^2 \theta - d^2}} \tag{1}$$

for $\phi(\theta)$, where d is an integration constant.

The solutions of this differential equation are great circles. To show this, find the θ-parametrization of a great circle, and verify that it solves Eq.(1), proceeding as follows.

(d) A great circle is the intersection of the sphere with a plane through the origin, i.e. the set of all points in $\mathbb{R}^3$ satisfying the relations $ax^1 + bx^2 + cx^3 = 0$ and $(x^1)^2 + (x^2)^2 + (x^3)^2 = 1$. Show that in spherical coordinates, this leads to a relation between ϕ and θ of the form

$$\sin(\phi - \phi_0) = \alpha \cot \theta, \tag{2}$$

and find ϕ_0 and α as functions of the constants a, b, c defining the plane.

(e) Solve Eq. (2) for $\phi(\theta)$ and verify that this function satisfies Eq.(1), thereby establishing that great circles are indeed geodesics of the unit sphere.

P.C9 Calculus of complex functions (p. 338)

P.C9.1 Holomorphic functions (p. 338)

ᴇC9.1.1 Cauchy–Riemann equations (p. 339)

Write the following functions of $z = x + iy$ and $\bar{z} = x - iy$ in the form $f(x, y) = u(x, y) + iv(x, y)$ and explicitly check if the Cauchy–Riemann equations are satisfied. Which of these functions are analytic in z?

(a) $f(z) = e^z$,   (b) $f(z) = \bar{z}^2$.

ᴘC9.1.2 Cauchy–Riemann equations (p. 339)

Investigate, using the Cauchy–Riemann equations, which of the following functions are analytic in $z = x + iy$, and if so, in which domain in $\mathbb{C}$. Check your conclusions by attempting to express each function in terms of z and $\bar{z}$.

(a) $f(x, y) = (x^3 - 3xy^2) + i(3x^2 y - y^3)$.

(b) $f(x, y) = xy + i\frac{1}{2}y^2$.

(c) $f(x, y) = \dfrac{x - iy}{x^2 + y^2}$.

(d) $\left.\begin{array}{l} f_+(x, y) \\ f_-(x, y) \end{array}\right\} = e^x[x\cos y \pm y\sin y] + ie^x[x\sin y \mp y\cos y]$.

P.C9.2 Complex integration (p. 341)

ᴇC9.2.1 Cauchy's theorem (p. 345)

The function $f(z) = e^z$ for $z \in \mathbb{C}$ is analytic. Cauchy's theorem then states that (a) closed contour integrals over simply connected domains are zero, and (b) contour integrals between two points are independent of the chosen contour. Check these claims explicitly by calculating the following contour integrals:

(a) $I_{\gamma_R} = \oint_{\gamma_R} dz\, f(z)$, along the circle γ_R with radius R about the origin $z = 0$;

(b) $I_{\gamma_i} = \int_{\gamma_i} dz\, f(z)$, between the points $z_0 = 0$ and $z_1 = 1 - i$, along (i) the straight line $\gamma_1 : z(t) = (1 - i)t$, and (ii) the curve $\gamma_2 : z(t) = t^3 - it$, with $t \in (0, 1)$. Calculate explicitly the difference $F(z_1) - F(z_0)$, where $F(z)$ is the anti-derivative of $f(z)$.

ᴘC9.2.2 Cauchy's theorem (p. 345)

Compute the contour integral $I_{\gamma_i} = \int_{\gamma_i} dz\, (z - i)^2$ explicitly along the following contours, γ_i, and explain your answers with reference to Cauchy's theorem.

(a) γ_1 is the straight line from $z_0 = 0$ to $z_1 = 1$, γ_2 is the line from $z_1 = 1$ to $z_2 = i$, and γ_3 is the line from $z_2 = i$ to $z_0 = 0$. What is $I_{\gamma_1} + I_{\gamma_2} + I_{\gamma_3}$? Explain your answer.

(b) γ_4 is the quarter-circle with radius 1 from z_1 to z_2. Is there a connection between I_{γ_4} and the integrals from (a)?

P.C9.3 Singularities (p. 345)

ᴇC9.3.1 Laurent series, residues (p. 348)

Let $p(z)$ be a polynomial of order $k \geq 0$ on $\mathbb{C}$ that does not have any zeros at z_0, then $f_m(z) = \frac{p(z)}{(z-z_0)^m}$ (with $m \geq 1$) is an analytic function on $\mathbb{C} \backslash z_0$, with a pole of order m at z_0.

(a) Show, using a Taylor series of $p(z)$ about z_0, that the Laurent series of $f_m(z)$ about z_0 has the following form:

$$f_m(z) = \sum_{n=-m}^{k-m} \frac{p^{(n+m)}(z_0)}{(n+m)!} (z-z_0)^n, \quad \text{with} \quad p^{(n)}(z_0) = \frac{d^n}{dz^n} p(z)\bigg|_{z=z_0}.$$

(b) For $f_m(z) = \frac{z^3}{(z-2)^m}$, find the Laurent series about the pole at $z_0 = 2$.

(c) Find the residues of $f_m(z) = \frac{z^3}{(z-2)^m}$ about the pole $z_0 = 2$ for $m = 1, 2, 3, 4$ and 5, using the formula $\mathrm{Res}(f, z_0) = \lim_{z \to z_0} \frac{1}{(m-1)!} \frac{d^{m-1}}{dz^{m-1}} \big[(z-z_0)^m f(z) \big]$.
[Check your results: Are the residues from (c) consistent with the Laurent series of (b)?]

ᴘC9.3.2 Laurent series, residues (p. 348)

For each of the following functions, determine their poles, as well as the residues using the residue formula. Then find the Laurent series about each pole using an appropriately chosen Taylor series.

(a) $\dfrac{2z^3 - 3z^2}{(z-2)^3}$,   (b) $\dfrac{1}{(z-1)(z-3)}$,   (c) $\dfrac{\ln z}{(z-5)^2}$,   (d) $\dfrac{e^{\pi z}}{(z-i)^m}$ with $m \geq 1$.

Hint: The Laurent series of a function of the form $f(z) = g(z)/(z-z_0)^m$, with $g(z)$ analytic in some neighborhood of z_0, follows from the Taylor series of $g(z)$ about z_0.
[Check your results: The constant terms [coefficient of $(z-z_0)^0$] in the Laurent series are for (a) 2, (b) $-\frac{1}{4}$ for the poles at $z_0 = 1$ and 3, (c) $-\frac{1}{25}$, (d) $-\pi^m/m!$. Further check: do the residues match the coefficients of $(z-z_0)^{-1}$ for each Laurent series?]

P.C9.4 Residue theorem (p. 348)

ᴇC9.4.1 Circular contours, residue theorem (p. 350)

(a) Calculate the integrals $I_+^{(k)} = \oint_{k \text{ times: } |z|=R} \frac{dz}{z}$ and $I_-^{(k)} = \oint_{k \text{ times: } |z|=R} \frac{dz}{z}$,

where $I_+^{(k)}$ (resp. $I_-^{(k)}$) involves a circular contour with radius R, winding around the origin k times in the mathematically positive (negative) direction, i.e. anti-clockwise (clockwise). Do not use the residue theorem; rather calculate the integral directly using the parametrization $z(\phi) = R e^{i\phi}$ and a suitable choice of integration interval for ϕ.

Use the residue theorem to calculate the following closed contour integrals in the complex plane, for $0 < a \in \mathbb{R}$:

(b) $I_1(a) = \oint_{|z|=\frac{1}{2}} dz\, g(z)$, $I_2(a) = \oint_{2\ \text{times:}\ |z|=2} dz\, g(z)$,   with $g(z) = \dfrac{e^{iaz}}{z^2 + 1}$;

(c) $I_3(a) = \oint_{|z|=4} dz f(z)$,   with $f(z) = \dfrac{z}{z^3 + (ai - 6)z^2 + (9 - 6ai)z + 9ai}$.

Hint: One of the poles of $f(z)$ is at $z_1 = -ai$.

$\left[\text{Check your results: (b) } I_2(\ln 2) = 3\pi,\ \text{(c) } I_3(1) = 0,\ I_3(6) = \frac{4\pi}{25}(1 + \frac{4}{3}i).\right]$

P.C9.4.2 Circular contours, residue theorem (p. 350)

Consider the function $f(z) = \dfrac{4z}{(z - a)(z + 1)^2}$, with $1 < a \in \mathbb{R}$.

(a)   Determine the residues of the function f at each of its poles.

Calculate the integral $I_{\gamma_i}(a) = \int_{\gamma_i} dz\, f(z)$ for the following integration contours:

(b) γ_1: a circle with radius $R = 1$ about $z = a$, traversed in the anti-clockwise direction;

(c) γ_2: a circle with radius $R = 1$ about $z = -1$, traversed in the clockwise direction;

(d) γ_3: a circle with radius $R = 2a$ about the origin, traversed in the anti-clockwise direction.

$\left[\text{Check your results: (b) } I_{\gamma_1}(2) = \frac{16}{9}\pi i,\ \text{(c) } I_{\gamma_2}(3) = \frac{3}{2}\pi i.\right]$

P.C9.4.3 Integrating by closing contour and using residue theorem (p. 351)

Calculate the following integral, with $a, b \in \mathbb{R}$, by closing the contour along a suitably chosen semicircle with radius $\to \infty$:

$$I(a, b) = \int_{-\infty}^{\infty} dx\, \frac{1}{x^2 - 2xa + a^2 + b^2}\, .\qquad \left[\text{Check your results: } I(-1, -2) = \frac{\pi}{2}.\right]$$

P.C9.4.4 Integrating by closing contour and using residue theorem (p. 351)

Calculate the following integrals (with $a, b \in \mathbb{R}$ and $a > 0$) by closing the contour with a semicircle of radius $\to \infty$ in the upper or lower complex half-planes (show that both choices give the same result!):

(a) $I(a, b) = \displaystyle\int_{-\infty}^{\infty} dx \frac{x}{(x^2 + b^2)(x - ia)}$,   (b) $I(a, b) = \displaystyle\int_{-\infty}^{\infty} dx \frac{x}{(x + ib)^2(x - ia)}$.

$\left[\text{Check your results: (a) } I(3, -2) = \frac{\pi}{5},\ \text{(b) } I(3, 2) = \frac{6\pi}{25}.\right]$

P.C9.4.5 Various integration contours, residue theorem (p. 351)

Consider the function $f(z) = \dfrac{z^2}{(z^2 + 4)(z^2 + a^2)}$, with $a \in \mathbb{R}$, $3 \le a < 4$.

(a)   Determine the residues of f at each of its poles.

Calculate the integral $I_{\gamma_i}(a) = \int_{\gamma_i} dz\, f(z)$ for the following integration contours:

(b) γ_1: a circle with radius $R = 1$ about the origin, traversed in the anti-clockwise direction;

(c) γ_2: a circle with radius $R = \frac{1}{2}$ about $z = 2i$, traversed in the anti-clockwise direction;

(d) γ_3: a circle with radius $R = 2$ about $z = 2i$, traversed in the clockwise direction;

(e) γ_4: the real axis, traversed in the positive direction.

$\left[\text{Check your results: (c) } I_{\gamma_2}(3) = -\frac{2\pi}{5}, \text{ (d) } I_{\gamma_3}(\frac{10}{3}) = -\frac{3\pi}{16}, \text{ (e) } I_{\gamma_4}(\frac{7}{2}) = \frac{2\pi}{11}.\right]$

P C9.4.6 Various integration contours, residue theorem (p. 351)

Consider the function $f(z) = \dfrac{1}{[z^2 - 2az + a^2 + \frac{1}{4}]^2 (4z^2 + 1)}$, with $1 < a \in \mathbb{R}$.

(a)   Determine the residues of the function f at each of its poles.

Calculate the integrals $I_{\gamma_i}(a) = \int_{\gamma_i} dz\, f(z)$ for the following integration contours:

(b) γ_1: a circle with radius $R = 1$ about $z_1 = 0$, traversed in the anti-clockwise direction;

(c) γ_2: a circle with radius $R = \frac{1}{\sqrt{2}}a$ about $z_2 = \frac{1}{2}a(1 - i)$, traversed in the clockwise direction;

(d) γ_3: a circle with radius $R = a + \frac{1}{2}$ about $z_3 = \frac{1}{2}a$, traversed in the anti-clockwise direction;

(e) γ_4: the line $z = x$, with $x \in (-\infty, \infty)$, traversed in the positive x-direction;

(f) γ_5: the line $z = \frac{1}{3}a + iy$, with $y \in (-\infty, \infty)$, traversed in the positive y-direction.

$\left[\text{Check your results: (b) } I_{\gamma_1}(2) = \frac{\pi i}{25}, \text{ (c) } I_{\gamma_2}(2) = \frac{7\pi}{25}, \text{ (e) } I_{\gamma_4}(3) = \frac{3\pi}{25}, \text{ (f) } I_{\gamma_5}(3) = \frac{\pi i}{150}.\right]$

E C9.4.7 Inverse Fourier transform via contour closure (p. 353)

(a)   The Green function defined by the equation $(d_t + a)G(t) = \delta(t)$ (with $0 < a \in \mathbb{R}$) has a corresponding Fourier transform given by $\tilde{G}(\omega) = (a - i\omega)^{-1}$. Show that the corresponding inverse Fourier transform yields the following result:

$$G(t) = \int_{-\infty}^{\infty} \frac{d\omega}{2\pi} \frac{e^{-i\omega t}}{a - i\omega} = \Theta(t)\, e^{-at}, \quad \text{with} \quad \Theta(t) = \begin{cases} 1 & \text{for } t > 0, \\ 0 & \text{for } t < 0. \end{cases}$$

(b)   The Fourier transform of the exponential function, $\tilde{L}(\omega) = \int_{-\infty}^{\infty} dt\, e^{i\omega t} e^{-a|t|} = \frac{2a}{\omega^2 + a^2}$ (with $0 < a \in \mathbb{R}$), is a Lorentz curve. Find the inverse Fourier transform $L(t) = \int_{-\infty}^{\infty} \frac{d\omega}{2\pi} e^{-i\omega t} \tilde{L}(\omega)$, by explicitly calculating the integral.

Hint: Calculate the integral for $t \neq 0$ as a contour integral, by closing the contour with a suitably chosen semicircle with radius $\to \infty$.

P C9.4.8 Inverse Fourier transform via contour closure: Green function of damped harmonic oscillator (p. 353)

The Green function of the damped harmonic oscillator is defined by the differential equation $(d_t^2 + 2\gamma d_t + \Omega^2)G(t) = \delta(t)$ (with Ω and γ real and positive). Its Fourier transform, defined by $G(t) = \int_{-\infty}^{\infty} \frac{d\omega}{2\pi} e^{-i\omega t} \tilde{G}(\omega)$, is given by $\tilde{G}(\omega) = (\Omega^2 - \omega^2 - 2\gamma i\omega)^{-1}$. Express the Green function in the form $G(t) = \int_{-\infty}^{\infty} dz\, f(z)$, and calculate the integral by closing the contour in the complex plane. You should proceed as follows:

(a) Find the residues of $f(z)$. Distinguish between the following cases:

(i) $\Omega > \gamma$ (underdamped), (ii) $\Omega = \gamma$ (critically damped) and (iii) $\Omega < \gamma$ (overdamped). *Hint:* (i) and (iii) each have two poles of first order; (ii) has only a single pole, but of second order.

(b) Calculate $G(t)$ by closing the contour with an appropriately chosen semicircle with radius $R \to \infty$ (again distinguishing between the different cases!). [Check your results for $G(t)$: (i) for $\Omega = 1$ and $\gamma \to 0$, $G(\pi/2) = 1$; (ii) for $\Omega = \gamma = 1$, $G(1) = e^{-1}$; (iii) for $\Omega = 4$ and $\gamma = 5$, $G(1/3) = \frac{1}{3}e^{-5/3}\sinh(1)$.]

V

VECTOR CALCULUS

The third part of this book introduces the mathematics of smooth structures in higher-dimensional spaces. Methodologically, this requires a synthesis of concepts of linear algebra and calculus introduced in the first two parts of the text. We discuss the mathematical description of curves, surfaces and more general geometric objects, and will learn how to characterize these structures from both a global and a local perspective. The tools to be introduced for this task include generalized coordinates, vector fields and various concepts building on the metric tensor. In the advanced chapters of Part V we will introduce fundamental elements of modern differential geometry. We will learn how to describe nontrivial geometric structures which are not embedded into some larger space – for example, our Universe is believed to be a four-dimensional "curved" structure not embedded in some "bigger" structure. To this end, we will introduce the concepts of differentiable manifolds and differential forms as powerful tools for the description of such structures. The usefulness of these differential forms in physics will be illustrated in the final chapter on the example of a modern formulation of electrodynamics from a geometrical perspective.

Introductory remarks

The objects of interest in physics are mostly defined in higher-dimensional spaces – think, for example, of the three-dimensional trajectories of planets, or the Universe itself. In the introduction to Part C we reasoned that physical laws specify how such objects respond to small variations of relevant system parameters, which in mathematical terms is differentiation. Differentiation requires a sense of "smoothness": from a sufficiently close-up perspective, the objects under consideration must look almost flat, or "linearizable". This very general line of reasoning indicates that much of the mathematics of physics requires the synthesis of the three disciplines geometry (objects in higher-dimensional space), calculus (small variations) and linear algebra (flatness). All subjects discussed in Part V have in common that they rely on this union. We will begin with the discussion of concrete tasks such as the description of smooth curves in space, the use of generalized coordinates, and the characterization of various types of fields. In the later chapters we will then take a more general perspective and introduce concepts of *differential geometry*, which provides the language required for the investigation of many subjects of modern physics.

V1   Curves

REMARK This chapter presumes knowledge of Chapters L2, L3, C1 and C2.

Curves in d-dimensional space play an important role in many areas of physics. This is true in particular in mechanics where they describe the motion of bodies through space. In this chapter we will discuss the mathematical definition of curves and introduce quantities describing them. We will learn how to differentiate and integrate with reference to curves.

V1.1 Definition

vector-valued function

A curve can be imagined as a smooth line in d-dimensional space. To define this in mathematical terms we need the concept of a **vector-valued function**. This is a function

$$\mathbf{r} : I \rightarrow \mathbb{R}^d, \qquad t \mapsto \mathbf{r}(t), \tag{V1}$$

smoothly assigning a vector $\mathbf{r}(t)$ to the parameter variable t. Here, $I \subset \mathbb{R}$ is some interval, which is taken to be $I = (0, 1)$ unless otherwise stated. It is often useful to interpret t as a time-like variable, in which case $\mathbf{r}(t)$ is a time-dependent position vector describing the motion of a point through $\mathbb{R}^d$. One may introduce an orthonormal basis, $\{\mathbf{e}_j\}$, to parametrize $\mathbb{R}^d$ in Cartesian coordinates and represent $\mathbf{r}(t)$ in terms of a coordinate vector $\mathbf{x}(t) = (x^1(t), \ldots, x^d(t))^{\mathrm{T}}$. This representation is described by d real functions, $x^j(t)$, of the parameter t, which means that the mathematical description of vector-valued functions is not more difficult than that of scalar functions.

curve

The **curve**, γ, corresponding to the function $\mathbf{r}$ is defined as the image of $\mathbf{r}$,

$$\gamma \equiv \{\mathbf{r}(t) | t \in I\}, \tag{V2}$$

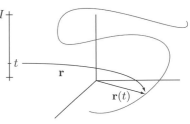

where the term "image" is defined as in Section L1.1. A graphical representation of γ in $\mathbb{R}^d$ is obtained by plotting the terminal points of the vectors $\mathbf{r}(t)$ for all t (see figure).

The function $\mathbf{r}(t)$ defines a *parametrization of the curve γ*. A parametrization contains more information than the curve itself. It describes not only its shape but also *how* the point $\mathbf{r}(t)$ moves (quickly/slowly?) along the curve as a function of t. For example, a

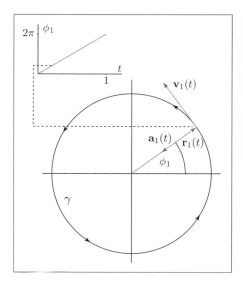 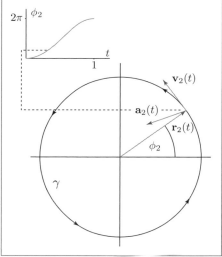

Fig. V1   Schematic depiction of the two parametrizations (V3) of the unit circle. Left: the angle $\phi_1(t)$ increases uniformly with time. Right: the angle $\phi_2(t)$ increases non-uniformly with time. The difference shows in velocity and acceleration vectors, as discussed in the text.

curve might be realized as a penciled route connecting two locations on a road map. A parametrization $\mathbf{r}(t)$ could then be used to describe a traveller's journey between those points.

Any curve has infinitely many distinct parametrizations. For example, the unit circle is a closed curve in $\mathbb{R}^2$ and the two functions $\mathbf{r}_1(t)$ and $\mathbf{r}_2(t)$,

$$\mathbf{r}_i(t) = \begin{pmatrix} \cos(\phi_i(t)) \\ \sin(\phi_i(t)) \end{pmatrix}; \qquad \phi_1(t) = 2\pi t, \qquad \phi_2(t) = \pi[1 - \cos(\pi t)], \qquad (\text{V3})$$

are two different parametrizations of it. To verify this, note that $(r_i^1)^2 + (r_i^2)^2 = 1$ lies on the unit circle. In either parametrization $\phi_i(t)$ increases monotonically from 0 to 2π (see Fig. V1) which means that the unit circle is fully covered.

INFO Sometimes, language not distinguishing between a *curve and its parametrization* is used and a parametrization $\mathbf{r}(t)$ is called a curve. This should not cause confusion as long as the difference between the two objects is kept in mind: a curve is an invariant geometric object in space and its parametrizations are different languages describing it.

V1.2 Curve velocity

velocity   The point $\mathbf{r}(t)$ moves through space at a certain **velocity**. Velocity is a vectorial quantity, $\mathbf{v}$, whose direction is tangent to the curve and whose magnitude measures the speed of the motion (see Fig. V1, where the different lengths of the velocity vector indicate that the two parametrizations traverse the curve at different speeds).

Mathematically, the velocity, $\mathbf{v}(t)$, of the curve at time t is defined as

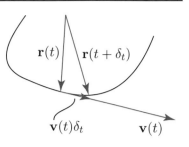

$$\mathbf{v}(t) = \lim_{\delta_t \to 0} \frac{\mathbf{r}(t + \delta_t) - \mathbf{r}(t)}{\delta_t} \equiv \frac{d\mathbf{r}(t)}{dt} \equiv \dot{\mathbf{r}}(t). \quad (V4)$$

The differential quotient expresses the intuitive understanding that the velocity describes the distance between two nearby points, $\mathbf{r}(t + \delta_t) - \mathbf{r}(t)$, in relation to their time difference, δ_t (see figure).[1] Inserting the component representation $\mathbf{r}(t) = \mathbf{e}_j x^j(t)$ into Eq. (V4), we obtain $\mathbf{v} = \dot{\mathbf{r}} = \mathbf{e}_j \dot{x}^j \equiv \mathbf{e}_j v^j$. This means that the derivative acting on the vector $\mathbf{r}$ is computed component-wise:

$$v^j(t) = \frac{dx^j(t)}{dt} \equiv \dot{x}^j(t).$$

To see that $\mathbf{v}$ is tangent to the curve, rewrite (V4) as

$$\mathbf{r}(t + \delta_t) \simeq \mathbf{r}(t) + \delta_t \mathbf{v}(t).$$

Varying δ_t at fixed t we obtain a tangent, i.e. a straight line $\mathbf{r}(t) + \delta_t \mathbf{v}(t)$, which for small δ_t is a good approximation to the curve near $\mathbf{r}(t)$.

INFO In the following we will often need to differentiate vector-valued functions. All these derivatives can be reduced to ordinary derivatives acting on the components of vectors. However, one should aim to keep the notation compact and avoid component representations where possible. Useful **vector differentiation rules** include:

vector differentiation rules

$$d_t(\mathbf{r} + \mathbf{s}) = d_t\mathbf{r} + d_t\mathbf{s},$$
$$d_t(a\mathbf{r}) = (d_t a)\,\mathbf{r} + a(d_t\mathbf{r}),$$
$$d_t(\mathbf{r} \cdot \mathbf{s}) = (d_t\mathbf{r}) \cdot \mathbf{s} + \mathbf{r} \cdot (d_t\mathbf{s}),$$
$$d_t(\mathbf{r} \times \mathbf{s}) = (d_t\mathbf{r}) \times \mathbf{s} + \mathbf{r} \times (d_t\mathbf{s}) \qquad \text{(for } d = 3\text{)}. \qquad (V5)$$

For the sake of notational transparency, we have omitted the time argument here, $\mathbf{r}(t) = \mathbf{r}$, etc. All these relations can be verified by applying standard rules of differentiation to the component representation of vectors, $\mathbf{r}(t) = \mathbf{e}_j r^j(t)$, remembering that the Cartesian basis vectors are time-independent, $\dot{\mathbf{e}}_j = \mathbf{0}$.

As an example, consider a circular curve parametrized by a vector, $\mathbf{r}(t)$, of fixed norm $\mathbf{r} \cdot \mathbf{r} = l^2 =$ const. Differentiation in time yields

$$0 = d_t l^2 = d_t(\mathbf{r} \cdot \mathbf{r}) = (d_t\mathbf{r}) \cdot \mathbf{r} + \mathbf{r} \cdot (d_t\mathbf{r}) = 2\mathbf{r} \cdot (d_t\mathbf{r}) = 2\mathbf{r} \cdot \mathbf{v}, \qquad (V6)$$

where $\mathbf{v} = d_t\mathbf{r}$ is the curve velocity. This shows that for a circular curve, the velocity vector is always perpendicular to the position vector.

curve acceleration

If the velocity along the curve changes in time, the motion is subject to **acceleration**. Acceleration is a vectorial quantity defined as the rate of change of velocity,

$$\mathbf{a}(t) \equiv \lim_{\delta_t \to 0} \frac{\mathbf{v}(t + \delta_t) - \mathbf{v}(t)}{\delta_t} = \dot{\mathbf{v}}(t) \equiv \ddot{\mathbf{r}}(t), \qquad (V7)$$

or in component notation, $a^j(t) = \dot{v}^j(t) = \ddot{r}^j(t)$.

---

[1] The dot notation, as in $\dot{\mathbf{r}}$ or $\dot{r}^j$, is customary for derivatives in a time-like parameter. However, one does not write $\dot{f}(x) = \frac{df}{dx}$ if x parametrizes length.

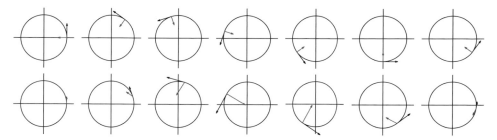

The velocity vectors (tangential arrows) and acceleration vectors (inward-pointing arrows) for the two parametrizations
$\mathbf{r}_1(t)$ (upper row) and $\mathbf{r}_2(t)$ (lower row) of a circle defined in Eq. (V3), sketched at several different times.

EXAMPLE Consider the two curves $\mathbf{r}_1(t)$ and $\mathbf{r}_2(t)$ defined in Eq. (V3). Their velocity and acceleration
vectors read

$$\mathbf{v}_i(t) = \begin{pmatrix} -\sin(\phi_i(t)) \\ \cos(\phi_i(t)) \end{pmatrix} \dot{\phi}_i(t), \qquad \dot{\phi}_1(t) = 2\pi, \quad \dot{\phi}_2(t) = \pi^2 \sin(\pi t),$$

$$\mathbf{a}_i(t) = -\begin{pmatrix} \cos(\phi_i(t)) \\ \sin(\phi_i(t)) \end{pmatrix} (\dot{\phi}_i(t))^2 + \begin{pmatrix} -\sin(\phi_i(t)) \\ \cos(\phi_i(t)) \end{pmatrix} \ddot{\phi}_i(t), \quad \ddot{\phi}_1(t) = 0, \quad \ddot{\phi}_2(t) = \pi^3 \cos(\pi t).$$

A series of "snapshots" of these vectors at different times is shown in Fig. V2. Notice that
$\mathbf{v}_i \cdot \mathbf{r}_i = 0$, which was proved generally in Eq. (V6). Also notice that the speed is constant
along the first curve, $\|\mathbf{v}_1(t)\| = 2\pi$, but changes along the second, $\|\mathbf{v}_2(t)\| = \pi^2 |\sin(\pi t)|$. Cor-
respondingly, the acceleration vector for the first curve is directed towards the center ($\ddot{\phi}_1 = 0$
implies $\mathbf{a}_1 \parallel -\mathbf{r}_1$), whereas the acceleration vector for the second curve has a tangential component ($\ddot{\phi}_2 \neq 0$ implies $\mathbf{a}_2 \cdot \mathbf{v}_2 \neq 0$), which acts to increase or decrease the speed along the curve
($\rightarrow$ V1.2.1-2).

INFO We conclude this section on curve velocity with a mathematical subtlety: in this text, we will
always parametrize curves by *open*[2] parameter intervals, such as (0, 1). This is done to guarantee the
global differentiability of the function $\mathbf{r}(t)$. For a closed interval, [0, 1], the parametrization would not
be differentiable[3] at $t = 0$ or $t = 1$, and this would lead to unwanted side effects, both mathematical
and physical. (For example, the derivative needed for the velocity, $\dot{\mathbf{r}}$, is not defined *at* the end points.)
For a curve defined on an open interval, (0, 1), the end points $\mathbf{r}(0)$ and $\mathbf{r}(1)$ are formally excluded
from the curve. In practice, this omission is not of relevance, because for any continuous curve the
endpoints can always be *defined* as limits, e.g. $\mathbf{r}(1) = \lim_{t \to 1} \mathbf{r}(t)$ and $\dot{\mathbf{r}}(1) = \lim_{t \to 1} \dot{\mathbf{r}}(t)$.

V1.3 Curve length

An important characteristic of a curve, γ, is its length. An estimate for the curve length
may be obtained by approximating the curve as a concatenation of many short, straight-line
segments (see figure). The length of each segment can be computed using the Euclidean

---

[2] The concept of an "open interval" is explained in Section L1.3, p. 17.
[3] The definition of differentiability requires the existence of the differential quotient (C1), $[f(t + \delta_t) - f(t)]/\delta_t$,
     irrespective of the sign of the incremental parameter, δ_t. This condition is violated at the boundary of a closed
     interval. For example, if $f(t)$ is defined on [0, 1], then $f(1 + \delta_t)$ is not defined for positive δ_t.

scalar product on $\mathbb{R}^d$. In the limit of an infinitely fine discretization, their sum converges to the length of the curve.

To make this prescription quantitative, we need a parametrization of the curve, $\mathbf{r} : (0, 1) \to \mathbb{R}^d$, $t \mapsto \mathbf{r}(t)$. We divide the parameter interval $(0, 1)$ into $N \gg 1$ subintervals of width $\delta_t = 1/N$, bounded by the parameter values $t_\ell = \ell\,\delta_t$, $\ell = 0, \ldots, N$. Each difference vector $\mathbf{r}(t_\ell + \delta_t) - \mathbf{r}(t_\ell)$ then defines a line segment approximately tangent to the curve. Adding the lengths of these segments,

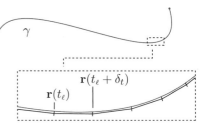

$$L_{\delta_t} \equiv \sum_{\ell=0}^{N-1} \|\mathbf{r}(t_\ell + \delta_t) - \mathbf{r}(t_\ell)\|,$$

an estimate of the length of the curve is obtained. For small δ_t, one may approximate $\|\mathbf{r}(t_\ell + \delta_t) - \mathbf{r}(t_\ell)\| \simeq \|\delta_t\,\dot{\mathbf{r}}(t_\ell)\| = \delta_t\|\dot{\mathbf{r}}(t_\ell)\|$ and this yields

$$L \equiv \lim_{\delta_t \to 0} L_{\delta_t} = \lim_{\delta_t \to 0} \delta_t \sum_{\ell} \|\dot{\mathbf{r}}((t_\ell)\| = \int_0^1 dt\,\|\dot{\mathbf{r}}(t)\|,$$

where in the last step we recognized the appearance of a Riemann sum (see Eq. (C19)).

curve length Denoting the **length of a curve** γ by $L[\gamma]$, we thus have

$$\boxed{L[\gamma] = \int_0^1 dt\,\|\dot{\mathbf{r}}(t)\|.} \qquad (V8)$$

This definition has been obtained by a geometric construction and should therefore not depend on the choice of parametrization. To verify its parametrization invariance, consider an arbitrary smooth and monotonically increasing function, $t : (a, b) \to (0, 1)$, $s \mapsto t(s)$, such that the function $\mathbf{r}' : (a, b) \to \mathbb{R}^d$, $s \mapsto \mathbf{r}'(s) \equiv \mathbf{r}(t(s))$ defines a different parametrization of the same curve. Applying the length formula in the new parametrization yields

$$L[\gamma] = \int_a^b ds\,\|d_s\mathbf{r}'(s)\| = \int_a^b ds\,\|d_s\mathbf{r}(t(s))\| \overset{(C8)}{=} \int_a^b ds\,\|d_t\mathbf{r}(t)\big|_{t=t(s)}d_st(s)\|$$

$$= \int_a^b ds\,\frac{dt(s)}{ds}\,\|d_t\mathbf{r}(t)\big|_{t=t(s)}\| \overset{(C28)}{=} \int_0^1 dt\,\|d_t\mathbf{r}(t)\|, \qquad (V9)$$

where the chain rule was used at the end of the first line, and a variable substitution at the end of the second. This confirms that the formula (V8) for the curve length is indeed parametrization invariant.

EXERCISE Compute the length of the curve of each Eq. (V3) in the two parametrizations given there, and show that the circumference of the unit circle is obtained in each case.

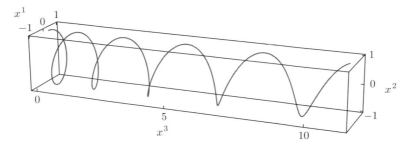

Fig. V3   The curve parametrized by Eq. (V10).

EXAMPLE Consider the curve γ shown in Fig. V3, with parametrization

$$\mathbf{r} : (0, 5) \to \mathbb{R}^3, \qquad t \mapsto \mathbf{r}(t) \equiv \left(\sin(2\pi t), \cos(2\pi t), \tfrac{2}{3} t^{3/2} \right)^{\mathsf{T}}. \tag{V10}$$

The velocity vector, $\frac{\mathrm{d}}{\mathrm{d}t} \mathbf{r}(t) = \left(2\pi \cos(2\pi t), -2\pi \sin(2\pi t), t^{1/2} \right)^{\mathsf{T}}$, has norm $\|\mathrm{d}_t \mathbf{r}(t)\| = \sqrt{4\pi^2 + t}$, and this integrates to

$$L[\gamma] = \int_0^5 \mathrm{d}t \, (4\pi^2 + t)^{1/2} = \tfrac{2}{3} (4\pi^2 + t)^{3/2} \Big|_0^5 = \tfrac{2}{3} \left[(4\pi^2 + 5)^{3/2} - (2\pi)^3 \right].$$

($\to$ V1.3.1-2)

INFO For a curve with parametrization $\mathbf{r} : (0, 1) \to \mathbb{R}$, $t \mapsto \mathbf{r}(t)$, consider the length, $s(t)$, of the curve segment corresponding to the partial interval $(0, t)$, $t \le 1$,

$$s(t) \equiv \int_0^t \mathrm{d}u \, \|\mathrm{d}_u \mathbf{r}(u)\|. \tag{V11}$$

The function $s(t)$ grows monotonically with t from 0 at $t = 0$ to $L[\gamma]$ as $t \to 1$. It defines a bijection, $s : (0, 1) \to (0, L[\gamma])$, and may therefore be inverted to yield a function, $t(s)$, assigning to each length $s \in (0, L[\gamma])$ the parameter $t(s)$ at which that length is reached. This observation suggests a parametrization of the curve by its own length function,

$$\mathbf{r}_L : (0, L[\gamma]) \to \mathbb{R}^d, \qquad s \mapsto \mathbf{r}_L(s) \equiv \mathbf{r}(t(s)).$$

This function is called the *natural parametrization* of the curve. In cases where it is evident that s refers to length, the subscript "L" is usually omitted, $\mathbf{r}_L(s) \equiv \mathbf{r}(s) \equiv \mathbf{r}(t(s))$.

The distinguishing feature of the natural parametrization is that its curve velocity, $\mathbf{v}(s) = \mathrm{d}_s \mathbf{r}(s)$, has unit magnitude:

$$\|\mathrm{d}_s \mathbf{r}(s)\| = \|\mathrm{d}_s \mathbf{r}(t(s))\| = \|\mathrm{d}_t \mathbf{r}(t)|_{t(s)}\| \frac{\mathrm{d}t(s)}{\mathrm{d}s} \overset{\text{(C9)}}{=} \frac{\|\mathrm{d}_t \mathbf{r}(t)|_{t(s)}\|}{\frac{\mathrm{d}s(t)}{\mathrm{d}t}\big|_{t=s(t)}} \overset{\text{(V11)}}{=} \frac{\|\mathrm{d}_t \mathbf{r}(t)|_{t(s)}\|}{\|\mathrm{d}_t \mathbf{r}(t)|_{t(s)}\|} = 1.$$

This in turn implies the orthogonality, $\mathbf{a}(s) \perp \mathbf{v}(s)$, between the curve velocity and the acceleration vector $\mathbf{a}(s) = \mathrm{d}_s \mathbf{v}(s)$, see Eq. (V6).

Try to develop some intuition for why in the natural parametrization, $\mathbf{r}(s)$, the curve is traversed at a uniform velocity. If in some other parametrization, $\mathbf{r}(t)$, a curve segment is traversed very quickly (or slowly), what does this mean for the rate of change of the length function $s(t)$, and what is the consequence for $\mathbf{v}(s)$? Compute the natural parametrization of the curve (V10). ($\to$ V1.3.3-4)

V1.4 Line integral

In physics, curves often appear as integration domains for a class of integrals known as *line integrals*. The idea of the line integral is best motivated via an application. In Eq. (L39) we defined the work done when a body subject to a constant force, $\mathbf{F}$, is moved along a straight path, $\mathbf{s}$, as $W = \mathbf{F} \cdot \mathbf{s}$. More generally, however, the force may *vary* along the path, and the path itself need not be straight (see the figure, where the path follows the curve γ).

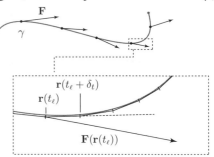

The work done under these generalized conditions is computed by straightforward adaption of the previous construction for the curve length: let $\mathbf{r}(t)$, $t \in (0,1)$, be a parametrization of the path, γ, and $\mathbf{F}(\mathbf{r}(t))$ be the force acting at the point $\mathbf{r}(t)$. To determine the work done, we divide the curve into $N = \delta_t^{-1} \gg 1$ segments, $\mathbf{s}_\ell = \mathbf{r}(t_\ell + \delta_t) - \mathbf{r}(t_\ell)$, where the discretization is defined as in Section V1.3. The work along each of these (straight) segments is given by $W_\ell = \mathbf{s}_\ell \cdot \mathbf{F}_\ell$ with $\mathbf{F}_\ell = \mathbf{F}(\mathbf{r}(t_\ell))$,[4] i.e.

$$W_\ell = [\mathbf{r}(t_\ell + \delta_t) - \mathbf{r}(t_\ell)] \cdot \mathbf{F}(\mathbf{r}(t_\ell)) \simeq \delta_t \, \dot{\mathbf{r}}(t_\ell) \cdot \mathbf{F}(\mathbf{r}(t_\ell)).$$

The total work along the path is obtained by summation over all segments, and in the limit $\delta_t \to 0$ of an infinitely fine segmentation one obtains

$$W \equiv \lim_{\delta_t \to 0} \sum_\ell W_\ell = \lim_{\delta_t \to 0} \delta_t \sum_\ell \dot{\mathbf{r}}(t_\ell) \cdot \mathbf{F}(\mathbf{r}(t_\ell)) = \int_0^1 dt \, \dot{\mathbf{r}}(t) \cdot \mathbf{F}(\mathbf{r}(t)) \equiv \int_\gamma d\mathbf{r} \cdot \mathbf{F}.$$

The last expression is symbolic notation for the line integral of the force along the curve; it is defined by the integral over time given in the second-last expression.

line integral   The construction above is an example of the **line integral** of a general vector-valued function, $\mathbf{f} : \gamma \to \mathbb{R}^d$, $\mathbf{r} \mapsto \mathbf{f}(\mathbf{r})$, defined on a curve γ in $\mathbb{R}^d$. The line integral is built according to the following procedure:

1. Parametrize the curve by a vector-valued function $\mathbf{r} : (0,1) \to \mathbb{R}^d$, $t \mapsto \mathbf{r}(t)$.

2. Construct the real-valued function $\dot{\mathbf{r}}(t) \cdot \mathbf{f}(\mathbf{r}(t))$.

3. Integrate that function over the domain of the curve parameter,

$$\boxed{\int_\gamma d\mathbf{r} \cdot \mathbf{f} \equiv \int_0^1 dt \, \dot{\mathbf{r}}(t) \cdot \mathbf{f}(\mathbf{r}(t)).} \tag{V12}$$

Although the construction makes reference to a particular parametrization, the result is *independent of the choice of parametrization*, and of the choice of the parametrization interval. (Proceed as in Section V1.3 to convince yourself that this is so.) This implies, for

---

[4] For sufficiently small δ_t and a smooth force function, $\mathbf{F}(\mathbf{r}(t)) \simeq \mathbf{F}(\mathbf{r}(t + \delta_t))$, the specific choice of the "readout point" at which the force is evaluated within each discretization segment is not of importance.

example, that the work invested to pull a body along a curve does not depend on the speed at which the curve is traversed. ($\rightarrow$ V1.4.1-2)

EXAMPLE Consider a vector-valued function,

$$\mathbf{f}\colon \mathbb{R}^2 \to \mathbb{R}^2, \qquad \mathbf{r} = \begin{pmatrix} x^1 \\ x^2 \end{pmatrix} \mapsto \mathbf{f}(\mathbf{r}) = \begin{pmatrix} x^2 \\ -x^1 \end{pmatrix},$$

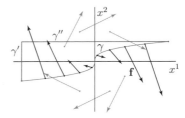

and a two-dimensional curve, γ, parametrized as

$$\mathbf{r}\colon (-1,1) \to \mathbb{R}^2, \qquad t \mapsto \mathbf{r}(t) = \begin{pmatrix} x^1(t) \\ x^2(t) \end{pmatrix} = \begin{pmatrix} t^3 \\ 2t \end{pmatrix}.$$

To compute the line integral, $\int_\gamma d\mathbf{r} \cdot \mathbf{f}$, we determine the curve velocity, $\dot{\mathbf{r}}(t)$, and evaluate the vector function along the curve, $\mathbf{f}(\mathbf{r}(t))$,

$$\dot{\mathbf{r}}(t) = \begin{pmatrix} 3t^2 \\ 2 \end{pmatrix}, \qquad \mathbf{f}(\mathbf{r}(t)) = \begin{pmatrix} x^2(t) \\ -x^1(t) \end{pmatrix} = \begin{pmatrix} 2t \\ -t^3 \end{pmatrix}.$$

Their scalar product, $\dot{\mathbf{r}}(t) \cdot \mathbf{f}(\mathbf{r}(t)) = 6t^3 - 2t^3 = 4t^3$, integrated over the curve parameter, t, yields

$$\int_\gamma d\mathbf{r} \cdot \mathbf{f} = \int_{-1}^{1} dt\, \dot{\mathbf{r}}(t) \cdot \mathbf{f}(\mathbf{r}(t)) = \int_{-1}^{1} dt\, 4t^3 = 0.$$

For the present choice of $\mathbf{f}$ and γ, the integrand is an antisymmetric function of t, hence the integral vanishes.[5] This antisymmetry reflects the "vortex-like" winding of the vectors $\mathbf{f}$ around the origin, as indicated in the figure. If $\mathbf{f}$ is interpreted as a force and the line integral as the work performed by it, contributions where $\mathbf{f}$ acts "along" the integration path cancel with those where it acts "opposite" to it. (For every point on the path for which the projection of $\mathbf{f}$ onto $\dot{\mathbf{r}}$ is positive, there is another point where it is negative, but with the same magnitude.)

Now consider another path connecting the initial and final points of γ, say $\gamma' \cup \gamma''$, as shown in the figure. Compute the line integral over this path as the sum of the line integrals over γ' and γ'', respectively. Show that the integral does not vanish. This result demonstrates that the work done against a force along a path between two points does, in general, depend on the shape of that path.

V1.5 Summary and outlook

In this chapter we introduced the most elementary class of smooth structures to be used in this part of the text, smooth curves. We learned how to parametrize curves through a time-like "coordinate", to perform derivatives of quantities defined for curves, and to integrate along them. All these concepts are very important in applications – smooth curves are pervasive in the description of all kinds of motion, not just in classical mechanics but also in other fields of physics. At the same time, the concepts of coordinate descriptions, differentiation and integration introduced here will all recur in the discussion of higher-dimensional geometric structures. We start this generalization in the next chapter and will eventually see (Chapter V4) that curves are one-dimensional members of a larger class of geometric structures which can all be described within one coherent framework.

---

[5] The integral $\int_{-a}^{a} dt\, g(t)$ vanishes if $g(t) = -g(-t)$ (why?).

Curvilinear coordinates

REMARK This chapter presumes knowledge of Chapters L3 and C3.

In Part L we argued that the Euclidean space $\mathbb{R}^d$ is best spanned by a fixed orthonormal basis. However, there are circumstances where it is preferable to abandon this principle and represent the vicinity of each point $\mathbf{r} \in \mathbb{R}^d$ by its own individual basis. To understand why this may be a better choice, consider the curve

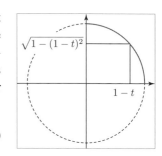

$$\mathbf{r}(t) = \left(1 - t, \sqrt{1 - (1 - t)^2}\right)^{\mathrm{T}}, \qquad t \in (0, 1). \qquad (V13)$$

At first sight, it may not be obvious that $\mathbf{r}(t)$ describes a quadrant of the unit circle, $S^1 = \{\mathbf{r} \in \mathbb{R}^2 \,|\, \|\mathbf{r}\| = 1\}$, i.e. a set of points with unit distance to the origin. The problem is that due to the square root in the second component, this representation "breaks the symmetry" between the 1- and 2-coordinates. However, the circle itself is symmetric under an exchange of these two coordinates. The representation (V13) thus has less symmetry than the object it describes, and consequently is inconvenient to work with.

It is always advisable to describe physical systems in a language reflecting the full symmetry of the problem. The coordinates used in the example above do not appropriately describe the rotational symmetry and we should seek alternative descriptions.

V2.1 Polar coordinates

Let us begin our discussion with a simple case study defined in two-dimensional space. The coordinates, $\mathbf{x} \equiv (x, y)^{\mathrm{T}}$, describing a point $\mathbf{r} = \mathbf{e}_x x + \mathbf{e}_y y$ in a fixed orthonormal basis, $\{\mathbf{e}_x, \mathbf{e}_y\}$, are called **Cartesian coordinates**. Above, we saw that Cartesian coordinates are not well suited to the description of problems with rotational symmetry. In such cases it is more natural to describe $\mathbf{r}$ by a pair of **polar coordinates**, $\mathbf{y} \equiv (\rho, \phi)^{\mathrm{T}}$, where ρ is the distance of $\mathbf{r}$ from the origin and ϕ the angle enclosed with a fixed reference direction, say $\mathbf{e}_x$ (see Fig. V4). The point $\mathbf{r}$ can now be represented in two different ways as

$$\mathbf{r} = \mathbf{e}_x x + \mathbf{e}_y y = \mathbf{e}_x \rho \cos\phi + \mathbf{e}_y \rho \sin\phi, \qquad (V14)$$

Cartesian coordinates

polar coordinates

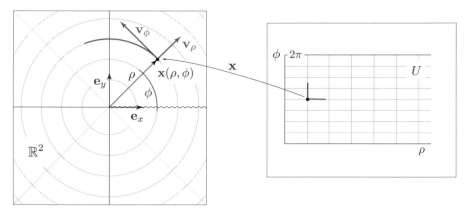

Fig. V4   The construction of polar coordinates. Each coordinate pair, $(\rho, \phi)^{\mathrm{T}} \in U$, describes a point, $\mathbf{x}(\rho, \phi) \in \mathbb{R}^2$, in the Cartesian coordinate plane. The concentric and radial lines are lines of constant radius ρ or angle ϕ, respectively. Tangent to these lines are the vectors of the coordinate basis discussed in Section V2.2 (see Eq. (V20)). The wriggly line denotes the non-negative real axis, $\mathbb{R}_0^+ \times \{0\} = \{(x, y)^{\mathrm{T}} | x \geq 0, y = 0\}$, which is not covered by the coordinate map.

where the second representation follows from the geometric definition of the polar coordinates. The correspondence between the two "languages", Cartesian and polar, is **transformation between Cartesian and polar coordinates** established by the *coordinate transformation*

$$\mathbf{x} : U \to U_{\mathrm{C}}, \qquad \mathbf{y} = \begin{pmatrix} \rho \\ \phi \end{pmatrix} \mapsto \mathbf{x}(\mathbf{y}) = \begin{pmatrix} x(\rho, \phi) \\ y(\rho, \phi) \end{pmatrix} = \begin{pmatrix} \rho \cos \phi \\ \rho \sin \phi \end{pmatrix}, \tag{V15a}$$

where $U = \mathbb{R}^+ \times (0, 2\pi) \ni (\rho, \phi)$ is the domain of definition of the polar coordinates, and $U_{\mathrm{C}} = \mathbb{R}^2 \backslash (\mathbb{R}_0^+ \times \{0\}) \ni (x, y)$ that of the Cartesian coordinates parametrized by them. The map $\mathbf{y} \mapsto \mathbf{x}(\mathbf{y})$ expresses the Cartesian coordinates through polar coordinates and in this sense defines a transformation from the Cartesian to the polar language (although the map arrow seems to indicate the opposite). The inverse map,

$$\mathbf{y} : U_{\mathrm{C}} \to U, \qquad \mathbf{x} = \begin{pmatrix} x \\ y \end{pmatrix} \mapsto \mathbf{y}(\mathbf{x}) = \begin{pmatrix} \rho(x, y) \\ \phi(x, y) \end{pmatrix} = \begin{pmatrix} \sqrt{x^2 + y^2} \\ \arctan(y/x) \end{pmatrix}, \tag{V15b}$$

describes the reverse transformation from polar to Cartesian coordinates.

A good way to visualize the polar coordinates is to plot the vectors $\mathbf{r}(\rho, \phi)$ as a function of ρ at fixed values of ϕ, and of ϕ at fixed values of ρ, respectively (see Fig. V4). This generates a system of *coordinate lines* in the form of a spider web. This visualization reveals the principal *advantage of polar coordinates*: they optimally describe situations with rotational symmetry. In polar coordinates, points at a fixed distance, R, from the origin have a simple representation, $(\rho, \phi)^{\mathrm{T}} = (R, \phi)^{\mathrm{T}}$. This is more natural than the Cartesian description as pairs, $(x, y)^{\mathrm{T}}$, subject to the constraint $\sqrt{(x)^2 + (y)^2} = R$. As we saw above, calculations in Cartesian coordinates involve cumbersome square roots and are more complicated and less intuitive.

integration
over open
coordinate
domains

INFO Notice that in Eq. (V15) both variables $\rho \in \mathbb{R}^+ \equiv (0, \infty)$ and $\phi \in (0, 2\pi)$ are defined on *open* intervals. The motivation for categorically choosing open coordinate domains is that in this way the transformation maps become globally differentiable, including at the boundaries of the domains.[3] Global differentiability will be required by the definition of various coordinate-dependent quantities and is considered a condition of highest priority. At the same time, the openness of the domains comes at a price: for example, the image of the map $\mathbf{x}(\mathbf{y})$ does not cover the entire plane, $\mathbb{R}^2$, but excludes the origin and points of the form $(\mathbf{e}_x \rho \cos(\phi) + \mathbf{e}_y \rho \sin(\phi))_{\phi=0,2\pi} = \mathbf{e}_x \rho + \mathbf{e}_y 0$. The union of all excluded points thus equals the non-negative real axis, $\mathbb{R}_0^+ \times \{0\} = \{(x, y)^{\mathrm{T}} | x \geq 0, y = 0\}$ (see the wriggly line in Fig. V4), and so the image of the coordinate transformation covers only a subset, $U_C \equiv \mathbb{R}^2 \setminus (\mathbb{R}_0^+ \times \{0\})$, of the Cartesian plane.

In this text we will exclusively work with open coordinate intervals. However, as we will see, the resulting exclusion of sets of lowered dimensionality – a half-line excluded from a plane, etc. – does not really limit the practical utility of the formalism.[6] For completeness we mention that parts of the physics literature sacrifice the condition of differentiability in exchange for a more extended (but never complete) coverage. For example, if the ϕ-coordinate domain is extended to the semi-open interval $[0, 2\pi)$, the positive real axis is included. However, the origin of the plane remains excluded because there the ϕ coordinate is not defined. For a discussion of coordinate representations providing full coverage, we refer to Section V4.1.

V2.2 Coordinate basis and local basis

REMARK In this section, various general structures pertaining to the concept of coordinates are introduced. We consider a coordinate map, $\mathbf{r}(\mathbf{y})$, involving a set of d generalized coordinates, y^j, defined in an open subset $U \subset \mathbb{R}^d$. The image, $\mathbf{r}(U) \equiv M$, of the coordinate map contains the points $\mathbf{r}(\mathbf{y})$. In this section, we assume M to be contained in a d-dimensional vector space equipped with an orthonormal basis. In this case, $\mathbf{r}(\mathbf{y})$ can be represented by a Cartesian vector, $\mathbf{x}(\mathbf{y}) = (x^1(\mathbf{y}), \ldots, x^d(\mathbf{y}))^{\mathrm{T}}$, whose components are smooth functions of the coordinates. (The assumption that M is contained in a vector space will be relaxed in Chapter V4, when coordinate representations of more general geometric structures are considered.)

Coordinates are defined to simplify the solution of problems possessing certain types of symmetries. This includes problems involving vectorial objects whose description requires vector space bases. Each coordinate system indeed defines a "coordinate basis" optimally suited to computations in this system. A key feature of this basis is that it varies from point to point. In this way, each point, $\mathbf{r} = \mathbf{r}(\mathbf{y})$, is equipped with d basis vectors tailored to the representation of vectors describing points in its neighborhood. The discussion of coordinate bases is the subject of the present section.

Coordinate lines and coordinate basis

curvilinear
coordinates

The polar coordinates defined in Section V2.1 are a system of **curvilinear coordinates**. Formally, coordinates are smooth invertible maps,

$$\mathbf{r} : U \to M, \qquad \mathbf{y} \mapsto \mathbf{r}(\mathbf{y}), \qquad (V16)$$

---

[6] For example, if $f(\rho, \phi)$ is a smooth function, its values at an excluded point are defined as limits, $f(\rho, 0) \equiv \lim_{\phi \to 0} f(\rho, \phi)$. In this sense, the coordinate representation defines a full coverage of the plane and for practical purposes the exclusion of a lower-dimensional set is not a major issue.

from an open subset $U \subset \mathbb{R}^d$ onto a set $M \equiv \mathbf{r}(U)$. In this chapter, we assume M to be contained in a d-dimensional vector space. Thanks to this "embedding", image points $\mathbf{r}(\mathbf{y})$ can be expanded as $\mathbf{r}(\mathbf{y}) = \mathbf{e}_a x^a(\mathbf{y})$ in an orthonormal basis $\{\mathbf{e}_a\}$ of the ambient space.[7] The component vector $\mathbf{x}(\mathbf{y}) = (x^1(\mathbf{y}), \ldots, x^d(\mathbf{y}))^{\mathrm{T}} \in \mathbb{R}^d$ then serves as Cartesian representation of $\mathbf{r}(\mathbf{y})$. (In the parlance of Part L of this text, the relation between $\mathbf{r}$ and $\mathbf{x}$ is like that between an "abstract vector", $\hat{\mathbf{x}} \in V$, and its component representation, $\mathbf{x} \in \mathbb{R}^d$. However, such notation is uncommon in the present context and will not be used here.) The set $U_C \equiv \mathbf{x}(U)$ containing all points $\mathbf{x}(\mathbf{y})$ lies open in $\mathbb{R}^d$ and defines the Cartesian image of the coordinate map. For example, in the discussion above, U was the domain of polar coordinates, and U_C the two-dimensional plane excluding the non-negative x-axis.

coordinate lines   Coordinates may be visualized through **coordinate lines**. A coordinate line representing the jth coordinate is generated by varying y^j while keeping all remaining coordinates fixed. This defines a *curve* in M,

$$\mathbf{r}_j : I_j \to M, \qquad y \mapsto \mathbf{r}_j(y) \equiv \mathbf{r}(y^1, \ldots, y, \ldots, y^d), \qquad \text{(V17)}$$

where the argument y takes the place of y^j and I_j is an interval of y-values such that $(y^1, \ldots, y, \ldots, y^d)^{\mathrm{T}}$ lies in U. In this way, each fixed choice for the remaining coordinates, $y^{i \neq j}$, specifies its own j-coordinate line in M. For example, Fig. V4 shows lines $\mathbf{r}_\rho$ and $\mathbf{r}_\phi$ in the Cartesian coordinate plane traced out by varying the polar coordinates ρ and ϕ, respectively.

Every point $\mathbf{r}(\mathbf{y}) \in M$ lies at the intersection of d coordinate lines, $\mathbf{r}_j(y)$, with $j = 1, \ldots, d$. For each coordinate line $\mathbf{r}_j(y)$ we can obtain a tangent vector at this point by computing the corresponding curve velocity, defined as[8]

$$(\mathbf{v}_j)_{\mathbf{r}} \equiv \mathbf{v}_{j,\mathbf{r}} = \frac{\mathrm{d}}{\mathrm{d}y} \mathbf{r}_j(y) \overset{\text{(V17)}}{=} \frac{\partial}{\partial y^j} \mathbf{r}(\mathbf{y}). \qquad \text{(V18)}$$

Since the directions of the coordinate lines generally vary with $\mathbf{r}$, the vectors $\mathbf{v}_{j,\mathbf{r}}$, too, depend on the base point $\mathbf{r}$ and a subscript will occasionally be used to emphasize this dependence. However, in both the physics and mathematics literature this subscript is not indicated unless necessary and we will follow this convention.

Since the map $\mathbf{y} \mapsto \mathbf{r}(\mathbf{y})$ is defined to be invertible, the d vectors $\mathbf{v}_{j,\mathbf{r}}$ are all linearly inde-
coordinate basis   pendent and form a basis of $\mathbb{R}^d$.[9] They define the **coordinate basis** of the $\mathbf{y}$-coordinates at $\mathbf{r}$. Their Cartesian representation has the form

$$\mathbf{v}_j(\mathbf{y}) = \mathbf{e}_a \frac{\partial x^a(\mathbf{y})}{\partial y^j}. \qquad \text{(V19)}$$

---

[7] The index $a = 1, \ldots, d$ on x^a is used for better distinguishability from the index j labeling the coordinates y^j.

[8] Here, the assumption that M is embedded in a *vector space* is of essential importance. The derivative $\mathrm{d}_y \mathbf{r}_j = \lim_{\delta \to 0} \delta^{-1}(\mathbf{r}_j(y + \delta) - \mathbf{r}_j(y))$ is computed from vector differences of nearby points $\mathbf{r}$ and such differences would not be defined without an underlying vector space structure.

[9] If the tangent vectors were linearly dependent, a nontrivial linear combination, $\mathbf{v}_j b^j = \partial_{y^j} \mathbf{r} \, b^j = \mathbf{0}$, would exist. In this case, a variation of the coordinates from $\mathbf{y}$ to a close-by configuration $\mathbf{y} + \delta \mathbf{b}$ would leave the image points unchanged: $\mathbf{r}(\mathbf{y} + \delta \mathbf{b}) - \mathbf{r}(\mathbf{y}) \simeq \delta(\partial_{y^j} \mathbf{r} \, b^j) = \mathbf{0}$. This, however, would be in conflict with the assumed injectivity of the coordinate description.

For example, in *polar coordinates*, the differentiation of $\mathbf{r}(\rho, \phi)$ in Eq. (V14) in the coordinates ρ and ϕ yields the polar coordinate basis vectors (illustrated in Fig. V4):

$$\mathbf{v}_\rho = \mathbf{e}_x \cos\phi + \mathbf{e}_y \sin\phi,$$
$$\mathbf{v}_\phi = \rho(-\mathbf{e}_x \sin\phi + \mathbf{e}_y \cos\phi). \tag{V20}$$

Note how these vectors depend on the coordinates $\mathbf{y} = (\rho, \phi)^{\mathrm{T}}$, and therefore vary from point to point, $\mathbf{r}(\mathbf{y})$.

Local metric tensor and local basis

By construction, the coordinate basis vectors point in the direction of the coordinate lines. This makes them optimally suited to describe problems in the representation defined by the **y**-coordinates. However, they are not in general normalized nor pairwise orthogonal. For example, the norm of the vector $\mathbf{v}_\phi$ in Eq. (V20) is obtained as $\|\mathbf{v}_{\phi,\mathbf{r}}\| = \rho$, and varies with the radial coordinate. This is natural, because the variation $\partial_\phi \mathbf{r} = \mathbf{v}_\phi$ of $\mathbf{r}$ under variations of the angle is larger the farther the point $\mathbf{r}$ is away from the origin (think about this point).

metric tensor in coordinate basis   The general geometric relation between the basis vectors is described by the **metric tensor**, Eq. (L57),

$$g_{ij,\mathbf{r}} \equiv \langle \mathbf{v}_{i,\mathbf{r}}, \mathbf{v}_{j,\mathbf{r}} \rangle, \tag{V21}$$

where $\langle \ , \ \rangle$ is the standard scalar product in $\mathbb{R}^d$. For example, the *metric in polar coordinates* is given by

$$g_{\rho\rho} = 1, \qquad g_{\phi\phi} = \rho^2, \qquad g_{\rho\phi} = g_{\phi\rho} = 0. \tag{V22}$$

Since the off-diagonal elements vanish, the basis vectors are orthogonal, $\langle \mathbf{v}_\rho, \mathbf{v}_\phi \rangle = 0$. For later reference, we note that the components of the inverse metric, defined in Eq. (L62), are given by

$$g^{\rho\rho} = 1, \qquad g^{\phi\phi} = \rho^{-2}, \qquad g^{\rho\phi} = g^{\phi\rho} = 0. \tag{V23}$$

local basis   In the physics community it is customary to *normalize* the vectors of the coordinate basis and in this way pass to a basis called the **local basis**. The vectors of the local basis are defined as

$$\mathbf{e}_{j,\mathbf{r}} = \frac{\mathbf{v}_{j,\mathbf{r}}}{\sqrt{g_{jj}}} = \frac{\mathbf{v}_{j,\mathbf{r}}}{\|\mathbf{v}_{j,\mathbf{r}}\|}, \tag{V24}$$

and frequently denoted by the letter $\mathbf{e}$. In this case, they are distinguished from the vectors of the Cartesian basis, $\{\mathbf{e}_a\}$, solely by their index. For example the *local basis in polar* **local basis in polar coordinates** *coordinates*, obtained from (V24), (V22) and (V20), is given by

$$\mathbf{e}_\rho = \mathbf{v}_\rho = \mathbf{e}_x \cos\phi + \mathbf{e}_y \sin\phi,$$
$$\mathbf{e}_\phi = \frac{1}{\rho}\mathbf{v}_\phi = -\mathbf{e}_x \sin\phi + \mathbf{e}_y \cos\phi, \tag{V25}$$

and $\mathbf{r} = \mathbf{e}_\rho \rho$. Although the normalization of the basis vectors is sometimes convenient, it comes with a price tag: the normalization factors, $(g_{jj}(\mathbf{y}))^{-1/2}$, are functions of $\mathbf{y}$. Their presence generally complicates calculations, and these complications can become quite painful in computations involving coordinate derivatives. The coordinate basis actually is an example of a basis where the *lack* of orthnormality is an intuitive feature and at the same time simplifies all calculations. This fact is well recognized by the mathematics community which favors working with this basis. In physics and engineering, however, the local basis is more frequently used. In this chapter on standard coordinate systems we will pay tribute to this convention and emphasize the local basis. However, later in the text, the coordinate basis will come to play a more prominent role. In either case, the two bases differ only by the normalization convention, Eq. (V24).

As an example of a derivative operation involving local basis vectors, consider the partial derivatives

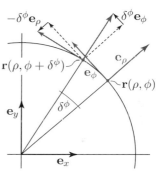

$$\partial_\rho \mathbf{e}_\rho = 0, \qquad \partial_\phi \mathbf{e}_\rho = \mathbf{e}_\phi,$$
$$\partial_\rho \mathbf{e}_\phi = 0, \qquad \partial_\phi \mathbf{e}_\phi = -\mathbf{e}_\rho, \qquad (V26)$$

which follow directly from the definition Eq. (V25). These derivatives describe how the vectors vary under variations of coordinates (see the figure): variations in ρ leave both basis vectors unchanged, while variations in ϕ change $\mathbf{e}_\rho$ in a direction parallel to $\mathbf{e}_\phi$, and $\mathbf{e}_\phi$ in a direction anti-parallel to $\mathbf{e}_\rho$.

In applications such derivatives appear, e.g., when the local basis is employed to describe *physical motion along curves*. As an example, consider a two-dimensional curve with parametrization $\mathbf{r}(t)$. The expansion of this vector in the Cartesian and the local basis assumes the form

$$\mathbf{r}(t) = \mathbf{e}_x x(t) + \mathbf{e}_y y(t) = \mathbf{e}_\rho(t)\, \rho(t) = \mathbf{e}_{\rho,\mathbf{r}(t)}\, \rho(t), \qquad (V27)$$

where the subscript $\mathbf{r}(t)$ in $\mathbf{e}_{\rho,\mathbf{r}(t)}$ emphasizes that the basis vector may change under variations of the base point $\mathbf{r}(t)$. Notice, however, that at all times the curve is described by a vector, $\mathbf{r}$, connecting the origin with a point on the curve. This is why the expansion of $\mathbf{r}$ contains the vector $\mathbf{e}_\rho$, but not $\mathbf{e}_\phi$.

curve velocity in polar coordinates   Application of the product rule to the computation of the curve velocity yields $d_t \mathbf{r}(t) = \mathbf{e}_\rho(t)\, d_t(\rho(t)) + d_t(\mathbf{e}_\rho(t))\, \rho(t)$, or

$$\dot{\mathbf{r}} = \mathbf{e}_\rho\, \dot\rho + \dot{\mathbf{e}}_\rho\, \rho,$$

in a more compact notation. It remains to compute the time derivative $\dot{\mathbf{e}}_\rho$. The time dependence of a local basis vector $\mathbf{e}_j(t)$ originates in the time dependence of the coordinates at which it is evaluated: $\mathbf{e}_j(t) = \mathbf{e}_{j,\mathbf{r}(t)} = \mathbf{e}_{j,\mathbf{r}(\rho(t),\phi(t))}$. The chain rule of calculus, Eq. (C41) (with the identifications $\mathbf{f} \to \mathbf{e}_j$, $\mathbf{g} \to \mathbf{y} = (\rho,\phi)^{\mathrm{T}}$, $x \to t$), thus yields

$$\dot{\mathbf{e}}_\rho = (\partial_\rho \mathbf{e}_\rho)\, \dot\rho + (\partial_\phi \mathbf{e}_\rho)\, \dot\phi \overset{(V26)}{=} \mathbf{e}_\phi\, \dot\phi, \qquad \dot{\mathbf{e}}_\phi = (\partial_\rho \mathbf{e}_\phi)\, \dot\rho + (\partial_\phi \mathbf{e}_\phi)\, \dot\phi \overset{(V26)}{=} -\mathbf{e}_\rho\, \dot\phi. \qquad (V28)$$

We conclude that the curve velocity may be written as

$$\dot{\mathbf{r}} = \mathbf{e}_\rho\, \dot\rho + \mathbf{e}_\phi\, \rho\, \dot\phi. \qquad (V29)$$

As one would expect, the components of the velocity vector are determined by the rates of change of the coordinates, $\dot\rho$ and $\dot\phi$. The effect of variations in ϕ grows with the separation,

ρ, of the position vector from the origin. This explains why the angular derivative appears in the combination $\rho\dot{\phi}$. Notice that in the coordinate basis, the curve velocity would assume the simpler form, $\dot{\mathbf{r}} = \mathbf{v}_\rho\dot{\rho} + \mathbf{v}_\phi\dot{\phi}$. Here, the scaling with ρ is included in the definition of the basis vector $\mathbf{v}_\phi$, and this leads to simpler formulas.

acceleration in polar coordinates

INFO   Occasionally, it becomes necessary to compute higher-order derivatives of curves represented in curvilinear coordinates. As an example, consider the *curve acceleration in polar coordinates*. Differentiating $\dot{\mathbf{r}}$ in time and using Eq. (V28) once more, we obtain

$$
\begin{aligned}
\mathbf{a} = \; & \mathrm{d}_t(\dot{\mathbf{r}}) \overset{(V29)}{=} \mathrm{d}_t(\mathbf{e}_\rho\,\dot{\rho} + \mathbf{e}_\phi\,\rho\,\dot{\phi}) = \dot{\mathbf{e}}_\rho\,\dot{\rho} + \mathbf{e}_\rho\,\ddot{\rho} + \dot{\mathbf{e}}_\phi\,\rho\,\dot{\phi} + \mathbf{e}_\phi\,\dot{\rho}\,\dot{\phi} + \mathbf{e}_\phi\,\rho\,\ddot{\phi} \\
& \overset{(V28)}{=} (\mathbf{e}_\phi\dot{\phi})\,\dot{\rho} + \mathbf{e}_\rho\,\ddot{\rho} + (-\mathbf{e}_\rho\dot{\phi})\,\rho\,\dot{\phi} + \mathbf{e}_\phi\,\dot{\rho}\,\dot{\phi} + \mathbf{e}_\phi\,\rho\,\ddot{\phi} \\
= \; & \mathbf{e}_\rho\,(\ddot{\rho} - \rho\,\dot{\phi}^2) + \mathbf{e}_\phi\,(2\dot{\rho}\,\dot{\phi} + \rho\,\ddot{\phi}).
\end{aligned}
$$

The different contributions to the acceleration vector are best understood by considering particular types of motion. For example, a circular motion at constant radius, $\dot{\rho} = \ddot{\rho} = 0$, generally leads to acceleration in the ϕ-direction, $\mathbf{e}_\phi\ddot{\phi}\rho$. However, even if the motion is not accelerated in the ϕ-direction, $\ddot{\phi} = 0$, one does have acceleration in the ρ-direction, $-\mathbf{e}_\rho\dot{\phi}^2$. An observer moving along the curve feels this acceleration as a centrifugal force in the opposite direction.

The derivation of the acceleration vector shows that computations with curvilinear coordinates require care. A useful safety check is to monitor the *physical dimensions* of terms at all stages of the computation. For example, the vector $\mathbf{r}$ has physical dimension "length". It is customary to keep track of this by writing $[\mathbf{r}] = $ length, where the square brackets are a shorthand for "dimension of". Each time derivative divides this dimension by one dimension of time, i.e. $[\frac{\mathrm{d}\mathbf{r}}{\mathrm{d}t}] = $ length/time and $[\mathbf{a}] = [\frac{\mathrm{d}^2\mathbf{r}}{\mathrm{d}t^2}] = $ length/time^2. The different terms contributing to the acceleration all have this dimension.

EXAMPLE   As an example illustrating the usefulness of curvilinear coordinates, consider the *work done by a rotationally symmetric force field*, $\mathbf{F}(\mathbf{r}) = f\mathbf{e}_\phi$, for $\mathbf{r} \neq \mathbf{0}$, along a spiral path, γ, in the plane. The force is colinear to $\mathbf{e}_\phi$, and has constant strength, $f = $ const. Consider a path parametrized as $\mathbf{y}(t) = (\rho(t), \phi(t))^\mathrm{T} = (\frac{tR}{t_0}, 2\pi\frac{t}{t_0})^\mathrm{T}$, with $t \in (0, t_0)$. It describes a spiral winding once around the origin and reaching a radial distance R at the final time, t_0 (see the figure for a plot with radial coordinates in units of R). Substituting this parametrization into Eq. (V29), we obtain $\dot{\mathbf{r}} = \frac{R}{t_0}(\mathbf{e}_\rho + \mathbf{e}_\phi 2\pi\frac{t}{t_0})$, and $\dot{\mathbf{r}} \cdot \mathbf{F} = \frac{2\pi Rf t}{t_0^2}$. This yields the work integral

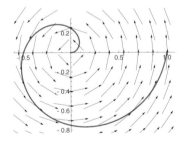

$$
W = \int_\gamma \mathrm{d}\mathbf{r} \cdot \mathbf{F} = \int_0^{t_0} \mathrm{d}t\,\dot{\mathbf{r}} \cdot \mathbf{F} = \int_0^{t_0} \mathrm{d}t\,\frac{2\pi Rf t}{t_0^2} = \pi Rf.
$$

The positive sign indicates that the force does positive work along a path with which it is approximately aligned. Try to formulate the same calculation in Cartesian coordinates and confirm that it becomes technically more complicated and less intuitive.

For further examples of line integrals using curvilinear coordinates, see problems V2.3.5-8.

V2.3 Cylindrical and spherical coordinates

Physical problems frequently possess symmetries under rotations. Low-dimensional examples include symmetry under rotations around a point in $\mathbb{R}^2$ or $\mathbb{R}^3$, or rotation around a fixed rotation axis in $\mathbb{R}^3$. In higher dimensions, more complex rotation symmetries can be realized. For all these cases, tailor-made curvilinear coordinate systems have been devised. The most prominent ones, illustrated in Fig. V5, are the polar coordinates discussed above, and the three-dimensional cylindrical and spherical coordinate systems, to be introduced next.

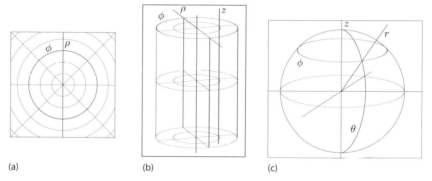

(a)   (b)   (c)

Fig. V5   Coordinate lines of the three most frequently used curvilinear coordinate systems. (a) Two-dimensional polar coordinates, (b) cylindrical coordinates, and (c) spherical coordinates (sometimes called three-dimensional polar coordinates).

Cylindrical coordinates

Cylindrical coordinates are designed to describe problems which are rotationally symmetric around a fixed *axis*. For example, the axially symmetric magnetic field generated by a straight, current-carrying wire is conveniently described in cylindrical coordinates. The construction of cylindrical coordinates is shown in the figure: following standard conventions, we define the z-axis of a Cartesian coordinate system as the symmetry axis of the problem. We next pick two directions orthogonal to it and

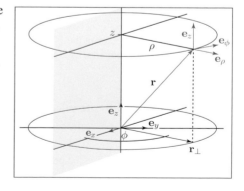

each other, to define a three-dimensional Cartesian system of coordinates (x, y, z). A vector $\mathbf{r}$ may then be decomposed as $\mathbf{r} = \mathbf{e}_z z + \mathbf{r}_\perp$, where $\mathbf{r}_\perp \perp \mathbf{e}_z$ lies in the xy-plane. Parametrizing this plane by polar coordinates, (ρ, ϕ), a representation $\mathbf{r}_\perp = \mathbf{e}_x \rho \cos\phi + \mathbf{e}_y \rho \sin\phi$ for the transverse component, and the representation

$$\mathbf{r}(\rho, \phi, z) = \mathbf{e}_x \, \rho \cos \phi + \mathbf{e}_y \, \rho \sin \phi + \mathbf{e}_z \, z \qquad \text{(V30)}$$

cylindrical coordinates

for the three-dimensional vector are obtained. The transformation map expressing the Cartesian coordinates $\mathbf{x} = (x, y, z)^T$ through **cylindrical coordinates** $\mathbf{y} = (\rho, \phi, z)^T$ is defined as

$$\mathbf{x}: U \to U_C, \qquad \mathbf{y} = \begin{pmatrix} \rho \\ \phi \\ z \end{pmatrix} \mapsto \mathbf{x}(\mathbf{y}) = \begin{pmatrix} x(\rho, \phi, z) \\ y(\rho, \phi, z) \\ z(\rho, \phi, z) \end{pmatrix} = \begin{pmatrix} \rho \cos \phi \\ \rho \sin \phi \\ z \end{pmatrix}, \qquad \text{(V31a)}$$

and the inverse map is

$$\mathbf{y}: U_C \to U, \qquad \mathbf{x} = \begin{pmatrix} x \\ y \\ z \end{pmatrix} \mapsto \mathbf{y}(\mathbf{x}) = \begin{pmatrix} \rho(x, y, z) \\ \phi(x, y, z) \\ z(x, y, z) \end{pmatrix} = \begin{pmatrix} \sqrt{x^2 + y^2} \\ \arctan(\frac{y}{x}) \\ z \end{pmatrix}. \qquad \text{(V31b)}$$

The domain of cylindrical coordinates, $U \equiv \mathbb{R}^+ \times (0, 2\pi) \times \mathbb{R} \ni (\rho, \phi, z)$, contains the open intervals $(0, \infty) = \mathbb{R}^+$ and $(0, 2\pi)$, for the radial and the angular coordinate, respectively. The image of the map then excludes the z-axis and all points, $(x, y, z)^T = (\rho, 0, z)^T$, having a positive x-coordinate and vanishing y-coordinate. The union of all excluded points defines the half-plane, $\mathbb{R}_0^+ \times \{0\} \times \mathbb{R}$ (see the shaded plane in the figure above) and so the image of the coordinate transformation is limited to $U_C \equiv \mathbb{R}^3 \backslash (\mathbb{R}_0^+ \times \{0\} \times \mathbb{R})$. ($\to$ V2.3.1-2)

Equations (V19), (V21) and (V24) may now be applied to compute the *coordinate basis vectors for cylindrical coordinates*, $\{\mathbf{v}_\rho, \mathbf{v}_\phi, \mathbf{v}_z\}$, the metric tensor, and the *local basis vectors*, $\{\mathbf{e}_\rho, \mathbf{e}_\phi, \mathbf{e}_z\}$. One finds ($\to$ V2.3.3)

coordinate basis and metric of cylindrical coordinates

$$\begin{aligned}
\mathbf{v}_\rho &= \mathbf{e}_x \cos \phi + \mathbf{e}_y \sin \phi, & g_{\rho\rho} &= 1, & \mathbf{e}_\rho &= \mathbf{v}_\rho, \\
\mathbf{v}_\phi &= \rho(-\mathbf{e}_x \sin \phi + \mathbf{e}_y \cos \phi), & g_{\phi\phi} &= \rho^2, & \mathbf{e}_\phi &= \frac{1}{\rho}\mathbf{v}_\phi, \\
\mathbf{v}_z &= \mathbf{e}_z, & g_{zz} &= 1, & \mathbf{e}_z &= \mathbf{e}_z.
\end{aligned} \qquad \text{(V32)}$$

EXERCISE Verify the *orthogonality of the cylindrical coordinate basis*, $g_{ij} = 0$ for $i \neq j$. Also verify that the local basis defines a right-handed system, $\mathbf{e}_\rho \times \mathbf{e}_\phi = \mathbf{e}_z$. The local basis comprises a fixed unit vector in the z-direction, $\mathbf{e}_z$, and the local basis, $\{\mathbf{e}_\rho, \mathbf{e}_\phi\}$, of a polar system spanning the plane perpendicular to z. Note that the point $\mathbf{r}$ is represented as

$$\mathbf{r} \overset{(V30)}{=} \mathbf{e}_\rho \, \rho + \mathbf{e}_z z. \qquad \text{(V33)}$$

From here, the velocity of a time-dependent vector $\mathbf{r}(t)$ is obtained as

$$\dot{\mathbf{r}} \overset{(V29)}{=} \mathbf{e}_\rho \, \dot\rho + \mathbf{e}_\phi \, \rho \, \dot\phi + \mathbf{e}_z \dot{z}. \qquad \text{(V34)}$$

Proceed as in the previous case of polar coordinates to obtain the acceleration, $\ddot{\mathbf{r}}$. ($\to$ V2.3.3)

EXAMPLE Let us take another look at the spiral curve defined in (V10). In cylindrical coordinates, it is represented as $(\rho(t), \phi(t), z(t))^T = (1, 2\pi t, (2/3)t^{3/2})^T$. We may now use the representation (V34) to compute its velocity as $\dot{\mathbf{r}}(t) = \mathbf{e}_\phi \, 2\pi + \mathbf{e}_z \, t^{1/2}$. This equation describes how the velocity vector winds around the z-axis while building up a growing z-component. The norm of this vector is obtained as $\|\dot{\mathbf{r}}\| = \sqrt{(2\pi)^2 + t}$, in agreement with the previous analysis after Eq. (V10). Note that the representation in cylindrical coordinates does a better job at exposing the geometry of the curve.

EXAMPLE As another example, consider a force $\mathbf{F} = f\mathbf{e}_\phi + c\,\mathbf{e}_z$ similar to that discussed on p. 420, but now with a constant component, c, in the z-direction. We aim to compute the work done along a three-dimensional spiral, which winds once around the origin in time t_0. At the final point it has reached a separation R from the central axis and a height Z in the z-direction (see the figure, where ρ and z are indicated in units of R and Z, respectively). In cylindrical coordinates, this curve can be parametrized as $\mathbf{y}(t) = (\rho(t), \phi(t), z(t))^{\mathrm{T}} = (R\frac{t}{t_0}, 2\pi\frac{t}{t_0}, Z\frac{t}{t_0})^{\mathrm{T}}$. Eq. (V34) yields

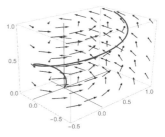

$\dot{\mathbf{r}} = \frac{R}{t_0}(\mathbf{e}_\rho + 2\pi\frac{t}{t_0}\mathbf{e}_\phi) + \frac{Z}{t_0}\mathbf{e}_z$, hence $\dot{\mathbf{r}} \cdot \mathbf{F} = \frac{2\pi Rft}{t_0^2} + \frac{Zc}{t_0}$. The work integral is then computed as

$$W = \int_\gamma d\mathbf{r} \cdot \mathbf{F} = \int_0^{t_0} dt\, \dot{\mathbf{r}} \cdot \mathbf{F} = \int_0^{t_0} dt \left(\frac{2\pi Rft}{t_0^2} + \frac{Zc}{t_0} \right) = \pi Rf + Zc \,.$$

Spherical coordinates

Spherical coordinates are used to describe problems possessing rotational symmetry around a fixed point in Euclidean space. We choose this point as the origin of a Cartesian coordinate system. In cases where the problem possesses a particular axis of interest,[10] this axis is commonly chosen as the z-axis. The x- and y-axes then span the "equatorial plane", as indicated in the figure. A point $\mathbf{r}$ is now described by the following three numbers: its **polar** distance, r, from the origin, the **polar angle**, θ, enclosed by the vector $\mathbf{r}$ and the z-axis, and **azimuthal** the **azimuthal angle**, ϕ, enclosed by the x-**angle** axis and the projection of $\mathbf{r}$ onto the equatorial plane.

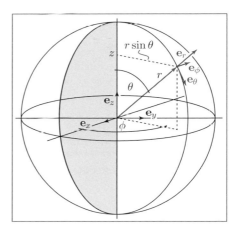

Elementary geometry shows that the z-component of the vector $\mathbf{r}$ is given by $r\cos\theta$, where $\theta \in (0, \pi)$. Similarly, the x- and y-components are given by $r\sin\theta\cos\phi$ and $r\sin\theta\sin\phi$, respectively. This leads to the representation

$$\boxed{\mathbf{r}(r, \theta, \phi) = \mathbf{e}_x\, r\sin\theta\cos\phi + \mathbf{e}_y r\sin\theta\sin\phi + \mathbf{e}_z r\cos\theta \,.} \qquad \text{(V35)}$$

spherical The map expressing the Cartesian coordinates, $\mathbf{x} = (x, y, z)^{\mathrm{T}}$, of $\mathbf{r}$ through **spherical**
coordinates **coordinates**, $\mathbf{y} = (r, \theta, \phi)^{\mathrm{T}}$, is thus given by

---

[10] For the description of a planet, this could be the axis connecting its magnetic north and south poles.

$$\mathbf{x} : U \to U_C, \quad \mathbf{y} = \begin{pmatrix} r \\ \theta \\ \phi \end{pmatrix} \mapsto \mathbf{x}(\mathbf{y}) = \begin{pmatrix} x(r,\theta,\phi) \\ y(r,\theta,\phi) \\ z(r,\theta,\phi) \end{pmatrix} = \begin{pmatrix} r \sin\theta \cos\phi \\ r \sin\theta \sin\phi \\ r \cos\theta \end{pmatrix}, \tag{V36a}$$

and the inverse map is

$$\mathbf{y} : U_C \to U, \quad \mathbf{x} = \begin{pmatrix} x \\ y \\ z \end{pmatrix} \mapsto \mathbf{y}(\mathbf{x}) = \begin{pmatrix} r(x,y,z) \\ \theta(x,y,z) \\ \phi(x,y,z) \end{pmatrix} = \begin{pmatrix} \sqrt{x^2+y^2+z^2} \\ \arccos\left(\frac{z}{\sqrt{x^2+y^2+z^2}}\right) \\ \arctan(y/x) \end{pmatrix}. \tag{V36b}$$

The coordinate domain, $U \equiv \mathbb{R}^+ \times (0,\pi) \times (0,2\pi) \ni (r,\theta,\phi)$, contains the open intervals $\mathbb{R}^+ = (0,\infty)$, $(0,\pi)$ and $(0,2\pi)$ for the radial, polar and azimuthal variable, respectively. As with cylindrical coordinates, the image of the map $\mathbf{x}(\mathbf{y})$ excludes the half-plane $\mathbb{R}_0^+ \times \{0\} \times \mathbb{R}$, and so is given by $U_C \equiv \mathbb{R}^3 \backslash (\mathbb{R}_0^+ \times \{0\} \times \mathbb{R})$. ($\to$ V2.3.1-2)

coordinate
basis and
metric of
spherical
coordinates

EXERCISE Verify that the *coordinate basis vectors for spherical coordinates*, $\{\mathbf{v}_r, \mathbf{v}_\theta, \mathbf{v}_\phi\}$, the metric tensor, and the *local basis vectors*, $\{\mathbf{e}_r, \mathbf{e}_\theta, \mathbf{e}_\phi\}$, are given by, respectively, ($\to$ V2.3.4)

$$\mathbf{v}_r = \mathbf{e}_x \sin\theta \cos\phi + \mathbf{e}_y \sin\theta \sin\phi + \mathbf{e}_z \cos\theta, \qquad g_{rr} = 1, \qquad \mathbf{e}_r = \mathbf{e}_r,$$

$$\mathbf{v}_\theta = r(\mathbf{e}_x \cos\theta \cos\phi + \mathbf{e}_y \cos\theta \sin\phi - \mathbf{e}_z \sin\theta), \qquad g_{\theta\theta} = r^2, \qquad \mathbf{e}_\theta = \frac{1}{r}\mathbf{v}_\theta, \tag{V37}$$

$$\mathbf{v}_\phi = r\sin\theta\,(-\mathbf{e}_x \sin\phi + \mathbf{e}_y \cos\phi), \qquad g_{\phi\phi} = r^2 \sin^2\theta, \qquad \mathbf{e}_\phi = \frac{1}{r\sin\theta}\mathbf{v}_\phi,$$

with $g_{i \neq j} = 0$. Again the off-diagonal elements of the metric tensor vanish, implying that the local basis vectors are mutually orthogonal. Verify that they form a right-handed system, $\mathbf{e}_r \times \mathbf{e}_\theta = \mathbf{e}_\phi$. In the local basis, the point $\mathbf{r}$ has the simple representation

$$\mathbf{r} \overset{(V35)}{=} \mathbf{e}_r\, r. \tag{V38}$$

Verify that the velocity of a time-dependent vector $\mathbf{r}(t)$, obtained by computing the time derivative of this expression, is given by

$$\dot{\mathbf{r}} = \mathbf{e}_r \dot{r} + \mathbf{e}_\theta\, r\dot{\theta} + \mathbf{e}_\phi\, r\dot{\phi} \sin\theta. \tag{V39}$$

Also compute the acceleration, $\ddot{\mathbf{r}}$, in spherical coordinates. ($\to$ V2.3.4)

INFO Curvilinear coordinates are tailored to describe structures of dimension d embedded in spaces of higher dimension, $n > d$. For example, the *surface of a sphere* of unit radius, S^2, is a $(d = 2)$-dimensional object embedded in $(n = 3)$-dimensional space. It can be parametrized by keeping the radius of the sphere, $r = 1$, fixed and letting the two angular coordinates θ and ϕ run through their domain of definition. In this way, a representation

$$(0,\pi) \times (0,2\pi) \to S^2 \subset \mathbb{R}^3, \qquad (\theta,\phi) \mapsto \mathbf{r}(1,\theta,\phi),$$

is obtained. In mathematical terminology, generalized "surfaces" defined in this way are called *differentiable manifolds*. Differentiable manifolds are a concept of great importance in various areas of physics and mathematics and we will discuss them in more detail in Chapter V4.

V2.4   A general perspective of coordinates

In retrospect, all the coordinate systems discussed above defined different "languages" for the description of geometrically defined points $\mathbf{r}$ in a set M. In this chapter, we assumed M to be embedded in a vector space whose dimension equals that of the coordinate domain, d. However, one frequently needs to consider more general settings. For example, the surface of a sphere in three-dimensional space is described by two coordinates. In this case, $d = 2$, while $\mathbf{r}$ is realized as a three-component vector, i.e. $M \subset \mathbb{R}^3$. In Chapter V4, even the assumption that M is contained in a vector space will be abandoned.

curvilinear coordinates   Irrespective of the concrete realization of the image set, M, a system of **curvilinear coordinates** is an invertible map,

$$\mathbf{r} : U \to M, \qquad \mathbf{y} \mapsto \mathbf{r}(\mathbf{y}) = \mathbf{r}(y^1, \dots, y^d), \tag{V40a}$$

between a coordinate domain $U \subset \mathbb{R}^d$, open in $\mathbb{R}^d$, and its image, M. The inverse map assigns to each point $\mathbf{r} \in M$ its coordinates $\mathbf{y}(\mathbf{r}) \in U$ (see Fig. V6):

$$\mathbf{y} : M \to U, \qquad \mathbf{r} \mapsto \mathbf{y}(\mathbf{r}) = (y^1(\mathbf{r}), \dots, y^d(\mathbf{r}))^{\mathrm{T}}. \tag{V40b}$$

Now consider another coordinate system for the same set M,[11]

$$\mathbf{r} : U' \to M, \qquad \mathbf{y}' \mapsto \mathbf{r}(\mathbf{y}') = \mathbf{r}(y'^1, \dots, y'^d). \tag{V41}$$

Its inverse assigns to points $\mathbf{r} \in M$ coordinates $\mathbf{y}'(\mathbf{r})$ which in general are different from $\mathbf{y}(\mathbf{r})$. For example, $\mathbf{y}(\mathbf{r}) = (\rho, \phi)^{\mathrm{T}}$ might be polar coordinates and $\mathbf{y}'(\mathbf{r}) \equiv \mathbf{x}(\mathbf{r}) = (x^1, x^2)^{\mathrm{T}}$ Cartesian coordinates of a point $\mathbf{r}$, as in Section V2.1 above.

To each $\mathbf{y}'$ we can now assign $\mathbf{y}(\mathbf{y}')$, defined such that $\mathbf{r}(\mathbf{y}') = \mathbf{r}(\mathbf{y}(\mathbf{y}'))$ represent the same point (see Fig. V6.) Formally, this correspondence between coordinates is described by a *coordinate transformation*, $\mathbf{y} : U' \to U$, $\mathbf{y}' \mapsto \mathbf{y}(\mathbf{y}')$, which is a map between the two coordinate domains. The image $\mathbf{y}(\mathbf{y}')$ is obtained by computing the $\mathbf{y}$-coordinates, $\mathbf{y}(\mathbf{r}(\mathbf{y}'))$,

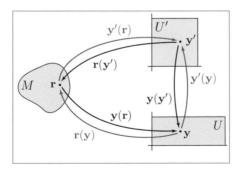

Fig. V6   Schematic of the coordinate representation of a subset $M \subset \mathbb{R}^n$ through different coordinate systems, $\mathbf{y}$ and $\mathbf{y}'$, and the transformations between them.

[11] It is customary to denote different coordinate systems by same symbol, $\mathbf{r} : U \to M$ and $\mathbf{r} : U' \to M$. Their distinction is indicated by the argument, $\mathbf{r}(\mathbf{y})$ vs. $\mathbf{r}(\mathbf{y}')$.

of the point $\mathbf{r}(\mathbf{y}')$. All coordinate transformations are invertible and smooth, which means
diffeomorphic that they are **diffeomorphic maps** between the coordinate domains. Equation (V15) and
map the other transformation formulas discussed earlier in this chapter were realizations of such
maps for the particular case where $\mathbf{y}' = \mathbf{x}$ were Cartesian coordinates.

As mentioned above, one may think of coordinates as *different languages* $(\mathbf{y}, \mathbf{y}', \dots)$
describing the same objects $(\mathbf{r})$. Coordinate transformations are the unique translations
between languages. As a rule, concrete mathematical calculations (differentiation, integration, etc.) are generally performed in a specific coordinate language, and not in M. In
Chapter V4 we will discuss cases where the operations of calculus are not even defined
in M. We will then explain how concepts relying on differentiation – coordinate bases,
curve velocities, etc. – are all defined with reference to coordinates, and differentiability
in M is never required. For the moment we just note that it generally pays to carefully
distinguish between geometric objects in M and the coordinates describing them, including in situations where this distinction may seem pedantic. For example, when working in
$M = \mathbb{R}^2$ it is easy to forget the distinction between points $\mathbf{r}$ and their Cartesian representations $\mathbf{x} = (x^1, x^2)^{\mathsf{T}}$. However, even then it is well to remember that there is a difference. An
early training of this view helps to avoid confusion in cases where the distinction becomes
crucial (the discipline of general relativity being a prominent example).

V2.5   Local coordinate bases and linear algebra

REMARK This section reinterprets the concepts introduced above from a linear algebraic perspective.
It introduces methodology that will facilitate our discussion of various geometric structures. The
section requires familiarity with matrices, in particular the transformation matrices between different
bases (Section L5.6), and with determinants (Chapter L6). We will refer back to this section later in
the text, however, it can be skipped at first reading if time is short.

Jacobi matrix

From the perspective of linear algebra, Eq. (V19) defines a linear transformation, $\mathbf{v}_j = \mathbf{e}_a J^a_{\ j}$, between the two bases $\{\mathbf{v}_j\}$ and $\{\mathbf{e}_a\}$. The transformation matrix has the elements

$$J^a_{\ j} \equiv \left(\frac{\partial \mathbf{x}}{\partial \mathbf{y}}\right)^a_j \equiv \frac{\partial x^a(\mathbf{y})}{\partial y^j}. \tag{V42}$$

Jacobi This matrix is called the **Jacobi matrix**
matrix of the map $\mathbf{y} \mapsto \mathbf{x}(\mathbf{y})$ and often denoted
as $J \equiv \frac{\partial \mathbf{x}}{\partial \mathbf{y}} \equiv \frac{\partial(x^1,\dots,x^d)}{\partial(y^1,\dots,y^d)}$. Its matrix elements are generally nonlinear functions of
the coordinates y^j. However, when considered at a fixed point $\mathbf{x}$, at "frozen values" of the coordinates $\mathbf{y}$, the numbers $J^a_{\ j}$
define a fixed $d \times d$ matrix, and all the
concepts of linear algebra apply to it.

> **Carl Gustav Jacob Jacobi**
> **1804–1851**
> German mathematician who made funda-
> mental contributions to various areas of
> mathematics, including the theory of special
> functions, number theory and differential
> equations. Jacobi was the first professor of
> Jewish descent to be appointed at a German university (Königsberg).

Specifically, the *inverse of the Jacobi matrix* is given by

$$(J^{-1})^j{}_a = \left(\frac{\partial \mathbf{y}}{\partial \mathbf{x}}\right)^j{}_a \equiv \frac{\partial y^j(\mathbf{x})}{\partial x^a}, \tag{V43}$$

where the curvilinear coordinates $\mathbf{y} = \mathbf{y}(\mathbf{x})$ are now interpreted as functions of the Cartesian coordinates. That these matrices are inverse to each other can be checked as

$$J^a{}_j (J^{-1})^j{}_b = \frac{\partial x^a(\mathbf{y})}{\partial y^j} \frac{\partial y^j(\mathbf{x})}{\partial x^b} = \frac{\partial x^a}{\partial x^b} = \delta^a{}_b,$$

where the chain rule Eq. (C42) was used. The inverse matrix can then be used to switch from the coordinate basis back to the Cartesian basis as $\mathbf{e}_a = \mathbf{v}_j (J^{-1})^j{}_a$, or

$$\mathbf{e}_a = \mathbf{v}_j(\mathbf{y}(\mathbf{x})) \frac{\partial y^j(\mathbf{x})}{\partial x^a}, \tag{V44}$$

in a more explicit notation. Here, both factors on the right vary as functions of $\mathbf{x}$. However, they do so in such a way that their product yields the constant Cartesian basis vectors $\mathbf{e}_a$.

Now let $\mathbf{u}(\mathbf{x})$ be a generic vector at the point $\mathbf{x} = \mathbf{x}(\mathbf{y})$. Depending on which basis is used, it affords the two representations $\mathbf{u} = \mathbf{e}_a u^a(\mathbf{x}) = \mathbf{v}_j(\mathbf{y}) u^j(\mathbf{y})$, where the choice of subscript, a vs. j, indicates which basis is referred to. From linear algebra we know that the corresponding *change between vector components* is given by the transformation matrix as $u^a = J^a{}_j u^j$, and $u^j = (J^{-1})^j{}_a u^a$, or

vector trans-
formations
via Jacobi
matrix

$$u^a(\mathbf{x}) = \left.\frac{\partial x^a(\mathbf{y})}{\partial y^j} u^j(\mathbf{y})\right|_{\mathbf{y}=\mathbf{y}(\mathbf{x})}, \qquad u^j(\mathbf{y}) = \left.\frac{\partial y^j(\mathbf{x})}{\partial x^a} u^a(\mathbf{x})\right|_{\mathbf{x}=\mathbf{x}(\mathbf{y})}. \tag{V45}$$

In the first equation, the specification $\mathbf{y} = \mathbf{y}(\mathbf{x})$ emphasizes that the expression on the right emerges as a function of the generalized coordinates, $\mathbf{y}$. However, when featuring in a coordinate change to Cartesian coordinates, $\mathbf{x}$, all $\mathbf{y}$-dependences should be expressed through $\mathbf{x}$ via the unique correspondence $\mathbf{y} = \mathbf{y}(\mathbf{x})$. A similar statement holds for the second equality. In the next section, we illustrate these abstract statements on a simple example.

Example: Jacobi matrix of polar coordinates

The *Jacobi matrix* of the polar coordinate system is readily obtained from Eq. (V15a) as

$$J = \frac{\partial(x, y)}{\partial(\rho, \phi)} = \begin{pmatrix} \frac{\partial x}{\partial \rho} & \frac{\partial x}{\partial \phi} \\ \frac{\partial y}{\partial \rho} & \frac{\partial y}{\partial \phi} \end{pmatrix} = \begin{pmatrix} \cos\phi & -\rho\sin\phi \\ \sin\phi & \rho\cos\phi \end{pmatrix}, \tag{V46}$$

where $J = J^a{}_j$ now carries the indices $a = x, y$ and $j = \rho, \phi$. Inserting the elements of J into $\mathbf{v}_j = \mathbf{e}_a J^a{}_j$, we confirm Eq. (V20) for the transformation between the basis vectors $\mathbf{v}_\rho, \mathbf{v}_\phi$ and $\mathbf{e}_x, \mathbf{e}_y$. Compared to the "direct" way of computing $\mathbf{v}_\rho = \partial_\rho \mathbf{x}$ and $\mathbf{v}_\phi = \partial_\phi \mathbf{x}$ from Eq. (V14), the Jacobi matrix simply offers an alternative way of doing the bookkeeping of coordinate derivatives. The corresponding **Jacobi determinant** is given by $\det(J) = \rho$.

The transformation in the reverse direction is described by the inverse Jacobi matrix as

$$J^{-1} = \frac{\partial(\rho, \phi)}{\partial(x, y)} = \begin{pmatrix} \frac{\partial \rho}{\partial x} & \frac{\partial \rho}{\partial y} \\ \frac{\partial \phi}{\partial x} & \frac{\partial \phi}{\partial y} \end{pmatrix} = \begin{pmatrix} \frac{x}{(x^2+y^2)^{1/2}} & \frac{y}{(x^2+y^2)^{1/2}} \\ -\frac{y}{x^2+y^2} & \frac{x}{x^2+y^2} \end{pmatrix}, \tag{V47}$$

where Eq. (V15b) was used to compute the derivatives. Equation (V44) then yields

$$\mathbf{e}_x = \mathbf{v}_\rho \frac{x}{(x^2 + y^2)^{1/2}} - \mathbf{v}_\phi \frac{y}{x^2 + y^2},$$

$$\mathbf{e}_y = \mathbf{v}_\rho \frac{y}{(x^2 + y^2)^{1/2}} + \mathbf{v}_\phi \frac{x}{x^2 + y^2},$$

where this time everything is expressed in the $\mathbf{x} = (x, y)^T$ language. As an exercise, use Eq. (V20) to confirm that the right-hand side does not dependent on the coordinates $(x, y)^T$, as required by the coordinate independence of the Cartesian basis vectors.

For a vector with Cartesian representation $\mathbf{u}(\mathbf{x}) = \mathbf{e}_a u^a(\mathbf{x})$, the polar coordinate representation $\mathbf{u}(\mathbf{x}) = \mathbf{v}_j(\mathbf{y}) u^j(\mathbf{y})$, is given by

$$\begin{pmatrix} u^\rho \\ u^\phi \end{pmatrix} (\mathbf{y}) \overset{(V45)}{=} J^{-1}(\mathbf{x}) \begin{pmatrix} u^x \\ u^y \end{pmatrix} (\mathbf{x}) \Bigg|_{\mathbf{x}=\mathbf{x}(\mathbf{y})}, \tag{V48}$$

where $\mathbf{y} = (\rho, \phi)^T$. The specification $\mathbf{x} = \mathbf{x}(\mathbf{y})$ emphasizes that the expression on the right is initially obtained as a function of Cartesian coordinates. In a second step, these have to be expressed through polar coordinates via Eq. (V15a). As an example, consider the vector with Cartesian representation $\mathbf{u} = (-y\mathbf{e}_x + x\mathbf{e}_y)(x^2 + y^2)^{-1/2}$, or $u^x = (x^2 + y^2)^{-1/2}(-y)$ and $u^y = (x^2 + y^2)^{-1/2}x$. Application of the matrix J^{-1} to this vector according to Eq. (V48) yields $u^\rho = 0$ and $u^\phi = (x^2 + y^2)^{-1/2}$. In a final step we express this through polar coordinates, $(x^2 + y^2)^{1/2} = \rho$, to obtain $\mathbf{u} = 0\mathbf{v}_\rho + \rho^{-1}\mathbf{v}_\phi = \mathbf{e}_\phi$.

EXERCISE Compute the Jacobi matrices for the transformations $\mathbf{y} \mapsto \mathbf{x}(\mathbf{y})$ between Cartesian and cylindrical or spherical coordinates. ($\rightarrow$ V2.5.1-2) Verify that their determinants are given by

$$\det\left(\frac{\partial(x^1, x^2, x^3)}{\partial(\rho, \phi, z)}\right) = \rho, \qquad \det\left(\frac{\partial(x^1, x^2, x^3)}{\partial(r, \theta, \phi)}\right) = r^2 \sin\theta. \tag{V49}$$

Jacobi matrix and metric

It is often useful to express the metric g_{ij} defined by the coordinate basis vectors in terms of the Jacobi matrix. To this end, we note that $g_{ij} \overset{(V21)}{=} (v_i)^a \delta_{ab}(v_j)^b$. The expansion $\mathbf{v}_j = \mathbf{e}_a J^a{}_j$ implies $(v_j)^a = J^a{}_j$, so that we obtain $g_{ij} = J^a{}_i \delta_{ab} J^b{}_j = (J^T){}_i{}^a J^a{}_j = (J^T J)_{ij}$, where the definition (L113) of the transpose of a matrix was used. This leads to a relation *connecting the metric with the Jacobi matrix*:

metric tensor from Jacobi matrix

$$\boxed{g_{ij} = (J^T J)_{ij}.} \tag{V50}$$

Although a direct computation of the metric from the scalar products may be more economical, Eq. (V50) plays an important role in various contexts, notably in integration theory (see the discussion following Eq. (C62) in Chapter C4 on multidimensional integration). As an exercise, apply relation (V50) to the polar Jacobi matrix to verify that the metric (V22) is obtained.

V2.6 Summary and outlook

In this chapter, coordinate representations alternative to the Cartesian coordinates defined by orthonormal bases were introduced. Coordinates were defined as smooth and bijective maps from a coordinate domain, U, onto a target domain, M. In general, M can be chosen from a large set of mathematical structures. However, in this chapter, we did not take the most general perspective and assumed M to be embedded in a vector space equipped with an orthonormal basis. We saw that the introduction of coordinates then led to various induced definitions: coordinate lines were obtained as curves in M by variation of just one coordinate, y^j. The tangents to these curves, $\mathbf{v}_j = \partial_j \mathbf{r}(\mathbf{y})$, defined a system of vectors, which could be employed as a powerful alternative to the Cartesian basis. We reasoned that these coordinate bases are tailored to computations in the $\mathbf{y}$-coordinates, and saw that they generally vary from point to point, $\mathbf{r}(\mathbf{y})$. These structures were made concrete for three very important coordinate systems, the polar, cylindrical and spherical coordinates. Besides the coordinate basis, we introduced the unit-normalized local basis as a workhorse for computations in these systems. (We also reasoned that the popularity of this basis in the physics and engineering culture is likely an artifact of tradition. At any rate, the normalization of coordinate basis vectors can lead to cumbersome expressions and does not always pay.)

In a later section of the chapter we interpreted coordinates from a more general viewpoint and reasoned that the embedding of M in a vector space actually was not essential. This generalization will become important in Chapters V4 to V7, when coordinate languages are applied to the description of more general geometric structures.

Physical information is often encoded in functions mapping d-dimensional space into image domains such as the real or complex numbers, the vector spaces $\mathbb{R}^n$ or $\mathbb{C}^n$, groups, or other mathematical structures. When they appear in physical contexts, such functions **field** are called **fields**.[12] Before defining fields in formal terms, let us give some examples of their use in physics.

▷ Local variations in the temperature in a volume $U \subset \mathbb{R}^3$ can be described by a *scalar field in three-dimensional space*, $T : U \to \mathbb{R}, \mathbf{r} \mapsto T(\mathbf{r})$, assigning to points $\mathbf{r}$ the local temperature, $T(\mathbf{r})$. Fields taking values in $\mathbb{R}$ or $\mathbb{C}$ are generally called real or complex **scalar field**   **scalar fields**, respectively.

▷ If the temperature profile depends on time, a *time-dependent scalar field*, $T : I \times U \to \mathbb{R}, (t, \mathbf{r})^\mathrm{T} \mapsto T(t, \mathbf{r})$, is required to describe the temperature variations over an interval of time, $I \subset \mathbb{R}$. This field is defined on a subset of four-dimensional space-time, $I \times U \in \mathbb{R} \times \mathbb{R}^3$.

▷ The flow of a fluid of homogeneous density is described by a field, $(t, \mathbf{r})^\mathrm{T} \mapsto \mathbf{v}(t, \mathbf{r}) \in \mathbb{R}^3$, where $\mathbf{v}$ describes the speed and direction of the flow at time t at space point $\mathbf{r}$. This is a three-dimensional *vector field* defined on four-dimensional space-time.

▷ The state of a *ferromagnet* is described by a field assigning to space-time points $(t, \mathbf{r})^\mathrm{T}$ a unit-normalized vector, $\hat{\mathbf{n}}(t, \mathbf{r}) \in \mathbb{R}^3$, describing the local magnetization of the material. For example, a ferromagnetic phase is distinguished by an approximately homogeneous magnetization, $\hat{\mathbf{n}}(t, \mathbf{r}) \simeq \text{const.}$, which generates the macroscopic magnetic field characteristic for magnetic substances. A unit-normalized vector, $\|\hat{\mathbf{n}}\| = 1$, may be identified with a point on the sphere of unit radius, S^2, so the field $\hat{\mathbf{n}} : \mathbb{R} \times \mathbb{R}^3 \to S^2$ maps space-time onto the unit sphere.

V3.1 Definition of fields

Mathematically, a *field* is a smooth map,

$$\mathbf{F} : M \to L, \qquad \mathbf{r} \mapsto \mathbf{F}(\mathbf{r}), \tag{V51}$$

base assigning to points $\mathbf{r}$ of the **base manifold**, M, values $\mathbf{F}(\mathbf{r})$ in the **target manifold**, L. We **manifold** assume that $M \subset \mathbb{R}^d$ and $L \subset \mathbb{R}^n$ are open subsets of d- and n-dimensional vector spaces, **target**
manifold

12   These fields are unrelated to the (number-)fields of mathematics defined in Section L1.3.

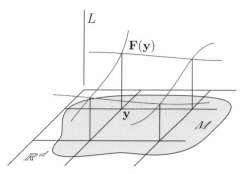

Fig. V7   Schematic illustration of the concept of a field as a map, **F**, from a base manifold M to a target manifold L.

respectively.[13] If M is parametrized by a system of generalized coordinates, the arguments $\mathbf{r} = \mathbf{r}(\mathbf{y})$ of the field can be parametrized by coordinate vectors, $\mathbf{y} = (y^1, \ldots, y^d)^T$. Likewise, a system of coordinates on L implies a coordinate representation for $\mathbf{F}$. Throughout this introductory section we will not rigorously discriminate between points $\mathbf{r}$ and their coordinate vectors $\mathbf{y}$. Likewise, we will identify $\mathbf{F} = (F^1, \ldots, F^n)^T$ with its coordinate representation. This leads to descriptions of the field as

$$\mathbf{F} : M \to L, \qquad \mathbf{y} = (y^1, \ldots, y^d)^T \mapsto \mathbf{F}(\mathbf{y}) = (F^1, \ldots, F^n)(\mathbf{y}), \qquad (V52)$$

schematically illustrated in Fig. V7. We finally note that, as in previous chapters, we reserve the symbol $\mathbf{x}$ for *Cartesian coordinates*. For example, a point in $(3+1)$-dimensional space-time has Cartesian representations as $\mathbf{x} = (x^0, x^1, x^2, x^3)^T$, where $x^0 = t$ parametrizes time, and $x^{1,2,3}$ are the three coordinates of a spatial vector. If $n = 1$, then $\mathbf{F} = F$ is a real-valued function, and fields of this type are called *real scalar fields*. For $L = \mathbb{C} \cong \mathbb{R}^2$ we have a *complex scalar field*. If $L = \mathbb{R}^{n>1}$ we speak of a *vector field*.[14] Finally, L may be a genuine subset of $\mathbb{R}^n$, such as the sphere $S^2 \subset \mathbb{R}^3$ mentioned above. The general concept of all these maps is illustrated in Fig. V7, where the shaded gray area represents the base manifold M and thick lines symbolize the image values $\mathbf{F}(\mathbf{y})$.

visualization of fields   Fields can be visualized in a variety of ways, as illustrated in Fig. V8 for three different examples. A scalar field in two-dimensional space is a "surface" floating over a plane (left panel) whose height relative to the plane is the field value. A field of two-dimensional vectors in two-dimensional space, $\mathbf{F}(\mathbf{r})$, can be represented by attaching the vectors $\mathbf{F}(\mathbf{r})$ to the base points $\mathbf{r}$. This leads to a swarm of arrows, as indicated in the center panel of the figure. ($\to$ V3.1.1-2) Such representations are often used to describe, e.g., distributions of current flow in the oceans. Likewise, three-dimensional vectors in three-dimensional space may be represented in terms of three-dimensional visualizations (right panel), depicting, e.g., the flow of a fluid in a vessel. However, the figure also illustrates how visual representations of three-dimensional vector fields tend to lack clarity.

In the rest of this chapter, we will learn how to describe fields in quantitative terms.

---

[13] Situations where this setting is too narrow are addressed in Section V4.1. In that section, the terminology "manifold" for the domain and the codomain of the field map will be explained.

[14] Formally, a complex *scalar* field, $L \cong \mathbb{R}^2$, may be identified with a two-dimensional real *vector* field. However, in the complex case, additional conditions discussed in Chapter C9 are generally imposed. This means that a complex scalar field has more structure than a generic two-dimensional real vector field, and that the two classes should be distinguished.

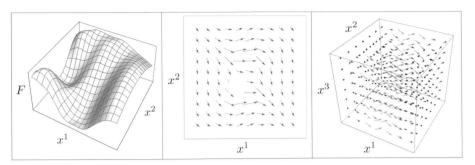

Fig. V8   Visualization of a scalar field in two dimensions, $n = 1, d = 2$ (left), a two-dimensional vector field in two dimensions, $n = 2, d = 2$ (center), and a three-dimensional vector field in three dimensions, $n = 3, d = 3$ (right).

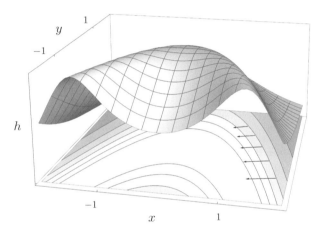

Fig. V9   Representation of the field $h(\mathbf{x})$ of Eq. (V53) as a surface floating over the xy-plane. The bottom of the graph contains a contour plot, showing contour lines along which the function remains constant. Dense contour lines indicate strong function changes. The arrows point in the direction in which the function increases most steeply, and their length indicates the magnitude of the increase.

V3.2 Scalar fields

Scalar fields are real-valued functions depending on more than one argument, $f = f(\mathbf{y}) = f(y^1, \ldots, y^d)$.[15] As an example, consider the two-dimensional field,

$$h : \mathbb{R}^2 \to \mathbb{R}, \qquad \mathbf{x} = (x, y)^{\mathrm{T}} \mapsto h(\mathbf{x}) = \frac{1}{(x^2 + y)^2 + c}, \qquad (V53)$$

where c is a positive constant. As mentioned above, such fields may be visualized as surfaces of local height, $h(\mathbf{x})$, over the two-dimensional plane (see Fig. V9). The surface analogy suggests an alternative graphical representation of the field. For a number

---

[15] Although an ordinary function $f(y)$ of one variable ($d = 1$) may also be considered as a field, the term is usually restricted to functions with base manifolds of dimension $d \geq 2$.

contour lines of constant values, $h_n \equiv a \cdot n$, $n \in \mathbb{Z}$, $a = $ const., **contour lines** along which the function remains at a constant value, $h(\mathbf{x}) = h_n$, are drawn in the (x, y)-plane. This yields the **contour plot** shown in the bottom of the figure. The contour plot indicates changes in the function through the density of contour lines. Widely spaced or densely spaced lines are indicative of shallow or steep function changes, respectively. Contour representations are frequently used in geographic maps where they indicate the altitude levels of the charted territory.

Total differential (Cartesian coordinates)

One frequently needs to describe how field values, $f(\mathbf{r})$, change under small variations of the argument, from $\mathbf{r}$ to $\mathbf{r} + \delta\mathbf{u}$, where $\mathbf{u}$ is a vector in the direction in which the change is monitored and δ an (infinitesimally) small variation parameter. This information is **total** provided by a quantity known as the **total dif-** **differential** **ferential** of the field. For a given point $\mathbf{r} \in M$, the total differential, $df_\mathbf{r}$, is a linear map which acts on vectors $\mathbf{u}$ to produce a number describing a rate of change of f in the direction of $\mathbf{u}$.[16] It is defined as

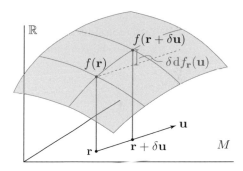

$$df_\mathbf{r} : \mathbb{R}^d \to \mathbb{R}, \qquad \mathbf{u} \mapsto df_\mathbf{r}(\mathbf{u}) \equiv \lim_{\delta \to 0} \frac{1}{\delta}\big(f(\mathbf{r} + \delta\mathbf{u}) - f(\mathbf{r})\big). \qquad (V54)$$

The construction shows that

> the action of the total differential df on a vector $\mathbf{u}$ equals the directional derivative of f along $\mathbf{u}$.

An argument similar to that proving Eq. (C37) shows that the differential $df_\mathbf{r}$ at fixed $\mathbf{r}$ is *linear* in its argument,[17] $(a, b \in \mathbb{R},\ \mathbf{u}, \mathbf{w} \in \mathbb{R}^d)$

$$df_\mathbf{r}(a\mathbf{u} + b\mathbf{w}) = a\, df_\mathbf{r}(\mathbf{u}) + b\, df_\mathbf{r}(\mathbf{w}). \qquad (V55)$$

For a fixed but small value of δ, Eq. (V54) implies $f(\mathbf{r} + \delta\mathbf{u}) - f(\mathbf{r}) \simeq \delta\, df_\mathbf{r}(\mathbf{u}) \overset{(V55)}{=} \delta df_\mathbf{r}(\delta\mathbf{u})$. This leads to a formula analogous to the one-dimensional Eq. (C2),

$$f(\mathbf{r} + \delta\mathbf{u}) \simeq f(\mathbf{r}) + df_\mathbf{r}(\delta\mathbf{u}), \qquad (V56)$$

---

[16] Notice that $\mathbf{u}$ need not necessarily lie in M. For example, if the field is defined on the disk of unit radius, $M = D \subset \mathbb{R}^2$, we may consider $\mathbf{u} = (100, 100)^\mathrm{T}$ and $\mathbf{r} = (0, 0)^\mathrm{T}$. For infinitesimal δ, $\mathbf{r} + \delta\mathbf{u}$ then lies in D although $\mathbf{u}$ does not.

[17] To see this explicitly, evaluate the differential on a sum of two vectors as
$$df(\mathbf{u} + \mathbf{w}) \equiv \lim_{\delta \to 0} \delta^{-1}\big(f(\mathbf{r} + \delta\mathbf{u} + \delta\mathbf{w}) - f(\mathbf{r} + \delta\mathbf{w}) + f(\mathbf{r} + \delta\mathbf{w}) - f(\mathbf{r})\big) = df(\mathbf{u}) + df(\mathbf{w}).$$
Here, we inserted $0 = -f(\mathbf{r} + \delta\mathbf{w}) + f(\mathbf{r} + \delta\mathbf{w})$ and in the second step noted that the offset $\delta\mathbf{w}$ in the difference $f(\mathbf{r} + \delta\mathbf{u} + \delta\mathbf{w}) - f(\mathbf{r} + \delta\mathbf{w})$ vanishes in the limit $\delta \to 0$.

showing how the differential linearly approximates f near $\mathbf{r}$. We finally note that it is customary to use *notation omitting the position argument*, and write $d_{\mathbf{r}}f \equiv df$ in cases where the identity of $\mathbf{r}$ is evident from the context.

The definition of the total differential does not make reference to a coordinate system. It remains somewhat abstract unless coordinates are applied to convert it into an expression containing concrete derivatives applied to the function f. In the following, we discuss the *total differential in Cartesian coordinates*, which is the form in which it often appears in the early physics curriculum. In the next subsection the generalization to arbitrary coordinates is presented.

total differential in Cartesian coordinates

In Cartesian coordinates, the argument, $\mathbf{r}$ can be identified with a component vector, $\mathbf{x} = (x^1, \ldots, x^d)^{\mathrm{T}}$, and $f(\mathbf{x})$ becomes a function of d real variables. The action of the total differential on a component-increment vector, $\mathbf{u}$, given by $d_{\mathbf{x}}(\mathbf{u}) = \delta^{-1}\big[f(\mathbf{x} + \delta\mathbf{u}) - f(\mathbf{x})\big]$, can then be computed by the chain rule, Eq. (C37), as

$$df_{\mathbf{x}}(\mathbf{u}) = \frac{\partial f(\mathbf{x})}{\partial x^k} u^k. \tag{V57}$$

In Cartesian coordinates the directional derivative thus assumes a particularly simple form. Specifically, its evaluation on the jth basis vector, $\mathbf{e}_j$, equals the partial derivative

$$df_{\mathbf{x}}(\mathbf{e}_j) = \frac{\partial f(\mathbf{x})}{\partial x^j}. \tag{V58}$$

Conversely, it is sometimes useful to consider the total differential of the ith coordinate function, x^i. This function maps a vector onto its ith component, $x^i(\mathbf{u}) = u^i$, and so the differential is obtained as $dx^i(\mathbf{u}) = u^i$, or $dx^i(\mathbf{e}_j) = \delta^i_j$ in the particular case of a basis vector. This representation may be substituted into (V57) to obtain a commonly used representation of the total differential, $df(\mathbf{u}) = \partial_k f \, dx^k(\mathbf{u})$. Holding for arbitrary arguments, $\mathbf{u}$, this formula implies an identity for the differentials themselves,

$$df = \frac{\partial f(\mathbf{x})}{\partial x^k} \, dx^k. \tag{V59}$$

Notice how the structure of this relation resembles that of the chain rule used to derive it.

EXERCISE Show that the total differential of the function defined in Eq. (V53) is given by

$$dh = -\frac{2(x^2 + y)}{((x^2 + y)^2 + c)^2}(2x dx + dy),$$

and that of the function $\phi(x, y) = \arctan(y/x)$ by $d\phi = \dfrac{x dy - y dx}{x^2 + y^2}$.

total differential in thermodynamics

INFO The total differential plays an important role in both mathematics and physics, where it is applied to describe the changes of functions in convenient and versatile ways. As an example from *thermodynamics* (where the use of total differentials is pervasive), let $p(V, T)$ be the pressure of a gas in a container of volume V at temperature T. We now ask for the change in pressure under small variations, δV and δT, of volume and temperature, respectively. To answer this question, we set up the total differential,

$$dp \stackrel{(V59)}{=} \frac{\partial p(V,T)}{\partial V} dV + \frac{\partial p(V,T)}{\partial T} dT. \tag{V60}$$

The corresponding change in pressure, given by $\delta p_{(V,T)} \equiv p(V + \delta V, T + \delta T) - p(V,T) \stackrel{(V56)}{\simeq} dp_{V,T}(\delta V, \delta T)$, can now conveniently be represented as[18]

$$\delta p_{(V,T)} = dp_{(V,T)}(\delta V, \delta T) = \frac{\partial p(V,T)}{\partial V} \delta V + \frac{\partial p(V,T)}{\partial T} \delta T.$$

For example, for an *ideal gas*, Clapeyron's formula states that $pV = nRT$, where n is the amount of substance (in units of moles) and R the Avogadro constant. This may be represented as $p(V,T) = nRT/V$, with $\partial_V p = -nR \frac{T}{V^2}$ and $\partial_T p = nR \frac{1}{V}$, so that

$$\delta p_{(V,T)} = dp_{(V,T)}(\delta V, \delta T) = nR \left(-\frac{T\delta V}{V^2} + \frac{\delta T}{V} \right). \tag{V61}$$

Regrettably, physics parlance tends to describe the differential itself (and not the arguments on which it acts) as a "small quantity". For example, one frequently finds formulations such as "let $dp = \partial_V p \, dV + \partial_T p \, dT$ be the variation of pressure, dp, under small variations of volume and temperature, dV and dT, respectively". From a mathematical perspective, this is nonsense – dp, dV, dT are linear maps which cannot be "small" – but this abuse of language is quite pervasive in physics. However, if it is kept in mind that the formulation actually refers to the action of the differential dp on a small argument, $(\delta V, \delta T)^T$, yielding $\delta p = \partial_V p \, \delta V + \partial_T p \, \delta T$, confusion can be avoided.

Total differential (general coordinates)

REMARK This subsection generalizes the concept of the total differential to arbitrary coordinates. In this form, it will be required in later sections when we discuss operations of vector calculus in curvilinear coordinates. Although the generalization is instructive and does not require significant formal effort, the section can be skipped at first reading if time is short.

The generalization of the total differential to arbitrary coordinates is easy and straightforward – provided a suitable basis is used. We consider a coordinate representation, $\mathbf{y} \mapsto \mathbf{r}(\mathbf{y})$, and ask for changes of the function $f(\mathbf{y}) \equiv f(\mathbf{r}(\mathbf{y}))$ under variations of the jth coordinate from y^j to $y^j + \delta$, or $\mathbf{y} \to \mathbf{y} + \delta \mathbf{e}_j$. On the one hand, $\delta^{-1}(f(\mathbf{y} + \delta \mathbf{e}_j) - f(\mathbf{y})) \simeq \partial_{y^j} f(\mathbf{y})$, by definition of the partial derivative (C34). On the other hand, the variation of the coordinate vector changes the argument from $\mathbf{r}(\mathbf{y})$ to $\mathbf{r}(\mathbf{y} + \delta \mathbf{e}_j) \simeq \mathbf{r}(\mathbf{y}) + \delta \mathbf{v}_j$, where $\mathbf{v}_j = \partial_{y^j} \mathbf{r}(\mathbf{y})$ is the jth vector of the coordinate basis. The variation of the argument thus leads to the difference $\delta^{-1}(f(\mathbf{r} + \delta \mathbf{v}_j) - f(\mathbf{r})) \simeq df(\mathbf{v}_j)$. Equating the two expressions, we obtain the identification

total differential in general coordinates

$$\boxed{df_{\mathbf{y}}(\mathbf{v}_j) = \partial_{y^j} f(\mathbf{y}).} \tag{V62}$$

This intuitive formula states that the rate of change of a function along the jth coordinate vector equals the partial derivative of the function's coordinate representation. In the

---

[18] In the last step, the total differentials dV and dT occurring in Eq. (V60) are applied to the vector $(\delta V, \delta T)^T$, yielding δV and δT, respectively.

following, the differential will appear frequently, and we often use shorthand notations such as $df_{\mathbf{r}(\mathbf{y})} \equiv df_{\mathbf{r}} \equiv df_{\mathbf{y}} \equiv df$, where the last representation leaves the dependence of the differential on the coordinates implicit. For a general vector expanded in the coordinate basis, $\mathbf{u} = \mathbf{v}_j u^j$, the linearity of the differential, $df(\mathbf{u}) = df(\mathbf{v}_j u^j) = df(\mathbf{v}_j) u^j$, and $u^j = dy^j(\mathbf{u})$, then implies the representation

$$df_{\mathbf{y}} = \frac{\partial f(\mathbf{y})}{\partial y^j} dy^j . \qquad\qquad (V63)$$

This equation generalizes the Cartesian formula (V59) to arbitrary coordinates. The fact that it looks the same is owed to the usage of the coordinate basis. If the local basis had been used instead, the presence of normalization factors, distinguishing local and coordinate basis vectors, would complicate the right-hand side of the formula.

EXAMPLE   The function $\phi = \arctan(y/x)$ considered in the exercise of the previous section is the angular variable of the polar coordinate system, Eq. (V15b). This means that in polar coordinates it becomes a coordinate function, with simple differential $d\phi$.

Gradient (Cartesian coordinates)

The local change of a scalar field, f, can be described by a quantity which contains the same information as the differential but is sometimes considered to be more "geometrical" (although this view can be disputed). This **gradient field**,

gradient field

$$\nabla f : M \to \mathbb{R}^d, \qquad \mathbf{r} \mapsto \nabla f_{\mathbf{r}}, \qquad\qquad (V64)$$

assigns to each $\mathbf{r} \in M$ a d-dimensional vector, $\nabla f_{\mathbf{r}}$. As with the differential, the $\mathbf{r}$-dependence is frequently left implicit and the shorthand notation ∇f used.

The gradient field is defined by the implicit condition that for each $\mathbf{u} \in \mathbb{R}^d$,

$$df_{\mathbf{r}}(\mathbf{u}) = \langle \nabla f_{\mathbf{r}}, \mathbf{u} \rangle , \qquad\qquad (V65)$$

where $\langle \ , \ \rangle$ is the standard inner product of $\mathbb{R}^d$: the scalar product of the gradient, ∇f, with any vector $\mathbf{u}$ must be equal to the differential df evaluated on that vector.

Geometric intuition for the structure of this vector field follows from the observation that $df(\mathbf{u})$ assumes its largest values along those directions for which $\mathbf{u}$ points in the direction of the largest variation of f. By definition (V65), $\langle \nabla f, \mathbf{u} \rangle$, too, will be largest for these directions of $\mathbf{u}$. This implies that ∇f is aligned with the directions of strongest increase:

The gradient field ∇f points in the direction of the strongest increase of f.

Temporarily denoting this direction by the unit vector $\mathbf{e}$, we may write $\nabla f = \mathbf{e} \|\nabla f\|$. To understand the meaning of the norm, $\|\nabla f\|$, of the gradient, we substitute $\mathbf{u} = \nabla f$ into the defining equation (V65), and obtain $\|\nabla f\| \, df(\mathbf{e}) \overset{(V55)}{=} df(\nabla f) \overset{(V65)}{=} \langle \nabla f, \nabla f \rangle = \|\nabla f\|^2$, where in the first equality the linearity of the differential was used. Division by $\|\nabla f\|$ leads to

$\|\nabla f\| = df(\mathbf{e})$. This shows that the norm of the gradient equals the value of the differential in the direction of maximal increase, $df(\mathbf{e})$, of the function:

> The norm of the gradient equals the differential $df(\mathbf{e})$ along a unit vector pointing in the direction of maximal increase.

gradient in Cartesian coordinates

To obtain a concrete representation we need coordinates, and as in the previous section we first consider the **gradient in Cartesian coordinates**, $\nabla f = \mathbf{e}_j \nabla f^j$. We know that the jth component of a vector in Cartesian coordinates is obtained as $u^j = u_j = \langle \mathbf{u}, \mathbf{e}_j \rangle$, where the notation reflects that for the standard scalar product, with metric $g_{ij} = \delta_{ij}$, the co- or contravariant positioning of indices is irrelevant. For the gradient, this identification yields $\nabla f^j = \nabla f_j = \langle \nabla f, \mathbf{e}_j \rangle \overset{(V65)}{=} df(\mathbf{e}_j) \overset{(V58)}{=} \partial_j f \equiv \partial^j f$. In the physics literature, this result is often expressed as

$$\nabla f = \mathbf{e}_j \nabla f^j = \begin{pmatrix} \partial^1 f \\ \vdots \\ \partial^d f \end{pmatrix}. \tag{V66}$$

In this representation, ∇f is a column vector containing the partial derivatives $\partial^j f = \partial_j f$ as components. A frequently met alternative interpretation considers ∇f as the action of a **gradient operator** (also called **nabla operator**), ∇, on the function f. The nabla operator is formally[19] defined as

nabla (∇) operator

$$\nabla \equiv \mathbf{e}_j \partial^j = \begin{pmatrix} \partial^1 \\ \vdots \\ \partial^d \end{pmatrix}. \tag{V67}$$

It is a d-component vector and at the same time a "differential operator" in the sense that it acts on the function to its right according to Eq. (V66), $f \mapsto \nabla f$. The denotations "nabla f" and "gradient f" are synonyms.

EXAMPLE As an example, consider the field $h(\mathbf{x})$ of Eq. (V53), shown in Fig. V9. Evaluating its partial derivatives, we find

$$\nabla h_{(x,y)} = \begin{pmatrix} \partial_x h(x,y) \\ \partial_y h(x,y) \end{pmatrix} = -\frac{2(x^2+y)}{[(x^2+y)^2+c]^2} \begin{pmatrix} 2x \\ 1 \end{pmatrix}. \tag{V68}$$

In the contour plot at the bottom of that figure, a few arrows representing this gradient field indicate how it stands orthogonal to the contour lines of the function (and thus points in a direction of maximal variation). For example, for any point on the line $x = 0$, the vector $(2x, 1)^T$ points in the y-direction. At $(x, y)^T = (0,0)^T$ the gradient vanishes, reflecting the absence of a direction of increase, i.e. this point is a local maximum. ($\rightarrow$ V3.2.1-4)

---

[19] One may make this interpretation rigorous and consider the nabla operator as an element of the linear space of differential operators acting on function spaces. However, this interpretation is of limited practical usefulness and we will not discuss it here.

Gradient (general coordinates)

A representation of the gradient, $\nabla f = \mathbf{v}_i \nabla f^i$, in the coordinate basis of a general coordinate system is obtained by a slight generalization of the procedure above. Insertion of the basis vector $\mathbf{v}_j$ as an argument into Eq. (V65) yields $df(\mathbf{v}_j) \stackrel{(V62)}{=} \partial_{y^j} f$ for the the left-hand side and $\langle \nabla f, \mathbf{v}_j \rangle = \langle \mathbf{v}_i \nabla f^i, \mathbf{v}_j \rangle \stackrel{(V21)}{=} \nabla f^i g_{ij}$ for the right-hand side. Applying the index lowering convention Eq. (L60), $\nabla f_j \equiv \nabla f^i g_{ij}$, leads to the identification

$$\boxed{(\nabla f_{\mathbf{r}(\mathbf{y})})_j = \partial_{y^j} f(\mathbf{y}).} \tag{V69}$$

We conclude that in the coordinate basis, the *covariant components of the gradient*, ∇f_j, are simply given by partial derivatives of f. The *contravariant components* are obtained by raising the index using Eq. (L63) and Eq. (L62) for the inverse metric tensor,

$$\nabla f^i = g^{ij} \nabla f_j = g^{ij} \partial_{y^j} f. \tag{V70}$$

EXAMPLE The covariant components of the *gradient in polar coordinates* are given by $\nabla f_\rho = \partial_\rho f$ and $\nabla f_\phi = \partial_\phi f$. From the metric, $g = \mathrm{diag}(1, \rho^2)$, the elements of the likewise diagonal inverse metric are obtained as $g^{\rho\rho} = 1$ and $g^{\phi\phi} = \rho^{-2}$. This yields $\nabla f^\rho = \partial_\rho f$ and $\nabla f^\phi = \rho^{-2}\partial_\phi f$, as well as the expansion

gradient in polar coordinates

$$\nabla f = \mathbf{v}_\rho \, \partial_\rho f + \mathbf{v}_\phi \frac{1}{\rho^2} \partial_\phi f. \tag{V71}$$

In *other bases* than the coordinate basis, different and generally more complicated expressions are obtained. For example, in systems with a diagonal metric tensor, $g = \mathrm{diag}(g_{11}, \ldots, g_{dd})$ (such as the standard systems polar, cylindrical, spherical), the unit normalization of the local basis vectors implies, $\mathbf{e}_j = \sqrt{g_{ii}}^{-1} \mathbf{v}_j$, and

$$\nabla f = \mathbf{v}_j \, g^{jj} \partial_{y^j} f = \mathbf{e}_j \frac{1}{\sqrt{g_{jj}}} \partial_{y^j} f. \tag{V72}$$

Although the right-most version of this formula is the one most frequently encountered in the physics literature, the presence of square-root normalization factors is an awkward feature of the local basis representation. From a geometric perspective, it is more natural to work in the un-normalized coordinate basis ("center representation"), or to avoid the usage of contravariant components, in which case the simple covariant formula Eq. (V69) suffices to describe the gradient.

Notice however, that in most applications the gradient appears as a vector in an inner product, as in $\langle \nabla f, \mathbf{u} \rangle = df(\mathbf{u})$, when the derivative of f in the direction of $\mathbf{u}$ is computed. In such cases, only the covariant components are required,

$$\langle \nabla f, \mathbf{u} \rangle = \nabla f_i u^i = (\partial_{y^i} f) u^i. \tag{V73}$$

INFO The gradient vector is often used to describe the *rate of change of a function along a curve*, $t \mapsto \mathbf{r}(t) = \mathbf{r}(\mathbf{y}(t))$, where in the second equality a parametrization in coordinates $\mathbf{y}$ is assumed. Application of the chain rule (C39) then yields

$$\frac{df(\mathbf{y}(t))}{dt} = \partial_{y^j} f(\mathbf{y}(t)) \frac{dy^j(t)}{dt} \overset{\text{(V69)}}{=} \nabla f_j \, \dot{y}^j = \nabla f_{\mathbf{r}(\mathbf{y}(t))} \cdot \dot{\mathbf{r}}(\mathbf{y}(t)). \tag{V74}$$

This formula states that the rate at which f changes along the curve is determined by its slope in the direction of the curve velocity.

gradient in spherical coordinates

EXERCISE Show that for the spherical coordinate system (r, θ, ϕ), with diagonal metric $g_{rr} = 1$, $g_{\theta\theta} = r^2$, $g_{\phi\phi} = r^2 \sin^2\theta$, the **gradient in the spherical coordinate basis** is given by

$$\nabla f^r = \partial_r f, \qquad \nabla f^\theta = \frac{1}{r^2} \partial_\theta f, \qquad \nabla f^\phi = \frac{1}{r^2 \sin^2\theta} \partial_\phi f. \tag{V75}$$

contour surfaces

INFO It is sometimes useful to characterize the geometry of the gradient vector in terms of **contour surfaces**. Contour surfaces generalize the contour lines describing two-dimensional functions, $f : \mathbb{R}^2 \to \mathbb{R}$, to functions in arbitrary dimensions, $f : \mathbb{R}^d \to \mathbb{R}$. A contour surface, $S_c \equiv \{\mathbf{r}' \in \mathbb{R}^d | f(\mathbf{r}') = c\}$, $c \in \mathbb{R}$ is defined as the set of all points $\mathbf{r}'$ for which f assumes the fixed value c. Every point $\mathbf{r}$ lies on such a contour surface, namely the surface $S_{f(\mathbf{r})}$ defined by all points on which f assumes the value $f(\mathbf{r})$.

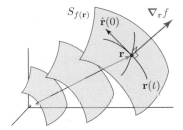

For $d=2$, these "surfaces" reduce to the contour lines illustrated in Fig. V9. In higher dimensions, the condition $f(\mathbf{r}') = f(\mathbf{r})$ is one real equation constraining a set of d variables, say the coordinates $\mathbf{y} = (y^1, \ldots, y^d)^\mathsf{T}$ of $\mathbf{r}'$. This constraint can be applied to express one of them, say $y^d = y^d(y^1, \ldots, y^{d-1})$, through $d-1$ free variables, $y^1, \ldots, y^{d-1}$, which may then be interpreted as coordinates parametrizing the surface. This shows that the contour surfaces are $(d-1)$-dimensional objects in d-dimensional space. (Exceptions to this rule occur at extremal points of the function f. For example, the contour surface corresponding to a global maximum, $f(\mathbf{r}_{\max}) = f_{\max}$, is just a single point, $S_{f_{\max}} = \{\mathbf{r}_{\max}\}$.)

The geometry of these surfaces is equivalently described by the statement that

> the gradient vector $\nabla_\mathbf{r} f$ stands perpendicular to the contour surface $S_{f(\mathbf{r})}$ through $\mathbf{r}$.

To see this, let $\mathbf{r}(t)$ be a curve in the contour surface $S_{f(\mathbf{r})}$ and running through $\mathbf{r}$ at $t = 0$. The function f then remains stationary along the curve, i.e. $\frac{d}{dt} f(\mathbf{r}(t)) = 0$. Application of Eq. (V74) at $t = 0$ thus yields $0 = \nabla f_{\mathbf{r}(t)} \cdot \dot{\mathbf{r}}(t)\big|_{t=0} = \nabla f_\mathbf{r} \cdot \dot{\mathbf{r}}(0)$. From this we conclude that the gradient vector is perpendicular to the tangent vectors of arbitrary curves in the surface. It is therefore perpendicular to the surface as such.

V3.3 Extrema of functions with constraints

REMARK This section introduces the method of Lagrange multipliers for finding extrema of functions in the presence of constraints. This technique is routinely required in the second year of the physics curriculum. First-year readers short of time may skip the section at first reading.

In applications, one often needs to find the extrema of functions, $f : \mathbb{R}^d \to \mathbb{R}, \mathbf{x} \mapsto f(\mathbf{x})$, subject to one or several constraints of the type $g(\mathbf{x}) = 0$. As an example, consider the linear function $f(x, y) = 10 + \frac{1}{2}x - y$, shown as the slanted plane in the figure. This function does not have an extremum. However, let us now ask for the maximal value of the function if a constraint, $g(x, y) \equiv x^2 + y^2 - 4 = 0$, is imposed. Geometrically, one now seeks its largest or smallest value on the intersection of the slanted plane with a cylinder of radius $2 = \sqrt{4}$. Such points evidently exist, demonstrating that the constrained function does have extrema.

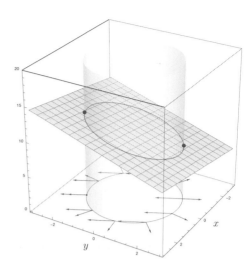

How does one solve problems of this type? An obvious strategy is to use the algebraic equation $g(\mathbf{x}) = 0$ to express one of the variables through the others, $x^1 = G(x^2, \ldots, x^d)$. Substitution into f then leads to a new function, $F(x^2, \ldots, x^d) = f(G(x^2, \ldots, x^d), x^2, \ldots, x^d)$, which may be varied without constraints. In the example above, one might solve for $y = \pm\sqrt{4 - x^2}$, to obtain $F(x) = 10 + \frac{1}{2}x \mp \sqrt{4 - x^2}$ as input for an unconstrained problem of one fewer variable. If more than one constraint equation is imposed, equally many variables can be eliminated in this way. In practice, however, this strategy is often avoided: first, the algebraic solution of constraint equations can be ambiguous, as in the example above. Second, the elimination approach often leads to algebraically cumbersome equations obscuring the symmetries of a problem.

In such cases, the *method of Lagrange multipliers* lends itself as a powerful alternative. We first formulate the solution strategy, and then discuss the rationale behind it.

▷ Define the function $F(\mathbf{x}, \lambda) = f(\mathbf{x}) - \lambda g(\mathbf{x})$, where the scalar parameter λ is called a **Lagrange multiplier**.

Lagrange multiplier

▷ Find the extremum of the function F by partial differentiation in all variables, including λ:

$$\partial_{x^j} F(\mathbf{x}, \lambda) = 0, \qquad \partial_\lambda F(\mathbf{x}, \lambda) = 0. \tag{V76}$$

▷ This leads to $d + 1$ algebraic equations which are solved for the $d + 1$ variables $(x^1, \ldots, x^d, \lambda)$. Each solution, $\mathbf{x} = (x^1, \ldots, x^d)$, specifies an extremal point of the constrained problem.

If there are more than one constraint equations, $g_1(\mathbf{x}) = \cdots = g_l(\mathbf{x}) = 0$, with $l < d$, the procedure is generalized to several Lagrange multipliers, $F = f - \sum_{i=1}^l \lambda_i g_i$.

For example, with $F(x, y) = 10 + \frac{1}{2}x - y - \lambda(x^2 + y^2 - 4)$, Eqs. (V76) yield

$$\tfrac{1}{2} - 2\lambda x = 0, \qquad -1 - 2\lambda y = 0, \qquad x^2 + y^2 - 4 = 0.$$

Elimination of λ from the first two equations yields $y = -2x$. In combination with the third equation we obtain $(x, y) = \pm \frac{2}{\sqrt{5}}(1, -2)$ for the extrema. Substitution of these values into the function shows that the configuration with positive x-value is a maximum, the other a minimum.

That the Lagrange multiplier method identifies the extrema of constrained functions is proven in lecture courses in mathematics and we here restrict ourselves to a heuristic discussion. First note that the condition $\partial_\lambda(f(\mathbf{x}) - \lambda g(\mathbf{x})) = 0$ simply enforces the constraint $g(\mathbf{x}) = 0$. The other partial derivatives require $\partial_j f = \lambda \partial_j g$. This is equivalent to the condition, $\nabla f = \lambda \nabla g$, that the gradients of the functions f and g be parallel. Above, we reasoned that the gradient ∇g is perpendicular to the contour surfaces of g, i.e. the surfaces in which g does not vary. Now assume that ∇f was not parallel to ∇g. In this case, the vector ∇f would have a non-vanishing component lying within g's contour surface. This would mean that the value of f could be changed – increased or decreased – without changing the constraint. By contrast, if $\nabla f \parallel \nabla g$, there is no freedom to alter f without violating the constraint, implying that an extremum has been reached. The idea is illustrated at the bottom of the figure where the arrows perpendicular to the circle indicate the gradient of the constraint, and the others that of the function. At the extrema they are parallel. The same idea underlies the general proof of the Lagrange multiplier method. ($\rightarrow$ V3.3.1-6)

V3.4 Gradient fields

REMARK In this section, we consider a class of fields known as gradient fields. These fields appear very early in the the physics curriculum, and in these early appearances Cartesian coordinates are generally used. However, much of the content of this section does not depend on the specific choice of coordinates and we therefore keep the discussion general.

As mentioned above, vector fields can be imagined as "swarms" of vectors in space. They often contain universal features which determine the structure of the swarm at large scales. For example, the field describing the steady flow of a fluid in a cylinder resembles a regular "stream" of vectors, see Fig. V10(a). The electric field created by charged particles contains regions from which vectors emanate, Fig. V10(b). The current field of a fluid with vorticity flow contains centers around which the vectors rotate, Fig. V10(c). Other fields may show all these features at once.

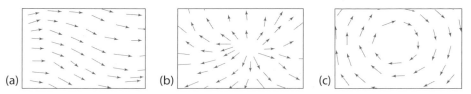

Fig. V10   (a) Source-free and rotation-free vector field. (b) Vector field containing a source. (c) Vector field with vorticity.

In the following three subsections, we introduce concepts to describe the universal features of vector fields. We begin with a discussion of the simplest type, fields describing regular flow without vorticity. In the subsequent two subsections, we then learn how to identify whether a vector field contains sources or centers of rotation, respectively.

Definition of gradient fields

Gradient fields are the simplest type of vector field. Their defining feature is that they can be represented as the gradient of a function. Given a smooth vector field, $\mathbf{u} : M \subset \mathbb{R}^d \to \mathbb{R}^d$, $\mathbf{r} \mapsto \mathbf{u}(\mathbf{r})$, a natural first question to ask is whether $\mathbf{u}$ is a **gradient field**: does there exist a smooth scalar field, $\varphi(\mathbf{r})$, such that $\mathbf{u}(\mathbf{r}) = \nabla \varphi_{\mathbf{r}}$? If so, the function φ is called the **potential** of the field.[20] Notice that for a gradient field, the target space $\mathbb{R}^d$ and the base manifold M must have the same dimension, e.g. a field of planar vectors, $\mathbf{u}(\mathbf{r}) \in \mathbb{R}^2$, defined in three-dimensional space, $\mathbf{r} \in \mathbb{R}^3$, cannot be a gradient field.

potential

If $\mathbf{u}$ is a gradient field, its (covariant) components can be written as $u_i = \partial_i \varphi = \partial_i \varphi$. Now compute the difference of "mixed" derivatives,

$$\partial_i u_j - \partial_j u_i = \partial_i \partial_j \varphi - \partial_j \partial_i \varphi = 0, \tag{V77}$$

where the last equality follows from the commutativity of partial derivatives acting on smooth functions, Eq. (C35). Equation (V77) represents a *necessary* condition for $\mathbf{u}$ to be a gradient field, meaning that if a field fails the condition, it cannot be a gradient field. However, the condition is not *sufficient*, i.e. there exist fields for which Eq. (V77) holds although they are not gradient fields. This point, and the extension of (V77) to a full criterion for ensuring that a vector field is a gradient field, will be discussed on p. 444 below.

Gradient fields play an important role in *physics*. For example, physical forces are described by vector fields, $\mathbf{F}$, where the vector $\mathbf{F}(\mathbf{r})$ represents the force acting at point $\mathbf{r}$. As discussed in Section V1.4, the work done by a force along a path γ is described by the line integral

$$W[\gamma] = \int_\gamma d\mathbf{r} \cdot \mathbf{F}.$$

Many forces occurring in physics have the property that no work is done if a body is moved along a *closed* path (see figure below), $\oint_\gamma d\mathbf{r} \cdot \mathbf{F} = 0$, where the standard symbol, $\oint$, is used to indicate a *line integral along a closed path*. This happens if the work done against the force along some portions of the path balances the work done by the force along others. A vector field whose line integral along an arbitrary closed path vanishes is called a **conservative vector field**. Gravitational forces, electrostatic forces and several others are examples of conservative force fields.

conservative vector field

An equivalent way of expressing the conservativeness of a force field is to say that its line integral along *any* path connecting two points, $\mathbf{r}'$ and $\mathbf{r}$, is independent of the choice of path. To see this, consider two different curves, γ_1 and γ_2, both running from $\mathbf{r}'$ to $\mathbf{r}$ (see the dashed lines in the figure). Let $-\gamma_2$ denote the curve γ_2 traversed in reverse order,

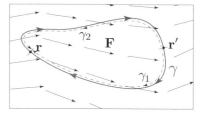

[20] In physics, it is customary to include a minus sign in the definition of the potential, writing $\mathbf{u} = -\nabla \varphi$.

from $\mathbf{r}$ to $\mathbf{r}'$. The line integral along $-\gamma_2$ is the negative[21] of that along γ_2, $W[-\gamma_2] = -W[\gamma_2]$. The curves γ_1 and $-\gamma_2$ may now be combined to a single closed curve, γ. The integral along γ is the sum of the integrals along the two segments, $W[\gamma] = W[\gamma_1] + W[-\gamma_2] = W[\gamma_1] - W[\gamma_2]$. If the field is conservative, then $W[\gamma] = 0$, which implies $W[\gamma_1] = W[\gamma_2]$. This shows that the value of the line integral is indeed independent of the choice of path.

criteria for gradient fields   The above construction shows that if the line integral of a vector field along any closed path vanishes, then its line integral along paths connecting two points are independent of the choice of the connecting paths. It is straightforward to show (do it!) that the converse is also true, so that we have the equivalence:

$$\forall\, \mathbf{r}', \mathbf{r} \in M : \quad \int_{\gamma_{\mathbf{r}'\to\mathbf{r}}} d\mathbf{r} \cdot \mathbf{u} \quad \text{is independent of } \gamma \quad \Longleftarrow\!\!\Longrightarrow \quad \oint_{\gamma} d\mathbf{r} \cdot \mathbf{u} = 0 . \tag{V78}$$

It turns out that these two conditions are equivalent to that of $\mathbf{u}$ being a gradient field:

$$\forall\, \mathbf{r}', \mathbf{r} \in M : \quad \int_{\gamma_{\mathbf{r}'\to\mathbf{r}}} d\mathbf{r} \cdot \mathbf{u} \quad \text{is independent of } \gamma \quad \Longleftrightarrow \quad \mathbf{u} \text{ is a gradient field.} \tag{V79}$$

This is a nontrivial statement whose two different directions, $\Leftarrow$ and $\Rightarrow$, need to be proven separately. First assume that $\mathbf{u}(\mathbf{r}) = \nabla\varphi_{\mathbf{r}}$ is a gradient field. Let $\gamma_{\mathbf{r}'\to\mathbf{r}}$ be a curve parametrized by $\mathbf{r} : (0,t) \to \mathbb{R}^d$, $s \mapsto \mathbf{r}(s)$, with $\mathbf{r}(0) = \mathbf{r}'$, $\mathbf{r}(t) = \mathbf{r}$. The line integral of the gradient field is then computed as

$$\int_{\gamma_{\mathbf{r}'\to\mathbf{r}}} d\mathbf{r} \cdot \mathbf{u} = \int_0^t ds\, \frac{d\mathbf{r}(s)}{ds} \cdot \nabla\varphi_{\mathbf{r}(s)} \overset{(V74)}{=} \int_0^t ds\, \frac{d\varphi(\mathbf{r}(s))}{ds}$$

$$= \varphi(\mathbf{r}(t)) - \varphi(\mathbf{r}(0)) = \varphi(\mathbf{r}) - \varphi(\mathbf{r}').$$

This expression depends on $\mathbf{r}'$ and $\mathbf{r}$, but not on the choice of the connecting path $\gamma_{\mathbf{r}'\to\mathbf{r}}$. Conversely, assume that the line integral between any two points $\mathbf{r}', \mathbf{r} \in M$ is independent of the connecting path $\gamma_{\mathbf{r}'\to\mathbf{r}}$. Pick a fixed point $\mathbf{r}'$ and define the function

$$\varphi(\mathbf{r}) = \int_{\gamma_{\mathbf{r}'\to\mathbf{r}}} d\mathbf{r} \cdot \mathbf{u} . \tag{V80}$$

This is a valid definition of a function of $\mathbf{r}$ because φ does not depend on the choice of $\gamma_{\mathbf{r}'\to\mathbf{r}}$. Using the same path parametrization as above, $\varphi(\mathbf{r}(t))$ may be represented as

$$\varphi(\mathbf{r}(t)) = \int_{\gamma_{\mathbf{r}'\to\mathbf{r}(t)}} d\mathbf{r} \cdot \mathbf{u} = \int_0^t ds\, \frac{d\mathbf{r}}{ds} \cdot \mathbf{u}(\mathbf{r}(s)) .$$

Differentiating this expression in t, and applying the chain rule (V74) to its left side, the fundamental theorem of calculus Eq. (C20) to the right side, we obtain

$$\dot{\varphi}(\mathbf{r}(t)) = \dot{\mathbf{r}}(t) \cdot \nabla\varphi_{\mathbf{r}(t)} = \dot{\mathbf{r}}(t) \cdot \mathbf{u}(\mathbf{r}(t)). \tag{V81}$$

---

[21] To verify this statement from the definition of the line integral, use the fact that if $\mathbf{r}(t)$, $t \in (0,1)$ is a parametrization of γ_2, then the reverse path, $-\gamma_2$, can be parametrized by $\mathbf{r}(1-t)$, $t \in (0,1)$.

This relation must hold regardless of the choices of $\dot{\mathbf{r}}(t)$ and $\mathbf{r}(t) \equiv \mathbf{r}$. This requires $\mathbf{u}(\mathbf{r}) = \nabla\varphi_{\mathbf{r}}$, showing that $\mathbf{u}$ is a gradient field. The construction also provides a constructive method for computing the potential, through (V80). Notice that the potential of a vector field is defined only up to a constant: since $\varphi(\mathbf{r}) + c$, with $c \in \mathbb{R}$, is a potential too. The freedom to add a constant reflects the arbitrariness of the starting point $\mathbf{r}'$ in the definition. Different starting points lead to potentials differing from φ only by a constant (why?).

EXAMPLE Consider the vector field $\mathbf{u} : \mathbb{R}^2 \to \mathbb{R}^2$, $\mathbf{r} \mapsto \mathbf{u}(\mathbf{r}) = (y, x)^{\mathrm{T}}$, where a Cartesian coordinate representation, $\mathbf{r}(\mathbf{x}) = \mathbf{x} = (x, y)^{\mathrm{T}}$, is implied. It satisfies $\partial_x u^y = \partial_y u^x$, and so might be a gradient field (recall that $u^i = u_i$ in Cartesian systems). To check for the existence of a potential function, consider the line integral along a straight line from the origin to $\mathbf{x}$, parametrized as $\mathbf{x}(t) = (tx, ty)^{\mathrm{T}}$. This defines $\varphi(\mathbf{x}) \stackrel{(\text{V80})}{=} \int_0^1 \mathrm{d}t\, \dot{\mathbf{x}}(t) \cdot \mathbf{u}(\mathbf{x}(t)) = \int_0^1 \mathrm{d}t\,(x \cdot (ty) + y \cdot (tx)) = \int_0^1 \mathrm{d}t\, 2xyt = xy$. Computing the gradient of this function, we indeed obtain $\nabla\varphi = \nabla(xy) = (y, x)^{\mathrm{T}} = \mathbf{u}$, confirming that $\mathbf{u}$ is a gradient field throughout its domain of definition. ($\to$ V3.4.1-2)

A topological criterion for gradient fields

Above, we argued that Eq. (V77) must necessarily hold if $\mathbf{u}$ is a gradient field. Whether or not this condition is *sufficient* depends on the "topology" (see the info section below) of the domain, M, on which $\mathbf{u}$ is defined.

A vector field $\mathbf{u} : M \to \mathbb{R}^d$, is a gradient field if its components obey the condition

$$\partial_i u_j - \partial_j u_i = 0, \qquad i = 1, \ldots, d, \qquad (\text{V82})$$

and its domain of definition, M, is simply connected.

connected set   Here, the attribute "simply connected" is defined as follows: a set M is called **connected** if any two of its elements can be connected by a continuous path in M. It is **simply connected** if it is connected and any closed path in M can be contracted to a trivial point-like path. For example, the set shown in the left panel of Fig. V11 is not connected because it contains pairs of points which cannot be connected by a path inside the set. The set shown in the center panel is connected but not simply connected: closed curves winding around the hole cannot be shrunk to point-like curves. The set shown in the right panel is connected and simply connected.

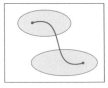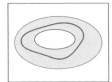

Fig. V11   Left panel: a disconnected subset of $\mathbb{R}^2$. Middle: a connected, but not simply connected, set. Right: a simply connected set.

topology INFO The connectedness of a subset is an example of a *topological criterion*. **Topology** is the discipline of mathematics addressing structures which do not change under continuous deformation. Topological features of a set remain invariant unless it is subjected to discontinuous operations such as tearing or gluing. (In topology, the terms "continuous", "tearing" and "gluing" are all defined in precise ways closely reflecting the daily-life interpretation of the words.) For example, a simply connected set remains simply connected unless a hole is drilled into it, the latter being a discontinuous operation. Topological features do not depend on geometric distances, angles or related geometric structures. However, they can often be described through integer-valued "topological invariants". For example, the number of holes contained in a surface defines an invariant.

Topological structures are of interest to physics because they are the most "universal" (detail-independent) features a system can have. For example, the vortex shown in Fig. V10(c) is an example of a topological structure in a two-dimensional vector field – it cannot be removed by any continuous operation and the number of times the field rotates around the center (once in the figure) is an example of a topological invariant. For their explanation of the role played by vortices in two-dimensional materials, J. M. Kosterlitz and D. J. Thouless were awarded the 2016 Nobel prize in physics (shared with D. Haldane).

The proof showing that simple connectedness and Eq. (V82) jointly define a criterion for gradient fields is beyond the scope of this text. However, the following counter-example demonstrates that the vanishing of the mixed derivative Eq. (V82) by itself is insufficient to establish gradientness.

EXAMPLE Consider the vector field

$$\mathbf{B}: M = \mathbb{R}^2\backslash\{(0,0)\} \mapsto \mathbb{R}^2, \qquad \begin{pmatrix} x \\ y \end{pmatrix} \mapsto \frac{1}{x^2+y^2}\begin{pmatrix} -y \\ x \end{pmatrix}, \qquad (V83)$$

defined in the "punctured plane", $\mathbb{R}^2\backslash\{(0,0)\}$. The origin $\{(0,0)\}$ needs to be excluded from the domain of definition, M, because the denominator in Eq. (V83) diverges at this point. Due to the exclusion of the origin, M is non-simply connected: paths winding around the origin cannot be contracted to a trivial point-like path.

example of a The vector field $\mathbf{B}$ rotates around the origin, as indicated in the fig-
non-gradient ure. In physics, the *magnetic fields* generated by currents display such
field winding behavior. For example, an infinitely long, straight wire carrying a uniform current creates a magnetic field which in the plane perpendicular to the wire behaves like our $\mathbf{B}$.

It is straightforward to verify that $\partial_x B_y - \partial_y B_x = 0$ everywhere in M, so Eq. (V77) holds. However, $\mathbf{B}$ is not a gradient field. To see why, compute the line integral along a circular path of radius R around the origin. Parametrizing this path as $\mathbf{x}(t) = R(\cos t, \sin t)^{\mathrm{T}}$, with $t\epsilon[0,1]$, we have $\dot{\mathbf{x}}(t) = R(-\sin t, \cos t)^{\mathrm{T}}$ and $\mathbf{B}(\mathbf{x}(t)) = (1/R)(-\sin t, \cos t)^{\mathrm{T}}$ along the path. Hence

$$\int_\gamma d\mathbf{r} \cdot \mathbf{B} = \int_0^{2\pi} dt\, \dot{\mathbf{r}}(t) \cdot \mathbf{B}(\mathbf{r}(t)) = \int_0^{2\pi} dt\, R\frac{1}{R}[\sin^2 t + \cos^2 t] = 2\pi. \qquad (V84)$$

We have thus identified a closed path along which $\mathbf{B}$ integrates to a non-vanishing result. According to Eqs. (V78) and (V79), this means that $\mathbf{B}$ cannot be a gradient field on M.

It is instructive to modify the above setup slightly so as to convert $\mathbf{B}$ into a proper gradient field. Consider a restricted domain of definition, $\tilde{M}$, obtained from M by removing the positive real axis (indicated by a wavy line in the figure) $\tilde{M} = \mathbb{R}^2\backslash(\mathbb{R}_0^+ \times \{0\})$. The restricted set does not contain "holes" and is simply connected, any path in $\tilde{M}$ can be contracted to a point. We also note that $\tilde{M}$ coincides with the image domain of a polar coordinate system, $\mathbf{y} = (\rho, \phi)^{\mathrm{T}}$, Eq. (V15). It is straightforward to verify (do it!) that in the polar coordinate basis the field has the simple representation $\mathbf{B} = \mathbf{v}_\phi \frac{1}{\rho^2}$, making its winding character visible. This equals the gradient, Eq. (V71), of the coordinate function $\varphi(\rho, \phi) \equiv \phi$, given by $\nabla\phi = (\mathbf{v}_\rho \partial_\rho + \mathbf{v}_\phi \frac{1}{\rho^2} \partial_\phi)\phi = \mathbf{v}_\phi \frac{1}{\rho^2}$. We conclude that on $\tilde{M}$ the field $\mathbf{B}$ is a gradient field, $\mathbf{B} = \nabla\varphi$.

Notice that the non-vanishing of the line integral along curves surrounding the origin is not in contradiction to this result. The closed curve considered above is not contained in $\tilde{M}$ because it intersects the "forbidden line". It can therefore not be applied for a gradient test criterion. What is contained in $\tilde{M}$ is the cut curve, $\tilde{\gamma}$, obtained by restriction of the parameter interval of t to the open interval $(0, 2\pi)$, so that the point $(0, 1)^\mathsf{T}$ is excluded. The integral of the continuous vector field $\mathbf{B}$ along the cut curve equals that along the closed curve, $\int_{\tilde{\gamma}} d\mathbf{r} \cdot \mathbf{B} = 2\pi R$. The fact that this does not vanish is not in conflict with the gradient property of $\mathbf{B}$ on $\tilde{M}$, since the cut curve $\tilde{\gamma}$ is not closed. ($\rightarrow$ V3.4.3-4)

gradient field criteria summarized

We conclude by summarizing the *criteria for a vector field being a gradient field*. A vector field, $\mathbf{u}$, is a gradient field

▷ if there exists a potential, φ, such that $\mathbf{u} = \nabla\varphi$, or

▷ if its line integral around any closed loop, γ, in its domain of definition vanishes, $\oint_\gamma d\mathbf{r} \cdot \mathbf{u} = 0$, or

▷ if the line integral between any two points in its domain of definition does not depend on the choice of the path connecting these two points, or

▷ if the criterion (V82) holds.

All four conditions are equivalent. Depending on the context one or another may be the most convenient to verify.

V3.5 Sources of vector fields

REMARK Requires Chapter C4. Readers who have not read that chapter in full may consult the first subsection of Section C4.4 for a definition of the integral of functions over two-dimensional surfaces embedded in three-dimensional space.

In the beginning of Section V3.4 we remarked that a vector field may contain "sources" from which vectors appear to emanate. In this section we introduce concepts to detect the presence of sources and to characterize them by quantitative measures.

Divergence (Cartesian coordinates)

divergence (in Cartesian coordinates)

Let $\mathbf{u} : M \subset \mathbb{R}^d \rightarrow \mathbb{R}^d$, $\mathbf{x} \mapsto \mathbf{u}(\mathbf{x})$ be a smooth d-component vector field in d-dimensional space, where a Cartesian parametrization of the argument domain, $\mathbf{u}(\mathbf{r}) = \mathbf{u}(\mathbf{r}(\mathbf{x})) \equiv \mathbf{u}(\mathbf{x})$, is assumed. The **divergence** of $\mathbf{u}$ is a scalar field defined as

$$\text{div }\mathbf{u} \equiv \boldsymbol{\nabla} \cdot \mathbf{u} : M \to \mathbb{R}, \quad \mathbf{x} \mapsto (\text{div }\mathbf{u})(\mathbf{x}) = (\boldsymbol{\nabla} \cdot \mathbf{u})(\mathbf{x}) \equiv \sum_{i=1}^{d} \partial_i u^i(\mathbf{x}), \qquad \text{(V85)}$$

where u^i are the Cartesian components of $\mathbf{u} = \mathbf{e}_i u^i$. The generalization of the divergence to curvilinear coordinate systems is discussed on p. 451 below.

EXAMPLE The divergence of the field $\mathbf{u}(x, y, z) = (xy, yz, zx)^\mathrm{T}$ is given by $\boldsymbol{\nabla} \cdot \mathbf{u} = \partial_x(xy) + \partial_y(yz) + \partial_z(zx) = y + z + x.$ ($\to$ V3.5.1-2)

The divergence of a vector field characterizes the source content of the field. To understand this statement, consider a vector field containing a point source, as in Fig. V10(b). The figure shows how the x-component u^x grows in the x-direction ($\partial_x u^x > 0$) and the y-component grows in the y-direction ($\partial_y u^y > 0$). According to Eq. (V85), the divergence div $\mathbf{u}$ will thus assume positive values and in this way indicate the presence of a source. It is a local probe in the sense that it describes $\mathbf{u}(\mathbf{x})$ through derivatives taken *at* $\mathbf{x}$. In the next section we characterize sources from a complementary global perspective, and then connect the global and local views to obtain a powerful unified picture.

Surface integrals of vector fields

REMARK While the definition (V85) defines the divergence of a vector field in general dimensions, this and the following sections focus on the three-dimensional case, $d = 3$, which is most important to early applications of the divergence in the physics curriculum.

Consider a two-dimensional surface, $S \subset M$, embedded in a three-dimensional domain M, and a vector field $\mathbf{u}$ defined on M. For a given point $\mathbf{r} \in S$ one may imagine $\mathbf{u}(\mathbf{r})$ as a vector specifying the direction and velocity of a liquid streaming through the surface. If S is closed and the total amount of outward flow is positive/negative then the vector field must contain sources/sinks inside S. The non-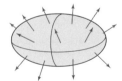vanishing of the flux through closed surfaces is a measure for the presence of sources, where a sink is interpreted as a "negative source".

To make this statement quantitative, we need to determine the flux of vector fields through surfaces via a suitably defined integral. We start by giving a surface an orientation. The *orientation of a surface* is a convention identifying one of its faces as the "outside" and the opposite face as the "inside". Not all surfaces admit such an assignment (see info section below), however most of practical relevance do. For example, the closed surface of a sphere has a natural out- and inside, and for a planar sheet one may define either side as the outside.

orientable
surface

Möbius strip

INFO A surface is called **orientable** if it permits the global assignment of an "inside" and an "outside" face. Planes, spheres, cylinders and most other surfaces are orientable in this sense. By contrast, the **Möbius strip** (see figure) is an example of a non-orientable surface. Inspection of the figure shows that it has just one face, not two. However it it not entirely obvious how to translate this observation into mathematical language. One of several equivalent ideas is as follows: a surface is non-orientable if it is possible to continuously deform a closed curve traversed in anti-clockwise direction into one which is traversed clockwise. The figure illustrates how this is achieved for the Möbius strip. By contrast, the direction of traversal of a closed curve on a sphere cannot be altered. For further discussion of orientability, see p. 490.

Consider a small surface element δS at $\mathbf{r}$, with outward-pointing unit normal vector, $\mathbf{n} = \mathbf{n}(\mathbf{r})$. The flux, $\delta\Phi$, of a vector field $\mathbf{u}$ through δS is defined as its component, $\mathbf{u}\cdot\mathbf{n}$, normal to δS, multiplied by the geometric area $|\delta S|$ of the element, $\delta\Phi \equiv |\delta S|\,(\mathbf{n}\cdot\mathbf{u})$. Concrete expressions for these quantities may be obtained by parametrizing the surface as $\mathbf{r} : U \subset \mathbb{R}^2 \to S, \mathbf{y} \mapsto \mathbf{r}(\mathbf{y})$, in terms of a two-component coordinate vector, $\mathbf{y} \equiv (y^1, y^2)^{\mathrm{T}}$. For a coordinate domain discretized via a rectangular grid with spacings δ^1 and δ^2, the induced surface element δS at $\mathbf{r}(\mathbf{y})$ is then spanned by the two vectors $\delta^i \partial_{y^i}\mathbf{r}(\mathbf{y}) \equiv \delta^i \mathbf{v}_i$ ($i = 1, 2$, no summation). It has geometric area $|\delta S| = \delta^1\delta^2 \|\mathbf{v}_1 \times \mathbf{v}_2\|$, and $\mathbf{n} \equiv \frac{\mathbf{v}_1 \times \mathbf{v}_2}{\|\mathbf{v}_1 \times \mathbf{v}_2\|}$ is its normal unit vector. We thus obtain the

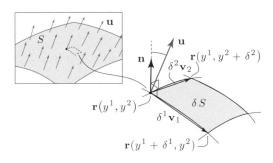

flux   **flux through the surface element as**[22]

$$\delta\Phi = |\delta S|\,(\mathbf{n}\cdot\mathbf{u}) = \delta^1\delta^2(\mathbf{v}_1 \times \mathbf{v}_2)\cdot\mathbf{u}. \tag{V86}$$

Note that $\delta\Phi = \delta\Phi(\mathbf{r})$ depends on the point $\mathbf{r}$ through the choice of the corner point used in its construction. There is arbitrariness in this choice: a different corner, or any other point in δS, might have been picked as a reference point. However, all these different choices, $\mathbf{r}'$, are close to each other in the sense that $\|\mathbf{r} - \mathbf{r}'\| = \mathcal{O}(\delta)$ is small in the coordinate increments $\delta^{1,2}$. Since all $\mathbf{r}$-dependent functions entering the construction are smooth, differences $\delta\Phi(\mathbf{r}) - \delta\Phi(\mathbf{r}')$ between fluxes obtained using different reference points within δS are therefore small by factors of δ compared to the fluxes

local approx-
imation

$\delta\Phi(\mathbf{r}), \delta\Phi(\mathbf{r}')$ themselves.[23] **Local approximations**, using the arbitrariness of readout coordinates within small geometric structures, are used frequently throughout this are text.

---

oriented
surface
element

[22] In the physics literature $\delta\mathbf{S} \equiv |\delta S|\,\mathbf{n} = \delta^1\delta^2(\mathbf{v}_1 \times \mathbf{v}_2)$ is often called an **oriented surface element**. The denotation emphasizes that $\delta\mathbf{S}$ contains information on both the geometric area and the orientation of the surface element.

[23] While $\delta\Phi(\mathbf{r}) \sim \mathcal{O}(\delta^2)$ is quadratic in δ, the differences $\delta\Phi(\mathbf{r}) - \delta\Phi(\mathbf{r}') \sim \mathcal{O}(\delta^3)$ are cubic.

flux of vector
field
surface
integral
The **flux of u through the surface**, Φ_S, is obtained as the Riemann sum over all surface elements. This defines the *vector field surface integral*, or just **surface integral** or **flux integral**, as

$$\Phi_S \equiv \int_S d\mathbf{S} \cdot \mathbf{u} \equiv \int_U dy^1 dy^2 \left(\partial_{y^1} \mathbf{r} \times \partial_{y^2} \mathbf{r} \right)\big|_{\mathbf{r}=\mathbf{r(y)}} \cdot \mathbf{u(r(y))}. \qquad (V87)$$

$\int_S d\mathbf{S} \cdot \mathbf{u}$ is a formal symbol, whose concrete meaning is given by the integral on the right.

Gauss's theorem

The discussion above introduced two criteria for the presence of sources in a vector field: a non-vanishing flux through a closed surface, and a non-vanishing divergence at the sources of a vector field, respectively. There must be a connection between them.

Johann Carl Friedrich Gauss 1777–1855
German astronomer, physicist and mathematician. Gauss made breakthrough contributions to a wide spectrum of mathematical disciplines. He worked on non-Euclidean geometry, algebra, the theory of special functions, and may be regarded the founding father of modern statistics.
Gauss was one of the last "universal scholars". His interests extended beyond the boundaries of physics and mathematics into the realms of geography, literature, cartography and other fields of science.

To establish this relation we consider an infinitesimal, box-shaped volume element, δV, inside the domain of definition of a vector field, $\mathbf{u}$, see Fig. V12. Let us calculate the outward flux of $\mathbf{u}$ over the surface, δS, of the box. Choosing coordinates as indicated in the figure, $\mathbf{e}_z$ is a unit vector normal to the top surface of the box and its opposite, $-\mathbf{e}_z$, is normal to the bottom surface. Since the variations of the vector field over the extensions of the box are very weak, the individual integrals over faces are accurately described by the local approximation, Eq. (V86). For example, the sum of the integrals over the infinitesimal top and bottom faces is given by

$$\delta\Phi_{\text{top}} + \delta\Phi_{\text{bot}} \simeq \delta^x \delta^y \left[\mathbf{e}_z \cdot \mathbf{u}(x, y, z + \delta^z) - \mathbf{e}_z \cdot \mathbf{u}(x, y, z) \right] \simeq \delta^x \delta^y \delta^z \partial_z u^z(x, y, z). \quad (V88)$$

Here, the differences in the values of the z-coordinate, although of $\mathcal{O}(\delta^z)$, do matter because a difference, $\mathbf{u}(\ldots, z+\delta^z) - \mathbf{u}(\ldots, z)$, vanishing in the limit $\delta^z \to 0$ is considered. However, after this difference has been evaluated (by Eq. (C38)) to obtain a result of $\mathcal{O}(\delta^z)$, the

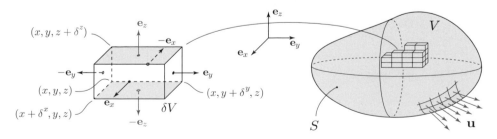

Fig. V12   On the derivation of Gauss's theorem.

choice of the coordinates x, y within the box faces becomes irrelevant, due to the local approximation principle discussed above.

Adding analogous contributions from the other two pairs of surfaces one obtains

$$\int_{\delta S} d\mathbf{S} \cdot \mathbf{u} \simeq \delta^x \delta^y \delta^z (\partial_x u^x + \partial_y u^y + \partial_z u^z) \overset{(V85)}{=} \delta^x \delta^y \delta^z (\boldsymbol{\nabla} \cdot \mathbf{u}) \simeq \int_{\delta V} dV \, \boldsymbol{\nabla} \cdot \mathbf{u}. \qquad \text{(V89)}$$

For the second step we noted that the contributions of the three pairs of surfaces combine to give the divergence of $\mathbf{u}$, and for the third that the near-constancy of the vector field within the box δV makes the term $\delta^x \delta^y \delta^z (\boldsymbol{\nabla} \cdot \mathbf{u})$ approximately equal to the volume integral of the divergence over the box. The result (V89) confirms that the flux of a vector field through a closed surface is caused by a non-vanishing divergence inside the surface.

Although this correspondence has been established only for the case of a small box, a simple argument shows that it carries over to arbitrarily shaped volumes, as in Fig. V12. Imagine a volume V filled up by a large number of infinitesimal boxes, δV_ℓ. The volume integral of $\boldsymbol{\nabla} \cdot \mathbf{u}$ over the full volume V can be expressed as the sum of box volume integrals, and to each of these Eq. (V89) may be applied,

$$\int_V dV \, \boldsymbol{\nabla} \cdot \mathbf{u} \simeq \sum_\ell \int_{\delta V_\ell} dV \, (\boldsymbol{\nabla} \cdot \mathbf{u}) = \sum_\ell \int_{\delta S_\ell} d\mathbf{S} \cdot \mathbf{u} \simeq \int_S d\mathbf{S} \cdot \mathbf{u}. \qquad \text{(V90)}$$

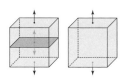

In the last step we observed that the sum of the flux integrals over the surfaces of the boxes approximately equals the flux integral over the outer hull or surface, S, of the full volume. To understand why, consider the sum of the flux integrals over just two adjacent boxes. The contributions to the integral from the touching faces cancel,[24] and the sum of the integrals equals the integral over the surface of the combined volume. By extension, the integral over a stack of boxes equals the integral over its outer hull, and this fact was used in Eq. (V90). In the limit of an infinitely fine decomposition, the approximate equalities become exact.

Summarizing, we have found that the volume integral of a divergence equals the flux integral over the volume's outer hull,

$$\boxed{\int_V dV \, \boldsymbol{\nabla} \cdot \mathbf{u} = \int_S d\mathbf{S} \cdot \mathbf{u},} \qquad \text{(V91)}$$

Gauss's theorem   a result known as **Gauss's theorem**. ($\rightarrow$ V3.5.3-8) The theorem states that a non-vanishing flux integral over a closed surface reflects a non-vanishing vector field divergence inside the surface. This correspondence is of importance in a large number of physical contexts.

INFO   The sources of vector fields describing physical quantities generally have a clearly defined **Gauss's law**   physical meaning. For example, **Gauss's law** of electromagnetism[25] states that

$$\boldsymbol{\nabla} \cdot \mathbf{E} = 4\pi \rho, \qquad \text{(V92)}$$

[24] The touching faces have equal area and normal vectors in opposite directions. The surface integrals are thus of equal magnitude but opposite sign.

[25] Although Gauss's law of electromagnetism assumes a form similar to Gauss's theorem of vector analysis, they are distinct results relying on different reasoning. That Gauss obtained breakthrough contributions in both physics and mathematics testifies to the extent of his scientific horizon.

where $\mathbf{E}$ is the electric field, and ρ the charge density, i.e. the amount of charge per unit volume, and we used CGS units (cf. footnote 23, Section C5.5). Integrating Eq. (V92) over a volume V bounded by a surface S and using Gauss's theorem (V91), one obtains an equivalent formulation of Gauss's law,

$$\int_S \mathrm{d}\mathbf{S} \cdot \mathbf{E} = 4\pi Q, \qquad (V93)$$

where $Q = \int_V \mathrm{d}V \rho$ is the charge contained in the volume V. Equations (V92) and (V93) are called the differential and the integral formulations of Gauss's law, respectively. Both state that electric fields are created by electric charge. While Eq. (V93) may be the more intuitive formulation, the differential representation Eq. (V92) does not make reference to a specific volume, which makes it a more versatile tool in applications.

electrostatic potential　　Another statement of electrostatics is that the electric field, $\mathbf{E}$, generated by a static charge distribution is a gradient field, $\mathbf{E} = -\nabla \varphi$, where φ is the **electrostatic potential**. Gauss's law then assumes the form $\nabla \cdot \nabla \varphi = -4\pi \rho$, known as the **Poisson equation**.

Divergence in general coordinates

Equation (V85) describes the divergence of a vector field in Cartesian coordinates. At the same time, the presence of sources is a universal feature not depending on coordinates. The derivation of Gauss's theorem contains the clue to a coordinate-invariant definition of the local divergence. To this end, let δV be an infinitesimal test volume of unspecified shape, with outer surface δS and geometric volume $|\delta V|$. Then, $\int_{\delta S} \mathrm{d}\mathbf{S} \cdot \mathbf{u} \overset{(V89)}{=} \int_{\delta V} \mathrm{d}V \operatorname{div} \mathbf{u} \simeq |\delta V| \operatorname{div} \mathbf{u}$, so that we may define the divergence as the outward flux per unit volume,

$$\boxed{\operatorname{div} \mathbf{u} \equiv \lim_{|\delta V| \to 0} \frac{1}{|\delta V|} \int_{\delta S} \mathrm{d}\mathbf{S} \cdot \mathbf{u}.} \qquad (V94)$$

coordinate-invariant definition of divergence　　This definition may now be applied to derive a formula for the *divergence in general coordinates*. Assume the volume of interest, V, to be described by a coordinate representation $\mathbf{r}(\mathbf{y})$. An infinitesimal cuboid in U, with corner points $\mathbf{y}$ and $\mathbf{y} + \mathbf{e}_i \delta^i$ (with $i = 1, 2, 3$, no summation), then maps onto a distorted cuboid, δV, as shown in the figure. The edges of δV are approximately defined by the vectors $\delta^i \partial_{y^i} \mathbf{r}(\mathbf{y}) = \delta^i \mathbf{v}_i(\mathbf{y})$, evaluated at the appropriate corners.

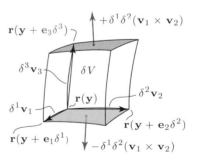

Our goal is to compute the outward flux $\int_{\delta S} \mathrm{d}\mathbf{S} \cdot \mathbf{u}$ through the surface of δV. In analogy to Eq. (V88), the top and bottom faces yield a contribution

$$\delta\Phi_{\text{top}} + \delta\Phi_{\text{bot}} = \left(|\delta S|\, \mathbf{n} \cdot \mathbf{u}\right)_{\text{top}} + \left(|\delta S|\, \mathbf{n} \cdot \mathbf{u}\right)_{\text{bottom}},$$

where the Cartesian expressions for $\delta^x \delta^y \mathbf{e}_z \cdot \mathbf{u}$ of (V88) need to be replaced by the triple products, $\delta^1 \delta^2 (\mathbf{v}_1 \times \mathbf{v}_2) \cdot \mathbf{u}$, of Eq. (V86). This leads to

$$\delta^1 \delta^2 \Big[((\mathbf{v}_1 \times \mathbf{v}_2) \cdot \mathbf{u})(\mathbf{y} + \mathbf{e}_3 \delta^3) - ((\mathbf{v}_1 \times \mathbf{v}_2) \cdot \mathbf{u})(\mathbf{y}) \Big] \simeq \delta^1 \delta^2 \delta^3 \partial_{y^3} \Big[((\mathbf{v}_1 \times \mathbf{v}_2) \cdot \mathbf{u})(\mathbf{y}) \Big],$$

where $((\mathbf{v}_1 \times \mathbf{v}_2) \cdot \mathbf{u})(\mathbf{y})$ is a shorthand for $(\mathbf{v}_1(\mathbf{y}) \times \mathbf{v}_2(\mathbf{y})) \cdot \mathbf{u}(\mathbf{y})$. The right-hand side can be simplified by expanding $\mathbf{u} = \mathbf{v}_i u^i$ in the coordinate basis. Then only the three-component survives in the triple product, yielding[26] $(\mathbf{v}_1 \times \mathbf{v}_2) \cdot \mathbf{u} = (\mathbf{v}_1 \times \mathbf{v}_2) \cdot \mathbf{v}_3 \, u^3 = \sqrt{g} \, u^3$, where in the last step we used Eq. (L181) to express the triple product of coordinate basis vectors through the determinant, $g \equiv \det(g)$, of the metric tensor.[27] The contributions from the top and bottom faces thus combine to $\delta^1 \delta^2 \delta^3 \partial_{y^3} (\sqrt{g} u^3)$.

Adding the contributions from the remaining pairs of surfaces, the total flux is obtained as $\int_{\delta S} d\mathbf{S} \cdot \mathbf{u} = \delta^1 \delta^2 \delta^3 \sum_i \partial_{y^i} (\sqrt{g} \, u^i)$. Since the volume of the distorted cuboid is given by $|\delta V| \simeq \delta^1 \delta^2 \delta^3 |(\mathbf{v}_1 \times \mathbf{v}_2) \cdot \mathbf{v}_3| = \delta^1 \delta^2 \delta^3 \sqrt{g}$, the flux can be expressed as $\frac{|\delta V|}{\sqrt{g}} \sum_i \partial_{y^i} (\sqrt{g} \, u^i)$. Division by $|\delta V|$ and comparison with Eq. (V94) finally lead to[28]

$$\operatorname{div} \mathbf{u} = \frac{1}{\sqrt{g}} \sum_i \partial_{y^i} (\sqrt{g} \, u^i) \tag{V95}$$

in general coordinates

for the **divergence in a general coordinate basis**. For example, in **spherical coordinates** where Eq. (C69) yields $\sqrt{g} = r^2 \sin\theta$, one obtains[29]

$$\operatorname{div} \mathbf{u} = \frac{1}{r^2} \partial_r (r^2 \, u^r) + \frac{1}{\sin\theta} \partial_\theta (\sin\theta \, u^\theta) + \partial_\phi u^\phi. \tag{V96}$$

EXAMPLE Consider a radially symmetric charge distribution described by a density $\rho(r)$. Represented in the spherical coordinate basis, the electric field generated by it will have the radial form $\mathbf{E} = \mathbf{v}_r E^r(r)$, where the absence of components in the θ and ϕ directions reflects the rotational symmetry of the distribution. For the same reason, the strength of the field, $E^r(r)$, depends on only the radial coordinate. A formula for the field strength can be obtained by application of Gauss's law, Eq. (V93), to a ball, B, of radius r centered around the charge. The charge contained in the ball is obtained as

$$Q(r) = \int_B dV \, \rho = 4\pi \int_0^r ds \, s^2 \rho(s).$$

The left-hand side of Gauss's law contains an integral over the surface, ∂B, of the ball, which is a sphere of radius r. Computing this integral by Eq. (V87), with $\partial_\theta \mathbf{r} = \mathbf{v}_\theta$ and $\partial_\phi \mathbf{r} = \mathbf{v}_\phi$, we obtain

$$\int_{\partial B} d\mathbf{S} \cdot \mathbf{E} = \int_0^\pi d\theta \int_0^{2\pi} d\phi \, (\mathbf{v}_\theta \times \mathbf{v}_\phi) \cdot \mathbf{v}_r E^r(r) = \int_0^\pi d\theta \int_0^{2\pi} d\phi \, r^2 \sin\theta \, E^r(r) = 4\pi r^2 E^r(r),$$

where $(\mathbf{v}_\theta \times \mathbf{v}_\phi) \cdot \mathbf{v}_r \overset{\text{(L181)}}{=} \sqrt{g} \overset{\text{(C69)}}{=} r^2 \sin\theta$ was used. We insert these two results into Eq. (V93) to obtain

$$E^r(r) = \frac{Q(r)}{r^2} = \frac{4\pi}{r^2} \int_0^r ds \, s^2 \rho(s). \tag{V97}$$

---

[26] We assume a right-handed coordinate system, so that the triple product $(\mathbf{v}_1 \times \mathbf{v}_2) \cdot \mathbf{v}_3$ is positive.

[27] We here assume $g > 0$. The more general case of indefinite metrics is addressed in Chapter V6.

[28] Although we derived the result under the assumption of positive orientation, it is of general validity. For example, a transformation $y^1 \to -y^1$ changing the orientation via the flip of just one coordinate leads to sign changes in both u^i and ∂_{y^i} canceling each other.

[29] For the *local* basis of spherical coordinates, the corresponding formula is given in Eq. (V122).

Coulomb field
If r lies outside the support of the charge distribution (so that $\rho(s) = 0$ for $s \geq r$), the charge $Q(r)$ within the ball equals the total charge of the distribution, Q, so that $E^r(r) = Q/r^2$. This is the **electric Coulomb field** generated by a charge Q.

Finally, one may use Eq. (V96) to compute the divergence of this field, $\text{div}(\mathbf{E}) = \frac{1}{r^2}\partial_r(r^2 E^r) = \frac{1}{r^2}4\pi\partial_r\int_0^r ds\, s^2\rho(s) = 4\pi\rho(r)$, and confirm that it indeed satisfies the differential form of Gauss's law, Eq. (V92).

Laplace operator

In applications, one is often interested in the source content of gradient fields, $\nabla \cdot \nabla\varphi$. The combination of two differential operations defines the second-order differential operator

$$\Delta = \nabla \cdot \nabla, \qquad \text{(V98)}$$

> **Pierre Simon Laplace**
> **1749–1827**
> French mathematician, physicist and astronomer. Important contributions include a five-volume work on celestial mechanics, the development of Bayesian statistics (of paramount importance to current-day statistical analysis in all sciences), the formulation of the Laplace equation and the Laplace transform, and many others. Laplace is remembered as one of the greatest scientists of all time, sometimes referred to as the "French Newton".

Laplace operator
known as the **Laplace operator** or Laplacian. In Cartesian coordinates it acts as $\Delta f = \sum_i \partial_i(\partial_i f)$, such that $\Delta = \sum_i \partial_i^2$. The Laplacian governs many important equations in physics, including (f, ρ are functions depending on space and/or time, v is a velocity)

$$\Delta f = 0 \qquad \textbf{Laplace equation,}$$
$$\Delta f = \rho \qquad \textbf{Poisson equation,}$$
$$v^2 \Delta f - \partial_t^2 f = 0 \qquad \textbf{wave equation.}$$

Laplace operator in general coordinates
It is also a building block of the Schrödinger equation of quantum mechanics. An expression for the **Laplace operator in general coordinates** is obtained by applying the covariant divergence operation, Eq. (V95), to the gradient vector,

$$\Delta f = \text{div}(\nabla f) = \frac{1}{\sqrt{g}}\sum_i \partial_{y^i}(\sqrt{g}\,\nabla f^i).$$

Using Eq. (V70), we obtain the generalized representation of the Laplace operator:

$$\Delta f = \frac{1}{\sqrt{g}}\sum_{ij} \partial_{y^i}(\sqrt{g}\,g^{ij}\partial_{y^j}f). \qquad \text{(V99)}$$

For example, the **Laplace operator in spherical coordinates** is given by

$$\Delta f = \frac{1}{r^2} \partial_r \left(r^2 \partial_r f \right) + \frac{1}{r^2 \sin \theta} \partial_\theta \left(\sin \theta \, \partial_\theta f \right) + \frac{1}{r^2 \sin^2 \theta} \partial_\phi^2 f, \qquad \text{(V100)}$$

and this formula is heavily used in electrodynamics and quantum mechanics.

V3.6 Circulation of vector fields

Centers of circulation as in Fig. V10(c) can be detected by methods similar to those introduced in the previous section. The defining characteristic of a vector field $\mathbf{u}$ with circulation is that there exist closed curves around which the line integrals $\oint_\gamma d\mathbf{r} \cdot \mathbf{u}$ assume non-vanishing values. These line integrals play a role similar to that of the surface integrals in the previous section. Their non-vanishing is equivalent to the presence of a differential quantity, the "curl" of a vector field.

Curl (Cartesian coordinates)

curl (in Cartesian coordinates)

Let $\mathbf{u} : M \subset \mathbb{R}^3 \to \mathbb{R}^3, \mathbf{x} \mapsto \mathbf{u}(\mathbf{x})$ be a three-dimensional vector field in three-dimensional space, where a Cartesian parametrization of the argument domain, $\mathbf{u}(\mathbf{r}) = \mathbf{u}(\mathbf{r}(\mathbf{x})) \equiv \mathbf{u}(\mathbf{x})$, is assumed. Its **curl**[30] is a vector field, defined as

$$\boxed{\text{curl } \mathbf{u} \equiv \nabla \times \mathbf{u} : M \to \mathbb{R}^3, \quad \mathbf{x} \mapsto (\text{curl } \mathbf{u})(\mathbf{x}) = (\nabla \times \mathbf{u})(\mathbf{x}) \equiv \mathbf{e}_i \epsilon^{ijk} \partial_j u^k \big|_{\mathbf{x}}.} \qquad \text{(V101)}$$

Notice the erratic appearance of two upstairs k-indices on the right-hand side. As usual this signals that elements of the metric tensor are required to promote the formula to one valid in arbitrary coordinates. The extension is discussed on p. 457 below.

A few comments: the symbol $\nabla \times \mathbf{u}$ can be interpreted as the cross product of the vector-differential operator $\nabla = (\partial_1, \partial_2, \partial_3)^\mathsf{T}$ with the argument vector field, $\mathbf{u}$, ($\to$ V3.6.1-2)

$$\nabla \times \mathbf{u} = \begin{pmatrix} \partial_2 u^3 - \partial_3 u^2 \\ \partial_3 u^1 - \partial_1 u^3 \\ \partial_1 u^2 - \partial_2 u^1 \end{pmatrix}. \qquad \text{(V102)}$$

To get the gist of this expression, consider the 3-component, $(\nabla \times \mathbf{u})^3 = \partial_1 u^2 - \partial_2 u^1$. A positive curl means that u^2 has a tendency to grow in the 1-direction, and u^1 to diminish in the 2-direction. This is what one expects from a vector field with circulation in the 1–2-plane (see the figure, where vertical and horizontal bars indicate how u^1 and u^2 change with position). As in the previous section, we have identified two criteria for the presence of circulation, a global one (the non-vanishing of closed-loop

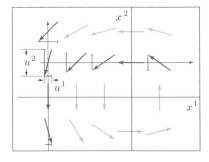

---

[30] The notations rot $\mathbf{u} \equiv$ curl $\mathbf{u}$ are equivalent and equally popular in the literature.

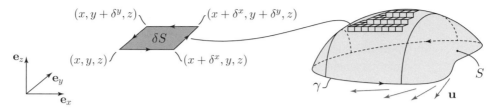

Fig. V13   On the derivation of Stokes's theorem. For a discussion, see the text.

line integrals) and a local one (the non-vanishing of the curl). It remains to establish the connection between the two.

Stokes's theorem

Consider a small rectangle, δS, in the xy-plane of a three-dimensional coordinate system, as in Fig. V13. We aim to compute the line integral, $\oint_{\gamma_{\delta S}} d\mathbf{r} \cdot \mathbf{u}$, of a smooth vector field $\mathbf{u}$ around its rectangular boundary, $\gamma_{\delta S}$. The integral is done in the mathematically positive orientation,

Sir George Gabriel Stokes 1819–1903
Irish physicist and mathematician. Stokes worked on the theory of hydrodynamics, wave propagation, and the theory of sound. He was the first to understand the principles of fluorescence.

counter-clockwise relative to the z-axis. The integration path comprises two pairs of parallel edges. These edges are traversed in opposite directions, and we expect the respective integrals to almost cancel against each other. Specifically, the integral along the pair of edges parallel to the x-axis yields

$$\int_{\text{front}} d\mathbf{r} \cdot \mathbf{u} + \int_{\text{back}} d\mathbf{r} \cdot \mathbf{u} \simeq \delta^x \mathbf{e}_x \cdot \mathbf{u}(x, y, z) - \delta^x \mathbf{e}_x \cdot \mathbf{u}(x, y + \delta^y, z)$$

$$\simeq -\delta^x \delta^y \partial_y u^x(x, y, z). \tag{V103}$$

Once more, the local approximation principle discussed on p. 448 justifies the first approximation. In the second step, the difference of the vector field components u^x on the opposing edges is evaluated to linear order in the small difference of their y-coordinates. Adding to this expression the sum of integrals, $\int_{\text{left}} + \int_{\text{right}}$, of the other pair of edges,[31] we obtain

$$\oint_{\gamma_{\delta S}} d\mathbf{r} \cdot \mathbf{u} \simeq \delta^x \delta^y (-\partial_y u^x + \partial_x u^y)(x, y, z) \overset{(\text{V101})}{=} \delta^x \delta^y (\nabla \times \mathbf{u}) \cdot \mathbf{e}_z \simeq \int_{\delta S} d\mathbf{S} \cdot (\nabla \times \mathbf{u}), \tag{V104}$$

where in the second step we noted that the contributions of the two pairs of edges combine to give the z-component of the curl of $\mathbf{u}$. In the final step we observed that the third expression is approximately equal to the area integral of $\nabla \times \mathbf{u}$ over δS, since the field is near-constant over the infinitesimal extent of the surface element. The orientation is defined such that the positively oriented traversal of the line integral corresponds to a surface normal vector pointing in the *positive* z-direction (give this point some consideration).

---

[31] This contribution has opposite sign, since $\delta^y \mathbf{e}_y \cdot \mathbf{u}(x + \delta^x, y, z) - \delta^y \mathbf{e}_y \cdot \mathbf{u}(x, y, z)$, expanded in δ^x, yields $\delta^x \delta^y \partial_x u^y(x, y, z)$.

As with Gauss's theorem, the construction is straightforwardly generalized to arbitrarily shaped extended integration domains: consider the curve γ shown in Fig. V13. Let $S \subset M$ be an arbitrary surface bounded by γ. Imagine an approximate "tiling" of S by infinitesimal, rectangular surface elements, δS_ℓ, as indicated in the figure. Summing over all tiles and using Eq. (V104) for each we then obtain (in analogy to Eq. (V90))

$$\int_S d\mathbf{S} \cdot (\nabla \times \mathbf{u}) \simeq \sum_\ell \int_{\delta S_\ell} d\mathbf{S} \cdot (\nabla \times \mathbf{u}) = \sum_\ell \oint_{\gamma_{\delta S_\ell}} d\mathbf{r} \cdot \mathbf{u} \simeq \oint_\gamma d\mathbf{r} \cdot \mathbf{u}. \qquad (V105)$$

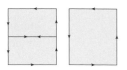

In the crucial last step we observed that the sum of the line integrals around all rectangles approximately equals the line integral over the boundary, γ, of the surface S. To understand this, consider the sum of the line integrals over just two adjacent rectangles. The contributions to the integral from their touching edges cancel and only the integral over the outer perimeter remains. By the same argument, the sum of line integrals around a set of rectangles tiling a more complex surface equals the line integral along the perimeter of the surface. In Eq. (V105) that path approximately equals γ. For vanishingly small tile sizes, the approximate equalities become exact.

Summarizing, we have found that the flux integral of a curl equals the line integral around the boundary of the surface,

$$\boxed{\int_S d\mathbf{S} \cdot (\nabla \times \mathbf{u}) = \oint_\gamma d\mathbf{r} \cdot \mathbf{u},} \qquad (V106)$$

Stokes's theorem

a result known as **Stokes's theorem**. ($\rightarrow$ V3.6.3-6) It states that non-vanishing line integrals around closed curves correspond to a non-vanishing vector field curl within surfaces bounded by the curves. We repeat that the choice of these surfaces is arbitrary, as long as they remain within the domain of definition of the vector field.

Curl (general coordinates)

As with its sources, the circulation of a vector field must be describable in an coordinate-invariant manner. We thus seek a coordinate-invariant definition of the curl. To this end, we apply Stokes's theorem to an infinitesimal surface element δS with normal unit vector $\mathbf{n}$. Since $\oint_{\gamma_{\delta S}} d\mathbf{r} \cdot \mathbf{u} \overset{(V104)}{=} \int_{\delta S} d\mathbf{S} \cdot (\operatorname{curl}\mathbf{u}) \simeq |\delta S| \, \mathbf{n} \cdot (\operatorname{curl}\mathbf{u})$, we may define the curl as the "circulation" per unit area,

coordinate-invariant definition of curl

$$\boxed{\mathbf{n} \cdot (\operatorname{curl}\mathbf{u}) \equiv \lim_{|\delta S| \to 0} \frac{1}{|\delta S|} \oint_{\gamma_{\delta S}} d\mathbf{r} \cdot \mathbf{u}.} \qquad (V107)$$

This definition can now be applied to derive a formula for the *curl in general coordinates*. Assuming a representation $\mathbf{r}(\mathbf{y})$ where $\mathbf{y} = (y^1, y^2, y^3)^T$ are the coordinates relative to a right-handed system, an infinitesimal coordinate rectangle with corner points $\mathbf{y}$ and $\mathbf{y} + \mathbf{e}_i \delta^i$ ($i = 1, 2$, no summation) defines a distorted rectangle, δS, as shown in the figure.

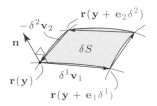

The edges of δS are approximately defined by the vectors $\delta^i \partial_{y^i} \mathbf{r}(\mathbf{y}) = \delta^i \mathbf{v}_i(\mathbf{y})$, its geometric area is $|\delta S| = \delta^1 \delta^2 \|\mathbf{v}_1 \times \mathbf{v}_2\|$, and $\mathbf{n} \equiv \frac{\mathbf{v}_1 \times \mathbf{v}_2}{\|\mathbf{v}_1 \times \mathbf{v}_2\|}$ the normal unit vector. Rearranging Eq. (V107), we thus obtain

$$\oint_{\gamma_{\delta S}} d\mathbf{r} \cdot \mathbf{u} \simeq |\delta S| \, \mathbf{n} \cdot (\text{curl } \mathbf{u}) \simeq \delta^1 \delta^2 (\mathbf{v}_1 \times \mathbf{v}_2) \cdot (\text{curl } \mathbf{u}) . \tag{V108}$$

We expand the curl in the coordinate basis, $\text{curl } \mathbf{u} = \mathbf{v}_i (\text{curl } \mathbf{u})^i$, use $(\mathbf{v}_1 \times \mathbf{v}_2) \cdot \mathbf{v}_{1,2} = 0$ and recall Eq. (L181), $(\mathbf{v}_1 \times \mathbf{v}_2) \cdot \mathbf{v}_3 = \sqrt{g}$, to represent the right-hand side through

$$(\mathbf{v}_1 \times \mathbf{v}_2) \cdot \mathbf{v}_3 \, (\text{curl } \mathbf{u})^3 = \sqrt{g} \, (\text{curl } \mathbf{u})^3 .$$

Turning to the left side of Eq. (V108), the contributions to the line integral from two opposing edges of the small coordinate rectangle can be obtained in a manner analogous to Eq. (V103). For the edges aligned in the y^1 direction we obtain the contribution $\delta^1 (\mathbf{v}_1 \cdot \mathbf{u})(\mathbf{y}) - \delta^1 (\mathbf{v}_1 \cdot \mathbf{u})(\mathbf{y} + \delta^2 \mathbf{e}_2) \simeq -\delta^1 \delta^2 \partial_{y^2} (\mathbf{v}_1 \cdot \mathbf{u})(\mathbf{y})$. Adding the analogous contribution from the edges parallel to y^2 we obtain the line integral is given by $\delta^1 \delta^2$ times

$$\partial_{y^1}(\mathbf{v}_2 \cdot \mathbf{u}) - \partial_{y^2}(\mathbf{v}_1 \cdot \mathbf{u}) = \epsilon^{3jk} \partial_{y^j} (\mathbf{v}_k \cdot \mathbf{u}) = \epsilon^{3jk} \partial_{y^j} (\mathbf{v}_k \cdot \mathbf{v}_l u^l) = \epsilon^{3jk} \partial_{y^j} (g_{kl} u^l) .$$

We used here the antisymmetric Levi–Civita symbol,[32] $\epsilon^{3jk} a_j b_k = a_1 b_2 - a_2 b_1$, to represent the alternating sum, in the second step expanded $\mathbf{u}$ in the coordinate basis, and in the third used $\mathbf{v}_k \cdot \mathbf{v}_l = g_{kl}$. Equating the expressions derived above for the right- and left-hand sides of Eq. (V108), we arrive at the conclusion $g^{1/2} (\text{curl } \mathbf{u})^3 = \epsilon^{3jk} \partial_{y^j} (g_{kl} u^l)$. Analogous relations hold for the 1- and 2-components, so that we have the representation

$$\boxed{(\text{curl } \mathbf{u})^i = \frac{1}{\sqrt{g}} \epsilon^{ijk} \partial_{y^j} (g_{kl} u^l)} \tag{V109}$$

curl in general coordinates

for the components of the **curl in a general coordinate basis**. For Cartesian coordinates, with $g_{kl} = \delta_{kl}$ and $g = 1$, the formula reduces to Eq. (V101). Explicit representations in other coordinate systems are somewhat more complicated. For example, a short calculation (try it!) shows that *the curl in spherical coordinates* is given by[33]

$$(\text{curl } \mathbf{u})^r = \frac{1}{\sin \theta} \left[\partial_\theta \left(\sin^2 \theta u^\phi \right) - \partial_\phi u^\theta \right],$$

$$(\text{curl } \mathbf{u})^\theta = \frac{1}{r^2} \left[\frac{1}{\sin \theta} \partial_\phi u^r - \sin \theta \partial_r (r^2 u^\phi) \right],$$

$$(\text{curl } \mathbf{u})^\phi = \frac{1}{r^2 \sin \theta} \left[\partial_r (r^2 u^\theta) - \partial_\theta u^r \right]. \tag{V110}$$

INFO The curl of physically relevant vector fields generally carries *physical significance* itself. For example, **Ampère's law** states that the line integral of the magnetic field, $\mathbf{B}$, along a closed loop, γ,

Ampère's law

---

[32] Recall that for the Levi–Civita symbol upper and lower indices are identical, $\epsilon^{ijk} \overset{(L7)}{=} \epsilon_{ijk}$, and not related to each other by index raising and lowering via g. (See also footnote 55 in Section L6.4.)

[33] For the *local* basis of spherical coordinates, the corresponding formula is given in Eq. (V122).

is proportional to the electric current I_S flowing through *any* area, S, bounded by that loop.[34] In CGS units, this equality is expressed by the formula

$$\oint_\gamma d\mathbf{r} \cdot \mathbf{B} = \frac{4\pi}{c} \int_S d\mathbf{S} \cdot \mathbf{j} \equiv \frac{4\pi}{c} I_S, \tag{V111}$$

current density
where c is the speed of light, and the current is represented as a surface integral over the **current density**, $\mathbf{j}$. (In fact, the current density is *defined* through the condition that its integral over any surface equals the physical current flowing through that surface.) Stokes's theorem states that

$$\oint_\gamma d\mathbf{r} \cdot \mathbf{B} = \int_S d\mathbf{S} \cdot (\nabla \times \mathbf{B}). \tag{V112}$$

Comparing Eqs. (V111) and (V112), we obtain the *differential form of Ampère's law*,

$$\nabla \times \mathbf{B} = \frac{4\pi}{c} \mathbf{j}, \tag{V113}$$

stating that the circulation of the magnetic field is caused by a flow of current.

curl in cylindrical coordinates
EXERCISE Use Eq. (V109) to verify that the **curl in cylindrical coordinates**, Eq. (V32), expressed in the coordinate basis, has components[35]

$$(\nabla \times \mathbf{u})^\rho = \frac{1}{\rho} \partial_\phi u^z - \rho \partial_z u^\phi,$$

$$(\nabla \times \mathbf{u})^\phi = \frac{1}{\rho} \left[\partial_z u^\rho - \partial_\rho u^z \right],$$

$$(\nabla \times \mathbf{u})^z = \frac{1}{\rho} \left[\partial_\rho (\rho^2 u^\phi) - \partial_\phi u^\rho \right]. \tag{V114}$$

Consider an infinite cylindrical wire centered on the z-axis, as shown in the figure, carrying a radially dependent current profile, $\mathbf{j} = \mathbf{v}_z j^z(\rho)$. (For example, $j^z(\rho) = c$ for $0 < \rho \le R$, and zero otherwise, would describe a wire of radius R and constant current density, $|\mathbf{j}| = c$.) Let D be a disk of radius ρ in the xy-plane centered on the axis of the wire. Confirm that the current flowing through D, $I = \int_D d\mathbf{S} \cdot \mathbf{j}$, is given by $I = 2\pi \int_0^\rho ds\, s j^z(s)$.

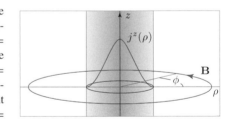

The magnetic field generated by this current distribution winds circularly around the axis of the wire, $\mathbf{B} = \mathbf{v}_\phi B^\phi(\rho)$, where the translational symmetry of the problem in the z-direction and the rotational symmetry in the ϕ-direction imply that the coefficient $B^\phi(\rho)$ depends on only the radial coordinate. Compute the line integral along the disk's boundary, ∂D, a circle of radius ρ, and confirm that $\int_{\partial D} d\mathbf{r} \cdot \mathbf{B} = 2\pi \rho^2 B^\phi(\rho)$. According to Eq. (V111) this must be proportional to the current, so that we have the identification

$$B^\phi(\rho) = \frac{4\pi}{c\rho^2} \int_0^\rho ds\, s j^z(s).$$

Compute the curl of this magnetic field to confirm that the current density satisfies Eq. (V113).

---

[34] We assume here time-independent current flow. If it is time-dependent, the situation becomes more complicated, cf. the discussion in Chapter V7.

[35] For the *local* basis of spherical coordinates, the corresponding formula is given in Eq. (V121).

Table V1 Parallels between the theorems of Gauss and Stokes.

	addresses	local	global	correspondence
Gauss's theorem	sources	$\nabla \cdot \mathbf{u}$	$\int_S d\mathbf{S} \cdot \mathbf{u}$	$\int_V dV \, \nabla \cdot \mathbf{u} = \int_S d\mathbf{S} \cdot \mathbf{u}$
Stokes's theorem	circulation	$\nabla \times \mathbf{u}$	$\oint_\gamma d\mathbf{r} \cdot \mathbf{u}$	$\int_S d\mathbf{S} \cdot (\nabla \times \mathbf{u}) = \oint_\gamma d\mathbf{r} \cdot \mathbf{u}$

integral theorems of vector analysis

We conclude by highlighting a number of striking parallels between Stokes's and Gauss's theorems, summarized in Table V1. These **integral theorems of vector analysis** are special cases of a more general mathematical relation, the *generalized Stokes's theorem*. The basic identity $\int_a^b dx \frac{df}{dx} = f(b) - f(a)$ is another variant of that theorem. Like the higher-dimensional variants of Stokes's theorem, it establishes a connection between the behavior of a function on the "boundary" $\{a, b\}$, and the integral of the derivative of that function over the interior $[a, b]$. The general version of Stokes's theorem will be discussed in Section V5.4 (see Eq. (V191) there) after the required mathematical terminology has been introduced.

V3.7 Practical aspects of three-dimensional vector calculus

The vector calculus operations gradient, divergence and curl play an important role in various introductory physics lecture courses. For example, the laws of electrodynamics are all formulated through vector differential operations acting on vector fields. One is often interested in situations displaying a high level of symmetry – cylindrically symmetric wires, or spherically symmetric charge distributions – in which case non-Cartesian coordinates are applied. In this section, which largely serves reference purposes, we summarize various useful vector calculus identities. We also translate from representations in the coordinate basis to the local basis, which is more frequently used in the physics literature.

Nabla identities

In the preceding sections we discussed the three operations gradient, divergence and curl, individually. However, in practice, they are often combined into higher-order differential operators acting on one field. We already mentioned the Laplacian, i.e. the divergence of the gradient, $\nabla \cdot \nabla f = \nabla^2 f$ in Section V3.5. Important two-derivative *operations involving the curl* include

gradient has no curl:	$\nabla \times (\nabla f) = \mathbf{0},$	(V115a)
curl has no divergence:	$\nabla \cdot (\nabla \times \mathbf{u}) = 0,$	(V115b)
curl of curl:	$\nabla \times (\nabla \times \mathbf{u}) = \nabla(\nabla \cdot \mathbf{u}) - \nabla^2 \mathbf{u}.$	(V115c)

The first identity states that a gradient field has no curl, the second that the curl of a vector field has no divergence. The third expresses the curl of the curl through the gradient of the divergence and the Laplacian of the field components. These identities are easy to prove in Cartesian coordinates using the Levi–Civita tensor and the interchangeability of two-fold partial derivatives. ($\to$ V3.7.1-2)

Other frequently encountered expressions contain the gradient, divergence or curl acting on *products of fields*. Applying the product rule, it is straightforward to derive the following identities: ($\to$ V3.7.3-4)

$$\nabla\,(fg) = f\,(\nabla g) + g\,(\nabla f)\,, \tag{V116a}$$

$$\nabla\,(\mathbf{u}\cdot\mathbf{w}) = \mathbf{u}\times(\nabla\times\mathbf{w}) + \mathbf{w}\times(\nabla\times\mathbf{u}) + (\mathbf{u}\cdot\nabla)\,\mathbf{w} + (\mathbf{w}\cdot\nabla)\,\mathbf{u}, \tag{V116b}$$

$$\nabla\cdot(f\mathbf{u}) = f\,(\nabla\cdot\mathbf{u}) + \mathbf{u}\cdot(\nabla f)\,, \tag{V116c}$$

$$\nabla\cdot(\mathbf{u}\times\mathbf{w}) = \mathbf{w}\cdot(\nabla\times\mathbf{u}) - \mathbf{u}\cdot(\nabla\times\mathbf{w})\,, \tag{V116d}$$

$$\nabla\times(f\mathbf{u}) = f\,(\nabla\times\mathbf{u}) - \mathbf{u}\times(\nabla f)\,, \tag{V116e}$$

$$\nabla\times(\mathbf{u}\times\mathbf{w}) = (\mathbf{w}\cdot\nabla)\,\mathbf{u} - (\mathbf{u}\cdot\nabla)\,\mathbf{w} + \mathbf{u}\,(\nabla\cdot\mathbf{w}) - \mathbf{w}\,(\nabla\cdot\mathbf{u})\,, \tag{V116f}$$

Gradient, divergence and curl in local basis

Above, we represented the vector derivatives in coordinate bases to obtain expressions that were naturally covariant and no more complicated than the formulas in Cartesian coordinates. However, the physics community prefers to work in the *local basis*, $\mathbf{e}_i = \mathbf{v}_i/\|\mathbf{v}_i\|$, with its unit length basis vectors. For reference purposes, we here summarize the most important vector analysis operations in that representation. The spherical and cylindrical systems in which these formulas are generally applied are special in that they have a diagonal, positive definite metric, $g_{ij} = g_{ii}\delta_{ij} = \|\mathbf{v}_i\|^2\delta_{ij}$, so that $\mathbf{v}_i = \sqrt{g_{ii}}\,\mathbf{e}_i$. We assume this property throughout below and distinguish between coordinate and local basis expansion coefficients by using a tilde symbol for the former, $\mathbf{u} = \sum_i \mathbf{v}_i\tilde{u}^i = \sum_i \mathbf{e}_i u^i$, implying $u^i = \sqrt{g_{ii}}\,\tilde{u}^i$. The triple appearance of the same fixed index i on the right-hand side shows that the covariant form of pairwise index contractions will be lost in the local basis representation.

gradient The components of the **gradient** in the local basis are obtained as $\nabla f^i = \sqrt{g_{ii}}\,\widetilde{\nabla}f^i$. Using $\widetilde{\nabla}f^i \overset{\text{(V70)}}{=} g^{ij}\partial_{y^j}f$ and $g^{ij} = \delta^{ij}/g_{ii}$ we get

$$\nabla f^i = \frac{1}{\sqrt{g_{ii}}}\partial_{y^i}f\,. \tag{V117}$$

divergence The **divergence** is obtained from Eq. (V95) via the replacement of u^i by $\tilde{u}^i = u^i/\sqrt{g_{ii}}$:

$$\nabla\cdot\mathbf{u} = \frac{1}{\sqrt{g}}\sum_i \partial_{y^i}\left(\sqrt{\frac{g}{g_{ii}}}\,u^i\right). \tag{V118}$$

curl Similarly, the components of the **curl**, Eq. (V109), are transcribed to the local basis as

$$(\nabla\times\mathbf{u})^i = \sqrt{\frac{g_{ii}}{g}}\sum_{jk}\epsilon^{ijk}\partial_{y^j}\left(\sqrt{g_{kk}}\,u^k\right). \tag{V119}$$

Laplace operator

Finally, the **Laplace operator**, $\Delta = \nabla \cdot \nabla$, makes no reference to vector components and hence retains the form given by Eq. (V99):

$$\Delta f = \frac{1}{\sqrt{g}} \sum_i \partial_{y^i} \left(\frac{\sqrt{g}}{g_{ii}} \partial_i f \right). \tag{V120}$$

The representations of these formulas in cylindrical and spherical coordinates are frequently needed in courses of theoretical physics and are summarized here for the convenience of the reader.[36] ($\rightarrow$ V3.7.5-13)

Cylindrical coordinates: ($\sqrt{g_{\rho\rho}} = 1$, $\sqrt{g_{\phi\phi}} = \rho$, $\sqrt{g_{zz}} = 1$, $\sqrt{g} = \rho$),

gradient: $\nabla f = \mathbf{e}_\rho \, \partial_\rho f + \mathbf{e}_\phi \, \frac{1}{\rho} \partial_\phi f + \mathbf{e}_z \, \partial_z f,$

divergence: $\nabla \cdot \mathbf{u} = \frac{1}{\rho} \partial_\rho(\rho u^\rho) + \frac{1}{\rho} \partial_\phi u^\phi + \partial_z u^z,$

curl: $\nabla \times \mathbf{u} = \mathbf{e}_\rho \left[\frac{1}{\rho} \partial_\phi u^z - \partial_z u^\phi \right] + \mathbf{e}_\phi \left[\partial_z u^\rho - \partial_\rho u^z \right]$

$$+ \, \mathbf{e}_z \frac{1}{\rho} \left[\partial_\rho(\rho u^\phi) - \partial_\phi u^\rho \right],$$

Laplacian: $\Delta f = \frac{1}{\rho} \partial_\rho(\rho \, \partial_\rho f) + \frac{1}{\rho^2} \partial_\phi^2 f + \partial_z^2 f. \tag{V121}$

Spherical coordinates: ($\sqrt{g_{rr}} = 1$, $\sqrt{g_{\theta\theta}} = r$, $\sqrt{g_{\phi\phi}} = r\sin\theta$, $\sqrt{g} = r^2 \sin\theta$),

gradient: $\nabla f = \mathbf{e}_r \partial_r f + \mathbf{e}_\theta \, \frac{1}{r} \partial_\theta f + \mathbf{e}_\phi \, \frac{1}{r\sin\theta} \partial_\phi f,$

divergence: $\nabla \cdot \mathbf{u} = \frac{1}{r^2} \partial_r(r^2 u^r) + \frac{1}{r\sin\theta} \partial_\theta(\sin\theta \, u^\theta) + \frac{1}{r\sin\theta} \partial_\phi u^\phi,$

curl: $\nabla \times \mathbf{u} = \mathbf{e}_r \frac{1}{r\sin\theta} \left[\partial_\theta(\sin\theta \, u^\phi) - \partial_\phi u^\theta \right] + \mathbf{e}_\theta \frac{1}{r} \left[\frac{1}{\sin\theta} \partial_\phi u^r - \partial_r(r u^\phi) \right]$

$$+ \, \mathbf{e}_\phi \frac{1}{r} \left[\partial_r(r u^\theta) - \partial_\theta u^r \right],$$

Laplacian: $\Delta f = \frac{1}{r^2} \partial_r(r^2 \, \partial_r f) + \frac{1}{r^2 \sin\theta} \partial_\theta(\sin\theta \, \partial_\theta f) + \frac{1}{r^2 \sin^2\theta} \partial_\phi^2 f. \tag{V122}$

V3.8 Summary and outlook

In this chapter, we introduced the very important concept of fields. Fields are physically motivated maps between an argument domain generally interpreted as physical space, or space-time, and an image domain locally identifiable with $\mathbb{R}^n$. In this chapter emphasis was put on the fields most frequently occuring in the early physics curriculum, scalar fields, $n = 1$, and (low-dimensional) vector fields, $n = 2, 3$. We discussed how fields can be visualized and classified according to universal criteria such as the presence of scalar potentials, sources, or circulation centers. Each of these criteria had a local representation

---

[36] In deriving these formulas ($\rightarrow$ V3.7.5-6), notice that when ∂_{y^i} acts on $\sqrt{g}$ or $\sqrt{g_{ii}}$, factors independent of the variable y^i can be pulled upfront. For example, if $\sqrt{g} = r^2 \sin\theta$, then $\partial_r \sqrt{g} = \sin\theta \, \partial_r r^2$, $\partial_\phi \sqrt{g} = r^2 \sin\theta \, \partial_\phi$.

in terms of suitably designed derivative operations, and a global one in terms of integral diagnostics. We saw that the global and the local approach are related via integral theorems, and how both have their respective strengths and weaknesses. In the next chapter we add more geometric substance to our discussion. Differentiable manifolds will be introduced as generalizations of the geometric structures discussed so far. Combined with the machinery of calculus, this will lead to a powerful framework, naturally accommodating the vector calculus operations discussed so far as special cases.

REMARK This chapter requires familiarity with Chapter L10. As in that chapter, we abandon the boldface notation for vectors, e.g. we write r instead of $\mathbf{r}$. Coordinate vectors will mostly be denoted by $y \in \mathbb{R}^d$ and their components by y^j. If no confusion is possible, a fixed coordinate system y has been chosen, and different points r, p, q on a manifold are under consideration, we will label their coordinates as $r^j \equiv y^j(r), p^j, q^j$, etc. In this way, the introduction of ever new symbols for coordinate vectors is avoided. However, this notation must be used with due care. Throughout the chapter we will frequently differentiate curves $y(t)$ at $t = 0$ and use the abbreviated notation $d_t y(0) \equiv d_t\big|_{t=0} y(t)$ for this.

Smooth geometric structures define the arena for the formulation of various fields of physics and mathematics. Many of these structures are easy to understand on an intuitive level; consider empty space, a two-dimensional sphere, or a circle as examples. Others can be more challenging to visualize – a sphere in five-dimensional space – or may have no obvious visualization at all. For example, we have seen that the set of all rotations of three-dimensional space defines a group. There is a sense of smoothness in this set, because any rotation can be generated by a continuous deformation of any other rotation, and clearly the set of rotations is geometrical. Yet it may not be evident how to conceptualize its geometry in any obvious way.

differential geometry In this chapter, we introduce the foundations of **differential geometry**, a comprehensive framework to understand smooth geometric objects in a unified fashion. The gateway into differential geometry is the realization that all the structures alluded to above look locally, but in general not globally, like a flat space in $\mathbb{R}^d$. A circle looks locally like a segment of a line, i.e. a subset of $\mathbb{R}^1$. Seen as a whole, however, it is different from a line, and this difference between the global and the local level is of defining importance to its geometric description. Likewise, the set of rotations of three-dimensional space may be parametrized by three rotation angles, and this set of angles defines a three-dimensional cuboid, a subset of $\mathbb{R}^3$. Globally, however, there is a difference between the set of rotations and a cuboid, etc.

Building on our earlier discussion of curvilinear coordinates, we introduce here the framework for describing smooth geometric structures both locally and globally. In the next chapter we will then introduce *differential forms* as a key tool to work with these objects in mathematical and physical contexts. The concepts introduced in these chapters define the foundations of *differential geometry*, a field of mathematics that is becoming increasingly important in modern theoretical physics.[37]

[37] For in-depth introductions to differential geometry emphasizing applications in physics we refer to the books *Analysis, Manifolds and Physics, Part I: Basics* (revised edition) by Yvonne Choquet-Bruhat and Cecile

V4.1 Differentiable manifolds

differentiable manifold

The overarching mathematical terminology for "smooth geometric structure" is a **differentiable manifold**, or just manifold for short. Prominent examples of manifolds are spheres, tori, balls, smooth curves and various group structures, to name but a few. In this section we elaborate on the local and the global description of such objects and show how to advance from one to the other.

The local and the global perspective of manifolds

A reasonable first attempt to introduce manifolds would be to define a manifold as the image, M, of a *diffeomorphic* (infinitely often differentiable, and bijective) map $r : U \to M$, $y \mapsto r(y)$, where $U \subset \mathbb{R}^d$ is an *open* coordinate domain in $\mathbb{R}^d$. This approach introduces what we will later understand as the local view; it provides a one-to-one identification between M and an open subset of $\mathbb{R}^d$. A few examples of manifolds for which this local perspective is sufficient are shown in the figure.

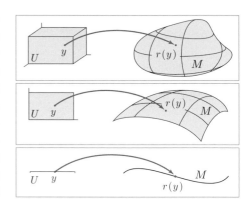

For others, however, it falls short of a complete representation. As a simple example, consider the $(d = 1)$-dimensional circle S^1 of unit radius. If the sole intention is to perform integral operations, then S^1 may be described as a circular curve parametrized on an open interval in such a way that its end points (almost) touch (see Fig. V14). In other cases, however, we need to describe a circle as what it is, a closed curve. For example, this distinction played a key role in our discussion of gradient fields on non-simply connected domains on p. 445. The closed circle cannot be represented as the diffeomorphic image of an open, i.e. end-point-excluding, parameter interval. In the next subsection we explain how augmented coordinate descriptions can be designed to obtain a global definition, including objects that cannot be represented as images of single coordinate maps.

However, before turning to this generalization let us mention a second aspect key to the understanding of manifolds: without even noticing, the circle is often understood as a circle embedded in $\mathbb{R}^2$, such as a circle drawn on a piece of paper. However, circles may be realized in different ways. Think of a circle in $\mathbb{R}^3$, or in a space of even higher dimensionality; the set of rotations around a fixed axis defines a circle where the rotation angle is a one-dimensional coordinate and a full rotation leads back to the origin; we saw that the set of unit-modular complex numbers $\exp(i\phi)$, $\phi \in \mathbb{R}$, is a circle in the complex plane, etc. These are examples of circles embedded in different sets. Now suppose we defined the circle as the diffeomorphic image of a map, $U \to S^1 \subset X$, where X is one of

DeWitt-Morette, North Holland; and *Geometry, Topology and Physics* (second edition), Graduate Student Series in Physics, by Mikio Nakahara, Taylor Frances.

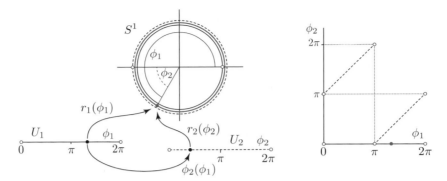

Fig. V14   Coordinate representation of a circle using two coordinate maps. Discussion, see text.

these sets. This approach would make reference to differentiability in the ambient set, X. In this way, X would become an essential part of the definition, and different realizations of X would lead to different definitions. A good definition should avoid such excess baggage and describe the circle as a stand-alone object, without reference to embeddings.

In the next subsection we introduce a definition of manifolds tailored to handle the complications addressed above. This definition will provide the foundation on which all further discussion is based.

Definition of manifolds

Our discussion above indicated that more than one coordinate map may be required to parametrize general manifolds M. The situation is illustrated in Fig. V14 on the example of the unit circle, $M = S^1$, embedded in $\mathbb{R}^2$ for graphical representability. One coordinate, $\phi_1 \in U_1 \equiv (0, 2\pi)$, is chosen to parametrize the circle as $r_1(\phi_1) \equiv (\cos\phi_1, \sin\phi_1)^{\mathrm{T}}$. This map does not reach the point $(1, 0)^{\mathrm{T}} \in S^1$, because U_1 is an open interval and the end-points 0 and 2π are excluded. However, a second coordinate, $\phi_2 \in U_2 \equiv (0, 2\pi)$,[38] may now be engaged to parametrize a different portion of S^1 as $r_2(\phi_2) \equiv (\cos(\phi_2 - \pi), \sin(\phi_2 - \pi))^{\mathrm{T}}$. This map does reach the previously excluded point, since $(1, 0)^{\mathrm{T}} = r(\phi_2 = \pi)$; however, it excludes the point $(-1, 0)^{\mathrm{T}}$. We conclude that each point on the circle is included in at least one of the images, $r_1(U_1)$ or $r_2(U_2)$, of the coordinate maps, and the majority, i.e., all except $(\pm 1, 0)^{\mathrm{T}}$, lie in both images.

Let $M_{12} \equiv r_1(U_1) \cap r_2(U_2) = S^1 \setminus \{(1, 0)^{\mathrm{T}}, (-1, 0)^{\mathrm{T}}\}$ be the intersection of the two coordinate images, i.e. the set of points reached by both maps. For each $r \in M_{12}$, we have two coordinate representations, $r = r_1(\phi_1) = r_2(\phi_2)$. This defines a map, $r_1^{-1}(M_{12}) \to r_2^{-1}(M_{12})$, $\phi_1 \mapsto \phi_2(\phi_1) = r_2^{-1}(r_1(\phi_1))$, assigning to each coordinate ϕ_1 the corresponding coordinate ϕ_2. By construction, the "transition map", $r_2^{-1} \circ r_1$, is a diffeomorphic map from $r_1^{-1}(M_{12})$ to $r_2^{-1}(M_{12})$. It defines the change from the r_1 to the r_2 language in the description of S^1.

---

[38] Although the two coordinate intervals $U_1 = (0, 2\pi)$ and $U_2 = (0, 2\pi)$ are identical, it is best to think of them as separate domains and label them differently.

EXERCISE Verify that $r_1^{-1}(M_{12}) = (0, 2\pi) \setminus \{\pi\} = (0, \pi) \cup (\pi, 2\pi)$. Show that on this union of intervals, $\phi_2(\phi_1) = \phi_1 + \pi$ for $\phi_1 < \pi$ and $\phi_2 = \phi_1 - \pi$ for $\phi_1 > \pi$. The switch between the two branches occurs at the limiting point between two almost touching open intervals. However, this point is excluded from the definition of the transition map. Within their respective domains of definition the maps $\phi_2 = \phi_1 \pm \pi$ are trivially invertible and infinitely often differentiable, i.e. they are diffeomorphisms. Notice that no part of the construction requires differentiability in the space $\mathbb{R}^2$ in which the manifold S^1 is embedded.

The circle as discussed above serves as a role model for the general definition of differentiable manifolds. Referring for a rigorous discussion to the literature, a *differentiable d-dimensional manifold*, M, is a set covered by the combined image, $M = \bigcup_{a=1}^{k} r_a(U_a)$, of k coordinate maps, $r_a : U_a \to M, y \mapsto r_a(y)$, where U_a are open coordinate domains in $\mathbb{R}^d$.[39] Each individual coordinate map, r_a, defines a **chart**, (r_a, U_a), of the manifold and a collection of charts, $\{(r_1, U_1), (r_2, U_2), \ldots, (r_k, U_k)\}$, fully covering the manifold is an **atlas**. In the above example, we parametrized $M = S^1$ through an atlas of two charts, $(r_1, (0, 2\pi))$ and $(r_2, (0, 2\pi))$.

> **chart**

> **atlas**

The maps $r_a : U_a \to M_a \subset M$ between the coordinate domains and their images, $r_a(U_a) \equiv M_a \subset M$, are required to be invertible and continuous.[40] We introduce their inverses, $y_a \equiv r_a^{-1}$, as $y_a : M_a \to U_a, r \mapsto y_a(r) = r_a^{-1}(r)$. The images M_a of different charts in M generally overlap. Where this happens the intersections, $M_{ab} \equiv M_a \cap M_b$, define regions with more than one coordinate representation, $r = r_a(y_a) = r_b(y_b)$ (see Fig. V15 for an illustration with two charts). The crucial condition of smoothness now is that the **transition functions**,

> **transition functions**

$$y_b \circ r_a : y_a(M_{ab}) \to y_b(M_{ab}), \qquad y_a \mapsto y_b(y_a) \equiv y_b(r_a(y_a)), \qquad \text{(V123)}$$

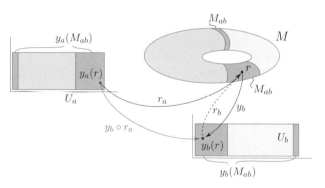

Fig. V15   On the coverage of manifolds by multiple charts. Discussion, see text.

[39] To avoid confusion between the vector-valued maps, r_a, and covariant components, r_i, we index the former with early Latin letters, $a, b, \ldots$.

[40] A technical remark: if M is embedded in a vector space such as $\mathbb{R}^n$, the continuity of the maps r_a may be defined using the mathematical structure of that embedding space. However, even if this is not the case it is still possible to define continuity, provided M does possess the mathematical structure of a *topological space*. Although a discussion of topological spaces is beyond the scope of this text, it is worth noting that the definition of manifolds requires comparatively little mathematical structure. For further discussion we refer to textbooks on differential geometry.

must be diffeomorphic maps between their domains of definition, $y_a(M_{ab})$ and $y_b(M_{ab})$. The definition of the unit circle required only one such function, $\phi_2(\phi_1)$, one overlap region, $M_{12} \subset S^1$, and a single intersection domain, $(0, \pi) \cup (\pi, 2\pi)$. If the above condition is met, M is a differentiable manifold. Notice that the differentiability criterion applies to maps $y_b(y_a)$ between different coordinates, but not to the maps $r(y_a)$ from coordinate domains into M. This is an important feature, which makes it possible to define manifolds which need not be embedded in $\mathbb{R}^n$. ($\to$ V4.1.1-2)

Submanifolds

submanifold

In applications we often consider curves in two- or three-dimensional structures, surfaces in volumes, or other lower-dimensional objects embedded in structures of higher dimension. The overarching terminology for them is **submanifolds**. Submanifolds are manifolds embedded in larger manifolds. To be precise, $L \subset M$ is a c-dimensional submanifold of the d-dimensional manifold M if for each point $r \in L$ there exists a chart, (r, U), of M, and a c-dimensional subdomain, $U \cap \mathbb{R}^c$, such that for constant $(a^{c+1}, \dots, a^d)$ in the complementary subdomain, $U \cap \mathbb{R}^{d-c}$,

$$q : U \cap \mathbb{R}^c \to L, \quad (x^1, \dots, x^c) \mapsto q(x^1, \dots, x^c) = r(x^1, \dots, x^c, a^{c+1}, \dots, a^d) \qquad \text{(V124)}$$

defines a chart of L. This definition states that a chart of L is obtained by taking a suitably chosen chart of M and constraining some of its coordinates. For example, the circle is a submanifold of the two-dimensional plane. The latter may be locally parametrized by a chart with two polar coordinates, (ρ, ϕ). If we freeze one of them, $\rho = R$, we obtain a chart of the circle with just one coordinate, ϕ. It is defined through the image of constrained coordinate configurations, (R, ϕ), of the two-dimensional chart.

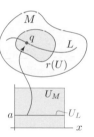

Note that submanifolds may have the same dimension as their host manifolds, $c = d$. For example, the unit-disk is a submanifold of two-dimensional space. In this case, the coordinates $(\rho, \phi) \in U = (0, 1) \times (0, 2\pi)$ define a local chart of both the embedding manifold $\mathbb{R}^2$ and the included disk, and no coordinates need to be frozen. The opposite extreme are zero-dimensional submanifolds, $c = 0$, which are just points in M.

The definition of submanifolds also encompasses the previous concept of *embedded manifolds*. To say that a manifold is embedded in $\mathbb{R}^n$ means that it is a submanifold of the manifold $\mathbb{R}^n$. The above example illustrates this for the embedding of the circle in $\mathbb{R}^2$.

INFO *How important are multi-chart representations of manifolds in practice?* The answer depends on the context. As discussed previously, single-chart coverages provide approximate coverings of manifolds, up to subsets of lower dimensionality. For example, the description of a circle by polar coordinates misses an isolated point, and the description of the two-sphere by spherical coordinates misses a line connecting the north and south poles along a great half-circle, see Section V2.1. Incomplete or local[41] coverages of this type are sufficient to compute integrals over manifolds (see

---

[41] A remark on a possibly disconcerting wording convention: single charts of manifolds are often said to provide a local description, even if "local" means everything-except-one-point. The rationale for this is that mathematicians define the attribute "local" as "non-global".

discussion on p. 243 in Section C4.2), or to describe their local geometric structure. In such cases, no extension to a complete coordinate description is required.

By contrast, full coverages provided by multi-chart atlases and their transition functions become important when global or topological aspects play a role. (For example, a Möbius strip is locally diffeomorphic to a rectangular strip, but is globally different from it. This difference shows in the transition functions mediating between a minimal atlas of two charts parametrizing the Möbius strip; think about this point.)

In either context, *differential forms* will emerge as powerful tools to describe the geometry of manifolds, and that of physically relevant structures defined on them. In the following we emphasize the local perspective, and mostly work with single charts. The extension to a global framework is left for lecture courses in topology or differential geometry, or advanced courses in theoretical physics.

V4.2   Tangent space

idea of tangent space

Much as smooth functions look locally linear, smooth manifolds look locally flat, as indicated schematically in Fig. V16. For example, the surface of Earth is a sphere, but looks locally planar. The planes locally approximating a d-dimensional manifold are d-dimensional vector spaces, called *tangent spaces*. In the example of Earth, this is easy to imagine just by staring at one's feet: the neighborhood of each geographic location is locally approximated by a two-dimensional plane in the embedding three-dimensional space. These planes differ from point to point, and it may be intuitively plausible that the infinite collection of all of them, a set called the "tangent bundle" and to be defined in the next subsection, contains exhaustive information on the geometric structure of Earth's surface.

Tangent spaces and the tangent bundle play an important role in the geometry of manifolds. In the following, we will learn how to describe them, both intuitively and in more formal terms. Our approach will encompass manifolds not embedded in a host vector space, and for which the visual picture of tangent spaces may be not quite as obvious as in the example of Earth.

Smooth functions on manifolds

The principal tools required for the description of tangent space (and, in fact, that of any other geometric structure on manifolds, too) are *smooth functions*, $f : M \to \mathbb{R}$. Although the concept of a smooth, real-valued function seems innocent, it may not be entirely evident

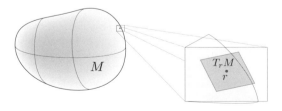

Fig. V16   The setup considered in this section. A d-dimensional manifold, M, is a smooth object, which implies its local flatness. The vicinity of any point r in M is approximated by the d-dimensional tangent vector space, $T_r M$, to r at M.

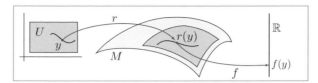

Fig. V17   A function on a manifold $f : M \to \mathbb{R}, r \mapsto f(r)$ is smooth if its coordinate representations $f : U \to \mathbb{R}$, $y \mapsto f(y) \equiv f(r(y))$ are smooth functions from $U \subset \mathbb{R}$ into the real numbers. Likewise, a curve in a manifold is smooth if its coordinate representation is a smooth curve.

how smoothness is defined in the present context. We agreed to abstain from differentiating (the principal operation required to detect smoothness) in M, and so the smoothness of functions must be defined with reference to a coordinate system, see Fig. V17: a function $f : M \to \mathbb{R}, r \mapsto f(r)$ is smooth if its coordinate representation, $f(r(y))$, is a smooth function of the coordinates y. More precisely, given a coordinate map $y : U \to M \ni r$ enclosing the point r, its composition with f, defined as $f : U \to \mathbb{R}, y \mapsto f(r(y)) \equiv f(y)$, must be a smooth function from $U \subset \mathbb{R}^d$ into the real numbers. Following common conventions we denote $f : M \to \mathbb{R}$ and its coordinate representation, $f : U \to \mathbb{R}$, by the same symbol.[42]

The smoothness of other objects defined on M is defined in similar ways. For example, a curve $r : I \to M, t \mapsto r(t)$ is smooth if its coordinate representation, $r : I \to U, t \mapsto y(t) \equiv y(r(t))$, is smooth. One may think of functions, curves, etc. as absolute objects, and of different coordinate representations as descriptions of these objects in different languages. Mathematical calculations require a language description and are therefore performed in coordinates. A substitution $f(y') = f(y(y'))$ defines the passage from the language $f(y)$ to $f(y')$ describing the object $f(r)$.

INFO   The definition of smoothness via coordinate representations is favored by fundamental and practical considerations. To understand why, consider the two-sphere $M = S^2 \subset \mathbb{R}^3$, embedded in $\mathbb{R}^3$. Let $f : S^2 \to \mathbb{R}$ be a function defined on the sphere in such a manner that an extension into the embedding space is not possible or meaningful. For example f might describe a two-dimensional surface mass distribution present on the sphere, but not in surrounding space. In this case, f is not differentiable in $\mathbb{R}^3$. (If x is on the sphere, then $x + \delta e_j$ is not, no matter how small δ, and the difference quotient, $\delta^{-1}(f(x + \delta e_j) - f(x))$, is not defined.) By contrast, the partial derivatives $\partial_{y^j} f(x(y))$ defined in a two-dimensional spherical coordinate representation of the sphere do not have this problem.

Tangent vectors as equivalence classes of curves

Let $r(t)$ be a smooth curve in M passing through the point r at $t = 0$, i.e. $r(0) = r$. The velocity vector at $t = 0$,

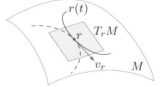

$$v_r \equiv \mathrm{d}_t r(0) \equiv \mathrm{d}_t\big|_{t=0} r(t), \qquad \text{(V125)}$$

[42] Since the transition maps between different coordinates are smooth, the smoothness of $f(y)$ implies that of $f(y') \equiv f(y(y'))$. It is therefore sufficient to run the smoothness test in an arbitrary coordinate system.

is tangent to the curve at r and therefore tangent to M. Intuitively, one may consider the tangent space at r as the set of all tangent velocity vectors constructed in this way.

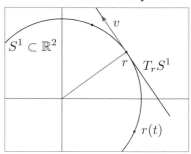

This is almost, but not quite, how the tangent space is actually defined. A drawback of Eq. (V125) is that it breaks the promise to not differentiate in M: the derivative $d_t r(t) = \lim_{\delta \to 0}(r(t + \delta) - r(t))$ can only be built if the manifold is embedded in a vector space, $M \subset \mathbb{R}^n$, and differences of vectors $r(t)$ are taken within that space. As a consequence, the tangent space to a circle, for example, would depend on whether the circle is embedded in $\mathbb{R}^2$ (as in the figure) or in $\mathbb{R}^3$, and this is a complication we should avoid.

curves defining tangent vectors
As a first step towards a better definition we note that different curves passing through r can have the same tangent vector. For example, the dashed curve indicated in the figure next to Eq. (V125) will have the same tangent vector as $r(t)$ if both are traversed at the same speed. This observation motivates the following construction: let $f : M \to \mathbb{R}$ be a smooth function on M (or at least in a neighborhood containing r). The *directional derivative* of f along $r(t)$ is then defined as $d_t f(r(0))$, i.e. the ordinary derivative of the real function $t \mapsto f(r(t))$ evaluated at $t = 0$. Two curves $r(t)$ and $r'(t)$ are called equivalent, $r(t) \sim r'(t)$, at $r = r(0) = r'(0)$ if they have the same directional derivative, $d_t f(r(0)) = d_t f(r'(0))$, for any f. The set of all equivalent curves defines an *equivalence class of curves*, $[r(t)]$, and this equivalence class in turn defines a tangent vector denoted as $\hat{v}_r$.

This definition may seem abstract, but in fact is very geometric. It associates tangent vectors with "bushels" of curves in M touching at a point r where they are all traversed in the same direction and at the same speed. Intuitively, the vector $\hat{v}_r$ is the common tangent vector of these curves. The d components representing $\hat{v}_r$ in a system of coordinates, $y = (y^1, \dots, y^d)^T$, for the manifold are defined as

$$v_r^i = d_t y^i(r(0)). \tag{V126}$$

According to this definition, the components of the tangent vector are derivatives of the coordinate functions y^i along any of $\hat{v}_r$'s representing curves.

As always, the coordinate representation, $v_r \equiv (v_r^1, \dots, v_r^d)^T$, of $\hat{v}_r$ is specific to a coordinate system. However, the components do not depend on the representing curve used to compute them: the functions $y^i(r)$ are smooth and by virtue of the equivalence relation any other curve, $r'(t) \in [r(t)]$, will give the same value, $d_t y^i(r(0)) = d_t y^i(r'(0))$. Also notice that the definition (V126) generalizes the naive definition (V125). That formula is recovered if M is embedded in a vector space $\mathbb{R}^d$ of equal dimensions. In this case, $r = \{r^i\}$ may be identified with a Cartesian coordinate representation of $\mathbb{R}^d$. For these coordinates, $y^i = r^i$, and $v_r^i = d_t r^i(t)$ is the coordinate representation of the tangent vector according to both Eqs. (V125) and (V126).

EXERCISE Consider $M = \mathbb{R}^2$ and the point $r = (R, 0)^T$ in a Cartesian coordinate system. Show that the curves $r_1(t) = R(\cos(t), \sin(t))^T$, $r_2(t) = R(2 - \cos(t), \sin(t))^T$ and $r_3(t) = R(1, t)^T$ all have

$r_a(0) = r$ and identical derivatives for smooth functions, $d_t f(r_a(0))$. They therefore lie in the same equivalence class. Verify that the components of the tangent vector in Cartesian coordinates are given by $(v^1, v^2)^T = (0, R)^T$, and in polar coordinates by $(v^r, v^\phi)^T = (0, 1)^T$. Why is the curve $r_4(t) = R(\cos(2t), \sin(2t))^T$ not equivalent to the curve $r_1(t)$, although it looks identical when sketched on paper?

Tangent vectors as directional derivative operators

The constructions above illustrate how tangent vectors and the curves representing them categorically appear in connection with directional derivatives of functions $f : M \rightarrow \mathbb{R}$. Indeed, equivalence classes of curves are defined for the sole purpose of taking of derivatives along their representatives. This suggests another way of thinking of a tangent vector

vectors as $\hat{v}_r$: it is a *differential operator*, $\partial_{v,r}$, acting on functions as $f \mapsto \partial_{v,r} f$, where $\partial_{v,r}$ is the
derivatives derivative of f at r along a representative in the class $[r(t)]$ defining $\hat{v}_r$:

$$\partial_{v,r} : f \mapsto \partial_{v,r} f \equiv d_t f(r(0)). \tag{V127}$$

The two definitions of tangent vectors, as equivalence classes of curves or as derivative operators, are equivalent and used interchangeably throughout. Both definitions may require some getting used to but after some practice turn out to be intuitive. Reflecting their meaning, we will use a (standard) derivative notation for tangent vectors, $\hat{v}_r = \partial_{v,r}$, where the subscript v is a reference to the vector $\hat{v}_r$ representing a class of curves.

Two more *remarks on notation*: the notation $\partial_{v,r}$ emphasizes that the definition of a tangent vector is specific to its base point, r. However, this reference is usually omitted and the slimmer notation $\partial_{v,r} \rightarrow \partial_v$ used instead (much as coordinate basis vectors, $\mathbf{v}_{j,r}$, were denoted as $\mathbf{v}_j$ in Section V2.2). Second, the object ∂_v is not an ordinary partial derivative. The definition of partial derivatives requires coordinates, whereas the definition of the action of ∂_v does not. However, we will see that formulas featuring ∂_v mimic ones involving partial derivatives, and this may be the origin of the ∂_v-notation.

In the new notation, the components v^i_r representing a tangent vector in a coordinate system $\{y^i\}$ assume the form

$$\boxed{v^i_r = \partial_{v,r} y^i(r) = d_t y^i(r(0)),} \tag{V128}$$

where the first representation stands for the application of the vector differential operator to a coordinate function, and the second makes the definition of this derivative explicit. In the next section, we will apply the above understanding of tangent vectors to describe the geometry of tangent space.

Tangent space bases

tangent The set of tangent vectors, $\partial_{v,r}$, defines a vector space, the **tangent space** to M at r,
space

$$T_r M \equiv \{\partial_{v,r} | \partial_{v,r} \text{ tangent to } M \text{ at } r\}. \tag{V129}$$

For concrete calculations in these spaces we need bases, and these are best constructed with reference to coordinate systems. Recall that the definition of tangent vectors uses curves

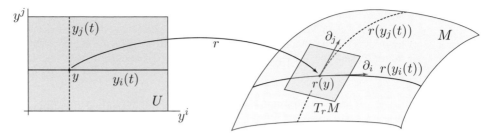

Fig. V18   Coordinate lines $y_j(t)$ in U define coordinate lines, $r(y_j(t))$, in M, whose tangent vectors, ∂_j, define a holonomic basis for $T_r M$. If M is embedded in $\mathbb{R}^n$, the n-dimensional representatives of ∂_j are the coordinate basis vectors $\mathbf{v}_j$ of Section V2.4.

$r(t)$ in M, and all these have coordinate representations, $y(t) = y(r(t))$, in U. Now consider the special curves $y_j(t) = y + e_j t$ in U, where y is the coordinate representation of the base point $r(y) = r$ and e_j is the jth standard unit vector in U. The images, $r(y_j(t))$, of these curves define coordinate lines in M (see Fig. V18). This construction generalizes the concept of coordinate lines introduced in Section V2.2 for manifolds embedded in $\mathbb{R}^2$ or $\mathbb{R}^3$ (see Eq. (V17), or Fig. V4).

The tangent vector defined by $y_j(t)$ is denoted by ∂_j:

$$\partial_j : f \mapsto \partial_j f \equiv d_t f(r(y_j(0))) \equiv d_t f(y_j(0)), \tag{V130}$$

where $f(y) \equiv f(r(y))$ is the coordinate representation of f. Observe that ∂_j acts on functions as a partial derivative in the y^j-direction: $\partial_j f = d_t f(y_j(0)) = d_t\big|_{t=0} f(y + e_j t) = \partial_j f(y)$, where ∂_j on the left side is the vector differential operator, and on the right side a standard partial derivative in y^j. This equality motivates the notation ∂_j for the coordinate tangent vector.

The *coordinate representation* of ∂_j is $e_j = (0, ..., 1, ..., 0)^{\mathsf{T}}$, the jth standard vector. This is seen by computing its ith component, $(\partial_j)^i$, via Eq. (V128): application of ∂_j to the ith coordinate function, $y^i = y^i(r)$, yields

$$(\partial_j)^i = \partial_j(y^i(r)) = d_t\big|_{t=0} y^i(y + e_j t) = \delta^i{}_j, \tag{V131}$$

which is the jth partial derivative of y^i in U or, equivalently, the derivative of the coordinate function y^i of M along the jth coordinate curve.

tangent
space basis   The d vectors $\{\partial_j\}$ define a **basis of tangent space**. To understand why, consider the tentative expansion $\partial_v = v^j \partial_j$ of a generic tangent vector in the vectors ∂_j. Its coefficients are obtained by applying both sides to a coordinate function, y^i. On the left, this yields $\partial_v y^i$, and on the right, $(v^j \partial_j)(y^i) = v^j(\partial_j y^i) \overset{(V131)}{=} v^i$. This gives $v^i = \partial_v y^i$, and defines the expansion as

$$\boxed{\partial_v = v^j \partial_j, \qquad v^j = \partial_v y^j,} \tag{V132}$$

through the derivatives of coordinate functions along v. Notice that the notation ∂_v is consistent with the interpretation of ∂_j as partial derivatives in the j-direction: the relation

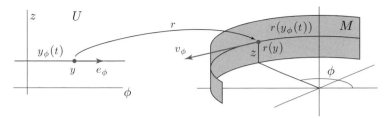

Fig. V19   A two-dimensional cylinder $M \subset \mathbb{R}^3$ of unit radius is locally described by cylindrical coordinates, $r(y)$, with coordinate vectors $y = (\phi, z)^T \equiv \phi e_\phi + z e_z$, where $e_\phi = (1,0)^T$, $e_z \equiv (0,1)^T$. (Do not confuse e_ϕ, a vector in the coordinate domain, with the local basis vectors $\mathbf{e}_\phi$ of Section V2.2.) The tangent vector ∂_ϕ at $r(y)$ is generated by the curve $y_\phi(t) = (\phi + t, z)^T = y + t e_\phi$ (see Eq. (V131)). In the embedding space, ∂_ϕ is a three-component vector, v_ϕ. In a Cartesian representation, $(x^1, x^2, x^3)^T$, its components are obtained by differentiation of the coordinate functions $x^i(\phi, z)$ along $y_\phi(t)$, i.e. $(v_\phi)^i = d_t x^i(y_\phi(0)) = \partial_\phi x^i(\phi, z)$. In this way one obtains $v_\phi = (-\sin\phi, \cos\phi, 0)^T$ for the Cartesian representation of ∂_ϕ in $\mathbb{R}^3$. This equals the vector $\mathbf{v}_\phi$ introduced in Section V2.2 as one of the elements of the cylindrical coordinate basis, see Eq. (V32).

$$\partial_v f(y) = v^j \partial_j f(y) = \frac{\partial f(y)}{\partial y^j} (\partial_v y^j) \tag{V133}$$

can be understood either as the expansion of ∂_v in the basis ∂_j, or, equivalently, as the chain rule applied to the directional derivative of the function $f(y)$.

holonomic   Bases $\{\partial_j\}$ defined with reference to a local coordinate system are called **holonomic**
basis   **bases** of tangent spaces. They are the preferred bases for the geometric description of manifolds.

INFO Although the construction above does not require an embedding, it is worthwhile discussing the particular case of *manifolds embedded in a vector space*, $\mathbb{R}^n$. Notice that n need not coincide with the manifold dimension, d (think of a ($d = 2$)-dimensional surface embedded in $\mathbb{R}^{(n=3)}$, etc.). In either case, the tangent space, $T_r M$, to a point $r \subset M$ is now a d-dimensional subspace of $\mathbb{R}^n$. In a Cartesian basis of $\mathbb{R}^n$, points r have n-component coordinate representations, $(x^1, \ldots, x^n) \in M$, where $x^i(y)$ are functions of M's local coordinates.

This setup leads to another representation of tangent vectors: a vector $\partial_v \in T_r M$ defines the coordinate vector $v = (v^1, \ldots, v^n)^T$, with components $v^i \equiv \partial_v x^i = d_t x^i(y(0)) \equiv d_t x^i(0)$, where $x^i(y(t)) = x^i(t)$ is the Cartesian description of a curve $r(t)$ representing ∂_v. Each tangent vector is thus represented by an n-component vector, $v \equiv e_j v^j$, in the embedding space $\mathbb{R}^n$. This construction leads back to the intuitive formula (V125) discussed in the beginning of the section. Specifically, the vectors of the holonomic basis ∂_j now have the coordinate representation

$$v_j = d_t r(y_j(0)) = \frac{\partial r(y)}{\partial y^j}. \tag{V134}$$

They equal the *coordinate basis* vectors introduced in Section V2.2 (there denoted in boldface fonts as $\mathbf{v}_j$). Figure V19 recapitulates this connection on the example of cylindrical coordinates. ($\rightarrow$ V4.2.1-2)

Now consider a situation where the manifold and its embedding space have equal dimensionality, $d = n$, and the former is an open subset of the latter, $M \subset \mathbb{R}^d$ (think of a two-dimensional disk in $\mathbb{R}^2$, etc.). Although the manifold itself covers only a subset of the embedding space, the tangent space at any point is equal to it, $T_r M = \mathbb{R}^d$. The reason for this is that vectors tangent to curves in M traversed at high speed may assume arbitrarily large values and realize any vector in $\mathbb{R}^d$. If

$\mathbb{R}^d$ is spanned by a Cartesian basis, $\{e_j\}$, both points r in the manifold and tangent vectors ∂_v are represented by d-component coordinate vectors, $e_j x^j$ and $v = e_j v^j$, respectively. In this case, the distinction between points r and tangent vectors ∂_v is easily forgotten. However, such conceptual sloppiness can be counterproductive and lead to confusion. In the present context it is best to avoid it.

Change of coordinates

Finally, let us explore how vector components, $v^j(y)$, change under a *change of coordinates* $y \mapsto y'(y)$, where y and y' are two systems with finite overlap on M. The components of a tangent vector,[43] $\partial_v = v^j(y)\partial_{y^j} = v^i(y')\partial_{y'^i}$, in these two systems are obtained by application of the vector to the respective coordinate functions, $v^j(y) = \partial_v y^j$ and $v^i(y') = \partial_v y'^i$. Given the components $v^j(y)$, the components $v^i(y')$ are obtained by application of the chain rule: $v^i(y') = \partial_v y'^i = \partial_v y'^i(y) = \frac{\partial y'^i(y)}{\partial y^j}\partial_v y^j = \frac{\partial y'^i(y)}{\partial y^j}v^j(y)$. The right-hand side of this expression yields the components $v^i(y'(y))$ expressed as functions of y. However, one may express all coordinates $y = y(y')$ through y' to obtain the representation

$$v^i(y') = \left.\frac{\partial y'^i(y)}{\partial y^j}v^j(y)\right|_{y=y(y')}. \tag{V135}$$

Note how Eq. (V135) generalizes the earlier result Eq. (V45). That relation described how the components of vectors defined on a subset $M \subset \mathbb{R}^d$ change if one switches from a Cartesian to a general coordinate basis. Equation (V135) describes the change of components for general transformations between holonomic bases. It is also used to transcribe vector components from one coordinate language to another in cases where a manifold cannot be covered by a single chart, see Fig. V15. In either case, the **Jacobi matrix**, $J^i_{\ j} = \frac{\partial y'^i}{\partial y^j}$, and its inverse, $(J^{-1})^j_{\ i} = \frac{\partial y^j}{\partial y'^i}$, are the central ingredients of the transformation formulas.

coordinate transformation of vector fields   To summarize, under coordinate transformations vector field expansions behave as

$$\partial_v = v^j(y)\partial_{y^j} = \left[v^j(y)\frac{\partial y'^i}{\partial y^j}\right]_{y=y(y')}\partial_{y'^i} = v^i(y')\partial_{y'^i}. \tag{V136}$$

Notice the structural similarity to the chain rule for a partial derivative $\partial_{y^j} = \frac{\partial y'^i}{\partial y^j}\partial_{y'^i}$ acting on a function $f(y'(y))$. ($\to$ V4.2.3-4)

To *summarize*, we have seen how to construct vector spaces $T_r M$ describing the neighborhoods of points r in a manifold. The vectors of these spaces are in one-to-one relation

---

[43] A word on *notational conventions*: vector fields can be expanded in different coordinate bases as $v = v^j(x)\partial_{x^j} = v^j(y)\partial_{y^j} = v^j(z)\partial_{z^j},\dots$. In this text, we will use a convention where the argument of the coefficient functions in $v^j(y)\partial_{y^j}$ indicates which coordinate system they refer to. It is important to keep this point in mind when coordinate changes are discussed. For example, in the relation $v^i(y'(y)) = \frac{\partial y'^i}{\partial y^j}v^j(y)$, the y-coefficient functions $v^j(y)$ should be carefully distinguished from the functions $v^i(y'(y))$, which are the y'-coefficient functions expressed through y-coordinates. Although expressions such as $v^i(y'(y))$ may be unwieldy, the systematic use of this notation avoids errors often occurring in more implicit formulations of coordinate changes.

to equivalence classes of curves, $[r(t)]$, traversing r in the same direction and at the same speed. Tangent vectors are pointers along these curves. However, in most applications it is better to consider them as directional derivatives, ∂_v, monitoring how smooth functions $f(r)$ change along these curves. The different pictures are equivalent, as follows from the representation of tangent vectors through their components in a system of coordinates, y. We learned how a system of coordinates, y^i, defines basis vectors, ∂_j, for the tangent spaces, and how generic tangent vectors ∂_v can be expanded in terms of these. Finally, we emphasized that the tangent space construction does not require the embedding of M in a vector space.

Tangent bundle and vector fields

The discussion of the previous section applied to tangent spaces of individual points, r. Although these spaces all have the same dimensionality, d, they differ from point to point (much like differently oriented planes in $\mathbb{R}^3$ are different two-dimensional vector spaces). **tangent** This motivates the introduction of a container set, the **tangent bundle**,
bundle

$$TM \equiv \bigcup_{r \in M} T_r M, \tag{V137}$$

as the formal union of all tangent spaces. The most important function of the tangent bundle **on manifold** is that it accommodates the vector fields of a manifold. A **vector field** ∂_v is a smooth map,

$$\partial_v : M \to TM, \qquad r \mapsto \partial_{v,r}, \tag{V138}$$

assigning a vector $\partial_{v,r}$ to each point r. The set of all vector fields on M is often denoted vect(M). If M is embedded in a vector space $\mathbb{R}^n$, this definition coincides with our earlier understanding of vector fields (think about this point).

A set of vector fields, $\{\partial_{v_j}\}$, $j = 1, \ldots, d$, such that for all r the vectors $\{\partial_{v_j,r}\}$ are a basis of $T_r M$, is **frame** called a **frame** of M. Most important in practice are **holonomic** **holonomic frames** defined by the vectors $\{(\partial_j)_r\}$ of a **frames** coordinate system (see figure). In these, vector fields ∂_v are expanded as $\partial_v = v^j \partial_j$, where the d compo-

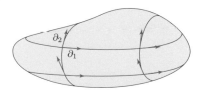

nents v^j are now smooth functions on the manifold, $v^j : M \to \mathbb{R}$, $r \mapsto v_r^j$. Their values are obtained by taking the directional derivatives, $v_r^j = \partial_{v,r} y^j$, of the coordinate functions according to the prescriptions discussed in the previous subsection (see Eq. (V132)).

flow   INFO   Vector fields on manifolds are often defined through **flows on the manifold**. A flow, Φ : $I \times X \mapsto M$, $(t, r) \mapsto \Phi_t(r)$, $I \subset \mathbb{R}$, $X \subset M$, is a family of curves on the manifold. They are defined such that $\Phi_t(r)$ is a curve in M passing through r at time $t = 0$, $\Phi_0(r) = r$. The pictorial analogy is that of a liquid streaming along the manifold. For example, the coordinate lines of the jth coordinate of a local chart define a flow through $\Phi_t(r) = r(y + e_j t)$. Likewise, the flow of a system of n differential equations introduced in Section C7.6 defines a system of curves where the domain of definition of the equations, $X \subset \mathbb{R}^n$, assumes the role of the manifold.

For each $r \in X$, $\Phi_t(r)$ is a representative curve for a vector tangent to the flow line at $t = 0$. Denoting this vector by $\partial_{v,r} \in T_r M$, a vector field, ∂_v, is defined through the set, $\{\partial_{v,r}\}$, of all flow vectors. For example, the ∂_j are the vector fields generated by the flow of the jth coordinate.

As an example, let $M = S^2$ be the unit sphere, locally parametrized by spherical coordinates, $y = (\theta, \phi)^T$. Consider the flow locally defined through $y(\Phi_t(r)) = (\theta + at, \phi + t)^T$, $a \in \mathbb{R}$. This is a system of spiraling curves on the sphere, as shown in the figure for $a = 0.2$. Its vector field has the coordinate representation $v_r^\theta = \partial_t \Phi_t^\theta\big|_{t=0}(r) = a$ and analogously 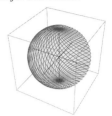 $v_r^\phi = 1$. The expansion of ∂_v in the holonomic basis thus reads $\partial_v = a\partial_\theta + \partial_\phi$.

V4.3 Summary and outlook

In this chapter we introduced elementary concepts of differential geometry. Differentiable manifolds were introduced as the geometric objects generalizing the concept of smooth surfaces. We emphasized a view in which manifolds need not be embedded into a larger space, and concrete calculations are performed in local coordinate systems. We reasoned that this approach leads to a cleaner and more universal description of manifolds (remember the definition of a circle independent of specific embeddings of the circle), but is also required by physics. For example, the Universe is a four-dimensional space-time manifold which is not a vector space, and not embedded into a larger space. Its mathematical description requires the concepts introduced in this chapter.

In the second part of the chapter we defined tangent vectors and vector spaces, and in this way introduced elements of linear algebra to the description of manifolds. In the next chapter we will extend the formalism to one integrating geometry, calculus and linear algebra. This unified framework will be the basis for the subsequent application of differential geometry in physics.

V5   Alternating differential forms

In this chapter we introduce differential forms as a tool to describe the geometry of manifolds, from both a mathematical and a physical perspective. Our discussion so far has shown how manifolds afford local linearizations – the system of tangent spaces. The focus was on the construction of the elements of these spaces, the tangent vectors, and on their description in terms of coordinates. We now take an essential next step and add elements of multilinear algebra to the framework. Adapting the concepts introduced in Chapter L10 for general vector spaces to the bundle of tangent spaces, we will introduce dual vectors and tensors of higher degree on manifolds.

These constructions synthesize elements of multilinear algebra and of calculus. Much as tensors generalize the concept of vectors, tensor fields emerge as generalizations of vector fields. Specifically, the field generalization of alternating multilinear forms defines a class of tensor fields called "differential forms". They play an important role in calculus and differential geometry, and are becoming increasingly important as a tool of modern physics. Indeed, many physical objects traditionally described as vector fields – force fields, vector potentials (sic), magnetic fluxes, etc. – are in fact differential forms. It is increasingly recognized that a language using differential forms provides cleaner and more natural descriptions of the physical phenomena related to these objects. Some of these applications will be touched upon in the final chapter of Part V where we outline how classical electrodynamics is formulated in the language of differential forms, and how this leads to an approach more powerful and physical than the traditional vector approach.

V5.1   Cotangent space and differential one-forms

In Section L10.2 we introduced the duals of vectors as linear maps of vectors into the reals. In the following, we will introduce the duals of tangent vectors and apply them to the geometric description of manifolds.

Cotangent space

cotangent space
The dual space of a tangent space, $T_r M$, is called the **cotangent space**, $T_r M^*$, to M at r. Its elements,

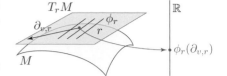

$$\phi_r : T_r M \to \mathbb{R}, \quad \partial_{v,r} \mapsto \phi_r(\partial_{v,r}), \qquad \text{(V139)}$$

are linear maps of tangent vectors into the real numbers and are called *one-forms*. In the figure, a form ϕ_r is represented by a pattern of parallel lines, as in Fig. L21(b). The definition of both tangent vectors and their duals is specific to a base point, as indicated by the subscript, r. However, this subscript is mostly suppressed in the notation, $\partial_{v,r} \to \partial_v$ and $\phi_r \to \phi$.

cotangent bundle
In analogy to the tangent bundle, we define the **cotangent bundle** of the manifold,

$$TM^* \equiv \bigcup_{r \in M} T_r M^*, \qquad \text{(V140)}$$

as the union of its cotangent spaces. Much as a vector field, $\partial_v : M \to TM$, $r \mapsto \partial_{v,r}$ is a map assigning vectors, $\partial_{v,r}$, to points, a **differential one-form** or just one-form, is a map,

one-form

$$\phi : M \to TM^*, \qquad r \mapsto \phi_r, \qquad \text{(V141)}$$

assigning dual vectors, ϕ_r, to points. The assignment is smooth in the sense that the action of the form on a vector field yields a smooth function, $\phi_r(\partial_{v,r})$, of r. The set of all one-forms on M is denoted as $\Lambda^1(M)$.

The definition above states that forms can be interpreted as maps,

$$\phi : \text{vect}(M) \to \Lambda^0(M), \qquad \partial_v \mapsto \phi(\partial_v), \qquad \text{(V142)}$$

zero-form
sending vector fields to functions, $\phi(\partial_v) : M \to \mathbb{R}$, $r \mapsto \phi_r(\partial_{v,r})$. In the present context, functions are frequently called "**zero-forms**", and the set of all functions is denoted by $\Lambda^0(M)$. The rationale behind this notation will become evident a little further on.

Note that the two equations, Eqs. (V141) and (V142), emphasize different aspects of the differential form. The first defines the differential form as map from points to forms, while the second defines it via its action on vector fields. These are two faces of the same coin and it makes sense to use the same symbol ϕ in both equations (although, strictly speaking, slightly different maps are described.)

Differential of functions

Consider a function, $f : \mathbb{R}^d \mapsto \mathbb{R}$, defined in $\mathbb{R}^d$. In Section V3.4 we introduced the total differential, df_x, as a map acting on vectors v as $df_x(v) = \partial_v f \equiv d_t|_{t=0} f(x + tv)$. This definition is readily generalized to the differential of functions on manifolds. We introduce **differential** the **differential of a function** $f : M \to \mathbb{R}$ as

$$df_r : T_r M \to \mathbb{R}, \qquad \partial_{v,r} \mapsto df_r(\partial_{v,r}) \equiv (\partial_v f)(r) = d_t f(r(0)). \qquad \text{(V143)}$$

This equation describes the local action of df_r on tangent space vectors $\partial_{v,r}$, where the directional derivative appearing on the right is the same as that in Eq. (V127). The corresponding *differential one-form*, df, extends the definition to an action on entire vector fields as

$$df : \text{vect}(M) \rightarrow \Lambda^0(M), \qquad \partial_v \mapsto df(\partial_v) = \partial_v f, \qquad \text{(V144)}$$

where $(df(\partial_v))_r = df_r(\partial_{v,r}) = (\partial_v f)(r)$ is defined locally as usual. Note that the definition $df(\partial_v) = \partial_v f$ given in Eq. (V143) can be read in two ways. We may consider $df(\partial_v)$ as the action of a function on argument vectors via the differential. Previously, in Eq. (V127), the same expression was interpreted as $\partial_v f$, i.e. the action of a vector on argument functions. In either interpretation the directional derivative of the function along the vector is taken. The differential is linear in its arguments, $df_r(\partial_u + \partial_v) = df_r(\partial_u) + df_r(\partial_v)$, and thus df is identified as a proper differential form.

EXAMPLE Consider the *height function*, $\cos\theta$, defined in the local domain of a spherical coordinate coordinate system, $y = (\theta, \phi)^\mathsf{T}$, on a sphere. The action of its differential on the coordinate vector fields is given by $d\cos\theta(\partial_\theta) = \frac{\partial}{\partial\theta}\cos\theta = -\sin\theta$ and $d\cos\theta(\partial_\phi) = \frac{\partial}{\partial\phi}\cos\theta = 0$. What is the geometric interpretation of these formulas?

Coordinate bases of cotangent space

Consider the differentials dy^i of a set of coordinate functions, y^i. These differentials define the dual basis of the holonomic basis ∂_j, and hence form a *basis of cotangent space*. To see how, consider the action of dy^i on ∂_j. This is the derivative of the ith coordinate in the direction of the jth coordinate basis vector,

$$dy^i(\partial_j) = \partial_j y^i = \delta^i{}_j, \qquad \text{(V145)}$$

which is the defining relation for a dual basis, see Eq. (L253). Arbitrary forms $\phi \in \Lambda^1(M)$ can be expanded in the dual basis as $\phi = \phi_i dy^i$. As with general dual basis expansions (Eq. (L255)), the components of the form are found by evaluating it on basis vectors, $\phi_i = \phi(\partial_i)$. Thus

$$\phi = \phi_i dy^i, \qquad \phi_i = \phi(\partial_i), \qquad \text{(V146)}$$

coordinate representation

where the components, $\phi_i = \phi_i(r)$, are functions defined in the domain of the coordinate chart. Expansions of this type are called **coordinate representations of forms**.

If the form is the differential of a function, $\phi = df$, its components are obtained as $df(\partial_i) = \partial_i f = \partial_{y^i} f(y)$. The expansion then assumes the form

$$df = \frac{\partial f(y)}{\partial y^i} dy^i, \qquad \text{(V147)}$$

and generalizes Eq. (V63) of Section V3.2 on the total differential of a function to the context of general differentiable manifolds.

As with tangent vectors, one often needs to *change coordinates in the representations of forms*. Consider a coordinate change expressing y' through y-coordinates via the map $y \mapsto y'(y)$, and the expansions of a form, $\phi = \phi_i(y')dy'^i = \phi_j(y)dy^j$, in these two systems.[44] The component functions in the y-system, $\phi_j(y)$, are obtained from those of the y'-system as

$\phi_j(y) = \phi(\partial_{y^j}) = (\phi_i(y')dy'^i)(\partial_{y^j}) \overset{(V144)}{=} \phi_i(y')\frac{\partial y'^i}{\partial y^j}$. Expressing the right-hand side through y-coordinates we are led to

$$\phi = \phi_i(y')\,dy'^i = \left[\phi_i(y')\frac{\partial y'^i}{\partial y^j}\right]_{y'=y'(y)} dy^j = \phi_j(y)\,dy^j. \tag{V148}$$

This formula is the twin of Eq. (V136) for the change of representation of vector fields: whereas the components of a vector field transform contravariantly with the Jacobi matrix, $J^i{}_j = \frac{\partial y'^i}{\partial y^j}$, as $v^i(y') = J^i{}_j v^j(y)$ (see Eq. (V135)), the components of the form transform covariantly with the inverse Jacobi matrix, $\phi_i(y') = \phi_j(y)(J^{-1})^j{}_i$. These transformation properties leave the action of a form on a vector invariant, $\phi(\partial_v) = \phi_i(y')v^i(y') = \phi_j(y)v^j(y)$, as they should.

EXAMPLE Let $U = \mathbb{R}^+ \times (0, 2\pi)$ be the polar coordinate domain of the slit plane, $M = \mathbb{R}^2\backslash(\mathbb{R}_0^+ \times \{0\})$. Consider the differential form $d\phi$ generated by the angular coordinate function ϕ. We now switch to a parametrization of M by Cartesian coordinates, $x = (x^1, x^2)^T$, with associated holonomic frame $\{\partial_1, \partial_2\}$. The coordinate representations of these vectors are identical to the standard Cartesian basis vectors, $v_j \overset{(V134)}{=} e_j$, and their dual vectors are the differential forms dx^i. They form a basis of TM^*, and hence there is an expansion $d\phi = \phi_1 dx^1 + \phi_1 dx^2$, with coordinate-dependent coefficients $\phi_i(x)$. To identify the latter, we have to evaluate $\phi_i = d\phi(\partial_i) = \partial_i\phi$ for $\phi(x) = \arctan(x^2/x^1)$. This gives $\phi_1 = -x^2/((x^1)^2 + (x^2)^2)$ and $\phi_2 = x^1/((x^1)^2 + (x^2)^2)$, or

$$d\phi = \frac{-x^2 dx^1 + x^1 dx^2}{(x^1)^2 + (x^2)^2}. \tag{V149}$$

($\to$ V5.1.1-2)

INFO Consider a point particle moving on a manifold M under the influence of a potential function, $\varphi : M \to \mathbb{R}$. This function describes how the energy of the particle varies with its position. We define the **force** acting on the particle as the differential one-form,

force

$$f = -d\varphi = -\frac{\partial\varphi(y)}{\partial y^i}dy^i, \tag{V150}$$

where y is a coordinate vector and $\varphi(y) = \varphi(r(y))$, as usual. To understand the meaning of this form, let $r_1 = r(y)$ and $r_2 = r(y + \Delta)$ be two points infinitesimally close to each other on the manifold, where $\Delta = \delta^j e_j$ is the displacement in the coordinate domain, and e_j are basis vectors in the y-space. The infinitesimal tangent vector $\partial_\Delta \equiv \delta^j \partial_j$ in $T_{r_1}M$ then is a vector with components δ^j, pointing from r_1 towards r_2. Equivalently, ∂_Δ may be considered as the tangent vector of a curve connecting r_1 and r_2 on the manifold. The energy difference $\delta\varphi \equiv \varphi(r_2) - \varphi(r_1)$ can be expressed as

$$\delta\varphi = \varphi(y + \Delta) - \varphi(y) \simeq \frac{\partial\varphi(y)}{\partial y^j}\delta^j \overset{(V145)}{=} \left(\frac{\partial\varphi(y)}{\partial y^i}dy^i\right)(\delta^j \partial_j) = -f(\partial_\Delta). \tag{V151}$$

---

[44] As for vectors (see footnote 6 of Section V4.2), the argument of a coefficient function indicates which coordinate systems it refers to.

This shows how the potential difference between nearby points equals the (negative of the) force form applied to the tangent vector, ∂_Δ, describing their separation. The definition (potential difference) = (force form acting on vector) differs from the traditional view where force is a vector. Representing this vector in a boldface notation $\mathbf{f}$, potential differences between points separated by a small vector $\mathbf{\Delta}$ are scalar products, $\delta\varphi = -\langle \mathbf{f}, \mathbf{\Delta} \rangle$. We will return to the discussion of the differences between the two formulations in Chapter V6.

Finally, a comment on *potentially confusing notation*: in the physics literature formulations like "let $d\varphi = \varphi(r_2) - \varphi(r_1)$ be the small potential difference between nearby points" are frequently used. Here, we express the same difference as $d\varphi(\partial_\Delta)$. The symbol $d\varphi$ represents a differential form which is not "small" in any sense. The smallness is in the argument vector, ∂_Δ. What the physics formulation actually refers to is the differential of the function, $d\varphi$, applied to a small argument vector. If one is aware of this interpretation, confusion can be avoided.

V5.2 Pushforward and pullback

At this stage, we have introduced the basic structures used to describe the geometry of manifolds, coordinates, tangent and cotangent space, vector fields and forms. We now take the next step to consider maps on manifolds. This will include maps $F : M \to L$ establishing connections between different manifolds whose dimensions, $\dim(M) \equiv d$ and $\dim(L) \equiv c$, need not even be the same, or maps $F : M \to M$ describing structures on a single manifold. An example with $d = c$ is a coordinate map, $F \equiv r : U \to L$, $y \mapsto r(y)$, of the manifold L, where $M \equiv U$ is the coordinate domain. Other examples include flows on a manifold, $F = \Phi_t : M \to M$, $r \mapsto \Phi_t(r)$ (see discussion in Section V4.2), real-valued functions, $F : M \to L \equiv \mathbb{R}$, $r \mapsto F(r)$, and many more.

pushforward and pullback idea
Our present goal is to understand how vector fields and forms behave under the action of a map. The general situation is illustrated in Fig. V20: given a map

$$F : M \to L, \qquad r \mapsto q(r), \tag{V152}$$

we can "push forward" a vector field $\partial_v \in \text{vect}(M)$ to a vector field $F_*\partial_v \in \text{vect}(L)$ on the image domain. Likewise, a differential form $\phi \subset \Lambda^1(L)$, defined on the image of F, can be "pulled back" to a differential form, $F^*\phi \in \Lambda^1(M)$, on M.

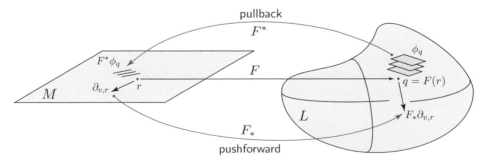

Fig. V20   Pushforward of vectors and pullback of forms: under pushforward, vectors $\partial_{v,r}$ on M map to vectors $F_*\partial_{v,r}$ on L. Forms ϕ_q on L get pulled back to forms $F^*\phi_q$ on M. Here the 1-form ϕ_q on the three-dimensional manifold L is visualized via a system of planes, and its pullback, the 1-form $F^*\phi_q$ on the two-dimensional M, via lines, see Section L10.5.

Pushforward

The *idea of pushforward*[45] is easily explained: assume that a vector $\partial_{v,r} \in T_r M$ describes the separation between two nearby points on the manifold M (such as the separation of two neigh-

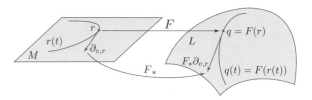

boring points on a curve). Under F these points get mapped to points which now are close in L. The image of $\partial_{v,r}$ under pushforward is the vector $F_* \partial_{v,r} \in T_q L$ describing the separation between these image points.

    The definition is made quantitative by representing $\partial_{v,r}$ through a curve $r(t)$. The **pushforward**, $F_* \partial_{v,r}$, is then represented by the curve $q(t) \equiv F(r(t))$. Pushforward is linear in its arguments,[46] $F_*(\partial_u + \partial_v) = F_*(\partial_u) + F_*(\partial_v)$, and so one usually writes $F_* \partial_v$ (as with linear maps of vector spaces). For later reference, we note that the pushforward of a vector under a *composite map*, $E \circ F : M \to K$, where $E : L \to K$, acts as

pushforward

$$(E \circ F)_* \partial_v = E_*(F_* \partial_v). \tag{V153}$$

Vectors get pushed in two steps, first by F_* from M to the intermediate manifold, L, and from there by E_* on to the final destination, K.

    Component representations of the pushforward vector, $F_* \partial_v$, are computed in the usual way: pick a coordinate system x on L and in this way define a holonomic basis ∂_{x^i}. The coefficients of the expansion $F_* \partial_v = (F_* \partial_v)^i \partial_{x^i}$ are then obtained by applying $F_* \partial_v$ to the coordinate functions, $(F_* \partial_v)^i = (F_* \partial_v) x^i = \mathrm{d}_t x^i(q(0)) = \mathrm{d}_t x^i(F(r(0))) \equiv \mathrm{d}_t F^i(r(0))$, where the shorthand $F^i = x^i(F)$ was used for the coordinate components of F. One often wants to express these components in terms of those of the vector $\partial_v = v^j \partial_{y^j}$, where a coordinate system y on M is assumed. This can be done by representing the map F itself in a coordinate language as $F(y) = F(r(y))$. Curves $r(t) = r(y(t))$ representing ∂_v then have coordinate representations $y(t)$, and $v^j = \partial_v y^j = \mathrm{d}_t y^j(0)$, as before. In this way, the components of the pushforward vector are obtained as $(F_* \partial_v)^i = \mathrm{d}_t F^i(y(0)) = \frac{\partial F^i(y(0))}{\partial y^j} \mathrm{d}_t y^j(0) = \frac{\partial F^i(y)}{\partial y^j} v^j$, and we have the expansion[47]

$$F_* \partial_v = \partial_{x^i} \left[\frac{\partial F^i(y)}{\partial y^j} v^j \right]_{y=y(x)}. \tag{V154}$$

The architecture of this formula underpins the meaning of pushforward as discussed. The **Jacobi matrix** $\frac{\partial F^i(y)}{\partial y^j}$, of dimension $c \times d$ (not necessarily square!) describes how the

Jacobi matrix

---

[45] The operation is commonly denoted as "pushforward", although "push-forward" or "push forward" would seem to be the correct ways of writing.

[46] Intuitively, the linearity of this map should be evident. It is made explicit by the coordinate representation below.

[47] Here, ∂_{x^i} is a vector, *not* a partial derivative. We write it to the left to emphasize the structure of the indices that are pairwise contracted.

coordinates $F^i(y)$ change under infinitesimal variations of their arguments y, and $\frac{\partial F^i}{\partial y^j} v^j$ is the change in F^i corresponding to the change in y described by the displacement vector ∂_v.

If $\partial_v : M \to TM$, $r \mapsto \partial_{v,r}$ is a *vector field*, the pushforward $F_* : TM \to TL$ may be applied to all vectors $\partial_{v,r}$. If, and only if, F is injective, this yields a vector field, $F_*\partial_v$, defined on the image $F(M) \subset L$. The injectivity condition is essential, for if F maps different points to the same image, $F(r_1) = F(r_2) = q$, a given vector field ∂_v defines different image tangent vectors, $F_*\partial_{v,r_1}$ and $F_*\partial_{v,r_2}$, in the tangent space T_qL. In this case, the image $F_*\partial_v$ does not represent a well-defined field. However, the *pushforward operation defined by injective maps F*,

$$F_* : \text{vect}(M) \to \text{vect}(F(M)), \qquad \partial_v \mapsto F_*\partial_v, \qquad (\text{V155})$$

is a map from the space of vector fields on M to those on $F(M)$.

pushforward
examples Various mathematical operations routinely applied in physics are pushforwards, although they are usually not understood in this way. Let us illustrate this point on two *examples*.

▷ Consider a *curve*, $\gamma : I \to L \subset \mathbb{R}^c$, $t \mapsto q(t)$. In this case F is the geometric image of the curve and the manifold $M = I$ is just an open one-dimensional interval. It makes little sense to distinguish between I and its coordinate representation,[48] and the curve parameter t may be considered as an element of I and as a coordinate at the same time.

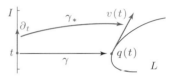

The manifold I has a one-dimensional tangent space, T_tI, spanned by a vector ∂_t. We may represent this vector via the representing curve $t(s) = t + s$, such that the single component of ∂_t is obtained as $d_s|_{s=0}(t + s) = 1$. Its pushforward, $\gamma_*\partial_t$, has the coordinate representation $d_s|_{s=0} q(t + s) = d_t q(t) = v(t)$, which is just the *curve velocity* in L. We observe that the tangent vectors of curves are the pushforwards of the unit vectors tangent to the parameter interval.

▷ Consider $M = L = \mathbb{R}^d$ and let $F = A$ be a *linear map*. As in the previous example, we do not distinguish between $M = \mathbb{R}^d$ and its (Cartesian) coordinate representation. The map then has the coordinate representation $y \mapsto y' = A(y) \equiv Ay$, with $y'^i = A^i_j y^j$ and Jacobi matrix $J^i_j = \frac{\partial y'^i}{\partial y^j} = A^i_j$. This demonstrates that the Jacobi matrix of a linear map equals the map itself. The pushforward of tangent vectors under a linear map, $A_*\partial_v \equiv \partial_{v'}$, have the coordinate representation $v'^i = A^i_j v^j$, as follows from Eq. (V154). As expected, they transform contravariantly under the map.

As a physical example, consider $M = L = \mathbb{R}^4$ and let $F = \Lambda$ be a *Lorentz transformation*, i.e. an element of the special orthogonal group, $O(1, 3)$, of linear maps Minkowski
metric leaving the **Minkowski metric**, $\eta = \text{diag}(1, -1, -1, -1)$, invariant, $\Lambda^T \eta \Lambda = \eta$ (see

[48] Formally, one may consider I parametrized by a "coordinate domain", $U = I$, where $U \ni y(t) = t \in I$ is an identity coordinate assignment. However, this formal overhead is excessive, unless different parametrizations, $s(t)$, of the same interval I are considered. This illustrates how in some cases one does not want to distinguish between a manifold and its coordinate representation. (However, think how a different choice of parametrization, $t \mapsto s(t)$, may be considered as a change of coordinates of I, and discuss how the formulas below change if the different "coordinate" s is used to parametrize I.)

Eq. (L211)). The Lorentz transformations, $\Lambda : \mathbb{R}^4 \to \mathbb{R}^4$, are linear and the above discussion implies that $v'^\mu = \Lambda^\mu{}_\nu v^\nu$ under the pushforward by these maps. In the parlance of special relativity it is said that space-time vectors v^μ transform contravariantly under Lorentz transformations. (Relativity is a field of physics where the distinction between co- and contravariance is widely acknowledged.)

Pullback

The reciprocal operation, pullback F^*, acts on forms, and it works in the reverse direction: a form $\phi \in \Lambda^1(F(M))$ defined on the image of $F(M) \subset L$ is "pulled back" to a form $F^*\phi \in \Lambda^1(M)$ on the pre-image M. The action is defined as in Section L10.8 for single vector spaces:

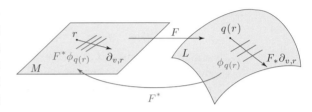

$$(F^*\phi_{q(r)})_r(\partial_{v,r}) \equiv \phi_{q(r)}(F_*\partial_{v,r}), \qquad (\text{V}156)$$

meaning that the value of the form $(F^*\phi_{q(r)})_r$ on a vector $\partial_{v,r} \in T_rM$ is found by acting with the form $\phi_{q(r)}$ on the pushforward of $\partial_{v,r}$. This operation is linear, $F^*\phi(\partial_u + \partial_v) = F^*\phi(\partial_u) + F^*\phi(\partial_v)$, and hence defines a one-form. Since it is evident that pullback connects forms at $q(r)$ with forms at r, reference to the base points are usually omitted, $F^*\phi_{q(r)} \to F^*\phi$. The extension of the pointwise pullback to the full image of F yields a map between differential forms,

$$F^* : \Lambda^1(F(M)) \to \Lambda^1(M), \qquad \phi \mapsto F^*\phi.$$

The pullback of a form by a *composite map*, $E \circ F : M \to K$, where $E : L \to K$, acts as

$$(E \circ F)^*\omega = F^*(E^*\omega), \qquad (\text{V}157)$$

so that ω is first pulled by E^* from the final destination, K, to the intermediary L, and from there by F^* back to M. This feature is a direct consequence of the definition.

Let us now find the *coordinate representation* of the pullback. Given coordinate representations $r(y)$ for M and $q(x)$ for L, a form ϕ and its pullback $F^*\phi$ can be expanded in their respective cotangent spaces as $\phi = \phi_i \, dx^i$ and $F^*\phi = (F^*\phi)_j \, dy^j$, respectively. We seek to establish a relation between the components $\phi_i(x)$ and $(F^*\phi)_j(y)$. To this end, we recall from Eq. (V146) that the components of ϕ are found by letting it act on L's tangent basis vectors, $\phi_i = \phi(\partial_{x^i})$. Likewise, the components of the pullback form, $F^*\phi$, are obtained by applying it to M's tangent basis vectors, $(F^*\phi)_j = (F^*\phi)(\partial_{y^j}) \overset{(\text{V}156)}{=} \phi(F_*\partial_{y^j})$. We expand the argument vector as $F_*\partial_{y^j} = (F_*\partial_{y^j})^i \partial_{x^i} \overset{(\text{V}154)}{=} \frac{\partial F^i}{\partial y^k}(\partial_{y^j})^k \partial_{x^i} \overset{(\text{V}131)}{=} \frac{\partial F^i}{\partial y^j} \partial_{x^i}$ to obtain $(F^*\phi)_j = \phi(\frac{\partial F^i}{\partial y^j} \partial_{x^i}) = \frac{\partial F^i}{\partial y^j} \phi_i$. This shows that the expansion of the pullback form is given by

$$F^*\phi = \phi_i(x(y))\frac{\partial F^i}{\partial y^j}\,dy^j\,. \tag{V158}$$

Note the structural similarity of this formula to Eq. (V154) for the pushforward. Again, the Jacobian matrix is the central building block. The difference is that this time it is contracted with the covariant components, ϕ_i, of the argument form, whereas for the pushforward the contraction was with the contravariant components, v^j, of the argument vector.

pullback   Let us discuss a few *examples of pullback operations*.
examples

▷ Consider the force one-form, $f = f_i\,dq^i$, defined in the vicinity of a *curve* $\gamma : I \to \mathbb{R}^d$, $t \mapsto q(t)$. (Here, $F \equiv \gamma$ plays the role of the map, $M = I$ is its one-dimensional argument manifold, $L = \mathbb{R}^d$ the image manifold, and we do not distinguish between these sets and their coordinate representations.) The form f may be pulled back to a form γ^*f defined on the parameter interval I. According to Eq. (V158), this form is given by

$$\gamma^*f = f_i\frac{dq^i}{dt}\,dt. \tag{V159}$$

Notice how this expression resembles that appearing in Eq. (V12) in the computation of the line integral of a force. There, the symbol dt denotes an integration measure, while here it is a one-form. We will explain the correspondence between the two expressions in Section V5.4 when we discuss the integration of forms.

▷ As in the analogous pushforward example, consider $M = L = \mathbb{R}^d$, let $F = A$ be a *linear map* with coordinate representation $y'^i = A^i{}_j y^j$ and ϕ' be a form on L. According to Eq. (V158), the pullback form, $\phi \equiv A^*\phi'$, on M has the coordinate representation $\phi_j = \phi'_i A^i{}_j$. This demonstrates that the components of forms transform covariantly under linear maps.

EXERCISE Discuss in what sense the pullback of linear algebra (Section L10.8) is a special case of the pullback defined here.

INFO Let us briefly address a subtlety concerning the *pushforward, pullback and the coordinate representations of forms and vectors*. Consider the differential of a coordinate function, dy^i. We can think of this form in two different ways: the first is as the differential dy^i of a coordinate function $y^i : M \mapsto \mathbb{R}$, where $r : U \to M, y \mapsto r(y)$ is a coordinate map (see figure) and we forget for a moment that coordinate functions are generally only defined on subsets $r(U) \subset M$ of the manifold. In this view, dy^i is a differential form on the tangent bundle

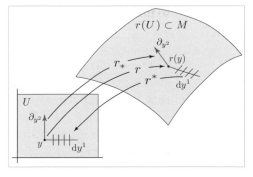

TM^*, and this is the interpretation emphasized above. Alternatively, we may consider the differential dy^i of the function $y^i : U \mapsto \mathbb{R}, y \mapsto y^i$, defined on the coordinate domain itself. The function y^i is now trivial, it just projects a vector $y \in U$ to its ith coordinate. Nevertheless, it is a valid function and $\mathrm{d}y^i$ is a valid differential form, now defined on the cotangent bundle of the coordinate domain, TU^*. The two definitions are closely related and generally denoted by the same symbol, $\mathrm{d}y^i$. For example, the symbol $\mathrm{d}\theta$ may refer to either a differential form defined in the tangent bundle of a two-sphere, or a differential form in the domain of spherical coordinates, (θ, ϕ), of the sphere. However, there are subtle differences, and it is sometimes worthwhile keeping in mind that $\mathrm{d}y^i$ on T^*M and T^*U really are two different forms. To see this in more detail, let us, for once, distinguish between objects defined on U and M by a subscript. Mathematically, $\mathrm{d}y^i_U$ is the pullback of $\mathrm{d}y^i_M$ under the map $r(y)$. This follows from the fact that if ∂_{v_U} is a vector in U with generating curve $y(t)$, we have $\mathrm{d}y^i_U(\partial_{v_U}) = \mathrm{d}_t y^i_U(y(0)) = \mathrm{d}_t y^i_M(r(y(0))) = \mathrm{d}y^i_M(r_*\partial_{v_U})$, where in the last equality we noted that the pushforward vector of ∂_{v_U} is generated by the curve $r(y(t))$.

pushforward and pullback vs. coordinate representations

The situation with coordinate basis vectors is similar. On the tangent bundle, we have defined $\partial_{y^j} \equiv \partial_{y^j,M}$ as vectors generated by the curves $r(y + te_j)$. These curves correspond to $y + te_j$ in the coordinate domain, where they generate vectors, $\partial_{y^j} \equiv \partial_{y^j,U}$, of TU commonly denoted by the same symbol. In U, ∂_{y_j} acts on functions as the ordinary partial derivative, $\partial_{y_j}f(y) \equiv \mathrm{d}_t\big|_{t=0}f(y + te_j)$. The partner vector $\partial_{y^j,M}$ in TM is generated by the curve $r(y + te_j)$ and hence is just the pushforward of $\partial_{y^j,U}$ under the coordinate map. Combining the pushforward and pullback relation between the vectors and forms, we obtain identical actions in U and M, $\delta^i{}_j = \mathrm{d}y^i_U(\partial_{y^j,U}) = r^*\mathrm{d}y^i_M(\partial_{y^j,U}) = \mathrm{d}y^i_M(r_*\partial_{y^j,U}) = \mathrm{d}y^i_M(\partial_{y^j,M})$, etc.

This bivariate picture implies alternative interpretations of various operations, notably those relating to coordinate changes. For example, Eq. (V148) was interpreted as a change of basis in the cotangent bundle TU^*. Alternatively, one may view $\phi_U \equiv \phi_{i,U}(y)\mathrm{d}y^i_U$ as a form in the U-domain of y-coordinates, and $\phi_{U'} \equiv \phi_{i,U'}(y')\mathrm{d}y'^i_{U'}$ as the same form represented in the U'-domain of y'-coordinates. The two are related as $\phi_U = (y' \circ y^{-1})^*\phi_{U'}$, i.e. one is the pullback of the other under the coordinate change $y' \circ y^{-1} : U \to U', y \mapsto y'(y)$. The two interpretations are fully consistent with each other, one may verify that the pullback formula (V158) applied to $F = (y' \circ y^{-1})$ is identical to the basis change formula Eq. (V148). As an exercise, discuss the analogous situation with vectors, and clarify how (V136) is obtained from (V154) if one interpretation is changed for the other.

Generally speaking, the form/vector-expanded-on-manifold view may be somewhat more geometrical, while the form/vector-expanded-in-coordinate domain is defined relative to open subsets of $\mathbb{R}^d$, and hence closer to calculus. It is often useful to switch between the two interpretations or to think about an operation in both ways at the same time.

Pushforward and pullback summary

To summarize, given a map $F : M \to L, r \mapsto q(r)$, pushforward F_* sends vectors $\partial_v \in TM$ to vectors $F_*\partial_v \in TL$, and pullback F^* sends forms $\phi \in TL^*$ to forms $F^*\phi \in TM^*$. If coordinates are used to represent elements of the manifolds as $r(y) \in M$ and $q(x) \in L$, respectively, and F defines a map of coordinates as $F : y \mapsto x(y)$, the two operations are described as

$$\boxed{\begin{aligned} F_*(v^j\partial_{y^j}) &= \partial_{x^i}\frac{\partial x^i}{\partial y^j}v^j, \\ F^*(\phi_i\mathrm{d}x^i) &= \phi_i\frac{\partial x^i}{\partial y^j}\mathrm{d}y^j. \end{aligned}} \qquad (V160)$$

The structure of these formulas is easy to remember: simply contract the matrix elements of the Jacobian $\frac{\partial x^i}{\partial y^j}$ with the components of either v^j or ϕ_i. ($\to$ V5.2.1-2)

V5.3   Forms of higher degree

REMARK Recapitulate Section L10.4.

In Section L10.4 we introduced alternating p-forms in vector spaces as antisymmetric tensors acting on p argument vectors. Differential p-forms are forms of degree p on manifolds, much as differential one-forms were one-forms on manifolds.

   In differential geometry, the degree of a differential form reflects the dimensionality of the geometric objects it describes. Specifically, one-forms are tailored to the representation of one-dimensional structures. For example, the force form describes the pairing of a directed force with a single displacement vector to a scalar work. Similarly, we will see that one-forms can be integrated over one-dimensional curves, etc. In this section, we generalize to differential p-forms and discuss how these objects represent higher-dimensional structures.

   A p-form, ω_r, defined with reference to a point r is an antisymmetric map,

$$\omega_r : \otimes^p(T_r M) \to \mathbb{R}, \qquad ((\partial_{v_1})_r, ..., (\partial_{v_p})_r) \mapsto \omega_r((\partial_{v_1})_r, ..., (\partial_{v_p})_r),$$

sending p tangent vectors to a number. The map is antisymmetric, i.e. the exchange of any two argument vectors yields a minus sign. We denote the set of all p-forms by $\Lambda^p(T_r M)$.

p-form   **Differential p-forms** are smooth extensions of pointwise-defined p-forms to forms defined on all of M,

$$\omega : \otimes^p \mathrm{vect}(M) \to \Lambda^0(M), \qquad (\partial_{v_1}, ..., \partial_{v_p}) \mapsto \omega(\partial_{v_1}, ..., \partial_{v_p}), \tag{V161}$$

where the function $\omega(\partial_{v_1}, ..., \partial_{v_p}) \in \Lambda^0(M)$ is defined through the pointwise construction, $\omega(\partial_{v_1}, ..., \partial_{v_p})(r) = \omega_r((\partial_{v_1})_r, ..., (\partial_{v_p})_r)$. The space spanned by all these forms is denoted by $\Lambda^p(M)$ (although $\Lambda^p(TM)$ would be a more accurate notation).

   For a tangent space of dimension $d = \dim(T_r M) = \dim(M)$, the space $\Lambda^p(T_r M)$ has dimension $\binom{d}{p}$ (see Section L10.4, p. 158), hence $\binom{d}{p}$ coordinate functions are required to specify a differential p-form. (We will see soon how these functions are obtained.) Since a d-dimensional vector space cannot support forms of degree higher than d, we have $\Lambda^{p>d}(M) = \{0\}$, the null space.

Wedge product

All operations introduced in Chapter L10 for p-forms of single vector spaces can be generalized to manifolds by smooth extension of pointwise definitions on manifolds. As an important example, consider the wedge product introduced in Section L10.6. The wedge-multiplication of two forms, $\phi_r \in \Lambda^p(T_r M)$ and $\psi_r \in \Lambda^q(T_r M)$, yields $\phi_r \wedge \psi_r \in \Lambda^{p+q}(T_r M)$, where the action of the product form on $p+q$ vectors, $\partial_{v_1}, ..., \partial_{v_{p+q}}$, is defined by Eq. (L272). Specifically, the product of two one-forms, $\phi_r, \psi_r \in \Lambda^1(T_r M)$, acts on vectors as

$$(\phi_r \wedge \psi_r)(\partial_u, \partial_v) = \phi_r(\partial_u)\psi_r(\partial_v) - \phi_r(\partial_v)\psi_r(\partial_u).$$

**wedge
product**

The extension of this operation defines the **wedge product (exterior product) of
differential forms**,

$$\wedge : \Lambda^p(M) \otimes \Lambda^q(M) \to \Lambda^{p+q}(M), \qquad (\phi, \psi) \mapsto \phi \wedge \psi, \qquad \text{(V162)}$$

where $\phi \wedge \psi$ acts on pairs of vectors fields, $(\phi \wedge \psi)(\partial_u, \partial_v)$, through the above point-wise
operation.

Coordinate representation

The wedge product is the key to the hierarchical construction of forms of higher degree,
and their coordinate representations. Consider a system of coordinate forms, $\{dy^i\}$, on a
d-dimensional manifold M. The expansion of a general form, $\omega \in \Lambda^p(M)$ reads as (see
Eq. (L276))

$$\omega = \frac{1}{p!}\omega_{i_1,\dots,i_p}(y) \bigwedge_{a=1}^p dy^{i_a}, \qquad \text{(V163)}$$

where $\bigwedge_{a=1}^p dy^{i_a} \equiv dy^{i_1} \wedge \dots \wedge dy^{i_p}$ is shorthand notation for a p-fold wedge product, and
the components $\omega_{i_1,\dots,i_p}(y)$, are smooth functions of the coordinates, antisymmetric in the
indices. These functions are obtained from the action of the form on the basis vector fields,

$$\omega_{i_1,\dots,i_p} = \omega(\partial_{i_1}, \dots, \partial_{i_p}). \qquad \text{(V164)}$$

As with forms in single vector spaces (see Eq. (L269)), it is sometimes convenient to switch
to an ordered sum over indices, $\omega = \sum_{i_1 < \dots < i_p} \omega_{i_1,\dots,i_p}(y) \bigwedge_{a=1}^p dy^{i_a}$.

**coordinate
transforma-
tion of
general form**

The change of representation under a *change of coordinates*, $y \mapsto y'(y)$, is obtained by
letting the y'-representation of the form act on the basis vectors ∂_{y^j}. Using $dy'^i(\partial_{y^j}) = \frac{\partial y'^i(y)}{\partial y^j}$, this yields

$$\omega = \frac{1}{p!}\omega_{i_1,\dots,i_p}(y') \bigwedge_{a=1}^p dy'^{i_a} = \frac{1}{p!}\left[\omega_{i_1,\dots,i_p}(y') \frac{\partial y'^{i_1}}{\partial y^{j_1}} \cdots \frac{\partial y'^{i_p}}{\partial y^{j_p}} \right]_{y'=y'(y)} \bigwedge_{b=1}^p dy^{j_b}, \qquad \text{(V165)}$$

in generalization of the one-form transformation rule (V148).

area form

EXAMPLE In $M = \mathbb{R}^2$ consider the Cartesian **area form**, $\omega = dx^1 \wedge dx^2$, a top-form defined
by a single component $\omega_{12} = 1$. It is called an area form because applied to a pair of vectors,
$\omega(\partial_u, \partial_v) = dx^1(\partial_u)dx^2(\partial_v) - dx^2(\partial_u)dx^1(\partial_v) = u^1 v^2 - u^2 v^1$, it yields the oriented area spanned by
the arguments. The component $\omega_{\rho,\phi}$ defining the expansion of ω in polar coordinates is given by

$$\omega_{\rho,\phi} \stackrel{\text{(V164)}}{=} \omega(\partial_\rho, \partial_\phi) = dx^1(\partial_\rho)dx^2(\partial_\phi) - dx^2(\partial_\rho)dx^1(\partial_\phi)$$

$$= (\partial_\rho x^1)(\partial_\phi x^2) - (\partial_\rho x^2)(\partial_\phi x^1) \stackrel{\text{(V15a)}}{=} \rho,$$

and so we have

$$\omega = \rho \, d\rho \wedge d\phi. \qquad \text{(V166)}$$

More generally, its representation in general coordinates, (y^1, y^2) reads

$$\omega = \left(\frac{\partial x^1}{\partial y^1} \frac{\partial x^2}{\partial y^2} - \frac{\partial x^2}{\partial y^1} \frac{\partial x^1}{\partial y^2} \right) dy^1 \wedge dy^2, \tag{V167}$$

where the expression in parentheses is the area of the parallelogram spanned by the Cartesian representation, $v_j = \partial_{y^j} x$, of the coordinate basis vectors. Alternatively, the coefficient may be interpreted as $\det\left(\frac{\partial x}{\partial y}\right)$, where $\left(\frac{\partial x}{\partial y}\right)$ is the Jacobian of the coordinate transformation $y \mapsto x(y)$. We will revisit this connection in our discussion of top-forms. ($\to$ V5.3.1-2)

INFO   The wedge product frequently appears in the construction of differential forms of physical significance. As an example, consider a situation where a large number of particles are confined to a region of space, V. Add a time axis to obtain a four-dimensional manifold, $M = \mathbb{R} \times V$, parametrized by Cartesian coordinates as $x = (x^0, x^1, x^2, x^3)$, where $x^{1,2,3}$ are coordinates of V, $x^0 \equiv ct$ is a time-like coordinate, and c a characteristic velocity (such as the speed of light).[49]

The *density* of particles, $\rho(x)$, is a space-time dependent function defined such that the number of particles in an infinitesimal cubical box of spatial volume $\delta^1 \delta^2 \delta^3$, at x, is given by $\rho(x)\delta^1 \delta^2 \delta^3$. In form language this is described by a *density form*,[50] whose application to the three vectors spanning the box yields the number of particles in it. This defines a three-form

$$\rho = j_{123} \, dx^1 \wedge dx^2 \wedge dx^3,$$

with $j_{123}(x) \equiv \rho(x)$. Applied to three infinitesimal tangent vectors, $\partial_{\Delta_i} = \delta^i \partial_i$ (no summation), this form yields $j_{123} \, dx^1 \wedge dx^2 \wedge dx^3 (\partial_{\Delta_1}, \partial_{\Delta_2}, \partial_{\Delta_3}) = j_{123}(x)\delta^1 \delta^2 \delta^3 = \rho(x)\delta^1 \delta^2 \delta^3$, as required. (Of course the construction is not specific to cubical boxes. Show that the application of the density form to three generic, positively oriented, infinitesimal vectors, $\partial_{\delta u}, \partial_{\delta v}, \partial_{\delta w}$, with Cartesian representations $\delta u, \delta v, \delta w$, yields j_{123} times the geometric volume of the parallelepiped spanned by these vectors. Maybe recapitulate Section L10.5 on the geometric meaning of volume forms.)

A slight modification of the construction describes the flow of current. The *current density* in the 3-direction, $j_{012}(x)$, is a function defined such that $j_{012}(x)\delta^0 \delta^1 \delta^2$ equals the number of particles passing within time $\delta_t = \delta^0/c$ through a rectangular area element lying within the $1-2$-plane at x and having area $\delta^1 \delta^2$. This equals the number of particles contained in a space-time box of spatial area $\delta^1 \delta^2$, temporal extension δ^0, and volume $\delta^0 \delta^1 \delta^2$, see Fig. V21. In analogy to the density form, we thus define a *current form*,

$$j = j_{012} \, dx^0 \wedge dx^1 \wedge dx^2,$$

whose application to the tangent vectors $\partial_{\Delta_0}, \partial_{\Delta_1}, \partial_{\Delta_2}$ yields $j_{012}(x)\delta^0 \delta^1 \delta^2$.

Forms with weight functions j_{023} and j_{031}, describing current through the $2-3$ and $3-1$ planes, respectively, are defined analogously. The application of the current form $\frac{1}{2} j_{0ij} \, dx^0 \wedge dx^i \wedge dx^j$ to an argument $(\partial_{\Delta_0}, \partial_{\delta v}, \partial_{\delta w})$, where δv and δw are spatial vectors with Cartesian components δv^i and δw^i, yields $j_{0ij} \delta^0 \delta v^i \delta w^j$. (Verify this result. In doing so, keep in mind that form-coefficients are antisymmetric in indices and that summations over Latin indices are spatial $1 \leq i, j \leq 3$.) This equals the number of particles flowing in time δ^0 through the area element spanned by δv and δw.

---

[49] The factor c is included to give all four coordinates, x^μ, $\mu = 0, 1, 2, 3$, the dimension of length. While many texts set $c = 1$ for simplicity, this factor is physically important, and we will keep it.

[50] There is a caveat here: consider an inversion of all coordinates, $x^i \to -x^i$, or other orientation non-preserving transformations. While this should not affect the number of particles per unit volume, $\rho = j_{123} \, dx^1 \wedge dx^2 \wedge dx^3$

twisted form   will change sign. In mathematics, the situation is dealt with via the introduction of **twisted alternating forms**. In essence, these are differential forms augmented with a sense of orientation. Under an orientation-changing coordinate transformation, twisted forms are defined to not change sign. Twisted forms of highest degree are

density   called **densities**. As long as one is working in one fixed orientation class, twisted forms do not differ from conventional forms, and this is the setting assumed here.

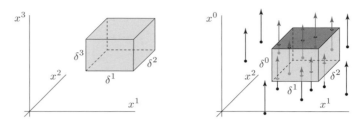

Fig. V21   On the definition of density and current flow via differential forms. Discussion, see text.

current   The full information on both density and currents is contained in the general **current three-form**,
three-form

$$j = \tfrac{1}{3!} j_{\mu\nu\sigma}\, dx^{\mu} \wedge dx^{\nu} \wedge dx^{\sigma}, \tag{V168}$$

where $0 \leq \mu, \nu \leq 3$ now run over all indices. This form contains four component functions, j_{123} for the particle density, and j_{0ij} for the current, respectively, all of the same physical dimension, $[\text{length}]^{-3}$. While the current form has been defined for a Cartesian system, Eq. (V168) can be expressed in terms of another coordinate system using the transformation $y \mapsto x(y)$. In an info section on p. 511 we will discuss how the present approach relates to the traditional description of current in terms of vector fields.

Top-dimensional forms and orientation

top-form   Highest-degree or **top-dimensional forms** are described by a single component function. On a d-dimensional manifold, their coordinate representation (V163) assumes the simple form[51]

$$\omega = \omega(y)\, dy^{1} \wedge \ldots \wedge dy^{d}, \tag{V169}$$

with a single *weight function*, $\omega(y)$. Top-dimensional forms play an important role in the global characterization of manifolds and in integration theory, see Section V5.4. They also

orientation   describe the **orientation of a manifold**.
of manifold   The heuristic meaning of orientation is that it gives a manifold an inside and an outside.
Möbius strip This is not always possible, the best known example being the **Möbius strip** (see p. 448). A manifold is orientable if, and only if, a top-dimensional form with globally non-vanishing weight function $\omega(y)$ exists. The condition $\omega(y) \neq 0$ is independent of the chosen coordinate system: under a coordinate change $y \mapsto y'(y)$, the form (V169) changes according to (V165), which for $p = d$ simplifies to

$$\omega = \omega(y') \bigwedge_{a=1}^{d} dy'^{a} = \omega(y'(y)) \bigwedge_{a=1}^{d} \frac{\partial y'^{a}}{\partial y^{i_a}} dy^{i_a}$$

$$= \omega(y'(y))\, \epsilon^{i_1,\ldots,i_d} \bigwedge_{a=1}^{d} \frac{\partial y'^{a}}{\partial y^{i_a}}\, dy^{a} \stackrel{(\text{L.159})}{=} \det\left(\frac{\partial y'}{\partial y}\right) \omega(y'(y)) \bigwedge_{a=1}^{d} dy^{a}. \tag{V170}$$

---

[51] Top-forms on higher-dimensional manifolds are often denoted by ω and we will adopt this convention here.

In the second line we ordered the forms under the wedge product in ascending order, and kept track of the ensuing sign factor by the Levi–Civita symbol $\epsilon^{i_1,\ldots,i_n}$ defined in Eq. (L7). Combination with the derivative factors leads to the appearance of the determinant of the Jacobi matrix, $\frac{\partial y'}{\partial y}$, in the final expression and the definition of the y-representation[52]

$$\omega(y) = \omega(y'(y)) \det\left(\frac{\partial y'}{\partial y}\right). \tag{V171}$$

(Jacobi) matrices describing coordinate changes have maximal rank and non-vanishing determinants. If $\omega(y')$ is non-vanishing, $\omega(y)$ will therefore be non-vanishing, too.

Orientable manifolds can be covered by coordinate charts $\{y^i\}$, $\{y'^i\}$, ... with positive weights $\omega(y)$, $\omega(y'),\ldots > 0$.[53] The Jacobi matrices of transformations between these coordinates have positive determinants and at the same time describe the change of tangent space bases, $\partial_{y^j} = \frac{\partial y'^i}{\partial y^j}\partial_{y'^i}$ (see Eq. (V136)). This means that these bases are all equally oriented. The Möbius strip introduced on p. 448 is an example of a manifold lacking orientability. Attempts to extend a basis to one covering its full tangent always lead to orientation conflicts, see the figure.

volume form **EXAMPLE** Consider the **volume form**, $\omega = dx^1 \wedge dx^2 \wedge dx^3$, of three-dimensional space $\mathbb{R}^3$. Under the transformation $y \mapsto x(y)$ expressing Cartesian through spherical coordinates, it transforms to $\omega = \det\left(\frac{\partial x}{\partial y}\right) dr \wedge d\theta \wedge d\phi$, with Jacobi determinant $\det\left(\frac{\partial x}{\partial y}\right) \overset{(V49)}{=} \rho^2 \sin\theta$. Thus, the *volume form in spherical coordinates* is given by

$$\omega = r^2 dr \wedge \sin\theta\, d\theta \wedge d\phi. \tag{V172}$$

stereographic coordinates **EXERCISE** The **stereographic coordinates** $z = (z^1, z^2)^{\mathrm{T}}$ are a coordinate system of the unit sphere S^2. As indicated in Fig. V22, they are defined by projecting points r on the sphere onto points q lying in a plane through the equator. (Sometimes, a plane tangent to the south pole is used as a projection plane instead.) In this way, the surface of the sphere gets mapped to a plane. Projections of this type find applications in differential geometry, but also in cartography.

stereographic coordinates Apply geometric reasoning to verify that the map function expressing spherical through stereographic coordinates, $z \mapsto y(z)$, is given by[54]

$$y : \mathbb{R}^2 \setminus (\mathbb{R}_0^+ \times \{0\}) \longrightarrow (0, \pi) \times (0, 2\pi),$$

$$z = \begin{pmatrix} z^1 \\ z^2 \end{pmatrix} \longmapsto y(z) = \begin{pmatrix} \theta(z) \\ \phi(z) \end{pmatrix} = \begin{pmatrix} 2\arctan(1/\rho) \\ \arctan(z^2/z^1) \end{pmatrix}, \quad \rho \equiv \sqrt{(z^1)^2 + (z^2)^2}. \tag{V173}$$

---

[52] The transformation of the area two-form, Eq. (V167), was a specific realization of this result.

[53] If a coordinate system has $\omega(y) < 0$, the inversion of one coordinate, $y^1 \to -y^1$, flips the sign of $\omega(y)$.

[54] The projection excludes the positive z^1-axis, $\{z^1 > 0, z^2 = 0\}$, corresponding to the set $\{\rho > 0, \phi = 0\}$, analogous to the exclusion of the positive x^1-axis for polar coordinates. The points $z = (0,0)^{\mathrm{T}}$ and "$z = \infty$" (i.e. the horizon of infinitely large z-values, $\rho = \infty$), corresponding to $\theta = \pi$ and $\theta = 0$, are also excluded.

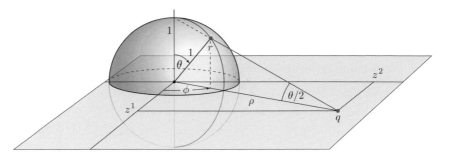

Fig. V22 Stereographic projection of the sphere. To each point r on the sphere we assign a point q in a plane through the equator by shining a light ray from the north pole through r. The projection point q is defined as the intersection of the light ray with the plane. Further discussion, see text.

Also verify that $\sin(\theta) = \frac{2\rho}{1+\rho^2}$. The inverse map, $y \mapsto z(y)$, is

$$z : (0, \pi) \times (0, 2\pi) \longrightarrow \mathbb{R}^2 \setminus (\mathbb{R}_0^+ \times \{0\}),$$

$$y = \begin{pmatrix} \theta \\ \phi \end{pmatrix} \longmapsto z(y) = \begin{pmatrix} z^1(y) \\ z^2(y) \end{pmatrix} = \begin{pmatrix} \rho \cos \phi \\ \rho \sin \phi \end{pmatrix}, \qquad \rho = \frac{1}{\tan(\theta/2)}. \qquad \text{(V174)}$$

Notice that positive increments in the θ-coordinate lead from north to south and are represented by negative increments in the stereographic coordinates.

area form   On the sphere, we define the spherical **area form** or **solid-angle form**

$$\omega \equiv \sin \theta \, \mathrm{d}\theta \wedge \mathrm{d}\phi. \qquad \text{(V175)}$$

The meaning of this form is explained in Chapter V6, where we show that it computes area elements on the unit sphere from pairs of tangent vectors. Use Eq. (V171) for the transformation $z \mapsto y(z)$ to verify that the stereographic representation of the area form is given by

$$\omega = w(z) \, \mathrm{d}z^1 \wedge \mathrm{d}z^2, \qquad w(z) = \frac{-4}{\left(1 + (z^1)^2 + (z^2)^2\right)^2}. \qquad \text{(V176)}$$

The negative sign of the weight function, $w(z)$, is inherited from the Jacobi determinant, $\det(\frac{\partial y}{\partial z})$; it implies that the stereographic projection is an orientation-changing map.[55] Observe how stereographic projection maps small areas around the north pole on the sphere, where θ is small, onto large regions in the outer plane, where ρ is large. This explains why the magnitude, $|w(z)|$, of the weight function decreases for large values of the z-coordinates: large values of z represent only small geometric areas on the sphere.
($\rightarrow$ V5.3.3-4)

Pullback

The pullback of forms $\omega \in \Lambda^p(L)$ under a map $F : M \to L$ is computed by generalization of the rule for one-forms:

$$\boxed{F^*\omega(\partial_{v_1}, \ldots, \partial_{v_p}) \equiv \omega(F_*\partial_{v_1}, \ldots, F_*\partial_{v_p}),} \qquad \text{(V177)}$$

---

[55] An orientation-preserving variant of the map can be obtained by coordinate exchange, $z^1 \leftrightarrow z^2$. However, the current version of the map defines an overall more natural coordinate assignment.

where the left-hand side defines $F^*\omega \in \Lambda^p(M)$. Let

$$\omega = \frac{1}{p!} \sum_{i_1,\dots,i_p=1}^{c} \omega_{i_1,\dots,i_p}(x) \bigwedge_{a=1}^{p} \mathrm{d}x^{i_a}$$

be a p-form on L expanded in local coordinates, $(x^1, \dots, x^c)$. The *coordinate representation of the pullback*, $F^*\omega$, to M is found as in the case of one-forms, see Eq. (V158): application of $F^*\omega$ to the basis vectors, $\partial_j \equiv \partial_{y^j}$, on M yields the components as $(F^*\omega)_{j_1,\dots,j_d} = F^*\omega(\partial_{y^{j_1}}, \dots, \partial_{y^{j_d}}) = \omega(F_*\partial_{y^{j_1}}, \dots, F_*\partial_{y^{j_d}})$. The pushforward $F_*\partial_{y^j} = \frac{\partial F^i}{\partial y^j}\partial_{x^i}$, then leads

pullback of general form

to the expansion

$$F^*\omega = \frac{1}{p!} \sum_{j_1,\dots,j_p=1}^{d} \sum_{i_1,\dots,i_p=1}^{c} \omega_{i_1,\dots,i_p}(x(y)) \frac{\partial F^{i_1}}{\partial y^{j_1}} \cdots \frac{\partial F^{i_p}}{\partial y^{j_p}} \bigwedge_{a=1}^{p} \mathrm{d}y^{j_a}. \qquad (V178)$$

The formula shows that the pullback of the constituent one-forms, $F^*\mathrm{d}x^i = \frac{\partial F^i}{\partial y^j}\mathrm{d}y^j$, defines that of the p-form.

Exterior derivative

So far, we did little more than extending pointwise-defined operations of multilinear algebra to the tangent bundle of manifolds. The operation to be introduced next is different in that it essentially relies on the presence of differentiable structures on manifolds: the

exterior derivative

exterior derivative, d, describes how forms change along specified directions on the manifold. As a directional operation, it requires a p-form, ϕ, and a tangent vector, ∂_v, as arguments. The exterior derivative, $\mathrm{d}\phi$, of ϕ then acts on $(p + 1)$ vectors, comprising the arguments of the p-form, and the directional argument, ∂_v. It is defined to be linear and antisymmetric in its arguments, and thus represents a $(p + 1)$-form.

The exterior derivative establishes relations between forms of neighboring degree, p, and $p + 1$. Its importance is reflected in the fact that the theory of differential forms is

exterior calculus

sometimes called **exterior calculus**. For example, the exterior derivative maps one-forms probing changes of quantities under directional increments to two-forms whose arguments are area elements. Below, we will see how these connections play an important role both in the geometry of manifolds and in physics.

The exterior derivative is an operator,

$$\mathrm{d} : \Lambda^p(M) \to \Lambda^{p+1}(M), \qquad \phi \mapsto \mathrm{d}\phi, \qquad (V179)$$

defined to obey the following conditions:

▷ d is *linear*: for $a, b \in \mathbb{R}$,

$$\mathrm{d}(a\phi + b\psi) = a\,\mathrm{d}\phi + b\,\mathrm{d}\psi;$$

graded Leibniz rule

▷ d obeys the **graded Leibniz rule**

$$\mathrm{d}(\phi \wedge \psi) = \mathrm{d}\phi \wedge \psi + (-)^{\mathrm{degree}(\phi)}\phi \wedge (\mathrm{d}\psi);$$

▷ d is *nilpotent*,

$$\mathrm{d} \circ \mathrm{d} = 0;$$

▷ when acting on a zero-form, $\phi \in \Lambda^0(M)$, i.e. a function, d acts as the *differential* discussed in Section V5.1, $\mathrm{d}\phi = \frac{\partial \phi}{\partial y^i}\mathrm{d}y^i$.

These conditions uniquely specify d. This is best seen by letting d act on the coordinate representation of a form, $\phi = \frac{1}{p!}\phi_{i_1,\ldots,i_p}\bigwedge_{a=1}^{p}\mathrm{d}y^{i_a}$. Application of the Leibniz rule simply yields $\mathrm{d}(\phi_{i_1,\ldots,i_p}\bigwedge_{a=1}^{p}\mathrm{d}y^{i_a}) = (\mathrm{d}\phi_{i_1,\ldots,i_p}) \wedge \bigwedge_{a=1}^{p}\mathrm{d}y^{i_a}$, since $\mathrm{d}\bigwedge_{a=1}^{p}\mathrm{d}y^{i_a} = 0$ by

coordinate representation nilpotency, $\mathrm{d}^2 = 0$. The expansion $\mathrm{d}\phi_{i_1,\ldots,i_p} \overset{(V147)}{=} \partial_j\phi_{i_1,\ldots,i_p}\mathrm{d}y^j$ then yields the **coordinate representation of the exterior derivative**

$$\boxed{\mathrm{d}\left(\frac{1}{p!}\phi_{i_1,\ldots,i_p}\bigwedge_{a=1}^{p}\mathrm{d}y^{i_a}\right) = \frac{1}{p!}\partial_j\phi_{i_1,\ldots,i_p}\,\mathrm{d}y^j\wedge\bigwedge_{a=1}^{p}\mathrm{d}y^{i_a}.} \qquad (V180)$$

Note that $\mathrm{d}\phi$ contains the derivatives of the component functions $\partial_j\phi_{i_1,\ldots,i_p}$. This underpins how d monitors the rate of change of ϕ along the manifold, different from the pointwise-defined operations discussed previously. A form that is stationary under the exterior

closed derivative, $\mathrm{d}\phi = 0$, is called a **closed form**. Conversely, a form which is the derivative

exact of another one, $\phi = \mathrm{d}\kappa$, is called an **exact form**. Every exact form is closed, $\mathrm{d}(\mathrm{d}\kappa) = 0$, but the reverse is not necessarily true. The classification of forms which are closed but not exact is a beautiful problem of mathematics, known under the name de Rham homology.

EXAMPLE In $M = \mathbb{R}^3$ consider the one-form $\lambda = \lambda_1\mathrm{d}x^1 + \lambda_2\mathrm{d}x^2 + \lambda_3\mathrm{d}x^3$. Application of Eq. (V180) yields

$$\mathrm{d}\lambda = (\partial_1\lambda_2 - \partial_2\lambda_1)\,\mathrm{d}x^1\wedge\mathrm{d}x^2 + (\partial_2\lambda_3 - \partial_3\lambda_2)\,\mathrm{d}x^2\wedge\mathrm{d}x^3 + (\partial_3\lambda_1 - \partial_1\lambda_3)\,\mathrm{d}x^3\wedge\mathrm{d}x^1. \quad (V181)$$

In $M = \mathbb{R}^3$ consider the two-form $\phi = \phi_1\,\mathrm{d}x^2\wedge\mathrm{d}x^3 + \phi_2\,\mathrm{d}x^3\wedge\mathrm{d}x^1 + \phi_3\,\mathrm{d}x^1\wedge\mathrm{d}x^2$. Then

$$\mathrm{d}\phi = (\partial_1\phi_1 + \partial_2\phi_2 + \partial_3\phi_3)\,\mathrm{d}x^1\wedge\mathrm{d}x^2\wedge\mathrm{d}x^3. \qquad (V182)$$

The components of $\mathrm{d}\lambda$ and $\mathrm{d}\phi$ resemble those of the curl and divergence operations of vector calculus, respectively. We will substantiate this observation in Section V6.3. ($\rightarrow$ V5.3.5-6)

It is straightforward (do it!) to compute the pullback, $F^*(\mathrm{d}\phi)$, of the exterior derivative Eq. (V180) via the pullback formula (V178). The same formulas determine the exterior derivative, $\mathrm{d}(F^*\phi)$, of the pullback of the form. The results agree, proving that *pullback and exterior derivative commute*:

$$\boxed{\mathrm{d}F^*\phi = F^*\,\mathrm{d}\phi.} \qquad (V183)$$

This identity can be applied to compute combinations of pullbacks and derivatives in the order which is most economical.

area form **EXAMPLE** The **area form** in polar coordinates, $\omega = \rho\,\mathrm{d}\rho \wedge \mathrm{d}\phi$ (see Eq. (V166)), is an exact form, expressible as the exterior derivative, $\omega = \mathrm{d}\kappa$, of the one-form $\kappa = \frac{1}{2}\rho^2\mathrm{d}\phi$. Now consider the pullback y^* to Cartesian coordinates, $x = (x^1, x^2)^T$, under the map $y : x \mapsto y(x) = (\theta, \phi)^T(x)$.

Equation (V183) implies the relation $y^*\omega = y^*\mathrm{d}\kappa = \mathrm{d}y^*\kappa$, i.e. we may either first differentiate κ and then pull back, $y^*\mathrm{d}\kappa$, or first pull back and then differentiate in Cartesian coordinates, $\mathrm{d}y^*\kappa$. Verify that both yield $\mathrm{d}x^1 \wedge \mathrm{d}x^2$. ($\rightarrow$ V5.3.7-8)

EXERCISE Consider the area two-form on the sphere, $\omega = \sin\theta\,\mathrm{d}\theta \wedge \mathrm{d}\phi$ (see Eq. (V175)). It can be represented (check!) as the exterior derivative of a one-form, $\omega = \mathrm{d}\kappa$, where $\kappa = -\cos\theta\mathrm{d}\phi$. Apply Eq. (V173) to verify that its pullback, $y^*\kappa$, to the stereographic plane (see Exercise on p. 491) reads

$$y^*\kappa = \frac{1-\rho^2}{\left(1+\rho^2\right)\rho^2}(z^2\mathrm{d}z^1 - z^1\mathrm{d}z^2), \qquad \rho^2 = (z^1)^2 + (z^2)^2.$$

Now compute $\mathrm{d}(y^*\kappa)$ to check that it coincides with $y^*\omega \overset{(V183)}{=} y^*(\mathrm{d}\kappa)$, as given by Eq. (V176). ($\rightarrow$ V5.3.9-10)

V5.4 Integration of forms

In previous chapters, we have introduced different types of integrals, beginning with the elementary one-dimensional integral of functions, then moving on to higher-dimensional integrals of functions and integrals of vector fields. Some of these integrals required the construction of specific area and volume elements (see Chapter C4) or the presence of a metric.[56] Looking at the situation at large, one may get the impression that each particular environment requires its own custom-made integral.

In this section we will see that differential forms can be applied to construct a more transparent and unified approach to integration. (i) In mathematics, integration is almost always defined as an integration of forms. (ii) The natural objects to integrate over d-dimensional manifolds are d-dimensional forms, one-forms over curves, two-forms over surfaces, three-forms over volumes, etc. (iii) No metric is required to define these integrals, and (iv) all variants of integrals introduced previously are special cases[57] of the unified concept.

Integration of one-dimensional forms in one-dimensional space

Consider a one-dimensional form, $\phi = \phi(y)\mathrm{d}y$, defined on an interval $I = (a,b)$. We define its integral over I as the ordinary Riemann integral of the weight function $\phi(y)$ over I:

$$\int_I \phi = \int_I \phi(y)\mathrm{d}y \equiv \int_a^b \phi(y)\mathrm{d}y. \tag{V184}$$

Thus the basis form, $\mathrm{d}y$, is formally replaced by the Riemann integration measure, also denoted $\mathrm{d}y$. For example, if $\phi = y\mathrm{d}y$, and $I = (0,1)$, then $\int_I \phi = \int_0^1 y\mathrm{d}y = \frac{1}{2}$. To understand the rationale behind this identification, consider the interval $I = (a,b)$ as a

---

[56] The integral of a vector field over a surface in three-dimensional space involved a vector field normal to that surface, see Eq. (V87). The definition of the normal field requires a metric.

[57] In cases where the identification of a traditional integral with a differential form integral looks awkward, the former is to blame. This happens, e.g., when the traditional integral requires the presence of a metric but the form integral does not.

Fig. V23   On the identification of the integral of one-forms with the Riemann integral.

straight, one-dimensional curve, partitioned into N infinitesimal segments $(y_\ell, y_{\ell+1})$, with $y_\ell = \ell\delta$ and $\delta = (b-a)/N$. To each segment we associate a vector, $\partial_{\Delta,\ell} = \delta\,\partial_y$, in the tangent space $T_{y_\ell}I$. This vector is defined to connect the segment's endpoints: its single component in the y-coordinate system is given by δ, and $y_\ell + \delta = y_{\ell+1}$. In this way the curve becomes a concatenation of infinitesimal tangent vectors, see Fig. V23. The application of the form ϕ to each vector yields an infinitesimal number, $\phi(\partial_{\Delta,\ell}) = \phi(y_\ell)\mathrm{d}y(\partial_{\Delta,\ell}) = \phi(y_\ell)\delta$, and it is natural to define its integral over I as the sum of these,

$$\int_I \phi \equiv \lim_{\delta\to 0} \sum_\ell \phi(\partial_{\Delta,\ell}) = \lim_{\delta\to 0} \delta \sum_\ell \phi(y_\ell).$$

This is a Riemann sum, which shows the identity of the form integral and the Riemann integral.

As a *physics example*, consider a force form $\phi = f$, and let the tangent vector $\partial_{\delta r,\ell}$ represent an infinitesimal displacement along a path. The sum of the works done along the displacements, $\sum_\ell f(\partial_{\delta r,\ell}) \xrightarrow{\delta\to 0} \int_I f$, defines the work along the path.

Integration of top-dimensional forms

Let ω be a *top-dimensional* form on a d-dimensional manifold M, e.g. a one-form on a line, a two-form on a surface, or a three-form in space, etc. We assume the manifold to be almost fully[58] covered by coordinates $r : U \to M$, $y \mapsto r(y)$. The form ω then affords the coordinate representation $\omega = \omega(y)\,\mathrm{d}y^1 \wedge \ldots \wedge \mathrm{d}y^d$.

integral of top-form on manifold   Following the logic of the previous section, it is natural to define the integral of ω over M as

$$\int_M \omega = \int_M \omega(y)\,\mathrm{d}y^1 \wedge\cdots\wedge \mathrm{d}y^d \equiv \int_U \omega(y)\,\mathrm{d}y^1 \wedge\cdots\wedge \mathrm{d}y^d \equiv \int_U \omega(y)\,\mathrm{d}y^1 \ldots \mathrm{d}y^d,$$

$$(\text{V}185)$$

where the individual representations emphasize different aspects of the integral: the first is the expression to be defined, and in the second we use the expansion of the form in coordinates. In the third, the integration of the form over M is identified with the integral of the pullback of the form to the coordinate domain (see discussion in the info section on p. 485 for added motivation for this identification). Finally, the integral of the form over $U \subset \mathbb{R}^d$ is identified with a multidimensional Riemann integral, where the role of

---

[58] Here, as always in integration theory the exclusion of lower-dimensional sets – points on a circle, lines on a sphere, etc. – is tolerated. As discussed in the info section on p. 243, such deficiencies do not affect integrals. In cases where no single coordinate system suffices to represent M even under these conditions, the different coordinate domains of a covering atlas need to be treated separately.

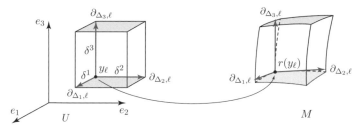

Fig. V24   The identification of the integral of d-forms with d-dimensional Riemann integrals, illustrated for $d = 3$.

the wedge product of coordinate forms is taken by the d-dimensional volume element, $dy^1 \wedge dy^2 \wedge \ldots \wedge dy^d \to dy^1 dy^2 \ldots dy^d$.

The latter interpretation of the *coordinate volume form as a Riemann volume element* extends the reasoning of the previous discussion of line integrals to higher dimensions: consider U partitioned into a large number of cuboids with infinitesimal side lengths δ^j ($j = 1, \ldots, d$) at coordinates y_ℓ (see Fig. V24, left, and Section C4.3 on volume integrals). Their edges are represented by tangent vectors, $\partial_{\Delta_j, \ell} = \delta^j \partial_{j, \ell}$ (no summation), in $T_{y_\ell} U$, and their volume is $(dy^1 \wedge \ldots \wedge dy^d)(\partial_{\Delta_1, \ell}, \ldots, \partial_{\Delta_d, \ell}) = \delta^1 \ldots \delta^d$. The integral of the form in the coordinate domain is defined as the limiting value of the sum of these expressions weighted by the component of the form, $\sum_\ell \omega(y_\ell) \delta^1 \ldots \delta^d$. In the limit $\delta^i \to 0$ this becomes $\int_U \omega(y) \, dy^1 \ldots dy^d$.

Finally, note that the distorted volume elements discussed in Chapter C4 (see Fig. C7, and several other related figures in that chapter) are spanned by the pushforwards of the tangent vectors $\partial_{\Delta_j} = \delta^j \partial_{y^j}$ from the coordinate domain TU to the tangent bundle TM (see Fig. V24, right). As discussed in the info section on p. 485, the tangent vectors in TU and TM are denoted by the same symbol, and so are the corresponding coordinate forms, dy^j. The coordinate form $dy^1 \wedge \ldots \wedge dy^d$ defined in TU^* or TM^* acts identically on the corresponding vectors. This identity is expressed by the second equality in Eq. (V185), which identifies the integral of ω over M with that of the integral of the pullback of ω to U.

EXAMPLE Consider the integral of the spherical area form, $\omega = \sin(\theta) d\theta \wedge d\phi$, over the unit sphere, $M = S^2$. In spherical coordinates, the integral formula gives

$$\int_{S^2} \sin(\theta) d\theta \wedge d\phi \equiv \int_{U=(0,\pi)\times(0,2\pi)} \sin(\theta) d\theta \wedge d\phi = \int_0^\pi \sin\theta \, d\theta \int_0^{2\pi} d\phi = 4\pi. \qquad \text{(V186)}$$

Changes of coordinates and general integral transforms

The above definition of integrals is independent of the choice of coordinates. Under a change, $y \mapsto y'(y)$, expressing y' through y coordinates, the representation of the form changes as in Eq. (V171), and the integral formula becomes

$$\int_{U'} \omega(y') dy'^1 \ldots dy'^d = \int_{U'} \omega(y') dy'^1 \wedge \ldots \wedge dy'^d$$
$$= \int_U \omega(y'(y)) \det\left(\frac{\partial y'}{\partial y}\right) dy^1 \wedge \ldots \wedge dy^d = \int_U \omega(y'(y)) \det\left(\frac{\partial y'}{\partial y}\right) dy^1 \ldots dy^d,$$

where the equality of the outermost expressions establishes the consistency with variable changes in Riemann integrals, see Eq. (C77). The inner equality connects the integrals of forms in different coordinate representations. It is instructive to read $\int_U \omega(y)\mathrm{d}y^1 \wedge \ldots \wedge \mathrm{d}y^d$ as the integral of the pullback of $\omega(y')\mathrm{d}y'^1 \wedge \ldots \wedge \mathrm{d}y'^d$ from U' to U.

The latter interpretation suggests a more general *connection between integrals of forms over different integration domains*: consider two orientable manifolds of equal dimension, M and L, represented by coordinate maps $r : U \to M$, $y \mapsto r(y)$ and $q : T \to L$, $x \mapsto q(x)$, respectively. Let $F : M \to L$ be an orientation-preserving diffeomorphism between the manifolds, i.e. a map whose coordinate representa-

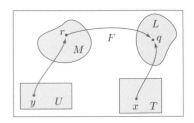

tions, $y \mapsto x(y) \equiv F(y)$, are diffeomorphic maps between coordinate domains and have positive Jacobi matrix, $\det\left(\frac{\partial F}{\partial y}\right) > 0$. For a top-form ω, defined on L, we then have the important formula

pullback of form-integrals

$$\int_M F^*\omega = \int_L \omega. \tag{V187}$$

This identity follows from the interpretation of the map $F \circ r : U \to L$, $y \mapsto F(r(y))$ as a coordinate representation of L, i.e. a system of coordinates alternative to x. The definition (V185), applied to this system, then states that $\int_L \omega = \int_U (F \circ r)^*\omega = \int_U r^*(F^*\omega)$, where the composition property of pullback, Eq. (V157), was used. $F^*\omega$ is a top-form on M, and the integral of its pullback $r^*(F^*\omega)$ to U, again according to Eq. (V185), equals the integral of this form over its ambient manifold M, $\int_U r^*(F^*\omega) = \int_M F^*\omega$. Combining these equalities, we obtain the statement (V187). As a side remark, notice how non-technical this proof is: it is very general, and at the same time does not involve complicated formulas. This combination of generality and structural transparency is a hallmark of differential form integration theory.

EXERCISE Consider the spherical area form on the two-sphere S^2, $\omega = \sin(\theta)\mathrm{d}\theta \wedge \mathrm{d}\phi$. Its pullback, $y^*\omega$, to the stereographic plane, $\mathbb{R}^2$, is given by Eq. (V176), where $z \mapsto y(z)$ is the map expressing spherical through stereographic coordinates, Eq. (V173). Show that ($\to$ V5.4.1-2)

$$\int_{S^2} \omega = \int_{\mathbb{R}^2} y^*\omega = 4\pi. \tag{V188}$$

Integration of lower-rank forms

Finally, let us discuss how to integrate forms of degree d, defined on manifolds, L, of higher dimension, $c \equiv n > d$, over d-dimensional submanifolds, $M \subset L$. Without much loss of generality, we assume the embedding manifold to be an open subset $L \subset \mathbb{R}^n$, although this identification will not be essential throughout. For example, we may consider the integration of a one-form in three-dimensional space ($d = 1$, $n = 3$) over a curve; or the integration of a two-form in three-dimensional space ($d = 2$, $n = 3$) over a surface.

The definition of such integrals follows from the previous constructions. Let M be parametrized by a coordinate map $F : U \to M$. A d-form, ϕ, defined on the n-dimensional manifold L, is then integrated over M by pulling it back to U and integrating there,

$$\int_M \phi \equiv \int_U F^*\phi. \tag{V189}$$

This makes sense, because the d-form $F^*\phi$, defined on the d-dimensional domain U, is top-dimensional and therefore integrable via (V185).

For example, consider a one-form, $\phi = \phi_i(x)dx^i$, defined in an n-dimensional manifold L with coordinates x. To compute its integral over a *curve*, $\gamma = M$, one parametrizes the latter by a coordinate, $r : I \to L, t \mapsto r(t)$. We then have $r^*\phi \overset{(V158)}{=} \phi_i(r(t))\dot{r}^i(t)dt$ and hence

$$\int_\gamma \phi = \int_I \phi_i(r(t))\, \dot{r}^i(t)\, dt. \tag{V190}$$

In this formula we recognize the familiar structure of a line integral, see Eq. (V12). The difference between the two expressions is that Eq. (V190) contains a form acting on the velocity vector, $\phi(\partial_{\dot{r}}) = \phi_i \dot{r}^i$, whereas the traditional line integral contains the scalar product of a vector with the velocity vector, $\mathbf{f} \cdot \dot{\mathbf{r}} = f^j g_{ji} \dot{r}^i$. We reason that the form-variant of the line integral is the more natural one. It does not require the excess baggage of a scalar product, and just sums the work accumulated by action of the force one-form, $\phi - f$, on infinitesimal path segments.

EXAMPLE Consider the work done against the force $f = x^2 dx^1 - x^1 dx^2$ (Cartesian coordinates in $\mathbb{R}^3$) along one revolution of a spiral curve, γ, with parametrization $r(t) = (\cos t, \sin t, ct)^T$. This force form has components $f_1 = x^2, f_2 = -x^1$ and $f_3 = 0$, hence Eq. (V190) yields

$$\int_\gamma f = \int_0^{2\pi} \left[x^2(t)\dot{x}^1(t) - x^1(t)\dot{x}^2(t) \right] dt = \int_0^{2\pi} [(\sin t)(-\sin t) - \cos t(\cos t)]\, dt = -2\pi.$$

EXERCISE A uniform current density in the z-direction of three-dimensional space is described by the differential two-form $j = j_0\, dx^1 \wedge dx^2$. Find the current flowing through (a) the northern hemisphere of a sphere of radius R, and (b) the full sphere. First try to find the answers by qualitative reasoning, then verify them by integration of j. ($\to$ V5.4.3-4 (a-c))

Stokes's theorem

In Sections V3.5 and V3.6 we discussed various integral identities of the form (integral of X over the $(d-1)$-dimensional boundary of Y) = (integral of derivative of X over the d-dimensional interior of Y). The fundamental law of calculus, $f(b) - f(a) = \int_a^b dx f'(x)$, too, is of this form if we interpret $\{a, b\}$ as the boundary of the interval $[a, b]$. The common structure of these identities suggests that a general principle is behind the scene.

Indeed, all integral identities mentioned above are special cases of one master identity, the **general Stokes's theorem**. It states that the integral of a $(d-1)$-form, ϕ, over the $(d-1)$-dimensional boundary, ∂M, of a manifold equals the integral of $d\phi$ over the manifold itself,

general
Stokes's
theorem

$$\boxed{\int_{\partial M} \phi = \int_M d\phi.}$$ (V191)

Notice the beauty of this formula: an expression containing just a few symbols subsumes all the different variants of boundary–bulk integral identities discussed previously.

INFO The *proof of Stokes's theorem* is similar to that of the more specialized integral theorems of Gauss and Stokes discussed in Chapter V3: let $r : U \to M, y \to r(y)$ be a coordinate representation of M. The domain boundary, ∂U, then parametrizes the boundary ∂M. The definition of form integrals means that Stokes's theorem assumes the form

$$\int_{\partial U} \omega = \int_U d\omega,$$ (V192)

where $\omega = r^*\phi$ is the pullback of ϕ to U, and on the r.h.s. we used the commutativity of pullback and exterior derivative, $r^* d\phi = dr^*\phi = d\omega$. Without loss of generality, we assume U to be a cuboid in $\mathbb{R}^d$ (see the figure for a two-dimensional representation). Following the strategy of the more specialized proofs discussed earlier, we partition U into many infinitesimal cuboids, U_ℓ, with boundary extensions $\delta^j, j = 1, ..., d$. The boundary integral over ∂U can then be written as $\int_{\partial U} \omega = \sum_\ell \int_{\partial U_\ell} \omega$, where we used the fact that all

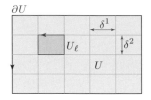

contributions from integrals over internal boundaries (the light gray lines in the figure) of the U_ℓs cancel, because the touching boundaries of adjacent cuboids have opposite orientations. At the same time, we have $\int_U d\omega = \sum_\ell \int_{U_\ell} d\omega$, and so the proof of the theorem reduces to the the verification of $\int_{\partial U_\ell} \omega = \int_{U_\ell} d\omega$ for infinitesimal domains. Let us expand the form ω with its $(d-1)$ independent components as

$$\omega = \omega_{2,...,d}\, dy^2 \wedge ... \wedge dy^d + \cdots + \omega_{1,...,d-1}\, dy^1 \wedge ... \wedge dy^{d-1}.$$

The above discussion of form integrals implies that the integral of ω over the pair of faces of cuboid U_ℓ normal to the 1-direction is given by $[\omega_{2,...,d}(y^1 + \delta^1, y^2, ..., y^d) - \omega_{2...d}(y^1, y^2, ..., y^d)]\delta^2...\delta^d \simeq \partial_1 \omega_{2,...,d}(y_\ell)\delta^1\delta^2...\delta^d$. Adding to this the contribution of the $(d-1)$ remaining pairs of faces we obtain $\int_{\partial U_\ell} \omega = [\partial_1 \omega_{2,...,d} + \cdots + \partial_d \omega_{1,...,d-1}]_\ell\, \delta^1...\delta^d$, which is equal to $\int_{U_\ell} d\omega$, the integral of the exterior derivative, $d\omega$, of the form over the interior of the cuboid U_ℓ. This proves Stokes's theorem for each infinitesimal cuboid and, upon extension, for the full integration domain.

As an example, consider the integral of a one-form, $\phi = \phi_i dy^i$, over a closed curve, γ. The curve is the boundary, $\gamma = \partial S$, of some area S and application of Stokes's theorem, $\oint_\gamma \phi = \int_S d\phi$, implies the equality of the coordinate integrals,

$$\oint_\gamma \phi_i dy^i = \int_S \partial_j \phi_i\, dy^j \wedge dy^i.$$ (V193)

The left-hand side is a line integral of the form Eq. (V190). On the right, the antisymmetric components, $\partial_j \phi_i - \partial_i \phi_j$, of the two-form have a structure reminiscent of the curl of a vector field. Equation (V193) thus assumes a role similar to that of Stokes's theorem of vector analysis, Eq. (V106). ($\to$ V5.4.3 (d)) In Chapter V6 we discuss how inner products may be applied to establish the precise connection between the general Stokes's theorem and those of vector analysis. For the moment, we just note that the general theorem works in all dimensions, and does not require an inner product.

EXERCISE Write down a three-dimensional version of Stokes's theorem, i.e. one for a two-form with expansion $\phi = \frac{1}{2}\phi_{ij}\mathrm{d}y^i \wedge \mathrm{d}y^j$, defined on a two-dimensional surface, $S = \partial V$, bounding a three-dimensional volume, V. Try to identify structures reminiscent of *Gauss's theorem*. How many different components does the two-form have, and how many has its exterior derivative? Do you see index structures similar to those appearing in the vector field divergence? Can you suggest the construction of a vector field from the components of the form such that the three-dimensional Stokes's theorem assumes a form similar to the familiar Gauss's theorem? These questions will be answered in detail in Section V6.3, however, it is instructive to speculate a little in advance.

EXAMPLE Stokes's theorem can be applied to compute the volume of an object via a surface integral over its bounding surface. To illustrate this statement in $\mathbb{R}^3$, consider a ball, B_R, of radius $r = R$. Its volume is obtained by computing the integral of the volume form in spherical coordinates, $\omega \stackrel{(\text{V172})}{=} r^2 \mathrm{d}r \wedge \sin(\theta)\mathrm{d}\theta \wedge \mathrm{d}\phi$, over the ball:

$$V_R = \int_{B_R} \omega \stackrel{(\text{V185})}{=} \int_0^R r^2 \mathrm{d}r \int_0^\pi \sin\theta\,\mathrm{d}\theta \int_0^{2\pi} \mathrm{d}\phi = \frac{1}{3}4\pi R^3 \,.$$

Alternatively, one may note that the volume form is exact, $\omega = \mathrm{d}\phi$, where $\phi = \frac{1}{3}r^3 \sin\theta\,\mathrm{d}\theta \wedge \mathrm{d}\phi$. According to Stokes's theorem, the integral of ϕ over the surface, ∂B_R, a sphere of radius R, yields an alternative expression for the volume:

$$V_R = \int_{B_R} \mathrm{d}\phi \stackrel{(\text{V191})}{=} \int_{\partial B_r} \phi \stackrel{(\text{V185})}{=} \frac{1}{3}R^3 \int_0^\pi \sin\theta\,\mathrm{d}\theta \int_0^{2\pi} \mathrm{d}\phi = \frac{1}{3}4\pi R^3 .$$

V5.5 Summary and outlook

In this chapter, we introduced differential forms as fields of antisymmetric tensors on manifolds. Within the framework of multilinear algebra, forms are the covariant partners of contravariant vectors. The same is true concerning vector fields and differential forms on manifolds. Specifically, we saw that all objects of multilinear algebra could be lifted to manifolds where they now defined smooth fields. The exterior derivative was introduced as a tool to describe the variation of differential forms on manifolds. And we defined the integral of forms, reasoning that these integrals are simpler and more universal than the various specialized integrals discussed in previous parts of the text. Establishing a connection between forms of neighboring degree, the exterior derivative was key to the formulation of a generalized variant of Stokes's theorem which we argued (but not yet showed) subsumes all integral theorems discussed in earlier chapters.

In the next chapter we introduce the concept of metrics on manifolds. This will put us in a position to measure distances and angles on geometric structures, and to connect in more concrete ways to the operations of vector analysis in their traditional formulation.

V6   Riemannian differential geometry

Beginning with Section L3.3, metric structures have appeared several times in the text and were discussed from different perspectives. In this chapter, we introduce the metric into the framework of differential geometry, thereby revealing its full geometric significance. Specifically, we will discuss how the metric describes geometric structures through lengths, angles and curvature, how it defines a bridge between forms and vectors, and also between forms of different degrees, and how it defines a unique top-dimensional form whose integral determines the volume of manifolds.

V6.1   Definition of the metric on a manifold

metric   A **metric** on a d-dimensional manifold M is a covariant tensor field of second degree,

$$g : M \to TM^* \otimes TM^*, \qquad r \mapsto g_r, \tag{V194}$$

smoothly assigning to each point r in M a bilinear form, g_r, on the tangent space $T_r M$. A bilinear form g_r takes two tangent vectors, $\partial_u, \partial_v \in T_r M$, as arguments to produce a number, $g_r(\partial_u, \partial_v)$. The extension to the manifold therefore defines a map

$$g : \mathrm{vect}(M) \times \mathrm{vect}(M) \to \Lambda^0(M), \qquad (\partial_u, \partial_v) \mapsto g(\partial_u, \partial_v), \tag{V195}$$

sending two vector fields to a function, with $g(\partial_u, \partial_v)_r = g_r(\partial_{u,r}, \partial_{v,r})$. As with the two interpretations of forms (see Eqs. (V141) and (V142)), the definitions above emphasize slightly different aspects of the metric.

The crucial difference distinguishing a metric from a two-form is that it is symmetric in its arguments. More specifically, g is required to be

▷   *symmetric*, $\forall \partial_u, \partial_v \in T_r M$, $g_r(\partial_u, \partial_v) = g_r(\partial_v, \partial_u)$, and

▷   *non-degenerate*, if $g_r(\partial_v, \partial_u) = 0 \ \forall \partial_u \in T_r M$, then $\partial_v = 0$.

Coordinate representation

A metric may be expanded in the basis of coordinate forms, $\{dy^i\}$, as

$$\boxed{g = g_{ij}(y)\, dy^i \otimes dy^j \equiv g_{ij}(y)\, dy^i\, dy^j,} \tag{V196}$$

metric tensor   where $g_{ij} = g_{ij}(y)$ are the components of the **metric tensor**. The second representation, omitting the tensor product sign, is not particularly clean but widely used in the physics literature. In Cartesian coordinates the *standard metric of* $\mathbb{R}^d$ has the form $g_{ij} = \delta_{ij}$, i.e.

$$g = dx^1 \otimes dx^1 + \cdots + dx^d \otimes dx^d, \qquad (V197)$$

where no reference to the metric tensor is made (resulting in an awkward two-index-upstairs summation). In the coordinate representation the metric acts on vector fields, $\partial_u = u^i \partial_i$ and $\partial_v = v^j \partial_j$, as

$$g(\partial_u, \partial_v) = u^i g_{ij} v^j.$$

Under a *coordinate change* $y \mapsto y'(y)$, expressing y' through y, with Jacobi matrix $J^i{}_k = \frac{\partial y'^i}{\partial y^k}$, it changes as a covariant tensor of second degree, or bilinear form:

$$g = g_{ij}(y')\, dy'^i \otimes dy'^j = g_{ij}(y'(y))\, J^i_k J^j_l\, dy^k \otimes dy^l \equiv g_{kl}(y)\, dy^k \otimes dy^l. \qquad (V198)$$

coordinate change of metric tensor   This implies the *covariant change of metric coefficients*,

$$g_{kl}(y) \equiv g_{ij}(y')\, J^i_k J^j_l \Big|_{y'=y'(y)}, \qquad (V199)$$

where the notation emphasizes that all y'-dependences on the right need to be expressed through y. In the literature, (V199) sometimes represented in a non-index notation as

$$g(y) = J^{\mathrm{T}} g(y'(y)) J. \qquad (V200)$$

Equation (V198) implies an important relation between the determinants, Eq. (L177), of the metric in the unprimed and primed systems. Applying arguments similar to those proving the determinant law $\det(AB) = \det(A)\det(B)$ (see Eq. (L170)), one obtains the factorization

$$\det\left(g(y)\right) = \left(\det\left(g(y')\right)\det J^2\right)_{y'=y'(y)}, \qquad (V201)$$

(Prove this relation in the two-dimensional case.) For an orientable manifold, $\det(J)$ can be chosen positive and Eq. (V201) implies the relation

$$\boxed{\sqrt{|g(y)|} = \sqrt{|g(y'(y))|}\, \det(J).} \qquad (V202)$$

EXAMPLE Consider the standard metric of two-dimensional space in Cartesian coordinates, $g = dx^1 \otimes dx^1 + dx^2 \otimes dx^2$. The polar representation, $x^1 = \rho\cos\phi$ and $x^2 = \rho\sin\phi$, implies $dx^1 = d(\rho\cos\phi) = \cos\phi\, d\rho - \rho\sin\phi\, d\phi$, and similarly $dx^2 = \sin\phi\, d\rho + \rho\cos\phi\, d\phi$. Substituting these expressions into the metric, we obtain the *standard metric of $\mathbb{R}^2$ in polar coordinates*,

$$g = d\rho \otimes d\rho + \rho^2\, d\phi \otimes d\phi.$$

Notice that the components $g_{\rho\rho} = 1$ and $g_{\phi\phi} = \rho^2$ appearing here agree with those defined in Eq. (V22) of Section V2.2. There, we had defined the metric through inner products taken between coordinate basis vectors, $g_{ij} = g(\mathbf{v}_i, \mathbf{v}_j) = \langle \mathbf{v}_i, \mathbf{v}_j \rangle$. In the present notation this equals $g_{ij} = g(\partial_i, \partial_j)$. This formula isolates the components g_{ij} in the expansion (V196), showing the equivalence of the formulations. Let us also verify Eq. (V202): For the map $y \mapsto x(y)$ expressing Cartesian through polar coordinates, $\sqrt{|g(y)|} = \sqrt{g_{\rho\rho}g_{\phi\phi}} = \rho$ is consistent with $\sqrt{|g(x(y))|}\, \det\left(\frac{\partial x}{\partial y}\right) \overset{(V46)}{=} 1 \cdot \rho$.

EXERCISE Show that the *standard metric of* $\mathbb{R}^3$ *in spherical coordinates*, (r, θ, ϕ), is given by

$$g = dr \otimes dr + r^2 d\theta \otimes d\theta + r^2 \sin^2 \theta \, d\phi \otimes d\phi. \tag{V203}$$

Check that this result is consistent with Eq. (V202). ($\to$ V6.1.1-2)

Orthonormal representation

A metric introduces concepts such as lengths and angles, etc., to tangent space. For example, the *norm of tangent vectors*, $\|\partial_v\| = \sqrt{g(\partial_v, \partial_v)}$, defines normalized vectors, $\partial_n \equiv \frac{\partial_v}{\|\partial_v\|}$, and by extension normalized vector fields. A basis, $\{\partial_{e_i}\}$, of normalized and mutually orthogonal vector fields, with $g(\partial_{e_i}, \partial_{e_j}) = \eta_{ij}$, is called an **orthonormal basis**. For a positive definite metric, orthonormal bases have $\eta_{ij} = \delta_{ij}$, while for a metric with **signature** $(p, d - p)$, they have $\eta_{ii} = 1$ for $i = 1, ..., p$ and $\eta_{ii} = -1$ for $i = p + 1, ..., d$, with $\delta_{i \neq j} = 0$. In such bases the metric has the locally orthonormal form

$$g(x) \equiv \eta = \eta_{ij} dx^i \otimes dx^j = \sum_i \eta_{ii} dx^i \otimes dx^i. \tag{V204}$$

Here and throughout, coordinates orthonormal in this sense are denoted by x^i. The notation generalizes the concept of Cartesian coordinates to positive indefinite metrics.

Manifolds equipped with a *positive definite* metric, $p = d$, are called **Riemannian manifolds**. Manifolds with non-positive metrics, $p \neq d$, are called **pseudo-Riemannian**. For example, the Minkowski metric of special relativity (see info section on p. 113) defines pseudo-Riemannian space-time with signature $(1, 3)$. Coordinate transformations preserving the form of the metric have Jacobians constrained by $\eta = J^T \eta J$ in the abbreviated notation of (V200) (see discussion of Eq. (L211) on p. 113).

As in finite vector spaces, it is often convenient to work in orthonormal representations, obtained via orthonormalization. For example, the holonomic basis vectors ∂_r, ∂_θ and ∂_ϕ of spherical coordinates have norms 1, r and $r \sin \theta$. The *curvilinear local bases* discussed in Section V2.2 are orthonormal bases obtained by normalization of such bases. Specifically, for spherical coordinates, $\partial_{e_r} = \partial_r$, $\partial_{e_\theta} = r^{-1} \partial_\theta$ and $\partial_{e_\phi} = (r \sin \theta)^{-1} \partial_\phi$. However, we have seen that such normalization makes the operations of vector analysis more complicated than in the un-normalized coordinate basis. Whether to normalize or not depends on the context and needs to be decided on a case-by-case basis.

EXERCISE Compute the pullback of the spherical coordinate metric, $g = d\theta \otimes d\theta + \sin^2 \theta \, d\phi \otimes d\phi$, to stereographic coordinates, $z = (z^1, z^2)$, introduced in Eq. (V173). Show that it takes the form $g = s(z) \sum_i dz^i \otimes dz^i$, and determine the scale factor $s(z)$. Since this metric tensor is proportional to but not equal to $\mathbb{1}$, the holonomic frame $(\partial_{z^1}, \partial_{z^2})$ is orthogonal but not orthonormal. How does the scale factor behave near the poles of the sphere? Give a geometric interpretation of this behavior. ($\to$ V6.1.3-4)

INFO The metric is key to the description of the geometry of manifolds. For example, it defines the local **curvature of surfaces** in three-dimensional space. To this end, one picks a point $r \in S$, constructs a vector normal to S (how is this done?) and then considers the set of all planes containing

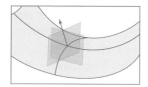  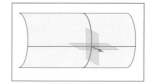

Fig. V25

On the definition of the Gaussian curvature of surfaces S in $\mathbb{R}^3$. For a given point $r \in S$, consider the set of curves defined by the intersection of all planes containing the normal to S at r. The Gaussian curvature, $\kappa = \kappa_{\min}\kappa_{\max}$, is the product of the minimum and maximum curvatures, $\kappa_{\min}$ and $\kappa_{\max}$, of these curves. Left/center/right: surface of negative/positive/vanishing curvature.

that normal, see Fig. V25. The intersection of any such plane with the surface defines a curve in S, and each of these curves has a local curvature, κ, at r. (This curvature is defined as the inverse of the radius of a circle locally approximating the curve at r.) Denote the minimum and maximum of these curvatures over the set of all curves through r by $\kappa_{\min}$ and $\kappa_{\max}$, respectively. The product of

Gaussian curvature

these two numbers, $K \equiv \kappa_{\min}\kappa_{\max}$, defines the **Gaussian curvature** at r. The Gaussian curvature is defined to be positive or negative depending on whether the two curves with extremal curvature bend in the same or in opposite directions. Figure V25 shows situations with negative, positive and vanishing curvature. For example, a cylinder, although different from a plane, is a surface of globally vanishing curvature since $\kappa_{\min} = 0$ everywhere. Generic surfaces have regions of positive and negative curvature separated by lines of vanishing curvature.

flat manifold

A manifold with globally vanishing curvature is called a **flat manifold**. Flatness implies the existence of coordinate systems with constant metric tensor, $g_{ij} = \eta_{ij}$. The coordinate lines of these coordinates are curves of vanishing curvature, and they define the vanishing Gaussian curvature of the manifold. For example, the vectors defined by coordinates (ϕ, z) for a cylinder of unit radius along the z-axis in $\mathbb{R}^3$ read (see Eq. (V134)) $\partial_\phi = -\sin(\phi)\partial_1 + \cos(\phi)\partial_2$, and $\partial_z = \partial_3$, where $\partial_{1,2,3}$ are Cartesian coordinate vectors in $\mathbb{R}^3$. The 2×2 diagonal metric tensor defined by these vectors is a unit matrix $g_{\phi\phi} = g_{zz} = 1$, demonstrating the flatness of the cylinder. Note that the requirement of a global representation of this form is essential here. On a curved manifold local transformations with representation $g_{r,ij} = \eta_{r,ij}$ can be found at any point r. However, it is not possible to extend them to a globally constant metric.

V6.2 Volume form and Hodge star

In Section L10.9 we showed how a metric defines two powerful algebraic structures in the space of alternating forms of a vector space, a canonical volume form and the Hodge star. Locally defined for individual vector spaces, these structures immediately carry over to manifolds.

Volume form

A metric provides the means to determine the volume of manifolds, such as the length of curves, the area of surfaces, the volume of three-dimensional structures, etc. Recall that the good objects to integrate on a d-dimensional manifold are d-forms. For a coordinate transformation $y \mapsto y'(y)$, the y- and y'-coordinate representations of a d-form have weight functions related by $\omega(y) = \omega(y'(y)) \det(\frac{\partial y'}{\partial y})$, see Eq. (V171). On a general manifold, the Jacobian of the coordinate transformation can be an arbitrary invertible matrix, hence the weight can change arbitrarily: there does not exist a canonical choice for the top-form.

The situation is different on (orientable) manifolds with a metric. In Section L10.9 we **volume form** showed that in this case the **volume form**,

$$\omega = |g(y)|^{1/2} \mathrm{d}y^1 \wedge \ldots \wedge \mathrm{d}y^d, \tag{V205}$$

plays a distinguished role, in that it remains invariant under transformations $y \mapsto y'(y)$, **volume of** because the basis forms transform linearly, $\mathrm{d}y'^i = \frac{\partial y'^i}{\partial y^j}\mathrm{d}y^j$ (see Eq. (L283)). The **volume** **manifold** **of a manifold** is defined as the integral over this form,

$$\mathrm{vol}(M) \equiv \int_M \omega. \tag{V206}$$

For a standard scalar product and Cartesian coordinates, $g = 1$ and $\omega = \mathrm{d}x^1 \wedge \ldots \wedge \mathrm{d}x^d$, the integral reduces to the standard volume integral discussed in Chapter C4. In cases where the integral does not exist – think of the two-dimensional plane, $M = \mathbb{R}^2$ – the manifold has undetermined or infinite volume. For a general coordinate system, the volume integral becomes

$$\mathrm{vol}(M) = \int_M |g(y)|^{1/2}\,\mathrm{d}y^1 \wedge \ldots \wedge \mathrm{d}y^d = \int_U |g(y)|^{1/2}\,\mathrm{d}y^1 \ldots \mathrm{d}y^d, \tag{V207}$$

where the second equality represents the integral as a pullback to a y-coordinate domain (see discussion on p. 485).

EXAMPLE In a Cartesian basis, the standard metric in $\mathbb{R}^3$ is given by (V197). It has $\det(g) = 1$ and the volume form is $\omega = \mathrm{d}x^1 \wedge \mathrm{d}x^2 \wedge \mathrm{d}x^3$. In spherical coordinates, the metric tensor is given by (V203). We now have $\sqrt{g} = r^2 \sin\theta$, hence the *spherical representation of the volume form* reads $\omega = r^2 \sin\theta\,\mathrm{d}r \wedge \mathrm{d}\theta \wedge \mathrm{d}\phi$. This agrees with the expression in Eq. (V172), derived from the Cartesian volume form via a coordinate transformation. The weight function, $r^2 \sin\theta$, matches the scale factor in the spherical integration volume element $\mathrm{d}V$, Eq. (C66), discussed in Section C4.3.

The *two-sphere* can be locally identified with the coordinate domain $U(\theta, \phi)$, having the spherical metric $g_{\theta\theta} = 1, g_{\phi\phi} = \sin^2\theta$. This means that its volume form, or in view of the two- **area form** dimensionality better to say its **area form**, is given by $\omega = \sin\theta\mathrm{d}\theta \wedge \mathrm{d}\phi$. We had considered this form before (see Eq. (V175)) and saw that its full integration indeed yields the area of the sphere, 4π.

Hodge star

In Section L10.9 we defined the Hodge star as a map between p-forms and $(d - p)$-forms of a d-dimensional vector space. Locally applied to the p-forms, ϕ_r of the tangent space **Hodge star** T_rM, it defines the **Hodge star** as a map between $\Lambda^p(M)$ and $\Lambda^{d-p}(M)$ as

$$
\boxed{
\begin{aligned}
& * : \Lambda^p(M) \to \Lambda^{d-p}(M), \qquad \phi \mapsto *\phi, \\
& *\phi = * \left(\frac{1}{p!} \phi_{i_1,\ldots,i_p} \bigwedge_{a=1}^{p} dy^{i_a} \right) = \frac{\sqrt{|g|}}{p!(d-p)!} \phi^{j_1,\ldots,j_p} \epsilon_{j_1,\ldots,j_p,j_{p+1},\ldots,j_d} \bigwedge_{b=p+1}^{d} dy^{j_b},
\end{aligned}
}
$$

$$\text{(V208)}$$

or

$$
\boxed{ (*\phi)_{j_{p+1},\ldots,j_d} = \frac{1}{p!} |g|^{1/2} \phi_{i_1,\ldots,i_p} g^{i_1 j_1} \cdots g^{i_p j_p} \epsilon_{j_1,\ldots,j_p,j_{p+1},\ldots,j_d}. }
\qquad \text{(V209)}
$$

We recapitulate from Section L10.9 that the Hodge star operation is self-involutory up to a sign factor,

$$
** \phi = \text{sgn}(g)(-)^{p(d-p)} \phi. \qquad \text{(V210)}
$$

EXAMPLE Consider three-dimensional space parametrized by spherical coordinates. The inverse metric is given by $g^{rr} = 1$, $g^{\theta\theta} = r^{-2}$, $g^{\phi\phi} = (r \sin\theta)^{-2}$. Show that ($\to$ V6.2.1-2)

$$
*dr = r^2 \sin\theta \, d\theta \wedge d\phi, \qquad *d\theta = \sin\theta \, d\phi \wedge dr, \qquad *d\phi = (\sin\theta)^{-1} \, dr \wedge d\theta. \qquad \text{(V211)}
$$

Note that the two-form $*dr$ equals the area form of the sphere. Connections of this type enter our discussion below of vector analysis in form language.

V6.3   Vectors vs. one-forms vs. two-forms in $\mathbb{R}^3$

In $\mathbb{R}^3$, there exist three distinct types of three-component tensors:[59]

$$
\begin{aligned}
\text{vectors:} \qquad & \partial_v = v^1 \partial_1 + v^2 \partial_2 + v^3 \partial_3, \\
\text{one-forms:} \qquad & \lambda = \lambda_1 \, dy^1 + \lambda_2 \, dy^2 + \lambda_3 \, dy^3, \\
\text{two-forms:} \qquad & \phi = \phi_{12} \, dy^1 \wedge dy^2 + \phi_{23} \, dy^2 \wedge dy^3 + \phi_{31} \, dy^3 \wedge dy^1. \qquad \text{(V212)}
\end{aligned}
$$

They all have a place in physics. Indeed, we saw above that forces, currents, densities and other physical quantities are best described as differential forms. This is in contrast to traditional teaching which indiscriminately treats all three-component fields in $\mathbb{R}^3$ as vector fields.

The translation between the two languages requires a metric. We will see that operations which assume a simple form in differential form language become cluttered by elements of the metric tensor required for the passage between the two formulations. For example, the coordinate-invariant formulation of Stokes's theorem, Eq. (V192), translates to Stokes's and Gauss's theorems of vector analysis. The differential operations $\nabla \times, \nabla \cdot$ appearing in these theorems assume a relatively complicated form in the traditional formulation, especially if curvilinear coordinates are used.

This section discusses the translation between forms and vectors and its ramifications in the operations of vector calculus. We will return to the subject from a physical

---

[59] All operations discussed in this section are defined locally. They work equally for vectors and forms of a vector space, and for vector fields and differential forms of a manifold.

perspective in Chapter V7 where we compare the two approaches in their application to electrodynamics.

from vectors to one-forms

The translation between forms and vectors requires the conversion of vectors to one-forms, of one-forms to two-forms, and of vectors to two-forms. The *passage between vectors and one-forms* is achieved via the isomorphism $J : V \to V^*$,[60] between vector spaces and their duals (cf. Section L10.2). Recall that to a vector, $v \in V$, one may assign a dual vector, $J(v) \in V^*$, by requiring that (see Eq. (L258))

$$g(v, u) = J(v)u, \qquad \forall u \in V. \tag{V213}$$

In component language this condition is written as $(J(v))_i \overset{(\text{L259})}{=} v^j g_{ji} \equiv v_i$, and $(J^{-1}(J(v)))^j \overset{(\text{L260})}{=} v_i g^{ij} \equiv v^j$ for the inverse map, where the components of the dual vector and the vector are distinguished by the co- and contravariant positioning of the indices, respectively. In the same way a vector field, $\partial_v = v^j \partial_{y^j}$, defines a differential one-form, $\lambda = J(\partial_v)$, with components $\lambda_i = (J(\partial_v)_i) = v^j g_{ji} = v_i$. The inverse map acts on one-forms, $\lambda = \lambda_i dy^i$, to yield vectors $J^{-1}(\lambda)$, with components $\lambda^j = (J^{-1}(\lambda))^j = \lambda_i g^{ij}$. In either case the positioning of the indices determines whether a contravariant vector or a covariant form is at hand. The action of J on one-forms and vectors is summarized by

$$J(v^j \partial_j) = v^j g_{ji} dy^i, \qquad J^{-1}(\lambda_i dy^i) = \lambda_i g^{ij} \partial_j. \tag{V214}$$

from one-forms to two-forms

The *passage between one-forms and two-forms* is defined by the $d = 3$ version of the Hodge star. To a one-form, λ, the Hodge star assigns a two-form, $\phi \equiv *\lambda$, with component representation (see Eq. (V209)) $\phi_{jk} = |g|^{1/2} \lambda_l g^{li} \epsilon_{ijk}$. From a two-form, ϕ, we may pass back to a one-form by applying the Hodge star once more, $\lambda = *\phi$. In components, $\lambda_i = \frac{1}{2} |g|^{1/2} \phi_{jk} g^{jj'} g^{kk'} \epsilon_{j'k'i} = \frac{1}{2} \text{sgn}(g) |g|^{-1/2} \phi_{jk} \epsilon^{jkl} g_{li}$, where we used $\epsilon_{j'k'i} = \epsilon_{j'k'l'} g^{l'l} g_{li}$ and in the second step evoked Eq. (L177), $g^{ij'} g^{kk'} g^{ll'} \epsilon_{j'k'l'} = \epsilon^{jkl} g^{-1}$, to convert the antisymmetrized product of three metric components into a determinant.

from two-forms to vectors

Finally, we may pass *from two-forms to vectors* in a two-step transformation, which first maps a two-form, ϕ, to a one-form, $*\phi$, and that to a vector, $J^{-1}(*\phi)$. The component representation is given by $v^i = (*\phi)_{i'} g^{i'i} = \frac{1}{2} \text{sgn}(g) |g|^{-1/2} \phi_{jk} \epsilon^{jki}$, and for the inverse operation, $\phi = *J(\partial_v)$, the components are $\phi_{jk} = |g|^{1/2} v^{i'} g_{i'l} g^{li} \epsilon_{ijk} = |g|^{1/2} v^i \epsilon_{ijk}$.

An overview of all these operations is given in Table V2. We next apply them to translate from form to vector representations of various differential operations.

gradient

Gradient: The gradient of a function, f, is the vector field associated with the exterior derivative df. It is defined as

$$\nabla f \equiv J^{-1}(df), \tag{V215}$$

and its components are given by

$$\boxed{\nabla f^i = g^{ij} \partial_j f.} \tag{V216}$$

This representation coincides with the earlier definition of the gradient of a function in curvilinear coordinates, Eq. (V70). As discussed in Section V3.2, the gradient vector field points in the direction of the steepest ascent of the function f, and to describe this geometric

---

[60] Do not be confused by the usage of the same symbol, J, for the isomorphism $V \to V^*$ and the Jacobian of maps. The two objects are unrelated.

Table V2 Invariants involving forms and vector fields in a three-dimensional metric space. Top: conversion rules. Bottom: derivative operations. The table works for metrics of both positive and negative signature, $\text{sgn}(g) = \pm 1$.

operation	invariant	components
0-form $\rightarrow$ 3-form	$\omega = *f$	$\omega_{ijk} = |g|^{1/2} f \epsilon_{ijk}$
3-form $\rightarrow$ 0-form	$f = *\omega$	$f = \text{sgn}(g)|g|^{-1/2} \frac{1}{3!} \omega_{ijk} \epsilon^{ijk}$
vector $\rightarrow$ 1-form	$\lambda = J \partial_v$	$\lambda_i = v^j g_{ji}$
1-form $\rightarrow$ vector	$\partial_v = J^{-1} \lambda$	$v^j = \lambda_i g^{ij}$
1-form $\rightarrow$ 2-form	$\phi = *\lambda$	$\phi_{jk} = |g|^{1/2} \lambda_l g^{li} \epsilon_{ijk}$
2-form $\rightarrow$ 1-form	$\lambda = *\phi$	$\lambda_i = \frac{1}{2!} \text{sgn}(g)|g|^{-1/2} \phi_{jk} \epsilon^{jkl} g_{li}$
2-form $\rightarrow$ vector	$\partial_v = J^{-1} * \phi$	$v^i = \frac{1}{2!} \text{sgn}(g)|g|^{-1/2} \phi_{jk} \epsilon^{jki}$
vector $\rightarrow$ 2-form	$\phi = *J \partial_v$	$\phi_{jk} = |g|^{1/2} v^i \epsilon_{ijk}$
gradient	$J^{-1}(df)$	$\nabla f = g^{ij} \partial_j f$
divergence	$\text{sgn}(g)(*d * J) \partial_v$	$\nabla \cdot \partial_v = |g|^{-1/2} \partial_i \left(|g|^{1/2} v^i \right)$
curl	$(J^{-1} * dJ) \partial_v$	$(\nabla \times \partial_v)^i = \text{sgn}(g)|g|^{-1/2} \epsilon^{ijk} \partial_j (g_{kl} v^l)$
Laplacian	$\text{sgn}(g)(*d * d)f$	$\Delta f = \nabla \cdot \nabla f = |g|^{-1/2} \partial_i (|g|^{1/2} g^{ij} \partial_j f)$

orientation a metric is required. Within the framework of forms, the exterior derivative df describes the variation of f and no reference to a metric is made.

divergence **Divergence**: The divergence is a first-order derivative operation mapping vector fields to functions. This is reminiscent of the exterior derivative of two-forms (cf. Eq. (V182)), which, likewise, converts forms with three coefficients to ones with only one coefficient. The correspondence is made explicit by first mapping the argument vector field to a two-form, $\phi = *J(\partial_v)$. An exterior derivative converts the two- to a three-form, $d\phi = d(*J\partial_v)$, which then is mapped by the Hodge star to a zero-form. The succession of the three operations,

$$\nabla \cdot \partial_v \equiv \text{sgn}(g)(*d * J)\partial_v, \tag{V217}$$

maps a vector field to a function and defines the *divergence*.[61] Note that unlike all previous definitions of the divergence as a derivative operation, no coordinates are required. The factor $\text{sgn}(g)$ is included to ensure equivalence to the previous definition (V95). Indeed, using Table V2 to apply the succession of operations to a vector field with components v^i, one obtains the coordinate representation (verify!)

$$\boxed{\nabla \cdot \partial_v = |g|^{-1/2} \partial_i \left(|g|^{1/2} v^i \right).} \tag{V218}$$

---

[61] The factor $\text{sgn}(g)$ is part of the definition; it cancels the $\text{sgn}(g)$ involved in the three-to-zero-form Hodge map, so that the resulting coordinate representation of the divergence, Eq. (V218), contains no sign factors.

curl **Curl**: The curl is a first-order derivative operation mapping three-component vector fields to three-component vector fields. It resembles the exterior derivative acting on (three-component) one-forms to (three-component) two-forms (cf. Eq. (V181)). This suggests the identification

$$\nabla \times \partial_v = (J^{-1} * dJ)\partial_v, \tag{V219}$$

whereby a vector field ∂_v is successively converted to a one-form by J, to a two-form by d, to a one-form by $*$, and finally back to a vector field by J^{-1}. Application of the relations of Table V2 confirms that this operation is equivalent to the previous definition, Eq. (V109):[62]

$$(\nabla \times \partial_v)^i = \text{sgn}(g)|g|^{-1/2}\epsilon^{ijk}\partial_j\left(g_{kl}v^l\right). \tag{V220}$$

Laplacian **Laplacian**: The Laplacian is a second-order derivative operation turning functions into functions.[63] The invariant succession of steps achieving this operation reads (0-form) $\overset{d}{\to}$ (1-form) $\overset{*}{\to}$ (2-form) $\overset{d}{\to}$ (3-form) $\overset{*}{\to}$ (0-form), or $\Delta \equiv \text{sgn}(g) * d * d$, where the signum of the metric is included as part of the definition. This can be written as

$$\Delta = \text{sgn}(g)(*d*J)(J^{-1}d) = \nabla \cdot \nabla. \tag{V221}$$

Indeed, the coordinate representation, obtained by straightforward composition of Eqs. (V216) and (V218), reads

$$\Delta = |g|^{-1/2}\partial_i\left(|g|^{1/2}g^{ij}\partial_j\right), \tag{V222}$$

reproducing Eq. (V99). For the example of the Laplacian in spherical coordinates, see Eq. (V122).

EXERCISE Apply the formulas derived above for the example of spherical coordinates (r,θ,λ) with $g_{rr} = 1$, $g_{\theta\theta} = r^2$, $g_{\phi\phi} = r^2\sin^2\theta$. Check that the previously derived relations (V75), (V96) and (V110) for the gradient, divergence and curl are reproduced.

INFO The usefulness of the Hodge star $*$ and the metric isomorphism J is not limited to three-dimensional situations. As an example, consider *physical force* which, regardless of dimension, the present approach describes by a one-form, f, see discussion on p. 480. In physics, the same quantity is represented by a vector, ∂_f, and $\partial_f = J^{-1}f$ translates between the two descriptions. This relationship follows directly from the definitions: in form language, the work performed by the force along a small displacement vector, ∂_Δ, equals the force form acting on the vector, $f(\partial_\Delta)$. Within the traditional approach, the same quantity is the inner product of the force vector and the displacement vector, $g(\partial_f, \partial_\Delta)$. Comparing the two expressions, $f_j(\partial_\Delta)^j = (\partial_f)^i g_{ij}(\partial_\Delta)^j$, we realize that $f_j = (\partial_f)^i g_{ij}$, i.e. ∂_f and f are related by the isomorphism J. For a conservative force, there exists a *potential function*, φ, such that $f = -d\varphi$. In this case, $f(\partial_\Delta) = -d\varphi(\partial_\Delta) = -\partial_\Delta(\varphi)$, and the work performed

---

[62] In our previous discussion we considered a positive metric, $\text{sgn}(g) = 1$. The extra sign appearing in Eq. (V220) ultimately is a matter of definition. It is convenient because it cancels the sign in $g_{kl} = \eta_{kl} = -1$ appearing in a negative definite metric. This situation is realized, e.g., in the theory of special relativity, where three-dimensional real space is represented by a metric of negative signature.

[63] There exists a generalized variant of the Laplacian acting on forms of arbitrary degree. However, we will not discuss this extension here.

by the force between nearby points equals minus the change in the potential function φ. The force vector, ∂_f, corresponding to $-\mathrm{d}\varphi$ is given by $J^{-1}(-\mathrm{d}\varphi) \overset{(V215)}{=} -\nabla\varphi$, and we recover the statement that conservative forces are the gradients of functions. However, the representation by forms is more natural and direct: physical forces are measured (see discussion on p. 152) as the work required to move a test particle between nearby points in space. This is a linear assignment, (separation in space) $\mapsto$ (number), and precisely what a differential form does. The construction of a vector "pointing in the direction of the force" requires a metric and complicates matters. For example, in any coordinate system, the exterior derivative of conservative force potentials assumes the simple form, $f = -\mathrm{d}\varphi = -\partial_i\varphi\,\mathrm{d}y^i$. By contrast, gradient vector fields, $\partial_f = (-\nabla\varphi)^j\,\partial_j = -(\partial_i\varphi)g^{ij}\partial_j$, contain derivatives mixed with elements of the metric, complicating the mathematics.

current
three-form   As a four-dimensional example, consider the **current three-form** defined in the info section on p. 489. In the context of special relativity, current is described by a *four-component current vector field*, ∂_j (here, the subscript refers to current, j, and is not an index). Its spatial components, $j^i(x)$, are the components of a three-dimensional current density vector "pointing in the direction of current flow", and its zeroth component, $j^0(x) \equiv c\rho(x)$, is the particle density scaled by the speed of light, c, so that all four components have dimensions (area $\cdot$ time)$^{-1}$. The dependence of $j^\mu(x)$ on the space-time argument $x = (x^0, x^1, x^2, x^3)^{\mathsf{T}} = (ct, x^1, x^2, x^3)^{\mathsf{T}}$ describes spatio-temporal variations of these quantities.

One may guess on formal grounds how the connection between the current three-form (V168) and the current vector, ∂_j, is established: an application of the Hodge star transforms the three-form into a $4 - 3 = 1$-form, and a subsequent application of cJ^{-1} maps the latter to a vector.[64] For the space-time manifold with its signature $(1, 3)$ metric, g, with $\det(g) = -1$, these operations are given by

$$\partial_j \equiv cJ^{-1}(*j) \overset{(V208)}{=} cJ^{-1}\left(\tfrac{1}{3!}j^{\mu\nu\sigma}\epsilon_{\mu\nu\sigma\rho}\mathrm{d}y^\rho\right) \overset{(V214)}{=} \tfrac{c}{3!}j^{\mu\nu\sigma}\epsilon_{\mu\nu\sigma\rho}g^{\rho\tau}\partial_\tau$$

$$= \tfrac{c}{3!}j_{\lambda\kappa\delta}g^{\lambda\mu}g^{\kappa\nu}g^{\delta\sigma}g^{\tau\sigma}\epsilon_{\mu\nu\sigma\rho}\partial_\tau \overset{(L159)}{=} -\tfrac{c}{3!}j_{\lambda\kappa\delta}\epsilon^{\lambda\kappa\delta\tau}\partial_\tau \equiv j^\tau\partial_\tau, \quad (V223)$$

where in the last step we used $\det(g^{-1}) = -1$. Using $\epsilon^{\lambda\kappa\delta\tau} = -\epsilon^{\tau\lambda\kappa\delta}$ we identify the components of the current vector field through the right-hand side of the equation as

$$j^\tau = \tfrac{c}{3!}\epsilon^{\tau\lambda\kappa\delta}j_{\lambda\kappa\delta}. \qquad (V224)$$

This construction confirms the physical expectation that the current vector and the current form are closely related. Specifically, the zeroth component, $c\rho = j^0 = \tfrac{c}{3!}\epsilon^{0\lambda\kappa\delta}j_{\lambda\kappa\delta} = c\epsilon^{0123}j_{123} = cj_{123}$, is given by the weight function of the density form, which we argued on p. 489 describes the particle density. The spatial component, $j^3 = \tfrac{c}{3!}\epsilon^{3\lambda\kappa\delta}j_{\lambda\kappa\delta} = c\epsilon^{3012}j_{012} = -cj_{012}$, defines the current flow in the 3-direction, and up to the factor $-c$ equals the weight function j_{012} of the three-form. Treating the other spatial directions in the same way, we find that j^i and j_{0jk} are related through

$$j^i = -\tfrac{c}{2}j_{0jk}\epsilon^{jki}, \qquad j_{0jk} = -c\epsilon_{jki}j^i. \qquad (V225)$$

This shows how the current flow in the i-direction (vector language) is related to the current flow through area elements perpendicular to that direction (form language).

As with the physical force discussed above, the definition of current as "the number of particles flowing through an area per unit time" is more natural than that via a "vector pointing in the direction of current flow". This is because actual detectors measure flow by counting numbers of particles per unit area. While the form-definition of the current describes just this, the translation between experiment and current-is-a-vector approach requires the excess baggage of a metric. We will return to this point in the next chapter when we discuss current flow in the context of electrodynamics.

---

[64] The factor c is needed for dimensional reasons, since the components of the three-form j have dimension volume^{-1}.

V6.4   Case study: metric structures in general relativity

Above, we have stressed that one should leave room for manifolds not embedded in an ambient vector space. The most important example motivating this view is the Universe itself. Since Einstein, we understand the Universe as a four-dimensional space-time manifold, equipped with a nontrivial metric. We do not believe in the existence of a larger vector space containing this structure. The justification of this view is a subject beyond the scope of this text. However, we have now developed sufficient proficiency with manifolds that it is tempting to discuss a few of its consequences.

equivalence principle

The starting point of general relativity is several observations subsumed under the name **equivalence principles**. For example, Newton's second law states that $\mathbf{F} = m_I\mathbf{a}$, the acceleration of a test body is proportional to the strength of an applied force, where the proportionality constant defines the (inertial) mass of the body. At the same time, we know that the gravitational force experienced by a body, $\mathbf{F} = m_G\mathbf{g}$, is proportional to a constant vector $\mathbf{g}$ in the direction of the gravitational field, where the proportionality constant defines the (gravitational) mass of the body. Empirically, gravitational and inertial mass are found to be identical, $m_G = m_I$. This means that in a perfectly homogeneous gravitational field all bodies experience the same gravitational acceleration, $\mathbf{a} = \mathbf{g}$. The situation is very different from that with other fundamental forces where, for example, the acceleration of a body by the force of an electric field, $\mathbf{a} = qm_I^{-1}\mathbf{E}$, depends on its inertial mass. The close connection between inertial and gravitational forces has been illuminated by various of Einstein's famous thought experiments. For example, an observer inside a constantly accelerated elevator in empty space will not be able to tell the difference from an elevator at rest subject to a gravitational field.

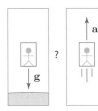

After a decade of thought, Einstein formulated a theory of gravitation which explains the peculiar features of gravitation in a way that is radically geometrical. (In fact the term *geometrodynamics* has been introduced as a synonym for general relativity.) The main principles of the new approach can be roughly summarized as follows:

Albert Einstein 1879–1955
German theoretical physicist who developed the theory of relativity, and made several other breakthrough contributions to theoretical physics. Relativity and quantum mechanics define the two fundamental pillars of modern physics, and Einstein derived the former more or less single-handedly. Being of Jewish origin, he did not return to Germany from a visit to the United States in 1933.

▷   There is no such thing as a fundamental gravitational force between bodies. (Let's recall that Newtonian mechanics axiomatically presumed the presence of a force, but never addressed its origins.)

▷   Rather, all bodies in the universe move freely along the shortest possible connections through a curved space-time. Space-time is considered a four-dimensional manifold whose curvature is described by a metric.

▷ Mass distributions in the Universe cause this curvature and in this way affect the motion of other bodies.

For example, in the view of general relativity, Earth does not act on bodies via a central gravitational force. Rather, it causes a distortion in the curvature of space time, which makes them move freely towards Earth's center. What we experience as weight is the hindrance of this motion due to Earth's finite solid extension. Perhaps this is a good point to once more stress that the proper place for introductions to this subject are lecture courses in relativity.

gravitational redshift

One of many observable predictions of the equivalence principles is the phenomenon of **gravitational redshift**. Suppose Alice stands on a tower of height h and sends a sequence of light signals, at constant intervals τ_A, to Bob, who is at height 0. What will be the intervals, τ_B, at which Bob receives the signals? According to the equivalence principle, a situation in which Alice and Bob are at rest and experience a gravitational force with gravitational constant g is equivalent to one in which they move through empty space with constant acceleration $a = g$. Let us look at the situation from the second perspective.

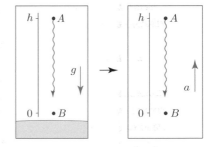

Q1 Assuming coordinates such that at the time $t = 0$ when the first signal is sent, Alice and Bob are at rest,[65] calculate the time at which Bob will receive the first signal. When Alice sends her second signal, at $t = \tau_A$, she and Bob have accelerated to a finite velocity (which one?) relative to the origin of the coordinate system. Assume that this velocity is negligibly small compared to the speed of light, c. Under this assumption, show that

$$\tau_B \simeq \left(1 - \frac{gh}{c^2}\right)\tau_A. \tag{V226}$$

(*Hints:* Some work can be saved by economic definitions of variables. Set up equations defining the times, t_1 and t_2, at which Bob receives the first and second signal respectively. Substitute $t_2 = t_1 + \tau_B$, and use the equation for t_1 to simplify the second equation. Neglect velocities of $\mathcal{O}(t_1, t_2, \tau_A, \tau_B)g$ compared to c.)

The time Bob has to wait between signals is the difference between the emission times. Now consider the specific case where τ_A is the time, $2\pi/\omega$, between two consecutive passages of the maximum amplitude of an electromagnetic wave. For Bob, this $\simeq (2\pi/\omega)(1 - gh/c^2)$, which means that he observes light of higher frequency, $\omega' = \omega(1 - gh/c^2)^{-1} \simeq \omega(1 + gh/c^2)$, than Alice. This phenomenon was predicted by Einstein as gravitational redshift. Light "falling" in a gravitational field is shifted to frequencies that increase with the "fall distance" h. Conversely, this means a shift to longer frequencies (for visible light, towards red) at higher altitudes. Gravitational redshift has been observed for various astronomical bodies, including white dwarfs, the Sun, and galaxy clusters. For the

---

[65] This means that the space-time coordinate system is "co-moving" such that at the initial time its spatial origin coincides with the position of Bob and moves with Bob's instantaneous velocity.

first terrestrial measurement, see R. Pound, G. Rebka, Apparent Weight of Photons, *Phys. Rev. Lett.* **4**, 337 (1960).

A1 At time t, Bob's position has moved to $x_B(t) = \frac{1}{2}gt^2$. The tip of the light ray is at $x_L(t) = h - ct$. The two positions coincide at $x_B(t_1) = x_L(t_1)$, which defines t_1 as $h - ct_1 = \frac{1}{2}gt_1^2$. At times, t, after the emission of the second signal, Bob's position continues to increase as $x_B(t)$, while the second light ray is at Alice's position at the time of emission, $h + \frac{1}{2}g\tau_A^2$, minus the traversed distance since emission, $x_L(t) = h + \frac{1}{2}g\tau_A^2 - c(t - \tau_A)$. We substitute $t_2 = t_1 + \tau_B$ in the defining equation, $x_B(t_2) = x_L(t_2)$, to obtain $\frac{1}{2}g(t_1 + \tau_B)^2 = h + \frac{1}{2}g\tau_A^2 - c(t_1 + \tau_B - \tau_A)$. Using the equation for t_1 and neglecting terms $\frac{1}{2}g\tau_{A,B}^2$, we obtain Eq. (V226).

Now let us examine the situation from the curved space-time perspective. To this end, we need to import a few facts from general relativity: the **Einstein equations** describe how space and time curve in response to the presence of mass distributions. In mathematical terms this means that the four-dimensional space-time

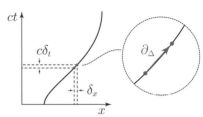

Einstein equations

manifold carries a metric, g, different from the Minkowski metric. The metric assigns to vectors, ∂_v, tangent to the space-time manifold, a value $g(\partial_v, \partial_v)$. We need to develop some intuition for the meaning of this quantity. To this end, consider a two-dimensional cartoon

world line

of space-time, and a **world line** of a particle moving through it. Physically, a world line is a sequence of space-time events defining the motion of a body through space. Mathematically, it is a curve described by the coordinate representation $x^\mu = (ct, x^i(t))$, where t is the time coordinate of an arbitrary coordinate system, and $x^i(t)$ are the spatial coordinates describing the trajectory. Importantly, both t and x^i do not have absolute physical reality.[66] Consider two close points on a world line and characterize their separation by a small vector, ∂_Δ, with local coordinate representation $(c\delta_t, \delta_x)$. For example, the world line might be that of a car moving on a highway. Observed from a system in which the highway is at rest, the two events are described by coordinates $(ct, x(t))^T$ and $(c(t + \delta_t), x(t + \delta_t))^T$. In this case, ∂_Δ has the coordinate representation $(c\delta_t, \dot{x}(t)\delta_t)^T$. The two events may be observed from any coordinate system, including a co-moving system in which they occur at the same spatial coordinate. For example, a coordinate system whose spatial origin is locked to the car has this property. In this system, ∂_Δ has the representation $(c\delta_{t'}, 0)$, where t' is the time coordinate of the chosen co-moving system. While this coordinate does not have absolute meaning, the norm $(c\delta_\tau)^2 \equiv \|\partial_\Delta\|^2 = g(\partial_\Delta, \partial_\Delta) = g_{00}(c\delta_{t'})^2$ defines the

proper time

proper time, or "wristwatch time", δ_τ, separating the events in the co-moving system. This quantity is invariant in the sense that a rescaled choice of time coordinate will change t' and g_{00} in such a way that δ_τ remains invariant. The equation $(c\delta_\tau)^2 = g(\partial_\Delta, \partial_\Delta)$ tells us that

---

[66] Einstein's theory famously predicts the relativity of time. For example, two events observed at coinciding times in one system need not be simultaneous in a different system. Never make the mistake of assigning physical reality to coordinates in relativity. For further discussion, we refer to specialized courses.

> the space-time separation of two events taking place in close spatio-temporal vicinity of a point x is characterized by a tangent vector ∂_Δ. The proper time between the events, $\delta_\tau = \sqrt{g_x(\partial_\Delta, \partial_\Delta)}/c$, is the time passing between them in a system in which ∂_Δ has only temporal components and the events take place at the same point in space.

Q2 Recapitulate the invariance of $g(\partial_v, \partial_v) = (\partial_v)^\mu g_{\mu\nu}(\partial_v)^\nu$ under coordinate changes. Discuss the freedom in the choice of different co-moving frames. Is the choice of the time axis of the co-moving system unique? And that of the spatial axes? Discuss what happens if different units of time, t, on the time axis are chosen. Why do δ_t and g_{00} change in such a way that δ_τ remains invariant?

We can now discuss the interpretation of gravitational redshift in a mindset which does not make reference to gravitational forces. Instead, we adopt Einstein's view in which Earth creates a tiny local distortion of the metric of the Universe. That modified metric follows from Schwarzschild's solution of the Einstein

Karl Schwarzschild 1873–1916
German theoretical physicist known for the first nontrivial solution of the Einstein equations. Schwarzschild derived his solution in 1915, the year in which Einstein introduced general relativity. Shortly after, he died of a medical condition developed during his service at the eastern front in the Great War.

equations in the presence of spherically symmetric (non-rotating) bodies of mass M, a setup well applicable to stars or planets. In the space outside these bodies, the **Schwarzschild** **metric** is given by

Schwarzschild metric

$$g = \left(1 - \frac{r_s}{r}\right) c^2 \mathrm{d}t \otimes \mathrm{d}t - \left(1 - \frac{r_s}{r}\right)^{-1} \mathrm{d}r \otimes \mathrm{d}r - r^2 \mathrm{d}\theta \otimes \mathrm{d}\theta - r^2 \sin^2\theta \, \mathrm{d}\phi \otimes \mathrm{d}\phi.$$

$$(\text{V227})$$

Here, $r_s = 2GM/c^2$ is the *Schwarzschild radius* defined in terms of the body's mass, and Newton's **gravitational constant**, G. We recall that in Newtonian mechanics two bodies of mass m_1 and m_2 separated by a distance r exert a force of magnitude $F = Gm_1m_2/r$ on each other. With $G \simeq 6.674 \cdot 10^{-11} \mathrm{m}^3\mathrm{kg}^{-1}\mathrm{s}^{-2}$ one finds that the Schwarzschild radius of "conventional" astronomic bodies (i.e. excluding black holes or neutron stars) is much smaller than their extension. For example, the Sun has a Schwarzschild radius of about 3 km and Earth one of a few mm.

gravitational constant

Q3 For radii r much larger than r_s the Schwarzschild metric approaches the Minkowski metric introduced on p. 46, where the spatial part is represented in spherical coordinates, see Eq. (V203). We aim to understand the tiny effects of the metric distortion for values of r comparable to the radius of Earth, r_E. Focusing on the radial and the temporal coordinates, let us define $x \equiv r - r_E$, and temporarily forget about the angular part of the metric. Show that the resulting two-dimensional metric defined in the space of coordinates, $(t,x)^T$, has the form

$$g_{ct,x} \simeq \left(1 - \kappa + \frac{2gx}{c^2}\right) \mathrm{d}t \otimes \mathrm{d}t - \left(1 + \kappa - \frac{2gx}{c^2}\right) \mathrm{d}x \otimes \mathrm{d}x, \qquad (\text{V228})$$

where $\kappa = r_s/r_E$ and the subscript on $g_{ct,x}$ indicates that the metric varies as a function of the space-time coordinates (ct, x) (although in the present case it is independent of time, and static).

A3 Substitution of $r = r_E + x$ in Eq. (V227) leads to the expression $r_s/r \simeq (r_s/r_E)(1 - x/r_E) = \kappa - 2GM/(c^2 r_E^2)x$, where $\kappa = r_s/r_E$. Now observe that in Newtonian mechanics a body of mass m on Earth's surface experiences the gravitational force $GMm/r_E^2 \equiv gm$. This leads to the identification $r_s/r \simeq \kappa - 2gx/c^2$. The second term in the metric (V228) is obtained via the approximation $(1 - r_s/r)^{-1} \simeq 1 + r_s/r$.

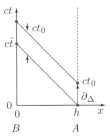

We may now discuss what happens in Alice and Bob's experiment from the metric perspective. To this end, consider the world lines of the light rays involved in the experiment. The first ray is emitted at time 0 and at $x = h$. It then moves towards Earth along an approximately[67] straight line before hitting ground at some later time, $\tilde{t}$. The second ray is emitted shortly after the first, with delay t_0, and hits ground at time $\tilde{t} + t_0$. That its delay at emission and receipt has the same value, t_0, follows from the fact that the metric is purely static. The world lines of the two light rays therefore look identical, up to a time shift by t_0. Crucially, t_0 has no intrinsic physical meaning, it is just a coordinate.

Q4 The proper delay times between the second and first signals, observed by Alice at emission as τ_A and by Bob at receipt as τ_B, each in their respective rest frames, do have invariant meaning. In either case, the delay is described by the same translation vector, ∂_Δ, with coordinate representation $(ct_0, 0)^T$ (see figure). Nevertheless, Alice and Bob find different proper delay times, τ_A and τ_B, for this translation vector, because proper times depend on the metric, which in turn depends on position due to gravity. Use the Schwarzschild metric to compute the proper delay times and show that $\tau_B/\tau_A \simeq 1 - \frac{gh}{c^2}$, in agreement with the result (V226) from our earlier treatment of the redshift phenomenon.

A4 To find the proper delay times τ_A and τ_B, we evaluate the metric on the temporal translation vector ∂_Δ at the space-time coordinates of Alice and Bob. In Alice's case, this yields $(c\tau_A)^2 = g_{(0,h)}(\partial_\Delta, \partial_\Delta) = \left(1 - \kappa + \frac{2gh}{c^2}\right)(ct_0)^2$, and for Bob $(c\tau_B)^2 = g_{(c\tilde{t},0)}(\partial_\Delta, \partial_\Delta) = (1 - \kappa)(ct_0)^2$. The ratio is given by $\tau_B/\tau_A = (1 - \kappa)^{1/2}/(1 - \kappa + \frac{2gh}{c^2})^{1/2} \simeq 1 - \frac{gh}{c^2}$.

We have thus obtained gravitational redshift from a construction that makes no reference to forces, nor the (equivalent) concept of accelerated motion. The difference in proper times is rather attributed to the bending of space-time by Earth, as predicted by the Schwarzschild solution. Other phenomena predicted on the same basis include the bending of light rays by mass distributions, or the deviations of planetary motion from purely ellipsoidal ones. All these have been observed experimentally, and are triumphs of general relativity.

---

[67] Einstein's theory predicts that light in the presence of massive bodies moves along bent trajectories, but that is another story.

V6.5　Summary and outlook

In this chapter we took the crucial step of equipping manifolds with a metric. From linear algebra, we know that a metric is at the origin of most structures associated with geometry. However, whereas in a single vector space geometry begins and ends with the measurement of lengths and angles, the situation on manifolds is richer. Here, the metric determines the shape of geometric objects, their curvature and their volume. A metric defines a canonical integral, it establishes connections between forms of different degrees, and between forms and vectors. The latter aspect was key to establishing the connection between the traditional approach to vector analysis and the one building on differential forms. Specifically, we discussed how form operations have vector analogues, the latter generally being more complicated than the former. Finally, we hinted at the importance of the joint concept of abstract manifolds (no embedding) and metric structures in general relativity. In the next and final chapter, we continue our discussion of differential geometry in physics and illustrate how it affords a clean and intuitive description of electromagnetism.

REMARK   This chapter assumes familiarity of the reader with the laws of electromagnetism at the level of an undergraduate experimental physics course. We will frequently compare vectors with their associated one- or two-forms. In such cases, the two objects are distinguished by plain symbols (E) for forms and underlined symbols ($\underline{E}$) for vectors. As usual, the positioning of indices, E^i and E_i, identifies the components of vectors and forms, respectively.

Throughout, the signature $(1,3)$ metric of space-time plays a distinguished role. We will work in orthogonal frames where the metric assumes the Minkowski form, given by $\eta = \mathrm{diag}(1,-1,-1,-1)$, and denote coordinates by the symbols x^μ. (While the usage of orthonormal bases is not essential for the form representation of the theory, it facilitates the comparison to the traditional vector approach.)

The chapter relies on the full machinery developed in Chapters V4 to V6. We frequently state identities requiring a brief justification. For example, if we say that the components of the magnetic field vector B^i relate to those of the field form as $B^i = -\frac{1}{2}\epsilon^{ijk}B_{jk}$, readers should verify the statement and test their knowledge. In this final chapter, we do not add explicit exhortations such as (check!).

In this chapter we will illustrate the usefulness of exterior calculus in physics on the example of electromagnetism. The modern theory of electromagnetism was formulated in the nineteenth century by *Maxwell*, then in the language of vectors. Below, we will review the physical principles of Maxwell theory and then formulate them in the language of forms.

James Clerk Maxwell 1831–1879
Scottish theoretical physicist and mathematician. Amongst many other achievements, he is credited with the formulation of the theory of electromagnetism, synthesizing all previous unrelated experiments and equations of electricity, magnetism and optics into a consistent theory. (He is also known for creating the first true color photograph in 1861.)

Importantly, this reformulation goes beyond a mere change of language. The theory of electromagnetism is intimately related to that of relativity. It was the first *relativistically invariant theory*, i.e. a theory compatible with coordinate transformations respecting the constancy of the speed of light. As we saw above, the metric of space-time is a concept central to Einstein's theory of relativity. Within special relativity, the Lorentz transformations stabilizing the Minkowski metric in the sense of Eq. (L211) describe how coordinates change between different systems, and in the later extension of general relativity the metric describes how space-time acquires curvature due to the presence of masses. A proper treatment of the metric is therefore essential in relativistic theories, and this is where the traditional description of electromagnetism fails.

Within the traditional approach, all physical quantities relevant to electromagnetism – electric and magnetic fields, and currents – are treated as vectors. However, we reasoned above that the true identity of a physical quantity follows from a measurement protocol.

In the case of electromagnetic fields and currents these protocols define not vectors, but differential forms in space-time. As discussed in previous chapters the translation to vectors requires the metric, and complicates the mathematics. More serious is the fact that the vector formulation obscures the role of the metric in a relativistically invariant context – it is difficult to tell whether a metric tensor appearing in a formula is required for physical reasons or only as a translational tool. This way of teaching electrodynamics has not changed in the last 100 years, likely due to social inertia. However, we hope that the exposition below demonstrates that an alternative formulation, more in line with modern developments in physics, is possible and actually more intuitive than the traditional one.

V7.1 The ingredients of electrodynamics

REMARK In this section we introduce electrodynamics, starting with the non-invariant approach in which differential forms are defined in three-dimensional space. Exterior derivatives in three-dimensional space will be denoted by d_s, in distinction from d acting in four-dimensional space-time. Similarly, differential forms occurring in both representations are labeled as ρ_s and j_s (three-dimensional), in distinction from ρ and j (four-dimensional).

In this section we introduce the essential ingredients of the theory of electromagnetism – space-time, charges and currents, and electromagnetic fields – from both a physical and a mathematical perspective. This will set the stage for the discussion of the next section where these building blocks are put in relation to each other.

Space-time

Minkowski space-time Classical electrodynamics is defined in flat **Minkowski space-time**, the manifold $M = \mathbb{R}^4$ equipped with the signature $(1,3)$ metric, $g = \eta = \mathrm{diag}(1,-1,-1,-1)$. As usual in this context, we employ a coordinate notation $x = x^\mu = (x^0, x^1, x^2, x^3)^\mathrm{T} \equiv (ct, x^1, x^2, x^3)^\mathrm{T}$, where $x^0 = ct$, t is time, c the speed of light, and x^i, $i = 1, 2, 3$ an orthonormal system of spatial coordinates. The theory defined on this manifold describes electromagnetism in the context of special relativity. As a side remark we note that the extension to general relativity and the presence of gravitation requires the generalization of M to a curved space-time with a non-constant signature $(1,3)$ metric. Although this represents a major complication, the concept of Riemannian manifolds introduced in Chapter V6 defines the proper mathematical framework to handle that situation.

Within both the traditional and the exterior calculus approach we distinguish between a *relativistically invariant and a non-invariant representation of electrodynamics*. These standard denotations are not ideal because electrodynamics is compatible with the laws of relativity no matter how it is represented. The difference is that the invariant formulation treats space and time on equal footing and is formulated on $(\mathbb{R}^4, \eta)$, space-time with the Minkowski metric. The non-invariant formulation splits $\mathbb{R}^4 = \mathbb{R} \times \mathbb{R}^3$ into a one-dimensional time axis and three-dimensional space. In this setting, time enters as a parameter and $\mathbb{R}^3$ assumes the role of the manifold with a negative definite signature $(0, 3)$. Space-time points are denoted by (t, x), where $x = \{x^i\}$ denotes a three-dimensional spatial coordinate vector (not to be confused with the four-dimensional $x = \{x^\mu\}$ of the invariant formulation). For example, the electric field, $\underline{E}(t, x)$, is a three-dimensional vector field on $\mathbb{R}^3$, depending on time, t, as a parameter, and the spatial coordinate x. (Within the invariant formulation, the electric field is absorbed into a different object, the field-strength tensor, to be discussed in Section V7.4 below.)

Following the historical sequence of developments, in Sections V7.1 to V7.3 we use the non-invariant formulation to discuss the experimental observations motivating the laws of electrodynamics, and their formulation through the Lorentz force law and the Maxwell equations. Finally, in Section V7.4, we discuss the invariant formulations of these equations.

Charges and currents

Charges and currents are the sources of electromagnetic fields, and their presence is described by a non-vanishing current three-form, j, Eq. (V168). Charge and currents densities are tied to the presence of matter – an electron carrying a negative unit-charge, for example – and for this reason the components of the current-form are sometimes called *matter fields*. This denotation emphasizes the difference from the electromagnetic fields, E, D, B, H, which are immaterial.

electric charge   *Electric charge* is represented by a three-form, ρ_s, defined such that the integral

$$Q = \int_V \rho_s$$

equals the charge, Q, contained in a volume V. *Electric charge is conserved.* If the charge density decreases in a region of space, $\partial_t Q < 0$, current must flow out of it, see the figure. This current **current** is represented by a **current two-form**,[68] j_s, defined such that **two-form**

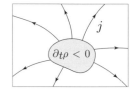

$$I = \int_S j_s \qquad (V229)$$

equals the electric current flowing through a surface, S. Here, it is assumed that S carries an orientation determining the sign of the current. Specifically, for closed surfaces, $S = \partial V$, defined as boundaries of volumes, the conventions are such that a diminishing of charge inside the volume, $d_t Q < 0$, is compensated for by a positive outward current, **continuity** $I = -d_t Q > 0$. The balance is expressed by the **integral form of the continuity equation**, **equation**

$$d_t \int_V \rho_s + \int_{\partial V} j_s = 0. \qquad (V230)$$

For a static choice of the volume, $d_t \int_V \rho_s = \int_V \partial_t \rho_s$, and application of Stokes's theorem yields $\int_V (\partial_t \rho_s + d_s j_s) = 0$. Valid regardless of the choice of V, this implies the **differential form of the continuity equation**,

---

[68]   For its relation to the invariant current three-form introduced on p. 489, see Section V7.4.

$$\boxed{\partial_t \rho_s + d_s j_s = 0.} \tag{V231}$$

How does one pass from the differential form representation of current conservation to the *traditional vector language*? A scalar charge density function, $\rho = \rho_{123}$, is defined as the coefficient of the charge form, $\rho_s = \rho\, dx^1 \wedge dx^2 \wedge dx^3$. With this identification, $Q = \int_V \rho_s = \int_V \rho\, dV$ assumes the form of a conventional volume integral over ρ, as expected. The three-dimensional Hodge star may be applied to pass from the three-form ρ_s to a scalar. Using Eq. (V208) or Table V2 and noting that for the Minkowski metric $\rho^{123} = -\rho_{123} = -\rho$, we find that this scalar is $*\rho_s = -\rho$. Turning to the current, application of the rules of Section V6.3 shows that $J^{-1} * j_s$ converts the current two-form into a vector. We expect this vector to define the traditional current density vector, $\underline{j}$, up to an as yet unidentified sign. This sign is determined by acting on the continuity equation with $-*$,

$$\underbrace{(-*\partial_t \rho_s)}_{\partial_t \rho} + \underbrace{(-*d_s *J)}_{\text{div}}\underbrace{(-J^{-1} * j_s)}_{\underline{j}} = 0,$$

where in the second term we inserted $\mathbb{1} \overset{\text{(V210)}}{=} -** = -*JJ^{-1}*$ and noted that div $\overset{\text{(V217)}}{=} -*d_s *J$. The minus sign in the divergence arises because $\text{sgn}(g) = \text{sgn}(\eta) = \text{sgn diag}(-1,-1,-1) = -1$. Compatibility with the traditional form of the continuity relation, $\partial_t \rho + \text{div}\,\underline{j} = 0$, is established via the definition

$$\underline{j} = -J^{-1} * j_s. \tag{V232}$$

current density vector The components of the **current vector** $\underline{j}$ are thus given by

$$j^i = \tfrac{1}{2}\epsilon^{ijk} j_{s,jk}, \qquad j_{s,jk} = \epsilon_{jki} j^i. \tag{V233}$$

Notice that the metric, and with it various pesky factors of $\text{sgn}(g) = -1$, enter the conversion between the current two-form and its vector representation via the Hodge star. This is an example of a metric dependence required solely by the passage from the metric-free form representation to the traditional language.

Electromagnetic fields

Electromagnetism addresses the mutual influence of charged matter and electromagnetic fields. The non-invariant formulation of the traditional approach defines four fields, $\underline{E}, \underline{D}, \underline{B}, \underline{H}$, all three-component vector fields in space-time. Conceptually, these are vector fields defined on the tangent bundle of space, $\mathbb{R}^3$, and following the conventions of Chapter V4 they should be denoted as ∂_E, etc. However, this notation is so alien to the traditional approach that we rather use $\underline{E}$ throughout. Likewise, we write $\underline{E} = E^i \underline{e}_i$ instead of $\partial_E = E^i \partial_{x^i}$ for component expansions. However, keep in mind that $\underline{E}$ remains a vector field in the sense of Chapter V4.

Within the exterior calculus approach, the four fields are differential forms of degree one (E, H) and two (D, B). The physical meaning of these forms is discussed in Section V7.3 after the fundamental laws of electromagnetism have been introduced. For the moment

Table V3   The four differential forms describing the electromagnetic fields within the framework of exterior calculus.

field	name	nature	degree	sector
E	electric field	electric	1	homogeneous
B	magnetic induction	magnetic	2	homogeneous
D	displacement field	electric	2	inhomogeneous
H	magnetic field	magnetic	1	inhomogeneous

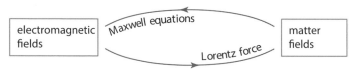

Fig. V26   Maxwell theory describes the creation of fields by charges and currents. Conversely, the Lorentz force describes the feedback of fields on charges via forces.

we just refer to Table V3, where various of their features are summarized. Specifically, the fields E, D are electric in nature, while B, H are magnetic. The fields E, B describe the influence of electromagnetic fields on charged matter, while D, H are created via the reciprocal influence of charged matter on fields. This distinction is labeled as homogeneous vs. inhomogeneous, indicating that D, H will be defined by differential equations with **the fields of electrody- namics** matter sources as inhomogeneities.

Having defined the constituents of the theory, we next discuss their interconnections via the laws of electromagnetism. These laws describe how charged matter is affected by electromagnetic fields through a force called the Lorentz force. A second group of laws, the inhomogeneous Maxwell equations, describes the inverse influence, the creation of fields by charges and currents (see Fig. V26). Third, the homogeneous Maxwell equations describe intrinsic relations between the fields. In the following we discuss these three groups of laws in turn.

V7.2   Laws of electrodynamics I: Lorentz force

A charged particle exposed to electric and/or magnetic fields experiences mechanical forces. In the following, we show how these forces define the electric and magnetic fields as one- and two-forms, respectively, and establish the connection to the traditional vector language.

Electric and magnetic field forms

An electric field acts on a test particle via an *electrostatic force* proportional to its charge. Mathematically, the force is described by a one-form, $F_{\mathrm{E}} \in \Lambda^1(\mathbb{R}^3)$. This force defines the **electric field** **electric field one-form** as force per charge, $E = F_E/q \in \Lambda^1(\mathbb{R}^3)$.

If a magnetic field is present and the particle is in motion, it experiences an additional *magnetic force*, F_M, proportional to the particle's charge and velocity. The magnetic induction field is defined through the work done against this force in a displacement process. To this end, consider a particle moving with velocity vector $\partial_v = \underline{v}$, and undergoing a small displacement by the vector $\partial_\Delta = \underline{\Delta}$.[69] Empirically, the work W required for this process is proportional to q, linear in $\underline{v}$ and $\underline{\Delta}$, and antisymmetric under exchange of these arguments. It is described by the application of a *magnetic induction two-form*, B, to these arguments, $W = -qB(\partial_v, \partial_\Delta)$. Recalling the definition of the inner derivative, Eq. (L278), the work relation may be written as $W = -q(i_{\partial_v}B)(\partial_\Delta)$, with the one-form $i_{\partial_v}B = B(\partial_v, \cdot)$. The one-form $F_M \equiv -qi_{\partial_v}B$ defines the velocity-dependent magnetic force, and its application to a displacement vector yields the magnetic work, $W = F_M(\partial_\Delta) = -qB(\partial_v, \partial_\Delta)$.

The total force acting on a particle in the presence of an electric and a magnetic field, **Lorentz force** $F_E + F_M$, is called the **Lorentz force**, and given by

$$F = q(E - i_v B).$$ (V234)

INFO It can be instructive to *visualize the electric field and magnetic induction forms*. Following the conventions defined in Fig. L21, the electric field, E, and the force form, F_E, are represented as a pattern of parallel planes, as in Fig. V27(a). The work, $E(\partial_\Delta)$, required for the displacement of a unit charge by $\partial_\Delta = \underline{\Delta}$ equals the number of planes intersected by the vector.

Similarly, the two-form B is a pattern of parallel lines. As indicated in Fig. V27(b), the readout, $B(\partial_v, \partial_\Delta)$, of the form on a pair of vectors, $\partial_v = \underline{v}$ and $\partial_\Delta = \underline{\Delta}$, is proportional to the number of flux lines piercing the area spanned by these vectors (the parallelogram indicated by thick black lines). The magnetic force one-form, F_M, then corresponds to the pattern of planes spanned by the vector $\underline{v}$ and the flux lines. This yields the work to be done by the force during a displacement, $F_M(\partial_\Delta) = -qB(\partial_v, \partial_\Delta)$, as the number of planes intersected by ∂_Δ.

Although such visualizations play no role in quantitative constructions, they may facilitate the qualitative understanding of a situation.

Traditional representation of fields, currents, and densities

The electromagnetic fields are traditionally described as three-component vector fields, $\underline{E}$, $\underline{D}$, $\underline{B}$, $\underline{H}$. These vectors are related to the field forms via the translation rules detailed in

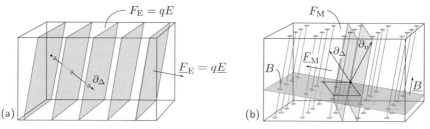

Fig. V27   On the visualization of the electric field one-form, E, the magnetic induction two-form, B, and the corresponding force forms F_E, F_M. The figure also indicates the corresponding vectors of the traditional approach.

[69] This setup models a current-carrying wire in which charge carriers move with drift velocity $\underline{v}$ along the wire axis. One measures the work required to displace the wire by $\underline{\Delta}$.

Table V4 Summary of conversion rules between electromagnetic forms and their scalar or vector representations.

scalar/vector	components	form	components
$\rho = - * \rho_s$	$\rho = \rho_{123}$	$\rho_s = *\rho$	$\rho_{123} = \rho$
$\underline{j} = -J^{-1} * j_s$	$j_s^i = \frac{1}{2}\epsilon^{ijk} j_{s,jk}$	$j_s = *J\underline{j}$	$j_{s,jk} = \epsilon_{jki} j_s^i$
$\underline{E} = J^{-1} E$	$E^i = -E_i$	$E = J\underline{E}$	$E_i = -E^i$
$\underline{B} = J^{-1} * B$	$B^i = -\frac{1}{2}\epsilon^{ijk} B_{jk}$	$B = - * J\underline{B}$	$B_{jk} = -\epsilon_{jki} B^i$
$\underline{H} = -J^{-1} H$	$H^i = H_i$	$H = -J\underline{H}$	$H_i = H^i$
$\underline{D} = -J^{-1} * D$	$D^i = \frac{1}{2}\epsilon^{ijk} D_{jk}$	$D = *J\underline{D}$	$D_{jk} = \epsilon_{jki} D^i$

Section V6.3. Specifically, the *electric field vector*, $\underline{E}$, is dual to the one-form E under the metric isomorphism, $\underline{E} \equiv \partial_E = J^{-1}(E)$, and similarly $\underline{F}_E \equiv \partial_{F_E} = J^{-1}(F_E)$. Notice that the metric enters the translation. This is important because the measurement of forces defines a form, rather than a vector, as discussed on p. 510. We thus encounter another case where the vectorial language introduces a fake dependence on the metric. Later, we will see that the metric is truly needed for physical reasons for only one purpose, namely to establish connections between the fields (E, B) and (D, H).

components of electric field　　In a *component language*, the electric field form and vector are represented as

$$E = E_i dx^i, \qquad \underline{E} = E^i \underline{e}_i, \qquad E^i = \eta^{ij} E_j = -E_i, \tag{V235a}$$

where $\underline{e}_i \equiv \partial_{x^i}$ are orthonormal coordinate basis vectors, and the components $E_i(t, x)$ and $E^i(t, x)$ may depend on time. The translation between components, $E^i = \eta^{ij} E_j = -E_i$, follows from Table V2, applied to the spatial part of the metric, $g_{ij} = \eta_{ij} = (-1)\delta_{ij}$. The value of the form acting on a displacement vector is now represented as $E_i \Delta^i = E^i \eta_{ij} \Delta^j$, i.e. the inner product, $g(\underline{E}, \underline{\Delta}) = \eta(\underline{E}, \underline{\Delta})$, of the vectors $\underline{E}$ and $\underline{\Delta}$. This is equivalent to the statement that the vector $\underline{E}$ stands perpendicular to the planes representing the form E (explain why!).

The *magnetic induction vector*, $\underline{B}$, corresponding to the two-form B is given by $\underline{B} = \partial_B = J^{-1}(*B)$. Since $** = -\mathbb{1}$, the reverse operation reads as $B = - * J\underline{B}$.

components of magnetic induction　　In a component representation we have (see Table V2)

$$B = \frac{1}{2} B_{jk} dx^j \wedge dx^k, \qquad \underline{B} = B^i \underline{e}_i, \qquad B^i = -\frac{1}{2}\epsilon^{ijk} B_{jk}, \qquad B_{jk} = -\epsilon_{jki} B^i. \tag{V235b}$$

As with our earlier discussion of the continuity relation, these translation rules are designed such that the form representation of the laws of electrodynamics matches the traditional form. For example, the magnetic force form, $F_M = -qi_{\partial_v} B = -qv^j B_{jk} dx^k$, is represented by the vector $\underline{F}_M \equiv \partial_{F_M} = J^{-1}(F_M) = -qv^j B_{jk} \eta^{ki} \underline{e}_i = -qv^j (-\epsilon_{jkl} B^l)(-\underline{e}_k) = q\underline{v} \times \underline{B}$, showing that the *Lorentz force in conventional language* is given by

traditional form Lorenz force

$$\underline{F} = q(\underline{E} + \underline{v} \times \underline{B}).$$

For later reference, we note that the two-form representing the displacement field, D, is converted to a *displacement field vector* using $\underline{D} \equiv -\partial_D = -J^{-1} * D$, with components $D^i = \frac{1}{2}\epsilon^{ijk} D_{jk}$. And the magnetic field one-form, H, is converted to a *magnetic field vector*

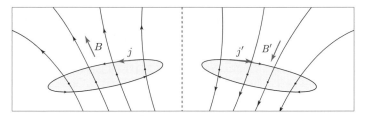

Fig. V28   The magnetic field, $B \propto \mathrm{curl}\, j$, generated by a current density vector j does not transform contravariantly and therefore is not a true vector. For example, under space reflection at a plane (indicated by the dashed line), its component perpendicular to that plane does not change sign. This makes its transformation behavior different from that of vectors.

as $\underline{H} \equiv -\partial_H = -J^{-1}H$, with components $H^i = H_i$. The translation between all forms and vectors relevant to our discussion is summarized in Table V4.

INFO   The conversion rules for the displacement vector, $\underline{D}$, and the current, j, from their respective two-forms differ by a minus sign from that of the magnetic induction vector, $\underline{B}$. Similarly, the conversion for the magnetic field vector, $\underline{H}$, differs by a sign from that for the electric field, $\underline{E}$. Formally, these sign differences are required to establish compatibility between the vectorial and the form representations of the Maxwell equations.

Physically, they reflect the fact that under *orientation-changing transformations*, such as reflections at a plane, j, $\underline{E}$ and $\underline{D}$ transform as conventional (or "polar") vectors, while $\underline{B}$ and $\underline{H}$ do not (Fig. V28). Acknowledging this difference, the latter are sometimes called "axial" vectors. For example, the magnetic force (polar) is related to the velocity (polar) and the magnetic induction (axial) as $\underline{F} \sim \underline{v} \times \underline{B}$. Under an inversion of space, $x^i \to -x^i$, both $\underline{F}$ and $\underline{v}$ change sign, enforcing that $\underline{B}$ must stay the same. In this chapter, we do not discuss the transformation of fields under orientation-changing transformations and use a right-handed basis, $\{\underline{e}_1, \underline{e}_2, \underline{e}_3\}$, for $\mathbb{R}^3$ throughout. In this setting the difference between polar and axial vectors shows up in the above minus sign.

V7.3   Laws of electrodynamics II: Maxwell equations

The Lorentz force describes the influence of fields on charged matter. Conversely, the Maxwell equations describe the creation of electric and magnetic fields by charges and currents. Maxwell's equations are a set of four first-order linear partial differential equations for the forms E, D, B, H. They can be grouped into two sets of two, where the first group, the homogeneous Maxwell equations, describe intrinsic connections between the fields E and B. The second group, the inhomogeneous equations, describe how the fields D and H are created by charges and currents as represented by the forms ρ_s and j_s. At this level the two groups are independent, and they do not depend on the metric. The latter is required for establishing connections between E and D and between B and H through (non-differential) equations supplementing the Maxwell equations. In vacuum, the connection between the one-form E and the two-form D is provided by the Hodge star, and likewise for H and B. These connections assume a trivial form if an orthonormal system is used. The situation becomes more complicated in extended media, where microscopic properties of the host medium affect the coupling relations. This complication, which is **macroscopic electro-dynamics** the subject of **macroscopic electrodynamics**, will not be addressed here, and we stay in vacuum throughout.

We next proceed to motivate Maxwell's equations from experimental observations of electrodynamics.

traditional form of Maxwell equations

INFO For later reference, we state the **Maxwell equations in their traditional form**, as taught in standard courses of electromagnetism. The theory is formulated in terms of the electric field $\underline{E}$, the magnetic induction $\underline{B}$, the electric displacement field $\underline{D}$ and the magnetic field $\underline{H}$. These fields are created by electric charge densities ρ and/or current densities $\underline{j}$. All fields depend on space and time, $\underline{E} = \underline{E}(x, t)$, etc.

The connection between matter and electromagnetic fields is established by the Maxwell equations,[70]

$$\underline{\nabla} \cdot \underline{D} = 4\pi\rho, \qquad\qquad\qquad\qquad \text{(Gauss)}$$

$$\underline{\nabla} \times \underline{H} - \tfrac{1}{c}\partial_t \underline{D} = \tfrac{4\pi}{c}\,\underline{j}, \qquad\qquad \text{(Ampère–Maxwell)}$$

$$\underline{\nabla} \cdot \underline{B} = 0, \qquad\qquad\qquad \text{(absence of magnetic sources)}$$

$$\underline{\nabla} \times \underline{E} + \tfrac{1}{c}\partial_t \underline{B} = 0, \qquad\qquad \text{(Faraday)} \qquad \text{(V236)}$$

where c is the speed of light. The first two equations are inhomogeneous equations, the last two homogeneous. These equations need to be augmented with relations connecting $\underline{D}$ with $\underline{E}$ and $\underline{H}$ with $\underline{B}$. In vacuum, the identification is particularly simple, $\underline{D} = \underline{E}, \underline{H} = \underline{B}$.

Homogeneous Maxwell equations

absence of magnetic charges

Empirically, there are no magnetic charges. In form language, the **absence of magnetic charges** means that the integration of the magnetic induction two-form, B, over the surface, ∂V, of any region in space, V, yields zero,

$$\int_{\partial V} B = 0, \qquad\qquad (\text{V237})$$

such that there is no net outward magnetic flux. The situation is illustrated in Fig. V29(a), where the two-form B is represented through closed field lines. By Stokes's theorem,

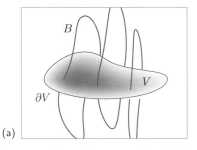 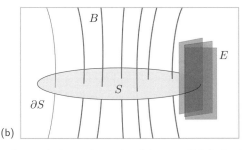

(a)   (b)

Fig. V29 The homogeneous laws of electromagnetism. (a) Absence of magnetic charges: integration of the magnetic induction two-form B over a closed surface equals zero; lines characterizing the form are closed loops, indicating that the field has no sources. (b) Law of induction: integration of the electric one-form E (indicated by shaded planes) over a closed curve ∂S bounding a surface S equals the integral of the negative time derivative, $-\partial_t B$, of the magnetic induction two-form over S.

---

[70] We here use CGS (centimeter, gram, second) units. In the alternative SI (meter, kilogram, second, ampere) system, the equations differ from the present ones by prefactors which are inessential for our discussion.

$\int_{\partial V} B = \int_V d_s B$. The vanishing of the integral for any volume requires the closedness of the magnetic induction form,

$$\boxed{d_s B = 0.}$$ (V238)

This is the first of the two homogeneous Maxwell equations. In vector language, it states that the magnetic induction vector field $\underline{B}$ has no sources, see the third of Eqs. (V236).

INFO It is straightforward to translate the form representation of the Maxwell equation (V238) to traditional language. Application of the Hodge star converts the equation for three-forms into one for scalar functions, $0 = *d_s B$. We now recall the representation (V217) of the divergence, div $= - * d_s * J$, to rewrite the equation as $0 = *d_s B = - * d_s * *B = - * d_s * JJ^{-1} * B = \underline{\nabla} \cdot \underline{B}$. Alternatively, the identities of Table V4 may be applied to translate between the two languages in a coordinate representation. ($\rightarrow$ V7.3.1 (a)).

Faraday law   The second homogeneous Maxwell equation expresses the **Faraday law of induction**: If a surface is threaded by a time-dependent magnetic induction, B, an electric field winding around the surface is induced (see Fig. V29(b)). Experiment shows that the integral of the field around the curve bounding the surface, $\int_{\partial S} E$ (i.e. the mechanical work required to drag a unit charge, $q = 1$, around the curve), equals the negative of $1/c$ times the time derivative of the magnetic flux, $\int_S B$, through the surface,

$$\int_{\partial S} E + \frac{1}{c} \frac{d}{dt} \int_S B = 0.$$

Applying Stokes's theorem to the first integral, and pulling the time derivative into the integral (the surface itself is static), we obtain $\int_S (d_s E + \frac{1}{c} \partial_t B) = 0.$[71] Once more, the validity of this relation for any integration domain implies the vanishing of the integrand,

$$\boxed{d_s E + \frac{1}{c} \partial_t B = 0.}$$ (V239)

This is the second homogeneous Maxwell equation.

INFO We translate Eq. (V239) to the *traditional form of the law of induction* by acting on it with $J^{-1}*$ to convert two-forms to vectors. Recalling that $J^{-1} * d_s J \overset{(V219)}{=}$ curl, we obtain $0 = J^{-1} * d_s JJ^{-1}E + \frac{1}{c} \partial_t J^{-1} *B = \underline{\nabla} \times \underline{E} + \frac{1}{c} \partial_t \underline{B} = 0$, thus recovering the fourth Maxwell equation (V236). ($\rightarrow$ V7.3.1 (b))

Inhomogeneous Maxwell equations

The inhomogeneous equations describe the creation of the fields D, H by charges and currents. **Gauss's law** (see footnote 14 of Chapter V3) states that electric charge is the source of the electric displacement field (see Fig. V30(a)): The integral $\int_{\partial V} D$ over the surface of a volume V equals 4π times the included charge, $Q = \int_V \rho_s$,

Gauss's law

---

[71] Recall that the time derivative in $\partial_t B = \partial_t \frac{1}{2} B_{ij}(x, t) dx^i \wedge dx^j$ acts on the time dependence of the form components as an ordinary partial derivative.

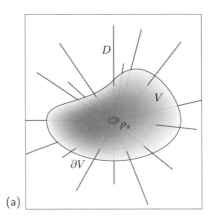
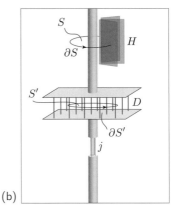

(a)   (b)

Fig. V30   The inhomogeneous laws of electromagnetism. (a) Law of Gauss: a charge density three-form, ρ_s, generates an electric displacement field, D. Integration of D over closed surfaces equals the integral of ρ_s over the enclosed volumes. (b) Law of Ampère–Maxwell: current flow described by a non-vanishing two-form, j, generates a magnetic field one-form, H, and/or an electric displacement field one-form, D. The integral of the sum $H - \frac{1}{c}\partial_t D$ over a closed curve bounding a surface (such as S or S' in the figure) equals the integral of j over that surface.

$$\int_{\partial V} D = 4\pi \int_V \rho_s.$$

Once more we apply Stokes's theorem, $\int_V (d_s D - 4\pi\rho_s) = 0$, to conclude that

$$\boxed{d_s D = 4\pi\rho_s.} \tag{V240}$$

Ampère–
Maxwell
law

Finally, the law of **Ampère–Maxwell** states that electric current creates a magnetic field, H, and/or a time-varying electric displacement field, D. The relation is such that (see Fig. V30(b)) the integral $\int_{\partial S} H$ over the closed boundary curve of a surface S equals the sum of $4\pi/c$ times the current flow through that surface, $4\pi I/c \overset{(V229)}{=} (4\pi/c)\int_S j_s$, and $1/c$ times the time derivative, $d_t \int_S D$, of the integral of D over the same surface:[72]

$$\int_{\partial S} H = \frac{4\pi}{c}\int_S j_s + \frac{1}{c}\int_S \partial_t D.$$

Again we apply Stokes's theorem to conclude that

$$\boxed{d_s H - \tfrac{1}{c}\partial_t D = \tfrac{4\pi}{c} j_s.} \tag{V241}$$

INFO As with the previous equations, application of the Hodge star leads from the form representation to the vectorial formulation (V236) ($\rightarrow$ V7.3.2):

---

[72] To understand the sum, recall that a time-varying current in a wire may be interrupted by the insertion of capacitor plates, as shown in the figure. While no electric current can flow between the plates, there may be a time-varying electric field, the displacement field. The displacement field and the electric current both contribute to the magnetic field winding around the setup.

$$0 = * \left[d_s(- * JJ^{-1}*)D - 4\pi\rho_s \right] = -\underline{\nabla} \cdot \underline{D} + 4\pi\rho,$$

$$0 = J^{-1} * \left[d_s(JJ^{-1})H - \tfrac{1}{c}\partial_t D - \tfrac{4\pi}{c} j_s \right] = -\underline{\nabla} \times \underline{H} + \tfrac{1}{c}\partial_t\underline{D} + \tfrac{4\pi}{c}\underline{j}.$$

V7.4 Invariant formulation

Above, we formulated Maxwell's equations in three-dimensional space, $\mathbb{R}^3$. Time entered as a parameter. In this section, we will treat space and time in a more egalitarian manner, and in this way arrive at a manifestly Lorentz invariant formulation in four-dimensional space-time, with metric signature $(1, 3)$. In this setting, the exterior derivative acts on forms as $d\phi = dt \wedge \partial_t\phi + d_s\phi$, or in an operator language,

$$d = dt \wedge \partial_t \mid d_s = dx^0 \wedge \partial_0 \mid d_s, \tag{V242}$$

where $x^0 = ct$. The Hodge star is now defined to act in $\Lambda(\mathbb{R}^4)$.

Invariant current form

Above, we described charge density and current by a *density three-form*, ρ_s, and a *current two-form*, j_s, respectively. Within the non-invariant approach, these two are combined to the current three-form

$$j = \rho_s - dt \wedge j_s. \tag{V243}$$

Comparison with the discussion on p. 489 shows that j is a form of dimension [charge], defined such that its evaluation on three spatial arguments $\partial_1, \partial_2, \partial_3$ equals the charge contained in the volume spanned by the three vectors. Likewise, the evaluation on $\partial_0, \partial_i, \partial_j$ equals the (negative of the) charge flowing in time $c\partial_0$ through the surface spanned by ∂_i and ∂_j. The sign is introduced such that the law of **current conservation** assumes a simple **continuity** form, $dj = (dt \wedge \partial_t + d_s)(\rho_s - dt \wedge j_s) = dt \wedge (\partial_t\rho_s + d_s j_s) = 0$, where Eq. (V231) was **equation** used. Current conservation is thus equivalent to the closedness of the current three-form,

$$\boxed{dj = 0.} \tag{V244}$$

Expanding the three-form as $j \overset{(V168)}{=} j_{123}\, dx^1 \wedge dx^2 \wedge dx^3 + \sum_{j<k} j_{0jk}\, dx^0 \wedge dx^j \wedge dx^k$, and comparing with $\rho_s \equiv \rho\, dx^1 \wedge dx^2 \wedge dx^3$ and $j_s \equiv \tfrac{1}{2}j_{s,jk}\, dx^j \wedge dx^k$, its components are identified as

$$j_{123} = \rho, \qquad j_{0jk} = -\tfrac{1}{c}j_{s,jk}. \tag{V245}$$

From the three-form, one may pass to a vector, $\partial_j = j^\mu\partial_\mu \equiv cJ^{-1} * j$, which we computed in Section V6.3. Their components are related by (V224), $j^0 = cj_{123}$ and $j^i \overset{(V225)}{=} -\tfrac{c}{2}j_{0jk}\epsilon^{jki}$. Combined with Eqs. (V245) and (V233), this yields

$$(j^0, j^1, j^2, j^3) \equiv (c\rho, j^1, j^2, j^3), \tag{V246}$$

Table V5 Summary of conversion rules between current forms and
current vector.

vector	non-invariant form	invariant form
$j^0 = c\rho$	ρ_{123}	j_{123}
j^i	$\frac{1}{2}\epsilon^{ijk}j_{s,jk}$	$-\frac{1}{2}c\,\epsilon^{ijk}j_{0jk}$

where j^i on the left should be interpreted as components of the four-vector ∂_j, obtained from the one-form $c*j$ via the canonical isomorphism J^{-1}, and on the right as components of the current density vector $\underline{j}$ of traditional teaching.

The continuity relation can be expressed as

$$0 = *\mathrm{d}j = *\tfrac{1}{3!}\partial_\mu j_{\lambda\kappa\delta}\,\mathrm{d}x^\mu \wedge \mathrm{d}x^\lambda \wedge \mathrm{d}x^\kappa \wedge \mathrm{d}x^\delta = -\tfrac{1}{3!}\partial_\mu j_{\lambda\kappa\delta}\epsilon^{\mu\lambda\kappa\delta} \stackrel{(V224)}{=} -\partial_\mu j^\mu,$$

which defines the representation of *current conservation in vector language*,

$$\partial_\mu j^\mu = 0. \tag{V247}$$

For the orientation of the reader, the connections between different representations of the current are summarized in Table V5.

Field-strength tensors

To proceed towards an invariant formulation of the Maxwell equations we now introduce the two-forms

$$F \equiv -\mathrm{d}x^0 \wedge E + B,$$
$$G \equiv \quad \mathrm{d}x^0 \wedge H + D. \tag{V248}$$

field-strength tensor

The form F is called the **field-strength tensor**, and G is called the **dual field-strength tensor**. F and G contain the forms appearing in the homogeneous and inhomogeneous Maxwell equations, respectively. At this point, the two objects are decoupled.

Acting on the field-strength tensor with the exterior derivative d, Eq. (V242), we obtain $\mathrm{d}F = (\mathrm{d}x^0 \wedge \partial_{x^0} + \mathrm{d}_s)(-\mathrm{d}x^0 \wedge E + B) = \mathrm{d}x^0 \wedge (\mathrm{d}_s E + \partial_0 B) + \mathrm{d}_s B$. The two terms vanish separately, as follows from the homogeneous Maxwell equations, Eqs. (V238) and (V239), and so we obtain the *closedness of the field-strength form*, $\mathrm{d}F = 0$.

In a similar manner, $\mathrm{d}G = \mathrm{d}x^0 \wedge (-\mathrm{d}_s H + \partial_0 D) + \mathrm{d}_s D = -\mathrm{d}x^0 \wedge \frac{4\pi}{c}j_s + 4\pi\rho_s = 4\pi j$, from the inhomogeneous Maxwell equations (V240) and (V241). Summarizing, we have

invariant form of Maxwell equations

arrived at the manifestly **space-time invariant form of the Maxwell equations**,

$$\boxed{\begin{aligned}\mathrm{d}F &= 0,\\ \mathrm{d}G &= 4\pi j.\end{aligned}} \tag{V249}$$

It is remarkable that the Maxwell equations, which describe such a diverse range of physical phenomena, can be represented this compactly.

INFO In a coordinate language, $F \equiv \frac{1}{2} F_{\mu\nu} dx^{\mu} \wedge dx^{\nu}$ and $G \equiv \frac{1}{2} G_{\mu\nu} dx^{\mu} \wedge dx^{\nu}$ are described by the covariant matrices

$$\{F_{\mu\nu}\} = \begin{pmatrix} 0 & E^1 & E^2 & E^3 \\ -E^1 & 0 & -B^3 & B^2 \\ -E^2 & B^3 & 0 & -B^1 \\ -E^3 & -B^2 & B^1 & 0 \end{pmatrix}, \quad \{G_{\mu\nu}\} = \begin{pmatrix} 0 & H^1 & H^2 & H^3 \\ -H^1 & 0 & D^3 & -D^2 \\ -H^2 & -D^3 & 0 & D^1 \\ -H^3 & D^2 & -D^1 & 0 \end{pmatrix}, \quad \text{(V250)}$$

where Table V3 was used to represent form components by components of the vector fields ($\rightarrow$ V7.4.1-2). In the coordinate language, the Maxwell equations (V249) take the form ($\rightarrow$ V7.4.3-4 (a,b))

$$\partial_\alpha F_{\mu\nu} + \partial_\mu F_{\nu\alpha} + \partial_\nu F_{\alpha\mu} = 0,$$
$$\partial_\alpha G_{\mu\nu} + \partial_\mu G_{\nu\alpha} + \partial_\nu G_{\alpha\mu} = 4\pi j_{\alpha\mu\nu}. \quad \text{(V251)}$$

The representations (V250) are frequently found in the literature, including in the traditional formulations of electrodynamics. In that context, $F_{\mu\nu}$ and $G_{\mu\nu}$ are defined as formal containers of field amplitudes previously introduced as vectors. This amounts to the implicit concession that the relativistic invariance of electrodynamics is at odds with the vectorial formulation.

Electrodynamics and Minkowski metric

So far, the fields E, B and D, H entering the homogeneous and the inhomogeneous Maxwell equations, respectively, remained decoupled. The connection between them follows from the empirical observation that in vacuum and in an orthonormal coordinate system, the components of the electric field equal those of the displacement field, $\underline{E} = \underline{D}$, and the components of the magnetic field those of the magnetic induction, $\underline{B} = \underline{H}$. Translated into form language using Table V4, this implies

$$D_{jk} = -E_i \epsilon_{ijk}, \qquad B_{jk} = -H_i \epsilon_{ijk}. \quad \text{(V252)}$$

The structure of these equations suggests that a Hodge star is at work, and indeed

$$D = *E, \qquad B = *H,$$

empolying the three-dimensional Hodge star. As a consistency check, note that conversion to vector field language correctly recovers the above-mentioned empirical relations, $\underline{D} = -J^{-1}(*D) = J^{-1}(E) = \underline{E}$ and $\underline{B} = J^{-1}(*B) = -J^{-1}(H) = \underline{H}$.

These relations also imply a connection between F and G, via the four-dimensional Hodge star. The latter acts on coordinate two-forms as

$$*(dx^\mu \wedge dx^\nu) = \frac{1}{2} \eta^{\mu\alpha} \eta^{\nu\beta} \epsilon_{\alpha\beta\gamma\delta} dx^\gamma \wedge dx^\delta. \quad \text{(V253)}$$

Using this relation, we find that $*F = \frac{1}{2} E_i \epsilon_{ijk} dx^j \wedge dx^k + \frac{1}{2} B_{ij} \epsilon_{ijk} dx^0 \wedge dx^k \overset{\text{(V252)}}{=} -\frac{1}{2} D_{jk} dx^j \wedge dx^k - H_i dx^0 \wedge dx^i = -G$. In this way the important identification

$$\boxed{*F = -G} \quad \text{(V254)}$$

is established. It shows that F and G are related to each other by the self-involutory operation $*$ and justifies the denotation *dual* field-strength tensor for G. Equations (V249) and (V254) provide a complete description of the laws of classical electromagnetism. Although these equations are coordinate independent, the connections between the components of F and G assume a simple form only in systems with orthonormal metric, $g = \eta$, i.e. systems related to each other by Lorentz transformations (L211).[73]

INFO In coordinates, Eq. (V254) reads as $G^{\alpha\beta} = \frac{1}{2}\epsilon^{\alpha\beta\mu\nu}F_{\mu\nu}$. With this relation, the component formulation of the Maxwell equations, Eqs. (V251), can be expressed more compactly as

$$\partial_\alpha G^{\alpha\beta} = 0,$$
$$\partial_\alpha F^{\alpha\beta} = 4\pi j^\beta.$$

These equations, often found in traditional formulations of electrodynamics, are the coordinate representations of the invariant equations $J^{-1} * \mathrm{d}F = 0$ and $J^{-1} * (\mathrm{d}G - 4\pi j) = 0$. ($\rightarrow$ V7.4.3-4 (c,d)) Note that this representation of the theory requires a metric (via Hodge) while the Maxwell equations in pristine form, Eqs. (V249), are non-metric.

V7.5 Summary and outlook

This concludes our survey of electrodynamics in form language. Starting with a summary of experimental observations, we discussed the laws of electrodynamics first in the original non-invariant formulation, then in a manifestly Lorentz invariant form. Readers may wonder why we did not discuss the *traditional formulation* of the invariant approach to electrodynamics. The answer is that it does not really exist. The description of the theory by four three-component vector fields $\underline{E}, \underline{B}, \underline{D}, \underline{H}$ is tailored to the separate treatment of space and time and has no good invariant extension. Even in traditional teaching, it is standard to formulate the invariant approach via the field-strength tensors $F_{\mu\nu}$ and $G_{\mu\nu}$. They are introduced as 4×4 matrices containing the components of $\underline{E}$ and $\underline{B}$ or $\underline{D}$ and $\underline{H}$ as entries. In this way, differential forms are effectively introduced but not discussed as such. However, this does not change the fact that the Lorentz invariant formulation requires the introduction of forms and the abandoning of the all-is-vector paradigm.

In the physics literature, electrodynamics and relativity are almost always discussed in coordinates ($F_{\mu\nu}$) rather than in invariant language (F). There is a tendency to treat $F_{\mu\nu}$ as a formally defined object, and not to bother much about its underlying identity. Whether one prefers this pragmatic approach or one emphasizing the conceptual meaning of objects is a matter of taste and personal inclination. However, the importance of differential forms in physics is on the rise, and we speculate that future generations of physicists will require this beautiful concept as part of their standard portfolio.

Poincaré group

[73] More generally, shifts of the coordinate origin, $x \rightarrow x + a$, $a \in \mathbb{R}^4$ also preserve the form of the equations. The generalized set of coordinate transformations, $x' = \Lambda x + a$, $\Lambda \in O(1,3)$ defines the **Poincaré group**.

The problems come in odd–even-numbered pairs, labeled ε for "example" and ᵖ for "practice". Each example problem prepares the reader for tackling the subsequent practice problem. The solutions to the odd-numbered example problems are given in Chapter SV. A password-protected manual containing solutions of all even-numbered practice problems will be made available to instructors.

P.V1 Curves (p. 406)

P.V1.2 Curve velocity (p. 407)

εV1.2.1 Velocity and acceleration (p. 409)

Consider the curve $\gamma = \{\mathbf{r}(t) \,|\, t \in (0, 2\pi/\omega)\}$, $\mathbf{r}(t) = (aC(t), S(t))^T \in \mathbb{R}^2$, with $C(t) = \cos[\pi(1 - \cos \omega t)]$, $S(t) = \sin[\pi((1 - \cos \omega t)]$, and $0 < a, \omega \in \mathbb{R}$.

(a) Calculate the curve's velocity vector, $\dot{\mathbf{r}}(t)$, and its acceleration vector, $\ddot{\mathbf{r}}(t)$. Can $\mathbf{r}(t)$ be expressed in terms of $\dot{\mathbf{r}}(t)$ and $\ddot{\mathbf{r}}(t)$?

(b) Can you represent the curve without the parameter t using an equation? Do you recognize the curve? Sketch the curve for the case $a = 2$.

(c) Calculate $\mathbf{r}(t) \cdot \dot{\mathbf{r}}(t)$. For which values of a is $\mathbf{r}(t) \cdot \dot{\mathbf{r}}(t) = 0$ true for all t?

ᵖV1.2.2 Velocity and acceleration (p. 409)

Consider the curve $\gamma = \{\mathbf{r}(t) \,|\, t \in (-\infty, \infty)\}$, $\mathbf{r}(t) = (e^{-t^2}, ae^{t^2})^T \in \mathbb{R}^2$, with $0 < a \in \mathbb{R}$ ($0 < a < 1$ for (c)).

(a) Calculate the curve's velocity vector, $\dot{\mathbf{r}}(t)$, and its acceleration vector, $\ddot{\mathbf{r}}(t)$. Can $\mathbf{r}(t)$ be expressed in terms of $\dot{\mathbf{r}}(t)$ and $\ddot{\mathbf{r}}(t)$?

(b) Can you represent the curve without the parameter t using an equation? Do you recognize the curve? Sketch the curve for the case $a = 2$.

(c) Calculate $\mathbf{r}(t) \cdot \dot{\mathbf{r}}(t)$. Find the time, $t(a)$, for which $\mathbf{r}(t) \cdot \dot{\mathbf{r}}(t) = 0$ holds. [Check your result: $t(e^{-2}) = \pm 1$.]

P.V1.3 Curve length (p. 409)

εV1.3.1 Curve length (p. 411)

Compute the length of the curve $\gamma : (0, 1) \mapsto \mathbb{R}^2, t \mapsto \mathbf{r}(t) = (\frac{1}{2}t^2, \frac{1}{3}at^3)^T$, as a function of $a \in \mathbb{R}$. [Check your result: If $a = 2$, then $L[\gamma] = \frac{2}{27}(19^{3/2} - 1)$.]

pV1.3.2 Curve length (p. 411)

Compute the length of the curve $\gamma : (0, \tau) \mapsto \mathbb{R}^3$, $t \mapsto \mathbf{r}(t) = (t^4, t^6, t^6)^\mathsf{T}$, as a function of $\tau > 0$. [Check your result: If $\tau = 1$, then $L[\gamma] = \frac{1}{54}(22^{3/2} - 8)$.]

εV1.3.3 Natural parametrization of a curve (p. 411)

Consider the curve $\mathbf{r}(t) = (t - \sin t, 1 - \cos t)^\mathsf{T} \in \mathbb{R}^2$ for $t \in (0, 2\pi)$.

(a) Sketch the curve qualitatively.

(b) Determine its arc length, $s(t)$, in the time interval $(0, t)$. [Check your answer: $s(2\pi) = 8$.]

(c) Find the natural parametrization, $\mathbf{r}_L(s)$. [Check your answer: $\mathbf{r}_L(4) = (\pi, 2)^\mathsf{T}$.]

pV1.3.4 Natural parametrization of a curve (p. 411)

Consider the curve $\gamma = \{\mathbf{r}(t) \mid t \in (0, \tau)\}$, $\mathbf{r}(t) = e^{ct}(\cos \omega t, \sin \omega t)^\mathsf{T} \in \mathbb{R}^2$, with $c \in \mathbb{R}$.

(a) Sketch the curve for the case of $\tau = 8\pi/\omega$ and $c = 1/\tau$. [This information only applies to part (a), not for parts (b)–(e).]

(b) Calculate the magnitude of the curve velocity, $\|\dot{\mathbf{r}}(t)\|$.

(c) Calculate the arc length, $s(t)$, covered in the time interval $(0, t)$.

(d) Determine the natural parametrization, $\mathbf{r}_L(s)$.

(e) Check explicitly that $\left\| \frac{d\mathbf{r}_L}{ds} \right\| = 1$.

$\left[\text{Check your answer: For } c = \omega = \tau = 1: \text{(b) } \sqrt{2}e^t, \text{(c) } \sqrt{2}(e^t - 1), \text{(d) } \mathbf{r}_L(s) = [s/\sqrt{2} + 1]\left(\cos[\ln(s/\sqrt{2} + 1)], \sin[\ln(s/\sqrt{2} + 1)]\right)^\mathsf{T}.\right]$

P.V1.4 Line integral (p. 412)

εV1.4.1 Line integral: mountain hike (p. 413)

Two hikers want to hike from the point $\mathbf{r}_0 = (0, 0)^\mathsf{T}$ in the valley to a mountain hut at the point $\mathbf{r}_1 = (3, 3a)^\mathsf{T}$. Hiker 1 chooses the straight path from valley to hut, γ_1. Hiker 2 chooses a parabolic path, γ_2, via the mountain top at the apex of the parabola, at $\mathbf{r}_2 = (2, 4a)^\mathsf{T}$ (see figure). They are acted on by the force of gravity $\mathbf{F}_g = -10\,\mathbf{e}_y$, and a height-dependent wind force, $\mathbf{F}_w = -y^2\,\mathbf{e}_x$.

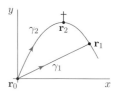

Find the work, $W[\gamma_i] = -\int_{\gamma_i} d\mathbf{r} \cdot \mathbf{F}$, performed by the hikers along γ_1 and γ_2, as functions of the parameter a. [Check your results: For $a = 1$ one finds $W[\gamma_1] = 39$, $W[\gamma_2] = 303/5$.]

pV1.4.2 Line integrals in Cartesian coordinates (p. 413)

Let $\mathbf{F}(\mathbf{r}) = (x^2, z, y)^\mathsf{T}$ be a three-dimensional vector field in Cartesian coordinates, with $\mathbf{r} = (x, y, z)^\mathsf{T}$. Calculate the line integral $\int_\gamma d\mathbf{r} \cdot \mathbf{F}$ along the following paths from $\mathbf{r}_0 \equiv (0, 0, 0)^\mathsf{T}$ to $\mathbf{r}_1 \equiv (0, -2, 1)^\mathsf{T}$.

(a) $\gamma_a = \gamma_1 \cup \gamma_2$ is the composite path consisting of γ_1, the straight line from $\mathbf{r}_0$ to $\mathbf{r}_2 \equiv (1, 1, 1)^\mathsf{T}$, and γ_2, the straight line from $\mathbf{r}_2$ to $\mathbf{r}_1$.

(b) γ_b is parametrized by $\mathbf{r}(t) = (\sin(\pi t), -2t^{1/2}, t^2)^T$, with $0 < t < 1$.

(c) γ_c is a parabola in the yz-plane with the form $z(y) = y^2 + \frac{3}{2}y$.

[Check your results: The sum of the answers from (a), (b) and (c) is -6.]

P.V2   Curvilinear coordinates (p. 414)

P.V2.3   Cylindrical and spherical coordinates (p. 421)

ₑV2.3.1 Coordinate transformations (p. 424)

Consider three points whose Cartesian coordinates, (x, y, z), are P_1: $(3, -2, 4)$, P_2: $(1, 1, 1)$ and P_3: $(-3, 0, -2)$. What is the representation of these three points in cylindrical coordinates, (ρ, ϕ, z), and in spherical coordinates, (r, θ, ϕ)? (Give the angles in radians.)

ₚV2.3.2 Coordinate transformations (p. 424)

The point P_1 has spherical coordinates $(r, \theta, \phi) = (2, \pi/6, 2\pi/3)$. What are its Cartesian and cylindrical coordinates, (x, y, z) and (ρ, ϕ, z), respectively? The point P_2 has cylindrical coordinates $(\rho, \phi, z) = (4, \pi/4, 2)$. What are its Cartesian and spherical coordinates? (Give the angles in radians.)

ₑV2.3.3 Cylindrical coordinates: velocity, kinetic energy, angular momentum (p. 422)

The relation between Cartesian and cylindrical coordinates is given by: $x = \rho \cos\phi$, $y = \rho \sin\phi$, $z = z$, with $\rho \in (0, \infty)$, $\phi \in (0, 2\pi)$, $z \in (-\infty, \infty)$.
Basis vectors: Construct the local basis vectors for cylindrical coordinates, $\{\mathbf{e}_{y_i}\} = \{\mathbf{e}_\rho, \mathbf{e}_\phi, \mathbf{e}_z\}$, and show explicitly that they have the following properties:
(a) $\mathbf{e}_{y_i} \cdot \mathbf{e}_{y_j} = \delta_{ij}$ and (b) $\mathbf{e}_{y_i} \times \mathbf{e}_{y_j} = \varepsilon_{ijk} \mathbf{e}_{y_k}$.
Physical quantities: Show that in cylindrical coordinates (c) the velocity vector, $\mathbf{v} = \frac{d}{dt}\mathbf{r}$, (d) the kinetic energy, $T = \frac{1}{2}mv^2$, and (e) the angular momentum, $\mathbf{L} = m(\mathbf{r} \times \mathbf{v})$, have the following forms:

$$\mathbf{v} = \dot\rho\,\mathbf{e}_\rho + \rho\dot\phi\,\mathbf{e}_\phi + \dot z\,\mathbf{e}_z, \qquad T = \frac{1}{2}m[\dot\rho^2 + \rho^2\dot\phi^2 + \dot z^2],$$
$$\mathbf{L} = m[-z\rho\dot\phi\,\mathbf{e}_\rho + (z\dot\rho - \rho\dot z)\,\mathbf{e}_\phi + \rho^2\dot\phi\,\mathbf{e}_z].$$

ₚV2.3.4 Spherical coordinates: velocity, kinetic energy, angular momentum (p. 424)

The relationship between Cartesian and spherical coordinates is given by: $x = r\sin\theta\cos\phi$, $y = r\sin\theta\sin\phi$, $z = r\cos\theta$, with $r \in (0, \infty)$, $\phi \in (0, 2\pi)$, $\theta \in (0, \pi)$.
Basis vectors: Construct the local basis vectors for spherical coordinates, $\{\mathbf{e}_{y_i}\} = \{\mathbf{e}_r, \mathbf{e}_\theta, \mathbf{e}_\phi\}$, and show explicitly that
(a) $\mathbf{e}_{y_i} \cdot \mathbf{e}_{y_j} = \delta_{ij}$ and (b) $\mathbf{e}_{y_i} \times \mathbf{e}_{y_j} = \varepsilon_{ijk} \mathbf{e}_{y_k}$.
Physical quantities: Show that in spherical coordinates (c) the velocity vector $\mathbf{v} = \frac{d}{dt}\mathbf{r}$, (d) the kinetic energy, $T = \frac{1}{2}mv^2$, and (e) the angular momentum, $\mathbf{L} = m(\mathbf{r} \times \mathbf{v})$, have the following forms:

$$\mathbf{v} = \dot r\,\mathbf{e}_r + r\dot\theta\,\mathbf{e}_\theta + r\dot\phi\sin\theta\,\mathbf{e}_\phi, \quad T = \frac{1}{2}m[\dot r^2 + r^2\dot\theta^2 + r^2\dot\phi^2\sin^2\theta], \quad \mathbf{L} = mr^2[\dot\theta\,\mathbf{e}_\phi - \dot\phi\sin\theta\,\mathbf{e}_\theta].$$

ₑV2.3.₅ Line integral in polar coordinates: spiral (p. 420)

The curve $\gamma_S = \{\mathbf{r}(\rho, \phi) \in \mathbb{R}^2 |\ \rho = R + \frac{1}{2\pi}\phi\Delta,\ \phi \in (0, 2\pi)\}$, with $0 < R, \Delta \in \mathbb{R}$, describes a spiral path in two dimensions, parametrized using polar coordinates.

(a)   Sketch the spiral path γ_S and calculate the line integral $W_1[\gamma_S] = \int_{\gamma_S} d\mathbf{r}\cdot\mathbf{F}_1$ of the field $\mathbf{F}_1 = \mathbf{e}_\phi$ along γ_S. [Check your result: If $R = \Delta = 1$, then $W_1[\gamma] = 3\pi$.]

(b)   Calculate the line integral $W_2[\gamma] = \int_\gamma d\mathbf{r} \cdot \mathbf{F}_2$ of the field $\mathbf{F}_2 = \mathbf{e}_x$ along the straight path γ_G from the point $(R, 0)^\mathsf{T}$ to the point $(R + \Delta, 0)^\mathsf{T}$, and also along the spiral path γ_S. Are the results related? Explain!

ₚV2.3.₆ Line integral in Cartesian and spherical coordinates (p. 420)

Consider the vector field $\mathbf{F} = (0, 0, fz)^\mathsf{T}$, with $f \in \mathbb{R}$. Compute the line integral $W[\gamma] = \int_\gamma d\mathbf{r} \cdot \mathbf{F}$ from $\mathbf{a} = (1, 0, 0)^\mathsf{T}$ to $\mathbf{b} = (0, 0, 1)^\mathsf{T}$ explicitly along the following two paths.

(a) γ_1: a straight line. [Check your result: If $f = 2$, then $W[\gamma_1] = 1$.]

(b) γ_2: a segment of a circle with radius $R = 1$ centered at the origin. Use spherical coordinates. [Check your result: If $f = 3$, then $W[\gamma_2] = \frac{3}{2}$.]

ₑV2.3.₇ Line integral in spherical coordinates: satellite in orbit (p. 420)

A satellite travels along an unusual trajectory γ that circles the north–south axis of the Earth as it travels from a point high above the north pole to a point high above the south pole. In spherical coordinates, the trajectory is given by $r(t) = r_0$, $\theta(t) = \omega_1 t$, $\phi(t) = \omega_2 t$, with $t \in (0, \pi/\omega_1)$. Due to the rotation of the Earth, there is a wind in the upper atmosphere, exerting a force $\mathbf{F} = -F_0 \sin\theta\, \mathbf{e}_\phi$ on the satellite.

(a)   Make a qualitative sketch of the orbit, for $\omega_2 = 20\omega_1$. How many times does the path circle around the north–south axis?

(b)   What is the velocity vector $\dot{\mathbf{r}}$ written in spherical coordinates?

(c)   Give the length $L[\gamma]$ of the orbit in terms of an integral. (You are not required to solve it.)

(d)   Use the line integral $W[\gamma] = \int_\gamma d\mathbf{r} \cdot \mathbf{F}$, to compute the work performed against the wind by the satellite along its orbit. [Check your result: If $F_0 = r_0 = \omega_1 = \omega_2 = 1$, then $W[\gamma] = -\frac{\pi}{2}$.]

ₚV2.3.₈ Line integrals in cylindrical coordinates: bathtub drain (p. 420)

A soap bubble travels along a spiral-shaped path γ towards the drain of a bathtub. In cylindrical coordinates the path is given by $\rho(t) = \rho_0 e^{-t/\tau}$, $\phi(t) = \omega t$, $z(t) = z_0 e^{-t/\tau}$, with $\rho_0 > \rho_d$ and $t \in [0, t_d]$, where ρ_d is the drain radius and $t_d = \tau \ln(\rho_0/\rho_d)$ the time at which the bubble reaches the drain.

(a)   Make a qualitative sketch of the path (e.g. for $\omega = 6\pi/\tau$ in $\rho_0 = 10\rho_A$).

(b)   What is the velocity vector $\mathbf{v} = \dot{\mathbf{r}}$ in cylindrical coordinates? What is the magnitude of the final velocity, i.e $v_d = \|\mathbf{v}(t_d)\|$?

(c)   Show that the length of the path is given by $L[\gamma] = \tau v_d\ (\rho_0/\rho_d - 1)$.

(d) Using the line integral $W[\gamma] = \int_\gamma d\mathbf{r} \cdot \mathbf{F}$, find the work done by gravity $\mathbf{F} = -mg\mathbf{e}_z$ along the path of the soap bubble. Give a physical interpretation for this result!

[Check your results: If $\tau = 2/\omega$, $z_0 = 2\rho_0$ and $\rho_d = \rho_0/3$, then: (b) $v_d = \rho_0/\tau$, (c) $L = 2\rho_0$, (d) $W[\gamma] = mg\rho_04/3$.]

P.V2.5 Local coordinate bases and linear algebra (p. 426)

εV2.5.1 Jacobian determinant for cylindrical coordinates (p. 428)

(a) Compute the Jacobian matrix, $\frac{\partial(x,y,z)}{\partial(\rho,\phi,z)}$, for the transformation expressing Cartesian through cylindrical coordinates.

(b) Compute the Jacobi matrix, $J^{-1} = \frac{\partial(\rho,\phi,z)}{\partial(x,y,z)}$, for the inverse transformation expressing cylindrical through Cartesian coordinates. [Check your result: Verify that $JJ^{-1} = \mathbb{1}$.]

(c) Compute the Jacobi determinants $\det(J)$ and $\det(J^{-1})$. [Check your results: Does their product equal 1?]

pV2.5.2 Jacobian determinant for spherical coordinates (p. 428)

(a) Compute the Jacobian matrix, $J = \frac{\partial(x,y,z)}{\partial(r,\theta,\phi)}$, for the transformation expressing Cartesian through spherical coordinates.

(b) Compute the Jacobi matrix, $J^{-1} = \frac{\partial(r,\theta,\phi)}{\partial(x,y,z)}$, for the inverse transformation expressing spherical through Cartesian coordinates. [Check your result: Verify that $JJ^{-1} = \mathbb{1}$.]

(c) Compute the Jacobi determinants $\det(J)$ and $\det(J^{-1})$. [Check your results: Does their product equal 1?]

P.V3   Fields (p. 430)

P.V3.1   Definition of fields (p. 430)

εV3.1.1 Sketching a vector field (p. 431)

Sketch the following vector fields in two dimensions, with $\mathbf{r} = (x, y)^\mathsf{T}$.

(a) $\mathbf{u} : \mathbb{R}^2 \to \mathbb{R}^2$, $\mathbf{r} \mapsto \mathbf{u}(\mathbf{r}) = (\cos y, 0)^\mathsf{T}$.

(b) $\mathbf{w} : \mathbb{R}^2 \to \mathbb{R}^2$, $\mathbf{r} \mapsto \mathbf{w}(\mathbf{r}) = \dfrac{1}{\sqrt{x^2 + y^2}}(x, -y)^\mathsf{T}$.

For several points $\mathbf{r}$ in the domain of the vector field map (e.g. $\mathbf{u}$), the sketch should depict the corresponding vectors, $\mathbf{u}(\mathbf{r})$, from the codomain of the map. For a chosen point $\mathbf{r}$ one draws an arrow with midpoint at $\mathbf{r}$, whose direction and length represents the vector $\mathbf{u}(\mathbf{r})$. The unit of length may be chosen differently for vectors, $\mathbf{r}$, from the domain and vectors, $\mathbf{u}(\mathbf{r})$, from the codomain, in order to avoid arrows from overlapping and to obtain an uncluttered figure (e.g. by drawing unit vectors $\hat{\mathbf{u}}(\mathbf{r})$ shorter than unit vectors $\hat{\mathbf{r}}$). Indeed, for the visual depiction of codomain vectors usually only their directions and *relative* lengths are of interest, not their absolute lengths.

P.V3.1.2 Sketching a vector field (p. 431)

Sketch the following vector fields in two dimensions:

(a) $\mathbf{u}(x, y) = (\cos x, 0)^{\mathrm{T}}$,   (b) $\mathbf{w}(x, y) = (2y, -x)^{\mathrm{T}}$.

P.V3.2 Scalar fields (p. 432)

P.V3.2.1 Gradient of a mountainside (p. 437)

A hiker encounters a mountainside (as shown in the figure) whose height is given by the function $h(\mathbf{r}) = \frac{x}{r} + 1$, with $\mathbf{r} = (x, y)^{\mathrm{T}}$ and $r = \sqrt{x^2 + y^2}$. Describe the topography of the slope by answering the following questions. Make use of the properties of the gradient vector $\nabla h_{\mathbf{r}}$.

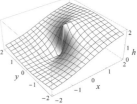

(a) Calculate the gradient, $\nabla h_{\mathbf{r}}$, and the total differential, $dh_{\mathbf{r}}(\mathbf{n})$, for the vector $\mathbf{n} = (n_x, n_y)^{\mathrm{T}}$.

(b) The hiker is at the point $\mathbf{r} = (x, y)^{\mathrm{T}}$. In which direction does the mountainside increase most steeply?

(c) In which direction do the contour lines run at this point?

(d) Sketch a contour plot of the mountainside. Also draw the gradient vectors $\nabla h_{\mathbf{r}}$ at the points $\mathbf{r}_1 = (-1, 1)^{\mathrm{T}}, \mathbf{r}_2 = (0, \sqrt{2})^{\mathrm{T}}$ and $\mathbf{r}_3 = (1, 1)^{\mathrm{T}}$.

(e) Is there a contour line in the positive quadrant $(x, y \geq 0)$ such that $x = y$? If so, at what height does it occur?

(f) Find an equation describing the contour line at height $h(\mathbf{r}) = H$ in the positive quadrant $(x, y \geq 0)$.

(g) Where is the mountainside least steep? What is its height at that position?

(h) Where is the mountainside at its steepest? Describe, in detail, how its topography close to that point depends on x and y.

P.V3.2.2 Gradient of a valley (p. 437)

A hiker encounters a valley as shown in the figure. The height of the valley is described by the equation $h(\mathbf{r}) = e^{xy}$, with $\mathbf{r} = (x, y)^{\mathrm{T}}$. Describe the topography of the valley by answering the following questions. Make use of the properties of the gradient vector $\nabla h_{\mathbf{r}}$.

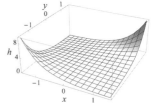

(a) Calculate the gradient $\nabla h_{\mathbf{r}}$ and the total differential $dh_{\mathbf{r}}(\mathbf{n})$ for the vector $\mathbf{n} = (n_x, n_y)^{\mathrm{T}}$.

(b) The hiker stands at the point $\mathbf{r} = (x, y)^{\mathrm{T}}$. In which direction does the slope of the valley increase most steeply?

(c) In which direction do the contour lines run at this point?

(d) Sketch a figure containing the contour plot of the side of the valley. Also draw the gradient vectors $\nabla h_{\mathbf{r}}$ at the points $\mathbf{r}_1 = \frac{1}{\sqrt{2}}(-1, 1)^{\mathrm{T}}, \mathbf{r}_2 = (0, 1)^{\mathrm{T}}$ and $\mathbf{r}_3 = \frac{1}{\sqrt{2}}(1, 1)^{\mathrm{T}}$.

(e) Obtain an equation for the contour line at a height $h(\mathbf{r}) = H(> 0)$.

(f)   At what point is the valley least steep? What is its height at this point?

(g)   At a distance of $r = \|\mathbf{r}\|$ from the origin, where is the valley at its steepest?

ₑV3.2.3   Gradient of $\ln(1/r)$ (p. 437)

Consider the scalar field $\varphi(\mathbf{r}) = \ln\left(r^{-1}\right)$, where $r = \sqrt{x^2 + y^2 + z^2}$. At which spatial points does $\|\nabla\varphi\| = 1$ hold?

ₚV3.2.4   Gradient of $\varphi(r)$ (p. 437)

(a)   For $\mathbf{r} \in \mathbb{R}^3$ and $r = \sqrt{x^2 + y^2 + z^2} = \|\mathbf{r}\|$, compute ∇r and ∇r^2.

(b)   Let $\varphi(r)$ be a general, twice differentiable function of r. Calculate $\nabla\varphi(r)$ in terms of $\varphi'(r)$, the first derivative of φ with respect to r.

P.V3.3   Extrema of functions with constraints (p. 439)

ₑV3.3.1   Minimal distance to hyperbola (p. 441)

Consider the hyperbola $x^2 + 8xy + 7y^2 = 5c^2$ (with constant c). Use a Lagrange multiplier to find the point on this hyperbola minimizing the distance, d, to the origin. Also find the minimal distance as function of c. [Check your result: If $c = 3\sqrt{5}$, then $d_{\min} = 5$.]
Hint: Minimize d^2 rather than d; the calculation is simpler, but the result the same.

ₚV3.3.2   Minimal area of open box with specified volume (p. 441)

A rectangular box with five faces, open at the top, and with specified volume, V, is to be constructed using a minimal amount of paper. Use a Lagrange multiplier to find the area, A, of paper needed for the minimal construction as a function of V. [Check your result: If $V = \frac{1}{2}\mathrm{m}^3$, then $A = 3\mathrm{m}^2$.]

ₑV3.3.3   Intersecting planes: minimal distance to origin (p. 441)

Consider the line of intersection of the two planes defined by the equations $x + y + z = 1$ and $x - y + 2z = 2$, respectively. Use Lagrange multipliers to find the point on this line lying closest to the origin. [Check your result: Its distance to the origin is $\sqrt{5/7}$.]

ₚV3.3.4   Maximal volume of box enclosed in ellipsoid (p. 441)

Consider the ellipsoid defined by $\frac{x^2}{a^2} + \frac{y^2}{b^2} + \frac{z^2}{c^2} = 1$. Also consider a rectangular box whose corners lie on the surface of the ellipsoid and whose edges are parallel to the ellipsoid's symmetry axes. Let $P = (x_p, y_p, z_p)^{\mathrm{T}}$ denote that corner of the box that lies in the positive quadrant ($x_p > 0$, $y_p > 0$, $z_p > 0$). How should this corner be chosen to maximize the volume of the box? What is the value of the maximal volume?

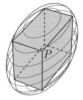

Hint: Maximize the volume $V(x, y, z) = 8xyz$ of a box having a corner at $(x, y, z)^{\mathrm{T}}$, under the constraint that this point lies on the ellipsoid.
[Check your result: If $a = \frac{1}{2}$, $b = 3$, $c = \sqrt{3}$, then $V_{\max} = 4$.]

 εV3.3.5 Entropy maximization subject to constraints (p. 441)

This problem and the next illustrate the use of Lagrange multipliers for a textbook topic from quantum statistical physics. For an in-depth discussion of the concepts mentioned below, refer to lecture courses in quantum physics and statistical physics.

Suppose a quantum system can be in any one of M possible states, $j = 1, \ldots, M$, with a probability p_j of being in the state j. The sum of these probabilities, $P = \sum_j p_j$, is fixed at $P = 1$. (Here, and in the following, $\sum_j$ stands for $\sum_{j=1}^{M}$.) When the system is in the quantum state j, the system has energy E_j and particle number N_j. In quantum statistical physics, the **entropy**, S, and **average energy**, E, of the system are defined as:

$$S = -\sum_j p_j \ln p_j , \qquad E = \sum_j E_j p_j . \tag{1}$$

Show that maximizing the entropy $S(\{p_j\})$ with respect to the probabilities p_j, subject to the constraints set out below, leads to the following forms for the p_js.

(a) If $P = 1$ is the only constraint, the entropy is maximal when all probabilities are equal, i.e. $p_j = 1/M$.

(b) If the constraint $P = 1$ is augmented by a second constraint, namely that the average energy has a specified value, $E = \sum_j E_j p_j$, the entropy is maximal when the probabilities p_j depend exponentially on the energies E_j as $p_j = Z^{-1} e^{-\beta E_j}$ (this is the Boltzmann distribution), where $Z = \sum_j e^{-\beta E_j}$ and $\beta > 0$ is a real constant.

Remarks: Z is known as the **partition function** of the system. In statistical physics, it is known that β is inversely proportional to the temperature, $\beta = 1/(k_B T)$, where the **Boltzmann constant**, k_B, is a universal constant. The average energy of the system, given by $E = \sum_j E_j p_j = \sum_j E_j e^{-\beta E_j}/Z$, is therefore governed by temperature: when T increases, E increases as well. In the limit $T \gg \max(E_j)$ we have $p_j = 1/M$, just as in (a), i.e. then all states are equally likely. In the limit of $T = 0$, p_j is nonzero only if E_j equals the lowest energy in the spectrum. If there is only a single state with lowest energy (a "non-degenerate ground state"), say with index $i = 1$, we have $p_j = \delta_{i1}$, i.e. at zero temperature the system is in the ground state with certainty.

ₚV3.3.6 Entropy maximization subject to constraints, continued (p. 441)

Consider the same setup as in the previous problem. Show that maximizing the entropy with respect to the probabilities p_j, subject to the three constraints of $P = 1$, specified average energy $E = \sum_j p_j E_j$, and specified average particle number, $N = \sum_j p_j N_j$, leads to probabilities of the form $p_j = Z^{-1} e^{-\beta(E_j - \mu N_j)}$, where $Z = \sum_j e^{-\beta(E_j - \mu N_j)}$ and $\beta > 0$ and μ are constants. Here Z is known as the **grand-canonical partition function**, and the constant μ, referred to as the **chemical potential**, regulates the average number of particles.

P.V3.4 Gradient fields (p. 441)

εV3.4.1 Potential of a vector field (p. 444)

Consider a vector field $\mathbf{u} : \mathbb{R}^3 \to \mathbb{R}^3$, $\mathbf{r} \mapsto \mathbf{u}(\mathbf{r}) = \left(2xy + z^3, x^2, 3xz^2\right)^{\mathrm{T}}$.

(a) Calculate the line integral $I_1 = \int_{\gamma_1} d\mathbf{r} \cdot \mathbf{u}(\mathbf{r})$ from $\mathbf{0} = (0,0,0)^{\mathrm{T}}$ to $\mathbf{b} = (1,1,1)^{\mathrm{T}}$, along the path $\gamma_1 = \{\mathbf{r}(t) = (t,t,t)^{\mathrm{T}} \mid 0 < t < 1\}$.

(b) Does the line integral depend on the shape of the path?

(c) Calculate the potential $\varphi(\mathbf{r})$ of the vector field $\mathbf{u}(\mathbf{r})$, using the line integral, $\varphi(\mathbf{r}) = \int_{\gamma_\mathbf{r}} d\mathbf{r} \cdot \mathbf{u}(\mathbf{r})$, along a suitably parametrized path $\gamma_\mathbf{r}$ from $\mathbf{0}$ to $\mathbf{r} = (x, y, z)^T$.

(d) Consistency check: Verify by explicit calculation that your result for $\varphi(\mathbf{r})$ satisfies the equation $\nabla \varphi(\mathbf{r}) = \mathbf{u}(\mathbf{r})$.

(e) Calculate the integral I_1 from part (a) over the vector field by considering the difference in potential $\varphi(\mathbf{r})$ (the anti-derivative!) at the integration limits $\mathbf{b}$ and $\mathbf{0}$. Consistency check: Do you obtain the same result as in part (a) of the exercise?

ₚV3.4.2 Line integral of a vector field (p. 444)

Compute the line integral $W[\gamma] = \int_\gamma d\mathbf{r} \cdot \mathbf{u}$ of the vector field $\mathbf{u}(\mathbf{r}) = (xe^{yz}, ye^{xz}, ze^{xy})^T$ along the straight line γ from the point $\mathbf{0} = (0,0,0)^T$ to the point $\mathbf{b} = b(1,2,1)^T$, with $b \in \mathbb{R}$. [Check your result: For $b^2 = \ln 2$, $W[\gamma] = 7/2$.] Does the line integral depend on the path taken?

ₑV3.4.3 Line integral of magnetic field of a current-carrying conductor (p. 446)

This problem illustrates that $\partial_i B^j - \partial_j B^i = 0$ does not necessarily imply $\oint d\mathbf{r} \cdot \mathbf{B} = 0$.
The magnetic field of an infinitely long current-carrying conductor has the form

$$\mathbf{B}(\mathbf{r}) = \frac{c}{x^2 + y^2} \begin{pmatrix} -y \\ x \\ 0 \end{pmatrix}.$$

(a) Show that $\partial_i B^j - \partial_j B^i = 0$ holds if $\sqrt{x^2 + y^2} \neq 0$.

(b) Compute the line integral $W[\gamma_C] = \int_{\gamma_C} d\mathbf{r} \cdot \mathbf{B}$ for the closed path along the circle C with radius R around the origin, $\gamma_C = \{\mathbf{r}(t) = R(\cos t, \sin t, 0)^T | t \in [0, 2\pi]\}$.

(c) Compute the line integral $W[\gamma_R] = \int_{\gamma_R} d\mathbf{r} \cdot \mathbf{B}$ for the closed path γ_R along the edges of the rectangle with corners $(1,0,0)^T$, $(2,0,0)^T$, $(2,3,0)^T$ and $(1,3,0)^T$.

(d) Are your results from (a) to (c) consistent with each other? Explain!

ₚV3.4.4 Line integral of vector field on non-simply connected domain (p. 446)

Consider the vector field

$$\mathbf{B}(\mathbf{r}) = \frac{1}{(x^2 + y^2)^2} \begin{pmatrix} -yx^n \\ x^{n+1} \\ 0 \end{pmatrix}.$$

(a) For what value of the exponent n does $\partial_i B^j - \partial_j B^i = 0$ hold, if $\sqrt{x^2 + y^2} \neq 0$?

In the following questions, use the value of n found in (a).

(b) Compute the line integral $W[\gamma_C] = \oint_{\gamma_C} d\mathbf{r} \cdot \mathbf{B}$ for the closed path along the circle C with radius R around the origin, $\gamma_C = \{\mathbf{r}(t) = R(\cos t, \sin t, 0)^T | t \in [0, 2\pi]\}$.

(c) What is the value of the line integral $W[\gamma_T] = \oint_{\gamma_T} d\mathbf{r} \cdot \mathbf{B}$ for the closed path γ_T along the edges of the triangle with corners $(-1, -1, 0)^T$, $(1, -1, 0)^T$ and $(a, 1, 0)^T$, with $a \in \mathbb{R}$? Sketch the result as function of $a \in [-2, 2]$. *Hint:* You may write down the result without a calculation, but should offer a justification for it.

P.V3.5 Sources of vector fields (p. 446)

ᴇV3.5.1 Divergence (p. 447)

(a) Compute the divergence, $\nabla \cdot \mathbf{u}$, of the vector field $\mathbf{u} : \mathbb{R}^3 \to \mathbb{R}^3$, $\mathbf{u}(\mathbf{r}) = (xyz, y^2, z^3)^{\mathsf{T}}$.
[Check your results: If $\mathbf{r} = (1, 1, 1)^{\mathsf{T}}$, then $\nabla \cdot \mathbf{u} = 6$.]

(b) Let $\mathbf{a} \in \mathbb{R}^3$ be a constant vector and $f : \mathbb{R}^3 \to \mathbb{R}$, $\mathbf{r} \mapsto f(r)$ a scalar function of $r = \|\mathbf{r}\|$. Show that

$$\nabla \cdot [\mathbf{a}f(r)] = \frac{\mathbf{r} \cdot \mathbf{a}}{r} f'(r).$$

Rule of thumb: ∇ acting on $f(r)$ generates $\hat{\mathbf{r}} = \mathbf{r}/r$ times the derivative, $f'(r)$.

ᴘV3.5.2 Divergence (p. 447)

(a) Compute the divergence, $\nabla \cdot \mathbf{u}$, of the vector field

$$\mathbf{u} : \mathbb{R}^3 \to \mathbb{R}^3, \quad \mathbf{u}(\mathbf{r}) = (xyz, z^2 y^2, z^3 y)^{\mathsf{T}}.$$

[Check your results: If $\mathbf{r} = (1, 1, 1)^{\mathsf{T}}$, then $\nabla \cdot \mathbf{u} = 6$.]

(b) Let $\mathbf{a}$ and $\mathbf{b}$ be constant vectors in $\mathbb{R}^3$. Show that $\nabla \cdot [(\mathbf{a} \cdot \mathbf{r})\mathbf{b}] = \mathbf{a} \cdot \mathbf{b}$.
Rule of thumb: ∇ "kills" the $\mathbf{r}$ in a way that generates another meaningful scalar product.

ᴇV3.5.3 Gauss's theorem – cube (Cartesian coordinates) (p. 450)

Consider the cube C, defined by $x \in (0, a)$, $y \in (0, a)$, $z \in (0, a)$, and the vector field $\mathbf{u}(\mathbf{r}) = (x^2, y^2, z^2)^{\mathsf{T}}$. Compute its outward flux, $\Phi = \int_S d\mathbf{S} \cdot \mathbf{u}$, through the cube's surface, $S \equiv \partial C$, in two ways:

(a) directly as a surface integral;

(b) as a volume integral via Gauss's theorem.

[Check your result: If $a = 2$, then $\Phi = 48$.]

ᴘV3.5.4 Gauss's theorem – cube (Cartesian coordinates) (p. 450)

Consider the cuboid C, defined by $x \in (0, a)$, $y \in (0, b)$, $z \in (0, c)$, and the vector field $\mathbf{u}(\mathbf{r}) = (\frac{1}{2}x^2 + x^2 y, \frac{1}{2}x^2 y^2, 0)^{\mathsf{T}}$. Compute its outward flux, $\Phi = \int_S d\mathbf{S} \cdot \mathbf{u}$, through the cube's surface, $S \equiv \partial C$, in two ways:

(a) directly as a surface integral; and

(b) as a volume integral via Gauss's theorem.

[Check your results: If $a = 2$, $b = 3$, $c = \frac{1}{2}$, then $\Phi = 18$.]

ᴇV3.5.5 Computing volume of barrel using Gauss's theorem (p. 450)

Consider a three-dimensional body bounded by a surface S. One method of computing its volume, V, is to express the latter as a flux integral over S by evoking Gauss's theorem for a vector field, $\mathbf{u}$, satisfying $\nabla \cdot \mathbf{u} = 1$:

$$V = \int_V dV = \int_V dV \, \boldsymbol{\nabla} \cdot \mathbf{u} \overset{\text{Gauss}}{=} \int_S d\mathbf{S} \cdot \mathbf{u} \, .$$

Use this method with $\mathbf{u} = \frac{1}{2}(x, y, 0)^{\mathsf{T}}$ to compute, in cylindrical coordinates, the volume of

(a)   a cylinder with height h and radius R, and

(b)   a cylindrical barrel with height h and z-dependent radius, $\rho(z) = R[1 + a\sin(\pi z/h)]^{1/2}$, with $z \in (0, h)$ and $a > 0$. [Check your result: If $a = \pi/4$, then $V = \frac{3}{2}\pi R^2 h$.]

pV3.5.6 Computing volume of grooved ball using Gauss's theorem (p. 450)

The volume of a body can be computed using a surface integral, $V = \int_S d\mathbf{S} \cdot \frac{1}{3}\mathbf{r}$, over the body's surface, S (see problem V3.5.5). Use this method to compute, in spherical coordinates,

(a)   the volume, V, of a ball with radius R, and

(b)   the volume, $V(\epsilon, n)$, of a "grooved ball", whose ϕ-dependent radius is described by the function $r(\phi) = R[1 + \epsilon\sin(n\phi)]^{2/3}$, where $1 \le n \in \mathbb{N}$ determines the number of grooves and $\epsilon < 1$ their depth. [Check your result: $V(\frac{1}{4}, 4) = \frac{33}{32}V(0, 0)$.]

εV3.5.7 Flux integral: flux of vector field through surface with cylindrical symmetry (p. 450)

Consider a cylinder centered at the origin, with length $2h$ and radius R. A point charge Q at the origin causes an electric field of the form $\mathbf{E}(\mathbf{r}) = E_0 \mathbf{r}/r^3$, with $E_0 = Q$. Find the outward flux, $\Phi_C = \Phi_T + \Phi_B + \Phi_W$, of this field through the entire surface of the cylinder, by computing the flux (a) through the top, Φ_T, and bottom, Φ_B, as well as (b) through the side wall, Φ_W. [Check your results: $\Phi_C = Q/\varepsilon_0$.]

pV3.5.8 Flux integral: flux of vector field through surface with cylindrical symmetry (p. 450)

Consider a solid of revolution, rotationally symmetric about the z-axis, with z-dependent radius $\rho(z) = \mathrm{e}^{-az}$ for $z \in (0, 1)$. It is bounded by the surface $S = S_W \cup S_B \cup S_T$, with side wall, bottom and top defined by:

$$S_W = \{(x, y, z) \in \mathbb{R}^3 : x^2 + y^2 = \mathrm{e}^{-2az}, z \ge 0\} \, ,$$

$$S_B = \{(x, y, z) \in \mathbb{R}^3 : x^2 + y^2 \le 1, z = 0\} \, ,$$

$$S_T = \{(x, y, z) \in \mathbb{R}^3 : x^2 + y^2 \le \mathrm{e}^{-2a}, z = 1\} \, .$$

Compute the outward flux, $\Phi = \int_S d\mathbf{S} \cdot \mathbf{u} = \Phi_W + \Phi_B + \Phi_T$, of the vector field $\mathbf{u}(x, y, z) = (x, y, -2z)$ through the surface S. [Check your results: If $a = 1$, then $\Phi_W = -\Phi_T = 2\pi\mathrm{e}^{-2}$.]

P.V3.6   Circulation of vector fields (p. 454)

εV3.6.1 Curl (p. 454)

(a)   Compute the curl, $\boldsymbol{\nabla} \times \mathbf{u}$, of the vector field, $\mathbf{u} : \mathbb{R}^3 \to \mathbb{R}^3$, $\mathbf{u}(\mathbf{r}) = (xyz, y^2, z^2)^{\mathsf{T}}$.
[Check your results: If $\mathbf{r} = (3, 2, 1)^{\mathsf{T}}$, then $\boldsymbol{\nabla} \times \mathbf{u} = (0, 6, -3)^{\mathsf{T}}$.]

(b)   Let $\mathbf{a} \in \mathbb{R}^3$ be a constant vector and $f : \mathbb{R}^3 \to \mathbb{R}, \mathbf{r} \mapsto f(r)$ a scalar function of $r = \|\mathbf{r}\|$. Show that

$$\nabla \times [\mathbf{a}f(r)] = \frac{\mathbf{r} \times \mathbf{a}}{r} f'(r).$$

Rule of thumb: ∇ acting on $f(r)$ generates $\hat{\mathbf{r}} = \mathbf{r}/r$ times the derivative, $f'(r)$.

ₚV3.6.2 Curl (p. 454)

(a)   Compute the curl, $\nabla \times \mathbf{u}$, of the vector field $\mathbf{u} : \mathbb{R}^3 \to \mathbb{R}^3$, $\mathbf{u}(\mathbf{r}) = (xyz, y^2z^2, xyz^3)^{\mathrm{T}}$.
      [Check your result: If $\mathbf{r} = (3, 2, 1)^{\mathrm{T}}$, then $\nabla \times \mathbf{u} = (-5, 4, -3)^{\mathrm{T}}$.]

(b)   Let $\mathbf{a}$ and $\mathbf{b}$ be constant vectors in $\mathbb{R}^3$. Show that $\nabla \times [(\mathbf{a} \cdot \mathbf{r})\mathbf{b}] = \mathbf{a} \times \mathbf{b}$.
      Rule of thumb: ∇ "kills" the $\mathbf{r}$ in a way that generates another meaningful vector product.

ₑV3.6.3 Stokes's theorem – cuboid (Cartesian coordinates) (p. 456)

Consider the cube C, defined by $x \in (0, a)$, $y \in (0, a)$, $z \in (0, a)$, and the vector field $\mathbf{w}(\mathbf{r}) = (-y^2, x^2, 0)^{\mathrm{T}}$. Compute the outward flux of its curl, $\Phi = \int_S d\mathbf{S} \cdot (\nabla \times \mathbf{w})$, through the surface $S \equiv \partial C \backslash \text{top}$, consisting of all faces of the cube except the top one at $z = a$, in two ways:

(a)   directly as a surface integral;

(b)   as a line integral via Stokes's theorem.

[Check your result: If $a = 2$, then $\Phi = -16$.]

ₚV3.6.4 Stokes's theorem – cuboid (Cartesian coordinates) (p. 456)

Consider the cuboid C, defined by $x \in (0, a)$, $y \in (0, b)$, $z \in (0, c)$, and the vector field $\mathbf{w}(\mathbf{r}) = \frac{1}{2}(yz^2, -xz^2, 0)^{\mathrm{T}}$. Compute the outward flux of its curl, $\Phi = \int_S d\mathbf{S} \cdot (\nabla \times \mathbf{w})$, through the surface $S \equiv \partial C \backslash \text{top}$, consisting of all faces of the cube except the top one at $z = c$, in two ways:

(a)   directly as a surface integral;

(b)   as a line integral via Stokes's theorem.

[Check your results: If $a = 2, b = 3, c = \frac{1}{2}$, then $\Phi = \frac{3}{2}$.]

ₑV3.6.5 Stokes's theorem – magnetic dipole (spherical coordinates) (p. 456)

Every magnetic field can be represented as $\mathbf{B} = \nabla \times \mathbf{A}$, where the vector field $\mathbf{A}$ is known as the **vector potential** of the field. For a magnetic dipole,

$$\mathbf{A} = \frac{1}{c} \frac{\mathbf{m} \times \mathbf{r}}{r^3}, \qquad \mathbf{B} = \frac{1}{c} \frac{3\mathbf{r}(\mathbf{m} \cdot \mathbf{r}) - \mathbf{m}r^2}{r^5},$$

where c is the speed of light. Let the constant dipole moment $\mathbf{m}$ be oriented in the z-direction, $\mathbf{m} = \mathbf{e}_z m$. Let H be a hemisphere with radius R, oriented with base surface in the xy-plane, symmetry axis along the positive z-axis and "north pole" on the latter. Compute the flux integral of the magnetic field through this hemisphere, $\Phi_H = \int_H d\mathbf{S} \cdot \mathbf{B}$, in two different ways:

(a)   directly, using spherical coordinates;

(b)   use $\mathbf{B} = \nabla \times \mathbf{A}$ and Stokes's theorem to express Φ as a line integral of $\mathbf{A}$ over the boundary of the surface of H, and evaluate the line integral.

ₚV3.6.₆ Stokes's theorem – cylinder (cylindrical coordinates) (p. 456)

Consider a cylinder, C, with radius R and height aR^2, centered on the z-axis, with base in the xy-plane, and the vector field $\mathbf{u} = \frac{x^2+y^2}{z}(-y, x, 0)^T$. Compute the flux of its curl, $\Phi_T = \int_T d\mathbf{S} \cdot (\nabla \times \mathbf{u})$, through the top face, T, of the cylinder in two different ways:

(a)   directly, using cylindrical coordinates; and

(b)   by using Stokes's theorem to express Φ_T as a line integral of $\mathbf{u}$ over the boundary, ∂T, of the cylinder top, and then computing the integral.

P.V3.7   Practical aspects of three-dimensional vector calculus (p. 459)

ₑV3.7.₁ Curl of gradient field (p. 460)

Let $f : \mathbb{R}^3 \ > \ \mathbb{R}$ be a smooth scalar field. Show that the curl of its gradient vanishes:

$$\nabla \times (\nabla f) = \mathbf{0}.$$

Recommendation: Use Cartesian coordinates, for which contra- and covariant components are equal, $\partial^i = \partial_i$, and write all indices downstairs.

ₚV3.7.₂ Derivatives of curl of vector field (p. 460)

Let $\mathbf{u} : \mathbb{R}^3 \to \mathbb{R}^3$ be a smooth vector field. Show that the following identities hold:

(a) $\nabla \cdot (\nabla \times \mathbf{u}) = 0$;   (b) $\nabla \times (\nabla \times \mathbf{u}) = \nabla(\nabla \cdot \mathbf{u}) - \nabla^2\mathbf{u}$.
       Recommendation: Use Cartesian coordinates and write all indices downstairs.

(c)   Check both identities for the field $\mathbf{u}(x, y, z) = (x^2yz, xy^2z, xyz^2)^T$.

ₑV3.7.₃ Nabla identities (p. 460)

(a)   Consider the scalar fields $f(x, y, z) = ze^{-x^2}$ and $g(x, y, z) = yz^{-1}$, and the vector fields $\mathbf{u}(x, y, z) = \mathbf{e}_x x^2 y$ and $\mathbf{w}(x, y, z) = (x^2 + y^3)\mathbf{e}_x$. Compute $\nabla f, \nabla g, \nabla^2 f, \nabla^2 g, \nabla \cdot \mathbf{u}, \nabla \times \mathbf{u}, \nabla \cdot \mathbf{w}$, $\nabla \times \mathbf{w}$. [Check your results: At the point $(x, y, z)^T = (1, 1, 1)^T$, we have $\nabla f = (-2e^{-1}, 0, e^{-1})^T$, $\nabla g = (0, 1, -1)^T$, $\nabla^2 f = \frac{2}{e}$, $\nabla^2 g = 2$, $\nabla \cdot \mathbf{u} = 2$, $\nabla \times \mathbf{u} = -\mathbf{e}_z$, $\nabla \cdot \mathbf{w} = 2$, $\nabla \times \mathbf{w} = -3\mathbf{e}_z$.]

(b)   Prove the following identities for *general* smooth scalar and vector fields, $f(x, y, z)$, $g(x, y, z)$ and $\mathbf{u}(x, y, z)$, $\mathbf{w}(x, y, z)$. Do *not* represent $\mathbf{u}$, $\mathbf{w}$ and ∇ as column vectors; instead use index notation. *Recommendation:* Use Cartesian coordinates and write all indices downstairs.

   (i) $\nabla (fg) = f(\nabla g) + g(\nabla f)$;

   (ii) $\nabla (\mathbf{u} \cdot \mathbf{w}) = \mathbf{u} \times (\nabla \times \mathbf{w}) + \mathbf{w} \times (\nabla \times \mathbf{u}) + (\mathbf{u} \cdot \nabla)\mathbf{w} + (\mathbf{w} \cdot \nabla)\mathbf{u}$;

   (iii) $\nabla \cdot (f\mathbf{u}) = f(\nabla \cdot \mathbf{u}) + \mathbf{u} \cdot (\nabla f)$.

(c)   Check the identities from (b) explicitly for the fields given in (a). [Check your results: At the point $(x, y, z)^T = (1, -1, 1)^T$, we have $\nabla (fg) = e^{-1}(2, 1, 0)^T$, $\nabla (\mathbf{u} \cdot \mathbf{w}) = (-2, -3, 0)^T$, $\nabla \cdot (f\mathbf{u}) = 0$.]

pV3.7.4 Nabla identities (p. 460)

(a) Consider the scalar field $f(x, y, z) = y^{-1} \cos z$ and two vector fields, $\mathbf{u}(x, y, z) = \left(-y, x, z^2\right)^{\mathrm{T}}$ and $\mathbf{w}(x, y, z) = (x, 0, 1)^{\mathrm{T}}$. Compute ∇f, $\nabla^2 f$, $\nabla \cdot \mathbf{u}$, $\nabla \times \mathbf{u}$, $\nabla \cdot \mathbf{w}$, $\nabla \times \mathbf{w}$. [Check your results: At the point $(x, y, z)^{\mathrm{T}} = (1, 1, 0)^{\mathrm{T}}$, $\nabla f = -\mathbf{e}_y$, $\nabla^2 f = 1$, $\nabla \cdot \mathbf{u} = 0$, $\nabla \times \mathbf{u} = 2\mathbf{e}_z$, $\nabla \cdot \mathbf{w} = 1$, $\nabla \times \mathbf{w} = \mathbf{0}$.]

(b) Prove the following identities for *general* smooth scalar and vector fields $f(x, y, z)$, $\mathbf{u}(x, y, z)$ and $\mathbf{w}(x, y, z)$. Do *not* represent $\mathbf{u}$, $\mathbf{w}$ and ∇ as column vectors; instead use index notation. *Recommendation:* Use Cartesian coordinates and write all indices downstairs.

(i) $\nabla \cdot (\mathbf{u} \times \mathbf{w}) = \mathbf{w} \cdot (\nabla \times \mathbf{u}) - \mathbf{u} \cdot (\nabla \times \mathbf{w})$;

(ii) $\nabla \times (f\mathbf{u}) = f (\nabla \times \mathbf{u}) - \mathbf{u} \times (\nabla f)$;

(iii) $\nabla \times (\mathbf{u} \times \mathbf{w}) = (\mathbf{w} \cdot \nabla) \mathbf{u} - (\mathbf{u} \cdot \nabla) \mathbf{w} + \mathbf{u} (\nabla \cdot \mathbf{w}) - \mathbf{w} (\nabla \cdot \mathbf{u})$.

(c) Check the identities from (b) explicitly for the fields given in (a).
[Check your results: At the point $(x, y, z)^{\mathrm{T}} = (1, 1, 0)^{\mathrm{T}}$: $\nabla \cdot (\mathbf{u} \times \mathbf{w}) = 2$, $\nabla \times (f\mathbf{u}) = (0, 0, 1)^{\mathrm{T}}$, $\nabla \times (\mathbf{u} \times \mathbf{w}) = (0, 2, 0)^{\mathrm{T}}$.]

εV3.7.5 Gradient, divergence, curl, Laplace in cylindrical coordinates (p. 461)

We consider a curvilinear *orthogonal* coordinate system with coordinates $\mathbf{y} = (y^1, y^2, y^3)^{\mathrm{T}} \equiv (\eta, \mu, \nu)^{\mathrm{T}}$, position vector $\mathbf{r}(\mathbf{y}) = \mathbf{r}(\eta, \mu, \nu)$ and coordinate basis vectors $\partial_\eta \mathbf{r} = \mathbf{e}_\eta n_\eta$, $\partial_\mu \mathbf{r} = \mathbf{e}_\mu n_\mu$, $\partial_\nu \mathbf{r} = \mathbf{e}_\nu n_\nu$, with $\|\mathbf{e}_j\| = 1$ and normalization factors n_η, n_μ, n_ν (i.e. no summations over η, μ and ν here!). Furthermore, let $f(\mathbf{r})$ be a scalar field and $\mathbf{u}(\mathbf{r}) = \mathbf{e}_\eta u^\eta + \mathbf{e}_\mu u^\mu + \mathbf{e}_\nu u^\nu$ a vector field, expressed in the *local basis*. Then, the gradient, divergence, curl and Laplace operator are given by

$$\nabla f = \mathbf{e}_\eta \frac{1}{n_\eta} \partial_\eta f + \overset{\eta \curvearrowright \mu}{\underset{\nu}{\circlearrowleft}} + \overset{\eta \curvearrowright \mu}{\underset{\nu}{\circlearrowleft}},$$

$$\nabla \cdot \mathbf{u} = \frac{1}{n_\eta n_\mu n_\nu} \partial_\eta \left(n_\mu n_\nu u^\eta\right) + \overset{\eta \curvearrowright \mu}{\underset{\nu}{\circlearrowleft}} + \overset{\eta \curvearrowright \mu}{\underset{\nu}{\circlearrowleft}},$$

$$\nabla \times \mathbf{u} = \mathbf{e}_\eta \frac{1}{n_\mu n_\nu} \left[\partial_\mu \left(n_\nu u^\nu\right) - \partial_\nu \left(n_\mu u^\mu\right)\right] + \overset{\eta \curvearrowright \mu}{\underset{\nu}{\circlearrowleft}} + \overset{\eta \curvearrowright \mu}{\underset{\nu}{\circlearrowleft}},$$

$$\nabla^2 f = \nabla \cdot (\nabla f) = \frac{1}{n_\eta n_\mu n_\nu} \partial_\eta \left(\frac{n_\mu n_\nu}{n_\eta} \partial_\eta f\right) + \overset{\eta \curvearrowright \mu}{\underset{\nu}{\circlearrowleft}} + \overset{\eta \curvearrowright \mu}{\underset{\nu}{\circlearrowleft}},$$

where circles with three arrows denote cyclical permutation of indices.
Now consider the cylindrical coordinates defined by $\mathbf{r}(\rho, \phi, z) = (\rho \cos \phi, \rho \sin \phi, z)^{\mathrm{T}}$.

(a) Write down formulas for $\mathbf{e}_\rho$, $\mathbf{e}_\phi$, $\mathbf{e}_z$ and n_ρ, n_ϕ, n_z.

Starting from the general formulas given above, find explicit formulas for

(b) ∇f,   (c) $\nabla \cdot \mathbf{u}$,   (d) $\nabla \times \mathbf{u}$,   (e) $\nabla^2 f$.

(f) Verify explicitly that $\nabla \times (\nabla f) = \mathbf{0}$, using the given formulas for the gradient and curl in general curvilinear coordinates η, μ, ν (i.e. not specifically cylindrical coordinates).

(g) Use cylindrical coordinates to compute ∇f, $\nabla \cdot \mathbf{u}$, $\nabla \times \mathbf{u}$ and $\nabla^2 f$ for the fields $f(\mathbf{r}) = \|\mathbf{r}\|^2$ and $\mathbf{u}(\mathbf{r}) = (x, y, 2z)^{\mathrm{T}}$. [Check your results: If $\mathbf{r} = (1, 1, 1)^{\mathrm{T}}$, then $\nabla f = (2, 2, 2)^{\mathrm{T}}$, $\nabla \cdot \mathbf{u} = 4$, $\nabla \times \mathbf{u} = \mathbf{0}$ and $\nabla^2 f = 6$.]

$_P$V3.7.$_6$ Gradient, divergence, curl, Laplace in spherical coordinates $_{(p.\ 461)}$

Consider a curvilinear *orthogonal* coordinate system with coordinates $\mathbf{y} = (y^1, y^2, y^3)^T \equiv (\eta, \mu, \nu)^T$, position vector $\mathbf{r}(\mathbf{y}) = \mathbf{r}(\eta, \mu, \nu)$ and coordinate basis vectors $\partial_\eta \mathbf{r} = \mathbf{e}_\eta n_\eta$, $\partial_\mu \mathbf{r} = \mathbf{e}_\mu n_\mu$, $\partial_\nu \mathbf{r} = \mathbf{e}_\nu n_\nu$, with $\|\mathbf{e}_j\| = 1$. Furthermore, $f(\mathbf{r})$ is a scalar field and $\mathbf{u}(\mathbf{r}) = \mathbf{e}_\eta u^\eta + \mathbf{e}_\mu u^\mu + \mathbf{e}_\nu u^\nu$ is a vector field, expressed in the local basis. Then, the gradient, divergence, curl and Laplace operator are given by

$$\nabla f = \mathbf{e}_\eta \frac{1}{n_\eta} \partial_\eta f \;+\; \overset{\eta\curvearrowright\mu}{\underset{\nu}{\circlearrowleft}} \;+\; \overset{\eta\curvearrowleft\mu}{\underset{\nu}{\circlearrowright}},$$

$$\nabla \cdot \mathbf{u} = \frac{1}{n_\eta n_\mu n_\nu} \partial_\eta \left(n_\mu n_\nu u^\eta \right) \;+\; \overset{\eta\curvearrowright\mu}{\underset{\nu}{\circlearrowleft}} \;+\; \overset{\eta\curvearrowleft\mu}{\underset{\nu}{\circlearrowright}},$$

$$\nabla \times \mathbf{u} = \mathbf{e}_\eta \frac{1}{n_\mu n_\nu} \left[\partial_\mu \left(n_\nu u^\nu \right) - \partial_\nu \left(n_\mu u^\mu \right) \right] \;+\; \overset{\eta\curvearrowright\mu}{\underset{\nu}{\circlearrowleft}} \;+\; \overset{\eta\curvearrowleft\mu}{\underset{\nu}{\circlearrowright}},$$

$$\nabla^2 f = \nabla \cdot (\nabla f) = \frac{1}{n_\eta n_\mu n_\nu} \partial_\eta \left(\frac{n_\mu n_\nu}{n_\eta} \partial_\eta f \right) \;+\; \overset{\eta\curvearrowright\mu}{\underset{\nu}{\circlearrowleft}} \;+\; \overset{\eta\curvearrowleft\mu}{\underset{\nu}{\circlearrowright}}.$$

Consider the spherical coordinates defined by $\mathbf{r}(r, \theta, \phi) = (r \sin\theta \cos\phi, r \sin\theta \sin\phi, r \cos\theta)^T$.

(a)   Write down formulas for $\mathbf{e}_r$, $\mathbf{e}_\theta$, $\mathbf{e}_\phi$ and n_r, n_θ, n_ϕ.

Starting from the general formulas given above, find an explicit formula for

(b) ∇f,   (c) $\nabla \cdot \mathbf{u}$,   (d) $\nabla \times \mathbf{u}$,   (e) $\nabla^2 f$.

(f)   Verify explicitly that $\nabla \cdot (\nabla \times \mathbf{u}) = 0$, using the above formulas for the divergence and the curl for general curvilinear coordinates η, μ, ν (i.e. not specifically spherical coordinates).

(g)   Use spherical coordinates to compute ∇f, $\nabla \cdot \mathbf{u}$, $\nabla \times \mathbf{u}$ and $\nabla^2 f$ for the fields $f(\mathbf{r}) = \|\mathbf{r}\|^2$ and $\mathbf{u}(\mathbf{r}) = (0, 0, z)^T$. [Check your results: If $\mathbf{r} = (1, 1, 1)^T$, then $\nabla f = (2, 2, 2)^T$, $\nabla \cdot \mathbf{u} = 1$, $\nabla \times \mathbf{u} = \mathbf{0}$ and $\nabla^2 f = 6$.]

$_E$V3.7.$_7$ Gradient, divergence, curl (spherical coordinates) $_{(p.\ 461)}$

Consider the scalar field $f(\mathbf{r}) = \frac{1}{r}$ and the vector field $\mathbf{u}(\mathbf{r}) = (e^{-r/a}/r)\mathbf{r}$, with $\mathbf{r} = (x, y, z)^T$ and $r = \sqrt{x^2 + y^2 + z^2}$. Calculate ∇f, $\nabla \cdot \mathbf{u}$, $\nabla \times \mathbf{u}$ and $\nabla^2 f$ explicitly for $r > 0$,

(a)   in Cartesian coordinates;   (b)   in spherical coordinates.

Verify that your results from (a) and (b) are consistent with one another.

$_P$V3.7.$_8$ Gradient, divergence, curl (cylindrical coordinates) $_{(p.\ 461)}$

Consider the scalar field $f(\mathbf{r}) = z(x^2 + y^2)$ and the vector field $\mathbf{u}(\mathbf{r}) = (zx, zy, 0)^T$. Calculate ∇f, $\nabla \cdot \mathbf{u}$, $\nabla \times \mathbf{u}$ and $\nabla^2 f$ explicitly in

(a)   Cartesian coordinates;   (b)   cylindrical coordinates.

Verify that your results from (a) and (b) are consistent with one another.

ₑV3.7.9 Gauss's theorem – cylinder (cylindrical coordinates) (p. 461)

Consider a vector field, $\mathbf{u}$, defined in cylindrical coordinates by $\mathbf{u}(\mathbf{r}) = \mathbf{e}_\rho z\rho$, and a cylindrical volume, V, defined by $\rho \in (0, R)$, $\phi \in (0, 2\pi)$, $z \in (0, H)$.

(a) Compute the divergence of the vector field $\mathbf{u}$ in cylindrical coordinates.

Compute the flux, Φ, of the vector field $\mathbf{u}$ through the surface, S, of the cylindrical volume V, via two methods:

(b) by calculating the surface integral, $\Phi = \int_S d\mathbf{S} \cdot \mathbf{u}$, explicitly;

(c) by using Gauss's theorem to convert the flux integral to a volume integral of $\nabla \cdot \mathbf{u}$ and then computing the volume integral explicitly.

ₚV3.7.10 Gauss's theorem – wedge ring (spherical coordinates) (p. 461)

Consider the "wedge-ring", W, which is shaded grey in the sketch. This shape can be expressed in spherical coordinates by the conditions $r \in (0, R)$ and $\theta \in (\pi/3, 2\pi/3)$. (Such a ring-like object, with wedge-shaped inner profile and rounded outer profile, is constructed from a sphere with radius R, by removing a double cone centred on the z-axis with apex angle $\pi/3$.) Compute the outward flux, Φ_W, of the vector field $\mathbf{u}(\mathbf{r}) = \mathbf{e}_r r^2$ through the surface, ∂W, of the wedge-ring, in two different ways.

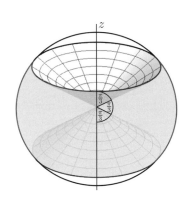

(a) Compute the flux integral, $\Phi_W = \int_{\partial W} d\mathbf{S} \cdot \mathbf{u}$. [Check your result: If $R = \frac{1}{2}$, then $\Phi_W = \frac{\pi}{8}$.]

(b) Use Gauss's theorem to convert the flux integral into a volume integral of the divergence $\nabla \cdot \mathbf{u}$, and compute the volume integral explicitly. *Hint:* In the local basis of spherical coordinates,

$$\nabla \cdot \mathbf{u} = \frac{1}{r^2} \partial_r \left(r^2 u^r \right) + \frac{1}{r \sin \theta} \partial_\theta \left(\sin \theta u^\theta \right) + \frac{1}{r \sin \theta} \partial_\phi u^\phi.$$

(b) For the vector field $\mathbf{w}(\mathbf{r}) = -\mathbf{e}_\theta \cos \theta$, calculate the outward flux, $\tilde{\Phi}_W = \int_{\partial W} d\mathbf{S} \cdot \mathbf{w}$, through the surface of the wedge-ring, either directly or by using Gauss's theorem. [Check your result: If $R = \frac{1}{\sqrt{3}}$, then $\tilde{\Phi}_W = \frac{\pi}{\sqrt{12}}$.]

ₑV3.7.11 Stokes's theorem – magnetic field of infinite current-carrying wire (cylindrical coordinates) (p. 461)

Let an infinitely long, infinitesimally thin conductor be oriented along the z-axis and carry a current I. It generates a magnetic field of the following form:

$$\mathbf{B}(\mathbf{r}) = \frac{2I}{c} \frac{1}{x^2 + y^2} \begin{pmatrix} -y \\ x \\ 0 \end{pmatrix} = \mathbf{e}_\phi \frac{2I}{c} \frac{1}{\rho}, \qquad \text{for} \quad \rho = \sqrt{x^2 + y^2} > 0.$$

Calculate the divergence and rotation of $\mathbf{B}(\mathbf{r})$ explicitly for $\rho > 0$, using

(a) Cartesian coordinates; and

(b) cylindrical coordinates. [Compare your results from (a) and (b)!]

(c)   Use cylindrical coordinates to compute the line integral, $\oint_\gamma d\mathbf{r} \cdot \mathbf{B}$, of the magnetic field along the edge, γ, of a circular disk, D, with radius $R > 0$, centered on the z-axis, and oriented parallel to the xy-plane.

(d)   Use Stokes's theorem and the result from (c) to compute the flux integral, $\int_D d\mathbf{S} \cdot (\nabla \times \mathbf{B})$, of the curl of the magnetic field over the disk D prescribed in (c).

(e)   Use your results for $\nabla \times \mathbf{B}$ from (a) and (d) to argue that the curl of the field is proportional to a two-dimensional δ-function, $\nabla \times \mathbf{B} = \mathbf{e}_z\, C\delta(x)\delta(y)$. Find the constant C. [*Hint*: The two-dimensional δ-function is normalized such that $\int_D dS\, \delta(x)\delta(y) = 1$ for the area integral over any surface D which lies parallel to the xy-plane and intersects the z-axis.]

(f)   Write the result obtained in (e) in the form $\nabla \times \mathbf{B} = \frac{4\pi}{c}\, \mathbf{j}(\mathbf{r})$ and determine $\mathbf{j}(\mathbf{r})$. This equation is Ampère's law (one of the Maxwell equations), where $\mathbf{j}(\mathbf{r})$ is the current density. Can you give a physical interpretation of your result for $\mathbf{j}(\mathbf{r})$?

ₚV3.7.12 Gauss's theorem – electrical field of a point charge (spherical coordinates) (p. 461)

The electric field of a point charge Q at the origin has the form

$$\mathbf{E}(\mathbf{r}) = \frac{Q}{r^3}\mathbf{r} = \mathbf{e}_r\frac{Q}{r^2}, \qquad \text{with} \quad r > 0, \qquad r = \sqrt{x^2 + y^2 + z^2}\,.$$

Calculate the divergence and the curl of $\mathbf{E}(\mathbf{r})$ explicitly for $r > 0$, using

(a)   Cartesian coordinates; and

(b)   spherical coordinates. [Compare your results from (a) and (b)!]

(c)   Use spherical coordinates to compute the flux, $\Phi_S = \int_S d\mathbf{S} \cdot \mathbf{E}$, of the electric field through a sphere, S, with radius $R > 0$, centered at the origin.

(d)   Use Gauss's theorem and the result from (c) to compute the integral, $\int_V dV(\nabla \cdot \mathbf{E})$, over the volume, V, enclosed by the sphere S described in (c).

(e)   Use your results for $\nabla \cdot \mathbf{E}$ from (a) and (d) to argue that the divergence of the field is proportional to a three-dimensional δ-function, i.e. has the form $\nabla \cdot \mathbf{E} = C\,\delta^{(3)}(\mathbf{r})$. Find the constant C. [*Hint:* The normalization of $\delta^{(3)}(\mathbf{r}) = \delta(x)\delta(y)\delta(z)$ is given by the volume integral $\int_V dV\, \delta^{(3)}(\mathbf{r}) = 1$, for any volume, V, that contains the origin.]

(f)   Write your result from (e) in the form $\nabla \cdot \mathbf{E} = 4\pi\rho(\mathbf{r})$, and determine $\rho(\mathbf{r})$. This equation is (the physical) Gauss's law (one of the Maxwell equations), where $\rho(\mathbf{r})$ is the charge density. Can you interpret your result in terms of $\rho(\mathbf{r})$?

ₑV3.7.13 Gauss's theorem – electrical dipole potential (spherical coordinates) (p. 461)

The potential of an electric dipole with dipole moment $\mathbf{p} = p\mathbf{e}_z$ is given by

$$\Phi(\mathbf{r}) = \frac{\mathbf{p} \cdot \mathbf{r}}{r^3} = \frac{pz}{r^3}.$$

(a)   Calculate the electric field, $\mathbf{E} = -\nabla\Phi(\mathbf{r})$, explicitly in Cartesian coordinates.

(b)   Represent $\Phi(\mathbf{r})$ in spherical coordinates and calculate the electric field explicitly in spherical coordinates. Compare this result with the result obtained in (a).
        Hint: $\mathbf{e}_z = \cos\theta\,\mathbf{e}_r - \sin\theta\,\mathbf{e}_\theta$.

(c)   Calculate the divergence and the curl of the electric field explicitly in Cartesian coordinates.

(d) Calculate the divergence and the curl of the electric field explicitly in spherical coordinates. [Compare the results obtained in (b) and (c)!]

(e) According to the (physical) law of Gauss we have $\int_S d\mathbf{S} \cdot \mathbf{E} = 4\pi Q$, where Q is the total charge contained within the volume of S. Now consider a sphere, S, of radius R, centered at the origin. Calculate Q by performing the flux integral over the sphere. Does your result for Q make physical sense? Explain!

(f) Now compute the flux integral in an alternative manner: convert it via the (mathematical) theorem of Gauss into a volume integral over $\nabla \cdot \mathbf{E}$, and evaluate this integral using the result from (d). Comment on the behavior of the integrand at $r = 0$.

P.V4   Introductory concepts of differential geometry (p. 463)

P.V4.1   Differentiable manifolds (p. 464)

ε V4.1.1   Four-chart atlas for $S^1 \subset \mathbb{R}^2$ (p. 467)

Consider the unit circle, S^1, embedded in $\mathbb{R}^2$, $S^1 = \{(x_1, x_2)|(x^1)^2 + (x^2)^2 = 1\} \subset \mathbb{R}^2$. One possible atlas for S^1 is provided by a collection of four coordinate maps or charts, $\{r_{i,\pm}\}$ (with $i \in \{1, 2\}$), each parametrizing a different half-circle, $S_{i,\pm} \equiv \{(x_1, x_2) \in S^1 | x^i \gtrless 0\} \subset S^1$. These coordinate maps are defined such that their inverses, $r_{i,\pm}^{-1}$, map $S_{i,\pm}^1$ to its projection on the axis where $x^i = 0$, so that it is parametrized by the portion $U_{i,\pm} = (-1, 1)$ of that axis:

$$r_{i,\pm}^{-1}: S_{i,\pm} \to U_{i,\pm}, \qquad r_{1,\pm}^{-1}: (x^1, x^2)^{\mathrm{T}} \mapsto x^2, \qquad r_{2,\pm}^{-1}: (x^1, x^2)^{\mathrm{T}} \mapsto x^1.$$

(a) Construct the maps $r_{i,\pm}$, i.e. give formulas for the images of x^2 and x^1 under $r_{1,\pm}$ and $r_{2,\pm}$, respectively. Convince yourself that these charts indeed provide an atlas for S^1.

(b) Now consider the quarter-circle $S_{-,+} = S_{1,-} \cap S_{2,+}$ and construct the transition function $r_{1,-}^{-1} \circ r_{2,+}$ relating its two alternative coordinate representations (i.e. express x^2 as function of x^1). Use sketches to elucidate how this transition function acts. [Check your result: $r_{1,-}^{-1} \circ r_{2,+}$ maps the point $x^1 = -\frac{1}{2} \in U_{2,+}$ to the point $x^2 = \frac{\sqrt{3}}{2} \in U_{1,-}$.]

(c) Similarly, construct the transition function $r_{2,-}^{-1} \circ r_{1,-}$ relating the two coordinate representations of the quarter-circle $S_{-,-} = S_{2,-} \cap S_{1,-}$.

ρ V4.1.2   Six-chart atlas for $S^2 \subset \mathbb{R}^3$ (p. 467)

Consider the unit sphere, S^2, embedded in $\mathbb{R}^3$, $S^2 = \{(x_1, x_2, x_3)|(x^1)^2 + (x^2)^2 + (x^3)^2 = 1\}$. One possible atlas for S is provided by a collection of six coordinate maps or charts, $\{r_{i,\pm}\}$ (with $i \in \{1, 2, 3\}$), each parametrizing a different half-sphere, $S_{i,\pm} \equiv \{(x^1, x^2, x^3) \in S^2 | \pm x^i > 0\} \subset S^2$. These maps are defined such that their inverses, $r_{i,\pm}^{-1}$, map $S_{i,\pm}^1$ to its projection onto the plane where $x^i = 0$, so that it is parametrized by a disk, $U_{i,\pm}$, lying within that plane:

$$r_{i,\pm}^{-1}: S_{i,\pm} \to U_{i,\pm}, \qquad r_{1,\pm}^{-1}: (x^1, x^2, x^3) \mapsto (x^2, x^3),$$

$$r_{2,\pm}^{-1} : (x^1, x^2, x^3) \mapsto (x^1, x^3),$$
$$r_{3,\pm}^{-1} : (x^1, x^2, x^3) \mapsto (x^1, x^2).$$

The figure shows some examples:

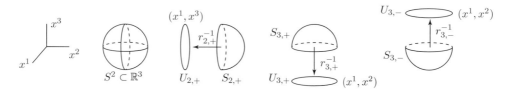

(a) Construct the maps $r_{i,\pm}$. Convince yourself that they indeed provide an atlas for S^2.

(b) Now consider the quarter-sphere $S_{3,-;2,+} = S_{3,-} \cap S_{2,+}$ for which $x^3 < 0$, $x^2 > 0$. Construct the transition function $r_{3,-}^{-1} \circ r_{2,+}$ relating the two coordinate representations of this quarter-sphere determined via $r_{3,-}$ and $r_{2,+}$. Use sketches to elucidate how this transition function is defined. [Check your result: $r_{3,-}^{-1} \circ r_{2,+}$ maps the point $(x^1, x^3)^\mathsf{T} = (\frac{2}{3}, \frac{2}{3})^\mathsf{T} \in U_{2,+}$ to the point $(x^1, x^2)^\mathsf{T} = (\frac{2}{3}, \frac{1}{3})^\mathsf{T} \in U_{3,-}$.]

P.V4.2 Tangent space (p. 468)

ℰV4.2.1 Tangent vectors on a paraboloidal manifold (p. 473)

Consider the manifold $M = \{r(\rho, \phi, z) \,|\, \rho > 0, \phi \in (0, 2\pi), z = \frac{1}{2}\rho^2\} \subset \mathbb{R}^3$, describing the surface of a paraboloid in cylindrical coordinates. Let $y : U \to M$, $y \mapsto r(y)$ be a parametrization of M in terms of $y = (\rho, \phi)^\mathsf{T} \equiv \rho\, e_\rho + \phi\, e_\phi$ ($e_\rho = (1,0)^\mathsf{T}$ and $e_\phi = (0,1)^\mathsf{T}$ are canonical basis vectors of U), and denote the Cartesian components of $r(y)$ by $(r^1, r^2, r^3)^\mathsf{T} = (\rho \cos\phi, \rho \sin\phi, \frac{1}{2}\rho^2)^\mathsf{T}$.

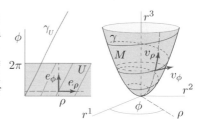

(a) Let $\{\partial_\rho, \partial_\phi\}$ denote the holonomic basis of this manifold at some point $r(y)$. Find the two-dimensional representations, v_ρ and v_ϕ, of these vectors in U, and their corresponding three-dimensional representations in the embedding space $\mathbb{R}^3$. Indicate the relation between all these vectors in sketches of U and $M \subset \mathbb{R}^3$. [Check your results: At $y = (2, \pi)^\mathsf{T}$, the $\mathbb{R}^3$-representations read $v_\rho = (-1, 0, 2)^\mathsf{T}$, $v_\phi = (0, -2, 0)^\mathsf{T}$.]

(b) Let $\gamma : (0,1) \to M$, $t \mapsto r(t)$ describe a spiralling curve in the manifold M, with coordinate representation $\gamma_U : (0,1) \to U$, $t \mapsto y(t) = (Rt, \omega t \bmod 2\pi)^\mathsf{T}$. (The angle ϕ increases linearly with t, but angles differing by multiples of 2π are identified with each other.) Let $\partial_u \equiv \partial_{u,r}$ denote the tangent vector to this curve at point r. Find $(u^\rho, u^\phi)^\mathsf{T}$, its two-dimensional representation in U, and $(u^1, u^2, u^3)^\mathsf{T}$, its three-dimensional representation in $\mathbb{R}^3$, and express these in terms of the basis vectors constructed in (a). [Check your results: If $R = 3$, $\omega = \pi$, then at $t = 1$ we have $(u^1, u^2, u^3)^\mathsf{T} = (-3, -3\pi, 9)^\mathsf{T}$.]

(c) Let $f : M \to \mathbb{R}$, $r \mapsto f(r) = r^1 + r^2 + r^3$ be a function defined on M. Compute its directional derivative, $\partial_u f$, along the spiral curve, as a function of t. [Check your result: If $R = 2$, $w = \frac{\pi}{2}$, then at $t = 1$ we have $\partial_u f = 6 - \pi$.]

ₚV4.2.2 Tangent vectors on the sphere S^2 (p. 473)

Consider the manifold $M = \{r(R, \theta, \phi)|\theta \in (0, \pi), \phi \in (0, 2\pi)\} \subset \mathbb{R}^3$ describing a sphere of radius R in spherical coordinates. Let $y : U \to M$, $y \mapsto r(y)$ be a parametrization of M in terms of $y = (\theta, \phi)^{\mathrm{T}} \equiv e_\theta + \phi\, e_\phi$ (where $e_\theta = (1, 0)^{\mathrm{T}}$ and $e_\phi = (0, 1)^{\mathrm{T}}$ are canonical basis vectors of U), and denote the Cartesian components of $r(y)$ by $(r^1, r^2, r^3)^{\mathrm{T}} = R(\cos\phi\sin\theta, \sin\phi\sin\theta, \cos\theta)^{\mathrm{T}}$.

(a) Let $\{\partial_\theta, \partial_\phi\}$ denote the holonomic basis of this manifold at some point $r(y)$. Find the two-dimensional representations, v_θ and v_ϕ, of these vectors in U, and their corresponding three-dimensional representations in the embedding space $\mathbb{R}^3$. Indicate the relation between all these vectors in sketches of U and $M \subset \mathbb{R}^3$. [Check your results: If $R = 2$, and $y = (\frac{\pi}{4}, \frac{\pi}{4})^{\mathrm{T}}$ the $\mathbb{R}^3$ representations read $v_\theta = (1, 1, -\sqrt{2})^{\mathrm{T}}, v_\phi = (-1, 1, 0)^{\mathrm{T}}$.]

(b) Let $\gamma : (0, 1) \to M$, $t \mapsto r(t)$ describe a spiralling curve in the manifold M, with coordinate representation $\gamma_U : (0, 1) \to U$, $t \mapsto y(t) = (\pi t, 2\pi n t \bmod 2\pi)^{\mathrm{T}}$. (The angle ϕ increases linearly with t, but angles differing by multiples of 2π are identified with each other.) Let $\partial_u \equiv \partial_{u,r}$ denote the tangent vector to this curve at point r. Find $(u^\theta, u^\phi)^{\mathrm{T}}$, its two-dimensional representation in U, and $(u^1, u^2, u^3)^{\mathrm{T}}$, its three-dimensional representation in $\mathbb{R}^3$. [Check your results: If $R = 2$, $n = \frac{3}{2}$, then at time $t = \frac{3}{4}$ we have $(u^1, u^2, u^3)^{\mathrm{T}} = \pi(-4, 2, -\sqrt{2})^{\mathrm{T}}$.]

(c) Let $f : M \to \mathbb{R}$, $r \mapsto f(r) = r^1 - r^2$ be a function defined on M. Compute its directional derivative, $\partial_u f$, along the spiral curve, as a function of t. [Check your result: If $R = 2$, $n = \frac{3}{2}$, then at $t = \frac{3}{4}$ we have $\partial_u f = -6\pi$.]

ₑV4.2.3 Holonomic basis for hyperbolic coordinates (p. 474)

The upper right quadrant, $\mathbb{R}^{+2}$, of the two-dimensional plane can be parametrized by **hyperbolic coordinates**, $y = (\rho, \alpha)^{\mathrm{T}}$, related to the Cartesian coordinates, $x = (x^1, x^2)^{\mathrm{T}}$, via $x^1 = \rho\, e^\alpha$, $x^2 = \rho\, e^{-\alpha}$, where the set $U = \{(\rho, \alpha)|\rho > 0\} = \mathbb{R}^+ \times \mathbb{R}$ defines the coordinate domain. The inverse relations read $\rho = \sqrt{x^1 x^2}$ and $\alpha = \ln\sqrt{x^1/x^2}$, hence the coordinate lines of fixed ρ or α are hyperbolae or straight lines, respectively.

Let $\{\partial_1, \partial_2\}$ and $\{\partial_\rho, \partial_\alpha\}$ denote the holonomic basis vectors of the Cartesian and hyperbolic coordinate systems, respectively. Express one in terms of the other:

(a) Find the Cartesian components, $(v_{yj})^i$, of the hyperbolic basis vectors, $\partial_{yj} = (v_{yj})^i \partial_{xi}$.

(b) Find the hyperbolic components, $(v_{xi})^j$, of the Cartesian basis vectors, $\partial_{xi} = (v_{xi})^j \partial_{yj}$.

(c) Construct the Jacobi matrix J for the transformation $y \mapsto x(y)$ from the components found in (a), and its inverse, J^{-1}, from those found in (b), and verify that $JJ^{-1} = \mathbb{1}$.

$\left[\text{Check your results: (a) if } (x^1, x^2) = (4, 1) \text{ then } ((v_\rho)^1, (v_\rho)^2)^{\mathrm{T}} = (2, \frac{1}{2})^{\mathrm{T}}, ((v_\alpha)^1, (v_\alpha)^2)^{\mathrm{T}} = (4, -1)^{\mathrm{T}}; \text{ (b) if } (\rho, \alpha) = (2, \ln 2), \text{ then } ((v_1)^\rho, (v_1)^\alpha)^{\mathrm{T}} = (\frac{1}{4}, \frac{1}{8})^{\mathrm{T}}, ((v_2)^\rho, (v_2)^\alpha)^{\mathrm{T}} = (1, -\frac{1}{2})^{\mathrm{T}}.\right]$

ₚV4.2.4 Holonomic basis for generalized polar coordinates (p. 474)

The plane $\mathbb{R}^2$ can be parametrized by **generalized polar coordinates**, $y = (\mu, \phi)^{\mathrm{T}}$, related to the Cartesian coordinates, $x = (x^1, x^2)^{\mathrm{T}}$, via $x^1 = \mu a \cos\phi$, $x^2 = \mu b \sin\phi$ (with $a, b \in \mathbb{R}^+$), where the set $U = \{(\mu, \phi)|\mu \in \mathbb{R}^+, \phi \in (0, 2\pi)\}$ defines the coordinate domain. The inverse relations read $\mu = [(x^1/a)^2 + (x^2/b)^2]^{1/2}$ and $\phi = \arctan(\frac{x^2 a}{x^1 b})$, thus coordinate lines of fixed μ or ϕ are ellipses or straight lines, respectively.

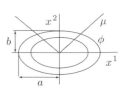

Let $\{\partial_1, \partial_2\}$ and $\{\partial_\mu, \partial_\phi\}$ denote the holonomic basis vectors of the Cartesian and generalized polar coordinate systems, respectively. Express one in terms of the other:

(a) Find the Cartesian components, $(v_{yj})^i$, of the generalized polar basis vectors, $\partial_{yj} = (v_{yj})^i \partial_{x^i}$.

(b) Find the generalized polar components, $(v_{xi})^j$, of the Cartesian basis vectors, $\partial_{x^i} = (v_{x^i})^j \partial_{yj}$.

(c) Construct the Jacobi matrix J for the transformation $y \mapsto x(y)$ from the components found in (a), and its inverse, J^{-1}, from those found in (b), and verify that $JJ^{-1} = \mathbb{1}$.

$\Big[$Check your results for $a = 2$, $b = 4$: (a) if $(x^1, x^2) = (8, 12)$ then $((v_\mu)^1, (v_\mu)^2)^{\mathrm{T}} = (\tfrac{8}{5}, \tfrac{12}{5})^{\mathrm{T}}$ and $((v_\phi)^1, (v_\phi)^2)^{\mathrm{T}} = (-6, 16)^{\mathrm{T}}$; (b) if $(\mu, \phi) = (3, \tfrac{\pi}{6})$, then $((v_1)^\mu, (v_1)^\phi)^{\mathrm{T}} = (\tfrac{\sqrt{3}}{4}, -\tfrac{1}{12})^{\mathrm{T}}$ and $((v_2)^\mu, (v_2)^\phi)^{\mathrm{T}} = (\tfrac{1}{8}, \tfrac{1}{8\sqrt{3}})^{\mathrm{T}}.\Big]$

P.V5 Alternating differential forms (p. 477)

P.V5.1 Cotangent space and differential one-forms (p. 477)

ᴇV5.1.1 Differential of a function in Cartesian and polar coordinates (p. 480)

Consider the differential, df, of the function $f : \mathbb{R}^2 \to \mathbb{R}$, $x = (x^1, x^2)^{\mathrm{T}} \mapsto f(x) = x^1 x^2$.

(a) Find the coefficients, $f_i(x)$, of the differential written in the form $df = f_i(x) dx^i$.

(b) Evaluate the action of df on the vector $\partial_u = x^1(\partial_1 + \partial_2)$.

(c) Express df in polar coordinates, $y = (\rho, \phi)^{\mathrm{T}}$, i.e. find the coefficients, $f_i(y)$, of $df = f_i(y) dy^i$.

(d) Evaluate the action of df on the vector $\partial_u = \partial_\rho - \partial_\phi$.

(e) Visualize the forms dx^1, dx^2, $d\rho$ and $d\phi$ by making a schematic sketch for each.

[Check your results: At $(x^1, x^2)^{\mathrm{T}} = (2, 1)^{\mathrm{T}}$ we have (a) $(f_1, f_2) = (1, 2)$; (b) $df(\partial_u) = 6$. At $(\rho, \phi)^{\mathrm{T}} = (2, \tfrac{\pi}{6})^{\mathrm{T}}$ we have (c) $(f_\rho, f_\phi) = (\sqrt{3}, 2)$; (d) $df(\partial_u) = \sqrt{3} - 2$.]

ᴘV5.1.2 Differential of a function in Cartesian and spherical coordinates (p. 480)

Consider the differential, df, of the function $f : \mathbb{R}^3 \to \mathbb{R}$, $x = (x^1, x^2, x^3)^{\mathrm{T}} \mapsto f(x) = \tfrac{1}{2}(x^1 x^1 + x^2 x^2)$.

(a) Find the coefficients, $f_i(x)$, of the differential written in the form $df = f_i(x) dx^i$.

(b) Evaluate the action of df on the vector $\partial_u = x^1 \partial_1 - x^2 \partial_2 + \partial_3$.

(c) Express df in spherical coordinates, $y = (r, \theta, \phi)^{\mathrm{T}}$, i.e. find the coefficients, $f_i(y)$, of $df = f_i(y) dy^i$.

(d) Evaluate the action of df on the vector $\partial_u = \partial_r - \partial_\theta + \partial_\phi$.

[Check your results: At $(x^1, x^2, x^3)^{\mathrm{T}} = (3, 1, 2)^{\mathrm{T}}$ we have (a) $(f_1, f_2, f_3) = (3, 1, 0)$; (b) $df(\partial_u) = 8$. At $(r, \theta, \phi)^{\mathrm{T}} = (4, \tfrac{\pi}{6}, \tfrac{\pi}{8})^{\mathrm{T}}$ we have (c) $(f_r, f_\theta, f_\phi) = (1, 4, 0)$; (d) $df(\partial_u) = -3$.]

P.V5.2 Pushforward and pullback (p. 481)

ℰV5.2.1 Pushforward and pullback: generalized polar coordinates (p. 486)

Consider the map $F : y \mapsto x(y)$ expressing the Cartesian coordinates, $x = (x^1, x^2)^{\mathsf{T}}$, in $\mathbb{R}^2$ through generalized polar coordinates, $y = (\mu, \phi)^{\mathsf{T}}$, through

$$x^1 = \mu a \cos \phi, \qquad x^2 = \mu b \sin \phi, \qquad (a, b \in \mathbb{R}).$$

(a) Compute the Jacobian matrix of the coordinate transformation F.

(b) Compute the pushforward of the vector fields ∂_μ and ∂_ϕ from generalized polar to Cartesian coordinates.

(c) Compute the pullback of the forms dx^1 and dx^2 from Cartesian to generalized polar coordinates.

(d) Evaluate the action of the form $\lambda = x^2 \, dx^1 - x^1 \, dx^2$ on the vector fields $\mu \partial_\mu$ and ∂_ϕ in two ways: (i) by pushforward of the vectors to Cartesian coordinates; and (ii) by pullback of the form to generalized polar coordinates.

[Check your results: If $a = 2$, $b = 3$, then at $(x^1, x^2)^{\mathsf{T}} = (6, 12)^{\mathsf{T}}$ we have (b) $F_* \partial_\mu = \frac{1}{5}(6 \, \partial_{x^1} + 12 \, \partial_{x^2})$, $F_* \partial_\phi = -8 \, \partial_{x^1} + 9 \, \partial_{x^2}$, and (d) $\lambda(F_* \mu \partial_\mu) = 0$, $\lambda(F_* \partial_\phi) = -150$; and at $(\mu, \phi)^{\mathsf{T}} = (5, \frac{\pi}{3})^{\mathsf{T}}$ we have (c) $F^* dx^1 = d\mu - 5\sqrt{3} \, d\phi$, $F^* dx^2 = \frac{3}{2}\sqrt{3} \, d\mu + \frac{15}{2} \, d\phi$.]

ₚV5.2.2 Pushforward and pullback: hyperbolic coordinates (p. 486)

Consider the map $F : y \mapsto x(y)$ expressing the Cartesian coordinates, $x = (x^1, x^2)^{\mathsf{T}}$, in $\mathbb{R}^2$ in terms of hyperbolic coordinates $y = (\rho, \alpha)^{\mathsf{T}}$, through

$$x^1 = \rho \, e^\alpha, \qquad x^2 = \rho \, e^{-\alpha}.$$

(a) Compute the Jacobian matrix of this coordinate transformation.

(b) Compute the pushforward of the vector fields ∂_ρ and ∂_α from hyperbolic to Cartesian coordinates.

(c) Compute the pullback of the forms dx^1 and dx^2 from Cartesian to hyperbolic coordinates.

(d) Evaluate the action of the form $\lambda = x^1 \, dx^1 + x^2 \, dx^2$ on the vector fields $\rho \partial_\rho$ and ∂_α in two ways: (i) by pushforward of the vector to Cartesian coordinates; (ii) by pullback of the form to hyperbolic coordinates.

[Check your results: At $(x^1, x^2)^{\mathsf{T}} = (1, 4)^{\mathsf{T}}$ we have (b) $F_* \partial_\rho = \frac{1}{2}\partial_{x^1} + 2\partial_{x^2}$, $F_* \partial_\alpha = \partial_{x^1} - 4\partial_{x^2}$; and (d) $\lambda(F_* \rho \partial_\rho) = 17$, $\lambda(F_* \partial_\alpha) = -15$; and at $(\rho, \alpha)^{\mathsf{T}} = (3, \ln 3)^{\mathsf{T}}$ we have (c) $F^* dx^1 = 3 \, d\rho + 9 \, d\alpha$, $F^* dx^2 = \frac{1}{3} \, d\rho - d\alpha$.]

P.V5.3 Forms of higher degree (p. 487)

ℰV5.3.1 Wedge product in Cartesian and polar coordinates (p. 489)

Consider two one-forms, $\lambda, \eta \in \Lambda^1(\mathbb{R}^2)$, defined in Cartesian coordinates, $x = (x^1, x^2)^{\mathsf{T}}$, by

$$\lambda = 2x^1 x^2 \, dx^1 + ((x^1)^2 + (x^2)^2) \, dx^2 \quad \text{and} \quad \eta = e^{x^1 x^2} dx^1 - dx^2.$$

(a) Construct their wedge product, $\lambda \wedge \eta \in \Lambda^2(\mathbb{R}^2)$.

(b)   Compute its action, $(\lambda \wedge \eta)(\partial_u, \partial_v)$, on the pair of vectors $\partial_u = x^1 \partial_1 + x^2 \partial_2$ and $\partial_v = \partial_1 - \partial_2$.

(c)   Express $\lambda \wedge \eta$ in polar coordinates, $y = (\rho, \phi)^T$.

[Check your results: At $(x^1, x^2)^T = (2, 3)^T$ we have (a): $\lambda \wedge \eta = -(12 + 13e^6)$; (b): $(\lambda \wedge \eta)(\partial_u, \partial_v) = 60 + 65e^6$; and at $(\rho, \phi)^T = (2, \frac{\pi}{4})^T$ we have (c): $\lambda \wedge \eta = 8(1 + e^2)\, d\rho \wedge d\phi$.]

ᴘV5.3.2 Wedge product in Cartesian and cylindrical coordinates (p. 489)

Consider two one-forms, $\lambda, \eta \in \Lambda^1(\mathbb{R}^3)$, defined in Cartesian coordinates, $x = (x^1, x^2)^T$, by

$$\lambda = x^3 dx^1 + x^1 dx^2 + x^2 dx^3 \quad \text{and} \quad \eta = x^2 dx^1 + x^3 dx^2 + x^1 dx^3.$$

(a)   Construct their wedge product, $\lambda \wedge \eta \in \Lambda^2(\mathbb{R}^3)$.

(b)   Compute its action, $(\lambda \wedge \eta)(\partial_u, \partial_v)$, on the pair of vectors $\partial_u = x^1 \partial_1 + x^2 \partial_2 + x^3 \partial_3$ and $\partial_v = \partial_1 + \partial_2 + \partial_3$.

(c)   Express $\lambda \wedge \eta$ in cylindrical coordinates, $y = (\rho, \phi, z)^T$.

[Check your results: At $(x^1, x^2, x^3)^T = (1, 2, 3)^T$ we have (a) $\lambda \wedge \eta = 7\, dx^1 \wedge dx^2 - 5\, dx^2 \wedge dx^3 + dx^3 \wedge dx^1$, (b) $(\lambda \wedge \eta)(\partial_u, \partial_v) = 0$; and at $(\rho, \phi, z)^T = (2, \frac{\pi}{4}, 3)^T$ we have (c) $\lambda \wedge \eta = 14\, d\rho \wedge d\phi + (4\sqrt{2} - 12)\, d\phi \wedge dz + 0\, dz \wedge d\rho$.]

ᴇV5.3.3 Stereographic projection of spherical area form (p. 492)

Let the unit sphere S^2 be parametrized by spherical coordinates, $y = (\theta, \phi)^T \in (0, \pi) \times (0, 2\pi)$, with spherical area two-form $\omega = \sin\theta\, d\theta \wedge d\phi$. Let $z \mapsto y(z)$ be the transformation expressing spherical through **stereographic coordinates**, $z = (z^1, z^2)^T \in \mathbb{R}^2$ (see figure on p. 491), with

$$\theta(z) = 2\arctan(1/\rho), \quad \phi(z) = \arctan(z^2/z^1), \quad \text{with} \quad \rho(z) = \sqrt{(z^1)^2 + (z^2)^2}.$$

These relations imply $\sin[\frac{1}{2}\theta(z)] = \frac{1}{\sqrt{1+\rho^2}}$ and $\sin\theta(z) = \frac{2\rho}{1+\rho^2}$. The inverse transformation is

$$z^1(y) = \rho\cos\phi, \quad z^2(y) = \rho\sin\phi, \quad \text{with} \quad \rho(y) = \frac{1}{\tan(\theta/2)}.$$

(a)   Compute the Jacobian matrix, $J(z) = \frac{\partial y}{\partial z}$, its inverse, $J^{-1}(y) = \frac{\partial z}{\partial y}$, and their determinants.

(b)   Show that when the spherical area two-form is expressed through stereographic coordinates as $\omega = \omega(z)\, dz^1 \wedge dz^2$, the weight function has the form $\omega(z) = \frac{-4}{(1+\rho^2)^2}$.

(c)   Compute the action of the two-form ω on the pair of vectors $(\partial_\theta, \partial_\phi)$, using both the y- and z-representations of ω.

$\Big[$Check your results: (a) does JJ^{-1} equal $\mathbb{1}$? If $(z^1, z^2) = (4, 3)^T$ then (b) $\omega(z^1, z^2) = -\frac{1}{169}$, (c) $\omega(\partial_\theta, \partial_\phi) = \frac{5}{13}.\Big]$

ᵨV5.3.4 Mercator projection: spherical area form (p. 492)

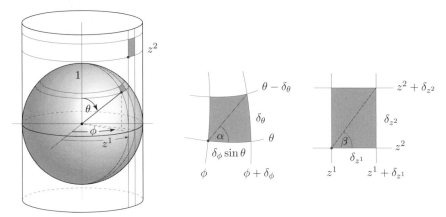

Let the unit sphere, S^2, be parametrized by spherical coordinates, $y = (\theta, \phi)^{\mathrm{T}}$, with spherical area two-form $\omega = \sin\theta\, d\theta \wedge d\phi$. Let $z \mapsto y(z)$ be the transformation expressing spherical through **Mercator coordinates**, $z = (z^1, z^2)^{\mathrm{T}}$:

$$y : (0, 2\pi) \times \mathbb{R} \longrightarrow (0, \pi) \times (0, 2\pi),$$

$$z = \begin{pmatrix} z^1 \\ z^2 \end{pmatrix} \longmapsto y(z) = \begin{pmatrix} \theta(z) \\ \phi(z) \end{pmatrix} = \begin{pmatrix} \pi - 2\arctan(e^{z^2}) \\ z^1 \end{pmatrix}. \tag{1}$$

The inverse map, $y \mapsto z(y)$, is

$$z : (0, \pi) \times (0, 2\pi) \longrightarrow (0, 2\pi) \times \mathbb{R},$$

$$y = \begin{pmatrix} \theta \\ \phi \end{pmatrix} \longmapsto z(y) = \begin{pmatrix} z^1(y) \\ z^2(y) \end{pmatrix} = \begin{pmatrix} \phi \\ \ln[\tan(\tfrac{1}{2}(\pi - \theta))] \end{pmatrix}. \tag{2}$$

The Mercator projection is widely used for two-dimensional maps of the globe. It maps the unit sphere onto a cylinder wrapped around its equator, with $z^1 = \phi$, and $z^2(\theta)$ defined in such a way that *local angles are preserved*. This condition means that in the above figures, showing two infinitesimal area elements in spherical and Mercator coordinates, respectively, $\tan\alpha = \tan\beta$ must hold. (This makes Mercator coordinates well suited for navigation purposes: a line of "constant course" on the sphere, making a fixed angle α with the meridians, is a straight line on a Mercator map, making the same angle with the lines of constant z^2.) The angle-preserving condition implies $-\dfrac{\delta_\theta}{\delta_\phi \sin\theta} = \dfrac{\delta_{z^2}}{\delta_{z^1}}$ (with a minus sign since increasing θ corresponds to decreasing z^2). With $\delta_{z^1} = \delta_\phi$ this leads to

$$\frac{dz^2}{d\theta} = -\frac{1}{\sin\theta}, \tag{3}$$

a differential equation for $z^2(\theta)$, to be solved subject to the boundary condition $z^2(\pi/2) = 0$.

(a)   Verify that Eq. (2) solves Eq. (3).

(b)   Find the values of $z^2(\pi/2)$, $z^2(0)$ and $z^2(\pi)$.

(c)   Explain intuitively why Mercator coordinates distort the size of objects as the latitude increases from the equator to the poles.

(d)   Show that Eq. (1) is equivalent to the relation $\sin\theta = \operatorname{sech}(z^2)$, and that this in turn implies $\cos\theta = \tanh(z^2)$. Show that differentiating the latter w.r.t. θ leads to Eq. (3).

(e) Compute the Jacobian matrix, $J(z) = \frac{\partial y}{\partial z}$, its inverse, $J^{-1}(y) = \frac{\partial z}{\partial y}$, and their determinants.

(f) Show that when the spherical area two-form is expressed through Mercator coordinates as $\omega = \omega(z)\,dz^1 \wedge dz^2$, the weight function has the form $\omega(z) = (\mathrm{sech}(z^2))^2$.

(g) Compute the action of the two-form ω on the pair of vectors $(\partial_\theta, \partial_\phi)$, using both the y- and z-representations of ω.

$\Big[$Check your results: (a) does JJ^{-1} equal $\mathbb{1}$? If $(z^1, z^2) = (\ln 2, \ln 3)^{\mathrm{T}}$ then (f) $\omega(z^1, z^2) = \frac{9}{25}$, (g) $\omega(\partial_\theta, \partial_\phi) = \frac{3}{5}.\Big]$

εV5.3.5 Exterior derivative (p. 494)

Compute the exterior derivative of the following forms:

(a) $f = x^1 x^2 \in \Lambda^0(\mathbb{R}^2)$,

(b) $\lambda = x^2 dx^1 + x^1 x^2 dx^2 \in \Lambda^1(\mathbb{R}^2)$,

(c) $\phi = x^2 x^3 dx^1 \wedge dx^2 + x^3 x^1 dx^2 \wedge dx^3 \in \Lambda^2(\mathbb{R}^3)$.

[Check your results: If $x^1 = 1, x^2 = 2, x^3 = 3$ then (a) $df = 2\,dx^1 + dx^2$, (b) $d\lambda = dx^1 \wedge dx^2$, (c) $d\phi = 5\,dx^1 \wedge dx^2 \wedge dx^3$.]

ₚV5.3.6 Exterior derivative (p. 494)

Compute the exterior derivative of the following forms:

(a) $f = x^1 x^2 + x^2 x^3 \in \Lambda^0(\mathbb{R}^3)$,

(b) $\lambda = x^1 x^2 dx^1 + x^2 x^3 dx^2 + x^1 x^2 dx^3 \in \Lambda^1(\mathbb{R}^3)$,

(c) $\omega = x^1 x^2 x^3 (dx^1 \wedge dx^2 + dx^2 \wedge dx^3 + dx^1 \wedge dx^3) \in \Lambda^2(\mathbb{R}^3)$.

[Check your results: If $x^1 = 1, x^2 = 2, x^3 = 3$ then (a) $df = 2\,dx^1 + 4dx^2 + 2dx^3$, (b) $d\lambda = -dx^1 \wedge dx^2 + 2dx^1 \wedge dx^3 - dx^2 \wedge dx^3$, (c) $d\omega = 5\,dx^1 \wedge dx^2 \wedge dx^3$.]

εV5.3.7 Pullback of polar area form to Cartesian coordinates in $\mathbb{R}^2$ (p. 495)

Consider the map $x \mapsto y(x)$ expressing polar coordinates, $y = (\rho, \phi)^{\mathrm{T}}$, through Cartesian coordinates, $x = (x^1, x^2)^{\mathrm{T}}$ in $\mathbb{R}^2$. The polar area form is given by $\omega = \rho d\rho \wedge d\phi$.

(a) Show that the pullback of ω to Cartesian coordinates is given by $y^*\omega = dx^1 \wedge dx^2$.

(b) The polar area form is exact, since it can also be expressed as $\omega = d\kappa$, with $\kappa = \frac{1}{2}\rho\,d\phi$. Use this fact to compute its pullback via $y^*\omega = y^* d\kappa = dy^*\kappa$, i.e. first compute the pullback $y^*\kappa$, then its exterior derivative $dy^*\kappa$.

ₚV5.3.8 Pullback of spherical area form to Cartesian coordinates in $\mathbb{R}^3$ (solid-angle form) (p. 495)

Consider the map $x \mapsto y(x)$ expressing spherical coordinates, $y = (r, \theta, \phi)^{\mathrm{T}}$, through Cartesian coordinates, $x = (x^1, x^2, x^3)^{\mathrm{T}}$ in $\mathbb{R}^3$. The spherical area form is given by $\omega = \sin\theta d\theta \wedge d\phi$.

(a) Show that the pullback of ω to Cartesian coordinates is given by $y^*\omega = \frac{1}{2}\frac{1}{r^3}\epsilon_{ijk}x^i dx^j \wedge dx^k$.

(b) The spherical area form is exact, since it can also be expressed as $\omega = d\kappa$, with $\kappa = -\cos\theta\,d\phi$. Use this fact to compute its pullback via $y^*\omega = y^* d\kappa = dy^*\kappa$, i.e. first compute the pullback $y^*\kappa$, then its exterior derivative $dy^*\kappa$.

Remark: The spherical area form constructs area elements on a sphere from pairs of tangent vectors. Since such area elements subtend corresponding solid angles in three-dimensional space, the pull-back $y^*\omega$ to $\mathbb{R}^3$ is also called the "**solid-angle form**". Integrating this form over an arbitrary surface in $\mathbb{R}^3$ yields the solid angle subtended by that surface.

εV5.3.9 Pullback of spherical area form from spherical to stereographic coordinates (p. 495)

(a) On the unit sphere, S^2, described using spherical coordinates, $y = (\theta, \phi)^T$, the area form is $\omega = \sin\theta \, d\theta \wedge d\phi$. Show that it is exact, $\omega = d\kappa$, with $\kappa = -\cos\theta \, d\phi$.

(b) Consider the transformation, $z \mapsto y(z)$, expressing spherical coordinates through stereographic coordinates, $(z^1, z^2)^T \in \mathbb{R}^2$, as (see problem V5.3.3)

$$\theta(z) = 2\arctan(\tfrac{1}{\rho}), \quad \phi(z) = \arctan(\tfrac{z^2}{z^1}), \quad \text{with } \rho(z) = \sqrt{(z^1)^2 + (z^2)^2}, \quad \sin\theta(z) = \tfrac{2\rho}{1+\rho^2}.$$

Compute the pullback of ω from spherical to stereographic coordinates, using $y^*\omega = y^* d\kappa = dy^*\kappa$, i.e. first compute the pullback $y^*\kappa$, then its exterior derivative $dy^*\kappa$. Does the result for the weight function, $\omega(z)$, agree with that found in problem V5.3.3?

$\Big[$Check your results: At the point $(z^1, z^2) = (2, 3)$, the components of the pullback form, $y^*\kappa = \kappa_i(z)\, dz^i$, take the values $(\kappa_1, \kappa_2) = \tfrac{6}{91}(2, -3)$ and the weight function equals $\omega(z) = \tfrac{-1}{49}$.$\Big]$

ρV5.3.10 Pullback of spherical area form from spherical to Mercator coordinates (p. 495)

The area form in the spherical coordinates, $y = (\theta, \phi)^T$, is exact, in that it can be written as $\omega = d\kappa$, $\kappa = -\cos\theta \, d\phi$. Consider the transformation $y : z \mapsto y(z)$, expressing spherical through Mercator coordinates, $z = (z^1, z^2)^T$, as

$$\cos(\theta(z^2)) = \tanh(z^2), \qquad \phi(z) = z^1$$

(see problem V5.3.4) Compute the pullback of ω from spherical to Mercator coordinates, using $y^*\omega = y^* d\kappa = dy^*\kappa$. Does the result for the weight function, $\omega(z)$, agree with that found in problem V5.3.4?

P.V5.4 Integration of forms (p. 495)

εV5.4.1 Stereographic coordinates: computing area of sphere (p. 498)

Let $y : \mathbb{R}^2 \to S^2$, $z \mapsto y(z)$, be the map expressing the spherical coordinates, $y = (\theta, \phi)^T$, for the unit sphere through stereographic coordinates, $z = (z^1, z^2)^T \in U = \mathbb{R}^2 \backslash(\{0\} \times \mathbb{R}_0^+)$ (see Problem V5.3.3). Compute the area of the unit sphere by integrating $y^*\omega = |\omega(z)| \, dz^1 \wedge dz^2$, the pullback of the spherical area form to stereographic coordinates, over their domain of definition. The weight function is $\omega(z) = \tfrac{-4}{(1+\rho^2)^2}$, where $\rho = \sqrt{(z^1)^2 + (z^2)^2}$.

ρV5.4.2 Mercator coordinates: computing area of sphere (p. 498)

Let $y : U \to S^2$, $z \mapsto y(z)$, be the map expressing the spherical coordinates, $y = (\theta, \phi)^T$, for the unit sphere through Mercator coordinates, $z = (z^1, z^2)^T \in U = (0, 2\pi) \times \mathbb{R}$ (see Problem V5.3.4). Compute the area of the unit sphere by integrating $y^*\omega = \omega(z) \, dz^1 \wedge dz^2$, the pullback of the spherical area form to Mercator coordinates, over their domain of definition. The weight function is $\omega(z) = (\text{sech}(z^2))^2$.

εV5.4.3 **Pullback of current form from Cartesian to spherical coordinates** (p. 499)

Consider the current form, $j = j_0 \, dx^1 \wedge dx^2 \in \Lambda^2(\mathbb{R}^3)$, describing a uniform current, with current density j_0, flowing in the x^3-direction. Let $x : y \mapsto x(y)$ be a parametrization of a sphere, S_R, of radius R, expressing the Cartesian coordinates $x = (x^1, x^2, x^3)^{\mathsf{T}}$ through the spherical coordinates $y = (\theta, \phi)^{\mathsf{T}}$.

(a) Compute the pullback, x^*j, of the current form onto the sphere S_R.

(b) Use x^*j to compute the current, $I = \int x^*j$, flowing through (i) the upper half-sphere, S_R^+, and (ii) the full sphere, S_R.

(c) Repeat the calculations from (b), this time using a suitable flux integral, $I = \int d\mathbf{S} \cdot \mathbf{j}$, where now the current density is represented by the vector $\mathbf{j} = j_0 \mathbf{e}_z$.

(d) The current form is exact, being expressible as $j = d\lambda$, with $\lambda = x^1 dx^2$. Use Stokes's theorem to compute the current through the upper half-sphere as $I_{S_R^+} = \int_M j = \int_M d\lambda = \int_{\partial M} \lambda$.

ρV5.4.4 **Pullback of current form from Cartesian to polar coordinates** (p. 499)

Consider the current form, $j = j_0 \, dx^1 \wedge dx^2 \in \Lambda^2(\mathbb{R}^3)$, describing a uniform current, with current density j_0, flowing in the x^3-direction. Let $x : y \mapsto x(y)$ be a parametrization of a cone, C, of height h and base radius R, centered on the x^3-axis, expressing the Cartesian coordinates $x = (x^1, x^2, x^3)^{\mathsf{T}}$ through the polar coordinates $y = (\rho, \phi)^{\mathsf{T}}$.

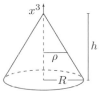

(a) Compute the pullback, x^*j, of the current form onto the cone.

(b) Use x^*j to compute the current, $I_C = \int x^*j$, flowing through the slanted surface of the cone.

(c) Repeat the calculation from (b), this time using a suitable flux integral, $I_C = \int_C d\mathbf{S} \cdot \mathbf{j}$, where now the current density is represented by the vector $\mathbf{j} = j_0 \mathbf{e}_z$.

P.V6   Riemannian differential geometry (p. 502)

P.V6.1   Definition of the metric on a manifold (p. 502)

εV6.1.1 **Standard metric of $\mathbb{R}^3$ in cylindrical coordinates** (p. 504)

Express the standard metric of $\mathbb{R}^3$, $g = \sum_{i=1}^3 dx^i \otimes dx^i$, in terms of cylindrical coordinates.

ρV6.1.2 **Standard metric of $\mathbb{R}^3$ in spherical coordinates** (p. 504)

Express the standard metric of $\mathbb{R}^3$, $g = \sum_{i=1}^3 dx^i \otimes dx^i$, in terms of spherical coordinates.

εV6.1.3 **Standard metric of unit sphere in stereographic coordinates** (p. 504)

On the unit sphere, the standard metric has the form $g = d\theta \otimes d\theta + \sin^2\theta \, d\phi \otimes d\phi$ in spherical coordinates. Express it through stereographic coordinates (defined in Problem V5.3.3) as $g = g_{kl}(z) dz^k \otimes dz^l$. Show that the metric tensor is proportional to the unit tensor, $g_{kl}(z) = s(z)\delta_{kl}$, and find the proportionality factor, $s(z)$. What is its value at points corresponding to the north and south poles of the sphere?

[Check your results: At $(z^1, z^2) = (1, 3)$ we have $s(z) = \frac{4}{121}$.]

Remark: Since the metric tensor $g_{kl}(z)$ is diagonal in stereographic coordinates, they form an orthogonal coordinate system. Moreover, it is proportional to $\mathbb{1}$, hence the coordinate system is "isotropic" – the z^1 and z^2 directions are locally equivalent. This means that the stereographic projection is a "conformal" (angle-preserving) map – angles between tangent vectors with the same base point remain unchanged when mapped from the sphere to the stereographic plane. Moreover, the distance between two nearby points on the plane is proportional to that on the sphere, irrespective of their relative orientation. However, the metric does not define an ortho*normal* system, since the proportionality factor $s(z)$ depends on position. Moreover, it does so in a manner that treats the northern and southern hemispheres asymmetrically.

ₚV6.1.4 Standard metric of unit sphere in Mercator coordinates (p. 504)

Express the standard metric on the unit sphere, $g = d\theta \otimes d\theta + \sin^2\theta \, d\phi \otimes d\phi$, in terms of Mercator coordinates (defined in Problem V5.3.4) as $g = g_{kl}(z)dz^k \otimes dz^l$. Discuss the properties of the metric tensor $g_{kl}(z)$ and explain why they make Mercator coordinates well-suited for navigation purposes.

[Check your results: At $(z^1, z^2) = (\ln 2, \ln 3)$ we have $g_{z^1 z^1} = g_{z^2 z^2} = \frac{9}{25}$.]

P.V6.2   Volume form and Hodge star (p. 505)

ₑV6.2.1 Hodge duals of basis one-forms in spherical coordinates (p. 507)

Consider $\mathbb{R}^3$ parametrized by spherical coordinates, with metric $g_{rr} = 1$, $g_{\theta\theta} = r$, $g_{\phi\phi} = r\sin\theta$. Compute the action of the Hodge star on the basis one-forms dr, $d\theta$ and $d\phi$.

ₚV6.2.2 Hodge duals of all basis forms in spherical coordinates (p. 507)

Consider $\mathbb{R}^3$ parametrized by spherical coordinates, with metric $g_{rr} = 1$, $g_{\theta\theta} = r$, $g_{\phi\phi} = r\sin\theta$. Compute the action of the Hodge star on all the basis forms of $\Lambda(\mathbb{R}^3)$,

$$1, \quad dr, \quad d\theta, \quad d\phi, \quad dr \wedge d\theta, \quad d\theta \wedge d\phi, \quad d\phi \wedge dr, \quad dr \wedge d\theta \wedge d\phi.$$

Verify that for this metric, $**$ acts like the identity map.

P.V7   Differential forms and electrodynamics (p. 518)

P.V7.3   Laws of electrodynamics II: Maxwell equations (p. 525)

ₑV7.3.1 Homogeneous Maxwell equations: form-to-traditional transcription (p. 527)

Use the Hodge star to transcribe the homogeneous Maxwell equations,

(a) $d_s B = 0$, 　　　　　　　　　　　　(b) $d_s E + \frac{1}{c}\partial_t B = 0$,

to differential equations for the vector fields $\underline{E} = J^{-1}E$ and $\underline{B} = J^{-1}*B$. Perform this conversion in two ways: (i) using component-free language, and (ii) using the explicit component representations of the field forms, $E = E_i dx^i$ and $B = \frac{1}{2}B_{jk}dx^j \wedge dx^k$.

Hint: In preparation for (ii), recapitulate why the following relations hold:

$$J^{-1}dx^i = -\underline{e}_i, \qquad J^{-1} * dx^j \wedge dx^k = -\epsilon^{jki}\underline{e}_i, \qquad *dx^i \wedge dx^j \wedge dx^k = -\epsilon^{ijk}.$$

ₚV7.3.2 Inhomogeneous Maxwell equations: form-to-traditional transcription (p. 528)

Use the Hodge star to transcribe the inhomogeneous Maxwell equations,

(a) $d_s D = 4\pi \rho_s,$ (b) $d_s H - \frac{1}{c}\partial_t D = \frac{4\pi}{c}j_s,$

to differential equations relating the vector fields $\underline{H} = -J^{-1}H$ and $\underline{D} = -J^{-1}*D$ to the charge and current densities, $\rho = -*\rho_s$ and $j = -J^{-1}*j_s$. Perform this conversion in two ways: (i) using component-free language, and (ii) using the explicit component representations of the forms involved, $H = H_i dx^i, D = \frac{1}{2}D_{jk}dx^j \wedge dx^k, \rho_s = \rho\,dx^1 \wedge dx^2 \wedge dx^3, j_s = \frac{1}{2}j_{jk}dx^j \wedge dx^k.$

P.V7.4 Invariant formulation (p. 579)

ₑV7.4.1 Field-strength tensor (p. 531)

Expand the field-strength tensor, $F = -dx^0 \wedge E + B$, in components as $F = \frac{1}{2}F_{\mu\nu}dx^\mu \wedge dx^\nu$, and find $\{F_{\mu\nu}\}$ in terms of the components, E^i and B^i, of the vector fields $\underline{E} = J^{-1}E$ and $\underline{B} = J^{-1}*B$.

ₚV7.4.2 Dual field-strength tensor (p. 531)

Expand the dual field-strength tensor, $G = dx^0 \wedge H + D$, in components as $G \equiv \frac{1}{2}G_{\mu\nu}dx^\mu \wedge dx^\nu$, and find $\{G_{\mu\nu}\}$ in terms of the components, H^i and D^i, of the vector fields $\underline{H} = -J^{-1}H$ and $\underline{D} = -J^{-1}*D$.

ₑV7.4.3 Homogeneous Maxwell equations: four-vector notation (p. 532)

(a) Show that the equation $dF = 0$, expressed in terms of the components of $F = \frac{1}{2}F_{\mu\nu}dx^\mu \wedge dx^\nu$, reads as

$$0 = \partial_\alpha F_{\mu\nu} + \partial_\mu F_{\nu\alpha} + \partial_\nu F_{\alpha\mu}. \qquad (1)$$

(b) Show that Eq. (1) is equivalent to the homogeneous Maxwell equations for the vector fields $\underline{E}$ and $\underline{B}$.

(c) Equation (1) can be converted to a four-vector equation using the four-dimensional Hodge star, $J^{-1} * dF = 0$. Show that this leads to

$$\partial_\alpha G^{\mu\nu} = 0, \qquad \text{with} \qquad G^{\alpha\beta} = \frac{1}{2}\epsilon^{\alpha\beta\mu\nu}F_{\mu\nu}, \qquad (2)$$

where $G^{\alpha\beta}$ are the contravariant components of the dual field-strength tensor, defined via $G = -*F$.

(d) Verify explicitly that Eq. (2) leads to Eq. (1).

ₚV7.4.4 Homogeneous Maxwell equations: four-vector notation (p. 532)

(a) Show that the equation $dG = 4\pi j$, expressed through the components of $G = \frac{1}{2}G_{\mu\nu}dx^\mu \wedge dx^\nu$, reads

$$\partial_\alpha G_{\mu\nu} + \partial_\mu G_{\nu\alpha} + \partial_\nu G_{\alpha\mu} = 4\pi j_{\alpha\mu\nu}. \qquad (1)$$

(b)   Show that Eq. (1) is equivalent to the inhomogeneous Maxwell equations relating the vector fields $\underline{D}$ and $\underline{H}$ to the charge and current densities, ρ and $\underline{j}$.

(c)   Equation (1) can be converted to a four-vector equation using $J^{-1} *(\mathrm{d}G - 4\pi j) = 0$. Show that this leads to

$$\partial_\alpha F^{\alpha\beta} = 4\pi j^\beta, \qquad \text{with} \qquad F^{\alpha\beta} = \tfrac{1}{2}\epsilon^{\alpha\beta\mu\nu} G_{\mu\nu}, \tag{2}$$

where $F^{\alpha\beta}$ are the contravariant components of the field-strength tensor, defined via $F = *G$.

(d)   Verify explicitly that Eq. (2) leads to Eq. (1).

S

SOLUTIONS

The fourth and final part of this book contains detailed solutions to all odd-numbered problems and to all case studies presented in the preceding three parts. A password-protected manual containing solutions of all even-numbered problems will be made available to instructors.

Solutions: Linear Algebra

S.L1 Mathematics before numbers (p. 3)

S.L1.1 Sets and maps (p. 3)

ᴇL1.1.1 Composition of maps (p. 7)

(a) Since A maps $\mathbb{Z}$ to $\mathbb{Z}$ and B maps $\mathbb{Z}$ to $\mathbb{N}_0$, it follows that $C = B \circ A$ maps $\mathbb{Z}$ to $\mathbb{N}_0$. The image of n is $C(n) = B(A(n)) = B(n+1) = |n+1|$. To summarize:

$$\boxed{C : \mathbb{Z} \to \mathbb{N}_0, \quad n \mapsto C(n) = |n+1|.}$$

(b) A, B and C are all surjective. A is also injective and bijective. B is not injective, because any positive $n \in \mathbb{N}_0$ is the image of *two* points in $\mathbb{Z}$, $B(n) = B(-n) = n$. Consequently, B is not bijective either. It follows that C, too, is not injective and thus not bijective.

S.L1.2 Groups (p. 7)

ᴇL1.2.1 The group $\mathbb{Z}_2$ (p. 8)

(a) The composition table implies the following properties.

+	0	1
0	0	1
1	1	0

(i) Closure: the result of any possible addition is listed in the table and belongs to the set $\{0, 1\}$. ✓

(i) Associativity:

$$(1 \boldsymbol{+} 0) \boldsymbol{+} 0 = 1 \boldsymbol{+} 0 = 1 \overset{?}{=} 1 \boldsymbol{+} (0 \boldsymbol{+} 0) = 1 \boldsymbol{+} 0 = 1 \checkmark$$
$$(0 \boldsymbol{+} 1) \boldsymbol{+} 0 = 1 \boldsymbol{+} 0 = 1 \overset{?}{=} 0 \boldsymbol{+} (1 \boldsymbol{+} 0) = 0 \boldsymbol{+} 1 = 1 \checkmark$$
$$(1 \boldsymbol{+} 1) \boldsymbol{+} 0 = 0 \boldsymbol{+} 0 = 0 \overset{?}{=} 1 \boldsymbol{+} (1 \boldsymbol{+} 0) = 1 \boldsymbol{+} 1 = 0 \checkmark$$
$$(1 \boldsymbol{+} 0) \boldsymbol{+} 1 = 1 \boldsymbol{+} 1 = 0 \overset{?}{=} 1 \boldsymbol{+} (0 \boldsymbol{+} 1) = 1 \boldsymbol{+} 1 = 0 \checkmark$$
$$(0 \boldsymbol{+} 1) \boldsymbol{+} 1 = 1 \boldsymbol{+} 1 = 0 \overset{?}{=} 0 \boldsymbol{+} (1 \boldsymbol{+} 1) = 0 \boldsymbol{+} 0 = 0 \checkmark$$
$$(0 \boldsymbol{+} 0) \boldsymbol{+} 1 = 0 \boldsymbol{+} 1 = 1 \overset{?}{=} 0 \boldsymbol{+} (0 \boldsymbol{+} 1) = 0 \boldsymbol{+} 1 = 1 \checkmark$$

(ii) The neutral element is 0, since adding it yields no change: $0 \boldsymbol{+} 0 = 0, 0 \boldsymbol{+} 1 = 1$.

(iii) For every element in the group, there is exactly one inverse, since every row of the table contains exactly one 0.

(iv) The group is abelian since the table is symmetric with respect to the diagonal.

(b) The group ($\{+1, -1\}$, ·), with standard multiplication as group operation, is isomorphic to $\mathbb{Z}_2$, since their composition tables have the same
structure if we identify $+1$ with 0 and -1 with 1.

·	+1	−1
+1	+1	−1
−1	−1	+1

ɛL1.2.3 Group of discrete translations in one dimension (p. 8)

(a) Consider the group axioms.

(i) Closure: the integers are closed under usual addition: $m, n \in \mathbb{Z} \Rightarrow m+n \in \mathbb{Z}$. All $x, y \in \mathbb{G}$
are integer multiples of λ, hence there exist integers $n_x, n_y \in \mathbb{Z}$ such that $x = \lambda n_x, y = \lambda n_y$.
It follows that $T(x, y) = x + y = \lambda \cdot n_x + \lambda \cdot n_y = \lambda \cdot (n_x + n_y) \in \lambda \cdot \mathbb{Z} = \mathbb{G}$. ✓

(ii) Associativity: the usual addition rule for real numbers is associative: $a, b, c \in \mathbb{R} \Rightarrow (a + b) + c = a + (b + c)$. For $x, y, z \in \mathbb{G}$ we therefore have $T(T(x, y), z) = T(x + y, z) = (x + y) + z = x + (y + z) = T(x, y + z) = T(x, T(y, z))$. ✓

(iii) Neutral element: the neutral element is $0 = \lambda \cdot 0 \in \mathbb{G}$: For all $x \in \mathbb{G}$ we have: $T(x, 0) = x + 0 = x$. ✓

(iv) Inverse element: the inverse element of $n \in \mathbb{Z}$ is $-n \in \mathbb{Z}$. Thus the inverse of $x = \lambda \cdot n \in \mathbb{G}$
is $-x \equiv \lambda \cdot (-n) \in \mathbb{G}$, since $T(x, -x) = \lambda \cdot n + \lambda \cdot (-n) = \lambda \cdot (n + (-n)) = \lambda \cdot 0 = 0$. ✓

(v) Commutativity (for the group to be abelian): for all $x, y \in \mathbb{G}$ we have $T(x, y) = x + y = y + x = T(y, x)$, since the usual addition of real numbers is commutative. ✓

Since $(\mathbb{G}, T)$ satisfies properties (i)–(v), it is an abelian group. ✓
Remark: For $\lambda = 1$, the group $(\mathbb{G}, T)$ is identical to $(\mathbb{Z}, +)$.

(b) The group axioms of $(\mathbb{T}, +)$ follow directly from those of $(\mathbb{G}, T)$.

(i) Closure: $\mathcal{T}_x, \mathcal{T}_y \in \mathbb{T} \Rightarrow \mathcal{T}_x + \mathcal{T}_y = \mathcal{T}_{T(x,y)} \in \mathbb{T}$, since if $x, y \in \mathbb{G}$, then $T(x, y) \in \mathbb{G}$ [see (a)]. ✓

(ii) Associativity: for $\mathcal{T}_x, \mathcal{T}_y, \mathcal{T}_z \in \mathbb{T}$ we have: $(\mathcal{T}_x + \mathcal{T}_y) + \mathcal{T}_z = \mathcal{T}_{T(x,y)} + \mathcal{T}_z = \mathcal{T}_{T(T(x,y),z)} \overset{(a)}{=} \mathcal{T}_{T(x,T(y,z))} = \mathcal{T}_x + \mathcal{T}_{T(y,z)} = \mathcal{T}_x + (\mathcal{T}_y + \mathcal{T}_z)$. ✓

(iii) Neutral element: the neutral element is $\mathcal{T}_0 \in \mathbb{T}$: For all $\mathcal{T}_x \in \mathbb{T}$ we have: $\mathcal{T}_x + \mathcal{T}_0 = \mathcal{T}_{T(x,0)} = \mathcal{T}_{x+0} = \mathcal{T}_x$. ✓

(iv) Inverse element: the inverse element of $\mathcal{T}_x \in \mathbb{T}$ is $\mathcal{T}_{-x} \in \mathbb{T}$, where $-x$ is the inverse
element of $x \in \mathbb{G}$ with respect to T, since $\mathcal{T}_x + \mathcal{T}_{-x} = \mathcal{T}_{T(x,-x)} = \mathcal{T}_{x+(-x)} = \mathcal{T}_0$. ✓

(v) Commutativity (for the group to be abelian): for all $x, y \in \mathbb{G}$ we have $\mathcal{T}_x + \mathcal{T}_y = \mathcal{T}_{T(x,y)} = \mathcal{T}_{T(y,x)} = \mathcal{T}_y + \mathcal{T}_x$, since the composition rule T in $\mathbb{G}$ is commutative. ✓

Since $(\mathbb{T}, +)$ satisfies properties (i)–(v), it is an abelian group. ✓

ɛL1.2.5 The permutation group S_3 (p. 11)

(a) The entries of the composition table can found by evaluating the image of 123 under P followed
by P'. For example $123 \overset{[213]}{\longmapsto} 213 \overset{[321]}{\longmapsto} 231$, hence $[321] \circ [213] = [231]$.

$P' \circ P$	[123]	[231]	[312]	[213]	[321]	[132]
[123]	[123]	[231]	[312]	[213]	[321]	[132]
[231]	[231]	[312]	[123]	[321]	[132]	[213]
[312]	[312]	[123]	[231]	[132]	[213]	[321]
[213]	[213]	[132]	[321]	[123]	[312]	[231]
[321]	[321]	[213]	[132]	[231]	[123]	[312]
[132]	[132]	[321]	[213]	[312]	[231]	[123]

(b)   The neutral element is the permutation that "does nothing", [123]. Each element has a unique inverse, since every row and column contains the neutral element exactly once.

(c)   The composition table is not symmetric, $P' \circ P \neq P \circ P'$, hence S_3 is *not* an abelian group. For example, $[312] \circ [213] = [132]$, whereas $[213] \circ [312] = [321]$.

S.L1.3 Fields (p. 12)

ɛL1.3.1 Complex numbers – elementary computations (p. 15)

For $z_1 = 12 + 5i$, $z_2 = -3 + 2i$ and $z_3 = a - ib$ $(a, b \in \mathbb{R})$ we find:

(a) $\bar{z}_1 = 12 - 5i$,

(b) $z_1 + z_2 = 12 + (-3) + (5 + 2)i = 9 + 7i$,

(c) $z_1 + \bar{z}_3 = 12 + a + (5 + b)i$,

(d) $z_1 z_2 = 12 \cdot (-3) - 5 \cdot 2 + i[5 \cdot (-3) + 12 \cdot 2] = -46 + 9i$,

(e) $\bar{z}_1 z_3 = 12 \cdot a - (-5) \cdot (-b) + i[(-5) \cdot a + 12 \cdot (-b)] = 12a - 5b - i(5a + 12b)$,

(f) $\dfrac{z_1}{z_2} = \dfrac{12 + 5i}{-3 + 2i} = \dfrac{(12 + 5i)(-3 - 2i)}{(-3 + 2i)(-3 - 2i)} = \dfrac{-36 + 10 + i(-15 - 24)}{9 + 4} = -2 - 3i$,

(g) $|z_1| = \sqrt{z_1 \bar{z}_1} = \sqrt{144 + 25} = 13$,

(h) $|z_1 + z_2| = \sqrt{9^2 + 7^2} = \sqrt{130}$,

(i) $a z_2 + 3 z_3 = (-3a + 2ai) + (3a - 3bi) = i(2a - 3b)$,

   $|a z_2 + 3 z_3| = \sqrt{i(2a - 3b) \cdot (-i)(2a - 3b)} = \sqrt{(2a - 3b)^2} = |2a - 3b|$.

ɛL1.3.3 Algebraic manipulations with complex numbers (p. 15)

(a) $z + \bar{z} = x + iy + x - iy = \boxed{2x} = 2\mathrm{Re}(z)$,

(b) $z - \bar{z} = x + iy - (x - iy) = \boxed{i2y} = i2\mathrm{Im}(z)$,

(c) $z \cdot \bar{z} = (x + iy)(x - iy) = \boxed{x^2 + y^2}$,

(d) $\dfrac{z}{\bar{z}} \overset{(c)}{=} \dfrac{z \cdot z}{\bar{z} \cdot z} = \dfrac{(x + iy)^2}{x^2 + y^2} = \boxed{\dfrac{x^2 - y^2}{x^2 + y^2} + i\dfrac{2xy}{x^2 + y^2}}$,

(e) $\dfrac{1}{z} + \dfrac{1}{\bar{z}} = \dfrac{\bar{z} + z}{z \cdot \bar{z}} \overset{(a),(c)}{=} \boxed{\dfrac{2x}{x^2 + y^2}}$,

(f) $\quad \dfrac{1}{z} - \dfrac{1}{\bar{z}} = \dfrac{\bar{z} - z}{z \cdot \bar{z}} \overset{(b),(c)}{=} \boxed{i\dfrac{(-2y)}{x^2 + y^2}},$

(g) $\quad z^2 + z = (x + iy)^2 + (x + iy) = \boxed{(x^2 - y^2 + x) + i(2xy + y)},$

(h) $\quad z^3 = (x + iy)^3 = (x^3 + 3x^2iy + 3x(iy)^2 + (iy)^3 = \boxed{(x^3 - 3xy^2) + i(3x^2y - y^3)}.$

ɛL1.3.5 Multiplying complex numbers – geometrical interpretation (p. 16)

(a) With $z_j = (\rho_j \cos\phi_j, \rho_j \sin\phi_j)$ and the given trigonometric identities, we have

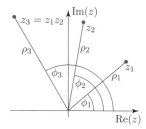

$$\begin{aligned}
z_3 = z_1 z_2 &= \rho_1(\cos\phi_1 + i\sin\phi_1)\rho_2(\cos\phi_2 + i\sin\phi_2) \\
&= \rho_1\rho_2\big[(\cos\phi_1\cos\phi_2 - \sin\phi_1\sin\phi_2) \\
&\quad + i\,(\sin\phi_1\cos\phi_2 + \cos\phi_1\sin\phi_2)\big] \\
&= \rho_1\rho_2\big[\cos(\phi_1 + \phi_2) + i\sin(\phi_1 + \phi_2)\big] \\
&\equiv \rho_3\big[\cos\phi_3 + i\sin\phi_3\big]
\end{aligned}$$

We read off: $\rho_3 = \rho_1\rho_2$, $\phi_3 = (\phi_1 + \phi_2)\,\mathrm{mod}(2\pi)$. ✓

(b) The complex number $z = x + iy$ is represented in the complex plane by the Cartesian coordinates $z \mapsto (x, y)$, or the polar coordinates $\rho = |z| = \sqrt{x^2 + y^2}$, $\phi = \arg(z) = \arctan\big(\frac{y}{x}\big)$. The latter formula determines ϕ only modulo π; to uniquely fix $\phi \in [0, 2\pi)$, we identify the quadrant containing the point (x, y).

$$z_1 = \sqrt{3} + i \mapsto (\sqrt{3}, 1), \qquad \rho_1 = \sqrt{3+1} = 2, \qquad \phi_1 = \arctan\Big(\tfrac{1}{\sqrt{3}}\Big) = \tfrac{\pi}{6},$$

$$z_2 = -2 + 2\sqrt{3}i \mapsto (-2, 2\sqrt{3}), \qquad \rho_2 = \sqrt{12+4} = 4, \qquad \phi_2 = \arctan\Big(\tfrac{-2\sqrt{3}}{2}\Big) = \tfrac{2\pi}{3},$$

$$z_3 = z_1 z_2 = (\sqrt{3} + i)(-2 + 2\sqrt{3}i) \qquad \rho_3 = \sqrt{16\cdot 3 + 16} = 8, \qquad \phi_3 = \arctan\Big(\tfrac{4}{-4\sqrt{3}}\Big) = \tfrac{5\pi}{6},$$

$$= -4\sqrt{3} + 4i \mapsto (-4\sqrt{3}, 4),$$

$$z_4 = \tfrac{1}{z_1} = \tfrac{1}{\sqrt{3}+i} = \tfrac{(\sqrt{3}-i)}{(\sqrt{3}+i)(\sqrt{3}-i)} \qquad \rho_4 = \tfrac{1}{4}\sqrt{3+1} = \tfrac{1}{2}, \qquad \phi_4 = \arctan\Big(\tfrac{-1/4}{\sqrt{3}/4}\Big) = \tfrac{11\pi}{6},$$

$$= \tfrac{\sqrt{3}}{4} - \tfrac{1}{4}i \mapsto (\tfrac{\sqrt{3}}{4}, -\tfrac{1}{4}),$$

$$z_5 = \bar{z}_1 = \sqrt{3} - i \mapsto (\sqrt{3}, -1), \qquad \rho_5 = \sqrt{3+1} = 2, \qquad \phi_5 = \arctan\Big(\tfrac{-1}{\sqrt{3}}\Big) = \tfrac{11\pi}{6}.$$

As expected, we find:

$\rho_3 = \rho_1\rho_2,$

$\phi_3 = \phi_1 + \phi_2,$

$\rho_4 = 1/\rho_1,$

$\phi_4 = -\phi_1\,\mathrm{mod}(2\pi),$

$\rho_5 = \rho_1,$

$\phi_5 = -\phi_1\,\mathrm{mod}(2\pi).$

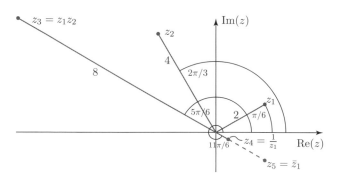

ₑL1.3.7 Field axioms for $\mathbb{F}_4$ (p. 12)

Multiplication table: the first row and column must contain only zeros, since $x \cdot 0 = 0 \cdot x = 0$ for each element x in the field. (Reason: $x \cdot y = x \cdot (y + 0) = x \cdot y + x \cdot 0$, hence $x \cdot 0 = 0$.) The second row and column follow from $1 \cdot x = x$. The remaining four entries must be arranged symmetrically to ensure commutativity. Let us begin by specifying the inverses of a and b, by entering a 1 in each of their columns.

·	0	1	a	b
0	0	0	0	0
1	0	1	a	b
a	0	a		
b	0	b		

Suppose that in each column the 1 appears *on* the diagonal; then the table's symmetry requires the remaining off-diagonal entries to contain the same element x, i.e. $a \cdot b = x = b \cdot a$. This implies the relations

$$a \cdot x = a \cdot (a \cdot b) \overset{\text{assoc.}}{=} (a \cdot a) \cdot b = 1 \cdot b = b \qquad \Leftrightarrow \qquad a \cdot x = b,$$

$$b \cdot x = b \cdot (b \cdot a) \overset{\text{assoc.}}{=} (b \cdot b) \cdot a = 1 \cdot a = a \qquad \Leftrightarrow \qquad b \cdot x = a,$$

·	0	1	a	b
0	0	0	0	0
1	0	1	a	b
a	0	a	1	x
b	0	b	x	1

(contradiction)

which lead to a contradiction, since it is easily verified that no element $x \in \{0, 1, a, b\}$ can satisfy both equations.

Therefore the 1 must appear as off-diagonal entry in each column, $a \cdot b = 1 = b \cdot a$. Finally, the two remaining diagonal entries need to be filled with a or b. If $a \cdot a = a$, then associativity would again lead to a contradiction: $1 = b \cdot a = b \cdot (a \cdot a) \overset{\text{assoc.}}{=} (b \cdot a) \cdot a = 1 \cdot a = a$. Thus $a \cdot a = b$ must hold, and analogously, $b \cdot b = a$.

·	0	1	a	b
0	0	0	0	0
1	0	1	a	b
a	0	a	b	1
b	0	b	1	a

Addition table: to ensure commutativity, the table must be symmetric. Each element has an additive inverse, hence each row and column must contain exactly one 0. Consider the option $a + 1 = 0$. Then distributivity would yield $0 = b \cdot 0 = b \cdot (a + 1) \overset{\text{distr.}}{=} b \cdot a + b \cdot 1 = 1 + b$, hence 1 would have two additive inverses (*a and b*), contradicting the field axioms. Analogously, the option $b + 1 = 0$ would likewise lead to a contradiction. Thus, the only remaining option, $1 + 1 = 0$, must hold. The rest of the addition table follows from analogous arguments.

+	0	1	a	b
0	0	1	a	b
1	1	0	b	a
a	a	b	0	1
b	b	a	1	0

S.L2 Vector spaces (p. 19)

S.L2.3 Vector spaces: examples (p. 25)

ₑL2.3.1 Vector space axioms: rational numbers (p. 28)

(a) First, we show that $(\mathbb{Q}^2, +)$ forms an abelian group.

    (i) Closure holds by definition. ✓

    (ii) Associativity:

$$\left[\begin{pmatrix} x^1 \\ x^2 \end{pmatrix} + \begin{pmatrix} y^1 \\ y^2 \end{pmatrix} \right] + \begin{pmatrix} z^1 \\ z^2 \end{pmatrix} = \begin{pmatrix} x^1 + y^1 \\ x^2 + y^2 \end{pmatrix} + \begin{pmatrix} z^1 \\ z^2 \end{pmatrix}$$

$$= \begin{pmatrix} x^1 + y^1 + z^1 \\ x^2 + y^2 + z^2 \end{pmatrix} = \begin{pmatrix} x^1 \\ x^2 \end{pmatrix} + \begin{pmatrix} y^1 + z^1 \\ y^2 + z^2 \end{pmatrix}$$

$$= \begin{pmatrix} x^1 \\ x^2 \end{pmatrix} + \left[\begin{pmatrix} y^1 \\ y^2 \end{pmatrix} + \begin{pmatrix} z^1 \\ z^2 \end{pmatrix} \right]. \checkmark$$

(iii) Neutral element: $\begin{pmatrix} 0 \\ 0 \end{pmatrix}$ is the neutral element. ✓

(iv) Additive inverse: $\begin{pmatrix} -x^1 \\ -x^2 \end{pmatrix} \in \mathbb{Q}^2$ is the additive inverse of $\begin{pmatrix} x^1 \\ x^2 \end{pmatrix} \in \mathbb{Q}^2$. ✓

(v) Commutativity: follows (component-wise) from the commutativity of $\mathbb{Q}$. ✓

Second, we show that scalar multiplication, $\cdot$, likewise has the properties required for $(\mathbb{Q}^2, +, \cdot)$ to form a vector space. Since the product of two rational numbers is always rational, $\left(\dfrac{p_1}{q_1} \cdot \dfrac{p_2}{q_2} = \dfrac{(p_1 p_2)}{(q_1 q_2)} \right)$, closure holds by definition. Moreover:

(vi) multiplication of a sum of scalars and a vector is distributive:

$$(\lambda + \mu) \cdot \begin{pmatrix} x^1 \\ x^2 \end{pmatrix} = \begin{pmatrix} (\lambda + \mu)x^1 \\ (\lambda + \mu)x^2 \end{pmatrix} = \begin{pmatrix} \lambda x^1 + \mu x^1 \\ \lambda x^2 + \mu x^2 \end{pmatrix} = \lambda \begin{pmatrix} x^1 \\ x^2 \end{pmatrix} + \mu \begin{pmatrix} x^1 \\ x^2 \end{pmatrix}; \; ✓$$

(vii) multiplication of a scalar and a sum of vectors is distributive:

$$\lambda \cdot \left[\begin{pmatrix} x^1 \\ x^2 \end{pmatrix} + \begin{pmatrix} y^1 \\ y^2 \end{pmatrix} \right] = \begin{pmatrix} \lambda x^1 + \lambda y^1 \\ \lambda x^2 + \lambda y^2 \end{pmatrix} = \lambda \begin{pmatrix} x^1 \\ x^2 \end{pmatrix} + \lambda \begin{pmatrix} y^1 \\ y^2 \end{pmatrix}; \; ✓$$

(viii) multiplication of a product of scalars and a vector is associative:

$$(\lambda \mu) \cdot \begin{pmatrix} x^1 \\ x^2 \end{pmatrix} = \begin{pmatrix} \lambda \mu x^1 \\ \lambda \mu x^2 \end{pmatrix} = \lambda \left[\mu \cdot \begin{pmatrix} x^1 \\ x^2 \end{pmatrix} \right]; \; ✓$$

(ix) neutral element: $1 \cdot \begin{pmatrix} x^1 \\ x^2 \end{pmatrix} = \begin{pmatrix} x^1 \\ x^2 \end{pmatrix}$. ✓

Therefore, the triple $(\mathbb{Q}^2, +, \cdot)$ represents a $\mathbb{Q}$-vector space.

(b) The set of integers $\mathbb{Z}$ does not form a field, since a multiplicative inverse $a^{-1} \in \mathbb{Z}$ does not exist for each $a \in \mathbb{Z} \setminus \{0\}$ (e.g. the equation $2 \cdot a = 1$ has no solution within the integers). Hence, it is also *not* possible to construct any vector space over the integers.

εL2.3.3 Vector space of real functions (p. 26)

We have to verify that all the axioms for a vector space are satisfied. First, $(F, +)$ indeed has all the properties of an abelian group.

(i) Closure holds by definition: adding two functions from F again yields a function in F. ✓

(ii,v) Associativity and commutativity follow trivially from the corresponding properties of $\mathbb{R}$. For example associativity:

$$\left[f + [g + h] \right](x) = f(x) + [g + h](x) = f(x) + \left(g(x) + h(x) \right)$$
$$= \left(f(x) + g(x) \right) + h(x) = [f + g](x) + h(x) = \left[[f + g] + h \right](x). \; ✓$$

(iii) The neutral element is the null function, defined by $f_{\text{null}} : x \mapsto f_{\text{null}}(x) \equiv 0$, since $f + f_{\text{null}} : x \mapsto f(x) + f_{\text{null}}(x) = f(x) + 0 = f(x)$. ✓

(iv) The additive inverse of f is $-f$, defined by $-f : x \mapsto (-f)(x) \equiv -f(x)$, since $f + (-f) : x \mapsto f(x) + (-f(x)) = 0$. ✓

Moreover, multiplication of any function with a scalar also has all the properties required for $(F, +, \cdot)$ to be a vector space. Closure holds per definition. Furthermore:

(vi) multiplication of a sum of scalars and a function is distributive:

$$[(\gamma + \lambda) \cdot f](x) = (\gamma + \lambda)f(x) = \gamma f(x) + \lambda f(x) = [\gamma \cdot f](x) + [\lambda \cdot f](x)$$
$$= [\gamma \cdot f + \lambda \cdot f](x) \, ; \, \checkmark$$

(vii) multiplication of a scalar and a sum of functions is distributive:

$$[\lambda \cdot (f + g)](x) = \lambda \Big([f + g](x)\Big) = \lambda \Big(f(x) + g(x)\Big) = \lambda f(x) + \lambda g(x)$$
$$= [\lambda \cdot f](x) + [\lambda \cdot g](x) = [\lambda \cdot f + \lambda \cdot g](x) \, ; \, \checkmark$$

(viii) multiplication of a product of scalars and a function is associative:

$$[(\gamma \lambda) \cdot f](x) = (\gamma \lambda)f(x) = \gamma \Big(\lambda f(x)\Big) = \gamma [\lambda \cdot f](x) = [\gamma \cdot (\lambda \cdot f)](x) \, ; \, \checkmark$$

(ix) neutral element: $[1 \cdot f](x) = 1f(x) = f(x). \, \checkmark$

Therefore, the triple $(F, +, \cdot)$ is an $\mathbb{R}$-vector space.

εL2.3.5 Vector space with unusual composition rule – addition (p. 28)

First, we show that $(V_a, +)$ forms an abelian group.

(i)   Closure holds by definition. $\checkmark$

(ii)   Associativity: $(\mathbf{v}_x + \mathbf{v}_y) + \mathbf{v}_z = \mathbf{v}_{x+y+a} + \mathbf{v}_z = \mathbf{v}_{(x+y+a)+z+a} = \mathbf{v}_{x+y+z+2a}$
$$= \mathbf{v}_{x+(y+z+a)+a} = \mathbf{v}_x + \mathbf{v}_{y+z+a} = \mathbf{v}_x + (\mathbf{v}_y + \mathbf{v}_z). \, \checkmark$$

(iii) Neutral element: $\mathbf{v}_x + \mathbf{v}_{-a} = \mathbf{v}_{x+(-a)+a} = \mathbf{v}_x , \quad \Rightarrow \quad \mathbf{0} = \mathbf{v}_{-a}. \, \checkmark$

(iv)   Additive inverse: $\mathbf{v}_x + \mathbf{v}_{-x-2a} = \mathbf{v}_{x+(-x-2a)+a} = \mathbf{v}_{-a} = \mathbf{0}, \, \Rightarrow \, -\mathbf{v}_x = \mathbf{v}_{-x-2a}. \, \checkmark$

(v)   Commutativity: $\mathbf{v}_x + \mathbf{v}_y = \mathbf{v}_{x+y+a} = \mathbf{v}_{y+x+a} = \mathbf{v}_y + \mathbf{v}_x. \, \checkmark$

Second, we show that scalar multiplication, $\cdot$, likewise has the properties required for $(V_a, +, \cdot)$ to form a vector space. Closure holds by definition. Moreover:

(vi) multiplication of a sum of scalars and a vector is distributive:

$$(\gamma + \lambda) \cdot \mathbf{v}_x = \mathbf{v}_{(\gamma+\lambda)x + a(\gamma+\lambda-1)} = \mathbf{v}_{\gamma x + a(\gamma-1)+\lambda x + a(\lambda-1)+a}$$
$$= \mathbf{v}_{\gamma x + a(\gamma-1)} + \mathbf{v}_{\lambda x + a(\lambda-1)} = \gamma \cdot \mathbf{v}_x + \lambda \cdot \mathbf{v}_x \, ; \, \checkmark$$

(vii) multiplication of a scalar and a sum of vectors is distributive:

$$\lambda \cdot (\mathbf{v}_x + \mathbf{v}_y) = \lambda \cdot \mathbf{v}_{x+y+a} = \mathbf{v}_{\lambda(x+y+a)+a(\lambda-1)} = \mathbf{v}_{\lambda x + a(\lambda-1)+\lambda y + a(\lambda-1)+a}$$
$$= \mathbf{v}_{\lambda x + a(\lambda-1)} + \mathbf{v}_{\lambda y + a(\lambda-1)} = \lambda \cdot \mathbf{v}_x + \lambda \cdot \mathbf{v}_y \, ; \, \checkmark$$

(viii) multiplication of a product of scalars and a vector is associative:

$$(\gamma \lambda) \cdot \mathbf{v}_x = \mathbf{v}_{(\gamma\lambda)x + a(\gamma\lambda-1)} = \mathbf{v}_{\gamma(\lambda x + a(\lambda-1))+a(\gamma-1)} = \gamma \cdot \mathbf{v}_{\lambda x + a(\lambda-1)} = \gamma \cdot (\lambda \cdot \mathbf{v}_x) \, ; \, \checkmark$$

(ix)   neutral element: $1 \cdot \mathbf{v}_x = \mathbf{v}_{x+a(1-1)} = \mathbf{v}_x$. ✓

Therefore, the triple $(V_a, +, \cdot)$ represents an $\mathbb{R}$-vector space.

S.L2.4 Basis and dimension (p. 28)

ɛL2.4.1 Linear independence (p. 30)

(a)   The three vectors are linearly independent if and only if the only solution to the equation

$$\mathbf{0} = a^1 \mathbf{v}_1 + a^2 \mathbf{v}_2 + a^3 \mathbf{v}_3 = a^1 \begin{pmatrix} 0 \\ 1 \\ 2 \end{pmatrix} + a^2 \begin{pmatrix} 1 \\ -1 \\ 1 \end{pmatrix} + a^3 \begin{pmatrix} 2 \\ -1 \\ 4 \end{pmatrix}, \quad \text{with} \quad a^j \in \mathbb{R}, \qquad (1)$$

is the trivial one, $a^1 = a^2 = a^3 = 0$. The vector equation (1) yields a system of three equations, (i)–(iii), one for each of the three components of (1), which we solve as follows:

(i) $0a^1 + 1a^2 + 2a^3 = 0$ $\overset{(i)}{\Rightarrow}$ (iv) $\boxed{a^2 = -2a^3}$;

(ii) $1a^1 - 1a^2 - 1a^3 = 0$ $\overset{\text{(iv) in (ii)}}{\Rightarrow}$ (v) $\boxed{a^1 = -a^3}$;

(iii) $2a^1 + 1a^2 + 4a^3 = 0$ $\overset{\text{(iv,v) in (iii)}}{\Rightarrow}$ (vi) $0 = 0$.

(i) yields (iv): $a^2 = -2a^3$. (iv) inserted into (ii) yields (v): $a^1 = -a^3$. Inserting (iv) and (v) into (iii) yields no new information. There are thus infinitely many nontrivial solutions (one for every value of $a^3 \in \mathbb{R}$), hence $\mathbf{v}_1$, $\mathbf{v}_2$ and $\mathbf{v}_3$ are $\boxed{\text{not}}$ linearly independent.

(b)   The desired vector $\mathbf{v}_2' = (x, y, z)^{\mathsf{T}}$ should be linearly independent from $\mathbf{v}_1$ and $\mathbf{v}_3$, i.e. its components x, y and z should be chosen such that the equation $\mathbf{0} = a^1 \mathbf{v}_1 + a^2 \mathbf{v}_2' + a^3 \mathbf{v}_3$ has no nontrivial solution, i.e. that it implies $a^1 = a^2 = a^3 = 0$.

(i) $0a^1 + xa^2 + 2a^3 = 0$ $\overset{(i)}{\Rightarrow}$ (iv)   choose $\boxed{x = 0}$, then $a^3 = 0$.

(ii) $1a^1 + ya^2 - 1a^3 = 0$ $\overset{\text{(iv) in (ii)}}{\Rightarrow}$ (v)   choose $\boxed{y = 0}$, then $a^1 = 0$.

(iii) $2a^1 + za^2 + 4a^3 = 0$ $\overset{\text{(iv),(v) in (iii)}}{\Rightarrow}$ (vi)   choose $\boxed{z = 1}$, then $a^2 = 0$.

(i) yields (iv): $2a^3 = -xa^2$; to enforce $a^3 = 0$ we choose $x = 0$. (iv) inserted into (ii) yields (v): $a^1 = -ya^2$; to enforce $a^1 = 0$ we choose $y = 0$. (iv), (v) inserted into (iii) yields $za^2 = 0$; to enforce $a^2 = 0$ we choose $z = 1$. Thus $\boxed{\mathbf{v}_2' = (0, 0, 1)^{\mathsf{T}}}$ is a choice for which $\mathbf{v}_1$, $\mathbf{v}_2'$ are $\mathbf{v}_3$ linearly independent. This choice is not unique – there are infinitely many alternatives; one of them, e.g. is $\mathbf{v}_2' = (0, 1, 0)^{\mathsf{T}}$.

ɛL2.4.3 Einstein summation convention (p. 32)

(a) $a_i b^i = b^j a_j$ is $\boxed{\text{true}}$, since i and j are dummy variables which are summed over, hence we may rename as we please:

$$a_i b^i = \sum_{i=1}^{2} a_i b^i = a_1 b^1 + a_2 b^2 = b^1 a_1 + b^2 a_2 = \sum_{j=1}^{2} b^j a_j = b^j a_j. ✓$$

(b) $a_i \delta^i{}_j b^j = a_k b^k$ is $\boxed{\text{true}}$, since $\delta^i{}_j$ is nonzero only for $i = j$, in which case it equals 1:

$$a_i \delta^i{}_j b^j = a_1 \underbrace{(\delta^1{}_1)}_{=1} b^1 + a_1 \underbrace{(\delta^1{}_2)}_{=0} b^2 + a_2 \underbrace{(\delta^2{}_1)}_{=0} b^1 + a_2 \underbrace{(\delta^2{}_2)}_{=1} b^2 = a_1 b^1 + a_2 b^2 = a_k b^k. ✓$$

(c) $a_i b^j a_j b^k \overset{?}{=} a_k b^l a_l b^i$ is $\boxed{\text{false}}$, since the indices i and k are *not* repeated, i.e. they are not summed over and hence may not be renamed. For example, for $i = 1$ and $k = 2$ the left-hand side, $a_1(b^1 a_1 + b^2 a_2)b^2$, clearly differs from the right-hand side, $a_2(b^1 a_1 + b^2 a_2)b^1$.

(d) $a_1 a_i b^1 b^i + b^2 a_j a_2 b^j = (a_i b^i)^2$ is $\boxed{\text{true}}$, since multiplication is associative and commutative and we may rename dummy indices as we please:

$$a_1 a_i b^1 b^i + b^2 a_j a_2 b^j = a_1 b^1 a_i b^i + a_2 b^2 a_i b^i = (a_1 b^1 + a_2 b^2)(a_i b^i) = (a_j b^j)(a_i b^i) = (a_i b^i)^2. \checkmark$$

In practice, the arguments illustrated above need not be written out explicitly. Relations such as (a), (b) and (d) may be simply written down without further discussion.

S.L3   Euclidean geometry (p. 38)

S.L3.2   Normalization and orthogonality (p. 11)

εL3.2.1 Angle, orthogonal decomposition (p. 42)

(a) $\cos(\angle(\mathbf{a}, \mathbf{b})) = \dfrac{\mathbf{a} \cdot \mathbf{b}}{\|\mathbf{a}\|\|\mathbf{b}\|} = \dfrac{3 \cdot 7 + 4 \cdot 1}{\sqrt{9 + 16} \cdot \sqrt{49 + 1}} = \dfrac{1}{\sqrt{2}} \quad \Rightarrow \quad \angle(\mathbf{a}, \mathbf{b}) = \boxed{\dfrac{\pi}{4}}$.

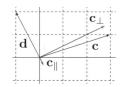

(b) $\mathbf{c}_{\parallel} = \dfrac{(\mathbf{c} \cdot \mathbf{d})\mathbf{d}}{\|\mathbf{d}\|^2} = \dfrac{3 \cdot (-1) + 1 \cdot 2}{1 + 4} \begin{pmatrix} -1 \\ 2 \end{pmatrix} = \boxed{\dfrac{1}{5} \begin{pmatrix} 1 \\ -2 \end{pmatrix}}$,

$\mathbf{c}_{\perp} = \mathbf{c} - \mathbf{c}_{\parallel} = \begin{pmatrix} 3 \\ 1 \end{pmatrix} - \dfrac{1}{5} \begin{pmatrix} 1 \\ -2 \end{pmatrix} = \boxed{\dfrac{7}{5} \begin{pmatrix} 2 \\ 1 \end{pmatrix}}$.

Consistency check: $\mathbf{c}_{\perp} \cdot \mathbf{c}_{\parallel} = \dfrac{1}{25}(1 \cdot 14 - 2 \cdot 7) = 0$. $\checkmark$

S.L3.3   Inner product spaces (p. 46)

εL3.3.1 Inner product for vector space of continuous functions (p. 47)

(a) All the defining properties of an inner product are satisfied:

(i) Symmetric: $\langle f, g \rangle = \displaystyle\int_I \mathrm{d}x f(x) g(x) = \int_I \mathrm{d}x\, g(x) f(x) = \langle g, f \rangle . \checkmark$

(ii,iii) Linear: $\langle \lambda \cdot f + g, h \rangle = \displaystyle\int_I \mathrm{d}x\, (\lambda f(x) + g(x)) h(x)$

$$= \lambda \int_I \mathrm{d}x f(x) h(x) + \int_I \mathrm{d}x\, g(x) h(x) = \lambda \langle f, h \rangle + \langle g, h \rangle . \checkmark$$

(iv) Positive semi-definite: $\langle f, f \rangle = \displaystyle\int_I \mathrm{d}x f^2(x) \geq 0$.

Since the integrand is everywhere ≥ 0, the integral is also ≥ 0. $\checkmark$ Moreover, since f is continuous, the integral can equal 0 if and only if the integrand f^2 vanishes everywhere. Therefore $f(x) = 0$, i.e. f is the zero function. $\checkmark$

Optional: mathematical justification for the last statement: Suppose $f \neq 0$, then there exists an $x_0 \in I$ such that $(f(x_0))^2 \neq 0$. Since f is continuous, $(f(x))^2$ is non-zero in some neighbourhood of x_0, i.e., there exists a $\delta > 0$, such that for all $|x_0 - x| < \delta$, $|(f(x))^2| > \frac{1}{2}|(f(x_0))^2|$. Thus the integral must be larger than zero; e.g. we can find a lower bound as follows: $\langle f, f \rangle = \int_I dx \, (f(x))^2 \geq \int_{x_0-\delta}^{x_0+\delta} dx \, (f(x))^2 > \int_{x_0-\delta}^{x_0+\delta} dx \, \frac{1}{2}(f(x_0))^2 = \delta f(x_0)^2 > 0$. Furthermore, for $f(x) \equiv 0$, we have $\langle f, f \rangle = \int_I dx \, (f(x))^2 = \int_I dx \, 0 = 0$. ✓

(b)
$$\langle f_1, f_2 \rangle = \int_{-1}^{1} dx f_1(x) f_2(x) = \int_{-1}^{1} dx \, \sin\left(\frac{x}{\pi}\right) \cos\left(\frac{x}{\pi}\right) = \boxed{0},$$

because the integrand is antisymmetric. Thus the two functions are 'orthogonal' to each other. Explicitly, the substitution $u = \sin(x/\pi)$, $du = dx \cos(x/\pi)/\pi$ gives:

$$\int_{-1}^{1} dx \, \sin\left(\frac{x}{\pi}\right) \cos\left(\frac{x}{\pi}\right) = \pi \int_{-\sin\left(\frac{1}{\pi}\right)}^{\sin\left(\frac{1}{\pi}\right)} du \, u = \frac{\pi u^2}{2} \Bigg|_{-\sin\left(\frac{1}{\pi}\right)}^{\sin\left(\frac{1}{\pi}\right)} = 0 .$$

ɛL3.3.3 Projection onto an orthonormal basis (p. 51)

(a)
$$\langle \mathbf{e}_1', \mathbf{e}_1' \rangle = \frac{1}{2}\big[1 \cdot 1 + 1 \cdot 1\big] = 1, \qquad\qquad \langle \mathbf{e}_1', \mathbf{e}_2' \rangle = \frac{1}{2}\big[1 \cdot 1 + (-1) \cdot 1\big] = 0.$$
$$\langle \mathbf{e}_2', \mathbf{e}_2' \rangle = \frac{1}{2}\big[1 \cdot 1 + (-1) \cdot (-1)\big] = 1.$$

The two vectors are normalized and orthogonal to each other, $\boxed{\langle \mathbf{e}_i', \mathbf{e}_j' \rangle = \delta_{ij}}$, therefore they form an orthonormal basis of $\mathbb{R}^2$. ✓

(b) Since the vectors $\{\mathbf{e}_1', \mathbf{e}_2'\}$ form an orthonormal basis, the component w^i of the vector $\mathbf{w} = (-2, 3)^{\mathrm{T}} = \mathbf{e}_i' w^i$ with respect to this basis is given by the projection $w^i = \langle \mathbf{e}'^i, \mathbf{w} \rangle$ (with $\mathbf{e}'^i = \mathbf{e}_i'$):

$$w^1 = \langle \mathbf{e}'^1, \mathbf{w} \rangle = \frac{1}{\sqrt{2}}\big[1 \cdot (-2) + 1 \cdot 3\big] = \boxed{\frac{1}{\sqrt{2}}},$$

$$w^2 = \langle \mathbf{e}'^2, \mathbf{w} \rangle = \frac{1}{\sqrt{2}}\big[1 \cdot (-2) - 1 \cdot 3\big] = \boxed{-\frac{5}{\sqrt{2}}}.$$

ɛL3.3.5 Gram–Schmidt orthonormalization (p. 46)

Strategy: iterative orthogonalization and normalization, starting from $\mathbf{v}_{1,\perp} = \mathbf{v}_1$.

Starting vector:	$\mathbf{v}_{1,\perp} = \mathbf{v}_1 = (1, -2, 1)^{\mathrm{T}}$.
Normalizing $\mathbf{v}_{1,\perp}$.	$\mathbf{e}_1' = \dfrac{\mathbf{v}_{1,\perp}}{\|\mathbf{v}_{1,\perp}\|} = \boxed{\dfrac{1}{\sqrt{6}}(1, -2, 1)^{\mathrm{T}}} = \mathbf{e}'^1$.
Orthogonalizing $\mathbf{v}_2$:	$\mathbf{v}_{2,\perp} = \mathbf{v}_2 - \mathbf{e}_1' \langle \mathbf{e}'^1, \mathbf{v}_2 \rangle = (1, 1, 1)^{\mathrm{T}} - \mathbf{e}_1'(0)$.
Normalizing $\mathbf{v}_{2,\perp}$:	$\mathbf{e}_2' = \dfrac{\mathbf{v}_{2,\perp}}{\|\mathbf{v}_{2,\perp}\|} = \boxed{\dfrac{1}{\sqrt{3}}(1, 1, 1)^{\mathrm{T}}} = \mathbf{e}'^2$.
Orthogonalizing $\mathbf{v}_3$:	$\mathbf{v}_{3,\perp} = \mathbf{v}_3 - \mathbf{e}_1' \langle \mathbf{e}'^1, \mathbf{v}_3 \rangle - \mathbf{e}_2' \langle \mathbf{e}'^2, \mathbf{v}_3 \rangle$
	$= (0, 1, 2)^{\mathrm{T}} - \mathbf{e}_1'(0) - \dfrac{1}{\sqrt{3}}(1, 1, 1)^{\mathrm{T}}\left(3\dfrac{1}{\sqrt{3}}\right) = (-1, 0, 1)^{\mathrm{T}}$.
Normalizing $\mathbf{v}_{3,\perp}$:	$\mathbf{e}_3' = \dfrac{\mathbf{v}_{3,\perp}}{\|\mathbf{v}_{3,\perp}\|} = \boxed{\dfrac{1}{\sqrt{2}}(-1, 0, 1)^{\mathrm{T}}} = \mathbf{e}'^3$.

ₑL3.3.7 Non-orthonormal basis vectors and metric (p. 50)

(a) $\hat{\mathbf{v}}_1 = \begin{pmatrix} 2 \\ 0 \end{pmatrix}$, $\hat{\mathbf{v}}_2 = \begin{pmatrix} 1 \\ 1 \end{pmatrix}$; $\Rightarrow$ $\hat{\mathbf{e}}_1 = \begin{pmatrix} 1 \\ 0 \end{pmatrix} = \boxed{\tfrac{1}{2}\hat{\mathbf{v}}_1}$, $\hat{\mathbf{e}}_2 = \begin{pmatrix} 0 \\ 1 \end{pmatrix} = \boxed{-\tfrac{1}{2}\hat{\mathbf{v}}_1 + \hat{\mathbf{v}}_2}$.

The vectors $\hat{\mathbf{v}}_1$ and $\hat{\mathbf{v}}_2$ form a basis, because both standard basis vectors $\hat{\mathbf{e}}_1$ and $\hat{\mathbf{e}}_2$ can be written in terms of them.

(b)   A representation of the vectors $\hat{\mathbf{x}}$ and $\hat{\mathbf{y}}$ as column vectors in the standard basis of $\mathbb{R}^2$ can be found as follows:

$$\hat{\mathbf{x}} = \hat{\mathbf{v}}_1 x^1 + \hat{\mathbf{v}}_2 x^2, \quad x^1 = 3, \ x^2 = -4 \quad \Rightarrow \quad \hat{\mathbf{x}} = \begin{pmatrix} 2 \\ 0 \end{pmatrix} 3 + \begin{pmatrix} 1 \\ 1 \end{pmatrix}(-4) = \boxed{\begin{pmatrix} 2 \\ -4 \end{pmatrix}}.$$

$$\hat{\mathbf{y}} = \hat{\mathbf{v}}_1 y^1 + \hat{\mathbf{v}}_2 y^2, \quad y^1 = -1, \ y^2 = 3 \quad \Rightarrow \quad \hat{\mathbf{y}} = \begin{pmatrix} 2 \\ 0 \end{pmatrix}(-1) + \begin{pmatrix} 1 \\ 1 \end{pmatrix} 3 = \boxed{\begin{pmatrix} 1 \\ 3 \end{pmatrix}}.$$

Scalar product: $\langle \hat{\mathbf{x}}, \hat{\mathbf{y}} \rangle_{\mathbb{R}^2} = \begin{pmatrix} 2 \\ -4 \end{pmatrix} \begin{pmatrix} 1 \\ 3 \end{pmatrix} = 2 \cdot 1 + (-4) \cdot 3 = \boxed{10}$.

(c) $g_{11} = \langle \hat{\mathbf{v}}_1, \hat{\mathbf{v}}_1 \rangle_{\mathbb{R}^2} = \boxed{4}$, $g_{12} = \langle \hat{\mathbf{v}}_1, \hat{\mathbf{v}}_2 \rangle_{\mathbb{R}^2} = \boxed{2}$,

                $g_{21} = \langle \hat{\mathbf{v}}_2, \hat{\mathbf{v}}_1 \rangle_{\mathbb{R}^2} = \boxed{2}$, $g_{22} = \langle \hat{\mathbf{v}}_2, \hat{\mathbf{v}}_2 \rangle_{\mathbb{R}^2} = \boxed{2}$.

(d) $\langle \hat{\mathbf{x}}, \hat{\mathbf{y}} \rangle_{\mathbb{R}^2} = \langle \mathbf{x}, \mathbf{y} \rangle_g = x^i g_{ij} y^j$

$$= 3 \cdot 4 \cdot (-1) + 3 \cdot 2 \cdot 3 + (-4) \cdot 2 \cdot (-1) + (-4) \cdot 2 \cdot 3 = \boxed{-10}. \ \checkmark \ [= \text{(b)}]$$

S.L4   Vector product (p. 55)

S.L4.2   Algebraic formulation (p. 58)

ₑL4.2.1 Elementary computations with vectors (p. 59)

Using $\mathbf{a} = (4, 3, 1)^T$ and $\mathbf{b} = (1, -1, 1)^T$, we obtain:

(a) $\|\mathbf{b}\| = \sqrt{1 + 1 + 1} = \boxed{\sqrt{3}}$,

            $\mathbf{a} - \mathbf{b} = (4 - 1, 3 - (-1), 1 - 1)^T = \boxed{(3, 4, 0)^T}$,

            $\mathbf{a} \cdot \mathbf{b} = 4 \cdot 1 + 3 \cdot (-1) + 1 \cdot 1 = \boxed{2}$,

$$\mathbf{a} \times \mathbf{b} = \begin{pmatrix} 4 \\ 3 \\ 1 \end{pmatrix} \times \begin{pmatrix} 1 \\ -1 \\ 1 \end{pmatrix} = \begin{pmatrix} 3 \cdot 1 - 1 \cdot (-1) \\ 1 \cdot 1 - 4 \cdot 1 \\ 4 \cdot (-1) - 3 \cdot 1 \end{pmatrix} = \boxed{\begin{pmatrix} 4 \\ -3 \\ -7 \end{pmatrix}}.$$

(b) $\mathbf{a}_\| = \dfrac{\mathbf{a} \cdot \mathbf{b}}{\|\mathbf{b}\|^2} \mathbf{b} = \dfrac{2}{3}\mathbf{b} = \boxed{\dfrac{2}{3}(1, -1, 1)^T}$,

            $\mathbf{a}_\perp = \mathbf{a} - \mathbf{a}_\| = (4, 3, 1)^T - (2/3, -2/3, 2/3)^T = \boxed{(10/3, 11/3, 1/3)^T}$.

(c) $\mathbf{a}_\| \cdot \mathbf{b} = \dfrac{2}{3}\mathbf{b} \cdot \mathbf{b} = \dfrac{2}{3} \cdot 3 = \boxed{2} = \mathbf{a} \cdot \mathbf{b}, \ \checkmark$

            $\mathbf{a}_\perp \cdot \mathbf{b} = \dfrac{10}{3} - \dfrac{11}{3} + \dfrac{1}{3} = \boxed{0}, \ \checkmark$

$$\mathbf{a}_{\parallel} \times \mathbf{b} = \frac{2}{3} \begin{pmatrix} 1 \\ -1 \\ 1 \end{pmatrix} \times \begin{pmatrix} 1 \\ -1 \\ 1 \end{pmatrix} = \frac{2}{3} \begin{pmatrix} (-1) \cdot 1 - 1 \cdot (-1) \\ 1 \cdot 1 - 1 \cdot 1 \\ 1 \cdot (-1) - (-1) \cdot 1 \end{pmatrix} = \boxed{\begin{pmatrix} 0 \\ 0 \\ 0 \end{pmatrix}}, \checkmark$$

$$\mathbf{a}_{\perp} \times \mathbf{b} = \frac{1}{3} \begin{pmatrix} 10 \\ 11 \\ 1 \end{pmatrix} \times \begin{pmatrix} 1 \\ -1 \\ 1 \end{pmatrix} = \frac{1}{3} \begin{pmatrix} 11 + 1 \\ 1 - 10 \\ -10 - 11 \end{pmatrix} = \boxed{\begin{pmatrix} 4 \\ -3 \\ -7 \end{pmatrix}} = \mathbf{a} \times \mathbf{b}. \checkmark$$

As expected, we have: $\mathbf{a}_{\parallel} \cdot \mathbf{b} = \mathbf{a} \cdot \mathbf{b}$, $\mathbf{a}_{\perp} \cdot \mathbf{b} = 0$, $\mathbf{a}_{\parallel} \times \mathbf{b} = \mathbf{0}$ and $\mathbf{a}_{\perp} \times \mathbf{b} = \mathbf{a} \times \mathbf{b}$. $\checkmark$

εL4.2.3 Levi–Civita tensor (p. 59)

(a) $a^i b^j \epsilon_{ij2} = -a^k \epsilon_{k2l} b^l$ is $\boxed{\text{true}}$. Indeed, writing out both sides explicitly, we find:

$$a^i b^j \epsilon_{ij2} = a^1 b^3 \epsilon_{132} + a^3 b^1 \epsilon_{312} = a^3 b^1 - a^1 b^3,$$
$$-a^k \epsilon_{k2l} b^l = -a^1 \epsilon_{123} b^3 - a^3 \epsilon_{321} b^1 = a^3 b^1 - a^1 b^3.$$

More compactly, we can bring the r.h.s. into the form of the l.h.s. by relabeling summation indices and using the antisymmetry of the ϵ tensor: $-a^k \epsilon_{k2l} b^l = -a^i b^j \epsilon_{i2j} = a^i b^j \epsilon_{ij2}$.

For the next two subproblems, we use the identity $\epsilon_{ijk} \epsilon_{mnk} = \delta_{im}\delta_{jn} - \delta_{in}\delta_{jm}$. To be able to apply it, it might be necessary to cyclically rearrange indices on one of the Levi–Civita factors.

(b) $$\epsilon_{1ik} \epsilon_{kj1} = \epsilon_{1ik} \epsilon_{j1k} = \boxed{\delta_{1j}\delta_{i1} - \delta_{11}\delta_{ij}} = \begin{cases} -1 & \text{if } i = j \in \{2, 3\}, \\ 0 & \text{otherwise.} \end{cases}$$

Note: For $i = j = 1$, the delta functions yield $\delta_{1j}\delta_{i1} - \delta_{11}\delta_{ij} = 1 \cdot 1 - 1 \cdot 1 = 0$.
As a check, we write out the k-sum explicitly: $\epsilon_{1ik} \epsilon_{kj1} = \epsilon_{1i1}\epsilon_{1j1} + \epsilon_{1i2}\epsilon_{2j1} + \epsilon_{1i3}\epsilon_{3j1}$. The first term vanishes, because two indices on ϵ are equal. The second term is nonzero only for $i = j = 3$, in which case it yields $\epsilon_{132}\epsilon_{231} = (-1) \cdot (+1) = -1$. The third term is nonzero only for $i = j = 2$ in which case it yields $\epsilon_{123}\epsilon_{321} = (+1) \cdot (-1) = -1$.

(c) $$\epsilon_{1ik} \epsilon_{kj2} = \epsilon_{1ik} \epsilon_{j2k} = \delta_{1j}\delta_{i2} - \delta_{12}\delta_{ij} = \boxed{\delta_{1j}\delta_{i2}} = \begin{cases} 1 & \text{if } i = 2 \text{ and } j = 1, \\ 0 & \text{otherwise.} \end{cases}$$

As a check, we write out the k-sum explicitly: $\epsilon_{1ik} \epsilon_{2kj} = \epsilon_{1i1}\epsilon_{21j} + \epsilon_{1i2}\epsilon_{22j} + \epsilon_{1i3}\epsilon_{23j}$. The first and second terms vanish, since they contain ϵ-factors on which two indices are equal. The third term is nonzero only if $i = 2$ and $j = 1$, in which case it yields $\epsilon_{123}\epsilon_{231} = 1$.

S.L4.3 Further properties of the vector product (p. 60)

εL4.3.1 Grassmann identity (BAC-CAB) and Jacobi identity (p. 61)

(a) Consider the kth component of $\mathbf{a} \times (\mathbf{b} \times \mathbf{c})$ in an orthonormal basis, for $k \in \{1, 2, 3\}$:

$$[\mathbf{a} \times (\mathbf{b} \times \mathbf{c})]^k \overset{(i)}{=} a^i [\mathbf{b} \times \mathbf{c}]^j \epsilon_{ijk} \overset{(ii)}{=} a^i b^m c^n \epsilon_{mnj} \epsilon_{kij}$$

$$\overset{(iii)}{=} a^i b^m c^n (\delta_{mk}\delta_{ni} - \delta_{mi}\delta_{nk}) \overset{(iv)}{=} a^i b^k c^i - a^i b^i c^k \overset{(v)}{=} b^k (\mathbf{a} \cdot \mathbf{c}) - c^k (\mathbf{a} \cdot \mathbf{b}).$$

Explanation: We (i), (ii) employed the Levi–Civita representation of the cross product; (iii) used the identity from the hint to perform the sum over the repeated index j in the product of two Levi–Civita tensors; (iv) performed the sums on the repeated indices m and n, exploiting the

Kronecker-δs; and (v) identified the remaining sums on i as scalar products. As a guide for the eye, we used horizontal brackets ("contractions") to indicate which repeated indices will be summed over in the next step.

(b) $\mathbf{a} \times (\mathbf{b} \times \mathbf{c}) + \mathbf{b} \times (\mathbf{c} \times \mathbf{a}) + \mathbf{c} \times (\mathbf{a} \times \mathbf{b})$

$$\overset{\text{Grassmann}}{=} \left[\mathbf{b}(\mathbf{a} \cdot \mathbf{c}) - \mathbf{c}(\mathbf{a} \cdot \mathbf{b})\right] + \left[\mathbf{c}(\mathbf{b} \cdot \mathbf{a}) - \mathbf{a}(\mathbf{b} \cdot \mathbf{c})\right] + \left[\mathbf{a}(\mathbf{c} \cdot \mathbf{b}) - \mathbf{b}(\mathbf{c} \cdot \mathbf{a})\right] = \boxed{\mathbf{0}}. \checkmark$$

(c) $\mathbf{a} = (1, 1, 2)^{\mathrm{T}}, \mathbf{b} = (3, 2, 0)^{\mathrm{T}}, \mathbf{c} = (2, 1, 1)^{\mathrm{T}}.$ $\mathbf{a} \cdot \mathbf{c} = \boxed{5},$ $\mathbf{a} \cdot \mathbf{b} = \boxed{5}.$

$$\mathbf{b} \times \mathbf{c} = \begin{pmatrix} 3 \\ 2 \\ 0 \end{pmatrix} \times \begin{pmatrix} 2 \\ 1 \\ 1 \end{pmatrix} = \begin{pmatrix} 2 \\ -3 \\ -1 \end{pmatrix}, \qquad \mathbf{a} \times (\mathbf{b} \times \mathbf{c}) = \begin{pmatrix} 1 \\ 1 \\ 2 \end{pmatrix} \times \begin{pmatrix} 2 \\ -3 \\ -1 \end{pmatrix} = \boxed{\begin{pmatrix} 5 \\ 5 \\ -5 \end{pmatrix}}.$$

$$\mathbf{a} \times \mathbf{b} = \begin{pmatrix} 1 \\ 1 \\ 2 \end{pmatrix} \times \begin{pmatrix} 3 \\ 2 \\ 0 \end{pmatrix} = \begin{pmatrix} -4 \\ 6 \\ -1 \end{pmatrix}, \qquad \mathbf{c} \times (\mathbf{a} \times \mathbf{b}) = \begin{pmatrix} 2 \\ 1 \\ 1 \end{pmatrix} \times \begin{pmatrix} -4 \\ 6 \\ -1 \end{pmatrix} = \boxed{\begin{pmatrix} -7 \\ -2 \\ -16 \end{pmatrix}}.$$

$$\mathbf{c} \times \mathbf{a} = \begin{pmatrix} 2 \\ 1 \\ 1 \end{pmatrix} \times \begin{pmatrix} 1 \\ 1 \\ 2 \end{pmatrix} = \begin{pmatrix} 1 \\ -3 \\ 1 \end{pmatrix}, \qquad \mathbf{b} \times (\mathbf{c} \times \mathbf{a}) = \begin{pmatrix} 3 \\ 2 \\ 0 \end{pmatrix} \times \begin{pmatrix} 1 \\ -3 \\ 1 \end{pmatrix} = \boxed{\begin{pmatrix} 2 \\ -3 \\ -11 \end{pmatrix}}.$$

$$\mathbf{b}(\mathbf{a} \cdot \mathbf{c}) - \mathbf{c}(\mathbf{a} \cdot \mathbf{b}) = 5 \begin{pmatrix} 3 \\ 2 \\ 0 \end{pmatrix} - 5 \begin{pmatrix} 2 \\ 1 \\ 1 \end{pmatrix} = \boxed{\begin{pmatrix} 5 \\ 5 \\ -5 \end{pmatrix}} = \mathbf{a} \times (\mathbf{b} \times \mathbf{c}). \checkmark$$

$$\mathbf{a} \times (\mathbf{b} \times \mathbf{c}) + \mathbf{b} \times (\mathbf{c} \times \mathbf{a}) + \mathbf{c} \times (\mathbf{a} \times \mathbf{b}) = \begin{pmatrix} 5 \\ 5 \\ -5 \end{pmatrix} + \begin{pmatrix} -7 \\ -2 \\ 16 \end{pmatrix} + \begin{pmatrix} 2 \\ -3 \\ -11 \end{pmatrix} = \boxed{\mathbf{0}}. \checkmark$$

εL4.3.3 Scalar triple product (p. 61)

(a)
$$S(y) = \mathbf{v}_1 \cdot (\mathbf{v}_2 \times \mathbf{v}_3) = \begin{pmatrix} 1 \\ 0 \\ 2 \end{pmatrix} \cdot \left(\begin{pmatrix} 3 \\ 2 \\ 1 \end{pmatrix} \times \begin{pmatrix} -1 \\ -2 \\ y \end{pmatrix} \right)$$

$$= \begin{pmatrix} 1 \\ 0 \\ 2 \end{pmatrix} \cdot \begin{pmatrix} 2y + 2 \\ -1 - 3y \\ -4 \end{pmatrix} = (2y + 2) + 0 - 8 = 2y - 6.$$

(b)
$$\mathbf{0} = a^1 \mathbf{v}_1 + a^2 \mathbf{v}_2 + a^3 \mathbf{v}_3 = a^1 \begin{pmatrix} 1 \\ 0 \\ 2 \end{pmatrix} + a^2 \begin{pmatrix} 3 \\ 2 \\ 1 \end{pmatrix} + a^3 \begin{pmatrix} -1 \\ -2 \\ y \end{pmatrix}, \quad \text{with} \quad a^j \in \mathbb{R}.$$

This vector equation yields a system of three equations which we solve as follows:

(i) $1a^1 + 3a^2 - 1a^3 = 0$ $\overset{(ii)}{\Rightarrow}$ (iv) $a^3 = a^2;$

(ii) $0a^1 + 2a^2 - 2a^3 = 0$ $\overset{\text{(iv) in (i)}}{\Rightarrow}$ (v) $a^1 = -2a^2;$

(iii) $2a^1 + 1a^2 + ya^3 = 0.$ $\overset{\text{(iv),(v) in (iii)}}{\Rightarrow}$ (vi) $a^2(-4 + 1 + y) = 0.$

(ii) yields (iv): $a^3 = a^2$. Inserting (iv) into (i) yields (v): $a^1 = -2a^2$. Inserting (iv) and (v) into (iii) yields (vi): $a^2(y - 3) = 0$. For $y \neq 3$ we have $0 \overset{(vi)}{=} a^2 \overset{(v)}{=} a^1 \overset{(iv)}{=} a^3$, thus the vectors are linearly independent. For $\boxed{y = 3}$, however, (vi) yields $0 = 0$, hence it does not fix the value of a^2. There are then infinitely many nontrivial solutions (one for every value of $a^2 \in \mathbb{R}$), hence $\mathbf{v}_1, \mathbf{v}_2$ and $\mathbf{v}_3$ are linearly dependent.

(c)   For $y = 3$ we have $S(3) \overset{(a)}{=} 2 \cdot 3 - 6 = \boxed{0}$, hence the volume of the parallelepiped spanned by the three vectors vanishes. Therefore they all lie in the same plane in $\mathbb{R}^3$ and thus are linearly dependent, as found in (b).

Remark: This example illustrates the following general fact: three vectors in $\mathbb{R}^3$ are linearly dependent if and only if their scalar triple product vanishes.

S.L5 Linear Maps (p. 63)

S.L5.1 Linear maps (p. 63)

ɛL5.1.1 Checking linearity (p. 65)

The map is linear if $F(v+w) = 2(v+w)+B$ is equal to $F(v)+F(w) = 2v+B+2w+B = 2(v+w)+2B$. This requires $\boxed{B = 0}$.

S.L5.3 Matrix multiplication (p. 69)

ɛL5.3.1 Matrix multiplication (p. 70)

The matrix product AB is defined only when A has the same number of columns as B has rows. The possible products of two of the matrices P, Q and R therefore are:

$$PQ = \begin{pmatrix} 4 & -3 & 1 \\ 2 & 2 & -4 \end{pmatrix} \begin{pmatrix} 3 & 0 & 1 \\ 1 & 2 & 5 \\ 1 & -6 & -1 \end{pmatrix} = \begin{pmatrix} 10 & -12 & -12 \\ 4 & 28 & 16 \end{pmatrix}, \quad PR = \begin{pmatrix} 4 & -3 & 1 \\ 2 & 2 & -4 \end{pmatrix} \begin{pmatrix} 3 & 0 \\ 1 & 2 \\ 1 & -6 \end{pmatrix} = \begin{pmatrix} 10 & -12 \\ 4 & 28 \end{pmatrix},$$

$$QR = \begin{pmatrix} 3 & 0 & 1 \\ 1 & 2 & 5 \\ 1 & -6 & -1 \end{pmatrix} \begin{pmatrix} 3 & 0 \\ 1 & 2 \\ 1 & -6 \end{pmatrix} = \begin{pmatrix} 10 & -6 \\ 10 & -26 \\ -4 & -6 \end{pmatrix}, \quad RP = \begin{pmatrix} 3 & 0 \\ 1 & 2 \\ 1 & -6 \end{pmatrix} \begin{pmatrix} 4 & -3 & 1 \\ 2 & 2 & -4 \end{pmatrix} = \begin{pmatrix} 12 & -9 & 3 \\ 8 & 1 & -7 \\ -8 & -15 & 25 \end{pmatrix},$$

$$QQ = \begin{pmatrix} 3 & 0 & 1 \\ 1 & 2 & 5 \\ 1 & -6 & -1 \end{pmatrix} \begin{pmatrix} 3 & 0 & 1 \\ 1 & 2 & 5 \\ 1 & -6 & -1 \end{pmatrix} = \begin{pmatrix} 10 & -6 & 2 \\ 10 & -26 & 6 \\ -4 & -6 & -28 \end{pmatrix}.$$

ɛL5.3.3 Spin-$\frac{1}{2}$ matrices (p. 70)

(a)
$$\mathbf{S}^2 = \tfrac{1}{4}\begin{pmatrix} 0 & 1 \\ 1 & 0 \end{pmatrix}\begin{pmatrix} 0 & 1 \\ 1 & 0 \end{pmatrix} + \tfrac{1}{4}\begin{pmatrix} 0 & -i \\ i & 0 \end{pmatrix}\begin{pmatrix} 0 & -i \\ i & 0 \end{pmatrix} + \tfrac{1}{4}\begin{pmatrix} 1 & 0 \\ 0 & -1 \end{pmatrix}\begin{pmatrix} 1 & 0 \\ 0 & -1 \end{pmatrix}$$

$$= \tfrac{1}{4}\begin{pmatrix} 1 & 0 \\ 0 & 1 \end{pmatrix} + \tfrac{1}{4}\begin{pmatrix} 1 & 0 \\ 0 & 1 \end{pmatrix} + \tfrac{1}{4}\begin{pmatrix} 1 & 0 \\ 0 & 1 \end{pmatrix} = \tfrac{3}{4}\begin{pmatrix} 1 & 0 \\ 0 & 1 \end{pmatrix} = \boxed{\tfrac{3}{4}\cdot\mathbb{1}}.$$

This has the form $\mathbf{S}^2 = s(s+1)\mathbb{1}$, with $s \overset{\checkmark}{=} \boxed{\tfrac{1}{2}}$.

(b)
$$[S_x, S_y] = \tfrac{1}{4}\left[\begin{pmatrix} 0 & 1 \\ 1 & 0 \end{pmatrix}\begin{pmatrix} 0 & -i \\ i & 0 \end{pmatrix} - \begin{pmatrix} 0 & -i \\ i & 0 \end{pmatrix}\begin{pmatrix} 0 & 1 \\ 1 & 0 \end{pmatrix}\right]$$

$$= \tfrac{1}{4}\left[\begin{pmatrix} i & 0 \\ 0 & -i \end{pmatrix} - \begin{pmatrix} -i & 0 \\ 0 & i \end{pmatrix}\right] = \tfrac{1}{2}\begin{pmatrix} i & 0 \\ 0 & -i \end{pmatrix} = \boxed{iS_z} \overset{\checkmark}{=} i\epsilon_{xyz}S_z.$$

$$[S_y, S_z] = \tfrac{1}{4}\left[\begin{pmatrix} 0 & -i \\ i & 0 \end{pmatrix}\begin{pmatrix} 1 & 0 \\ 0 & -1 \end{pmatrix} - \begin{pmatrix} 1 & 0 \\ 0 & -1 \end{pmatrix}\begin{pmatrix} 0 & -i \\ i & 0 \end{pmatrix}\right]$$

$$= \tfrac{1}{4}\left[\begin{pmatrix} 0 & i \\ i & 0 \end{pmatrix} - \begin{pmatrix} 0 & -i \\ -i & 0 \end{pmatrix}\right] = \tfrac{1}{2}\begin{pmatrix} 0 & i \\ i & 0 \end{pmatrix} = \boxed{iS_x} \overset{\checkmark}{=} i\epsilon_{yzx}S_x.$$

$$[S_z, S_x] = \tfrac{1}{4}\left[\begin{pmatrix} 1 & 0 \\ 0 & -1 \end{pmatrix}\begin{pmatrix} 0 & 1 \\ 1 & 0 \end{pmatrix} - \begin{pmatrix} 0 & 1 \\ 1 & 0 \end{pmatrix}\begin{pmatrix} 1 & 0 \\ 0 & -1 \end{pmatrix}\right]$$

$$= \tfrac{1}{4}\left[\begin{pmatrix} 0 & 1 \\ -1 & 0 \end{pmatrix} - \begin{pmatrix} 0 & -1 \\ 1 & 0 \end{pmatrix}\right] = \tfrac{1}{2}\begin{pmatrix} 0 & 1 \\ -1 & 0 \end{pmatrix} = \boxed{\mathrm{i}S_y} \overset{\checkmark}{=} \mathrm{i}\epsilon_{zxy}S_y\,.$$

Commutators are antisymmetric, hence $[S_y, S_x] = -[S_y, S_x] = -\mathrm{i}S_z \overset{\checkmark}{=} \mathrm{i}\epsilon_{yxz}S_z$, and $[S_x, S_x] = 0 \overset{\checkmark}{=} \mathrm{i}\epsilon_{xxk}S_k$, etc. Clearly the spin-$\tfrac{1}{2}$ matrices do satisfy the SU(2) algebra.

EL5.3.5 Matrix multiplication (p. 70)

(a)
$$A = \begin{pmatrix} 0 & 0 & 0 \\ a_1 & a_2 & a_3 \\ 0 & 0 & 0 \end{pmatrix}, \quad B = \begin{pmatrix} b_1 & 0 & 0 \\ 0 & b_2 & 0 \\ 0 & 0 & b_3 \end{pmatrix}, \quad AB = \boxed{\begin{pmatrix} 0 & 0 & 0 \\ a_1b_1 & a_2b_2 & a_3b_3 \\ 0 & 0 & 0 \end{pmatrix}}.$$

(b) $\quad (AB)^i{}_j = \sum_k A^i{}_k B^k{}_j = \sum_k a_k \delta^i{}_m b_k \delta^k{}_j = \boxed{\delta^i{}_m\, a_j b_j}\,.$

$$AB = \overset{m\text{th row}}{\longrightarrow} \begin{pmatrix} 0 & \cdots & 0 \\ \vdots & & \vdots \\ a_1 & \cdots & a_N \\ \vdots & & \vdots \\ 0 & \cdots & 0 \end{pmatrix} \begin{pmatrix} b_1 & 0 & \cdots & 0 \\ 0 & b_2 & \cdots & 0 \\ \vdots & & \ddots & \vdots \\ 0 & \cdots & \cdots & b_N \end{pmatrix} = \boxed{\begin{pmatrix} 0 & \cdots & 0 \\ \vdots & & \vdots \\ a_1 b_1 & \cdots & a_N b_N \\ \vdots & & \vdots \\ 0 & \cdots & 0 \end{pmatrix}}.$$

S.L5.4   The inverse of a matrix (p. 72)

EL5.4.1 Gaussian elimination and matrix inversion (p. 78)

(a) We bring the augmented matrix into the desired form by taking linear combinations of rows. We indicate these using square brackets which refer to the rows of the previous augmented matrix. For example, in the first step below we make the replacements $[1] \to [1]$, $[2] \to \tfrac{1}{2}(3[2] - 2[1])$ and $[3] \to 6[3] + 2[1]$. It is advisable to keep the mental arithmetic simple by avoiding the occurrence of fractions until the left side has been brought into row echelon form (only zeros on one side of the diagonal).

$[1]:$	3	2	-1	$\vert$	1		$[1]:$	3	2	-1	$\vert$	1
$[2]:$	2	-2	4	$\vert$	-2	$\longrightarrow$	$\tfrac{1}{2}(3[2]-2[1]):$	0	-5	7	$\vert$	-4
$[3]:$	-1	$\tfrac{1}{2}$	-1	$\vert$	0		$6[3]+2[1]:$	0	7	-8	$\vert$	2

$\tfrac{1}{3}(5[1]+2[2]):$	5	0	3	$\vert$	-1		$\tfrac{1}{5}([1]-3[3]):$	1	0	0	$\vert$	1
$-[2]:$	0	5	-7	$\vert$	4	$\longrightarrow$	$\tfrac{1}{5}([2]+7[3]):$	0	1	0	$\vert$	-2
$\tfrac{1}{9}(7[2]+5[3]):$	0	0	1	$\vert$	-2		$[3]:$	0	0	1	$\vert$	-2

Hence the solution to the system of equations is: $\mathbf{x} = (x^1, x^2, x^3)^{\mathrm{T}} = \boxed{(1, -2, -2)^{\mathrm{T}}}$.

(b) If the last equation is removed, we obtain after the second step (see above):

$[1]:$	3	2	-1	$\vert$	1	$\longrightarrow$	$[1]:$	3	2	-1	$\vert$	1
$[2]:$	2	-2	4	$\vert$	-2		$\tfrac{1}{2}(3[2]-2[1]):$	0	-5	7	$\vert$	-4

The system is now underdetermined, since there are more unknowns than equations. The solution thus depends on a free parameter, which we call $x^3 = \lambda$. We now complete the system with a corresponding third line and then bring the left side into diagonal form:

$$
\begin{array}{c}
[1]: \\
[2]: \\
[3]:
\end{array}
\left[
\begin{array}{ccc|c}
3 & 2 & -1 & 1 \\
0 & -5 & 7 & -4 \\
0 & 0 & 1 & \lambda
\end{array}
\right]
\longrightarrow
\begin{array}{c}
\frac{1}{5}[\frac{1}{3}(5[1]+2[2])-3[3]]: \\
-\frac{1}{5}([2]-7[3]): \\
[3]:
\end{array}
\left[
\begin{array}{ccc|c}
1 & 0 & 0 & -\frac{1}{5}(1+3\lambda) \\
0 & 1 & 0 & \frac{1}{5}(4+7\lambda) \\
0 & 0 & 1 & \lambda
\end{array}
\right]
$$

There are evidently infinitely many solutions, $\mathbf{x} = \boxed{(-\frac{1}{5}-\frac{3}{5}\lambda, \frac{4}{5}+\frac{7}{5}\lambda, \lambda)^{\mathrm{T}}}$. They lie along a straight line in $\mathbb{R}^3$, parametrized by λ.

(c) If the last equation is replaced by $-x^1 + \frac{2}{7}x^2 - x^3 = 0$, we obtain:

$$
\begin{array}{c}
[1]: \\
[2]: \\
[3]:
\end{array}
\left[
\begin{array}{ccc|c}
3 & 2 & -1 & 1 \\
2 & -2 & 4 & -2 \\
-1 & \frac{2}{7} & -1 & 0
\end{array}
\right]
\longrightarrow
\begin{array}{c}
[1]: \\
\frac{1}{2}(3[2]-2[1]): \\
6[3]+2[1]:
\end{array}
\left[
\begin{array}{ccc|c}
3 & 2 & -1 & 1 \\
0 & -5 & 7 & -4 \\
0 & \frac{40}{7} & -8 & 2
\end{array}
\right]
$$

$$
\begin{array}{c}
\frac{1}{3}(5[1]+2[2]): \\
-[2]: \\
8[2]+7[3]:
\end{array}
\left[
\begin{array}{ccc|c}
5 & 0 & 3 & -1 \\
0 & 5 & -7 & 4 \\
0 & 0 & 0 & -18
\end{array}
\right]
$$

The last equation reads $0x^1 + 0x^2 + 0x^3 = -18$, which is a logical contradiction. In this case, this system of equations thus has $\boxed{\text{no solution}}$.

(d) We set up an augmented matrix containing the unit matrix on the right. Using the same steps as in (a), we manipulate it until the unit matrix appears on the left:

$$
\begin{array}{c}
[1]: \\
[2]: \\
[3]:
\end{array}
\left[
\begin{array}{ccc|ccc}
3 & 2 & -1 & 1 & 0 & 0 \\
2 & -2 & 4 & 0 & 1 & 0 \\
-1 & \frac{1}{2} & -1 & 0 & 0 & 1
\end{array}
\right]
\longrightarrow
\begin{array}{c}
[1]: \\
\frac{1}{2}(3[2]-2[1]): \\
6[3]+2[1]:
\end{array}
\left[
\begin{array}{ccc|ccc}
3 & 2 & -1 & 1 & 0 & 0 \\
0 & -5 & 7 & -1 & \frac{3}{2} & 0 \\
0 & 7 & -8 & 2 & 0 & 6
\end{array}
\right]
$$

$$
\begin{array}{c}
\frac{1}{3}(5[1]+2[2]): \\
-[2]: \\
\frac{1}{9}(7[2]+5[3]):
\end{array}
\left[
\begin{array}{ccc|ccc}
5 & 0 & 3 & 1 & 1 & 0 \\
0 & 5 & -7 & 1 & -\frac{3}{2} & 0 \\
0 & 0 & 1 & \frac{1}{3} & \frac{7}{6} & \frac{10}{3}
\end{array}
\right]
\longrightarrow
\begin{array}{c}
\frac{1}{5}([1]-3[3]): \\
\frac{1}{5}([2]+7[3]): \\
[3]:
\end{array}
\left[
\begin{array}{ccc|ccc}
1 & 0 & 0 & 0 & -\frac{1}{2} & -2 \\
0 & 1 & 0 & \frac{2}{3} & \frac{4}{3} & \frac{14}{3} \\
0 & 0 & 1 & \frac{1}{3} & \frac{7}{6} & \frac{10}{3}
\end{array}
\right]
$$

The right side of the final augmented matrix contains the columns of the inverse matrix:

$$
A^{-1} = \boxed{\frac{1}{3}\begin{pmatrix} 0 & -\frac{3}{2} & -6 \\ 2 & 4 & 14 \\ 1 & \frac{7}{2} & 10 \end{pmatrix}}. \quad
\text{Check:}\quad \mathbf{x} = A^{-1}\mathbf{b} = \frac{1}{3}\begin{pmatrix} 0 & -\frac{3}{2} & -6 \\ 2 & 4 & 14 \\ 1 & \frac{7}{2} & 10 \end{pmatrix}\begin{pmatrix} 1 \\ -2 \\ 0 \end{pmatrix} = \begin{pmatrix} 1 \\ -2 \\ -2 \end{pmatrix}.\ \checkmark
$$

εL5.4.3 Matrix inversion (p. 78)

(a) The inverse of $M_2 = \begin{pmatrix} 1+m & 0 \\ 1 & m \end{pmatrix}$ follows from the formula $\begin{pmatrix} a & b \\ c & d \end{pmatrix} = \frac{1}{ad-bc}\begin{pmatrix} d & -b \\ -c & a \end{pmatrix}$:

$$
M_2^{-1} = \boxed{\begin{pmatrix} \frac{1}{1+m} & 0 \\ -\frac{1}{m(1+m)} & \frac{1}{m} \end{pmatrix}}. \quad
\text{Check:}\quad \begin{pmatrix} \frac{1}{1+m} & 0 \\ -\frac{1}{m(1+m)} & \frac{1}{m} \end{pmatrix}\cdot\begin{pmatrix} 1+m & 0 \\ 1 & m \end{pmatrix} = \begin{pmatrix} 1 & 0 \\ 0 & 1 \end{pmatrix}.\ \checkmark
$$

We compute the inverse of $M_3 = \begin{pmatrix} 1+m & 0 & 0 \\ 1 & m & 0 \\ 1 & 0 & m \end{pmatrix}$ using Gaussian elimination:

$$
\begin{array}{cccc|ccc}
[1]: & 1+m & 0 & 0 & 1 & 0 & 0 \\
[2]: & 1 & m & 0 & 0 & 1 & 0 \\
[3]: & 1 & 0 & m & 0 & 0 & 1
\end{array}
\longrightarrow
\begin{array}{ccc|ccc}
\frac{1}{1+m}[1]: & 1 & 0 & 0 & \frac{1}{1+m} & 0 & 0 \\
\frac{1}{m}([2]-\frac{1}{1+m}[1]): & 0 & 1 & 0 & -\frac{1}{m(1+m)} & \frac{1}{m} & 0 \\
\frac{1}{m}([3]-\frac{1}{1+m}[1]): & 0 & 0 & 1 & -\frac{1}{m(1+m)} & 0 & \frac{1}{m}
\end{array}
$$

The right side of the augmented matrix gives the inverse matrix M_3^{-1}:

$$
M_3^{-1} = \boxed{\begin{pmatrix} \frac{1}{1+m} & 0 & 0 \\ -\frac{1}{m(1+m)} & \frac{1}{m} & 0 \\ -\frac{1}{m(1+m)} & 0 & \frac{1}{m} \end{pmatrix}}. \quad \text{Check:} \quad \begin{pmatrix} \frac{1}{1+m} & 0 & 0 \\ -\frac{1}{m(1+m)} & \frac{1}{m} & 0 \\ -\frac{1}{m(1+m)} & 0 & \frac{1}{m} \end{pmatrix}\begin{pmatrix} 1+m & 0 & 0 \\ 1 & m & 0 \\ 1 & 0 & m \end{pmatrix} \overset{\checkmark}{=} \begin{pmatrix} 1 & 0 & 0 \\ 0 & 1 & 0 \\ 0 & 0 & 1 \end{pmatrix}.
$$

(b) The results for M_2^{-1} and M_3^{-1} have the following properties: the first diagonal element equals $\frac{1}{1+m}$, the remaining diagonal elements equal $\frac{1}{m}$, and the remaining elements of the first column equal $-\frac{1}{m(1+m)}$. The checks performed in (a) illustrate why these properties are needed. We thus formulate the following ansatz for the form of M_n^{-1} for a general n:

$$
M_n^{-1} = \begin{pmatrix} \frac{1}{1+m} & 0 & 0 & \cdots & 0 \\ -\frac{1}{m(1+m)} & \frac{1}{m} & 0 & \ddots & \vdots \\ -\frac{1}{m(1+m)} & 0 & \frac{1}{m} & \ddots & 0 \\ \vdots & \vdots & \ddots & \ddots & 0 \\ -\frac{1}{m(1+m)} & 0 & \cdots & 0 & \frac{1}{m} \end{pmatrix}.
$$

Now let us check our ansatz explicitly: does $M_n^{-1} M_n = \mathbb{1}$ hold?

$$
M_n^{-1} \cdot M_n = \begin{pmatrix} \frac{1}{1+m} & 0 & 0 & \cdots & 0 \\ -\frac{1}{m(1+m)} & \frac{1}{m} & 0 & \ddots & \vdots \\ -\frac{1}{m(1+m)} & 0 & \frac{1}{m} & \ddots & 0 \\ \vdots & \vdots & \ddots & \ddots & 0 \\ -\frac{1}{m(1+m)} & 0 & \cdots & 0 & \frac{1}{m} \end{pmatrix}\begin{pmatrix} 1+m & 0 & 0 & \cdots & 0 \\ 1 & m & 0 & \ddots & \vdots \\ 1 & 0 & m & \ddots & 0 \\ \vdots & \vdots & \ddots & \ddots & 0 \\ 1 & 0 & \cdots & 0 & m \end{pmatrix}
$$

$$
= \begin{pmatrix} \frac{1+m}{1+m} & 0 & 0 & \cdots & 0 \\ -\frac{1+m}{m(1+m)}+\frac{1}{m} & \frac{m}{m} & 0 & \ddots & \vdots \\ -\frac{1+m}{m(1+m)}+\frac{1}{m} & 0 & \frac{m}{m} & \ddots & 0 \\ \vdots & \vdots & \ddots & \ddots & 0 \\ -\frac{1+m}{m(1+m)}+\frac{1}{m} & 0 & \cdots & 0 & \frac{m}{m} \end{pmatrix} = \begin{pmatrix} 1 & 0 & 0 & \cdots & 0 \\ 0 & 1 & 0 & \ddots & \vdots \\ 0 & 0 & & \ddots & 0 \\ \vdots & \vdots & \ddots & \ddots & 0 \\ 0 & 0 & \cdots & 0 & 1 \end{pmatrix}. \checkmark
$$

(c) Alternative formulation using index notation: $(M_n^{-1})^i{}_j = \boxed{\frac{1}{m}\left(\delta^i{}_j - \frac{1}{1+m}\delta^1{}_j\right)}$.

$$
(M_n^{-1} \cdot M_n)^i{}_j = \sum_l (M_n^{-1})^i{}_l (M_n)^l{}_j = \sum_l \left(\frac{1}{m}\delta^i{}_l - \frac{1}{m(1+m)}\delta^1{}_l\right)\left(m\delta^l{}_j + \delta^1{}_j\right)
$$

$$
= \delta^i{}_j + \frac{1}{m}\delta^1{}_j - \frac{1}{1+m}\delta^i{}_j - \frac{1}{m(1+m)}\delta^1{}_j = \delta^i{}_j + \delta^1{}_j\frac{1+m-m-1}{m(1+m)} = \boxed{\delta^i{}_j}. \checkmark
$$

S.L5.5   General linear maps and matrices (p. 78)

ɛL5.5.1 Two-dimensional rotation matrices (p. 79)

(a)   For $R_\theta : \mathbf{e}_j \mapsto \mathbf{e}'_j = \mathbf{e}_i(R_\theta)^i{}_j$ the image vector $\mathbf{e}'_j$ yields column j of the rotation matrix:

$$R_\theta : \quad \begin{pmatrix} 1 \\ 0 \end{pmatrix} \mapsto \begin{pmatrix} \cos\theta \\ \sin\theta \end{pmatrix}, \quad \begin{pmatrix} 0 \\ 1 \end{pmatrix} \mapsto \begin{pmatrix} -\sin\theta \\ \cos\theta \end{pmatrix} \quad \Rightarrow \quad \boxed{R_\theta = \begin{pmatrix} \cos\theta & -\sin\theta \\ \sin\theta & \cos\theta \end{pmatrix}}.$$

(b)   For $\theta_1 = 0, \theta_2 = \frac{\pi}{4}, \theta_3 = \pi/2$ and $\theta_4 = \pi$ we have:

$$R_0 = \begin{pmatrix} 1 & 0 \\ 0 & 1 \end{pmatrix}, \qquad R_0\mathbf{a} = \mathbf{a}, \qquad R_0\mathbf{b} = \mathbf{b}.$$

$$R_{\frac{\pi}{4}} = \frac{1}{\sqrt{2}}\begin{pmatrix} 1 & -1 \\ 1 & 1 \end{pmatrix}, \quad R_{\frac{\pi}{4}}\mathbf{a} = \frac{1}{\sqrt{2}}\begin{pmatrix} 1 \\ 1 \end{pmatrix}, \quad R_{\frac{\pi}{4}}\mathbf{b} = \frac{1}{\sqrt{2}}\begin{pmatrix} -1 \\ 1 \end{pmatrix}.$$

$$R_{\frac{\pi}{2}} = \begin{pmatrix} 0 & -1 \\ 1 & 0 \end{pmatrix}, \qquad R_{\frac{\pi}{2}}\mathbf{a} = \begin{pmatrix} 0 \\ 1 \end{pmatrix}, \qquad R_{\frac{\pi}{2}}\mathbf{b} = \begin{pmatrix} -1 \\ 0 \end{pmatrix}.$$

$$R_\pi = \begin{pmatrix} -1 & 0 \\ 0 & -1 \end{pmatrix}, \qquad R_\pi\mathbf{a} = \begin{pmatrix} -1 \\ 0 \end{pmatrix}, \qquad R_\pi\mathbf{b} = \begin{pmatrix} 0 \\ -1 \end{pmatrix}.$$

(c)   Using the addition theorems, we readily obtain:

$$R_\theta R_\phi = \begin{pmatrix} \cos\theta & -\sin\theta \\ \sin\theta & \cos\theta \end{pmatrix}\begin{pmatrix} \cos\phi & -\sin\phi \\ \sin\phi & \cos\phi \end{pmatrix}$$

$$= \begin{pmatrix} \cos\theta\cos\phi - \sin\theta\sin\phi & -\cos\theta\sin\phi - \sin\theta\cos\phi \\ \sin\theta\cos\phi + \cos\theta\sin\phi & -\sin\theta\sin\phi + \cos\theta\cos\phi \end{pmatrix} = \begin{pmatrix} \cos(\theta+\phi) & -\sin(\theta+\phi) \\ \sin(\theta+\phi) & \cos(\theta+\phi) \end{pmatrix} = R_{\theta+\phi}. \checkmark$$

(d)   Rotating the vector $\mathbf{r} = (x, y)^\mathrm{T}$ by the angle θ leaves its length unchanged:

$$R_\theta\mathbf{r} = \begin{pmatrix} x\cos\theta - y\sin\theta \\ x\sin\theta + y\cos\theta \end{pmatrix}$$

$$\|R_\theta\mathbf{r}\| = \sqrt{(x^2+y^2)(\cos^2\theta+\sin^2\theta)+(2\cos\theta\sin\theta)(xy-xy)} = \sqrt{(x^2+y^2)} = \|\mathbf{r}\|. \checkmark$$

S.L5.6   Matrices describing coordinate changes (p. 80)

ɛL5.6.1 Basis transformations and linear maps in $\mathbb{E}^2$ (p. 84)

(a)   The relation $\hat{\mathbf{v}}_j = \hat{\mathbf{v}}'_i T^i{}_j$ between old and new bases yields the transformation matrix T:

$$\begin{aligned} \hat{\mathbf{v}}_1 &= \tfrac{3}{4}\hat{\mathbf{v}}'_1 + \tfrac{1}{3}\hat{\mathbf{v}}'_2 \equiv \hat{\mathbf{v}}'_1 T^1{}_1 + \hat{\mathbf{v}}'_2 T^2{}_1 \\ \hat{\mathbf{v}}_2 &= -\tfrac{1}{8}\hat{\mathbf{v}}'_1 + \tfrac{1}{2}\hat{\mathbf{v}}'_2 \equiv \hat{\mathbf{v}}'_1 T^1{}_2 + \hat{\mathbf{v}}'_2 T^2{}_2 \end{aligned} \quad \Rightarrow \quad T = \begin{pmatrix} T^1{}_1 & T^1{}_2 \\ T^2{}_1 & T^2{}_2 \end{pmatrix} = \boxed{\begin{pmatrix} \tfrac{3}{4} & -\tfrac{1}{8} \\ \tfrac{1}{3} & \tfrac{1}{2} \end{pmatrix}}.$$

(b)   Using T^{-1} and $\hat{\mathbf{v}}'_i = \hat{\mathbf{v}}_j (T^{-1})^j{}_i$ we can write the new basis in terms of the old:

$$T^{-1} = \frac{1}{\det T}\begin{pmatrix} T^2{}_2 & -T^1{}_2 \\ -T^2{}_1 & T^1{}_1 \end{pmatrix} = \boxed{\frac{12}{5}\begin{pmatrix} \tfrac{1}{2} & \tfrac{1}{8} \\ -\tfrac{1}{3} & \tfrac{3}{4} \end{pmatrix}} \equiv \begin{pmatrix} (T^{-1})^1{}_1 & (T^{-1})^1{}_2 \\ (T^{-1})^2{}_1 & (T^{-1})^2{}_2 \end{pmatrix}.$$

$$\hat{\mathbf{v}}'_1 = \hat{\mathbf{v}}_1 (T^{-1})^1{}_1 + \hat{\mathbf{v}}_2 (T^{-1})^2{}_1 = \boxed{\tfrac{6}{5}\hat{\mathbf{v}}_1 - \tfrac{4}{5}\hat{\mathbf{v}}_2}.$$

$$\hat{\mathbf{v}}'_2 = \hat{\mathbf{v}}_1 (T^{-1})^1{}_2 + \hat{\mathbf{v}}_2 (T^{-1})^2{}_2 = \boxed{\tfrac{3}{10}\hat{\mathbf{v}}_1 + \tfrac{9}{5}\hat{\mathbf{v}}_2}.$$

Alternatively, these relations can be derived by solving the equations for $\hat{\mathbf{v}}_1$ and $\hat{\mathbf{v}}_2$ to give $\hat{\mathbf{v}}_1'$ and $\hat{\mathbf{v}}_2'$. (This is equivalent to finding T^{-1}.)

(c) The components of $\hat{\mathbf{x}} = \hat{\mathbf{v}}_j x^j = \hat{\mathbf{v}}_i' x'^i$ in the old and new bases, $\mathbf{x} = (x^1, x^2)^\mathrm{T}$ and $\mathbf{x}' = (x'^1, x'^2)^\mathrm{T}$ respectively, are related by $x'^i = T^i{}_j x^j$:

$$\mathbf{x} = \begin{pmatrix} 1 \\ 2 \end{pmatrix}, \quad \mathbf{x}' = T\mathbf{x} = \frac{1}{24}\begin{pmatrix} 18 & -3 \\ 8 & 12 \end{pmatrix}\begin{pmatrix} 1 \\ 2 \end{pmatrix} = \boxed{\begin{pmatrix} \frac{1}{2} \\ \frac{4}{3} \end{pmatrix}}, \quad \Rightarrow \quad \boxed{\hat{\mathbf{x}} = \hat{\mathbf{v}}_1 + 2\hat{\mathbf{v}}_2 = \tfrac{1}{2}\hat{\mathbf{v}}_1' + \tfrac{4}{3}\hat{\mathbf{v}}_2'}.$$

(d) The components of $\hat{\mathbf{y}} = \hat{\mathbf{v}}_i' y'^i = \hat{\mathbf{v}}_j y^j$ in the new and old bases, $\mathbf{y}' = (y'^1, y'^2)^\mathrm{T}$ and $\mathbf{y} = (y^1, y^2)^\mathrm{T}$ respectively, are related by $y^j = (T^{-1})^j{}_i y'^i$:

$$\mathbf{y}' = \begin{pmatrix} \frac{3}{4} \\ \frac{1}{3} \end{pmatrix}, \quad \mathbf{y} = T^{-1}\mathbf{y}' = \frac{1}{10}\begin{pmatrix} 12 & 3 \\ -8 & 18 \end{pmatrix}\begin{pmatrix} \frac{3}{4} \\ \frac{1}{3} \end{pmatrix} = \boxed{\begin{pmatrix} 1 \\ 0 \end{pmatrix}}, \quad \Rightarrow \quad \boxed{\hat{\mathbf{y}} = \tfrac{3}{4}\hat{\mathbf{v}}_1' + \tfrac{1}{3}\hat{\mathbf{v}}_2' = \hat{\mathbf{v}}_1}.$$

(e) The matrix representation A' of the map $\hat{A}$ in the new basis describes its action on that basis: the image of basis vector j, written as $\hat{\mathbf{v}}_j' \overset{\hat{A}}{\mapsto} \hat{\mathbf{v}}_i' A'^i{}_j$, yields column j of A':

$$\begin{aligned} \hat{\mathbf{v}}_1' &\mapsto 2\hat{\mathbf{v}}_1' + 0\hat{\mathbf{v}}_2' \equiv \hat{\mathbf{v}}_1' A'^1{}_1 + \hat{\mathbf{v}}_2' A'^2{}_1 \\ \hat{\mathbf{v}}_2' &\mapsto 0\hat{\mathbf{v}}_1' + 1\hat{\mathbf{v}}_2' \equiv \hat{\mathbf{v}}_1' A'^1{}_2 + \hat{\mathbf{v}}_2' A'^2{}_2 \end{aligned} \quad \Rightarrow \quad A' = \begin{pmatrix} A'^1{}_1 & A'^1{}_2 \\ A'^2{}_1 & A'^2{}_2 \end{pmatrix} = \boxed{\begin{pmatrix} 2 & 0 \\ 0 & 1 \end{pmatrix}}.$$

The basis transformation T now yields the matrix representation A of $\hat{A}$ in the old basis:

$$A' = TAT^{-1} \Rightarrow A = T^{-1}A'T = \frac{1}{10}\begin{pmatrix} 12 & 3 \\ -8 & 18 \end{pmatrix}\begin{pmatrix} 2 & 0 \\ 0 & 1 \end{pmatrix}\frac{1}{24}\begin{pmatrix} 18 & -3 \\ 8 & 12 \end{pmatrix} = \boxed{\frac{1}{10}\begin{pmatrix} 19 & -\frac{3}{2} \\ -6 & 11 \end{pmatrix}}.$$

(f) For $\hat{\mathbf{x}} \overset{\hat{A}}{\mapsto} \hat{\mathbf{z}}$, the components $\hat{\mathbf{z}}$ are obtained by matrix multiplying the components of $\hat{\mathbf{x}}$ with the matrix representation of $\hat{A}$, in either the new or the old basis:

$$\mathbf{z}' = A'\mathbf{x}' = \begin{pmatrix} 2 & 0 \\ 0 & 1 \end{pmatrix}\begin{pmatrix} \frac{1}{2} \\ \frac{4}{3} \end{pmatrix} = \boxed{\begin{pmatrix} 1 \\ \frac{4}{3} \end{pmatrix}}, \quad \mathbf{z} = A\mathbf{x} = \frac{1}{20}\begin{pmatrix} 38 & -3 \\ -12 & 22 \end{pmatrix}\begin{pmatrix} 1 \\ 2 \end{pmatrix} = \boxed{\frac{8}{5}\begin{pmatrix} 1 \\ 1 \end{pmatrix}}.$$

The results for $\mathbf{z}'$ and $\mathbf{z}$ are consistent, $T\mathbf{z} = \frac{1}{24}\begin{pmatrix} 18 & -3 \\ 8 & 12 \end{pmatrix}\frac{8}{5}\begin{pmatrix} 1 \\ 1 \end{pmatrix} = \begin{pmatrix} 1 \\ \frac{4}{3} \end{pmatrix} = \mathbf{z}'$. ✓

(g) The component representation of the standard basis of $\mathbb{E}^2$ is $\hat{\mathbf{e}}_1 = (1,0)^\mathrm{T}$ and $\hat{\mathbf{e}}_2 = (0,1)^\mathrm{T}$. Once the old basis has been specified by making the choice $\hat{\mathbf{v}}_1 = 3\hat{\mathbf{e}}_1 + \hat{\mathbf{e}}_2 = (3,1)^\mathrm{T}$ and $\hat{\mathbf{v}}_2 = -\tfrac{1}{2}\hat{\mathbf{e}}_1 + \tfrac{3}{2}\hat{\mathbf{e}}_2 = (-\tfrac{1}{2},\tfrac{3}{2})^\mathrm{T}$, that also fixes the new basis, as well as $\hat{\mathbf{x}}$ and $\hat{\mathbf{z}}$. The components of $\hat{\mathbf{x}}$ and $\hat{\mathbf{z}}$ in the standard basis of $\mathbb{E}^2$ can be computed via either the old or the new basis. In the standard basis we obtain the following representation:

(h)

$$\hat{\mathbf{v}}'_1 = \hat{v}_j (T^{-1})^j{}_1 = \frac{6}{5}\begin{pmatrix} 3 \\ 1 \end{pmatrix} - \frac{4}{5}\begin{pmatrix} -\frac{1}{2} \\ \frac{3}{2} \end{pmatrix} = \boxed{\begin{pmatrix} 4 \\ 0 \end{pmatrix}}.$$

$$\hat{\mathbf{v}}'_2 = \hat{v}_j (T^{-1})^j{}_2 = \frac{3}{10}\begin{pmatrix} 3 \\ 1 \end{pmatrix} + \frac{9}{5}\begin{pmatrix} -\frac{1}{2} \\ \frac{3}{2} \end{pmatrix} = \boxed{\begin{pmatrix} 0 \\ 3 \end{pmatrix}}.$$

$$\hat{\mathbf{x}} = \hat{v}_j x^j = 1\begin{pmatrix} 3 \\ 1 \end{pmatrix} + 2\begin{pmatrix} -\frac{1}{2} \\ \frac{3}{2} \end{pmatrix} = \boxed{\begin{pmatrix} 2 \\ 4 \end{pmatrix}}, \qquad \hat{\mathbf{x}} = \hat{v}'_i x'^i = \frac{1}{2}\begin{pmatrix} 4 \\ 0 \end{pmatrix} + \frac{4}{3}\begin{pmatrix} 0 \\ 3 \end{pmatrix} = \boxed{\begin{pmatrix} 2 \\ 4 \end{pmatrix}}. \checkmark$$

$$\hat{\mathbf{z}} = \hat{v}_j z^j = \frac{8}{5}\begin{pmatrix} 3 \\ 1 \end{pmatrix} + \frac{8}{5}\begin{pmatrix} -\frac{1}{2} \\ \frac{3}{2} \end{pmatrix} = \boxed{\begin{pmatrix} 4 \\ 4 \end{pmatrix}}, \qquad \hat{\mathbf{z}} = \hat{v}'_i z'^i = 1\begin{pmatrix} 4 \\ 0 \end{pmatrix} + \frac{4}{3}\begin{pmatrix} 0 \\ 3 \end{pmatrix} = \boxed{\begin{pmatrix} 4 \\ 4 \end{pmatrix}}. \checkmark$$

By comparing $\hat{\mathbf{x}}$ and $\hat{\mathbf{z}}$ we see that $\hat{A}$ stretches the $\hat{\mathbf{e}}_1$ direction by a factor 2.

ᴇL5.6.3 Basis transformations (p. 84)

(a) The effect of the map A on the standard basis, $\mathbf{e}_j \overset{A}{\mapsto} \mathbf{A}_j \equiv \mathbf{e}_i A^i{}_j$, gives the column vectors of the matrix representation $A = (\mathbf{A}_1, \mathbf{A}_2, \mathbf{A}_3)$. In this case, we have:

$$\begin{pmatrix} 1 \\ 0 \\ 0 \end{pmatrix} \overset{A}{\mapsto} \begin{pmatrix} \cos\theta_3 \\ \sin\theta_3 \\ 0 \end{pmatrix}, \quad \begin{pmatrix} 0 \\ 1 \\ 0 \end{pmatrix} \overset{A}{\mapsto} \begin{pmatrix} -\sin\theta_3 \\ \cos\theta_3 \\ 0 \end{pmatrix}, \quad \begin{pmatrix} 0 \\ 0 \\ 1 \end{pmatrix} \overset{A}{\mapsto} \begin{pmatrix} 0 \\ 0 \\ 1 \end{pmatrix}.$$

For the angle $\theta_3 = \pi$ we use the compact notation $\cos\theta_3 = \sin\theta_3 = \frac{1}{\sqrt{2}} \equiv s$. Thus:

$$A = \begin{pmatrix} s & -s & 0 \\ s & s & 0 \\ 0 & 0 & 1 \end{pmatrix}, \qquad B = \begin{pmatrix} 3 & 0 & 0 \\ 0 & 1 & 0 \\ 0 & 0 & 1 \end{pmatrix}, \qquad C = \begin{pmatrix} 0 & 0 & 1 \\ 0 & 1 & 0 \\ -1 & 0 & 0 \end{pmatrix}.$$

(b)
$$\mathbf{y} = B\mathbf{x} = \begin{pmatrix} 3 & 0 & 0 \\ 0 & 1 & 0 \\ 0 & 0 & 1 \end{pmatrix}\begin{pmatrix} 1 \\ 1 \\ 1 \end{pmatrix} = \boxed{\begin{pmatrix} 3 \\ 1 \\ 1 \end{pmatrix}}.$$

(c) The composite map D is found by straightforward matrix multiplication:

$$D = CBA = \begin{pmatrix} 0 & 0 & 1 \\ 0 & 1 & 0 \\ -1 & 0 & 0 \end{pmatrix}\begin{pmatrix} 3 & 0 & 0 \\ 0 & 1 & 0 \\ 0 & 0 & 1 \end{pmatrix} A = \begin{pmatrix} 0 & 0 & 1 \\ 0 & 1 & 0 \\ -3 & 0 & 1 \end{pmatrix}\begin{pmatrix} s & -s & 0 \\ s & s & 0 \\ 0 & 0 & 1 \end{pmatrix} = \begin{pmatrix} 0 & 0 & 1 \\ s & s & 0 \\ -3s & 3s & 0 \end{pmatrix},$$

$$\mathbf{z} = D\mathbf{x} = \begin{pmatrix} 0 & 0 & 1 \\ s & s & 0 \\ -3s & 3s & 0 \end{pmatrix}\begin{pmatrix} 1 \\ 1 \\ 1 \end{pmatrix} = \boxed{\begin{pmatrix} 1 \\ 2s \\ 0 \end{pmatrix}}.$$

(d) On the one hand we have $\mathbf{e}_j \overset{A}{\mapsto} \mathbf{e}'_j$, with $\mathbf{e}'_j = \mathbf{e}_i A^i{}_j$, because the image of the standard basis vector $\mathbf{e}_j$ under the mapping A, written in the standard basis, is given by the column vector j of the matrix A, with components $A^i{}_j$. The inverse relationship is given by $\mathbf{e}_j = \mathbf{e}'_i (A^{-1})^i{}_j$. On the other hand we have $\mathbf{e}_j = \mathbf{e}'_i T^i{}_j$, by the definition of the transformation matrix. It follows that $\boxed{T = A^{-1}}$. Using the fact that A is a rotation matrix, we know that: $A^{-1}(\theta_3) = A(-\theta_3)$. Therefore:

$$
T = \begin{pmatrix} s & s & 0 \\ -s & s & 0 \\ 0 & 0 & 1 \end{pmatrix}.
$$

(e) $\quad \mathbf{x}' = T\mathbf{x} = \begin{pmatrix} s & s & 0 \\ -s & s & 0 \\ 0 & 0 & 1 \end{pmatrix} \begin{pmatrix} 1 \\ 1 \\ 1 \end{pmatrix} = \boxed{\begin{pmatrix} 2s \\ 0 \\ 1 \end{pmatrix}}, \quad \mathbf{y}' = T\mathbf{y} = \begin{pmatrix} s & s & 0 \\ -s & s & 0 \\ 0 & 0 & 1 \end{pmatrix} \begin{pmatrix} 3 \\ 1 \\ 1 \end{pmatrix} = \boxed{\begin{pmatrix} 4s \\ -2s \\ 1 \end{pmatrix}}.$

(f)   The transformed matrix B' can be found by matrix multiplication:

$$
B' = TBT^{-1} = A^{-1}BA = \begin{pmatrix} s & s & 0 \\ -s & s & 0 \\ 0 & 0 & 1 \end{pmatrix} \begin{pmatrix} 3 & 0 & 0 \\ 0 & 1 & 0 \\ 0 & 0 & 1 \end{pmatrix} A = \begin{pmatrix} 3s & s & 0 \\ -3s & s & 0 \\ 0 & 0 & 1 \end{pmatrix} \begin{pmatrix} s & -s & 0 \\ s & s & 0 \\ 0 & 0 & 1 \end{pmatrix}
$$

$$
= \begin{pmatrix} 4s^2 & -2s^2 & 0 \\ -2s^2 & 4s^2 & 0 \\ 0 & 0 & 1 \end{pmatrix} = \boxed{\begin{pmatrix} 2 & -1 & 0 \\ -1 & 2 & 0 \\ 0 & 0 & 1 \end{pmatrix}}.
$$

$$
\mathbf{y}' = T\mathbf{y} = TB\mathbf{x} = \underbrace{TBT^{-1}}_{B} \underbrace{T\mathbf{x}}_{\mathbf{x}'} = B'\mathbf{x}' = \begin{pmatrix} 2 & -1 & 0 \\ -1 & 2 & 0 \\ 0 & 0 & 1 \end{pmatrix} \begin{pmatrix} 2s \\ 0 \\ 1 \end{pmatrix} = \boxed{\begin{pmatrix} 4s \\ -2s \\ 1 \end{pmatrix}} \; [= (e)\checkmark].
$$

S.L6 Determinants (p. 86)

S.L6.2 Computing determinants (p. 88)

L6.2.1 Computing determinants (p. 88)

We expand the determinant along the indicated row or column:

$$
\det A = \begin{vmatrix} 2 & 1 \\ 5 & -3 \end{vmatrix} = 2 \cdot (-3) - 5 \cdot 1 = \boxed{-11}.
$$

$$
\det B = \begin{vmatrix} 3 & 2 & 1 \\ 4 & -3 & 1 \\ 2 & -1 & 1 \end{vmatrix} \overset{\text{column 3}}{=} \begin{vmatrix} 4 & -3 \\ 2 & -1 \end{vmatrix} - \begin{vmatrix} 3 & 2 \\ 2 & -1 \end{vmatrix} + \begin{vmatrix} 3 & 2 \\ 4 & -3 \end{vmatrix} = \boxed{-8}.
$$

$$
\det C = \begin{vmatrix} a & a & a & 0 \\ a & 0 & 0 & b \\ 0 & 0 & b & b \\ a & b & b & 0 \end{vmatrix} \overset{\text{row 2}}{=} -a \begin{vmatrix} a & a & 0 \\ 0 & b & b \\ b & b & 0 \end{vmatrix} + b \begin{vmatrix} a & a & a \\ 0 & 0 & b \\ a & b & b \end{vmatrix}
$$

$$
= -a \left[a \begin{vmatrix} b & b \\ b & 0 \end{vmatrix} - a \begin{vmatrix} 0 & b \\ b & 0 \end{vmatrix} \right] + b \left[0 + 0 - b \begin{vmatrix} a & a \\ a & b \end{vmatrix} \right]
$$

$$
= -a^2(-b^2) + a^2(-b^2) - b^2(ab - a^2) = \boxed{a^2 b^2 - ab^3}.
$$

S.L7 Matrix diagonalization (p. 98)

S.L7.3 Matrix diagonalization (p. 102)

ₑL7.3.₁ Matrix diagonalization (p. 103)

(a) The zeros of the characteristic polynomial yield the eigenvalues:

Char. polynomial:
$$0 \overset{!}{=} \det(A - \lambda\mathbb{1}) = \begin{vmatrix} -1-\lambda & 6 \\ -2 & 6-\lambda \end{vmatrix} = (-1-\lambda)(6-\lambda) + 12$$
$$= \lambda^2 - 5\lambda + 6 = (\lambda - 2)(\lambda - 3).$$

Eigenvalues: $\lambda_1 = \boxed{2}$, $\lambda_2 = \boxed{3}$.

Checks: $\lambda_1 + \lambda_2 = \boxed{5} \overset{\checkmark}{=} \operatorname{Tr} A = -1 + 6$, $\lambda_1\lambda_2 = \boxed{6} \overset{\checkmark}{=} \det A = -1 \cdot 6 - (-2) \cdot 6.$

Eigenvectors:

$\lambda_1 = 2$:
$$\mathbf{0} \overset{!}{=} (A - \lambda_1\mathbb{1})\mathbf{v}_1 = \begin{pmatrix} -3 & 6 \\ -2 & 4 \end{pmatrix}\mathbf{v}_1 \quad \Rightarrow \quad \mathbf{v}_1 = \boxed{\begin{pmatrix} 2 \\ 1 \end{pmatrix}};$$

$\lambda_2 = 3$:
$$\mathbf{0} \overset{!}{=} (A - \lambda_2\mathbb{1})\mathbf{v}_2 = \begin{pmatrix} -4 & 6 \\ -2 & 3 \end{pmatrix}\mathbf{v}_2 \quad \Rightarrow \quad \mathbf{v}_2 = \boxed{\begin{pmatrix} 3 \\ 2 \end{pmatrix}}.$$

Explicitly: the two rows of the matrix $(A - \lambda_j\mathbb{1})$ are proportional to each other (as expected, since the determinant of this matrix equals zero). Thus both rows yield the same information about the eigenvector $\mathbf{v}_j$. For $\mathbf{v}_1 = (v^1{}_1, v^2{}_1)^{\mathrm{T}}$ we have $-3v^1{}_1 + 6v^2{}_1 = 0$, implying $v^2{}_1 = \frac{1}{2}v^1{}_1$, thus it has the form $\mathbf{v}_1 = a_1(2, 1)^{\mathrm{T}}$. Similarly one finds $\mathbf{v}_2 = a_2(3, 2)^{\mathrm{T}}$. The prefactors a_1 and a_2 are *not* fixed by the eigenvalue equation, since if $\mathbf{v}_j$ satisfies $(A - \lambda_j\mathbb{1})\mathbf{v}_j = 0$, the same is true for $a_j\mathbf{v}_j$, with $a_j \in \mathbb{R}$. If one desires the eigenvectors to be normalized, the normalization condition $\|\mathbf{v}_j\| = 1$ fixes the absolute value of the prefactor, $|a_j|$. However, that is not the case here, hence we may choose the prefactor as we please – here we take $a_1 = a_2 = 1$.

The similarity transformation T contains the eigenvectors as columns; its inverse follows via the inversion formula for 2×2 matrices, $\begin{pmatrix} a & b \\ c & d \end{pmatrix}^{-1} = \frac{1}{ad-bc}\begin{pmatrix} d & -b \\ -c & a \end{pmatrix}$:

Sim-tr.:
$$T = (\mathbf{v}_1, \mathbf{v}_2) = \boxed{\begin{pmatrix} 2 & 3 \\ 1 & 2 \end{pmatrix}}, \quad T^{-1} = \frac{1}{2\cdot 2 - 3\cdot 1}\begin{pmatrix} 2 & -3 \\ -1 & 2 \end{pmatrix} = \boxed{\begin{pmatrix} 2 & -3 \\ -1 & 2 \end{pmatrix}}.$$

Check:
$$T^{-1}AT = T^{-1}\begin{pmatrix} -1 & 6 \\ -2 & 6 \end{pmatrix}\begin{pmatrix} 2 & 3 \\ 1 & 2 \end{pmatrix} = \begin{pmatrix} 2 & -3 \\ -1 & 2 \end{pmatrix}\begin{pmatrix} 4 & 9 \\ 2 & 6 \end{pmatrix} = \boxed{\begin{pmatrix} 2 & 0 \\ 0 & 3 \end{pmatrix}} \overset{\checkmark}{=} \begin{pmatrix} \lambda_1 & 0 \\ 0 & \lambda_2 \end{pmatrix}.$$

Conversely:
$$TDT^{-1} = T\begin{pmatrix} 2 & 0 \\ 0 & 3 \end{pmatrix}\begin{pmatrix} 2 & -3 \\ -1 & 2 \end{pmatrix} = \begin{pmatrix} 2 & 3 \\ 1 & 2 \end{pmatrix}\begin{pmatrix} 4 & -6 \\ -3 & 6 \end{pmatrix} = \boxed{\begin{pmatrix} -1 & 6 \\ -2 & 6 \end{pmatrix}} \overset{\checkmark}{=} A.$$

The converse check is a bit more convenient, since D involves zeros whereas A does not.

(b) The zeros of the characteristic polynomial yield the eigenvalues:

Char. polynomial:
$$0 \overset{!}{=} \det(A - \lambda\mathbb{1}) = \begin{vmatrix} -\mathrm{i}-\lambda & 0 \\ 2 & \mathrm{i}-\lambda \end{vmatrix} = (-\mathrm{i}-\lambda)(\mathrm{i}-\lambda).$$

Eigenvalues: $\lambda_1 = \boxed{+\mathrm{i}}$, $\lambda_2 = \boxed{-\mathrm{i}}$.

Checks: $\lambda_1 + \lambda_2 = \boxed{0} \overset{\checkmark}{=} \operatorname{Tr} A = \mathrm{i} + (-\mathrm{i})$, $\lambda_1\lambda_2 = \boxed{1} \overset{\checkmark}{=} \det A = -\mathrm{i} \cdot \mathrm{i} - 2 \cdot 0.$

Eigenvectors:

$$\lambda_1 = +\mathrm{i}: \quad \mathbf{0} \overset{!}{=} (A - \lambda_1 \mathbb{1})\mathbf{v}_1 = \begin{pmatrix} -2\mathrm{i} & 0 \\ 2 & 0 \end{pmatrix}\mathbf{v}_1 \quad \Rightarrow \quad \boxed{\mathbf{v}_1 = a_1 \begin{pmatrix} 0 \\ 1 \end{pmatrix}};$$

$$\lambda_2 = -\mathrm{i}: \quad \mathbf{0} \overset{!}{=} (A - \lambda_2 \mathbb{1})\mathbf{v}_2 = \begin{pmatrix} 0 & 0 \\ 2 & 2\mathrm{i} \end{pmatrix}\mathbf{v}_2 \quad \Rightarrow \quad \boxed{\mathbf{v}_2 = a_2 \begin{pmatrix} 1 \\ \mathrm{i} \end{pmatrix}}.$$

Explicitly: for the eigenvector $\mathbf{v}_1 = (v^1{}_1, v^2{}_1)^{\mathrm{T}}$ we have $-2\mathrm{i}v^1{}_1 + 0 v^2{}_1 = 0$, implying $v^1{}_1 = 0$, thus it has the form $\mathbf{v}_1 = a_1(0,1)^{\mathrm{T}}$. Similarly one finds $\mathbf{v}_2 = a_2(1,\mathrm{i})^{\mathrm{T}}$. The prefactors are complex numbers and thus have the general form $a_j = |a_j|(\cos\phi_j + \mathrm{i}\sin\phi_j)$, with norm $|a_j|$ and phase ϕ_j. For illustration purposes, let us choose the eigenvectors to have norm $\|\mathbf{v}_j\| = 1$. This fixes the norm $|a_j|$ of each prefactor, but not its phase, which we may choose as we please. We here choose $\phi_j = 0$, hence $a_1 = 1$ and $a_2 = \frac{1}{\sqrt{2}}$.

The similarity transformation T contains the eigenvectors as columns; its inverse follows via the inversion formula for 2×2 matrices, which holds also for complex matrices:

$$\text{Sim-tr.:} \qquad T = (\mathbf{v}_1, \mathbf{v}_2) = \boxed{\begin{pmatrix} 0 & \frac{1}{\sqrt{2}} \\ 1 & \frac{\mathrm{i}}{\sqrt{2}} \end{pmatrix}}, \quad T^{-1} = \frac{1}{\frac{0\cdot\mathrm{i}}{\sqrt{2}} - \frac{1\cdot 1}{\sqrt{2}}} \begin{pmatrix} \frac{\mathrm{i}}{\sqrt{2}} & \frac{-1}{\sqrt{2}} \\ -1 & 0 \end{pmatrix} = \boxed{\begin{pmatrix} -\mathrm{i} & 1 \\ \sqrt{2} & 0 \end{pmatrix}}.$$

To compute TDT^{-1}, we exploit the zeros in $D = \begin{pmatrix} \mathrm{i} & 0 \\ 0 & -\mathrm{i} \end{pmatrix}$ to directly write down DT^{-1}.

$$\text{Check:} \qquad TDT^{-1} = \begin{pmatrix} 0 & \frac{1}{\sqrt{2}} \\ 1 & \frac{\mathrm{i}}{\sqrt{2}} \end{pmatrix}\begin{pmatrix} \mathrm{i}\cdot(-\mathrm{i}) & \mathrm{i}\cdot 1 \\ -\mathrm{i}\cdot\sqrt{2} & -\mathrm{i}\cdot 0 \end{pmatrix} = \boxed{\begin{pmatrix} -\mathrm{i} & 0 \\ 2 & \mathrm{i} \end{pmatrix}} \overset{\checkmark}{=} A.$$

Remark: Our decision to use normalized eigenvectors has yielded some square roots. Using unnormalized eigenvectors avoids these and leads to simpler expressions, e.g.:

$$\tilde{T} = \begin{pmatrix} 0 & 1 \\ 1 & \mathrm{i} \end{pmatrix}, \quad \tilde{T}^{-1} = \begin{pmatrix} -\mathrm{i} & 1 \\ 1 & 0 \end{pmatrix}, \quad \tilde{T}D\tilde{T}^{-1} = \begin{pmatrix} 0 & 1 \\ 1 & \mathrm{i} \end{pmatrix}\begin{pmatrix} \mathrm{i}\cdot(-\mathrm{i}) & \mathrm{i}\cdot 1 \\ -\mathrm{i}\cdot 1 & -\mathrm{i}\cdot 0 \end{pmatrix} = \begin{pmatrix} -\mathrm{i} & 0 \\ 2 & \mathrm{i} \end{pmatrix} \overset{\checkmark}{=} A.$$

(c)   The zeros of the characteristic polynomial yield the eigenvalues:

$$\text{Char. polynomial:} \qquad 0 \overset{!}{=} \det(A - \lambda\mathbb{1}) = \begin{vmatrix} 1-\lambda & 0 & -1 \\ 0 & 2\mathrm{i}-\lambda & 0 \\ 1 & 0 & 1-\lambda \end{vmatrix} = (1-\lambda)^2(2\mathrm{i}-\lambda) + (2\mathrm{i}-\lambda)$$

$$= (2\mathrm{i}-\lambda)(2 - 2\lambda + \lambda^2) = (2\mathrm{i}-\lambda)(1+\mathrm{i}-\lambda)(1-\mathrm{i}-\lambda).$$

Eigenvalues: $\lambda_1 = \boxed{2\mathrm{i}}, \lambda_2 = \boxed{1+\mathrm{i}}, \lambda_3 = \boxed{1-\mathrm{i}}$.

Checks: $\lambda_1 + \lambda_2 + \lambda_3 = \boxed{2+2\mathrm{i}} \overset{\checkmark}{=} \operatorname{Tr} A = 1 + 2\mathrm{i} + 1,$

$$\lambda_1\lambda_2\lambda_3 = \boxed{4\mathrm{i}} \overset{\checkmark}{=} \det A = 1\cdot(2\mathrm{i}\cdot 1) + 1\cdot(-1)(2\mathrm{i})(-1).$$

Eigenvectors:

$$\lambda_1 = 2\mathrm{i}: \qquad \mathbf{0} \overset{!}{=} (A - \lambda_1\mathbb{1})\mathbf{v}_1 = \begin{pmatrix} 1-2\mathrm{i} & 0 & -1 \\ 0 & 0 & 0 \\ 1 & 0 & 1-2\mathrm{i} \end{pmatrix}\mathbf{v}_1 \quad \Rightarrow \quad \boxed{\mathbf{v}_1 = a_1 \begin{pmatrix} 0 \\ 1 \\ 0 \end{pmatrix}};$$

$$\lambda_2 = 1+\mathrm{i}: \quad \mathbf{0} \overset{!}{=} (A - \lambda_2\mathbb{1})\mathbf{v}_2 = \begin{pmatrix} -\mathrm{i} & 0 & -1 \\ 0 & -1+\mathrm{i} & 0 \\ 1 & 0 & -\mathrm{i} \end{pmatrix}\mathbf{v}_2 \quad \Rightarrow \quad \boxed{\mathbf{v}_2 = a_2 \begin{pmatrix} 1 \\ 0 \\ -\mathrm{i} \end{pmatrix}};$$

$$\lambda_3 = 1 - i: \quad \mathbf{0} \overset{!}{=} (A - \lambda_3 \mathbb{1})\mathbf{v}_3 = \begin{pmatrix} i & 0 & -1 \\ 0 & -1+3i & 0 \\ 1 & 0 & i \end{pmatrix} \mathbf{v}_3 \quad \Rightarrow \quad \mathbf{v}_3 = \boxed{a_3 \begin{pmatrix} 1 \\ 0 \\ i \end{pmatrix}}.$$

Explicitly: for the eigenvector $\mathbf{v}_1 = (v^1{}_1, v^2{}_1, v^3{}_1)^T$, the first row of the matrix $(A - \lambda_1 \mathbb{1})$ yields the condition $(1 - 2i)v^1{}_1 - 1v^3{}_1 = 0$, while the third row yields $v^1{}_1 + (1 - 2i)v^3{}_1 = 0$; this directly implies $v^1{}_1 = v^3{}_1 = 0$. The component $v^2{}_1$ can be chosen arbitrarily, hence $\mathbf{v}_1 = a_1(0, 1, 0)^T$. For the eigenvector $\mathbf{v}_2 = (v^1{}_2, v^2{}_2, v^3{}_2)^T$, the second row of $(A - \lambda_2 \mathbb{1})$ yields $(-1+i)v^2{}_2 = 0$, implying $v^2{}_2 = 0$. The first and third rows are multiples of each other and yield $-iv^1{}_2 - v^3{}_2 = 0$, which can be satisfied, for example, by choosing $v^1{}_2 = a_2$ and $v^3{}_2 = -ia_2$. The third eigenvector is found analogously to the second. The prefactors $a_j \in \mathbb{C}$ are not fixed by the eigenvalue equation. For definiteness, let us choose the first nonzero component of each eigenvector equal to unity and hence take $a_j = 1$.

The similarity transformation T diagonalizing A contains the eigenvectors as columns:

$$T = (\mathbf{v}_1, \mathbf{v}_2, \mathbf{v}_3) = \boxed{\begin{pmatrix} 0 & 1 & 1 \\ 1 & 0 & 0 \\ 0 & -i & i \end{pmatrix}}. \text{ We find its inverse using Gaussian elimination:}$$

$$
\begin{array}{llll}
[1]: & 0 \ \ 1 \ \ 1 & \Big| & 1 \ \ 0 \ \ 0 \\
[2]: & 1 \ \ 0 \ \ 0 & \Big| & 0 \ \ 1 \ \ 0 \\
[3]: & 0 \ -i \ \ i & \Big| & 0 \ \ 0 \ \ 1
\end{array}
\longrightarrow
\begin{array}{llll}
[2]: & 1 \ \ 0 \ \ 0 & \Big| & 0 \ \ 1 \ \ 0 \\
\frac{1}{2}([1]+i[3]): & 0 \ \ 1 \ \ 0 & \Big| & \frac{1}{2} \ \ 0 \ \ \frac{i}{2} \\
\frac{1}{2}([1]-i[3]): & 0 \ \ 0 \ \ 1 & \Big| & \frac{1}{2} \ \ 0 \ \ \frac{-i}{2}
\end{array}
$$

Hence $T^{-1} = \boxed{\begin{pmatrix} 0 & 1 & 0 \\ \frac{1}{2} & 0 & \frac{i}{2} \\ \frac{1}{2} & 0 & \frac{-i}{2} \end{pmatrix}}$. Finally, let us check whether TDT^{-1} reproduces A.

Check: $\quad TDT^{-1} = \begin{pmatrix} 0 & 1 & 1 \\ 1 & 0 & 0 \\ 0 & -i & i \end{pmatrix} \begin{pmatrix} 2i \cdot 0 & 2i \cdot 1 & 2i \cdot 0 \\ (1+i) \cdot \frac{1}{2} & (1+i) \cdot 0 & (1+i) \cdot \frac{i}{2} \\ (1-i) \cdot \frac{1}{2} & (1-i) \cdot 0 & (1-i) \cdot \frac{-i}{2} \end{pmatrix} = \boxed{\begin{pmatrix} 1 & 0 & -1 \\ 0 & 2i & 0 \\ 1 & 0 & 1 \end{pmatrix}} \overset{\checkmark}{=} A.$

eL7.3.3 Diagonalizing a matrix that depends on a variable (p. 103)

Char. poly.: $\quad 0 \overset{!}{=} \det(A - \lambda \mathbb{1}) = \begin{vmatrix} x-\lambda & 1 & 0 \\ 1 & 2-\lambda & 1 \\ 3-x & -1 & 3-\lambda \end{vmatrix}$

$$= (x-\lambda)(2-\lambda)(3-\lambda) + (3-x) + (x-\lambda) - (3-\lambda)$$

$$= (x-\lambda)(2-\lambda)(3-\lambda).$$

Eigenvalues: $\quad \boxed{\lambda_1 = x, \quad \lambda_2 = 2, \quad \lambda_3 = 3.}$

Checks: $\quad \lambda_1 + \lambda_2 + \lambda_3 = x + 5 \overset{\checkmark}{=} \operatorname{Tr} A,$

$$\lambda_1 \lambda_2 \lambda_3 = 6x \overset{\checkmark}{=} \det A = x \cdot (2 \cdot 3 - (-1) \cdot 1) - 1 \cdot (1 \cdot 3 - (3-x) \cdot 1).$$

Eigenvectors:

$$0 \overset{!}{=} (A - \lambda_1 \mathbb{1})\mathbf{v}_1 = \begin{pmatrix} 0 & 1 & 0 \\ 1 & 2-x & 1 \\ 3-x & -1 & 3-x \end{pmatrix} \mathbf{v}_1 \overset{\text{Gauss}}{\longrightarrow} \boxed{\mathbf{v}_1 = a_1 \begin{pmatrix} 1 \\ 0 \\ -1 \end{pmatrix}, \ |a_1| = \frac{1}{\sqrt{2}}};$$

$$0 \overset{!}{=} (A - \lambda_2 \mathbb{1})\mathbf{v}_2 = \begin{pmatrix} x-2 & 1 & 0 \\ 1 & 0 & 1 \\ 3-x & -1 & 1 \end{pmatrix}\mathbf{v}_2 \qquad \overset{\text{Gauss}}{\longrightarrow} \qquad \boxed{\mathbf{v}_2 = a_2 \begin{pmatrix} 1 \\ 2-x \\ -1 \end{pmatrix}, \quad |a_2| = \frac{1}{\sqrt{6-4x+x^2}}} \ ;$$

$$0 \overset{!}{=} (A - \lambda_3 \mathbb{1})\mathbf{v}_3 = \begin{pmatrix} x-3 & 1 & 0 \\ 1 & -1 & 1 \\ 3-x & -1 & 0 \end{pmatrix}\mathbf{v}_3 \qquad \overset{\text{Gauss}}{\longrightarrow} \qquad \boxed{\mathbf{v}_3 = a_3 \begin{pmatrix} 1 \\ 3-x \\ 2-x \end{pmatrix}, \quad |a_3| = \frac{1}{\sqrt{14-10x+2x^2}}} \ .$$

εL7.3.5 Degenerate eigenvalue problem (p. 105)

The zeros of the characteristic polynomial yield the eigenvalues:

$$0 \overset{!}{=} \det(A - \lambda \mathbb{1}) = \begin{vmatrix} 2-\lambda & -1 & 2 \\ -1 & 2-\lambda & -2 \\ 2 & -2 & 5-\lambda \end{vmatrix}$$

$$\overset{\text{(i)}}{=} (2-\lambda)\big[(2-\lambda)(5-\lambda)-4\big] - (-1)\big[(-1)(5-\lambda)+4\big] + 2\big[2-(2-\lambda)2\big]$$

$$\overset{\text{(ii)}}{=} (2-\lambda)\big[\lambda^2 - 7\lambda + 6\big] + 5(\lambda - 1) \overset{\text{(iii)}}{=} (2-\lambda)\big[(\lambda-1)(\lambda-6)\big] + 5(\lambda-1)$$

$$\overset{\text{(iv)}}{=} (\lambda-1)\big[(2-\lambda)(\lambda-6)+5\big] \overset{\text{(v)}}{=} -(\lambda-1)(\lambda-1)(\lambda-7).$$

Remarks: (i) We calculate the determinant using the Laplace expansion along the first column, and (ii) then simplify. (iii) From the hint that $\lambda = 1$ is an eigenvalue, we know that $\det(A - \lambda\mathbb{1})$, and thus the square bracket, too, must contain a factor of $(\lambda - 1)$. (iv) We evaluate this bracket and factorize (v) again, using the quadratic formula for example:

(v): $(2-\lambda)(\lambda-6)+5 = -\lambda^2 + 8\lambda - 7 = -(\lambda-1)(\lambda-7)$, since $\frac{-8\pm\sqrt{64-28}}{-2} = 4 \mp 3 = \begin{cases} 1 \\ 7 \end{cases}$.

Alternatively (if the factorization is not apparent): (iii′) completely multiply out $\det(A-\lambda\mathbb{1})$, (iv′) then use polynomial division to factorize out the factor $(\lambda - 1)$, and (v′) factorize the residual quadratic polynomial as in step (v) above:

$$\det(A - \lambda\mathbb{1}) \overset{\text{(iii′)}}{=} -\lambda^3 + 9\lambda^2 - 15\lambda + 7 \overset{\text{(iv′)}}{=} (\lambda - 1)(-\lambda^2 + 8\lambda - 7)$$

$$\overset{\text{(v′)}}{=} -(\lambda - 1)(\lambda - 1)(\lambda - 7).$$

(iv′) Polynomial division:

$$
\begin{array}{l}
\big(-\lambda^3 + 9\lambda^2 - 15\lambda + 7 \big)\,/\,(\lambda - 1) = -\lambda^2 + 8\lambda - 7 \\
\underline{\lambda^3 \ -\lambda^2} \\
\qquad 8\lambda^2 - 15\lambda \\
\qquad \underline{-8\lambda^2 \ +8\lambda} \\
\qquad\qquad -7\lambda + 7 \\
\qquad\qquad \underline{7\lambda - 7} \\
\qquad\qquad\qquad 0
\end{array}
$$

Eigenvalues: $\lambda_1 = \lambda_2 = \boxed{1}$, $\lambda_3 = \boxed{7}$. The eigenvalue 1 is two-fold degenerate.

Checks: $\lambda_1 + \lambda_2 + \lambda_3 = \boxed{9} \overset{\checkmark}{=} \operatorname{Tr} A$, $\quad \lambda_1\lambda_2\lambda_3 = \boxed{7} \overset{\checkmark}{=} \det A$.

Determination of the normalized eigenvector $\mathbf{v}_3$ of the non-degenerate eigenvalue $\lambda_3 = 7$:

$$\lambda_3 = 7 : \qquad 0 \overset{!}{=} (A - \lambda_3\mathbb{1})\mathbf{v}_3 = \begin{pmatrix} -5 & -1 & 2 \\ -1 & -5 & -2 \\ 2 & -2 & -2 \end{pmatrix}\mathbf{v}_3 \qquad \overset{\text{Gauss}}{\Longrightarrow} \qquad \boxed{\mathbf{v}_3 = \frac{1}{\sqrt{6}}\begin{pmatrix} 1 \\ -1 \\ 2 \end{pmatrix}}.$$

Details of the Gauss method:

$$
\begin{array}{ccc|c}
v^1_3 & v^2_3 & v^3_3 & \\
\hline
-5 & -1 & 2 & 0 \\
-1 & -5 & -2 & 0 \\
2 & -2 & -2 & 0
\end{array}
\;\Rightarrow\;
\begin{array}{l}
-[1]: \\
\frac{1}{12}([1]-5[2]): \\
-\frac{1}{6}([3]+2[2]):
\end{array}
\begin{array}{ccc|c}
v^1_3 & v^2_3 & v^3_3 & \\
\hline
5 & 1 & -2 & 0 \\
0 & 2 & 1 & 0 \\
0 & 2 & 1 & 0
\end{array}
\;\Rightarrow\;
\begin{array}{l}
\frac{1}{10}(2[1]-[2]): \\
\frac{1}{2}[2]: \\
{[2]}-[3]:
\end{array}
\begin{array}{ccc|c}
v^1_3 & v^2_3 & v^3_3 & \\
\hline
1 & 0 & -\frac{1}{2} & 0 \\
0 & 1 & \frac{1}{2} & 0 \\
0 & 0 & 0 & 0
\end{array}
$$

The system on the right gives two relations between the components of $\mathbf{v}_3 = (v^1_3, v^2_3, v^3_3)^{\mathrm{T}}$, namely. $v^1_3 - \frac{1}{2}v^3_3 = 0$ and $v^2_3 + \frac{1}{2}v^3_3 = 0$. Since the third row contains only zeros, the eigenvector is determined (as expected) only up to a prefactor $a_3 \in \mathbb{C}$, which can be freely chosen: $\mathbf{v}_3 = a_3(1, -1, 2)^{\mathrm{T}}$. The normalization condition $\|\mathbf{v}_3\| = 1$ implies that $a_3 = \pm\frac{1}{\sqrt{6}}$; here we select the positive sign (the negative sign would be equally legitimate).

Determination of the normalized eigenvectors $\mathbf{v}_{1,2}$ of the degenerate eigenvalue $\lambda_{1,2} = 1$:

$$
\lambda_1 = \lambda_2 = 1: \qquad\qquad \mathbf{0} \overset{!}{=} (A - \lambda_j \mathbb{1})\mathbf{v}_j = \begin{pmatrix} 1 & -1 & 2 \\ -1 & 1 & -2 \\ 2 & -2 & 4 \end{pmatrix} \mathbf{v}_j.
$$

All three rows are proportional to each other, $[3] = 2[1] = -2[2]$. In case one does not notice this immediately and uses Gaussian elimination, one is led to the same conclusion:

$$
\begin{array}{ccc|c}
v^1_j & v^2_j & v^3_j & \\
\hline
1 & -1 & 2 & 0 \\
-1 & 1 & -2 & 0 \\
2 & -2 & 4 & 0
\end{array}
\;\Rightarrow\;
\begin{array}{l}
[1]: \\
([1]+[2]): \\
(2[1]-[3]):
\end{array}
\begin{array}{ccc|c}
v^1_j & v^2_j & v^3_j & \\
\hline
1 & -1 & 2 & 0 \\
0 & 0 & 0 & 0 \\
0 & 0 & 0 & 0
\end{array}
$$

The augmented matrix on the right contains nothing but zeros in both its second and third rows. This is a direct consequence of the fact that on the left, the second and third rows are both proportional to the first. Since only one row is nontrivial, we obtain only *one* relation between the components of $\mathbf{v}_j = (v^1_j, v^2_j, v^3_j)^{\mathrm{T}}$, namely $v^1_j - v^2_j + 2v^3_j = 0$. Therefore, we can choose two components of $\mathbf{v}_j$ freely and thereby construct two linearly independent eigenvectors, e.g., $\mathbf{v}_1 = (1, 1, 0)^{\mathrm{T}}$ and $\mathbf{v}_2 = (0, 2, 1)^{\mathrm{T}}$. We could normalize them, too, but choose to postpone this to the next and final step, namely orthonormalizing all eigenvectors.

The eigenvectors of different eigenvalues are already orthogonal, $\langle \mathbf{v}_j, \mathbf{v}_3 \rangle = 0$ for $j = 1, 2$. (This is no coincidence; as explained in Chapter L8, it follows from the fact that $A = A^{\mathrm{T}}$.) Thus it suffices to orthonormalize the degenerate eigenvectors $\mathbf{v}_1$ and $\mathbf{v}_2$. We use the Gram–Schmidt procedure, with

$$
\mathbf{v}'_1 = \mathbf{v}_1/\|\mathbf{v}_1\| = \boxed{\frac{1}{\sqrt{2}}(1, 1, 0)^{\mathrm{T}}} \quad \text{and}
$$

$$
\mathbf{v}'_{2,\perp} = \mathbf{v}_2 - \mathbf{v}'_1 \langle \mathbf{v}'_1, \mathbf{v}_2 \rangle = \begin{pmatrix} 0 \\ 2 \\ 1 \end{pmatrix} - \frac{1}{\sqrt{2}}\begin{pmatrix} 1 \\ 1 \\ 0 \end{pmatrix}\frac{2}{\sqrt{2}} = \begin{pmatrix} -1 \\ 1 \\ 1 \end{pmatrix}, \quad \Rightarrow \quad \mathbf{v}'_2 = \frac{\mathbf{v}'_{2,\perp}}{\|\mathbf{v}'_{2,\perp}\|} = \boxed{\frac{1}{\sqrt{3}}\begin{pmatrix} -1 \\ 1 \\ 1 \end{pmatrix}}.
$$

The vectors $\{\mathbf{v}'_1, \mathbf{v}'_2, \mathbf{v}_3\}$ form an orthonormal basis of $\mathbb{R}^3$. We use them as columns for T. Inverting the latter, e.g. using Gaussian elimination, we find that $T^{-1} = T^{\mathrm{T}}$:

$$
\text{Sim. tr.:} \qquad T = (\mathbf{v}'_1, \mathbf{v}'_2, \mathbf{v}_3) = \boxed{\begin{pmatrix} \frac{1}{\sqrt{2}} & \frac{-1}{\sqrt{3}} & \frac{1}{\sqrt{6}} \\ \frac{1}{\sqrt{2}} & \frac{1}{\sqrt{3}} & \frac{-1}{\sqrt{6}} \\ 0 & \frac{1}{\sqrt{3}} & \frac{2}{\sqrt{6}} \end{pmatrix}}, \quad T^{-1} = \boxed{\begin{pmatrix} \frac{1}{\sqrt{2}} & \frac{1}{\sqrt{2}} & 0 \\ \frac{-1}{\sqrt{3}} & \frac{1}{\sqrt{3}} & \frac{1}{\sqrt{3}} \\ \frac{1}{\sqrt{6}} & \frac{-1}{\sqrt{6}} & \frac{2}{\sqrt{6}} \end{pmatrix}} = \begin{pmatrix} \mathbf{v}'^{\mathrm{T}}_1 \\ \mathbf{v}'^{\mathrm{T}}_2 \\ \mathbf{v}^{\mathrm{T}}_3 \end{pmatrix} = T^{\mathrm{T}}.
$$

Check: $TDT^{-1} = \begin{pmatrix} \frac{1}{\sqrt{2}} & \frac{-1}{\sqrt{3}} & \frac{1}{\sqrt{6}} \\ \frac{1}{\sqrt{2}} & \frac{1}{\sqrt{3}} & \frac{-1}{\sqrt{6}} \\ 0 & \frac{1}{\sqrt{3}} & \frac{2}{\sqrt{6}} \end{pmatrix} \begin{pmatrix} 1 \cdot \frac{1}{\sqrt{2}} & 1 \cdot \frac{1}{\sqrt{2}} & 1 \cdot 0 \\ 1 \cdot \frac{-1}{\sqrt{3}} & 1 \cdot \frac{1}{\sqrt{3}} & 1 \cdot \frac{1}{\sqrt{3}} \\ 7 \cdot \frac{1}{\sqrt{6}} & 7 \cdot \frac{-1}{\sqrt{6}} & 7 \cdot \frac{2}{\sqrt{6}} \end{pmatrix} = \boxed{\begin{pmatrix} 2 & -1 & 2 \\ -1 & 2 & -2 \\ 2 & -2 & 5 \end{pmatrix}} \overset{\checkmark}{=} A.$

The fact that $T^{-1} = T^{\mathsf{T}}$ is no coincidence. As explained in Chapter L8, this property follows from the orthonormality of the eigenvectors forming the columns of T.

S.L7.4 Functions of matrices (p. 107)

εL7.4.1 Functions of matrices (p. 108)

(a) For $A = \begin{pmatrix} 0 & a \\ 0 & 0 \end{pmatrix}$ we have $A^2 = 0$, thus the Taylor series for e^A contains only two terms:

$$e^A = A^0 + A = \mathbb{1} + A = \boxed{\begin{pmatrix} 1 & a \\ 0 & 1 \end{pmatrix}}.$$

(b) We seek e^A, with $A = \theta\tilde{\sigma}$, $\tilde{\sigma} = \begin{pmatrix} 0 & -1 \\ 1 & 0 \end{pmatrix}$. The matrix $\tilde{\sigma}$ has the following properties:

$$\tilde{\sigma}^2 = \begin{pmatrix} 0 & -1 \\ 1 & 0 \end{pmatrix}\begin{pmatrix} 0 & -1 \\ 1 & 0 \end{pmatrix} = -\mathbb{1}, \quad \tilde{\sigma}^{2m} = (\tilde{\sigma}^2)^m = (-1)^m\mathbb{1}, \quad \tilde{\sigma}^{2m+1} = \tilde{\sigma}(\tilde{\sigma}^2)^m = (-1)^m\tilde{\sigma}.$$

Therefore: $e^A = \sum_{l=0}^{\infty} \frac{1}{l!}A^l = \sum_{m=0}^{\infty} \frac{1}{(2m)!}\theta^{2m} \underbrace{\tilde{\sigma}^{2n}}_{(-1)^m\mathbb{1}} + \sum_{m=0}^{\infty} \frac{1}{(2m+1)!}\theta^{2m+1} \underbrace{\tilde{\sigma}^{2m+1}}_{(-1)^m\tilde{\sigma}}$

$$= \mathbb{1}\cos\theta + \tilde{\sigma}\sin\theta = \boxed{\begin{pmatrix} \cos\theta & -\sin\theta \\ \sin\theta & \cos\theta \end{pmatrix}}.$$

Remark: This matrix describes a rotation by the angle θ in $\mathbb{R}^2$. Evidently $e^{\theta\tilde{\sigma}}$ is an exponential representation of such a rotation matrix.

(c) First equality:

$$\boxed{f(A)} - \sum_{l=0}^{\infty} c_l (\underbrace{TDT^{-1}}_{A})^l \overset{(i)}{=} \sum_{l=0}^{\infty} c_l TD^l T^{-1} = T\Big(\sum_{l=0}^{\infty} c_l D^l\Big)T^{-1} = \boxed{Tf(D)T^{-1}}.$$

(i) For the third step we used the following relation (holding for any matrix A, hence also for D):

$$\boxed{(TAT^{-1})^l} = (TA\underbrace{T^{-1})(T}_{=\mathbb{1}}A\underbrace{T^{-1})(T}_{=\mathbb{1}}AT^{-1})\cdots(TAT^{-1}) = \boxed{TA^l T^{-1}}.$$

Second equality:

$$\boxed{[f(D)]_{ij}} = \Big[\sum_{l=0}^{\infty} c_l D^l\Big]_{ij} = \sum_{l=0}^{\infty} c_l [D^l]_{ij} \overset{(ii)}{=} \sum_{l=0}^{\infty} c_l \lambda_i^l \delta_{ij} = \delta_{ij}\sum_{l=0}^{\infty} c_l \lambda_i^l = \boxed{\delta_{ij}f(\lambda_i)}.$$

(ii) For the third step we used the fact that the lth power of a diagonal matrix D is diagonal too, with $D^l = \text{diag}(\lambda_1^l, \ldots, \lambda_n^l)$.

(d) Given: e^A, with $A = \theta\begin{pmatrix} 0 & -1 \\ 1 & 0 \end{pmatrix}$. We begin by diagonalizing A.

Char. polynomial: $0 \overset{!}{=} \det(A - \lambda\mathbb{1}) = \lambda^2 + \theta^2 \Rightarrow$ Eigenvalues: $\boxed{\lambda_\pm = \pm i\theta}$.

Checks: $\qquad\qquad\qquad \lambda_+ + \lambda_- = 0 \overset{\checkmark}{=} \mathrm{Tr}\,A\,, \quad \lambda_+\lambda_- = \theta^2 \overset{\checkmark}{=} \det A\,.$

Normalized eigenvectors: $\quad \mathbf{0} \overset{!}{=} (A - \lambda_\pm \mathbb{1})\mathbf{v}_\pm \;\Rightarrow\; \mathbf{v}_\pm = \frac{1}{\sqrt{2}}\begin{pmatrix} 1 \\ \mp i \end{pmatrix}.$

Similarity transf.: $\qquad\qquad T = (\mathbf{v}_+, \mathbf{v}_-) = \frac{1}{\sqrt{2}}\begin{pmatrix} 1 & 1 \\ -i & i \end{pmatrix}, \qquad T^{-1} = \frac{1}{\sqrt{2}}\begin{pmatrix} 1 & i \\ 1 & -i \end{pmatrix}.$

$e^A = T e^D T^{-1}:$

$$e^A = T \begin{pmatrix} e^{i\theta} & 0 \\ 0 & e^{-i\theta} \end{pmatrix} \frac{1}{\sqrt{2}}\begin{pmatrix} 1 & i \\ 1 & -i \end{pmatrix} = \frac{1}{2}\begin{pmatrix} 1 & 1 \\ -i & i \end{pmatrix}\begin{pmatrix} e^{i\theta} & ie^{i\theta} \\ e^{-i\theta} & -ie^{-i\theta} \end{pmatrix}$$

$$= \frac{1}{2}\begin{pmatrix} e^{i\theta} + e^{-i\theta} & ie^{i\theta} - ie^{-i\theta} \\ -ie^{i\theta} + ie^{-i\theta} & e^{i\theta} + e^{-i\theta} \end{pmatrix} = \boxed{\begin{pmatrix} \cos\theta & -\sin\theta \\ \sin\theta & \cos\theta \end{pmatrix}}.$$

This agrees with the result from (b).

S.L8 Unitarity and Hermiticity (p. 109)

S.L8.1 Unitarity and orthogonality (p. 109)

L8.1.1 Orthogonal and unitary matrices (p. 111)

(a) The real matrix A is orthogonal, since

$$A A^\mathsf{T} = \begin{pmatrix} \sin\theta & \cos\theta \\ -\cos\theta & \sin\theta \end{pmatrix}\begin{pmatrix} \sin\theta & -\cos\theta \\ \cos\theta & \sin\theta \end{pmatrix} = \mathbb{1}.$$

The complex matrix B is not unitary, since

$$B B^\dagger = \frac{1}{1-i}\frac{1}{1+i}\begin{pmatrix} 2 & 1+i & 0 \\ 1+i & -1 & 1 \\ 0 & 2 & i \end{pmatrix}\begin{pmatrix} 2 & 1-i & 0 \\ 1-i & -1 & 2 \\ 0 & 1 & -i \end{pmatrix} = \frac{1}{2}\begin{pmatrix} 6 & 1-3i & 2+2i \\ 1+3i & 4 & -2-i \\ 2-2i & -2+i & 5 \end{pmatrix} \neq \mathbb{1}.$$

(b) $\qquad\qquad \mathbf{x} = (1,2)^\mathsf{T}, \qquad \mathbf{a} = A\mathbf{x} = (\sin\theta + 2\cos\theta, -\cos\theta + 2\sin\theta)^\mathsf{T},$

$$\|\mathbf{x}\| = \boxed{\sqrt{5}}, \qquad \|\mathbf{a}\| = \sqrt{5\sin^2\theta + 5\cos^2\theta + \sin\theta\cos\theta(4-4)} = \boxed{\sqrt{5}}.$$

Since A is orthogonal, the norm is conserved.

(c) $\qquad\qquad \mathbf{y} = (1,2,i)^\mathsf{T}, \qquad\qquad \mathbf{b} = B\mathbf{y} = \frac{1}{1-i}(2 + 2(1+i), (1+i) - 2 + i, 4 + i^2)^\mathsf{T}$

$$\|\mathbf{y}\| = \sqrt{1+4+1} = \boxed{\sqrt{6}}, \qquad \|\mathbf{b}\| = \sqrt{\tfrac{1}{2}[(16+4) + (1+4) + (9)]} = \boxed{\sqrt{17}}.$$

Since B is not unitary, the norm is not conserved.

S.L8.2 Hermiticity and symmetry (p. 117)

L8.2.1 Diagonalizing symmetric or Hermitian matrices (p. 119)

(a) The zeros of the characteristic polynomial yield the eigenvalues:

Char. polynomial: $\qquad 0 \overset{!}{=} \det(A - \lambda\mathbb{1}) = \begin{vmatrix} 3-\lambda & -4 \\ -4 & -3-\lambda \end{vmatrix} = (3-\lambda)(-3-\lambda) - 16$

$$= \lambda^2 - 25 = (\lambda - 5)(\lambda + 5).$$

Eigenvalues: $\lambda_1 = \boxed{5}$, $\lambda_2 = \boxed{-5}$.

Checks: $\lambda_1 + \lambda_2 = \boxed{0} \stackrel{\checkmark}{=} \mathrm{Tr}\, A = 3 - 3$, $\lambda_1\lambda_2 = \boxed{-25} \stackrel{\checkmark}{=} \det A = 3\cdot(-3) - (-4)^2$.

Eigenvectors:

$\lambda_1 = 15$: $\mathbf{0} \stackrel{!}{=} (A - \lambda_1 \mathbb{1})\mathbf{v}_1 = \begin{pmatrix} -2 & -4 \\ -4 & -8 \end{pmatrix}\mathbf{v}_1 \quad \Rightarrow \quad \mathbf{v}_1 = \boxed{\dfrac{1}{\sqrt{5}}\begin{pmatrix} 2 \\ -1 \end{pmatrix}}.$

$\lambda_2 = -5$: $\mathbf{0} \stackrel{!}{=} (A - \lambda_2 \mathbb{1})\mathbf{v}_2 = \begin{pmatrix} 8 & -4 \\ -4 & 2 \end{pmatrix}\mathbf{v}_2 \quad \Rightarrow \quad \mathbf{v}_2 = \boxed{\dfrac{1}{\sqrt{5}}\begin{pmatrix} 1 \\ 2 \end{pmatrix}}.$

Explicitly: for the eigenvector $\mathbf{v}_1 = (v^1{}_1, v^2{}_1)^{\mathrm{T}}$ we have $-2v^1{}_1 - 4v^2{}_1 = 0$, hence it has the form $\mathbf{v}_1 = a_1(2, -1)^{\mathrm{T}}$. Analogously one finds $\mathbf{v}_2 = a_2(1, 2)^{\mathrm{T}}$. To ensure the orthogonality of the similarity transformation T containing the orthonormalized eigenvectors as column vectors, we normalize them as $\|\mathbf{v}_j\| = 1$. This fixes each prefactor a_j up to a sign. We choose $a_1 = a_2 = \frac{1}{\sqrt{5}}$, but other signs, e.g. $a_1 = -a_2 = \frac{1}{\sqrt{5}}$, would work, too.

Sim.-tr.: $T = (\mathbf{v}_1, \mathbf{v}_2) = \boxed{\dfrac{1}{\sqrt{5}}\begin{pmatrix} 2 & 1 \\ -1 & 2 \end{pmatrix}}, \quad T^{-1} = T^{\mathrm{T}} = \begin{pmatrix} \mathbf{v}_1^{\mathrm{T}} \\ \mathbf{v}_2^{\mathrm{T}} \end{pmatrix} = \boxed{\dfrac{1}{\sqrt{5}}\begin{pmatrix} 2 & -1 \\ 1 & 2 \end{pmatrix}}.$

Check: $TDT^{-1} = \dfrac{1}{\sqrt{5}}\begin{pmatrix} 2 & 1 \\ -1 & 2 \end{pmatrix} \dfrac{1}{\sqrt{5}}\begin{pmatrix} 5\cdot 2 & 5\cdot(-1) \\ -5\cdot 1 & -5\cdot 2 \end{pmatrix} = \boxed{\begin{pmatrix} 3 & -4 \\ -4 & -3 \end{pmatrix}} \stackrel{\checkmark}{=} A.$

Remark: Of course it is also possible to construct the similarity transformation T using eigenvectors that are *not* normalized. However then its inverse, $\tilde{T}^{-1}$, will not equal $\tilde{T}^{\mathrm{T}}$, but will have to be found in an additional step using the appropriate inversion formula. For example, if we choose the prefactors above as $a_1 = 1$ and $a_2 = 2$, thus taking $\tilde{\mathbf{v}}_1 = (2, -1)^{\mathrm{T}}$ and $\tilde{\mathbf{v}}_2 = (2, 4)^{\mathrm{T}}$ as eigenvectors, we obtain:

Sim.-tr.: $\tilde{T} = (\tilde{\mathbf{v}}_1, \tilde{\mathbf{v}}_2) = \boxed{\begin{pmatrix} 2 & 2 \\ -1 & 4 \end{pmatrix}}, \quad \tilde{T}^{-1} = \boxed{\dfrac{1}{10}\begin{pmatrix} 4 & -2 \\ 1 & 2 \end{pmatrix}}.$

Since the columns of T and $\tilde{T}$, and the rows of T^{-1} and $\tilde{T}^{-1}$, only differ by prefactors, the check here works analogously to the one above:

Check: $\tilde{T}D\tilde{T}^{-1} = \begin{pmatrix} 2 & 2 \\ -1 & 4 \end{pmatrix} \dfrac{1}{10}\begin{pmatrix} 5\cdot 4 & 5\cdot(-2) \\ -5\cdot 1 & -5\cdot 2 \end{pmatrix} = \boxed{\begin{pmatrix} 3 & -4 \\ -4 & -3 \end{pmatrix}} \stackrel{\checkmark}{=} A.$

(b)   The zeros of the characteristic polynomial yield the eigenvalues:

Char. polynomial: $0 \stackrel{!}{=} \det(A - \lambda \mathbb{1}) = \begin{vmatrix} 1 - \lambda & i \\ -i & 1 - \lambda \end{vmatrix} = (1 - \lambda)^2 - 1 = \lambda(\lambda - 2).$

Eigenvalues: $\lambda_1 = \boxed{0}$, $\lambda_2 = \boxed{2}$.

Checks: $\lambda_1 + \lambda_2 = \boxed{2} \stackrel{\checkmark}{=} \mathrm{Tr}\, A = 1 + 1$, $\lambda_1\lambda_2 = \boxed{0} \stackrel{\checkmark}{=} \det A = 1 - i(-i).$

Eigenvectors:

$\lambda_1 = 0$: $\mathbf{0} \stackrel{!}{=} (A - \lambda_1 \mathbb{1})\mathbf{v}_1 = \begin{pmatrix} 1 & i \\ -i & 1 \end{pmatrix}\mathbf{v}_1 \quad \Rightarrow \quad \mathbf{v}_1 = \boxed{\dfrac{1}{\sqrt{2}}\begin{pmatrix} 1 \\ i \end{pmatrix}}.$

$\lambda_2 = 2$: $\mathbf{0} \stackrel{!}{=} (A - \lambda_2 \mathbb{1})\mathbf{v}_2 = \begin{pmatrix} -1 & i \\ -i & -1 \end{pmatrix}\mathbf{v}_2 \quad \Rightarrow \quad \mathbf{v}_2 = \boxed{\dfrac{1}{\sqrt{2}}\begin{pmatrix} 1 \\ -i \end{pmatrix}}.$

Explicitly: for the eigenvector $\mathbf{v}_1 = (v^1{}_1, v^2{}_1)^{\mathrm{T}}$ we have $v^1{}_1 + iv^2{}_1 = 0$, hence it has the form $\mathbf{v}_1 = a_1(1, i)^{\mathrm{T}}$. Similarly one finds $\mathbf{v}_2 = a_2(1, -i)^{\mathrm{T}}$. To ensure the unitarity of the similarity transformation T containing the orthonormalized eigenvectors as column vectors, we normalize them as $\|\mathbf{v}_j\| = 1$. This fixes each prefactor a_j up to a phase. We choose $a_1 = a_2 = \frac{1}{\sqrt{2}}$, but other phase choices, e.g. $a_1 = -a_2 = \frac{i}{\sqrt{2}}$, would work, too.

Sim-tr.: $\qquad T = (\mathbf{v}_1, \mathbf{v}_2) = \boxed{\dfrac{1}{\sqrt{2}} \begin{pmatrix} 1 & 1 \\ i & -i \end{pmatrix}}, \quad T^{-1} = T^\dagger = \begin{pmatrix} \mathbf{v}_1^\dagger \\ \mathbf{v}_2^\dagger \end{pmatrix} = \boxed{\dfrac{1}{\sqrt{2}} \begin{pmatrix} 1 & -i \\ 1 & i \end{pmatrix}}.$

Check: $\quad TDT^{-1} = \dfrac{1}{\sqrt{2}} \begin{pmatrix} 1 & 1 \\ i & -i \end{pmatrix} \dfrac{1}{\sqrt{2}} \begin{pmatrix} 0\cdot 1 & 0\cdot(-i) \\ 2\cdot 1 & 2\cdot i \end{pmatrix} = \boxed{\begin{pmatrix} 1 & i \\ -i & 1 \end{pmatrix}} \overset{\checkmark}{=} A.$

(c) The zeros of the characteristic polynomial yield the eigenvalues:

Char. polynomial: $\qquad 0 \overset{!}{=} \det(A - \lambda\mathbb{1}) = \begin{vmatrix} 1-\lambda & 0 & -i \\ 0 & 1-\lambda & 0 \\ i & 0 & 1-\lambda \end{vmatrix} = (1-\lambda)^3 - (1-\lambda)$

$$= (1-\lambda)(\lambda^2 - 2\lambda) = (1-\lambda)\lambda(\lambda - 2).$$

Eigenvalues: $\quad \lambda_1 = \boxed{1}, \lambda_2 = \boxed{0}, \lambda_3 = \boxed{2}.$

Checks: $\lambda_1 + \lambda_2 + \lambda_3 = \boxed{2} \overset{\checkmark}{=} \mathrm{Tr}\, A = 1 + 1 + 1, \quad \lambda_1\lambda_2\lambda_3 = \boxed{0} \overset{\checkmark}{=} \det A = 1\cdot 1\cdot 1 - i(-i).$

Eigenvectors:

$\lambda_1 = 1:\qquad 0 \overset{!}{=} (A - \lambda_1\mathbb{1})\mathbf{v}_1 = \begin{pmatrix} 0 & 0 & -i \\ 0 & 0 & 0 \\ i & 0 & 0 \end{pmatrix}\mathbf{v}_1 \qquad \Rightarrow \quad \mathbf{v}_1 = \boxed{\begin{pmatrix} 0 \\ 1 \\ 0 \end{pmatrix}};$

$\lambda_2 = 0:\qquad 0 \overset{!}{=} (A - \lambda_2\mathbb{1})\mathbf{v}_2 = \begin{pmatrix} 1 & 0 & -i \\ 0 & 1 & 0 \\ i & 0 & 1 \end{pmatrix}\mathbf{v}_2 \qquad \Rightarrow \quad \mathbf{v}_2 = \boxed{\dfrac{1}{\sqrt{2}}\begin{pmatrix} 1 \\ 0 \\ -i \end{pmatrix}};$

$\lambda_3 = 2:\qquad 0 \overset{!}{=} (A - \lambda_3\mathbb{1})\mathbf{v}_3 = \begin{pmatrix} -1 & 0 & -i \\ 0 & -1 & 0 \\ i & 0 & -1 \end{pmatrix}\mathbf{v}_3 \qquad \Rightarrow \quad \mathbf{v}_3 = \boxed{\dfrac{1}{\sqrt{2}}\begin{pmatrix} 1 \\ 0 \\ i \end{pmatrix}}.$

To ensure the unitarity of T, we have normalized all eigenvectors as $\|\mathbf{v}_j\| = 1$.

Sim. tr.: $T = (\mathbf{v}_1, \mathbf{v}_2, \mathbf{v}_3) = \boxed{\begin{pmatrix} 0 & \frac{1}{\sqrt{2}} & \frac{1}{\sqrt{2}} \\ 1 & 0 & 0 \\ 0 & \frac{-i}{\sqrt{2}} & \frac{i}{\sqrt{2}} \end{pmatrix}}, \quad T^{-1} = T^\dagger = \begin{pmatrix} \mathbf{v}_1^\dagger \\ \mathbf{v}_2^\dagger \\ \mathbf{v}_3^\dagger \end{pmatrix} = \boxed{\begin{pmatrix} 0 & 1 & 0 \\ \frac{1}{\sqrt{2}} & 0 & \frac{i}{\sqrt{2}} \\ \frac{1}{\sqrt{2}} & 0 & \frac{-i}{\sqrt{2}} \end{pmatrix}}.$

Check: $TDT^{-1} = \begin{pmatrix} 0 & \frac{1}{\sqrt{2}} & \frac{1}{\sqrt{2}} \\ 1 & 0 & 0 \\ 0 & \frac{-i}{\sqrt{2}} & \frac{i}{\sqrt{2}} \end{pmatrix}\begin{pmatrix} 1\cdot 0 & 1\cdot 1 & 1\cdot 0 \\ 0\cdot\frac{1}{\sqrt{2}} & 0\cdot\frac{1}{\sqrt{2}} & 0\cdot\frac{i}{\sqrt{2}} \\ 2\cdot\frac{1}{\sqrt{2}} & 2\cdot 0 & 2\cdot\frac{-i}{\sqrt{2}} \end{pmatrix} = \begin{pmatrix} 1 & 0 & -i \\ 0 & 1 & 0 \\ i & 0 & 1 \end{pmatrix} \overset{\checkmark}{=} A.$

εL8.2.3 Spin-$\frac{1}{2}$ matrices: eigenvalues and eigenvectors (p. 119)

For S_x we obtain:

Char. pol.: $\qquad 0 \overset{!}{=} \det(S_x - \lambda\mathbb{1}) = \begin{vmatrix} -\lambda & \frac{1}{2} \\ \frac{1}{2} & -\lambda \end{vmatrix} = \lambda^2 - \frac{1}{4} = (\lambda - \frac{1}{2})(\lambda + \frac{1}{2}).$

Eigenvalues: $\qquad \boxed{\lambda_{x,1} = \frac{1}{2}, \quad \lambda_{x,2} = -\frac{1}{2}.}$

Checks: $\lambda_{x,1} + \lambda_{x,2} = 0 \overset{\checkmark}{=} \operatorname{Tr} S_x$, $\lambda_{x,1}\lambda_{x,2} = -\frac{1}{4} \overset{\checkmark}{=} \det S_x$.

Eigenvectors $\mathbf{v}_{x,a}$:

$$0 \overset{!}{=} (S_x - \lambda_{x,1}\mathbb{1})\mathbf{v}_{x,1} = \frac{1}{2}\begin{pmatrix} -1 & 1 \\ 1 & -1 \end{pmatrix}\mathbf{v}_{x,1} \qquad \Rightarrow \qquad \boxed{\mathbf{v}_{x,1} = \frac{1}{\sqrt{2}}\begin{pmatrix} 1 \\ 1 \end{pmatrix}},$$

$$0 \overset{!}{=} (S_x - \lambda_{x,2}\mathbb{1})\mathbf{v}_{x,2} = \frac{1}{2}\begin{pmatrix} 1 & 1 \\ 1 & 1 \end{pmatrix}\mathbf{v}_{x,2} \qquad \Rightarrow \qquad \boxed{\mathbf{v}_{x,2} = \frac{1}{\sqrt{2}}\begin{pmatrix} 1 \\ -1 \end{pmatrix}}.$$

For S_y we obtain:

Char. pol.: $$0 \overset{!}{=} \det(S_y - \lambda\mathbb{1}) = \begin{vmatrix} -\lambda & -\frac{i}{2} \\ \frac{i}{2} & -\lambda \end{vmatrix} = \lambda^2 - \frac{1}{4} = (\lambda - \tfrac{1}{2})(\lambda + \tfrac{1}{2}).$$

Eigenvalues: $\boxed{\lambda_{y,1} = \frac{1}{2}, \quad \lambda_{y,2} = -\frac{1}{2}}.$

Checks: $\lambda_{y,1} + \lambda_{y,2} = 0 \overset{\checkmark}{=} \operatorname{Tr} S_y$, $\lambda_{y,1}\lambda_{y,2} = -\frac{1}{4} \overset{\checkmark}{=} \det S_y$.

Eigenvectors $\mathbf{v}_{y,a}$:

$$0 \overset{!}{=} (S_y - \lambda_{y,1}\mathbb{1})\mathbf{v}_{y,1} = \frac{1}{2}\begin{pmatrix} -1 & -i \\ i & -1 \end{pmatrix}\mathbf{v}_{y,1} \qquad \Rightarrow \qquad \boxed{\mathbf{v}_{y,1} = \frac{1}{\sqrt{2}}\begin{pmatrix} 1 \\ i \end{pmatrix}},$$

$$0 \overset{!}{=} (S_y - \lambda_{y,2}\mathbb{1})\mathbf{v}_{y,2} = \frac{1}{2}\begin{pmatrix} 1 & -i \\ i & 1 \end{pmatrix}\mathbf{v}_{y,2} \qquad \Rightarrow \qquad \boxed{\mathbf{v}_{y,2} = \frac{1}{\sqrt{2}}\begin{pmatrix} 1 \\ i \end{pmatrix}}.$$

For S_z we obtain:

Char. pol.: $$0 \overset{!}{=} \det(S_z - \lambda\mathbb{1}) = \begin{vmatrix} \frac{1}{2} - \lambda & 0 \\ 0 & -\frac{1}{2} - \lambda \end{vmatrix} = -(\tfrac{1}{2} - \lambda)(\tfrac{1}{2} + \lambda).$$

Eigenvalues: $\boxed{\lambda_{z,1} = \frac{1}{2}, \quad \lambda_{z,2} = -\frac{1}{2}}.$

Checks: $\lambda_{z,1} + \lambda_{z,2} = 0 \overset{\checkmark}{=} \operatorname{Tr} S_z$, $\lambda_{z,1}\lambda_{z,2} = -\frac{1}{4} \overset{\checkmark}{=} \det S_z$.

Eigenvectors $\mathbf{v}_{z,a}$:

$$0 \overset{!}{=} (S_z - \lambda_{z,1}\mathbb{1})\mathbf{v}_{z,1} = \frac{1}{2}\begin{pmatrix} 0 & 0 \\ 0 & 1 \end{pmatrix}\mathbf{v}_{z,1} \qquad \Rightarrow \qquad \boxed{\mathbf{v}_{z,1} = \begin{pmatrix} 1 \\ 0 \end{pmatrix}},$$

$$0 \overset{!}{=} (S_z - \lambda_{z,2}\mathbb{1})\mathbf{v}_{z,2} = \frac{1}{2}\begin{pmatrix} 1 & 0 \\ 0 & 0 \end{pmatrix}\mathbf{v}_{z,2} \qquad \Rightarrow \qquad \boxed{\mathbf{v}_{z,2} = \begin{pmatrix} 0 \\ 1 \end{pmatrix}}.$$

EL8.2.5 Inertia tensor (p. 156)

Point masses: $m_1 = 4$ at $\mathbf{r}_1 = (1,0,0)^{\mathrm{T}}$; $m_2 = M$ at $\mathbf{r}_2 = (0,1,2)^{\mathrm{T}}$; $m_3 = 1$ at $\mathbf{r}_3 = (0,4,1)^{\mathrm{T}}$.

$$\tilde{I}_{ij} = \sum_a m_a\left(\delta_{ij}\mathbf{r}_a^2 - r^i{}_a r^j{}_a\right) \quad \Rightarrow \quad \tilde{I} = \sum_a m_a \begin{pmatrix} \mathbf{r}_a^2 - r^1{}_a r^1{}_a & -r^1{}_a r^2{}_a & -r^1{}_a r^3{}_a \\ -r^2{}_a r^1{}_a & \mathbf{r}_a^2 - r^2{}_a r^2{}_a & -r^2{}_a r^3{}_a \\ -r^3{}_a r^1{}_a & -r^3{}_a r^2{}_a & \mathbf{r}_a^2 - r^3{}_a r^3{}_a \end{pmatrix}.$$

$$\tilde{I} = 4\cdot\begin{pmatrix} 1-1 & 0 & 0 \\ 0 & 1-0 & 0 \\ 0 & 0 & 1-0 \end{pmatrix} + M\cdot\begin{pmatrix} 5-0 & 0 & 0 \\ 0 & 5-1 & -2 \\ 0 & -2 & 5-4 \end{pmatrix} + 1\cdot\begin{pmatrix} 17-0 & 0 & 0 \\ 0 & 17-16 & -4 \\ 0 & -4 & 17-1 \end{pmatrix}$$

$$= \boxed{\begin{pmatrix} 5M+17 & 0 & 0 \\ 0 & 4M+5 & -2(M+2) \\ 0 & -2(M+2) & M+20 \end{pmatrix}}.$$

The zeros of the characteristic polynomial yield the moments of inertia (eigenvalues):

$$0 \overset{!}{=} \det(\tilde{I} - \lambda \mathbb{1}) = \begin{vmatrix} 5M + 17 - \lambda & 0 & 0 \\ 0 & 4M + 5 - \lambda & -2(M+2) \\ 0 & -2(M+2) & M + 20 - \lambda \end{vmatrix}$$

$$= (5M + 17 - \lambda)\left[(4M + 5 - \lambda)(M + 20 - \lambda) - 4(M+2)^2 \right]$$

$$= (5M + 17 - \lambda)\left[\lambda^2 - 5(M+5)\lambda + 69M + 84 \right].$$

Moments of inertia:

$$\lambda_1 = \boxed{5M + 17},$$

$$\lambda_{2,3} = \tfrac{1}{2}\left[5(M+5) \pm \sqrt{25(M+5)^2 - 4(69M + 84)} \right]$$

$$= \boxed{\tfrac{1}{2}\left[5(M+5) \pm \sqrt{25M^2 - 26M + 289} \right]}.$$

For $M = 5$: $\boxed{\lambda_1 = 42}$, $\lambda_{2,3} = \tfrac{1}{2}[50 \pm \sqrt{784}] = 25 \pm 14$, hence $\boxed{\lambda_2 = 39}$, $\boxed{\lambda_3 = 11}$.

S.L8.3 Relation between Hermitian and unitary matrices (p. 120)

L8.3.1 Exponential representation of two-dimensional rotation matrix (p. 122)

(a) We use the product decomposition $R_\theta = \left[R_{\theta/m}\right]^m$. For $m \gg 1$, $\theta/m \ll 1$ we have $\cos(\theta/m) = 1 + \mathcal{O}((\theta/m)^2)$ and $\sin(\theta/m) = \theta/m + \mathcal{O}((\theta/m)^3)$. Therefore

$$R_{\theta/m} = \begin{pmatrix} \cos\theta & -\sin\theta \\ \sin\theta & \cos\theta \end{pmatrix} = \begin{pmatrix} 1 & -\frac{\theta}{m} \\ \frac{\theta}{m} & 1 \end{pmatrix} + \mathcal{O}((\tfrac{\theta}{m})^2) = \mathbb{1} + \tfrac{\theta}{m}\tilde{\sigma} + \mathcal{O}((\tfrac{\theta}{m})^2), \quad \tilde{\sigma} = \boxed{\begin{pmatrix} 0 & -1 \\ 1 & 0 \end{pmatrix}}.$$

(b) The identity $\lim_{m\to\infty}[1 + \tfrac{x}{m}]^m = e^x$ now yields an exponential representation of R_θ:

$$R_\theta = \lim_{m\to\infty}\left[R_{\theta/m}\right]^m = \lim_{m\to\infty}\left[\mathbb{1} + \tfrac{\theta}{m}\tilde{\sigma}\right]^m = \boxed{e^{\theta\tilde{\sigma}}}.$$

S.L10 Multilinear algebra (p. 147)

S.L10.1 Direct sum and direct product of vector spaces (p. 147)

L10.1.1 Direct sum and direct product (p. 149)

We are given the two vectors in $\mathbb{R}^2$, $u = (2, 1)^T$, $v = (3, -1)^T$.

(a) $u \oplus v = (e_i, 0)u^i + (0, e_i)v^i = (e_i u^i, e_i v^i) = (u, v) = \boxed{(2, 1, 3, -1)^T}$.

$v \oplus u = (v, u) = \boxed{(3, -1, 2, 1)^T}$.

(b) $u \otimes v = e_i \otimes e_j u^i v^j = (e_i u^i) \otimes (e_j v^j) = (2e_1 + e_2) \otimes (3e_1 - e_2)$

$= 2 \cdot 3 e_1 \otimes e_1 + 2 \cdot (-1)e_1 \otimes e_2 + 1 \cdot 3e_2 \otimes e_1 + 1 \cdot (-1)e_2 \otimes e_2$

$= \boxed{6e_1 \otimes e_1 - 2e_1 \otimes e_2 + 3e_2 \otimes e_1 - e_2 \otimes e_2}.$

$$v \otimes u = e_i \otimes e_j v^i u^j = (e_i v^i) \otimes (e_j u^j) = (3e_1 - e_2) \otimes (2e_1 + e_2)$$
$$= 3 \cdot 2e_1 \otimes e_1 + 3 \cdot 1e_1 \otimes e_2 + (-1) \cdot 2e_2 \otimes e_1 + (-1) \cdot 1e_2 \otimes e_2$$
$$= \boxed{6e_1 \otimes e_1 + 3e_1 \otimes e_2 - 2e_2 \otimes e_1 - e_2 \otimes e_2}.$$

S.L10.2 Dual space (p. 150)

ʟ10.2.1 Dual vector in $\mathbb{R}^{2*}$: height of slanted plane (p. 151)

(a) Given $x_1 = \binom{2}{1}$ and $x_2 = \binom{5}{3}$, the standard basis of $\mathbb{R}^2$ can be expressed as:
$e_1 = 3x_1 - x_2$ and $e_2 = -5x_1 + 2x_2$. Therefore:

$$h_1 = h(e_1) = 3h(x_1) - h(x_2) = 9 - a, \ \ h_2 = h(e_2) = -5h(x_1) + 2h(x_2) = -15 + 2a.$$

Hence the plane is specified by the dual vector $h = (h_1, h_2) = \boxed{(9-a, -15+2a)}$.

(b) The height at $x_3 = \binom{7}{3} = 7e_1 + 3e_2$ is $h(x_3) = 7h(e_1) + 3h(e_2) = \boxed{18 - a}$.

ʟ10.2.3 Dual vectors in $\mathbb{R}^{2*}$ (p. 151)

(a) The standard basis vectors, expressed through $x_1 = \binom{a}{1}$ and $x_2 = \binom{a}{2}$, read $e_1 = (2x_1 - x_2)/a$, $e_2 = -x_1 + x_2$. Hence:

$$w(e_1) = (2w(x_1) - w(x_2))/a = (2 \cdot 1 - (-1))/a = \boxed{3/a},$$
$$w(e_2) = -w(x_1) + w(x_2) = -1 + (-1) = \boxed{-2}.$$

(b) $w(y) = w(e_1)2 + w(e_2)b = \boxed{6/a - 2b}$.

(c) If the given vectors x_1 and x_2 are expanded as $x_j \equiv e_i(x_j)^i \equiv e_i X^i{}_j$, the components of x_j give the jth column of X, and $e_j = x_i(X^{-1})^i{}_j$. Given the images, $w(x_i)$, of the map when acting on the vectors x_i, we can find the images of the basis vectors as $w(e_j) = \boxed{w(x_i)(X^{-1})^i{}_j}$, and of the vector $y = e_j y^j$ as $w(y) = w(e_j)y^j = \boxed{w(x_i)(X^{-1})^i{}_j y^j}$.

For (a,b) we have $X = \binom{a\ a}{1\ 2}$ and $y = \binom{2}{b}$. With $X^{-1} = \binom{2/a\ \ -1}{-1/a\ \ \ 1}$ we obtain:

$$(w(1), w(2)) = (1, -1)\begin{pmatrix} 2/a & -1 \\ -1/a & 1 \end{pmatrix} = \boxed{(3/a, -2)}, \checkmark$$
$$w(y) = w(e_j)y^j = (3/a, -2)\begin{pmatrix} 2 \\ b \end{pmatrix} = \boxed{6/a - 2b}. \checkmark$$

ʟ10.2.5 Basis transformation for vectors and dual vectors (p. 152)

Given a basis transformation $e'_j = e_i(T^{-1})^i{}_j$, the dual basis transform as $e'^i = T^i{}_j e^j$. The components of e'_j give the jth column of T^{-1}; the ith row of T gives the components of e'^i:

$$e'_1 = \binom{2}{3}, \ e'_2 = \binom{3}{5} \ \Rightarrow \ T^{-1} = \binom{2\ 3}{3\ 5}, \ T = \begin{pmatrix} 5 & -3 \\ -3 & 2 \end{pmatrix} \ \Rightarrow \ e'^1 = \boxed{(5, -3)}, \ e'^2 = \boxed{(-3, 2)}.$$

εL10.2.7 Canonical map between vectors and dual vectors via metric (p. 154)

(a) The components of a vector, $u \in V$, and its dual, $J(u) \in V^*$, are related by $J(u)_j = u^i g_{ij} \equiv u_j$. The vector $u = e_1 + a\,e_2 = e_i u^i$ has the component representation $\begin{pmatrix} u^1 \\ u^2 \end{pmatrix} = \begin{pmatrix} 1 \\ a \end{pmatrix}$, thus its dual, $J(u) = u_j e^j$, has the component representation

$$(u_1, u_2) = (1, a) \begin{pmatrix} 2 & 3 \\ 3 & 5 \end{pmatrix} = (2 + 3a, 3 + 5a),$$

and is given by $J(u) = \boxed{e^1(2 + 3a) + e^2(3 + 5a)}$.

Notice that index-lowering, $u_j = u^i g_{ij}$, can be performed via "matrix multiplication", arranged such that u_j equals the contraction of the u^is with the jth column of g_{ij}.

(b) The components of a dual vector, $w \in V^*$, and its dual, $J^{-1}(w) \in V$, are related by $(J^{-1}(w))^i = g^{ij} w_j \equiv w^i$. The inverse metric reads $g^{ij} = \begin{pmatrix} 5 & -3 \\ -3 & 2 \end{pmatrix}$. The dual vector $w = e^1 - 2a\,e^2 = w_j e^j$ has component representation $(w_1, w_2) = (1, -2a)$, thus its dual, $J^{-1}(w) = e_i w^i$, has the component representation

$$\begin{pmatrix} w^1 \\ w^2 \end{pmatrix} = \begin{pmatrix} 5 & -3 \\ -3 & 2 \end{pmatrix} \begin{pmatrix} 1 \\ -2a \end{pmatrix} = \begin{pmatrix} 5 + 6a \\ -3 - 4a \end{pmatrix},$$

and is given by $J^{-1}(w) = \boxed{e_1(5 + 6a) + e_2(-3 - 4a)}$.

Notice that index raising, $w^i = g^{ij} w_j$, can be performed via "matrix multiplication", arranged such that w^i equals the contraction of the w_js with the ith row of g^{ij}.

(c) $J(u)v = [e^1(2 + 3a) + e^2(3 + 5a)][2e_1 - e_2] = (2 + 3a)2 - (3 + 5a) = \boxed{1 + a}$, where we used $e^i e_j = \delta^i{}_j$. This agrees with

$$g(u, v) = u^i g_{ij} v^j = (1, a) \begin{pmatrix} 2 & 3 \\ 3 & 5 \end{pmatrix} \begin{pmatrix} 2 \\ -1 \end{pmatrix} = 1 + a, \checkmark$$

as it should, because by definition, $J(u)v = g(u, v)$.

S.L10.3   Tensors (p. 154)

εL10.3.1 Linear transformations in tensor spaces (p. 155)

(a) The map $A = \frac{1}{\sqrt{2}} \begin{pmatrix} 1 & 1 \\ 1 & -1 \end{pmatrix}$ acts on the basis vectors as $e_1 \mapsto Ae_1 = e_k A^k{}_1 = \frac{1}{\sqrt{2}}(e_1 + e_2)$, $e_2 \mapsto Ae_2 = e_k A^k{}_2 = \frac{1}{\sqrt{2}}(e_1 - e_2)$. The image of the tensor $t_+ = \frac{1}{\sqrt{2}}(e_1 \otimes e_1 + e_2 \otimes e_2)$ therefore is:

$$t'_+ = At_+ = \frac{1}{\sqrt{2}}\left[\frac{1}{\sqrt{2}}(e_1 + e_2) \otimes \frac{1}{\sqrt{2}}(e_1 + e_2) + \frac{1}{\sqrt{2}}(e_1 - e_2) \otimes \frac{1}{\sqrt{2}}(e_1 - e_2)\right]$$

$$= \frac{1}{\sqrt{2}}\frac{1}{2}[(1 + 1)(e_1 \otimes e_1) + (1 - 1)(e_1 \otimes e_2) + (1 - 1)(e_2 \otimes e_1) + (1 + 1)(e_2 \otimes e_2)]$$

$$= \frac{1}{\sqrt{2}}(e_1 \otimes e_1 + e_2 \otimes e_2) = \boxed{t_+}.$$

(b) Similarly, the image of $t_- = \frac{1}{\sqrt{2}}(e_1 \otimes e_2 - e_2 \otimes e_1)$ is:

$$t'_- = At_- = \frac{1}{\sqrt{2}} \cdot \frac{1}{2}[(e_1 + e_2) \otimes (e_1 - e_2) - (e_1 - e_2) \otimes (e_1 + e_2)]$$

$$= \frac{1}{\sqrt{2}} \frac{1}{2}[(1-1)(e_1 \otimes e_1) + (-1-1)(e_1 \otimes e_2) + (1+1)(e_2 \otimes e_1) + (-1+1)(e_2 \otimes e_2)]$$

$$= \frac{1}{\sqrt{2}}(-e_1 \otimes e_2 + e_2 \otimes e_1) = \boxed{-t_-}.$$

(c) The image of a general tensor, $t = e_i \otimes e_j t^{ij}$, under a general map A is: $t' = (Ae_i) \otimes (Ae_j)t^{ij} = e_k A^k{}_i e_l A^l{}_j t^{ij} \equiv e_k e_l t'^{kl} \Rightarrow \boxed{t'^{kl} = A^k{}_i A^l{}_j t^{ij}}$. The tensor will remain invariant if the coefficient remain unchanged, i.e. $t^{kl} = A^k{}_i A^l{}_j t^{ij} = A^k{}_i t^{ij} (A^T)^l{}_j$, or $t' A t A^T$, in matrix notation.

(d) If $t^{ij} = \delta^{ij}$, this condition reads $\delta^{kl} = A^k{}_i A^l{}_j \delta^{ij} = A^k{}_i A^l{}_i = A^k{}_i (A^T)_i{}^l = (AA^T)^{kl}$, i.e $\boxed{AA^T = \mathbb{1}}$, so A must be an orthogonal matrix.

εL10.3.3 Tensors in $T^1{}_2(V)$ (p. 155)

For $u = 3e_1 + ae_2$, $v = be_1 - e_2$, $w = 5e^1 - 2ce^2$ and $t = e_2 \otimes e^1 \otimes e^2 \in T^1{}_2(\mathbb{R}^2)$, we find:

(a) $t(w; u, v) = e_2(w)e^1(u)e^2(v) = w_2 u^1 v^2 = (-2c) \cdot 3 \cdot (-1) = \boxed{6c}$,

(b) $t(w; ., u) = w_2 e^1 u^2 = (-2c) \cdot e^1 \cdot a = \boxed{-2ace^1}$,

(c) $t(.; v, u) = e_2 v^1 u^2 = e_2 \cdot b \cdot a = \boxed{abe_2}$.

S.L10.4 Alternating forms (p. 157)

εL10.4.1 Three-form in $\mathbb{R}^3$ (p. 158)

(a) $\omega = \sum_{P \in S_3} \text{sgn}(P) e^{P1} \otimes e^{P2} \otimes e^{P3}$

$$= \boxed{e^1 \otimes e^2 \otimes e^3 - e^1 \otimes e^3 \otimes e^2 + e^2 \otimes e^3 \otimes e^1 - e^2 \otimes e^1 \otimes e^3 + e^3 \otimes e^1 \otimes e^2 - e^3 \otimes e^2 \otimes e^1}.$$

(b) $\phi = \omega(., v, w) = \boxed{e^1(v^2 w^3 - v^3 w^2) + e^2(v^3 w^1 - v^1 w^3) + e^3(v^1 w^2 - v^2 w^1)}$.

$\omega(u, v, w) = \phi(u) = \boxed{u^1 v^2 w^3 - u^1 v^3 w^2 + u^2 v^3 w^1 - u^2 v^1 w^3 + u^3 v^1 w^2 - u^3 v^2 w^1}$.

The components of $\phi \equiv \phi_i e^i$, given by $\phi_i = \boxed{\epsilon_{ijk} v^j v^k}$, are those of the cross product of the vectors v and w, while $\omega(u, v, w) = \det(u, v, w) = u \cdot (v \times w)$ equals the determinant of the matrix containing the vectors u, v, w as columns, which equals their triple product.

(c) For $u = (1, 1, a)^T$, $v = (1, a, 1)^T$, $w = (a, 1, 1)^T$, we have

$$\phi = \boxed{e^1(a-1) + e^2(a-1) + e^3(1-a^2)}.$$

$\omega(u, v, w) = \phi(u) = 1 \cdot (a-1) + 1 \cdot (a-1) + a \cdot (1-a^2) = \boxed{-a^3 + 3a - 2}$.

S.L10.6 Wedge product (p. 161)

ₜL10.6.1 Alternating forms in $\Lambda^2(\mathbb{R}^3)$ (p. 162)

Given $u = 3e_1 - ae_2$ and $v = 2e_1 + be_3$ in $\mathbb{R}^3$, we have:

(a) $(e^1 \wedge e^3)(u, v) = (e^1 \otimes e^3 - e^3 \otimes e^1)(u, v) = u^1 v^3 - u^3 v^1 = 3 \cdot b - 0 \cdot 2 = \boxed{3b}$,

(b) $(e^1 \wedge e^3)(v, u) = v^1 u^3 - v^3 u^1 = 2 \cdot 0 - b \cdot 3 = \boxed{-3b}$,

(c) $(e^3 \wedge e^2)(u, u) = u^3 u^2 - u^2 u^3 = \boxed{0}$,

(d) $(e^2 \wedge e^1)(u, v) = u^2 v^1 - u^1 v^2 = (-a) \cdot 2 - 3 \cdot 0 = \boxed{-2a}$,

(e) $(e^1 \wedge e^2 \wedge e^3) = \sum_P \text{sgn}(P)\, e^{P1} \otimes e^{P2} \otimes e^{P3}$

$$= \boxed{e^1 \otimes e^2 \otimes e^3 - e^1 \otimes e^3 \otimes e^2 + e^2 \otimes e^3 \otimes e^1 - e^2 \otimes e^1 \otimes e^3 + e^3 \otimes e^1 \otimes e^2 - e^3 \otimes e^2 \otimes e^1}.$$

ₜL10.6.3 Wedge products in the Grassmann algebra $\Lambda(\mathbb{R}^4)$ (p. 162)

Using $e^i \wedge e^j = -e^j \wedge e^i$ (implying $e^i \wedge e^i = 0$), we obtain:

(a) $\phi_B \wedge \phi_A = e^4 \wedge e^3 = \boxed{-e^3 \wedge e^4}$,

(b) $\phi_A \wedge \phi_C = e^3 \wedge (e^3 \wedge e^1 + e^2 \wedge e^4) = \boxed{-e^2 \wedge e^3 \wedge e^4}$,

(c) $\phi_A \wedge \phi_D = e^3 \wedge (e^1 \wedge e^2 \wedge e^3) = \boxed{0}$,

(d) $\phi_B \wedge \phi_D = e^4 \wedge (e^1 \wedge e^2 \wedge e^3) = \boxed{-e^1 \wedge e^2 \wedge e^3 \wedge e^4}$.

S.L10.7 Inner derivative (p. 162)

ₜL10.7.1 Inner derivative of two-form (p. 163)

$$i_u\phi = (e^1 \wedge e^2 + e^2 \wedge e^3 + e^3 \wedge e^1)(u, .)$$
$$= (e^1 \otimes e^2 - e^2 \otimes e^1 + e^2 \otimes e^3 - e^3 \otimes e^2 + e^3 \otimes e^1 - e^1 \otimes e^3)(u, .)$$
$$= u^1 e^2 - u^2 e^1 + u^2 e^3 - u^3 e^2 + u^3 e^1 - u^1 e^3 = \boxed{e^1(u^3 - u^2) + e^2(u^1 - u^3) + e^3(u^2 - u^1)}.$$

For $u = e_1 - ae_2$, we thus obtain $i_u\phi = \boxed{ae^1 + e^2 + (-a - 1)e^3}$.

S.L10.8 Pullback (p. 163)

ₜL10.8.1 Pullback from $\mathbb{R}^2$ to $\mathbb{R}^2$ (p. 165)

(a) The pullback of a one-form, $\phi = \phi_i e^i \in \Lambda^1(\mathbb{R}^2)$, by F is defined as $F^*(\phi) = F^*(\phi_i e^i) = \phi_i F^i{}_j e^j$. For $\phi = e^1 - e^2 = \phi_i e^i$, with $(\phi_1, \phi_2) = (1, -1)$, and $F = \begin{pmatrix} 2 & a \\ a & 1 \end{pmatrix}$ this yields:

$$F^*(\phi) = \phi_i F^i{}_1 e^1 + \phi_i F^i{}_2 e^2 = (1 \cdot 2 - 1 \cdot a)e^1 + (1 \cdot a - 1 \cdot 1)e^2 = \boxed{(2 - a)e^1 + (a - 1)e^2}.$$

(b) The pullback of a two-form, $\omega = \omega_{i_1 i_2} e^{i_1} \wedge e^{i_2} \in \Lambda^2(\mathbb{R}^2)$ by F is defined as

$$F^*(\omega) = F^*(\omega_{i_1 i_2} e^{i_1} \wedge e^{i_2}) = \omega_{i_1 i_2} F^{i_1}{}_{j_1} F^{i_2}{}_{j_2} e^{j_1} \wedge e^{j_2}$$
$$= \omega_{i_1 i_2} (F^{i_1}{}_1 F^{i_2}{}_2 - F^{i_1}{}_2 F^{i_2}{}_1) e^1 \wedge e^2.$$

For $\omega = e^1 \wedge e^2$ and $F = \begin{pmatrix} 2 & a \\ a & 1 \end{pmatrix}$ this yields:

$$F^*\omega = (F^1{}_1 F^2{}_2 - F^1{}_2 F^2{}_1) e^1 \wedge e^2 = (2 \cdot 1 - a \cdot a) e^1 \wedge e^2 = \boxed{(2 - a^2) e^1 \wedge e^2}.$$

S.L10.9   Metric structures (p. 165)

ɛL10.9.1 Hodge duals on a three-dimensional vector space (p. 167)

(a) In $\mathbb{R}^3$, the Hodge star acts as follows on zero-, one-, two- and three-forms:

$$1 = \tfrac{1}{0!} \mapsto \quad *1 = \tfrac{1}{3!}(*1)_{ijk} e^i \wedge e^j \wedge e^k, \quad (*1)_{ijk} = \tfrac{\sqrt{|g|}}{0!} \epsilon_{ijk},$$

$$\lambda = \tfrac{1}{1!} \lambda_i e^i \mapsto \quad *\lambda = \tfrac{1}{2!}(*\lambda)_{jk} e^j \wedge e^k, \quad (*\lambda)_{jk} = \tfrac{\sqrt{|g|}}{1!} \lambda_{i'} g^{i'i} \epsilon_{ijk},$$

$$\phi = \tfrac{1}{2!} \phi_{jk} e^j \wedge e^k \mapsto \quad *\phi = \tfrac{1}{1!}(*\phi)_l e^l, \quad (*\phi)_l = \tfrac{\sqrt{|g|}}{2!} \phi_{j'k'} g^{j'j} g^{k'k} \epsilon_{jkl},$$

$$\omega = \tfrac{1}{3!} \omega_{ijk} e^i \wedge e^j \wedge e^k \mapsto \quad *\omega = \tfrac{1}{0!}(*\omega)1, \quad (*\omega) = \tfrac{\sqrt{|g|}}{3!} \omega_{i'j'k'} g^{i'i} g^{j'j} g^{k'k} \epsilon_{jkl}.$$

(b) For the standard metric $g_{ij} = \delta_{ij}$, the inverse is likewise trivial, $g^{ij} = \delta^{ij}$, and $\sqrt{|g|} = 1$. Thus we have the following simplifications:

$$(*1)_{ijk} = \boxed{\epsilon_{ijk}}, \quad (*\lambda)_{jk} = \boxed{\lambda_i \epsilon_{ijk}}, \quad (*\phi)_l = \boxed{\tfrac{1}{2!}\phi_{jk}\epsilon_{jkl}}, \quad (*\omega) = \boxed{\tfrac{1}{3!}\omega_{ijk}\epsilon_{ijk}}. \quad (1)$$

Now exploit the antisymmetry of wedge products and form components to simplify sums over repeated indices, e.g. $\tfrac{1}{2!}\epsilon_{1ij} e^i \wedge e^j = \tfrac{1}{2!}(\epsilon_{123} e^2 \wedge e^3 + \epsilon_{132} e^3 \wedge e^2) = e^2 \wedge e^3$, and $\tfrac{1}{2!}\phi_{jk}\epsilon_{jk3} = \tfrac{1}{2!}(\phi_{12}\epsilon_{123} + \phi_{21}\epsilon_{213}) = \phi_{12}$:

$$*1 = \boxed{e^1 \wedge e^2 \wedge e^3}, \qquad *\lambda = \boxed{\lambda_1(e^2 \wedge e^3) + \lambda_2(e^3 \wedge e^1) + \lambda_3(e^1 \wedge e^2)},$$

$$*\phi = \boxed{\phi_{23}\, e^1 + \phi_{31}\, e^2 + \phi_{12}\, e^3}, \qquad *\omega = \boxed{\omega_{123}}.$$

(c) Since $*\lambda$ is a two-form, with components $(*\lambda)_{jk} \overset{(1)}{=} \lambda_i \epsilon_{ijk}$, its Hodge dual is given by

$$(**\lambda)_l \overset{(1)}{=} \tfrac{1}{2!}(*\lambda)_{jk}\epsilon_{jkl} = \tfrac{1}{2!}(\lambda_i\epsilon_{ijk})\epsilon_{jkl} = \lambda_i\delta_{il} = \lambda_l, \quad \Rightarrow \quad \boxed{**\lambda = \lambda}.$$

Since $*\phi$ is a one-form, with components $(*\phi)_i \overset{(1)}{=} \tfrac{1}{2!}\phi_{j'k'}\epsilon_{j'k'i}$, its Hodge dual is given by

$$(**\phi)_{jk} \overset{(1)}{=} (*\phi)_i \epsilon_{ijk} = (\tfrac{1}{2!}\phi_{j'k'}\epsilon_{j'k'i})\epsilon_{ijk} = \tfrac{1}{2!}(\phi_{jk} - \phi_{kj}) = \phi_{jk}, \quad \Rightarrow \quad \boxed{**\phi = \phi}.$$

(d) Now consider a general metric. We know from (a) that $(*\lambda)_{j'k'} = \sqrt{|g|}\lambda_i g^{ii'}\epsilon_{i'j'k'}$, and that $(**\lambda)_l = \tfrac{1}{2!}\sqrt{|g|}(*\lambda)_{j'k'} g^{j'j} g^{k'k}\epsilon_{jkl}$. Combining these, we obtain

$$(**\lambda)_l \overset{(i)}{=} \tfrac{\sqrt{|g|}}{2!}\left(\sqrt{|g|}\lambda_i g^{ii'}\epsilon_{i'j'k'}\right)g^{j'j}g^{k'k}\epsilon_{ljk}(-)^{1\cdot(3-1)} \overset{(ii)}{=} \tfrac{|g|}{g}\lambda_i\left[\tfrac{1}{2!}\epsilon^{ijk}\epsilon_{ljk}\right] \overset{(iii)}{=} \boxed{\mathrm{sgn}(g)\lambda_l}.$$

This shows that $**\lambda = \text{sgn}(g)\lambda$. ✓ For step (i), we used $\epsilon_{jkl} = \epsilon_{ljk}(-)^{1\cdot(3-1)}$, where the sign arises from permuting one 1-form index, l, past $(3-1)$ indices, jk. (For a p-form in d dimensions, the analogous step includes permuting p indices past $(d-p)$ ones, generating a sign $(-)^{p(d-p)}$.) For step (ii), we used the definition of the metric determinant, $g^{ii'}g^{jj'}g^{kk'}\epsilon_{i'j'k'} = \epsilon^{ijk}\det(g^{-1})$, and the shorthand $\det(g^{-1}) = g^{-1}$. For step (iii), we noted that for a given value of the index l, nonzero contributions arise from the $\epsilon \cdot \epsilon$ factor only if j,k differ from both l and i, i.e. only for $i = l$. There are 2! choices for jk, which both have the same sign, hence $\frac{1}{2!}\epsilon^{ijk}\epsilon_{ljk} = \delta^i{}_l$. (For example, if $l = 3$, then the nonzero contributions come from $jk = 12$ or 21, with $i = 3$.)

Similarly, we know from (a) that $(*\phi)_{i'} = \frac{\sqrt{|g|}}{2!}\phi_{jk}g^{jj'}g^{kk'}\epsilon_{j'k'i'}$, and that $(**\phi)_{mn} = \sqrt{|g|}(*\phi)_{i'}g^{i'i}\epsilon_{imn}$. Combining these, we obtain

$$(**\phi)_{mn} \overset{(i)}{=} \frac{|g|}{2!}\left(\phi_{jk}g^{jj'}g^{kk'}\epsilon_{j'k'i'}\right)g^{i'i}\epsilon_{mni}(-)^{1\cdot(3-1)} \overset{(ii)}{=} \frac{|g|}{g}\phi_{jk}\left[\frac{1}{2!}\epsilon^{jki}\epsilon_{mni}\right]$$
$$\overset{(iii)}{=} \boxed{\text{sgn}(g)\phi_{mn}}.$$

This shows that $**\phi = \text{sgn}(g)\,\phi$. ✓ The justification for steps (i) and (ii) is analogous to that given above. For step (iii), we noted that for a given choice of indices mn, nonzero contributions arise from the $\epsilon \cdot \epsilon$ factor only if i differs from both mn and jk, i.e. if either $jk = mn$ or $kj = mn$. Since $\phi_{jk}\epsilon^{jkl}$ is symmetric in jk, both cases contribute with the same sign. This sign follows from the particular choice $jk = mn$, for which $\epsilon \cdot \epsilon = 1$.

(e) For $(g_{ij}) = \text{diag}(-1,-1,-1)$, we have $g = \det g = -1$ and $(g^{ij}) = \text{diag}(-1,-1,-1)$, so the simplifications from (a) read

$$(*1)_{ijk} = \boxed{\epsilon_{ijk}}, \quad (*\lambda)_{jk} = \boxed{-\lambda_i\epsilon_{ijk}}, \quad (*\phi)_l = \boxed{\frac{1}{2!}\phi_{jk}\epsilon_{jkl}}, \quad (*\omega) = \boxed{-\frac{1}{3!}\omega_{ijk}\epsilon_{ijk}}.$$

Hence $*1$ and $*\phi$ are the same as in (a), whereas $*\lambda$ and $*\omega$ pick up an extra sign.

EL10.9.3 Coordinate invariance of the Hodge star operation: $p = 1, n = 3$ (p. 167)

Starting from $*e^i$, we (i) transform the argument form into the primed system using $e^i = (T^{-1})^i{}_k e'^k$, (ii) apply the $*$-operation using its primed version, $*e'^k = \frac{1}{2!}\sqrt{|g'|}g'^{kl_1}\epsilon_{l_1l_2l_3}e'^{l_2}\wedge e'^{l_3}$, and (iii) transform back to the unprimed basis, using $\sqrt{|g'|} = \sqrt{|g|}\det T^{-1}$ and $g'^{kl} = T^k{}_i T^l{}_j g^{ij}$, which we will need in the form $(T^{-1})^i{}_k g'^{kl} = T^l{}_j g^{ij}$:

$$*e^i \overset{(i)}{=} *(T^{-1})^i{}_k e'^k \overset{(ii)}{=} \frac{1}{2!}\sqrt{|g|}(T^{-1})^i{}_k g'^{kl_1}\epsilon_{l_1l_2l_3}e'^{l_2}\wedge e'^{l_3}$$
$$\overset{(iii)}{=} \frac{1}{2!}\sqrt{|g'|}\det T^{-1}(T^{l_1}{}_{j_1}g^{ij_1})\epsilon_{l_1l_2l_3}(T^{l_2}{}_{j_2}e^{j_2})\wedge(T^{l_3}{}_{j_3}e^{j_3})$$
$$= \boxed{\frac{1}{2!}\sqrt{|g|}g^{ij_1}\epsilon_{j_1j_2j_3}e^{j_2}\wedge e^{j_3}}.$$

For the last equality, we noted that the three-fold antisymmetric product of transformation matrix elements, $T^{l_1}{}_{j_1}T^{l_2}{}_{j_2}T^{l_3}{}_{j_3}\epsilon_{l_1l_2l_3} = \det T\epsilon_{j_1j_2j_3}$, yields a determinant, which cancels the $\det T^{-1}$ in the prefactor. Comparing the initial and final expressions, we have recovered the unprimed version of the Hodge operation from its primed version. This establishes its invariance.

Solutions: Calculus

S.C1 Differentiation of one-dimensional functions (p. 208)

S.C1.1 Definition of differentiability (p. 208)

ᴇC1.1.1 Differentiation of polynomials (p. 211)

(a) $f'(x) = 9x^2 + 2,$ $\qquad\qquad$ $f''(x) = 18x.$

(b) $f'(x) = 4x^3 - 4x,$ $\qquad\qquad$ $f''(x) = 12x^2 - 4.$

S.C1.2 Differentiation rules (p. 212)

ᴇC1.2.1 Product rule and chain rule (p. 213)

Using $\sin' x = \cos x$, $\cos' x = -\sin x$, and the product and chain rules, we obtain:

(a) $f'(x) = \sin x + x \cos x,$

(b) $f'(x) = -\sin[\pi(x^2 + x)]\pi(2x + 1),$

(c) $f'(x) = \dfrac{2x}{(7 - x^2)^2},$

(d) $f'(x) = \dfrac{1}{x+1} - \dfrac{x-1}{(x+1)^2} = \dfrac{2}{(x+1)^2}.$

S.C1.3 Derivatives of selected functions (p. 213)

ᴇC1.3.1 Differentiation of trigonometric functions (p. 214)

Using $\frac{d}{dx}\sin x = \cos x$, $\frac{d}{dx}\cos x = -\sin x$, and $\sin^2 x + \cos^2 x = 1$, we readily find

(a) $\dfrac{d}{dx}\tan x = \dfrac{d}{dx}\dfrac{\sin x}{\cos x} = \dfrac{\cos x}{\cos x} + \dfrac{\sin^2 x}{\cos^2 x} = \boxed{1 + \tan^2 x}$, ✓

$\qquad\qquad = \dfrac{\cos^2 x + \sin^2 x}{\cos^2 x} = \dfrac{1}{\cos^2 x} = \boxed{\sec^2 x}$. ✓

(b) $\dfrac{d}{dx}\cot x = \dfrac{d}{dx}\dfrac{\cos x}{\sin x} = -\dfrac{\sin x}{\sin x} - \dfrac{\cos^2 x}{\sin^2 x} = \boxed{-1 - \cot^2 x}$, ✓

$\qquad\qquad = \dfrac{-\sin^2 x - \cos^2 x}{\sin^2 x} = -\dfrac{1}{\sin^2 x} = \boxed{-\csc^2 x}$. ✓

ₜC1.3.3 Differentiation of inverse trigonometric functions (p. 214)

The trigonometric functions $f = \sin$, $\cos$ and $\tan$ are all periodic, hence their inverses, $f^{-1} = \arcsin$, arccos and arctan, each have infinitely many branches, one for each x-domain of f on which a bijection can be defined. On any given branch, the slope of f^{-1} has the same sign as the slope of f. We consider representative examples of such branches, and for each case compute the derivative of f^{-1} using $(f^{-1})'(x) = \dfrac{1}{f'(y)|_{y=f^{-1}(x)}}$.

(a)   arcsin is the inverse function of sin, with $\sin(\arcsin x) = x$. We consider two branches of $\arcsin x$, with slopes of opposite sign.
I. The function $\sin\colon (-\tfrac{1}{2}\pi, \tfrac{1}{2}\pi) \to (-1, 1)$ has positive slope, $\sin' x = \cos x$, and inverse $\arcsin\colon (-1, 1) \to (-\tfrac{1}{2}\pi, \tfrac{1}{2}\pi)$.
II. The function $\sin\colon (\tfrac{1}{2}\pi, \tfrac{3}{2}\pi) \to (1, -1)$ has negative slope, $\sin' x = \cos x$, and inverse $\arcsin\colon (-1, 1) \to (\tfrac{3}{2}\pi, \tfrac{1}{2}\pi)$.
Using upper/lower signs for branches I/II, we obtain

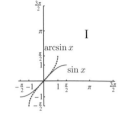

$$\arcsin' x = \frac{1}{\sin'(y)|_{y=\arcsin x}} = \frac{1}{\cos(\arcsin x)}$$

$$= \frac{\pm 1}{\sqrt{1 - \sin^2(\arcsin x)}} = \boxed{\frac{\pm 1}{\sqrt{1 - x^2}}}.$$

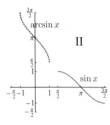

Unless stated otherwise, the notation arcsin refers to branch I.

(b)   arccos is the inverse function of cos, with $\cos(\arccos x) = x$. We consider two branches of arccos, with slopes of opposite sign.
I. The function $\cos\colon (0, \pi) \to (1, -1)$ has negative slope, $\cos' x = -\sin x$, and inverse $\arccos\colon (-1, 1) \to (\pi, 0)$.
II. The function $\cos x\colon (-\pi, 0) \to (-1, 1)$ has positive slope, $\cos' x = -\sin x$, and inverse $\arccos\colon (-1, 1) \to (-\pi, 0)$.
Using upper/lower signs for branches I/II, we obtain

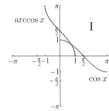

$$\arccos' x = \frac{1}{\cos'(y)|_{y=\arccos x}} = \frac{-1}{\sin(\arccos x)}$$

$$= \frac{\mp 1}{\sqrt{1 - \cos^2(\arccos x)}} = \boxed{\frac{\mp 1}{\sqrt{1 - x^2}}}.$$

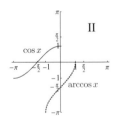

Unless stated otherwise, the notation arccos refers to branch I.

(c)   arctan is the inverse function of tan, with $\tan(\arctan x) = x$. The slope of tan, given by $\tan' x = \sec^2 x$, is positive for every branch. We consider only the branch centered on zero, $\tan\colon (-\tfrac{\pi}{2}, \tfrac{\pi}{2}) \to \mathbb{R}$, with inverse $\arctan\colon \mathbb{R} \to (-\tfrac{\pi}{2}, \tfrac{\pi}{2})$:

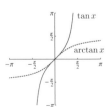

$$\arctan' x = \frac{1}{\tan'(y)|_{y=\arctan x}} = \frac{1}{\sec^2(\arctan x)}$$

$$= \frac{1}{1 + \tan^2(\arctan x)} = \boxed{\frac{1}{1 + x^2}}.$$

ɛC1.3.5 **Differentiation of powers, exponentials, logarithms** (p. 214)

(a) $f'(x) = \dfrac{1}{2\sqrt{2x^3}}$,

(b) $f'(x) = \dfrac{1}{2}\dfrac{1}{x^{1/2}(x+1)^{1/2}} - \dfrac{1}{2}\dfrac{x^{1/2}}{(x+1)^{3/2}} = \dfrac{1}{2}\dfrac{1}{x^{1/2}(x+1)^{3/2}}$,

(c) $f'(x) = e^x(2x-1)$,

(d) $f'(x) = \dfrac{d}{dx}e^{\ln 3^x} = \dfrac{d}{dx}e^{x\ln 3} = e^{x\ln 3}\ln 3 = 3^x\ln 3$,

(e) $f'(x) = \ln x + \dfrac{x}{x} = \ln x + 1$,

(f) $f'(x) = \ln(9x^2) + x\dfrac{1}{9x^2}18x = \ln(9x^2) + 2$.

ɛC1.3.7 **L'Hôpital's rule** (p. 214)

For (a,b) we may apply L'Hôpital's rule in the form $\lim_{x\to x_0}\dfrac{f(x)}{g(x)} = \lim_{x\to x_0}\dfrac{f'(x)}{g'(x)}$, since the given functions f and g both vanish at the limiting point x_0, whereas f' and g' are finite there.

(a) $\lim\limits_{x\to 1}\dfrac{x^2+(a-1)x-a}{x^2+2x-3} = \lim\limits_{x\to 1}\dfrac{2x+(a-1)}{2x+2} = \dfrac{2+(a-1)}{2+2} = \boxed{\dfrac{a+1}{4}}$.

(b) $\lim\limits_{x\to 0}\dfrac{\sin(ax)}{x+ax^2} = \lim\limits_{x\to 0}\dfrac{a\cos(ax)}{1+2ax} = \boxed{a}$.

(c) We use $\lim_{x\to 0}\dfrac{f(x)}{g(x)} = \lim_{x\to 0}\dfrac{f'(x)}{g'(x)} = \lim_{x\to 0}\dfrac{f''(x)}{g''(x)}$, since not only f and g but also f' and g' vanish at $x=0$.

$\lim\limits_{x\to 0}\dfrac{1-\cos(ax)}{\sin^2 x} = \lim\limits_{x\to 0}\dfrac{a\sin(ax)}{2\sin x\cos x} = \lim\limits_{x\to 0}\dfrac{a^2\cos(ax)}{2[\cos^2 x - \sin^2 x]} = \boxed{\dfrac{a^2}{2}}$.

(d) We use L'Hôpital's rule three times, since $f^{(n)}(0)$ and $g^{(n)}(0)$ all vanish for $n=0,1,2$:

$\lim\limits_{x\to 0}\dfrac{x^3}{\sin(ax)-ax} = \lim\limits_{x\to 0}\dfrac{3x^2}{a\cos(ax)-a} = \lim\limits_{x\to 0}\dfrac{6x}{-a^2\sin(ax)} = \lim\limits_{x\to 0}\dfrac{6}{-a^3\cos(ax)} = \boxed{\dfrac{-6}{a^3}}$.

(e) The naive answer, $\lim_{x\to 0}(x\ln x)\overset{?}{=}0\cdot\infty$, is ill-defined, hence we evoke L'Hôpital's rule for the case $\lim_{x\to 0}|f(x)| = \lim_{x\to 0}|g(x)| = \infty$, with $f(x)=\ln x$ and $g(x)=x^{-1}$:

$\lim\limits_{x\to 0}(x\ln x) = \lim\limits_{x\to 0}\dfrac{\ln x}{x^{-1}} = \lim\limits_{x\to 0}\dfrac{x^{-1}}{-x^{-2}} = \lim\limits_{x\to 0}-x = \boxed{0}$.

S.C2   Integration of one-dimensional functions (p. 216)

S.C2.2   One-dimensional integration (p. 218)

ɛC2.2.1 **Elementary integrals** (p. 220)

(a) $I(x) = \displaystyle\int^x_1 dy(2y^3 - 2y + 3) = \left[\tfrac{1}{2}y^4 - y^2 + 3y\right]^x_1 = \boxed{\tfrac{1}{2}x^4 - x^2 + 3x - \tfrac{5}{2}}$.

(b) $I(x) = \displaystyle\int^x_0 dy\,e^{3y} = \left[\tfrac{1}{3}e^{3y}\right]^x_0 = \boxed{\tfrac{1}{3}(e^{3x} - 1)}$.

S.C2.3 Integration rules (p. 222)

C2.3.1 Integration by parts (p. 223)

(a) $I(z) = \int_0^z dx\, \overset{u}{x}\, \overset{v'}{e^{2x}} = \left[\overset{u}{x}\, \overset{v}{\tfrac{1}{2}e^{2x}}\right]_0^z - \int_0^z dx\, \overset{u'}{1}\cdot \overset{v}{\tfrac{1}{2}e^{2x}} = \boxed{\tfrac{1}{2}ze^{2z} - \tfrac{1}{4}(e^{2z}-1)}$,

$I'(z) = \left[\tfrac{1}{2}(1+2z) - \tfrac{1}{4}2\right]e^{2z} \overset{\checkmark}{=} ze^{2z}$, $I(\tfrac{1}{2}) \overset{\checkmark}{=} \tfrac{1}{4}$.

Note the cancellation pattern: $I' = u'v + uv' - u'v = uv'$. [Similarly for (c), (d).]

(b) $I(z) = \int_0^z dx\, \overset{u}{x^2}\, \overset{v'}{e^{2x}} = \left[\overset{u}{x^2}\, \overset{v}{\tfrac{1}{2}e^{2x}}\right]_0^z - \int_0^z dx\, \overset{u'}{2x}\, \overset{v}{\tfrac{1}{2}e^{2x}}$.

The integral on the right can be done by integrating by parts a second time, see (a):

$I(z) \overset{(a)}{=} \boxed{\tfrac{1}{2}z^2 e^{2z} - \tfrac{1}{2}ze^{2z} + \tfrac{1}{4}\left[e^{2z}-1\right]}$,

$I'(z) = \left[\tfrac{1}{2}(2z+2z^2) - \tfrac{1}{2}(1+2z) + \tfrac{1}{4}2\right]e^{2z} \overset{\checkmark}{=} z^2 e^{2z}$, $I(\tfrac{1}{2}) \overset{\checkmark}{=} \tfrac{e}{8} - \tfrac{1}{4}$.

Since we integrated by parts twice, I' yields more involved cancellations than for (a).

(c) $I(z) = \int_0^z dx\, \overset{u}{(\ln x)}\cdot \overset{v'}{1} = \left[\overset{u}{(\ln x)}\,\overset{v}{x}\right]_0^z - \int_0^z dx\, \overset{u'}{\tfrac{1}{x}}\,\overset{v}{x} = \boxed{(\ln z)z - z}$,

$I'(z) = \tfrac{1}{z}z + \ln z - 1 \overset{\checkmark}{=} \ln z$, $I(1) \overset{\checkmark}{=} -1$.

(d) $I(z) = \int_0^z dx\, \overset{u}{(\ln x)}\cdot \overset{v'}{\tfrac{1}{\sqrt{x}}} = \left[\overset{u}{(\ln x)}\,\overset{v}{2\sqrt{x}}\right]_0^z - \int_0^z dx\, \overset{u'}{\tfrac{1}{x}}\,\overset{v}{2\sqrt{x}} = \boxed{(\ln z)2\sqrt{z} - 4\sqrt{z}}$.

To evaluate $\left[\ln(x)\sqrt{x}\right]_{x=0}$, we used the rule of L'Hôpital ($\to$ C1.3.7-8):

$$\left[(\ln x)\sqrt{x}\right]_{x=0} = \lim_{x\to 0}\frac{\ln x}{x^{-1/2}} = \lim_{x\to 0}\frac{\frac{d}{dx}\ln x}{\frac{d}{dx}x^{-1/2}} = \lim_{x\to 0}\frac{x^{-1}}{-\tfrac{1}{2}x^{-3/2}} = \lim_{x\to 0}\left[-2x^{1/2}\right] = \boxed{0}.$$

Thus the divergence of $\ln(x)$ for $x \to 0$ is so slow that $\sqrt{x}$ suppresses it.

$I'(z) = 2\left[\tfrac{1}{z}\sqrt{z} + (\ln z)\tfrac{1}{2}\tfrac{1}{\sqrt{z}}\right] - 4\tfrac{1}{2}\tfrac{1}{\sqrt{z}} \overset{\checkmark}{=} (\ln z)\tfrac{1}{\sqrt{z}}$, $I(1) \overset{\checkmark}{=} -4$.

(e) $I(z) = \int_0^z dx\, \overset{u}{\sin x}\, \overset{v'}{\sin x} = \left[\overset{u}{\sin x}\,\overset{v}{(-\cos x)}\right]_0^z - \int_0^z dx\, \underset{\sin^2 x - 1}{\underbrace{\overset{u'}{\cos x}\,\overset{v}{(-\cos x)}}}$.

Re-express the integral on the right in terms of $I(z)$,

$I(z) = -\sin z\cos z - I(z) + \int_0^z dx\, 1$,   and solve for $I(z)$:

$I(z) = \boxed{\tfrac{1}{2}(-\sin z\cos z + z)}$,

$$I'(z) = \tfrac{1}{2}(-\cos^2 z + \sin^2 z + 1) \overset{\checkmark}{=} \sin^2 z, \qquad\qquad\qquad I(\pi) \overset{\checkmark}{=} \tfrac{\pi}{2}.$$

(f) $\displaystyle I(z) = \int_0^z dx\ \overset{u}{\sin^3 x}\ \overset{v'}{\sin x} = \Big[\overset{u}{\sin^3 x}\ \overset{v}{(-\cos x)}\Big]_0^z - \int_0^z dx\ \Big(\overset{u'}{3\sin^2 x\ \cos x}\Big)\overset{v}{(-\cos x)}.$

$$\underbrace{}_{\sin^2 x - 1}$$

Re-express the integral on the right in terms of $I(z)$,

$$I(z) = -\sin^3 z \cos z - 3\Big[I(z) - \int_0^z dx\ \sin^2 x\Big], \qquad \text{solve for } I(z), \text{ and use (e):}$$

$$I(z) \overset{(e)}{=} \boxed{\tfrac{1}{4}\Big[-\sin^3 z \cos z + \tfrac{3}{2}(-\sin z \cos z + z)\Big]},$$

$$I'(z) = \tfrac{1}{4}\Big[-3\sin^2 z\ \underbrace{\cos^2 z}_{1-\sin^2 z} + \sin^4 z + \tfrac{3}{2}(-\cos^2 z + \sin^2 z + 1)\Big] \overset{\checkmark}{=} \sin^4 z, \qquad I(\pi) \overset{\checkmark}{=} \tfrac{3\pi}{8}.$$

ℰC2.3.3 Integration by substitution (p. 224)

(a) $\displaystyle I(z) = \int_0^z dx\ x\cos(x^2 + \pi) \qquad \Big[y(x) = x^2,\ dy = 2x\,dx\Big]$

$$= \tfrac{1}{2}\int_{y(0)}^{y(z)} dy\ \cos(y+\pi) = \tfrac{1}{2}\sin(y+\pi)\Big|_0^{z^2} = \boxed{\tfrac{1}{2}\sin(z^2 + \pi)},$$

$$I'(z) = \tfrac{1}{2}\cos(z^2+\pi)\tfrac{d}{dz}z^2 \overset{\checkmark}{=} \cos(z^2+\pi)\,z, \qquad\qquad I\Big(\sqrt{\tfrac{\pi}{2}}\Big) \overset{\checkmark}{=} -\tfrac{1}{2}.$$

(b) $\displaystyle I(z) = \int_0^z dx\ \sin^3 x \cos x \qquad \Big[y(x) = \sin x,\ dy = \cos x\,dx\Big]$

$$= \int_{y(0)}^{y(z)} dy\ y^3 = \tfrac{1}{4}y^4\Big|_0^{\sin z} = \boxed{\tfrac{1}{4}\sin^4 z},$$

$$I'(z) = \sin^3 z \tfrac{d}{dz}\sin z \overset{\checkmark}{=} \sin^3 z \cos z, \qquad\qquad I\Big(\tfrac{\pi}{4}\Big) \overset{\checkmark}{=} \tfrac{1}{16}.$$

(c) $\displaystyle I(z) = \int_0^z dx\ \sin^3 x = \int_0^z dx\ \sin x\big[1 - \cos^2 x\big] \qquad \Big[y(x) = \cos x,\ dy = -\sin x\,dx\Big]$

$$= -\int_{y(0)}^{y(z)} dy\ (1 - y^2) = -(y - \tfrac{1}{3}y^3)\Big|_{\cos 0}^{\cos z} = \boxed{-\cos z + \tfrac{1}{3}\cos^3 z + \tfrac{2}{3}},$$

$$I'(z) = \sin z + \cos^2 z(-\sin z) = \sin z(1 - \cos^2 z) \overset{\checkmark}{=} \sin^3 z, \qquad I\Big(\tfrac{\pi}{3}\Big) \overset{\checkmark}{=} \tfrac{5}{24}.$$

(d) $\displaystyle I(z) = \int_0^z dx\ \cosh^3 x = \int_0^z dx\ \cosh x\big[1 + \sinh^2 x\big]\ \Big[y(x) = \sinh x,\ dy = \cosh x\,dx\Big]$

$$= \int_{y(0)}^{y(z)} dy\ (1 + y^2) = (y + \tfrac{1}{3}y^3)\Big|_0^{\sinh z} = \boxed{\sinh z + \tfrac{1}{3}\sinh^3 z},$$

$$I'(z) = \cosh z + \sinh^2 z \cosh z = \cosh z(1 + \sinh^2 z) \overset{\checkmark}{=} \cosh^3 z, \qquad I(\ln 2) \overset{\checkmark}{=} \tfrac{57}{64}.$$

(e) $\displaystyle I(z) = \int_0^z dx\ \sqrt{1 + \ln(x+1)}\ \tfrac{1}{x+1} \qquad \Big[y(x) = \ln(x+1),\ dy = \tfrac{1}{1+x}\,dx\Big]$

$$= \int_{y(0)}^{y(z)} dy\ \sqrt{1+y} = \tfrac{2}{3}(1+y)^{3/2}\Big|_0^{\ln(z+1)} = \boxed{\tfrac{2}{3}\Big[(1 + \ln(z+1))^{3/2} - 1\Big]},$$

$$I'(z) = (1 + \ln(z+1))^{1/2}\tfrac{d}{dz}\ln(z+1) \overset{\checkmark}{=} \sqrt{1 + \ln(z+1)}\ \tfrac{1}{z+1}, \qquad I(e^3 - 1) \overset{\checkmark}{=} \tfrac{14}{3}.$$

(f) $\displaystyle I(z) = \int_0^z dx\, x^3 e^{-x^4} \qquad \left[y(x) = x^4,\; dy = 4x^3\, dx \right]$

$\displaystyle \quad = \frac{1}{4} \int_{y(0)}^{y(z)} dy\, e^{-y} = -\frac{1}{4} e^{-y} \Big|_0^{z^4} = \boxed{\frac{1}{4}\left[1 - e^{-z^4} \right]},$

$\displaystyle I'(z) = \frac{1}{4} e^{-z^4} \frac{d}{dz} z^4 \overset{\checkmark}{=} e^{-z^4} z^3, \qquad\qquad\qquad\qquad\qquad I(\sqrt[4]{\ln 2}) \overset{\checkmark}{=} \frac{1}{8}.$

C2.3.5 $\sqrt{1-x^2}$ integrals by trigonometric substitution (p. 224)

(a) Since $\frac{d}{dx} \arcsin x = \frac{1}{\sqrt{1-x^2}}$, the primitive function of the integrand is known, and we may conclude immediately that $I(z) = [\arcsin x]_0^z = \arcsin z$.

Equivalently, we may compute the integral using the substitution $x = \sin y$, with $dx = dy \frac{dx}{dy} = dy \sin' y = dy \cos y$ and $\sqrt{1-x^2} = \sqrt{1-\sin^2 y} = \cos y$. The new integration boundaries are found by evaluating $y = \arcsin x$ at $x = 0$ and $x = z$:

$\displaystyle I(z) = \int_0^z dx\, \frac{1}{\sqrt{1-x^2}} = \int_{\arcsin 0}^{\arcsin z} dy \cos y \frac{1}{\cos y} = \int_0^{\arcsin z} dy = \boxed{\arcsin z}.$

Check result: $I\left(\frac{1}{\sqrt{2}}\right) = \arcsin\left(\frac{1}{\sqrt{2}}\right) = \frac{\pi}{4}$, since $\sin\left(\frac{\pi}{4}\right) = \frac{1}{\sqrt{2}}$. ✓

General check: $\displaystyle \frac{dI(z)}{dz} = \frac{d}{dz} \arcsin z = \boxed{\frac{1}{\sqrt{1-z^2}}}.$ ✓

(b) We substitute $x = \frac{1}{a} \sin y$, with $dx = dy \frac{dx}{dy} = dy \frac{1}{a} \cos y$ and $\sqrt{1-a^2 x^2} = \cos y$:

$\displaystyle I(z) = \int_0^z dx\, \sqrt{1-a^2 x^2} = \frac{1}{a} \int_{\arcsin 0}^{\arcsin(az)} dy \cos y \cos y \equiv \frac{1}{a} \tilde{I}(b).$

We compute the $\cos^2 y$ integral, with upper limit $b = \arcsin(az)$, by integrating by parts, with $u = \cos y$, $v = \sin y$, $u' = -\sin y$, $v' = \cos y$:

$\displaystyle \tilde{I}(b) = \int_0^b dy\, \underset{u}{\cos y} \underset{v'}{\cos y} \overset{uv - \int u'v}{=} [\cos y \sin y]_0^b - \int_0^b dy\, \underbrace{[-\sin y]\sin y}_{\cos^2 y - 1}$

$\displaystyle \qquad = b + \cos b \sin b - \tilde{I}(b)$

$\displaystyle \Rightarrow \tilde{I}(b) = \frac{1}{2}[b + \sin b \cos b] = \frac{1}{2}\left[b + \sin b \sqrt{1-\sin^2 b} \right].$

We expressed the r.h.s. through sin, because the argument of $\tilde{I}(b)$ is $b = \arcsin(az)$.

$\displaystyle \Rightarrow I(z) = \frac{1}{a} \tilde{I}(\arcsin(az)) = \boxed{\frac{1}{2a}\left[\arcsin(az) + az\sqrt{1-a^2 z^2} \right]}.$

Check result: for $a = \frac{1}{2}$, $I(\sqrt{2}) = \arcsin\left(\frac{1}{\sqrt{2}}\right) + \frac{1}{\sqrt{2}}\sqrt{1-\frac{1}{2}} = \frac{\pi}{4} + \frac{1}{2}$. ✓

General check: $\displaystyle \frac{dI(z)}{dz} \overset{(a)}{=} \frac{1}{2}\left[\frac{1}{\sqrt{1-a^2 z^2}} + \sqrt{1-a^2 z^2} + az \frac{-az}{\sqrt{1-a^2 z^2}} \right] = \sqrt{1-a^2 z^2}.$ ✓

ₑC2.3.7 $1/(1 - x^2)$ integrals by hyperbolic substitution (p. 224)

(a)   Since $\frac{d}{dx}\operatorname{arctanh} x = \frac{1}{1-x^2}$, the primitive function of the integrand is known, and we may conclude immediately that $I(z) = [\operatorname{arctanh} x]_0^z = \operatorname{arctanh} z$.

Equivalently, we may compute the integral using the substitution $x = \tanh y$, with $dx = dy\frac{dx}{dy} = dy \tanh' y = dy \operatorname{sech}^2 y$ and $1 - x^2 = 1 - \tanh^2 y = \operatorname{sech}^2 y$. The new integration boundaries are found by evaluating $y = \operatorname{arctanh} x$ at $x = 0$ and $x = z$:

$$I(z) = \int_0^z dx\, \frac{1}{1-x^2} = \int_{\operatorname{arctanh} 0}^{\operatorname{arctanh} z} dy\, \operatorname{sech}^2 y\, \frac{1}{\operatorname{sech}^2 y} = \int_0^{\operatorname{arctanh} z} dy\, 1 = \boxed{\operatorname{arctanh} z}\,.$$

Check result: $I(\frac{3}{5}) = \operatorname{arctanh}(\frac{3}{5}) = \ln 2$, since $\tanh(\ln 2) = \frac{e^{\ln 2} - e^{-\ln 2}}{e^{\ln 2} + e^{-\ln 2}} = \frac{2 - 1/2}{2 + 1/2} \overset{\checkmark}{=} \frac{3}{5}$.

General check: $\dfrac{dI(z)}{dz} = \dfrac{d}{dz}\operatorname{arctanh} z = \boxed{\dfrac{1}{1-z^2}}\cdot\ \checkmark$

(b)   We substitute $x = \frac{1}{a}\tanh y$, with $dx = dy\frac{dx}{dy} = dy\frac{1}{a}\operatorname{sech}^2 y$ and $1 - a^2x^2 = \operatorname{sech}^2 y$.

$$I(z) = \int_0^z dx\, \frac{1}{(1 - a^2x^2)^2} = \frac{1}{a}\int_{\operatorname{arctanh} 0}^{\operatorname{arctanh}(az)} dy\, \frac{\operatorname{sech}^2 y}{\operatorname{sech}^4 y} = \frac{1}{a}\int_0^{\operatorname{arctanh}(az)} dy\, \cosh^2 y \equiv \frac{1}{a}\tilde{I}(b).$$

We compute the $\cosh^2 y$ integral, with upper limit $b = \operatorname{arctanh}(az)$, by integrating by parts, with $u = \cosh y$, $v = \sinh y$, $u' = \sinh y$, $v' = \cosh y$:

$$\tilde{I}(b) = \int_0^b dy\, \overset{u}{\cosh y}\,\overset{v'}{\cosh y} \overset{uv - \int u'v}{=} \left[\cosh y\, \sinh y\right]_0^b - \int_0^b dy\, \underbrace{\sinh y\, \sinh y}_{\cosh^2 y - 1}$$

$$= b + \cosh b\, \sinh b - \tilde{I}(b)$$

$$\Rightarrow \tilde{I}(b) = \frac{1}{2}\left[b + \sinh b\, \cosh b\right] = \frac{1}{2}\left[b + \frac{\tanh b}{1 - \tanh^2 b}\right].$$

We expressed the r.h.s. through $\tanh$, using $\sinh\cosh = \tanh/\operatorname{sech}^2 = \tanh/(1 - \tanh^2)$, because the argument of $\tilde{I}(b)$ is $b = \operatorname{arctanh}(az)$.

$$\Rightarrow I(z) = \frac{1}{a}\tilde{I}(\operatorname{arctanh}(az)) = \boxed{\frac{1}{2a}\left[\operatorname{arctanh}(az) + \frac{az}{1 - a^2z^2}\right]}.$$

Check result: for $a = 3$, we have $I(\frac{1}{5}) = \frac{1}{6}\left[\operatorname{arctanh}(\frac{3}{5}) + \frac{3/5}{1 - (3/5)^2}\right] = \frac{1}{6}\ln 2 + \frac{5}{32}$. $\checkmark$

General check: $\dfrac{dI(z)}{dz} = \dfrac{1}{2}\left[\dfrac{1}{1 - a^2z^2} + \dfrac{(1 - a^2z^2) + 2a^2z^2}{(1 - a^2z^2)^2}\right] = \dfrac{1}{(1 - a^2z^2)^2}\cdot\ \checkmark$

S.C2.4   Practical remarks on one-dimensional integration (p. 225)

ₑC2.4.1   Partial fraction decomposition (p. 226)

(a)   The integral has the form $I(z) = \int_0^z dx\, f(x)$, with

$$f(x) = \frac{3}{(x + 1)(x - 2)}\,. \tag{1}$$

To compute it using a partial fraction decomposition, we make the ansatz

$$f(x) = \frac{A}{x+1} + \frac{B}{x-2} \,, \tag{2}$$

and determine the coefficients A and B by writing (2) in the form (1):

$$f(x) = \frac{A(x-2) + B(x+1)}{(x+1)(x-2)} = \frac{(A+B)x - 2A + B}{(x+1)(x-2)} \,. \tag{3}$$

Comparing coefficients in the numerators of (3) and (1) we obtain:

$$A + B = 0 \qquad\qquad \Rightarrow \qquad\qquad A = -B \,, \tag{4}$$

$$-2A + B = 3 \qquad\qquad \overset{(4)}{\Rightarrow} \qquad\qquad -3A = 3 \quad \Rightarrow \quad A = -1, \ B = 1 \,. \tag{5}$$

Now we insert the coefficients from (5) into (2) and integrate:

$$I(z) = \int_0^z dx f(x) = \int_0^z dx \left[\frac{-1}{x+1} + \frac{1}{x-2} \right] = \left[-\ln|x+1| + \ln|x-2| \right]_0^z$$

$$= \boxed{\ln \left| \frac{1 - \frac{1}{2}z}{1+z} \right|} \,.$$

Remark: The form of the ansatz (2), as well as the coefficients A and B, follow from the asymptotic behavior of the function $f(x)$ at its singularities, $x = -1$ and $x = 2$, respectively:

$$x = -1 + \delta : \qquad f(-1+\delta) = \frac{3}{(-1+\delta+1)(-1+\delta-2)} \overset{\delta \to 0}{\longrightarrow} -\frac{1}{\delta} + \mathcal{O}(\delta^0) \,, \tag{6}$$

$$x = 2 + \delta : \qquad f(2+\delta) = \frac{3}{(2+\delta+1)(2+\delta-2)} \overset{\delta \to 0}{\longrightarrow} \frac{1}{\delta} + \mathcal{O}(\delta^0) \,. \tag{7}$$

Equations (6) and (7) directly imply that $A = -1$ and $B = 1$, because these are the only values for which ansatz (2) shows the same asymptotic behavior as the function (1) at its singularities.

(b) The integral has the form $I(z) = \int_0^z dx f(x)$, with

$$f(x) = \frac{3x}{(x+1)^2(x-2)} \,. \tag{8}$$

To compute it using a partial fraction decomposition, we make the ansatz

$$f(x) = \frac{A}{x+1} + \frac{B}{(x+1)^2} + \frac{C}{x-2} \,, \tag{9}$$

and determine the coefficients A, B and C by bringing (9) into the form (8):

$$f(x) = \frac{A(x+1)(x-2) + B(x-2) + C(x+1)^2}{(x+1)^2(x-2)}$$

$$= \frac{A(x^2 - x - 2) + Bx - 2B + C(x^2 + 2x + 1)}{(x+1)^2(x-2)}$$

$$= \frac{(A+C)x^2 + (-A+B+2C)x - 2A - 2B + C}{(x+1)^2(x-2)} \,. \tag{10}$$

Comparing coefficients in the numerators of (10) and (8), we obtain:

$$A + C = 0 \qquad\qquad \Rightarrow \qquad\qquad A = -C \,, \tag{11}$$

$$-A + B + 2C = 3 \quad \overset{(11)}{\Rightarrow} \quad B + 3C = 3 \Rightarrow B = 3 - 3C, \tag{12}$$

$$-2A - 2B + C = 0 \quad \overset{(11)}{\Rightarrow} \quad 3C - 2B = 0 \tag{13}$$

$$\overset{(12,13)}{\Rightarrow} \quad 3C - 2(3 - 3C) = 0 \Rightarrow C = \tfrac{2}{3}, B = 1, A = -\tfrac{2}{3}. \tag{14}$$

Now we insert the coefficients from (14) into (9) and integrate:

$$I(z) = \int_0^z f(x) = \int_0^z dx \left[-\frac{2}{3} \frac{1}{x+1} + \frac{1}{(x+1)^2} + \frac{2}{3} \frac{1}{x-2} \right]$$

$$= \left[-\frac{2}{3} \ln|x+1| - \frac{1}{x+1} + \frac{2}{3} \ln|x-2| \right]_0^z = \boxed{\frac{2}{3} \ln \left| \frac{1 - \frac{1}{2}z}{1+z} \right| + \frac{z}{z+1}}.$$

Remark: The form of the ansatz (9), as well as the coefficients A, B and C, follow from the asymptotic behavior of the function $f(x)$ at its singularities, $x = -1$ and $x = 2$, respectively:

$$x = -1 + \delta: \quad f(-1 + \delta) = \frac{3(-1 + \delta)}{(-1 + \delta + 1)^2(-1 + \delta - 2)} = \frac{-3(1 - \delta)}{\delta^2(-3)(1 - \frac{1}{3}\delta)} \tag{15}$$

$$\overset{\delta \to 0}{\longrightarrow} \frac{(1 - \delta)(1 + \frac{1}{3}\delta + \mathcal{O}(\delta^2))}{\delta^2} = \frac{1}{\delta^2} - \frac{2}{3\delta} + \mathcal{O}(\delta^0). \tag{16}$$

$$x = 2 + \delta: \quad f(2 + \delta) = \frac{3(2 + \delta)}{(2 + \delta + 1)^2(2 + \delta - 2)}$$

$$\overset{\delta \to 0}{\longrightarrow} \frac{6(1 + \mathcal{O}(\delta^1))}{3^2\delta} = \frac{2}{3\delta} + \mathcal{O}(\delta^0). \tag{17}$$

For the step from (15) to (16) we used $\frac{1}{1 - \frac{1}{3}\delta} = 1 + \frac{1}{3}\delta + \mathcal{O}(\delta^2)$. [This follows from $(1 - \frac{1}{3}\delta)(1 + \frac{1}{3}\delta) = 1 + \mathcal{O}(\delta^2)$, or more generally, from the first two terms of the geometric series for $\frac{1}{1 - \frac{1}{3}\delta}$, see Section C5.1, Eq. (C85).] Equations (16) and (17) directly imply that $A = -\frac{2}{3}$, $B = 1$ and $C = \frac{2}{3}$, because these are the only values for which ansatz (9) shows the same asymptotic behavior as the function (8) at its singularities.

S.C2.4.3 Elementary Gaussian integrals (p. 227)

(a) We compute I using polar coordinates: $x = \rho \cos\phi$, $y = \rho \sin\phi$, $dxdy = \rho \, d\rho \, d\phi$.

$$I = \int_{-\infty}^{+\infty} dxdy \, e^{-(x^2+y^2)} = \int_0^{2\pi} d\phi \int_0^\infty d\rho \, \rho \, e^{-\rho^2} = 2\pi \left[-\frac{1}{2} e^{-\rho^2} \right]_0^\infty = \boxed{\pi}.$$

(b) In the two-dimensional integral I, the x and y integrals are independent and factorize:

$$I = \int_{-\infty}^{+\infty} dxdy \, e^{-(x^2+y^2)} = \left[\int_{-\infty}^\infty dx \, e^{-x^2} \right] \left[\int_{-\infty}^\infty dy \, e^{-y^2} \right] = \underbrace{\left[\int_{-\infty}^\infty dx \, e^{-x^2} \right]^2}_{I_0(1)} = [I_0(1)]^2.$$

$$I_0(1) = +\sqrt{I} = \boxed{\sqrt{\pi}}. \qquad \text{(Sign: } I_0(1) \text{ is positive since the integrand } e^{-x^2} > 0.\text{)}$$

The required Gaussian integral $I_0(a)$ is obtained using the substitution $y = \sqrt{a}x$:

$$I_0(a) = \int_{-\infty}^\infty dx \, e^{-ax^2} \overset{y=\sqrt{a}x}{=} \int_{-\infty}^\infty dy \frac{1}{\sqrt{a}} e^{-y^2} = \boxed{\sqrt{\frac{\pi}{a}}}.$$

(c) Completing the square in the exponent yields:

$$-ax^2 + bx = -a\left[x^2 - \frac{b}{a}x\right] = -a\left[x^2 - 2\left(\frac{b}{2a}\right)x + \left(\frac{b}{2a}\right)^2 - \left(\frac{b}{2a}\right)^2\right]$$

$$= -a\left[x - \frac{b}{2a}\right]^2 + \frac{b^2}{4a} = -ay^2 + D, \quad \text{with } C = \frac{b}{2a}, D = \frac{b^2}{4a}, y = x - C.$$

$$I_1(a,b) = \int_{-\infty}^{\infty} dx\, e^{-ax^2+bx} = \int_{-\infty}^{\infty} dy\, e^{-ay^2+D} = I_0(a)e^D = \boxed{\sqrt{\frac{\pi}{a}}\,e^{\frac{b^2}{4a}}}.$$

§C2.4.5 **Definite exponential integrals of the form** $\int_0^{\infty} dx\, x^n e^{-ax}$ (p. 227)

Below, I_n stands for $I_n(a)$, i.e. the a dependence of the integral will not be indicated explicitly.

(a) Repeated partial integration gives:

$$I_n = \int_0^{\infty} dx\, x^n e^{-ax},$$

$$I_0 = \int_0^{\infty} dx\, e^{-ax} = -\frac{e^{-ax}}{a}\Big|_0^{\infty} = \frac{1}{a},$$

$$I_1 = \int_0^{\infty} dx\, \overset{u}{x}\,\overset{v'}{e^{-ax}} \overset{uv - \int u'v}{=} \left[x\left(-\frac{e^{-ax}}{a}\right)\right]_0^{\infty} - \int_0^{\infty} dx\, 1\left(-\frac{e^{-ax}}{a}\right) = 0 + \frac{1}{a}I_0 = \frac{1}{a^2},$$

$$I_2 = \int_0^{\infty} dx\, \overset{u}{x^2}\,\overset{v'}{e^{-ax}} \overset{uv - \int u'v}{=} \left[x^2\left(-\frac{e^{-ax}}{a}\right)\right]_0^{\infty} - \int_0^{\infty} dx\, 2x\left(-\frac{e^{-ax}}{a}\right) = 0 + \frac{2}{a}I_1 = \frac{2}{a^3},$$

$$\cdots$$

$$I_n = \int_0^{\infty} dx\, \overset{u}{x^n}\,\overset{v'}{e^{-ax}} \overset{uv - \int u'v}{=} x^n\left[-\frac{1}{a}e^{-ax}\right]_0^{\infty} - \int_0^{\infty} dx\, nx^{n-1}\left(-\frac{1}{a}e^{-ax}\right) = \frac{n}{a}I_{n-1}.$$

The resulting pattern is:

$$I_n = \frac{n}{a}I_{n-1} = \frac{n}{a}\frac{n-1}{a}I_{n-2} = \frac{n}{a}\frac{n-1}{a}\frac{n-2}{a}I_{n-3} = \frac{n}{a}\frac{n-1}{a}\frac{n-2}{a}\cdots\frac{2}{a}\frac{1}{a}\frac{1}{a} = \boxed{\frac{n!}{a^{n+1}}}.$$

(b) Repeated differentiation gives (with $(-)^n \equiv (-1)^n$):

$$I_0 = \int_0^{\infty} dx\, e^{-ax} \qquad\qquad \Rightarrow \quad I_0 = -\frac{e^{-ax}}{a}\Big|_0^{\infty} = \frac{1}{a},$$

$$\frac{dI_0}{da} = \int_0^{\infty} dx\,(-x)e^{-ax} = -I_1 \qquad \Rightarrow \quad I_1 = (-)^1\frac{d}{da}\frac{1}{a} = \frac{1}{a^2},$$

$$\frac{d^2I_0}{da^2} = \int_0^{\infty} dx\,(-x)^2 e^{-ax} = (-)^2 I_2 \qquad \Rightarrow \quad I_2 = (-)^2\frac{d^2}{da^2}\frac{1}{a} = -\frac{d}{da}\frac{1}{a^2} = \frac{2\cdot 1}{a^3},$$

$$\frac{d^3I_0}{da^3} = \int_0^{\infty} dx\,(-x)^3 e^{-ax} = (-)^3 I_3 \qquad \Rightarrow \quad I_3 = (-)^3\frac{d^3}{da^3}\frac{1}{a} = -\frac{d}{da}\frac{2\cdot 1}{a^3} = \frac{3\cdot 2\cdot 1}{a^4}.$$

The resulting pattern is:

$$\frac{d^n I_0}{da^n} = \int_0^{\infty} dx\,(-x)^n e^{-an} = \boxed{(-)^n I_n} \quad \Rightarrow \quad I_n = (-)^n\frac{d^n}{da^n}\frac{1}{a} = -\frac{d}{da}\frac{(n-1)!}{a^n} = \boxed{\frac{n!}{a^{n+1}}}.$$

S.C3 Partial differentiation (p. 229)

S.C3.1 Partial derivative (p. 229)

ᴇC3.1.1 Partial derivatives (p. 230)

(a) $f(x, y) = x^2 y^3 - 2xy,$ $\partial_x f(x, y) = \boxed{2xy^3 - 2y},$ $\partial_y f(x, y) = \boxed{3x^2 y^2 - 2x}.$

(b) $f(x, y) = \sin[xe^{2y}],$ $\partial_x f(x, y) = \boxed{\cos[xe^{2y}]e^{2y}},$ $\partial_y f(x, y) = \boxed{\cos[xe^{2y}]2xe^{2y}}.$

S.C3.2 Multiple partial derivatives (p. 230)

ᴇC3.2.1 Partial derivates of first and second order (p. 231)

$$\partial_x r = \partial_x \sqrt{x^2 + y^2} = \frac{1}{2} \frac{2x}{\sqrt{x^2 + y^2}} = \frac{x}{r}, \quad \text{similarly:} \quad \partial_y r = \frac{y}{r},$$

$$\partial_x f(\mathbf{r}) = \partial_x \frac{x}{r} = \frac{1}{r} - x \cdot \frac{1}{r^2} \cdot \frac{x}{r} = \frac{r^2 - x^2}{r^3} = \frac{y^2}{r^3},$$

$$\partial_y f(\mathbf{r}) = \partial_y \frac{x}{r} = -x \cdot \frac{1}{r^2} \cdot \frac{y}{r} = -\frac{xy}{r^3},$$

$$\partial_{y,x}^2 f(\mathbf{r}) = \partial_y \left(\frac{y^2}{r^3} \right) = \frac{2y}{r^3} - y^2 \cdot \frac{3}{r^4} \cdot \frac{y}{r} = \frac{2yr^2 - 3y^3}{r^5} = \frac{2yx^2 - y^3}{r^5},$$

$$\partial_{x,y}^2 f(\mathbf{r}) = \partial_x \left(-\frac{xy}{r^3} \right) = -\frac{y}{r^3} + xy \cdot \frac{3}{r^4} \cdot \frac{x}{r} = \frac{-yr^2 + 3x^2 y}{r^5} = \frac{2yx^2 - y^3}{r^5},$$

$$\partial_x^2 f(\mathbf{r}) = \partial_x \left(\frac{y^2}{r^3} \right) = -y^2 \frac{3}{r^4} \cdot \frac{x}{r} = -\frac{3xy^2}{r^5},$$

$$\partial_y^2 f(\mathbf{r}) = \partial_y \left(-\frac{xy}{r^3} \right) = -\frac{x}{r^3} + xy \cdot \frac{3}{r^4} \cdot \frac{y}{r} = \frac{-xr^2 + 3xy^2}{r^5} = \frac{2xy^2 - x^3}{r^5}.$$

S.C3.3 Chain rule for functions of several variables (p. 231)

ᴇC3.3.1 Chain rule for functions of two variables (p. 235)

(a) Direct computation of $f(\mathbf{g}(\mathbf{x}))$ and its partial derivatives yields:

$$f(\mathbf{g}(\mathbf{x})) = \|\mathbf{g}(\mathbf{x})\| = \left\| \left(\ln x^2, 3 \ln x^1 \right)^{\mathsf{T}} \right\| = \boxed{\ln^2 x^2 + 9 \ln^2 x^1}.$$

$$\partial_{x^1} f(\mathbf{g}(\mathbf{x})) = \partial_{x^1} \left[\ln^2 x^2 + 9 \ln^2 x^1 \right] = \left[0 + 9(2 \ln x^1) \partial_{x^1} \ln x^1 \right] = \boxed{\frac{18 \ln x^1}{x^1}},$$

$$\partial_{x^2} f(\mathbf{g}(\mathbf{x})) = \partial_{x^2} \left[\ln^2 x^2 + 9 \ln^2 x^1 \right] = \left[(2 \ln x^2) \partial_{x^2} \ln x^2 + 0 \right] = \boxed{\frac{2 \ln x^2}{x^2}}.$$

(b)   The chain rule, $\partial_{x^k} f(\mathbf{g}(\mathbf{x})) = \sum_j \left[\partial_{y^j} f(\mathbf{y}) \right]_{\mathbf{y}=\mathbf{g}(\mathbf{x})} \partial_{x^k} g^j(\mathbf{x})$, yields the same results:

$$\partial_{x^1} f(\mathbf{g}(\mathbf{x})) = \left[\partial_{y^1} f(\mathbf{y}) \right]_{\mathbf{y}=\mathbf{g}(\mathbf{x})} \cdot \partial_{x^1} g^1(\mathbf{x}) + \left[\partial_{y^2} f(\mathbf{y}) \right]_{\mathbf{y}=\mathbf{g}(\mathbf{x})} \cdot \partial_{x^1} g^2(\mathbf{x})$$

$$= \left[\partial_{y^1} \left[(y^1)^2 + (y^2)^2 \right] \right]_{\mathbf{y}=\mathbf{g}(\mathbf{x})} \cdot \partial_{x^1} \left[\ln x^2 \right] + \left[\partial_{y^2} \left[(y^1)^2 + (y^2)^2 \right] \right]_{\mathbf{y}=\mathbf{g}(\mathbf{x})} \cdot \partial_{x^1} \left[3 \ln x^1 \right]$$

$$= 0 + \left[2y^2 \right]_{\mathbf{y}=\mathbf{g}(\mathbf{x})} \frac{3}{x^1} = 2g^2(\mathbf{x}) \cdot \frac{3}{x^1} = 2(3 \ln x^1) \frac{3}{x^1} = \boxed{\frac{18 \ln x^1}{x^1}} \overset{\checkmark}{=} \text{(a)}.$$

$$\partial_{x^2} f(\mathbf{g}(\mathbf{x})) = \left[\partial_{y^1} f(\mathbf{y}) \right]_{\mathbf{y}=\mathbf{g}(\mathbf{x})} \cdot \partial_{x^2} g^1(\mathbf{x}) + \left[\partial_{y^2} f(\mathbf{y}) \right]_{\mathbf{y}=\mathbf{g}(\mathbf{x})} \cdot \partial_{x^2} g^2(\mathbf{x})$$

$$= \left[\partial_{y^1} \left[(y^1)^2 + (y^2)^2 \right] \right]_{\mathbf{y}=\mathbf{g}(\mathbf{x})} \cdot \partial_{x^2} \left[\ln x^2 \right] + \left[\partial_{y^2} \left[(y^1)^2 + (y^2)^2 \right] \right]_{\mathbf{y}=\mathbf{g}(\mathbf{x})} \cdot \partial_{x^2} \left[3 \ln x^1 \right]$$

$$= \left[2y^1 \right]_{\mathbf{y}=\mathbf{g}(\mathbf{x})} \cdot \frac{1}{x^2} + 0 = 2g^1(\mathbf{x}) \cdot \frac{1}{x^2} = 2 \ln x^2 \frac{1}{x^2} = \boxed{\frac{2 \ln x^2}{x^2}} \overset{\checkmark}{=} \text{(a)}.$$

In both (a) and (b), computing the partial derivatives involves first differentiating the sum of squares, then differentiating logarithms. The chain rule route (b) simply makes this notationally somewhat more explicit than the direct route (a).

S.C4   Multidimensional integration (p. 238)

S.C4.1   Cartesian area and volume integrals (p. 238)

εC4.1.1   Fubini's theorem (p. 240)

(a) $I(a) = \int_0^a dx \int_0^1 dy\, x(x^2+y)^{\frac{1}{2}} = \int_0^a dx\, \frac{2}{3} \left[x(x^2+y)^{\frac{3}{2}} \right] \Big|_{y=0}^{y=1} = \int_0^a dx\, \frac{2}{3} \left[x(x^2+1)^{\frac{3}{2}} - x^4 \right]$

$\qquad = \frac{2}{3} \left[\frac{2}{5} \frac{1}{2} (x^2+1)^{\frac{5}{2}} - \frac{1}{5} x^5 \right] \Big|_{x=0}^{x=a} = \boxed{\frac{2}{15} \left[(a^2+1)^{\frac{5}{2}} - 1 - a^5 \right]}.$

(b) $I(a) = \int_0^1 dy \int_0^a dx\, x(x^2+y)^{\frac{1}{2}} = \int_0^1 dy\, \frac{2}{3} \frac{1}{2} \left[(x^2+y)^{\frac{3}{2}} \right] \Big|_{x=0}^{x=a} = \int_0^1 dy\, \frac{1}{3} \left[(a^2+y)^{\frac{3}{2}} - y^{\frac{3}{2}} \right]$

$\qquad = \frac{1}{3} \frac{2}{5} \left[(a^2+y)^{\frac{5}{2}} - y^{\frac{5}{2}} \right] \Big|_{y=0}^{y=1} = \boxed{\frac{2}{15} \left[(a^2+1)^{\frac{5}{2}} - 1 - a^5 \right]}.$

εC4.1.3   Violation of Fubini's theorem (p. 241)

(a)   First, we calculate the derivatives given in the hint:

$$\frac{\partial}{\partial y} \frac{y}{x^2+y^2} = \frac{1}{x^2+y^2} - \frac{2y^2}{(x^2+y^2)^2} = \frac{x^2-y^2}{(x^2+y^2)^2} = f(x,y) \overset{\text{analogously}}{=} -\frac{\partial}{\partial x} \frac{x}{x^2+y^2} \cdot \checkmark$$

This equation yields the anti-derivatives of $f(x,y)$ for integrating over either y or x:

$$I_A(a) = \int_a^1 dx \int_0^1 dy\, \frac{\partial}{\partial y} \frac{y}{x^2+y^2} = \int_a^1 dx\, \frac{1}{x^2+1} = \arctan x \Big|_a^1 = \boxed{\frac{\pi}{4} - \arctan a}.$$

$$I_B(a) = -\int_0^1 dy \int_a^1 dx \frac{\partial}{\partial x} \frac{x}{x^2+y^2} = -\int_0^1 dy \left[\frac{1}{1+y^2} - \frac{a}{a^2+y^2} \right]$$

$$= -\left[\arctan y - \arctan \frac{y}{a} \right]_0^1 = -\frac{\pi}{4} + \arctan \frac{1}{a} = \boxed{\frac{\pi}{4} - \arctan a} \overset{\checkmark}{=} I_A(a).$$

For the last step, we used the identity $\arctan \frac{1}{a} = \frac{\pi}{2} - \arctan a$.

(b) For $a = 0$, the calculation of I_A is analogous to that of (a) and gives $I_A(0) = \boxed{\frac{\pi}{4}}$. The calculation of I_B, however, changes significantly – the contribution of the lower x-integration boundary yields zero *before* the y-integral is carried out:

$$I_B(0) = -\int_0^1 dy \int_0^1 dx \frac{\partial}{\partial x} \frac{x}{x^2+y^2} = -\int_0^1 dy \left[\frac{1}{1+y^2} - 0 \right] = -\left[\arctan y \right]_0^1 = \boxed{-\frac{\pi}{4}}.$$

We therefore have $\boxed{I_{A,B}(a \to 0) = I_A(0) = -I_B(0)}$.

(c) We split the integration domain into two parts, $R_0 = R_0^+ \cup R_0^-$, in a way that ensures $f \gtrless 0$ on $R_0^\pm$, by choosing $R_0^+ = \{0 \le y \le x \le 1\}$ and $R_0^- = \{0 \le x \le y \le 1\}$:

$$I_0^+ = \int_0^1 dx \int_0^x dy\, f(x,y) = \int_0^1 dx \int_0^x dy \frac{\partial}{\partial y} \frac{y}{x^2+y^2} = \int_0^1 dx \frac{x}{x^2+x^2} = \frac{1}{2} \ln \left(\frac{1}{0} \right) = \boxed{\infty}.$$

$$I_0^- = \int_0^1 dy \int_0^y dx\, f(x,y) = \int_0^1 dy \int_0^y dx \frac{\partial}{\partial x} \frac{-x}{x^2+y^2} = \int_0^1 dy \frac{-y}{y^2+y^2} = -\frac{1}{2} \ln \left(\frac{1}{0} \right) = \boxed{-\infty}.$$

(d) We split the integration domain as $R_\delta = R_\delta^+ \cup R_\delta^-$, such that $I_\delta = I_\delta^+ + I_\delta^-$, with $I_\delta^\pm = \int_{R_\delta^\pm} dx\, dy\, f(x,y)$. These two integrals can be computed in a manner similar to (c), except that the integration domain for the outer integrals is changed from $(0,1)$ to $(\delta, 1)$. This yields

$$I_\delta^+ = \int_\delta^1 dx \int_0^x dy\, f(x,y) = \frac{1}{2} \ln \left(\frac{1}{\delta} \right), \qquad I_\delta^- = \int_\delta^1 dy \int_0^y dx\, f(x,y) = -\frac{1}{2} \ln \left(\frac{1}{\delta} \right).$$

We obtain $I_\delta = \boxed{0}$ for any $\delta < 1$, i.e. the limit $\delta \to 0$ is well-defined. The reason is that the removed square S_δ is symmetric w.r.t. $x \leftrightarrow y$, so that it regularizes the divergence in a way that ensures perfect cancellations between positive and negative contributions.

_EC4.1.5 Two-dimensional integration (Cartesian coordinates) (p. 241)

$$I(a) = \int_G dx\, dy\, f(x,y) = \int_0^1 dy \int_1^{a-y} dx\, xy = \int_0^1 dy\, y \left[\frac{1}{2} x^2 \right]_1^{a-y} = \frac{1}{2} \int_0^1 dy\, y \left[(a-y)^2 - 1 \right]$$

$$= \frac{1}{2} \int_0^1 dy \left[y^3 - ay^2 + (a^2-1)y \right] = \left[\frac{1}{8} y^4 - \frac{1}{3} ay^3 + \frac{1}{4}(a^2-1)y^2 \right]_0^1 = \boxed{-\frac{1}{8} - \frac{1}{3}a + \frac{1}{4}a^2}.$$

₅C4.1.7 Area enclosed by curves (Cartesian coordinates) (p. 241)

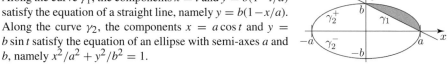

(a)   Along the curve γ_1, the components $x = t$ and $y = b(1-t/a)$ satisfy the equation of a straight line, namely $y = b(1-x/a)$.
Along the curve γ_2, the components $x = a\cos t$ and $y = b\sin t$ satisfy the equation of an ellipse with semi-axes a and b, namely $x^2/a^2 + y^2/b^2 = 1$.

(b)   When computing areas it is useful to parametrize curves by one of the coordinates of $\mathbb{R}^2$. Here we use x as the curve parameter (y would work equally well) and parametrize the upper and lower branches of the ellipse, $y_2^{\pm}$, by $y_2^{\pm}(x) = \pm b\sqrt{1-x^2/a^2}$, with $|x| < a$. Its area is then described by $-a < x < a$ and $y_2^-(x) < y < y_2^+(x)$:

$$S(a,b) = \int_{-a}^{a} dx \int_{y_2^-(x)}^{y_2^+(x)} dy\, 1 = \int_{-a}^{a} dx\left[y_2^+(x) - y_2^-(x)\right] = 2\int_{-a}^{a} dx\, b\sqrt{1-x^2/a^2}$$

$$\overset{x\,=\,a\sin u}{=} 2ab\int_{-\pi/2}^{\pi/2} du\cos^2 u \overset{\text{part. int.}}{=} 2ab\tfrac{1}{2}\left[u + \sin u\cos u\right]_{-\pi/2}^{\pi/2} = \boxed{\pi ab}.$$

(c)   We parametrize the straight line γ_1 by $y_1(x) = b(1-x/a)$. According to the figure, it intersects the ellipse γ_2 at the points $(a,0)^{\mathrm{T}}$ and $(0,b)^{\mathrm{T}}$. This is consistent with the following algebraic argument. Since $y_1(x) \geq 0$ for $x \leq a$, the straight line γ_1 intersects only the positive branch y_2^+ of the ellipse, namely when

$$0 = y_1(x) - y_2^+(x) = b(1-x/a) - b\sqrt{1-\tfrac{x^2}{a^2}}, \quad \Rightarrow \quad x = a \text{ or } x = 0.$$

The desired area (shaded in sketch) is thus described by $0 < x < a$ and $y_1(x) < y < y_2^+(x)$:

$$A(a,b) = \int_0^a dx \int_{y_1(x)}^{y_2^+(x)} dy\, 1 = \int_0^a dx\left[y_2^+(x) - y_1(x)\right] = \int_0^a dx\, b\left[\sqrt{1-\tfrac{x^2}{a^2}} - \left(1-\tfrac{x}{a}\right)\right]$$

$$\overset{(a)}{=} \tfrac{1}{4}\pi ab - b\left[x - \tfrac{1}{2}\tfrac{x^2}{a}\right]_0^a = \boxed{ab\left(\tfrac{1}{4}\pi - \tfrac{1}{2}\right)}.$$

Geometric consideration: the desired area is a quarter of the area of the ellipse, namely $\tfrac{1}{4}\pi ab$, minus the area of a triangle with base a and height b, namely $\tfrac{1}{2}ab$.

₅C4.1.9 Area integral for volume of a pyramid (Cartesian coordinates) (p. 241)

(a)   The plane E intersects the three axes at $x = a$, $y = b$ and $z = c$. Therefore the pyramid has base area $A = \tfrac{1}{2}ab$, height $h = c$ and volume $V = \tfrac{1}{3}Ah = \boxed{\tfrac{1}{6}abc}$.

(b)   The diagonal of the pyramid's base area in the xy-plane is described by $y_D(x) = b(1-x/a)$, and the base area itself by $0 \leq x \leq a$ and $0 \leq y \leq y_D(x)$. The pyramid's height above the base area is $h(x,y) = c(1 - x/a - y/b)$. Thus the volume is:

$$V = \int_A dxdy\, h(x,y) = \int_0^a dx \int_0^{y_D(x)} dy\, c\left(1 - \tfrac{x}{a} - \tfrac{y}{b}\right) = c\int_0^a dx\left[y\left(1-\tfrac{x}{a}\right) - \tfrac{1}{2}\tfrac{y^2}{b}\right]_0^{y_D(x)}$$

$$= c\int_0^a dx\left[b\left(1-\tfrac{x}{a}\right)^2 - \tfrac{1}{2}b^2\left(1-\tfrac{x}{a}\right)^2\tfrac{1}{b}\right] = \tfrac{1}{2}cb\left(-\tfrac{1}{3}a\right)\left[\left(1-\tfrac{x}{a}\right)^3\right]_0^a = \boxed{\tfrac{1}{6}abc}.\ \checkmark$$

S.C4.2   Curvilinear area integrals (p. 242)

ᴇC4.2.1   Area of an ellipse (generalized polar coordinates) (p. 246)

(a)   In generalized polar coordinates, the area element is given by:

$$dS = d\mu \, d\phi \, \|\partial_\mu \mathbf{r} \times \partial_\phi \mathbf{r}\|.$$

The integration measure can be calculated by using $\mathbf{r} = \mu a \cos\phi \, \mathbf{e}_1 + \mu b \sin\phi \, \mathbf{e}_2$:

$$\|\partial_\mu \mathbf{r} \times \partial_\phi \mathbf{r}\| = \left\| \begin{pmatrix} a\cos\phi \\ b\sin\phi \\ 0 \end{pmatrix} \times \begin{pmatrix} -a\mu\sin\phi \\ b\mu\cos\phi \\ 0 \end{pmatrix} \right\| = ab\,\mu(\cos^2\phi + \sin^2\phi) = \boxed{ab\,\mu}.$$

The integration limits now correspond to $0 < \mu < \infty, 0 < \phi < 2\pi$. Thus,

$$I = ab \int_0^\infty d\mu\,\mu \int_0^{2\pi} d\phi\, f(\mu^2) = \boxed{2\pi ab \int_0^\infty d\mu\,\mu f(\mu^2)}.$$

(b)   In Cartesian coordinates, an ellipse is defined by $(x/a)^2 + (y/b)^2 = 1$, and by $\mu^2 = 1$ in generalized polar coordinates. Using the latter, its area can be computed as follows:

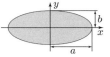

$$A_E = \int_{(x/a)^2+(y/b)^2\leq 1} dx\,dy = \int_{\mathbb{R}^2} dx\,dy f\left((x/a)^2 + (y/b)^2\right) \quad \text{with } f(\mu^2) = \begin{cases} 1 & \text{for } 0 < \mu^2 \leq 1, \\ 0 & \text{for } \mu^2 > 1, \end{cases}$$

$$= \int_0^{2\pi} d\phi \int_0^\infty d\mu\, ab\,\mu f(\mu^2) = 2\pi ab \int_0^1 d\mu\,\mu = 2\pi ab \frac{1}{2} = \boxed{\pi ab}.$$

S.C4.3   Curvilinear volume integrals (p. 249)

ᴇC4.3.1   Volume and moment of inertia (cylindrical coordinates) (p. 251)

In cylindrical coordinates, $x = \rho\cos\phi$, $y = \rho\sin\phi$, z, the volume element is $dV = d\phi\,dz\,d\rho\,\rho$ and the perpendicular distance to the z axis is $d_\perp^2 = x^2 + y^2 = \rho^2$. The homogeneous body F has density $\rho_0 = M/V_F$. The integration region is $0 < \phi < 2\pi$, $H \leq z \leq 2H$, $0 < \rho \leq az$, hence the integral over ρ has a z-dependent upper limit and must be computed before the integral over z.

(a) $V_F(a) = \int_K dV = \int_0^{2\pi} d\phi \int_H^{2H} dz \int_0^{az} d\rho\,\rho = 2\pi \int_H^{2H} dz \left[\frac{\rho^2}{2}\right]_0^{az} = \pi a^2 \int_H^{2H} dz\,z^2$

$$= \pi a^2 \left[\frac{z^3}{3}\right]_H^{2H} = \boxed{\frac{7\pi a^2}{3} H^3}.$$

(b) $I_F(a) = \int_K dV\,\rho_0\,\rho^2 = \rho_0 \int_0^{2\pi} d\phi \int_H^{2H} dz \int_0^{az} d\rho\,\rho^3 = 2\pi \int_H^{2H} dz \left[\frac{\rho^4}{4}\right]_0^{az}$

$$= \rho_0 \frac{\pi a^4}{2} \int_H^{2H} dz\,z^4 = \rho_0 \frac{\pi a^4}{2} \left[\frac{z^5}{5}\right]_H^{2H} = \rho_0 \frac{31\pi a^4}{10} H^5 = \boxed{\frac{93a^2}{70} MH^2}.$$

E4.3.3 Volume of a buoy (spherical coordinates) (p. 251)

(a) In spherical coordinates, $x = r \sin\theta \cos\phi$, $y = r \sin\theta \sin\phi$, $z = r \cos\theta$, the volume element is $dV = r^2 \sin\theta \, dr d\theta d\phi$. The inequality for z can be written as $r \cos\theta \geq a\sqrt{r^2 \sin^2\theta(\cos^2\phi + \sin^2\phi)} = ar\sin\theta$, or $1/a \geq \tan\theta \Rightarrow \boxed{\theta \leq \arctan(1/a) \equiv \widetilde{\theta}}$.

(b)
$$V(R,a) = \int_0^R dr\, r^2 \int_0^{\widetilde{\theta}} d\theta\, \sin\theta \int_0^{2\pi} d\phi = \left[\tfrac{1}{3}r^3\right]_0^R \left[-\cos\theta\right]_0^{\widetilde{\theta}} \cdot 2\pi$$

$$= \boxed{\frac{2\pi}{3}R^3 \left[1 - \frac{a}{\sqrt{1+a^2}}\right]} \qquad \text{since} \quad \cos\widetilde{\theta} = a/\sqrt{1+a^2}.$$

E4.3.5 Wave functions of two-dimensional harmonic oscillator (polar coordinates) (p. 251)

(a) The area integral O factorizes into a radial integral P and an angular integral $\widetilde{P}$.

Area integral:
$$O^{mm'}_{nn'} = \int_{\mathbb{R}^2} dS\, \overline{\Psi}_{nm}(\mathbf{r})\Psi_{n'm'}(\mathbf{r})$$
$$= \int_0^\infty d\rho\, \rho \int_0^{2\pi} d\phi \left(\overline{R}_{n|m|}(\rho)\overline{Z}_m(\phi)\right)\left(R_{n'|m'|}(\rho)Z_{m'}(\phi)\right)$$
$$= P^{|m||m'|}_{nn'}\widetilde{P}^{mm'}.$$

Radial integral:
$$P^{|m||m'|}_{nn'} = \int_0^\infty d\rho\, \rho\, R_{n|m|}(\rho)R_{n'|m'|}(\rho) \quad \text{(since } \overline{R} = R).$$

Angular integral:
$$\widetilde{P}^{mm'} = \int_0^{2\pi} d\phi\, \overline{Z}_m(\phi)Z_{m'}(\rho).$$

(b) The angular integral over the functions $Z_m(\phi) = \frac{1}{\sqrt{2\pi}}e^{im\phi}$ can be calculated easily for arbitrary m and m'. To evaluate the case $m \neq m'$ we use the Euler identity, $e^{i2\pi k} = 1$ if $k \in \mathbb{Z}$:

$$\widetilde{P}^{mm'} = \frac{1}{2\pi}\int_0^{2\pi} d\phi\, e^{i(-m+m')\phi}$$

$$= \left. \begin{cases} \dfrac{1}{2\pi}\displaystyle\int_0^{2\pi} d\phi\, e^0 = 1 & \text{for } m = m' \\[3mm] \dfrac{1}{2\pi}\dfrac{e^{i2\pi(m'-m)} - 1}{i(m'-m)} \stackrel{\text{Euler}}{=} \dfrac{1}{2\pi}\dfrac{1-1}{i(m'-m)} = 0 & \text{for } m \neq m' \end{cases} \right\} = \boxed{\delta_{mm'}} \cdot \checkmark$$

(c) According to (b), $O^{mm'}_{nn'} \propto \delta_{mm'}$, hence only radial integrals of the form $P^{mm}_{nn'}$ are of interest here. We compute them using the substitution $x = \rho^2$, $dx = 2\rho\, d\rho$, and $\int_0^\infty dx\, x^n e^{-x} = n!$:

$$P^{00}_{00} = \int_0^\infty d\rho\, \rho \left[\sqrt{2}e^{-\rho^2/2}\right]^2 = \int_0^\infty dx\, e^{-x} = \boxed{1},$$

$$P^{11}_{11} = \int_0^\infty d\rho\, \rho \left[\sqrt{2}\rho e^{-\rho^2/2}\right]^2 = \int_0^\infty dx\, x e^{-x} = \boxed{1},$$

$$P^{22}_{22} = \int_0^\infty d\rho\, \rho \left[\rho^2 e^{-\rho^2/2}\right]^2 = \int_0^\infty dx\, \tfrac{1}{2}x^2 e^{-x} = \tfrac{1}{2}\cdot 2 = \boxed{1},$$

$$P^{00}_{22} = \int_0^\infty d\rho\, \rho \left[\sqrt{2}(\rho^2-1)e^{-\rho^2/2}\right]^2 = \int_0^\infty dx\,(x^2-2x+1)e^{-x} = (2-2\cdot1+1) = \boxed{1}.$$

$$P_{20}^{00} = \int_0^\infty d\rho\, \rho\sqrt{2}(\rho^2-1)e^{-\rho^2/2}\sqrt{2}e^{-\rho^2/2} = \int_0^\infty dx\,(x-1)e^{-x} = 1-1 = \boxed{0}\,.$$

It follows that $O_{00}^{00} = O_{11}^{11} = O_{22}^{22} = O_{22}^{00} = \boxed{1}$ and $O_{20}^{00} = \boxed{0}$. ✓

S.C4.4 Curvilinear integration in arbitrary dimensions (p. 252)

C4.4.1 Surface integral: area of a sphere (p. 254)

(a)   We use $\mathbf{r}(x,y) = (x,y,\sqrt{R^2-x^2-y^2})^{\mathrm{T}}$ to parametrize the upper half-sphere by Cartesian coordinates on the disk D. Computing the coordinate basis vectors, we find

$$\partial_x\mathbf{r} = \left(1,0,\frac{-x}{\sqrt{R^2-x^2-y^2}}\right)^{\mathrm{T}}, \qquad \partial_y\mathbf{r} = \left(0,1,\frac{-y}{\sqrt{R^2-x^2-y^2}}\right)^{\mathrm{T}}.$$

$$\|\partial_x\mathbf{r} \times \partial_y\mathbf{r}\| = \left[\|\partial_x\mathbf{r}\|^2\,\|\partial_x\mathbf{r}\|^2 - \|\partial_x\mathbf{r}\cdot\partial_y\mathbf{r}\|^2\right]^{1/2}$$

$$= \left[\left(1+\frac{x^2}{R^2-x^2-y^2}\right)\left(1+\frac{y^2}{R^2-x^2-y^2}\right) - \frac{x^2y^2}{(R^2-x^2-y^2)^2}\right]^{1/2} = \frac{R}{\sqrt{R^2-x^2-y^2}}\,.$$

To compute the area of the sphere, $A_S = 2A_{S_+} = 2\int_D dxdy\,\|\partial_x\mathbf{r} \times \partial_y\mathbf{r}\|$, we integrate over the disk D defined by the inequalities $|x| < R$ and $y \le \sqrt{R^2-x^2}$:

$$A_S = \int_{-R}^R dx \int_{-\sqrt{R^2-x^2}}^{\sqrt{R^2-x^2}} dy\,\frac{2R}{\sqrt{R^2-x^2-y^2}} = \int_{-R}^R dx \int_{-\sqrt{R^2-x^2}}^{\sqrt{R^2-x^2}} dy\,\frac{2R}{\sqrt{R^2-x^2}}\,\frac{1}{\sqrt{1-\frac{y^2}{R^2-x^2}}}\,.$$

Using the substitution $t = \frac{y}{\sqrt{R^2-x^2}}$, with $t \in (-1,1)$, we obtain

$$A_S = \int_{-R}^R dx \int_{-1}^1 dt\,\frac{2R}{\sqrt{1-t^2}} = 2R\int_{-R}^R dx\,\big[\arcsin t\big]_{-1}^1 = 2R(2R)\left[\frac{\pi}{2}-\left(-\frac{\pi}{2}\right)\right] = \boxed{4\pi R^2}\,.$$

(b)   We use $\mathbf{r}(\theta,\phi) = R(\sin\theta\cos\phi, \sin\theta\sin\phi, -\sin\theta)\mathrm{T}$ to parametrize the sphere by spherical coordinates. Computing the coordinate basis vectors, we find

$$\partial_\theta\mathbf{r} = R(\cos\theta\cos\phi, \cos\theta\sin\phi, \cos\theta)^{\mathrm{T}}, \qquad \partial_\phi\mathbf{r} = R(-\sin\theta\sin\phi, \sin\theta\cos\phi, 0)^{\mathrm{T}},$$

$$\|\partial_\theta\mathbf{r} \times \partial_\phi\mathbf{r}\| = \left[\|\partial_\theta\mathbf{r}\|^2\,\|\partial_\phi\mathbf{r}\|^2 - \|\partial_\theta\mathbf{r}\cdot\partial_\phi\mathbf{r}\|^2\right]^{1/2} = \left[R^4-0\right]^{1/2} = R^2\,.$$

To compute the area of the sphere, we integrate over the full domain $U = (0,\pi) \times (0,2\pi)$:

$$A_S = \int_S dS = \int_U d\theta d\phi\,\|\partial_\theta\mathbf{r} \times \partial_\phi\mathbf{r}\| = R^2\int_0^{2\pi} d\phi \int_0^\pi d\theta\,\sin\theta = -2\pi R^2\big[\cos\theta\big]_0^\pi = \boxed{4\pi R^2}\,.$$

C4.4.3 Volume and surface integral: parabolic solid of revolution (p. 254)

(a)   In cylindrical coordinates the paraboloid P is defined by $z(\rho) = \rho^2$, or equivalently, $\rho(z) = \sqrt{z}$. Its volume is:

$$V = \int_P dV = \int_0^{2\pi} d\phi \int_0^{z_{max}} dz \int_0^{\rho(z)} d\tilde\rho\,\tilde\rho = 2\pi \int_0^{z_{max}} dz\,\frac{z}{2} = \boxed{\frac{\pi}{2}z_{max}^2}\,.$$

(b)   The area, A, of the curved part, C, of the surface of P is calculated as follows:

Parametrization of C: $\mathbf{r}(\phi, z) = \rho\,\mathbf{e}_\rho + z\,\mathbf{e}_z$, with $\rho = \rho(z) = \sqrt{z}$.

Tangent vectors: $\partial_\phi \mathbf{r} = \rho\,\mathbf{e}_\phi$, $\partial_z \mathbf{r} = \rho'\,\mathbf{e}_\rho + \mathbf{e}_z$, with $\rho' = \partial_z\rho(z) = \frac{1}{2\sqrt{z}}$.

This gives: $\|\partial_\phi \mathbf{r} \times \partial_z \mathbf{r}\| = \left[(\partial_\phi \mathbf{r})^2(\partial_z \mathbf{r})^2 - \partial_\phi \mathbf{r} \cdot \partial_z \mathbf{r}\right]^{\frac{1}{2}}$

$$= \left[\rho^2\left(\rho'^2 + 1\right) - 0\right]^{\frac{1}{2}} = \sqrt{z}\left[\frac{1}{(2\sqrt{z})^2} + 1\right]^{\frac{1}{2}} = \left[\tfrac{1}{4} + z\right]^{\frac{1}{2}}.$$

Area of C: $A = \displaystyle\int_C \mathrm{d}S = \int_C \mathrm{d}\phi\,\mathrm{d}z\,\|\partial_\phi \mathbf{r} \times \partial_z \mathbf{r}\| = \int_0^{2\pi} \mathrm{d}\phi \int_0^{z_{\max}} \mathrm{d}z\,\left[\tfrac{1}{4} + z\right]^{\frac{1}{2}}$

$$= 2\pi\tfrac{2}{3}\left[\tfrac{1}{4} + z\right]^{\frac{3}{2}}\Big|_0^{z_{\max}} = \boxed{\tfrac{\pi}{6}\left[(1 + 4z_{\max})^{\frac{3}{2}} - 1\right]}.$$

εC4.4.5 **Surface area of a circular cone** (p. 254)

We place the tip of the cone at the origin, as shown in the sketch. We adopt polar coordinates, $(\rho, \phi)^{\mathrm{T}}$, defined on the domain $U = (0, R) \times (0, 2\pi)$, such that $x = \rho\cos\phi$, $y = \rho\sin\phi$. On the conical surface, $z = b\rho$ with $b = h/R$, hence it can be parametrized as

$$\mathbf{r} : U \to S_C, \quad (\rho, \phi)^{\mathrm{T}} \mapsto \mathbf{r}(\rho, \phi) = \begin{pmatrix} \rho\cos\phi \\ \rho\sin\phi \\ \rho b \end{pmatrix}.$$

The conical area, $A_C = \displaystyle\int_{S_C} \mathrm{d}S = \int_U \mathrm{d}\rho\,\mathrm{d}\phi\,\|\partial_\rho \mathbf{r} \times \partial_\phi \mathbf{r}\|$, can thus be computed as follows:

$$\|\partial_\rho \mathbf{r} \times \partial_\phi \mathbf{r}\| = \left\|\begin{pmatrix} \cos\phi \\ \sin\phi \\ b \end{pmatrix} \times \begin{pmatrix} -\rho\sin\phi \\ \rho\cos\phi \\ 0 \end{pmatrix}\right\| = \rho\left\|\begin{pmatrix} b\cos\phi \\ -b\sin\phi \\ 1 \end{pmatrix}\right\| = \rho\sqrt{b^2 + 1}.$$

$$A_C = \int_0^R \mathrm{d}\rho\,\rho \int_0^{2\pi} \mathrm{d}\phi\sqrt{b^2 + 1} = \tfrac{1}{2}R^2(2\pi)\sqrt{b^2 + 1} = \boxed{\pi R\sqrt{h^2 + R^2}}.$$

S.C4.5   Changes of variables in higher-dimensional integration (p. 256)

εC4.5.1 Variable transformation for two-dimensional integral (p. 257)

(a)   The variable transformation $x = \tfrac{1}{2}(X + Y)$, $y = \tfrac{1}{2}(X - Y)$ has inverse $X = x + y$, $Y = x - y$. The corresponding Jacobi matrices are

$$J = \frac{\partial(x,y)}{\partial(X,Y)} = \begin{pmatrix} \frac{\partial x}{\partial X} & \frac{\partial x}{\partial Y} \\ \frac{\partial y}{\partial X} & \frac{\partial y}{\partial Y} \end{pmatrix} = \boxed{\begin{pmatrix} \frac{1}{2} & \frac{1}{2} \\ \frac{1}{2} & -\frac{1}{2} \end{pmatrix}}, \quad J^{-1} = \frac{\partial(X,Y)}{\partial(x,y)} = \begin{pmatrix} \frac{\partial X}{\partial x} & \frac{\partial X}{\partial y} \\ \frac{\partial Y}{\partial x} & \frac{\partial Y}{\partial y} \end{pmatrix} = \boxed{\begin{pmatrix} 1 & 1 \\ 1 & -1 \end{pmatrix}},$$

with determinants $\det J = \boxed{-\tfrac{1}{2}}$ and $\det J^{-1} = \boxed{-2}$.

(b) The integration ranges for X and Y must be chosen such that the inequalities defining the square S are respected. The minimal and maximal values of $X = x + y$ are 0 and 2, but those of $Y = x - y$ depend on X. For $X \in (0, 1)$, the Y-domain is bounded from above and below by the lines $x = 0$ and $y = 0$ respectively, i.e. by $Y = -X$ and $Y = X$. For $X \in (1, 2)$, the Y-domain is bounded from above and below by the lines $y = 1$ and $x = 1$ respectively, i.e. by $Y = X - 2$ and $Y = 2 - X$.

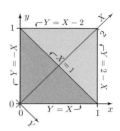

$$I_1 = \int_S dx\,dy = \int_S dX\,dY \left| \frac{\partial(x,y)}{\partial(X,Y)} \right| = \int_0^1 dX \int_{-X}^X dY \tfrac{1}{2} + \int_1^2 dX \int_{X-2}^{2-X} dY \tfrac{1}{2}$$

$$= \tfrac{1}{2} \int_0^1 dX\, 2X + \tfrac{1}{2} \int_1^2 dX\, 2(2 - X) = \left[\tfrac{1}{2} X^2 \right]_0^1 + \left[2X - \tfrac{1}{2} X^2 \right]_1^2 = \boxed{1}. \checkmark$$

(c) To integrate over the triangle T instead of the square S, we proceed as in (b), but restrict X to the range $(0, 1)$, since $y < 1 - x$ implies $X < 1$.

$$I_2(n) = \int_T dx\,dy\,|x - y|^n = \int_T dX\,dY \left| \frac{\partial(x,y)}{\partial(X,Y)} \right| |Y|^n = \int_0^1 dX \int_{-X}^X dY \tfrac{1}{2} |Y|^n$$

$$= \tfrac{1}{2} \int_0^1 dX\, 2 \int_0^X dY\, Y^n = \int_0^1 dX \frac{1}{n+1} X^{n+1} = \boxed{\frac{1}{(n+1)(n+2)}}.$$

C4.5.3 **Three-dimensional Gaussian integral via linear transformation** (p. 257)

Using the substitution $u = a(x + y)$, $v = b(z - y)$, $w = c(x - z)$ gives

$$I = \int_{\mathbb{R}^3} dx\,dy\,dz\, e^{-\left[a^2(x+y)^2 + b^2(z-y)^2 + c^2(x-z)^2 \right]} = \int_{\mathbb{R}^3} du\,dv\,dw \left| \frac{\partial(x,y,z)}{\partial(u,v,w)} \right| e^{-\left[u^2 + v^2 + w^2 \right]}.$$

Because x, y and z are all integrated over $(-\infty, \infty)$, the same holds for u, v and w. Furthermore, because u, v and w are given as functions of x, y and z, the Jacobian determinant is calculated simply using the formula:

$$J = \left| \det \begin{pmatrix} \frac{\partial u}{\partial x} & \frac{\partial u}{\partial y} & \frac{\partial u}{\partial z} \\ \frac{\partial v}{\partial x} & \frac{\partial v}{\partial y} & \frac{\partial v}{\partial z} \\ \frac{\partial w}{\partial x} & \frac{\partial w}{\partial y} & \frac{\partial w}{\partial z} \end{pmatrix}^{-1} \right| = \left| \det \begin{pmatrix} a & a & 0 \\ 0 & b & -b \\ c & 0 & -c \end{pmatrix}^{-1} \right| = \left| abc(-1 - 1 + 0) \right|^{-1} = \boxed{\frac{1}{2abc}}.$$

Note: The Jacobian determinant is always positive, because it is given by the *magnitude* of the determinant of the matrix of partial derivatives.

Note: Alternatively the Jacobian determinant can also be calculated using $J = \left| \frac{\partial(x,y,z)}{\partial(u,v,w)} \right|$. For this, however, we must first invert the transformation (i.e. using Gaussian elimination), and determine x, y and z as functions of u, v and w. This gives $x = \tfrac{1}{2}(u/a + v/b + w/c)$, $y = \tfrac{1}{2}(u/a - v/b - w/c)$, $z = \tfrac{1}{2}(u/a + v/b - w/c)$, and

$$J = \left| \det \begin{pmatrix} \frac{\partial x}{\partial u} & \frac{\partial x}{\partial v} & \frac{\partial x}{\partial w} \\ \frac{\partial y}{\partial u} & \frac{\partial y}{\partial v} & \frac{\partial y}{\partial w} \\ \frac{\partial z}{\partial u} & \frac{\partial z}{\partial v} & \frac{\partial z}{\partial w} \end{pmatrix} \right| = \frac{1}{2^3} \left| \det \begin{pmatrix} 1/a & 1/b & 1/c \\ 1/a & -1/b & -1/c \\ 1/a & 1/b & -1/c \end{pmatrix} \right| = \frac{1}{2^3 abc} \left| 2 - (-2) + 0 \right| = \boxed{\frac{1}{2abc}}.$$

Because in this case the Jacobian is independent of u, v and w, the integral $\int du\,dv\,dw$ decomposes into three independent Gaussian integrals: $\boxed{I = \sqrt{\pi^3}/(2abc)}$.

εC4.5.5 Multidimensional Gaussian integrals (p. 257)

(a) We write the exponent as $\mathbf{x}^T A \mathbf{x}$, with $\mathbf{x} = (x, y)^T$ and a symmetric matrix A, with $A^1{}_2 = A^2{}_1$:

$$\mathbf{x}^T A \mathbf{x} = \begin{pmatrix} x & y \end{pmatrix} \begin{pmatrix} A^1{}_1 & A^1{}_2 \\ A^2{}_1 & A^2{}_2 \end{pmatrix} \begin{pmatrix} x \\ y \end{pmatrix} = A^1{}_1 x^2 + A^1{}_2 xy + A^2{}_1 yx + A^2{}_2 y^2$$

$$\overset{!}{=} (a+3)x^2 + 2(a-3)xy + (a+3)y^2 .$$

We therefore identify

$$A^1{}_1 = A^2{}_2 = a+3, \quad A^1{}_2 = A^2{}_1 = a - 3, \quad \Rightarrow \quad A = \begin{pmatrix} a+3 & a-3 \\ a-3 & a+3 \end{pmatrix}.$$

Eigenvalues: $0 \overset{!}{=} \det(A - \lambda \mathbb{1}) = (2a - \lambda)(6 - \lambda) \quad \Rightarrow \quad \lambda_1 = \boxed{2a}, \; \lambda_2 = \boxed{6}.$

Checks: $\lambda_1 + \lambda_2 = 2a + 6 \overset{\checkmark}{=} \operatorname{Tr} A, \quad \lambda_1 \lambda_2 = 12a \overset{\checkmark}{=} \det A = (a+3)^2 - (a-3)^2 .$

Normalized eigenvectors:

$$\lambda_1 = 2a : \quad 0 \overset{!}{=} (A - \lambda_1 \mathbb{1}) \mathbf{v}_1 = \begin{pmatrix} -a+3 & a-3 \\ a-3 & -a+3 \end{pmatrix} \mathbf{v}_1 \quad \Rightarrow \quad \mathbf{v}_1 = \boxed{\frac{1}{\sqrt{2}} \begin{pmatrix} 1 \\ 1 \end{pmatrix}},$$

$$\lambda_2 = 6 : \quad 0 \overset{!}{=} (A - \lambda_2 \mathbb{1}) \mathbf{v}_2 = \begin{pmatrix} a-3 & a-3 \\ a-3 & a-3 \end{pmatrix} \mathbf{v}_2 \quad \Rightarrow \quad \mathbf{v}_2 = \boxed{\frac{1}{\sqrt{2}} \begin{pmatrix} 1 \\ -1 \end{pmatrix}}.$$

There therefore exists a transformation T such that $A = TDT^T$, with $D = \boxed{\begin{pmatrix} 2a & 0 \\ 0 & 6 \end{pmatrix}}$. The columns of this transformation T are the normalized eigenvectors of A:

$$T = \boxed{\frac{1}{\sqrt{2}} \begin{pmatrix} 1 & 1 \\ 1 & -1 \end{pmatrix}}.$$

With $\tilde{\mathbf{x}} = (\tilde{x}, \tilde{y})^T = T^T \mathbf{x}$, the exponent thus takes the following form:

$$(a+3)x^2 + 2(a-3)xy + (a+3)y^2 = \mathbf{x}^T A \mathbf{x} = \mathbf{x}^T T D T^T \mathbf{x} = \tilde{\mathbf{x}}^T D \tilde{\mathbf{x}}$$

$$= \begin{pmatrix} \tilde{x} & \tilde{y} \end{pmatrix} \begin{pmatrix} 2a & 0 \\ 0 & 6 \end{pmatrix} \begin{pmatrix} \tilde{x} \\ \tilde{y} \end{pmatrix} = 2a\tilde{x}^2 + 6\tilde{y}^2 .$$

(b) Since $T^{-1} = T^T$, the inverse of the relation $\tilde{\mathbf{x}} = T^T \mathbf{x}$ is $\mathbf{x} = T\tilde{\mathbf{x}}$. Explicitly:

$$x = T^1{}_1 \tilde{x} + T^1{}_2 \tilde{y} = \frac{1}{\sqrt{2}} (\tilde{x} + \tilde{y}), \quad y = T^2{}_1 \tilde{x} + T^2{}_2 \tilde{y} = \frac{1}{\sqrt{2}} (\tilde{x} - \tilde{y}).$$

The Jacobian determinant equals 1, since T is orthogonal. Explicitly:

$$J = \left| \det \begin{pmatrix} \frac{\partial x}{\partial \tilde{x}} & \frac{\partial x}{\partial \tilde{y}} \\ \frac{\partial y}{\partial \tilde{x}} & \frac{\partial y}{\partial \tilde{y}} \end{pmatrix} \right| = \left| \det \begin{pmatrix} T^1{}_1 & T^1{}_2 \\ T^2{}_1 & T^2{}_2 \end{pmatrix} \right| = \left| \det \frac{1}{\sqrt{2}} \begin{pmatrix} 1 & 1 \\ 1 & -1 \end{pmatrix} \right| = \left| \frac{1 - (-1)}{\sqrt{2}^2} \right| = \boxed{1}.$$

(c) We now compute $I(a)$, using the Gaussian integral $\int_{-\infty}^{\infty} dx\, e^{-b^2 x^2} = \frac{\sqrt{\pi}}{b}$ for the last step:

$$I(a) = \int_{\mathbb{R}^2} dx\, dy\, e^{-[(a+3)x^2 + 2(a-3)xy + (a+3)y^2]} = \int_{\mathbb{R}^2} dx\, dy\, e^{-\mathbf{x}^T A \mathbf{x}} = \int_{\mathbb{R}^2} d\tilde{x}\, d\tilde{y}\, J\, e^{-\tilde{\mathbf{x}}^T D \tilde{\mathbf{x}}}$$

$$= \int_{\mathbb{R}^2} d\tilde{x}\, d\tilde{y}\, e^{-(2a\tilde{x}^2 + 6\tilde{y}^2)} = \int_{-\infty}^{\infty} d\tilde{x}\, e^{-2a\tilde{x}^2} \int_{-\infty}^{\infty} d\tilde{y}\, e^{-6\tilde{y}^2} = \frac{\sqrt{\pi}}{\sqrt{2a}} \frac{\sqrt{\pi}}{\sqrt{6}} = \boxed{\frac{\pi}{2\sqrt{3}a}}.$$

The result has the expected form: $I = \sqrt{\frac{\pi^n}{\det A}}$, with $n = 2$ and $\det A = \lambda_1 \lambda_2 = 12a$.

S.C5 Taylor series (p. 259)

S.C5.2 Complex Taylor series (p. 263)

C5.2.1 Addition theorems for sine and cosine (p. 264)

On the one hand: $e^{i(a+b)} = \cos(a+b) + i\sin(a+b)$ (Euler formula).   (1)

On the other hand: $e^{ia}e^{ib} = (\cos a + i\sin a)(\cos b + i\sin b)$

$$= \cos a \cos b - \sin a \sin b + i(\sin a \cos b + \cos a \sin b).$$ (2)

If a and b are real, a comparison of the real and imaginary parts of $e^{i(a+b)} = e^{ia}e^{ib}$ yields:

(a) $\mathrm{Re}\,(1) = \mathrm{Re}\,(2)$: $\Rightarrow$ $\boxed{\cos(a+b) = \cos a \cos b - \sin a \sin b}$,

(b) $\mathrm{Im}\,(1) = \mathrm{Im}\,(2)$: $\Rightarrow$ $\boxed{\sin(a+b) = \sin a \cos b + \cos a \sin b}$.

If a and b are complex, this argument is not applicable. However, the sought identities follow with almost no additional effort from the upper or lower sign versions of the identity $e^{i(a+b)} \pm e^{-i(a+b)} = e^{ia}e^{ib} \pm e^{-ia}e^{-ib}$.

S.C5.3 Finite-order expansion (p. 266)

C5.3.1 Taylor series (p. 266)

(a)   Method 1: Use the known formula for the geometric series, $\frac{1}{1-x} = \sum_{n=0}^{\infty} x^n$ (for $|x| < 1$), as well as the series expansion for sine, $\sin x = x - \frac{x^3}{6} + \mathcal{O}(x^5)$:

$$f(x) = \frac{1}{1-\sin x} = \frac{1}{1-[x+\frac{1}{6}x^3 + \mathcal{O}(x^5)]}$$

$$= 1 + \left[x - \frac{1}{6}x^3\right] + \left[x - \frac{1}{6}x^3\right]^2 + \left[x - \frac{1}{6}x^3\right]^3 + \left[x - \frac{1}{6}x^3\right]^4 + \mathcal{O}(x^5)$$

$$= 1 + x - \frac{1}{6}x^3 + x^2 - \frac{1}{3}x^4 + x^3 + x^4 + \mathcal{O}(x^5)$$

$$= \boxed{1 + x + x^2 + \frac{5}{6}x^3 + \frac{2}{3}x^4 + \mathcal{O}(x^5)} .$$

Method 2: Determine the Taylor coefficients via successive derivatives:

$f(x) = \frac{1}{1-\sin x}$, $\Rightarrow$ $f(0) = \boxed{1}$,

$f^{(1)}(x) = \frac{\cos(x)}{(1-\sin(x))^2}$ $\Rightarrow$ $f^{(1)}(0) = \boxed{1}$,

$f^{(2)}(x) = \frac{2\cos^2(x)}{(1-\sin(x))^3} - \frac{\sin(x)}{(1-\sin(x))^2}$ $\Rightarrow$ $f^{(2)}(0) = \boxed{2}$,

$f^{(3)}(x) = \frac{6\cos^3(x)}{(1-\sin(x))^4} - \frac{6\sin(x)\cos(x)}{(1-\sin(x))^3} - \frac{\cos(x)}{(1-\sin(x))^2}$ $\Rightarrow$ $f^{(3)}(0) = \boxed{5}$,

$f^{(4)}(x) = \frac{24\cos^4(x)}{(1-\sin(x))^5} - \frac{36\sin(x)\cos^2(x)}{(1-\sin(x))^4} + \frac{6\sin^2(x)}{(1-\sin(x))^3}$

$\qquad - \frac{8\cos^2(x)}{(1-\sin(x))^3} + \frac{\sin(x)}{(1-\sin(x))^2}$ $\Rightarrow$ $f^{(4)}(0) = \boxed{16}$,

$$\Rightarrow \quad f(x) = f(0) + f^{(1)}(0)x + \tfrac{1}{2!}f^{(2)}(0)x^2 + \tfrac{1}{3!}f^{(3)}(0)x^3 + \tfrac{1}{4!}f^{(4)}(0)x^4 + \mathcal{O}\left(x^5\right)$$

$$= \boxed{1 + x + x^2 + \tfrac{5}{6}x^3 + \tfrac{2}{3}x^4 + \mathcal{O}(x^5)}.$$

Remark: This example shows that taking successive derivatives of products or quotients is rather tedious, because each application of the product rule generates additional terms. If the series expansions of the involved factors are known, it is simpler to use these, as shown in Method 1.

(b) Method 1: Use the substitution $x = 1 + y$ and use the known series expansion of the logarithm, $\ln(1 + y) = -\sum_{k=0}^{\infty} \frac{(-y)^{k+1}}{k+1}$, as well as the sine function [see (a)]:

$$g(x) = \sin(\ln(x)) = \sin(\ln(1 + y)) = \sin\left(-\left(-y + \tfrac{1}{2}y^2 + \mathcal{O}(y^3)\right)\right)$$

$$= y - \tfrac{1}{2}y^2 + \mathcal{O}(y^3) \overset{y=x-1}{=} \boxed{(x - 1) - \tfrac{1}{2}(x - 1)^2 + \mathcal{O}\left((x - 1)^3\right)}.$$

Method 2: Determine the Taylor coefficients via successive derivatives:

$$g(x) = \sin(\ln(x)), \qquad\qquad\qquad \Rightarrow \quad g(1) = \sin(\ln(1)) = \boxed{0},$$

$$g^{(1)}(x) = \cos(\ln(x))\frac{1}{x}, \qquad\qquad \Rightarrow \quad g^{(1)}(1) = \cos(0) = \boxed{1},$$

$$g^{(2)}(x) = -\sin(\ln(x))\frac{1}{x^2} - \cos(\ln(x))\frac{1}{x^2}, \qquad \Rightarrow \quad g^{(2)}(1) = \boxed{-1},$$

$$\Rightarrow \quad g(x) = g(1) + g^{(1)}(1)(x - 1) + \tfrac{1}{2!}g^{(2)}(1)(x - 1)^2 + \mathcal{O}\left((x - 1)^3\right)$$

$$= \boxed{(x - 1) - \tfrac{1}{2}(x - 1)^2 + \mathcal{O}\left((x - 1)^3\right)}.$$

(c) Method 1: Use the Taylor expansions for the exponential and cosine functions:

$$h(x) = e^{\cos x} = e^{\left[1 - \frac{1}{2}x^2 + \mathcal{O}(x^4)\right]} = e^1 e^{-\left[\frac{1}{2}x^2 + \mathcal{O}(x^4)\right]} = \boxed{e\left[1 - \tfrac{1}{2}x^2 + \mathcal{O}(x^4)\right]}.$$

Factoring out an e^1 before Taylor expanding is necessary in order to be able to use the well-known series expansion of the exponential function in a form (namely $e^{-\frac{1}{2}x^2}$) in which its argument vanishes in the limit $x \to 0$, so that the series can be truncated after a few terms. If instead $e^{1-\frac{1}{2}x^2}$ is expanded in powers of its full argument, $(1 - \tfrac{1}{2}x^2)$, the complete Taylor series with infinitely many terms is needed to recover the correct result for $x \to 0$, namely e^1:

$$e^{1-\frac{1}{2}x^2} = \sum_{l=0}^{\infty} \frac{1}{l!}(1 - \tfrac{1}{2}x^2)^l \overset{x=0}{\longrightarrow} \sum_{l=0}^{\infty} \frac{1}{l!} = e^1.$$

To recover the second term in the Taylor expansion of $h(x)$ in this way, namely $-\tfrac{1}{2}x^2$, the binomial theorem has to be used: $(1 + y)^l = \sum_{k=0}^{l} \frac{l! y^k}{k!(l-k)!}$:

$$e^{1-\frac{1}{2}x^2} = \sum_{l=0}^{\infty} \frac{1}{l!}(1 - \tfrac{1}{2}x^2)^l = \sum_{l=0}^{\infty} \frac{1}{l!} \sum_{k=0}^{l} \frac{l!(-\frac{1}{2}x^2)^k}{k!(l-k)!} = \sum_{l=0}^{\infty} \frac{1}{l!}\left[1 + l(-\tfrac{1}{2}x^2) + \mathcal{O}(x^4)\right]$$

$$= \sum_{l=0}^{\infty} \frac{1}{l!} + \sum_{l=1}^{\infty} \frac{1}{(l-1)!}(-\tfrac{1}{2}x^2) + \mathcal{O}(x^4) = e^1 - e^1 \tfrac{1}{2}x^2 + \mathcal{O}(x^4). \checkmark$$

Clearly this approach is much more tedious than directly expanding $e^{-\frac{1}{2}x^2}$!
Method 2: Determine the Taylor coefficients via successive derivatives:

$$h(x) = e^{\cos x}, \qquad\qquad \Rightarrow \quad h(0) = \boxed{e^1},$$

$$h^{(1)}(x) = -\sin x\, e^{\cos x}, \qquad\qquad \Rightarrow \quad h^{(1)}(0) = \boxed{0},$$

$$h^{(2)}(x) = -\cos x\, e^{\cos x} + (-\sin x)^2\, e^{\cos x}, \qquad \Rightarrow \quad h^{(2)}(0) = \boxed{-e^1},$$

$$\Rightarrow \quad h(x) = h(0) + h^{(1)}(0)x + \tfrac{1}{2}h^{(2)}(0)\,x^2 + \mathcal{O}\left(x^3\right) = \boxed{e - e\tfrac{1}{2}x^2 + \mathcal{O}(x^3)}.$$

The question whether or not to factor out an e^1, discussed above for Method 1, does not arise at all for Method 2. The strategy of iteratively taking derivatives and setting $x = 0$ automatically finds the right answer. In this sense Method 2 is in general easier to apply, since no subtleties need to be considered.

S.C5.4 Solving equations by Taylor expansion (p. 267)

C5.4.1 Series expansion for iteratively solving an equation (p. 268)

Equation to be solved: $\qquad\qquad 0 = e^{y-1} + \epsilon y - 1\,.$ $\qquad\qquad\qquad$ (1)

Series ansatz: $\qquad\qquad y(\epsilon) = y_0 + y_1\epsilon + \tfrac{1}{2!}y_2\epsilon^2 + \mathcal{O}(\epsilon^3),$ with $y_n \equiv y^{(n)}(0)\,.$ $\qquad$ (2)

(a) Method 1: Expansion of equation. We insert (2) into (1), expand to order $\mathcal{O}(\epsilon^2)$ using $e^{z+c} = e^c\left[1 + z + \tfrac{1}{2!}z^2\right] + \mathcal{O}(x^3)$, with $z = y_1\epsilon + \tfrac{1}{2}y_2\epsilon^2 + \mathcal{O}(\epsilon^3)$, and collect powers of ϵ:

$$0 = e^{\left(y_0 + y_1\epsilon + \frac{1}{2}y_2\epsilon^2 + \mathcal{O}(\epsilon^3)\right) - 1} + \epsilon\left(y_0 + y_1\epsilon + \mathcal{O}(\epsilon^2)\right) - 1$$

$$= e^{y_0-1}\left[1 + \left(y_1\epsilon + \tfrac{1}{2}y_2\epsilon^2 + \mathcal{O}(\epsilon^3)\right) + \tfrac{1}{2!}\left(y_1\epsilon + \mathcal{O}(\epsilon^2)\right)^2\right] + y_0\epsilon + y_1\epsilon^2 - 1 + \mathcal{O}(\epsilon^3)$$

$$= \left[e^{y_0-1} - 1\right] + \left[e^{y_0-1}y_1 + y_0\right]\epsilon + \left[e^{y_0-1}\tfrac{1}{2}(y_2 + y_1^2) + y_1\right]\epsilon^2 + \mathcal{O}(\epsilon^3)$$

The coefficients of each ϵ^n must vanish, yielding a hierarchy of equations for y_n:

$$\epsilon^0: \quad 0 = e^{y_0-1} - 1, \qquad\qquad\qquad\qquad \Rightarrow \qquad y_0 = \boxed{1},$$

$$\epsilon^1: \quad 0 = e^{y_0-1}y_1 + y_0 = y_1 + 1 \qquad\qquad \Rightarrow \qquad y_1 = \boxed{-1},$$

$$\epsilon^2: \quad 0 = e^{y_0-1}\tfrac{1}{2}(y_2 + y_1^2) + y_1 = \tfrac{1}{2}(y_2 + 1) - 1 \qquad \Rightarrow \qquad y_2 = \boxed{1}.$$

The desired solution thus has the form: $\quad y(\epsilon) = \boxed{1 - \epsilon + \tfrac{1}{2}\epsilon^2 + \mathcal{O}(\epsilon^3)}.$

(b) Method 2: Repeated differentiation. We write the given equation in the form $0 = \mathcal{F}(y(\epsilon), \epsilon) \equiv F(\epsilon)$ and set up the hierarchy $0 = d_\epsilon^n F(\epsilon)|_{\epsilon=0}$, with $y(0) = y_0$, $y'(0) = y_1$ and $y''(0) = y_2$:

$$F: \qquad 0 = e^{y-1} + \epsilon y - 1 \qquad\qquad \overset{\epsilon=0}{\Rightarrow} \qquad 0 = e^{y_0-1} - 1,$$

$$d_\epsilon F: \qquad 0 = e^{y-1}y' + y + \epsilon y' \qquad\qquad \overset{\epsilon=0}{\Rightarrow} \qquad 0 = e^{y_0-1}y_1 + y_0,$$

$$d_\epsilon^2 F: \qquad 0 = e^{y-1}(y'^2 + y'') + y' + \epsilon y'' + y' \qquad \overset{\epsilon=0}{\Rightarrow} \qquad 0 = e^{y_0-1}(y_1^2 + y_2) + 2y_1.$$

We obtain the same hierarchy as in (a), but with somewhat less tedium. By solving each equation in the hierarchy as it becomes available, that information can be used while setting up the next equation. This would have simplified the equations on the right to

$$0 = e^{y_0-1} - 1 \Rightarrow y_0 = \boxed{1}, \quad 0 = y_1 + 1 \Rightarrow y_1 = \boxed{-1}, \quad 0 = (1 + y_2) - 2 \Rightarrow y_2 = \boxed{1}.$$

εC5.4.3 Series expansion of inverse function (p. 268)

We discuss two equivalent methods for determining the coefficients y_n. Method 1: expansion of the equation of interest in powers of x. Method 2: repeated differentiation.

(a) $\ln(x)$ is the inverse of e^x. We are looking for the series expansion of $y(x) \equiv \ln(1 + x)$ in powers of x. This function fulfills the defining equation

$$e^{y(x)} = 1 + x. \tag{1}$$

Series ansatz: $\qquad y(x) = y_0 + y_1 x + \frac{1}{2!}y_2 x^2 + \mathcal{O}(x^3), \quad \text{with } y_n \equiv y^{(n)}(0). \tag{2}$

Method 1: expansion of equation. We insert (2) into (1), expand to order $\mathcal{O}(x^2)$ using $e^{z+c} = e^c[1 + z + \frac{1}{2!}z^2] + \mathcal{O}(z^3)$, with $z = y_1 x + \frac{1}{2}y_2 x^2$, and collect powers of x:

$$0 = e^{y(x)} - x - 1 = e^{y_0 + y_1 x + \frac{1}{2}y_2 x^2 + \mathcal{O}(x^3)} - x - 1$$
$$= e^{y_0}\left[1 + \left(y_1 x + \frac{1}{2}y_2 x^2\right) + \frac{1}{2!}(y_1 x)^2\right] - x - 1 + \mathcal{O}(x^3)$$
$$= (e^{y_0} - 1) + (e^{y_0}(y_1) - 1)x + \left[e^{y_0}(\frac{1}{2}y_2 + \frac{1}{2!}y_1)\right]x^2 + \mathcal{O}(x^3).$$

The coefficients of each x^n must vanish, yielding a hierarchy of equations for y_n. Solving them successively yields, for x^0: $0 = e^{y_0} - 1$, hence $y_0 = \boxed{0}$; for x^1: $0 = e^{y_0}y_1 - 1$, hence $y_1 = \boxed{1}$; for x^2: $0 = e^{y_0}\frac{1}{2}(y_2 + y_1^2)$, hence $y_2 = \boxed{-1}$. This gives the solution:

$$\ln(1 + x) \equiv y(x) \overset{(2)}{=} \boxed{x - \frac{1}{2}x^2 + \mathcal{O}(x^3)}.$$

Method 2: repeated differentiation. We write Eq. (1) in the form $0 = \mathcal{F}(y(x), x) \equiv F(x)$, set up the hierarchy $0 = d_x^n F(x)|_{x=0}$, and solve it successively for the y_ns:

$F:$	$0 = e^{y(x)} - x + 1$	$\overset{x=0}{\Rightarrow}$	$0 = e^{y_0} - 1$	$\Rightarrow$	$y_0 = y(0) = \boxed{0},$
$d_x F:$	$0 = y'e^y - 1$	$\overset{x=0}{\Rightarrow}$	$0 = y_1 e^{y_0} - 1$	$\Rightarrow$	$y_1 = y'(0) = \boxed{1},$
$d_x^2 F:$	$0 = \left[(y')^2 + y''\right]e^{y(0)}$	$\overset{x=0}{\Rightarrow}$	$0 = \left[1^2 + y_2\right]e^0$	$\Rightarrow$	$y_2 = y''(0) = \boxed{-1}.$

We obtain the same hierarchy as with method 1, but in somewhat simpler fashion. ✓

(b) $\frac{\ln(x)}{\ln(2)}$ is the inverse of 2^x. We are looking for the series expansion of $y(x) \equiv 2^x$ in powers of x. This function fulfills the defining equation:

$$\frac{\ln(y(x))}{\ln(2)} = x. \tag{3}$$

Series ansatz: $\qquad y(x) = y_0 + y_1 x + \frac{1}{2!}y_2 x^2 + \mathcal{O}(x^3), \quad \text{with } y_n \equiv y^{(n)}(0). \tag{4}$

Method 1: expansion of Eq. (3), using $\ln(1+z) = z - \frac{1}{2}z^2 + \mathcal{O}(z^3)$:

$$0 = \ln\left(y_0 + y_1 x + \frac{1}{2}y_2 x^2 + \mathcal{O}(x^3)\right) - x\ln(2)$$

$$= \ln(y_0) + \ln\left(1 + \frac{y_1}{y_0}x + \frac{1}{2}\frac{y_2}{y_0}x^2 - x\ln(2)\mathcal{O}(x^3)\right)$$

$$= \ln(y_0) + \left(\frac{y_1}{y_0}x + \frac{1}{2}\frac{y_2}{y_0}x^2 + \mathcal{O}(x^3)\right) - \frac{1}{2}\left(\frac{y_1}{y_0}x + \frac{1}{2}\frac{y_2}{y_0}x^2 - x\ln(2)\mathcal{O}(x^3)\right)^2$$

$$= \ln(y_0) + \left[\frac{y_1}{y_0} - \ln(2)\right]x + \left[\frac{1}{2}\frac{y_2}{y_0}x^2 - \frac{y_1^2}{y_0^2}\right]x^2 + \mathcal{O}(x^3).$$

The coefficients of each x^n must vanish, yielding a hierarchy of equations for y_n. Solving them successively yields, for x^0: $0 = \ln(y_0)$, thus $y_0 = \boxed{1}$; for x^1: $0 = \frac{y_1}{y_0} - \ln(2)$, thus $y_1 = \boxed{\ln(2)}$; for x^2: $0 = e^{y_0}\frac{1}{2y_0^2}(y_1^2 - y_2)$, thus $y_2 = \boxed{\ln^2(2)}$. This gives the solution:

$$2^x \equiv y(x) \overset{(4)}{=} \boxed{1 + \ln(2)x + \frac{1}{2}\ln^2(2)x^2 + \mathcal{O}(x^3)}.$$

Method 2: repeated differentiation of Eq. (3), written in the form $0 = \mathcal{F}(y(x),x) \equiv F(x)$:

$$F: \quad 0 = \frac{\ln(y(x))}{\ln(2)} - x \quad \overset{x=0}{\Rightarrow} \quad 0 = \ln y_0 \quad \Rightarrow \quad y_0 = y(0) = \boxed{1},$$

$$d_x F: \quad 0 = \frac{y'}{\ln(2)y} - 1 \quad \overset{x=0}{\Rightarrow} \quad 0 = \frac{y_1}{\ln(2)y_0} - 1 \quad \Rightarrow \quad y_1 = y'(0) = \boxed{\ln(2)},$$

$$d_x^2 F: \quad 0 = \frac{1}{\ln(2)}\left[\frac{y''}{y} - \frac{(y')^2}{y^2}\right] \quad \overset{x=0}{\Rightarrow} \quad 0 = \frac{y_2}{y_0} - \frac{(y_1)^2}{y_0^2} \quad \Rightarrow \quad y_2 = y''(0) = \boxed{\ln^2(2)}.$$

S.C5.5 Higher-dimensional Taylor series (p. 269)

C5.5.1 Taylor expansions in two dimensions (p. 269)

(a) Multiplication of the series expansion of the exponential and cosine functions:

$$g(x,y) = e^x \cos(x + 2y)$$

$$= \left[1 + x + \frac{1}{2}x^2 + \mathcal{O}(x^3)\right]\left[1 - \frac{1}{2}(x+2y)^2 + \mathcal{O}(x^3, y^3, x^2 y, xy^2)\right]$$

$$= \left[1 + x + \frac{1}{2}x^2 + \mathcal{O}(x^3)\right]\left[1 - \frac{1}{2}x^2 - 2y^2 - 2xy + \mathcal{O}(x^3, y^3, x^2 y, xy^2)\right]$$

$$= 1 - \frac{1}{2}x^2 - 2y^2 - 2xy + x + \frac{1}{2}x^2 + \mathcal{O}(x^3, y^3, x^2 y, xy^2)$$

$$= \boxed{1 + x - 2y^2 - 2xy + \mathcal{O}(x^3, y^3, x^2 y, xy^2)}.$$

(b) The Taylor series of a function of two variables (including up to second order) reads:

$$g(x,y) = \left[1 + (x\partial_{\tilde{x}} + y\partial_{\tilde{y}}) + \frac{1}{2}(x\partial_{\tilde{x}} + y\partial_{\tilde{y}})(x\partial_{\tilde{x}} + y\partial_{\tilde{y}})\right]g(\tilde{x},\tilde{y})\Big|_{\tilde{x}=\tilde{y}=0}$$

$$+ \mathcal{O}(x^3, y^3, x^2 y, xy^2)$$

$$= \left[1 + x\partial_x + y\partial_y + \frac{1}{2}x^2\partial_x^2 + \frac{1}{2}y^2\partial_y^2 + xy\partial_x\partial_y\right]g(0,0) + \mathcal{O}(x^3, y^3, x^2 y, xy^2).$$

Notation: $\partial_i g(0,0) \equiv \partial_i g(x,y)|_{x=y=0}$, i.e. take the derivative first, then set $x = y = 0$.

$$g(x,y) = e^x \cos(x+2y), \qquad\qquad\qquad \Rightarrow \qquad g(0,0) = \boxed{1},$$

$$\partial_x g(x,y) = e^x \cos(x+2y) - e^x \sin(x+2y),$$

$$= g(x,y) - e^x \sin(x+2y), \qquad\qquad \Rightarrow \qquad \partial_x g(0,0) = \boxed{1},$$

$$\partial_x^2 g(x,y) = \partial_x g(x,y) - e^x \sin(x+2y) - e^x \cos(x+2y)$$

$$= -2e^x \sin(x+2y), \qquad\qquad \Rightarrow \qquad \partial_x^2 g(0,0) = \boxed{0},$$

$$\partial_y g(x,y) = -2e^x \sin(x+2y), \qquad\qquad \Rightarrow \qquad \partial_y g(0,0) = \boxed{0},$$

$$\partial_y^2 g(x,y) = -4e^x \cos(x+2y), \qquad\qquad \Rightarrow \qquad \partial_y^2 g(0,0) = \boxed{-4},$$

$$\partial_x \partial_y g(x,y) = -2e^x \sin(x+2y) - 2e^x \cos(x+2y) \qquad \Rightarrow \qquad \partial_x \partial_y g(0,0) = \boxed{-2},$$

$$\partial_y \partial_x g(x,y) = \partial_x \partial_y g(x,y) \qquad\qquad\qquad \Rightarrow \qquad \partial_y \partial_x g(0,0) = \boxed{-2},$$

$$\Rightarrow \quad g(x,y) = \boxed{1 + x - 2y^2 - 2xy + \mathcal{O}(x^3, y^3, x^2 y, xy^2)}.$$

S.C6 Fourier calculus (p. 271)

S.C6.1 The δ-function (p. 272)

C6.1.1 Integrals with δ-function (p. 277)

(a) $\quad I_1(a) = \displaystyle\int_{-\infty}^{\infty} dx\, \delta(x-\pi) \sin(ax) = \boxed{\sin(a\pi)}$.

(b) $\quad I_2(a) = \displaystyle\int_{\mathbb{R}^3} dx^1 dx^2 dx^3\, \delta(\mathbf{x}-\mathbf{y}) \|\mathbf{x}\|^2 = \|\mathbf{y}\|^2 = \boxed{a^2 + 1^2 + 2^2}$.

(c) $\quad I_3(a) = \displaystyle\int_0^a dx\, \frac{\delta(x-\pi)}{a + \cos^2(x/2)} = \begin{cases} \frac{1}{a+\cos^2(\pi/2)} = \frac{1}{a} & \text{for } a > \pi \\ \frac{1}{2a} & \text{for } a = \pi \\ 0 & \text{for } a < \pi \end{cases}$.

The δ peak at $x = \pi$ lies inside the domain of integration $[0,a]$ if $a > \pi$, and outside if $a < \pi$. The case $a = \pi$ yields half the value of the case $a > \pi$. (The latter statement follows from representing the δ-function as a series of ever sharper, symmetric, normalized peaks: if one does not integrate over the full peak, but only up to its maximum, only half its weight contributes, hence $\int_{-\infty}^0 dx\, \delta(x) = \frac{1}{2}$ and $\int_0^\infty dx\, \delta(x) = \frac{1}{2}$.)

(d) $\quad I_4(a) = \displaystyle\int_0^3 dx\, \delta(x^2 - 6x + 8)\sqrt{e^{ax}} = \int_0^3 dx \left[\frac{\delta(x-4)}{2} + \frac{\delta(x-2)}{2} \right] \sqrt{e^{xa}} = \frac{\sqrt{e^{2a}}}{2} = \boxed{\frac{e^a}{2}}$.

We use $\delta(g(x)) = \sum_i \frac{\delta(x-x_i)}{|g'(x_i)|}$, where x_i are the zeros of g. For $g(x) = x^2 - 6x + 8$ the zeros are $x_1 = 4$ and $x_2 = 2$. At these points the absolute value of $g'(x) = 2x - 6$ takes the values $|g'(x_1)| = 2$ and $|g'(x_2)| = 2$. Since x_1 lies outside the integration domain $[0,3]$, only x_2 contributes to the integral.

(e) $\quad I_5(a) = \displaystyle\int_{\mathbb{R}^2} dx^1 dx^2\, \delta(\mathbf{x}-a\mathbf{y})\, \mathbf{x} \cdot \mathbf{y} = a\mathbf{y} \cdot \mathbf{y} = a(1 + 3^2) = \boxed{10a}$.

ᴇC6.1.3 **Lorentz representation of the Dirac delta function** (p. 277)

We have to verify that in the limit $\epsilon \to 0$, $\delta^\epsilon(x)$ possesses the defining properties of the Dirac delta function, namely:

(i) $\delta(0) = \infty$,   (ii) $\delta(x \neq 0) = 0$ (i.e. peak width = 0),   (iii) $\int_{-\infty}^{\infty} dx\, \delta(x) = 1$.

Lorentz peak: $\delta^\epsilon(x) = \delta^\epsilon(x) = \dfrac{\epsilon/\pi}{x^2 + \epsilon^2}$.

(i) Height: $\delta^\epsilon(0) = \dfrac{1}{\pi \epsilon} \overset{\epsilon \to 0}{\longrightarrow} \boxed{\infty}$. ✓

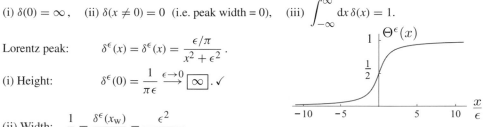

(ii) Width: $\dfrac{1}{2} = \dfrac{\delta^\epsilon(x_{\mathrm{w}})}{\delta^\epsilon(0)} = \dfrac{\epsilon^2}{x_{\mathrm{w}}^2 + \epsilon^2}$

    $\Rightarrow \quad x_{\mathrm{w}} = \epsilon \overset{\epsilon \to 0}{\longrightarrow} \boxed{0}$. ✓

   $\delta^\epsilon(x \neq 0) \overset{\epsilon \ll x}{\longrightarrow} \dfrac{\epsilon/\pi}{x^2}\left[1 + \mathcal{O}(\epsilon^2/x^2)\right] \overset{\epsilon \to 0}{\longrightarrow} \boxed{0}$. ✓

(iii) Weight, computed by using the substitution:

$$x = \epsilon \tan y = \epsilon \frac{\sin y}{\cos y},$$

$$\frac{dx}{dy} = \frac{\epsilon}{\cos^2 y}, \quad x^2 + \epsilon^2 = \frac{\epsilon^2}{\cos^2 y}.$$

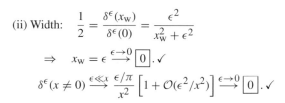

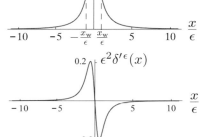

$$\int_{-\infty}^{\infty} dx\, \delta^\epsilon(x) = \int_{-\infty}^{\infty} dx\, \frac{\epsilon/\pi}{x^2 + \epsilon^2} \overset{x = \epsilon \tan y}{=} \int_{-\frac{\pi}{2}}^{\frac{\pi}{2}} dy\, \frac{dx}{dy}\, \frac{\epsilon/\pi}{\epsilon^2/\cos^2 y} = \frac{1}{\pi} \int_{-\frac{\pi}{2}}^{\frac{\pi}{2}} dy = \boxed{1} . ✓$$

Thus $\delta^\epsilon(x)$ does possess the defining properties (i)–(iii) of a Dirac δ-function. ✓

(iv) Step: $\Theta^\epsilon(x) = \int_{-\infty}^{x} dx'\, \delta^\epsilon(x') = \frac{1}{\pi} \int_{-\frac{\pi}{2}}^{y(x)} dy' = \frac{1}{\pi} y' \Big|_{-\frac{\pi}{2}}^{y(x)} = \frac{1}{\pi}\left[\arctan(x/\epsilon) + \frac{\pi}{2}\right]$

$$= \boxed{\frac{1}{2}\left[\frac{2}{\pi} \arctan(x/\epsilon) + 1\right]} \overset{\epsilon \to 0}{\longrightarrow} \begin{cases} 1 & \text{for } x > 0, \\ \frac{1}{2} & \text{for } x = 0, \\ 0 & \text{for } x < 0. \end{cases}$$

(v) Derivative: $\delta'^\epsilon(x) = \dfrac{d}{dx} \dfrac{\epsilon/\pi}{x^2 + \epsilon^2} = \boxed{-\dfrac{2x\epsilon/\pi}{(x^2 + \epsilon^2)^2}}$.

ᴇC6.1.5 **Series representation of the coth function** (p. 280)

For $y \in \mathbb{R}^+$, the variable $\omega \equiv e^{-y}$ satisfies the inequality $|\omega| < 1$, hence the geometric series $\sum_{n=0}^{\infty} \omega^n = 1/(1-\omega)$ converges.

(a) $\displaystyle\sum_{n=0}^{\infty} e^{-y(n+1/2)} = e^{-y/2} \sum_{n=0}^{\infty} w^n = \frac{e^{-y/2}}{1-w} = \frac{e^{-y/2}}{1-e^{-y}} = \frac{1}{e^{y/2} - e^{-y/2}} = \boxed{\dfrac{1}{2\sinh(y/2)}}$.

(b) $\sum_{n=0}^{\infty}(-1)^n e^{-y(n+1/2)} = e^{-y/2}\sum_{n=0}^{\infty}(-w)^n = \dfrac{e^{-y/2}}{1+w} = \dfrac{e^{-y/2}}{1+e^{-y}} = \dfrac{1}{e^{y/2}+e^{-y/2}} = \boxed{\dfrac{1}{2\cosh(y/2)}}$.

(c) $\sum_{n\in\mathbb{Z}} e^{-y|n|} = \sum_{n\geq 0} e^{-yn} + \sum_{n\leq 0} e^{yn} - 1 = 2\sum_{n\geq 0} e^{-yn} - 1 = 2\sum_{n\geq 0} w^n - 1 = 2\dfrac{1}{1-w} - 1$

$\qquad = \dfrac{1+w}{1-w} = \dfrac{1+e^{-y}}{1-e^{-y}} = \dfrac{e^{-y/2}(e^{y/2}+e^{-y/2})}{e^{-y/2}(e^{y/2}-e^{-y/2})} = \dfrac{\cosh(y/2)}{\sinh(y/2)} = \boxed{\coth(y/2)}$.

S.C6.2   Fourier series (p. 277)

C6.2.1   Fourier series of the sawtooth function (p. 285)

The sawtooth function: $f(x) = x$ for $-\pi < x < \pi$, is periodic with period $L = 2\pi$. The Fourier series is defined by: $f(x) = \frac{1}{L}\sum_{k=-\infty}^{\infty} e^{ikx}\tilde{f}_k$, with $k = 2\pi n/L = n$, for $n \in \mathbb{Z}$.

$n \neq 0:\qquad \tilde{f}_n = \int_{-\pi}^{\pi} dx\, f(x)e^{-inx} = \int_{-\pi}^{\pi} dx\, x e^{-inx} \overset{\text{part. int.}}{=} -\dfrac{x}{in}e^{-inx}\Big|_{-\pi}^{\pi} + \int_{-\pi}^{\pi} dx\, \dfrac{1}{in}e^{-inx}$

$\qquad\qquad = \left[\dfrac{i}{n}xe^{-inx} + \dfrac{1}{n^2}e^{-inx}\right]_{-\pi}^{\pi} = \boxed{\dfrac{2i\pi}{n}(-1)^n}$.

$n = 0:\qquad \tilde{f}_0 = \int_{-\pi}^{\pi} dx\, x e^0 = \int_{-\pi}^{\pi} dx\, x = \boxed{0}$.

Thus the Fourier series has the following form:

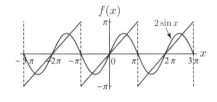

$f(x) = \dfrac{1}{2\pi}\sum_{n=-\infty}^{\infty} e^{inx}\tilde{f}_n = \boxed{\sum_{\substack{n=-\infty \\ n\neq 0}}^{\infty} \dfrac{i}{n}(-1)^n e^{inx}}$.

Remark: Evidently $\tilde{f}_n = -\tilde{f}_{-n}$ (a consequence of the fact that the function $f(x)$ is antisymmetric). The Fourier series can therefore also be rewritten as a sine series (not asked for here), by using $e^{inx} - e^{-inx} = 2i\sin(nx)$:

$$f(x) = \boxed{\sum_{n=1}^{\infty} \dfrac{2}{n}(-1)^{n+1}\sin(nx)}.$$

C6.2.3   Cosine series (p. 278)

(a)   Inserting the Fourier ansatz $f(x) = \frac{1}{L}\sum_{k'} e^{ik'x}\tilde{f}_{k'}$ into the formula for the Fourier coefficient yields:

$$\int_{-\frac{L}{2}}^{\frac{L}{2}} dx\, e^{-ikx}f(x) = \sum_{k'}\tilde{f}_{k'}\underbrace{\dfrac{1}{L}\int_{-\frac{L}{2}}^{\frac{L}{2}} dx\, e^{i(k'-k)x}}_{\equiv I_{kk'} = \delta_{kk'}} = \boxed{\tilde{f}_k}. \checkmark \qquad\qquad (1)$$

The orthonormality of the Fourier modes, $I_{kk'} = \delta_{kk'}$, is seen as follows:

$$I_{k=k'} = \dfrac{1}{L}\int_{-\frac{L}{2}}^{\frac{L}{2}} dx = \boxed{1}.$$

$$I_{k \neq k'} = \frac{1}{L} \int_{-\frac{L}{2}}^{\frac{L}{2}} dx\, e^{i(k'-k)x} = \frac{e^{i(k'-k)\frac{L}{2}} - e^{-i(k'-k)\frac{L}{2}}}{i(k'-k)}$$

$$= \frac{e^{i(k'-k)\frac{L}{2}}}{i(k'-k)} \underbrace{\left(1 - e^{-i(k'-k)L}\right)}_{=1-e^{-i\frac{2\pi}{L}(n'-n)L}=0} = \boxed{0}.$$

(b) For an even function, we have $f(x) = f(-x)$. Therefore, we can write $\tilde{f}_k$ as follows, using the substitution $x \to -x$ for the part of the integral where $x \in [-L/2, 0]$:

$$\tilde{f}_k = \int_{-\frac{L}{2}}^{\frac{L}{2}} dx\, e^{-ikx} f(x) = \int_0^{\frac{L}{2}} dx\, e^{-ikx} f(x) + \int_{-\frac{L}{2}}^0 dx\, e^{-ikx} f(x)$$

$$= \int_0^{\frac{L}{2}} dx \left[e^{-ikx} f(x) + e^{+ikx} \underbrace{f(-x)}_{-f(x)} \right] = \boxed{2 \int_0^{\frac{L}{2}} dx\, \cos(kx) f(x)}. \qquad (2)$$

Since $\cos(kx)$ is an even function in k, we have $\tilde{f}_k = \tilde{f}_{-k}$. It follows that:

$$f(x) = \frac{1}{L} \sum_k e^{ikx} \tilde{f}_k = \frac{\tilde{f}_0}{L} + \frac{1}{L} \sum_{k>0} \underbrace{\left(e^{ikx} \tilde{f}_k + e^{-ikx} \tilde{f}_{-k} \right)}_{= \tilde{f}_k} = \frac{1}{L} \tilde{f}_0 + \frac{1}{L} \sum_{k>0} \tilde{f}_k 2 \cos(kx)$$

$$\equiv \boxed{\frac{1}{2} a_0 + \sum_{k>0} a_k \cos(kx)}, \qquad \text{with} \qquad a_k \equiv \frac{2}{L} \tilde{f}_k \overset{(2)}{=} \boxed{\frac{4}{L} \int_0^{\frac{L}{2}} dx\, \cos(kx) f(x)}. \qquad (3)$$

(c) Cosine coefficients via (3), where only terms with $k \geq 0$ occur:

$$k = 0: \quad a_0 \overset{(3)}{=} \frac{4}{L} \int_0^{\frac{L}{2}} dx\, \underbrace{\cos(0)}_{=1} f(x) = \frac{4}{L} \int_0^{\frac{L}{4}} dx\, 1 + \frac{4}{L} \int_{\frac{L}{4}}^{\frac{L}{2}} dx\, (-1) = \boxed{0}, \qquad (4)$$

$$k > 0: \quad a_k \overset{(3)}{=} \frac{4}{L} \int_0^{\frac{L}{2}} dx\, \cos(kx) f(x) = \frac{4}{L} \int_0^{\frac{L}{4}} dx\, \cos(kx) - \frac{4}{L} \int_{\frac{L}{4}}^{\frac{L}{2}} dx\, \cos(kx)$$

$$= \boxed{\frac{4}{Lk} \left[2 \sin(kL/4) - \sin(kL/2) \right]}. \qquad (5)$$

In comparison, calculating the Fourier coefficients via (1) is a bit more cumbersome:

$$k = 0: \quad \tilde{f}_0 \overset{(1)}{=} \int_{-\frac{L}{2}}^{\frac{L}{2}} dx\, e^0 f(x) = \int_{-\frac{L}{4}}^{\frac{L}{4}} dx - \int_{-\frac{L}{2}}^{-\frac{L}{4}} dx - \int_{\frac{L}{4}}^{\frac{L}{2}} dx = \boxed{0} \quad [\overset{(4)}{=} \tfrac{L}{2} a_0 \checkmark], \qquad (6)$$

$$k \neq 0: \quad \tilde{f}_k \overset{(1)}{=} \int_{-\frac{L}{2}}^{\frac{L}{2}} dx\, e^{-ikx} f(x) = \int_{-\frac{L}{4}}^{\frac{L}{4}} dx\, e^{-ikx} - \int_{-\frac{L}{2}}^{-\frac{L}{4}} dx\, e^{-ikx} - \int_{\frac{L}{4}}^{\frac{L}{2}} dx\, e^{-ikx}$$

$$= \frac{1}{-ik} \left[\left(e^{-ikL/4} - e^{ikL/4} \right) - \left(e^{ikL/4} - e^{ikL/2} \right) - \left(e^{-ikL/2} - e^{-ikL/4} \right) \right]$$

$$= \boxed{\frac{2}{k} \left[2 \sin(kL/4) - \sin(kL/2) \right]} \qquad [\overset{(5)}{=} \tfrac{L}{2} a_k \checkmark]. \qquad (7)$$

Now set $0 \neq k = 2\pi n/L$ in (5), with $n \in \mathbb{Z}$:

$$a_k = \frac{2}{L}\tilde{f}_k \overset{(5)}{=} \frac{2}{\pi n}\left[2\sin(\pi n/2) - \underbrace{\sin(\pi n)}_{=0}\right] = \begin{cases} 0 & \text{for } 0 \neq n = 2m, \\ \dfrac{4}{\pi(2m+1)}(-1)^m & \text{for } n = 2m+1, \end{cases}$$

with $m \in \mathbb{N}_0$. Therefore the cosine representation (3) of $f(x)$ has the following form:

$$f(x) \overset{(3)}{=} \frac{4}{\pi}\sum_{m\geq 0}\frac{(-1)^m}{2m+1}\cos\left(\frac{2\pi(2m+1)x}{L}\right)$$

$$= \frac{4}{\pi}\left[\cos\left(\frac{2\pi x}{L}\right) - \frac{1}{3}\cos\left(\frac{6\pi x}{L}\right) + \frac{1}{5}\cos\left(\frac{10\pi x}{L}\right) + \cdots\right]$$

The sketch shows the function $f(x)$ and the approximation thereof that arises from the first three terms of the cosine series.

S.C6.3　Fourier transform (p. 285)

ₑC6.3.1　Properties of Fourier transformations (p. 288)

(a)　Fourier transform of $f(x-a)$, rewritten using the substitution $\bar{x} = x - a$:

$$\int_{-\infty}^{\infty}dx\,e^{-ikx}f(x-a) = \int_{-\infty}^{\infty}d\bar{x}\,e^{-ik(\bar{x}+a)}f(\bar{x}) = e^{-ika}\int_{-\infty}^{\infty}d\bar{x}\,e^{-ik\bar{x}}f(\bar{x}) = \boxed{e^{-ika}\tilde{f}(k)}.$$

(b)　Fourier transform of $f(ax)$, rewritten using the substitution $\bar{x} = ax$:

$$\int_{-\infty}^{\infty}dx\,e^{-ikx}f(ax) = \int_{-\infty}^{\infty}d\bar{x}\left|\frac{dx}{d\bar{x}}\right|e^{-i\frac{k}{a}\bar{x}}f(\bar{x}) = \frac{1}{|a|}\int_{-\infty}^{\infty}d\bar{x}\,e^{-i\frac{k}{a}\bar{x}}f(\bar{x}) = \boxed{\frac{1}{|a|}\tilde{f}(k/a)}.$$

ₑC6.3.3　Fourier transformation of a Gauss peak (p. 290)

Normalized Gaussian:　　　$g^{[\sigma]}(x) = \dfrac{1}{\sqrt{2\pi}\sigma}e^{-x^2/2\sigma^2}$,　　　$\displaystyle\int_{-\infty}^{\infty}dx\,g^{[\sigma]}(x) = 1.$　　　(1)

Fourier transformation:　　　$\tilde{g}_k^{[\sigma]} = \displaystyle\int_{-\infty}^{\infty}dx\,e^{-ikx}g^{[\sigma]}(x) = \int_{-\infty}^{\infty}dx\,\frac{1}{\sqrt{2\pi}\sigma}e^{-\frac{1}{2\sigma^2}(x^2+2\sigma^2ikx)}.$

Completing the square:　　　$\left(x^2 + 2\sigma^2 ikx\right) = \left(x+\sigma^2 ik\right)^2 + \sigma^4 k^2 \overset{\bar{x}=x+\sigma^2 ik}{=} \bar{x}^2 + \sigma^4 k^2.$

$$\tilde{g}_k^{[\sigma]} \overset{dx=d\bar{x}}{=} \int_{-\infty}^{\infty}d\bar{x}\,\frac{1}{\sqrt{2\pi}\sigma}e^{-\frac{\bar{x}^2}{2\sigma^2}}e^{-\frac{\sigma^4 k^2}{2\sigma^2}} = e^{-\sigma^2 k^2/2}\underbrace{\int_{-\infty}^{\infty}d\bar{x}\,g^{[\sigma]}(\bar{x})}_{\overset{(1)}{=}1} = \boxed{e^{-\sigma^2 k^2/2}}.$$

Remark: The Fourier transform of a Gaussian distribution with width σ is a Gaussian of width $1/\sigma$. This is a good example of Fourier reciprocity; the Fourier transform of a narrow distribution is a wide distribution and vice versa.

C6.3.5 Parseval's identity and convolution (p. 291)

(a) Explicit computation of the integral:

$$\int_{-\pi}^{\pi} dx \bar{f}(x)\, g(x) = \int_{-\pi}^{\pi} dx\, x \sin(x) = \left[-x \cos(x) + \sin(x) \right]_{-\pi}^{\pi} = \pi + \pi = \boxed{2\pi}. \qquad (1)$$

Summation of Fourier coefficients: the sawtooth function $f(x)$ has period $L = 2\pi$, hence its Fourier series has the form $f(x) = \frac{1}{L} \sum_k e^{ikx} \tilde{f}_k = \frac{1}{2\pi} \sum_n e^{inx} \tilde{f}_n$, with $k = 2\pi n/L = n \in \mathbb{Z}$. The same is true for the sine function $g(x) = \sin(x)$. Their Fourier coefficients are known to have the following form:

$$\tilde{f}_n = \frac{2\pi i (-1)^n}{n} \quad \text{for } n \neq 0, \quad \text{and} \quad \tilde{f}_0 = 0.$$

$$\tilde{g}_n = \frac{\pi}{i}(\delta_{n,1} - \delta_{n,-1}), \quad \text{since} \quad g(x) = \sin(x) = \frac{1}{2i}\left(e^{ix} - e^{-ix} \right) = \frac{1}{2\pi} \sum_n e^{inx} \tilde{g}_n.$$

Parseval's identity yields:

$$\int_{-\pi}^{\pi} dx \bar{f}(x)\, g(x) = \frac{1}{L} \sum_n \bar{\tilde{f}}_n \tilde{g}_n = \frac{1}{2\pi}\left[\bar{\tilde{f}}_1 \tilde{g}_1 + \bar{\tilde{f}}_{-1} \tilde{g}_{-1} \right]$$

$$= \frac{1}{2\pi}\left[\frac{(-2\pi i)(-1)^{+1}}{(+1)} \cdot \frac{\pi}{i} + \frac{(-2\pi i)(-1)^{-1}}{(-1)} \cdot \frac{(-\pi)}{i} \right] = \boxed{2\pi} \overset{\checkmark}{=} (1).$$

(b) Special case of Parseval's identity : $\int_{-\pi}^{\pi} dx\, |f(x)|^2 = \frac{1}{2\pi} \sum_n |\tilde{f}_n|^2.$

On the one hand:

$$\int_{-\pi}^{\pi} dx\, |f(x)|^2 = \int_{-\pi}^{\pi} dx\, x^2 = \frac{1}{3}x^3 \Big|_{-\pi}^{\pi} = \boxed{\frac{2\pi^3}{3}}. \qquad (2)$$

On the other hand:

$$\frac{1}{2\pi} \sum_{n=-\infty}^{\infty} |\tilde{f}_n|^2 = \frac{2}{2\pi} \sum_{n=1}^{\infty} \frac{(2\pi)^2}{n^2} = \boxed{4\pi \sum_{n=1}^{\infty} \frac{1}{n^2}}. \qquad (3)$$

Parseval: (2) = (3)

$$\boxed{\sum_{n=1}^{\infty} \frac{1}{n^2} = \frac{\pi^2}{6}}. \qquad (4)$$

(c) Direct computation of the convolution integral: since f and g are periodic, with the same period, the domain of integration can be chosen to be an arbitrary interval of length equal to this period. We here choose $(-\pi, \pi)$ and thus compute the following convolution integral:

$$(f * g)(x) = \int_{-\pi}^{\pi} dx'\, f(x - x')g(x') = \int_{-\pi}^{\pi} dx'\, g(x - x')f(x') = (g * f)(x). \qquad (5)$$

The formulas on the right express the fact that the convolution of two functions is commutative. (That can be seen, e.g., in Fourier space, since the convolution theorem gives $\widetilde{(f * g)}_k = \tilde{f}_k \tilde{g}_k = \tilde{g}_k \tilde{f}_k = \widetilde{(g * f)}_k$. Also see Eq. (7).) In the present case it is simplest to use the form on the right (the form on the left is discussed further below):

$$(g * f)(x) \overset{(5),\text{right}}{=} \int_{-\pi}^{\pi} dx'\, \sin(x - x')\, x' = \left[x' \cos(x - x') + \sin(x - x') \right]_{-\pi}^{\pi} \qquad (6a)$$

$$= \pi \cos(x - \pi) + \pi \cos(x + \pi) + \sin(x - \pi) - \sin(x + \pi) = \boxed{-2\pi \cos x}, \qquad (6b)$$

since $\cos(x - \pi) = \cos(x + \pi) = -\cos(x)$ and $\sin(x - \pi) = \sin(x + \pi)$.

Alternative computation via summation of Fourier coefficients:

$$(f * g)(x) = \frac{1}{2\pi} \sum_{n=-\infty}^{\infty} \tilde{f}_n \tilde{g}_n e^{inx} = \frac{1}{2\pi} \sum_{\substack{n=-\infty \\ n\neq 0}}^{\infty} \frac{2\pi i \, (-1)^n}{n} \left(\frac{\pi}{i}\right) \left[\delta_{n,1} - \delta_{n,-1}\right] e^{inx}$$

$$= \pi \left[\frac{(-1)^{+1}}{(+1)} e^{-ix} - \frac{(-1)^{-1}}{(-1)} e^{-ix}\right] = 2\pi \left[-\frac{1}{2} e^{ix} - \frac{1}{2} e^{-ix}\right] = \boxed{-2\pi \cos(x)} \overset{\checkmark}{=} \text{(6b)}.$$

Remark: It is instructive to perform the direct computation of the convolution integral also using the left expression in Eq. (5). The functions occurring therein are defined as follows, e.g. for $x \in (0, \pi)$ (an analogous discussion holds for $x \in (-\pi, 0)$):

$$g(x') = \sin(x') \qquad \text{for} \quad x' \in (-\pi, \pi) \quad \Rightarrow \quad -\pi < x' < \pi, \qquad \text{(I)}$$

$$f(x - x') = \begin{cases} x - x' & \text{for} \quad x - x' \in (-\pi, \pi) \quad \Rightarrow \quad x - \pi < x' < x + \pi, \qquad \text{(II)} \\ x - x' - 2\pi & \text{for} \quad x - x' \in (\pi, 3\pi) \quad \Rightarrow \quad x - 3\pi < x' < x - \pi. \qquad \text{(III)} \end{cases}$$

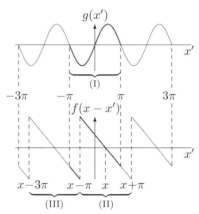

When x' traverses the domain of integration $(-\pi, \pi)$, $g(x')$ is described by a single formula, (I), throughout the entire domain, whereas for $f(x - x')$ two cases have to be distinguished: since this function exhibits a discontinuity when its argument $x - x'$ passes the point π, we need formula (II) for $x - \pi < x' < \pi$, but formula (III) for $-\pi < x' < x - \pi$. [(III) is the "periodic continuation" of (II), shifted by one period.] Two possible strategies for dealing with this situation are either (i) to shift the domain of integration, or (ii) to split it into two.

(i) Shifting the domain of integration: to avoid the discontinuity the domain of integration can be shifted from $(-\pi, \pi)$ to $(-\pi + x, \pi + x)$ (which is allowed, since the integrand is periodic). This yields (via the substitution $x'' = x - x'$)

$$(f * g)(x) = \int_{x - \pi}^{x + \pi} dx' f(x - x') g(x') \overset{x'' = x - x'}{=} \int_{-\pi}^{\pi} dx'' g(x - x'') f(x'') = (g * f)(x), \qquad \text{(7)}$$

thus reproducing Eq. (5)$_{\text{right}}$, which constitutes another proof of the fact that a convolution is commutative. For the current example, (5)$_{\text{right}}$ has the advantage compared to (5)$_{\text{left}}$ that not only g but also f are each described by only a *single* formula throughout the entire domain of integration, $g(x - x') = \sin(x - x')$ and $f(x') = x'$. Consequently, in the above calculation from (6a) to (6b) it was not necessary to distinguish two cases separately.

(ii) Splitting the domain of integration: alternatively, the discontinuity can be accounted for by accordingly splitting the domain $(-\pi, \pi) = (-\pi, x - \pi) \cup [x - \pi, \pi)$ into two:

$$(f * g)(x) \overset{\text{(5)}_{\text{left}}}{=} \int_{-\pi}^{\pi} dx' f(x - x') g(x') = \int_{-\pi}^{x - \pi} dx' f(x - x') g(x') + \int_{x - \pi}^{\pi} dx' f(x - x') g(x')$$

$$= \int_{-\pi}^{x - \pi} dx' \underbrace{(x - x' - 2\pi)}_{\text{(III)}} \underbrace{\sin(x')}_{\text{(I)}} + \int_{x - \pi}^{\pi} dx' \underbrace{(x - x')}_{\text{(II)}} \underbrace{\sin(x')}_{\text{(I)}}$$

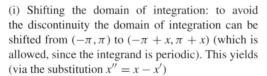

$$= \int_{-\pi}^{\pi} dx'\,(x - x')\sin(x') + (-2\pi)\int_{-\pi}^{x-\pi} dx'\,\sin x'$$

$$= \left[-x\cos x' + x'\cos x' - \sin x'\right]_{-\pi}^{\pi} + 2\pi\left[\cos x'\right]_{-\pi}^{x-\pi}$$

$$= 0 - 2\pi + 0 + 2\pi[\cos(x-\pi) + 1] = \boxed{-2\pi\cos x}. \tag{8}$$

For the current example the computation using strategy (i), shifting the domain of integration (or equivalently, exploiting the commutativity of convolutions), is simpler than using strategy (ii), splitting this domain. However, if *both* functions f and g are defined via periodic continuation and exhibit discontinuities, strategy (i) is not helpful: the domain of integration will then contain a discontinuity for both $(f * g)$ and $(g * f)$, so that splitting the domain of integration is unavoidable.

ɛC6.3.7 Poisson summation formula (p. 297)

(a) We multiply the completeness relation with $f(y/L)$ and integrate over $x = y/L$:

Completeness:
$$\frac{1}{L}\sum_{n\in\mathbb{Z}} e^{-i2\pi ny/L} = \sum_{m\in\mathbb{Z}}\delta(y - Lm) = \frac{1}{L}\sum_{m\in\mathbb{Z}}\delta\left(\frac{y}{L} - m\right); \tag{1}$$

$$\int dx f(x)(1): \qquad \sum_{n\in\mathbb{Z}}\int_{-\infty}^{\infty} dx f(x)e^{-i2\pi nx} = \sum_{m\in\mathbb{Z}}\int_{-\infty}^{\infty} dx f(x)\delta(x - m) = \sum_{m\in\mathbb{Z}} f(m).$$

The left integral corresponds to the Fourier transform, $\tilde{f}(k) = \int_{-\infty}^{\infty} dx f(x)e^{-ikx}$, with $k = 2\pi n$. Thus we obtain:

$$\boxed{\sum_{n\in\mathbb{Z}}\tilde{f}(2\pi n) = \sum_{m\in\mathbb{Z}} f(m)}. \tag{2}$$

(b) The Fourier transform of the function $f(x) = e^{-a|x|}$ has the following form:

$$\tilde{f}(k) = \int_{-\infty}^{\infty} dx\, e^{-ikx} f(x) = \int_{-\infty}^{\infty} dx\, e^{-(ikx+a|x|)} = \frac{2a}{k^2 + a^2}.$$

The Poisson summation formula (2) then gives:

$$\boxed{\sum_{n\in\mathbb{Z}}\frac{2a}{(2\pi n^2 + a^2)}} = \sum_{m\in\mathbb{Z}} e^{-a|m|} = 2\sum_{m=0}^{\infty} e^{-am} - 1$$

$$= \frac{2}{1 - e^{-a}} - 1 = \frac{1 + e^{-a}}{1 - e^{-a}} = \frac{e^{a/2} + e^{-a/2}}{e^{a/2} - e^{-a/2}} = \boxed{\coth(a/2)}.$$

S.C6.4   Case study: Frequency comb for high-precision measurements (p. 296)

A1: We insert the Fourier series for $p(t)$ into the formula for the Fourier transform of $p(t)$:

$$\tilde{p}(\omega) = \int_{-\infty}^{\infty} dt\, e^{i\omega t} p(t) = \frac{1}{\tau}\sum_{m}\underbrace{\int_{-\infty}^{\infty} dt\, e^{i\omega t}e^{-i\omega_m t}}_{2\pi\delta(\omega-\omega_m)}\tilde{p}_m = \boxed{\omega_r\sum_{m}\tilde{p}_m\delta(\omega - \omega_m)},$$

with $\omega_r = 2\pi/\tau$. We now clearly see that $\tilde{p}(\omega)$ is a periodic frequency comb of δ-functions, whose weights are fixed by the coefficients $\tilde{p}_m$ of the Fourier series.

A2: We insert the Fourier representation, $f(t) = \int_{-\infty}^{\infty} \frac{d\omega}{2\pi} e^{-i\omega t} \tilde{f}(\omega)$, into the definition of $p(t)$ and then perform the substitution $\omega = y\omega_r$ (implying $\omega\tau = 2\pi y$):

$$p(t) = \sum_n f(t - n\tau) = \sum_n \int_{-\infty}^{\infty} \frac{d\omega}{2\pi} e^{-i\omega(t-n\tau)} \tilde{f}(\omega)$$

$$\overset{\omega=y\omega_r}{=} \sum_n \int_{-\infty}^{\infty} dy\, e^{i2\pi yn} \underbrace{\left[e^{-iy\omega_r t} \tfrac{1}{\tau} \tilde{f}(y\omega_r) \right]}_{\equiv F(y)} = \sum_n \tilde{F}(2\pi n) \overset{\text{(Poisson)}}{=} \sum_m F(m).$$

Here we have defined the function $F(y) = e^{-iy\omega_r t}\tfrac{1}{\tau}\tilde{f}(y\omega_r)$, with Fourier transform $\tilde{F}(k)$, and used the Poisson summation formula ($\to$ C6.3.7). Using $\omega_m = m\omega_r = 2\pi m/\tau$, we thus obtain:

$$p(t) = \sum_m F(m) = \frac{1}{\tau} \sum_m e^{-im\omega_r t} \underbrace{\tilde{f}(m\omega_r)}_{\equiv \tilde{p}_m} \overset{\omega_m=m\omega_r}{=} \frac{1}{\tau} \sum_m e^{-i\omega_m t} \tilde{p}_m \quad \text{with} \quad \boxed{\tilde{p}_m = \tilde{f}(\omega_m)}.$$

The middle term has the form of a discrete Fourier series, from which we can read off the discrete Fourier coefficients $\tilde{p}_m$ of $p(t)$. They are clearly given by $\tilde{p}_m = \tilde{f}(\omega_m)$, and correspond to the Fourier transform of $f(t)$ evaluated at the discrete frequencies ω_m.

A3: From **A1** and **A2** we directly obtain the following form for the Fourier transform of $p(t)$:

Fourier spectrum: $\qquad \tilde{p}(\omega) \overset{\text{A1}}{=} \omega_r \sum_m \tilde{p}_m \delta(\omega - \omega_m) \overset{\text{A2}}{=} \boxed{\omega_r \sum_m \tilde{f}(\omega_m)\delta(\omega - \omega_m)}.$

For a series of Gaussian functions, $p_G(t) = \sum_n f_G(t - n\tau)$, the envelope of the frequency comb, $\tilde{f}_G(\omega)$, has the form of a Gaussian, too ($\to$ C6.3.3):

Envelope: $\qquad \tilde{f}_G(\omega) = \int_{-\infty}^{\infty} dt\, e^{i\omega t} \frac{1}{\sqrt{2\pi T^2}} e^{-\frac{t^2}{2T^2}} = \boxed{e^{-\frac{1}{2}T^2\omega^2}}.$

A4: The Fourier transform of $E(t) = e^{-i\omega_c t} p(t)$, to be denoted $\tilde{E}(\omega)$, is the same as that of $p(t)$, except that the frequency argument is shifted by ω_c:

$$\tilde{E}(\omega) = \int_{-\infty}^{\infty} dt\, e^{i\omega t} E(t) = \int_{-\infty}^{\infty} \frac{d\omega'}{2\pi} \tilde{p}(\omega') \underbrace{\int_{-\infty}^{\infty} dt\, e^{i(\omega-\omega_c-\omega')t}}_{2\pi\delta(\omega-\omega_c-\omega')} = \tilde{p}(\omega - \omega_c)$$

$$\overset{\text{A3}}{=} \frac{2\pi}{\tau} \sum_m \tilde{f}(\omega_m)\delta(\omega - \omega_m - \omega_c) \overset{m=n-N}{=} \boxed{\frac{2\pi}{\tau} \sum_n \tilde{f}(\omega_{n-N})\delta(\omega - \omega_n - \omega_{\text{off}})}.$$

For the last step we used $\omega_c = N\omega_r + \omega_{\text{off}}$ and renamed the summation index, $m = n - N$, such that $\omega_m + \omega_c = \omega_n + \omega_{\text{off}}$. Thus $\tilde{E}(\omega)$ forms an "offset-shifted" frequency comb, whose peaks relative to the Fourier frequencies ω_n are shifted by the offset frequency ω_{off}. The "center" of the comb lies at the frequency where $\tilde{f}(\omega_{n-N})$ is maximal, i.e. at $n = N$, with frequency $\omega_N \simeq \omega_c$.

A5: We begin with the definition of the Fourier transform of $p_\gamma(t)$:

Definition:
$$\tilde{p}_\gamma(\omega) = \int_{-\infty}^{\infty} dt\, e^{i\omega t} p_\gamma(t) = \int_{-\infty}^{\infty} dt\, e^{i\omega t} \sum_n f(t - n\tau) e^{-|n|\tau\gamma}$$

$t' = t - n\tau :$
$$= \underbrace{\sum_n e^{in\tau\omega} e^{-|n|\tau\gamma}}_{\equiv S^{[\gamma,\omega_r]}(\omega)} \underbrace{\int_{-\infty}^{\infty} dt'\, e^{i\omega t'} f(t')}_{=\tilde{f}(\omega)} . \tag{1}$$

The sum
$$S^{[\gamma,\omega_r]}(\omega) \equiv \sum_{n\in\mathbb{Z}} e^{in\tau\omega} e^{-|n|\tau\gamma} \overset{\tau=2\pi/\omega_r}{=} \sum_{n\in\mathbb{Z}} e^{i2\pi n\omega/\omega_r} e^{-2\pi|n|\gamma/\omega_r} \tag{2}$$

has the same form as a damped sum over Fourier modes,
$$S^{[\epsilon,L]}(x) \equiv \sum_{k\in\frac{2\pi}{L}\mathbb{Z}} e^{ikx-\epsilon|k|} = \sum_{n\in\mathbb{Z}} e^{i2\pi nx/L} e^{-2\pi|n|\epsilon/L} . \tag{3}$$

The latter can be summed using geometric series in the variables $e^{-2\pi(\epsilon\mp ix)/L}$ ($\to$ C6.1.6):
$$S^{[\epsilon,L]}(x) = \frac{1 - e^{-4\pi\epsilon/L}}{1 + e^{-4\pi\epsilon/L} - 2e^{-2\pi\epsilon/L}\cos(2\pi x/L)} \simeq L \sum_{m\in\mathbb{Z}} \delta_{\text{LP}}^{[\epsilon]}(x - mL) . \tag{4}$$

The result is a periodic sequence of peaks at the positions $x \simeq mL$, each with the form of a Lorentzian peak (LP), $\delta_{\text{LP}}^{[\epsilon]}(x) = \frac{\epsilon/\pi}{x^2+\epsilon^2}$ for $x, \epsilon \ll L$. Using the association $x \mapsto \omega$, $\epsilon \mapsto \gamma$ and $L \mapsto \omega_r$ we obtain:
$$S^{[\gamma,\omega_r]}(\omega) \overset{(2,4)}{=} \omega_r \sum_{m\in\mathbb{Z}} \delta_{\text{LP}}^{[\gamma]}(\omega - m\omega_r) \tag{5}$$

and
$$\tilde{p}_\gamma(\omega) \overset{(1,5)}{=} \boxed{\omega_r \sum_{m\in\mathbb{Z}} \delta_{\text{LP}}^{[\gamma]}(\omega - \omega_m)\tilde{f}(\omega)} .$$

Thus the spectrum of a series of periodic pulses, truncated beyond $|n| \lesssim 1/(\tau\gamma)$, corresponds to a frequency comb with Lorentz-broadened peaks as teeth, each with width $\simeq \gamma$.

S.C7 Differential equations (p. 300)

S.C7.2 Separable differential equations (p. 302)

C7.2.1 Separation of variables (p. 304)

(a) The autonomous differential equation $\dot{x} = x^2$ can be solved by separation of variables and subsequent integration.

$$\frac{dx}{dt} = x^2 \quad\Rightarrow\quad \int_{x(t_0)}^{x(t)} \frac{d\tilde{x}}{\tilde{x}^2} = \int_{t_0}^{t} d\tilde{t} \quad\Rightarrow\quad -\frac{1}{x(t)} + \frac{1}{x(t_0)} = t - t_0 .$$

Initial condition (i) is $x(0) = 1$: $\Rightarrow$ $-\dfrac{1}{x(t)} + 1 = t$ $\Rightarrow$ $\boxed{x(t) = \dfrac{1}{1-t}}$,

where the solution is defined on the interval $(-\infty, 1)$.

Initial condition (ii) is $x(2) = -1$: $\Rightarrow$ $-\dfrac{1}{x(t)} - 1 = t - 2$ $\Rightarrow$ $\boxed{x(t) = \dfrac{1}{1-t}}$,

where the solution is defined on the interval $(1, \infty)$.

(b) Graphical analysis of the equation $\dot{x} = x^2$.

(i) For all $x \neq 0$, $x^2 > 0$ and also $\dot{x} > 0$, i.e the curve increases monotonically. (ii) For $x = 1$, we have $\dot{x} = 1$, which fixes the slope at $x = 1$. For $x \to \pm\infty$, we have $\dot{x} \to \infty$. This suggests that there is a value for t where the curve diverges. According to the solution to the differential equation, this happens at $t = 1$.

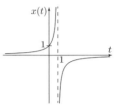

C7.2.3 Separation of variables: barometric formula (p. 304)

With a linear temperature gradient, $T(x) = T_0 - b(x - x_0)$, separation of variables yields:

$$\frac{dp(x)}{dx} = -\alpha \frac{p(x)}{T(x)} = -\alpha \frac{p(x)}{T_0 + bx_0 - bx} \quad \Rightarrow \quad \int_{p_0}^{p} \frac{d\tilde{p}}{\tilde{p}} = -\alpha \int_{x_0}^{x} d\tilde{x} \frac{1}{T_0 + bx_0 - b\tilde{x}}$$

$$\Rightarrow \quad \ln \frac{p(x)}{p_0} = \frac{\alpha}{b} \ln \frac{T_0 + bx_0 - bx}{T_0} = \frac{\alpha}{b} \ln\left(\frac{T(x)}{T_0}\right) \quad \Rightarrow \quad \boxed{\frac{p(x)}{p_0} = \left(\frac{T(x)}{T_0}\right)^{\frac{\alpha}{b}}}.$$

This result is the so-called barometric height formula.

C7.2.5 Substitution and separation of variables (p. 304)

(a) By inserting the substitution $y = ux$ and $y' = u'x + u$ into the differential equation $y' = f(y/x)$, we obtain a separable equation,

$$u'x + u = f(u) \quad \Rightarrow \quad \boxed{\frac{du}{dx} = \frac{f(u) - u}{x}},$$

since on the right-hand side of the boxed equation the u and x dependencies factorize.

(b) The equation $xy' = 2y + x$ is not directly separable, but it can be made separable by rearranging and substituting $y = ux$:

$$y' = 2y/x + 1 = f(y/x) \overset{y=ux}{=} f(u), \quad \text{with} \quad f(u) = 2u + 1.$$

$$\frac{du}{dx} = \frac{f(u) - u}{x} = \frac{2u + 1 - u}{x} = \frac{u + 1}{x} \quad \Rightarrow \quad \int_{u_0}^{u} d\tilde{u} \frac{1}{\tilde{u} + 1} = \int_{x_0}^{x} d\tilde{x} \frac{1}{\tilde{x}}$$

$$\Rightarrow \quad \ln \frac{u + 1}{u_0 + 1} = \ln \frac{x}{x_0} \quad \Rightarrow \quad \frac{u + 1}{u_0 + 1} = \frac{x}{x_0}.$$

Initial condition: $y(x_0 = 1) = 0$ $\Rightarrow$ $u_0 = u(x_0 = 1) = 0$ $\Rightarrow$ $u + 1 = x$.
Solution: $u(x) = x - 1$ $\Rightarrow$ $\boxed{y(x) = ux = (x - 1)x}$.   Initial condition: $y(1) = 0$. ✓

S.C7.3   Linear first-order differential equations (p. 306)

ᴇC7.3.₁ Inhomogeneous linear differential equation: variation of constant (p. 307)

(a)   The general solution to the DEQ $\dot{x}_h(t) + 2x_h(t) = 0$ is: $\boxed{x_h(t) = x_h(0)e^{-2t}}$.
      (We find this by recognizing the integral, or by separating variables.)

(b)   Variation of constants: ansatz: $x_p(t) = \tilde{c}(t)x_h(t) = c(t)e^{-2t}$.
      Insert this in the DEQ with $t_0 = 0$, $c(0) = 0$:

$$t = \dot{x}_p(t) + 2x_p(t) = [\dot{c}(t) - 2c(t) + 2c(t)]\,e^{-2t} = \dot{c}(t)e^{-2t} \;\Rightarrow\; \dot{c}(t) = te^{2t}$$

$$\Rightarrow \quad c(t) = \int_0^t d\tilde{t}\,\tilde{t}e^{2\tilde{t}} \overset{\text{P.I.}}{=} \tfrac{1}{2}\tilde{t}e^{2\tilde{t}}\Big|_0^t - \int_0^t d\tilde{t}\,\tfrac{1}{2}e^{2\tilde{t}} = \tfrac{1}{2}te^{2t} - \tfrac{1}{4}e^{2\tilde{t}}\Big|_0^t = \tfrac{1}{2}te^{2t} - \tfrac{1}{4}e^{2t} + \tfrac{1}{4}$$

$$\Rightarrow \quad x_p(t) = c(t)e^{-2t} = \boxed{\tfrac{1}{2}t - \tfrac{1}{4} + \tfrac{1}{4}e^{-2t}}\,.$$

Initial condition is $x(0) = 0 \quad\Rightarrow\quad x_h(0) = 0$:

$$x(t) = x_h(t) + x_p(t) = x_h(0)e^{-2t} + \tfrac{1}{2}t - \tfrac{1}{4} + \tfrac{1}{4}e^{-2t}, \quad\Rightarrow\quad \boxed{x(t) = \tfrac{1}{2}t - \tfrac{1}{4} + \tfrac{1}{4}e^{-2t}}\,.$$

S.C7.4   Systems of linear first-order differential equations (p. 308)

ᴇC7.4.₁ Linear homogeneous differential equation with constant coefficients (p. 309)

Differential equation:
$$\dot{\mathbf{x}}(t) = A \cdot \mathbf{x}(t)\,, \quad \text{with} \quad A = \frac{1}{5}\begin{pmatrix} 3 & -4 \\ -4 & -3 \end{pmatrix}.$$

General exponential ansatz:
$$\mathbf{x}(t) = \sum_j \mathbf{v}_j e^{\lambda_j t} c^j\,, \quad \text{with} \quad A\mathbf{v}_j = \lambda_j \mathbf{v}_j\,.$$

Characteristic polynomial:
$$0 = \det(A - \lambda\mathbb{1}) = \tfrac{1}{25}\left[(3 - 5\lambda)(-3 - 5\lambda) - 16\right]$$
$$= \tfrac{1}{25}(\lambda^2 - 1)$$

Eigenvalues:
$$\lambda^2 = 1 \quad\Rightarrow\quad \lambda_\pm = \pm 1.$$

Checks:
$$\lambda_+ + \lambda_- = 0 \overset{\checkmark}{=} \operatorname{Tr}A\,,$$
$$\lambda_+\lambda_- = -1 \overset{\checkmark}{=} \det A = \left(\tfrac{1}{5}\right)^2 (3 \cdot (-3) - 4 \cdot 4)\,.$$

Eigenvectors:
$\lambda_+ = 1$:
$$\mathbf{0} = (A - \lambda_+\mathbb{1})\mathbf{v}_+ = \tfrac{1}{5}\begin{pmatrix} -2 & -4 \\ -4 & -8 \end{pmatrix}\mathbf{v}_+ \qquad \Rightarrow \mathbf{v}_+ = \frac{1}{\sqrt{5}}\begin{pmatrix} 2 \\ -1 \end{pmatrix},$$

$\lambda_- = -1$:
$$\mathbf{0} = (A - \lambda_-\mathbb{1})\mathbf{v}_- = \tfrac{1}{5}\begin{pmatrix} 8 & -4 \\ -4 & 2 \end{pmatrix}\mathbf{v}_- \qquad \Rightarrow \mathbf{v}_- = \frac{1}{\sqrt{5}}\begin{pmatrix} 1 \\ 2 \end{pmatrix}.$$

Similarity transformation for the diagonalization:
$$T = (\mathbf{v}_+, \mathbf{v}_-) = \frac{1}{\sqrt{5}}\begin{pmatrix} 2 & 1 \\ -1 & 2 \end{pmatrix}, \quad T^{-1} = T^{\mathrm{T}} = \frac{1}{\sqrt{5}}\begin{pmatrix} 2 & -1 \\ 1 & 2 \end{pmatrix}.$$

Determination of the initial condition:
$$\mathbf{x}(0) = \sum_i \mathbf{v}_i c^i \quad\Rightarrow\quad \mathbf{c} = T^{\mathrm{T}} \cdot \mathbf{x}(0).$$

[Explanation:
$$x^j(0) = v^j{}_i c^i = T^j{}_i c^i \quad\Rightarrow\quad c^i = (T^{-1})^i{}_j x^j(0)$$

$$\mathbf{c} = T^{-1}\mathbf{x}(0) = \frac{1}{\sqrt{5}}\begin{pmatrix} 2 & -1 \\ 1 & 2 \end{pmatrix}\begin{pmatrix} 2 \\ 1 \end{pmatrix} = \frac{1}{\sqrt{5}}\begin{pmatrix} 3 \\ 4 \end{pmatrix} \equiv \begin{pmatrix} c^+ \\ c^- \end{pmatrix}.\Bigg]$$

Solution: $\mathbf{x}(t) = \mathbf{v}_+ e^{\lambda_+ t} c^+ + \mathbf{v}_- e^{\lambda_- t} c^- = \frac{1}{\sqrt{5}}\begin{pmatrix} 2 \\ -1 \end{pmatrix} \cdot \frac{3}{\sqrt{5}} e^{1 \cdot t} + \frac{1}{\sqrt{5}}\begin{pmatrix} 1 \\ 2 \end{pmatrix} \cdot \frac{4}{\sqrt{5}} e^{-1 \cdot t}$

$$= \boxed{\frac{3}{5}\begin{pmatrix} 2 \\ -1 \end{pmatrix} e^t + \frac{4}{5}\begin{pmatrix} 1 \\ 2 \end{pmatrix} e^{-t}}.$$

Check: $\mathbf{x}(0) = \frac{3}{5}\begin{pmatrix} 2 \\ -1 \end{pmatrix} + \frac{4}{5}\begin{pmatrix} 1 \\ 2 \end{pmatrix} = \frac{1}{5}\begin{pmatrix} 10 \\ 5 \end{pmatrix} \overset{\checkmark}{=} \begin{pmatrix} 2 \\ 1 \end{pmatrix}.$

Check: $\dot{\mathbf{x}} \overset{?}{=} A\mathbf{x}$

$$\dot{\mathbf{x}} = \frac{3}{5}\begin{pmatrix} 2 \\ -1 \end{pmatrix} e^t - \frac{4}{5}\begin{pmatrix} 1 \\ 2 \end{pmatrix} e^{-t}$$

$$A \cdot \mathbf{x} = \frac{1}{5}\begin{pmatrix} 3 & -4 \\ -4 & -3 \end{pmatrix}\left[\frac{3}{5}\begin{pmatrix} 2 \\ -1 \end{pmatrix} e^t + \frac{4}{5}\begin{pmatrix} 1 \\ 2 \end{pmatrix} e^{-t}\right]$$

$$= \frac{1}{5} \cdot \frac{3}{5}\begin{pmatrix} 10 \\ -5 \end{pmatrix} e^t + \frac{1}{5} \cdot \frac{4}{5}\begin{pmatrix} -5 \\ -10 \end{pmatrix} e^{-t} = \frac{3}{5}\begin{pmatrix} 2 \\ -1 \end{pmatrix} e^t - \frac{4}{5}\begin{pmatrix} 1 \\ 2 \end{pmatrix} e^{-t} \overset{\checkmark}{=} \dot{\mathbf{x}}.$$

εC7.4.3 **System of linear differential equations with non-diagonizable matrix** (p. 312)

(a) Characteristic polynomial:

$$0 \overset{!}{=} \det(A - \lambda I) = (\tfrac{1}{3})^3 \begin{vmatrix} 7-3\lambda & 2 & 0 \\ 0 & 4-3\lambda & -1 \\ 2 & 0 & 4-3\lambda \end{vmatrix} = (\tfrac{1}{3})^3\left[(7-3\lambda)(4-3\lambda)^2 - 4\right]$$

$$= -\lambda^3 + 5\lambda^2 - 8\lambda + 4 = -(\lambda-1)(\lambda-2)^2.$$

Therefore there is a simple zero, $\lambda_1 = \boxed{1}$, and a double zero, $\lambda_2 = \lambda_3 = \boxed{2}$.

Checks: $\lambda_1 + \lambda_2 + \lambda_3 = 5 \overset{\checkmark}{=} \operatorname{Tr} A = \tfrac{1}{3}(7+4+4),$

$$\lambda_1 \lambda_2 \lambda_3 = 4 \overset{\checkmark}{=} \det A = \left(\tfrac{1}{3}\right)^3 (7 \cdot 4 \cdot 4 + 2 \cdot (-2)).$$

(b) We begin by finding the eigenvector $\mathbf{v}_1$ for λ_1.

$$\lambda_1 = 1: \qquad \mathbf{0} \overset{!}{=} (A - \lambda_1 \mathbb{1})\mathbf{v}_1 = \frac{1}{3}\begin{pmatrix} 4 & 2 & 0 \\ 0 & 1 & -1 \\ 2 & 0 & 1 \end{pmatrix}\mathbf{v}_1, \qquad \Rightarrow \qquad \mathbf{v}_1 = \boxed{\frac{1}{3}\begin{pmatrix} -1 \\ 2 \\ 2 \end{pmatrix}}.$$

The eigenvector $\mathbf{v}_1$ can be written down by inspection. It is nevertheless instructive to also determine it using Gaussian elimination:

[1] :	4	2	0		0		
[2] :	0	1	−1		0		
[3] :	2	0	1		0		

$\longrightarrow$

[1] :	4	2	0		0
[2] :	0	1	−1		0
[1] − 2([2] + [3]) :	0	0	0		0

$\checkmark$

[1] :	4	2	0		0
[2] :	0	1	−1		0
[3] :	0	0	1		α

$\longrightarrow$

$\frac{1}{4}$[1] − $\frac{1}{2}$([2] + [3]) :	1	0	0		$-\frac{1}{2}\alpha$
[2] + [3] :	0	1	0		α
[3] :	0	0	1		α

Since $(A - \lambda_1 \mathbb{1}) = 0$, the rows of the extended matrix are not linearly independent, hence the second system yields a row that contains only zeros. Thus $\mathbf{v}_1$ involves one free parameter, which we chose as $v^3_{\,1} = \alpha$ in the third system. By now taking, for example, $\alpha = \frac{2}{3}$, we obtain $\mathbf{v}_1 = \frac{1}{3}(-1, 2, 2)^T$.

Next we consider the eigenspace of λ_2.

$$\lambda_2 = 2: \qquad \mathbf{0} \overset{!}{=} (A - \lambda_2 \mathbb{1})\mathbf{v}_2 = \frac{1}{3}\begin{pmatrix} 1 & 2 & 0 \\ 0 & -2 & -1 \\ 2 & 0 & -2 \end{pmatrix}\mathbf{v}_2, \qquad \Rightarrow \qquad \mathbf{v}_2 = \boxed{\frac{1}{3}\begin{pmatrix} 2 \\ -1 \\ 2 \end{pmatrix}}.$$

$\mathbf{v}_2$, too, can be written down by inspection. But since λ_2 is a two-fold zero, the question immediately arises whether there exists another eigenvector associated with λ_2, linearly independent of $\mathbf{v}_2$. To clarify this, we solve the above system by Gaussian elimination:

[1]:	1	2	0	|	0		[1]:	1	2	0	|	0
[2]:	0	−2	−1	|	0		−[2]:	0	2	1	|	0
[3]:	2	0	−2	|	0		2([1]+[2])−[3]:	0	0	0	|	0

$\longrightarrow$

[1]:	1	2	0	|	0		[1]−[2]+[3]:	1	0	0	|	α
[2]:	0	2	1	|	0		$\frac{1}{2}$([2]−[3]):	0	1	0	|	$-\frac{1}{2}\alpha$
[3]:	0	0	1	|	α		[3]:	0	0	1	|	α

$\longrightarrow$

Since $(A - \lambda_2 \mathbb{1}) = 0$, the rows of the extended matrix are not linearly independent. But although λ_2 is a *two*-fold zero of the characteristic polynomial, Gaussian elimination here yields only *one* row containing purely zeros, hence $\mathbf{v}_2$ involves only one free parameter. (In the third system we chose it as $v^3_{\,2} = \alpha$, and in the end used $\alpha = \frac{2}{3}$ to obtain $\mathbf{v}_2 = \frac{1}{3}(2, -1, 2)^T$.) Therefore the degenerate eigenvalue λ_2 has only *one* eigenvector, i.e. the eigenspace of λ_2 is only one-dimensional, just as the eigenspace of λ_1. Hence the matrix A is not diagonalizable. (For a diagonalizable matrix a two-fold zero would yield an extended matrix containing *two* rows consisting purely of zeros. The solution would then involve *two* independent parameters, so that it would be possible to construct *two* linearly independent eigenvectors, $\mathbf{v}_2$ and $\mathbf{v}_3$, both with eigenvalue λ_2.)

(c) To determine $\mathbf{v}_3$, we use Gaussian elimination to solve the equation $(A - \lambda_2 \mathbb{1})\mathbf{v}_3 = \mathbf{v}_2$:

[1]:	1	2	0	|	2		[1]:	1	2	0	|	2
[2]:	0	−2	−1	|	−1		−[2]:	0	2	1	|	1
[3]:	2	0	−2	|	2		2[1]+2[2]−[3]:	0	0	0	|	0

$\longrightarrow$

[1]:	1	2	0	|	2		[1]−2[2]+[3]:	1	0	0	|	$1+\alpha$
[2]:	0	2	1	|	1		$\frac{1}{2}$([2]−[3]):	0	1	0	|	$\frac{1}{2}(1-\alpha)$
[3]:	0	0	1	|	α		[3]:	0	0	1	|	α

$\longrightarrow$

The second system contains a row consisting purely of zeros, thus the solution has a free parameter; in the third system we choose it as $v^3_{\,3} = \alpha$. Setting, for example, $\alpha = -\frac{1}{3}$, we obtain $\mathbf{v}_3 = \frac{1}{3}(2, 2, -1)^T$. This choice is particularly convenient, since then $\mathbf{v}_1$, $\mathbf{v}_2$ and $\mathbf{v}_3$ form an orthonormal system. It yields the following similarity transformation T:

$$\boxed{T = (\mathbf{v}_1, \mathbf{v}_2, \mathbf{v}_3) = \frac{1}{3}\begin{pmatrix} -1 & 2 & 2 \\ 2 & -1 & 2 \\ 2 & 2 & -1 \end{pmatrix} = T^T = T^{-1}}.$$

(d) For the coefficient $\mathbf{c}(0)$ we obtain:

$$\mathbf{c}(0) = T^{-1}\mathbf{x}(0) = \frac{1}{3}\begin{pmatrix} -1 & 2 & 2 \\ 2 & -1 & 2 \\ 2 & 2 & -1 \end{pmatrix}\begin{pmatrix} 1 \\ 1 \\ 1 \end{pmatrix} = \boxed{\begin{pmatrix} 1 \\ 1 \\ 1 \end{pmatrix}}.$$

Thus the solution of the differential equation is:

$$\mathbf{x}(t) = c^1(0)e^t\mathbf{v}_1 + \left(c^2(0) + tc^3(0)\right)e^{2t}\mathbf{v}_2 + c^3(0)e^{2t}\mathbf{v}_3$$

$$= \boxed{\frac{1}{3}e^t\begin{pmatrix} -1 \\ 2 \\ 2 \end{pmatrix} + \frac{1}{3}(1+t)e^{2t}\begin{pmatrix} 2 \\ -1 \\ 2 \end{pmatrix} + \frac{1}{3}e^{2t}\begin{pmatrix} 2 \\ 2 \\ -1 \end{pmatrix}}.$$

(e) Explicit check:

$$\dot{\mathbf{x}}(t) = \frac{1}{3}e^t\begin{pmatrix} -1 \\ 2 \\ 2 \end{pmatrix} + \frac{2}{3}(1+t)e^{2t}\begin{pmatrix} 2 \\ -1 \\ 2 \end{pmatrix} + \frac{1}{3}e^{2t}\begin{pmatrix} 2 \\ -1 \\ 2 \end{pmatrix} + \frac{2}{3}e^{2t}\begin{pmatrix} 2 \\ 2 \\ -1 \end{pmatrix}$$

$$= \frac{1}{3}e^t\begin{pmatrix} -1 \\ 2 \\ 2 \end{pmatrix} + \frac{2}{3}(1+t)e^{2t}\begin{pmatrix} 2 \\ -1 \\ 2 \end{pmatrix} + e^{2t}\begin{pmatrix} 2 \\ 1 \\ 0 \end{pmatrix}.$$

On the other hand:

$$A\mathbf{x}(t) = \frac{1}{3}e^t\begin{pmatrix} -1 \\ 2 \\ 2 \end{pmatrix} + \frac{2}{3}(1+t)e^{2t}\begin{pmatrix} 2 \\ -1 \\ 2 \end{pmatrix} + \frac{1}{9}e^{2t}\begin{pmatrix} 7 & 2 & 0 \\ 0 & 4 & -1 \\ 2 & 0 & 4 \end{pmatrix}\begin{pmatrix} 2 \\ 2 \\ -1 \end{pmatrix}$$

$$= \frac{1}{3}e^t\begin{pmatrix} -1 \\ 2 \\ 2 \end{pmatrix} + \frac{2}{3}(1+t)e^{2t}\begin{pmatrix} 2 \\ -1 \\ 2 \end{pmatrix} + e^{2t}\begin{pmatrix} 2 \\ 1 \\ 0 \end{pmatrix} \overset{\checkmark}{=} \dot{\mathbf{x}}(t).$$

ɛC7.4.5 **Coupled oscillations of two point masses** (p. 313)

(a) Equations of motion:

in matrix form:
$$\begin{pmatrix} \ddot{x}^1 \\ \ddot{x}^2 \end{pmatrix} = -\underbrace{\begin{pmatrix} \dfrac{K_1+K_{12}}{m_1} & -\dfrac{K_{12}}{m_1} \\ -\dfrac{K_{12}}{m_2} & \dfrac{K_2+K_{12}}{m_2} \end{pmatrix}}_{\equiv A}\begin{pmatrix} x^1 \\ x^2 \end{pmatrix},$$

Compact notation: $\ddot{\mathbf{x}} = -A\mathbf{x}$.   (1)

(b) Transformation into an eigenvalue problem:

Ansatz for solution: $\mathbf{x}(t) = \mathbf{v}\cos(\omega t).$ (2)

Differentiating the ansatz twice: $\ddot{\mathbf{x}}(t) = -\omega^2\mathbf{v}\cos(\omega t).$ (3)

Inserting (2), (3) in (1): $-\omega^2\mathbf{v}\cos(\omega t) = -A\mathbf{v}\cos(\omega t).$

Eigenvalue equation: $A\mathbf{v} = \omega^2\mathbf{v}.$ (4)

(c) For $m_1 = m_2$, $K_2 = m_1\Omega^2$, $K_1 = 4K_2$ and $K_{12} = 2K_2$, we have: $\boxed{A = \Omega^2\begin{pmatrix} 6 & -2 \\ -2 & 3 \end{pmatrix}}$

$\dfrac{1}{\Omega^2}\cdot$[eigenvalue equation (4)]: $\dfrac{1}{\Omega^2}A\mathbf{v} \overset{(4)}{=} \dfrac{\omega^2}{\Omega^2}\mathbf{v} = \lambda\mathbf{v},$ with $\lambda \equiv (\omega/\Omega)^2.$ (5)

Determination of the eigenvalues λ_j of the matrix $\frac{1}{\Omega^2}A$:

Characteristic polynomial:
$$0 \overset{!}{=} \det\begin{pmatrix} 6-\lambda & -2 \\ -2 & 3-\lambda \end{pmatrix} = (6-\lambda)(3-\lambda) - 4$$
$$= \lambda^2 - 9\lambda + 14 = (\lambda - 7)(\lambda - 2).$$

Eigenvalues:
$$\lambda_1 = \boxed{2}, \quad \lambda_2 = \boxed{7}. \tag{6}$$

Eigenvectors $\mathbf{v}_j$: $(A - \lambda_j \mathbb{1})\,\mathbf{v}_j = \mathbf{0}$.

$\lambda_1 = 2$: $\begin{pmatrix} 4 & -2 \\ -2 & 1 \end{pmatrix}\mathbf{v}_1 = \mathbf{0}$ $\Rightarrow$ $\mathbf{v}_1 = \boxed{\dfrac{1}{\sqrt{5}}\begin{pmatrix} 1 \\ 2 \end{pmatrix}}$. $\tag{7}$

$\lambda_2 = 7$: $\begin{pmatrix} -1 & -2 \\ -2 & -4 \end{pmatrix}\mathbf{v}_2 = \mathbf{0}$ $\Rightarrow$ $\mathbf{v}_2 = \boxed{\dfrac{1}{\sqrt{5}}\begin{pmatrix} 2 \\ -1 \end{pmatrix}}$. $\tag{8}$

Eigenfrequency $\omega_j \overset{(5)}{=} \sqrt{\lambda_j}\Omega$: $\omega_1 \overset{(6)}{=} \boxed{\sqrt{2}\Omega}, \quad \omega_2 \overset{(6)}{=} \boxed{\sqrt{7}\Omega}.$

Same-phase eigenmode: $\mathbf{x}_1(t) \overset{(2)}{=} \mathbf{v}_1 \cos(\omega_1 t) \overset{(7)}{=} \boxed{\dfrac{1}{\sqrt{5}}\begin{pmatrix} 1 \\ 2 \end{pmatrix}\cos(\sqrt{2}\Omega t)}.$

Out-of-phase eigenmode: $\mathbf{x}_2(t) \overset{(2)}{=} \mathbf{v}_2 \cos(\omega_2 t) \overset{(8)}{=} \boxed{\dfrac{1}{\sqrt{5}}\begin{pmatrix} 2 \\ -1 \end{pmatrix}\cos(\sqrt{7}\Omega t)}.$

(d)   For the eigenmode $\mathbf{x}_1(t)$ (left sketch), both the masses swing in phase, and for $\mathbf{x}_2(t)$ (right sketch), both the masses oscillate out of phase. The latter requires stronger expansion and compression of the springs. Therefore the out-of-phase mode costs more energy and thus has a higher frequency than the same-phase mode. The schematic sketch below illustrates the positions of the point masses at time $t = 0$ and the thick arrow illustrates their velocities a small time (e.g. a quarter period) later.

Remark: Both the masses at time $t = 0$ are displaced by $x^1{}_0$ and $x^2{}_0$ respectively, and the subsequent oscillation is a superposition of both the eigenmodes, $\mathbf{x}(t) = \sum_j c^j \mathbf{x}_j(t)$, whose coefficients c^j are fixed by the initial displacement $\mathbf{x}_0 = (x^1{}_0, x^2{}_0)^T$, with $\mathbf{x}_0 = \sum_j c^j \mathbf{v}_j$.

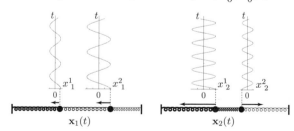

$\mathbf{x}_1(t)$ $\mathbf{x}_2(t)$

ɛC7.4.7 Inhomogeneous linear differential equation of second order: driven overdamped harmonic oscillator (p. 313)

(a)   Simplification of matrix equation:
$$\ddot{x} + 2\gamma\dot{x} + \Omega^2 x = f_A(t), \quad x(0) = 0, \quad \dot{x}(0) = 1, \tag{1}$$

can be written as a first-order matrix DEQ, using $\mathbf{x} \equiv (x, \dot{x})^T = (x^1, x^2)^T$ and $\ddot{x} = \dot{x}^2$.

New variables: $\dot{x} = \dot{x}^1 = x^2, \quad \ddot{x} = \dot{x}^2 = -\Omega^2 x^1 - 2\gamma x^2 + f_A(t).$

Matrix form:
$$\underbrace{\begin{pmatrix} \dot{x}^1 \\ \dot{x}^2 \end{pmatrix}}_{\dot{\mathbf{x}}} = \underbrace{\begin{pmatrix} 0 & 1 \\ -\Omega^2 & -2\gamma \end{pmatrix}}_{A} \underbrace{\begin{pmatrix} x^1 \\ x^2 \end{pmatrix}}_{\mathbf{x}} + \underbrace{f_A(t) \begin{pmatrix} 0 \\ 1 \end{pmatrix}}_{\mathbf{b}(t)}.$$

Compact notation: $\dot{\mathbf{x}} = A \cdot \mathbf{x} + \mathbf{b}(t)$. $\hspace{3cm}$ (2)

Initial values: $\mathbf{x}_0 = \mathbf{x}(0) = (x(0), \dot{x}(0))^\mathsf{T} = (0, 1)^\mathsf{T}$.

(b) Homogeneous solution.
We first determine the eigenvalues λ_j and eigenvectors $\mathbf{v}_j$ ($j = +, -$) of A:

$$0 \overset{!}{=} \det(A - \lambda \mathbb{1}) = \begin{vmatrix} -\lambda & 1 \\ -\Omega^2 & -2\gamma - \lambda \end{vmatrix} = \lambda(2\gamma + \lambda) + \Omega^2 = \lambda^2 + 2\gamma\lambda + \Omega^2. \hspace{1cm} (3)$$

Eigenvalues (for $\gamma > \Omega$): $\boxed{\lambda_\pm = -\gamma \pm \gamma_r, \text{ with } \gamma_r \equiv \sqrt{\gamma^2 - \Omega^2} \text{ real}}$. $\hspace{1cm}$ (4)

Checks: $\lambda_+ + \lambda_- = -2\gamma \overset{\checkmark}{=} \mathrm{Tr}\,A, \quad \lambda_+\lambda_- = \gamma^2 - \gamma_r^2 = \Omega^2 \overset{\checkmark}{=} \det A$.

Eigenvectors: $\mathbf{0} = (A - \lambda_j \mathbb{1})\mathbf{v}_j = \begin{pmatrix} -\lambda_j & 1 \\ -\Omega^2 & -2\gamma - \lambda_j \end{pmatrix} \mathbf{v}_j \quad \Rightarrow \quad \boxed{\mathbf{v}_j = \begin{pmatrix} 1 \\ \lambda_j \end{pmatrix}}$.

This works since $-\Omega^2 \cdot 1 + (-2\gamma - \lambda_j)\lambda_j \overset{(3)}{=} 0$ if λ_j is an eigenvalue.

Since $\mathbf{x}_j(t) = \mathbf{v}_j e^{\lambda_j t}$ satisfies the homogeneous equation $\dot{\mathbf{x}}_j = A \cdot \mathbf{x}_j$, the first component of $\mathbf{x}_j(t)$, i.e. $x_j(t) = e^{\lambda_j t}$, fulfills the DEQ $(1)|_{f_A(t)=0}$. Check that this is the case:

$$(\mathrm{d}_t^2 + 2\gamma \mathrm{d}_t + \Omega^2)e^{\lambda_j t} = \left(\gamma_j^2 + 2\gamma\lambda_j + \Omega^2\right)e^{\lambda_j t} \overset{(3)}{=} 0. \quad \checkmark \hspace{1cm} (5)$$

The most general form of the homogeneous solution is $\mathbf{x}_h(t) = \sum_j c_h^j \mathbf{x}_j(t)$. For a given initial value $\mathbf{x}_0$, the coefficient vector $\mathbf{c}_h = (c_h^1, c_h^2)^\mathsf{T}$ is fixed by $\mathbf{x}_h(0) = \sum_j \mathbf{v}_j c_h^j = \mathbf{x}_0$, or in matrix notation, $T\mathbf{c}_h = \mathbf{x}_0$, where the matrix $T = \{v^i_{\;j}\}$ has the eigenvectors $\mathbf{v}_j$ as columns, i.e. $T = (\mathbf{v}_1, \mathbf{v}_2)$:

$$T = \begin{pmatrix} 1 & 1 \\ \lambda_+ & \lambda_1 \end{pmatrix}, \quad T^{-1} = \frac{1}{\lambda_- - \lambda_+}\begin{pmatrix} \lambda_- & -1 \\ -\lambda_+ & 1 \end{pmatrix} \overset{(4)}{=} -\frac{1}{2\gamma_r}\begin{pmatrix} \lambda_- & -1 \\ -\lambda_+ & 1 \end{pmatrix}, \hspace{1cm} (6)$$

$$\mathbf{c}_h = T^{-1}\mathbf{x}_0 \Rightarrow \begin{pmatrix} c_h^+ \\ c_h^- \end{pmatrix} = -\frac{1}{2\gamma_r}\begin{pmatrix} \lambda_- & -1 \\ -\lambda_+ & 1 \end{pmatrix}\begin{pmatrix} 0 \\ 1 \end{pmatrix} = \frac{1}{2\gamma_r}\begin{pmatrix} 1 \\ -1 \end{pmatrix} \Rightarrow \boxed{c_h^\pm = \pm\frac{1}{2\gamma_r}}.$$

The homogeneous solution of the matrix DEQ, $(2)|_{\mathbf{b}(t)=0}$, is thus

$$\mathbf{x}_h(t) = \sum_j c_h^j \mathbf{x}_j(t) = \frac{1}{2\gamma_r}\left[e^{\lambda_+ t}\begin{pmatrix} 1 \\ \lambda_+ \end{pmatrix} - e^{\lambda_- t}\begin{pmatrix} 1 \\ \lambda_- \end{pmatrix}\right],$$

and the homogeneous solution of the initial second-order DEQ, $(1)|_{f_A(t)=0}$, is

$$x_h(t) = x_h^1(t) = \boxed{\frac{1}{2\gamma_r}\left[e^{\lambda_+ t} - e^{\lambda_- t}\right]} = \frac{e^{-\gamma t}}{\gamma_r}\sinh(\gamma_r t).$$

Check that $x_h(t)$ has the required properties (not really necessary, since all relevant properties have already been checked above, but nevertheless instructive):

$$(d_t^2 + 2\gamma d_t + \Omega^2)x_h = \frac{1}{2\gamma_r}\left[\left(\gamma_+^2 + 2\gamma\gamma_+ + \Omega^2\right)e^{\gamma_+ t} - \left(\gamma_-^2 + 2\gamma\gamma_- + \Omega^2\right)e^{\gamma_- t}\right] \overset{(3)}{=} 0.\ \checkmark$$

Initial value: $x_h(0) = 0,\quad \dot{x}_h(0) = 1.\ \checkmark$ (7)

(c)   Particular solution: The method of variation of constants looks for a particular solution of the matrix DEQ (2) of the form $\mathbf{x}_p(t) = \sum_j c_p^j(t)\mathbf{x}_j(t)$, with $c_p^j(t)$ chosen such that

$$\sum_j \dot{c}_p^j(t)\mathbf{x}_j(t) = \mathbf{b}(t). \tag{8}$$

A solution to (8), with $c^j(0) = 0$ (and therefore $\mathbf{x}_p(0) = \mathbf{0}$), is given by

$$c_p^j(t) = \int_0^t d\tilde{t}\, \tilde{b}^j(\tilde{t})\, e^{-\lambda_j \tilde{t}}, \tag{9}$$

where the $\tilde{b}^j(t)$ originate from the decomposition of $\mathbf{b}(t) = \sum_j \mathbf{v}_j \tilde{b}^j(t)$ into eigenvectors. In components, $b^i(t) = v^i_{\ j}\tilde{b}^j(t)$, and in matrix notation, $\mathbf{b}(t) = T\tilde{\mathbf{b}}(t)$, $\tilde{\mathbf{b}}(t) = T^{-1}\mathbf{b}(t)$:

$$\begin{pmatrix}\tilde{b}^+(t) \\ \tilde{b}^-(t)\end{pmatrix} \overset{(6)}{=} -\frac{1}{2\gamma_r}\begin{pmatrix}\lambda_- & -1 \\ -\lambda_+ & 1\end{pmatrix}f_A(t)\begin{pmatrix}0 \\ 1\end{pmatrix} = \frac{f_A(t)}{2\gamma_r}\begin{pmatrix}1 \\ -1\end{pmatrix} \Rightarrow \boxed{\tilde{b}^\pm(t) = \pm\frac{f_A(t)}{2\gamma_r}}.$$

For the given driving function, we have $f_A(t) = f_A$ for $t \geq 0$, and therefore we obtain:

$$c_p^\pm(t) \overset{(9)}{=} \pm\frac{f_A}{2\gamma_r}\int_0^t d\tilde{t}\, e^{-\lambda_\pm \tilde{t}} = \boxed{\pm\frac{f_A}{2\gamma_r\lambda_\pm}\left[1 - e^{-\lambda_\pm t}\right]}.$$

Check initial value: $c_p^j(0) \overset{\checkmark}{=} 0$. Check that (8) holds:

$$\sum_j \dot{c}_p^j(t)\mathbf{x}_j(t) \overset{(8)?}{=} \frac{f_A}{2\gamma_r}\left[\frac{\lambda_+ e^{-\lambda_+ t}}{\lambda_+}\begin{pmatrix}1 \\ \lambda_+\end{pmatrix}e^{\lambda_+ t} - \frac{\lambda_- e^{-\lambda_- t}}{\lambda_-}\begin{pmatrix}1 \\ \lambda_-\end{pmatrix}e^{\lambda_- t}\right] = f_A\begin{pmatrix}0 \\ 1\end{pmatrix} \overset{\checkmark}{=} \mathbf{b}(t).$$

The desired particular solution for $t > 0$ is therefore given by:

$$\mathbf{x}_p(t) = \sum_j c_p^j(t)\mathbf{v}_j e^{\lambda_j t} = \frac{f_A}{2\gamma_r}\left[\frac{e^{\lambda_+ t} - 1}{\lambda_+}\begin{pmatrix}1 \\ \lambda_+\end{pmatrix} - \frac{e^{\lambda_- t} - 1}{\lambda_-}\begin{pmatrix}1 \\ \lambda_-\end{pmatrix}\right],$$

$$x_p(t) = x_p^1(t) = \boxed{\frac{f_A}{2\gamma_r}\left[\frac{e^{\lambda_+ t} - 1}{\lambda_+} - \frac{e^{\lambda_- t} - 1}{\lambda_-}\right]}.$$

Check that $x_p(t)$ has the required properties (not really necessary, since all relevant properties have already been checked above, but nevertheless instructive):

$$(d_t^2 + 2\gamma d_t + \Omega^2)x_p(t)$$

$$= \frac{f_A}{2\gamma_r}\left[\underbrace{\left(\lambda_+^2 + 2\gamma\lambda_+ + \Omega^2\right)}_{\overset{(4)}{=}0}\frac{e^{\lambda_+ t}}{\lambda_+} - \underbrace{\left(\lambda_-^2 + 2\gamma\lambda_- + \Omega^2\right)}_{\overset{(4)}{=}0}\frac{e^{\lambda_- t}}{\lambda_-} - \Omega^2\left(\frac{1}{\lambda_+} - \frac{1}{\lambda_-}\right)\right]$$

$$= -\Omega^2\frac{f_A}{2\gamma_r}\left[\frac{\lambda_- - \lambda_+}{\lambda_-\lambda_+}\right] \overset{(4)}{=} \Omega^2\frac{f_A}{2\gamma_r}\frac{2\gamma_r}{\Omega^2} = f_A.\ \checkmark$$

Initial value: $x_p(0) = 0,\quad \dot{x}_p(0) = 0.\ \ \checkmark$ (10)

(d) Qualitative discussion: According to (7) and (10), the solution $x(t) = x_h(t) + x_p(t)$ of the inhomogeneous DEQ (1) has the required initial values $x(0) = 0$ and $\dot{x}(0) = 1$. Since $\lambda_\pm < 0$, the long time limit $t \to \infty$ is determined by the constant contribution of $x_p(t)$ alone:

$$x_p(t) \overset{t \to \infty}{\longrightarrow} x_p(\infty) = -\frac{f_A}{2\gamma_r}\left[\frac{1}{\lambda_+} - \frac{1}{\lambda_-}\right] \overset{(4)}{=} \boxed{\frac{f_A}{\Omega^2}}.$$

For $t > 0$, the driving force $F_A = mf_A$ is time-independent. Therefore, it leads to a constant shift in the equilibrium position from 0 to $x_p(\infty) = f_A/\Omega^2$. For this shift, the restoring force of the harmonic oscillator, $F_R = -m\Omega^2 x_p(\infty)$, cancels the driving force exactly, i.e. $F_A + F_R = 0$.

Concerning the sketch, for $f_A < 0$: according to the given initial condition, $x(t)$ initially increases for small times (starting from 0), attains a maximum [at $\dot{x}_p(t) = 0$, where $e^{\lambda_+ t} = e^{\lambda_- t}$ i.e. at $t = 1/(2\gamma_r)$], and thereafter tends to the long time limit from above, $x_p(\infty) < 0$.

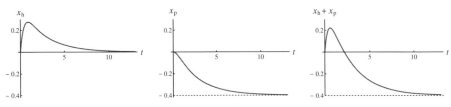

Parameters for the sketch: $\Omega = 1$, $\gamma = 1.5$, $f_A = -0.4$.

S.C7.5   Linear higher-order differential equations (p. 308)

₣C7.5.1   Green function of $(d_t + a)$ (p. 319)

(a) We first recall two properties of the δ-function. Firstly, it corresponds to the derivative of the Θ-function, $d_t\Theta(t) = \delta(t)$. Secondly, for an arbitrary function $b(t)$, the relation $\delta(t)b(t) = \delta(t)b(0)$ holds.

Now we verify the validity of the given ansatz for the Green function.

Ansatz: $\qquad\qquad\qquad\qquad\qquad\qquad G(t) = \Theta(t)x_h(t).$ (1)

Homogeneous solution satisfies $\qquad\qquad \hat{L}(t)x_h(t) = 0,$ (2)

with initial condition $\qquad\qquad\qquad\qquad x_h(0) = 1.$ (3)

Thus we have:
$$\begin{aligned}
d_t\big[\Theta(t)x_h(t)\big] &= \big[d_t\Theta(t)\big]x_h(t) + \Theta(t)d_t x_h(t) \\
&= \delta(t)x_h(t) + \Theta(t)d_t x_h(t), \\
&= \delta(t)\underbrace{x_h(0)}_{\overset{(3)}{=}1} + \Theta(t)d_t x_h(t).
\end{aligned}$$
       (4)

$$\Rightarrow \qquad \hat{L}(t)G(t) \overset{(1)}{=} (d_t + a)\big[\Theta(t)x_h(t)\big] \overset{(4)}{=} \delta(t) + \underbrace{\Theta(t)d_t x_h(t) + a\Theta(t)x_h(t)}_{\Theta(t)\underbrace{\left[\hat{L}(t)x_h(t)\right]}_{\overset{(2)}{=}0}} = \delta(t). \quad \checkmark$$

(b) The homogeneous equation $(d_t + a)x_h(t) = 0$, with initial condition $x_h(0) = 1$, has the solution $x_h = e^{-at}$. Consequently, the Green function is:

$$\boxed{G(t) \overset{(1)}{=} \Theta(t)e^{-at}}.$$
       (5)

(c)   Fourier integral: $\tilde{G}(\omega) = \int_{-\infty}^{\infty} dt\, e^{i\omega t} G(t) \overset{(5)}{=} \int_{-\infty}^{\infty} dt\, \Theta(t) e^{(i\omega-a)t} = \int_{0}^{\infty} dt\, e^{(i\omega-a)t}$

$[e^{(i\omega-a)\infty} = 0,\ \text{since } a > 0.]$ $= \dfrac{1}{i\omega - a} \left[e^{(i\omega-a)t} \right]_{0}^{\infty} = \boxed{\dfrac{1}{a - i\omega}}.$ (6)

(d)   Consistency check:

Defining Eq. for $G(t)$: $\hat{L}(t) G(t) = \delta(t),$ with $\hat{L}(t) = d_t + a.$ (7)

Fourier transform: $\tilde{L}(-i\omega)\tilde{G}(\omega) = 1$ with $\tilde{L}(-i\omega) = -i\omega + a.$ (8)

(8) solved for $\tilde{G}(\omega)$: $\tilde{G}(\omega) = \dfrac{1}{\tilde{L}(-i\omega)} = \boxed{\dfrac{1}{-i\omega + a}}.$ $[= (6)\ \checkmark].$ (9)

The step from (7) to (8) is justified as follows:

Fourier representations: $G(t) = \int_{-\infty}^{\infty} \dfrac{d\omega}{2\pi} e^{-i\omega t} \tilde{G}(\omega),$ $\delta(t) = \int_{-\infty}^{\infty} \dfrac{d\omega}{2\pi} e^{-i\omega t}.$ (10)

Inserting (10) into (7): $\int_{-\infty}^{\infty} \dfrac{d\omega}{2\pi} \underbrace{\left[\hat{L}(t) \right] e^{-i\omega t}}_{= \tilde{L}(-i\omega) e^{-i\omega t}} \tilde{G}(\omega) = \int_{-\infty}^{\infty} \dfrac{d\omega}{2\pi} e^{-i\omega t}.$

Therefore: $\tilde{L}(-i\omega)\tilde{G}(\omega) = 1$ $\left[\text{agrees with (8)}\ \checkmark \right].$

(e)   To solve the DEQ: $\hat{L}(t) x(t) = f(t),$ with $f(t) = e^{2at}.$

Solution ansatz: $x(t) = \int_{-\infty}^{\infty} du\, G(t-u) f(u) \overset{s=t-u}{=} \int_{-\infty}^{\infty} ds\, G(s) f(t-s).$ (11)

The substitution simplifies the argument of G. Using it is not essential, but advisable for the current example, since here G is a more complicated function than f.

$$x(t) \overset{(5),(11)}{=} \int_{-\infty}^{\infty} ds\, \Theta(s) e^{-as} e^{2a(t-s)} = e^{2at} \int_{0}^{\infty} ds\, e^{-3as} = \boxed{\dfrac{1}{3a} e^{2at}}.$$ (12)

The solution may be verified via explicit insertion into the differential equation:

$$\hat{L}(t) x(t) \overset{(12)}{=} (d_t + a) \dfrac{1}{3a} e^{2at} = \dfrac{1}{3a} (2a + a) e^{2at} = e^{2at}. \quad \checkmark$$

S.C7.6   General higher-order differential equations (p. 313)

ₑC7.6.1 Field lines in two dimensions (p. 324)

Along a field line, $\mathbf{r}(t)$, we have $\dot{\mathbf{r}} \parallel \mathbf{u}(\mathbf{r})$, i.e. $(\dot{x}, \dot{y})^T \parallel (-ay, x)^T.$

DEQ for field lines: $\dfrac{dy}{dx} = \dfrac{\dot{y}}{\dot{x}} = \dfrac{x}{-ay}$

Separation of variables: $\int_{y_0}^{y} d\tilde{y}(-a\tilde{y}) = \int_{x_0}^{x} d\tilde{x}\tilde{x}$

Integrate: $-\tfrac{1}{2}a\left(y^2 - y_0^2\right) = \tfrac{1}{2}\left(x^2 - x_0^2\right)$

Rearrange: $\boxed{x^2 + ay^2 = x_0^2 + ay_0^2}.$

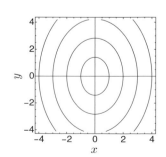

For $a > 0$, this equation describes an ellipse passing through the point $(x_0, y_0)^T$. Different choices for the latter correspond to different initial conditions for the DEQ, yielding different field lines. The sketch shows those obtained for $a = \frac{1}{2}$ by choosing $x_0 \in \{1, 2, 3, 4\}$ and $y_0 = 0$.

S.C7.7 Linearizing differential equations (p. 326)

ₑC7.7.₁ Fixed points of a differential equation in one dimension (p. 327)

Differential equation: $\dot{x} = f_\lambda(x) = (x^2 - \lambda)^2 - \lambda^2$.

(a) Fixed points: $f_\lambda(x^*) = 0 \Rightarrow (x^*)^2 = \pm\lambda + \lambda$.

    (i)   For $\lambda \le 0$ there is a single fixed point:

$$x_0^* = \boxed{0}.$$

    (ii) For $\lambda > 0$ there are three fixed points:

$$x_-^* = \boxed{-\sqrt{2\lambda}}, \quad x_0^* = \boxed{0}, \quad x_+^* = \boxed{\sqrt{2\lambda}}.$$

(c) The stability of a fixed point is determined by the sign of $\dot{x} = f_\lambda(x)$ directly to the left and right of the fixed point, i.e. at $x = x^* \mp \epsilon$ (with $\epsilon \to 0^+$).

(b)

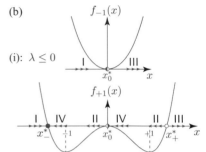

(i): $\lambda \le 0$

(ii): $\lambda > 0$

Left of x^*:   for $\dot{x} = f_\lambda(x^* - \epsilon)$ $\begin{cases} > 0,\ x(t) \text{ increases} & \Rightarrow \text{ flows towards } x^*, \quad\quad (I) \\ < 0,\ x(t) \text{ decreases} & \Rightarrow \text{ flows away from } x^*. \quad (II) \end{cases}$

Right of x^*:   for $\dot{x} = f_\lambda(x^* + \epsilon)$ $\begin{cases} > 0\ x(t),\ \text{increases} & \Rightarrow \text{ flows away from } x^*, \quad (III) \\ < 0\ x(t),\ \text{decreases} & \Rightarrow \text{ flows towards } x^*. \quad\quad (IV) \end{cases}$

Via a graphical analysis (see sketch) we find that:

(i) $\lambda \le 0$: x_0^* is $\boxed{\text{semistable}}$ (see I, III).

(ii) x_0^* $\boxed{\text{semistable}}$ (see II, IV); x_+^* $\boxed{\text{unstable}}$ (see II, III).

ₑC7.7.₃ Stability analysis in two dimensions (p. 327)

(a) Expressed in vector notation, the differential equation reads:

$$\begin{pmatrix} \dot{x} \\ \dot{y} \end{pmatrix} = \begin{pmatrix} 2x^2 - xy \\ c(1-x) \end{pmatrix} \quad \Rightarrow \quad \dot{\mathbf{x}} = \mathbf{f}(\mathbf{x}), \quad \text{with} \quad \mathbf{x} = \begin{pmatrix} x \\ y \end{pmatrix}, \quad \mathbf{f}(x) = \begin{pmatrix} 2x^2 - xy \\ c(1-x) \end{pmatrix}$$

Fixed point: $\mathbf{f}(\mathbf{x}^*) = \mathbf{0} \Rightarrow x^* = 1,\ y^* = 2x^* = 2,$ $\boxed{\mathbf{x}^* = \begin{pmatrix} 1 \\ 2 \end{pmatrix}}$.

(b) Explicit insertion of $\mathbf{x} = \mathbf{x}^* + \boldsymbol{\eta} = \begin{pmatrix} 1 + \eta^1 \\ 2 + \eta^2 \end{pmatrix}$ into the DEQ.

DEQ: $\dot{\mathbf{x}} = \dot{\boldsymbol{\eta}} = \begin{pmatrix} 2\left(1 + \eta^1\right)^2 - \left(1 + \eta^1\right)\left(2 + \eta^2\right) \\ c\left(1 - \left(1 + \eta^1\right)\right) \end{pmatrix},$

linear order: $= \begin{pmatrix} 2\eta^1 - \eta^2 \\ -c\eta^1 \end{pmatrix} = \begin{pmatrix} 2 & -1 \\ -c & 0 \end{pmatrix} \begin{pmatrix} \eta^1 \\ \eta^2 \end{pmatrix} = A\boldsymbol{\eta}.$

(c) $\left(\dfrac{\partial f^i}{\partial x^j}\right) = \begin{pmatrix} \frac{\partial f^1}{\partial x} & \frac{\partial f^1}{\partial y} \\ \frac{\partial f^2}{\partial x} & \frac{\partial f^2}{\partial y} \end{pmatrix} = \begin{pmatrix} 4x-y & -x \\ -c & 0 \end{pmatrix}$ $\Rightarrow$ $\left(\dfrac{\partial f^i}{\partial x^j}\right)\Big|_{\mathbf{x}=\mathbf{x}^*} = \boxed{\begin{pmatrix} 2 & -1 \\ -c & 0 \end{pmatrix}} = A.\checkmark$

(d) Determination of the eigenvalues and eigenvectors of A.

Eigenvalues: $\quad 0 \overset{!}{=} \det(A - \lambda \mathbb{1}) = \begin{vmatrix} 2-\lambda & -1 \\ -c & -\lambda \end{vmatrix} = (2-\lambda)(-\lambda) - c = \lambda^2 - 2\lambda - c$

$$\lambda_\pm = 1 \pm \tfrac{1}{2}\sqrt{4+4c} = \boxed{1 \pm \sqrt{1+c}}\,.$$

Eigenvectors: $\quad \mathbf{0} \overset{!}{=} (A - \lambda_\pm \mathbb{1})\mathbf{v}_\pm = \begin{pmatrix} 2-\lambda_\pm & -1 \\ -c & -\lambda_\pm \end{pmatrix}\mathbf{v}_\pm$ $\Rightarrow$ $\mathbf{v}_\pm = \boxed{\begin{pmatrix} \lambda_\pm \\ -c \end{pmatrix}}\,.$

(e) For short times, the time-dependence of a displacement in the $\mathbf{v}_\pm$-direction is given by $\eta_\pm(t) = \eta_\pm(0)e^{\lambda_\pm t}$. The fixed point is unstable in the $\mathbf{v}_+$-direction, since the eigenvalue λ_+ is strictly positive (we set $c > 0$). The fixed point is stable in the $\mathbf{v}_-$-direction (since $\lambda_- < 0$). The characteristic time scale, for which the displacement $\eta_\pm(t)$ grows or shrinks respectively, is given by $\tau_\pm = |\lambda_\pm|^{-1}$.

Check your results: If $c = 3$, then $\lambda_\pm = 1 \pm 2 = \boxed{\left\{\begin{smallmatrix} 3 \\ -1 \end{smallmatrix}\right\}}$, $\mathbf{v}_+ = \boxed{\begin{pmatrix} 3 \\ -3 \end{pmatrix}}$, $\mathbf{v}_- = \boxed{\begin{pmatrix} -1 \\ -3 \end{pmatrix}}$.

S.C8   Functional calculus (p. 330)

S.C8.3   Euler–Lagrange equations (p. 333)

ₑC8.3.1 Snell's law (p. 335)

(a) Along the curve $\mathbf{r}(x) = (x, y(x))^\mathsf{T}$, the curve velocity is given by $\mathrm{d}_x\mathbf{r}(x) = (1, y'(x))^\mathsf{T}$, hence the length functional takes the form

$$\int_I \mathrm{d}x \|\mathrm{d}_x\mathbf{r}(x)\| = \int_I \mathrm{d}x\sqrt{1 + y'^2(x)} = \int_I \mathrm{d}x L(y(x), y'(x)) \equiv L[y].$$

Its Lagrangian, $L(y, y') = \boxed{\sqrt{1 + y'^2}}$, depends only on y', not on y.

For a geometric interpretation, note that an increment δ_x causes an increase in arc length of

$$\delta_s = \|\mathbf{r}(x + \delta_x) - \mathbf{r}(x)\| = \left[\delta_x^2 + (y(x+\delta_x) - y(x))^2\right]^{1/2} = \delta_x\left[1 + y'^2(x)\right]^{1/2}.$$

Summing up the δ_s increments by integrating over x yields the above functional $L[y]$.

(b) The corresponding Euler–Lagrange equation is

$$\frac{\mathrm{d}}{\mathrm{d}x}\frac{\partial L}{\partial y'} = \frac{\partial L}{\partial y} \qquad \Rightarrow \qquad \boxed{\frac{\mathrm{d}}{\mathrm{d}x}\frac{y'}{\sqrt{1+y'^2}} = 0}\,.$$

Its solution, $y'(x) = 0$, describes a line with constant slope, i.e. a straight line, as expected.

(c) An increment, δ_x, in the curve parameter leads to an increment, δ_t, in the traversal time $t(s(x))$ given by

$$\delta_t = \frac{\mathrm{d}t(s(x))}{\mathrm{d}x}\delta_x = \frac{\mathrm{d}t(s)}{\mathrm{d}s}\Big|_{s=s(x)}\frac{\mathrm{d}s(x)}{\mathrm{d}x}\delta_x = \frac{1}{v(\mathbf{r}(x))}\sqrt{1 + y'^2(x)}\,\delta_x.$$

Here $\frac{ds}{dt} = v(\mathbf{r})$ is the speed of light in the medium at point $\mathbf{r}$, and we used $\frac{ds}{dx} = \frac{d}{dx}\int_a^x d\bar{x}\sqrt{1+y'^2(\bar{x})} = \sqrt{1+y'^2(x)}$. Using $v(\mathbf{r}) = \frac{c}{n(\mathbf{r})}$ and integrating over x to sum up all time increments yields a functional for the traversal time of a given path:

$$t[\mathbf{r}] = \frac{1}{c}\int_a^b dx\, n(x, y(x))\sqrt{1+y'^2(x)}.$$

(d) The refraction index, $n(x) = n_A\Theta(-x) + n_B\Theta(x)$, is independent of y, hence the traversal time functional has the form $t[y] = \int_a^b dx L(y, y', x)$, where the function $L = \frac{1}{c}n(x)\sqrt{1+y'^2(x)}$ does not depend on y. Its extremal paths satisfy the Euler–Lagrange equation

$$\frac{d}{dx}\frac{\partial L}{\partial y'} = \frac{\partial L}{\partial y} \qquad \Rightarrow \qquad \frac{d}{dx}\left[n(x)\frac{y'(x)}{\sqrt{1+y'^2(x)}}\right] = 0,$$

implying that $C = n(x)y'(x)/\sqrt{1+y'^2(x)}$ is constant along the entire path. Since $n(x)$ is constant but different on the left and right, this condition is fulfilled by having paths of constant but different slopes on the left and right, $y'(x) = m_A$ and $y'(x) = m_B$, with the slopes related by

$$n_A\frac{m_A}{\sqrt{1+m_A^2}} = n_B\frac{m_B}{\sqrt{1+m_B^2}}.$$

The slope on the left, m_A, determines the angle of incidence, α, as follows (analogously for β), see figure:

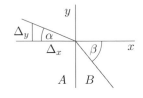

$$\frac{m_A}{\sqrt{1+m_A^2}} = \frac{\Delta y/\Delta x}{\sqrt{1+(\Delta y/\Delta x)^2}} = \frac{\Delta y}{\sqrt{\Delta_x^2 + \Delta_y^2}} = \sin\alpha$$

$$\Rightarrow \qquad \boxed{\frac{n_A}{n_B} = \frac{\sin\beta}{\sin\alpha}}.$$

ɛC8.3.3 Length functional in two dimensions, in polar coordinates (p. 337)

(a) Along the curve $\mathbf{r}(\rho) = (\rho\cos\phi(\rho), \rho\sin\phi(\rho))^T$, the curve velocity,

$$d_\rho\mathbf{r} = (\cos\phi - \rho\,\phi'\sin\phi, \sin\phi + \rho\,\phi'\cos\phi)^T, \qquad \text{with} \qquad \phi' = \frac{d\phi}{d\rho},$$

has magnitude $\|\dot{\mathbf{r}}\| = \sqrt{1+\rho^2\phi'^2}$. Hence the length functional takes the form

$$L[\phi] = \int_I d\rho\,\|d_\rho\mathbf{r}(\rho)\| = \int_I d\rho\, L(\phi(\rho), \phi'(\rho), \rho), \quad \text{with} \quad L(\phi, \phi', \rho) = \boxed{\sqrt{1+\rho^2\phi'^2}}.$$

(b) Since L has no explicit ϕ-dependence, the Euler–Lagrange equation,

$$d_\rho\big[\partial_{\phi'}L\big] = \partial_\phi L \qquad \Rightarrow \qquad \boxed{d_\rho\left[\frac{\rho^2\phi'}{\sqrt{1+\rho^2\phi'^2}}\right] = 0},$$

can be trivially integrated to obtain a conservation law, $\frac{\rho^2\phi'}{\sqrt{1+\rho^2\phi'^2}} = \rho_0 = \text{const}$. By solving this equation for ϕ', we obtain a differential equation for $\phi(\rho)$:

$$\rho^4\phi'^2 = \rho_0^2(1+\rho^2\phi'^2) \qquad \Rightarrow \qquad \boxed{\phi' = \frac{\rho_0}{\rho\sqrt{\rho^2-\rho_0^2}}}. \qquad (1)$$

(c) The straight-line relation, $\cos(\phi - \phi_0) = \rho_0/\rho$, implies $\phi(\rho) = \rho_0 + \arccos(\rho_0/\rho)$. Thus

$$\phi'(\rho) = \frac{1}{\sqrt{1 - \rho_0^2/\rho^2}} \frac{\rho_0}{\rho^2} = \boxed{\frac{\rho_0}{\rho^2\sqrt{\rho^2 - \rho_0^2}}},$$

in accord with Eq. (1). Thus straight lines indeed minimize the length functional on $\mathbb{R}^2$.

(d) Now switch to a different parametrization, where both ρ and ϕ depend on time. Along the curve $\mathbf{r}(t) = (\rho(t)\cos\phi(t), \rho(t)\sin\phi(t))^{\mathsf{T}}$, the velocity, $\dot{\mathbf{r}} = (\dot\rho\cos\phi - \rho\dot\phi\sin\phi, \dot\rho\sin\phi + \rho\dot\phi\cos\phi)^{\mathsf{T}}$ has magnitude $\|\dot{\mathbf{r}}\| = (\dot\rho^2 + \rho^2\dot\phi^2)^{1/2}$. Hence the length functional reads as

$$L[\mathbf{y}] = \int_I dt\, \|\dot{\mathbf{r}}(t)\| = \int_I dt\, L(\mathbf{y}(t), \dot{\mathbf{y}}(t)), \quad \text{with} \quad L(\mathbf{y}, \dot{\mathbf{y}}) = \boxed{(\dot\rho^2 + \rho^2\dot\phi^2)^{1/2}}.$$

(e) The Lagrangian does not depend on ϕ, and the Euler–Lagrange equations take the form

$$\frac{d}{dt}\frac{\partial L}{\partial \dot\rho} = \frac{\partial L}{\partial \rho} \quad\Rightarrow\quad \boxed{\frac{d}{dt}\frac{\dot\rho}{L} = \frac{\rho\dot\phi^2}{L}}, \tag{2}$$

$$\frac{d}{dt}\frac{\partial L}{\partial \dot\phi} = \frac{\partial L}{\partial \phi} \quad\Rightarrow\quad \boxed{\frac{d}{dt}\frac{\rho^2\dot\phi}{L} = 0}. \tag{3}$$

(f) A straight line is defined by $\cos(\phi(t) - \phi_0) = \rho_0/\rho(t)$. Differentiating w.r.t. t, we find that the radial and angular velocities satisfy the relation $\sin(\phi - \phi_0)\dot\phi = \rho_0\,\dot\rho/\rho^2$. If the Lagrangian function is evaluated while assuming that these relations hold, it simplifies to

$$L = \rho\dot\phi\sqrt{(\tfrac{\rho}{\rho_0})^2\sin^2(\phi - \phi_0) + 1} = \rho\dot\phi\sqrt{\tan^2(\phi - \phi_0) + 1} = \frac{\rho\dot\phi}{\cos(\phi - \phi_0)} = \boxed{\frac{\rho^2\dot\phi}{\rho_0}}. \tag{4}$$

It follows that $\rho^2\dot\phi/L = \rho_0 = \text{const.}$, hence Eq. (3) is satisfied. Equation (2) holds, too, since

$$\frac{d}{dt}\frac{\dot\rho}{L} = \frac{d}{dt}\frac{\rho_0\dot\rho}{\rho^2\dot\phi} = \frac{d}{dt}\sin(\phi - \phi_0) = \cos(\phi - \phi_0)\dot\phi \overset{(4)}{=} \frac{\rho\dot\phi^2}{L}. \checkmark$$

Hence straight lines indeed extremize the length functional, as expected.

S.C9 Calculus of complex functions (p. 338)

S.C9.1 Holomorphic functions (p. 338)

S.C9.1.1 Cauchy–Riemann equations (p. 339)

(a) $e^z = e^{x+iy} = e^x e^{iy} = e^x\cos y + ie^x\sin y, \quad\Rightarrow\quad u(x, y) = e^x\cos y, \; v(x, y) = e^x\sin y$.

$$\partial_x u = \boxed{e^x\cos y}, \qquad \partial_y v = \boxed{e^x\cos y}, \qquad\Rightarrow\qquad \partial_x u = \partial_y v. \checkmark$$

$$\partial_y u = \boxed{-e^x\sin y}, \qquad \partial_x v = \boxed{e^x\sin y}, \qquad\Rightarrow\qquad \partial_y u = -\partial_x v. \checkmark$$

The Cauchy–Riemann equations are satisfied. This was to be expected, because e^z depends on x and y via the combined variable $z = x + iy$.

(b)　$\bar{z}^2 = (x - iy)^2 = (x^2 - y^2) - i2xy$,　$\Rightarrow$　$u(x, y) = x^2 - y^2$, $v(x, y) = -2xy$.

$$\partial_x u = \boxed{2x}, \qquad \partial_y v = \boxed{-2x}, \qquad\qquad \Rightarrow \qquad \partial_x u \neq \partial_y v . \checkmark$$

$$\partial_y u = \boxed{-2y}, \qquad \partial_x v = \boxed{-2y}, \qquad\qquad \Rightarrow \qquad \partial_y u \neq -\partial_x v . \checkmark$$

The Cauchy–Riemann equations are not satisfied. This was to be expected, because $\bar{z}^2$ does not depend on x and y via the combined variable $z = x + iy$, but rather depends on $\bar{z} = x - iy$.

S.C9.2　Complex integration (p. 341)

ɛC9.2.1　Cauchy's theorem (p. 345)

Given: the analytic function $f(z) = e^z$, with anti-derivative $F(z) = e^z$. A contour integral of this function along a contour $z(t)$, with $t \in (0, 1)$, has the form:

$$I_\gamma = \int_\gamma dz\, f(z) = \int_0^1 dt\, \frac{dz(t)}{dt} f(z(t)) = \int_{z(0)}^{z(1)} dz\, F'(z) = F(z(1)) - F(z(0)).$$

(a)　Parametrization of the circle: $\gamma_R : z(t) = R e^{i2\pi t}$, with $t \in (0, 1)$.

$$I_{\gamma_R} = \oint_{\gamma_R} dz f(z) = \int_0^1 dt\, \frac{dz}{dt} f(z(t)) = \int_0^1 dt\, (i2\pi R e^{i2\pi t}) e^{R e^{i2\pi t}}$$

$$= \left[e^{R e^{i2\pi t}} \right]_0^1 = e^{R e^{2\pi i}} - e^{R e^0} = e^{R \cdot 1} - e^{R \cdot 1} = \boxed{0}. \text{ [as expected } \checkmark\text{]}$$

(b)　(i) Parametrization of the line: $\gamma_1 : z(t) = (1 - i)t$, with $t \in (0, 1)$.

$$I_{\gamma_1} = \int_0^1 dt\, \frac{dz}{dt} f(z(t)) = \int_0^1 dt\, (1-i) e^{(1-i)t} = \left[e^{(1-i)t} \right]_0^1 = \boxed{e^{1-i} - 1}.$$

(ii) Parametrization of the curve: $\gamma_2 : z(t) = t^3 - it$, with $t \in (0, 1)$.

$$I_{\gamma_2} = \int_0^1 dt\, \frac{dz}{dt} f(z(t)) = \int_0^1 dt\, (3t^2 - i) e^{t^3 - it} = \left[e^{t^3 - it} \right]_0^1 = \boxed{e^{1-i} - 1}.$$

As expected, we obtain $I_{\gamma_1} = I_{\gamma_2} = e^{1-i} - 1 = F(z_1) - F(z_0)$. $\checkmark$

S.C9.3　Singularities (p. 345)

ɛC9.3.1　Laurent series, residues (p. 348)

(a)　The Taylor series of $p(z)$ about z_0 reads: $p(z) = \sum_{\bar{n}=0}^k \frac{p^{(\bar{n})}(z_0)}{\bar{n}!} (z - z_0)^{\bar{n}}$. Consequently, the Laurent series of $f_m(z)$ about z_0 has the following form:

$$f_m(z) = \frac{p(z)}{(z - z_0)^m} = \sum_{\bar{n}=0}^k \frac{p^{(\bar{n})}(z_0)}{\bar{n}!} (z - z_0)^{\bar{n}-m} \overset{n = \bar{n} - m}{=\!=} \boxed{\sum_{n=-m}^{k-m} \frac{p^{(n+m)}(z_0)}{(n + m)!} (z - z_0)^n}.$$

(b) The Laurent series of $f_m(z) = \frac{z^3}{(z-2)^m}$ about $z_0 = 2$ follows from the Taylor series of $p(z) = z^3$ about z_0. With $p^{(1)}(z) = 3z^2$, $p^{(2)}(z) = 6z$ and $p^{(3)}(z) = 6$ we obtain:

$$p(z) = \sum_{\bar{n}=0}^{3} \frac{p^{(\bar{n})}(2)}{\bar{n}!}(z-2)^{\bar{n}} = 2^3 + 3 \cdot 2^2(z-2) + \tfrac{1}{2!}6 \cdot 2(z-2)^2 + \tfrac{1}{3!}6(z-2)^3 .$$

$$f_m(z) = \frac{p(z)}{(z-2)^m} = \boxed{8(z-2)^{-m} + 12(z-2)^{1-m} + 6(z-2)^{2-m} + (z-2)^{3-m}} .$$

(c) The residues of $f_m(z)$ at $z_0 = 2$ (pole of order m) read:

$$\text{Res}(f_m, 2) = \lim_{z \to 2} \frac{1}{(m-1)!} \frac{d^{m-1}}{dz^{m-1}}\left[(z-2)^m f_m(z)\right] = \lim_{z \to z_0} \frac{1}{(m-1)!} \frac{d^{m-1}}{dz^{m-1}}\left[z^3\right] ;$$

$$\text{Res}(f_1, 2) = \lim_{z \to 2} z^3 = \boxed{8} ; \qquad\qquad \text{Res}(f_2, 2) = \lim_{z \to 2} \frac{d}{dz} z^3 = \lim_{z \to 2} 3z^2 = \boxed{12} ;$$

$$\text{Res}(f_3, 2) = \lim_{z \to 2} \frac{1}{2!} \frac{d^2}{dz^2} z^3 = \lim_{z \to 2} \frac{3!}{2!} z = \boxed{6} ; \qquad \text{Res}(f_4, 2) = \lim_{z \to 2} \frac{1}{3!} \frac{d^3}{dz^3} z^3 = \lim_{z \to 2} \frac{3!}{3!} = \boxed{1} ;$$

$$\text{Res}(f_m, 2) = \lim_{z \to 2} \frac{1}{(m-1)!} \frac{d^{m-1}}{dz^{m-1}} z^3 = \boxed{0} \quad \text{for} \quad m \geq 5 .$$

As expected, the residue $\text{Res}(f_m, 2)$ corresponds to the coefficient of $(z-2)^{-1}$ in the Laurent series of $f_m(z)$ presented in (b). ✓

S.C9.4 Residue theorem (p. 348)

C9.4.1 Circular contours, residue theorem (p. 350)

(a) With the Parametrization $z(\phi) = Re^{i\phi}$ and $\phi \in (0, \pm k2\pi)$ for $I_{\pm}^{(k)}$ we obtain:

$$\left.\begin{array}{l} I_+^{(k)} = \oint_{|z|=R} \frac{dz}{z} \\[2ex] I_-^{(k)} = \oint_{|z|=R} \frac{dz}{z} \end{array}\right\} = \int_0^{\pm k2\pi} d\phi \, \frac{dz(\phi)}{d\phi} \frac{1}{z(\phi)} = \int_0^{\pm k2\pi} d\phi \, \frac{iRe^{i\phi}}{Re^{i\phi}} = \boxed{\pm k2\pi i} .$$

Note: This result holds in general: when $g(z)$ has an isolated pole at z_0, and the contour γ_k circles this pole k times in the mathematically positive (resp. negative) direction, then $\oint_{\gamma_k} dz \, g(z) = \pm k2\pi i \, \text{Res}(g, z_0)$.

(b) The function $g(z)$ has two poles of order 1 at $z_\pm = \pm i$:

$$g(z) = \frac{e^{iaz}}{z^2 + 1} = \frac{e^{iz}}{(z-i)(z+i)} = \frac{e^{iz}}{(z-z_+)(z-z_-)} .$$

The corresponding residues are:

$$\text{Res}(g, z_\pm) = \lim_{z \to z_\pm}\left[(z - z_\pm) \frac{e^{iaz}}{(z-z_+)(z-z_-)}\right] = \frac{e^{iaz_\pm}}{(z_\pm - z_\mp)} = \boxed{\frac{e^{\mp a}}{\pm 2i}} .$$

For $I_1 = \oint_{|z|=\frac{1}{2}} dz \, g(z)$ both poles lie outside the integration contour, $\Rightarrow I_1 = \boxed{0}$.

For $I_2 = \oint_{2 \text{ times: } |z|=2} dz \, g(z)$ both poles lie inside the integration contour, hence

$$I_2(a) = -2 \cdot 2\pi i\left[\text{Res}(g, z_+) + \text{Res}(g, z_-)\right] = -4\pi i \frac{e^{-a} - e^a}{2i} = \boxed{4\pi \sinh a} .$$

(c) The poles of $f(z)$ are located at the zeros of the denominator. According to the hint, the denominator thus contains a factor of $(z + ai)$. Using polynomial division, we can factorize as follows:

$$z^3 + (ai - 6)z^2 + (9 - a6i)z + 9ai = (z + ai)(z^2 - 6z + 9) = (z + ai)(z - 3)^2 .$$

Consequently $f(z)$ has a pole of order 1 at $z_a = -ai$ and a pole of order 2 at $z_3 = 3$:

$$f(z) = \frac{z}{z^3 + (ai - 6)z^2 + (9 - a6i)z + 9ai} = \frac{z}{(z - z_a)(z - z_3)^2} .$$

The residues are thus:

$$\mathrm{Res}(f, z_a) = \lim_{z \to z_a} \left[(z - z_a)f(z) \right] = \frac{z_a}{(z_a - z_3)^2} = \boxed{\frac{-ai}{(-ai - 3)^2}} ;$$

$$\mathrm{Res}(f, z_3) = \lim_{z \to z_3} \frac{d}{dz} \left[(z - z_3)^2 g(z) \right] = \lim_{z \to z_3} \frac{d}{dz} \frac{z}{z - z_a} = \frac{(z_3 - z_a) - z_3}{(z_3 - z_a)^2} = \boxed{\frac{ai}{(3 + ai)^2}} .$$

For $a < 4$ the integration contour $|z| = 4$ encloses both poles, however for $a > 4$ it encloses only the pole at $z_2 = 3$. Hence:

$$I_3(a) = \oint_{|z|=4} dz\, f(z) = \begin{cases} 2\pi i \left[\mathrm{Res}(f, -ai) + \mathrm{Res}(f, 3) \right] = \boxed{0}, & \text{for } a < 4, \\[2mm] 2\pi i \left[\mathrm{Res}(f, 3) \right] = \boxed{-\dfrac{2\pi a}{(3 + ai)^2}}, & \text{for } a > 4. \end{cases}$$

That I_3 vanishes in the first case can also be seen as follows: for $a < 4$, both poles lie inside the integration contour, and hence the circular contour with radius $|z| = 4$ may be extended to a circle with radius $R \to \infty$ without crossing any poles. Because the integrand vanishes as $f(z) \sim z^{-2} \sim R^{-2}$ for large arguments, while the integration measure only grows proportionally to R, the integral vanishes:

$$I_3(a < 4) = \lim_{R \to \infty} \oint_{|z|=R} dz\, f(z) = \lim_{R \to \infty} \oint_{|z|=R} \frac{dz}{z} = \lim_{R \to \infty} \int_0^{2\pi} d\phi \frac{iRe^{i\phi}}{(iRe^{i\phi})^2} = \boxed{0}. \checkmark$$

Note: This is an example of the following general fact: if a contour integral $I_\gamma = \oint_\gamma dz\, f(z)$ encloses all poles of $f(z)$, then it may be extended to a circular contour with radius $R \to \infty$, without crossing any poles. The integral $I_\gamma = \lim_{R \to \infty} \oint dz\, f(z)$ will vanish as long as the integrand has the property $\lim_{|z| \to \infty} \left[zf(z) \right] = 0$. Hence it follows that for such functions, the sum of all the residues is always equal to zero – a useful consistency check! (This does not apply to part (b), because along the imaginary axis, where $z = iy$ and $f(iy) \propto e^{-ay}$, the limit $\lim_{y \to -\infty} \left[(iy)f(iy) \right] = \infty$.)

 εC9.4.3 Integrating by closing contour and using residue theorem (p. 351)

We regard the integral (with $a, b \in \mathbb{R}$) as a contour integral in the complex plane,

$$I(a, b) = \int_{-\infty}^{\infty} dx\, f(x) = \lim_{R \to \infty} \int_{\Gamma_0} dz\, f(z), \quad \text{with} \quad f(z) = \frac{1}{z^2 - 2za + a^2 + b^2},$$

where $\Gamma_0 : \{z(x) = x \mid x \in (-R, R)\}$ is a section of the real axis. The integrand has two poles of first order, at $z_\pm = a \pm i|b|$ in the upper/lower half-plane respectively:

$$f(z) = \frac{1}{(z - a - i|b|)(z - a + i|b|)} = \frac{1}{(z - z_+)(z - z_-)} .$$

Residues: $\mathrm{Res}(f, z_\pm) = \lim\limits_{z \to z_\pm}\left[(z - z_\pm)f(z)\right] = \lim\limits_{z \to z_\pm}\dfrac{1}{(z - z_\mp)} = \dfrac{1}{z_\pm - z_\mp} = \boxed{\dfrac{1}{\pm 2i|b|}}.$ (1)

In order to use the residue theorem, we require a closed contour. We therefore add to the contour Γ_0 a semicircle about the origin with radius R. We can insert this circle in either the upper or lower half-planes, denoting it by Γ_+ or Γ_-, respectively, since in both cases the integral along the semicircle, $\int_{\Gamma_\pm} dz\, f(z)$, vanishes in the limit $R \to \infty$. This can be seen as follows:

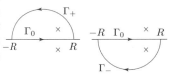

The parametrization $\Gamma_\pm$: $\{z(\phi) = Re^{i\phi} | \phi \in (0, \pm\pi)\}$ gives

$$\int_{\Gamma_\pm} dz\, f(z) = \int_0^{\pm\pi} d\phi\, \frac{dz(\phi)}{d\phi} f(z(\phi)) \xrightarrow{R \to \infty} \int_0^{\pm\pi} d\phi\, \left(iRe^{i\phi}\right)\frac{1}{(Re^{i\phi})^2} = 0.$$

Both of the two possible closed contours, $\Gamma_0 \cup \Gamma_+ = \curvearrowright$ and $\Gamma_0 \cup \Gamma_- = \curvearrowright$, encircle a single pole, z_+ and z_- respectively. They are traversed in the mathematically positive or negative directions respectively, which yields a plus or minus sign in the residue theorem, cancelling the sign from the poles and thus giving the same result:

$$I(a, b) = \lim_{R \to \infty} \left\{ \begin{array}{c} \displaystyle\int_{\curvearrowright} dz\, f(z) \\[2em] \displaystyle\int_{\curvearrowright} dz\, f(z) \end{array} \right\} = \pm 2\pi i\, \mathrm{Res}(f, z_\pm) \stackrel{(1)}{=} \pm 2\pi i\, \frac{1}{\pm 2i|b|} = \boxed{\frac{\pi}{|b|}}.$$

Note: The above strategy of enclosing a contour in the upper/lower half-planes is possible whenever $\lim_{|z| \to \infty}\left[zf(z)\right] = 0$. This applies, for example, to any rational function of the form $f(z) = p(z)/q(z)$, with polynomials $p(z)$ and $q(z)$ of degree n_p and n_q respectively, with $n_p \leq n_q - 2$.

εC9.4.5 Various integration contours, residue theorem (p. 351)

(a)   The function $f(z)$ has four poles of order 1, at $z_2^\pm = \pm 2i$ and at $z_a^\pm = \pm ai$:

$$f(z) = \frac{z^2}{(z^2 + 4)(z^2 + a^2)}$$
$$= \frac{z^2}{(z - z_2^+)(z - z_2^-)(z - z_a^+)(z - z_a^-)}.$$

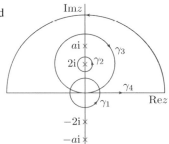

The associated residues read:

$$\mathrm{Res}(f, z_2^\pm) = \lim_{z \to z_2^\pm}\left[(z - z_2^\pm)f(z)\right] = \frac{(z_2^\pm)^2}{(z_2^\pm - z_2^\mp)((z_2^\pm)^2 + a^2)} = \frac{-4}{2(\pm 2i)(a^2 - 4)} = \frac{\pm i}{a^2 - 4};$$

$$\mathrm{Res}(f, z_a^\pm) = \lim_{z \to z_a^\pm}\left[(z - z_a^\pm)f(z)\right] = \frac{(z_a^\pm)^2}{((z_a^\pm)^2 + 4)(z_a^\pm - z_a^\mp)} = \frac{-a^2}{(a^2 - 4)2(\pm ai)} = \frac{\mp ia}{2(a^2 - 4)}.$$

(b)   The circular contour γ_1 does not encircle any poles, therefore $I_{\gamma_1} = \int_{\gamma_1} dz\, f(z) = \boxed{0}$.

γ_2 encloses the pole at z_2^+, and γ_3 encloses both the poles z_2^+ and z_a^+. Consequently, we have:

(c) $I_{\gamma_2} = \displaystyle\int_{\gamma_2} dz\, f(z) = 2\pi i\, \mathrm{Res}(f, z_2^+) = \boxed{-\dfrac{2\pi}{a^2 - 4}};$

(d) $I_{\gamma 3} = \displaystyle\int_{\gamma 3} dz\, f(z) = -2\pi i \big[\mathrm{Res}(f, z_2^+) + \mathrm{Res}(f, z_a^+)\big] = \boxed{\dfrac{2\pi(1 - \frac{1}{2}a)}{a^2 - 4}}.$

(e)   The contour γ_4 along the real axis can be calculated by closing it with a semicircle of radius $\to \infty$ in the upper or lower half-planes, since $\lim_{|z|\to\infty}[zf(z)] = 0$. We choose the upper half-plane, because it allows us to use the fact that $\curvearrowright$ encloses that same poles as γ_3. Because the traversal direction is reversed relative to $I_{\gamma 3}$, we have $I_{\gamma 4} = I_{\curvearrowright} = \boxed{-I_{\gamma 3}}$.

εC9.4.7 Inverse Fourier transform via contour closure (p. 353)

(a)   We consider the Fourier integral (with $t \neq 0$) as a contour integral in the complex plane:

$$G(t) = \int_{-\infty}^{\infty} \frac{d\omega}{2\pi} \frac{e^{-i\omega t}}{a - i\omega} = \lim_{R\to\infty} \int_{\Gamma_0} dz\, f(z), \quad \text{with} \quad f(z) = \frac{1}{2\pi} \frac{e^{-izt}}{a - iz} \quad (0 < a \in \mathbb{R}),$$

where $\Gamma_0 : \{z(x) = x \mid x \in (-R, R)\}$ is a section of the real axis. The integrand has a single pole of first order at $z_0 = -ia$, with residue

$$\mathrm{Res}(f, z_0) = \lim_{z\to z_0}\big[(z - z_0)f(z)\big] = \lim_{z\to -ia}\left[\frac{(z + ia)}{2\pi} \frac{e^{-izt}}{-i(z + ia)}\right] = \frac{e^{-at}}{-2\pi i}. \tag{1}$$

In order to use the residue theorem, we require a closed contour. To this end, we close the contour Γ_0 by a semicircle about the origin of radius R. Depending on the sign of t, we choose this semicircle to be in either the upper or lower half-plane, Γ_+ or Γ_-, respectively, in such a way that $\mathrm{Re}(-izt) < 0$, because that ensures that the semicircular integral $\int_{\Gamma_\pm} dz\, f(z)$ vanishes as $R \to \infty$.

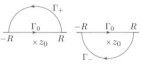

To see this in detail, let us use the parametrization $\Gamma_\pm : \{z(\phi) = Re^{i\phi} \mid \phi \in (0, \pm\pi)\}$:

$$\int_{\Gamma_\pm} dz\, f(z) = \int_0^{\pm\pi} d\phi\, \frac{dz(\phi)}{d\phi} f(z(\phi)) = \int_0^{\pm\pi} d\phi\, (iRe^{i\phi}) \frac{1}{2\pi} \frac{e^{-itR(\cos\phi + i\sin\phi)}}{a - iz}.$$

The behavior of the integrand in the limit $R \to \infty$ is determined by the factor

$$e^{tR\sin\phi} \xrightarrow{R\to\infty} \begin{cases} 0 & \text{for} \quad t\sin\phi < 0, \\ \infty & \text{for} \quad t\sin\phi > 0. \end{cases}$$

To ensure that the semicircular integral vanishes, we require the first case. Thus, for $t < 0$ or $t > 0$ we choose the contour such that $\sin\phi > 0$ or < 0 respectively, i.e. Γ_+ or Γ_-. The pole of $f(z)$ lies in the lower half plane, and so is not enclosed by the contour $\curvearrowright = \Gamma_0 \cup \Gamma_+$. It is, however, enclosed by the contour $\curvearrowleft = \Gamma_0 \cup \Gamma_-$ (which is traversed in the mathematically negative direction, and so picks up a negative sign in the residue theorem). Thus we conclude that:

$$\left.\begin{aligned} G(t < 0) &= \int_{\curvearrowright} dz\, f(z) = 0 \\ G(t > 0) &= \int_{\curvearrowleft} dz\, f(z) = -2\pi i\, \mathrm{Res}(f, z_0) \overset{(1)}{=} e^{-a|t|} \end{aligned}\right\} \Rightarrow G(t) = \boxed{\Theta(t)\, e^{-at}}. \checkmark$$

(b) We proceed in analogy to (a). The integral has the form $L(t) = \int_{-\infty}^{\infty} dz\, f(z)$, with

$$f(z) = \frac{e^{-izt}}{2\pi}\tilde{L}(z) = \frac{a\,e^{-izt}}{\pi(z^2 + a^2)} = \frac{a\,e^{-izt}}{\pi(z - z_+)(z - z_-)}.$$

Hence $f(z)$ has two poles of order 1, at $z_\pm = \pm ia$, with residue

$$\mathrm{Res}(f, z_\pm) = \lim_{z \to z_\pm}\left[(z - z_\pm)f(z)\right] = \frac{a\,e^{-iz_\pm t}}{\pi(z_\pm - z_\mp)} = \frac{a\,e^{\pm at}}{\pi(\pm 2ia)} = \frac{e^{\pm at}}{\pm 2\pi i}. \tag{2}$$

For $t < 0$ or $t > 0$ we close the contour in the upper or lower half-planes respectively:

$$\left.\begin{array}{ll}
L(t < 0) & = \displaystyle\int_{\frown} dz\, f(z) = 2\pi i\,\mathrm{Res}(f, z_+) \overset{(1)}{=} e^{+at} \\[2mm]
L(t > 0) & = \displaystyle\int_{\smile} dz\, f(z) = -2\pi i\,\mathrm{Res}(f, z_-) \overset{(1)}{=} e^{-at}
\end{array}\right\} \quad \Rightarrow \quad L(t) = \boxed{e^{-a|t|}}.\ \checkmark$$

As expected, the inverse Fourier transform of the Lorentz function has given us back the exponential function $e^{-a|t|}$, which was the starting point for our calculation.

.

S.V1 Curves (p. 406)

S.V1.2 Curve velocity (p. 407)

εV1.2.1 Velocity and acceleration (p. 409)

(a) It will be economical to use the compact notation $C(t) = \cos[\pi(1 - \cos\omega t)]$ and $S(t) = \sin[\pi((1 - \cos\omega t)]$, with $C^2 + S^2 = 1$, and derivatives $\dot{C} = -\omega\pi \sin(\omega t)S$, $\dot{S} = \omega\pi \sin(\omega t)C$. Then the curve, and the velocity and acceleration along it, take the following forms:

$$\mathbf{r}(t) = \left(aC, S\right)^{\mathrm{T}}, \qquad \dot{\mathbf{r}}(t) = \boxed{\omega\pi \sin(\omega t)\left(-aS, C\right)^{\mathrm{T}}},$$

$$\ddot{\mathbf{r}}(t) = \boxed{\omega^2\pi \cos(\omega t)\left(-aS, C\right)^{\mathrm{T}} - [\omega\pi \sin(\omega t)]^2\left(aC, S\right)^{\mathrm{T}}}$$

$$= \boxed{\omega \cot(\omega t)\dot{\mathbf{r}} - [\omega\pi \sin(\omega t)]^2 \mathbf{r}}.$$

(b) A parameter-free representation of the curve is $\boxed{(x/a)^2 + y^2 = 1}$. This traces out an ellipse. The sketch shows the case for $a = 2$.

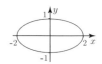

(c) $\mathbf{r}(t) \cdot \dot{\mathbf{r}}(t) = \boxed{\pi\omega \sin(\omega t)\, CS(a^2 - 1)}$. This vanishes when $\boxed{a = 1}$, in which case the curve describes a circle. For a circular trajectory, the velocity vector is perpendicular to the position vector at every point.

S.V1.3 Curve length (p. 409)

εV1.3.1 Curve length (p. 411)

For the curve $\mathbf{r}(t) = (\frac{1}{2}t^2, \frac{1}{3}at^3)^{\mathrm{T}}$, the curve velocity is $\dot{\mathbf{r}}(t) = (t, at^2)^{\mathrm{T}}$, with $\|\dot{\mathbf{r}}(t)\| = \sqrt{t^2 + a^2 t^4} = t\sqrt{1 + a^2 t^2}$. The curve length is

$$L[\gamma] = \int_0^1 dt\, \|\dot{\mathbf{r}}(t)\| = \int_0^1 dt\, t\sqrt{1 + a^2 t^2} = \frac{1}{3a^2}\left[(1 + a^2 t^2)^{3/2}\right]_0^1 = \frac{1}{3a^2}\left[(1 + a^2)^{3/2} - 1\right].$$

ₑV1.3.3 Natural parametrization of a curve (p. 411)

(a)

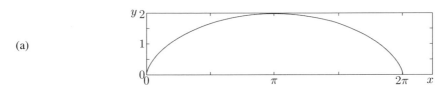

(b) $\mathbf{r}(t) = (t - \sin t, 1 - \cos t)^{\mathrm{T}}, \qquad \dot{\mathbf{r}}(t) = (1 - \cos t, \sin t)^{\mathrm{T}},$

$\|\dot{\mathbf{r}}(t)\| = \sqrt{(1 - \cos t)^2 + (\sin t)^2} = \sqrt{1 - 2 \cos t + 1} = 2|\sin(t/2)|.$

For the last step we used $\cos 2A = 1 - 2 \sin^2 A$. For $t < 2\pi$, $\sin(t/2) > 0$, so the modulus sign may be omitted:

$$s(t) = \int_0^t du \, \|\dot{\mathbf{r}}(u)\| = \int_0^t du \, 2 \sin{(u/2)} = -4 \cos{(u/2)} \Big|_0^t = \boxed{4 - 4 \cos{(t/2)}}.$$

(c) $t(s) = 2 \arccos{(1 - s/4)} = 2 \arccos \tilde{s}, \quad \text{with } \tilde{s} = 1 - s/4 \quad \text{[inverse function of (b)]}.$

$$\mathbf{r}_{\mathrm{L}}(s) = \boxed{\left(2 \arccos \tilde{s} - \sin{\left[2 \arccos \tilde{s}\right]}, 1 - \cos{\left[2 \arccos \tilde{s}\right]}\right)^{\mathrm{T}}}.$$

$$= \begin{pmatrix} 2 \arccos \tilde{s} - 2 \sin(\arccos \tilde{s}) \cos(\arccos \tilde{s}) \\ 1 - \cos^2(\arccos \tilde{s}) + \sin^2(\arccos \tilde{s}) \end{pmatrix} = \boxed{\begin{pmatrix} 2 \arccos \tilde{s} - 2\sqrt{1 - \tilde{s}^2}\,\tilde{s} \\ 1 - \tilde{s}^2 + 1 - \tilde{s}^2 \end{pmatrix}}.$$

S.V1.4   Line integral (p. 412)

ₑV1.4.1 Line integral: mountain hike (p. 413)

Strategy for the line integral $\int_\gamma d\mathbf{r} \cdot \mathbf{F} = \int_I dt \, \dot{\mathbf{r}}(t) \cdot \mathbf{F}(\mathbf{r}(t))$: find a parametrization $\mathbf{r}(t)$ of the curve, then determine $\dot{\mathbf{r}}(t)$, $\mathbf{F}(\mathbf{r}(t))$ and $\dot{\mathbf{r}}(t) \cdot \mathbf{F}(\mathbf{r}(t))$, then integrate.

Given: $\mathbf{r}_0 \equiv (0, 0)^{\mathrm{T}}$, $\mathbf{r}_1 \equiv (3, 3a)^{\mathrm{T}}$, $\mathbf{r}_2 \equiv (2, 4a)^{\mathrm{T}}$, $\mathbf{F}(\mathbf{r}) = \mathbf{F}_{\mathrm{g}} + \mathbf{F}_{\mathrm{w}} = (-y^2, -10)^{\mathrm{T}}$.
Hiker 1: path γ_1 is a straight line from $\mathbf{r}_0$ to $\mathbf{r}_1$ and hence has the form $y(x) = ax$. A possible parametrization, with $t = x \in (0, 3)$ as curve parameter, is:

γ_1: $\mathbf{r}(t) = (x(t), y(x))^{\mathrm{T}} = (t, at)^{\mathrm{T}},$

$\dot{\mathbf{r}}(t) = (1, a)^{\mathrm{T}},$

$\mathbf{F}(\mathbf{r}(t)) = (-y^2(t), -10)^{\mathrm{T}} = (-a^2 t^2, -10)^{\mathrm{T}},$

$\left[\dot{\mathbf{r}}(t) \cdot \mathbf{F}(\mathbf{r}(t))\right]_{\gamma_1} = -(a^2 t^2 + 10a),$

$W[\gamma_1] = -\int_{\gamma_1} d\mathbf{r} \cdot \mathbf{F} = \int_0^3 dt \, [a^2 t^2 + 10a] = \left[\tfrac{1}{3}a^2 t^3 + 10at\right]_0^3 = \boxed{9a^2 + 30a}.$

Hiker 2: path γ_2 is a parabola with apex $\mathbf{r}_2 = (2, 4a)^{\mathrm{T}}$ and has the form $y(x) = -k(x - 2)^2 + 4a$. Inserting $\mathbf{r}_0 = (0, 0)^{\mathrm{T}}$ or $\mathbf{r}_1 = (3, 3a)^{\mathrm{T}}$ yields the curvature, $k = a$. A possible parametrization, with $t = x \in (0, 3)$ as curve parameter, is:

$\gamma_2:$
$$\mathbf{r}(t) = (x(t), y(x))^{\mathrm{T}} = (t, -a(t-2)^2 + 4a)^{\mathrm{T}},$$
$$\dot{\mathbf{r}}(t) = (1, -2a(t-2))^{\mathrm{T}},$$
$$\mathbf{F}(\mathbf{r}(t)) = (-y^2(t), -10)^{\mathrm{T}} = (-[-a(t-2)^2 + 4a]^2, -10)^{\mathrm{T}},$$
$$[\dot{\mathbf{r}}(t) \cdot \mathbf{F}(\mathbf{r}(t))]_{\gamma_1} = -[-a(t-2)^2 + 4a]^2 + 20a(t-2),$$
$$W[\gamma_2] = -\int_{\gamma_2} d\mathbf{r} \cdot \mathbf{F} = \int_0^3 dt \left[[-a(t-2)^2 + 4a]^2 - 20a(t-2)\right]$$
$$= \int_0^3 dt \left[a^2(t-2)^4 - 8a^2(t-2)^2 + 16a^2 - 20at + 40a\right]$$
$$= \left[\tfrac{1}{5}a^2(t-2)^5 - \tfrac{8}{3}a^2(t-2)^3 - 10at^2 + (16a^2 + 40a)t\right]_0^3$$
$$= \left(\tfrac{1}{5} - \tfrac{8}{3} + 48 + \tfrac{32}{5} - \tfrac{64}{3}\right)a^2 + (-90+120)\,a = \boxed{\tfrac{153}{5}a^2 + 30a}.$$

S.V2 Curvilinear coordinates (p. 414)

S.V2.3 Cylindrical and spherical coordinates (p. 421)

ɛV2.3.1 Coordinate transformations (p. 424)

Cylindrical coord.: $\rho = \sqrt{x^2+y^2}$, $\phi = \arctan(y/x)+n_\phi\pi$, $z = z$.

Spherical coord.: $r = \sqrt{x^2+y^2+z^2}$, $\theta = \arccos(z/r)$, $\phi = \arctan(y/x)+n_\phi\pi$,

with $\theta \in (0, \pi)$, and $n_\phi \in \mathbb{Z}$ chosen such that $\phi \in (0, 2\pi)$ lies in the correct quadrant.

$P_1:$ $(x, y, z) = (3, -2, 4)$, $(\rho, \phi, z) = \boxed{(\sqrt{13}, 5.69, 4)}$, $(r, \theta, \phi) = \boxed{(\sqrt{29}, 0.73, 5.69)}$,

    $x > 0, y < 0 \;\Rightarrow\; \phi$ lies in the 4th quadrant $\;\Rightarrow\; \phi \in (3\pi/2, 2\pi)$,
    $\phi = \arctan(-2/3) + 2\pi \approx -0.59 + 6.28 = 5.69$ (equals $326°$),
    $\theta = \arccos(4/\sqrt{29}) \approx 0.73$ (equals $42°$).

$P_2:$ $(x, y, z) = (1, 1, 1)$, $(\rho, \phi, z) = \boxed{(\sqrt{2}, \pi/4, 1)}$, $(r, \theta, \phi) = \boxed{(\sqrt{3}, 0.96, \pi/4)}$,

    $x > 0, y > 0 \;\Rightarrow\; \phi$ lies in the 1st quadrant $\;\Rightarrow\; \phi \in (0, \pi/2)$,
    $\phi = \arctan(1/1) = \pi/4$ (equals $45°$),
    $\theta = \arccos(1/\sqrt{3}) \approx 0.96$ (equals $55°$).

$P_3:$ $(x, y, z) = (-3, 0, -2)$, $(\rho, \phi, z) = \boxed{(3, \pi, -2)}$, $(r, \theta, \phi) = \boxed{(\sqrt{13}, 2.16, \pi)}$,

    $x < 0, y = 0 \;\Rightarrow\; \phi$ lies on the negative x-axis $\;\Rightarrow \phi = \pi$ (equals $180°$),
    $\theta = \arccos(-2/\sqrt{13}) \approx 2.16$ (equals $124°$).

ɛV2.3.3 Cylindrical coordinates: velocity, kinetic energy, angular momentum (p. 422)

In terms of the cylindrical coordinates $y_1 = \rho$, $y_2 = \phi$, $y_3 = z$, the Cartesian coordinates $x_i = x_i(y_j)$ are given by $x_1 = x = \rho \cos\phi$, $x_2 = y = \rho \sin\phi$, $x_3 = z$. We also have $\mathbf{e}_{x_i} \cdot \mathbf{e}_{x_j} = \delta_{ij}$.

Position vector: $\mathbf{r} = x\,\mathbf{e}_x + y\,\mathbf{e}_y + z\,\mathbf{e}_z = \rho\cos\phi\,\mathbf{e}_x + \rho\sin\phi\,\mathbf{e}_y + z\mathbf{e}_z = \boxed{\rho\,\mathbf{e}_\rho + z\mathbf{e}_z}$.

(a) Construction of the local basis vectors: $\mathbf{v}_{y_i} = \partial\mathbf{r}/\partial y_i$, $v_{y_i} = \|\partial\mathbf{r}/\partial y_i\|$, $\mathbf{e}_{y_i} = \mathbf{v}_{y_i}/v_{y_i}$.

$\mathbf{v}_\rho = \cos\phi\,\mathbf{e}_x + \sin\phi\,\mathbf{e}_y$, $v_\rho = (\sin^2\phi + \cos^2\phi)^{\frac{1}{2}} = 1$, $\mathbf{e}_\rho = \boxed{\cos\phi\,\mathbf{e}_x + \sin\phi\,\mathbf{e}_y}$.

$\mathbf{v}_\phi = -\rho\sin\phi\,\mathbf{e}_x + \rho\cos\phi\,\mathbf{e}_y$, $v_\phi = (\rho^2\sin^2\phi + \rho^2\cos^2\phi)^{\frac{1}{2}} = \rho$, $\mathbf{e}_\phi = \boxed{-\sin\phi\,\mathbf{e}_x + \cos\phi\,\mathbf{e}_y}$.

$\mathbf{v}_z = \mathbf{e}_z$, $v_z = 1$, $\mathbf{e}_z = \boxed{\mathbf{e}_z}$.

Normalization is guaranteed by construction: $\mathbf{e}_\rho \cdot \mathbf{e}_\rho = \mathbf{e}_\phi \cdot \mathbf{e}_\phi = \mathbf{e}_z \cdot \mathbf{e}_z = \boxed{1}$.
Orthogonality:

$$\mathbf{e}_\rho \cdot \mathbf{e}_\phi = (\cos\phi\,\mathbf{e}_x + \sin\phi\,\mathbf{e}_y) \cdot (-\sin\phi\,\mathbf{e}_x + \cos\phi\,\mathbf{e}_y) = -\cos\phi\sin\phi + \sin\phi\cos\phi = \boxed{0} ,$$

$$\mathbf{e}_\rho \cdot \mathbf{e}_z = (\cos\phi\,\mathbf{e}_x + \sin\phi\,\mathbf{e}_y) \cdot \mathbf{e}_z = \boxed{0} , \quad \mathbf{e}_\phi \cdot \mathbf{e}_z = (-\sin\phi\,\mathbf{e}_x + \cos\phi\,\mathbf{e}_y) \cdot \mathbf{e}_z = \boxed{0} .$$

Hence: $\boxed{\mathbf{e}_{y_i} \cdot \mathbf{e}_{y_j} = \delta_{ij}}$. ✓

(b) Cross product: $\mathbf{e}_\rho \times \mathbf{e}_\rho = \mathbf{e}_\phi \times \mathbf{e}_\phi = \mathbf{e}_z \times \mathbf{e}_z = \boxed{0}$.

$$\mathbf{e}_\rho \times \mathbf{e}_\phi = (\cos\phi\,\mathbf{e}_x + \sin\phi\,\mathbf{e}_y) \times (-\sin\phi\,\mathbf{e}_x + \cos\phi\,\mathbf{e}_y) = \cos^2\phi\,\mathbf{e}_z - \sin^2\phi(-\mathbf{e}_z) = \boxed{\mathbf{e}_z} ,$$

$$\mathbf{e}_\phi \times \mathbf{e}_z = (-\sin\phi\,\mathbf{e}_x + \cos\phi\,\mathbf{e}_y) \times \mathbf{e}_z = -\sin\phi(-\mathbf{e}_y) + \cos\phi\,\mathbf{e}_x = \boxed{\mathbf{e}_\rho} ,$$

$$\mathbf{e}_z \times \mathbf{e}_\rho = \mathbf{e}_z \times (\cos\phi\,\mathbf{e}_x + \sin\phi\,\mathbf{e}_y) = \cos\phi\,\mathbf{e}_y + \sin\phi(-\mathbf{e}_x) = \boxed{\mathbf{e}_\phi} .$$

Hence: $\boxed{\mathbf{e}_{y_i} \times \mathbf{e}_{y_j} = \varepsilon_{ijk}\mathbf{e}_{y_k}}$. ✓

Remark: In three dimensions a set of *orthonormal* basis vectors "automatically" satisfies the cross product formula. Above, the explicit calculations show that the cross product is cyclic in its arguments. Alternatively, you can convince yourself of this using a sketch.

(c) $\mathbf{v} = \dfrac{\mathrm{d}}{\mathrm{d}t}\mathbf{r}(\rho,\phi,z) = \dot\rho\partial_\rho\mathbf{r} + \dot\phi\partial_\phi\mathbf{r} + \dot z\partial_z\mathbf{r} \overset{(a)}{=} \boxed{\dot\rho\,\mathbf{e}_\rho + \rho\dot\phi\,\mathbf{e}_\phi + \dot z\,\mathbf{e}_z}$.

(d) $T = \tfrac{1}{2}m\mathbf{v}^2 = \tfrac{1}{2}m(\dot\rho\,\mathbf{e}_\rho + \rho\dot\phi\,\mathbf{e}_\phi + \dot z\,\mathbf{e}_z)^2 = \boxed{\tfrac{1}{2}m[\dot\rho^2 + \rho^2\dot\phi^2 + \dot z^2]}$.

(e) $\mathbf{L} = m(\mathbf{r} \times \mathbf{v}) = m(\rho\,\mathbf{e}_\rho + z\mathbf{e}_z) \times (\dot\rho\,\mathbf{e}_\rho + \rho\dot\phi\,\mathbf{e}_\phi + \dot z\,\mathbf{e}_z)$

$\qquad = m[\rho^2\dot\phi\,(\mathbf{e}_\rho \times \mathbf{e}_\phi) + z\dot\rho(\mathbf{e}_z \times \mathbf{e}_\rho) + \rho\dot z(\mathbf{e}_\rho \times \mathbf{e}_z) + z\rho\dot\phi(\mathbf{e}_z \times \mathbf{e}_\phi)]$

$\qquad = \boxed{m[-z\rho\dot\phi\,\mathbf{e}_\rho + (z\dot\rho - \rho\dot z)\mathbf{e}_\phi + \rho^2\dot\phi\,\mathbf{e}_z]}$.

εV2.3.5 **Line integral in polar coordinates: spiral** (p. 420)

The spiral radius depends on the angle as $\rho(\phi) = R + \frac{1}{2\pi}\phi\Delta$.

(a) $\mathbf{r} = \rho\,\mathbf{e}_\rho$, $\partial_\phi\mathbf{r} = \partial_\phi\rho\,\mathbf{e}_\rho + \rho\,\mathbf{e}_\phi$, $\mathbf{F}_1(\mathbf{r}) = \mathbf{e}_\phi$.

$W_1[\gamma_S] = \displaystyle\int_0^{2\pi} \mathrm{d}\phi\,(\partial_\phi\mathbf{r})\cdot\mathbf{F}_1 = \int_0^{2\pi} \mathrm{d}\phi(\partial_\phi\rho\,\mathbf{e}_\rho + \rho\,\mathbf{e}_\phi)\cdot\mathbf{e}_\phi = \int_0^{2\pi} \mathrm{d}\phi\,\rho$

$\qquad = \displaystyle\int_0^{2\pi} \mathrm{d}\phi\,(R + \tfrac{1}{2\pi}\Delta\phi) = \left[R\phi + \tfrac{1}{4\pi}\Delta\phi^2\right]_0^{2\pi} = \boxed{2\pi R + \pi\Delta}$.

(b) Along the straight path γ_G we use Cartesian coordinates:

$\mathbf{r} = x\mathbf{e}_x$, $\partial_x\mathbf{r} = \mathbf{e}_x$, $\mathbf{F}_2(\mathbf{r}) = \mathbf{e}_x$.

$$W_2[\gamma_G] = \int_R^{R+\Delta} dx\,(\partial_x \mathbf{r})\cdot\mathbf{F}_2 = \int_R^{R+\Delta} dx\,\mathbf{e}_x\cdot\mathbf{e}_x = \int_R^{R+\Delta} dx = \boxed{\Delta}.$$

Along the spiral path γ_S we use polar coordinates, with $\mathbf{F}_2 = \mathbf{e}_x = \cos\phi\,\mathbf{e}_\rho - \sin\phi\,\mathbf{e}_\phi$.

$$W_2[\gamma_S] = \int_0^{2\pi} d\phi\,(\partial_\phi \mathbf{r})\cdot\mathbf{F}_2 = \int_0^{2\pi} d\phi(\partial_\phi \rho\,\mathbf{e}_\rho + \rho\,\mathbf{e}_\phi)\cdot(\cos\phi\,\mathbf{e}_\rho - \sin\phi\,\mathbf{e}_\phi)$$

$$= \int_0^{2\pi} d\phi\left[\tfrac{1}{2\pi}\Delta\cos\phi + (R + \tfrac{1}{2\pi}\phi\Delta)(-\sin\phi)\right]$$

$$= 0 + 0 - \tfrac{1}{2\pi}\Delta\int_0^{2\pi} d\phi\,\phi\sin\phi \overset{\text{part. int.}}{=} -\tfrac{1}{2\pi}\Delta(-2\pi) = \boxed{\Delta}.$$

Discussion: since $\mathbf{F}_2$ is a gradient field (with $\mathbf{F}_2 = \nabla x$), the value of a line integral depends on only the starting point and endpoint of its path. These are the same for γ_G and γ_S, hence $W[\gamma_G] = W[\gamma_S]$.

εV2.3.7 Line integral in spherical coordinates: satellite in orbit (p. 420)

(a) During the flight, $t_D = \pi/\omega_1$, θ varies linearly from 0 to $\omega_1 t_D = \pi$, and ϕ varies from 0 to $\omega_2 t_D = 10(2\pi)$. Therefore the spiral circles around the north–south axis $\boxed{10}$ times.

(b) $\mathbf{r}(t) = r(t)\,\mathbf{e}_r(t)$, with $r = r_0$, $\theta(t) = \omega_1 t$, $\phi(t) = \omega_2 t$.

$$\dot{\mathbf{r}} = \dot{r}\,\mathbf{e}_r + r\dot\theta\,\mathbf{e}_\theta + r\dot\phi\sin\theta\,\mathbf{e}_\phi = \boxed{r_0\omega_1\,\mathbf{e}_\theta + r_0\omega_2\sin(\omega_1 t)\,\mathbf{e}_\phi}.$$

(c) $$L[\gamma] = \int_0^{\pi/\omega_1} dt\,\|\mathbf{v}(t)\| = \boxed{\int_0^{\pi/\omega_1} dt\,r_0\sqrt{\omega_1^2 + \omega_2^2\sin^2(\omega_1 t)}}.$$

(d) $\mathbf{F} = -F_0\sin\theta\,\mathbf{e}_\phi$, $\dot{\mathbf{r}}(t)\cdot\mathbf{F}(\mathbf{r}(t)) = -F_0 r_0\omega_2\sin^2(\omega_1 t)$, since $\mathbf{e}_\theta\cdot\mathbf{e}_\phi = 0$, $\mathbf{e}_\phi\cdot\mathbf{e}_\phi = 1$.

$$W[\gamma] = \int_\gamma d\mathbf{r}\cdot\mathbf{F} = \int_0^{\pi/\omega_1} dt\,\dot{\mathbf{r}}(t)\cdot\mathbf{F}(\mathbf{r}(t)) = -F_0 r_0\omega_2\int_0^{\pi/\omega_1} dt\,\sin^2(\omega_1 t)$$

$$= -F_0 r_0\omega_2\frac{1}{2}\left[t - \frac{1}{\omega_1}\sin(\omega_1 t)\cos(\omega_1 t)\right]_0^{\pi/\omega_1} = \boxed{-F_0\pi r_0\frac{\omega_2}{2\omega_1}}.$$

S.V2.5 Local coordinate bases and linear algebra (p. 426)

εV2.5.1 Jacobian determinant for cylindrical coordinates (p. 428)

(a) Cylindrical coordinates: $\mathbf{x} = (x, y, z)^{\mathrm{T}} = (\rho\cos\phi, \rho\sin\phi, z)^{\mathrm{T}}$. Jacobi matrix for $\mathbf{y} \mapsto \mathbf{x}(\mathbf{y})$:

$$J = \frac{\partial(x, y, z)}{\partial(\rho, \phi, z)} = \begin{pmatrix} \frac{\partial x}{\partial\rho} & \frac{\partial x}{\partial\phi} & \frac{\partial x}{\partial z} \\ \frac{\partial y}{\partial\rho} & \frac{\partial y}{\partial\phi} & \frac{\partial y}{\partial z} \\ \frac{\partial z}{\partial\rho} & \frac{\partial z}{\partial\phi} & \frac{\partial z}{\partial z} \end{pmatrix} = \begin{pmatrix} \cos\phi & -\rho\sin\phi & 0 \\ \sin\phi & \rho\cos\phi & 0 \\ 0 & 0 & 1 \end{pmatrix}.$$

(b) Inverse transformation: $\mathbf{y} = (\rho, \phi, z)^{\mathrm{T}} = \left((x^2 + y^2)^{1/2}, \arctan\left(\frac{y}{x}\right), z\right)^{\mathrm{T}}$.

$$J^{-1} = \frac{\partial(\rho, \phi, z)}{\partial(x, y, z)} = \begin{pmatrix} \frac{\partial \rho}{\partial x} & \frac{\partial \rho}{\partial y} & \frac{\partial \rho}{\partial z} \\ \frac{\partial \phi}{\partial x} & \frac{\partial \phi}{\partial y} & \frac{\partial \phi}{\partial z} \\ \frac{\partial z}{\partial x} & \frac{\partial z}{\partial y} & \frac{\partial z}{\partial z} \end{pmatrix} = \boxed{\begin{pmatrix} \frac{x}{(x^2+y^2)^{1/2}} & \frac{y}{(x^2+y^2)^{1/2}} & 0 \\ -\frac{y}{x^2+y^2} & \frac{x}{x^2+y^2} & 0 \\ 0 & 0 & 1 \end{pmatrix}}$$

$$= \begin{pmatrix} \cos\phi & \sin\phi & 0 \\ -\frac{\sin\phi}{\rho} & \frac{\cos\phi}{\rho} & 0 \\ 0 & 0 & 1 \end{pmatrix}.$$

Check:

$$J \cdot J^{-1} = \begin{pmatrix} \cos\phi & -\rho\sin\phi & 0 \\ \sin\phi & \rho\cos\phi & 0 \\ 0 & 0 & 1 \end{pmatrix} \cdot \begin{pmatrix} \cos\phi & \sin\phi & 0 \\ -\frac{\sin\phi}{\rho} & \frac{\cos\phi}{\rho} & 0 \\ 0 & 0 & 1 \end{pmatrix} = \begin{pmatrix} 1 & 0 & 0 \\ 0 & 1 & 0 \\ 0 & 0 & 1 \end{pmatrix} = \mathbb{1} . \checkmark$$

(c) $\det(J) = \begin{vmatrix} \cos\phi & -\rho\sin\phi & 0 \\ \sin\phi & \rho\cos\phi & 0 \\ 0 & 0 & 1 \end{vmatrix} = \rho(\cos^2\phi + \sin^2\phi) = \boxed{\rho}.$

$$\det(J^{-1}) = \begin{vmatrix} \cos\phi & \sin\phi & 0 \\ -\frac{\sin\phi}{\rho} & \frac{\cos\phi}{\rho} & 0 \\ 0 & 0 & 1 \end{vmatrix} = \frac{1}{\rho}(\cos^2\phi + \sin^2\phi) = \boxed{\frac{1}{\rho}}.$$

Check: $\det(J) \cdot \det(J^{-1}) = 1 . \checkmark$

S.V3　Fields (p. 430)

S.V3.1　Definition of fields (p. 430)

εV3.1.1 Sketching a vector field (p. 431)

(a) The direction of the vector field $\mathbf{u}(\mathbf{r}) = (\cos y, 0)^T$ is always parallel to $\mathbf{e}_x$, independent of $\mathbf{r}$. For a fixed value of y the field has a fixed value, independent of x, depicted by arrows that all have the same length and direction. For a fixed value of x, the length and direction of the arrows change periodically with y, as $\cos(y)$. In particular, $\mathbf{u} = \mathbf{e}_x$ for $y = n2\pi$, $\mathbf{u} = -\mathbf{e}_x$ for $y = (n + \frac{1}{2})2\pi$, and $\mathbf{u} = \mathbf{0}$ for $y = (n + \frac{1}{2})\pi$, with $n \in \mathbb{Z}$.

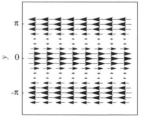

(b) The norm of the vector field $\mathbf{w}(\mathbf{r}) = (x^2 + y^2)^{-1/2}(x, -y)^T$ is independent of $\mathbf{r}$, $\|\mathbf{w}(\mathbf{r})\| = 1$, thus all arrows have the same length. On the x-axis we have $\mathbf{w}(\mathbf{r}) = \frac{x}{|x|}\mathbf{e}_x = \text{sign}(x)\mathbf{e}_x$, thus the arrows point outward (away from the origin). On the y-axis we have $\mathbf{w}(\mathbf{r}) = -\text{sign}(y)\mathbf{e}_y$, thus the arrows point inward (toward the origin). On the diagonal, $x = y$, we have $\mathbf{w}(\mathbf{r}) = \text{sign}(x)\frac{1}{\sqrt{2}}(1, -1)^T$, thus for $x > 0$ (or $x < 0$) all arrows point with slope -1 towards the bottom right (or the top left). Analogously for the other diagonal. Arrow directions between axes and diagonals follow by interpolation.

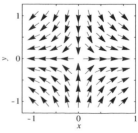

In both figures the axis labels refer to the units used for **r**-arrows from the domain of the map. The unit of length for arrows from the codomain has not been specified, hence only their direction and relative length carries any significance, not their absolute length. Moreover, the size of the arrow heads has been chosen proportional to the arrow length; this makes it visually clearer how the field strength varies.

S.V3.2 Scalar fields (p. 432)

εV3.2.1 Gradient of a mountainside (p. 437)

(a) The gradient and total differential are given by:

$$\nabla h_{\mathbf{r}} = \begin{pmatrix} \partial_x h \\ \partial_y h \end{pmatrix} = \begin{pmatrix} \dfrac{\sqrt{x^2+y^2}-x\cdot\dfrac{2x}{2\sqrt{x^2+y^2}}}{x^2+y^2} \\ \dfrac{-x\cdot\dfrac{2y}{2\sqrt{x^2+y^2}}}{x^2+y^2} \end{pmatrix} = \boxed{\dfrac{y}{r^3}\begin{pmatrix} y \\ -x \end{pmatrix}}.$$

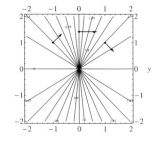

$$dh_{\mathbf{r}}(\mathbf{n}) = (\partial_x h)n_x + (\partial_y h)n_y = \boxed{\dfrac{y}{r^3}(yn_x - xn_y)}.$$

(b) The gradient vector, $\nabla h_{\mathbf{r}} = (y/r^3)(y, -x)^{\mathrm{T}}$, points in the direction of steepest increase; the unit vector parallel to it is $\hat{\mathbf{n}}_{\parallel} = \nabla h_{\mathbf{r}}/\|\nabla h_{\mathbf{r}}\| = \boxed{\mathrm{sign}(y)(y, -x)^{\mathrm{T}}/r}$.

(c) The contour lines at the point $\mathbf{r}$ are perpendicular to the gradient vector $\nabla h_{\mathbf{r}}$ and therefore run along the unit vector $\hat{\mathbf{n}}_{\perp} = \boxed{\mathrm{sign}(y)(x, y)^{\mathrm{T}}/r}$. (Verify that $dh_{\mathbf{r}}(\hat{\mathbf{n}}_{\perp}) = 0$, which confirms that h does not change along the direction of $\hat{\mathbf{n}}_{\perp}$.)

(d) The arrows with starting points $\mathbf{r}_1 = (-1, 1)^{\mathrm{T}}$, $\mathbf{r}_2 = (0, \sqrt{2})^{\mathrm{T}}$ and $\mathbf{r}_3 = (1, 1)^{\mathrm{T}}$ depict the vectors $\nabla h_{\mathbf{r}_1} = \boxed{2^{-3/2}(1, 1)^{\mathrm{T}}}$, $\nabla h_{\mathbf{r}_2} = \boxed{2^{-1/2}(1, 0)^{\mathrm{T}}}$ and $\nabla h_{\mathbf{r}_3} = \boxed{2^{-3/2}(1, -1)^{\mathrm{T}}}$, respectively.

(e) Yes: for $x = y \geq 0$, $h(x, x) = 1 + 1/\sqrt{2}$ is constant. Therefore, this defines a contour line which is at a height of $\boxed{1 + 1/\sqrt{2}}$.

(f) The contour line at a height of $h(\mathbf{r}) \equiv H$ is defined by the equation

$$H \equiv \dfrac{x}{\sqrt{x^2 + y^2}} + 1.$$

For a given value of x, we rearrange the equation, then square both sides and solve for y:

$$(H - 1)^2[x^2 + y^2] = x^2,$$

$$(H - 1)^2 y^2 = x^2[1 - (H - 1)^2] \qquad \Rightarrow \qquad y = \boxed{x\left[\dfrac{1}{(H - 1)^2} - 1\right]^{1/2}}.$$

Check: the contour at $y = x$ implies that $H = 1 + 1/\sqrt{2}$, and is therefore consistent with (iv).

(g) Regions where the mountainside is horizontally flat locally are given by the equation $\nabla h_{\mathbf{r}} = \mathbf{0}$. This is satisfied when $\boxed{y = 0 \text{ with } x \neq 0}$. The line $\{(x, 0)|x < 0\}$ defines the "valley", and is at a height of $\boxed{h=0}$. The line $\{(x,0)|x>0\}$ defines the "ridge" of the mountainside and is at a height of $\boxed{h=2}$.

(h) At the point $\boxed{\mathbf{r}=\mathbf{0}=(0,0)^{\mathrm{T}}}$, we find an infinitely steep, i.e. vertical "wall". This is evident from the fact that the gradient vector $\nabla h_{\mathbf{r}}$ is not well defined at that point, since it depends on the direction from which $\mathbf{r}$ approaches the point $\mathbf{0}$. For example, on the one hand $\lim_{x\to 0}[\nabla h_{(x,0)}]=(0,0)^{\mathrm{T}}$ (the valley and the ridges remain flat even for arbitrarily small $|x|$), while on the other hand, $\lim_{y\to 0}\left[\nabla h_{(0,y)}^{\mathrm{T}}\right]=\lim_{y\to 0}(1/|y|,0)=(\infty,0)^{\mathrm{T}}$ (the gradient in the x-direction along the line $x=0$ increases with decreasing values of $|y|$). Actually, the truly vertical part of the wall is infinitesimally narrow, because for all $\mathbf{r}\neq\mathbf{0}$, $\nabla h_{\mathbf{r}}$ is well-defined and finite.

εV3.2.3 Gradient of $\ln(1/r)$ (p. 437)

$$\varphi(\mathbf{r})=\ln\frac{1}{r},\quad r=\sqrt{x^2+y^2+z^2},$$

$$\nabla\varphi=\begin{pmatrix}\partial_x\\\partial_y\\\partial_z\end{pmatrix}\varphi=\varphi'(r)\begin{pmatrix}\partial_x\\\partial_y\\\partial_z\end{pmatrix}r=\frac{1}{r}\left[-\frac{1}{2}\right]\frac{1}{(x^2+y^2+z^2)^{1/2}}\begin{pmatrix}2x\\2y\\2z\end{pmatrix}=-\frac{1}{r^2}\begin{pmatrix}x\\y\\z\end{pmatrix}=-\frac{\hat{\mathbf{r}}}{r},$$

$$\|\nabla\varphi\|^2=\frac{1}{r^4}(x^2+y^2+z^2)=\frac{1}{r^2}\quad\Rightarrow\quad \|\nabla\varphi\|=1\quad\text{at}\quad\boxed{r=1}.$$

This describes a spherical surface around the origin, with radius 1.

S.V3.3 Extrema of functions with constraints (p. 439)

εV3.3.1 Minimal distance to hyperbola (p. 441)

We seek the point, (x,y), which minimizes the (squared) distance, $d^2=x^2+y^2$, under the constraint that it lies on the given hyperbola, $g(x,y)=x^2+8xy+7y^2-5c^2=0$. To this end we extremize the auxiliary function,

$$F(x,y,\lambda)=d(x,y)-\lambda g(x,y)=x^2+y^2-\lambda(x^2+8xy+7y^2-5c^2),$$

with Lagrange multiplier λ, with respect to all its variables:

$$0\overset{!}{=}\partial_x F=2x-2\lambda x-8\lambda y\qquad\Rightarrow\qquad x=\lambda(x+4y)\,;\tag{1}$$

$$0\overset{!}{=}\partial_y F=2y-8\lambda x-14\lambda y\qquad\Rightarrow\qquad y=\lambda(4x+7y)\,;\tag{2}$$

$$0\overset{!}{=}\partial_\lambda F=x^2+8xy+7y^2-5c^2\,.\tag{3}$$

Eliminating λ by forming the ratio of Eqs. (1) and (2), we obtain

$$\frac{x}{y}=\frac{x+4y}{4x+7y}\quad\Rightarrow\quad 4x^2+7xy=xy+4y^2\quad\Rightarrow\quad 4\left[(x/y)^2+\tfrac{3}{2}(x/y)-1\right]=0\,.$$

This equation has two possible solutions: (i) $x/y=-2$ and (ii) $x/y=\frac{1}{2}$. Inserted into Eq. (3), solution (i) yields $5c^2=y^2(2^2-8\cdot2+7)=-5$, which is a contradiction, since c^2 is positive. Solution (ii), though, yields a meaningful result:

$$5c^2=x^2(1+8\cdot2+2^2\cdot7)=45x^2,\quad\Rightarrow\quad x=\boxed{\pm\tfrac{1}{3}c},\quad y=\boxed{\pm\tfrac{2}{3}c}.$$

Therefore, the minimal distance is $d_{\min}=\sqrt{x^2+y^2}=\boxed{\tfrac{1}{3}\sqrt{5}c}$.

εV3.3.3 Intersecting planes: minimal distance to origin (p. 441)

We seek the extremum of the (squared) distance, $d^2(\mathbf{r}) = x^2 + y^2 + z^2$, of the point $\mathbf{r}$ to the origin, subject to the two constraints that $\mathbf{r}$ lies on both planes and hence satisfies $g_1(\mathbf{r}) = x + y + z - 1 = 0$ and $g_2(\mathbf{r}) = x - y + 2z - 2 = 0$. To this end, we extremize the auxiliary function

$$F(\mathbf{r}; \lambda_1, \lambda_2) = d^2(\mathbf{r}) - \lambda_1 g_1(\mathbf{r}) - \lambda_2 g_2(\mathbf{r})$$
$$= x^2 + y^2 + z^2 - \lambda_1 x - \lambda_1 y - \lambda_1 z + \lambda_1 - \lambda_2 x + \lambda_2 y - 2\lambda_2 z + 2\lambda_2 \,,$$

with Lagrange multipliers λ_1 and λ_2, with respect to all its variables. The conditions $\nabla F = 0$ and $\partial_{\lambda_1} F = 0$, $\partial_{\lambda_2} F = 0$ yield:

$$[\tilde{1}]: \quad 0 \overset{!}{=} \partial_x F = 2x - \lambda_1 - \lambda_2 \,,$$
$$[\tilde{2}]: \quad 0 \overset{!}{=} \partial_y F = 2y - \lambda_1 + \lambda_2 \,,$$
$$[\tilde{3}]: \quad 0 \overset{!}{=} \partial_y F = 2z - \lambda_1 - 2\lambda_2 \,.$$

$$[1]: \quad 0 \overset{!}{=} \partial_{\lambda_1} F = x + y + z - 1 \,,$$
$$[2]: \quad 0 \overset{!}{=} \partial_{\lambda_2} F = x - y + 2z - 2 \,.$$

We now use Gaussian elimination to eliminate λ_1 and λ_2 from equations $[\tilde{1}]$ to $[\tilde{3}]$:

	x	y	z	λ_1	λ_2	
$[\tilde{1}]$:	2	0	0	-1	-1	0
$[\tilde{2}]$:	0	2	0	-1	1	0
$[\tilde{3}]$:	0	0	2	-1	-2	0

$\longrightarrow$

	x	y	z	λ_1	λ_2	
$[\tilde{1}'] = \frac{1}{2}\left([\tilde{1}] + [\tilde{2}]\right)$:	1	1	0	-1	0	0
$[\tilde{2}'] = 2[\tilde{2}] + [\tilde{3}]$:	0	4	2	-3	0	0
$[\tilde{3}'] = 3[\tilde{1}'] - [\tilde{2}']$:	3	-1	-2	0	0	0

$[\tilde{3}']$ implies $3x - y - 2z = 0$. We solve this equation, together with the above equations $[1]$ and $[2]$, for x, y and z, again using Gaussian elimination:

	x	y	z	
$[1]$:	1	1	1	1
$[2]$:	1	-1	2	2
$[\tilde{3}']$:	3	-1	-2	0

$\longrightarrow$

	x	y	z	
$[1'] = [1] + [2]$:	2	0	3	3
$[2'] = [1] - [2]$:	0	2	-1	-1
$[\tilde{3}'] = \frac{1}{7}\left(3[1] - [\tilde{3}'] - 2[2']\right)$:	0	0	1	$\frac{5}{7}$

$\longrightarrow$

	x	y	z	
$\frac{1}{2}\left([1'] - 3[\tilde{3}']\right)$:	1	0	0	$\frac{3}{7}$
$\frac{1}{2}\left([2'] + [\tilde{3}']\right)$:	0	1	0	$-\frac{1}{7}$
$[\tilde{3}']$:	0	0	1	$\frac{5}{7}$

Thus the desired extremum lies at the point $\boxed{\mathbf{r} = (x, y, z)^{\mathrm{T}} = \tfrac{1}{7}(3, -1, 5)^{\mathrm{T}}}$.

εV3.3.5 Entropy maximization subject to constraints (p. 441)

(a) To minimize the entropy, $S(\{p_j\}) = -\sum_j p_j \ln p_j$, under the constraint of unit total probability, $g_1(\{p_j\}) = \sum_j p_j - 1 = 0$, we extremize the auxiliary function

$$F(\{p_j\}) = S(\{p_j\}) - \lambda_1 g_1(\{p_j\}) = -\sum_j p_j \ln p_j - \lambda_1 \Big(\sum_j p_j - 1\Big),$$

with Lagrange multiplier λ_1, with respect to all its variables:

$$0 \overset{!}{=} \partial_{p_j} F = -\ln p_j - 1 - \lambda_1 \quad \Rightarrow \quad \boxed{p_j = e^{-\lambda_1 - 1}} \,.$$

This implies that p_j does not depend on j: $p_j = p$. The constraints tell us that:

$$1 \overset{!}{=} \sum_{i=1}^{M} p_j = \sum_{i=1}^{M} p = Mp \quad \Rightarrow \quad \boxed{p = \frac{1}{M}} \,.$$

(b) To additionally impose the constraint of fixed average energy, $g_2(\{p_j\}) = \sum_j E_j p_j - E = 0$, we introduce a second Lagrange multiplier λ_2,

$$F(\{p_j\}) = S(\{p_j\}) - \lambda_1 g_1(\{p_j\}) - \lambda_2 g_2(\{p_j\})$$

$$= -\sum_j p_j \ln p_j - \lambda_1 \left(\sum_j p_j - 1\right) - \lambda_2 \left(\sum_j p_j E_j - E\right),$$

and extremize this auxiliary function with respect to all its variables:

$$0 \overset{!}{=} \partial_{p_j} F = -\ln p_j - 1 - \lambda_1 - \lambda_2 E_j \quad \Rightarrow \quad \boxed{p_j = e^{-\lambda_1 - 1} e^{-\lambda_2 E_j}}.$$

This implies that p_j depends exponentially on E_j. We define $\boxed{p_j = e^{-\beta E_j}/Z}$, with $e^{\lambda_1 + 1} \equiv Z$ and $\lambda_2 \equiv \beta$. Then the condition $\sum_j p_j = 1$ implies that the normalization constant has the form $\boxed{Z = \sum_j e^{-\beta E_j}}$.

S.V3.4   Gradient fields (p. 441)

εV3.4.1 Potential of a vector field (p. 444)

(a) Along the integration path $\gamma_1 = \{\mathbf{r}(t) = (t, t, t)^\mathrm{T} \mid 0 < t < 1\}$, we have $\frac{d\mathbf{r}}{dt} = (1, 1, 1)^\mathrm{T}$ for the velocity vector, and $\mathbf{u}(\mathbf{r}(t)) = (2t^2 + t^3, t^2, 3t^3)^\mathrm{T}$ for the vector field. Hence:

$$\int_{\gamma_1} d\mathbf{r} \cdot \mathbf{u}(\mathbf{r}) = \int_0^1 dt \, \frac{d\mathbf{r}}{dt} \cdot \mathbf{u}(\mathbf{r}(t)) = \int_0^1 dt \, (3t^2 + 4t^3) = \boxed{2}.$$

(b)
$$\mathbf{u}(\mathbf{r}) = \begin{pmatrix} u_x(\mathbf{r}) \\ u_y(\mathbf{r}) \\ u_z(\mathbf{r}) \end{pmatrix} = \begin{pmatrix} 2xy + z^3 \\ x^2 \\ 3xz^2 \end{pmatrix} \quad \Rightarrow \quad
\begin{aligned}
\partial_x u_y - \partial_y u_x &= -2x + 2x = \boxed{0}, \\
\partial_y u_z - \partial_z u_y &= 0 - 0 = \boxed{0}, \\
\partial_z u_x - \partial_x u_z &= -3z^2 + 3z^2 = \boxed{0}.
\end{aligned}$$

Therefore, $\partial_i u_j - \partial_j u_i = 0$ holds. Furthermore, the domain of $\mathbf{u}(\mathbf{r})$, which is $\mathbb{R}^3$, is simply connected. Therefore I_1 is independent of the path γ_1 between $\mathbf{0}$ and $\mathbf{b}$.

(c) Choose a suitable parametrization, e.g. $\gamma_{\mathbf{r}} = \{\mathbf{r}(t) = t\mathbf{r} = (tx, ty, tz) \mid 0 < t < 1\}$.

$$\mathbf{r}(t) = \big(x(t), y(t), z(t)\big)^\mathrm{T} = t\mathbf{r} = (tx, ty, tz)^\mathrm{T}, \quad \dot{\mathbf{r}}(t) = \mathbf{r} = (x, y, z)^\mathrm{T},$$

$$\mathbf{u}(\mathbf{r}(t)) = \Big(2x(t)y(t) + z^3(t), x^2(t), 3x(t)z^2(t)\Big)^\mathrm{T}$$

$$= \Big(2(tx)(ty) + (tz)^3, (tx)^2, 3(tx)(tz)^2\Big)^\mathrm{T},$$

$$\dot{\mathbf{r}}(t) \cdot \mathbf{u}(\mathbf{r}(t)) = 2x^2 y t^2 + xz^3 t^3 + yx^2 t^2 + 3xz^3 t^3 = 3x^2 y t^2 + 4xz^3 t^3,$$

$$\varphi(\mathbf{r}) = \int_{\gamma_{\mathbf{r}}} d\mathbf{r} \cdot \mathbf{u}(\mathbf{r}) = \int_0^1 dt \, \dot{\mathbf{r}}(t) \cdot \mathbf{u}(\mathbf{r}(t)) = \int_0^1 dt \, \big(3x^2 y t^2 + 4xz^3 t^3\big) = \boxed{x^2 y + xz^3}.$$

(d) Consistency check:

$$\nabla \varphi(\mathbf{r}) = \begin{pmatrix} \partial_x \\ \partial_y \\ \partial_z \end{pmatrix} (x^2 y + xz^3) = \boxed{\begin{pmatrix} 2xy + z^3 \\ x^2 \\ 3xz^2 \end{pmatrix}}. \qquad \text{Evidently} \quad \nabla \varphi(\mathbf{r}) \overset{\checkmark}{=} \mathbf{u}(\mathbf{r}).$$

(e)

$$I_1 = \int_{\gamma_1} d\mathbf{r} \cdot \mathbf{u}(\mathbf{r}) = \int_{\gamma_1} d\mathbf{r} \cdot \nabla\varphi(\mathbf{r}) = \varphi(\mathbf{b}) - \varphi(\mathbf{0}) = 1^2 \cdot 1 + 1 \cdot 1^3 - 0 = \boxed{2}.$$

This is in agreement with part (a) of the exercise! ✓

V3.4.3 Line integral of magnetic field of a current-carrying conductor (p. 446)

(a) For the given field, $\mathbf{B} = \frac{c}{x^2+y^2}(-y, x, 0)^{\mathsf{T}}$, the expression $\partial_i B^j - \partial_j B^i$ vanishes away from the z axis ($\sqrt{x^2 + y^2} \neq 0$):

$$\partial_z B^x - \partial_x B^z = \boxed{0}, \quad \partial_z B^y - \partial_y B^z = \boxed{0},$$

$$\partial_x B^y - \partial_y B^x = c\left[\frac{(x^2 + y^2) - 2x^2}{(x^2 + y^2)^2} - (-)\frac{(x^2 + y^2) - 2y^2}{(x^2 + y^2)^2} \right] = \boxed{0}.$$

(b) Along the circular path γ_C with radius R around the origin, with $t \in [0, 2\pi]$, one finds:

$$\mathbf{r}(t) = \big(x(t), y(t), z(t)\big)^{\mathsf{T}} = R(\cos(t), \sin(t), 0)^{\mathsf{T}}, \quad \dot{\mathbf{r}}(t) = R(-\sin(t), \cos(t), 0)^{\mathsf{T}},$$

$$\mathbf{B}(\mathbf{r}(t)) = \frac{c}{x(t)^2 + y(t)^2}(-y(t), x(t), 0)^{\mathsf{T}} = \frac{cR}{R^2}(-\sin(t), \cos(t), 0)^{\mathsf{T}},$$

$$\dot{\mathbf{r}}(t) \cdot \mathbf{B}(\mathbf{r}(t)) = c\big[\sin^2(t) + \cos^2(t)\big] = c,$$

$$W[\gamma_C] = \int_{\gamma_K} d\mathbf{r} \cdot \mathbf{B} = \int_0^{2\pi} dt\, \dot{\mathbf{r}}(t) \cdot \mathbf{B}(\mathbf{r}(t)) = \int_0^{2\pi} dt\, c = \boxed{2\pi c}.$$

(c) The line integral has four contributions, corresponding to the four edges of the rectangle:

$$\frac{1}{c}W[\gamma_R] = \int_1^2 dx \frac{0}{x^2 + 0^2} + \int_0^3 dy \frac{2}{2^2 + y^2} + \int_2^1 dx \frac{-3}{x^2 + 3^2} + \int_3^0 dy \frac{1}{1^2 + y^2}$$

$$\overset{\tilde{y} = \frac{y}{2}, \tilde{x} = \frac{x}{3}}{=} 0 + \int_0^{\frac{3}{2}} d\tilde{y} \frac{1}{1 + \tilde{y}^2} + \int_{\frac{1}{3}}^{\frac{2}{3}} d\tilde{x} \frac{1}{\tilde{x}^2 + 1^2} - \int_0^3 dy \frac{1}{1^2 + y^2}$$

$$= [\arctan(\tfrac{3}{2}) - \arctan(0)] + [\arctan(\tfrac{2}{3}) - \arctan(\tfrac{1}{3})] - [\arctan(3) - \arctan(0)]$$

$$= \tfrac{\pi}{2} - \tfrac{\pi}{2} = \boxed{0},$$

since $\arctan(\tfrac{A}{B}) + \arctan(\tfrac{B}{A}) = \tfrac{\pi}{2}$ holds for arbitrary positive numbers A and B.

(d) The line integral $W[\gamma] = \oint_\gamma d\mathbf{r} \cdot \mathbf{B}$ along a closed curve $\gamma \in \mathbb{R}^3$ vanishes if, and only if, there exists a domain U_γ with the following properties: (i) the domain U_γ encloses the entire curve ($\gamma \subset U_\gamma$); the relation $\partial_i B^j = \partial_j B^i$ holds throughout the domain U_γ; (iii) the domain U_γ is *simply connected*. In other words: the line integral vanishes if, and only if, it is possible to shrink the integration path down to a point, *without* leaving the domain on which $\partial_i B^j = \partial_j B^i$ holds.

For the present vector field $\mathbf{B}$, $\partial_i B^j = \partial_j B^i$ holds, according to (a), in a domain that is not simply connected: $\mathbb{R}^3$ without the z axis, $\mathbb{R}^3/\{(0, 0, z)^{\mathsf{T}} | z \in \mathbb{R}\}$ [or, for fixed $z = 0$, the xy-plane without the origin, $\mathbb{R}^2/(0, 0)^{\mathsf{T}}$]. Therefore the line integral $W[\gamma]$ along a closed curve γ vanishes if, and only if, the curve does *not* encircle the z axis. This is the case for the rectangular path γ_R of part (c), but not for the circular path γ_C of part (b).

The figure shows examples of two domains (shaded), a rectangle U_{γ_R} and a ring U_{γ_C}, that satisfy properties (i) and (ii): (i) both domains enclose the corresponding integration paths, γ_R or γ_C, respectively, and (ii) the curl of $\mathbf{B}$ vanishes throughout both domains. However, property (iii) holds only for the first domain: the rectangle U_{γ_R} is simply connected, the ring U_{γ_C} is not.

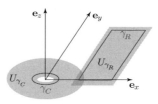

Conclusion: $\partial_i B^j = \partial_j B^i$ is a necessary, but not a sufficient condiction for $\oint d\mathbf{r} \cdot \mathbf{B} = 0$. The latter requires $\partial_i B^j = \partial_j B^i$ to hold on a *simply connected* domain.

S.V3.5 Sources of vector fields (p. 446)

ᵣV3.5.1 Divergence (p. 447)

(a) $\nabla \cdot \mathbf{u} = \partial_i u^i = \partial_x(xyz) + \partial_y(y^2) + \partial_z(z^3) = \boxed{yz + 2y + 3z^2}$.

(b) $\nabla \cdot [\mathbf{a}f(r)] = \partial_i[a^i f(r)] = a^i f'(r)\,\partial_i r = a^i \dfrac{r^i}{r} f'(r) = \boxed{\dfrac{\mathbf{r} \cdot \mathbf{a}}{r} f'(r)}$.

Here we used $\partial_i r = \partial_i (r^j r^j)^{1/2} = \frac{1}{2}(r^j r^j)^{-1/2}\partial_i(r^j r^j) = \frac{1}{2}r^{-1}2r^i = r^i/r$.

ᵣV3.5.3 Gauss's theorem – cube (Cartesian coordinates) (p. 450)

C is a cube with edge length a (see sketch). Let S_1 to S_6 be its 6 faces with normal vectors $\mathbf{n}_{1,2} = \pm\mathbf{e}_x$, $\mathbf{n}_{3,4} = \pm\mathbf{e}_y$, $\mathbf{n}_{5,6} = \pm\mathbf{e}_z$. We seek the flux, $\Phi = \int_S d\mathbf{S} \cdot \mathbf{u}$, of the vector field $\mathbf{u}(\mathbf{r}) = (x^2, y^2, z^2)^T$, through the cube's surface, $S \equiv \partial C \equiv S_1 \cup S_2 \cup \cdots \cup S_6$.

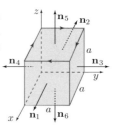

(a)   Direct calculation of $\Phi = \sum_{i=1}^{6} \Phi_i$, with $\Phi_i = \int_{S_i} d\mathbf{S} \cdot \mathbf{u}$, yields

$$\Phi_1 + \Phi_2 = \int_0^a dy \int_0^a dz \left[\underbrace{(\mathbf{n}_1 \cdot \mathbf{u})_{x=a}}_{a^2} + \underbrace{(\mathbf{n}_2 \cdot \mathbf{u})_{x=0}}_{0} \right] = a^4 .$$

Analogously: $\Phi_3 + \Phi_4 = \Phi_5 + \Phi_6 = a^4$. We thus obtain $\Phi = \boxed{3a^4}$.

(b)   Alternatively, using Gauss's theorem, $\Phi = \int_S d\mathbf{S} \cdot \mathbf{u} \overset{\text{Gauss}}{=} \int_C dV \nabla \cdot \mathbf{u}$, we obtain

$$\Phi = \int_0^a \int_0^a \int_0^a dx\, dy\, dz\, \nabla \cdot \mathbf{u} = \int_0^a \int_0^a \int_0^a dx\, dy\, dz\, (2x + 2y + 2z)$$

$$= [x^2]_0^a \cdot [y]_0^a \cdot [z]_0^a + [x]_0^a \cdot [y^2]_0^a \cdot [z]_0^a + [x]_0^a \cdot [y]_0^a \cdot [z^2]_0^a = \boxed{3a^4} \overset{\checkmark}{=} \text{(a)} .$$

ᵣV3.5.5 Computing volume of barrel using Gauss's theorem (p. 450)

(a)   For a cylinder, the surface, S, consists of a bottom, side wall, and top, $S = B \cup W \cup T$. Let $\mathbf{u}$ be a vector field with the property that $\nabla \cdot \mathbf{u} = 1$, then the volume V of the cylinder can be determined via the following flux integral through its surface.

Volume: $V = \displaystyle\int_V dV = \int_V dV(\nabla \cdot \mathbf{u}) = \int_S d\mathbf{S} \cdot \mathbf{u} = \Phi_B + \Phi_W + \Phi_T$.

In cylindrical coordinates, $\mathbf{r} = \mathbf{e}_\rho\,\rho + \mathbf{e}_z\,z$, the *outward*-facing surface elements for the three surfaces are given by:

bottom: $d\mathbf{S} = -\mathbf{e}_z\,\rho\,d\rho\,d\phi$, $z = 0$,

side wall: $d\mathbf{S} = \;\;\mathbf{e}_\rho\,R\,d\phi\,dz$, $\rho = R$,

top: $d\mathbf{S} = \;\;\mathbf{e}_z\,\rho\,d\rho\,d\phi$, $z = h$.

The field $\mathbf{u} = \frac{1}{2}(x, y, 0)^\mathsf{T}$ satisfies $\nabla\cdot\mathbf{u} = 1$ and in cylindrical coordinates takes the form $\mathbf{u} = \mathbf{e}_\rho\frac{1}{2}\rho$. Therefore $d\mathbf{S}\cdot\mathbf{u} = 0$ for the bottom and top, thus only the side wall contributes, with $d\mathbf{S}\cdot\mathbf{u} = (\mathbf{e}_\rho\,R\,d\phi\,dz)\cdot(\mathbf{e}_\rho\frac{1}{2}R) = d\phi\,dz\frac{1}{2}R^2$:

$$V = \Phi_W = \int_W d\mathbf{S}\cdot\mathbf{u} = \int_0^{2\pi} d\phi \int_0^h dz\,\tfrac{1}{2}R^2 = \boxed{\pi R^2 h}\,.\checkmark$$

Alternatively, albeit less elegantly, one could choose $\mathbf{u} = \frac{1}{3}\mathbf{r}$, for example, which also fulfills the condition $\nabla\cdot\mathbf{u} = 1$. Since $\mathbf{r} = \mathbf{e}_\rho\rho + \mathbf{e}_z z$, we get:

bottom: $\Phi_B = \dfrac{1}{3}\displaystyle\int_B d\mathbf{S}\cdot\mathbf{u} = -\dfrac{1}{3}\int_0^R \rho\,d\rho\int_0^{2\pi}d\phi\cdot 0 = 0$,

side wall: $\Phi_W = \dfrac{1}{3}\displaystyle\int_W d\mathbf{S}\cdot\mathbf{u} = \dfrac{1}{3}\int_0^{2\pi}d\phi\int_0^h dz\,R^2 = \tfrac{2}{3}\pi h R^2$,

top: $\Phi_T = \dfrac{1}{3}\displaystyle\int_T d\mathbf{S}\cdot\mathbf{u} = \dfrac{1}{3}\int_0^R \rho\,d\rho\int_0^{2\pi}d\phi\,h = \tfrac{1}{3}\tfrac{1}{2}R^2 2\pi h = \tfrac{1}{3}\pi h R^2$.

Combined: $V = \Phi_B + \Phi_W + \Phi_T = 0 + \tfrac{2}{3}\pi h R^2 + \tfrac{1}{3}\pi h R^2 = \boxed{\pi R^2 h}\,.\checkmark$

(b) For a barrel with z-dependent radius, $\rho(z)$, the surface of the side wall can still be parametrized using cylindrical coordinates: $S = \{\mathbf{r} = \mathbf{e}_\rho\rho(z) + \mathbf{e}_z z \mid \phi \in (0, 2\pi), z \in (0, h)\}$. The oriented surface element, $d\mathbf{S} = d\phi\,dz\,(\partial_\phi\mathbf{r}\times\partial_z\mathbf{r})$, then takes the form

$$d\mathbf{S} = d\phi\,dz\big(\mathbf{e}_\phi\rho(z)\times[\mathbf{e}_\rho\,d_z\rho(z) + \mathbf{e}_z]\big) = d\phi\,dz\big(-\mathbf{e}_z d_z\rho(z) + \mathbf{e}_\rho\big)\rho(z),$$

and for $\mathbf{u} = \frac{1}{2}(x, y, 0)^\mathsf{T} = \mathbf{e}_\rho\frac{1}{2}\rho$ the corresponding flux element is $d\mathbf{S}\cdot\mathbf{u} = d\phi\,dz\,\frac{1}{2}\rho^2(z)$. The volume of a barrel with radius $\rho(z) = R\big[1 + a\sin(\pi z/h)\big]^{1/2}$ thus is

$$V = \int_W d\mathbf{S}\cdot\mathbf{u} = \int_0^{2\pi}d\phi\int_0^h dz\,\tfrac{1}{2}\rho^2(z) = 2\pi\int_0^h dz\,\tfrac{1}{2}R^2\big[1 + a\sin(\pi z/h)\big]$$

$$= \pi R^2\Big[z - a\tfrac{h}{\pi}\cos(\pi z/h)\Big]_0^h = \boxed{\pi R^2 h\big[1 + \tfrac{2a}{\pi}\big]}\,.$$

ₑV3.5.7 Flux integral: flux of electric field through cylinder (p. 450)

We choose the symmetry axis of the cylinder to be the z-axis, and use cylindrical coordinates.

Position vector: $\mathbf{r} = \rho\,\mathbf{e}_\rho + z\,\mathbf{e}_z$, $r^2 = \rho^2 + z^2$.

Electric field: $\mathbf{E}(\mathbf{r}) = E_0\dfrac{\mathbf{r}}{r^3} = E_0\dfrac{\rho\,\mathbf{e}_\rho + z\,\mathbf{e}_z}{(\rho^2 + z^2)^{\frac{3}{2}}}$.

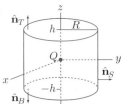

(a)   The calculations of the flux through both the top (T) and the bottom (B) are analogous; we thus treat them together, with the upper/lower sign referring to T/B:

$$\Phi_{T/B} = \int_0^{2\pi} d\phi \int_0^R d\rho\, \rho\, (\hat{\mathbf{n}}_{T/B} \cdot \mathbf{E})_{z=\pm h} = 2\pi \int_0^R d\rho\, \rho\, \frac{E_0 h}{(\rho^2 + h^2)^{\frac{3}{2}}} = -\left[\frac{2\pi E_0 h}{(\rho^2 + h^2)^{\frac{1}{2}}}\right]_0^R$$

$$= \boxed{2\pi E_0 h \left[\frac{1}{h} - \frac{1}{(R^2 + h^2)^{\frac{1}{2}}}\right]}.$$

The above form of the surface integral is obtained via the following considerations. At the top/bottom faces, we have $z = \pm h$ and $\mathbf{r}(\rho, \phi) = \rho\, \mathbf{e}_\rho \mp h \mathbf{e}_z$. Therefore these faces are spanned by the coordinate basis vectors $\partial_\rho \mathbf{r} = \mathbf{e}_\rho$ and $\partial_\phi \mathbf{r} = \rho \mathbf{e}_\phi$. Their normal vectors, $\partial_\rho \mathbf{r} \times \partial_\phi \mathbf{r} = \rho\, \mathbf{e}_z$, are parallel to $\mathbf{e}_z$, as expected. We choose the signs of the corresponding normalized vectors, $\hat{\mathbf{n}}_T$ and $\hat{\mathbf{n}}_B$ respectively, such that they point *outwards*, hence:

normal vector: $\hat{\mathbf{n}}_{T/B} = \pm\dfrac{\partial_\rho \mathbf{r} \times \partial_\phi \mathbf{r}}{\|\partial_\rho \mathbf{r} \times \partial_\phi \mathbf{r}\|} = \pm \mathbf{e}_z$, $\hat{\mathbf{n}}_{T/B} \cdot \mathbf{E}\big|_{z=\pm h} = E_0 \dfrac{(\pm 1)(\pm h)}{(\rho^2 + (\pm h)^2)^{\frac{3}{2}}}$;

surface element: $d\mathbf{S}_{T/B} = dS\, \hat{\mathbf{n}}_{T/B}$, $dS = d\rho\, d\phi\, \|\partial_\rho \mathbf{r} \times \partial_\phi \mathbf{r}\| = d\rho\, d\phi\, \rho$.

As expected, $dS = \|d\mathbf{S}\| = d\rho\, d\phi\, \rho$ corresponds to the surface element for polar coordinates.

(b)   The calculation of the flux through the side wall (W) proceeds as follows:

$$\Phi_W = \int_0^{2\pi} d\phi\, R \int_{-h}^h dz\, (\hat{\mathbf{n}}_W \cdot \mathbf{E})_{\rho=R} = 2\pi R \int_{-h}^h dz\, \frac{E_0 R}{(R^2 + z^2)^{\frac{3}{2}}} = \left[\frac{2\pi E_0 z}{(R^2 + z^2)^{\frac{1}{2}}}\right]_{-h}^h$$

$$= \boxed{\frac{4\pi E_0 h}{(R^2 + h^2)^{\frac{1}{2}}}}.$$

The above form of the surface integral results from the following considerations. Along the side face, we have $\rho = R$, $\mathbf{r}(\phi, z) = R\, \mathbf{e}_\rho + z \mathbf{e}_z$, with coordinate basis vectors $\partial_\phi \mathbf{r} = R \mathbf{e}_\phi$ and $\partial_z \mathbf{r} = \mathbf{e}_z$. Their normal vector, $\partial_\phi \mathbf{r} \times \partial_z \mathbf{r} = R \mathbf{e}_\rho$, is parallel to $\mathbf{e}_\rho$ and directed outwards, as expected. The corresponding normalized vector reads:

normal vector: $\hat{\mathbf{n}}_W = \dfrac{\partial_\phi \mathbf{r} \times \partial_z \mathbf{r}}{\|\partial_\phi \mathbf{r} \times \partial_z \mathbf{r}\|} = \mathbf{e}_\rho$, $\hat{\mathbf{n}}_S \cdot \mathbf{E}\big|_{\rho=R} = E_0 \dfrac{R}{(R^2 + z^2)^{\frac{3}{2}}}$;

surface element: $d\mathbf{S}_W = dS\, \hat{\mathbf{n}}_W$, $dS = d\phi\, dz\, \|\partial_\phi \mathbf{r} \times \partial_z \mathbf{r}\| = d\phi\, R\, dz$.

As expected, dS is the product of the line element for integration along a circle with radius R, $R\, d\phi$, and the line element in the z-direction, dz.

The integral for Φ_W may be calculated as follows (with $s = z/R$):

$$I = \int ds\, \frac{1}{(1 + s^2)^{\frac{3}{2}}} \quad \text{[Substitution: } s = \sinh y,\ ds = dy\cosh y,\ \sqrt{1 + s^2} = \cosh y.]$$

$$= \int dy\, \frac{\cosh y}{\cosh^3 y} = \int dy\, \frac{1}{\cosh^2 y} = \tanh y = \frac{\sinh y}{\cosh y} = \boxed{\frac{s}{(1 + s^2)^{\frac{1}{2}}}}.$$

Check: $\dfrac{d}{ds} \dfrac{s}{(1 + s^2)^{\frac{1}{2}}} = \dfrac{1}{(1 + s^2)^{\frac{1}{2}}} - \dfrac{s^2}{(1 + s^2)^{\frac{3}{2}}} = \dfrac{1 + s^2 - s^2}{(1 + s^2)^{\frac{3}{2}}} = \dfrac{1}{(1 + s^2)^{\frac{3}{2}}} \cdot \checkmark$

For the total outward flux, Φ_W cancels the second term in $\Phi_T + \Phi_B$, with the result:

$$\Phi_C = \Phi_T + \Phi_B + \Phi_W = \boxed{4\pi E_0} = \boxed{4\pi Q}.$$

This is an example of Gauss's law for electrostatics: the flux of an electric field through a closed surface, which encloses an electric charge Q_{tot}, is always equal to $\Phi = 4\pi Q_{\text{tot}}$.

S.V3.6   Circulation of vector fields (p. 454)

εV3.6.1 Curl (p. 454)

(a) $\displaystyle \mathbf{V} \times \mathbf{u} = \mathbf{e}_i \epsilon_{ijk} \partial_j u^k = \begin{pmatrix} \partial_y u^z - \partial_z u^y \\ \partial_z u^x - \partial_x u^z \\ \partial_x u^y - \partial_y u^x \end{pmatrix} = \begin{pmatrix} \partial_y(z^2) - \partial_z(y^2) \\ \partial_z(xyz) - \partial_x(z^2) \\ \partial_x(y^2) - \partial_y(xyz) \end{pmatrix} = \boxed{\begin{pmatrix} 0 \\ xy \\ -xz \end{pmatrix}}.$

(b)   We use Cartesian coordinates and put all indices downstairs:

$$\mathbf{V} \times [\mathbf{a}f(r)] = \mathbf{e}_i \epsilon_{ijk} \partial_j a_k f(r) = \mathbf{e}_i \epsilon_{ijk} \partial_j r a_k f'(r) = \mathbf{e}_i \epsilon_{ijk} \frac{r_j}{r} a_k f'(r) = \boxed{\frac{\mathbf{r} \times \mathbf{a}}{r} f'(r)}.$$

εV3.6.3 Stokes's theorem – cuboid (Cartesian coordinates) (p. 456)

C is a cube with edge length a (see sketch). Let S_1 to S_6 be its 6 faces with normal vectors $\mathbf{n}_{1,2} = \pm\mathbf{e}_x$, $\mathbf{n}_{3,4} = \pm\mathbf{e}_y$, $\mathbf{n}_{5,6} = \pm\mathbf{e}_z$. Given the vector field $\mathbf{w}(\mathbf{r}) = (-y^2, x^2, 0)^T$, we seek the flux of its curl, $\Phi = \int_S d\mathbf{S} \cdot (\mathbf{V} \times \mathbf{w})$, through the surface $S \equiv \partial C \backslash \text{top} = S_1 \cup S_2 \cup S_3 \cup S_4 \cup S_6$.

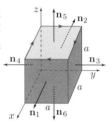

(a)   Direct calculation of $\Phi = \Phi_1 + \Phi_2 + \Phi_3 + \Phi_4 + \Phi_6$:

$$\mathbf{V} \times \mathbf{w} = (0, 0, 2x + 2y)^T = \mathbf{e}_z 2(x + y).$$

$$\Phi_i = \int_{S_i} d\mathbf{S}_i \cdot (\mathbf{V} \times \mathbf{w}) = \int_{S_i} dS_i \, (\mathbf{n}_i)_z \, 2(x + y).$$

For each face ($i = 1, 2, 3, 4$), the corresponding normal vector lies perpendicular to $\mathbf{e}_z$, therefore $(\mathbf{n}_i)_z = 0$. Consequently, they make no contribution to the flux: $\Phi_i = 0$. The flux through the bottom surface, S_6, thus gives the total flux:

$$\Phi = \Phi_6 = -\int_0^a dx \int_0^a dy \, 2(x + y) = -[x^2]_0^a [y]_0^a - [x]_0^a [y^2]_0^a = \boxed{-2a^3}.$$

(b)   Alternatively, using Stokes's theorem: $\displaystyle \Phi = \int_S d\mathbf{S} \cdot (\mathbf{V} \times \mathbf{w}) \overset{\text{Stokes}}{=} \oint_{\partial S} d\mathbf{r} \cdot \mathbf{w}.$

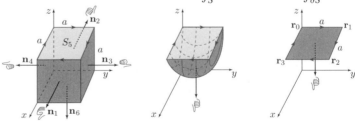

The five faces $S = S_1 \cup S_2 \cup S_3 \cup S_4 \cup S_6$ (coloured dark grey in the sketch) have the same boundary as the top face S_5 (coloured light grey), hence $\partial S = \partial S_5$. Because of the outward/downward orientation of the surface S, the line integral around the top face must be performed in the *clockwise* direction. This follows from the right-hand rule applied to the normal vector of the faces (left sketch). [Alternatively, one can consider a deformation of the surface S into the face S_5

(see middle and right sketches), follow how the normal vectors to the surface change during the deformation, and then apply the right-hand rule to the resulting normal vectors. Such a deformation is indeed legitimate, since via Stokes's theorem, the flux integral of the curl of a vector field only depends on the boundary of the surface, not on the rest of the shape.]

To calculate the line integral, let $\mathbf{r}_0 = (0, 0, a)^T$, $\mathbf{r}_1 = (0, a, a)^T$, $\mathbf{r}_2 = (a, a, a)^T$, $\mathbf{r}_3 = (a, 0, a)^T$ and parametrize the line segments by $t \in I = (0, a)$:

$$\gamma_1[\mathbf{r}_0 \to \mathbf{r}_1]: \qquad \mathbf{r}(t) = (0, t, a)^T, \qquad\qquad\qquad \dot{\mathbf{r}}(t) = (0, 1, 0)^T,$$
$$\mathbf{w}(\mathbf{r}(t)) = (-t^2, 0, 0)^T, \qquad [\dot{\mathbf{r}}(t) \cdot \mathbf{w}(\mathbf{r}(t))]_{\gamma_1} = 0.$$

$$\gamma_2[\mathbf{r}_1 \to \mathbf{r}_2]: \qquad \mathbf{r}(t) = (t, a, a)^T, \qquad\qquad\qquad \dot{\mathbf{r}}(t) = (1, 0, 0)^T,$$
$$\mathbf{w}(\mathbf{r}(t)) = (-a^2, t^2, 0)^T, \qquad [\dot{\mathbf{r}}(t) \cdot \mathbf{w}(\mathbf{r}(t))]_{\gamma_2} = -a^2.$$

$$\gamma_3[\mathbf{r}_2 \to \mathbf{r}_3]: \qquad \mathbf{r}(t) = (a, a - t, a)^T, \qquad\qquad \dot{\mathbf{r}}(t) = (0, -1, 0)^T,$$
$$\mathbf{w}(\mathbf{r}(t)) = (-(1 - t)^2, a^2, 0)^T, \qquad [\dot{\mathbf{r}}(t) \cdot \mathbf{w}(\mathbf{r}(t))]_{\gamma_3} = -a^2.$$

$$\gamma_4[\mathbf{r}_3 \to \mathbf{r}_0]: \qquad \mathbf{r}(t) = (a - t, 0, a)^T, \qquad\qquad \dot{\mathbf{r}}(t) = (-1, 0, 0)^T,$$
$$\mathbf{w}(\mathbf{r}(t)) = (0, (1 - t)^2, 0)^T, \qquad [\dot{\mathbf{r}}(t) \cdot \mathbf{w}(\mathbf{r}(t))]_{\gamma_4} = 0.$$

The line integral along the boundary, $\partial S = \gamma_1 \cup \gamma_2 \cup \gamma_3 \cup \gamma_4$, yields:

$$\Phi = \int_{\partial S} d\mathbf{r} \cdot \mathbf{w} = \int_0^a dt \left[0 - a^2 - a^2 + 0 \right] = \boxed{-2a^3} \overset{\checkmark}{=} (a).$$

EV3.6.5 **Stokes's theorem – magnetic dipole (spherical coordinates)** (p. 456)

(a) Magnetic field:
$$\mathbf{B} = \frac{1}{c} \frac{\mathbf{e}_r 3r(\mathbf{e}_z m \cdot \mathbf{e}_r r) - \mathbf{e}_z m r^2}{r^5}$$
$$= \frac{m}{c} \frac{1}{r^3} (\mathbf{e}_r 3(\mathbf{e}_z \cdot \mathbf{e}_r) - \mathbf{e}_z) \overset{\mathbf{e}_r \mathbf{e}_z = \cos\theta}{=} \frac{m}{cr^3} (\mathbf{e}_r 3\cos(\theta) - \mathbf{e}_z).$$

Surface element:
$$d\mathbf{S} = \mathbf{e}_r dS = \mathbf{e}_r \sin\theta R^2 d\phi \, d\theta.$$

Flux:
$$\Phi = \int_H d\mathbf{S} \cdot \mathbf{B} = \frac{m}{cR^3} \int_0^{2\pi} d\phi \int_0^{\pi/2} d\theta \, R^2 \sin(\theta)(\mathbf{e}_r 3\cos(\theta) - \mathbf{e}_z) \cdot \mathbf{e}_r$$
$$= \frac{m}{cR^3} 2\pi R^2 \underbrace{\int_0^{\pi/2} d\theta \, \sin(\theta) \cos(\theta)}_{\frac{1}{2}\sin^2(\theta)|_0^{\pi/2} = \frac{1}{2}} (3 - 1) = \boxed{\frac{2\pi m}{cR}}.$$

(b) By Stokes's theorem, $\int_H d\mathbf{S} \cdot \mathbf{B} = \oint_\gamma d\mathbf{r} \cdot \mathbf{A}$, the flux can be computed as a line integral of $\mathbf{A}$ along the hemisphere's boundary, a circle with radius R. We parametrize it in spherical coordinates as $\mathbf{r}(\phi) = \mathbf{e}_r R$ with $\theta = \pi/2$, $\phi \in (0, 2\pi)$, and $\frac{d\mathbf{r}}{d\phi} = \partial_\phi \mathbf{e}_r R = \mathbf{e}_\phi R$.

Vector field :
$$\mathbf{A} = \frac{1}{c} \frac{\mathbf{e}_z m \times \mathbf{e}_r r}{r^3} = \mathbf{e}_\phi \frac{1}{c} \frac{m}{r^2} \quad \Rightarrow \quad \frac{d\mathbf{r}}{d\phi} \cdot \mathbf{A} = R \frac{1}{c} \frac{m}{R^2} = \frac{m}{cR}.$$

Flux:
$$\Phi = \oint_\gamma d\mathbf{r} \cdot \mathbf{A} = \frac{m}{cR} \int_0^{2\pi} d\phi \, \frac{d\mathbf{r}}{d\phi} \cdot \mathbf{A} = \frac{m}{cR} \int_0^{2\pi} d\phi = \boxed{\frac{2\pi m}{cR}}. \checkmark$$

S.V3.7   Practical aspects of three-dimensional vector calculus (p. 459)

ɛV3.7.1 Curl of gradient field (p. 460)

The identity to be proven is a vector relation. We consider its ith component:

$$\left[\boldsymbol{\nabla}\times(\boldsymbol{\nabla}f)\right]_i = \epsilon_{ijk}\partial_j\partial_k f \overset{(i)}{=} -\epsilon_{ikj}\partial_j\partial_k f \overset{(ii)}{=} -\epsilon_{ijk}\partial_k\partial_j f \overset{(iii)}{=} -\epsilon_{ijk}\partial_j\partial_k f = -\left[\boldsymbol{\nabla}\times(\boldsymbol{\nabla}f)\right]_i .$$

For step (i) we used the antisymmetry of the Levi–Civita symbol under exchange of indices; for (ii) we relabeled summation indices as $j \leftrightarrow k$; for (iii) we used Schwarz's theorem to change the order of taking partial derivatives. We have thus shown that $\left[\boldsymbol{\nabla}\times(\boldsymbol{\nabla}f)\right]_i$ is equal to minus itself. Hence it must vanish, implying $\boxed{\boldsymbol{\nabla}\times(\boldsymbol{\nabla}f) = \mathbf{0}}$.

ɛV3.7.3 Nabla identities (p. 460)

(a)
$$\boldsymbol{\nabla}f = \begin{pmatrix}\partial_x\\\partial_y\\\partial_z\end{pmatrix} ze^{-x^2} = e^{-x^2}\boxed{\begin{pmatrix}-2xz\\0\\1\end{pmatrix}}, \quad \boldsymbol{\nabla}g = \begin{pmatrix}\partial_x\\\partial_y\\\partial_z\end{pmatrix} yz^{-1} = \boxed{\begin{pmatrix}0\\z^{-1}\\-yz^{-2}\end{pmatrix}}.$$

$$\boldsymbol{\nabla}^2 f = \left(\partial_x^2 + \partial_y^2 + \partial_z^2\right) ze^{-x^2} = \boxed{-2ze^{-x^2}\left(1 - 2x^2\right)}.$$

$$\boldsymbol{\nabla}^2 g = \left(\partial_x^2 + \partial_y^2 + \partial_z^2\right) yz^{-1} = \boxed{2yz^{-3}}.$$

$$\boldsymbol{\nabla}\cdot\mathbf{u} = \partial_x u_x + \partial_y u_y + \partial_z u_z = \partial_x(x^2 y) = \boxed{2xy}.$$

$$\boldsymbol{\nabla}\times\mathbf{u} = \begin{pmatrix}\partial_y u_z - \partial_z u_y\\\partial_z u_x - \partial_x u_z\\\partial_x u_y - \partial_y u_x\end{pmatrix} = \begin{pmatrix}\partial_y 0 - \partial_z 0\\\partial_z(x^2 y) - \partial_x 0\\\partial_x 0 - \partial_y(x^2 y)\end{pmatrix} = \begin{pmatrix}0\\0\\-x^2\end{pmatrix} = \boxed{-\mathbf{e}_z x^2}.$$

$$\boldsymbol{\nabla}\cdot\mathbf{w} = \partial_x w_x + \partial_y w_y + \partial_z w_z = \partial_x(x^2 + y^3) + \partial_y 0 + \partial_z 0 = \boxed{2x}.$$

$$\boldsymbol{\nabla}\times\mathbf{w} = \begin{pmatrix}\partial_y w_z - \partial_z w_y\\\partial_z w_x - \partial_x w_z\\\partial_x w_y - \partial_y w_x\end{pmatrix} = \begin{pmatrix}\partial_y 0 - \partial_z 0\\\partial_z(x^2 + y^3) - \partial_x 0\\\partial_x 0 - \partial_y(x^2 + y^3)\end{pmatrix} = \begin{pmatrix}0\\0\\-3y^2\end{pmatrix} = \boxed{-\mathbf{e}_z 3y^2}.$$

(b) Equations (i) and (ii) are vector equations, which we will consider for a specific component, say i. In contrast, (iii) is a scalar equation.

(i) $\left[\boldsymbol{\nabla}(fg)\right]_i = \partial_i(fg) = f(\partial_i g) + g(\partial_i f) = f(\boldsymbol{\nabla}g)_i + g(\boldsymbol{\nabla}f)_i \checkmark$

(ii) $\left[\mathbf{u}\times(\boldsymbol{\nabla}\times\mathbf{w}) + \mathbf{w}\times(\boldsymbol{\nabla}\times\mathbf{u}) + (\mathbf{u}\cdot\boldsymbol{\nabla})\mathbf{w} + (\mathbf{w}\cdot\boldsymbol{\nabla})\mathbf{u}\right]_i$

$= \epsilon_{ijk}u_j(\boldsymbol{\nabla}\times\mathbf{w})_k + \epsilon_{ijk}w_j(\boldsymbol{\nabla}\times\mathbf{u})_k + u_j\partial_j w_i + w_j\partial_j u_i$

$= \underbrace{\epsilon_{ijk}\epsilon_{klm}}_{=\delta_{il}\delta_{jm}-\delta_{im}\delta_{jl}}\left(u_j\partial_l w_m + w_j\partial_l u_m\right) + u_j\partial_j w_i + w_j\partial_j u_i$

$= u_j\partial_i w_j - u_j\partial_j w_i + w_j\partial_i u_j - w_j\partial_j u_i + u_j\partial_j w_i + w_j\partial_j u_i$

$= u_j\partial_i w_j + w_j\partial_i u_j = \partial_i\left(u_j w_j\right) = [\boldsymbol{\nabla}(\mathbf{u}\cdot\mathbf{w})]_i \checkmark$

(iii) $\boldsymbol{\nabla}\cdot(f\mathbf{u}) = \partial_i(f u_i) = f\partial_i u_i + u_i\partial_i f = f(\boldsymbol{\nabla}\cdot\mathbf{u}) + \mathbf{u}\cdot(\boldsymbol{\nabla}f) \checkmark$

(c)　Using the results from (a) we obtain:

(i)　　　$\mathbf{\nabla}(fg) = \mathbf{\nabla}(y e^{-x^2}) = \boxed{e^{-x^2}\begin{pmatrix} -2xy \\ 1 \\ 0 \end{pmatrix}}$,

$f(\mathbf{\nabla}g) + g(\mathbf{\nabla}f) = z e^{-x^2}\begin{pmatrix} 0 \\ z^{-1} \\ -yz^{-2} \end{pmatrix} + yz^{-1}e^{-x^2}\begin{pmatrix} -2xz \\ 0 \\ 1 \end{pmatrix} = \boxed{e^{-x^2}\begin{pmatrix} -2xy \\ 1 \\ 0 \end{pmatrix}} \overset{\checkmark}{=} \mathbf{\nabla}(fg);$

(ii)　　　$\mathbf{\nabla}(\mathbf{u}\cdot\mathbf{w}) = \begin{pmatrix} \partial_x \\ \partial_y \\ \partial_z \end{pmatrix}\left(x^2 y(x^2 + y^3) + 0 + 0 \right) = \boxed{\begin{pmatrix} 4x^3 y + 2xy^4 \\ x^4 + 4x^2 y^3 \\ 0 \end{pmatrix}}$,

$\mathbf{u}\times(\mathbf{\nabla}\times\mathbf{w}) + \mathbf{w}\times(\mathbf{\nabla}\times\mathbf{u}) + (\mathbf{u}\cdot\mathbf{\nabla})\mathbf{w} + (\mathbf{w}\cdot\mathbf{\nabla})\mathbf{u}$

$= \begin{pmatrix} x^2 y \\ 0 \\ 0 \end{pmatrix}\times\begin{pmatrix} 0 \\ 0 \\ -3y^2 \end{pmatrix} + \begin{pmatrix} x^2+y^3 \\ 0 \\ 0 \end{pmatrix}\times\begin{pmatrix} 0 \\ 0 \\ -x^2 \end{pmatrix} + x^2 y\, \partial_x\begin{pmatrix} x^2+y^3 \\ 0 \\ 0 \end{pmatrix} + (x^2+y^3)\partial_x\begin{pmatrix} x^2 y \\ 0 \\ 0 \end{pmatrix}$

$= \begin{pmatrix} 0 \\ 3x^2 y^3 \\ 0 \end{pmatrix} + \begin{pmatrix} 0 \\ x^4 + x^2 y^3 \\ 0 \end{pmatrix} + \begin{pmatrix} 2x^3 y \\ 0 \\ 0 \end{pmatrix} + \begin{pmatrix} 2x^3 y + 2xy^4 \\ 0 \\ 0 \end{pmatrix} = \boxed{\begin{pmatrix} 4x^3 y + 2xy^4 \\ x^4 + 4x^2 y^3 \\ 0 \end{pmatrix}} \overset{\checkmark}{=} \mathbf{\nabla}(\mathbf{u}\cdot\mathbf{w});$

(iii)　　　$\mathbf{\nabla}\cdot(f\mathbf{u}) = \mathbf{\nabla}\cdot\begin{pmatrix} ze^{-x^2}x^2 y \\ 0 \\ 0 \end{pmatrix} = \boxed{yze^{-x^2}\left(-2x^3 + 2x \right)}$,

$f(\mathbf{\nabla}\cdot\mathbf{u}) + \mathbf{u}\cdot(\mathbf{\nabla}f) = ze^{-x^2}(2xy) + \begin{pmatrix} x^2 y \\ 0 \\ 0 \end{pmatrix}\cdot\begin{pmatrix} -2xze^{-x^2} \\ 0 \\ 0 \end{pmatrix} = e^{-x^2}\left(2xyz - 2x^3 yz \right)$

$= \boxed{yze^{-x^2}\left(-2x^3 + 2x \right)} \overset{\checkmark}{=} \mathbf{\nabla}\cdot(f\mathbf{u})\ .$

ɛV3.7.5　Gradient, divergence, curl, Laplace in cylindrical coordinates (p. 461)

(a)　　　$\mathbf{r}(\rho,\phi,z) = (\rho\cos\phi,\ \rho\sin\phi,\ z)^{\mathrm{T}}\ .$

$\partial_\rho\mathbf{r} \equiv n_\rho\mathbf{e}_\rho,$　　with　　$n_\rho = 1,$　　　$\mathbf{e}_\rho = (\cos\phi,\ \sin\phi,\ 0)^{\mathrm{T}}.$

$\partial_\phi\mathbf{r} \equiv n_\phi\mathbf{e}_\phi,$　　with　　$n_\phi = \rho,$　　　$\mathbf{e}_\phi = (-\sin\phi,\ \cos\phi,\ 0)^{\mathrm{T}}.$

$\partial_z\mathbf{r} \equiv n_z\mathbf{e}_z,$　　with　　$n_z = 1,$　　　$\mathbf{e}_z = (0,\ 0,\ 1)^{\mathrm{T}}.$

(b)　　　$\mathbf{\nabla}f = \mathbf{e}_\rho\frac{1}{n_\rho}\partial_\rho f + \mathbf{e}_\phi\frac{1}{n_\phi}\partial_\phi f + \mathbf{e}_z\frac{1}{n_z}\partial_z f = \boxed{\mathbf{e}_\rho\partial_\rho f + \mathbf{e}_\phi\frac{1}{\rho}\partial_\phi f + \mathbf{e}_z\partial_z f}\ .$

(c)　　　$\mathbf{\nabla}\cdot\mathbf{u} = \frac{1}{n_\rho n_\phi n_z}\left[\partial_\rho\left(n_\phi n_z u^\rho \right) + \partial_\phi\left(n_z n_\rho u^\phi \right) + \partial_z\left(n_\rho n_\phi u^z \right) \right]$

$= \frac{1}{\rho}\left[\partial_\rho\left(\rho u^\rho \right) + \partial_\phi u^\phi + \partial_z\left(\rho u^z \right) \right] = \boxed{\frac{1}{\rho}\partial_\rho\left(\rho u^\rho \right) + \frac{1}{\rho}\partial_\phi u^\phi + \partial_z u^z}\ .$

(d) $\nabla \times \mathbf{u} = \mathbf{e}_\rho \dfrac{1}{n_\phi n_z}\Big[\partial_\phi\big(n_z u^z\big) - \partial_z\big(n_\phi u^\phi\big)\Big] + \mathbf{e}_\phi \dfrac{1}{n_z n_\rho}\Big[\partial_z\big(n_\rho u^\rho\big) - \partial_\rho\big(n_z u^z\big)\Big]$

$\qquad\qquad + \mathbf{e}_z \dfrac{1}{n_\rho n_\phi}\Big[\partial_\rho\big(n_\phi u^\phi\big) - \partial_\phi\big(n_\rho u^\rho\big)\Big]$

$\qquad = \mathbf{e}_\rho \dfrac{1}{\rho}\Big[\partial_\phi\big(u^z\big) - \partial_z\big(\rho u^\phi\big)\Big] + \mathbf{e}_\phi\Big[\partial_z\big(u^\rho\big) - \partial_\rho\big(u^z\big)\Big] + \mathbf{e}_z \dfrac{1}{\rho}\Big[\partial_\rho\big(\rho u^\phi\big) - \partial_\phi\big(u^\rho\big)\Big]$

$\qquad = \boxed{\mathbf{e}_\rho\Big[\dfrac{1}{\rho}\partial_\phi u^z - \partial_z u^\phi\Big] + \mathbf{e}_\phi\Big[\partial_z u^\rho - \partial_\rho u^z\Big] + \mathbf{e}_z \dfrac{1}{\rho}\Big[\partial_\rho\big(\rho u^\phi\big) - \partial_\phi u^\rho\Big]}$.

(e) $\nabla^2 f = \nabla \cdot \nabla f$

$\qquad = \dfrac{1}{\rho}\partial_\rho\big(\rho \partial_\rho f\big) + \dfrac{1}{\rho}\partial_\phi\Big(\dfrac{1}{\rho}\partial_\phi f\Big) + \partial_z\big(\partial_z f\big) = \boxed{\dfrac{1}{\rho}\partial_\rho\big(\rho\partial_\rho f\big) + \dfrac{1}{\rho^2}\partial_\phi^2 f + \partial_z^2 f}$.

(f) $\mathbf{D} \equiv \nabla f = \mathbf{e}_\eta \underbrace{\dfrac{1}{n_\eta}\partial_\eta f}_{\equiv D^\eta} + \mathbf{e}_\mu \underbrace{\dfrac{1}{n_\mu}\partial_\mu f}_{\equiv D^\mu} + \mathbf{e}_\nu \underbrace{\dfrac{1}{n_\nu}\partial_\nu f}_{\equiv D^\nu}$,

$\qquad \nabla \times \mathbf{D} = \mathbf{e}_\eta \dfrac{1}{n_\mu n_\nu}\Big[\partial_\mu\big(n_\nu D^\nu\big) - \partial_\nu\big(n_\mu D^\mu\big)\Big] + \;\overset{\eta\;\mu}{\underset{\nu}{\curvearrowright}}\; + \;\overset{\eta\;\mu}{\underset{\nu}{\circlearrowleft}}$,

$\qquad \nabla \times (\nabla f) = \mathbf{e}_\eta \dfrac{1}{n_\mu n_\nu}\Big[\partial_\mu\Big(n_\nu \dfrac{1}{n_\nu}\partial_\nu f\Big) - \partial_\nu\Big(n_\mu \dfrac{1}{n_\mu}\partial_\mu f\Big)\Big] + \;\overset{\eta\;\mu}{\underset{\nu}{\curvearrowright}}\; + \;\overset{\eta\;\mu}{\underset{\nu}{\circlearrowleft}}$

$\qquad = \mathbf{e}_\eta \dfrac{1}{n_\mu n_\nu}\Big[\partial_\mu \partial_\nu - \partial_\nu \partial_\mu\Big]f + \;\overset{\eta\;\mu}{\underset{\nu}{\curvearrowright}}\; + \;\overset{\eta\;\mu}{\underset{\nu}{\circlearrowleft}}$

$\qquad = \boxed{0}$ [using Schwarz's theorem] .   ✓

(g) In cylindrical coordinates, the fields read $f(\mathbf{r}) = \|\mathbf{r}\|^2 = \rho^2 + z^2$ and $\mathbf{u}(\mathbf{r}) = (x, y, 2z)^\mathsf{T} = (\rho\cos\phi, \rho\sin\phi, 2z)^\mathsf{T} = \mathbf{e}_\rho \rho + \mathbf{e}_z 2z = \mathbf{e}_\rho u^\rho + \mathbf{e}_\phi u^\phi + \mathbf{e}_z u^z$, with $u^\rho = \rho$, $u^\phi = 0$, $u^z = 2z$.

$\qquad \nabla f = \mathbf{e}_\rho \partial_\rho f + \mathbf{e}_\phi \dfrac{1}{\rho}\partial_\phi f + \mathbf{e}_z \partial_z f = \boxed{\mathbf{e}_\rho 2\rho + \mathbf{e}_z 2z}$.

$\qquad \nabla \cdot \mathbf{u} = \dfrac{1}{\rho}\partial_\rho\big(\rho u^\rho\big) + \dfrac{1}{\rho}\partial_\phi u^\phi + \partial_z u^z = \dfrac{1}{\rho}\partial_\rho\big(\rho^2\big) + \dfrac{1}{\rho}\partial_\phi(0) + \partial_z(2z) = 2 + 2 = \boxed{4}$.

$\qquad \nabla \times \mathbf{u} = \mathbf{e}_\rho\Big[\dfrac{1}{\rho}\partial_\phi u^z - \partial_z u^\phi\Big] + \mathbf{e}_\phi\Big[\partial_z u^\rho - \partial_\rho u^z\Big] + \mathbf{e}_z \dfrac{1}{\rho}\Big[\partial_\rho\big(\rho u^\phi\big) - \partial_\phi u^\rho\Big]$

$\qquad = \mathbf{e}_\rho\Big[\dfrac{1}{\rho}\partial_\phi(2z) - \partial_z(0)\Big] + \mathbf{e}_\phi\Big[\partial_z(0) - \partial_\rho(2z)\Big] + \mathbf{e}_z \dfrac{1}{\rho}\Big[\partial_\rho(\rho\cdot 0) - \partial_\phi(0)\Big] = \boxed{0}$.

$\qquad \nabla^2 f = \dfrac{1}{\rho}\partial_\rho\big(\rho\partial_\rho f\big) + \dfrac{1}{\rho^2}\partial_\phi^2 f + \partial_z^2 f = \dfrac{1}{\rho}\partial_\rho\big(\rho\cdot 2\rho\big) + 0 + 2 = 4 + 2 = \boxed{6}$.

ᴇV3.7.7 Gradient, divergence, curl (spherical coordinates) (p. 461)

(a)   We use Cartesian coordinates, $\mathbf{r} = (x, y, z)^\mathsf{T}$, with $r = \sqrt{x^2 + y^2 + z^2}$, and write all indices downstairs, denoting the components of $\mathbf{r}$ by $x_j = x, y, z$. Since both f and $\mathbf{u}$ depend on the radius r, it is advisable to compute some partial derivatives beforehand: The scalar field f

depends only on the radius, $f(\mathbf{r}) = 1/r$. Consequently, the partial derivative of $f(r(x, y, z))$ with respect to x_j can be computed as follows using the chain rule:

$$\partial_j r = \partial r / \partial x_j = x_j / r \qquad \text{with} \qquad x_j = x, y, z , \tag{1}$$

$$\partial_r f = \partial_r (1/r) = -1/r^2 , \tag{2}$$

$$\partial_j f = (\partial_r f)(\partial_j r) \overset{(1)}{=} (\partial_r f)(x_j/r) \overset{(2)}{=} -x_j/r^3 , \tag{3}$$

$$\nabla f \overset{(3)}{=} \boxed{-\mathbf{r}/r^3} . \tag{4}$$

The vector field $\mathbf{u} = (e^{-r/a}/r)\mathbf{r}$ has Cartesian components $u_j = R(r)x_j$, where

$$R = \frac{e^{-r/a}}{r} , \qquad \text{with} \qquad \partial_j R = (\partial_r R)(\partial_j r) = (\partial_r R)(x_j/r) , \tag{5}$$

$$\partial_r R = \left(-\frac{1}{r^2} - \frac{1}{ar} \right) e^{-r/a} = \left(-\frac{1}{r} - \frac{1}{a} \right) R . \tag{6}$$

In the following, we use Einstein summation notation (e.g. $x_j x_j = r^2$),

$$\nabla \cdot \mathbf{u} = \partial_j u_j = (\partial_j R)x_j + R(\partial_j x_j)$$

$$\overset{(5)}{=} (\partial_r R) \underbrace{(x_j/r)x_j}_{= r^2/r = r} + 3R \overset{(6)}{=} \left[-1 - \frac{r}{a} + 3 \right] R = \boxed{\left[2 - \frac{r}{a} \right] \frac{e^{-r/a}}{r}} .$$

$$\nabla \times \mathbf{u} = \partial_i u_j \varepsilon_{ijk} \mathbf{e}_k = \partial_i (R x_j) \varepsilon_{ijk} \mathbf{e}_k = \left[(\partial_i R)x_j + R(\partial_i x_j) \right] \varepsilon_{ijk} \mathbf{e}_k$$

$$= \left[(\partial_r R)(x_i/r)x_j + R\delta_{ij} \right] \varepsilon_{ijk} \mathbf{e}_k = \boxed{\mathbf{0}} \qquad \text{(using antisymmetry of } \varepsilon_{ijk}) .$$

$$\nabla^2 f = \partial_j [\partial_j f] \overset{(3)}{=} \partial_j \left[-x_j/r^3 \right] = -\left[(\partial_j x_j)/r^3 + x_j \partial_j (1/r^3) \right]$$

$$= -\left[3/r^3 + x_j \partial_r (1/r^3)(\partial_j r) \right] = -\left[3/r^3 + \underbrace{x_j (-3/r^4)(x_j/r)}_{-3r^2/r^5} \right] = \boxed{0} .$$

(b)　Spherical coordinates:

$$f(r, \theta, \phi) = 1/r, \qquad \mathbf{u}(r, \theta, \phi) = \mathbf{e}_r\, e^{-r/a}, \quad \Rightarrow \quad u^r = e^{-r/a}, \quad u^\theta = 0, \quad u^\phi = 0 .$$

$$\nabla f = \left(\mathbf{e}_r \partial_r + \mathbf{e}_\theta \frac{1}{r} \partial_\theta + \mathbf{e}_\phi \frac{1}{r \sin\theta} \partial_\phi \right) \frac{1}{r} = \boxed{-\mathbf{e}_r \frac{1}{r^2}} .$$

$$\nabla \cdot \mathbf{u} = \frac{1}{r^2} \partial_r \left(r^2 u^r \right) + \frac{1}{r \sin\theta} \partial_\theta \left(\sin\theta\, u^\theta \right) + \frac{1}{r \sin\theta} \partial_\phi u^\phi$$

$$= \frac{1}{r^2} \partial_r (r^2 e^{-r/a}) = \frac{1}{r^2} \left(2r - \frac{r^2}{a} \right) e^{-r/a} = \boxed{\left(2 - \frac{r}{a} \right) \frac{e^{-r/a}}{r}} .$$

$$\nabla \times \mathbf{u} = \mathbf{e}_r \frac{1}{r \sin\theta} \left(\partial_\theta \left(\sin\theta\, u^\phi \right) - \partial_\phi u^\theta \right) + \mathbf{e}_\theta \frac{1}{r} \left(\frac{1}{\sin\theta} \partial_\phi u^r - \partial_r (r u^\phi) \right)$$

$$+ \mathbf{e}_\phi \frac{1}{r} \left(\partial_r (r u^\theta) - \partial_\theta u^r \right) = \boxed{\mathbf{0}} .$$

$$\nabla^2 f = \frac{1}{r^2} \partial_r \left(r^2 \partial_r f \right) + \frac{1}{r^2 \sin\theta} \partial_\theta \left(\sin\theta\, \partial_\theta f \right) + \frac{1}{r^2 \sin^2\theta} \partial_\phi^2 f = \frac{1}{r^2} \partial_r \left(-r^2 \frac{1}{r^2} \right) = \boxed{0} .$$

As expected, Cartesian and spherical coordinates yield the same results, but the latter more elegantly exploit the fact that f depends only on r, and $\mathbf{u}$ only on r and $\mathbf{e}_r$.

ₑV3.7.9 Gauss's theorem – cylinder (cylindrical coordinates) (p. 461)

(a) We compute the divergence of the vector field $\mathbf{u} = \mathbf{e}_\rho \, z\rho$ in cylindrical coordinates:

$$\boldsymbol{\nabla} \cdot \mathbf{u} = \frac{1}{\rho} \partial_\rho \left(\rho u^\rho \right) + \frac{1}{\rho} \partial_\phi u^\phi + \partial_z u^z = \frac{1}{\rho} \partial_\rho \left(z\rho^2 \right) + 0 + 0 = \boxed{2z}.$$

(b) Only the side wall contributes to the flux, since at the top and bottom faces $\mathrm{d}\mathbf{S} \propto \mathbf{e}_z \perp \mathbf{u}$.

$$\Phi = \int_S \mathrm{d}\mathbf{S} \cdot \mathbf{u} = \int_{\text{side wall}} \mathrm{d}\mathbf{S} \cdot \mathbf{u} = \int_0^{2\pi} \mathrm{d}\phi \int_0^H \mathrm{d}z \, \left(\mathbf{e}_\rho R \right) \cdot \left(\mathbf{e}_\rho R z \right) = \boxed{\pi H^2 R^2}.$$

(c) $$\Phi = \int_S \mathrm{d}\mathbf{S} \cdot \mathbf{u} \overset{\text{Gauss}}{=} \int_V \mathrm{d}V \boldsymbol{\nabla} \cdot \mathbf{u} = \int_0^{2\pi} \mathrm{d}\phi \int_0^H \mathrm{d}z \int_0^R \mathrm{d}\rho \rho \, (2z) = \boxed{\pi H^2 R^2}.$$

ₑV3.7.11 Stokes's theorem – magnetic field of infinite current-carrying wire (cylindrical coordinates) (p. 461)

(a) Cartesian coordinates, with $\sqrt{x^2 + y^2} \neq 0$:

$$\mathbf{B} = \frac{2I}{c} \frac{1}{x^2 + y^2} \left(-y, x, 0 \right)^{\mathrm{T}},$$

$$\boldsymbol{\nabla} \cdot \mathbf{B} = \sum_i \partial_i B^i = \frac{2I}{c} \left(-\frac{2x(-y)}{(x^2+y^2)^2} - \frac{2yx}{(x^2+y^2)^2} \right) = \boxed{0},$$

$$\boldsymbol{\nabla} \times \mathbf{B} = \begin{pmatrix} \partial_y B^z - \partial_z B^y \\ \partial_z B^x - \partial_x B^z \\ \partial_x B^y - \partial_y B^x \end{pmatrix} = \frac{2I}{c} \begin{pmatrix} 0 \\ 0 \\ \frac{1}{x^2+y^2} - \frac{2x^2}{(x^2+y^2)^2} + \frac{1}{x^2+y^2} - \frac{2y^2}{(x^2+y^2)^2} \end{pmatrix} = \boxed{0}.$$

(b) Now we repeat the calculations in cylindrical coordinates, $\mathbf{r}(\rho, \phi, z)$, with $\rho > 0$:

$$\mathbf{B} = \mathbf{e}_\phi \frac{2I}{c} \frac{1}{\rho}, \quad \Rightarrow \quad B^\rho = B^z = 0, \quad B^\phi = \frac{2I}{c} \frac{1}{\rho}, \tag{1}$$

$$\boldsymbol{\nabla} \cdot \mathbf{B} = \frac{1}{\rho} \partial_\rho \left(\rho B^\rho \right) + \frac{1}{\rho} \partial_\phi B^\phi + \partial_z B^z = \boxed{0},$$

$$\boldsymbol{\nabla} \times \mathbf{B} = \mathbf{e}_\rho \left(\frac{1}{\rho} \partial_\phi B^z - \partial_z B^\phi \right) + \mathbf{e}_\phi \left(\partial_z B^\rho - \partial_\rho B^z \right) + \mathbf{e}_z \frac{1}{\rho} \left(\partial_\rho \left(\rho B^\phi \right) - \partial_\phi B^\rho \right)$$

$$= \mathbf{e}_\rho 0 + \mathbf{e}_\phi 0 + \mathbf{e}_z \frac{2I}{c} \frac{1}{\rho} \partial_\rho (1) = \boxed{0}.$$

The same results are obtained using Cartesian and cylindrical coordinates, but the latter more elegantly exploit the fact $\mathbf{B}$ depends only on ρ and $\mathbf{e}_\phi$.

(c) Parametrization of the path: $\mathbf{r}(\phi) = \mathbf{e}_\rho R$, with $\phi \in (0, 2\pi)$, and $\frac{\mathrm{d}\mathbf{r}(\phi)}{\mathrm{d}\phi} = \mathbf{e}_\phi R$.

$$\oint_\gamma \mathrm{d}\mathbf{r} \cdot \mathbf{B} = \int_0^{2\pi} \mathrm{d}\phi \frac{\mathrm{d}\mathbf{r}}{\mathrm{d}\phi} \cdot \mathbf{B}(\mathbf{r}) = \int_0^{2\pi} \mathrm{d}\phi \, R B^\phi \overset{(1)}{=} \int_0^{2\pi} \mathrm{d}\phi \, R \frac{2I}{cR} = 2\pi \frac{2I}{c} = \boxed{\frac{4\pi}{c} I}. \tag{2}$$

(d) Using Stokes's theorem, we see immediately that:

$$\int_D \mathrm{d}\mathbf{S} \cdot (\boldsymbol{\nabla} \times \mathbf{B}) \overset{\text{Stokes}}{=} \oint_\gamma \mathrm{d}\mathbf{r} \cdot \mathbf{B} \overset{(2)}{=} \boxed{\frac{4\pi}{c} I}. \tag{3}$$

(e)   On the one hand, it follows from (a) that $\nabla \times \mathbf{B} = 0$ for all spatial points with $\rho > 0$, i.e. for all points except those that lie directly on the z-axis. On the other hand, it follows from (d) that the flux integral of $\nabla \times \mathbf{B}$ does not vanish over the disk D, but rather is equal to $4\pi I/c$. This appears paradoxical at first: how can the flux integral of a vector field yield a finite value if the field apparently vanishes everywhere? However, notice that the calculation in part (a) does not apply for the case $\rho = 0$, hence we have no reason to conclude that the curl vanishes on the z-axis. The fact that the flux integral of $\nabla \times \mathbf{B}$ yields a finite value, although the integrand vanishes everywhere except at $\rho = 0$, tells us that $\nabla \times \mathbf{B}$ must be proportional to a two-dimensional δ-function, peaked at $\rho = 0$:

$$\nabla \times \mathbf{B} = \boxed{C\,\mathbf{e}_z\,\delta(x)\delta(y)}. \tag{4}$$

The direction of $\nabla \times \mathbf{B}$ is equal to $\mathbf{e}_z$ from symmetry arguments, since $\mathbf{B}$ points in the $\mathbf{e}_\phi$-direction, and $\mathbf{e}_z$ is the sole unit vector that stays orthogonal to $\mathbf{e}_\phi$ for all angles ϕ. The constant C can be determined as follows:

$$\frac{4\pi I}{c} \overset{(3)}{=} \int_D d\mathbf{S}\cdot(\nabla\times\mathbf{B}) \overset{(4)}{=} \int_D dS\,\mathbf{e}_z\cdot C\mathbf{e}_z\,\delta(x)\delta(y) = C\underbrace{\int_D dS\,\delta(x)\delta(y)}_{=1}, \Rightarrow \boxed{C = \tfrac{4\pi}{c}I}.$$

(f)   Inserting the above result for C into (4) yields $\nabla \times \mathbf{B} = \frac{4\pi}{c}\,\mathbf{j}(\mathbf{r})$, with $\mathbf{j}(\mathbf{r}) = \mathbf{e}_z\,I\delta(x)\delta(y)$. This corresponds to Ampère's law (one of the Maxwell equations), where $\mathbf{j}(\mathbf{r})$ is the current density of an infinitesimally thin conductor with current I along z-axis.

εV3.7.13 Gauss's theorem – electrical dipole potential (spherical coordinates) (p. 461)

(a)   We have: $\partial_i x_i = x_i/r$ with $x_i = x, y, z$.

$$\mathbf{E} = -\nabla\Phi = -\begin{pmatrix} \partial_x\Phi \\ \partial_y\Phi \\ \partial_z\Phi \end{pmatrix} = -p\begin{pmatrix} -\frac{3zx}{r^5} \\ -\frac{3zy}{r^5} \\ -\frac{3z^2}{r^5} + \frac{1}{r^3} \end{pmatrix} = \boxed{p\left(3\frac{z}{r^5}\mathbf{r} - \frac{1}{r^3}\mathbf{e}_z\right)}.$$

(b)   In spherical coordinates, the field takes on the form $\Phi(\mathbf{r}) = p\frac{\cos\theta}{r^2}$.

$$\mathbf{E} = -\nabla\Phi = -\left(\mathbf{e}_r\partial_r + \mathbf{e}_\theta\frac{1}{r}\partial_\theta + \mathbf{e}_\phi\frac{1}{r\sin\theta}\partial_\phi\right)\Phi = \boxed{p\left(\frac{2\cos\theta}{r^3}\mathbf{e}_r + \frac{\sin\theta}{r^3}\mathbf{e}_\theta\right)}.$$

Since $\mathbf{e}_z = \cos\theta\mathbf{e}_r - \sin\theta\mathbf{e}_\theta$, this corresponds to the Cartesian result:

$$\mathbf{E} = p\Big(3\underbrace{\frac{\cos\theta}{r^3}}_{z/r^4}\mathbf{e}_r - \frac{1}{r^3}\underbrace{(\cos\theta\mathbf{e}_r - \sin\theta\mathbf{e}_\theta)}_{\mathbf{e}_z}\Big)$$

(c)   Using Cartesian coordinates:

$$\nabla\cdot\mathbf{E} = \sum_i \frac{\partial}{\partial x_i}E_i = p\left(\frac{9z}{r^5} + \frac{3z}{r^5} - 15\frac{z}{r^6}\left(\frac{x^2}{r} + \frac{y^2}{r} + \frac{z^2}{r}\right) + \frac{3}{r^4}\frac{z}{r}\right) = \boxed{0}.$$

$$\nabla\times\mathbf{E} = p\begin{pmatrix} -\frac{15z^2y}{r^7} + \frac{3y}{r^5} - \frac{3y}{r^5} + \frac{15z^2y}{r^7} \\ \frac{3x}{r^5} - \frac{15z^2x}{r^7} + \frac{15z^2x}{r^7} - \frac{3x}{r^5} \\ -\frac{15zyx}{r^7} + \frac{15zxy}{r^7} \end{pmatrix} = \boxed{0}.$$

(d)   Using spherical coordinates:

$$\nabla \cdot \mathbf{E} = \frac{1}{r^2} \partial_r \left(r^2 E_r\right) + \frac{1}{r \sin \theta} \partial_\theta \left(\sin \theta E_\theta\right) + \frac{1}{r \sin \theta} \partial_\phi E_\phi$$

$$= p \left[\frac{1}{r^2} \partial_r \left(\frac{2 \cos \theta}{r}\right) + \frac{1}{r \sin \theta} \partial_\theta \left(\frac{\sin^2 \theta}{r^3}\right)\right] = p \left[\frac{-2 \cos \theta}{r^4} + \frac{2 \cos \theta}{r^4}\right] = \boxed{0}.$$

$$\nabla \times \mathbf{E} = \mathbf{e}_r \frac{1}{r \sin \theta} \left(\partial_\theta \left(E_\phi \sin \theta\right) - \partial_\phi E_\theta\right) + \mathbf{e}_\theta \frac{1}{r} \left(\frac{1}{\sin \theta} \partial_\phi E_r - \partial_r E_r\right)$$

$$+ \mathbf{e}_\phi \frac{1}{r} \left(\partial_r \left(r E_\theta\right) - \partial_\theta E_r\right)$$

$$= p \mathbf{e}_\phi \frac{1}{r} \left[\underbrace{\partial_r \left(\frac{\sin \theta}{r^2}\right)}_{-2\frac{\sin \theta}{r^3}} - \underbrace{\partial_\theta \left(\frac{2 \cos \theta}{r^3}\right)}_{-2\frac{\sin \theta}{r^3}}\right] = \boxed{0}.$$

(e)   The spherical surfaces S has area element $d\mathbf{S} = dS \, \mathbf{e}_r$, with $dS = d\phi \, d\theta \, \sin \theta R^2$:

$$\int_S d\mathbf{S} \cdot \mathbf{E} = \int_S dS \, \underbrace{\mathbf{e}_r \cdot \mathbf{E}(r = R)}_{E_r} = \frac{3 p R^2}{4\pi \varepsilon_0 R^3} \int_0^{2\pi} d\phi \underbrace{\int_0^\pi d\theta \sin \theta \cos \theta}_{0} = 0.$$

The flux integral yields zero, hence the (physical) law of Gauss implies that the total charge vanishes, $4\pi Q = \int_S d\mathbf{S} \cdot \mathbf{E} = 0$. This reflects the physical fact that the total charge of an electric dipole is equal to zero.

(f)   The (mathematical) theorem of Gauss gives

$$\int_S d\mathbf{S} \cdot \mathbf{E} = \int_{V_S} dV \nabla \cdot \mathbf{E} \overset{?}{=} 0.$$

Since $\nabla \cdot \mathbf{E} = 0$ for all $\mathbf{r} \neq \mathbf{0}$, it seems natural to conclude immediately that the integral $\int_{V_S} dV \nabla \cdot \mathbf{E}$ vanishes. But there is a subtlety – since the components of $\mathbf{E}$ diverge *at* the origin, the integral might have a finite value after all. That happens, e.g. for the potential of a point charge. Here, however, the integral indeed does yield zero, as demonstrates unambiguously by directly computing the flux integral over the sphere, as done in part (e).

S.V4   Introductory concepts of differential geometry (p. 463)

S.V4.1   Differentiable manifolds (p. 464)

ɛV4.1.1 Four-chart atlas for $S^1 \subset \mathbb{R}^2$ (p. 467)

(a)   The maps $r_{i,\pm} : U_{i,\pm} \to S_{i,\pm}$ act as follows:

$$r_{1,\pm}: \quad x^2 \mapsto (\pm\sqrt{1-(x^2)^2},x^2)^T,$$
$$r_{2,\pm}: \quad x^1 \mapsto (x^1,\pm\sqrt{1-(x^1)^2})^T.$$

(b) $S_{-,+} = S_{1,-} \cap S_{2,+}$ is the portion of the unit circle lying in the upper left quadrant, with $x^1 < 0$, $x^2 > 0$. It can be described by both $r_{2,+}$ and $r_{1,-}$ and the transition function translating between these two descriptions, mapping $U_{2,+}|_{x^1<0} \to U_{1,-}|_{x^2>0}$, is

$$r_{1,-}^{-1} \circ r_{2,+} : (-1,0) \to (0,1), \qquad x^1 \mapsto x^2(x^1) = \sqrt{1-(x^1)^2}.$$

(c) $S_{-,-} = S_{2,-} \cap S_{1,-}$, is the portion of the unit circle lying in the lower left quadrant, with $x^1 < 0, x^2 < 0$. It can be described by both $r_{1,-}$ and $r_{2,-}$ and the transition function translating between these two description, mapping $U_{1,-}|_{x^2<0} \to U_{2,-}|_{x^1<0}$, is

$$r_{2,-}^{-1} \circ r_{1,-} : (-1,0) \to (-1,0), \qquad x^2 \mapsto x^1(x^2) = -\sqrt{1-(x^2)^2}.$$

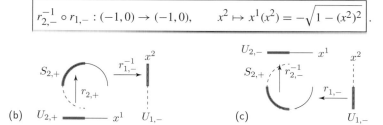

S.V4.2 Tangent space (p. 468)

ᴇV4.2.1 Tangent vectors on a paraboloidal manifold (p. 473)

(a) The holonomic basis is generated by the curves $y_\rho(t) = (\rho + t, \phi)^T = y + te_\rho$ and $y_\phi(t) = (\rho, \phi + t)^T = y + te_\phi$ in U. The corresponding basis vectors are represented as:

in U: $\qquad v_\rho = d_t y_\rho(t)\big|_{t=0} = e_\rho = \boxed{(1,0)^T}$,

$\qquad\qquad v_\phi = d_t y_\phi(t)\big|_{t=0} = e_\phi = \boxed{(0,1)^T}$;

in $\mathbb{R}^3$: $\qquad v_\rho = d_t r(y_\rho(t))\big|_{t=0} = \dfrac{\partial r(y)}{\partial \rho} = \boxed{(\cos\phi, \sin\phi, \rho)^T}$,

$\qquad\qquad v_\phi = d_t r(y_\phi(t))\big|_{t=0} = \dfrac{\partial r(y)}{\partial \phi} = \boxed{(-\rho\sin\phi, \rho\cos\phi, 0)^T}$.

(b) In U, the components $(u^\rho, u^\phi)^T$ of the tangent vector ∂_u at $y(t)$ are given by $u^i = \partial_u(y^i) = d_t y^i(t)$, the directional derivatives of the component functions, $y^i(t) \equiv y^i(y(t))$, along the curve representing the spiral in U, namely $y(t) = (y^\rho(t), y^\phi(t))^T = (Rt, \omega t)^T$:

$$u^\rho = d_t y^\rho = d_t(Rt) = \boxed{R}, \qquad u^\phi = d_t y^\phi = d_t(\omega t) = \boxed{\omega},$$
$$\Rightarrow \quad (u^\rho, u^\phi)^T = (R, \omega)^T \overset{(a)}{=} \boxed{e_\rho R + e_\phi\,\omega}.$$

Similarly, in $\mathbb{R}^3$ the Cartesian components $(u^1, u^2, u^3)^T$ of ∂_u are given by $u^i = \partial_u(r^i) = d_t r^i(t)$, the directional derivatives of the Cartesian components of the spiral curve in $M \subset \mathbb{R}^3$, $r(t) = (r^1(t), r^2(t), r^3(t))^T = (\rho(t)\cos\phi(t), \rho(t)\sin\phi(t), \tfrac{1}{2}\rho^2(t))^T$:

$$u^1 = d_t r^1(t) = d_t(Rt \cos \omega t) = \boxed{R \cos \omega t - Rt\omega \sin \omega t},$$

$$u^2 = d_t r^2(t) = d_t(Rt \sin \omega t) = \boxed{R \sin \omega t + Rt\omega \cos \omega t},$$

$$u^3 = d_t r^3(t) = d_t(\tfrac{1}{2}R^2 t^2) = \boxed{R^2 t},$$

$$\Rightarrow \quad (u^1, u^2, u^3)^T = R(\cos \omega t, \sin \omega t, Rt)^T + \omega(-Rt \sin \omega t, Rt \cos \omega t, 0)^T$$

$$= \left[R(\cos \phi, \sin \phi, \rho)^T + w(-\rho \sin \phi, \rho \cos \phi, 0)^T\right]_{r=r(t)}$$

$$\overset{(a)}{=} \boxed{[Rv_\rho + \omega v_\phi]_{r=r(t)}}.$$

Equivalently, but more compactly, we can obtain the above results as follows:

in U: $\qquad (u^\rho, u^\phi)^T = d_t y(t) = \dfrac{\partial y}{\partial \rho} \dfrac{d\rho(t)}{dt} + \dfrac{\partial y}{\partial \phi} \dfrac{d\phi(t)}{dt} \overset{(a)}{=} \boxed{e_\rho R + e_\phi \omega} = (R, \omega)^T;$

in $\mathbb{R}^3$: $\qquad (u^1, u^2, u^3)^T = d_t r(t) = \dfrac{\partial r}{\partial \rho} \dfrac{d\rho(t)}{dt} + \dfrac{\partial r}{\partial \phi} \dfrac{d\phi(t)}{dt} \overset{(a)}{=} \boxed{[v_\rho R + v_\phi \omega]_{r=r(t)}}.$

(c)   Along the spiral, the function $f(t) = r^1(t) + r^2(t) + r^3(t) \equiv f(y(t))$ has the form

$$f(t) = \left[\rho \cos \phi + \rho \sin \phi + \tfrac{1}{2}\rho^2\right]_{y=y(t)} = Rt \cos \omega t + Rt \sin \omega t + \tfrac{1}{2}R^2 t^2.$$

Its directional derivative along the spiral is

$$\partial_u f = d_t f(t) = \boxed{R(\cos \omega t + \sin \omega t + Rt) + \omega(-Rt \sin \omega t + Rt \cos \omega t)}.$$

Equivalently, but more compactly:

$$\partial_u f = u^j \partial_j f \overset{(b)}{=} u^\rho \partial_\rho f + u^\phi \partial_\phi f = \boxed{[R(\cos \phi + \sin \phi + \rho) + \omega(-\rho \sin \phi + \rho \cos \phi)]_{y=y(t)}}.$$

ₑV4.2.₃ Holonomic basis for hyperbolic coordinates (p. 474)

(a)   Under the coordinate transformation $y \mapsto x(y)$, expressing Cartesian through hyperbolic coordinates, the holonomic basis vectors transform as $\partial_{y^j} = \dfrac{\partial x^i}{\partial y^j} \partial_{x^i} \equiv (v_{y^j})^i \partial_{x_i}$. With $x^1 = \rho\, e^\alpha$, $x^2 = \rho\, e^{-\alpha}$, the Cartesian components of ∂_ρ and ∂_α thus read:

$$v_\rho \equiv ((v_\rho)^1, (v_\rho)^2)^T = \left(\tfrac{\partial x^1}{\partial \rho}, \tfrac{\partial x^2}{\partial \rho}\right)^T = (e^\alpha, e^{-\alpha})^T = \boxed{(\sqrt{x^1/x^2}, \sqrt{x^2/x^1})^T},$$

$$v_\alpha \equiv ((v_\alpha)^1, (v_\alpha)^2)^T = \left(\tfrac{\partial x^1}{\partial \alpha}, \tfrac{\partial x^2}{\partial \alpha}\right)^T = (\rho\, e^\alpha, -\rho\, e^{-\alpha})^T = \boxed{(x^1, -x^2)^T}.$$

(b)   The inverse relations read $\partial_{x^i} = \dfrac{\partial y^j}{\partial x^i} \partial_{y^j} \equiv (v_{x^i})^j \partial_{y^j}$. With $\rho = \sqrt{x^1 x^2}$ and $\alpha = \ln\sqrt{x^1/x^2}$, the hyperbolic components of ∂_1 and ∂_2 thus read:

$$v_1 \equiv ((v_1)^\rho, (v_1)^\alpha)^T = \left(\tfrac{\partial x^\rho}{\partial x^1}, \tfrac{\partial x^\alpha}{\partial x^1}\right)^T = \tfrac{1}{2}(\sqrt{x^2/x^1}, 1/x^1)^T = \boxed{\tfrac{1}{2}(e^{-\alpha}, e^{-\alpha}/\rho)^T},$$

$$v_2 \equiv ((v_2)^\rho, (v_2)^\alpha)^T = \left(\tfrac{\partial x^\rho}{\partial x^2}, \tfrac{\partial x^\alpha}{\partial x^2}\right)^T = \tfrac{1}{2}(\sqrt{x^1/x^2}, -1/x^2)^T = \boxed{\tfrac{1}{2}(e^\alpha, -e^\alpha/\rho)^T}.$$

(c)   The Jacobi matrix is defined as $J^i{}_j = \frac{\partial x^i}{\partial y^j} = (v_{yj})^i$, with inverse $(J^{-1})^j{}_i = \frac{\partial y^j}{\partial x^i} = (v_{xi})^j$:

$$JJ^{-1} = \begin{pmatrix} \frac{\partial x^1}{\partial \rho} & \frac{\partial x^1}{\partial \alpha} \\ \frac{\partial x^2}{\partial \rho} & \frac{\partial x^2}{\partial \alpha} \end{pmatrix} \begin{pmatrix} \frac{\partial \rho}{\partial x^1} & \frac{\partial \rho}{\partial x^2} \\ \frac{\partial \alpha}{\partial x^1} & \frac{\partial \alpha}{\partial x^2} \end{pmatrix} = \begin{pmatrix} e^\alpha & \rho\,e^\alpha \\ e^{-\alpha} & -\rho\,e^{-\alpha} \end{pmatrix} \frac{1}{2} \begin{pmatrix} e^{-\alpha} & e^\alpha \\ e^{-\alpha}/\rho & -e^\alpha/\rho \end{pmatrix} = \begin{pmatrix} 1 & 0 \\ 0 & 1 \end{pmatrix}. \checkmark$$

S.V5   Alternating differential forms (p. 477)

S.V5.1   Cotangent space and differential one-forms (p. 477)

εV5.1.1 Differential of a function in Cartesian and polar coordinates (p. 480)

(a)   The components of the differential, $df = f_i dx^i$, of the function $f(x) = x^1 x^2$ are given by
$f_i = df(\partial_i) = \frac{\partial}{\partial x^i} f$, hence $f_1 = \frac{\partial}{\partial x^1} f = \boxed{x^2}$, $f_2 = \frac{\partial}{\partial x^2} f = \boxed{x^1} \Rightarrow df = x^2 dx^1 + x^1 dx^2$.

(b)   Using $dx^i(u) = dx^i(u^j \partial_j) = u^j \partial_j x^i = u^j \delta_j{}^i = u^i$, we obtain:
$$df(\partial_u) = (x^2 dx^1 + x^1 dx^2)(x^1(\partial_1 + \partial_2)) = \boxed{x^2 x^1 + (x^1)^2}.$$

(c)   When expressed through polar coordinates using $x^1(y) = \rho\cos\phi$, $x^2(y) = \rho\sin\phi$, the function f reads : $f(x(y)) = x^1(y)x^2(y) = \rho^2 \cos\phi \sin\phi = \frac{1}{2}\rho^2 \sin 2\phi$. We obtain the coefficients of $df = f_i(y)dy^i$ in polar coordinates using $f_i(y) = f(\partial_{y^i}) = \frac{\partial}{\partial y^i} f(y)$:

$$f_\rho = \frac{\partial}{\partial \rho} f = \boxed{\rho \sin 2\phi}, \qquad f_\phi = \frac{\partial}{\partial \phi} f = \boxed{\rho^2 \cos 2\phi}.$$

Thus, in polar coordinates the form reads: $df = \boxed{\rho \sin 2\phi\, d\rho + \rho^2 \cos 2\phi\, d\phi}$.

Alternatively, we can use the transformation rule for the coefficients of forms. Under the transformation $y \mapsto x(y)$, expressing Cartesian through polar coordinates, we have:
$$f_i(x) = [f_j(y)(J^{-1})^j{}_i]_{y=y(x)} \Rightarrow f_j(y) = [f_i(x)J^i{}_j]_{x=x(y)}, \text{ with Jacobian}$$

$$J^i{}_j = \left(\frac{\partial x^i}{\partial y^j}\right) = \begin{pmatrix} \frac{\partial x^1}{\partial \rho} & \frac{\partial x^1}{\partial \phi} \\ \frac{\partial x^2}{\partial \rho} & \frac{\partial x^2}{\partial \phi} \end{pmatrix} = \begin{pmatrix} \cos\phi & -\rho\sin\phi \\ \sin\phi & \rho\cos\phi \end{pmatrix}.$$

With $f_1(x(y)) = x^2(y) = \rho\sin\phi$, $f_2(x(y)) = x^1(y) = \rho\cos\phi$, we obtain:

$$f_\rho(y) = [f_1(x)J^1{}_\rho + f_2(x)J^2{}_\rho]_{x=x(y)} = \rho\sin\phi\,\cos\phi + \rho\cos\phi\,\sin\phi = \boxed{\rho\sin 2\phi}, \checkmark$$

$$f_\phi(y) = [f_1(x)J^1{}_\phi + f_2(x)J^2{}_\phi]_{x=x(y)} = -\rho\sin\phi\,\rho\sin\phi + \rho\cos\phi\,\rho\cos\phi = \boxed{\rho^2 \cos 2\phi}. \checkmark$$

(d)   Using $dy^i(\partial_{y^j}) = \delta^i{}_j$ we obtain:

$$df(\partial_u) = (\rho\sin 2\phi\, d\rho + \rho^2 \cos 2\phi\, d\phi)(\partial_\rho - \partial_\phi) = \boxed{\rho\sin 2\phi - \rho^2 \cos 2\phi}.$$

(e)

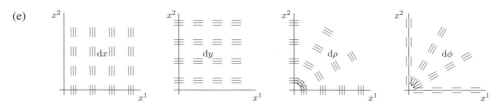

The figure uses triples of parallel lines to visualize the action of a form at a given point, follow-ing the conventions of Fig. L21 from Section L10.5: the triples visualizing df are perpendicular to the direction in which f changes most rapidly, i.e. parallel to lines of constant f. For $d\rho$ and $d\phi$, the orientation of these triples depends on position, since the holonomic basis vectors, ∂_ρ and ∂_ϕ, do, too.

S.V5.2 Pushforward and pullback (p. 481)

εV5.2.1 Pushforward and pullback: generalized polar coordinates (p. 486)

(a)   For $x^1 = \mu a \cos\phi$, $x^2 = \mu b \sin\phi$: $J^i{}_j = \left(\dfrac{\partial x^i}{\partial y^j} \right) = \begin{pmatrix} \frac{\partial x^1}{\partial \mu} & \frac{\partial x^1}{\partial \phi} \\ \frac{\partial x^2}{\partial \mu} & \frac{\partial x^2}{\partial \phi} \end{pmatrix} = \boxed{\begin{pmatrix} a\cos\phi & -\mu a \sin\phi \\ b\sin\phi & \mu b \cos\phi \end{pmatrix}}.$

(b)   The pushforward of a general vector, $\partial_u = \partial_{y^j} u^j$, is given by $F_*\partial_u = \partial_{x^i} J^i{}_j u^j$:

$$F_*\partial_\mu = \partial_{x^i} J^i{}_\mu = \partial_{x^1} a\cos\phi + \partial_{x^2} b\sin\phi = \boxed{\dfrac{\partial_{x^1} x^1 + \partial_{x^2} x^2}{\left[(x^1/a)^2 + (x^2/b)^2 \right]^{1/2}}},$$

$$F_*\partial_\phi = \partial_{x^i} J^i{}_\phi = -\partial_{x^1} \mu a \sin\phi + \partial_{x^2} \mu b \cos\phi = \boxed{-\partial_{x^1} \frac{a}{b} x^2 + \partial_{x^2} \frac{b}{a} x^1},$$

where we used $\mu = \left[(x^1/a)^2 + (x^2/b)^2 \right]^{1/2}$ to express the r.h.s. in terms of x. (Note that in these expressions ∂_{x^i} are not partial derivatives, but tangent vectors!)

(c)   The pullback of a general form, $\phi = \phi_i dx^i$, is given by $F^*\phi = \phi_i J^i{}_j dy^j$:

$$F^* dx^1 = J^1{}_j dy^j = \boxed{a\cos\phi\, d\mu - \mu a \sin\phi\, d\phi},$$
$$F^* dx^2 = J^2{}_j dy^j = \boxed{b\sin\phi\, d\mu + \mu b \cos\phi\, d\phi}.$$

(d)   (i)   Using pushforward of the vectors (with $F_*\mu\partial_\mu = \partial_{x^1} x^1 + \partial_{x^2} x^2$), we obtain

$$\lambda(F_*\mu\partial_\mu) \overset{(b)}{=} (x^2 dx^1 - x^1 dx^2)(\partial_{x^1} x^1 + \partial_{x^2} x^2) = x^2 x^1 - x^1 x^2 = \boxed{0},$$

$$\lambda(F_*\partial_\phi) \overset{(b)}{=} (x^2 dx^1 - x^1 dx^2)\left[-\partial_{x^1} \frac{a}{b} x^2 + \partial_{x^2} \frac{b}{a} x^1 \right] = \boxed{-\frac{a}{b}(x^2)^2 - \frac{b}{a}(x^1)^2}.$$

(ii)   Using pullback of the form $\lambda = x^2 dx^1 - x^1 dx^2$,

$$F^*\lambda \overset{(c)}{=} \mu b \sin\phi(a\cos\phi\, d\mu - \mu a \sin\phi\, d\phi) - \mu a \cos\phi(b\sin\phi\, d\mu + \mu b\cos\phi\, d\phi),$$
$$= -\mu^2 ab(\sin^2\phi + \cos^2\phi)d\phi = -\mu^2 ab\, d\phi$$

we obtain: $F^*\lambda(\mu\partial_\mu) = \boxed{0}$, ✓ $F^*\lambda(\partial_\phi) = \boxed{-\mu^2 ab} = -\frac{a}{b}(x^2)^2 - \frac{b}{a}(x^1)^2.$ ✓

S.V5.3 Forms of higher degree (p. 487)

εV5.3.1 Wedge product in Cartesian and polar coordinates (p. 489)

(a) $\lambda \wedge \eta = (2x^1 x^2 \, dx^1 + ((x^1)^2 + (x^2)^2) \, dx^2) \wedge (e^{x^1 x^2} \, dx^1 - dx^2)$

$$= \boxed{-\left[2x^1 x^2 + ((x^1)^2 + (x^2)^2)e^{x^1 x^2}\right] dx^1 \wedge dx^2}.$$

(b) For the vectors $\partial_u = x^1 \partial_1 + x^2 \partial_2$ and $\partial_v = \partial_1 - \partial_2$, with components $(u^1, u^2) = (x^1, x^2)^{\mathrm{T}}$ and $(v^1, v^2) = (1, -1)^{\mathrm{T}}$, we have $(dx^1 \wedge dx^2)(\partial_u, \partial_v) = u^1 v^2 - u^2 v^1 = x^1 \cdot (-1) - x^2 \cdot 1$.

$$\Rightarrow (\lambda \wedge \eta)(\partial_u, \partial_v) = \boxed{\left[2x^1 x^2 + ((x^1)^2 + (x^2)^2)e^{x^1 x^2}\right](x^1 + x^2)}.$$

(c) Under the coordinate transformation $y \mapsto x(y)$ expressing Cartesian through polar coordinates, $x^1 = \rho \cos\phi, x^2 = \rho \sin\phi$, we have:

$$dx^1 \wedge dx^2 = J^1{}_{j_1} J^2{}_{j_2} \, dy^{j_1} \wedge dy^{j_2} = (J^1{}_1 J^2{}_2 - J^1{}_2 J^2{}_1) \, dy^1 \wedge dy^2,$$

with Jacobian matrix $J^i{}_j = \left(\frac{\partial x^i}{\partial y^j}\right) = \begin{pmatrix} \cos\phi & -\rho\sin\phi \\ \sin\phi & \rho\cos\phi \end{pmatrix}$. Hence

$$\lambda \wedge \eta = \left[2x^1 x^2 + ((x^1)^2 + (x^2)^2)e^{x^1 x^2}\right]_{x=x(y)} (\rho\cos^2\phi + \rho\sin^2\phi) \, d\rho \wedge d\phi$$

$$= \boxed{\rho^3 \left[2\cos\phi\sin\phi + e^{\rho^2\cos\phi\sin\phi}\right] d\rho \wedge d\phi}.$$

εV5.3.3 Stereographic projection of spherical area form (p. 492)

(a) Given the transformation $\theta(z) = 2\arctan(1/\rho)$, $\phi(z) = \arctan(z^2/z^1)$, with $\rho = \sqrt{(z^1)^2 + (z^2)^2}$, the derivatives needed to construct the Jacobian matrix of the map $z \mapsto y(z)$ are:

$$\frac{\partial\rho}{\partial z^i} = \frac{\partial}{\partial z^i}\sqrt{(z^1)^2 + (z^2)^2} = \frac{z^i}{\rho}, \qquad \frac{\partial\theta}{\partial z^i} = \frac{2}{1 + 1/\rho^2}\left(\frac{-1}{\rho^2}\right)\frac{\partial\rho}{\partial z^i} = \frac{-2z^i}{(\rho^2+1)\rho},$$

$$\frac{\partial\phi}{\partial z^1} = \frac{1}{1 + (z^2/z^1)^2}\frac{(-z^2)}{(z^1)^2} = \frac{-z^2}{\rho^2}, \qquad \frac{\partial\phi}{\partial z^2} = \frac{1}{1 + (z^2/z^1)^2}\frac{1}{z^1} = \frac{z^1}{\rho^2}.$$

$$\Rightarrow J(z) = \frac{\partial y}{\partial z} = \begin{pmatrix} \frac{\partial\theta}{\partial z^1} & \frac{\partial\theta}{\partial z^2} \\ \frac{\partial\phi}{\partial z^1} & \frac{\partial\phi}{\partial z^2} \end{pmatrix} = \boxed{\begin{pmatrix} \frac{-2z^1}{(\rho^2+1)\rho} & \frac{-2z^2}{(\rho^2+1)\rho} \\ \frac{-z^2}{\rho^2} & \frac{z^1}{\rho^2} \end{pmatrix}}.$$

For the inverse transformation, $(z^1, z^2)^{\mathrm{T}} = (\rho\cos\phi, \rho\sin\phi)^{\mathrm{T}}$, with $\rho = \frac{1}{\tan(\theta/2)}$, we have:

$$\frac{\partial\rho}{\partial\theta} = -\frac{1}{2}\frac{1}{\tan^2(\theta/2)}\frac{1}{\cos^2(\theta/2)} = -\frac{1}{2\sin^2(\theta/2)} = -\frac{1}{2}(1 + \rho^2),$$

$$\Rightarrow J(y)^{-1} = \frac{\partial z}{\partial y} = \begin{pmatrix} \frac{\partial z^1}{\partial\theta} & \frac{\partial z^1}{\partial\phi} \\ \frac{\partial z^2}{\partial\theta} & \frac{\partial z^2}{\partial\phi} \end{pmatrix} = \boxed{\begin{pmatrix} \frac{-\cos\phi}{2\sin^2(\theta/2)} & \frac{-\sin\phi}{\tan(\theta/2)} \\ \frac{-\sin\phi}{2\sin^2(\theta/2)} & \frac{\cos\phi}{\tan(\theta/2)} \end{pmatrix}}.$$

Check: $JJ^{-1} = \begin{pmatrix} \frac{-2\cos\phi}{\rho^2+1} & \frac{-2\sin\phi}{\rho^2+1} \\ \frac{-\sin\phi}{\rho} & \frac{\cos\phi}{\rho} \end{pmatrix}\begin{pmatrix} -\frac{1}{2}(1+\rho^2)\cos\phi & -\rho\sin\phi \\ -\frac{1}{2}(1+\rho^2)\sin\phi & \rho\cos\phi \end{pmatrix} = \mathbb{1}. \checkmark$

The corresponding determinants are

$$\det J = \det\left(\frac{\partial y}{\partial z}\right) = -2\frac{(z^1)^2 + (z^2)^2}{(\rho^2 + 1)\rho^3} = \boxed{\frac{-2}{(\rho^2 + 1)\rho}}, \quad \det(J^{-1}) = \boxed{-\frac{1}{2}\rho(\rho^2 + 1)}.$$

Remark: The fact that these determinants are negative indicates that the stereographic and spherical holonomic bases, $\{\partial_{z^1}, \partial_{z^2}\}$ and $\{\partial_\theta, \partial_\phi\}$, have opposite relative orientation, hence the stereographic projection is an orientation-changing map.

(b) The spherical area form, $\omega = \sin\theta\, d\theta \wedge d\phi$, transforms as:

$$\omega = \sin(\theta(z))\det\left(\frac{\partial y}{\partial z}\right) dz^1 \wedge dz^2 = \frac{2\rho}{(1 + \rho^2)}\frac{(-2)}{(\rho^2 + 1)\rho}\, dz^1 \wedge dz^2 = \omega(z)\, dz^1 \wedge dz^2,$$

with weight function $\omega(z) = \boxed{\dfrac{-4}{(1 + \rho^2)^2}}$.

(c) Using the area form in spherical coordinates, we directly obtain $\omega(\partial_\theta, \partial_\phi) = \sin\theta(d\theta \wedge d\phi)(\partial_\theta, \partial_\phi) = \boxed{\sin\theta}$. To use stereographic coordinates, we first transform the vectors:

$$\partial_\theta = \frac{\partial z^i}{\partial\theta}\partial_{z^i} \overset{(a)}{=} -\frac{1}{2}(1 + \rho^2)\big[\cos\phi\, \partial_{z^1} + \sin\phi\, \partial_{z^2}\big],$$

$$\partial_\phi = \frac{\partial z^i}{\partial\phi}\partial_{z^i} \overset{(a)}{=} \big[-\rho\sin\phi\, \partial_{z^1} + \rho\cos\phi\, \partial_{z^2}\big].$$

The minus sign for ∂_θ reflects the fact that with increasing θ, both z^1 and z^2 decrease.

$$\omega(\partial_\theta, \partial_\phi) = \omega(z)(dz^1 \wedge dz^2)\left(\frac{\partial z^i}{\partial\theta}\partial_{z^i}, \frac{\partial z^j}{\partial\phi}\partial_{z^j}\right) = \omega(z)\left(\frac{\partial z^1}{\partial\theta}\frac{\partial z^2}{\partial\phi} - \frac{\partial z^1}{\partial\phi}\frac{\partial z^2}{\partial\theta}\right) = \omega(z)\det J^{-1}$$

$$\overset{(a,b)}{=} \frac{-4}{(1 + \rho^2)^2}\left(-\frac{1}{2}\rho\right)(\rho^2 + 1) = \boxed{\frac{2\rho}{1 + \rho^2}} = \sin\theta. \checkmark$$

Thus $\omega(\partial_\theta, \partial_\phi)$, the area of the surface element spanned by $(\partial_\theta, \partial_\phi)$, goes to zero as $\rho \to \infty$ or 0, as expected for $\theta \to 0$ or π, respectively.

εV5.3.5 Exterior derivative (p. 494)

When evaluating $d\phi = \partial_j(\phi_{i_1\ldots i_p})\, dx^j \wedge \prod_{a=1}^{p} d\wedge^{i_a}$, the sum over j can be restricted to index values not occurring in the wedge product, $j \notin \{i_1\ldots i_p\}$.

(a) $$df = d(x^1 x^2) = \frac{\partial}{\partial x^i}(x^1 x^2)dx^i = \boxed{x^2 dx^1 + x^1 dx^2}.$$

(b) $$d\lambda = d(x^2 dx^1 + x^1 x^2 dx^2) = \frac{\partial}{\partial x^i}(x^2)\, dx^i \wedge dx^1 + \frac{\partial}{\partial x^i}(x^1 x^2)\, dx^i \wedge dx^2$$

$$= \frac{\partial}{\partial x^2}(x^2)\, dx^2 \wedge dx^1 + \frac{\partial}{\partial x^1}(x^1 x^2)\, dx^1 \wedge dx^2 = \boxed{(-1 + x^2)dx^1 \wedge dx^2}.$$

(c) $$d\phi = d(x^2 x^3 dx^1 \wedge dx^2 + x^3 x^1 dx^2 \wedge dx^3)$$

$$= \frac{\partial}{\partial x^i}(x^2 x^3)\, dx^i \wedge dx^1 \wedge dx^2 + \frac{\partial}{\partial x^i}(x^3 x^1)\, dx^i \wedge dx^2 \wedge dx^3$$

$$= \boxed{(x^2 + x^3)dx^1 \wedge dx^2 \wedge dx^3}.$$

ɛV5.3.7 **Pullback of polar area form to Cartesian coordinates in $\mathbb{R}^2$** (p. 495)

(a)   The derivatives, $\frac{\partial y^j}{\partial x^i}$, needed for the pullback, $y^*\omega = \omega_j \frac{\partial_{y^j}}{\partial_{x^i}} dx^i$, are all elements of the Jacobi matrix for the transformation $x \mapsto y(x)$, defined as:

$$\rho = \sqrt{(x^1)^2 + (x^2)^2}, \qquad \phi = \arctan(\tfrac{x^2}{x^1}).$$

The Jacobi matrix is given by

$$J = \frac{\partial y}{\partial x} = \begin{pmatrix} \frac{\partial \rho}{\partial x^1} & \frac{\partial \rho}{\partial x^2} \\ \frac{\partial \phi}{\partial x^1} & \frac{\partial \phi}{\partial x^2} \end{pmatrix} = \begin{pmatrix} \frac{x^1}{\rho} & \frac{x^2}{\rho} \\ -\frac{x^2}{\rho^2} & \frac{x^1}{\rho^2} \end{pmatrix},$$

$$y^*\omega = y^*(\rho\, d\rho \wedge d\phi) = \rho(x)\left(\tfrac{\partial \rho}{\partial x^1} dx^1 + \tfrac{\partial \rho}{\partial x^2} dx^2\right) \wedge \left(\tfrac{\partial \phi}{\partial x^1} dx^1 + \tfrac{\partial \phi}{\partial x^2} dx^2\right)$$

$$= \rho(x)\left(\tfrac{\partial \rho}{\partial x^1}\tfrac{\partial \phi}{\partial x^2} - \tfrac{\partial \rho}{\partial x^2}\tfrac{\partial \phi}{\partial x^1}\right) dx^1 \wedge dx^2 = \rho(x)\left(\tfrac{x^1}{\rho}\tfrac{x^1}{\rho^2} - \tfrac{x^2}{\rho}\tfrac{(-x^2)}{\rho^2}\right) dx^1 \wedge dx^2 = \boxed{dx^1 \wedge dx^2}.$$

(b)   The pullback of the form $\kappa = \tfrac{1}{2}\rho^2 d\phi$ is given by

$$y^*\kappa = \tfrac{1}{2}\rho^2(x)\left(\tfrac{\partial \phi}{\partial x^1} dx^1 + \tfrac{\partial \phi}{\partial x^2} dx^2\right) = \tfrac{1}{2}\rho^2\left[-\tfrac{x^2}{\rho^2} dx^1 + \tfrac{x^1}{\rho^2} dx^2\right] = \boxed{\tfrac{1}{2}(-x^2 dx^1 + x^1 dx^2)}.$$

The exterior derivative of $y^*\kappa$ is $dy^*\kappa = \tfrac{1}{2}(-dx^2 \wedge dx^1 + dx^1 \wedge dx^2) = \boxed{dx^1 \wedge dx^2} \overset{\checkmark}{=} \omega$.

ɛV5.3.9 **Pullback of spherical area form from spherical to stereographic coordinates** (p. 495)

(a)   The exterior derivative of $\kappa = -\cos\theta d\phi$ is

$$d\kappa = \tfrac{\partial}{\partial y^i}(-\cos\theta)\, dy^i \wedge d\phi = \tfrac{\partial}{\partial\theta}(-\cos\theta)\, d\theta \wedge d\phi = \boxed{\sin\theta\, d\theta \wedge d\phi} = \omega.\ \checkmark$$

(b)   Using $\phi(z) = \arctan(z^2/z^1)$, we obtain:

$$y^*\kappa = y^*(-\cos\theta d\phi) = -\cos\theta(z)\tfrac{\partial\phi}{\partial z^i} dz^i = \tfrac{1-\rho^2}{1+\rho^2}\left[-\tfrac{z^2}{\rho^2} dz^1 + \tfrac{z^1}{\rho^2} dz^2\right]$$

$$\equiv \kappa_i\, dz^i, \quad \text{with } (\kappa_1, \kappa_2) = \boxed{\tfrac{1-\rho^2}{(1+\rho^2)\rho^2}(-z^2, z^1)}.$$

Here we used $\cos\theta = \pm\sqrt{1 - \sin^2\theta} = \pm\sqrt{1 - \left(\tfrac{2\rho}{1+\rho^2}\right)^2} = \pm\sqrt{\tfrac{(1-\rho^2)^2}{(1+\rho^2)^2}} = \tfrac{\rho^2-1}{1+\rho^2}$, and choose the sign such that $\cos\theta \gtrless 0$ for $\rho \gtrless 1$. Now we take the exterior derivative of the pullback:

$$dy^*\kappa = d(\kappa_i\, dz^i) = \left(\tfrac{\partial\kappa_i}{\partial z^j} dz^j\right) \wedge dz^i = (-\partial_{z^2}\kappa_1 + \partial_{z^1}\kappa_2)\, dz^1 \wedge dz^2 \equiv \omega(z)\, dz^1 \wedge dz^2,$$

where the weight function is given by

$$\omega(z) = (-\partial_{z^2}\kappa_1 + \partial_{z^1}\kappa_2) = \tfrac{1-\rho^2}{(1+\rho^2)\rho^2}\left[\tfrac{\partial z^2}{\partial z^2} + \tfrac{\partial z^1}{\partial z^1}\right] + \left(\tfrac{\partial}{\partial\rho}\tfrac{1-\rho^2}{(1+\rho^2)\rho^2}\right)\left[\tfrac{\partial\rho}{\partial z^2}(z^2) + \tfrac{\partial\rho}{\partial z^1}(z^1)\right]$$

$$= \tfrac{2(1-\rho^2)}{(1+\rho^2)\rho^2} + \tfrac{2(\rho^4 - 2\rho^2 - 1)}{(1+\rho^2)^2\rho^3}\rho = \boxed{\tfrac{-4}{(1+\rho^2)^2}}.$$

To obtain the second line above, we used

$$\tfrac{\partial}{\partial\rho}\tfrac{1-\rho^2}{(1+\rho^2)\rho^2} = \tfrac{-2\rho(1+\rho^2)\rho - 2(1-\rho^2)\rho^2 - 2(1-\rho^2)(1+\rho^2)}{(1+\rho^2)^2\rho^3} = \tfrac{2(\rho^4 - 2\rho^2 - 1)}{(1+\rho^2)^2\rho^3}, \text{ and } \tfrac{\partial\rho}{\partial z^i}z^i = \tfrac{z^i}{\rho}z^i = \rho.$$

As expected from $dy^*\kappa = y^*\omega$, our result for $\omega(z)$ agrees with that found in Problem V5.3.3.

S.V5.4 Integration of forms (p. 495)

εV5.4.1 Stereographic coordinates: computing area of sphere (p. 498)

The area of the unit sphere is given by an integral over the entire stereographic coordinate domain:

$$A_{S^2} = \int_{S^2} \omega = \int_{\mathbb{R}^2} y^* \omega = \int_{\mathbb{R}^2} |\omega(z)|\, dz^1 \wedge dz^2 = \int_{\mathbb{R}^2} |\omega(z)|\, dz^1 dz^2\,.$$

Since the stereographic projection is an orientation-changing map, the weight function, $\omega(z) = -\frac{4}{(1+\rho^2)^2}$, is negative, thus its absolute value is used to compute areas. Since this function depends only on the distance, $\rho = \sqrt{(z^1)^2 + (z^2)^2}$, from the origin within the stereographic plane, we perform the integral over $\mathbb{R}^2$ using polar coordinates, with $(z^1, z^2) = (\rho \cos \phi, \rho \sin \phi)$. Then

$$A_{S^2} = \int_0^{2\pi} d\phi \int_0^\infty d\rho\, \omega(\rho)\rho = 2\pi \int_0^\infty d\rho \frac{4\rho}{(1+\rho^2)^2} \overset{\eta = \rho^2}{=} 2\pi \int_0^\infty d\eta \frac{2}{(1+\eta)^2} = 4\pi \left[\frac{-1}{1+\eta}\right]_0^\infty = \boxed{4\pi}\,. \checkmark$$

εV5.4.3 Pullback of current form from Cartesian to spherical coordinates (p. 499)

(a) We parametrize the sphere using the map

$$x : U \to S_R, \quad y = (\theta, \phi)^{\mathrm{T}} \mapsto x(y) = (x^1, x^2, x^3)^{\mathrm{T}} = R(\cos \phi \sin \theta, \sin \phi \sin \theta, \cos \theta)^{\mathrm{T}},$$

with $U = (0, \pi) \times (0, 2\pi)$, and compute the pullback of the current form as follows:

$$x^* j = x^*(j_0\, dx^1 \wedge dx^2) = j_0 \frac{\partial x^1}{\partial y^i} \frac{\partial x^2}{\partial y^j} dy^i \wedge dy^j = j_0 \left(\frac{\partial x^1}{\partial \theta} \frac{\partial x^2}{\partial \phi} - \frac{\partial x^1}{\partial \phi} \frac{\partial x^2}{\partial \theta} \right) d\theta \wedge d\phi$$

$$= j_0 R^2 \left[(\cos \phi \cos \theta)(\cos \phi \sin \theta) - (-\sin \phi \sin \theta)(\sin \phi \cos \theta) \right] d\theta \wedge d\phi$$

$$= \boxed{j_0 R^2 \tfrac{1}{2} \sin 2\theta\, d\theta \wedge d\phi}\,.$$

(b) The current through the curved surface of the sphere is obtained by integrating the pullback form over the coordinate domain parametrizing it.

(i) The coordinate domain for the upper half-sphere is $U^+ = (0, \tfrac{1}{2}\pi) \times (0, 2\pi)$. The current flowing through it is

$$I_{S_R^+} = \int_{S_R^+} j = \int_{U^+} x^* j = \tfrac{1}{2} j_0 R^2 \int_{U^+} \sin 2\theta\, d\theta \wedge d\phi$$

$$= \tfrac{1}{2} j_0 R^2 \int_0^{2\pi} d\phi \int_0^{\pi/2} d\theta \sin 2\theta = \tfrac{1}{2} j_0 R^2 \cdot 2\pi \cdot \left[-\tfrac{1}{2} \cos 2\theta \right]_0^{\pi/2} = \boxed{j_0 \pi R^2}\,.$$

The result is just j_0 times the area of a disk of radius R, corresponding to the projection of the upper half-sphere onto the $x^1 x^2$-plane.

(ii) The integral over the full sphere, S_R, is proportional to

$$\int_0^\pi d\theta \sin 2\theta = \left[-\tfrac{1}{2} \cos 2\theta \right]_0^\pi = \boxed{0}\,.$$

(c) Alternatively, let us compute the flux integral of the current density, $\mathbf{j} = j_0 \mathbf{e}_z$, over a spherical surface with surface element $d\mathbf{S} = R^2 d\theta\, d\phi \sin \theta\, \mathbf{e}_r$. Since $\mathbf{e}_r \cdot \mathbf{e}_z = \cos \theta$, we obtain:

(i) $I_{S_R^+} = \int_{S_R^+} d\mathbf{S} \cdot \mathbf{j} = j_0 R^2 \int_0^{2\pi} d\phi \int_0^{\frac{\pi}{2}} d\theta \sin \theta\, \mathbf{e}_r \cdot \mathbf{e}_z = j_0 R^2 2\pi \int_0^{\frac{\pi}{2}} d\theta \sin \theta \cos \theta = \boxed{j_0 R^2 \pi}\,. \checkmark$

(ii) Similarly, integrating over the full sphere yields $I_{S_R} = \boxed{0}\,. \checkmark$

(d) Note that the x^3-coordinate, determining the curved shape of the sphere, does not enter the calculation in (b) at all. This means that the integral $I_M = \int_M j$ must yield the same result for any manifold M having the same boundary, $\partial M = \gamma$, as the upper half-sphere, namely a circle of radius R. This is a reflection of Stokes's theorem. The current form is exact, being expressible as $j = d\lambda$, with $\lambda = j_0 x^1 dx^2$. Its pullback to ∂M, using $\gamma : (0, 2\pi) \to \partial M$, $\phi \mapsto x(\phi) = R(\cos\phi, \sin\phi)$, reads:

$$\gamma^*\lambda = j_0 x^1(\phi)\frac{\partial x^2}{\partial\phi}d\phi = j_0(R\cos\phi)(R\cos\phi)d\phi = j_0 R^2 \cos^2\phi\, d\phi.$$

Using Stokes's theorem, we thus obtain:

$$I_M = \int_M j = \int_M d\lambda = \int_{\partial M} \lambda = j_0 R^2 \int_0^{2\pi} \cos^2\phi\, d\phi = \boxed{\pi j_0 R^2}.\ \checkmark$$

S.V6   Riemannian differential geometry (p. 502)

S.V6.1   Definition of the metric on a manifold (p. 502)

ᴇV6.1.1 Standard metric of $\mathbb{R}^3$ in cylindrical coordinates (p. 504)

In general, the metric transform as $g = g_{ij}(x)\, dx^i \otimes dx^j = g_{kl}(y)\, dx^k \otimes dx^l$, with $g_{kl}(y) = g_{ij}(x(y))J^i{}_k J^j{}_l$. For the Cartesian metric, $g_{ij}(x) = \delta_{ij}$, this reduces to $g_{kl}(y) = J^i{}_k J^i{}_l = (J^T J)_{kl}$. For the transformation $x \mapsto y(x)$ expressing Cartesian through cylindrical coordinates,

$$x^1 = \rho\cos\phi, \quad x^2 = \rho\sin\phi, \quad x^3 = z,$$

the Jacobi matrix has the form

$$J = \frac{\partial x}{\partial y} = \begin{pmatrix} \frac{\partial x^1}{\partial\rho} & \frac{\partial x^1}{\partial\phi} & \frac{\partial x^1}{\partial z} \\ \frac{\partial x^2}{\partial\rho} & \frac{\partial x^2}{\partial\phi} & \frac{\partial x^2}{\partial z} \\ \frac{\partial x^3}{\partial\rho} & \frac{\partial x^3}{\partial\phi} & \frac{\partial x^3}{\partial z} \end{pmatrix} = \begin{pmatrix} \cos\phi & -\rho\sin\phi & 0 \\ \sin\phi & \rho\cos\phi & 0 \\ 0 & 0 & 1 \end{pmatrix}.$$

$$\Rightarrow \quad J^T J = \begin{pmatrix} \cos\phi & \sin\phi & 0 \\ -\rho\sin\phi & \rho\cos\phi & 0 \\ 0 & 0 & 1 \end{pmatrix}\begin{pmatrix} \cos\phi & -\rho\sin\phi & 0 \\ \sin\phi & \rho\cos\phi & 0 \\ 0 & 0 & 1 \end{pmatrix} = \begin{pmatrix} 1 & 0 & 0 \\ 0 & \rho^2 & 0 \\ 0 & 0 & 1 \end{pmatrix}$$

$$\Rightarrow \quad g = \boxed{d\rho \otimes d\rho + \rho^2\, d\phi \otimes d\phi + dz \otimes dz}.$$

The metric is diagonal, hence cylindrical coordinates form an orthogonal coordinate system.

ᴇV6.1.3 Standard metric of unit sphere in stereographic coordinates (p. 504)

The metric transforms as $g = g_{ij}(y)\, dy^i \otimes dy^j = g_{kl}(z)\, dz^k \otimes dz^l$, with $g_{kl}(z) = g_{ij}(y(z))J^i{}_k J^j{}_l = (J^T g(y(z))J)_{kl}$. For the map $z \mapsto y(z)$, expressing spherical through stereographic coordinates,

$$\theta(z) = 2\arctan\left(\tfrac{1}{\rho}\right), \quad \phi(z) = \arctan\left(\tfrac{z^2}{z^1}\right), \quad \text{with } \rho(z) = \sqrt{(z^1)^2 + (z^2)^2}, \quad \sin\theta(z) = \frac{2\rho}{1+\rho^2},$$

the Jacobi matrix has the form (see Problem V5.3.3). With $g_{ij}(y) = \begin{pmatrix} 1 & 0 \\ 0 & \sin^2\theta \end{pmatrix}$, we obtain

$$J(z) = \frac{\partial y}{\partial z} = \begin{pmatrix} \frac{\partial \theta}{\partial z^1} & \frac{\partial \theta}{\partial z^2} \\ \frac{\partial \phi}{\partial z^1} & \frac{\partial \phi}{\partial z^2} \end{pmatrix} = \begin{pmatrix} \frac{-2z^1}{(\rho^2+1)\rho} & \frac{-2z^2}{(\rho^2+1)\rho} \\ \frac{-z^2}{\rho^2} & \frac{z^1}{\rho^2} \end{pmatrix}$$

$$\Rightarrow \quad J^T g(y(z)) J = \begin{pmatrix} \frac{-2z^1}{(\rho^2+1)\rho} & \frac{-z^2}{\rho^2} \\ \frac{-2z^2}{(\rho^2+1)\rho} & \frac{z^1}{\rho^2} \end{pmatrix} \begin{pmatrix} 1 & 0 \\ 0 & \frac{(2\rho)^2}{(1+\rho^2)^2} \end{pmatrix} \begin{pmatrix} \frac{-2z^1}{(\rho^2+1)\rho} & \frac{-2z^2}{(\rho^2+1)\rho} \\ \frac{-z^2}{\rho^2} & \frac{z^1}{\rho^2} \end{pmatrix}$$

$$= \frac{4}{(\rho^2+1)^2\rho^2} \begin{pmatrix} (z^1)^2 + (z^2)^2 & z^1 z^2 - z^2 z^1 \\ z^2 z^1 - z^1 z^2 & (z^2)^2 + (z^1)^2 \end{pmatrix} = \frac{4}{(\rho^2+1)^2} \mathbb{1}.$$

$$\Rightarrow \quad g = s(z)\big[dz^1 \otimes dz^1 + dz^2 \otimes dz^2\big], \qquad s(z) = \boxed{\frac{4}{(\rho^2+1)^2}}.$$

Near the north or south poles, where $\rho = \infty$ or 0, the scale factor $s(z)$ tends to 0 or 4, respectively. Its vanishing in the former case compensates for the fact that points near the north pole with small differences, δ_θ or $\delta\phi$, in latitude or longitude, lie very far apart on the stereographic plane. Near the south pole, however, this is not the case.

S.V6.2 Volume form and Hodge star (p. 505)

εV6.2.1 Hodge duals of basis one-forms in spherical coordinates (p. 507)

For a diagonal metric, the Hodge star operation acts on the basis one-forms in $\mathbb{R}^3$ as

$$*dy^i = \frac{1}{2!}\sqrt{|g|}g^{ii'}\epsilon_{i'jk}dy^j \wedge dy^k = \sqrt{|g|}g^{ii}dy^j \wedge dy^k,$$

where on the far right it is understood that ijk are related cyclically (i.e. jk are not summed over). For spherical coordinates, $y = (r,\theta,\phi)^T$, we have $\sqrt{|g|} = r^2 \sin\theta$ and inverse metric $g^{rr} = 1$, $g^{\theta\theta} = r^{-2}$, $g^{\phi\phi} = (r\sin\theta)^{-2}$:

$$*dr = \sqrt{|g|}g^{rr}d\theta \wedge d\phi = r^2 \sin\theta \cdot 1 \cdot d\theta \wedge d\phi = \boxed{r^2 \sin\theta\, d\theta \wedge d\phi},$$

$$*d\theta = \sqrt{|g|}g^{\theta\theta}d\phi \wedge dr = r^2 \sin\theta \cdot r^{-2} d\phi \wedge dr = \boxed{\sin\theta\, d\phi \wedge dr},$$

$$*d\phi = \sqrt{|g|}g^{\phi\phi}dr \wedge d\theta = r^2 \sin\theta \cdot (r\sin\theta)^{-2} dr \wedge d\theta = \boxed{(\sin\theta)^{-1}dr \wedge d\theta}.$$

S.V7 Differential forms and electrodynamics (p. 518)

S.V7.3 Laws of electrodynamics II: Maxwell equations (p. 525)

εV7.3.1 Homogeneous Maxwell equations: form-to-traditional transcription (p. 527)

In $\Lambda(\mathbb{R}^3)$, with metric $g_{ij} = \eta_{ij} = -\delta_{ij}\mathbb{1}$, we have

$$J^{-1}dx^i = \eta^{il}\underline{e}_l = -\underline{e}_i, \tag{1}$$

$$J^{-1}*dx^j \wedge dx^k = J^{-1}\eta^{jj'}\eta^{kk'}\epsilon_{j'k'i'}dx^{i'} = \eta^{jj'}\eta^{kk'}\epsilon_{j'k'i'}\eta^{i'i}\underline{e}_i = -\epsilon^{jki}\underline{e}_i, \tag{2}$$

$$*dx^i \wedge dx^j \wedge dx^k = \eta^{ii'}\eta^{jj'}\eta^{kk'}\epsilon_{i'j'k'} = -\epsilon^{ijk}, \tag{3}$$

The relation between the forms E, B and their vector representations $\underline{E}$, $\underline{B}$ read:

$$E = E_i dx^i, \qquad\qquad \underline{E} \equiv E^i \underline{e}_i \equiv J^{-1}E, \qquad\qquad E^i \overset{(1)}{\underline{\underline{\;}}} E_l \eta^{li} = -E_i \;; \qquad (4)$$

$$B = \tfrac{1}{2} B_{jk} dx^j \wedge dx^k, \qquad\qquad \underline{B} \equiv B^i \underline{e}_i \equiv J^{-1}*B, \qquad\qquad B^i \overset{(2)}{\underline{\underline{\;}}} -\tfrac{1}{2} B_{jk} \epsilon^{jki}. \qquad (5)$$

Inconsistent index positions arise on the far right of Eqs. (1) and (4), because once the sum on l has been performed, the notation no longer keeps track of the metric.

(a) (i) To convert the three-form equation $d_s B = 0$ to a scalar equation, we act on it with the Hodge star, insert factors of $\mathbb{1} = -**$ and $\mathbb{1} = JJ^{-1}$, and note that $-*d_s*J = \text{div}$ is the invariant formulation of the divergence:

$$0 = *d_s B = *d_s(-**)B = -*d_s*(JJ^{-1})*B$$

$$= \underbrace{(-*d_s*J)}_{\text{div}} \underbrace{(J^{-1}*B)}_{\underline{B}} = \boxed{\nabla \cdot \underline{B}}\,.$$

(ii) Expressed in terms of components, this strategy reads as follows:

$$0 = *d_s B = *\tfrac{1}{2}\partial_i B_{jk} dx^i \wedge dx^j \wedge dx^k \overset{(3)}{\underline{\underline{\;}}} -\tfrac{1}{2}\partial_i B_{jk}\epsilon^{ijk} \overset{(5)}{\underline{\underline{\;}}} \partial_i B^i = \boxed{\nabla \cdot \underline{B}}\,.$$

(b) (i) To convert the two-form equation $d_s E + \tfrac{1}{c}\partial_t B = 0$ to a vector field equation, we act on it with the two-form-to-vector conversion operation $J^{-1}*$, insert a factor $\mathbb{1} = JJ^{-1}$, and note that $J^{-1}*d_s J = \text{curl}$ is the invariant formulation of the curl:

$$0 = J^{-1}*(d_s E + \tfrac{1}{c}\partial_t B) = \underbrace{J^{-1}*d_s J}_{\text{curl}} \underbrace{J^{-1}E}_{\underline{E}} + \tfrac{1}{c}\partial_t \underbrace{J^{-1}*B}_{\underline{B}} = \boxed{\nabla \times \underline{E} + \tfrac{1}{c}\partial_t \underline{B}}\,.$$

$$0 = J^{-1}*(d_s E + \tfrac{1}{c}\partial_t B) = J^{-1}*(\partial_j E_k + \tfrac{1}{c}\partial_t \tfrac{1}{2}B_{jk}) dx^j \wedge dx^k$$

$$\overset{(2)}{\underline{\underline{\;}}} -(\partial_j E_k + \tfrac{1}{c}\partial_t \tfrac{1}{2}B_{jk})\epsilon^{jki}\underline{e}_i \overset{(4,5)}{\underline{\underline{\;}}} (\partial_j E^k \epsilon^{jki} + \tfrac{1}{c}\partial_t B^i)\underline{e}_i = \boxed{\nabla \times \underline{E} + \tfrac{1}{c}\partial_t \underline{B}}\,.$$

S.V7.4 Invariant formulation (p. 529)

ɛV7.4.1 Field-strength tensor (p. 531)

Expanding the field-strength tensor in components as

$$F = -E_i dx^0 \wedge dx^i + \tfrac{1}{2}B_{jk} dx^j \wedge dx^k \equiv \tfrac{1}{2}F_{\mu\nu} dx^\mu \wedge dx^\nu,$$

we identify $F_{0i} = -E_i = E^i$ and $F_{jk} = B_{jk} = -\epsilon_{jki}B^i$, hence

$$\{F_{\mu\nu}\} = \begin{pmatrix} 0 & E^1 & E^2 & E^3 \\ -E^1 & 0 & -B^3 & B^2 \\ -E^2 & B^3 & 0 & -B^1 \\ -E^3 & -B^2 & B^1 & 0 \end{pmatrix}.$$

(a)
$$0 = dF = \sum_{\alpha\mu\nu} \tfrac{1}{2}\partial_\alpha F_{\mu\nu}\, dx^\alpha \wedge dx^\mu \wedge dx^\nu$$

$$= \sum_{\alpha<\mu<\nu} (\partial_\alpha F_{\mu\nu} + \partial_\mu F_{\nu\alpha} + \partial_\nu F_{\alpha\mu})\, dx^\alpha \otimes dx^\mu \otimes dx^\nu$$

Here we rearranged all terms in order of ascending indices, exploiting the antisymmetry of $F_{\mu\nu}$ and of the wedge product to collect similiar terms. For every fixed choice of $\alpha < \mu < \nu$, the coefficient of $dx^\alpha \otimes dx^\mu \otimes dx^\mu$ must vanish, hence

$$\boxed{0 = \partial_\alpha F_{\mu\nu} + \partial_\mu F_{\nu\alpha} + \partial_\nu F_{\alpha\mu}}.$$

(b)   Using $F_{0i} = -F_{i0} = E^i$ and $F_{jk} = -\epsilon_{ijk}B^i$, we obtain:

$\alpha = 0,\ \mu = j,\ \nu = k$: $\qquad 0 = \partial_0 F_{jk} + \partial_j F_{k0} + \partial_k F_{0j} = -\epsilon_{jki}\partial_0 B^i - \partial_j E^k + \partial_k E^j$.

Contract with $-\tfrac{1}{2}\epsilon^{jkl}e_l$: $\Rightarrow$ $\boxed{0 = \tfrac{1}{c}\partial_t \underline{B} + \underline{\nabla} \times \underline{E}}.\ \checkmark$

$\alpha = 1,\ \mu = 2,\ \nu = 3$: $\qquad 0 = \partial_1 F_{23} + \partial_2 F_{31} + \partial_3 F_{12} = -(\partial_1 B^1 + \partial_2 B^2 + \partial_3 B^3)$

$\Rightarrow$ $\boxed{0 = \underline{\nabla} \cdot \underline{B}}.\ \checkmark$

(c)
$$0 = J^{-1}*dF = J^{-1}*\partial_\alpha \tfrac{1}{2}F_{\mu\nu}dx^\alpha \wedge dx^\mu \wedge dx^\nu$$

$$= J^{-1}\partial_\alpha \tfrac{1}{2}F_{\mu\nu}g^{\alpha\alpha'}g^{\mu\mu'}g^{\nu\nu'}\epsilon_{\alpha'\mu'\nu'\beta'}dx^{\beta'}$$

$$= \partial_\alpha \tfrac{1}{2}F_{\mu\nu}g^{\alpha\alpha'}g^{\mu\mu'}g^{\nu\nu'}\epsilon_{\alpha'\mu'\nu'\beta'}g^{\beta'\beta}\underline{e}_\beta = \partial_\alpha \tfrac{1}{2}F_{\mu\nu}\epsilon^{\alpha\mu\nu\beta}\underline{e}_\beta$$

$\Rightarrow$ $\boxed{0 = \partial_\alpha G^{\alpha\beta}}$,   with $G^{\alpha\beta} = \tfrac{1}{2}\epsilon^{\alpha\beta\mu\nu}F_{\mu\nu}$.   (1)

For the last step, we used the component representation of the relation $G = -*F$,

$$G = -*\tfrac{1}{2}F_{\mu\nu}dx^\mu \wedge dx^\nu = -\tfrac{1}{2}F_{\mu'\nu'}g^{\mu'\mu}g^{\nu'\nu}\epsilon_{\mu\nu\alpha\beta}dx^\alpha \wedge dx^\beta,$$

which implies $G_{\alpha\beta} = -\tfrac{1}{2}F^{\mu\nu}\epsilon_{\mu\nu\alpha\beta}$. Raising the indices of G and lowering those of F using a product of four gs produces another minus sign, hence $G^{\alpha\beta} = \tfrac{1}{2}\epsilon^{\alpha\beta\mu\nu}F_{\mu\nu}$.

(d)   Starting from (1), we obtain, for fixed β:

$$\partial_\alpha \tfrac{1}{2}\epsilon^{\alpha\mu\nu\beta}F_{\mu\nu} = 0.$$

Nonzero contributions arise only for all four indices different from each other. For example, for $\beta = 0$:

$$0 = \partial_1 \tfrac{1}{2}(F_{23} - F_{32}) + \partial_2 \tfrac{1}{2}(F_{31} - F_{13}) + \partial_3 \tfrac{1}{2}(F_{12} - F_{21}) = \partial_1 F_{23} + \partial_2 F_{31} + \partial_3 F_{12},$$

since $F_{\mu\nu} = -F_{\nu\mu}$, and similarly for $\beta = 1, 2, 3$. Hence we obtain:

$$0 = \boxed{\partial_\alpha F_{\mu\nu} + \partial_\mu F_{\nu\alpha} + \partial_\nu F_{\alpha\mu}}.\ \checkmark$$

This expression vanishes unless α, μ and ν are all different.

Index